Global Innovation Science Handbook

Praveen Gupta   Editor

Brett E. Trusko, Ph.D., MBA   Editor

New York   Chicago   San Francisco
Athens   London   Madrid
Mexico City   Milan   New Delhi
Singapore   Sydney   Toronto

2 3 4 5 6 7 8 9 0   DOW/DOW   1 9 8 7 6 5 4

ISBN 978-0-07-179270-7
MHID 0-07-179270-8

Sponsoring Editor
Judy Bass

Editing Supervisor
Stephen M. Smith

Production Supervisor
Pamela A. Pelton

Acquisitions Coordinator
Amy Stonebraker

Project Manager
Raghavi Khullar,
Cenveo® Publisher Services

Copy Editors
Namita Panda and Upendra Prasad,
Cenveo Publisher Services

Proofreaders
Linda Leggio and Connie Blazewicz

Indexer
Robert Swanson

Art Director, Cover
Jeff Weeks

Composition
Cenveo Publisher Services

Printed and bound by RR Donnelley.

I respectfully dedicate this book to my role models who have inspired me to continue their work. They are W. Edwards Deming, Peter F. Drucker, Robert F. Galvin, Eliyahu M. Goldratt, H. James Harrington, Kaoru Ishikawa, Robert S. Kaplan, Tom J. Peters, Genichi Taguchi, Walter A. Shewhart, and A. William (Bill) Wiggenhorn. I am indebted for their influence in my life.

—Praveen Gupta

I started on this book almost three years before it was completed. The authors of these chapters who were not my friends at the time became so over the year of recruitment and the year of writing and editing. These authors represent the best of innovation! Each of them contributed only a piece of this book, but nearly all of them could have written the entire book by themselves. To the authors, I am grateful.

I am also thankful for my family. When pulling together a project of this size, while holding down a full-time job, the time has to come from somewhere; and, unfortunately, giving up sleep entirely isn't really an option. Therefore, I dedicate the book to my incredibly supportive and lovely wife, Kirsten, and my three boys, Nikolas, Dominick, and Treyton, who all gave up time with their husband and father to allow me to do this.

—Brett E. Trusko

About the Editors

Praveen Gupta, MSEE, is a corporate executive and thought leader in excellence and innovation management for achieving profitable growth through Six Sigma, process excellence, business innovation, and performance measurements. He has led the development of innovation science since 2003. Mr. Gupta has published in many magazines and has authored many books on Six Sigma and innovation, including *Business Innovation in the 21st Century* and *The Innovation Solution*. He was the founding editor-in-chief of the *International Journal of Innovation Science*. Mr. Gupta has taught innovation classes at the University of Illinois and Illinois Institute of Technology. He is also the founding director of the Center for Innovation Science and Applications (CISA) at IIT Chicago. Mr. Gupta has recently relocated to Silicon Valley while continuing his research interest and contributions through CISA.

Brett E. Trusko, Ph.D., MBA, is an assistant professor at Texas A&M University, where he is also the director of the Health LINKS innovation lab, a 4000-ft^2 innovation center located in the world-famous Texas Medical Center. He is the president and CEO of the nonprofit International Association of Innovation Professionals (www.IAOIP.org), a global innovation standards development and certification organization. Dr. Trusko is the coauthor of *Improving Healthcare Quality and Cost with Six Sigma* and the current editor-in-chief of the *International Journal of Innovation Science*.

Contents

Section 3 **Creativity Tools**

Contributors

Jeremy Alexis   M.Des.; Director of Interdisciplinary Education, Senior Lecturer, Illinois Institute of Technology (CHAP. 31)

Fernando Almeida   (CHAP. 46)

Jørn Bang Andersen   M.Sc.; Director, Clareo Europe; Advisory Board Member, Kellogg Innovation Network (CHAP. 36)

Eduardo Armando   Ph.D.; Full Professor, FIA Business School, São Paulo, Brazil; Full Professor, University of São Paulo School of Economics, Business, and Accountancy (CHAP. 40)

Siva Balasubramanian   Ph.D.; Harold L. Stuart Professor of Marketing and Associate Dean, Stuart School of Business, Illinois Institute of Technology (CHAP. 48)

Billy R. Brocato   M.A., M.B.A.; Research Assistant, Texas A&M University (CHAP. 33)

Vern Burkhardt   M.A.; Director and Author, IdeaConnection Ltd. (CHAP. 23)

C. Robert Carlson   Ph.D.; Dean, School of Applied Technology, Illinois Institute of Technology (CHAP. 47)

Donald A. Coates   Ph.D., P.E.; Assistant Professor (retired), Kent State University; Director, New Paradigm Innovations, LLC (CHAP. 28)

David Conley   M.B.A.; President, Innovation Corp.; Managing Partner, PQR Group (CHAP. 17)

Cathie M. Currie   Ph.D.; Education-Workforce Strategist, Altshuller Institute for TRIZ Studies (CHAP. 20)

C. Anthony Di Benedetto   Ph.D., M.B.A.; Professor, Temple University (CHAPS. 30, 45)

Ellen Di Resta   M.S., M.B.A.; Strategic Marketing, Becton Dickinson; Author/Lecturer, Strategic Innovation and Design, Synaptics Group (CHAPS. 8, 38)

Ellen R. Domb   Ph.D.; PQR Group (CHAP. 24)

Rameshwar Dubey   Ph.D., FIIPE; Assistant Professor, Symbiosis Institute of Operations Management (CHAP. 39)

Thomas N. Duening   Ph.D.; El Pomar Chair in Business and Entrepreneurship and Director, Center for Entrepreneurship, University of Colorado at Colorado Springs (CHAP. 6)

Nina Fazio   B.S.; Copy Editor; International Journal of Innovation Science (CHAP. 42)

Cristina I. Fernandes   Ph.D.; Researcher, Research Centre in Business Sciences, University of Beira Interior, Portugal (CHAP. 35)

João J. M. Ferreira   Associate Professor, University of Beira Interior, Portugal; Researcher, Research Centre in Business Sciences, University of Beira Interior (CHAP. 35)

Helena Forsman   D.Sc.; Professor, University of Tampere School of Management (CHAP. 43)

Nikolaus Franke   Prof. Dr.; Head of the Institute for Entrepreneurship and Innovation, WU Vienna (Vienna University of Economics and Business) (CHAP. 19)

Lisa Friedman   Ph.D.; Cofounder, Enterprise Development Group (CHAPS. 2, 4)

Juan Vicente García Manjón   Ph.D., Professor of Business Innovation, Universidad Europea Miguel de Cervantes, Spain (CHAP. 26)

Brian Glassman   (CHAP. 22)

Michael Gorham   Ph.D., M.S.; Professor of Finance and Director of the IIT Stuart Center for Financial Innovation, Illinois Institute of Technology (CHAP. 48)

Richard B. Greene   (CHAP. 15)

Michael Grieves   (CHAP. 11)

Praveen Gupta   M.S.E.E.; Director, IIT Center for Innovation Science and Applications; Corporate Quality, Prysm, Inc. (CHAPS. 1, 7, 21, 25, 37, 47)

Laszlo Gyorffy   M.S.; President, Enterprise Development Group (CHAP. 4)

Herman Gyr   (CHAP. 2)

H. James Harrington   *Ph.D.; CEO, Harrington Institute* (CHAP. 13)

Jack Hipple   *B.S.Ch.E.; Principal, TRIZ and Engineering Training Services, LLC* (CHAP. 3)

Andria Long   *M.B.A.; Vice President, Innovation, Johnsonville* (CHAP. 41)

Heather McGowan   *B.F.A., M.B.A.; Strategic Advisor and Academic Entrepreneur, Becker College, Philadelphia University* (CHAP. 38)

Andrea Meyer   *M.S.; President, Working Knowledge®* (CHAP. 29)

José Monteiro   (CHAP. 46)

Joseph T. Nabor   *J.D.; Partner, Fitch Even Tabin & Flannery LLP; President, Chicago Intellectual* (CHAP. 44)

Steven G. Parmelee   *J.D.; Patent Attorney and Partner, Fitch Even Tabin & Flannery LLP* (CHAP. 44)

Mário L. Raposo   *Full Professor, University of Beira Interior, Portugal; Researcher, Research Centre in Business Sciences, University of Beira Interior* (CHAP. 35)

Arin N. Reeves   *J.D., Ph.D.; President, Nextions* (CHAP. 5)

Christopher S. Rollyson   *Managing Director, CSRA, Inc.; Founder, Chief Digital Officer, Chief Architect, Social Network Roadmap* (CHAP. 16)

José Santos   (CHAP. 46)

Suzanna Schmeelk   *Ed.D., M.S.; Network Security Research Scientist, LGS Bell Labs Innovations, LLC* (CHAPS. 49, 50)

Thomas Schmidt   *D.Sc.; Visiting Professor, Parsons The New School for Design* (CHAP. 33)

Victor Scholten   *Ph.D.; Assistant Professor, Delft University of Technology, The Netherlands* (CHAP. 27)

Fiona Schweitzer   *Ph.D.; Professor for Marketing and Market Research, University of Applied Sciences Upper Austria* (CHAP. 18)

Klaus Solberg Søilen   *B.A. Bus Admin (CLU); B.A. Pol Sci (CLU); M.B.A. (HEC); Dr. rer. pol. (LIPSIENSIS); FBS, Halmstad University* (CHAP. 10)

Melissa Sterry   *Design Scientist and Futurist, Bionic City* (CHAP. 9)

Serdal Temel   *Ph.D.; Senior Researcher, Ege University Science and Technology Centre* (CHAPS. 27, 43)

Larry R. Thompson   (CHAP. 14)

Maria B. Thompson   *M.S.; Senior Director of Innovation Strategy, Motorola Solutions, Inc.* (CHAP. 42)

Brett E. Trusko   *Ph.D., M.B.A.; President, International Association of Innovation Professionals; Assistant Professor, Texas A&M University* (CHAPS. 1, 12, 21, 37)

Rajesh Tyagi   *Ph.D., M.B.A.; Assistant Professor, HEC Montréal* (CHAP. 32)

Eduardo Vasconcellos   (CHAP. 40)

Navneet Vidyarthi   *Ph.D.; Assistant Professor, Concordia University* (CHAP. 32)

Frank Voehl   *Lean Six Sigma Grand Master Black Belt; Chancellor, Harrington Institute; Strategy Associates, Inc.* (CHAP. 13)

Abram Walton   *Ph.D.; Associate Professor, Technology Management & Innovation, Florida Institute of Technology* (CHAP. 22)

Margaret Morgan Weichert   *M.B.A.; Partner/Principal, Ernst & Young Advisory* (CHAP. 34)

Robert C. Wolcott   *Ph.D.; Cofounder and Executive Director, Kellogg Innovation Network, Kellogg School of Management; Senior Lecturer, Innovation & Entrepreneurship, Kellogg School of Management; Partner, Clareo Partners* (CHAP. 36)

Foreword

We have entered an era in which talent-driven innovation is the only sustainable source of competitive advantage for many companies and countries. Today, knowledge, information, and technology are widely distributed, increasingly commodities, and globally accessible. Those who know best what to do with these building blocks of innovation once they get them gain a competitive edge and reap the economic rewards.

As innovation takes center stage in the world's economies, it has never been more important to explore the innovator's craft—the innovative thinking and know-how applied by insightful and determined business leaders, brilliant scientists and engineers, master entrepreneurs, creative organizations, and champions of change. If we can turn innovation from an art practiced by a few to a science accessible to many, we can enable many more people and organizations to achieve their innovative potential.

Opportunities for innovators are booming around the globe. Digital technology, biotechnology, and nanotechnology are creating unparalleled prospects for innovation across numerous industries. Demand for innovations tailored to the cultural and socioeconomic attributes of rapidly growing emerging economies has created a wealth of markets. Big data analytics is providing powerful new tools for research and scientific discovery, as well as unprecedented insight and intelligence for businesses and analysts in a wide range of disciplines. In the hands of inventors, producers, and entrepreneurs, modeling and simulation are a force multiplier and innovation accelerator, offering an extraordinary opportunity to design products and services faster and to develop innovations that would otherwise be impossible. While new materials and digitization are poised to drive revolutionary innovation in manufacturing, additive manufacturing and 3D printing are democratizing access to the means of production, giving creators, designers, and entrepreneurs new power to innovate.

Innovation has expanded beyond science and technology embedded in hardware, products, and processes. It now includes innovations such as Web-based businesses, novel service delivery, new ways to connect with customers, new retail distribution and shopping systems, new media, and high-value lifestyle products and services. The "Internet of Things" is evolving rapidly, and, in the decade ahead, as many as 50 billion objects and devices could be connected, generating $14 trillion in economic value. Yes, innovation is wonder drugs—microchips with blazing speed, efficient thin-film solar cells, and nanoelectromechanical systems—but it is also the captivating invented world of Harry Potter, $1 music downloads, and $4 Caffè Vanilla Frappuccinos served at thousands of sleek, high-end coffee shops around the world.

At the same time, the pace of technological change is quickening, and innovation itself is changing. Hypercompetition and rapid technical change do not favor the time-consuming transfers and handoffs when innovations are developed in a traditional serial process. Innovation is becoming more multidisciplinary and often exceeds the scale and scope of single-discipline project models.

Praveen shared his vision for innovation science with my organization about eight years ago. In the 21st century, we need new rules and tools of innovation for faster and better solutions and higher return on investment. The growing number of opportunities for innovation means that we need to engage intellectually a much wider population. With countries and companies alike facing an "innovation imperative," it has never been more important to teach people and organizations how to out-imagine, out-create, and out-innovate. We must learn how best to manage the innovation process to achieve the speed, efficiency, cost-effectiveness, and outcomes that will deliver competitive advantage, economic prosperity, higher standards of living, and a better quality of life.

The insight and experiences captured by Brett and Praveen in the *Global Innovation Science Handbook*, whose chapters were authored by global experts, make an important contribution to more pervasive, manageable, and profitable innovations.

Deborah Wince-Smith
President, Council on Competitiveness

Acknowledgments

It is a humbling experience and an honor to be a coeditor of the *Global Innovation Science Handbook*, with contributions from dozens of experts around the world. It is also very appropriate for my exploration of innovation science to culminate in this handbook, a milestone in the maturity of the innovation field. On this journey I have had the opportunity to meet thousands of people in many countries, and the privilege to work with innovation experts and good people. It has been a great honor and pleasure to work with Brett Trusko, one of the best partners one could ask for on such a monumental project. This handbook would not have been possible without Brett.

My innovation journey started in 2003 after the dotcom bomb with my desire to rediscover myself for the next project. While driving my daughter Avanti to a science fair at the University of Illinois, Urbana-Champaign, in 2004, I asked her, "When do kids think creatively?" Half-asleep as she was, I heard, "Dad, let me sleep." I tried one more time, and then came the response, "When they are annoyed." Her simple reply became my motivation to accelerate development of an innovation framework to demystify innovation. Thanks to Avanti for being a catalyst in my innovation journey. I also would like to acknowledge my son Krishna for his commitment to excellence and hard work, innovatively and successfully, that helped me raise the bar and expectations for myself. As much as my kids might have been inspired by my work, my work is a reflection of the inspiration I derived from their aspirations.

There are numerous people who have contributed to my innovation knowledge and who believed in my ability to advance the innovation body of knowledge. I regret that many names will be omitted in this list. Thanks to Anoop Verma, the first person to recognize its potential while I was sketching the innovation framework on a piece of paper. My sincere thanks for valuable input to Paul Davis, Prof. Mohanbir Sawhney, Rajesh Tyagi, Arvind Srivastava, Jay Patel, Rajiv Khanna, Richard O'Brien, Alex Goncalves, Hans Hansen, Dr. James Harrington, Rajeev Jain, Abhai Johri, Lauri LaMantia, Wayne Rothschild, Justin Swindells, Joseph Nabor, Glen Nevogt, Marjorie Hook, Gina Jones, Beth Daley, Jan E. Droege (late and founder of the National Association of Idea Management), Cissy Pettenon, Mike Lippitz, Baber Inayat, Shellie Tate, Dan Pongetti, Joe Boggio, Randall Kempner, Joao Mendes, Jorge Teixeira, Alberto Casal, Veronica Perea Medina, Helena Santos, Tony Reyes, Charles Cooney, Ganesh Raman, Tarun Kumar, Rob Peters, Adam Hecktman, Scott Pfeiffer, and many more.

Special thanks to Prof. C. Robert Carlson, Dean, and Robert Anderson at IIT Chicago for accepting my first presentation and helping me launch my innovation journey and setting up the Center for Innovation Science and Applications. My absolutely highest regard for the late Robert W. Galvin, former Chairman and CEO of Motorola, for not only contributing a chapter to my first book, but also lending his name to support my innovation work. He has been an exemplary business leader in my life. I would like to express my personal gratitude to Deborah Wince-Smith, President of the Council on Competitiveness, for giving me the opportunity to share my innovation framework with her team in the early years of its development. I have been blessed to share one of my chapters with Dipak Jain, Dean of the Kellogg School of Management and INSEAD, and I appreciate that he wrote an excellent foreword for my first book, giving me confidence to continue my journey. My special thanks to Luis Reis, Chief Operating Officer at Sonae in Portugal, and Ray Mehra, President of R-Squared, for supporting my pursuit of innovation science. I salute Anita Raman, Bill Hughes, and Paul Bailey of Multi-Science for legitimizing my journey of innovation science by launching the *International Journal of Innovation Science*.

I owe all of my work to thousands of people around the world who have believed in me, including one in Jordan who said, "Praveen, I love your guts for not giving up." He was right. I could not give up for the last 10 years. Thanks to Judy Bass, Steve Chapman, and Philip Ruppel of McGraw-Hill Education's Professional Division for this booster as our innovation journey continues.

Praveen Gupta

I want to thank all the authors of this book for the time they each put into their chapters. More important, for never losing their composure as I constantly harassed them for their contributions. I would also like to thank Kylie Saunders for checking citations and "setting up" the chapters before my edits. She did the job that I truly did not want to do, and she did so with a smile and positive attitude. I would like to thank Praveen for contacting the people who wouldn't call me back. Judy Bass and Amy Stonebraker at McGraw-Hill met with me every month to coach me through the enormous undertaking. Their insight and guidance is what allowed me to organize and complete the project. Most of all, I would like to acknowledge Nina Fazio, editor extraordinaire! Nina is a coauthor of one of the chapters, and kindly offered to help me with the edits if I needed her. I took her up on this offer and she edited at least 15 chapters. All that she asked is that I help her build a copy-editing business (a true entrepreneur). So here goes: If you need a great copy editor, contact Nina Fazio!

Thanks to everyone!

Brett E. Trusko

Introduction

Innovation means success; there can be no failure in life when there is innovation. Sooner or later, someone will discover a solution to *any* problem. That is how the innovation journey began for Praveen when he began his search for something to do after a major dotcom disappointment. After some soul-searching and an experiment in optimizing Six Sigma using the theory of constraints, he was inspired to better understand innovative thinking in a more systematic way—if for no better reason than to be able to teach it. For Brett, innovation was just the logical path from clinical medicine to finance, to marketing, and finally to technology.

As a team, we were inspired to do something different. To us, this meant taking Six Sigma to the next level. Accordingly, Praveen developed the Six Sigma Business Scorecard, which immediately led to the need to expand the knowledge required for solving problems. Additional research in this area highlighted the opportunity to study innovation as a system rather than just be limited by the perception that innovation is a gift, secret sauce, or something magical and mysterious. Tom Kelley's book, *The Art of Innovation,* prompted us to explore the possibility that innovation could be a science, a methodology that could be formalized and approached as a set of tools and methods that innovators and innovation teams could put to good use. So, in 2004, we started the journey of "innovation science."

Praveen began a study of the great scientists and innovators of the last several hundred years to try to discover what made them special. He looked at scientists such as Einstein, Newton, Galileo, Edison, and Henry Ford. This led to theories about many of the questions he had about innovation, which in turn led to the theory of breakthrough innovation now known as Brinnovation™. A framework and methodology soon followed. After several years of evangelizing the concepts of "innovation science," Praveen began to notice that more and more people came to accept the idea of innovation as a science. Granted, there are still people who argue with us in online forums and blogs. There have been aggressive attacks on the idea of a certification in innovation, specifically because of the long-held belief that innovators are simply the "chosen," or that it is something that everyone is born with but loses as they age and socialize. In 2008, after a presentation to a group of businesspeople, a debate began about the science of innovation that led to the launching of the *International Journal of Innovation Science* (IJIS). We felt that this formally laid the foundation for innovation science, and began the process of slowly building a community of like-minded thinkers from around the world. Many prominent innovation experts and thought leaders have joined the innovation science movement through the IJIS, the business innovation networks, the Business Innovation Conference, and our new baby, "Innova-Con." All of this culminated in the formation of the International Association of Innovation Professionals (www.iaoip .org), which is focused on organizing the discipline of innovation and creating a certification process to demonstrate mastery of innovation skills. Engaging experts from different disciplines of innovation from around the world gave us the critical mass necessary for creating a body of knowledge that would help organizations develop in-house competencies in innovation methods. Thus, this *Global Innovation Science Handbook* was launched about 3 years ago as an attempt to formalize the field of innovation science through the body of knowledge and certification.

The *Global Innovation Science Handbook* (GISH) was created based on a defined body of knowledge that includes intent, methodology, tools, and measurements. The GISH was developed to challenge the popular paradigm that "learned" innovation was impossible, and that it was something in the realm of magic. We believed that, collectively, we all could make innovation a learned skill. The GISH is organized in a way such that any chapter can be independently read and utilized in the daily practice of innovation. Like creating new knowledge or innovative solutions, the "prior art" matters; similarly, we have attempted to learn from the "prior art" of innovation in this handbook.

Section 1 begins with the establishment of the framework for a successful innovation effort. This is the first step in cultivating an innovation competency within an organization or team. It includes a thoughtful determination to embark on innovation in a creative and purposeful way. This means that the organization must take a serious look at its current ability to innovate. Many innovation efforts fail because an organization is simply not set up to be successful at innovation. Too many organizations embark on the road to innovation, but do not eliminate constraining hierarchies and other cultural impediments to success. Finally, the strategy for innovation must be considered: how it will be developed, how it will be sustained.

Section 2 emphasizes the general concepts of innovation that don't fit well into the traditional discussion of innovation or, for that matter, a traditional business model. Examples of these concepts include personal and corporate creativity, innovation and neuroscience, association with bionics, benchmarking, and process versus practice innovation. While benchmarking potentially could have been included in Section 5, which covers methods, we chose to leave it here because of its broad applicability to all areas of innovation.

Section 3 highlights perhaps the most familiar but least understood part of innovation: creativity. Creativity is thought by many to be the essential ingredient, or the "secret sauce," of innovation. That ability that diminishes from the moment you are born until the day of your death. We all know the story: kids are born creative, but lose the ability to be creative through aging, formal education, personal and professional responsibilities, marriage, growing families—and something in the water? The authors in this section of the book would argue that all of these and the hundred other reasons you can come up with are all wrong. People never stop being creative.

Section 4 is best represented by a quote from the first author in this section: "Innovation is not for the timid." As Cathie Currie states in her chapter on managing the development of ideas and the importance of bringing the most valuable ideas from concept to reality, "Innovative thinkers boldly perceive opportunities in circumstances where conventional thinkers see few prospects. Innovators trawl among bountiful sources, whereas the less innovative find themselves fishing in an empty stream." In the chapter on quality of ideas, we share the process of creating, screening, and exploring ideas. In the chapter on managing development of innovation ideas, the authors discuss the process and methods behind evaluating ideas.

Section 5 is, to put it mildly, the "mother" of all sections. Many readers will have purchased this book just for this methodology section. Despite the 12 rich chapters that we've included, there are probably 24 other topics that our readers will tell us we should have included. As we developed the book, we understood that we were building a body of knowledge in a rapidly changing and evolving field. Therefore, we tried to include what was most appropriate at the time it was published. We have included a chapter on types of innovation in an effort to capture some of the less common methodologies. We've also included some chapters that people may not consider innovation. This section, taken as a whole, offers a perspective on innovation that will allow you to evaluate all other methodologies in a whole new light.

Section 6 explores the measurement of innovation and at first blush appears to be another one of those topics that can't really be addressed. After all, how can people know if they're successful in innovation given that most can't benchmark their current level of innovation competency? Unfortunately, this is exactly the state of many organizations. Unfortunately, the term *fuzzy*, as in fuzzy front end of innovation, has permeated the innovation community. Utilizing the MMI (measured and managed innovation) model, one chapter describes innovation radar built on 12 dimensions. Measuring all these dimensions allows the Nordic region to evaluate companies based on innovation capabilities. Another chapter highlights process-based measures of innovation and some business innovation indices we offer to clarify some of these measurements—if not for your organization, then at least to stimulate your thoughts.

Section 7 is similar to the methodology section, and if that was the "mother" of all sections, then deployment is surely the "father." The steps involved in the deployment of an innovation program are numerous. None of the steps can be ignored, though some may already have been taken somewhere along the way. This section is organized as a kind of life-cycle approach to innovation. Starting with inspiration, we move to strategy, followed by organizing, developing excellence, culture, a measurement system, protection of intellectual property, launching an innovative product, and, finally, the maturity model for sustaining a culture of innovation.

Section 8 includes case studies from the financial, nanotechnology, government, and education sectors. As organizations gain more experience in managing innovation and produce more success stories over the years, this section is expected to grow with each subsequent edition of the GISH.

The *Global Innovation Science Handbook* represents a leap in knowledge of innovation science with content from over 40 contributors from around the world. If professionals and educators can articulate their questions about innovation, the GISH should have answers for most if not all of them. As new questions arise, new editions will arrive with an up-to-date body of knowledge. This is our attempt to consolidate the collective wisdom of our peers and make innovation a learned skill for anyone who wants to become a better innovator. Innovation competency has become a critical necessity for individuals and organizations. This compilation of the innovation body of knowledge has led to the next step in recognizing competency in innovation, along with the International Association of Innovation Professionals to shepherd the way forward. If you believe in the advancement of innovation as a science, we strongly encourage you to join the association.

If you would like to share your questions or contribute to this ever-evolving body of knowledge for innovation through future editions of the *Global Innovation Science Handbook*, we would love to hear from you. There are many ways to reach us: You can find us on Twitter, LinkedIn, Facebook, and the association's website, and of course via good old e-mail and phone.

Happy innovating!

Preparing for Innovation

Establishing the framework for a successful innovation effort is the key first step to establishing an innovation competency or team. Specifically, this requires several things to be considered before innovation efforts even commence. Included in this is a decision to embark on innovation in a creative and purposeful way. This means that the organizations take a serious look at their ability to innovate. Many innovation efforts fail, because an organization is not "set up" to be successful at innovation. An honest and critical look at your ability to innovate can mean the difference between success and failure. A second consideration in this effort, and related to the first is to consider that beyond the organizational structure, there is a need to be culturally prepared to innovate. Too many organizations embark on the road to innovation, but do not eliminate hierarchies and other cultural impediments to success. As is discussed so often in innovation, fear of failure directly relates to fear of innovating. To create this culture one must take a serious look at the organizations' leadership. How many innovative organizations have lost their innovative competency when there is a leadership change? Finally, you must consider the strategy for innovation: how you develop it, how you maintain it.

Strategy

A basic tenet of business is to create a strategy, simply, a roadmap to get to where you want to be. Many believe that innovation is all about freedom, flexibility, and a devil–may-care attitude about the process. Yes, many of the legends seem to have found the magic formula to innovation by accident, but if you dig deep into the history of the person, or company that you are thinking about, you will find that most of what they accomplished was thoughtful and purposeful. The legend has it that the founder of FedEx, Fred Smith, was given a grade of C by his professor because the idea was not feasible. The story of Steve Jobs is rife with stories of how he forced his vision on his employees and the market—he had a plan to change the world and he worked hard on every one of his inventions. Sure, they may have been fuzzy at the inception, but by the time they were in design, they knew well what they were creating. This is the same with your innovation organization. Only by knowing what you would like to be can you become it, and you can only become it by creating a clear picture of what *it* might be.

The Organization's Capability

Creation of a blueprint of your innovation organization is perfectly descriptive. Just as you have a blueprint of a building before you build it, so too must you have a blueprint of an organization. This could also be called a framework, structure, or model. Anything that helps you with the concept of designing the organization that helps you become what you see yourself as in the future. Of course this can take many shapes and forms, just like a building. And just like a building there are certain things that are required such as electricity, heating, cooling, and light. In your innovate framework you need to understand your business (internal environment), which consists of the people, the structure, and the tasks you are trying to accomplish. You need a clear understanding of how the external environment affects and interacts with your organization. This applies to where you are now, and where you would like to be in the future. Of course bridging the past and the future is also a little like remodeling a building. What do we have, what can be kept, and what needs to be discarded?

Understanding the Culture in an Innovation Effort

If your organization is located in the heart of Palo Alto, the cultural considerations are entirely different than if your organization resides in Waco (no offense to Waco). Culture is easier where people believe in innovation and accept the start-up mentality and approach. If your organization is in Palo Alto your cultural issues may be entirely different from anywhere else. As an organization, you may, in fact, find that your best people are hard to keep because the innovation culture is so strong and entrenched in the everyday life of the majority of the population. You may find that you will be limited because you cannot keep good people around. On the flip side, in Waco you may find that it is difficult to remake the culture to be more like Palo Alto. All people are innovative at heart, the trick is to understand your culture and reshape it in a way that makes the most sense for your organization's strategy framework.

Leader's Innovation Commitment

A common problem with innovation is the true commitment of the leader to the process and the end goal. There are stories of great leaders that reflect individuals who are obnoxiously stubborn about where they see their organization and how it will make a difference in the world. Whether you are reflecting on Vince Lombardi or Steve Jobs, a great leader has several attributes. They have a clear vision of where they want the organization to be in the future and they move the organization in that direction. They advocate for the customer, and architect an organization to serve that customer. To accomplish their vision, they find the money to pay for it, while taking time to mentor and break down walls to working together. They act as a role model, networker, and most important, create a culture where this can all come together.

CHAPTER 1

Strategy for Innovation

Praveen Gupta and Brett E. Trusko

Strategy for innovation is a long-term commitment to sustained growth. It is a vision to achieve greater successes and create bigger opportunities for growth of stakeholder value. When it comes to innovation, the strategy may look into 5-, 10-, 20-, or 30-year or even longer time frames to create a path for an organization—the path that is checked periodically and the roadmap adjusted routinely. But there is a target, however fuzzy it may be. Strategy operates within the scope of the organization's vision or the so-called *fundamental strategy* of achieving sustained profitable growth.

In many corporations, strategies are formulated to increase profitability, and growth is an aspect of it besides cutting costs. These firms cut costs and have practically no resources for innovation. Innovation seems to be too risky and counterproductive to their cost-cutting efforts. Some firms develop strategies for growth and hope that new products will be profitable and make money for the company, and then the company ends up losing money with new products. We have learned that in order to sustain profitable growth, firms need to develop strategies for growth and ensure growth is profitable through excellence in execution. Successful execution of a strategy depends on many factors including clarity and feasibility, leadership commitment to make it work, buy-in from middle management through their participation in strategy development, incentives and recognition for successful execution of the strategy, establishing measures of success and accountable ownership for execution (execute or be executed!), aligning the organization for synchronizing cross-functional priorities, and ensuring necessary resources.

Basic Tenets of Business Success

To cut costs in an organization, the operations executive looks into reducing waste, improving processes, leaning operations, reducing floor space, and rightsizing headcount. Cost-cutting requires a lot of actions in the present, and a little thinking about the future. Planning for future growth requires a lot of thinking. Actually, when an entrepreneur starts a business, what is the first objective? Is it to make money or to get a customer? It is *always* to find the first customer and to make the sale.

Similarly, for a mature business, the first objective can be to grow sales, not to lose the entrepreneurial spirit. Once we have made a sale, we have identified new users that need or want our product or service. If we can deliver the product or service without any waste, we will make money. It has been said that the main purpose of a business is not to make money, but rather to make a product or service customers need and love to have. If we do it well, we will make money. So, making money is not the end of the business, it is the means and outcome of doing business. Sometimes, money is a very important ingredient for a business. If there is a major accident or a problem, money is the oxygen a business would need.

Having said that, we must drive growth in revenue and profit. Profit comes from reducing waste and revenue growth is achieved through new products or services. We have identified the following five tenets that every business should follow to optimize a business for its purpose and sustain profitable growth:

1. Stick to the fundamental strategy of any business: sustaining profitable growth. In other words, seek revenue growth first, and then deliver flawlessly. Sales mean business, and continuing business means sales growth.

2. Implement a sensible business scorecard that can prioritize strategies, and identify opportunities to reduce waste for higher profit margins, or innovate solutions for high-margin growth.

3. Strive for perfection in everything in the business. There should not be any hesitation in becoming very good or best-in-class. If you don't, somebody out there is taking business away from you. This is necessary to remain profitable.

4. Innovate continually for new sales opportunities. Maintain a portfolio of new products in development for immediate, short-term, and long-term and future. Plan to sustain revenue growth. Optimize new products for affordability and profit margin. Avoid substandard products.

5. Sustain profitable growth by creating a culture of process thinking and management system. Design business processes to achieve success targets.

Strategy Execution

Strategy execution is like going on a vacation. If the vacation is not planned well in detail, it can become more work than the work we are getting away from. So, we must first define a destination, a target, then a roadmap or an action plan; resources, fuel or things necessary to carry on vacation; and finally, people for whom the vacation is planned. Similarly, it is for the people—creating growth opportunities for everyone associated with an organization for which that strategy is designed.

If an organization tries to operate at peak performance, it is bound to crash and burn. Imagine running as fast as you can—let's say 10 miles an hour. How long and for what distance can you sustain that pace? Certainly not forever! Similarly, each organization must set an optimum level of operating conditions that yield the best, sustained results. Stop-and-go driving is no better than driving intelligently at a normal speed. Thus, for operating a business in normal operating conditions, one must understand and optimize the business' capabilities. Good execution requires leadership, management, processes, knowledge, measures, information, systems or rules, disciplined work, people, and continual communication.

Understanding the need for excellence in execution is essential to making innovation profitable. The strategy for execution begins with a good set of measures correlated to the fundamental strategy. An intelligent business scorecard identifies operational opportunities for improving profit and achieving growth. Opportunities for profit improvement are exploited using tools like Six Sigma and Lean principles. The increase in revenue growth is achieved through institutionalizing innovation principles using a phased deployment or a maturity model. The following describes the steps for executing an innovation strategy in an organization:

1. Establish a fundamental strategy to achieve sustained profitable growth (SPG).

2. Perform a gap assessment and identify areas for improvement.

3. Establish or emphasize vision, beliefs, goals, and mission.

4. Construct a business process flowchart and identify critical processes and their measures of effectiveness.

5. Build a business scorecard linking strategies, products, processes, operational targets, and resources.

6. Align organizational resources and establish responsibility to accomplish innovation objectives.

7. Measure performance, summarize data, display performance, and communicate expectations continually to all employees.

8. Question old management principles and tools for their suitability as influenced by the Internet, technology, and smart devices. Devise or utilize new management tools relevant to employees in the 21st century.

Developing the Innovation Strategy

The main purpose of innovation in business is profitable revenue growth through new products and services. Innovation can also be used for improving processes, or designing new processes. Primarily when leadership is considering creating a culture of innovation or institutionalizing innovation, the intent is to sustain profitable growth. The question is how much growth a company should plan for to determine how much innovation is necessary. Target market segments and revenue growth opportunities should be identified using

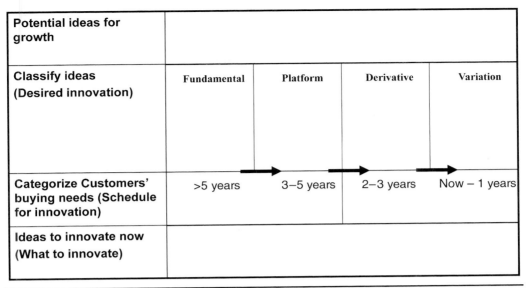

Potential ideas for growth				
Classify ideas (Desired innovation)	Fundamental	Platform	Derivative	Variation
Categorize Customers' buying needs (Schedule for innovation)	>5 years	3–5 years	2–3 years	Now – 1 years
Ideas to innovate now (What to innovate)				

FIGURE 1-1 Planning for a portfolio of innovations.

market research, focus groups, ethnography, or even co-locating with customers. These opportunities can be classified based on the time-to-market into four innovation strategies: variation, derivative, platform, and fundamental. Figure 1-1 provides a form to plan for a portfolio of innovations. The variation innovations are planned for near-term growth, derivative innovation for short-term growth, platform innovations for long-term growth, and fundamental innovations for the longer-term future. Variation and derivative innovation could be considered sustaining types of innovations, and platform and fundamental innovation offer more opportunities for becoming disruptive innovations.

Create a portfolio of innovations to meet short- and long-term growth requirements and allocate resources to all types of innovations. Some cost-conscious large corporations are trimming their R&D budget by focusing on innovations for the 3-year horizon. Foregoing fundamental innovations can mean sacrificing the future for large corporations. These organizations may choose an open innovation, mergers, or acquisition path to supplement disbanded in-house innovations. The extent of the research and development budget allocated to variation, derivative, platform, and fundamental innovations could be a ratio similar to 30 percent, 20 percent, 30 percent, and 20 percent, respectively. Another consideration to reduce risk with investment in innovation and R&D is to create a next-level breakdown of sources of innovations. Some innovations can be organic, developed in house, while others could be outsourced, open innovations, or crowd sourced. Figure 1-2 shows a breakdown by

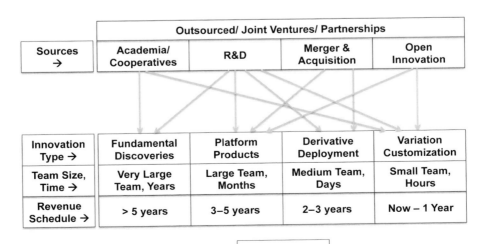

FIGURE 1-2 Sources of innovations for new products.

typical sources of innovation for new products. An organization can create its unique strategy combining types of innovation and its sources to create profitable innovations for market leadership in its industry.

Reducing Risks of Investment in Innovation

Considering the low success rate of new products, start-ups, venture-capital funded projects, and mergers and acquisitions there is tremendous risk associated with innovation. No wonder CEOs do not trust their innovation process because of associated risks with the investment, or uncertainties with the outcome. Reducing the risk of the innovation process requires a clear understanding of inputs, activities and outputs, and resources management. Below are some of the reasons that innovations fail:

1. Wrong products, nobody wants it. Ill-defined market input and customer insights
2. Not ready in time to launch during the market window
3. Marginal performance due to suboptimal design
4. "Me too" or minimal innovation, similar to its predecessor
5. Difficult to reproduce or problems in manufacturing
6. Too expensive to build, can't make money yet
7. Difficult for customers to buy
8. Customers do not like using it
9. Just can't make it work

There are many more that go wrong with new products. Many times the product does what the customer needs but it does not excite the customer. Capturing target user needs, wants, desires, and wishes is an important aspect of exciting the customer. The extent of innovation must be such that the customer must enjoy the innovative solution. Customers must be surprised and wowed repeatedly so they are inclined to share their experience with their friends and become loyal users. If users enjoy the innovative product or service, they will help it grow faster. If users dislike a product, they won't.

Marketing Innovations

In many companies, the marketing manager leads the innovation process. In the 21st century, the marketing process has changed dramatically. No longer does the marketing depend solely on the Marketing department; now "everyone" can become a marketing professional due to the advent of social media. Now anyone can reach out and introduce innovations to millions of customers almost instantly using social media engines such as Facebook and LinkedIn. Each major country has its own set of social media platforms.

Unfortunately, this approach can make a marketing manager cringe, especially when this communication methodology is given to a 20-something-year-old social media specialist with little experience in messaging strategy. As an example of marketing strategy and social media, Apple CEO Steve Jobs was notorious about controlling the marketing massage at his company, and Apple is generally seen as the poster child for innovation in the late 1990s until his death on October 5, 2011. In the Walter Isaacson memoir of Steve Jobs,[1] he outlines in great detail the control Jobs exercised over the timing, marketing, and message of Apple products, especially when it was a new product.

Regardless, this control seemed to be more of an exception than a rule with most companies. Although Jobs was allowed to get away with this kind of marketing control, the average company must consider how they market their innovations well before the innovation is released. This includes all aspects of the lead up to launch as well as the post launch. This also means that messages may need to change at a moment's notice. If a launch is not going as planned, social media may need to be coordinated so that the message can be adjusted. Additionally, in the modern world, messaging may come from many partners in the value chain. For example, a smartphone manufacturer with third-party developers constantly has issues with the release of the developer's kit. Release the kit too early and you run the risk of future upgrades being "leaked," but release too late and you run the risk that there will be no applications available for the product.

This now brings to light the involvement of the entire company in the marketing message and strategy. Engineers, production personnel, executives, and others must understand the marketing strategy. Without an understanding of what the company is trying to accomplish, a simple tweet or e-mail to a friend can become a leak of proprietary and competitive information. Unlike the old days, when the marketing department could control the entire message, marketing is becoming more of a company competency than ever before.

Launching the Innovation Initiative

Practicing innovation is an explicitly strategic initiative for some companies, while for others this practice is an implicit one. In order to launch the innovation initiative, the map of innovation must be understood and adopted. Innovation starts at the top, where the leader commits to sustained profitable growth rather than just "making money." Growth-driven leaders are interested in innovation. Sustained growth can be realized if a "growth office" is appointed and headed by a leader committed to ensuring continual growth through innovation.

Having assigned a clear responsibility for innovation, a good understanding of the innovation process helps in making progress. The four components of a good innovation process are resources, knowledge, play, and imagination. Without resources, innovation is not possible. Innovation does require investment of financial and intellectual resources. Once the resources are deployed, ideas for innovation are managed and commercialized as necessary to achieve a significant return on investment. The process continues from commercialization back to concepts to sustain and accelerate innovation.

Resources for Innovation

Resources to support the commitment to innovation are a critical first action in order to execute the strategy well. Resources can either be financial or human capital. Proper utilization of human capital offers many opportunities for accelerating the innovation of new products and services. Of course, human capital is an outcome of the financial investment in employees. In order to grow human capital, a corporation must invest in employees through training, exposure, and experience in various aspects of the business.

Human capital consisting of intellectual capability is often one of the least utilized resources. Since so much opportunity exists, any improvement in its usage will have an amazing impact on the outcome. If the human brain's average utilization doubled from about 5 percent to 10 percent, the world would be a very different place. In such a scenario, a continual stream of new products or services would result. Innovation on demand and mass customization for customers would become the new status quo. When a switch is made from primarily utilizing mechanical resources to primarily utilizing intellectual resources, then nothing is standard anymore; that is the impact of such a change.

Financial capital still plays a role in providing support to the innovation strategy. Organizations must invest in facility and policy upgrades to create an ambiance to spur creativity—an innovation room or physical space for experiments, and enough time to reflect and dig deeper into available mental resources. Organizations like to hire best-in-class people in all functions and keep them busy running around. People do not feel comfortable sitting in their offices thinking because someone, including a supervisor, may think they are not doing any work. Managers love to see highly intelligent people run around, but they rarely allow them to use their brains.

One of the investments in innovation must be similar to 3M's allowance of 15 percent of time for thinking, creativity, or learning without any accountability. Setting up an innovation room that provides facilities and resources for accelerated learning, such as a library with research material, knowledge management software, a laboratory for experimentation, resources for benchmarking, and a solitary space for deep thinking without any distraction, is an additional investment. More resources may be required for acquiring related new technology or tools to stretch organizational capability into new domains.

Organization Structure

Interestingly, every organization has a CEO, COO, and CFO, or their equivalents. Some organizations have a CTO, the Chief Technology Officer, for leading new product development. Otherwise, most of the corporate attention goes to managing shareholders' perceptions of the organization as well as operational performance through cost reduction by pressuring suppliers and counting beans. Too much focus is on profit and not enough focus is on growth. Money is spent on cutting costs—sometimes more than what is spent on developing new products and services. Employees' jobs in the longer term are dependent on future opportunities, while still maintaining excellence in the portfolio.

In a typical R&D function, most of the resources are spent on development and little on research. As a result, product performance is marginal and manufacturability is highly questionable. We quickly release new designs from the Design Engineering department to the Product Development department and wave good-bye. The Manufacturing department pays for the questionable designs due to lack of research and thoroughness in the design phase.

One of the reasons for marginal performance in manufacturing is attributed to the lack of defined targets for various parameters. Excellence in manufacturing is impossible without specifying targets for these parameters. In the absence of targets, the product is built to specification limits, and thus the product is produced with marginal performance.

Thus, organizations must be realigned for innovation and excellence. Without excellence, innovations will be wasted in marginal environments for performance. With excellence in operations, innovation can be accelerated because of faster and precise feedback from experiments. To create a structure for launching and sustaining the innovation initiative, organizations must have an innovation office or leader who can clearly bring economic sense to the design function and establish goals and processes to maximize use of human capital.

All innovation begins with an idea in someone's mind; thus excellence in idea management is a first critical step to building an innovative organization. The management process must be standardized, where every employee is expected to contribute ideas to create value at the activity, process, or product level. Starting with a goal of one idea per quarter per employee is a reasonable goal upon which to build. Eventually, a continual flow of employee ideas toward development of new products or services in response to demand from customers or the marketplace will likely become the status quo.

Communicating the Innovation Message

Another important aspect in effectively executing the innovation strategy is the story. Leadership must develop a coherent message demonstrating the need for continual innovation, the benefits of innovation, and the consequences for not innovating enough to keep up with customer demand. Consistency and constancy of the message are critical in generating employee interest across the organization, minimizing conflicts, and aligning organizational resources toward the common goal. Typically, the story may include a finding from the benchmarking study identifying opportunities to gain market share and improve, or even create a message to achieve fundamental business objectives.

In many companies, the innovation message is subtle through actions of certain individuals in executive positions. For example, the late Steve Jobs was known to drive and demand innovation at Apple.[1] At Google the work environment, also known as Googleplex, is committed to innovation and improving the "great" performance that promotes innovation at Google. Googleplex presents a unique lobby and hallway décor, office clusters, recreation, a café, and a snack room to promote free thinking and, combined with the goal, to develop innovative search-related products that promote continual innovation. The famous "15 percent investment in unaccountable time for employees to use" is a great way to communicate commitment to innovation.

Incentives and Controls

On a personal level, incentives have played a significant role in Mr. Gupta's desire to do things differently over the course of his career. The Silver Quill awards at Motorola given for publishing articles challenged him to write his first paper. Besides getting $100 per published page, the company also encouraged more writing by making the reward continual. In addition, filing any application for a patent at Motorola resulted in at least $300 in the early-to-mid-1980s, and the reward amount continued to grow as the application moved further down the patent filing process. In both cases, the reward was given for starting to do things differently.

Once the flavor of doing things differently is tasted, it becomes difficult to stop. Understanding the innovation process in the brain shows that a personal incentive for learning can be much more effective than simply the incentive of achieving a more innovative outcome. Without learning multiple subjects, accelerating innovation is difficult. Thus, some incentives for learning must be in place. Unfortunately, in tough economic times, learning incentives are the first to go at many companies, which only highlights operational problems leading to corporate troubles. Can anyone imagine turning around or growing a company without having employees learn and innovate in place? This scenario certainly sounds like a recipe for winding down business through continual cost-cutting and negative revenue growth.

Culture and Change

Culture and change are two nebulous aspects of a business. Culture is about how we interact with each other and how we make decisions. On a daily basis, these two acts depend on corporate values. Thus, defining the corporate values that stand up under dire circumstances is imperative. A methodology can become a strategy, a strategy can become the corporate values, and new values can become the culture. Thus, once the innovation strategy begins to be implemented, based on its success and widespread institutionalization, it can become a corporate value of doing things differently. Once the value is accepted and practiced by everyone, all employees simply do things differently. Thus, the innovation strategy can become part of the corporate culture.

One of the major topics within the realm of corporate leadership is "change." In one of his live seminars, Tom Peters asked a question, "How long does it take one person to change his mind?" He talked about managing resistance to change, how leadership gradually phases in new practices, employee resistance, and implementation that is fragmented due to resistance. Then, he answered his own question. He said the same people who resist at work actually change at home all the time when needed. When people are doing things in a certain way, and then asked to change their behavior, resistance will occur because of the unanswered question, "Why should one change?"

People change in no time if they know why they should adapt new practices, and how they would benefit from such changes. To accept change, employees must recognize the benefits of innovation through incentives or recognition. In any case, Tom Peters' answer for time needed to change a mind is "a moment." Once the decision is made, change is made in the mind and followed by the practice.

Figure 1-3, the Innovative Thinking Matrix, shows various aspects of an organization that impact the collective mind of an organization once the decision is made to institutionalize innovation. As reported earlier, an individual can change his/her mind any time; however, as a group of employees, different people choose different moments to make up their minds. If the leadership communicates the strategy, employee roles, expected practices, and desired outcome, more employees will make the decision to accept the innovation initiative faster. Then, leadership must walk the talk and encourage innovation practices while making decisions.

Business Aspects	Conventional Thinking	Innovative Thinking
Purpose of business	Make money	Create value, and make money
Customer Demand	Satisfy	See as a larger opportunity
Leadership	Manage for quarterly profits	Lead to build a business
Decision Making	React to fix	Respond to solve systemically
Goal Setting	Easy to achieve in short term goals	Challenging long term goals
Market Analysis	Limited external knowledge	Extensive benchmarking
Direction	Random and personal	Driven by vision and values
Profitable Growth	Profit or growth	Optimized profit and growth
Organizational Values	Competitive and negative	Collaborative and positive
Employee Learning	Hire and stale skilled employees	Build and renew employee skills
Innovation	Flash of a genius	Learned skill
Improvement	Incremental	Aggressive
Method of Innovation	Brain Storming	Well defined process
Innovators	Selected few	Everyone
Resources for Innovation	Allocated sporadically	Invested continually
Building Block of Innovation	Clusters of people	Networked Individual

FIGURE 1-3 Innovative thinking.

The aspects, as listed in Fig. 1-3, range from purpose of the business, customer demand, decision-making process, organizational values, employee learning, innovation method, and resources. The matrix highlights elements of the innovation process at the methodology and its context levels. For each aspect, examine the conventional thinking and the innovative thinking, and plan to move toward innovative thinking through policy, procedures, and practice.

For example, the main purpose of business is to make money. One can make money in many ways—legal or illegal. Thus, the purpose of a business must be clearly stated. One good definition of this purpose is "to provide value to customers by doing the right things efficiently and making money." Customers pay for value. A business cannot make money unless the customer pays. The market capitalization of the business depends on the long-term execution of the strategy. Any short-term manipulation of the stock through some short-term intent can at best be considered a gimmick. Such manipulation is not the purpose of a business.

One of the leadership decisions that drives innovation in an organization is to set aggressive goals for improvement and continual change or renewal. The definition of "aggressive" can be understood as a sufficient amount or rate of change that forces one to think and do differently. For example, if one decides to make 10 percent more money at the personal level, one would think of certain incremental activities, either work hours or a bonus, caused the 10 percent gain. However, if one decides to earn 50 percent more than the previous years, one starts thinking seriously and asks questions like "What else can I do?" The level of change that forces us to "do differently" beyond our comfort level is labeled aggressive. Aggressive is not the same as "pie in the sky." Aggressive means stretching current capability, resources, and thinking. Figure 1-3 can be used to assess corporate culture for the required changes in an organization.

Identifying Gaps

Launching the innovative initiative begins with an understanding of current practices. The objective is to identify strengths and weaknesses, build on strengths and build in the areas of weaknesses. This step leads to an action plan that can enable an organization to make progress. Many organizations already have similar diagnostics, matrices, or assessment tools. However, their adequacy is sometimes questioned because of the lack of a framework for the innovation process.

Once there is an understanding of the innovation process, assessing and establishing a baseline for an organization for elements of the innovation process is easier. Thus the assessment includes questions about strategy, leadership, process inputs, process activities, process outputs, and measures of innovation. At the early stage, one needs to highlight critical areas for change in order to realize an innovation-friendly organization.

To evaluate each aspect of an organization, one can simply assign a percent score based on the applicable approach, deployment, and results (see Fig. 1-4). For example, in assessing strategic commitment for growth through innovation, one can look for clearly stated and documented objectives, its institutionalization through tactics and processes, and outcomes in terms of continual leadership interest in growth through incomes. Considering these three elements with equal significance for evaluating each statement, one can assign a percent score. As for grading guidelines, one can consider a score of 0–20 as ad hoc, 21–40 as marginal, 41–60 as practiced sporadically, 61–80 as standardizing the practice, and 81–100 as achieving desired results. While assessing the organization's performance, one needs not to split hairs about the absolute score. The objective is rather relative significance in order to initiate some actions to start making progress. For benchmarking purposes, the overall average can be calculated for assessing future progress.

For many years, leaders wanted to have an initiative, such as "Make Money and Have Fun." Once the initiative was implemented, measuring how much money was made was easier, but measuring how much fun employees had while working in the organization was impossible. While creating a process for developing innovative ideas faster, the author realized that as people think in terms of good, crazy, stupid, and funny ideas, first it took longer, and then ideas became more innovative. Most importantly, a measure of having fun evolved from the process.

Thus, one way to know when employees are having fun while working at a corporation is to measure how many funny ideas are coming from employees, or how freely employees can present funny ideas without fear of retribution or ridicule. When employees are having fun, they can pretty much say whatever they want to improve the company performance; no idea is discouraged. Being an innovative organization, we need all the funny ideas employees can come up with to improve or develop products or services.

Item#	Aspects of Innovation	Score (%)
1	A strategic commitment has been made to drive growth through innovation.	
2	An executive has been assigned full time to lead innovation.	
3	A strategy has been executed to accelerate innovation.	
4	Sufficient resources have been committed to support innovation activities.	
5	Departmental goals have been established to develop innovative solutions at the process level.	
6	Leadership has established a prestigious award for an innovative solution that creates exceptional value.	
7	Leadership understands the innovation process, and actively promotes risk-taking and doing things differently.	
8	A process has been established to achieve excellence in managing employee ideas.	
9	All employees have been given access to the Internet for conducting research in real time.	
10	Employees are encouraged to rotate among various departments.	
11	Company has in-house library of industry and related books and journals, and has access to on-line research services.	
12	Continual learning is rewarded at all levels, and time is allowed for learning.	
13	There is a facility for employees to brainstorm, play or experiment to test their ideas.	
14	Employees are encouraged to 'think' for new ideas for improving processes, products and services.	
15	Measures related to CEO recognition, employee ideas, and revenue from new offerings have been established.	
16	Employees are free to give funny ideas, and are not afraid of failures.	
	Average =	
Legend	0 – 20 = Ad Hoc; 21 - 40 = Marginal; 41 - 60 = Practiced; 61 – 80 = Standardized; 81 – 100 = Proven	

FIGURE 1-4   Diagnostics of innovation.

Innovative Leadership

The success of a new strategy, without questioning its formulation, depends on how passionately the leader champions for its success. Innovation has been used either as a corporate "value" or a strategy to facilitate turnaround. In either case, the CEO or executive must believe in its intended outcome, successfully drive the organization by providing direction, resources, and support, and continually engaging employees through timely feedback and follow-up.

In many organizations, the leader focuses on profit and initiates cost-cutting measures, which may be necessary in the short term for a struggling company; however, in doing so the leader acts counter to innovative thinking. Success begins with a thought in the leader's mind and is achieved through leadership traits. Figure 1-5 organizes various leadership traits according to the process of innovation and lists corresponding approaches of an innovative leader. People do what their leaders do—not what they ask for. Successful leaders demonstrate these behaviors and thus set an example for others to follow.

Leadership Traits	Innovative Leader
Learning	Reads a lot about a variety of subjects; interacts with community groups, employees, customers, and suppliers
Listening	Listens well to all ideas for noise, and noise for ideas
Personal style	Takes risks and executes tasks well
Interaction with employees	Encourages doing things differently better
Interaction with customers	Listens to their needs, and accepts challenges
Interaction with suppliers	Demands partnership for innovative solutions
Interaction with shareholders	Seeks support for longer term performance
Giving feedback	Rewards successes, understands failures, and encourages experiments
Behavior	Presents himself as positively enthusiastic, energetic, and an exemplary person

FIGURE 1-5   Innovative leadership.

Making an Innovation Strategy Work

Lawrence G. Hrebiniak, in his book *Making Strategy Work*,[2] provides a template for leading effective execution and change. According to Hrebiniak, and the methodology used for Six Sigma projects, one might consider the following tactics for successful execution of the innovation strategy:

- Define a clear charter with cost and benefit analysis
- Identify stakeholders and utilize their influence
- Align organizational structure
- Develop a roadmap with clearly defined accountability
- Coordinate tasks and frequently share information
- Support and reinforce execution
- Manage change and culture
- Establish a process for sustaining innovation
- Reward success and inspire excellence
- Learn and adjust the strategy

A lot has been written about strategy execution; however, success depends on this ultimate factor: the desire of the leadership to make the strategy work. If a leader is committed to making innovation become an integral part of doing business, it will happen. Otherwise innovation will not happen.

Leading Innovation with a Sense of Urgency

Surveys consistently show that innovation is one of the top priorities of business executives except when it comes to allocating resources. A recent IBM Global CEO Study[3] shows that creativity in leadership has become a critical requirement for success in today's economy. In an interview published in *International Journal of Innovation Science*, Chris Galvin,[4] former Chairman and CEO of Motorola, redefined leadership as taking followers to a place never visited. Leaders must be creative—otherwise how would they create new growth opportunities!

Due to intense global competition from rising economies and reduced merger activity, the need for organic innovation has skyrocketed. The challenge then is to determine how most corporations can pursue innovation with a sense of urgency to achieve their business objectives. It requires a different perspective of business. The purpose of a business must be to provide value to its customers and to society. If a business serves customers well it will make money and profit. Thus the leadership must focus on serving more customers well, leading the business to profitable growth. Leadership is required for revenue growth and good management is necessary for profit. To create a sense of urgency for innovation, corporate leadership must commit to sustained profitable growth. The absence of such an approach indicates too much dependence on existing product mix, and/or the complacency caused by the absence of crisis at the leadership level, a poor understanding of the innovation process, ineffective corporate performance (including measures of innovation), or too much talk about innovation as an annual program.

In tough economic times, corporate leadership must also go back to the basic idea that community and business need each other. In order for business to succeed, it needs community support for employment and infrastructure at the least, and community needs business support for employment opportunities and new solutions. Sustained profitable growth creates a better community and business relationship that sours when a business purely focuses on profit without any focus on revenue growth.

References

1. Isaacson, W. *Steve Jobs*. Simon and Schuster, 2011.
2. Hrebiniak, L. *Making Strategy Work: Leading Effective Execution and Change*. Pearson Prentice Hall, 2005.
3. IBM Institute for Business Value. Connected Generation, 2012.
4. Gupta, P. "A CEO's Perspective—Making Innovation Work." *International Journal of Innovation Science*, 2 (1), 2010.

CHAPTER 2

Creating Your Innovation Blueprint: Assessing Current Capabilities and Building a Roadmap to the Future

Lisa Friedman and Herman Gyr

"All organizations are perfectly designed to get the results they get."

—Arthur Jones[1]

Today's business challenges demand that companies innovate at exponential speed while facing global competition from established competitors and start-ups alike. It's a new world: global, mobile, social, green, cost-pressured, and time sensitive. Everything is in flux at once—technologies, regulations, customers, competitors, partners, customer demands, and what the web has to say about it all. In this chaotic business climate, what can give innovators the edge they need to succeed?

By definition, innovation is risky—if we already knew what worked, it would not be innovation. Enterprises can increase their chance of success by having a clear and shared innovation strategy for their changing business environment, and by ensuring that key business and organization elements are all aligned to enhance innovation capability.

Back in the 1990s, Peter Drucker wrote that "All organizations—not just businesses—need one core competence: innovation."[2] Many years later, most organizations are working toward this goal, and for many, this is still an evolving work in progress.

Leaders now have many solutions and tools available, as you will see in the chapters following in this book. However, one key ingredient is often missing: an integrated overview of how these solutions fit together into a coherent roadmap that drives innovation forward.

Too often, people are encouraged to innovate while many business and organizational factors are lined up against them. Innovators get a mixed message. Leaders' words often tell them one thing, but the ongoing organization structure and business practices tell them something else. Their company may even set up an innovation space, create skunkworks teams, hold an innovation competition, or install innovation software; but underneath, the core way of running the company stays the same. Innovation becomes an add-on and is not built into the core DNA of the project team or company or network. Innovators can be clear where they're headed, but if they encounter obstacle after obstacle in their path, their progress will be slowed just at the time when speed is important.

Innovators' chance of success is dramatically increased when they have a coherent picture in their minds about where their industry is going and a shared view of what they want to create and why. Next, they are most likely to succeed when all the key business and organizational factors that influence their day-to-day work are aligned to support their innovation success. It's a tough global marketplace and innovators need every advantage they can get.

Two Key Innovation Challenges

In 1998, we published a book called *The Dynamic Enterprise: Tools for Turning Chaos into Strategy and Strategy into Action.*[3] These two challenges—turning chaos into strategy and strategy into action—are two of the most important tasks for innovators as well.

Turning Chaos into Shared Strategy—Seeing the Future

Industry after industry is in the midst of revolution as their products, services, and even business models are continually being reinvented. In this rapidly changing world, one of the most important guides to effective innovation is to clarify a compelling destination—to ensure you innovate the right things, to set your course toward a worthwhile future.

Setting your innovation destination requires two important skills. The ability to

1. Understand the emerging trends in your industry, to see where your industry is headed, regardless of whether you go there or not.

2. Create a roadmap to the future that all your people can share. What do you want to create in this emerging future? Who do you want to become?

If you say "full speed ahead!" to both a missionary and a warrior, they are going to go in different directions. The key to creating a clear and shared innovation strategy is for key stakeholders to see where the future is headed and to understand what they want to create, together, in this future.

Strategy into Action—Aligning the Key Drivers of Innovation

Turning this shared strategy into action also requires two important skills. The ability to

1. Identify the key elements needed to deliver innovation as a core competence in a team or throughout an enterprise.

2. Coordinate and align these elements to enable people to work together and to move in the same direction.

One risk is that organizations or teams implement one or more solutions or tools to enhance innovation, while ignoring other key elements. Other parts of the organization may even push strongly in the opposite direction. Sometimes the pushes in other directions are so strong that the intended innovation has no chance to succeed. A piecemeal approach to innovation can occasionally work because other elements of the organization are already naturally aligned to drive innovation forward. However, this is not always the case. The risks of failure are quite high when an organization tries to install a few new innovation practices without embedding these in an overall integrated commitment to innovation.

For a wonderful example of alignment, think of a retail department store at Christmas. Every element of the store is designed for a common purpose—to enhance holiday shopping. From the moment a customer walks through the door of the department store, it feels, sounds, looks, and sometimes even smells like Christmas. Some stores blast a burst of warm air near the entrances to give shoppers an instant sense of warmth. Holiday music plays in the background and there may be scented candles burning near the entrances to give off smells of cinnamon and hot cider, or the scent of pine needles of a Christmas tree. Holiday colors and decoration are everywhere. Special holiday products are displayed prominently. Extra staff is often hired and employees receive special coaching to enhance their holiday spirit. Even receipts may have holiday messages printed on them. Every element you can imagine has been designed to communicate that this is a time to celebrate—and to give presents. It's an annual exercise in alignment and shared purpose.

How can a team or organization use a similar kind of focus and commitment to ensure they can deliver innovation? The good news is that there are straightforward and common-sense ways to identify the key elements needed to drive innovation forward, and simple ways to check if these elements are aligned.

This chapter will describe an integrated system that identifies the key elements of a successful innovation strategy and show how they fit together. In fact, your innovation blueprint can bring these elements together to create one overall visual map to your intended

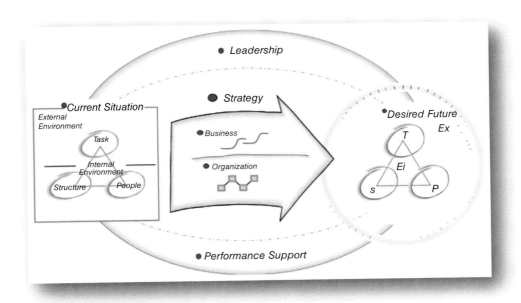

FIGURE **2-1**   Innovation blueprint.

future. Having an integrated overview also enables you to assess your current capabilities, to see which areas need the most attention to ensure you succeed.

This chapter should enable you to determine which of the following chapters in the book will be most essential reading for you personally and for your team, organization, or innovation community.

Benefits of an Innovation Blueprint

An innovation blueprint (see Fig. 2-1) can help people to be literally "on the same page." People working together can see a common vision of where their enterprise is headed and why. They can understand the kinds of innovation needed most. A blueprint can also help to ensure the organization as a whole is truly designed to deliver innovation rather than having only one group here or there trying for this goal. The blueprint can help to create a spirit and capability for "innovation everywhere."

An enterprise where all elements are aligned to enable innovation will be able to move more quickly to innovate at all levels—for small but urgently needed repairs, for ongoing continuous improvements, and for major business transformations.

An innovation blueprint enables

- Shared innovation strategy—people throughout a team have a shared view of where they are going and why, a common destination, and a shared understanding about where they most need to innovate.

- Organization alignment for innovation—ensuring elements throughout the enterprise are designed to support innovation, that there is a logical "innovation architecture" in place.

- Clear communication—everyone can literally "see" where the company is headed and what is required to get there.

- Monitoring and assessing progress—it becomes easy to assess current strengths and challenges, and to understand what still needs to be put in place to enable innovation.

Key Elements of an Innovation Blueprint

An *enterprise* is a business supported by an organization. Innovation in an enterprise can occur either in its business (e.g., in its connection to its marketplace or its products and services), or in its organization, (e.g., in its organization design, systems, and structures; people skills and working relationships; or in its internal culture and climate). Form (organization) almost always follows function (business). When business needs change or new opportunities open up in the marketplace, the organization needs to adapt to ensure it can enable the business to succeed under the changing conditions.

On the blueprint in Fig. 2-1, the top half of the map shows the business and the lower half shows the organization.

Business

The key business elements are the external environment and the core task of the business.

External Environment

An enterprise exists within a given external environment of multiple forces: customers; competitors; strategic partners; cultural, economic, and social trends; regulations; science and technology, resources, and new opportunities. At times, the external environment will be in flux, where industries as a whole are being redefined.

These large-scale business transformations belong in the environment external to the enterprise, because in a global economy, these shifts generally occur whether a particular business goes along with them or not. In the innovation mapping process we typically begin with assessing the external environment, as it is important to understand where your industry and customers are headed before clarifying the kinds of innovation you want your business to deliver.

Task

Task clarifies the specific value the business delivers to customers. The core task of the business is to create and deliver products or services that meet the customer demands and market opportunities present in the external environment. Task can also include the business model—how revenue is generated, and the brand—the way these products and services are positioned in the marketplace, what the business stands for. For example, while two companies might both offer household cleaning products, one might offer products sold in large volumes with basic packaging, designed for price-conscious customers. Another company might offer products with nontoxic ingredients sold in compostable containers, designed for customers concerned about health and the environment. Both companies sell similar products but their businesses might differ in significant ways.

These strategic business drivers then filter down to the more operational and tactical elements of task that describe how the work is accomplished; for example, work processes/work flow, quality standards, department and team goals, and core competencies needed.

All of these task elements together should provide a picture of the advantage this particular business has over its competitors in the external environment, and how its solutions deliver more value to its customers than competing solutions can. This part of the blueprint can also reveal areas of risk and opportunity, to show where business innovation would make the greatest impact.

Organization

The organization includes the elements that can be arranged to help deliver business success: the organization's structure, its people, and its internal environment or culture.

Structure

We think of the structure as the *riverbanks* that determine how the work flows, and how people accomplish their tasks. Elements of structure can include the organization chart (divisions, job descriptions, and reporting relationships), roles and responsibilities, locations and facilities, finance systems, information systems, communication systems, and human resource (HR) policies and practices—especially incentives and rewards. Even meeting norms can be engineered to influence how people work together.

People

This category includes the kinds of people needed and how they get along, how they work together. Elements related to people can include skills and talents that are needed, demographics, quality of the working relationships, teamwork, and the quality of communication and collaboration.

Internal Environment

This category includes how it feels to work inside the enterprise or the organization's culture. It can include vision and values, sense of identity, emotional climate, and levels of trust, spirit, or energy (or lack of these).

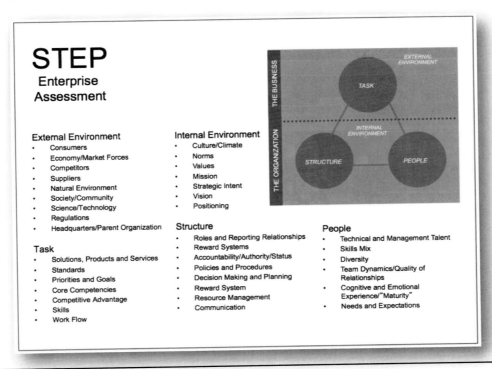

STEP
Enterprise Assessment

External Environment
- Consumers
- Economy/Market Forces
- Competitors
- Suppliers
- Natural Environment
- Society/Community
- Science/Technology
- Regulations
- Headquarters/Parent Organization

Task
- Solutions, Products and Services
- Standards
- Priorities and Goals
- Core Competencies
- Competitive Advantage
- Skills
- Work Flow

Internal Environment
- Culture/Climate
- Norms
- Values
- Mission
- Strategic Intent
- Vision
- Positioning

Structure
- Roles and Reporting Relationships
- Reward Systems
- Accountability/Authority/Status
- Policies and Procedures
- Decision Making and Planning
- Reward System
- Resource Management
- Communication

People
- Technical and Management Talent
- Skills Mix
- Diversity
- Team Dynamics/Quality of Relationships
- Cognitive and Emotional Experience/"Maturity"
- Needs and Expectations

FIGURE 2-2   STEP elements of the enterprise.

STEP

The acronym STEP—for Structure, Task, Environments (both internal and external), and People—helps to frame these elements as one interconnected system. See Fig. 2-2.

Alignment

An enterprise works best when all STEP elements are aligned. Whenever one element of STEP changes significantly, other elements typically need to shift as well to avoid conflicts and keep the enterprise running smoothly. As a brief example, imagine a business that was founded and grown over many years as part of the traditional Swiss watch industry. Many years back, its external environment was quite stable, in an industry where quality and tradition were highly valued. Customers wanted an elegant, refined, and reliable watch that they considered a fine piece of jewelry. In task, the key products were finely tuned, hand-crafted watches with a classic design, where styles changed little from year to year. Structure in the company was most often a traditional hierarchy that also changed very little over time. Jobs were stable and employees could expect predictable, routine work. The internal environment focused on high quality, elegance, attention to detail, and a timeless and comforting predictability.

Now consider how the watch industry began to change with the global emergence of inexpensive digital watches from Asia. Global market share for the classic watches described earlier dropped dramatically and jobs in the watch-making industry decreased to less than a third of what they had been.

This disruption gave rise to Swatch, a company that responded to these market changes by making inexpensive and colorful, yet still high-quality and Swiss-brand watches. This was a very different kind of enterprise. In the external environment, customers viewed their Swatch as a whimsical, affordable, fashion accessory, and they might buy multiple watches over time. In task, the product was entirely different from a traditional watch in nearly all respects: the materials, colors, production processes, marketing, and sales. A Swatch was an inexpensive and fun watch, where styles constantly changed. Special designs commemorated holidays or global sporting events. Given this entirely new business model, the organization's structure needed to change as well. The structure mirrored, and enabled, its more open and flexible business practices. More work was project-based, with jobs that might change from project to project. People in the organization needed to stay connected to current trends and open to outside influences and new ideas. People collaborated across the boundaries of the company to work with artists and well-known designers around the world. The internal environment was lively, emotional, quick-paced, curious, and open to change.

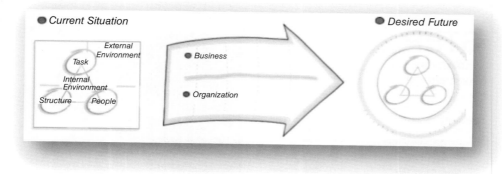

Figure 2-3   Current and future in the innovation blueprint.

This example shows how each element of STEP can be effectively realigned in response to disruptions in the external business environment. When market or customer demands shift and an enterprise changes its business in response, a very different kind of organization will typically be required as well.

Adding Current and Future to the Blueprint

Because today's global external business environment is changing so rapidly along so many fronts at once, the enterprise is dynamic; it's on the move. Various STEP elements of the enterprise are typically always under development, moving from the current state to the intended future. The innovation blueprint can be extremely helpful in capturing this movement. It shows both current STEP and future STEP (see Fig. 2-3). Having both the current innovation capability and the capabilities needed for the future in a single map helps to clarify the requirements for moving from one to the other.

Three Drivers of Innovation

The arches and the arrow in the blueprint (see Fig. 2-4) show the three groups of people who determine whether the enterprise truly moves from its current state to its intended future:

- Leaders—inspire, communicate, authorize, fund, and enable stakeholders to build the future. Leaders can be formal or informal, and can be internal or external; for example, sometimes community or government leaders, or leaders from strategic alliances, might be needed for success.

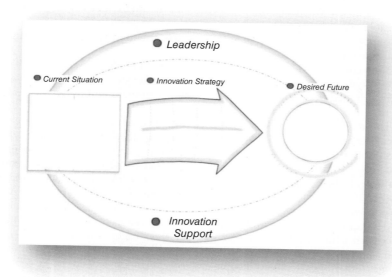

Figure 2-4   Leadership, stakeholders, and innovation support.

- Key stakeholders—employees, partners, suppliers, and sometimes even customers—who are the people who actually build the future.
- Support—the people or teams who coordinate changes; who bring in tools, training, and outside expertise as needed; and who generally keep complex changes on track.

While leaders educate, inspire, authorize, and fund stakeholders to build the future, people working on the innovation support team ensure that stakeholders are *capable* of building the future.

Using the Blueprint to Build the Innovation Capability of Your Enterprise

The blueprint is a useful tool to design key elements of the enterprise—both in the business and the organization—to enable innovation. The sections below will explore how the design of each element of STEP—connection to the external environment, business task, structure, people, and internal environment—all contribute to innovation capability.

Because the field of innovation itself is rapidly changing, these examples are meant to provide sample ideas and current best practices, and to provoke your thinking. Without a doubt, within a short period of time we will see a whole collection of new ways to enhance innovation. For example, many organizations now use online innovation platforms and mobile apps to capture and develop ideas across large groups of employees, partners, suppliers, and customers. However, the technology, design, multimedia capability, and ease of use of these platforms are improving by the month. Because innovation tools and best practices are themselves a target of innovation, we encourage you to use the examples below as thinking tools to provoke your own ideas about designing the key elements of your enterprise to promote innovation.

The innovation blueprint is a thinking tool for asking important questions, but it also can be drawn as an actual paper map, as a physical roadmap to the future (see Fig. 2-5).

An innovation blueprint can be of any size. Individuals and teams may choose to work with smaller maps (see Fig. 2-6). Digital versions can be drawn easily as well, and of course,

Figure 2-5   Example of an innovation blueprint wall map.

Figure 2-6   Individual and team blueprints.

given the nature of entrepreneurial ventures, even a blueprint drawn by hand on a paper napkin has been known to capture breakthrough ideas.

Planning for the Future and Assessing Current Capabilities

The sections below will give you ideas you can use to fill in the right-hand side of the innovation blueprint for the specific innovation enablers you want to build into your enterprise. A helpful early step is to decide how far into the future you want to look when you create your blueprint. Some work groups find that looking one year ahead seems quite far, while others set a three year-, five year, or even a longer "deep future" timeline. They understand that their enterprise will never look the way they predict, but thinking ahead helps them prepare for what arises in the actual future. We also have found that participants often place ideas into "the future" on their maps that actually already exist today—somewhere. These practices seem cutting edge at the moment and participants assume they will be more common in the future. A few groups invent a future no one has seen before and work to make it happen.

Each section below will also list questions you can use to assess your current innovation capabilities, to help identify the elements you need to develop most. These current capability questions are adapted from an online innovation assessment tool called Value STEP[4] that is used to gather input from large numbers of stakeholders throughout an enterprise. Very often, leaders and managers say these capabilities already exist while employees tell a very different story. Using a larger number of employees to assess current capabilities can also be helpful to assess where employees feel inspired, empowered, and enabled to innovate, and where they still need more help. Because innovation capability is rarely evenly distributed throughout an enterprise, an online survey can show areas that need the most attention or can help assess progress over time toward an innovation goal.

Innovation Blueprint: External Environment

External Environment—The True "Front End of Innovation"

Many discussions of the "front end of innovation" begin with idea generation or creativity. However, in a rapidly changing business environment, first understanding the emerging market and customer shifts that are taking place can be essential to ensure that the ideas generated are relevant, that they meet real needs. The risk of starting your innovation practices with the ideas themselves, without this earlier understanding of where markets and customers are heading, is that innovators generate ideas that are not really relevant for the times. They might generate ideas that are interesting, but that are not important in the world that is emerging.

It can be important to look for emerging trends in customers, competitors, suppliers, partners, technologies, government regulations that affect your enterprise, the economy, society and culture, and the natural environment. It can also be good to understand the general state of your industry and how you are positioned in it. For example, is your industry growing or shrinking? Is it being disrupted by new technology or social trends?

Staying Connected to Your Customers

In addition to looking out at the market in general, it is also important to understand customers at a deep level—what they need, the problems they would like to be solved, what is most important to them, and what they wish they had that you could provide better than anyone else. Innovation demands close customer knowledge, far beyond the overall demographics or statistics found in general marketing reports. In your innovation blueprint, include as much information as you can about customers' experience, what they value most, and how this is changing.

The External Environment Indicates the Kind of Innovation Needed

These market and customer questions can seem quite practical and commonsense, but in a rapidly changing business environment, the answers can be surprising and lead to unexpected conclusions about the kinds of innovation needed.

For example, consider a television network that found they were losing their younger audiences and knew they needed to be more innovative to entice them back. Leaders pushed their people to create more innovative programs that would appeal to young people, but their youth audience numbers kept dropping. As they took a deeper look into this situation, they realized that significant numbers of their younger audiences had not noticed the improved programming because they were no longer watching TV in their living rooms. Their customers had changed. Many members of their youth audience had moved to inter-active games, to spending time communicating with friends online, or to watching TV and shorter videos online—on whatever device they had on hand at the moment.

In addition, the broadcaster's competitors who offered online services to this audience were beginning to offer original TV programming as well. The situation was about to get worse. Companies that offered online games (e.g., Microsoft), subscriptions to rent movies and TV online (e.g., Netflix), online book and product sellers (e.g., Amazon), and online video services (e.g., YouTube) were all beginning to create their own original programs. These competitors had taken TV viewers' attention away from traditional living room televi-sion sets, and now that they had this attention, they were aiming to create high-quality programs of their own to offer through their online services and platforms.

In addition, every month, more mobile apps were available to enhance viewers' media experience by directing them to "second screen" services, which gave enhanced information about "first screen" shows. Some offered a chance to interact with other viewers while a show was still in progress.

These insights into changing customers, technologies, and competitors can lead to insights about the kinds of innovation that might be required for success. A TV broadcasting company that simply needs its people to improve the quality of traditional programs likely already has many of the innovation skills it needs. However, if the company needs to move into interactive online media to produce content in ways it has rarely done before, it will need to build business and organization practices that promote disruptive innovation. (See later sections on Task, Structure, People, and Internal Environment for specific ideas about how to design each element to promote disruptive innovation.)

Adding Industry Lifecycle to Your Blueprint

We are living in a unique moment in time, where many industries are undergoing major transformation, all at the same time. As our society becomes global, mobile, social, and has a growing emphasis on sustainability, industry after industry has to respond in distinctly new ways. These industries feel a common pressure to innovate, but clarifying the depth of the change in the external marketplace can help to clarify the depth of innovation being called for.

Lifecycle Curves

Sometimes so much improvement or change in a product or services occurs that it leads to a "state change." Think of warming ice that gets warmer and warmer, until it reaches a thresh-old where it melts and becomes water. It takes on a new form, with new qualities. If water continues to grow hotter and hotter, it reaches its boiling point, where it stops being water and turns into steam. The same kind of state change can happen with individual products and services, with businesses, with industries as a whole, and sometimes even with multiple industries transforming at the same time.

Consider the lifecycle curve of an individual product or service (see Fig. 2-7). The first curve dips down as the product requires initial investment before it "rounds the curve" and begins slow growth. If the product is successful in its marketplace, it enters a phase of rapid growth. Finally, over time, its growth may begin to slow, plateau, and even decline. While a product may disappear altogether, there is also the chance that it will be reinvented and transformed into a whole new form. Think of the horse and carriage giving way to cars. The core product was still transportation, but it was delivered within a whole new form, a "state change" into a whole new generation of transportation.

Figure 2-7   Lifecycle curves for a product, a business, an industry, or a shift in eras.[5]

Notice that the two curves overlap. In a transformational time, the mainstream majority is still on the traditional curve, while the second curve is growing quickly, signaling future growth.

This same kind of shift can take place in a business or even in an industry as a whole. Changes in technologies and products can first affect one company and then another, but as a new generation of products becomes successful, eventually all companies within the industry have to adopt the new innovation if they want to survive.

At unique moments, an even deeper shift occurs that affects all industries at once—a shift in eras. Think of the agricultural era giving way to the industrial era. Industry after industry was transformed at once, in parallel.

Today, we are at the cusp of another shift in eras. The world is moving from the Analog Era to the Digital Era and this change is having profound and wide-ranging effects. Information is now interactive, global, mobile, and social. The TV broadcasting company described earlier is not alone—it is hard to name an industry that does not feel this impact. What is the book of the future, or publishing company of the future, now that individuals' comments or recommendations can be shared so easily, or now that multimedia, and even interactive media, can be put into e-books? What is health care of the future, now that patients have increasing access to expert medical information online, to devices that track health information of all kinds, and to their own genetic data? What is the impact for hospitals and medical centers when patients can communicate with doctors by e-mail or video, or when they can communicate with other patients with similar symptoms or syndromes?[6] What is the store of the future? What is the next generation in education?

An important question to consider for your innovation blueprint: Is there a new curve, a next generation, emerging for your industry in the Digital era?

In a similar way, a shift to sustainability is also affecting a wide range of industries. Bill McDonough and Michael Braungart have labeled Cradle to Cradle principles for "remaking the way we make things" in any industry.[7] They talk about simultaneously enhancing the economy, the environment, and social equity. In this emerging era, business leaders and innovators are using these principles to ask: What are sustainable buildings, landscaping, or community design? What are green materials or manufacturing processes? What is the car of the future? What are home energy systems of the future? What is sustainably produced clothing or what is a sustainable food system? Industry after industry is looking for new answers.

For your innovation blueprint, consider: Is there a new curve, a next generation, emerging for your industry that uses sustainability principles to simultaneously create financial, environmental, and social value?

Lifecycle Curve Implications for Your Blueprint

If you find that your industry is the midst of being redefined, you will want to ensure that your innovation practices can move you beyond the daily repairs and improvements that any business needs. You would want to install practices and tools to enable disruptive, game-changing innovations that move you into the new era in your industry.

Many times, as the lifecycle curves show, the traditional products are still the mainstream business and have the largest customer base, even though it is clear that they will not be the products or services for many years into the future. In this case, your innovation practices will need to cover how to continue to grow your first curve—to improve quality or streamline operations in your current business, or to take your current products into new markets. At the same time, other parts of your innovation system would need to encourage and enable the creation of products and services that redefine businesses and transform marketplaces. A deep look into your external environment will tell you the kind of innovation practices you need to build.

Assessing Current Capabilities—External Environment

- Are you aware of the external trends that affect your business or industry? (Changes in technology, culture, economy, regulations, competitors, suppliers, partners, or customers?)

- Do you understand your customer's most important needs?

- Is your industry in the midst of a transformation? Do you need to innovate on both lifecycle curves at once?

Innovation Blueprint: Task

The Task section of your blueprint clarifies the business practices that foster innovation throughout your enterprise. Two of the most important areas to consider are

1. **A clear innovation strategy**—ensuring your people know the areas where innovation is needed most.

2. **An integrated innovation system**—that aligns and connects diverse innovation activities into an effective, continuous, end-to-end innovation process. A strong integrated innovation system creates a complete and consistent flow of activities from market and customer insights, to idea generation, optimizing value, selecting and funding the best, all the way through to implementation or commercialization.

Task: Big Picture Strategy

Innovation Strategy—Getting the Big Picture Right

Any area of business can be a target of innovation. A first step in enhancing your innovation capability is to determine where innovation is most essential. Do you need to improve the quality of your current products and services, invent completely new products and services, streamline operations, cut costs, move into new markets, find new ways to connect with your customers, reinvent your business model, or make any number of other changes your business might need? Different teams in your enterprise might use the same innovation practices and tools for very different goals, but it helps if these expectations are clear to all involved.

Creating a Balanced Portfolio of Innovations

When setting innovation strategy, it is helpful to create a portfolio of projects that balances timing, degree of risk, and amount of innovation stretch. Some projects may aim to deliver immediate benefits to current customers. Others might look further ahead and take on riskier projects meant for near-term future customers. The most ambitious but unpredictable projects might aim for blue-sky, high-risk projects that explore future possibilities, require a great deal of learning, and potentially invent entirely new business models.

Innovation Strategy: Bottom-Up and Top-Down, Inside-Out and Outside-In

Setting innovation strategy is often seen as a job of leaders, as direction setting to be led from the top. When leaders act as futurists, in touch with the emerging trends in the external environment, they may be the first to identify new areas where innovation is urgently needed. They often launch "innovation campaigns," where they outline strategic priorities, clarify the business case for a new area where innovation is needed, and communicate guidelines to innovators about how to develop and propose ideas and value propositions. Leaders issue an "invitation to innovate" within a given target area, and innovators know there is a sponsor waiting for their ideas and solutions.

In addition, in a dynamic enterprise, the wisdom of the crowd can help set the strategy as well. People working on the front lines are sometimes the first to spot emerging customer needs, advances in technology, potential business partners, or other information that dramatically influences the areas where innovation is needed, and suddenly possible. Sometimes employees discover important possibilities through their activities or relationships outside of the workplace that others would not have known to ask them (as employees' knowledge and connections are not limited by their current work roles). Other times, when given the chance, motivated customers suggest surprising offerings that can identify important needs that had been overlooked inside the company. Potential business partners may also contribute strong, creative ideas about future opportunities.

A truly innovative enterprise will find a way for any of their key stakeholders to bring game-changing insights and connections forward—even in areas where leaders had not yet asked for ideas. Many of these strategies and proposals may not be appropriate for any number of business reasons, but a few might be exactly the kind of opportunity that is needed, and that only someone from the front lines or outside the organization would have seen. One of the most famous examples of this is Tony Fadell, the inventor of both the iPod and the online music store that supported it. Fadell shopped his vision of a device that could launch a revolution in music to multiple companies—who all turned him down—until

Steve Jobs hired him to create the iPod at Apple.[8] This decision transformed Apple as much as it did the music industry. Not long after, the company changed its name from Apple Computer to Apple, Inc.—a name that could more easily encompass its new role as the world's largest music retailer.

Task: Creating an Integrated System for Turning Insight into Innovations

If you ask the people on your team or throughout your enterprise to describe their innovation system, could they do it? Would they all describe the same process? Do they use shared tools and methods that enable them to collaborate with each other?

Having an integrated system that aligns the many separate innovation activities within your team or enterprise has multiple benefits. An integrated innovation system ensures

- You have a complete "end-to-end" process for taking insights to ideas to strong value propositions to actions.
- Separate activities within this process are aligned. If innovation practices and tools are based on similar concepts, language, or graphics, users can understand them and put them to use more easily and quickly. Users can mix and match the best combination of practices for their specific need at the time.
- Innovators share common concepts, practices, language, and tools. They will be able to give feedback to each other more easily, to collaborate to strengthen the value of each other's ideas.

The complete set of innovation activities should guide innovators to explore market and customer opportunities, to generate breakthrough ideas, and to help them turn their ideas into strong value propositions and optimize the value of these. Once the value propositions are as strong as they can be, these can be presented and pitched in a clear process for selection and funding. Finally, innovators should be mentored and guided as needed through implementation or commercialization (see Fig. 2-8).

Explore Opportunities

In this phase, specific practices are put into place to ensure that innovators are connected to external market trends and changing customer needs and interests. Marketing departments often focus their attention on what currently exists and may not be looking forward as futurists. It can help to look outside your organization and think forward in time, to build relationships with

- Think tanks (e.g., Institute for the Future in Silicon Valley that create trends maps for a decade ahead[9])
- Start-ups whose products or services would impact your industry
- Start-up accelerators or incubators that bring together multiple start-ups, often from a focused industry
- Innovation conferences for your industry (vs. traditional conferences that often focus on improvements, rather than larger-scale transformations)
- Pitching competitions where entrepreneurs in your industry compete for funding
- Trend blogs or future-oriented publications

A number of larger companies are staying connected to the latest innovations in their industry by offering office space to local start-ups; for example, Turner opened an innovation accelerator called Media Camp that invests in and advises media start-ups. AOL offers space

| Explore | Generate | Optimize | Select | Mobilize |
| OPPORTUNITIES | BREAKTHROUGH IDEAS | VALUE | THE BEST | FOR RESULTS |

FIGURE 2-8 An integrated innovation process.

and mentors to start-ups at First Floor Labs in their Palo Alto office. The New York Times recently began TimeSpace, where they offer space and collaboration opportunities with digital media start-ups, and BBC Worldwide has a similar incubator, called BBC Worldwide Labs in London and is opening a similar group in New York.[10]

Connecting to customers is equally as important as connecting to emerging marketplace trends. Marketing departments gather customer data and may summarize this and make it available to innovators, but innovators need to live and breathe their customers' experience if they want to truly add value. The rise in social media is leading to an explosion of ways where enterprises can hear the "voice of the customer" directly from the customers themselves. Because websites, apps, and online communities of interest are changing so quickly, innovators can look for the most relevant ways to connect to customers as these methods and new tools evolve.

Generate Breakthrough Ideas

Once innovators have identified significant opportunities, they can generate ideas about potential breakthrough solutions and then choose the strongest ones to develop further. Several chapters in this book outline methods for enhancing creativity and for brainstorming. Idea generation can take place either in-person or using online idea and innovation platforms and apps, which are particularly helpful for communities where people are spread across many different locations.

One practice that is extremely useful in this phase is for innovators to turn their ideas into value propositions, which highlight the strongest potential value of an idea. There are many different forms of value propositions, such as Barnes et al.'s Value Proposition Builder,[11] Guy Kawasaki's ten questions,[12] SRI International's NABC—Need, Approach, Benefit, and Competition,[13] or Osterwalder and Pigneur's Business Model Canvas.[14]

The value proposition tool we use most often is CO-STAR,[15] which guides innovators to identify the essence of the customer and the business value inherent in their ideas (see Table 2-1).

A shared value proposition template helps people to communicate their ideas clearly and concisely to others. If team members or innovators across a large enterprise or network use a shared template, it makes it easier to see the range of ideas proposed and to see how they might be similar or different. They also have a shared "innovation language," which makes it easier to collaborate to improve each other's value propositions.

Optimize Value

Some innovators generate solutions and then immediately move to pitch these for support or approval. Most often, however, ideas are never at their best at the beginning. This phase emphasizes the discipline of turning fresh ideas into well thought out value propositions, and then getting feedback from others to make these value propositions as strong as they can be—before pitching for resources for funding.

Having a shared value proposition template enables innovators to gather focused feedback—either in person or online—to build group intelligence and expertise into their value propositions. For example, others may be able to identify additional *customer* groups who would benefit from the *solution*; a new technology or market insight that might open up whole new *opportunities*; key partners that might be helpful to have on the *team*; competing *alternatives* that the innovators may not have known, and so forth.

Prototyping is also helpful in this phase, as innovators can test out their ideas and get more realistic feedback from others. Anything that helps make the value proposition more tangible can be helpful—from drawings, figures, and diagrams, to mock-ups, enactments,

C	Your intended **Customers** and their most important unmet needs
O	The most significant **Opportunity** within your idea
S	Your **Solution** to capture the full potential of the opportunity
T	Your **Team** to deliver the solution
A	Your **Advantage** over competing alternatives
R	**Results** you expect to deliver: The **Rewards** to your customers, and the **Returns** or **Revenue** to your team, company, suppliers, partners, or funders

TABLE 2-1   The CO-STAR Value Proposition Template

and actual working prototypes. Eric Ries, the author of *The Lean Start-Up*,[16] advocates quickly creating a "Minimal Viable Product" as a test to further learning. He recommends launching several "MVPs" at once to learn customers' response, as part of the path of designing a successful product more quickly. While Ries writes about software products, trying out multiple minimum prototypes can often be helpful for learning about service or process innovations as well.

Working on each of the CO-STAR categories in more depth during the optimization phase also provides the foundation for a business plan. With each successive round of building and improving their value propositions, innovators can document the size or concerns of various customer segments, how to take a solution to market, how to engage suppliers, predicted costs for creating and marketing the solution, more detailed results or revenue projections, or how customers respond to prototypes or to early versions of the solution.

Selecting and Funding the Best

An integrated innovation system will also have clear methods for how competing ideas, value propositions, and business plans are evaluated and selected. Do ideas go to innovators' bosses and managers, to top executives, to a specially selected innovation team, to outside funders, or to the wisdom of the crowd in online systems that let users vote their favorites to the top of the list?

Improvement innovations may go to the boss or team of the department in question for evaluation or approval, while larger or more disruptive innovation projects might compete in a stage-gate, or multiple-phase, selection process. A phased selection process can take place over several months, where small teams with fresh ideas typically present to an internal innovation panel to receive small amounts of funding to further their work. There might be a middle phase where expanded teams present more developed value propositions and gain increased resources. In the final round, teams present prototypes or more complete business plans to an innovation panel that might include internal or external leaders and funders in the field, and winning teams gain commitment and resources to launch their project.

An increasingly popular practice condenses these phases into one concentrated week or even a weekend of work. In fast-paced, weeklong "innovation boot camps" or "start-up weekends," teams form and innovators quickly develop ideas into value propositions, mini-business plans, and working prototypes when possible. The events typically end with innovators pitching their ideas to top executives or other funders.

Just as innovators learn to pitch, leaders have to "catch," to give feedback to and evaluate pitches—which is often a new role. (See Chap. 4 on Leading Innovation for more detail on the role of leaders as internal venture capitalists.)

Mobilizing for Results/Commercialization

This is also a step that is often left out of the innovation process. Once ideas are funded, innovators or teams are often left on their own to implement their solutions—even when implementation or commercialization might involve business skills or connections the team does not possess. Innovators who can create the product or service with the best technology or the strongest customer value often still can use help navigating the business or financial or political realities required to bring their project to life. This is a place where leaders, funders, colleagues, mentors, or advisory boards can play important roles in making needed connections, in "barrier busting," or otherwise speeding the path to success. (See Chap. 4 on Leading Innovation for more details on leaders' roles in mentoring, networking, and in clearing the path for innovators to succeed.)

Assessing Current Capabilities—Task

Questions to ask to assess your current capabilities include

- Are you connected to your customers, and are you enabling your people to innovate in the areas that create the most compelling customer value?

- Are you innovating for all the areas of the lifecycle curves that apply to your business or industry (e.g., moving up the curve of your traditional business; launching new curves/new generations of your business)?

- Do your people share a common picture of your innovation strategy? Do they understand where innovation is needed most?

- Are you creating a balanced innovation portfolio?

- Do your people have shared innovation concepts, practices, tools, and languages?

- Do you have online as well as in-person innovation activities and tools?

- Do you have an integrated, end-to-end innovation system? (If so, what percent of your people know what it is?) Do you have specific practices and tools to explore opportunities, generate ideas, optimize value, select and fund the best, and launch or commercialize?

- Can you innovate across the boundaries of departments and of your enterprise? Are customers, suppliers, partners, or other interested stakeholders included in your innovation process?

S—Structure

Structure includes all those elements of the organization that leaders get to design: the organization chart, job descriptions and roles, your facilities and locations, all of your corporate functions (e.g., HR, finance, IT, marketing, etc.). Every element of the organization can be designed to enhance innovation.

Innovation Architecture

The most important element of organization design to ensure innovation capability is to focus on the innovation system itself. Think of the structural elements that support the practices designed in the Task section earlier. Some areas to consider are

- Creating official roles and jobs where innovation is the deliverable. Some enterprises hire a Chief Innovation Officer or VP of Innovation to oversee the innovation program as a whole.

- Some larger enterprises also create a team of innovation professionals to grow the capabilities throughout a larger enterprise. This function may have grown out of previous organization development or quality improvement roles, as innovation includes quality improvements to existing products and services, and then moves beyond this to also include the creation of new and potentially disruptive products and services. Also, borrowing roles and terminology from the quality movement, some innovation trainers, coaches, or mentors may be part-time "innovation green belts," while the full-time team members may be known as "innovation black belts."

- Building in a "futurist" function—putting people in roles to create the relationships or gather the information about industry trends and marketplace opportunities. For example, Swisscom, the largest telecom company in Switzerland, created a Silicon Valley Outpost where staff members gather information on new technology trends, create relationships with start-ups in the area, and host tours and trainings for leaders and project teams. Universal Music Group has an insight team that gathers marketplace trends and also connects with customers on a regular basis. Their online Speakerbox community brings together 3000 customers who have agreed to respond to questions about what is important to them that leaders or project teams want to ask.[17]

- Clarifying the roles needed for the innovation system described earlier. If innovators will be pitching their ideas, who will be "catching" these? How will ideas be captured and sorted (e.g., will individuals or teams do this, or will there be software that manages these tasks)? What innovation panels or juries will be set up to evaluate ideas and who will serve on these? Designing the innovation system itself, and the specific roles and timelines within it, is an important activity.

- Some enterprises create a calendar for campaigns, where a leader or project team will launch a campaign each month, to call for innovations in a particular area.

- Open innovation—innovation across the boundaries of an enterprise—is an important skill to have when much of the knowledge and expertise required for particular projects lie outside a single enterprise. Some groups begin their innovation efforts internally at first, and then move to friendly collegial external partners once they are comfortable with their system and practices. When this larger system is working well, they will then expand their innovation efforts to include a wider range of external partners and customers.

Organization Alignment for Innovation

Every element of the organization can be aligned to enhance innovation. As an experiment, ask a colleague to name any organization function in your enterprise, at random. Then think about how that function would be designed if you truly wanted to enable powerful innovation capability throughout your enterprise. Every function can be designed to drive toward innovation. Sample ideas include the following

Finance for Innovation

If budgets are set at the beginning of the year, it is often difficult to fund new ideas that are proposed in the middle of the budget cycle. One remedy is that Finance or business leaders set aside an "innovation bank" to create funding for new projects throughout the year. Also, Finance may ensure that innovation projects are measured and tracked by creating "innovation dashboards" to show how much investment was used, the number of ideas and value propositions created, and the value delivered. Finance can also collaborate with HR to determine financial incentives for innovators. For example, financial rewards can go back to the individual or team creating the idea, to anyone involved in the project overall, to the department or division whose innovation efforts led to the idea, and even to the company overall—to ensure that everyone helps to build the innovation culture.

HR for Innovation

There are HR practices that can promote innovation at each step in the "employee journey": recruiting, hiring, orientation, training and development, performance evaluation, career development and promotion, and even termination. For example, imagine how recruiting and hiring might be different based on a commitment to innovation. Tom Kelley, a partner in the innovative design firm IDEO, has described how he asks job candidates to describe an interesting or unusual product they have noticed recently. If a candidate seems skilled and capable, but then says, "I can't think of anything at the moment. I'll have to think about that one for a while," Kelley assumes that candidate may not be right for their company. However, if the individual lights up and talks with real passion about an interesting product with a unique or intriguing feature, Kelley would be very interested in that candidate. The company is committed to hiring for design expertise, but also for the observation skills and creative spirit needed to innovate.

Ensuring that new employee orientation includes training in the organization's innovation practices and tools gives the message that innovation is a key priority—as well as actually enabling new staff to join in ongoing innovation activities. However, if people are encouraged to innovate, but their performance evaluations later don't measure it or reward them for it, they learn very quickly that innovation is not really that important. It is invisible to management. They tend to put their time, attention, and resources on what is measured instead.

The choice of people who are promoted also sends an important message to the organization. If only employees who focused on traditional projects are promoted, other employees can get the message that it is good to focus on basics, to stay the course. If employees who took chances and tried out new ideas are promoted, particularly if some of their ideas didn't work well at the beginning, this gives a message to employees that it is OK to take risks and it's good to try something new.

HR may also lead the design of the innovation architecture, research and purchase innovation software and tools, and establish the training and development needed to ensure their people work together as skilled innovators. In addition, while leadership or product development teams sometimes lead innovation initiatives or "innovation turnarounds," in many organizations, these efforts are led by HR.

Facilities Design for Innovation

Facilities and workspace design can also encourage collaboration and innovation. Many enterprises are transforming meeting spaces from the traditional boardroom meeting spaces (with a long conference table and large heavy chairs placed in the middle of a long, narrow room), into larger, open innovation labs with movable furniture and whiteboards. These labs are designed for flexibility and interaction, and typically have planning and prototyping materials available as well to encourage experimentation and collaboration.

Sometimes facilities can shape the innovation expected in work that occurs inside the buildings. For example, a company interested in creating more environmentally responsible products may model sustainable innovation in their building design—through using green materials and building techniques, installing renewable energy, designing for energy and water conservation, promoting sustainable transportation, or landscaping with native plants and water flows.

Also, as discussed earlier, some companies offer space to start-ups, partly so their people can develop relationships with the companies with synergistic solutions, and also to enhance the entrepreneurial start-up spirit in their enterprise.

IT for Innovation

Information technologies to enhance innovation are evolving quickly. It seems that each month, there are updated methods and tools available, including

- Enterprise-wide online innovation platforms, where employees share their ideas and value propositions, where leaders launch innovation campaigns and competitions, and where people can discover similar ideas, and give each other feedback to use group intelligence to make their ideas better.
- Smaller-scale innovation apps for individuals and teams.
- Brainstorming software.
- Knowledge management systems, to help employees find who knows information they need. These are particularly helpful as many employees have knowledge far beyond their current job titles.
- Social media for enterprises add to multidisciplinary brainstorming, communication, and collaboration.
- Advances in online education systems contribute to the mix, as employees can more easily access learning that can help them enhance the skills needed for their projects.

Legal for Innovation

Many innovators view their legal department as the group that tells them why they can't try something new, when in fact, legal departments can add a great deal of value for innovators. Legal advisors can help to create agreements with partners, suppliers, or external funders that enable groups to innovate across boundaries. They can guide innovators through the process of checking patents or trademarks, and to secure intellectual property created. Legal teams can also provide the basic agreements needed in large open innovation networks or communities, to help these groups create ownership or reward arrangements that work for all involved.

Assessing Current Capabilities—Structure

Questions to ask to assess how well your current organization structure enables innovation include

- Do you have a clear innovation architecture to support your innovation process from insights through implementation? (For example, Is it clear who receives and reviews the ideas in each phase of the innovation process? Are there clear roles and timelines?)
- Is your organization aligned to enhance innovation? (Do HR, Finance, IT, Legal, and other functions support innovation?)
- Do your facilities offer space that promotes innovation, interaction, and collaboration?
- Do you have an "innovation bank" or finance system to fund new projects?

P—People

In reality, an "enterprise" doesn't innovate—it is the people working together inside an enterprise who innovate. There is no business strategy or organization design per se that is best for innovation. The innovation tasks and structure described so far only succeed when they enable the people of an enterprise to create the future together.

Informed Innovators

First, it's important that people throughout the enterprise understand the most important innovation challenges. Thinking back on the lifecycle curves in Fig. 2-7, it's important for people to understand whether they are being invited to create "repairs" or problem solving for current products, services, processes, or organizational issues. Or, are they being invited to incrementally improve current solutions, perhaps to expand these into new customer segments or new markets? Or do they need to "leap to the next curve," to create a whole new generation of solutions, disruptive solutions that might transform their business? A detailed "innovation brief" can ensure that people have the information they need to understand the challenge at hand.

Capable Innovators

Building on the idea of a shared innovation system described earlier, it is also important that people get training in using the system. Particularly when systems are simple and straight-forward, the training may be accomplished quite quickly. Some of the online innovation systems include training about using it with the software system or the app itself. This becomes part of the process of ensuring that people feel ready and able to join innovation campaigns, that they can bring their own best ideas forward and that they know how to support others' ideas as well.

Innovators Everywhere

It's equally important that people throughout an enterprise feel their contribution is important. In some companies innovation challenges are open to large numbers of people, as employees are invited to submit ideas to particular challenges. In one of our recent client projects, an administrative assistant came up with one of the winning new project ideas. In another example in the same group, employees used the Q+ online innovation platform to put CO-STAR value propositions online.[18] Employees throughout the organization could give feedback on each section of their value propositions. For the T section in one team's CO-STAR (team), an employee from a different department connected them to one of her personal friends who had recently founded a start-up with a technology that would be perfect for them. This new connection was the missing link that made the whole project possible and the two groups began to collaborate.

Where innovation is concerned, people throughout the enterprise may have the information or expertise needed to make a creative new project succeed. You never know who, or whose friends and colleagues, will have the missing piece of inspiration and expertise that can give a potential project the boost it needs to succeed.

Talents and Skills for the Future

As strategies and business models change, it is important that the enterprise keeps up with people whose skills and talents can create the new products and services most needed. Entirely new skills might be required. For example, think of book publishers creating content that can be read on e-readers, with the potential for videos or opportunities for interaction interwoven into the text. Publishers face decisions about whether these new skills become an essential part of their work and belong inside their organization, or whether to partner with existing experts. An innovative business strategy often leads to a new people strategy as well.

Innovation Champions

Innovation rarely follows a straight course, proceeding exactly as planned. True innovation requires one or more champions who care enough about the potential in a project that they will weather the ups and downs, and have the commitment to find a way to make a project succeed. Innovation champions, by definition, are trying to do something that has not been done before. There is often no clear path or simple answer about how something will work best. Champions are the ones who wake up at 3:00 am, with a sudden insight about how to succeed. Champions bring the personal passion for their projects, the commitment, and the entrepreneurial spirit to find a way around obstacles that will inevitably arise. Curt Carlson, the President and CEO of SRI, says, "No champion, no project, no exception."[19]

Collaborative Teams

Just as start-ups often begin with a team of three key founders—one technical or product leader, one business/financial leader, and one marketing leader who connects to customers—innovation project teams often find that they need all three of these skill sets to successfully move their idea through to implementation. Even internal innovation projects for streamlining processes or cutting costs will reach their innovation goals more quickly when innovators fully understand the technicalities of the process in question, the larger system of finances involved, and when they understand the process from their internal customer's perspective. Organizations where innovators can bring all of these points of view together can develop stronger solutions and implement them more quickly.

Working in a collaborative team is also an individual skill. People need to bring their own skills to their innovation projects, but also need to be able to integrate others' skills and knowledge into their work as well. Tom Kelley from IDEO describes how they look for "T-shaped people," those who have in-depth expertise (the vertical line of the T) and know how to work well with others (the horizontal part of the T).[20]

Assessing Current Capabilities—People

Questions to ask to assess how well your people are prepared to innovate include

- Do your people have the skills and talents needed to succeed?
- Do you have enough people willing and able to be innovation champions?
- Can your people work together in collaborative teams?
- Can your people collaborate across boundaries—of teams, departments, division, or with others outside the company?

Ei—Internal Environment

Some work environments help employees to feel excited about their work, full of ideas, and eager to contribute, while other environments do just the opposite. Internal environment or culture is not easy to design on its own, as the task, structure, and people elements of the enterprise influence the way it feels to the people working there. The internal environment or culture then in turn impacts how well each of the other elements can succeed. If innovation is promoted verbally, but all other actions reinforce department silos and internal competitiveness above customer value creation, people would likely feel frustrated or even cynical about innovation. Fewer of them would offer their ideas or participate in challenges and innovation campaigns. On the other hand, when leaders set out a clear story about where the industry is headed, about how the business or organization needs to change to be a leader in this new environment, and how important new ideas can be to this success, people are likely to take notice. If this is followed up with a clear innovation system, where everyone knows how to offer their ideas, how to work with others to create and improve value propositions, or know the rewards if they succeed, the whole innovation environment is starting to look more appealing. People get energized about the possibilities and get to see that innovation is fun and fulfilling. They get a sense of contributing to something worthwhile—for their customers, their team, their company, and for themselves.

One area where innovation cultures often get into trouble is when some teams are working on improving the current business or operations while others are working on more revolutionary future-oriented projects. Think of the two curves from Fig. 2-7. On days with their worst behavior, innovators working on one curve might not trust the people working on the other curve. People working to grow or improve the current business could easily think, "We have all the customers. We bring in all the funding and they just waste it. They keep trying these ideas and lots of them never work." And people working on the second curve, creating quite revolutionary products or services, or new ways of working, might say about their more traditional colleagues, "We're the future. They're just the old culture guard, trying to keep anything from changing." Conflict can easily arise between groups working on innovation at different levels or scales.

However, innovation flourishes where people are focused on customers and the challenges at hand, rather than spending their time caught up in a toxic or blaming internal culture. It's important to ensure that people understand that both types of innovation are typically needed, simultaneously. One group may be working on keeping current customers satisfied and giving them enough smaller improvements to keep them engaged. Their work funds and enables the innovators working on entirely new ways of operating. The new products or services or new ways of working may ultimately become the standard in the future, but at the time the innovations are first being created, both groups of innovators are needed. Innovation thrives when both groups appreciate each other and value each other's work. "They keep the store running, so we can experiment, to see what works for the future." And, "They are trying out new solutions, and someday we might all leap to the new curve they build." If innovators support each other, they all can focus on creating their unique form of customer value.

As our clients create innovation blueprints, there are often common themes about the culture they want to create. They aim for an internal environment with a lively and powerful innovation culture, where

- Anyone can innovate. We know that everything can be better—from our day-to-day operations to how we might revolutionize our industry. And we know how to join in the innovation efforts in our team or company.
- Innovation is visible. We can see the ideas that are being generated and can watch these improve over time. We get to see the evolution of a success.
- People are passionate about the work they do. People care.
- We have high standards. We want to be the best.
- We have an entrepreneurial spirit. People take initiative.
- Wild and crazy ideas are welcomed as a way of getting to really good ideas.
- People trust and respect each other.
- We help each other to make our ideas better. We contribute to each other's success.
- We ask lots of questions. We have a spirit of learning.
- We stay open and flexible. The best solution may be different than the one we first imagined.
- Our successes are rewarded and celebrated. Everyone gets to see them.

Assessing Current Capabilities—Internal Environment

- Is your environment filled with creativity and experimentation?
- Are your people energized and inspired? Do they want to innovate?
- Can anyone in your organization be an innovator?
- Do your people trust, respect, and support each other?
- Do your people want to be the best?

Three Drivers: Leadership, Stakeholder Engagement, and Innovation Support

Having a blueprint does not ensure the "building will get built." Too many times we see the blueprint in the drawer or the blueprint on the shelf. What ensures that the enterprise envisioned in the blueprint will come alive?

People take the actions that turn blueprints into reality. If an enterprise is committed to develop its innovation capability and creates a clear innovation blueprint, three groups of people are needed to take the next steps: leadership, stakeholders, and innovation support (Figs. 2-9, 2-10, and 2-11).

Leadership

Leaders ensure their people are willing and able to innovate. They inspire their people about what is possible and they authorize the funding and resources, the innovation systems, and the organization that can enable innovation throughout the enterprise. In addition, good innovation leaders encourage their people to step forward and provide their own inspiration. They know that their people will sometimes be the ones who step forward to show the future possibilities.

FIGURE 2-9   Leadership diagram.

The following chapters have many examples of leaders' roles in innovation. (See Chap. 4 about ten key roles leaders play in innovation). A blueprint without leadership is unlikely to be fully implemented. Furthermore, there are many kinds of leaders who often need to be involved—not only the formal executive leaders in an enterprise. There may be culture leaders, who people look to as role models. Sometimes industry leaders, or community or government leaders make a difference in inviting people to innovate. An important part of your own blueprint will be to identify all of the leaders who can contribute to enhancing innovation, to clarify what is needed from them, and to get them involved.

Assessing Current Capabilities—Leadership for Innovation

Questions to ask about innovation leadership include whether your leaders

- Treat innovation as a strategic priority? Do they take a stand for innovation?
- Communicate a compelling business case to your people about why innovation is essential?
- Ensure the systems, practices, tools, events, and support are in place to enhance innovation capability throughout their enterprise?
- Are role models for innovation?
- Coach and mentor innovators?
- Participate on innovation panels and juries, to evaluate and select the best projects to move forward?
- Celebrate and reward inspiring examples of innovation?

Stakeholder Engagement

While leaders inspire and authorize and support innovation, they rarely create all of it themselves. The stakeholders are the ones who innovate. It's important to find ways to engage key stakeholders, the potential innovators who will be the ones to build the future. Getting them engaged often first takes getting their attention, ensuring they have the means and methods to innovate, and then ensuring they can join in innovation activities.

Successful stakeholder engagement efforts typically include a mix of

- Kickoff events and materials, to publicize innovation programs.
- All-Hands events to engage innovators. Larger companies often conduct an "Innovation Road Trip," where leaders travel to different locations to emphasize the importance of the efforts and engage participants in innovation programs.
- Innovation training events, workshops, and/or online information, to explain needs and upcoming programs.
- Idea suggestion methods, in-person and online, where any employee can suggest ideas for improving products and processes.
- Innovation campaigns, where leaders invite ideas and value propositions for a targeted topic area. A campaign is quite interactive, with the expectation that people will give feedback to improve ideas, and later vote the best ones to the top of the list.
- Pitching competitions.
- Business plan competitions for larger venture ideas.

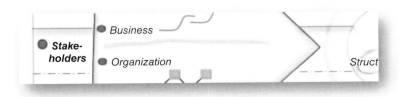

Figure 2-10 Stakeholder diagram.

- Innovation tours, connection to start-ups, connection to company innovation "outposts" if these exist (where company scouts create relationships with start-ups and external innovation resources).
- Communications to keep ideas in progress visible, to highlight winning pitches, and to celebrate successful projects and new ventures.

Assessing Current Capabilities—Stakeholder Engagement

Questions to ask include

- What percent of your people feel invited to innovate? Is this the percentage you want?
- Do people understand how to participate in innovation activities?
- Are people involved in improving current products, services, and processes, as well as creating breakthrough innovations?
- How many ideas are innovators submitting? How many people are engaged in commenting on and improving others' ideas, even if they don't submit ideas themselves?
- Are people satisfied with their ability to innovate in their day-to-day work and to participate in targeted innovation activities?

Innovation Support

This individual or team keeps innovation on track. If leaders set the direction, instill inspiration, and offer resources and funding, the support team ensures that people are *capable innovators*. They provide the skills, tools, events, and connection to outside experts as needed. They ensure there is a clear innovation architecture in place—the series of events where innovators scout out opportunities, generate ideas, build project teams, develop value propositions in-person or online, get feedback and iterate, and pitch in various rounds of development for time and resources. The innovation support team helps design this series of events and gets the word out about how to participate. They may work with IT to get an online innovation platform installed, or with Finance or HR to develop incentives and rewards. They handle the communications and events to keep everyone engaged. And they may work with outside resources as needed to keep up with the latest innovation practices or tools.

The team also monitors and coordinates innovation activities—and catches the inevitable breakdowns. What happens when too many ideas come in and it's hard to have the time to evaluate them all? What happens if there are too few ideas? This team catches any difficulty as it occurs, so they can fix it and ensure the innovation process keeps going strong. They provide rational, logistical support as well as emotional support. They may get back to employees whose ideas were not chosen, to make sure these individuals still feel appreciated and stay engaged. Or they may design the kinds of events and speakers that keep the energy high for innovation. Each group is different. This team finds the way to keep their particular group energized to search for and build creative new solutions.

Figure 2-11   Innovation support diagram.

Assessing Current Capabilities—Innovation Support

Questions to ask about Innovation Support include whether you have

- A team to build and monitor our innovation system?
- The knowledge and skills about how to innovate in your particular organization?
- The tools and practices to innovate?
- Innovation training as needed—for innovators and for leaders?
- A clear calendar for which innovation activities happen when?
- A way to monitor and measure your innovation process?
- A way to stay in touch with the emotional climate around innovation?
- A way to stay in touch with your innovators, to ensure they have what they need to create value?
- A way to continuously refine your innovation process to make it stronger?

Conclusion

Most people naturally have a desire to innovate, to contribute to making things better. Yet when large numbers of people throughout an enterprise or network want to innovate together, it can get chaotic pretty quickly. It helps to have a common map to align business strategy with innovation practices, tools, and resources; organization design; people's skills and talents; and an innovation culture. The innovation blueprint does exactly that. It also helps leaders and innovation support teams to define the critical roles they play to ensure innovation succeeds. The more elements of an organization are designed to enable innovation, the more likely it will be to occur consistently over time. The ultimate goal of the innovation blueprint is to ensure that people throughout an enterprise can create value together—that they can create a compelling future and can make a difference.

References

1. Hanna, D. *Designing Organizations for High Performance.* Reading, MA: Addison Wesley Longman, 1988.
2. Drucker, Peter F. *Management Challenges for the 21st Century.* New York, NY: HarperCollins Publishers, Inc., 1999, p. 119.
3. Friedman, Lisa and Gyr, Herman. *The Dynamic Enterprise: Tools for Turning Chaos into Strategy and Strategy into Action.* San Francisco, CA: Jossey-Bass Business and Management Series, John Wiley, 1998.
4. ValueSTEP Online Innovation Assessment. http://www.enterprisedevelop.com. Tools Section.
5. Friedman and Gyr. *The Dynamic Enterprise.* San Francisco, California: Jossey-Bass 1998, p. 152.
6. Friedman, Lisa, Gyr, Herman, and Gyr, Alex. "The Changing Patient in the Digital Era: A Typology for Guiding Innovation in Healthcare." *International Journal of Innovation Science,* December 2009.
7. McDonough, William and Braungart, Michael. *Cradle to Cradle: Remaking the Way We Make Things.* New York, NY: North Point Press, 2002.
8. Kalney, Leander. Inside Look at Birth of the iPod. *Wired.com,* 7-21-04. http://www.wired.com/gadgets/mac/news/2004/07/64286?currentPage=all.
9. Institute for the Future. http://www.iftf.org.
10. McKenzie, Hamish. Disruption from the Inside: Media Companies Seek Hope in Incubators. *Pando Daily,* Feb. 14, 2013. http://pandodaily.com/2013/02/14/disruption-from-the-inside-media-companies-seek-hope-in-incubators.
11. Barnes, Cindy, Blake, David, and Pinder, Helen. *Creating and Delivering Your Value Proposition: Managing Customer Experience for Profit.* Philadelphia, PA: Kogan Page Limited, 2009.
12. Kawasaki, Guy. *The Art of the Start: The Time-Tested, Battle-Hardened Guide for Anyone Starting Anything.* New York, NY: Penguin Group, Inc., 2004.
13. Carlson, Curt and Wilmot, William. *Innovation: The Five Disciplines for Creating What Customers Want.* New York, NY: Random House, 2006.

14. Osterwalder, Alexander and Pigneur, Yves. *Business Model Generation: A Handbook for Visionaries, Game Changers, and Challengers*. Hoboken, NJ: John Wiley & Sons, Inc. 2010.

15. Gyorffy, Laszlo and Friedman, Lisa. *Creating Value with CO-STAR: An Innovation Tool for Perfecting and Pitching Your Brilliant Idea*. Palo Alto, CA: Enterprise Development Group, Inc., Publishing, 2012.

16. Ries, Eric. *The Lean Start-Up: How Today's Entrepreneurs Use Continuous Innovation to Create Radically Successful Businesses*. New York, NY: Crown Business/Random House, Inc., 2011.

17. Bain, Robert. Face the Music. *Research Magazine: Research-Live.com*, September 2011. http://www.research-live.com/features/face-the-music/4006113.article.

18. Gyorffy and Friedman. *Creating Value with CO-STAR*, pp. 121–124.

19. Carlson, Curtis R. and Wilmot, William W. *Innovation: The Five Disciplines for Creating What Customers Want*. Random House Digital, Inc., 2006, p. 157.

20. Kelley, Tom. *The Ten Faces of Innovation: IDEO's Strategies for Beating the Devil's Advocate and Driving Creativity throughout Your Organization*. New York, NY: Doubleday, 2005, pp. 75–77.

CHAPTER 3

The Culture for Innovation

Jack Hipple

*"Your business plan is **what** you are, but culture is **who** you are."*
—Gary Kelly, Chairman/President/CEO, Southwest Airlines,
"The Southwest Culture: Our Secret Sauce,"
SW *Spirit* magazine, October 2012

Whaat is "culture"? Why is it important? Does your organization care about its culture? Do you know what it is? Does your organization understand the impact of a given culture? Why might it be desirable to change a culture? How does one go about changing culture? In the pages that follow, we will explore each of these questions.

What Is Culture?

Let's take the analogy of cooking. We can mix ingredients together, heat them, and get a meal. We add a sauce and it becomes so much more. Barbeque rib cook-offs are held around the country where the only difference between the end products is a "secret sauce" invented by someone decades ago and passed down from generation to generation. Coca-Cola and Pepsi-Cola have secret recipes whose total combination is known only by a few people in the company. People are absolutely passionate about liking one or the other and the choice is simply based on an immeasurable quality known as taste. Apple pie and turkey stuffing are other examples of combining the same ingredients in unique ways with "sauces."

When we mix together different kinds of people within an organization, what is the "sauce" that holds them together and gives an organization some measure of uniqueness? In an article by Gary Kelly from Southwest Airlines, he said that "Living the Southwest Way" involves three things:

1. Work hard

2. Have fun

3. Take care of each other

It is easy to see these things if you are a frequent Southwest flier. Why do the airline safety announcements have to be dull and boring? While the Federal Aviation Administration (FAA) requires certain specific things that must be told to passengers, it does not specify in what way they need to be said. Why does every Southwest flight have a different *style* of safety announcement? They are humorous, causing us to pay attention! Why is Southwest the only airline whose boarding gate desks are decorated with spooky stuff at Halloween and the gate agents are dressed up as witches? Is there something in other airlines' policy manuals that says you can't have fun? Why is Southwest the only airline whose baggage handlers brag about having fun handling bags (and not charging to do so!). Is there some rule that says this can't be done? Now one reaction to all of this is to think that Southwest is not a "professional" airline. If that's the case, why is Southwest the *only* airline that has had 25 years of uninterrupted profitability and no layoffs? The only airline which doesn't charge for bags and is not reliant on baggage fees for its existence? Why has it grown from serving only three cities in Texas to one of the largest airlines in the United States? Is what they do so

Question 1: How would you describe the culture in your organization?

FIGURE 3-1   Question 1.

What behaviors do you observe all the time?	What are the implications? What is unique? What does it indicate?

TABLE 3-1   Behavior Analysis

difficult to copy? So why do Southwest employees (and indirectly their customer) have fun? Because it's part of their *culture*. Culture is all about how people behave, treat each other, and treat customers.

So the very first question we need to ask is "what is our culture?" If we can't describe it in a few short sentences, as Southwest Airlines does, it's not culture, it is policy. It's a procedures manual. Policies tell people what to do, not how to do it. It needs to succinctly say what we do and who we are. If we don't know what the culture is, before we decide to analyze or change it, we need to find out what it is now. This *cannot* be a multiple choice, anonymous questionnaire sent out to all employees where we presume to know what some of the answers might be. It is not a mission or vision statement. Policies describe what we'd like to accomplish. It needs to be a blank box where we directly ask the question:

What is our culture? How would you describe it to an outsider who has never interacted with our organization or our employees before (Fig. 3-1)?

Here's a table to help you (Table 3-1).

Ask yourself also, where did these behaviors derive from? Are the reasons known? Are they still valid? A particular style of behavior within an organization may have been predicated on circumstances in the past that were perfectly valid at the time. In the case of Southwest, these behaviors have been found applicable and appropriate over a long period of time. This may or may not be the case in your organization. Think about it carefully!

Why Is Culture Important?

Culture is like glue that holds things together. Without it, things are things and people are people. Robots don't need culture to interact with others. They only need software programming and mechanical and electrical connections and maintenance once in a while. The robot does not have empathy or concern. It doesn't have fun unless it is programmed to do something mechanical that looks silly to an observer. We have all met and interacted with robots in our business encounters, haven't we? It's not a lot of fun interacting with a procedures manual, is it? Is it any more enjoyable to work with customer "service" agents whose first duty is to serve the organization they work for?

Now in the case of Southwest Airlines there was a deliberate decision made, at the time of its formation, as to what its culture was going to be. Not every organization is started that way. A small start-up company often takes on the personality and culture of its founders, who did all the original hiring. As the organization grows, there's a perceived need to develop procedures and policies to minimize variation and exceptions. The original founders frequently sell their early company to a venture capitalist who may not know or care anything about culture. They want to turn the investment into a quick profit for their stakeholders. Eventually the company may go public and the original roots are forever lost. During this process, however, people learn how to get things done in the most expedient way, independent of the high-level business decisions that are going on. The culture forms, slowly, subtly, and it's seldom written up in a policy manual. It's how they get things done.

Question 2: What is the nature and impact of your culture?

FIGURE 3-2 Question 2.

What are the internal effects of the culture you observe?	What are the external (customer/supplier) effects?

TABLE 3-2 Cultural Impact Analysis

It's how people are treated. It's how they deal with customers. Every organization has a "culture." It may be written or hung on a wall. Or it may be "just the way we work here" and never written down—just passed on informally in new employee training or with hand holding by senior people in the organization.

An organization's culture is predictive in the sense that it will frame the manner and method in which people behave, respond to problems, and deal with crises, customers, colleagues, and sudden changes in technology or the external business environment. (Fig. 3-2.)

Here's a table to help you (Table 3-2).

How would your culture deal with a major change in a government regulation affecting your business? How would your culture affect your organization dealing with a new competitive business situation? A blocking patent? Regardless of whether you believe the current culture is what is needed or appropriate, it is important to recognize the impact of what it is now and the potential impact of trying to change it. The situations to which a business or organization responds change all the time. The response of the organization and individuals within it will depend on its culture.

Does Your Organization Care about Its Culture?

Now this may seem like a silly question, especially if you have made a serious effort to understand it. But we have all seen many examples of employee surveys where the "surveyor" has no intention of doing anything with the answers. It's frequently just a way of making people feel good about being asked. Not saying anything about the results can be more damaging than never having asked the questions to start with. So before you begin this journey, ask yourself some very serious questions:

1. What will you do with the results? Will you share them openly? How? In a letter? Open forum? Lecture? How will you react and respond to what you hear in such a meeting?

2. What will be the consequences of asking the questions? Will the questions themselves carry some kind of unintended message? Whenever there is a corporate or organizational survey, most people will be looking for feedback in some form or another. What if the results are not what you expected and leave you feeling uncomfortable? What will you do?

3. What will be the reaction of various managers after hearing the truth? Will they welcome the input and be willing to adapt and change (if necessary)? Will they dismiss the results as not being accurate and reflective of the true culture in their part of the organization? How will you handle this?

4. If one of the reasons you are surveying about culture is to prepare the organization for an upcoming challenge, what if the results are not what you are expecting in terms of the organizational change that may be required? What will you do? How will you communicate this? To whom? What kind of plans do you have?

Question 3: Do we really care about the culture in ABC organization? Why? How do we demonstrate it?

FIGURE 3-3   Question 3.

Why do we care about culture? How do we demonstrate it?	How will we react to understanding the culture of the organization?

TABLE 3-3   Caring about Culture

5. What do you anticipate the culture assessment to show? What will be your *personal* reaction to a result showing something you don't expect? Will you be able to deal with the answers, especially if they are contrary to not only what you expected, but what you think is needed for some major organizational effort? What will you do? How will you figure out what to do?

Once we have started down this road, we can't turn back. We have to really want to know what peoples' perception of the culture is. Then we have to be ready to hear the truth and respond to it. We *really* have to care about the culture and its impact on our employees, our business, and our customers.

Now of course we could argue that, in some rare circumstances, culture doesn't matter all that much. Monopolistic companies and businesses that do not need to be customer responsive or capable of changing with conditions may be somewhat insulated from the need to understand culture and can get away with commanding what to do for a long period of time. But even the U.S. Postal Service, having seen its business whittled away by e-mail and other service providers such as UPS and FedEx (who provide the same *function* in a different way), and military organizations asked to fight in conditions requiring understanding of country and religious cultures, have learned that they cannot stick their heads in the sand and not be prepared to understand the local culture and its impact on the ability to gain the support of the local population.

Another thing to consider is the multiple levels of management involved. If the desire to understand an organizational culture is driven by the "C" level management, how well is the desire to understand culture accepted at lower levels? If it is perceived as a "this too shall pass" program of the month, there will be winks of support ("Just fill this out and get back to work…") and little buy-in. The effort to understand and utilize the understanding of culture must be thorough and complete. Otherwise the best intended effort will fail to take root and have the intended long-term benefits (Fig. 3-3).

Here's a table to help you (Table 3-3).

If someone asked you how you demonstrate concern for the culture in your organization, how would you respond?

How Do We Know and How Is It Measured?

We have now given serious thought as to what culture is and have decided that we really *are* interested in knowing what is within our organization. In addition, we have prepared ourselves, and all of our managers, to react to any survey results in a positive way.

We will progress in order of complexity in how we can assess the culture within an organization.

First, we can watch and listen. Now, of course, this involves getting out of the office and interacting with people. That's a given. Some simple observational questions and what they might imply with regard to culture:

1. In a meeting involving senior management, does anyone respond when the question "Does anyone have any questions?" is asked? Or is there total silence? When the meeting is over, is there a burst of hallway and lunchroom conversation that seems like it should have been discussed in the meeting? Might this demonstrate a top-down autocratic culture where people do only what they are told? Maybe that's okay if the senior leader is always right. Is that ever the case? This says something about the culture of "listening" within the organization.

2. At the beginning of a meeting, does the leader explicitly say that he or she is interested in questions or discussion at any time? If someone asks a question, does discussion actually occur or is a statement similar to "Let's talk about that off-line" or "Let's save that for later" made? Might that signify that the manager is more concerned about getting his or her point across rather than listening to input? Has an important decision already been made? If so, why not say so? Did we need to have a meeting to communicate a decision already made? If not at meetings, how is input solicited and taken into account?

3. Are off-line informal conversations around the coffeepot or water cooler observed where people are having some highly interactive discussions about an important technical or organizational issue? If so, what happens to those discussions if a more senior manager walks up to the group? Does the group stop talking or is the manager invited to join in the conversation?

4. Are managers walking around and listening? Where do they eat lunch? Is there an "open-door" policy? Are there "executive" parking spots? What do these things say about the hierarchical nature and culture of the organization? Now there may be legitimate reasons for some special places for certain people, but is it based on security concerns or just status?

5. Teamwork is such a buzzword today. It's used everywhere and anywhere. What does it mean in your organization? To you? To your employees? Are people enthused about being assigned to a "team?" Are teams appointed? Or do they form spontaneously with minimal management involvement? Does it seem that everyone has to approve everything? Is there support for individual initiative? Is an individual initiative that breaks the mold greeted with the expression, "you're not a team player?"

Now these kinds of observations show us the tip of the iceberg, but if we want to know what the culture is, we need to do some sort of psychological assessment of the individuals and teams. There are dozens of these types of assessment instruments and it is not appropriate to comment on all since there is no one who has familiarity with all of them. We will review two longtime used assessment instruments that have provided insights over the years in two key areas:

1. Social behavior relating to each other, concepts, and organizational styles

2. Problem-solving style

NOTE: *The following discussion uses examples related to two specific assessment instruments with which the author has had successful experience, but this does not imply an endorsement of these nor negativity about others. The author has no commercial interest in the sale or use of the assessment instruments discussed. Other assessment instruments can be used (that are in some cases "spin-offs" from the two discussed)—the important thing is to understand what we are measuring and why. Use whatever validated assessment instruments you want, but make sure they cover the two key issues—social style and problem-solving style.*

Organizations and teams that interface on a regular basis with external groups or companies need to be especially sensitive to the fact that styles and cultures can differ significantly.

Social Style

The most well-known of these types of assessment instruments is the Myers-Briggs, or MBTI. This measurement of social style preference is easily recognized by the use of the four letters commonly used to summarize the preferred style:

E/I, S/N, T/F, J/P, that is, ESTJ or INFP

There are 16 different possible combinations of these social style preferences and they are not equally distributed in the population.

This assessment has many competitive spin-offs developed by other companies trying to fine-tune both the assessment instrument in terms of simplicity, how the results are presented, and the confidence level of the results, but the output is in the same format. They all measure four different characteristics of an individual's *preferences* in style. Preference means a person's natural tendency to behave or relate in certain ways. It does not prevent them from behaving or relating in different ways when necessary. Preference does give an insight into the stress level that might result from being asked to behave or relate in a way that is not natural. The implication here is that if an organization is populated by a majority of individuals with certain preferences, and they are then asked to do or accomplish "something" that is not within their natural preferences, this transition may be frustrating to many. There will be organizational and individual stress, which may have an impact on achieving a goal. There may be significant friction within a team with different social styles. Let's take a look at the four measured behavior preferences. Remember these are *preferences*. Anyone can do something they don't like or behave in a way they are not comfortable for a period of time, but if asked to do this for a sustained period of time, significant stress will develop, both within the individual and in the relationships that the individual has with others. It is also possible for an individual's preference to be in the middle of these extremes. This characteristic makes it easy for such an individual to bridge the gap when working with others.

Extroversion/Introversion

What is an individual's natural/preferred inclination in this regard? Curling up with a book or making blind phone calls to generate leads and customers? Like to vocally contribute ideas and suggestions in a meeting (maybe overly so and dominate a meeting or conversation) or sit back and wait to be called on? Uncomfortable with public speaking? There is no right or wrong about either style. For example, someone could be comfortable speaking in front of a group, but not seek out the opportunity. As we said previously, people have preferences. This does not mean that an introvert cannot be a public speaker once in a while when they have to. It does not mean that an extrovert cannot be quiet when appropriate. This first behavior preference is indicated by an "E" (extrovert) or "I" (introvert). Though many individuals have strong preferences toward one or the other, an individual's preference can be anywhere along the continuum. There is no right or wrong about a preference, but we can easily imagine jobs and responsibilities that might require a stronger preference for one over the other. For example, a sales position requiring much conversation and communication is most likely more comfortable for an "E" than an "I." It is unlikely that we will find many "E" librarians. An "E" will tend to walk up to someone they don't know and introduce themselves and start a conversation (e.g., at a trade show or in a meeting), while an "I" will feel uncomfortable doing this and will wait to be introduced. They will not be comfortable with "cold calling." What is natural for an individual will most likely produce a situation where the individual will seek other employment or the stress caused will produce negative ill will in some form or another.

In terms of culture, we can imagine an organization whose primary focus is selling services will have a very extroverted culture, but the IT group (dealing with programs and computers that don't talk back) in that same company may be very introverted in their culture.

If we think about this behavior preference in terms of innovation, an extrovert will want to talk to a customer directly about their experiences while an introvert may feel that reading a summary of a customer interview or reading a literature article is adequate. An extrovert may see the need to talk with companies and individuals outside the business area of interest while an introvert may see most of the value in talking with people who have in-depth knowledge of the current business or technology.

One of the implications here is to think about major cultural shifts that an organization might need to deal with. An organization that has been very "inward" focused in terms of technology development and has decided to get out and talk to customers firsthand and watch what they do with products, may find this challenge difficult.

Sensor/Intuitor (S/N)

> **NOTE:** *When this characteristic is used, the capital letter "N" is used instead of the first letter in the word "Intuitor" since it has already been used in the first "Introversion/Extroversion" comparison.*

This is a measure of how individuals take in data and information, or assessing and evaluating a problem or situation. An "S" tends to focus on facts, data, and directly measurable things. An "N" is someone who will see more indirect and less tangible aspects of a problem or situation. A short exercise to assess this preference, without actually taking the assessment, is to ask someone how they would describe a leaf. If they say "green," "scaly," or some other easily measurable property, they are most likely an "S." Someone who says "light," "fluffy," or "reflects light" is most likely an "N."

Where does this come into the innovation equation? An "S" culture will focus on how a product or service is made and used from a practical standpoint. It will focus on prices and costs. An "N" culture will look at the impact of a product or service on someone's job or environment, how tailored it is, or how it affects other things which might be emotional in nature. For example, whether a product is red or blue may be analyzed by an "S" from a simple cost standpoint (which tinting resins are cheaper and easier to use?) whereas an "N" might also ask about preferences in colors and packaging. It is very easy, especially for technical people, to lose sight of the "soft" side of new products and how consumers perceive this aspect. If an organization is totally filled with "S" people, they may totally miss some "soft" or interface aspects of a new product or business. Conversely, an overly "N" organization may lose sight of the necessary questions and data in customer interviews that must be gathered and used to calculate costs of manufacturing, quality control, and reliability.

A current case to think about is in the context of "S" and "N" preferences for tablet, portable, and desk-based computers. Certainly there are cost and technical issues in designing and manufacturing these devices, but also the intangibles of how a consumer *interfaces* with such devices. A customer's desired preference for manipulating what is on the screen, accessing and moving icons, and enlarging and scrolling is something that can only be determined by actually watching what consumers do, not necessarily what they say they want to do.

Thinking/Feeling

This is the characterization of how an individual or group reacts to the "measuring" it does with the previous preference ("N" or "S"). In other words, once someone has assessed a product or opportunity need, what is the response? A "T" ("Thinker") will tend to consider facts, data, and actionable items in their actions and responses. They will focus on costs, timing, and manpower needs. The "F" ("Feeler") will be more concerned about the impact of the actions on the people side of the equation. For example, will the plan impact people's lives and schedules? Will it make life easier? In the delivery of a new product, the "T" will want to make sure it's delivered on time and the "F" may think to add a handwritten "Thank You" note to the new customer for helping launch the new product. It may be good to have both.

In a major new innovation effort or program, which of these characteristics is more important in your plans? What are you trying to change about your products? Services? Customer relations? Delivery mechanism? Are you planning to develop products and services that are closer to the end user? Do you know what their expectations are? Is your culture appropriate?

Judging/Perceiving

This aspect of behavior and culture relates to the nature and style of the actions resulting from data input and analysis. A "J" ("Judger") will want to see a critical path, schedules, close budget estimates, and commitments of who is going to do what and when. A "P" ("Perceiver") will fight to keep the door open for possibilities that may come about as a project or product develops and not lock in things so definitively. It will be harder to pin down a "P" as to dates, budgets, and commitments. They will see value in unrecognized opportunities that might come down the road.

In terms of innovation, a culture of "judging" (J) may result in an overly regimented product development process, resulting in a lack of awareness of customer or market changes during a long development process. A strong "perceiving" culture may miss deadlines for prototype or product deliveries, regulatory requirements, and other necessary items for a planned product launch.

There is no right or wrong about any of these characteristics or cultures. As we have discussed, there are benefits to all the different aspects of social behavior preferences. The key is to recognize the overriding cultural bias and when an addition or adjustment is necessary. For example, if an organization with a strong introversion culture is developing a new product that will have much more consumer interface and use than it is used to, it needs to recognize that hiring a market research firm to analyze consumer response may be required if this sensitivity is not in house. At the other extreme, if an excellent consumer product company ("N") faces an opportunity where costs are more important than in the past, it might need to pay more attention to details, costs, and project management issues.

This four-letter characterization of an individual is not unchangeable but is somewhat hard wired at an early age. Experiences can cause minor adjustments in preferences over time, but not usually dramatic swings. People who have strong preferences of a certain "type" like to associate with people like themselves resulting in an unrecognized culture that tends to propagate itself. If the business need or the customer base changes, it may not be possible for the organization to respond appropriately and not even recognize the reason why.

Problem-Solving Style

Problem solving is an inherent aspect of innovation in that a new innovation, regardless of whether it is a product or a service solves a problem that has not been solved before or replaces the current solution by significantly reducing cost, minimizing negative effects, reducing complexity, minimizing waste, or improving customer interface issues. Innovation solves problems.

To the extent to which an innovation effort is directed at problem solving, we need to ask, "what is the manner in which people solve problems?" It turns out there is a way to measure this, known as the Kirton KAI or Kirton Adaptive-Innovative Index. People with similar social styles can have different problem-solving styles. They may get along great in a social situation, but in a problem-solving meeting, friction between these individuals may be observed. Again, there are some other assessment instruments that measure this aspect of behavior and culture and what is presented here is one particular assessment instrument.

At one extreme of behavior preference are those described as "adapters." They tend to solve problems within the paradigm of knowledge they possess and with which they have experience and confidence. They also show a preference for "tried and true" solutions with minimal tendency for risk. In an idea generation session, they will tend to filter their ideas prior to expressing. This "filtration" will be based primarily on what they believe is possible. They will care about agendas and schedules. Their method of analysis and problem solving will be very visible and obvious.

At the other extreme are in Kirton's words, "innovators." (*Note:* This is an unfortunate choice of words as either problem-solving style preference can create innovation. It might be better to describe these individuals as pioneers, risk takers, or "out-of-the-box" individuals.) They will tend to throw out ideas for consideration without complete evaluation. They will care less about agendas and schedules. They will tend to associate elements from different areas and technologies to arrive at a solution. These mental connections will not be obvious to others until they are explained. Their process for innovation and idea generation is often not recognizable by anyone other than themselves. They will frequently have to "slow down" and explain where their ideas came from.

The KAI assessment is a number from 32 to 160 and the sum of three subscores discussed previously. The lower the profile, the more adaptive is the preferred problem-solving style. The higher the number, the more innovative or out of the box is the preferred style. The average of the world is about 90 and the "2 standard deviation" range is about 70 to 110. Profiles outside these ranges are very strongly adaptive or innovative, respectively.

This assessment is very hard wired at an early age within individuals. It is possible for someone to be "different" for a period of time, but the more the difference in expectations, the more stressful the journey. It is likely that an individual, required to "behave" in a way that is not natural for an extended period of time will seek other employment or, if staying with the organization trying to force the change, will become demoralized and stressed, affecting job performance. For example, if a high KAI person who enjoys and is hard wired to think out of the box and make disparate associations is asked to be a bank teller, they may be able to perform this function well for a short period of time, but if the time is extended, they will be thinking about ways to modify the process instead of complying with the rules.

Conversely, someone who is highly adaptive on the KAI spectrum is asked to lead a new idea generation session with little or no boundary conditions, they will, in a short while, become frustrated and uncomfortable dealing with uncertainty.

Let's take a look at these two extremes from an innovation culture standpoint. If the organization has a primarily adaptive profile (meaning that the majority of people in the organization have this characteristic), they will do an excellent job at product line extensions, but most likely miss a breakthrough that comes from a technology outside their normal realm of experience. They will also most likely only talk to their current customer base as a source of information about what is needed in the future, and only go to meetings and trade shows directly related to the current product and business. They may not talk to possible customers in a parallel universe. For example, auto manufacturers may talk to people who drive cars all the time to get information, but may not talk to people who fly most of the time. At the other extreme, strong innovators or pioneers will be very good at seeing and seeking out the nonobvious opportunities and make connections with other parallel universes, but may miss valuable short-range product extension opportunities. They may tend to confuse customers with a constantly changing set of questions and ideas. An adapter will enjoy reading current industry trade magazines, while an innovator or pioneer will enjoy reading about things unrelated to the current business or technology. Going to trade shows unrelated to the current business or technology will provide stimulus to such a person.

As with social style assessment, there are alternatives to the Kirton assessment, but they all measure these characteristics in some form.

These two characteristics, social style and problem-solving style, capture well the culture within an organization. In some organizations there may not be a dominating culture, but a mix of styles which the organization values and uses appropriately. What is important is to analyze and understand what the culture is, and if necessary, by department and function. Without this understanding, it is not possible to understand how culture affects innovation strengths and weaknesses. As mentioned previously, these behavioral preferences of individuals are fairly hard wired in someone's DNA, though there can be some subtle changes in the social style over time. This is important to understand as a dramatic change in the type of innovation required of an organization may require different social and problem-solving styles and the sudden demand for a change by a CEO will not necessarily produce the results desired. Stress will develop if a dramatic change in culture is required to achieve a totally different business objective. Prior to a CEO making major changes in business strategy, it is essential that the culture and the impact of planned organizational changes be understood. This will probably include revisions in where people are recruited, both in an academic sense as well as an industrial sense. There is nothing wrong with making a speech about new directions the organization may be heading, but to *demand* immediate agreement and buy-in may create organizational stress of such proportions that the goal has little chance of success.

Another point important to make is profiles that are in the middle. It is possible for individuals to be in the middle of the extremes on a KAI profile as well as on each of the individual social style designation letters. These types of individuals have the capability to bridge the gap between the two extremes of view and can greatly facilitate debates, discussions, and team meetings.

Before we leave this subject, there is an important item to discuss. These assessments (or any others you might use), given to individuals, whether in person by a certified provider of the assessments or on the web, is private information between the individual and the

Question 4: What is the culture inside the organization? How is it measured?

Figure 3-4 Question 4.

How do you plan to assess the culture within the organization?	How will you make sure that you measure both social and problem-solving styles?

TABLE 3-4   Measuring Culture

provider. This information is never shared with an employee's company or manager without the individual's permission. If these results are being used in a group setting, all participants must agree to have their information shared and in what manner.

Here's a table to help you (Table 3-4).

Understanding the Impact of Culture

Some of the implications of cultural differences have been covered in the previous section as we provided examples to illustrate the various social and problem-solving styles. We will now look at this issue in greater depth.

Let's start with the example of a company whose primary business has been a semi-monopoly for some time based on a strong patent portfolio. Some of the key patents are about to expire and competitors who have been waiting in the wings are preparing to enter the market. What might we expect the current organizational culture to be? There's a reasonable chance that this organization, over time (without conscious effort) has become very analogic in its problem solving and both sensing and judging in its behavior. Does this mean there are people in the organization that don't fit this profile? Of course! But the strategy and tone set by senior leaders will tend to be supportive of those of similar styles and tolerant and skeptical of those who are different. There may be a perceived need for other points of view, but the patience to tolerate these styles may be limited. A classic case is when a senior level corporate committee (probably composed primarily of "STJs") wants to review and possibly approve a new venture type of project and the summary is being presented by an "NP" who really wants to share the subtleties and possibilities of the proposal that go beyond just the facts. It is difficult for the "NP" to present the ideas in a concise form as there will be a natural desire to discuss side issues and possibilities. Coaching will be needed. Even organizations perceived to be very innovative have cultures and paradigms. There is an old story about 3M's considering entering the CD disk business, given its knowledge of some of the skills required to make these types of products. The project champion was given some advice by a veteran 3M'er—"Hold it parallel to the ground when discussing this opportunity. If it's thin and flat, they'll buy it." Think about nearly everything that 3M makes, from tapes to sandpaper. Even an innovative company like 3M may have a culture that is not recognized.

Think about the hot topic of today's business world and untold number of conferences—"Open Innovation." What does this mean? Generally, it describes an approach to innovation which admits that no organization knows it all and that working with outside groups, whether they be universities, competitors, or parallel universe technology companies, is necessary to compete in the long term. It recognizes that no one organization has enough resources to know everything. How would the various social and problem-solving type styles react to this concept? A strongly innovative KAI person would no doubt welcome such activities, but a strongly adaptive person might question the value of interacting with someone who was not familiar with the business or problem at hand. An extroverted version might welcome the opportunity to talk to unfamiliar people outside their industry whereas an introvert might not. However an introverted, strongly innovative KAI person might very well do a lot of reading and web browsing outside their normal work area.

What about collaborations and mergers? We have all seen examples of mergers negotiated by a few CEOs over lunch without giving the slightest thought to the cultures within their organizations. An agreement is signed and then the people deep down in the organizations have to begin to work together and quickly find out that their way of "doing business" is quite different. The execution of the merger falls apart. The ability to collaborate and work with others, both in an open innovation sense and a collaborative/merger sense is greatly affected by all the qualities previously discussed.

There are many lists of what is required within an organization to catalyze innovation. Here's one list that we can analyze:

1. Curiosity, courage, and risk
2. Positive contagion

3. Creating and nurturing collaborative teams

4. Rewards and recognition

5. Self-managing culture

Any other list can be analyzed in the same fashion by thinking through how various "types" will relate to the criteria judged to be important.

Curiosity—Assuming that this is a desirable trait, what would we expect of the various types of people we have discussed? It is likely that an innovative KAI profile person and an N/P social style person would have more natural curiosity? If we are expecting this and the people in the organization are "S" data driven, "J" schedule driven, and adaptive KAI people, this simply won't happen naturally. And it is unwise and unproductive from a morale standpoint for a manager to simply demand that such people become more curious. On the other hand, if there's a problem with internal accounting data resulting in an audit, these folks may be just who we want on the job. Curiosity is a natural trait that many people have. If there's been no need inside an organization for a high level of curiosity for a sustained period of time, we can't simply wave a wand and make it happen. We will have to hire some different types of people and/or hire some outsiders on a temporary basis to provide an example.

Positive contagion—We may all have images of what this means, but let's assume that it pertains to the ability of people to spread their innovation enthusiasm around the organization. Will this be easy for an introvert? Contagion normally also implies some level of emotion. This will be much more natural for an "F" social style who naturally interacts on that basis. Maybe our image of positive contagion is viral activity on the web. In this case, the extroversion/introversion difference may not matter. It is likely that a "P" social style person will also be more prone to this activity since they are thinking about possibilities more often than a "J" who is interested in closure.

Creating and nurturing collaborative teams—Teamwork is the buzzword of the last two decades. It requires the ability to interact with others, even those who may be quite different in personalities and styles. It is generally recognized that most innovation breakthroughs happen through a collaborative effort. Though there are certainly examples of lone inventors coming up with breakthroughs, this is the exception rather than the rule. Most problems requiring breakthrough innovation are more complex than in the past and require multidisciplinary teams to address. Even lone inventor examples give way to the complexity of skills required to manufacture, assemble, and sell the invention.

Teams always have a leader, either appointed or de facto. That leader needs to recognize that, in order to hear everyone's input, both extroverts and introverts need to be heard. Just because people don't participate and speak up does not mean they do not have ideas. They must be engaged through a proactive effort by the leader or facilitator.

Teams may also have deadlines to meet in terms of reports and deliverables. The "J's" will have no problem with this; the "P's" will. The "S's" and "T's" will be concerned about all the facts and data needed to support a recommendation to management; the "N's" and "F's" will want to consider the potential of decisions on the organization and will want to consider additional possibilities. A team leader, *if the profiles of the team members are known,* will be able to maximize the output of team. Occasionally this may be done by segregating issues of concern for a separate discussion that will not detract from meeting short-term goals.

Rewards and recognition—When a successful innovation has occurred and the decision is made to reward and recognize the individual or team, we often fall into the money pit trap, assuming that money is the primary motivator for everyone. Obviously there is some minimum level of monetary reward that people expect, but in the case of an extraordinary award, how does the idea of a travel award sound to a marketing person who is already on the road five days a week? To a strong introverted person, how much reinforcement is gained by a large party that requires a lot of socialization with others versus a quiet dinner out with their spouse? How would a "P" person respond to an opportunity to explore a new area with little management oversight for a period of time?

Self-managing culture—This characteristic implies that a culture that is not top-down driven is supportive of innovation. One could argue that a CEO with a grand vision and intelligence to support that vision (Steve Jobs?) might be counter to this trait, but it would be generally agreed that no one person knows everything about everything. So with that kind of exception, what does this concept mean in terms of culture? One interpretation might be that teams can self-organize with minimal management involvement to attack a problem brought in by anyone in the organization with the understanding that, at some point of significant funding, management approval at some level will be required.

Question 5: What is the impact of the culture in your organization?

FIGURE 3-5   Question 5.

What do you do now?	How does culture impact?

TABLE 3-5   Caring about Culture

How would you like to change what you do? How do you do it?	What culture changes are required?

TABLE 3-6   Changing Culture

Another interpretation might be that once a problem has been identified, certain individuals will *volunteer* to work on that project. If we think back to the individual characteristics discussed earlier, we can ask what kind of individuals might be comfortable with this sort of self-organizing structure and the freedom associated with it? Certainly individuals with strongly innovative KAI profiles will be very comfortable with this kind of framework, while more structured, adaptive KAI individuals will not be. There is no right or wrong about any of this—it is just a style *preference*. More structured social style preferences such as "S" and "T" will be looking for structure and guidance. Any of these kinds of people can work on a project that does not match their natural tendencies for some period of time, but the larger the differential between natural style and preference and the actual task and environment, the larger the stress level becomes.

You may have your own list of what's important in your innovation efforts. Take each bullet thought to be important and think about the various social and problem-solving style preferences and their ability to contribute positively to each.

Here are some tables to help you.

How Does One Go About Changing Culture?

Let's assume for the moment that there is a need to change the culture. Sports teams do this all the time by changing coaches and bringing in an entire new staff chosen by the new head coach. In the real commercial world, this is usually impractical. However, we have seen cases where a CEO is replaced or moved up and those individuals who did not get the top job leave for another opportunity and the new CEO (especially if hired from outside the company) hires a whole new immediate staff, many times from their previous employer. But we can only replace so many people easily and changes at the top allow different speeches to be made about how "things are going to change." It's never that easy.

An organization that realizes it needs to change its culture cannot simply fire everybody one day and instantly replace everyone with the "right" people the next day, even if it knew exactly what kind of people it needed. Culture change and adjustment takes time and patience.

These types of changes will be slow in coming and potentially disruptive to an organization, especially if major changes are perceived to be necessary to meet new business conditions. For example, the rise of the web as a commerce tool has replaced the need for many personal salespeople calling on customers who do not have special needs. It is unlikely that an extroverted salesperson who now needs to spend a significant amount of time behind a computer screen answering inquiries from people not seen or heard (nor who are golfing partners!) would make such adjustment easily even if the need for this transition was obvious to everyone. Similarly, the rise of the web in global team collaboration, replacing many long airline flights, changes the nature of the interaction in subtle ways. There is less impact from an outgoing personality unless a tool such as Skype is used.

Consider again the hot topic of the day—open innovation. This is a recognition that, in today's complex world, no one company has all the intelligence and technology needed to address all customers' problems. We now recognize the value of working with organizations outside our core area.

A highly innovative KAI person will be able to make disparate connections with outside technologies whereas a more adaptive KAI person will focus on connections and extensions within the current industry and may not see the value in connecting with parallel universes. A highly extroverted (E) and high KAI person will truly enjoy both the concept and activities typically involved in open innovation. If your organization is beginning to recognize the value in open innovation, but you have a history and culture of being very introspective, you may have an organization full of "S's" and "J's," and adaptive KAI profiles. Individuals with a strong history of involvement in Six Sigma programs are probably strong STJs in social style and more adaptive on the problem-solving style.

"Listening to the customer" is another buzzword phrase we hear. A highly extroverted person has trouble listening as they are normally talking over people. They will miss the subtleties of what someone is trying to say. A strong "S" person will pay more attention to the details of how a product is used and the nuances that cannot be captured in a survey. A high KAI person will be able to make associations between multiple customer needs not apparent to others. Matching of personal styles with the objective is important.

The last thing an organization should do is to make a speech discussing the need for new "behaviors" and then expecting the culture to change overnight. People with any kind of loyalty will try, but those who cannot will leave or retire. Since people learn best by example, here are some suggestions for dealing with this issue. These suggestions assume that individual style assessments have been given and people are willing to share in an open way within their teams, with their managers, and that managers are not using this information in a punitive way. These are big assumptions, but if there is not an open, fearless dialogue about this topic, the effort will not be effective.

1. Understand clearly the difference between the culture desired and the culture in place. Make a list of the pros and cons of each so that they are clearly understood. Make sure you understand why the culture that is in place is the way it is and how it developed.

2. Make sure you can clearly explain the need for change and why to anyone at any level in the organization.

3. In what places in the organization is one culture or behavior preference greater than another? Have various groups give workshops or "lunch and learns" about their culture and its impact as well as plusses and minuses. Is it possible to do some job rotations to expose people to different styles and cultures?

4. Consider having people put their style preferences on a display or desk title plaque for everyone to see, introducing a topic of conversation.

5. Identify the "bridgers" in the middle of some of the preferences (both social and problem solving) and ask them to proactively assist in any transition.

6. In group presentations, present examples of the strengths of each style and preference. Ask the groups to react and make suggestions as they relate to the business changes that are seen and required. Ask the groups to share examples they have seen where the various styles have made an impact in the organization.

7. If there is a large gap between what exists and what is desired, selectively hire outsiders to catalyze the change, but not in a destructive way ("we're here to change the way things are done around here").

8. Look for mismatches in styles and position requirements. For example, Six Sigma is more of a process monitoring effort followed by breakthrough idea generation. Match the skills with the personalities. Design for Six Sigma (DFSS) will appeal more to the "N's" and higher KAI people.

9. Increase the involvement with customers so that all the different styles can apply their strengths against the problem. Don't over screen the opportunity.

Do you understand the concept of culture?	Yes	No	If yes, how do you explain it to others? If not, why not? What will you do to better understand it?
Do you understand why culture is important?			If it is, what are you doing to understand it within your organization? If not, why do you think it is not? What would you do if you did?
Does your organization care about its culture?			If it does, what actions would an outsider see that would support that claim? If not, why not? What are the possible consequences of not caring? Have you seen any?
Are you measuring "culture?"			If so, how are you measuring? Why did you choose the methods or assessments that you did? Do the assessment instruments and methods measure what is important? How did you decide what was important?
Do you understand the impact of the culture within your organization? The impact of changing it?			What do your assessments tell you about this? What business and organizational changes do you see forthcoming? What does your analysis tell you about how your organization will react and handle it? What aspect of culture needs to be changed? How will you go about doing that? What will be the impact?

TABLE 3-7   Cultural Analysis Inventory

10. A major shift in internal culture that requires different styles, as opposed to using the existing styles more efficiently as the culture shifts, will come only from a slow change in the nature of the people hired. It will also come about to some extent by selective outplacement (there is no way to totally avoid this), and individuals volunteering to leave when the cultural change is too much of a stretch for them.

11. Recognize that the journey to change culture is a slow deliberate process. There are many scientifically based assessment tools that can assist, but only if used in the right way.

Summary

Culture change can be a necessary and important organizational journey that is occasionally necessary, both when organizations merge and as business conditions or the nature of an individual business changes. People, who *are* the culture, have some very hardwired behavior preferences that can be measured and utilized to assist in transitions. Understand them and use them appropriately.

Here is a summary table for use in assessing your understanding of culture and its impact on your organization and its business (see Table 3-7).

Bibliography

Keirsey, David. *Please Understand Me II.* Prometheus Nemesis Book Co., 1998.

Kirton, M.J. "Adaptors and Innovators—The Way People Approach Problems." *Planned Innovation*, 3, 1980: 51–54.

Kirton, M.J. "Adaptors and Innovators—Why New Initiatives Get Blocked." *Long Range Planning*, 7 (2), 1984: 137–144.

Yund, Jennifer. "The Relationship Between Foursight™ and Occupational Groups: Examining Creative Process Preferences for Chefs, Engineers, and Nurses." *Buffalo State College Department of Creative Studies Masters Thesis*, Dec. 2005.

http://www.kaicentre.com

http://www.buffalo-bcpi.com

http://en.wikipedia.org/wiki/Myers-Briggs_Type_Indicator

http://www.personalitydesk.com/learning-center

http://en.wikipedia.org/wiki/DISC_assessment

http://www.cpsb.com/assessments/view

CHAPTER 4

Leading Innovation: Ten Essential Roles for Harnessing the Creative Talent of Your Enterprise

Lisa Friedman and Laszlo Gyorffy

Introduction

People want to be innovative. Innovation appeals to an innate human desire to build a better future. In today's dynamic enterprise, employees naturally want to develop and implement new ideas. Companies have the opportunity to capture their employees' ingenuity by teaching skills and creating an environment that promotes and optimizes innovative behavior, then building the stage to let their superstars shine.

The goal should be for leaders to take their team and their organizations to a place where everyone, everywhere, is responsible for innovation everyday—whether as an idea generator, team member, facilitator, mentor, or innovation sponsor.

How does a leader start the process of fostering an environment that invites innovation and gets the most out of the ideas of their employees?

They do it by inspiring others about what is possible; by ensuring the importance of innovation is recognized by all; and by creating what may at first sound like an impossible paradox—establishing a discipline of innovation. Bringing this discipline to life and embedding it in the core DNA of an enterprise requires leaders to become good at playing multiple roles. There is no one-style-fits-all when it comes to inspiring and enabling innovation in an organization.

Over the years, 10 roles have stood out as defining characteristics of successful innovation leaders. This chapter will outline each role, define its core elements, offer examples, and end with questions to encourage you to reflect on your own capabilities. Collectively, these 10 roles offer a holistic leadership model for nurturing and harvesting the creative genius of your people.

A frequently quoted statistic about the connection between leadership and organizational creativity comes from Goran Ekvall, Professor Emeritus of Organizational Psychology at the University of Lund in Sweden.[1] Ekvall found that 67 percent of the variance in the creativity in an organization was directly related to leaders' behavior. That is, when employees were apathetic, indifferent, and disinterested in their work, there was a 67 percent chance this was related to leadership behavior. The inverse was also true: When people were producing creative solutions and products, when they found joy and meaning in their job, and when they were willing to invest energy in their work, 67 percent of the time, this was attributed to leaders' behavior as well.[2]

Playing the 10 Roles

We all wear different hats in our personal lives. On any given day, we might be a parent, a friend, a big brother or big sister, a chef, a spiritual advisor, a carpool driver, and/or a coach to the neighborhood sports team. Each of these roles carries a unique set of demands and requires very different actions. Living a fulfilling life often involves playing a number of diverse roles, with the ability to move seamlessly from one to the other. These roles may change over time and our capacity to do them well depends on our ability to evolve our thinking and skills.

It is easy to recognize the roles we play in our private lives, but which roles are most important when it comes to leading innovation?

One role differentiation that is important for innovation is the distinction between managing and leading. Both roles play a key part in business success, but innovation is largely a leadership activity. Management focuses on following orders, organizing work, budgeting time and resources, assigning the right people to the necessary tasks, and ensuring the job gets done. The focus is on efficiency, predictability, quality improvement for existing products and services, and effective control. You can see how this orientation and skill set can help at the back end of the innovation process where the primary focus is cost-effective, rapid implementation. But this kind of management is often a significant inhibiter at the fuzzy front end of innovation. Playing the role of leader requires building comfort with risk, seeing routes that others avoid as potential opportunities for advantage, happily testing new ideas, and finding ways around obstacles to accomplish complex tasks that might never have been done before. Leadership demands a future-oriented perspective that often challenges the very predictability that managers so keenly try to maintain.

Leading innovation requires 10 essential, and very different, roles:

1. **Futurist**: Looks toward the future, scouts new opportunities, and brings future possibilities out of the fog so that everyone can see them and their potential. Also enables people throughout the organization to discover the emerging trends that most impact their work.

2. **Direction setter**: Creates and communicates vision and business strategy in a compelling manner, and ensures innovation priorities are clear.

3. **Customer advocate**: Keeps the voice of the customer alive in the hearts, minds, and actions of innovators and teams.

4. **Architect**: Designs (or authorizes others to design) an end-to end, integrated innovation process, and also promotes organization design for innovation, where each function contributes to innovation capability.

5. **Venture capitalist**: Secures funding for innovation, evaluates and selects projects to receive resources, and guides implementation.

6. **Mentor**: Coaches and guides innovation champions and teams.

7. **Barrier buster**: Helps navigate political landmines and removes organizational obstacles.

8. **Networker**: Works across organizational boundaries to engage stakeholders, promotes connections across boundaries, and secures widespread support.

9. **Culture creator**: Ensures the spirit of innovation is understood, celebrated, and aligned with the strategy of the organization.

10. **Role model**: Provides a living example of innovation through attention and language, as well as through personal choices and actions. Key stakeholders often test the leader's words, to see if these are real. For innovation to move forward, the leader must pass these inevitable tests—to show that, yes, he or she is absolutely committed to innovation as essential to success.

1. Futurist

Here comes the future—ready or not! Industry after industry is finding itself in the middle of a revolution, where products and services, business models, competitors, potential partners, underlying technologies, and customer demands may all be changing at the same time.

Any part of your current business or organization can be a target of innovation, but today's rapidly evolving business environment often requires rethinking fundamental business strategies and looking ahead to the next generation of products or services. It's hard to imagine an industry that is not going through a revolution, which leaves the field wide open for companies determined to be winners in the emerging generation. (See the industry life-cycles discussion in Chap. 2 on Creating Your Innovation Blueprint.)

Playing the role of futurist requires leaders to become comfortable going beyond what they know, to be interested in discovering something new that has the potential to profoundly change their own enterprise. The role requires leaders to work with their teams to look in unexpected places for new and unanticipated trends. It means actively monitoring the external business environment to learn about emerging needs, technologies, competitors, and adjacent markets. It means being engaged with customers to see how the most forward-looking ones are beginning to use your products or services, or to notice what else they might be using instead. It means getting to know the start-up companies working in garages, labs, and innovation incubators all over the world, with plans to disrupt your industry. Mostly, it means being open to surprise. But in this case, surprises are filled with learning and innovation opportunities.

Too often we see innovation initiatives that completely skip this "futuring" aspect. They ignore the true front end of the innovation process and move directly to idea generation. Brainstorming based only on current conditions and customer needs runs the risk that the ideas created will not carry the company very far into the future. Generating lots of interesting solutions when you haven't uncovered significant new market and customer opportunities, when you haven't yet explored their dynamics and potential, will naturally limit your success when searching for breakthroughs.

There is also something very compelling about seeing the new ideas and processes that other innovators are creating. The future has a sense of energy and excitement to it. Looking ahead tends to add momentum to the idea generation process that follows it. Once you and your innovators connect to the immense creativity occurring in your field, new ideas typically flow naturally from what you learn.

The leader's role as a futurist involves

- Scouting trends that can impact the enterprise.
- Making these trends visible for others.
- Encouraging people to be on the lookout for important trends themselves and to bring important information back to their teams and leaders.
- Designing means for external trends to be brought into the organization on a regular basis.
- Building a future-oriented culture, a dynamic enterprise, where people understand that "we are on the move."

The future will never look the way we imagine it. In fact, much of what we foresee in the future already exists in today's leading edge companies. Some of the trends we see now will survive and thrive to become tomorrow's standards, while others will quietly disappear. We can't possibly predict the winners and losers, but the thinking we do to anticipate potential changes will prepare us for the actual future, as it becomes our new day-to-day reality. We'll be in touch with the customers, know the potential partners, understand where to find or acquire the new technologies, and have teams with background knowledge ready, who can quickly put their learning to use to create new solutions.

"Those that choose to build their present out of images of the past will miss the opportunities of the future."

—Winston Churchill

2. Direction Setter

As the Cheshire Cat told Alice, if you don't know where you're going, any road will take you there.[3] Innovation is a journey into the unknown and there are multiple paths and many options open to the leadership team. Once emerging trends are clear, leaders next need to set the strategy going forward: Who do you need to become in this emerging future? ("Who" you want to become asks more than simply "What" you want to become. "Who" goes beyond the products and services you decide to offer. What you offer, combined with your values, point of view, and what you stand for, all determine your identity as an enterprise. This tells your market and your customers who you intend to become in this emerging future.)

A future-driven business strategy defines the innovation agenda. What needs to be created that does not yet exist in the enterprise? What new products and services? What new business models or processes? And what changes in the organization will be needed to become the kind of enterprise that can succeed in the envisioned future?

In this role, leaders ensure their people have a clear direction that frames innovation priorities, so it is clear where fresh ideas and breakthrough projects are needed most. Leaders also secure funding for these priorities and make sure their innovators know resources are available for solutions that move the business in the intended direction. In addition, leaders help innovators know how far to stretch: Are incremental solutions needed that can be implemented simply and quickly or should innovators reach for bold, disruptive ideas that enable the organization to stake out new territory? Innovators can benefit from clear directions about business strategy, degree and timing of innovation needed, and how much risk to take, how far to stretch.

In our experience, the primary reason game-changing innovation fails is because time is not invested up front to take these actions, to align the organization behind a shared strategic blueprint for the future and a clear and widely shared innovation agenda. (See Chap. 2 for more detail.)

Creating a compelling vision of the future offers the organization's most creative people a sandbox in which they can operate. It pulls from their best thinking and offers motivation for them to follow through on whatever their innovation task. Direction setters communicate facts, set numerical targets, and use data to bolster their logic in setting goals or making their case for change. Then, because data alone rarely "moves" people, leaders expand their narrative and communicate the "why," to make the decisions meaningful. They bring together the compelling story about future potential that engages people more deeply and inspires them to bring their creative best.

Specifically, the direction setter

- **Brings focus**: Gets his or her organization aligned around the desired future and the innovation priorities that are demanded to achieve the goals.
- **Defines boundaries**: Sets stretch goals and concrete business targets that demand innovative solutions.
- **Generates energy**: The goal is to connect in a way that will inspire others to engage with the vision, and awaken the passion and imagination in each of those who elect to follow.
- **Encourages collaboration**: With a compelling future that stretches the capabilities of the enterprise, it becomes clear that no one can succeed on their own.

Consider the bold direction setting of Steve Jobs who was able to build support for Apple's dramatic leaps from computers to iPods and iTunes and then again to iPhones and mobile apps. The underlying design and technical expertise that makes these products distinctly Apple remained constant, but it was Jobs and his team that defined and led the company on a distinctly new path that allowed it to enjoy an unprecedented level of success—in a whole new industry. Apple is now the largest music retailer in the world. Let's not underestimate the courage of these direction setters. Today the path seems obvious, but back then, the outcomes of these choices were not guaranteed. Steve Jobs had the vision and fortitude to stand before the board and defend the opportunity to explore and fail, with the hope that they would explore and succeed.

Of all the 10 roles of innovation leaders, your ability to mobilize others to invent a positive future will have the biggest impact on your legacy as a leader. We often try to summarize this role in a way that captures the depth of its importance:

Creativity is thinking something different.

Invention is doing or producing something different.

Innovation is making a difference.

3. Customer Advocate

Playing the role of customer advocate means you are able to

- Stay informed about customer satisfaction and the competitiveness of your offerings.
- Bring the "voice of the customer" into decision making and idea generation.
- Set the expectation that creating customer value is everyone's job.
- Support staff to engage in open innovation with customers.

You want your employees to innovate with the customer in mind. Your team, unit, division, and even enterprise should know what is most important to your customers. They need to understand which problems, if solved, would have the greatest positive impact. These insights can become the inspirational sparks of your group's creative thinking. Strong innovation usually emerges from collaborative, hands-on or observational experiences that take place on the front lines with customers.

A midsize medical imaging manufacturer had been acquired by a larger company looking to establish a bigger footprint in the industry. The new parent company had its own sales force and regional account management structure, so the smaller firm lost the close contact it had previously had with its customers. Customer information was reduced to satisfaction studies and marketing reports. In the past, they had run end-user training at their plants, where engineers and designers got to meet and interact with customers on an ongoing basis. Customers were regulars in the cafeteria. Now the training was delivered at an educational center far from the community of product developers. Staff could go a year without seeing a customer and the results were beginning to show. The parent company was demanding that they cut costs in one of their key products, and the engineers started down their typical path of gathering the technical specialists needed to brainstorm cost-reduction ideas. However, this time the head of the division intervened. He had recently launched an innovation initiative and wanted to accomplish cost-cutting in light of a broader product strategy. He sensed the team needed greater exposure to the problem at hand. He encouraged them to visit hospitals and medical centers where their product was in use, to observe the experiences of doctors, lab technicians, nurses, and patients as they used the equipment. Given strict patient confidentiality requirements, it took a great deal of effort to get the permissions needed to gain access to actual patients using their equipment, but it was possible. Their observational exercise turned out to be quite eye opening.

The engineers learned much more than they had expected about their current product—both about aspects that seriously frustrated lab technicians and patients alike, and about features they had planned to improve that didn't matter to users. Observing customers helped them clarify how to simplify their products to create a better user experience. A few of the changes to create simplicity added cost, but many more of these changes decreased cost. The final result was a less expensive product that gave users value where they wanted it most.

The division head trusted his intuition that his team needed to reconnect to their customers and was rewarded with a much better product design at a lower cost. It is too easy to fall into the trap of assuming you know what customers need. Connecting with actual customers is an important step in developing products and services that create value.

In addition, in today's age of social media, there are more ways than ever before to connect with customers online. Customers increasingly rate and rank and review almost every kind of product or service. Customers ask each other questions and provide answers from their own experience. Innovation leaders find ways to ensure their people connect with customers online as well as in person.

For example, in 2003, the global software company SAP created an online customer community called SAP Community Network. A decade later, 2.5 million users worldwide communicate with each other across 430 subgroups that span multiple countries and languages. Users add up to 3000 new posts each day.[4] Many kinds of customers—software users, developers, consultants, mentors, and students—use the SAP Community Network to get help, share ideas, learn, innovate, and connect with others. Leaders can tap into this community at any time to understand customers' concerns and questions, and even their new product wishes. In 2011, SAP added Idea Place, where users submit ideas for product improvements or for entirely new products. In its first two years, this community produced 8700 ideas, 9000 comments, and 53,000 votes, which in turn led to over 600 implemented ideas.[5] From its beginning, SAP senior leaders envisioned this community, believed in its value, invested to nurture its growth, and encouraged employees throughout the company to use it to stay in touch with customers.

Jeff Bezos, the founder and CEO of Amazon, is well-known for advocating, "Start with the customer and work backwards."[6] Years ago, when Amazon sold primarily print books, company research showed that customers liked e-books—a new product that threatened Amazon's core business. Bezos took a strong stand that the company must not create a business strategy designed only to protect their existing business when their customers clearly wanted something else. He demanded his people do just the opposite—create what customers wanted and find a way to build a business around it. They did just that, and by 2011, e-books sales surpassed sales of print books.[7] In fact, by April 2011, Amazon became the world's largest e-book retailer.

Bezos consistently took a stand for customers even when it led to controversy inside the company. For example, Amazon developed a system to enable competitors to sell on its website, even when competitors offered cheaper prices. Each product page shows Amazon's price for new merchandise, other retailers offering the same products—often at cheaper prices, and even individuals selling used products at extremely lower prices. This decision cut into Amazon's sales, but it also brought more customers to the site and kept customers coming back to buy more products through Amazon overall. Now Amazon is known not only as an online product retailer, but as an open marketplace where customers can find full information on products they might want to buy. They can find product reviews—whether positive or negative, can post their own reviews, can interact with other customers, and can even sell their used products back to others. Amazon created their marketplace from the customers' point of view, and customers keep coming back.

When you give customers the chance, they will tell you what they value most. They will let you know in great detail what is broken and what is high on their wish list. Listen to them and work with them to make them happy. But don't rely only on them to create the future road map for your product or service. That's *your* role as direction setter and customer advocate, and the responsibility of your team. Customers often don't know what they don't know. Meeting the needs they identify for you may seem like the safest way to go, until a competitor arrives with a radically different and improved offering. Once your customers see what is possible, they will demand it. You don't want to be caught in the risky game of playing catch-up or trying to get out of the commodity trap. Part of being an effective customer advocate is to push the innovation envelope and stay out in front of your customers and competitors.

"If I had asked people what they wanted, they would have said a faster horse."

—Henry Ford[8]

Finally, the task of customer advocacy is even more complicated when applied to individuals who aren't yet your customers. A visionary innovation leader keeps the presence of future customers alive in the creative minds of his or her people. Think of Toyota when it released the Prius to an emerging green car-buying population. Toyota was targeting a new population—buyers who were distinctly different from the customers buying their traditional cars. In fact, Toyota's Prius helped to define this group's identity—to show how large a group they were. This shift into new market spaces required leaders to educate employees about the new customers they would be serving and the new solutions they would be providing.

4. Architect

"All organizations are perfectly designed to get the results they get."

—Arthur Jones[9]

Highly innovative enterprises are specifically designed and directed by their leaders to be capable of delivering increasingly better results, both from continuous improvements and from major breakthroughs that create new markets.

In a turbulent marketplace, when more complexity can add even more confusion, it is important to have a simple and straightforward innovation system. Shared and easy-to-use innovation practices and tools can enable everyone in a company to work together to develop ideas that deliver compelling customer value. This is not a time to rely on lone genius to carry the day. Yet few companies have developed a shared set of common innovation concepts, practices, and language that can enable them to collectively supercharge their efforts to grow their business, reduce costs, or outperform the competition.

An Integrated Innovation System

Leaders have the responsibility to ensure effective innovation architecture is created, understood, and widely shared throughout their enterprise. An integrated innovation system covers the full end-to-end innovation process, and ensures the practices and tools are aligned and flow easily from one to the other. Clear and shared innovation architecture ensures employees have the methods and tools to

- **Explore opportunities**: To discover emerging market trends and connect with customers to understand their needs

- **Generate breakthrough ideas**: To create and contribute their ideas

- **Optimize value**: How to develop their ideas into strong value propositions and business plans, how to work together with others to optimize the value of these over time

FIGURE 4-1   An Integrated innovation process.

- **Select and fund the best**: To understand the process for competing for approval and funding
- **Mobilize for results**: If their projects are selected, innovators understand where to get needed support for implementation or for taking their projects to market

(See Fig. 4-1. Also see Chap. 2 for more detail.)

Building Essential Innovation Skills

An effective innovation architecture includes training as needed, to ensure employees have the most essential innovation skills. If you are looking for ordinary people to come up with extraordinary ideas, they should be well versed in the fundamentals of innovation. You would not dream of mobilizing an army without running the troops through boot camp. Yet we send unarmed innovators into a battle for funding and management attention all the time. Innovators being asked to contribute their energy and imagination to assist the organization in its mission need the skills to

- Identify significant innovation opportunities
- Generate creative solutions
- Develop value propositions
- Communicate ideas in a clear, concise, and compelling manner
- Create cost-effective prototypes of their ideas
- Collaborate effectively with colleagues to rapidly improve the value of their ideas

Most employees possess some of these skills to some degree, but imagine the innovation potential if everyone in your enterprise was proficient at all of them. Not only would each person and his or her idea have a greater chance of success, but collectively whole new levels of teamwork would be unleashed through interactions with competent collaborators, mentors, and sponsors.

Innovation Tools

Too often the processes organizations use to pursue innovation can actually erode their capability to innovate. Systems built on stages and reviews can create a dry, bureaucratic process, limit the potential for breakthrough thinking, and deflect attention from customer experience. Other times, by limiting responsibility for innovation to a specific department, organizations underutilize the creative capabilities of other employees.

One example of a tool that cuts through the bureaucracy and enables employees at all levels to develop their innovative ideas is CO-STAR. By answering six basic, yet essential questions, CO-STAR can turn an inexact concept into a well-honed value proposition.

CO-STAR stands for Customer–Opportunity–Solution–Team–Advantage–and Results. These six elements are central to crystallizing the business value of an idea. It is a vehicle for enhancing innovative ideas through constructive dialogue and a process for focusing attention on the fundamentals that make an idea valuable.[10]

CO-STAR guides innovators to answer the following questions about their idea:

- Who are your intended **Customers** and what are their most important unmet needs?
- What is the full potential of the **Opportunity**?
- What is your proposed **Solution** for capturing the opportunity and satisfying customers?
- Who needs to be on your **Team** to ensure your solution's success?

- What is your competitive **Advantage** over alternatives?
- What **Results** will be achieved from your solution? (As well as the **Rewards** to your customers and the **Revenue** or **Return** to your team and company?)

There are always many more ideas than those that can provide important value. Every element of a CO-STAR value proposition is essential for an idea "to grow up" to deliver value. The continued insight and passion of an innovation champion and team is necessary to turn an initial idea into a value proposition worthy of investment.

Value propositions then become the new currency of innovation, rather than only raw ideas. Venture capitalists have been using methods like this for years to reduce risk and maximize returns on their investments.

Scale: Tapping the Genius of the Entire Enterprise

Social media and crowdsourcing have introduced a new era of collective brilliance. The latest technologies break down traditional silos and allow for open innovation with large numbers of employees, as well as with suppliers, alliance partners, and even customers.

Pioneering companies employ online innovation platforms to harness the genius of the group far beyond teams that meet in the same place or at the same time. For example, a web-based application "Q+ Innovation Management" enables innovators to share their ideas utilizing the CO-STAR framework online and to gather feedback from any group anywhere in the world. In addition, companies can run targeted innovation campaigns—issuing invitations to innovate within a specific area of importance (e.g., revenue growth, cost savings, operational efficiency, or product breakthroughs), with a two to three week window of time for collecting hundreds of ideas from employees. Participants submit value propositions online, as well as provide feedback and vote for ideas of others. "The wisdom of the crowd" drives the highest ranked value propositions to the top, along with comments or links to additional information and resources for each one. A real-time dashboard makes it clear how ideas, activities, and decisions are progressing through the pipeline.

Just as a good surgeon, engineer, or artist would insist on using the right equipment to get the best results, leaders now have powerful online innovation tools to support their disciplined practices. There is no longer any reason for companies not to scale their innovation activities to match the size of the problems they are tackling or the opportunities they are chasing.

Innovation Architecture for Quick Wins versus Big Game-Changing Ideas

Innovation is often slowed or inhibited altogether when new ideas go to managers, who then determine whether the ideas can fit into the next quarter's or next year's budgets, which are often already allotted for existing products and services. The budgeting process can slow the speed of putting any new ideas into practice. An effective innovation architecture offers alternative routes for approval, funding, and implementation for a range of projects of various size, scope, and disruptive potential:

- For ideas that fit within the boundaries of their own departments, managers can set aside an innovation budget for ideas they expect will arise between budget cycles.

- For ideas with an impact across departments or divisions that would still be relatively easy and quick to implement, leaders can set up a type of Quick Wins innovation panel. A small group of innovation leaders from various disciplines review ideas and give rapid responses for approval and resources.

- For big game-changing ideas, leaders often set up Game-Changer innovation panels, designed to evaluate larger and more disruptive proposals. These panels have innovation leaders from multiple disciplines or divisions of the organization, and may also bring in external experts as needed. These groups often fund proposals in sequential stages, with initial smaller amounts of funding that enable innovators to complete their business plans, prototypes, or initial customer testing, and larger amounts of funding available as projects show signs of success and need to resources to scale their solutions.

Aligning the Organization for Innovation

Leaders architect not only the innovation process itself, but each element of the organization that influences innovation as well. Organization designs that promote innovation include

- HR practices to reinforce innovation (e.g., recruiting, hiring, orientation, training, development, performance evaluations, and perhaps the most important—rewards and incentives).

- Finance ensures funds are available for developing ideas and innovation projects. Finance may also create the metrics by which innovation is monitored and evaluated and where successes can be proven.

- IT (information technology) supports online innovation, collaboration, and knowledge management tools.

- Legal offers IP (intellectual property) protection, as well as support for collaboration and partnering.

- The Facilities department designs environments for inspiring and enabling innovation and collaboration.

- Communications keeps innovation visible and celebrates successes.

Essentially any element of the organization can play its part in making innovation possible and probable. (See Chap. 2 for more detail.) In a large enterprise, each of these functions in the organization likely has its own set of leaders, so the Architect role in this case requires building a leadership coalition that fully commits to and enables innovation. Innovation architecture is a team sport.

5. Venture Capitalist

Venture capitalists (VCs) are good at finding, choosing, and bringing the creative ideas of others to market. They have good judgment about which ideas and value propositions will work. They can project how potential ideas may play out within the organization and in the marketplace. If innovators "pitch" their ideas and value propositions, leaders in the VC role are the ones who "catch"; they are the authorities who will decide the fate of the innovators' dreams.

Venture capitalists are disciplined risk takers. They understand that not every idea will develop into an innovative solution, bring the company new customers, or generate greater return from existing customers. They are looking for new and fresh ideas, but also for innovators who make a credible overall business case for why the organization should support their efforts. A tool such as CO-STAR (described earlier) or a similar system to evaluate ideas is extremely helpful in this case. VCs need a way to evaluate whether ideas meet an important customer need (for either internal or external customers), and whether the ideas respond to a significant market opportunity—an opportunity that goes far enough beyond current solutions to justify its unpredictability.

Inside companies and teams, innovation leaders in the role of venture capitalist often clarify the innovation challenge (the "CO" in CO-STAR). They typically go a step beyond the direction-setting role. Direction setters identify the opportunity in the marketplace and make it clearly visible, but the VC secures the funding and resources for projects, shows that these resources are available, and issues the invitation to innovate. Leaders in the VC role may create an innovation challenge or deliver an "innovation brief" in person, in writing, or through online video. The brief communicates the call for innovation, clarifies the resources available, and informs innovators about the process to develop and pitch their value propositions.

Playing the role of venture capitalist means you are able to

- Find and allocate time and funding to invest in innovation projects.

- Identify strategic areas or specialty fields to fund, while staying open to unexpected opportunities as well.

- Launch an innovation campaign, a call for innovation in a specific field.

- Communicate an innovation brief that gets people inspired about submitting their ideas.

- Determine the merits of an idea and its value to a customer.

- Evaluate the strength of a value proposition and business case, particularly compared to other alternatives.

- Determine a project's overall ROI, where "return on investment" includes financial as well as other potential kinds of returns.

- Mentor innovators once you have given their projects initial funding (see Mentor section below).

- Sponsor the implementation of big game-changing projects as well as incremental improvements.

Playing the role of VC also requires you to be comfortable establishing an innovation portfolio that works similarly to an effective stock portfolio. The portfolio should contain a mix of incremental improvements and revolutionary ideas for existing and new customers, and should balance risk across these ideas. Some ideas should offer a quicker and more certain return, some would offer a medium-range timeline or amount of risk, and a few more ambitious projects would be longer term and much less predictable. Any one idea might not succeed, but the portfolio at the end of the year should show a strong positive return on investment.

While professional VCs receive hundreds of business plans and hear hundreds of pitches every year filled with supposed world-changing ideas, most of them believe a great team in a big market can often figure out how to build a successful company. Many times, VCs initially invest in a start-up with a specific solution and business model, but in the course of growing the company and taking products to market, they and the start-up's founders discover that the market potential is not nearly as strong as they had predicted. In many instances, the business searches for and finds a completely new solution or business model. The founder team is able to "pivot," to launch a whole new version of the solution. The business that ultimately succeeds may barely resemble the initial business that was funded. As Eric Ries describes in his book *Lean Start-up*, an important part of the VC role is to recognize when it is time for a project to pivot and to test multiple alternative business directions quickly, to see if there is one that could ultimately succeed with customers.[11]

Still, there are times when a new venture will fail, and VCs know how to deal with this as well. Traditional business leaders often launch a small number of new projects and make every effort to ensure that all of them succeed. In fact, failure is too often not an option for a traditional business leader's pet project, no matter how negatively the market reacts. In contrast, VCs launch many start-ups and monitor all of them to see if they are achieving their expected milestones. When a new venture does not hit its numbers and a pivot is not possible, they cut losers quickly so they can fund the winners. VCs may make a number of initial investments and then rapidly reallocate resources when needed. The common summary is that for every 10 investments a VC firm makes

- Three to four are losers (−X)
- Five are "the living dead," which don't die out but don't grow either (1X)
- One to two are winners (superstars that produce 20X or more in return)

David Cowan, a successful VC with Bessemer Venture Partners who has 20 years of investing experience, says, "People are embarrassed to talk about their failures, but the truth is that if you don't have a lot of failures, then you're just not doing it right, because that means that you're not investing in risky ventures." He continues, "I believe failure *is* an option for entrepreneurs."[12] In fact, Cowan has invested in entrepreneurs whose last venture didn't succeed, after they made a strong case for what they learned and how they could put that learning to use to respond to a new and important opportunity.

While innovation leaders inside companies need venture investing skills, they also may have a wider range of success criteria than the average VC. Typical venture capitalists invest in a business to create a company profitable enough to sell or take public, so they can return funds to their investors with a sizable financial profit. Inside a company, the innovation success criteria are often more complex. Most companies aren't only bottom-line driven; innovation projects can claim a broader return on investment that includes elements such as strategic relationship building, brand development, or customer acquisition across the full range of a company's products or services. Instead of evaluating ideas based only on short-term financial gain for a single line of business, leaders can invest for reputation and status, or for strong connections with potential customers or strategic partners, if they think these will ultimately deliver financial gain for the company as a whole.

In addition, once VCs invest in start-up companies, they don't simply leave the innovators on their own to succeed or fail. Many VCs play a very active role on the board. They may bring in new staff or leaders. They may participate in the decisions about whether to stay the course or to pivot. They play an active role in ensuring their investments succeed. (See more detail in sections on Mentoring, Barrier Busting, and Networking).

6. Mentor

Mentors see themselves as people developers. They take a long-term view of their staff and see innovation projects as an opportunity to stretch their employees' capabilities and to help them achieve their aspirations. Depending on the scope of their ideas, innovators' careers can be significantly shaped by this journey.

Idea champions are likely to be high performers. They are people who are willing to go above and beyond the normal day job to do the right thing for the organization. They are often the rising stars in the organization and should be nurtured. The next time an employee walks into your office willing to champion a new idea, you might consider that he or she also likely has good management and leadership potential.

A mentor enables innovators to focus on their project results while also learning about themselves along the way. Leaders can mentor innovators through any of the practical business steps in the value creation process itself: connecting to emerging market trends and identifying the most significant opportunities, gaining customer insights, creating strong solutions and business models, connecting to others who can help them create compelling value propositions, learning how best to communicate and pitch their potential projects, and how to implement quickly once approved and funded. For example, a medical equipment manufacturer helped the engineers with new product ideas with their pitches. Many of the company's engineers were not comfortable presenting their big ideas to the senior staff, so a manager would work with them to develop the business case. Managers often copresented with the idea champion to the senior staff. This allowed the engineers to focus on the technical side of the equation where they felt most comfortable while gradually building their pitching and business skills.

Leaders can also encourage innovators to look inside themselves, to understand themselves as leaders or project champions. When is it time to stick to an idea, and when is it time to listen to others and change course? When is it time for individual vision and when is it essential to collaborate and build a strong team with a shared vision? Innovators face a wide mix of business strategy and interpersonal/team challenges on their road to success, and leaders who have experience in these areas can offer timely guidance to help innovators successfully navigate this multidimensional path.

Mentors don't simply give information and advice. They ask open-ended questions that force innovators to think about themselves and the business challenges they face. Asking what an innovation champion learned at each milestone is helpful to promote development. Identifying what worked, what didn't, and why, should be a regular topic of conversation. In addition, acknowledging that learning through failure is valuable goes a long way toward building trust, risk-taking, and fostering a climate of innovation.

Good mentors hold frequent development discussions, ensuring the dialogue is as much about the innovator as it is about the innovation they're developing. The time spent on mentoring varies from project to project, but is typically a matter of months rather than days. This time frame gives the mentor and idea champion time to explore the innovation journey together and develop opportunities for the learner to try out new capabilities. Mentors can act as a sounding board to test and explore new solutions and options, and give small pointers based on their understanding of the organization and experience with other innovation projects. Since it's also helpful when the mentor can coach the innovator through the tough challenges that inevitably show up, they need enough time working together to ensure they confront obstacles that will test the innovator's skills.

Playing the role of mentor means you are able to

- Use innovation efforts as an opportunity to develop innovators' capabilities and careers
- Coach an innovation champion and team through the entire innovation process
- Ask tough questions and allow innovators to struggle, without taking over the project
- Accelerate project team learning by encouraging experimentation, risk-taking, and iteration

A good mentor can cover the wide range of topics needed to enhance the growth of the individual and the results for the business.

7. Barrier Buster

Unless people understand why innovation is truly urgent, it often loses to the core business or is the first victim of budget cuts in the battle for resources. Companies often get overly protective of their current operations and lose sight of investing in the future. The core business is larger, is well established, is the center of influence, and can justify resources based on short-term financial results. Many times, leaders need to actively support a new idea, an innovator, or a team whose project they believe deserves to move forward.

Businesses that have been running for a while are typically quite capable of executing incremental changes that improve basic operations or add features to the existing product line. What they too often don't know how to do is generate and support radical ideas that might require more flexible ideation and execution processes.

Playing the role of barrier buster means you are able to

- Provide the necessary time, space, tools, and data for your staff to innovate
- Guide projects along the path of least resistance and avoid political pitfalls
- Adjust policies, procedures, and organization practices to facilitate new idea implementation
- Talk your peers through the fear, uncertainty, and doubt that often comes with change

Being a barrier buster requires you to be able to negotiate skillfully in tough situations with both internal and external groups. Innovation means change, and change can be quite disruptive and emotionally charged. Being able to gain concessions without damaging relationships is a valuable skill. Innovation leaders in these companies need to help new ideas mature and create paths of least resistance so projects can navigate the political, economic, and cultural obstacles that are likely to arise.

There are countless organizational barriers to innovation that cause it to be slow, inefficient, costly, risky, and frustrating. Being aware of some of the most typical obstacles can be helpful:

- The organization lacks the enterprise-wide methods (concepts, practices, tools, language, or skills) for innovation.
- There is not enough funding to form and facilitate innovation projects.
- The organization is overly consensus-oriented and any dissenting vote can bring an innovation project to a halt. Champions and sponsors give up or leave the company because it is too hard to get everyone onboard with ideas.
- The organization's relentless commitment to operational excellence prevents anything new and disruptive from being tried and tested. This is a classic example of a strength becoming a weakness.
- Past success has robbed the organization of its willingness to take risks. Leaders play it safe and settle for "me too" strategies just to keep up with the pack, rather than boldly investing in a better future.
- The organization lacks proper incentives for innovation. Idea champions are rarely recognized and rewarded for their efforts.
- People are overworked and simply don't have the bandwidth to take on their innovative ideas.
- Organization silos prevent cross-boundary collaboration and limit the scale, speed, and impact of innovation.

Barrier busters must be politically savvy to meet these kinds of challenges. They need sensitivity to know how the specific people and their organization are likely to react. Barrier busters help their idea champions or project teams maneuver through complex political situations effectively, because they can anticipate the organizational "landmines" and how to avoid them.

Barrier busters are also determined. They don't stop at the first signs of resistance, and refuse to accept "no" for an answer whenever there is hope for success. They are resourceful, looking for the support and resources wherever they can be found. Barrier busters know the difference between the market saying "no," and an organizational obstacle saying, "no." A leader might have learned from the VC role to let go of struggling projects, where customers don't respond as expected or where the market does not respond positively, in order to move the resources to fund innovation winners. However, as a barrier buster, this same leader knows that organizational protectiveness does not mean the project is struggling in the market. The barrier buster fights for the opportunity to let customers decide which product or service is the business of the future.

History is full of examples of innovators who were told their ideas would not work, but who ultimately found ways to find the support and resources they needed. Consider what would have happened if these innovators had not persisted in the face of obstacles:

"Man will never reach the moon regardless of all future scientific advances."
—Dr. Lee De Forest, "Father of Radio & Grandfather of Television."

"We don't like their sound, and guitar music is on the way out."
—Decca Recording Co. rejecting the Beatles, 1962.

"I think there is a world market for maybe five computers."
—Thomas Watson, chairman of IBM, 1943.

"The concept is interesting and well-formed, but in order to earn better than a 'C,' the idea must be feasible."
—A Yale University management professor in response to Fred Smith's paper proposing reliable overnight delivery service. (Smith went on to found FedEx.)

"Drill for oil? You mean drill into the ground to try and find oil? You're crazy!"
—Response from the drillers Edwin L. Drake tried to enlist in his project to drill for oil in 1859.[13]

8. Networker

A networker's primary task is to work across organizational boundaries to engage stakeholders and secure innovation support.

"No matter who you are, most of the smartest people work for someone else."
—Bill Joy, cofounder, Sun Microsystems[14]

No one is smart enough to solve really important problems alone. Being able to find, build, and test ideas through a network of diverse perspectives offers leaders, teams, and innovation projects the benefits of new insights and novel points of view, which increase the chance of uncovering real breakthroughs. A rich network of productive stakeholders offers both the "collective intelligence" needed to achieve breakthrough ideas and the means to implement them.

Networking is, at its core, relationship building and learning. Sometimes these relationships are tied directly to the development and implementation of new ideas and other times the focus is on gaining new perspectives on the world. The goal is to expand your network inside and outside your organization, to extend your sphere of influence, and to expand the expertise and diversity of your peer group and potential innovation partners.

Internal networking within your own organization is important, as many ideas and projects will cross the boundaries of your own area of the enterprise. Many ideas and projects can benefit from the expertise, skills, perspective, or even the contacts and working relationships that exist in other parts of the organization. Also, support from others whose work would change in response to an innovation can be crucial to getting buy-in and approval for projects with far-ranging impact.

Networking is "playing organizational politics" in a positive way. The most effective internal networkers are those people who consistently push for win-win outcomes. They work from a mind-set that making connections and reaching out to others will help all involved, including the company and its customers.

Breaking down organizational silos is vital to encouraging communication and building relationships across the enterprise. Effective networking supports the role of barrier buster, as trusted relationships can bridge the interdepartmental chasms that too often hinder the flow and support for bold ideas.

External networking is equally important. Getting out of the office to interact with other leaders, entrepreneurs, researchers, educators, artists, or thinkers from around the world enhances creativity and innovation skills in general. Expanded horizons often make it easier to see a larger solution set. Networkers go out of their way to meet people with different kinds of ideas and perspectives to extend their knowledge and expand their paradigms. For example, Starbucks founder Howard Shultz' travels through Europe allowed him meet coffee experts and café owners, and to crystallize his thinking about how Starbucks could bring a new coffee culture to the United States. Schultz was later able to draw on this community of coffee aficionados to guide his vision and to help him grow the business to unprecedented levels.[15]

Through these connections and experiences, leaders open themselves up to new possibilities and build their ability to embrace ideas that can be recombined in new ways. Successfully connecting seemingly unrelated questions, problems, or ideas from different people or fields, is an invaluable asset to a leader and his or her team.

"Point of view is worth an extra 80 IQ points."
—Alan Kay, Pioneering Computer Scientist[16]

In addition, in today's digital world, innovation leaders now have a wide range of choices of online tools to enhance networking. Business-oriented social media, online innovation platforms, and knowledge management systems enable leaders to search for individuals with the specific expertise, connections, or perspective they need. Internal online communities connect members within an enterprise and sometimes extend to include key partners, suppliers, or customers as well. Individuals can ask for information and expertise, for feedback on ideas and value propositions, for people with needed skills, for connections to individuals who could support or fund projects, and even to help find barrier busters when needed.

External online communities also now offer connections and discussion groups for linking individuals with similar interests across the globe (e.g., LinkedIn groups, industry associations, user groups, or specialty topic organizations). These communities exist for the express purpose of network building. They typically discourage individuals from simply trying to sell their own products or services, as they want to preserve the true working network. When there is a genuine question or request where members can help further a serious innovation effort, exchanges are generally quite lively as members enjoy the chance to contribute their insights and support.

Playing the role of networker means you are able to

- Build productive linkages and coalitions across organizational boundaries within your organization.
- Draw others in as joint sponsors for your teams' projects.
- Search outside the organization for innovative ideas and individuals who can help your projects.
- Maintain a strong professional network outside your organization.
- Build relationships with those who want to support innovations in your field.

9. Culture Creator

The culture creator's primary task is to ensure the spirit of the innovation process is understood, celebrated, and aligned with the strategy of the organization.

It is often said, "The soft stuff is the hard stuff." All of the appropriate innovation architecture and practices can be available, but ultimately, people have to want to use them.

Think back to Fig. 4-1, under the architect role, where there were five circles showing each of the key elements of innovation architecture. It is easy to predict the kind of team or organization culture that supports each of these circles.

A culture that encourages the active exploration of market and customer opportunities would most likely be forward-looking, curious, open to outside information, connected to customers and committed to delighting them, and connected to outside innovators as well.

A culture that leads people to generate breakthrough ideas would likely be creative, playful, and tolerant of crazy ideas—because they might lead to very good, very possible ideas. IDEO, the Palo Alto-based design firm known for innovation has its brainstorming rules written on meeting room walls:

- Defer judgment
- Encourage wild ideas
- Build on the ideas of others
- Stay focused on the topic
- One conversation at a time
- Go for quantity (e.g., an IDEO design team might try to brainstorm 101 ideas in an hour)[17]

Also, the more people who are engaged in generating new ideas, the more ideas there will be, which in turn means the more good ideas there will be that can be implemented. Swiss Post (the national postal service in Switzerland) developed an in-person and online platform called Post Idea, where employees could submit improvement ideas. They found that the program typically implemented seven percent of the ideas employees submitted. As more employees participated over time and submitted more ideas, the implementation percentage stayed approximately the same, meaning that the higher the number of ideas received, the more innovations were created over time.[18]

In addition, Swiss Post created an innovation fund for innovation, new business, and sustainability where employees could submit CO-STAR value propositions for creating new products and services, get collaborative feedback, and submit these for funding. In 2012, they supported 34 projects through this process, and new ventures are already being taken to market. Innovation leaders ensured that employees throughout the organization and key external stakeholders as well learned about the value this program created. As employees saw projects in new areas, such as advanced security systems for e-mails and digital signatures, or clean transportation programs (e.g., fuel cells for clean transportation on Post buses, or electric-bike sharing programs), people throughout the organization experienced the value their ideas were creating. Each new project launched and celebrated in turn triggered a wave of new ideas submitted.

The innovation phase of optimizing value is most effective in cultures where people clearly understand the innovation process and feel invited to participate in it. Communication, throughout the organization, repeated quite often, goes a long way to keeping innovation efforts visible and inviting. If innovation is seen as more than simply "luck or lone genius," individuals are more likely to want to participate. Then, if they have a clear process with simple and straightforward practices and online tools, they are more likely to know how to participate.

In addition, people are more likely to help each other improve their value propositions in cultures where collaboration and feedback are valued. A culture that elicits and appreciates feedback is actually quite rare. In many organizations, given a 1-hour meeting, innovators will pitch their ideas for 50 minutes and then ask for feedback in the few minutes left. When they get feedback, these innovators too often spend their time defending their ideas, deflecting any possible learning. In a culture that values feedback, innovators present in as short a time as possible and leave the majority of the meeting for feedback—which they then listen to intently. These innovators want to build their ideas and value propositions to be as strong as they can possibly be. They actively seek input from others and are continually open to learn how their product or service could be better.

Optimizing value also thrives in a culture that is open to experimentation, to trying things out to see how they work. Rapid prototyping, even with simple mock-ups in the beginning, leads to quick leaps in learning. The networker role is synergistic here, as innovators can call on their network to get valuable input and guidance.

The selecting and funding the best phase of innovation thrives best in organizations that are willing to take risks, and where leaders are willing to bet on ideas that might disrupt current products and services. If there is a call for innovation, but only the safe ideas that don't threaten current lines of business are the ones selected, innovators quickly get the message that innovation is a slogan, not a true demand.

Also, this phase works best where there is a culture of healthy competition and where decisions are considered to be fair. Whenever innovators sense that leaders play favorites, that they choose winning innovations to move forward to implementation based on company politics rather than market opportunity or potential customer value, motivation will drop pretty dramatically.

Innovation also thrives in cultures where "failure" is considered a natural part of innovation. Everyone knows that all ideas cannot be chosen in pitching competitions or online innovation campaigns. Effective innovation cultures honor all the individuals who participate, as winning ideas are often helped along the way by the creativity offered in surrounding solutions. Because innovators learn from each other, individuals are valued for their part in contributing to the overall "innovation ecosystem." In some organizations, this culture is proactively made visible, by offering rewards to winning innovators and teams, but also to the department where they work and to all who participate in the campaigns.

The "Mobilizing for Results and Commercialization" phase succeeds in cultures with an emphasis on implementation and delivery. Once projects are selected to move forward, these cultures bring all manner of resources to bear to help fledgling projects succeed. Mentors and networkers provide important supports here as well. These cultures emphasize that what counts is not only value imagined, but is "value delivered." Innovation leaders are those who ensure that projects move through all the phases—from the earliest insights, through testing and improving value, through more advanced development, and finally to full implementation or commercialization.

Culture Change

These paragraphs have described the culture that promotes innovation and makes it more likely to succeed. Yet many innovation leaders are hired into their roles precisely because this kind of culture does not exist in their organizations. How does a leader become a culture creator, who can transform a culture that inhibits innovation into one that helps it to flourish? A few of the early steps in playing the role of culture creator as a change agent include

- Create a safe, trusting, and collaborative environment.
- Establish common language, concepts, and practices for innovation.
- Encourage forward-looking curiosity.
- Be open to new ideas, even if they might cannibalize existing products and services.
- Recognize that risk-taking and failure is part of the innovation process.
- Tell inspiring stories and use symbols to reinforce the importance of innovation to your particular business. Make innovation compelling, both from a business perspective and from the personal viewpoint of individual innovators. People want to make a difference.
- Find ways to publicly celebrate the accomplishments of champions and teams.

Customizing Your Innovation Culture

It can help to link your innovation culture to your overall company culture. For example, when she was CEO of HP, Carly Fiorina linked their innovation culture to the founding values of the company, through the "Rules of the Garage." The rules were written across a photo of the one-car Palo Alto garage where Bill Hewlett and Dave Packard first began the company. Fiorina described her intent as a culture creator: "We have tried to capture the spirit of the original HP in what we call the Rules of the Garage. The garage is a special place for us. It represents that entrepreneurial, inventive spirit that is special about HP. The reason we wrote them down was to remind ourselves that this is what this place used to be about, it's what this place always needs to be about. Those soft things, those things that represent the soul and the spirit of the place, in the end, those are in many cases the most sustainable competitive advantage that you have."[19]

HP's Rules of the Garage include

- Believe you can change the world.
- Work quickly, keep the tools unlocked, work whenever possible.
- Know when to work alone and when to work together.
- Share tools, ideas. Trust your colleagues.
- No politics. No bureaucracy. (These are ridiculous in a garage.)
- The customer defines a job well done.
- Radical ideas are not bad ideas.
- Invent different ways of working.
- Make a contribution every day.
- If it doesn't contribute, it doesn't leave the garage.
- Believe that together we can do anything.
- Invent.[20]

10. Role Model

"Be the change that you wish to see in the world."

—Mahatma Gandhi

Innovators listen to what leaders say, but most of all, they watch what leaders do. Day-to-day actions and decisions show whether leaders are actually committed to innovation or whether it's simply a popular tagline that sounds good but has no real support.

Leaders who manage all of the other nine roles so far will have gone a very long way to showing their commitment to innovation. What is left to role model? There is one more very important aspect to innovation leadership. Beyond these other nine roles, people can sense when a leader has a real personal passion for innovation. Innovation is not easy. Almost any new idea and project eventually hits major obstacles or encounters people who say it could never work. Innovators need their own inner drive to continue through these difficult moments. Innovation champions are the ones who wake up at 3:00 am with the solution that could work—they find a way through. Leaders with a deep belief that innovation is essential and who care about the value it can deliver to their customers—and perhaps to society at large—help elicit that extra spark of inspiration that innovators need.

Being a role model means you are able to

- Convey a sense of passion for new ideas and inspire others to contribute.
- Easily utilize the tools and techniques of the innovation process to create value.
- Recognize and quickly adopt new ideas within your own team or business unit.
- Ask, "What can we learn?" when projects do not turn out as expected.
- Communicate and celebrate successes.

How a leader spends his or her time is always noticed by others. Innovation cannot be fully delegated to others—it will only be considered an important effort in a team or company when the leader is actively engaged. As an innovation leader, you might

- Convene special strategy sessions just for generating ideas and exploring options.
- Contribute ideas of your own, don't just sit back and judge the ideas of others.
- Have an "open office" policy where you invite people to share their ideas and value propositions.
- Offer balanced feedback on new ideas (e.g., what you like about the idea and what could be stronger).
- Approve experiments and encourage Lean prototyping.
- Be willing to collaborate with other units and promote your team members' ideas to others in the organization.
- Be inquisitive. Ask open-ended questions that challenge common wisdom (e.g., ask "Why?" "Why not?" "What if?" and "How could we?").
- Invite cross-disciplinary participation and encourage different perspectives.
- Admit when you are wrong and capture the learning from failed inventions.
- Assign some of your best people to innovation projects.
- Promote the best innovators in your staff.

One of the best role models of innovation leadership we know is Curt Carlson, President and CEO of SRI International. SRI is a large independent R&D organization that has created a wide range of well-known products that have transformed industries, including the computer mouse; Internet communications (SRI received the first transmission on Arpanet, the precursor to today's Internet); high-definition TV (HDTV); telerobotic surgery technology that led to the da Vinci robotic surgery systems now available in many operating rooms; and SIRI, the personal digital assistant that was recently acquired by Apple.[21] Carlson is a master of each of the roles of innovation leadership. If you follow him around during his day, you would see many interactions where he is asking people what they are working on and what they are most excited about. In addition, Carlson

- Built an integrated system of innovation and value creation inside SRI, with a clear architecture that people understand. Innovators share a common system of creating value propositions and have a number of collaborative sessions called "watering holes," where teams collaborate to help each other strengthen each other's value propositions. There are various levels of innovation boards, where teams can present for increasing amounts of funding first to validate their concepts, then to develop their solutions and business models, and ultimately to commercialize their products or services. If innovators want to tell Carlson about the technologies or products they are working on, he wants to hear a quick pitch about their value proposition. He doesn't want to hear only the idea or technology itself, but wants to hear the compelling customer value their idea would create.
- Models the use of this innovation system for his own leadership ideas as well. He creates value propositions for his ideas and asks others for input.
- Helps to create a strong feedback culture. Carlson often describes stories about how tough it is to succeed in a highly competitive global economy, and counsels that people need to make their value propositions as strong as they can possibly be, in order to have a chance for them to survive. He demonstrates this himself by listening

to others' comments about his own ideas and incorporating what he learns to revise and improve his own value propositions.

- Helps to find and engage outside experts to collaborate on projects.
- Helps to find additional external funders for larger projects and helps to launch new ventures.
- Gets genuinely excited about the value his people create. Carlson tells wonderful stories about what his people are inventing and how they are helping to solve some of the world's most important problems.

Being a role model is not always easy to do. Listen to the feedback of others to get a measure of your effectiveness. If you are involved in a 360-assessment process you will gain a clear snapshot of your performance as an innovation role model. You'll know you need to improve if you start receiving feedback that says you are too cautious and don't seek to be bold or different, prefer the tried and true, bring too narrow and tactical a perspective, don't connect with ideas outside your own area, use old solutions on new problems, or stifle the creativity of others. Being a good role model for innovation, or any other key leadership competence, means being conscious of your own behavior and recognizing where you are on your own developmental journey.

Overall, a positive attitude is infectious. If the leader is passionate and excited about the company's vision and the projects being pitched, others will catch the enthusiasm as well.

Conclusion

Leaders who can play all 10 of these roles will accelerate the development of a vibrant innovation culture that provides

- Infinite source of value-creating ideas
- Speed and agility of implementation
- Continuous performance improvement
- Employee engagement and satisfaction
- Delighted and loyal customers
- Competitive advantage

Companies looking to aggressively boost their innovation capabilities need leaders who recognize and embrace their role as innovation champions. They need a committed leadership team with an entrepreneurial mind-set. They need to be forward-looking and to inspire the organization about what is possible. They need to set direction, clarify priorities, and define the boundaries around the type of innovation required for the company to thrive. As they make the potential future visible, innovation leaders form a kind of "strategic sandbox" where entrepreneurs throughout the enterprise are invited to play, with the goal of creating compelling customer value.

Pioneering companies typically have a powerful, widely shared and strongly reinforced "stretch vision" of the intended future. Leaders act in full alignment with the intended future and model the necessary innovation practices. They embrace their natural optimism and become an appreciative audience for creative ideas, ask inspiring questions, encourage collaboration, free up resources, and clear paths through the bureaucracy. These behaviors are especially important for novel ideas that are harder to conceive and easier to kill.

There is no innovation without leadership. The best leaders know which role to play at which time to ensure a healthy pipeline of high-value ideas and a highly engaged workforce capable of inventing a better future.

Businesses don't fail—leaders do. Leaders who don't treat innovation as a priority simply cede opportunity to those who do. Effectively playing all 10 roles helps leaders ensure their teams consistently outthink, outcreate, and outperform the competition.

References

1. Ekvall, Goran. "Organizational Climate for Creativity and Innovation." *European Journal of Work and Organizational Psychology*, 5 (1), 1996, pp. 105–123.

2. Cabra, John, Talbot, Reginald, and Joniak, Andrew. "Potential Explanations of Climate Factors That Help and Hinder Workplace Creativity: A Case from Selected Colombian Companies." *Cuadernos de Administración—Bogotá,* 20 (33), 2007, p. 297.

3. Carrol, Lewis. *Alice in Wonderland.* 1865.

4. Wikipedia. SAP Community Network. http://en.wikipedia.org/wiki/SAP_Community_Network. Accessed April 16, 2013.

5. Yolton, Mark. Co-Innovating the Future with #SAP Idea Place. *About SCN,* March 15, 2013. http://scn.sap.com/community/about/blog/2013/03/15/co-innovating-the-future-with-sap-idea-place.

6. Bulygo, Zach. 12 Business Lessons You Can Learn from Amazon Founder and CEO Jeff Bezos. *KISSMetrics: A Blog about Analytics, Marketing, and Testing,* January 19, 2013. http://blog.kissmetrics.com/lessons-from-jeff-bezos.

7. Miller, Claire Cain and Bosman, Julie. E-Books Outsell Print Books at Amazon. *New York Times,* May 19, 2011. http://www.nytimes.com/2011/05/20/technology/20amazon.html?_r=0.

8. Ford, Bill. Transcript of Q4 2005 Ford Motor Company Earnings Conference Call, Ford Motor Company, January 23, 2006. Referenced in Quote Investigator, "My Customers Would Have Asked for a Faster Horse." http://quoteinvestigator.com/2011/07/28/ford-faster-horse.

9. Hanna, D. *Designing Organizations for High Performance.* Reading, MA: Addison Wesley Longman, 1988.

10. Gyorffy, Laszlo and Friedman, Lisa. *Creating Value with CO-STAR: An Innovation Tool for Perfecting and Pitching Your Brilliant Ideas.* Palo Alto, CA: Enterprise Development Group, Inc., Publishing, 2012.

11. Ries, Eric. *The Lean Start-up: How Today's Entrepreneurs Use Continuous Innovation to Create Radically Successful Businesses.* New York, NY: Crown Business, 2011.

12. Gage, Deborah. "The Venture Capital Secret." *The Wall Street Journal,* September 20, 2012.

13. Cerf, Christopher and Navasky, Victor. *The Experts Speak: The Definitive Compendium of Authoritative Misinformation.* New York, NY: Pantheon Books, 1984.

14. Joy, Bill. Quoted in Karim R. Lakhani and Jill A. Panetta. *Principles of Distributed Innovation.* The Berkman Center for Internet and Society at Harvard Law School, Research Publication No. 2007-7, October 2007.

15. Schultz, Howard and Yang, Dori Jones. *Pour Your Heart Into It: How Starbucks Built a Company One Cup at a Time.* New York, NY: Hyperion, 1997.

16. Kay, Alan. Talk at Creative Think Seminar, July 20, 1982. http://folklore.org/StoryView.py?project=Macintosh&story=Creative_Think.txt.

17. OpenIDEO. Tips on Brainstorming. *OpenIDEO blog.* Feb 23, 2011. http://www.openideo.com/fieldnotes/openideo-team-notes/seven-tips-on-better-brainstorming.

18. Caboussat, Pierre-Yves, Head of Innovation Management. "Overview: We Make Innovation Possible." *Swiss Post Innovation Management Annual Report, 2012.*

19. Knobel, Lance. "Rules of the Garage" Ensure Continued Innovation. *New Perspectives Quarterly.* Los Angeles Times Syndicate International, May 15, 2001. http://www.digitalnpq.org/global_services/global_ec_viewpoint/05-15-01.html.

20. Williams, Paul. Rules of the HP Garage. *Blogging Innovation,* January 2, 2010. http://www.business-strategy-innovation.com/2010/01/rules-of-hp-garage.html.

21. Carlson, Curtis, and Wilmott, William. *Innovation: The Five Disciplines for Creating What Customers Want.* New York, NY: Crown Business, 2006.

General Concepts

The general concepts of innovation don't fit well into the traditional outline of innovation. These areas do not fit well into the traditional business. Examples in this section are personal and corporate creativity, innovation and neuroscience, association with bionics, benchmarking, and process versus practice innovation. While benchmarking potentially could have been included in the methods section, we chose to leave it here because of its broad applicability to all areas of innovation.

All humans are born creative. It is in our DNA and this is what makes us different than most other animals. Sure, there is the occasional amazing trick a dog, cat, or monkey may perform. We see them all the time on YouTube. But for the most part, humans are the only ones that are creative in their problem solving nearly all the time. Some would argue Modern life might make us lazy as we depend more and more on technology; however, there are many bright lights to be seen in the way we adapt our modern technologies. How much of this is our creativity and how much is the creativity of others is debatable; however, I believe that a natural tendency toward innovation is human nature. In Dr. Aaron Reeves' chapter on personal creativity, she points out that innovation and creativity are not necessarily the same thing. Creativity, a mental precursor to innovation, is about imagination and ideas, whereas innovation is about actions and processes. She also points out that innovation requires creativity, but that creativity does not always lead to innovation. This is an excellent precursor to her book where we examine both creativity and innovation and try to distinguish between the two.

On the other hand, Tom Duening in his chapter addresses the concept of corporate creativity. Once again a distinction between creativity and innovation is established. An organization can be creative but not necessarily innovative. However, the chance that an organization will be innovative after demonstrating creativity is much better than for an individual. This is mainly due to the profit motive and the fact that there are definitely tangible rewards to innovation in organizations. The main premise in his chapter is that all employees in the company innovate, but the important difference from one company to the next is the velocity of its innovation. Velocity and innovation can only be done if the corporation is creative. He goes on to define corporate creativity and how to foster that creativity in your organization.

In our chapter on innovation and neuroscience, we distinguish between what is hard-wired into the human brain versus what is software. We readily admit that neither of us are neuroscientists, but remain very interested in the hard-wiring around innovation.

The following chapter by Ellen Di Resta deals with the question of neuroscience as software. What we mean is that neuroscience may shape us to be more innovative and creative as our lives progress. The accepted principle that creativity declines in old age is being challenged every day through the field of neuroscience. Some of the interesting components of this include the use of art that is discussed in this chapter and other places in the book to increase the ability to become a right-brained thinker. Also, neuroscience includes a better understanding of feelings and emotions, intuition, and getting in touch with your individual needs.

Another interesting area we want to explore in this book is the question of bionic association. To do so, we solicited a chapter from Melissa Sterry, a leading thinker in this field. It is Sterry's belief that mimicry is a great untapped innovation potential. For example, she points out that da Vinci, while considered a genius, did not feel the same way. When da Vinci faced a problem, he did not look within for the solution but instead looked to nature. By doing so, he was able to recreate the things that he saw and was credited with genius. She goes on to point out innovations in the human body (bionics) that replicate or improve upon the human condition. She also points out numerous other examples of innovation inspired by nature.

Innovation benchmarking, as pointed out by Klaus Solberg Søilen, is not just another benchmarking exercise, but primarily another approach to benchmarking. In other words, innovation benchmarking is used to identify the factors behind benchmark success. According to him, what makes innovation benchmarking so difficult is the fact that there are no factors or key performance indicators (KPIs) that specifically call out innovation and can be measured as such. He shows you how to find variables and utilize them to create your own benchmarking.

In the chapter on process, practice, and innovation, Michael Grieves calls our attention to the limitations of process. From an innovation perspective, process is designed specifically for incremental improvements, whereas practice is more appropriate for innovation. So, when we refer to best practices, we are referring to something entirely different than best processes.

Finally, in this section, we try once again to move into a space where businesspeople are uncomfortable. The roots of ethnography are in anthropology and the stereotype of the ethnographer is deep within the jungle. In fact, this is a rapidly changing field, finding a prominent place in innovation. Given that we are a global community, very few people can claim to be untouched by other cultures, religions, and ethnicities. In fact, even the lost tribes of old are becoming fewer and farther between. And with every interaction come new expectations and an appreciation for tolerance. Businesses that wish to compete globally must now consider the ramifications of their color, shape, and name. And although the *Chevy Nova in the Spanish-speaking country* story has been debunked, it is still a cautionary tale about knowing your audience. Imagine the power when you do market to people unlike yourself!

CHAPTER 5

Creating Creativity: Personal Creativity for Personal Productivity

Arin N. Reeves

Introduction

Although creativity and innovation are often used as interchangeable terms or meshed together as one concept, the difference between the two is an important one that actually helps us to understand each more fully. One way to differentiate between the two is to understand creativity as the mental precursor to innovation; creativity is about imagination and ideas where innovation is about action and process. The authors of the new book *Smart-Storming: The Game-Changing Process for Generating Bigger, Better Ideas*, Mitchell Rigie and Keith Harmeyer describe the difference between creativity and innovation as

> "Creativity is most often defined as the mental ability to conceptualize (imagine) new, unusual or unique ideas, to see the new connection between seemingly random or unrelated things. Innovation on the other hand, is defined as the process that transforms those forward-looking new ideas into real-world (commercial) products, services, or processes of enhanced value. The result of such a transformation can be incremental, evolutionary, or radical in its impact on the status quo."[1]

Innovation requires creativity, but creativity does not always lead to innovation. Organizations that seek innovative thinking need workplaces and talent development systems that foster creativity, and they also need process systems that can translate the creativity into innovation. Understanding these distinctions between creativity and innovation allows us to understand, learn, and maximize each more comprehensively.

Since creativity is separate from, albeit necessary to, innovation, individuals can develop and utilize their personal creativity capabilities regardless of whether their jobs and workplaces explicitly require or seek innovation. Furthermore, an IBM survey of more than 1500 Chief Executives from over 60 countries and 30 industries found that "chief executives believe that—more than rigor, management discipline, integrity, or even vision—successfully navigating an increasing complex world will require creativity."[2] Developing and using personal creativity in the workplace is no longer relegated to the "creative arts" or deemed as a "nice to have," developing and using personal creativity at work—regardless of the work, workplace, or industry—is quickly becoming a competitive differentiator in talent and a core competency for leadership whether or not it is fostered in the workplace where it is valued.

What Is Personal Creativity?

Thought leaders from many industries have defined creativity in a myriad of ways, and there is no universal way to define creativity that makes sense for all who are trying to understand it in the context of their own lives. The definition used in this exploration of creativity is based on the standard definition of creativity explored in the *Creativity Research*

Journal by researchers from The Torrance Creativity Center[3] as well as the definition set forth by Robert E. Franken in *Human Motivation.*[4]

> "Creativity is the ability to create *original* ideas, connections, alternatives, or possibilities that are *effective* in solving problems, communicating with others, and inspiring new and useful ideas in others."

Creativity requires both the unrestrained openness of originality and the "does it work" constraints of effectiveness. It is the ability to think of things that no one has thought of before or see connections between existing ideas that no one has seen before in a way that is inspirational, useful, and/or enjoyable. As the following quote suggests, creativity needs some constraint in order to be creative; otherwise, it could just be random, muddled, or even bizarre.

> *"Creativity is not simply originality and unlimited freedom. There is much more to it than that. Creativity also imposes restrictions. While it uses methods other than those of ordinary thinking, it must not be in disagreement with ordinary thinking—or rather, it must be something that, sooner or later, ordinary thinking will understand, accept, and appreciate. Otherwise, the result would be bizarre, not creative."*
>
> —Sylvano Arieti, *Creativity: The Magic Synthesis*

How Does Personal Creativity Work?

In order to explore the process of personal creativity, it is important to explicitly recognize that creativity is indeed a process, and it can be honed through practice. The process may feel intuitive and organic to some, and it may feel unnatural and uncomfortable for others, but the output from the process is creative regardless of how the process is internalized.

Based on the research we have conducted with dozens of individuals in a multitude of industries and professions, the process of generating original and effective ideas (see Fig. 5-1) requires

1. Diverse inputs/inclusive thinking
2. Context articulation
3. Divergent thinking
4. Convergent thinking

Diverse Inputs/Inclusive Thinking

While researchers have posited everything from neurobehavioral differences to cultural differences to differences in education levels and intelligence to explain the variations in individual creativity, recent research indicates that seeking and integrating diverse inputs into everyday thinking can simply and effectively increase overall personal creativity regardless of other factors.[5]

Diverse inputs from a multitude of cultures, ages, professions, languages, nationalities, geographic origins, educational backgrounds, interests, and so on, have the potential to enhance an individual's ability to generate new ideas and new connections between existing ideas as long as the individual thinks inclusively about integrating the diverse inputs into his everyday thinking. These diverse inputs can come from personal relationships, professional networks, learning sources, and other activities; the more these inputs are derived from natural networks (friends/close colleagues/organic interests) as opposed to unnatural networks (focus groups/networking events/forced interests), the more valuable they are in their ability to spark original and effective ideas and connections.

EXAMPLE: HopeLab Re-Mission is a video game created by HopeLab that has "significantly improved treatment adherence and indicators of cancer-related self-efficacy and knowledge in adolescents and young adults who were undergoing cancer therapy. The findings support current efforts to develop effective video game interventions for education and training in health care."

Figure 5-1 The process of generating effective ideas.

Pam Omidyar, the founder of HopeLab, is a self-described video game fanatic who also happened to work in an immunology laboratory. After a hard day of work in immunology, she would relax by playing video games with her husband, Pierre Omidyar, the software engineering whiz who founded eBay. So, what happens when you blend the diverse inputs of an immunology researcher, two video game fanatics, a software engineer, personal interests in helping children, and a passion for creativity? You get a carefully designed and enjoyable video game that children with cancer play because it is fun, and they want to play it! You get a carefully designed and enjoyable video game that makes it more likely that the children will understand their cancers better, adhere to their medical regimens with more regularity, and feel more in control of their ability to positively impact their fights against various cancers.

Any of the inputs by themselves would not have created the magic of Re-Mission. It was the diversity of inputs combined with Pam Omidyar's desire and ability to combine the various inputs into an original and effective idea that resulted in an *original* new way to help children fight cancer *effectively*.

The integration of diverse inputs into your thinking processes becomes even more critical when the communication of ideas crosses national and cultural boundaries as most business interactions do these days. Individuals who can integrate these diverse perspectives into their personal creativity are more informed, competitive, and effective than individuals who don't.

Researchers have discovered that the diversity of perspectives engaged in the understanding of creativity must be recognized in order to capture the variance in how creativity is understood, inspired, and executed.[6] Consistent with research that shows that Eastern cultures tend to think more in terms of "we" and "the collective" in contrast to the Western tendency of "I" and "the individual," research in Hong Kong demonstrated that Chinese people were more likely to define creativity in regard to the benefit to the collective whereas Westerners defined creativity in regard to individual ability or aesthetic.[7] Further, the lack of research on creativity in African or Latin American countries reminds us that there is much more that we have to learn in order to fully engage the creativity of people in the many cultures that contribute to the marketplace in which we communicate and compete.

Whether the objective is to be more personally creative in a global environment or have the opportunity to personally inspire greater creativity in a culturally diverse group of people, the entryway into a truly effective creative process begins with seeking out and integrating diverse perspectives into our thought processes.

Context Articulation

While diverse inputs prime us to be more personally creative, understanding the context in which we are attempting to be creative is the link between an "original idea" and an "original and effective" idea. Without clearly articulating the context in which the creativity is relevant, original ideas may be interesting but not very useful. In the HopeLab example, the connection between immunology (improving the immune system to fight cancer better) and video games made sense in the context of using video games to increase health outcomes for people who had both a love of video games and health challenges that could be addressed through the video games. The same technology in the context of a different audience and different challenges may seem more frivolous than creative.

In order for an idea to be both original and effective, there needs to be a clear and definite context that is articulated in which the idea can be relevant. Sometimes the creativity is a new idea in a particular context, and sometimes, creativity can be the transference of an idea from one context in an original and effective way into another context.

EXAMPLES: Newton's Apple and 3M's Post-It Note
Newton's Apple and the Importance of Context   Think of the story of Isaac Newton's creative epiphany about the theory of gravitation when he saw an apple fall from the tree. Newton's epiphany was original and effective because he contextualized his idea within the arena of physics, an arena in which he had studied and researched for years. The idea of "have you noticed an apple always falls to the ground instead of falling sideways or flying through the air" would have been quite less useful in the context of farming or medicine or painting.

3M's Post-It Note   In the 1970s, a 3M scientist named Spencer Silver was trying to create an adhesive that was stronger than others in the marketplace. Instead, he created a strangely weak adhesive that stuck paper to various surfaces but only semipermanently. The paper could be removed from the surface without leaving a residue and used several times without losing its adhesive qualities, but it was not strong enough to stick to the surface when pulled. Initially considered a failed product, Silver continued to work with it as a possible spray adhesive or surface for bulletin boards. Although 3M continued to reject his ideas for this project, he continued to present his ideas to his colleagues and friends.

Arthur Fry, another 3M scientist, sang in Silver's church and had been searching for a way to "bookmark" his hymn book without permanently marking the book. Remembering Silver's "low-tack" adhesive, Fry used the material to stick slips of paper as bookmarks that he removed after service. This concept of removable bookmarks garnered a lot of attention in and out of 3M, and both Silver and Fry are credited for their contributions to the now ever-present office supply, the Post-It Note.

Spencer Silver, the holder of over 20 U.S. patents, invented an adhesive that was a failure in one context and an original and effective breakthrough idea in a different context.

Divergent Thinking and Convergent Thinking

With diverse inputs informing the creative process and the context for creativity clearly artic-ulated, the next step in being creative is to engage in unrestrained divergent thinking before culling the product of the unrestrained thinking with convergent thinking. The power of orig-inal and effective personal creativity lies in that powerful and complex intersection between unrestrained thinking, constrained decision making, and the possible diverse relationships between the two in any particular context.

As early as the 1960s, J.P. Guilford theorized that creativity was comprised of both diver-gent thinking (coming up with many ideas/solutions to a problem) and convergent thinking (vetting the various ideas to identify the best workable solutions).[8] In the early to middle 1990s, this concept of unrestrained thinking and viable application was further explored in models like the Geneplore Model which includes a generative phase (a divergent thinking phase where an individual generates new ideas to a specific problem) and an exploratory phase (a convergent thinking phase where the individual evaluates the viability/functionality of the ideas within specific structural constraints).[9] Focusing on these ideas and theories within the context of specific arenas, industries, and organizations led to additional research stressing that new ideas, however vetted, needed to be actionable in a particular context for them to have value.[10]

The combination of divergent and convergent thinking allows for the unrestrained brainstorming of ideas without any limits, and this unstructured brainstorming funnels into the structured vetting process where the ideas are analyzed, vetted, and approved for even-tual use (see Fig. 5-2). In 2006, Marisa Mayer, now the CEO of Yahoo!, wrote the following in *BusinessWeek*:

"Constraints shape and focus problems and provide clear challenges to overcome. Creativity thrives best when constrained. But constraints must be balanced with a healthy disregard for the impossible. The creativity realized in this balance between constraint and disregard for the impossible is fueled by passion and leads to revolutionary change."[11]

While this model allows us to generally understand how we generate, vet, and translate new ideas for specific contexts, it doesn't help us understand how some people generate more

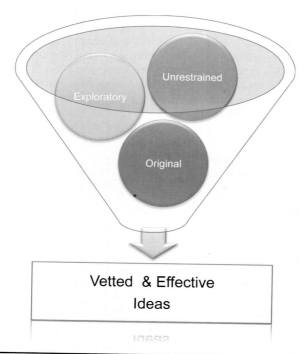

Figure 5-2 Idea funnel.

and better ideas (really different ideas) in the divergent thinking phase and how some people are better able to vet the divergent ideas to arrive at a convergent idea that really works.

When an individual begins with diverse inputs, the variety of perspectives informing his or her thinking lead to increased generation of different ideas in the divergent phase (the art of creativity) and better vetting in the convergent phase (the craft of creativity). This enhanced thinking occurs regardless of the context for the creativity, especially if the diversity of perspectives account for diversity within the specific context. As Donatella Versace, the Italian fashion designer known for her innovative fashion, asserted, "Creativity comes from a conflict of ideas." That essential conflict of ideas necessary for creativity requires an input of diverse perspectives.

Putting It All Together: A Case Study   A professional services firm and a large corporation teamed up to bid for the rights to develop an attractive tract of waterfront property in a large city. The decision-making body is made up of legislators, real estate specialists, environmentalists, and representatives of the surrounding communities. The bid guidelines have clearly articulated that all bidders need to address economic, environmental, and community revitalization issues in addition to the traditional architectural and construction matters.

Eager to leverage their best assets in the bid, the firm and the corporation submitted the names of the most experienced and renowned members of their organization as specialists for the project. These specialists brought the weight of their expertise and experience to bear in the divergent thinking process where the team brainstormed both traditional and innovative ideas for this project. Once they had completed this process, they began the task of vetting the various ideas in the convergent thinking process to identify the best ideas to include in the bid. As they focused on this "divergent to convergent based on the context" creativity process, they noticed that their ideas from the divergent thinking activity had not netted much diversity of thought. While most of the team agreed with the ideas that had been generated, the ease of the convergent thinking process was more about the lack of differences in ideas as opposed to creative consensus. In other words, it was easy for them to agree with each other because most of them had said the same thing.

As the bid development neared the end, one of the team leaders expressed dissatisfaction with the overall "innovative feel" of the bid. He invited a few people who were familiar with this bid selection process to give feedback on their bid and the accompanying presentation. The feedback that the team received was overwhelmingly mediocre bordering on negative.

What the leaders realized is that although they had been meticulous about fostering creativity in the process, there had not been enough diversity in the perspectives involved to result in the kind of breakthrough creativity they were seeking.

The leaders asked each of the people providing feedback if they could recommend independent contractors for the team who would add different perspectives to the process which resulted in the addition of 5 new team members (very diverse representatives of the environment movement, various community organizations, and political advocates) to the 10-person team.

The next iteration of the divergent idea generation process looked very different. It was filled with considerably more disagreement, conflict, and misunderstandings between people. Sorting out the communication and substantive issues in this process helped the team leaders recognize how much better the final product would be for having been born in inclusive thinking.

The team did end up winning the bid, and the independent contractors who had assisted with the process were brought on as permanent members of the team so that the creative process could continue to be productive.

In the case study above, the importance of including diverse (and contradictory/conflicting) perspectives is not just an added value to the overall creative process. Diverse perspectives are critical to the creative process actually working and can enhance or inhibit the effectiveness of the other steps in the process. That said, those diverse perspectives would have not had much value if they were not pulled together by a common context in which they needed to operate. Finally, with the diverse inputs and contexts in place, the process of gathering ideas without restraints and then vetting them with care allowed them to arrive at a project that both leveraged the individual and collective creativity of the group.

Common Barriers to Effective Personal Creativity

"The chief enemy of creativity is 'good' sense."

—Pablo Picasso

Even when the process for being more effectively creative is clear, the barriers that prevent full implementation of creativity abound. The barriers preventing personal creativity may seem to exist in the structure of an organization, a supervisor's rigid personality, crowded living space, a lack of creative mentors, and so on and so forth. While some of these external barriers may indeed be inhibiting some of us from maximizing our creative potential, the most common barriers to effective personal creativity are unconscious barriers that live and thrive in our own minds. We are born to be creative, to imagine new things, and create new connections between existing ideas; yet, we erode this innate ability slowly as we unwittingly allow the unconscious but real barriers to take root in our minds. Recognizing these barriers is the first step to removing their influence in limiting our personal creativity. Our research has found seven key barriers that prevent individuals from connecting with and developing their personal creativity.

Barrier 1: Our Perceived Definitions of Creativity   Most of us have been taught that creativity is the expression of artists, actors, musicians, and others who reside in the world of "the arts." When we conjure up an image of creativity, we see art and artists, and when we relegate creativity to existing in only those worlds, it is difficult to conjure up the internal personal drive and motivation to experiment with and experience creativity in our own lives. Challenging our own definitions of creativity and exploring new ways of defining and seeing creativity can be a great first step in rebuilding our relationship with our personal creativity. Consider these perspectives to assist you with rethinking and refining your own definition:

- Creativity is seeing what everyone else has seen and thinking what no one else has thought— Albert Einstein
- Making the simple complicated is commonplace; making the complicated simple, awesomely simple, that's creativity—Charles Mingus
- Creativity is piercing the mundane to find the marvelous—Bill Moyers
- Creativity is the power to connect the seemingly unconnected—William Plomer

Barrier 2: Our Presumed Uses for Creativity   Similar to the first barrier, this barrier also limits the ways in which we see creativity as applying to our lives. Even if we accept that we can personally benefit from creativity (regardless of our profession, our lives, or our interests), we still limit the ways in which creativity can be applicable in our lives. How would you answer the questions: "In what part of your life do you feel most creative?" and "In what part of your life do you feel the least creative?" What if you were asked to switch the two? That initial sense of discomfort or resistance to being creative in an area where you are not used to being creative (e.g., the route you take to work every morning) and less creative in an area where you are comfortable leaning on creativity (e.g., storytelling or creative problem solving) is your cognitive presumption for where creativity is necessary or warranted and where it is not. Creativity exercised in one area of your mind allows you to make new connections in a different part of your mind. Next time you are trying to solve a complex problem at work, change your commute or your breakfast routine and see what new ideas emerge to help you solve your problem.

Barrier 3: Overdependence on Knowledge

"Imagination is more important than knowledge."

—Albert Einstein

We are taught to value what we know, especially when we have expended time, financial and emotional resources to acquire the knowledge we have. When it comes to creativity, it is our rigid adherence to what we know that traps us from discovering what else could be.

As Ray Bradbury once said, "Don't think. Thinking is the enemy of creativity." While knowledge and experience are critical in the convergent thinking stage, reliance on knowledge can be debilitating in the divergent thinking stage.

Barrier 4: Our Experiences and Expertise

"Good judgment comes from experience and experience comes from bad judgment."

—Will Rogers

Our experiences and expertise can work well for us in the convergent thinking process, but they can limit our ability to think inclusively and generate the volume of ideas that we need in the divergent thinking process. As the Japanese proverb guides—"None of us is as smart as all of us." The ability to constantly seek opinions of people who do not share our experiences and expertise in order to learn and grow is part of the creative process.

There is a popular episode "The Opposite" from the hit comedy, *Seinfeld*, where George Costanza frustratedly says: "It all became very clear to me sitting out there today, that every decision I've ever made in my entire life has been wrong. My life is the complete opposite of everything I want it to be. Every instinct I have in every aspect of life, be it something to wear, something to eat. It's often wrong." Jerry Seinfeld, as one of his best friends, responds with, "If every instinct you have is wrong, then the opposite would have to be right." George goes on to great success with his "opposite" strategy and concludes: "A job with the New York Yankees! This has been the dream of my life ever since I was a child, and it's all happening because I'm completely ignoring every urge toward common sense and good judgment I've ever had." Letting go of past experiences can sometimes free us up to experience new realities, realities that foster our creative growth even if the shake-up does not end up being as profound as it was for George.

Barrier 5: Our Habits   By definition, our habits are behavior routines that we fall into unconsciously and repeat just as unconsciously. While some habits are good habits and take the pressure away from our conscious brain to think through every little thing in any given day, habits can also blind us to inputs that could be enhancing our art of creativity.

The Invisible Gorilla videos created by Christopher Chabris and Daniel Simons (www .theinvisiblegorilla.com) show how selective attention works. When viewers of the video are asked to watch and count how many times a group of people in a circle throw a basketball to each other, the majority of viewers do not see a very large black gorilla walk across the screen. Focusing on the basketball to count the passes causes us to tune out other critical details in the video. Habits are much like that basketball. Even if we are not focused on our habits consciously, our brains tune out everything that feels irrelevant to the execution of the habit, making us miss much that could enhance our creative thinking and output.

Barrier 6: Our Personal and Professional Relationship Networks   The more diverse our personal and professional networks, the more new ideas and new connections between old ideas we will generate. Steve Jobs observed that "Creativity is just connecting things. When you ask creative people how they did something, they feel a little guilty because they didn't really do it, they just saw something. It seemed obvious to them after a while. That's because they were able to connect experiences they've had and synthesize new things." What Jobs is describing is an organic convergent thinking process that occurs after a very rich divergent thinking session. The precursor to that rich divergent thinking session is a diverse personal and professional network from which various ideas from multiple perspectives float around in our minds until they suddenly gel as a creative answer to a vexing problem. Throughout his life, Jobs pointed to diverse inputs such as Zen Buddhism, calligraphy courses, and rock music as the inspiration for his creative thinking. It wasn't just these abstract concepts that inspired him; it was the people he met who were passionate about these topics—his calligraphy instructor who was passionate about beautiful writing made him think about how fonts could be so much better than they were on the old computers; his Zen Buddhist teachers who inspired him to think about how technology products need to mimic the simplicity of life; and his love for rock music and rock stars which drove him to build the first collective commercial platform through which to sell music, song by song.

"Ziba, a top innovation-consulting firm in Portland, maximizes the value of a diverse workforce. The company's 120 employees are from 18 different countries and speak 26 languages. According to Sohrab Vossoughi, the firm's founder and president, 'genetic diversity breeds creativity, much like it does with biology.'"[12]

Barrier 7: Our Fear of Failure   There is nothing worse for our personal creativity than our fear of failure, a fear that most of us have adopted as one of our deepest and most scary fears. Companies like Zappos and Google that invest in their employees' personal creativity foster that creativity by creating a culture of experimentation where a "failure" is not a failure but a success waiting to happen. One professional services firm in Chicago actually encourages their employees to fail at least once a quarter. They have to discuss their failures (and what they learned from their failures) in their quarterly evaluations.

Even if you are not in one of these uber creativity-nurturing organizations, understanding and defusing your own fear of failure will allow you to open up to the "Post-It" type mistakes that are actually successes waiting to happen.

Effective Strategies to Power Up Personal Creativity

While eliminating the barriers listed earlier will net you some success in enhancing your personal creativity, we have found the following proactive strategies to be especially effective in increasing the originality and effectiveness of your personal creativity as well as the frequency with which your creativity supports and strengthens your talents at work.

1. Trust yourself

    "Creativity makes a leap, then, looks to see where it is."

    —Mason Cooley

    The creative process begins with trusting yourself to be creative. Creative people have cultivated a deep trust in their own creativity, so they are willing to take leaps into the unknown and trust that they will land somewhere useful. In studying these individuals, we discovered that their trust in themselves developed from seeing their "failures" as "learning opportunities" which made them more willing instead of being afraid to take risks in the future.

    Trust yourself. Where can you take a leap of faith today knowing that wherever you land, there will be an opportunity to learn and grow there?

2. Open up

> *"One very important aspect of motivation is the willingness to stop and to look at things that no one else has bothered to look at. This simple process of focusing on things that are normally taken for granted is a powerful source of creativity."*
>
> —Edward de Bono

Our brain is inundated with so many inputs on a daily basis that our ability to focus requires that we shut out most of the details. Over a period of time, we create habits of shutting out the same things and letting in the same things over and over until our creativity can no longer be stoked because new inputs are not getting through. Opening ourselves up to inputs that we usually shut out allows our brains to make new connections and think new thoughts, both of which are critical to creativity.

3. Clean and organize

> *"Clean out a corner of your mind and creativity will instantly fill it."*
>
> —Dee Hock

Have you ever wondered why you feel the need to organize your desk or clean your house when you have an important project to do? Unfairly deemed to be a procrastination technique, the need to organize and clean is now known to be a cognitive clearing activity where you are literally making space in your brain to create new ideas and connections. So, clean away and see what fills the space you create.

4. Make mistakes

> *"Creativity is allowing yourself to make mistakes. Art is knowing which ones to keep."*
>
> —Scott Adams

Mistakes are not failures; they are moments for learning. Mistakes are also not errors; errors are careless whereas mistakes are experiments. People who make mistakes well know that mistakes are great for learning as long as you don't make the same mistake more than once. With every lesson that comes from a mistake, you become better informed and more creative.

If you are not in a work environment where mistakes are tolerated, allow yourself the luxury of at least one rough draft where you don't limit your thought process before you cull the final product to be free of mistakes.

5. Get angry

> *"A wonderful emotion to get things moving when one is stuck is anger. It was anger more than anything else that had set me off, roused me into productivity and creativity."*
>
> —Mary Garden

The power of anger is often misunderstood and misused. Anger is understood as a stressor, but anger can be a powerful motivator for creativity because anger opens us up and allows us to experience the world in a more intense way. Anger is the equivalent of a cognitive stimulant as long as you get angry at situations, not people. Anger with other people and yourself is draining, but general anger, especially anger aimed at a particular situation, is great for creativity.

6. Get enthusiastic

> *"Creativity is a natural extension of our enthusiasm."*
>
> —Earl Nightingale

If anger is a cognitive stimulant, enthusiasm is a super stimulant! Enthusiasm for anything—ideas, products, situations, people, and so on—gets creativity humming like nothing else. Enthusiasm for a topic makes you curious about the topic, which opens you up to new experiences, interesting people, and novel ideas.

Discover the things you are enthusiastic about and start your creative journey there!

7. Listen to your hunches

> *"A hunch is creativity trying to tell you something."*
>
> —Frank Capra

A hunch—or intuition—is often the organic convergent thinking process that occurs in your brain after you have taken in data from many sources. Stepping away from a problem allows this process to occur and rewards you with hunches when you return to the problem. Trust your hunches, but make sure that your hunches are occurring after taking in data. If you are getting hunches before you take in diverse inputs of data, chances are you abdicating your creativity to your unconscious preferences, so data, data, data, then, wait for the hunch. Now, listen to it.

8. **Subtract instead of adding**

"Creativity has more to do with the elimination of the inessential than with inventing something new."

—Helmut Jahn

Sometimes creativity is about paring things down to their simplest form. While creativity sometimes urges us to add improvements and embellishments to ideas and processes, creativity can also be the act of taking things away until ideas exist in their least complex guise.

Get creative about taking the unnecessary away from your work, your work environment, and from your work processes. When the inessential is gone, what new things can you learn about what is left?

9. **Move your body**

"All truly great thoughts are conceived while walking."

—Friedrich Nietzsche

Physical activity is especially useful in the transition between the divergent and convergent thinking phases. Once diverse inputs are used to conjure up unrestrained ideas and connections, transitioning to a constrained vetting process can be difficult. The frustration inherent in this transition can be greatly alleviated through physical movement such as taking a walk.

Get moving to get creative!

10. **Question your questions**

"The uncreative mind can spot wrong answers, but it takes a creative mind to spot wrong questions."

—Antony Jay

The right answer to a wrong question is no more helpful than a wrong answer to a right question. The ability to discern between a wrong question and a right question is crucial to the creativity process because the question is the starting point for the creative journey.

Questioning the questions you are trying to answer allows you to differentiate between relevant and irrelevant, important and unimportant, and effective and ineffective creative processes. What is a question that you are trying to answer right now that could be asked differently to trigger a very different thought process?

11. **Pump up the volume**

"To get a great idea, come up with lots of them."

—Thomas Edison

A greater volume of ideas increases the probability of getting a great idea, so pump up the volume of ideas. The best way to do this is to increase the diversity of inputs and create several divergent thinking opportunities before you transition to convergent thinking. Ask yourself questions to stir the divergent thinking process. Why do I want to solve this problem? If this problem was a rock star, who would it be? If the solution was a food, what food group do I think it would belong to? What ideas would a 5-year-old come up with to solve this problem? What ideas would a 75-year-old come up with to solve this problem? Use the above questions and create more of your own, the crazier the questions, the better your divergent thinking!

12. **Read read read**

"It seems to be one of the paradoxes of creativity that in order to think originally, we must familiarize ourselves with the ideas of others."

—George Kneller

All original ideas begin at the end of someone else's good idea. So, read a lot of things from different sources in different fields. If you work in the sciences, read some works in literary sources along with readings in your subject area. Creative sparks fly when different subjects crash together in unique ways. Create these crashes by reading other people's good ideas in different fields. Then, let your brain do its job, and it will make original and effective connections for you!

References

1. http://www.smartstorming.com/articles/the-relationship-between-creativity-and-innovation.
2. http://www-03.ibm.com/press/us/en/pressrelease/31670.wss.
3. Runco, Mark A. and Jaeger, Garrett J. "The Torrance Creativity Center. The Standard Definition of Creativity." *Creativity Research Journal*, 24 (1): 92–96, 2012.
4. Franken, Robert E. *Human Motivation*, 3rd ed. Pacific Grove, CA: Brooks/Cole Publishing Company, 1994.
5. Reeves, Arin N. *The Next IQ: The Next Level of Intelligence for 21st Century Leaders*. ABA Publishing, 2012.
6. Sternberg R.J. "Introduction" in Kaufman, J.C. and Sternberg, R.J. (eds.), *The International Handbook of Creativity*. Cambridge University, 2006.
7. Ibid.
8. Guilford, J.P. *The Nature of Human Intelligence*, 1967.
9. Ward, T.B. "What's Old about New Ideas" in Smith, Ward, and Finke. (eds.), *The Creative Cognition Approach*. London, UK: MIT Press, pp. 157–178, 1995.
10. Amabile, T.M. "How to Kill Creativity." *Harvard Business Review*, 1998.
11. http://www.businessweek.com/stories/2006-02-12/creativity-loves-constraints.
12. http://www.inc.com/articles/201106/josh-linkner-7-steps-to-a-culture-of-innovation.html.

CHAPTER 6

The Creative Corporation

Thomas N. Duening

Introduction

The modern era of global competition has introduced a new imperative for corporate innovation. Innovation is often cited today as a primary source of competitive advantage regardless of industry or national identity. A Google search on the terms "innovation imperative" returns no fewer than 60,000 hits.[1]

Clearly innovation is shaping the modern economy, but was it ever not so? Perhaps our generation suffers from the same myopic perspective as generations of the past—who tended also to think that *their* generation was one of profound innovation and change. The argument has been made that our generation is unique in that only recently have we developed global communication networks, global transportation systems, and massive online education systems. These and other factors have also served to make "place" less important as a factor in one's access to key knowledge, talent, and other resources essential to innovation. An innovation in one part of the world is now rapidly disseminated via the Internet and other channels to nearly every other place in the world. As such, it is difficult for any particular innovation to be sequestered or harbored by a single individual or firm. Once the genie is out of the bottle, so to speak, any other intrepid company or entrepreneur can leverage the same innovation for commercial purposes. Intellectual property protection can help build a "picket fence" around some innovations, but global competitive advantage is not as reliant on intellectual property as it once was.

Perhaps, then, what makes our era unique is not the *fact* of innovation being an important driver of competitive advantage, but rather that the *velocity* of innovation has increased. Arguably, this is the vital distinction between the modern era and its "innovation imperative," and the past where innovation was also important to commercial success (try to name an era where innovation was not important). Perhaps what matters most today is not *that* a company innovates, but rather that it increases the *velocity* of its innovation.

In order to compete in this rapidly evolving "innovation economy" it is no longer possible for organizations simply to "keep doing what they have always done." Past managerial imperatives—Six Sigma, Lean, Total Quality Management (TQM), and others—have limited value as contemporary guideposts. Managers need to improve operations to be sure, but if that is their *sole* focus they can be certain that they inexorably will improve themselves into irrelevance. Focusing laser-beam-like on doing better what the organization already does will not cut it in our fast-paced, globally competitive, innovation-based economy.

Innovation is defined in this chapter as "the ability to solve new or old problems in new ways." **Creativity** plays a role in innovation to the extent that innovation requires *insight* into how a problem can be solved in a new way. It is this "insight" that lies at the heart of the "creative moment." Elsewhere, the creative moment has been defined as the "eureka" or "ah-ha" moment.[2] You've probably experienced it yourself. You may have struggled with a problem for a long time and, when you may have least expected it, suddenly arrived at a solution that had not previously occurred to you. This "creative moment" probably followed hours, days, weeks, or even years of false starts, frustration, and dead-ends. Yet, the problem was solved. Creativity was displayed. What needs to be understood is what was happening during the hours, days, weeks, or years that preceded the creative moment. What was the organizational context of the "creating?" What were the thought processes that enabled the creative insight? What role do other people—for example, colleagues, managers, others—play in the creative process? That is what we will explore in this chapter.

As innovation and "corporate creativity" have become the predominant imperatives for modern organizations, the paucity of scholarly understanding of what creativity is and how it is to be accomplished begins to stand out. My research background is in entrepreneurship and so this chapter will endeavor to apply relevant concepts from that field to corporate creativity. This does not strike me as overreaching since entrepreneurs are businesspeople who often work on the frontiers of the economy and they often need to conjure their creative powers to deal with uncertainty and ambiguity. As it happens, scholarship in entrepreneurship has recently turned its attention to the processes that expert entrepreneurs actually use to solve the problems associated with launching and growing new ventures.

In the next section of this chapter we will explore a more in-depth definition of "corporate creativity" so that we can better grasp what we are trying to understand and promote. The ultimate aim of this chapter is to operationalize "corporate creativity" and to give readers tactics to use immediately in their workplaces. It would be beneficial toward this end to have at least a modest grasp of what we are talking about.

Next, we will examine some "best practices" that are associated with corporate creativity. There is no such thing as a "last word" on this topic, but it is helpful to know what some of the top scholars and practitioners of corporate creativity are thinking and prescribing.

The chapter then engages in a short discussion about how creativity manifests in society *writ large*. This may be one of the chief contributions of this chapter—it attempts to draw insights into corporate creativity from the most powerful creative engine humans have ever devised: free market economies. We will explore, in particular, Hayek's notions of the spontaneous order and abstract rules. It is difficult to find a social system that has produced more creativity, more *timely* creativity, and more prosperity than free market economies. This chapter assumes that the seminal ingredients of creativity are best evidenced in such free market systems. Although corporations are more goal-oriented than society *writ large,* we would be remiss if we didn't investigate this powerful creative engine for potential isomorphic links to corporate creativity.

This chapter suggests next that some of the concepts derived from the scholarship into how expert entrepreneurs create value amidst uncertainty are useful. We will explore key concepts and principles associated with expert entrepreneurship and how they can be applied within the boundaries of the corporation. As will be pointed out, creativity often means that the creator is functioning on the *frontier*. That is, the creator is testing ideas, concepts, products, and novelties that have not previously been tried or tested. Expert entrepreneurs do this as a matter of course, and it seems that insights into how *they* operate can provide insights into how corporate creators also might operate.

Finally, this chapter concludes with key takeaways that will guide managers and corporate leaders who want to nurture a more creative corporate culture. Ultimately, this chapter argues that the foundation of a creative corporate culture must include potent abstract rules that help produce what is referred to as a "bounded spontaneous order." Paradoxically, although creativity is desperately desired, it cannot be commanded. It can only be beckoned.

Defining Corporate Creativity

It is usual for reference books to define key terms by citing how those terms are defined in leading dictionaries. I'll spare you that here because creativity may be one of those terms that we all understand and know, but find it difficult to pin down into a simple or singular definition. It's like the famous definition of pornography: "I know it when I see it." Let's therefore begin by attempting to highlight some factors that seem to be part of any creative moment. I propose the following:

- Creativity is evidenced by novelty
- Creativity often involves combining well-known particulars in a new way
- Creativity usually occurs only after hard work or lengthy preparation
- Creativity often produces unusable or worthless outcomes
- Creativity can come from anyone and anywhere
- Creativity is more likely under duress, stress, or scarcity
- Creativity can involve luck and unpredictable contingencies

Let's take a deeper look at each of these factors.

Creativity is evidenced by novelty: It would be difficult to identify something as creative if it was not also novel. We tend to see ideas, products, designs, as creative if they are of a type that we haven't seen before. Of course, this does not mean that the creative idea isn't being concurrently developed elsewhere. In the global economy, it's likely that any creative output is either already being developed or is close to being developed somewhere else. Still, novelty is relative to people and markets, and there is no need to worry about "absolute novelty." What matters most in this innovation economy is to capture creative moments as they occur and leverage them for competitive advantage with deliberate speed. The old concept of "sustainable competitive advantage" is inexorably losing its relevance in the innovation economy. What matters more is continuous novelty creation, creativity capture, and leveraging creativity for what might be called "sustainable creative advantage" (SCA).*

Creativity often involves combining known particulars in new ways: Writers, artists, musicians, all work with limited tools, materials, and time to create new works. Writers have words, punctuation, and limited forms of media (although digital media has recently opened new vistas). Good writers combine words and punctuation marks—elements known to us all—in novel ways. We know a good writer when we read one, but that doesn't mean we can define, exactly, what makes them "creative." In fact, the nebulous character of the term has enabled the profession of "literary critic" and endless interpretations of the creative works of writers. Certainly there are writers (or other types of artists) who continue to be revered for their creative genius. But that doesn't mean that writing is finished or that there won't be new writers that will win accolades as creative geniuses. Knowing this, it's important to realize that corporate creativity may involve little more than recombining the resources that the organization already controls. With this understanding, the key is to help people become aware of the resources that are within their realm of control in order to enable creative recombination of the resource pool. How often have large companies lamented the organizational "silos" that inhibit cross-boundary communications and awareness? Creativity is usually not cited as a casualty of these silos, but given that creativity is often the result of the recombination of existing resources it is likely that corporate creativity suffers in organizations rife with silos.

Creativity usually occurs only after hard work or lengthy preparation: Many people assume that creative geniuses are gifted by genetics, God, good fortune, or something else outside their control. That is a myth that most creative people would repudiate. Such people are more likely to point to the hard work, discipline, practice, and years of effort that lies behind momentary, spontaneous acts of creation. Research into what it takes to develop expertise in nearly any field has identified "deliberate practice" as necessary. *Deliberate practice* is the term used to refer to the rehearsal of activities, collection of feedback on the quality of the performance of that activity, analysis of the feedback, and use of the feedback to improve on future performances.[3] Significantly, the research has determined that expertise development in nearly any field requires 10 years of deliberate practice.[4] So creativity, although manifested as a singular "ah-ha" moment is normally the result of deliberate practice and hard work. Von Mises wrote about the creative genius as follows:

> "Neither does the creative genius drive immediate gratification from his creative activities. Creating is for him agony and torment, a ceaseless excruciating struggle against internal and external obstacles; it consumes and crushes him. Such agonies are phenomena which have nothing in common with connotations generally attached to the notions of work and labor, production and success, breadwinning and enjoyment of life. The genius does not deliver to order. Men cannot improve the natural and social conditions which bring about the creator and his creation. But, of course, one can organize society in such a way that no room is left for pioneers and their path-breaking."[5]

Creativity can come from anyone and anywhere: This point resonates with the one immediately above. With enough hard work, practice, feedback, and rehearsal it is possible for anyone to develop creative capacity. Corporations employ legions of highly trained professionals who take their professions seriously and who have deliberately

---

*Although a search of the scholarly literature brings up 867 citations for "sustainable competitive advantage," there are zero citations for the phrases "sustainable creative advantage" and "sustainable innovation advantage." It seems that a sea of change is in order in the scholarly literature to respond to changes in the actual world of business and competition.

practiced and/or deliberately performed (a form or deliberate practice that refers to *in vivo* as opposed to *in vitro* practice and learning*). These individuals may be anywhere in the organization and potentially can produce creative outputs within their field of expertise. The key to enabling employee creativity is to ensure that everyone not only recognizes they are free to exercise their creativity, but also that they will be duly recognized and rewarded for their contributions. Too often employees are reluctant to express their creativity because someone else—often their immediate superior—will get the credit (or doesn't get the credit and retaliates against the subordinate for standing out). There is probably little more that an organization could do to stifle creativity than to reserve credit only for those with lofty titles and higher pay grades.

Creativity is more likely under duress, stress, or scarcity: In times of stress or scarcity, people often discover creative solutions that they would not have thought of under other circumstances. Anyone who has watched *Survivorman* on television has seen the creativity of its host, Les Stroud, as he struggles with survival in a wide range of environmental settings.[†] Key to the show is the disclosure at the beginning of the meager supplies that Mr. Stroud was left alone with in the wilderness. It is interesting to watch as he uses not only those supplies, but also those that he can scavenge, the clothes on his back, even his cameras for creative solutions to problems in a survival setting. While it would be difficult to recreate a survival setting in the corporate workplace, there clearly is value in creating an "us against the elements" mentality among the corporate team. Anyone who is a sports fan recognizes this mentality as a common, nigh unto cliché, sound bite for athletes and sports teams to motivate high performance. Keeping employees focused on key metrics, competitors, and global challenges may help generate a similar "survivalist" mentality that will help motivate creative efforts.

Creativity can involve luck and unpredictable contingencies: There is no denying the role that luck plays in nearly any achievement. Whether it's the ball bouncing the right way in an athletic event, the maestro finding the precisely talented musician to round out the orchestra, or finding that one person with whom one can build an entire life, luck is one among many causal factors. The experienced creative person recognizes this, grows comfortable with it, and learns to leverage every lucky stroke that comes his or her way. There is no "pride of ownership" or ego that requires credit for fortunate happenstance. Creative people "go with the flow" and don't worry about whether their success is based on luck or hard work. At the same time, recognizing that luck plays a role in anyone's ultimate creative success should engender both humility and gratitude. Humility is an underappreciated value in corporations. Most people confuse humility with deference or reticence. We are not talking here about creating a workplace full of shrinking violets—quite the contrary. Without significant achievement, there is nothing to be humble about. Gratitude is similar to humility except that the former is a social act. Demonstrating gratitude for the lucky circumstances that helped one enjoy the fruits of a creative moment helps to maintain the social equilibrium of the corporate culture. This is critical to cultivating a culture capable of competing based on sustainable creative advantage.

"Best Practices" in Corporate Creativity

"Creativity" is one of those "soft" concepts that many hard-nosed businesspeople look upon with skepticism—and rightly so. Creativity is often associated with "starving artists" and others who have no concern about the commercial potential of their creations. Businesses cannot afford to adopt that frame of mind, and they cannot afford overhead that doesn't somehow contribute to the organization's financial health. For example, one of the more often-cited techniques to nurture creativity within the corporation is referred to as "tinkering time." Google has a well-known "20 percent time" initiative where employees are encouraged to spend 20 percent of their time working on projects outside their job description.[6] However, research into the effectiveness of this approach indicates serious problems. The problem companies have found with "tinkering time" as a general policy is that most employees are not comfortable with the burdens and expectations of tinkering time. They

---

*Deliberate performance is similar to deliberate practice, except that the former occurs in the work setting. People become experts in their fields by using the same reflective techniques of deliberate practice, but they do so while actually performing in their assigned job roles.

[†]For more on this television program, see http://lesstroud.ca/survivorman/home.php.

find the requirement to "be creative" as an unwelcome burden on top of their regular job requirements.[7]

Creativity in the corporate context requires that it be channeled either to revenue-generating activities or to cost-reducing activities. That is to say, creativity within the organizational context cannot be free-flowing and without consequence. It must be disciplined, channeled, rule bound. We will take up the notion of rule-boundedness in the next section. Here, we will be concerned some "best practices" associated with nurturing a creative corporate culture.

Fortunately, there is ample research into factors that are vital to fostering a creative climate in the organizational setting. Ekvall's model of the creative climate identifies 10 factors that need to be present:

1. **Idea time:** People need time to think and to develop new ideas. Many organizational cultures frown on this as they tend to see "think time" as "idle time." Creative ideas need time to incubate. People may even need time to do some reading that helps them understand their creative ideas in more depth and to gauge the true novelty of what they are considering.

2. **Risk taking:** People need to be able to make decisions within acceptable risk boundaries. However, most organizations don't articulate risk boundaries clearly enough for people to feel comfortable in their decision making. Some organizations, such as Koch Industries, specify domains in which people have specific "decision rights" and they are empowered to follow their own thinking and instincts in taking risks within those boundaries. Repetitive success with specific decision rights helps an individual acquire additional rights over time and with continued success.[8]

3. **Challenge:** People need to feel challenged within their own areas of expertise. Csikszentmihalyi's work on the concept of "flow" indicates that people work at their peak when they are offered challenges that are achievable, but that require they stretch their talents to new levels.[9] Further, Duening and Ivancevich highlighted the need for the organizations' "Einsteins" to be challenged in a manner that helps them achieve the highest levels of Maslow's hierarchy of needs.[10]

4. **Freedom:** People need freedom (I prefer "autonomy" in the corporate context) to express their talents, make mistakes, learn from mistakes, and grow within their field of expertise. Some managers have a tendency toward too-soon interventions and tend to thwart the essential learning and unlearning that only comes from experience.

5. **Idea support:** People need to feel that their ideas are aligned with the interests of the organization. Organizations should establish clear objectives and clear contact points for individuals to discuss their ideas as they develop and to ensure that they are aligned with the long-term interests of the organization.

6. **Conflicts:** People need to be able to defend their creative ideas within the organization's "marketplace" of ideas. Managers should articulate that there are no "entitlements" within the organization, and that all ideas must live or die based on their relative merit within that marketplace. At the same time, people should be empowered with business case development and presentation skills that enable them adequately to articulate their creative insights within the context of the idea marketplace.

7. **Debates:** People should be engaged in debates about the merits of their creative ideas. Debating serves two primary purposes: (1) It forces the creator to think hard about possible objections/problems with their creative ideas; and (2) It introduces incremental improvements to ideas from those who are party to the debates.

8. **Playfulness/humor:** People often develop better ideas and more quickly when they are allowed to play with concepts in a nonthreatening way. Organizations that promote "stupid ideas" and allow them to live or die within the marketplace of internal conversation, conflict, and debate are more likely to surface those creative ideas that just may turn out to be the next "killer app."

9. **Trust/openness:** People function at their best within a context of trust and openness. This means that managers and others don't "sugarcoat" their responses to creative ideas, overstate their interest in ideas, or understate their interest. They also avoid any "not invented here" envy or jealousy that often sabotages relationships and the potential for the further development of good ideas.

10. **Dynamism/liveliness:** People today tend to want to minimize the distinction between work and play. Organizations that foster open spaces, free-flowing conversations, impromptu chat sessions, and favor a lively workplace "buzz" are the norm among today's most creative companies. Think Facebook, Google, eBay, and other titans of our time. They famously have created lively organizational environments that are equal parts work and play.

These 10 factors have been the subject of substantial research and there are some prescriptions about how to foster them within the organization and how to mix a recipe that is tailored to different types of organizations. For example, creativity within a steel mill will take on a different look and feel than creativity within an advertising agency. Under these two examples, clearly the relative importance and emphasis on these various factors will differ. As I pointed out elsewhere, the ingredients for organizational improvement are well known. What is less intensively researched and, thus, less well known is how to mix these ingredients into a recipe that is right for particular organizational settings.[11] The best that can be said right now is that managers must creatively mix these 10 ingredients and discover the recipe that is best for their particular organization.

Abstract Rules and Corporate Creativity

Life is governed by rules, not all of which are articulated or codified. Oftentimes we only have a vague notion that our actions are governed by rules, and we might even have difficulty answering a question about why we have taken specific actions. We tend to hush our voices in libraries. Why? We avoid standing directly next to someone else in an otherwise empty elevator. Why?

Myriad such "abstract rules" govern our mundane as well as our more important daily activities and decisions.[12] Abstract rules are those unarticulated, yet essential, guidelines, norms, traditions that people within a social setting tend to follow. Conversely, failure to follow the prevailing abstract rules leads to surprise, annoyance, or, in the worst case, ostracism. Abstract rules are unarticulated—which means they aren't explicitly written down or codified. They are essential in that most social systems could not function without them. For example, the busy sidewalks of New York could not function if pedestrians did not follow the abstract rule of trying to avoid direct, head-on collisions with other pedestrians. The abstract rule to avoid collisions can basically be worded as "stay to the right in the direction that you are walking and move further to the right if you are on a head on course to collide with someone walking in the opposite direction." Wayward pedestrians who violate this rule (a more common occurrence in our "text messaging" age) are a nuisance to others and a hazard to themselves.[13]

If we can accept the premise that social groups require abstract rules to function effectively, then the next question that arises is "Where do these rules come from?" We can turn to the work of Friedrich Hayek for an insightful and carefully analyzed answer to this question as it pertains to abstract rules that govern social activities in a society *writ large*.

Hayek's most trenchant analysis of the phenomenon of rule-governed social behavior identifies rule following as a phenomenon that lies between what he categorizes as "instinct" and "reason."[14] He points out that most of the rules we follow originally were based on instinctive activities that humans engaged in without thinking. Some of these behaviors worked, and some did not. Accepting the scientific understanding of humans as evolved beings that formerly lived primarily in nomadic groups of 100 or so members, it is easy to understand that such groups who tended to adopt the most useful behaviors and activities were more likely to survive and, potentially, to thrive.[15] Eventually, as language and writing evolved, humans developed the capacity to reflect on the activities and behaviors that worked. These behaviors and activities might have been codified into formal rules (such as the 10 Commandments) or they may simply have been passed on through oral traditions and stories.

The important point to note about this brief history of the origin of at least some of our more useful abstract rules is that they are not a result of *reasoning* about what works. Reason played no role in the *emergence* of the abstract rules. Abstract rules emerge out of people acting on "instinct" and creating new activities based on the conditions and contingencies under which they find themselves. Some of these new activities prove fruitful and are repeated and perhaps refined; others prove less useful (and occasionally fatal) and therefore are not passed on. Reason comes into play only *after* the new rules have emerged and proven to be useful.[16]

Reason is our capacity to articulate formerly abstract rules in the prevailing natural language. Once the rules are codified in language we become able to evaluate them, compare them to one another, and debate about whether the codified rule is aligned with what we think was the "intent" of those who established it. Reason enables continuous dialog about the utility and interpretation of our articulated rules, and how they might be improved to be aligned better with other codified and abstract rules that the social group accepts.[17]

Now we come to the crux of the issue that Hayek was addressing. Of course, the primary target of Hayek's analyses was the creativity stifling effect of attempts to centrally plan economic activities and outcomes. He wanted to demonstrate that central planning will always and everywhere be incomplete, and will always lead to social and economic stagnation because it is not *reason* that creates new rules to deal with new circumstances. Rather, Hayek asserts, new rules that lead to unexpected, novel human activity and outcomes emerge *between* instinct and reason. It is only *after* the new rules have emerged that human reason is engaged in the codification, modification, and comparative evaluation of the new rules with those that already exist.

Hayek's analysis can also be applied to the challenges of nurturing a creative corporate culture. It is important to remember that human adaptation to new circumstances (which requires creativity) cannot be achieved in advance through the application of reason. In other words, just as central planning of an economy cannot be successful in nurturing creativity because central planning inhibits creativity *on the frontier*; central planning also cannot be successful in nurturing creativity within the context of the corporation because that also occurs *on the frontier*. In short, creativity cannot be imposed by fiat or executive order, and it cannot be planned in advance. The creative corporation will only result from careful cultivation of the *conditions* under which creativity can *naturally emerge*.

We all know, however, that the corporation must have some codified rules that employees follow or it will fly apart. The notion of a corporation without formal, governing rules is not just an odd concept, it is literally undefined. Unlike a society, however, where many of the abstract rules that people follow are a result of many generations of trial and error learning leading to abstract rules and traditions that have proven to be useful, the average corporation doesn't have as long a history of such learning. So how can the corporation establish a culture bounded by abstract rules, but at the same time allows for creativity between instinct and reason that Hayek cites as the key to what he calls the "spontaneous order?"[18]

This is the challenge that lies at the heart of nurturing a creative corporate culture. The creativity that contributes to the long-term health and success of the corporation must be constrained within the boundaries of the abstract and formal rules of the corporation. But the corporation is nothing more than the current people that comprise its workforce. How do employees become acculturated to the abstract rules ("traditions") of the corporation? How do they develop an ability to work within the constraints of the rules that govern the daily activities of the corporation and yet preserve the potential to invent "on the frontier?"

The consequence of people acting freely in accord with abstract rules is referred to as "spontaneous order." The notion of something "spontaneous" occurring within the boundaries of the corporation may seem counterintuitive to most managers and corporate leaders, but it is the essence of creativity. If corporations want to increase their creative capacity they must allow for spontaneous order to emerge within the context of appropriate abstract rules. By definition, if spontaneity is stifled, creativity cannot and will not occur. If creativity were not spontaneous then it could be planned in advance and we would not have a need for this chapter or research that is attempting to understand and promote corporate creativity.

The spontaneous order is the term that Hayek uses to describe what he calls the "Open Society."[19] The astounding capacity for human beings to live freely and in relative harmony by following abstract rules is a recent discovery. Prior to the discovery of the power of individual freedom and the spontaneous order, human beings generally were governed by the whims of monarchs or dictators. The assumption behind this type of social structure is that everyone should pursue the same ends. That is, the "glory of the state" or the "glory of the monarchy" were deemed to be the supreme values to which all subjects must contribute and/or submit.

The advent of human liberty on the societal scale was the result of the revolutionaries in the United States and France, mostly in the 18th century. These revolutionaries ushered in a new era in which human beings became free to decide what is best for them as individuals or as family units within the boundaries of abstract rules. This new era of individual liberty precipitated an explosion of creativity, prosperity, and a standard of living that the world had never seen before.

The rise in the general standard of living was a direct result of unlocking the creative powers of every individual within society. Although these creative powers were unleashed to enable each person to pursue their own interests within the context of abstract rules, the unintended consequence was a flourishing of wealth for everyone participating in the free societies. The spontaneous order created by this unleashing of human creativity was not planned by anyone, and, importantly, it could not have been. There is no way a centralized planning authority could have acquired all of the knowledge that collectively resides within the minds of free individuals acting according to their own interests.[20]

As the spontaneous order of the Open Society is arguably the most powerful creative engine humans have ever stumbled upon, it seems likely there are lessons in the key elements of the Open Society for corporate creativity. Nurturing a culture that enables what we will call a "bounded spontaneous order" seems to be a matter of the prevailing abstract rules and inherent capacities of the employees. As we noted earlier, merely providing "tinkering time" is insufficient to promote corporate creativity. In the next section, we will explore some of the findings from the scholarship into expert entrepreneurs. It is argued that corporate creativity will be enhanced if employees are trained and allowed to practice some of the behaviors, attitudes, and principles of the expert entrepreneur to increase corporate creativity.

Effectuation and Corporate Creativity

The abstract rules that we have been discussing as the restraining component to unfettered creativity within the corporation has not been addressed by scholars. What are the elements of abstract rules that will enable someone to be a good employee, and yet empower them to experiment on the frontiers of their capacities and responsibilities? It is on this frontier that truly creative contributions to the corporation will be made.

Fortunately, relevant research into human exploration of the business frontier is available via scholarship into "expert entrepreneurship." Entrepreneurship is a relatively new topic of serious scholarly research, but significant results have begun to emerge. One of the more revealing lines of research in recent years looks at the way expert entrepreneurs actually think, decide, and take action. This line of research can be encapsulated under the concept of "effectuation."

Effectuation is defined as "taking action toward unpredictable future states using currently controlled resources and with imperfect knowledge about current circumstances."[21] In other words, scholarship into effectuation examines how entrepreneurs decide and act in the face of uncertainty. One of the hallmarks of effectual action is the application of an entirely different form of logic that is typically used by corporate leaders. Causal logic begins with a single clear goal in mind, and attempts to gather and deploy the resources that are required to achieve that singular goal. The objective of the corporate employee under causal logic would be to aggregate the means necessary to achieve the goal.

By way of contrast, effectual logic does not begin with a clear goal in mind. Most expert entrepreneurs will readily admit that the economic goals they've achieved often were unexpected at the time they launched their ventures. The expert entrepreneur acting under effectual logic begins with the means currently controlled, rather than the goal. Expert entrepreneurs leverage the resources they currently control to create any number of successful, alternative future outcomes—they are not constrained to pursuing a single, predetermined goal. The difference between causal and effectual logic is presented in Fig. 6-1.

Managerial or "causal" logic is depicted on the left side of the diagram. Note that causal logic assumes a well-defined "given goal." The challenge confronting the economic agent

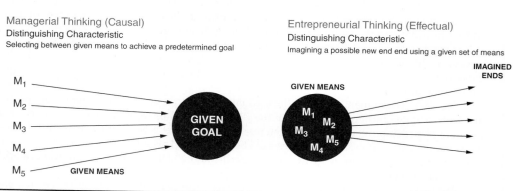

Figure 6-1 Causal versus effectual logic.[22]

under this logic is to gather the resources needed to pursue that singular goal. All measures of success and/or effectiveness are related to whether or not that goal was achieved.

The right side of this diagram depicts what is referred to as "effectual logic."[23] This is the form of logic that has been discovered to be used by expert entrepreneurs. As shown in the graphic, expert entrepreneurs begin with the means they currently control and then use those means to pursue economic opportunity. Effectual logic, in contrast to causal logic, does not focus on achieving any *particular* end. The expert entrepreneur realizes that pursuing economic opportunity is a discovery process, and that the ultimate end toward which one is tending is unknown in the present.[24]

Effectuation has been encapsulated into five fundamental principles. These principles have been revealed through investigation of expert entrepreneurs in a wide range of fields, from technology to restaurants, and from product-oriented to service-oriented. The five principles that have been explicated by Sarasvathy are[25]

1. The Bird in the Hand Principle
2. The Affordable Loss Principle
3. The Lemonade Principle
4. The Crazy Quilt Principle
5. The Pilot in the Plan Principle

The Bird in the Hand Principle

The Bird in the Hand Principle is the foundational principle of effectual logic. Effectual logic differs from causal logic in that it starts with a given set of means (resources) and deploys those means to achieve a wide range of potential goals, any one of which could be determined to be a "success." The Bird in the Hand Principle encourages economic agents to take stock of the resources they currently control, and to leverage these to create value.

For example, the expert entrepreneur starting a venture would begin by taking stock of the resources that are currently controlled and those that can readily be acquired. He or she would know from experience that the ends pursued are not well defined. The novice who has not learned this principle might be more inclined to hesitate in getting started as he or she attempts to acquire more knowledge about the market, set a clear goal, and then analyze whether it was possible to acquire the resources necessary to achieve that goal. The expert entrepreneur doesn't wait. The entrepreneur starts and relies on his or her experience and personal resourcefulness to persist until appropriate goals begin to emerge. The entrepreneur also seeks to gather additional resources as the venture grows to help accelerate that growth.

The Affordable Loss Principle

The Affordable Loss Principle stipulates that entrepreneurs risk no more than they are willing to lose. For example, expert entrepreneurs realize that by launching a new venture—and considering the time, energy, and general sacrifice that will be required to build it over time—they are thereby *not* doing something else that might work out better. Expert entrepreneurs have learned to minimize such opportunity costs by focusing on the opportunity they've chosen. Novice entrepreneurs, by way of contrast, may attempt to "hedge their bets" and try to minimize risk by attempting to pursue all of their opportunities at the same time. The expert, by focusing on a single opportunity, is willing to risk being wrong—something that would be difficult to do in most enterprise settings.

The Lemonade Principle

The Lemonade Principle is based on the old adage that goes "If life throws you lemons, make lemonade." In other words, make the best of the unexpected. Expert entrepreneurs, for example, have learned that predictability is reserved only for narrow domains in business and in life. Most of what happens as one is building a new venture could not have been predicted. The expert realizes that adaptability is a key characteristic for building successful ventures. Stubbornly sticking to original business models, or refusing to heed the market's desire for lower prices, better features, or something else is not effective. On the other hand, the expert entrepreneur has also learned that there must be valid reasons for changing business strategies. The reasons underlying a decision to change a venture's strategy are many, and the expert entrepreneur has learned that some are more potent evidence of the need for change than others. For example, if a single customer group demands lower prices the expert entrepreneur realizes that it's not possible to please everyone. On the other hand, if one's target customers consistently choose the lower-priced alternative offered by competitors it may be time to shift strategy.

The Crazy Quilt Principle

The Crazy Quilt Principle is based on the expert entrepreneur's strategy continuously to seek out people who may become valuable contributors to his or her venture. For example, many entrepreneurs learn the value of establishing knowledgeable advisory boards. Advisory boards can be as large or small as needed, but should include people who can add value to the venture. Adding value can be achieved in a number of ways, including contributing needed investment capital, providing insights about how a particular market works, opening doors to potential customers, and many others.

The Pilot in the Plane Principle

The Pilot in the Plane Principle is based on the concept of "control." The unique aspect of control under effectual logic is referred to as "nonpredictive control." The desire to control events in our environment seems to be a natural human tendency. Expert entrepreneurs believe they can determine their individual futures best by applying effectual logic to the resources they currently control. In other words, expert entrepreneurs focus on discovering how to create value with the resources they currently control.

These principles of effectual logic are simple enough to understand and learn, and they may have significant application within the corporation seeking to increase creativity. Imagine all employees, operating under appropriate abstract rules who are empowered also to act on the principles of expert entrepreneurs. That would create both a "bounded spontaneous order"—which is *essential* to a creative corporate culture—and address the problem with "tinkering time" by ensuring employees are equipped with the tools used by expert entrepreneurs to create economic value. Next, we will examine a few takeaways from the earlier discussions.

Conclusions and Action Items

Significant takeaways can be developed from this discussion only if you, the reader, have been convinced that the notions of "effectuation," "abstract rules," and "spontaneous order" have been developed sufficiently that harboring doubts is counterproductive. In order to move forward and make the necessary adjustments to nurture a more creative corporate culture, it is imperative that you thoroughly absorb and accept these discoveries and overcome any lingering predilections toward maintaining the status quo. There can be no half-measures toward this end. Employees will either be granted the freedom and appropriate empowerment to create a bounded spontaneous order, or they will not. The definition of creativity that we discussed earlier requires *absolutely* that people are given latitude to explore creative ideas on the frontier.

Of course, maintaining the status quo is the safe and traditional route followed by most large corporations. Careers often are made or broken by decisions to try something new. The concepts discussed in this chapter decidedly are new, and they only recently have been applied to the corporate setting. In this era where innovation has been cited as a key source of competitive advantage, it seems that corporations cannot afford to ignore the breaking research and insights into the factors associated with creativity. This chapter has explored both best practices, which can be imported immediately, as well as some leading-edge insights into creativity from the perspective of social systems, *writ large*.

So, given that you've been warned and given the reasonable assumption that you've read this chapter to learn something about how to begin to transform the creative capacities and culture of your organization, the following prescriptions are offered without reservation:

1. **Train employees in effectual logic:** Effectual logic opens up a whole new set of possibilities for creativity. Only those who have been trained to think in terms of multiple possible alternative successful outcomes will be able to conjure and use their personal creativity to deal with ambiguity and the unexpected. Goal setting is endemic to the corporate mind-set. In fact, goal setting is normally part of the annual review process to judge employee performance. As such, the causal-logic mind-set is deeply ingrained into most corporate employees. Only direct retraining in effectual logic and its principles can help people understand that non-predictive control of current resources is the *essence* of corporate creativity.

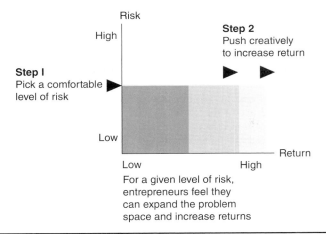

FIGURE 6-2 Entrepreneurial risk tolerance.

2. **Provide clear decision or resource rights:** Entrepreneurs using effectual logic and the entrepreneurial method begin with the resources they currently control. They then act creatively to manage risk and increase returns using these resources. Graphically, this is represented in Fig. 6-2.

   The expert entrepreneur acts on the Affordable Loss Principle and manages risk within the known level of tolerance. But knowing one's level of personal risk tolerance doesn't lead to inaction. For the entrepreneur, it is simply a function of understanding that to make something happen commercially requires creative extension of controlled resources toward generating returns.

3. **Reward unexpected results and positive returns:** There may be no management bromide that is more anathema to corporate creativity than the old "management by objectives" (MBO). MBO was developed by management guru Peter Drucker and that may in itself be reason enough that the approach will be exceedingly difficult to supplant. Managers love goal setting, and employees love knowing exactly what they need to do to receive their next raise or promotion. Still, it is possible that corporations have taken goal setting to an unhealthy extreme and it is time to step back to enable the spontaneous order to develop. The first step is to train employees in the art of effectual logic. The next step is to develop incentives that are directly tied to the results people achieve via creative use of the resources they control. Entrepreneurs endure hardship, stress, even failure all for the sake of controlling their destiny and pursuing their own rewards. These same factors—personal control and personal reward based on personally generated results—will motivate appropriately empowered employees to similar levels of effort.

4. **Accept failure as a step in the creative process:** Perhaps the most difficult management challenge for nurturing a creative corporate culture is to accept failure as a part of the creative process. Failure is common among entrepreneurs, and is generally regarded as a favorable thing if the failure occurred in the context of committed hard work and personal integrity. Entrepreneurs accept the unpredictable nature of the projects they undertake and are capable of cutting their losses and moving on to the next project in the event of a failure. Contrariwise, a well-known problem in management is the so-called "escalation of commitment" phenomenon that occurs when people feel a project is failing.[26] Rather than cutting their losses and accepting failure they tend to escalate commitment in the hope that they eventually will succeed. "Hope" as a strategy is not something that entrepreneurs choose to do. Hope also should not guide strategy within the corporation. Creativity naturally engenders failure. Corporations need to find a way to set clear guidelines for knowing when to accept failure, and when to forge ahead in the event that success is right around the corner. For the entrepreneur, failure is reached when the affordable loss they are willing to accept has been exceeded. Something similar needs to become a part of corporation projects. In a recent and very rare magazine interview, marketing guru Seth Godin said, "How many people go to work each day and say 'This might not work' and then go do it? The answer is, not enough."[27]

5. **Take charge of the "abstract rules":** As we discussed, no formal rules can be written to nurture corporate creativity. Corporate creativity must occur between instinct and reason, and it will be governed and channeled by "abstract rules." We all know what the abstract rules are in any particular workplace. Normally, within the corporate setting, it is evidenced by the phrase "that's the way we do things around here." The abstract rules that employees follow are created by what leaders do and say, not by the formal rules they create and/or enforce. Most of the formal rules in corporations are of the "thou shalt not" variety. They do not usually promote the kind of corporate creativity that we are seeking via this brief analysis. Only the abstract rules that enable creative pursuit of ends that are unpredictable in the present will promote the kind of hard work and toil that precedes genuine instances of creativity. It is vital that the organization have skilled managers that can promote a culture that empowers employees to use their creativity. Such promotion is a result of deeds, not directives. People cannot be directed to "be creative." Managers must be trained to enable the spontaneous order via their daily deeds that are examples of the abstract rules that apply to everyone within the organization.

In this chapter we have explored a definition of "creativity," examined some generally accepted "best practices," and ventured into new domains by linking Hayek's analysis of the greatest single engine of human creativity—free market economics—to the corporation. The recipe for nurturing a culture of creativity in the organization certainly includes some or perhaps all of the best practice ingredients that we examined here. However, this chapter also argued that simply importing these ingredients into the average corporate environment will be insufficient to nurture creativity. What is generally missing is an empowering framework where employees are not merely left alone to "be creative" (as in the "tinkering time" approach that has been shown generally to be ineffective), but actually empowered to be creative within the context of firm-specific abstract rules.

Employees can be empowered appropriately by training them in the effectual logic used by expert entrepreneurs. With this knowledge employees can look for creative ways to leverage existing resources toward multiple possible successful, yet unpredictable in the present, outcomes. This can also help prevent the common "escalation of commitment" toward a predetermined goal that may be unachievable.

In addition to preparing employees for creativity, managers and leaders can prepare themselves to be comfortable with an evolving spontaneous order. As a caveat, this chapter acknowledges that corporations cannot be as free-form as Hayek's Open Society, and that in the corporate setting a "bounded spontaneous order" is preferred instead. This can be managed through developing and promulgating the appropriate set of "abstract rules." Whether these are communicated via technical terminology such as "hurdle rates," or whether the abstract rules are communicated less directly via leadership actions is an important consideration. Formal rules are generally of the "thou shalt not" genre, and hardly conducive to promoting corporate creativity. Instead, employees need to have a firm idea of the resources they legitimately control, the boundaries within which these resources can be leveraged, and then be allowed to create a new spontaneous order—one that is *not* planned in advance.

Unfortunately, the formality of most corporate structures often preclude such spontaneity, but it would occur naturally under sufficiently well-known abstract rules if managers and leaders would allow it. Corporate leaders must get their arms around the prevailing corporate lore—"the way things are done around here"—in order to nudge the abstract rules toward a more favorable creative culture. Ultimately, those leaders who desire to nurture a more creative corporate culture must, ironically, learn to "let go" in order for a bounded spontaneous order to emerge. Sustainable creative advantage can be won, but it won't be won by directly ordering it or planning for it.

References

1. Search conducted on January 21, 2013 with search terms "innovation imperative" (quotation marks included).
2. Murphy, T. Eureka! *Forbes,* 133 (11): 218, 1984.
3. Day, D. "The Difficulties of Learning from Experience and the Need for Deliberate Practice." *Industrial & Organizational Psychology,* 3 (1): 41–44, 2010.
4. Ericcson, K.E., Prietula, M.J., and Cokely, E.T. "The Making of an Expert." *Harvard Business Review,* 85 (7/8): 114–121, 2007.
5. Von Mises, L. *Human Action: A Treatise on Economics.* New Haven, CT: Yale University Press, pp. 139–140, 1949.

6. Finkle, T.A. "Corporate Entrepreneurship and Innovation in Silicon Valley: The Case of Google, Inc." *Entrepreneurship: Theory & Practice*, 36 (4): 863–884, 2012.

7. Foege, A. *The Tinkerers: The Amateurs, DIYers, and Inventors Who Make America Great*. New York, NY: Perseus Books, 2013.

8. Koch, C.G. *The Science of Success: How Market-Based Management Built the World's Largest Private Company*. Hoboken, NJ: John Wiley & Sons, 2007.

9. Csikszentmihalyi, M. *Flow: The Psychology of Optimal Experience*. New York, NY: Harper & Row, 1990.

10. Duening, T.N. and Ivancevich , J.M. *Managing Einsteins: Leading Technical Workers in the Digital Age*. New York, NY: McGraw-Hill, 2001.

11. Duening, T.N. "Enterprise Process Innovation: The Ingredients Are Well Known, But What Is the Recipe?" *International Journal of Innovation and Technology Management*, 4 (1): 87–101, 2007.

12. Hayek, F. *New Studies in Philosophy, Politics, Economics, and the History of Ideas*. Chicago, IL: University of Chicago Press; Chapter 3, 1978.

13. Painter, K. "Texting, Music Put Distracted Pedestrians at Risk." *USA Today,* December 13, 2012.

14. Hayek, F. *The Fatal Conceit: The Errors of Socialism*. Chicago, IL: University of Chicago Press; Chapter 1, 1988.

15. Hayek, F. *Law, Legislation, and Liberty. Volume 1: Rules and Order*. Chicago, IL: University of Chicago Press; Chapter 1, 1973.

16. Hayek, F. *Individualism and Economic Order*. Chicago, IL: University of Chicago Press; Chapter 2, 1948.

17. Hayek, F. *Law, Legislation, and Liberty. Volume 2: The Mirage of Social Justice*. Chicago, IL: University of Chicago Press; Chapters 7 and 8, 1976.

18. Hayek, F. Op cit., p. 146, 1988.

19. Hayek, F. Op cit., p. 112, 1976.

20. Hayek, F. Op cit., p. 23, 1978.

21. Duening, T.N. and Stock, G.N. *The Entrepreneurial Method*. Dubuque, IA: Kendall Hunt Publishing, 2013.

22. Sarasvathy, S.D. From the Society for Effectual Action, http://www.effectuation.org; retrieved October 4, 2012.

23. Sarasvathy, S.D. and Venkataraman, S. "Entrepreneurship as Method: Open Questions for an Entrepreneurial Future." *Entrepreneurship: Theory & Practice*, 35 (1): 113–135, 2011.

24. Wiltbank, R., Dew, N., Read, S., and Sarasvathy, S.D. "What to Do Next? The Case for Non-predictive Strategy." *Strategic Management Journal*, 27 (10): 981–998, 2006.

25. Sarasvathy, S.D. *Effectual Entrepreneurship*. New York, NY: Routledge, 2011.

26. Staw, B.M. "The Escalation of Commitment to a Course of Action." *Academy of Management Review*, 6 (4): 577–587, 1981.

27. Brown, C. "The Guru Takes Flight." *Entrepreneur*, 41 (2): 32, 2013.

CHAPTER 7

Innovation Neuroscience Hardware

Praveen Gupta

Recently, the growing field of neuroscience has been examining more about the brain. A total understanding of the brain may be difficult to achieve; however, understanding the innovation process at a higher level may be a significant contribution to improving intellectual productivity in the business environment. This chapter at best creates a crude framework for raising awareness of the brain's role in institutionalizing innovation in organizations. Innovation does not occur in the brain somewhere, appearing to be random because of its frequency of occurrence. A little more understanding of the brain processes can lead to a significant improvement in the rate of innovation; improvement that is critical in the knowledge age.

Innovation has been considered an art for centuries. Innovators, however, believe they know how to repeatedly engage in it. For them, innovation is a science. Some people have dozens or even hundreds of patents. For them, they have mastered the art of innovation, or at least they have developed a science to their art, just like Thomas Edison, who decided to innovate and had a goal to file one patent per week. Edison ended up having 1063 patents in his name. For Edison, innovation was not an art; it became a science. What did these innovators do that can be learned by others?

Einstein's brain, though a little larger than normal, happened to be within acceptable statistical variations observed among human brains. Today, extensive work in the area of neuroscience is aimed at improving human health and building more intelligent machines for doing tasks in a similar way as humans do them. Businesses have stretched the physical limits of employees by improving productivity through tools such as computers and automation. As a result, however, people generate a lot of information without effectively using it.

The new methodologies such as Six Sigma require newer solutions faster. The required rate of improvement has been taken to a new and higher level. Sustaining growth and realizing profits require thinking differently. "Thinking" is the hardest task to do, though it should not be, as it is the core competency of humans when compared with other species in the environment. Thus understanding the brain and its functioning in the context of business innovation helps to accentuate usage of this practically unused faculty. Any incremental increase in the average utilization of the brain will cause tremendous change and innovation beyond the imagination in every aspect of life.

Innovation begins with a thought. An innovative thought requires focused thinking and capturing many thoughts as the ideas move through activity, process, product, and business stages. The brain is a building block of innovation, so understanding its functions, and increasing its utilization for accelerating innovation, can only help to further its innovative capability. Applying thinking to the innovation process, which is a somewhat unique approach to understanding it, is the first step in developing innovative solutions on demand.

What is seen in the universe is what is seen in the mind. Every idea or discovery occurs in someone's mind somewhere. In some respects, the human brain is almost as big as the universe itself (i.e., has almost as much capability or potential). In fact, each and every brain is equally powerful within some variation. Understanding the brain's involvement in the discovery process can increase the utilization of intellectual capacity.

The brain contains about 10 billion neurons (or nerve cells) and trillions of axons that facilitate connections with other neurons. Each neuron can have up to 10,000 connections (or synapses). The total number of possible neuron combinations is an astronomically high number (about 10^{80} connections). This number is equal to the number of positively charged particles in the whole known universe.[1]

The smartest person is essentially similar to Einstein, whose brain (people assume) had 100 percent of the connections made. A "dumb" individual can be considered someone with only one percent of the brain connections made. Yet even that one percent of brain connections has practically infinite potential. Many great innovators were explicitly deficient in some aspects—some physically and others mentally.

Every human can innovate and must understand the potential to innovate. This chapter, through mostly an overview of the brain and its correlation with the innovation process, demonstrates that the innovation process can be learned and must be learned to realize a person's innovative potential, compete through discovering new products and services, and assist in sustaining profitable growth.

Overview of Brain Anatomy

The human brain starts forming within the first 30 days after the conception of a human being. The typical brain looks something similar to the schematic shown in Fig. 7-1. (The outermost layer [i.e., the hard surface similar to leather] is called dura mater and protects the brain. Below dura mater is cerebrospinal fluid, which provides a soft protective cushion.

The inner layer is called pia mater and is similar to a crumbled sheet of crepe paper, which if extended is the size of a dinner table napkin. This sheet, which is called the cortex, consists of gyri and sulci and is divided into two asymmetrical halves called the left hemisphere and right hemisphere (see Fig. 7-2). Interestingly, the right side of the brain controls

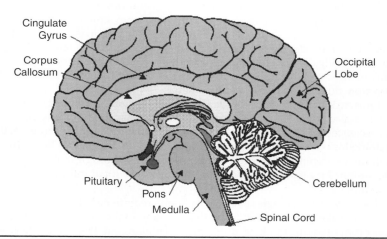

FIGURE 7-1 The brain—right down the middle. (*Source:* Reproduced with permission: Dr. Eric Chudler, *http://faculty.washington.edu/chudler/baw1.html.*)

Left Brain Functions	Right Brain Functions
Sequential Analysis	Holistic Functioning
Logical Interpretation	Comprehension of simultaneous multi-sensory input
Language, Mathematics	Visual and spatial capability
Reasoning	Coordinated complex functions such as dancing, singing, and gymnastics
Language Memory	Visual, spatial, and auditory memory

FIGURE 7-2 Brain hemispheres.

the left side of the body, and the left side of the brain controls the right side of the body. The two hemispheric sides of the cortex are connected through nerve fibers called the corpus callosum. Each hemisphere consists of four lobes, namely frontal (motor activities), parietal (image and recognition), occipital (vision), and temporal (hearing or timing). Each hemisphere, when viewed from the side, is divided into three sections named the forebrain, the midbrain, and the hindbrain.

The frontal lobe takes up half of the hemisphere, controls motor activities, and integrates emotion to convert thoughts into actions. The parietal lobe is the communication center for receiving and integrating all sensory inputs for creating information. The temporal lobe consists of the amygdala and hippocampus that are involved in learning, memory, and expression. The occipital lobe (the farthest of all) processes visual information.

The human nervous system includes central and peripheral elements. The brain and spinal cord form the central nervous system. The peripheral nervous system communicates to and with the central nervous system. In the central nervous system, the brain's three elements of hindbrain, midbrain, and forebrain are formed within a few days after conception.

The hindbrain produces the medulla, the cerebellum, and the pons. The medulla and pons are involved in controlling the physiological functions, like breathing or motor skills. The medulla is involved in the control of blood pressure, heart rate, and breathing and is located right above the spinal cord. The cerebellum is the interface between the higher brain and the muscles. The pons connects the hindbrain with the midbrain and the forebrain.

The forebrain develops structures like the diencephalon surrounded by the telencephalon (see Fig. 7-3). The diencephalon is the core of the forebrain, consisting of the thalamus and hypothalamus. The thalamus is responsible for managing the sensory information going to the telencephalon, and the hypothalamus is responsible for regulating physiological and biological functions. The telencephalon, or cerebrum, consists of the two cerebral hemispheres and is mainly responsible for sensory perception, learning, memory, and conscious behavior.[2]

The telencephalon includes the structures such as the thalamus, hypothalamus, hippocampus, and amygdala. The thalamus is located above the brain stem in the midbrain. The thalamus receives the information from various senses, transmits it to the cerebral cortex and other areas of the brain, and communicates signals from the cerebral cortex back to the spinal cord.

Brain Modules	Nucleus Structures	Key Functions
Telencephalon	Cerebral Cortex, Amygdala, Hippocampus, Basal Ganglia	Analyze sensory data, perform memory functions, learn new info, form thoughts and make decisions; plays a role in sense of smell, motivation and emotional behavior; two hippocampi, located on both sides of brain, involved in anger, fear, pleasure, formation and retention of memory for facts (database), and are motivators for problem solving; initiation and direction of voluntary movement, and balance movement
Dienephalon	Thalamus, Hypothalamus	Receives and integrates sensory inputs, and relays to cerebral cortex; control of body temperature, emotions, thirst and hunger
Mesencephalon	Substantia Nigra, Central Gray, Red Nuclieus	Transmitter of information, changes in metabolism, intellectual abilities including memory, judgment, abstract thinking; Spinal cord control, and auditory reflexes; Coordination of the instincts and emotions, reactions, and motion
Metencephalon	Pontine, Deep Cerebellar	Rapid Eye Movement (REM) sleep, subconscious thinking, dreaming, imagination, memory consolidation, comprehension; planning, predicting, and motor control
Myencephalon	Inferior Olive	Communicates sensory information and inputs from other brain nuclei and communicates sensory information to cerebellum

Figure 7-3 Brain functions.

The hypothalamus acts like a thermostat to control body temperature. If a person feels too hot, the hypothalamus sends signals to expand capillaries in the skin, causing blood to cool down. The hypothalamus is a small structure next to the thalamus and controls metabolic functions, body temperature, sexual arousal, thirst, hunger, and biological rhythms.

The hippocampus surrounds the thalamus and is divided into more than one section. Each hippocampus is connected to a structure called the amygdala, which is responsible for fear and fear memory. Blocking protein synthesis of the amygdala will block the formation of fear memory. The hippocampus and amygdala are responsible for processing and perceiving emotion and memory. The hippocampus evaluates or associates memories and forwards them from short-term memory to long-term memory in the cerebral cortex for permanent storage. Therefore, the hippocampus can be considered an important center of learning in the brain.

The grouping of the thalamus, hypothalamus, hippocampus, and amygdala is called the limbic system, which deals with physiological drives, instincts, and emotions. The limbic system is where the sensations of pleasure, pain, and anger are most keenly felt.

The cerebral cortex is a sheet of gray matter that is about 4 ± 2 mm, or the height of six business cards,[3] and makes up the outer layer of the brain. The cerebral cortex has bumps and grooves, normally called gyri or sulci (i.e., gyrus and sulcus for singular bump or groove), respectively. The cortex is full of neurons that develop connections, based on the information received throughout life. As a person learns, the neurons gain stimuli, grow dendrites, and connect with other neurons.

The cerebral cortex has several specific functional areas, and any area of the cerebral cortex not specified is used for association responsible for thoughts, judgment, humor, and behaviors. The specific functional areas of the cerebral cortex include the prefrontal cortex (for problem solving and complex thoughts), the motor cortex (for bodily movement), the somatosensory cortex (for processing multisensory input), the visual cortex (for detecting simple visual input), and the auditory cortex (for detection of sound quality). As highlighted earlier, besides these functional areas, association areas in the cerebral cortex are responsible for advanced capabilities in various areas.

The basal ganglia located at the base of the cerebral hemispheres receive input from the cerebral cortex through the striatum in the midbrain and forward it back to the cerebral cortex through the thalamus. Basically, the basal ganglia function to maintain the muscle tone needed to stabilize joints.

The brain stem controls basic functions of life, including muscle coordination. The brain stem in the midbrain processes visual and auditory information as well as the information between the higher brain and the spinal cord. The sensory information moving up the spinal system passes through the brain stem in the hindbrain area, where synapses are formed with brain stem neurons, called the reticular system. The reticular system controls bodily functions such as movement, coordination, and sensitivity to pain. Activity level in the reticular system affects the nerve systems, thus affecting the awakening or sleep state of the brain.

The cerebellum controls fine body movement, balance, and posture. The cerebellum is like a little brain inside the brain with its own cortex and even its own hemispheres. The cerebellum, though only about 10 percent of brain volume, contains about 50 percent of all the neurons in the brain The cerebellum receives commands from the somatomotor cerebral cortex and gives smoothness of motion and exactness of positioning through internal and external feedback comparisons.

The Innovation Process in the Brain

Medical researchers have mapped the brain inside out, and most of the functions are compartmentalized. The brain includes practically infinite numbers of neurons, which get excited by stimuli through the senses and produce electric potential. This electric potential is then converted into chemical reactions that transform into various proteins and other reactions. The role of Na^+, K^+, and Ca^+ ions in the functioning of neurons and the creation of synapses is well understood; however, what happens next is difficult to comprehend. While some sporadic phenomena can be explained, the root causes of those phenomena are often unknown.

In an attempt to understand the innovation process inside the brain, identifying certain parts associated with language and auditory interpretation is possible, but exactly how a person thinks and makes decisions is a mystery. However, everything that a person sees, does, or imagines does happen inside the brain. Actually, the brain makes it possible to have some knowledge of everything in the universe. If a person knew how to use the brain effectively, then everything discovered so far could have been discovered by one brain (given the recognized size of a person's brain).

The total absence of the brain occurs when the brain is absolutely nonfunctional or dead. Otherwise, the brain is so big in its potential, that significant damage would not render it dead. For example, the brain may have approximately billions of neurons, trillions of axons, and many more combinations of neurons and axons, making synapses. Synapses can be considered the information bits. The number of synapses can realistically approach infinity. A fraction of that capacity appears to be 10^{80}. A small fraction of such brain capacity is still enough to use the brain throughout life (or for thinking of innovative solutions).

Einstein said that nothing is invented; instead everything is discovered. If everything is discovered, the brain must see everything. Just how can a brain see everything? In other words, what is the process used by the brain to try out different combinations to produce innovative solutions? The brain is continually bombarded with information. What happens to it?

Look at electrical connections as an illustration. If a connection is broken, the circuit is literally dead. If the connection is made, the room is lighted and everyone is awakened. In the same way, the brain appears to be a set of connections that must be present in order to sustain proper functioning. Any brain function is a result of cumulative output of the input stimuli, be it visual, audio, touch, or taste.

Figure 7-4 shows a simplified version of the brain process. The cerebral cortex appears to be a collection of nerve cells called neurons. Brain neurons are available in practically limitless quantities to absorb sensory input for processing. Signals are received through various senses or activities. These signals are passed through nerve fibers to neurons in the cortex area for pattern formation.

Once the pattern is formed based on the purpose to use it or not to use it, the information that is compared in the association is in the cortex and is stored in its short-term or long-term memory bank. The cortex area and surrounding gray matter can absorb short- or long-term information in unlimited amounts. The short-term memory connections are elastic and reversible, while the long-term memory synapses are hardwired to last longer. Some patterns are permanently etched based on the extent of their strength, leading to virtually permanent connections or patterns. When the patterns are formed, they are evaluated or compared for signal generation and motor action.

The duration of the information storage is dependent on the elasticity of the material and the decision made to ignore or receive the material. The natural loss of information can be attributed to the property of the cortex material; however, how the decision is made in the brain to use or not to use the information is still unknown. Once the decision is made to use the information, processing of the information can be understood inside the brain. Humans have the capability to make such decisions without knowing the mechanism at this stage. Perhaps further research will reveal more secrets of the brain.

Once the information is imprinted on a group of neurons, a pattern is formed. The human brain continually forms patterns of signals it receives from various senses. Then the

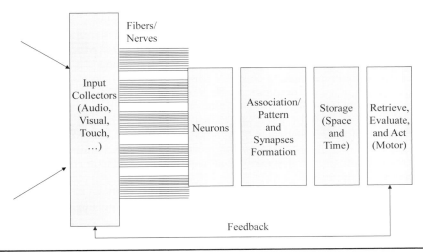

FIGURE 7-4   Mental process flow.

Elements of Innovation	Key Side	Key Brain Elements ...Accelerators
Gathering information	L	**Somasatory sensors, Thalamus, Hypothalamus** Receives and integrates sensory inputs, and relays to cerebral cortex, ... **Energy and comfort**
Learning, Comprehension	R	**Cerebral Cortex, Amygdala, Hippocampus** Analyze sensory data, perform memory functions, learn new info, form thoughts and make decisions,... **Motivation and Incentives**
Analysis, Questioning, Interpretation	L	**Substantia Nigra, Central Gray, Red Nuclieus** Transmitter of information, intellectual abilities including memory, judgment, abstract thinking, ... **Stamina and Time Management**
Association, Induction, Deduction	R	**Hippocampus** involved in formation and retention of declarative, spatial, or long-term memory for facts (database),... **Knowledge and Research**
Combinatorial Processing	L	**Stratium** responsible for procedural or short-term memory,... **Experiment and Play**
Extrapolation	R	**Pontine, Deep Cerebellar** causing subconscious thinking, dreaming, imagination, memory consolidation, comprehension,...**Rest and Reflect**
Formulation	LR	**Cerebellum, Ganglia** – Timing relationships, motor planning, predicting, and motor control,...**Evaluate and Select**

FIGURE 7-5 Innovation and brain functions. (*Source: http://www.benbest.com/science/anatmind/ anatmind.html*)[1]

decision to recognize patterns for comparisons is made. The question remaining is this: Which part of the brain makes this comparison and how?

The brain contains visual and auditory (i.e., somatosensory) association areas where the association among patterns is made. If the match is made (in the context to use this information), an electrical signal is generated, which leads to a chemical reaction producing certain tactical or facial movements or actions. When innovation is taking place, the existing patterns are reviewed in the memory. To discover a new or unique pattern, comparisons (through association) of existing patterns must occur. Then the person making those comparisons must determine the missing patterns.

Figure 7-5 represents implementation of the innovation process through brain functions. For example, gathering information through sensory inputs is processed through the thalamus and hypothalamus. The repetition and intensity of input determines the chemical signal strength in terms of its elements and temperature. The combination of input type, intensity, and temperature creates certain signals that determine people's behaviors through their tactical or facial expressions. The signal strength depends on the familiarity of the pattern, its repetition, and its intensity. When the patterns are associated or matched, a positive signal is generated. When the patterns are not recognized or missed, a negative signal is released.

Accordingly, the innovation process becomes a process of reviewing new or selected patterns in the cerebral cortex, evaluating them with existing patterns or potential patterns, and determining the missing or absent patterns. Such a process leads to signal strength for either generating ideas or questions. Thus, a thought or an idea is a feedback or observation of a pattern that is either present or absent in the cortex and then represented (or communicated) through tactile or facial expressions. Similarly, a question is an idea that is based on the nature of the resultant signal strength sent to Broca's and Wernicke's areas for asking questions. To ask questions, a person must first develop, understand, and speak the language. Without the development of language and speech faculties, a question is most likely going to be suppressed.

Once the generation of thoughts is understood, questions about the speed of thought may be asked. With the current understanding of brain functioning, the speed of thought thus becomes the rate at which patterns can be associated and compared in order to evaluate and generate new thoughts or new patterns. Given the number of neurons, axons, and synapses in the cortex, algorithms must be developed to speed up the evaluation process.

One way to speed up the speed of thought is to establish anchor locations in the brain. Anchors are long-term memory locations that have almost permanently etched information. These locations are formed through the repetitive use of the known pattern, which creates strong synapses that appear similar to cured synapses losing their elasticity. For example, children memorize tables that help them to perform mental math. Having the anchors in the memory, calculations are then expedited. Similarly, if certain patterns are anchored in the cortex, comparing patterns by association can be expedited. However, they may eventually be lost over a longer period of time.

Accelerating the Innovation Process in the Brain

Having understood the basic functioning of the brain, how to accelerate the innovation process becomes the next question. Acceleration can be achieved through creating patterns faster, thus speeding up the associations and evaluations. The creation of faster patterns requires continually gaining new experiences through somatosensory learning, as well as getting involved in more activities leading to increased input through the spinal cord. Creating associations and evaluating patterns faster requires anchors and the involvement of more neurons. These anchors and increased neural involvement can be achieved through speeding up thought and through the external excitement of neural activities.

Figure 7-5 captures the activity and ambience levels for improving various stages of the innovation process. For example, in order to improve information gathering through sensory inputs, enough energy and comfort must exist. To improve the analysis and questioning aspects of the innovation process, stamina and time management are required; to improve the association and evaluation processes, knowledge and rest to reflect or think are needed. To improve the entire innovation thinking process, understanding and developing knowledge anchors may be a key to doing so.

External excitement implies application of an external energy to stimulate certain brain structures associated with the innovation process. Accordingly, studies have been done to map the brain's wave patterns in its different states. Taylor Andrew Wilson, for example, has utilized the Brain Wave theory to create music that accelerates the brain's performance. In his book *The Mind Accelerator*,[4] Wilson identifies the Beta state (13–40 Hz) as associated with an attentive, conscious, and narrowly focused state of mind. The Alpha state (7–12 Hz) is associated with states of mind involving visualization, relaxation, and ingenuity. The Theta state (4–7 Hz) is associated with intuition, memory, and deep-thought states of mind.

According to Wilson's theory, to sustain various activities in the brain, music with certain brain waves is applied. For example, to investigate potential combinations or patterns, the selected brain-wave music will bring brain waves into a peak-performance frequency, thus capitalizing on the information-acquisition and knowledge-building efforts. To expand thinking in order to explore potential innovative solutions for generating significant change, therefore, the selected brain-wave music must trigger optimal waves of neural firing. Using Wilson's theory, such waves will thus deliver a focused and acute peak-performance brain state for evaluating and analyzing ideas, suggestions, and solutions.

Proper diet and exercise in order to fully utilize brain potential cannot be understated. Brain activity can actually be accelerated by increasing neuron activity levels through proper diet and exercise. Proper nutrition and exercise helps to maintain the chemical balance of the Na^+, K^+, and Ca^+ levels needed for reaching maximum potential in innovation activities.

Summary

We have learned that humans are born as a creative and innovative species, however the current education and work environment discourages people to utilize their creative potential and drive for innovation. It is a common belief that innovation is an art that is difficult to impart and manage, and people either have or do not have it. This chapter demonstrates that creativity and innovative thinking happen in people's brains, and if understood better could be taught for improving the efficiency of innovation. Analysis of the brain functions and correlating with innovation activities shows that individuals and organizations can take certain steps to harness human potential to create new knowledge and solutions. For example, in order to learn something new a person must be in a comfortable temperature or environment, and must have energy or be alert. Analyzing the opportunities and possible solutions require stamina and strength for maintaining commitment to innovation. Thinking of innovative ideas requires development of both, left and right brains. Thus, one of the main incentives to inspire innovation can be rewarding people in terms of opportunities to learn and develop new skills. After all, an active mind is an innovative mind. Finally, we have learned that all of us have practically equal capacity for innovation, thus everyone must be inspired and supported for innovation in their areas of interest or work.

References

1. Restak, R. *Brainscapes*. New York, NY: Hyperion, 1995.
2. Purveys, W., Sadava, D., Orians, G., and Heller, C. *Life—The Science of Biology*, 7th ed. Gordonsville, VA: W. H. Freeman Company, 2004.
3. Hawkins, J. and Blackeslee, S. *On Intelligence*. New York, NY: Times Books, 2004.
4. Wilson, T. *The Mind Accelerator*. Volition Thought House, 2004 at http://www.vth.biz /kb/html.php.

CHAPTER 8

Innovation and Neuroscience

Ellen Di Resta

Introduction

The astonishing pace of technological and macroeconomic change is requiring that companies innovate continuously. Most companies are good at optimizing their current product development processes continuously, but innovation is not the same as product development. Few would argue that, however, many organizations try to embed innovation into existing product development processes and expect different results. This seldom works. Innovation is doing something new that adds value to the organization. What is new can vary greatly from an upgrade to an existing offering, to something completely unknown. This chapter will focus on innovations that result in the creation of offerings that are not yet defined, and are completely new to the company.

Product development, on the other hand, is figuring out how a company will reliably produce an offering that has been defined. It is inherently impossible to create a completely new offering from within a process that requires a defined offering as a starting point. For this reason, innovation requires a completely different process and mind-set; one that embraces the ambiguity of the unknown, yet guides toward the definition of a new offering that can then be developed. Figure 8-1 shows how the innovation process feeds the product development process.

Once it is acknowledged that the innovation process should be separate from the product development process, the problem then becomes one of finding the right innovation process, people, and environment for the company. To be able to do this, we need to understand what type of work the process should facilitate. This highlights an important gap in the current understanding of innovation. Beyond knowing that we want an innovation program to result in the creation of new ideas that will help a company to achieve its goals, very little is known about how relevant new ideas can consistently be created.

As a result there are many random idea generation processes that pose as innovation methods. In these processes as many ideas as possible are generated with an emphasis on how "out of the box" they are perceived to be. Company leaders then vet the ideas based on current business needs, trends, or preferences. Opportunities that will catalyze new product portfolios to sustain company growth for a significant period of time are seldom generated as a result.

However, there are some innovation processes that have evolved to consistently yield opportunities that result in sustained market success. Past results may reveal consistent successes, yet the people who manage these processes often struggle when asked to identify how or why they are successful. Answers often point to factors that are generally consistent with the company's core business. For example, a market research team may point to the use of consumer insights as the key to success. These answers, however, do not adequately describe the source of success, as these stated factors could be found in both successful and unsuccessful innovation endeavors.

It has become clear that there is some type of intangible asset responsible for this success that the reasons above do not adequately capture. Since this intangible asset cannot be observed from the outside, I have turned my attention toward the inside. What drives the thinking of the people who are successful at creating relevant new ideas?

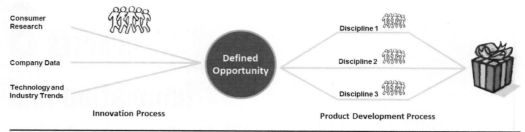

FIGURE 8-1 Innovation and product development processes.

To answer this question, I have surveyed the field of neuroscience searching for insight into the mental functions necessary for successful innovation programs. What I have found is that recent neuroscience research can now explain brain functions that are largely responsible for the creation of relevant new ideas.

Chapter Overview

This chapter will use the recent research in neuroscience to fill the gaps in understanding what type of work an innovation process needs to facilitate, identifying the qualifications of people who excel in innovation roles, and creating an environment to support their work. We will explore these topics in three ways.

First, we will look at fields that are currently applying neuroscience tools to their work. The fields of neuromarketing and neuroeconomics are currently using brain-imaging techniques to aid in market research and the product development process. We will also review the new field of neuroesthetics, which is beginning to define the mechanisms for how the arts communicate to us in ways of which we are currently unaware.

Second, we will focus on new research that presents promising insights into how we think about innovation. These insights illuminate the brain functions that are the source of new ideas, as well as those that let us know when we have found the right new idea. In this section we will also review an application of neuroscience research that discusses how three brain functions can prevent us from being innovative.

Finally, we will provide guidelines for the specific application of these insights to create a process, select talent, and create an environment conducive to successful innovation efforts.

Current Applications of Neuroscience

Neuroscience methods are being applied to help researchers to understand how consumers make decisions through the fields of neuromarketing and neuroeconomics. They are also being applied to understand cognitive responses to aesthetic experiences through the new field of neuroesthetics. We will explore the relevance of each to the innovation process.

Neuromarketing and Neuroeconomics

Neuromarketing and neuroeconomics are new related fields that are gaining popularity in the development process. In basic terms, they use neuroimaging techniques to enable the measurement of the brain activity of a respondent when exposed to different stimuli. Techniques such as functional magnetic resonance imaging (fMRI) allow researchers to see images of changes in blood flow, which correspond to activity levels in different areas of the brain. Neuromarketing is the application of these techniques to learn about consumer preferences, as the neuroimages are thought to represent a more accurate representation of preference than a respondent would self-report. Neuroeconomics is the translation of this information to consumer decision making more broadly. Professor Gregory Berns of Emory University is an expert in these fields, and suggests that the increased popularity is due to two main factors: First, it is possible that neuroimaging will become less expensive and faster than other marketing methods such as focus groups or large-based quantitative studies. Second, there is the hope that neuroimaging will provide market information that cannot currently be obtained through conventional methods.[1] These methodologies are also beginning to help companies get a better understanding of whether,

> **KEY QUESTION:** How could we develop neuromarketing and neuroeconomics tools to gain useful feedback before the ideas are fully formed?

and how, consumers will accept new offerings. As such, they are most often used in later stages of the product development process.

However, according to Berns, the most promising applications of neuroimaging methods may come before a product is even released, when it is just an idea being developed.[2] This new application can expand the use of neuromarketing and neuroeconomics beyond marketing functions, and assist in idea generation and innovation functions.

Neuroesthetics

Neuroesthetics is a relatively new field being pioneered by Professor Semir Zeki. Neuroesthetics uses neuroscience to explain and understand aesthetic experiences at the neurological level. It also uses art to understand something about the brain. Zeki describes neuroesthetics as the way that artists use their brains to communicate to our brains. He adds that this form of communication requires a unique and interesting intelligence that transcends most forms of communication

> **KEY QUESTION:** How do we apply neuroesthetics research to build "translation" capabilities we can use in creating new concepts in the innovation process?

that are explicitly recognized by the general public. Zeki acknowledges that while it is difficult to articulate this skill, it becomes obvious when it is there.[3]

I work with many people in the design field. When evaluating designers for suitability to innovation projects, I am looking for those who are able to communicate a product's benefits through the design of, or interaction with, a product. I am less concerned with typical hand skills that showcase drawing ability and technique. These people are able to deliberately elicit an intended response from a consumer through the design of the product or interaction. In one example, a designer created a prototype of a diaper for infants, intended to encourage a feeling of bonding between the mother and child as it was being applied. During an evaluation session with consumers, on more than one occasion the mothers started crying due to the warmth of feeling the product evoked, as its use deliberately mimicked a warm embrace.

When asked to describe the skills required of those with talent in this discipline, Zeki suggested that one of the most important functions is a high degree of lateral thinking (thinking that spans across disciplines) required to communicate in this way.[4] I call this skill "translation." In the example above, the designer was able to translate the feeling of bonding to a product experience. Neuroesthetics is such a new field that there is currently little formal adoption of its principles in the innovation or product development process.

Potential Applications of Neuroscience to Innovation

We will now turn our attention to recent research in neuroscience that has the potential to fill the knowledge gaps that make innovation such a mystery. Specifically, three topic areas can increase our understanding of how new ideas are created, how we know when these ideas are right, and the cognitive barriers to innovation.

The Genesis of New Ideas

Professor William Duggan of Columbia University has intensively studied the mechanism for how creative ideas happen in the human mind. In his book *Strategic Intuition* he outlines this mechanism, and provocatively debunks the notion that good ideas are the result of the experimental methods we all learned in school which have become the basis for many innovation and product development processes today. The experimental method is illustrated below in Fig. 8-2.

Troubled by the fact that the experimental method doesn't adequately describe how ideas are conceived, Duggan looked back through history and found the first writings on the scientific method by Roger Bacon in 1237. In reviewing Bacon's work, Duggan finds that the experimental method was only one part of the scientific method, which Bacon

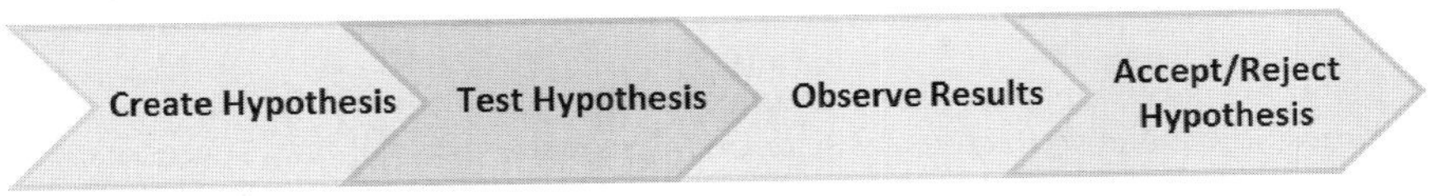

Create Hypothesis → Test Hypothesis → Observe Results → Accept/Reject Hypothesis

FIGURE 8-2 The experimental method.[5]

described as (1) look at the experiments of others, (2) look at your own experiments, (3) apply your reason, and only then devise experiments to test your reason via the experimental method. The scientific method includes the steps leading up to the creation of the idea, while the experimental method is the way to test the idea. It does not help us to create the idea in the first place. To return focus on the creation of the ideas, Duggan evolved Bacon's scientific method by combining it with more recent insights from the cognitive sciences to create what he calls strategic intuition.[6] Duggan compares strategic intuition to two other types of intuition (ordinary and expert) as outlined below:

- Ordinary intuition: This is a form of feeling, not thinking, that we call our gut instinct. We have all experienced this form of intuition, and know that it is sometimes correct, and sometimes incorrect as it is the result of feeling, not thinking. The next two forms of intuition are the result of thinking, not feeling.[7]

- Expert intuition: This is a form of rapid thinking where you jump to a conclusion when you recognize something familiar, as popularized in Malcolm Gladwell's book *Blink*. With expert intuition, the more experienced a person is in a given profession, the quicker and easier it is for them to recognize patterns that have occurred in the past, resulting in faster, more accurate problem solving.

- Strategic intuition: In contrast to expert intuition it is slow, but works well in new situations where there is less prior experience. The reason it is slow is that in new situations, the brain takes longer to make enough new connections to find a good answer. It is also holistic. Not only the idea is created, but its rationale for success is simultaneously connected. When the right connections are made, the flash of insight happens. The flash of insight is fast, but it may have taken the brain weeks to get there.[8]

It's obvious that expert intuition would fit well with the goals of the product development process, as product and performance metrics are well defined. People can become experts at solving similar types of problems over and over again. It also becomes obvious why strategic intuition would lend itself more strongly to the innovation process, as it does not require a proposed solution to evaluate. To the outsider, the ideas that result from strategic intuition may look random, however, they are not random at all. Key to understanding how strategic intuition works, it is necessary to first understand the concept of intelligent memory. To this end, Duggan cites the work of Barry Gordon, a neuroscientist who coined the term as the title of his 2003 book on the subject, and describes how intelligent memory works after an initial goal or problem is posed.

To paraphrase this concept, intelligent memory is the collection of examples from past experience that are stored on the shelves of the brain as you go through life. In reality there is no such thing as the right or left brain that separates analytical or intuitive thought. As the person takes in their experiences through their five senses, the brain parses these memories and stores them in different areas of the brain. Intelligent memory is a synthesis of analytic and intuitive thought. When a problem is posed, the brain connects the thought fragments that are relevant to the problem and pulls them together through an orchestration that happens as a single cognitive event.[9]

In the case of expert intuition, the recall of these memories draws upon the memories of a single person. Strategic intuition, however, requires that the memories of others are drawn upon as well, and the process facilitates the collection and synthesis of these diverse experiences. The strategic intuition process steps are outlined in Fig. 8-3 as follows:

- Seek out examples from a vast array of historical references.

- Presence of mind—The conscious freeing of all expectations and previous ideas of what you might do, including being open to modifying the initial goal. This step requires the mental discipline to free your thoughts to let the flash of insight come to you.

- The flash of insight itself—When the mind is free of initial expectations, then new ideas and connections can be made. This is also why the flash of insight often happens when we are not thinking about the problem at all.

- Resolution—You not only see what should be done, but you are ready to do it.[10]

In his book, Duggan cites examples of strategic intuition at the heart of many successes throughout history, from early scientific discoveries, to military successes, and to modern day businesses. The one commonality shared by these examples was the fact that the successful outcome could not be predicted beforehand. As he states "Many aspects of human

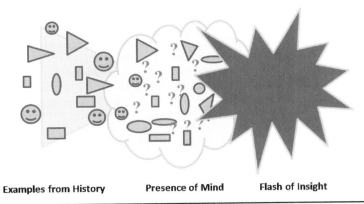

Examples from History   Presence of Mind   Flash of Insight

FIGURE 8-3   The strategic intuition process.

life may be predictable, but what comes together in somebody's mind will likely never be one of them."[11]

While the new solution may not be predictable, our intelligent memory ensures that it is not random. It does, however, require that a specific goal or problem is posed. This is why insights generated as a result of strategic intuition are more likely to be correct than the gut feeling that results from ordinary intuition, where a specific goal is not intentionally put forth. In strategic intuition, the brain may make new or interesting con-

> **KEY INSIGHT:** A successful innovation process should facilitate the creation of new ideas in a way that is consistent with how the brain creates them naturally.

nections, but these connections are associated with the focus of the original thought.[12]

The concept of strategic intuition is useful to understand how the brain generates new ideas, but it does not fully explain how we know when we've found the right one.

Knowing When You've Got the Right Idea

To shed some light on the question of how we know whether or not an idea is good, we can turn to the work of Dr. Robert Burton, MD, Associate Chief of the Department of Neurosciences at Mt. Zion—UCSF Hospital. In his book *On Being Certain*, he explores how it is that we know what we know. This work is an excellent complement to Duggan's work on strategic intuition, and an equally necessary part of an innovation process intended to create new ideas of value. To introduce the topic of how we know what we know, Burton describes what he calls the "Feeling of Knowing," which he defines as the feelings of certainty, rightness, conviction, and correctness about an idea or concepts. It is a form of metaknowledge, meaning knowledge about knowledge,[13] and he describes it in several ways. Among them are the following examples:

- The feeling that you know something, even if you can't remember what it is. For example, if you forget a name, you still know that you know the name, you just can't recall it.

- Burton uses an excellent example to describe the difference in how you feel both with and without a Feeling of Knowing where he asks that the paragraph below be read without skimming or skipping to the answer at the end:

   "A newspaper is better than a magazine. A seashore is a better place than the street. At first it is better to run than to walk. Even young children can enjoy it. Once successful, complications are minimal. Birds seldom get too close. Rain, however, soaks in very fast. If there are no complications, it can be very peaceful. If things break loose from it, however, you will not get a second chance."

At first the paragraph seems meaningless. You may try to sort through possible explanations, but when presented with a single word, kite, it all feels different. The paragraph has now been infused with a Feeling of Knowing.[14]

Burton describes intuition through the lens of the Feeling of Knowing, stating that intuition is the combination of unconscious thoughts with the Feeling of Knowing. Unconscious thoughts are thoughts that exist without an awareness that they are occurring. Burton suggests that without the ability to disassociate our thoughts from the awareness of them, the chaos would be overwhelming. We would not be able to concentrate on what we are doing in the moment if our brain did not silence less pressing, longer-range thoughts that occur at the same time.[15] For example, the brain may be processing a solution to a problem while you go about your daily activities. If you were aware of and focusing on every thought the mind ponders, you would not be able to focus on the other things that you need to do to get through the day. Solutions, after processing, only come into consciousness when they are completed.[16]

Burton presents two contrasting examples to illustrate the concept of unconscious thought and how it feels to us. In the first example, he describes a person who programs a computer to solve the problem of defining a possible drug formulation for a very rare disease. He then writes a second program that tells the computer to reject all solutions that do not have a very high likelihood of success. The computer begins working, and after a few years and the project is forgotten, the computer spits out a formula that it deems 99.999 percent likely to succeed. In this scenario, you don't feel as if the computer has done anything out of the ordinary.

In contrast, he gives an example of a novelist contemplating a multigenerational novel with a huge cast of characters. After a few months of working on it, the novelist becomes overwhelmed and instead focuses on a much simpler story; after a few years he all but forgets about the first novel. Then one day the novelist wakes up with the full idea of the first novel in his head, plot, characters, and all the difficult connections worked out. He's overwhelmed by the sense of rightness of the solution and the absence of effort in working it all out. In this scenario, the absence of awareness of the thoughts makes the experience feel "otherworldly." When an idea appears seemingly out of the blue, it doesn't feel intentional. On the other hand, the delay in the computer's processing speed doesn't dilute the belief in the computer's continued processing because we know it couldn't work without someone programming it with the original intention. "With our own thoughts, any significant delay between question and answer tends to strip the thought of a sense of being intentional. How the brain creates a sense of cause and effect isn't known, but the temporal relationship must be crucial. We must experience cause as preceding effect. The closer this proximity, the greater the feeling of intentionality."[17] This is similar to Duggan's description of strategic intuition. The flash of insight is fast, but may have taken the brain weeks to get there.[18]

With these thoughts in mind, it's easy to see how the idea generation process as it occurs in the mind could be thought to be random. Relevant ideas appear to pop up out of nowhere, which seems to make them no different than the randomly generated ideas from a day-long workshop or brainstorm. However, a good innovation process is anything but random. It requires a delicate balance that ensures the mental functions described by Duggan and Burton can happen, but also ensures that after the ideas are generated they must be useful to, and understood by, the broader organization.

The importance of setting clear goals for innovation also becomes evident, as this is what is needed to infuse the process with the intentionality necessary to know when the ideas are relevant. An interesting example of how this intentionality has been built into a business process can be found in the Disney process organization chart from 1943[19] (see Fig. 8-4). It shows that the first thing to happen at the top of the chart is the development of the story, which then drives everything. The fully developed story is then fed to the director who leads the development process of creating the movie, drawing upon all the production resources that would be typically found in a movie production studio at that time. What is interesting is that the organizational chart explicitly shows that it is the idea that drives the production. When organizations try to embed innovation processes into existing development processes, the opposite tends to happen. All too often ideas are driven by what can be accomplished within the existing process. This is fine if the goal is to incrementally improve what exists today. To create something new, as Disney has successfully done for decades, the creation of the right idea must come first.

> **KEY INSIGHT:** The brain will unconsciously filter ideas based on an intentionally posed problem. Therefore, problems whose solutions will be valuable to the organization should be explicitly posed as part of the innovation process.

The power of understanding the true genesis of new ideas, as well as the mechanism for knowing they are right, cannot be understated when we think about their application to an innovation process. In 2007, RTI International, a nonprofit technology research organization, summarized the results of three sources that analyzed the commercial success of new ideas. Among their findings was a commercialization success curve that showed one percent of new ideas resulted in a commercial success. (A commercial success was defined as a product that yielded greater returns than the company's investment in its development).[21] Clearly, there is no shortage of new ideas. It's also clear that organizations have not yet been able to consistently harness the power of the brain's natural ability to create relevant new ideas.

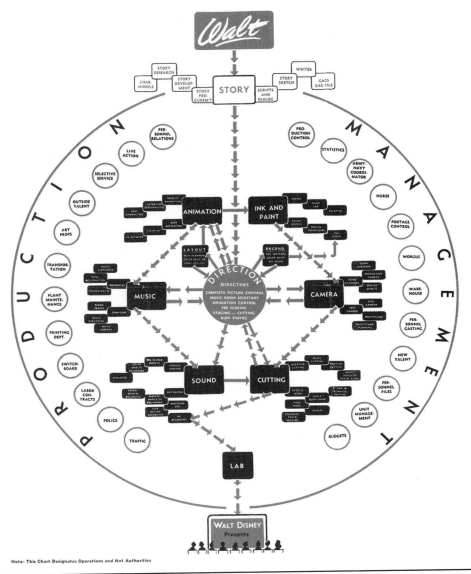

FIGURE 8-4   The Walt Disney Company process chart.[20]

Barriers to Innovative Thinking

The inability of most organizations to consistently create relevant new ideas begs the question that if the brain is naturally predisposed to create relevant new ideas, then why doesn't this happen more often? Neuroscience Professor Gregory Berns of Emory University explores this question in depth in his book *Iconoclast*, which is also his term for innovative people who think differently from the rest of us. His ideas about what it takes to create new ideas are similar to the work of Duggan and Burton. He then cites three cognitive barriers that keep people from becoming iconoclasts, which are as natural as the mechanisms that drive the creation of new ideas.[22] Essentially, the brain has competing mechanisms that must be overcome to create relevant new ideas. These barriers are perception, fear response, and social intelligence. We will look at each one separately.

Perception

According to Berns, the ability to perceive the world in new ways is the key to being able to imagine new possibilities and come up with new ideas. He suggests that the brain is efficient, and each time it executes a task, it looks for the most efficient way to do it. While this natural quest by the brain for efficiency is certainly conducive to the development of expert intuition, this tendency needs to be broken down intentionally or unintentionally in order to think more imaginative thoughts.[23] In one particularly literal example, Berns cites the work of glass artist Dale Chihuly. Chihuly invented the forms of glass sculpture he is now associated with, which include fanciful Dr. Seuss-like shapes that seem to defy gravity, and contain powerful color juxtapositions. Chihuly wears an eye patch due to a car accident in which he lost his eye and never regained his depth perception. Before the accident he never created such fanciful sculptures. The physical change in his perceptual ability resulted in this new perception that enables the work he creates today.[24]

Fear Response

Berns breaks fear down into two components; fear of uncertainty, and fear of public ridicule. This response encourages action without thought, as in the primal responses to external threats. When people feel exposed, uncertain, or as if they are being attacked or criticized, their fear distorts their perception, and they cannot imagine new possibilities. This does not mean that iconoclasts do not feel fear, but they are able to tame their natural stress response to fear.[25] For example, the innovation process requires questioning conventional wisdom, which brings up the uncertainty of the unknown. The presentation of these ideas brings with it the possibility that others in the organization will ridicule them. Those who cannot tame these fears will not allow themselves to follow through on the work required to innovate.

Social Intelligence

Berns acknowledges that ideas will not be valuable if others do not accept them. He suggests that the two things key to the acceptance of new ideas is their familiarity and their reputation. "The two go hand-in-hand. In order to sell one's ideas, one must create a positive reputation that will draw people toward something that is initially unfamiliar and potentially scary. Familiarity helps build one's reputation," which makes it necessary to network broadly and sell new ideas as part of the innovation process. Berns illustrated this idea by comparing Picasso's success while alive and Van Gogh's lack of success until after his death. Picasso circulated his ideas with wide circles of people who could promote his work. He also socialized the avant-garde expressions of his ideas to help people to understand them. In contrast, Van Gogh lived a reclusive life, his work unknown until well after his death when it was discovered and promoted by others.[26] This highlights the necessity to introduce new ideas to the organization as their value becomes more apparent to the innovation team.

In order to be successfully innovative, Berns suggests that one's perceptions must be able to be broken down to make new connections and create new ideas. Fear must be overcome to bring new ideas into being and social intelligence and networks must be developed to gain broad acceptance of the ideas. Therefore, it becomes apparent that the brain is equally adept at the skills necessary for innovation as it is at creating barriers to innovation. These competing instincts need to be consciously managed, which can be difficult as most organizations unintentionally reinforce these barriers.

We've Always Done It That Way

Over 100 years ago, the Industrial Revolution ushered in a new paradigm for how business operated.[27] The introduction of automation was a disruptive innovation that enabled mass production of goods that were much more affordable. The change was so profound that a paradigm was established that set the stage for how businesses compete to this day.

As noted by Richard Florida, the high school system was developed as a way to stamp out workers to feed the industrial machine.[28] New perspectives are rigorously worked out of the system, lack of conformity is met with public ridicule, and socialization reinforces the benefits of conformity to schedules and processes. Over the same time period, few resources have been devoted to developing the skills necessary to create new offerings or serve new customers.

> **KEY INSIGHT:** The key to business success for the last century reinforces the cognitive barriers to innovation. The innovation process must encourage thinking that bypasses these barriers.

Applying Neuroscience Insights to Innovation

The neuroscience research we have been discussing provides useful insights that can help us to understand the differences between those who work well with innovation, and those who do not. In the next section we will examine the process, talent selection, skill development, and environment conducive to innovation through the lens of this neuroscience research.

Innovation Process

We have explored the mechanisms for how the brain works to develop relevant new ideas, but innovation doesn't just happen by itself. We will turn our focus from the specific brain

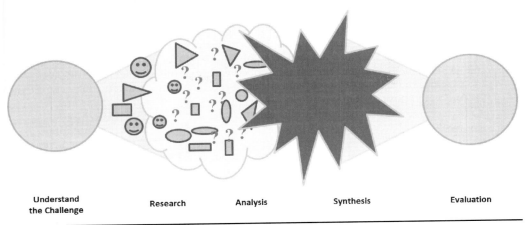

| Understand the Challenge | Research | Analysis | Synthesis | Evaluation |

FIGURE 8-5   The innovation process.

functions that are useful to innovation to looking at a process to apply these functions to innovation in a structured way. An ideal process should provide sufficient structure to keep random work from happening, yet still enable new ideas to be generated, developed, and broadly accepted. It should look less like the experimental method, and encourage the conditions for strategic intuition and the Feeling of Knowing. As shown in Fig. 8-5, we will look at each phase in the process and point out key applications of the research discussed earlier in the chapter.

Understand and Articulate the Challenge

As stated previously, a good innovation process does not encourage the random generation of ideas. It is important that the challenge is articulated such that it meets the goals and executional constraints of the larger organization. This part of the process is also necessary to plant the intentionality of the goal into the minds of the people on the project team, to guide their thinking toward relevant solutions. Without an articulation of the challenge, there would be no problem on which the brain could focus the processing of unconscious thoughts.

> **KEY APPLICATION:** Intentionally pose a problem and generate a broad base of examples from history.

Research

In this stage, it is important to conduct secondary research to get an initial understanding of market forces, and analogous problem types. Formal and informal leaders on the topic are found, and potential candidates are identified for primary research. Relevant methods for primary research are selected and primary research is conducted. This stage maps to Duggan's first stage of the strategic intuition process where we search the memories and collected examples of others. It serves as a way to build the team's knowledge base from which to draw insights.

Analysis

In this stage, the information collected in the research phase is broken down into basic elements. It's important in this phase to separate the elements from their initial context such that new associations can be made. Many different scenarios may be played out, critically reviewed, compared to analogs, and built into frameworks for future decision making. While it may not be readily apparent, this stage correlates to Duggan's presence of mind stage. A good critical analysis lets go of preconceived notions or outcomes relative to the solution, paving the way for the flash of insight to come. It also breaks down the current perceptions that Berns identifies as a barrier to innovation, allowing for new ideas to come from fresh perspectives.

> **KEY APPLICATION:** Create conditions to enable new perspectives, presence of mind, and allow time for unconscious processing.

Synthesis

In this stage the new idea is born. As Duggan suggests, the offering is envisioned in all aspects.[29] As with Burton's example of the paragraph suddenly becoming associated with the word kite, the new idea makes sense of the disparate idea fragments as they come together in their entirety. For example, if a new product idea is envisioned, then the way to reach the people motivated to buy it in the form of a new business model will also become evident. This new business model also lays the foundation for how to evaluate the new idea. In this stage the translation skills, as described earlier in this chapter, become particularly useful. Once the new idea is born, it needs to be represented such that the intent of the idea is represented through the design of the concept.

Evaluation

In this stage, the team needs to develop evaluation methods that are relevant to the new idea. Evaluating ideas from within the newly envisioned context for success creates different, yet relevant metrics to enable the larger organization to understand their value. This evaluation helps to check the logic, hone the idea to

> **KEY APPLICATION:** Socialize the new idea in terms the organization can understand.

increase relevance to either the market or organization, and provide criteria for the rest of the organization to follow. It will provide the validity that the idea is worth continued investment and support, socializing its value, and giving the organization the confidence to develop the idea into a market offering.

It can be seen that this process fully utilizes the ideas described by Duggan and Burton in terms of establishing the conditions for creating new ideas and ensuring they are relevant. However, it is very different from the process used in the product development process, as illustrated below in Fig. 8-6.

It's important to respect that each process is appropriate to meet different goals; the innovation process defines new opportunities and the product development process will produce a salable offering. A process will not work if it is staffed with people who are not well-suited to the type of work it necessitates. We will now focus on identifying people who display strong aptitude for the cognitive functions necessary for innovation, and can tame the cognitive barriers that work against it.

Talent Selection for Innovation

No one would argue that idea generation is considered a creative process, and it makes sense to include creativity in the list of skills that are necessary for the innovation process. There are three sources that are particularly relevant to creating the ideas and connections needed by the innovation process previously described.

The first is Professor Sarnoff Mednick. Back in 1962, Mednick described creativity in terms of associative thinking, and asserts that creative ideas must be relevant ideas.[30] Later work by Colin Martindale in 1998 states that creativity requires the simultaneous presence of intelligence, perseverance, unconventionality, and the ability to think in a particular manner. He states that while none of these traits are particularly rare, what is uncommon is finding all of these traits to be present in one person. He then goes on to discuss the fact that a creative idea must be both original and appropriate for the situation in which it occurs.[31] We can see that the idea of relevance is a key element of these definitions of creativity. They are in stark contrast to the popular idea that creativity may be defined as the ability to just make

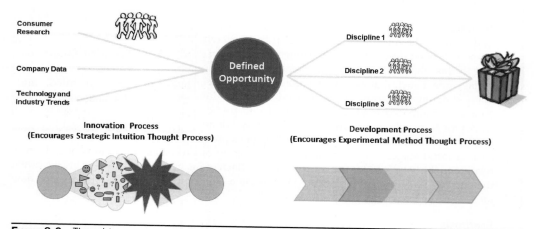

FIGURE 8-6 Thought processes applicable to innovation and product development.

up new ideas at random. This idea of relevance as part of the creative skill set as defined by cognitive psychologists is consistent with neuroscience research on how relevant new ideas are created.

For an equally consistent, yet slightly different perspective, we look to the work of Dr. Howard Gruber to understand more about the creative person's mind-set. Dr. Gruber was an expert in cognitive psychology and the founder of the Institute of Cognitive Studies. His focus is less on what creativity is, and more on who creative people are. He states that creative people show constant courage, not fearing to strike out in new directions, and perceiving situations or problems that would frighten the average person as an interesting challenge. Creative people don't just solve problems and settle back into stasis. To them "creative work begets new and fruitful problems; organizing matters so that this remains the case is part of what it means to lead a creative life."[32]

He also talks about how a creative life is nonhomeostatic, and that different modalities of thought are never separated by an unscalable wall.[33] This is an important trait relative to the innovation process, which often requires a translation between business goals, consumer needs, and the offerings that will satisfy them. A final point Gruber makes that is relevant to the innovation process is his statement that "creative people possess, and seek to possess, unique points of view or special perspectives on the world. Such points of view, in fact, are likely to distinguish the creative person more than any particular problem-solving ability."[34] This mind-set is consistent with Berns' idea that people suited to innovation have been able to break through the barriers of perception and fear to see the world in new ways.

> **KEY INSIGHT:** Creativity is less about making up new things, and more about the mind-set of creative people as they constantly seek relevant solutions to new problems.

A few years ago, I published some guidelines for how to recognize people who would likely be well-suited to innovation processes. While I didn't realize it at the time, these guidelines are intended to recognize evidence of the extent to which people possess both the skills and mind-set of creative people listed earlier. A paraphrased excerpt follows:

While looking for creative, multidisciplinary people is a good start, it is necessary to look at how people describe their work, and the ways in which they frame the examples they discuss.

Signs of a good fit for innovation projects

- A quest for holistic understanding of the goals they are pursuing.
- An intense interest in ensuring that they are solving the right problem when presented with a challenge, and may reframe the problem that was initially presented.
- They will modify processes or ways of working to ensure they are progressing toward the ultimate goal.
- They are very detail-oriented in defining criteria, a problem statement, or rationale leading to a conclusion, but less detail-oriented in executing a process or refining a solution.
- They can articulate the broad implications of their work, and the specific impacts to other groups in the organization.

Signs of a less than ideal fit for innovation projects

- They have diverse interests because they lack focus—They cannot weave a coherent story about why they have pursued each path.
- They are deeply interested in each discipline as a discrete pursuit. They may have high proficiency in each discipline, but they do not connect or translate between them.
- For people from creative disciplines, their primary focus may be on how well they can draw or write, and not on consciously translating the emotions and motivations to drive desired experiences.
- Alternatively, they may not be able to transcend their own preferences to translate an experience some else would value.[35]

These are skills that are not found on a typical resume, so we cannot expect many people to come to the organization fully prepared to work in the innovation process.

Developing Innovation Skills

Once we find people who exhibit traits that suggest they will do well in the innovation environment, it's important to understand how to encourage their development. It's also useful to understand the extent to which people can be trained in innovation at all. If so, what ideally needs to be taught? Harvard's disruptive innovation expert Clay Christensen put it this way in the HBS Working Knowledge Newsletter in May 2012.[36]

"I don't want to overstate the case. I think about 40 percent of people just are not going to be good at innovating regardless of what they do. And 5 percent are born with the instinct. The rest of us could learn what these innovators do if somebody would just crawl inside their brains and codify what to them is intuitive."

Jeffrey Dyer and Hal Gregersen collaborated with Christensen to codify what the rest of us could learn from successful innovators. In their book and articles they define what is called the innovator's DNA (iDNA). They describe five "discovery skills" that innovative entrepreneurs such as Jeff Bezos of Amazon.com, Scott Cook of Intuit, and many others possess. They describe innovative entrepreneurs as spending 50 percent more time on these discovery activities than executives with no track record for innovation. These skills are

- Associating—They are able to connect seemingly unrelated questions.
- Questioning—They constantly ask questions and challenge common wisdom.
- Observing—They scrutinize common phenomena, such as the behavior of others, as an anthropologist would.
- Experimenting—They actively try out new ideas by creating prototypes and launching pilots.
- Networking—They go out of their way to meet people with different perspectives to seek out new ideas and extend their own knowledge domains.[37]

At the end of the article they suggest that as innovators actively engage in these discovery skills, they become defined by them, and build confidence in their creative abilities. They then suggest that others can also become more creative by consciously engaging in these activities as well.[38] But how do we know which came first? Do these people pursue these activities because they are innovative people? Or did the pursuit of these activities make them more innovative? Is training people to perform these activities enough to encourage the creation of new, relevant ideas?

One clue to whether the cognitive skills necessary to innovate can be taught comes from research on neuroplasticity. Sharon Begley, the author of *Train Your Mind, Change Your Brain*, writes extensively about the topic. Neuroplasticity refers to changes in the structure and function of the brain—at the level of neural pathways and synapses—which are due to how people behave and think, to the environment they live in and the experiences they have, including bodily injury such as stroke and amputation.[39] The recognition of neuroplasticity has replaced the decades-old view that the adult brain is physiologically static, allowing neurobiologists to explore how the brain changes throughout life, what forces cause it to change, how extensive the changes can be and whether there are limits to neuroplasticity.[40] Begley cites the work of many scientists which debunk the idea that the adult brain is static. Most notably, she cites experiments showing that the brains of stroke patients can "remap" themselves so that a region that originally performed one function can take on a new one, substituting for the region knocked out by the stroke. If the regions of the brain could only perform a single function, as once believed, and could not adapt to perform new functions, the brain wouldn't be able to reorganize itself in these ways.[41]

On the other hand, if an ability has not been encoded anywhere in the brain through experience, or if a piece of knowledge has not been stored in it through learning, it cannot be put there artificially. People who have a lot of varied knowledge will be able to make more connections than those who have less knowledge; the challenge may be to focus on identifying what can foster the free-flowing thinking necessary to make the right connections.[42]

This gives credence that Dyer and Gregersen's suggestion that training people on the behaviors of innovators can enhance their ability to make the connections necessary to innovate. However, we have learned that the act of creating relevant new ideas requires specific conditions to ensure that the connections are relevant. Engaging in more associating, questioning, observing,

> **KEY INSIGHT:** The skills for innovation can be taught, but people must possess the right mind-set to use them effectively.

experimenting, and networking activities alone will not ensure that these conditions will be established.

Cultivating an Innovation Mind-Set

Duggan suggests that currently, the cultivation of these conditions is most likely self-taught through learning, work, or general life experiences. He rejects the idea of people being predisposed in one way or another, and suggests that however they got there, people who can grasp this mind-set can put his ideas for strategic intuition into practice. Duggan uses the example of Steve Jobs' early failures to illustrate that this mind-set is not typically natural, and can be self-taught. Although Jobs failed initially, he evolved to the point of being able to consistently succeed, resulting in a substantial increase in his hit rate.[43] Since expert intuition is the result of people becoming faster and more efficient as they gain experience, Jobs essentially developed expert intuition in executing the skills required for strategic intuition. As with any profession, exercising strategic intuition skills more frequently will build proficiency, resulting in greater efficiency in their use.[44]

For a different perspective, Brene Brown, an academic researcher who focuses on vulnerability and living what she calls a wholehearted life, presents an interesting twist to the idea of readiness and the ability to encourage the mind-set we have been discussing. She states that being open and vulnerable are the keys to everything of value at work—creativity, courage, and connection. Brown suggests that people hide behind their "work armor," because they are afraid to let their ideas out into the world to be criticized,[45] and that "there has never been a truly innovative idea put forward that was not laughed at."[46] In her TED talk, she described that in her role as a researcher she sought to understand material in an attempt to control and predict it; a mind-set that was undermining her ability to be open to truly new ideas.[47]

From Brown's perspective it is clear that the willingness to be open and vulnerable is a necessary component to breaking down the cognitive barriers to innovation discussed by Berns earlier in this chapter. This willingness to be open must be encouraged and supported, as this is what separates those who are successful from those who are not successful at innovation. If we are too afraid of being embarrassed or criticized, then we will not embrace the ambiguity or accept the new connections that are necessary to come up with a great idea. Brown suggests that it is not what you know, but who you are that needs to be brought to bear in accepting these types of challenges.[48] Through this lens, Christensen's comments about those who will never be able to innovate can be viewed differently. While there may be some people who are natural innovators, it is likely that they are also unafraid to embrace their own perspective, creativity, and ideas.

From this exploration we can conclude that the skills necessary to successfully innovate can certainly be taught. Not everyone will achieve the same levels of mastery in performing these skills, much like people have different levels of and proficiency in math, science, or the arts. However, we have also learned that while teaching skills is necessary, it is not sufficient to achieve innovation performance goals if the right mind-set cannot be cultivated.

> **KEY INSIGHT:** The right mind-set enables the effective use of innovation skills, by minimizing the impact of cognitive barriers to innovation.

Creating an Environment for a Successful Innovation Program

The extent to which the right mind-set can be taught has no clear answer. Current research suggests that it is more of a personal choice that people must make, whether consciously or unconsciously. This choice will enable them to be receptive to the external conditions that are conducive to their work. For those who are ready, the ideal environment will both physically and emotionally reinforce the process for encouraging new ideas and minimize the influence of the cognitive barriers. We will now turn our attention to these environmental factors.

Physical Environment

Work environments today vary from the ubiquitous office cube farm, to the uber-hip spaces that Google is notoriously known for, with many variations in between. If we are to encourage innovation it makes sense that the work environment be conducive to the activities that people in this line of work will need to perform.

In his book, *Where Good Ideas Come From*, Steven Johnson describes how the English coffeehouses in the 18th century were places that fertilized countless enlightenment-era innovations. They provided a shared environment where a diverse mix of distinct professions and passions overlapped.[49] I have found that a workspace suitable for innovation should encourage the stages of the innovation process, and would include:

- Spaces conducive to connection and interaction, both focused and nonfocused—These spaces encourage Duggan's first phase of the process, which is collecting examples from history. This can happen in gathering spaces intended for a specific project (the idea of a project war room), as well as more general gathering spaces (such as social kitchen/water cooler areas). This also helps to break down the socialization barrier, by creating spaces for like-minded people to find each other.

- Spaces conducive to quiet contemplation—These are spaces that provide the presence of mind to mull over all the new ideas that may have entered their heads, enabling people to clear their minds so they are open to receive new connections. They also give people a chance to intentionally work on the specific mental tasks associated with analyzing the problem.

- Spaces conducive to administrative tasks—These spaces may be the same as or different from the quiet contemplation spaces. For example, a quiet space setup like a library will not be as efficient as a desk for many day-to-day tasks.

> **KEY APPLICATION:** The physical space can and should encourage the interactions and contemplations necessary for relevant idea generation.

Emotional Environment

We will now shift our focus to the emotional environment conducive to innovation. This aspect is often the most difficult for an organization to get right, as it requires establishing a culture and reward systems that will encourage innovation without harming the environment that is conducive to product development and manufacturing processes. It is also the most important aspect in cultivating the right mind-set for innovation. In general, reward systems for innovation should be different, yet perceived as equally fair, to those used in the rest of the business. Most likely, the organization is already doing a good job of rewarding the efficiency of the product development and manufacturing teams.

> **KEY APPLICATION:** The emotional environment is often overlooked, but is where company leadership can have the strongest influence on innovation processes.

Innovation teams must be able to embrace ambiguity, and pursue some ideas that may not turn out to be correct. They need to be reassured that their job security will not be threatened for exhibiting behaviors that, to an outsider, may not appear to be as efficient as those in the product development teams. There is a big difference between pursuing an unknown path with less efficiency than a well-worn path, and slacking off or forging ahead without thinking. While it is important that an organization's leaders do not micromanage, it is important that they understand innovation well enough to know when the innovation team is doing great work, and to reward it accordingly. While there is no single formula for rewarding innovation, getting this level of understanding, involvement, and reward systems right will ensure a culture that will foster a productive mind-set for innovation.

Conclusion

Recent research in the field of neuroscience has illuminated the mechanisms for how the human brain creates relevant new ideas, the type of new ideas necessary for successful innovation. It has also illuminated the mechanisms that construct barriers to the creation of relevant new ideas. Understanding these mental functions has confirmed that the activities most critical for innovation are occurring inside the mind. The external tasks and activities that we observe can provide valuable input to these mental functions, but they are not the direct source of the ideas.

We have little control over an individual's desire or ability to overcome their cognitive barriers to innovation. For those who are ready, however, we can use this new knowledge to deliberately create programs that combine the right process, people, and environment to facilitate the creation of relevant new ideas. Since it is increasingly important that companies innovate in game-changing ways, an innovation program must be distinctly different from current product development programs. The focus for innovation professionals is to consistently apply new knowledge about these mental functions to their work, so that best practices for innovation can be established. Currently, there are no standards to guide the creation or selection of innovation programs. We can now remove the mystery behind why some innovation processes are successful and others are not, based on the extent to which they facilitate the mental functions that are the source of relevant new ideas, and minimize the mental barriers to creating these ideas.

Albert Einstein said that the definition of insanity is doing the same thing over and over and expecting different results. Companies can no longer incrementally improve existing product development processes and expect radically new results. Innovation professionals can no longer expect their clients to blindly trust that their processes will deliver the value their clients seek. Innovation is not random. The tools are now available to dramatically improve the innovation "hit rate." It's time to start learning how to use them.

References

1. Ariely, D. and Berns, G. Neuromarketing: The Hope and Hype of Neuroimaging in Business. *Nature Reviews*, 11, April 2010.
2. Ibid.
3. Zeki, S. Interview by the author, October 26, 2012.
4. Ibid.
5. Duggan, W. *Strategic Intuition*. Columbia University Press, p. 18, 2007.
6. Ibid., pp. 12–23.
7. Ibid., p. 2.
8. Ibid., pp. 58–59.
9. Ibid., pp. 33–35.
10. Ibid., pp. 58–59.
11. Ibid., p. 63.
12. Ibid., p. 33.
13. Burton, R. *On Being Certain*. St. Martin's Griffin, p. 3, 2008.
14. Ibid., p. 5.
15. Ibid., p. 130.
16. Ibid.
17. Ibid., p. 133.
18. Duggan, W. *Strategic Intuition*. Columbia University Press, 2007.
19. Hirasuna, D. *@Issue Journal*, August 7, 2009, V.01
20. Ibid.
21. Cope, J. "Commercialization Success Rates: A Brief Review." *RTI Tech Ventures Newsletter*, 4 (4), December 2007.
22. Berns, G. *Iconoclast*. Harvard Business School Press, p. 6, 2010.
23. Ibid., pp. 44–46.
24. Ibid., pp. 15–16.
25. Ibid., pp. 62–63.
26. Ibid., pp. 130–133.
27. Lucas, R., Jr. *Lectures on Economic Growth*. Cambridge, UK: Harvard University Press, pp. 109–110, 2002.
28. Florida, R. *The Rise of the Creative Class Revisited*, 10th ed. Basic Books, a member of The Perseus Books Group, 2012.
29. Duggan, W. *Strategic Intuition*. Columbia University Press, p. 5, 2007.
30. Mednick, S. "The Associative Basis of the Creative Process." *Psychological Review*, 69 (3), May 1962.
31. Martindale, C. The Biological Bases of Creativity; Chapter 7, p. 137. *Handbook of Creativity*, edited by Robert Sternberg, Cambridge University Press, 1st ed., November 13, 1998.
32. Gruber, H. "And the Bush Was Not Consumed. The Evolving Systems Approach to Creativity." In Sahan and Celia Modgli (eds). *Towards a Theory of Psychological Development*. Windsor, England: NFER, pp. 269–299.
33. Gruber, H. "On the Relation Between 'Aha Experiences' and the Construction of Ideas." *History of Science*, 19: 41–59, 1981.
34. Ibid.
35. Di Resta, E. "The Art of Translating Market Insight into Business Opportunities for Innovation." *International Journal of Innovation Science*, 2 (2), June 2010.
36. Emmons, G., Hanna, J., and Thompson, R. "Five Ways to Make Your Company More Innovative." Harvard Business School, Working Knowledge Newsletter, May 23, 2012.
37. Dyer, J., Gregersen, H., and Christensen, C. "The Innovator's DNA." *Harvard Business Review*, December 2009.

38. Ibid.
39. Pascual-Leone, A., Amedi, A., Fregni, F., and Merabet, L. "The Plastic Human Brain Cortex." *Annual Review of Neuroscience*, 28: 377–401, 2005. doi 10.1146/annurev.neuro.27.070203.144216.
40. Ibid.
41. Begley, S. *Train Your Mind, Change Your Brain*, by Mind Life Institute in arrangement with Ballantine, a division of Random House, Inc., 2007.
42. Begley, S. Interview with the author, October 18, 2012.
43. Duggan, W. Interview with the author, October 19, 2012.
44. Ibid.
45. Kellaway, L. "Time to Open Up at the Office." *Financial Times*, October 18, 2012.
46. Ibid.
47. Brown, B. *TED Talk*. TEDx Houston, 2010.
48. Ibid.
49. Johnson, S. *Where Good Ideas Come From*. Riverhead Books, p. 162, 2010.

CHAPTER 9

Biomimetics: Learning from Life

Melissa Sterry

3.8 Billion Years in the Making: An Introduction to Biomimetics

"Imitation is the sincerest form of flattery"

—Charles Caleb Colton

Mimicry is one of the most commonplace behaviors exhibited by living things. Many flora and fauna species use mimicry to merge with their surroundings, many as a means to hide from predators or prey. Humans engage in mimicry for many purposes, including, among other things, social bonding. The trait is inherent; we are born mimics, as expressed when a baby imitates a parent's facial expression. However, our mimetic inclinations are not solely for social purposes. Indeed, mimicry is a means by which we learn many skills, including reading, writing, art, sports, music, cooking, and more. By the time we reach early adulthood, most of us are adept copycats and much about our person will reflect this. Our accent will likely mimic that of our peer group, as will our choice of language, clothing, music, and interests. We continue to use mimicry throughout our lives, both consciously and unconsciously. For example, when attracted to someone we may consciously mimic their interests, such as take up a sport they pursue, while unconsciously mimicking their body language when in conversation with them.

Given that humans aren't the only primates to exhibit mimicry (for instance, orangutans use mimicry as a means to experience the emotions of others), it's arguably probable that mimicry was exhibited by our earliest ancestors. Equally clear is humans' long-held interest in the species about us, of which archeological records date back to Paleolithic times. Cave paintings, such as those at the Chauvet-Pont-d'Arc Cave in the Ardèche, Southern France, illustrate detailed observations of the form and behavior of animals. However, Paleolithic humans were not only watching the world about them, they were listening to it, too, which has led some academics to hypothesize that faunal sound mimesis may have been one of the earliest forms of language. But what else might our ancient ancestors have learned through faunal mimicry? The earliest known hominin technology comes in the form of rock fragments used for the cutting and scraping of bones 3.39 million years ago. What is not known is whether those tools were crafted by the hominin species *Australopithecus afarensis* or by one of our own *Homo* ancestors. What we do know is that tool use has been observed across the animal kingdom in both our closest relatives and in species far removed from us, such as octopuses, dolphins, and elephants.

We will never know which species was the very first to use a tool, nor in the event that species wasn't human, will we know if the species was contemporary, and therein observable, to the earliest hominid tool users. However, the mimesis of other life forms and the behaviors of those forms, both floral and faunal, have played an integral role in human development—socially, culturally, and technologically. Therein, our relationship with other species ought not be considered as a "them" and "us" scenario, but as symbiotic, wherein we are all part of a highly integrated, codependent system that involves both the exchange of resources and information. How much information? In 2009, I mused that given it's been at least 3.8 billion years in the making, "Life is the biggest R&D project on Earth."

"There are some four million different kinds of animals and plants in the world. Four million different solutions to the problems of staying alive."

—Sir David Attenborough

Biomimetics draws on humans' inherent mimicry skills and refers to the "transfer of ideas and analogues from biology to technology." The discipline became formally recognized in the 1950s, when biophysicist Otto Schmitt coined the term biomimetics. Also known as bionics, biomimicry, biognosis, and biomimesis, biomimetics has given rise to innumerable inventions, inspiring such innovation luminaries as Leonardo da Vinci and Buckminster Fuller. However ancient, though, its roots may be, biomimetics embodies the essence of leading-edge science in that to be effective, its practitioners must step outside of conventional scientific boundaries and adopt a hybrid approach. A fusion of science and technology, biomimetics is classified as a technoscience. Therein, by its very nature, the discipline promotes transdisciplinary thinking and the dissolution of disciplines.

In decades past, the joining of divergent dots was the sole preserve of the polymath. However, in this, the "Smart Age," the many, not the few, are endowed with such things as can enable them to see the infinite connections that bound the world about them. Science, once siloed, is now accessible to both the expert and novice alike. Not only that, but it's accessible in abundance. One-third of the world's population is online, 6 billion people have mobile-cellular subscriptions and over a billion people own a smartphone. Whether consciously or otherwise, we are all now part of The Internet of Things, as our actions are captured, in real time, by the information communications technology (ICT) about us. Top-down innovation is converging with bottom-up, wherein high tech such as nanoscale sensors and actuators, operating systems and supercomputers, smartphones and tablets, underpin low tech, in such forms as distributed computing, DIY apps and homemade hardware. The web has migrated from publishing to social to industrial platform. Data, once confined to linear constructs, now flows between people and places, individuals and organizations, living and inanimate, and the speed and scale at which that data flows is expanding exponentially, as is our capacity to use that data constructively.

The relationship between science and technology is an intimate one; advances in the former enable likewise in the latter, and so the cycle continues. However, this relationship is not exclusive, as recognized by those championing a migration from "STEM" to "STEAM." The arts—not least science fiction literature and film—have provided invaluable inspiration for scientists and technologists. "Artistic license" enables the likes of authors and film directors to explore ideas ahead of their time; to consider the context in which those ideas may evolve and how they might look and feel, without the need to understand their every facet. The expression "ideas have sex" coined by Matt Ridley applies. Many a scientific endeavor has inspired a novel, if not several, of which some were made into movies that drove interest in the hypothesis and in the professions and technologies associated with it, a case in point being Arthur C. Clarke and Stanley Kubrick's 1968 film *2001: A Space Odyssey*. In 2011, during a patent battle, Samsung claimed that a tablet concept depicted in Clarke and Kubrick's film invalidated Apple's design-related iPad patents. This wasn't the first instance in which patent ownership was challenged by artistic representations and likely will not be the last. Which brings us back to the most prolific biomimetician to date: painter, sculptor, architect, inventor, engineer, physicist, anatomist, botanist, zoologist, physical geographer, and set and costume designer Leonardo da Vinci.

"Human subtlety will never devise an invention more beautiful, more simple, or more direct than does nature because in her inventions nothing is lacking, and nothing is superfluous."

—Leonardo da Vinci

Giorgio Vasari wrote of da Vinci, "an artist of outstanding physical beauty, who displayed infinite grace in everything that he did and who cultivated his genius so brilliantly that all problems he studied he solved with ease." More recently it has been said that "he embodies the potential scope of a human being. He makes us confront what we have the potential to be." Emulating the achievements of a polymath, as accomplished as da Vinci, is no mean feat. Yet arguably, that feat has been made significantly easier, thanks to the web and that which it facilitates, such as instant access to information and to talent networks worldwide. Given his capacity for big thinking, how might da Vinci have used Big Data? A humanitarian with a love of nature, were he born in 1952 and not 1452, would he have seen potential in citizen science, as a means to upscale the rate at which environmental and social challenges are tackled? I suspect he would; I can picture a 21st century da Vinci embracing ICT—using a smartphone to capture footage of species and their behaviors; a tablet to sketch scientific, technological, and artistic concepts that he migrated to his desktop via the Cloud; a 3D printer to rapidly prototype his inventions; social media to communicate his many observations and thoughts on the world about him.

How might one mimic the mind-set of such a man? First, while we may consider da Vinci a genius, he did not: When confronted with problems he looked not within, but to nature for solutions. Second, da Vinci's perception of what was possible was not defined by public opinion, but by the limits of his seemingly endless imagination. Third, da Vinci embraced science, technology, and art with equal enthusiasm, integrating each into his innovation approach. While da Vinci sketched some several thousand visionary inventions, given that analysis indicates there is "only a 12 percent similarity between biology and technology in the principles which solutions to problems illustrate," were he alive today, he would surely feel there was still much work to be done. Thankfully, much is being done.

Biomimetics is one of the fastest growing research disciplines. Whereas less than 100 papers were published per year in the 1990s, several thousands were published per year in the first decade of this century. Converge humans' natural talent for mimicry with the volume of species from which we may draw inspiration and insight, with the potential embedded in emergent technologies to investigate those species in far greater detail than ever before, and you are looking at a discipline with nothing short of exponential potential. The scale and scope of innovation in biomimetics is such that this chapter will but scratch the surface. However, I hope to endow you with some insight that may inspire you to dig deeper and do so in the spirit of da Vinci.

We Have the Technology: Bionics for the Body

"We're getting a glimpse now of a new age, where we will carefully integrate technology with our very nature; an age in which you can't differentiate from our biology and the device; an age in which you cannot differentiate between what is human and not human and what is nature and what is not nature."
—Hugh Herr

Mimicry is just one of many human behaviors of which there is evidence over millennia. Body augmentation is another. Given that each of us comprises around 7 octillion atoms, most of which are between 7 and 13.7 billion years old, one could be forgiven for thinking we'd be happy with our lot. But not us, for we are a species that, above all others, likes to imagine upgraded versions of ourselves. While Martin Caidin's "Six Million Dollar" cyborg "Steve Austin" is the fictitious character most commonly associated with bionics, I turn to another, lesser-known superhero to assist me in expressing this point. Comic strip character "Vixen" is an African-American model-meets-zoomorphic and therianthropic member of the "Justice League of America": When she is not strutting down catwalks, she is saving lives and fighting evil—a feat she achieves by channeling the abilities of various animals and, upon occasion, those of her fellow superheroes. Vixen first appeared in Action Comics in July 1981, yet this mimetic superhero would not look at all out of place if found engraved in the walls of an ancient Egyptian or Sumerian temple. Since the dawn of civilization, humans have been envisaging superbeings with humanoid characteristics. Whether they are deities, comic strip heroes, or airbrushed supermodels, these superbeings personify our ideals of physicality, character, and intellect. While the attributes of such superbeings vary, each has one or more characteristics that makes them better than human, but as the 21st century unfolds, superhumans are no longer the stuff of fantasy.

In 1982, having been caught in a blizzard on Mount Washington for three nights in temperatures as low as −29°C, climber Hugh Herr (Fig. 9-1) lost both his lower legs to frostbite. Doctors told Herr he'd never climb again, but he had other ideas and crafted specialized prostheses adapted for climbing; prostheses that enabled him to grab hand- and footholds beyond the reach of a man with biological legs. Having augmented his own body to achieve remarkable feats, Herr set his sights on helping others. Today he heads the Biomechatronics research group at the MIT Media Lab and is founder and Chief Scientific Officer of Media Lab spin-off company iWalk. Dubbed the "Bionic Man," Herr is coholder of 14 patents related to assistive devices, including those for a computer-controlled artificial knee, an active ankle-foot orthosis, and the world's first powered ankle-foot prosthesis. Herr's work brings together developments in several fields of science and technology to make "intimate extensions of the human body—structurally, neurologically, and dynamically." Whereas prostheses were once inert and rather awkward objects, Herr's elegantly integrate the biological and the technological in biomimetic systems. Like their biological equivalent, Herr's prosthesis, such as the iWalk BIOM, are autoadaptive devices, able to anticipate and respond

Figure 9-1   Director of the Biomechatronics research group at MIT Media Lab and founder and Chief Scientific Officer at iWalk, Hugh Herr, PhD.

to environmental conditions, thanks to intuitive design systems built using sensors, synthetic muscle-like actuators, and onboard computers that use complex algorithms to mimic the body's own control mechanisms.

In the 20th century, the Moon landing was "one small step for man, one giant leap for mankind"; in the 21st century, the quote lends itself to the day-to-day breakthroughs being made in medical bionics. Herr envisages lightweight exoskeletons for all, that act in parallel with the lower limbs to transfer loads to the ground, thus reducing the metabolic demands of walking or running, pointing out that should such devices come to market "driving across town in a metal box with four wheels would be just absurd." The brand names of the latest generation of medical bionics give a clue as to the thinking that underpins them, iWalk, eLEGS (Ekso Bionics), and i-Limb (Touch Bionics). These are smart robotics capable of machine-to-machine (M2M) communications, devices with the capacity to send and receive data to any other connected device, natural or otherwise. Wearable technologies, such devices are already enabling what Ekso Bionics coin a brand new era of mobility in which we rethink physical limitations and augment human capacity. Ekso Bionics has been hailed one of the most innovative bionics businesses on the planet by the ilk of *Wired*, *Time*, and *Inc* magazines and, like Herr, the company sees wide-scale applications for bionic devices. "Test pilots" of its wearable robots range from people with spinal cord injuries to soldiers, the former empowered to walk again, the latter able to carry heavier loads without exerting the same load pressures on the body. Magnús Oddsson of Icelandic orthopedics company Össur thinks the next step in prosthetics could be "smart trousers" that "use the actuators that are already there, the muscles, and simply provided a new central controller."

"We are stardust, we are golden, we are billion year old carbon."

—Joni Mitchell

We are, to quote Carl Sagan, "star stuff," therein comprising atoms; positively charged protons, negatively charged electrons, and neutrally charged neutrons. In short, humans are biological power stations. When reduced to its lowest common denominator, our nervous system is an electronic information system that migrates data about our body. Nano-biointerfacing science is delivering some of the most game-changing developments in medical bionics—developments that are moving at a phenomenal pace. In 2009, a team of European Union scientists unveiled the "SmartHand"—a human-machine interface that wired a prosthetic hand directly to the existing nerve endings in the stump of a severed arm.

FIGURE 9-2   World's first implantable thought-controlled robotic arm, being developed by Chalmers researcher Max Ortiz Catalan. (*Source*: Oscar Mattsson)

The artificial hand not only enabled its user to undertake complex tasks such as writing, but also restored the sensation of feeling. Using osseointegration, in which a prosthesis is directly anchored to the skeleton, therein creating a direct interface between the prosthesis and the bone, Max Ortiz Catalan (Fig. 9-2) and his team at Chalmers University of Technology in Sweden are implanting electrodes directly on nerves and remaining muscles. The approach enables the bioelectric signals from the brain to the prosthesis to become more stable than they would be were the electrodes placed over the skin (Fig. 9-3). Thought control is enabled

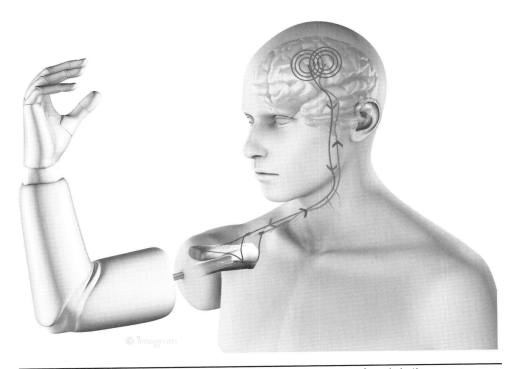

FIGURE 9-3.   Thought-controlled prosthesis, in which signals are transferred via the nerves through the arm stump and captured by electrodes, which transmit the signals through a titanium implant (OPRA implant system) to be decoded by the prosthetic arm. (*Photo: Integrum*)

by complex algorithms that decode the signals passing between the brain and prosthesis, mimicking a physiological system.

Today's brain-computer interfaces (BCI) are built on research dating back to 1875, when British physician Richard Caton published a paper in the *British Medical Journal* on his findings on electrical activity in the brains of rabbits and monkeys. In 2000, the sci-fi sounding "Thought Translation Device" developed by Niels Birbaumer and his team at the University of Tübingen's Institute of Medical Psychology and Behavioral Neurobiology, marked a significant leap forward, by allowing completely paralyzed users to write with their brains. But what of the future of BCI? Pondering "beyond state-of-the-art" neural interface technology is something Dr. Justin Sanchez and his team at the Neuroprosthetic Research Group (NRG) at the University of Miami do daily. NRG use the term "Renaissance Man" to describe the genre of talent they recruit for research projects an approach that mirrors that of smart city technology pioneers Living PlanIT.

What happens when we plug braincomputer interfaces into the smart city and the ilk of software platforms like Living PlanIT's Urban Operating System (UOS) that connect M2M devices, be those devices prosthetics, smartphones, buildings, or infrastructures? Ideas already demonstrated by Living PlanIT include the vital statistics of conference delegates being monitored by sensors in flying drones, which were then migrated to other devices via Wi-Fi. Sensors in a smart building could pick up such data, so that when an individual has vital organ failure, emergency services can be notified. Bioimplantronics—electronic devices implanted into biological organs—would facilitate even greater amounts of data being migrated through intelligent networks, meaning that when a patient arrived at Accident and Emergency (A&E), medics would already have a good idea what course of action to take.

"Hey mind reader,
*   You know what I feel, before I speak*
*   It always seems you gotta one up me"*

—VersaEmerge

Welcome to what could be called the "Symbionic Century"; a time when everything, including people, can be connected using electronic signals; integrated information communications technology that facilitates mutualistic interaction from the nano to the macro in scale. Symbiosis is one of the most fundamental features of living ecosystems; a relationship that benefits the parties—in some instances to such a degree that one cannot exist without another. A symbiotic relationship, at the landscape scale in which nanoscale and macroscale species are codependent, is found in forests. Fungus and the roots of vascular plants form a mutualistic relationship (mycorrhiza), which among other functions enables the mobilization of a range of nutrients across the forest floor. Complex systems such as this, along with advances in information communications technology, are inspiring a new paradigm in computing.

In February 2013, we took a step toward that paradigm when a team led by Miguel Nicolelis created "a sophisticated, direct communication linkage between brains." Nicolelis' team fitted two rats with brain-to-brain interfaces that enabled them to share sensory information, thus collaborate on simple tasks to earn rewards. Information transferred between the rats was not limited by geography; in one demonstration the team used the Internet to link the brains of a rat at Duke University in North Carolina, USA and another in Natal, Brazil. Taking the "organic computer" paradigm to an ecosystem level, cognitive psychologist and animal intelligence researcher Diana Reiss, musician Peter Gabriel, Director of MIT's Center of Bits and Atoms Neil Gershenfeld, and Internet pioneer and vice president and Chief Internet Evangelist for Google Vint Cert, announced at TED 2013 their plan to create an "interspecies Internet" that could connect humans with other highly intelligent species including dolphins, apes, and elephants.

"What was amazing to me was that [the animals] seemed a lot more adept at getting a handle on our language than we were at getting a handle on theirs," says Gabriel. "I work with a lot of musicians from around the world. Often we don't have any common language at all. We sit behind our instruments and it's a way to connect."

—Peter Gabriel

An interspecies Internet is a concept already envisaged in science fiction, in the form of the "Tree of Souls" in James Cameron's 2009 film *Avatar*; a film in which interdisciplinary practitioner Cameron, whose activities include film direction, deep-sea exploration, screenwriting, visual arts, editing and inventing, explored myriad emergent innovations in science, technology, and the arts. Gabriel's TED 2013 speech drew our attention to another genre of interspecies communication, music. Steven Spielberg's 1977 epic, *Close Encounters of the Third Kind*, famously featured a scene in which humans communicated with aliens by playing "5 tones"—enabling the parties to communicate using mathematical musical phrases. In that same scene, two ground technicians say lines that embody the essence of biomimetics: "Welcome to the first day of school fellas" and "It seems they are trying to teach us something." A biomimetician has the humility to admit that humans can learn from other species and perhaps in ways we cannot yet imagine. In his TED 2013 talk, Vint Cert showed an image from *E.T.*, saying that we are figuring out "what it means to communicate with something that's not a person." In a conversation I had with Living PlanIT CEO Steve Lewis, he expressed a similar sentiment that "one can imagine seeing what another species can see, what they can hear, how evolution has caused species to specialize and to interpret their environments, and this will, I suspect, challenge analytical models for many years to come."

In February 2012, a short film conceived by film director Ridley Scott and writer Damon Lindelof, was released as a viral Internet promotion for their movie Prometheus. Set in 2023, the film featured a character from Prometheus, Peter Weyland, giving a TED talk titled "We Are the Gods Now," a title that resonated with futurist Jason Silva. In November 2012, Silva gave a talk by the same name, in which he asked "when we reverse engineer life itself, when biology becomes the new canvas for our aesthetic design, what new forms of genius might come out of that?" When discussing what had led him to write "We Are the Gods Now," writer Lindelof had stated, "in really good science fiction, the line between the science and fiction is blurry. When I started attending TED, that line got even blurrier—I started hearing about ideas that were, in my own imagination, more far out than some of the science fiction I was seeing." Monkeys controlling robots with their minds—science fiction or fact? Fact. TEDMED 2012 saw Miguel Nicolelis presenting footage of an experiment in which a monkey, Aurora, controlled a monkey avatar with her thoughts via a BCI device, leaving her able to use her biological arms for other purposes. In 2011, Nicolelis went on *The Daily Show* and said that he would build a thought-controlled exoskeleton within 4 years. In February 2013, he said he's on track and plans to showcase the innovation at the Brazil World Cup in 2014, with the backing of the Brazilian government that awarded his research $20 million, and the approval of the secretary general of Fédération Internationale de Football Association (FIFA).

We All Have the Technology: DIY Bionics for All

"The only laws that could and did rule them were the natural laws."

—R. Buckminster Fuller

While fictitious cyborg the "Bionic Man" required the expertise of the "Office of Scientific Intelligence" and "Six Million Dollars" to build, in the real world, thanks to the Internet, and more recently to 3D printing, medical bionics is no longer the exclusive domain of those with a PhD and a laboratory at their disposal. In 2012, it was reported that Sun Jifa, a 51-year old from Guanmashan in Jilin province, northern China, had built himself a pair of DIY prosthetic arms. Jifa, a farmer, had lost his arms when a homemade bomb he was using for blast fishing detonated early. Without the funds to afford professional prosthetics, he decided to make his own from scrap metal at virtually no cost. While the task took him 8 years and aesthetically the result wouldn't look out of place in the dystopian action movie *Mad Max*, Jifa reported they enable him to undertake everyday tasks "just like anyone else."

In October 2012, in a keynote speech to 450 brand leaders in Russia, I stressed the imperative to embrace "Made at Home" and to experiment with 3D printing, DIY technology, and citizen science—because the next generation in design, manufacture, and retail will see consumers become makers and an integrated part of the research, development, and production process. The spirit of DIY technology is embodied in British start-up Raspberry Pi, of which the mission is to be a catalyst which encourages other companies to clone its approach, so that cheap, accessible, programmable computers are everywhere. Umpteen DIY technology magazines and forums now populate the Internet, each built on a sharing culture that sees expertise and tools shared freely, both locally and globally. Organizations such as Technology Will Save Us and Super/Collider encourage expert and novice alike to engage with science and technology, helping to break the boundaries that once ring-fenced active engagement in the fields.

The origins of 3D printing lay in the Italian Renaissance in the works of polymath Leon Battista Alberti (1404–1472). An author, artist, architect, poet, linguist, philosopher, humanist, and cryptographer, Alberti was to manufacturing what da Vinci was to science. Indeed, he could be credited as the "Father of 3D Printing." The pair shared much in common—both were dilettantes, both blurred the boundaries of STEAM subjects and both drew inspiration from nature. What relevance has 3D printing and DIY technology to biomimetics? One answer lies in the way ecosystems at both the local and the global level organize themselves. Take an ecosystem of your choice—pretty much anyone, anywhere, as long as the ecosystem hasn't been isolated for millennia; that is, it should not be located half a mile deep in a frozen lake in Antarctica, or in the very remotest depths of an underground cave system. Observe that system and you will find that, as exhibited in the forest scenario discussed earlier, while hierarchies will exist in such forms as food chains, flora and fauna species will exhibit a distinctly collaborative approach, a coalition of the willing. I will use ecosystem responses to natural hazards to illustrate this point.

On May 18, 1980, footage of one of the largest natural hazards in living memory aired live on television screens around the world. Millions of people watched as Mount St. Helens erupted, unleashing among other things, the largest terrestrial landslide in recorded history, an ash plume over 80,000-feet high and a lateral blast with an initial velocity of 220 mi/h that rapidly increased to 670 mi/h. Forty thousand acres of forest was leveled in an instant. The images that streamed about the planet were that of devastation and, by the looks of it, nothing in the blast zone could possibly have survived. In the aftermath of the eruption, an extensive study ensued to track the ecological recovery of the site. This past 33 years, ecologists have extensively documented every step in that recovery process—and in doing so have produced one of the most comprehensive ecosystem studies ever made. This study has revealed some surprising facts about the ecosystem's response to the eruption, which are similar to results I have seen in numerous other studies of ecosystem responses to hazards including hurricanes, wildfires, and flooding. These responses include pioneer species that either endure the hazard, or that rapidly repopulate a site and in doing so, increase its suitability for other less hardy species. For example, at Mount St. Helens the prairie lupine, a prodigious seed producer, was a pioneer species to arrive on the pumice plain. The species' mutualistic relationship with a root bacteria enables it to fix nitrogen, thus creating a microhabitat hospitable to several species of plants, which in turn helps support other life forms, such as insects and birds.

"Extreme events that are destructive for humans are often a blessing for natural populations"
—Hickey and Salas

Where humans see a glass half empty, some species see it half full, having evolved the capacity to turn a natural hazard into a resource opportunity. This isn't so surprising if we think back to the origins of life on Earth some 3.8 billion years ago, when conditions were decidedly hostile to life as we know it. Even today, an abundance of species still persist in extreme environments, some so extreme the species are known as extremophiles. The manufacturing paradigm of the Industrial Revolution—the paradigm by which all major global product brands still organize themselves, involves a top-down, resource-intensive structure, which comprises siloed specialist groups of suppliers, makers, distributors, retailers, and consumers. In this scenario we see a great many goods mass-produced in large industrial facilities to a relatively limited number of specifications, when compared with the size and the geographic distribution of their market. Historically, most goods manufacturers illustrated little in the way of regard for the conditions in which materials used in their goods were sourced; that is, showed no concern for the state of the ecosystem that ultimately supported their business model. Likewise, sweatshop labor underpinned this industrialized production model, wherein men, women and, in some instances, children worked long hours for little pay. Once goods exchanged hands, that is migrated from factory to store to consumer, the former passed responsibility for those goods to the latter until such time as the goods were no longer desired, at which point many went to landfills. The model was inequitable, wasteful, and ultimately unsustainable—a fact that has become abundantly clear.

Back to 3D printers and DIY technology, which together with ICT present the opportunity to adopt a radically different model to that of the Industrial Revolution, a model that embeds the principles of a natural ecosystem. Flora and fauna species use distributed data, data that come in the form of DNA. They use local resources to meet their material and energy needs, yet in many instances, have the capacity to migrate from place to place as environmental changes adjust the volume of resources available at different locations. Flora and fauna species process materials *in situ* and at ambient temperatures, materials that are generally part of closed-loop systems. While all life on Earth is derived from a common ancestor, not only is the design of each species unique, but also in many instances the design of individual members of that species is bespoke. Should such principles be applied to the medical bionics sector, one scenario we could see emerge by the year 2020, is described below.

Over the past 12 years, humanity has migrated from "take, make, waste" to a "closed-loop" economy (Fig. 9-4). Recycling is commonly undertaken at both the macro- and microscale. Early tools that bucked the former trend back in 2012 included the ilk of Filobot plastic filament maker that enabled users to recycle plastics in the home—materials that could then be used in 3D printers. "Do It Yourself" migrated to "Design It Yourself" and citizens now take a hands-on approach to product research and development, using platforms such as Constrvct, that enables users to customize garment designs online and to send those designs with their body measurements, thereby to receive bespoke fashions to their exact specifications delivered to their front door. More or less everyone, everywhere on the planet, has access to the technology that facilitates such customization, be it in his or her own home, or at service providers in their locality. Medical prosthesis suppliers have integrated such thinking into their operations.

Prosthesis users now specify their desired product, its design specifications (using both their own input and that of supplier algorithms), and their body measurements online. Users obtain their measurements using smart apps and personal 3D scanners. Some prostheses and prosthesis parts, that is, replacement parts, can be downloaded and 3D printed in the home. Whether DIY prostheses are made of recycled household materials, or medical grade products distributed by specialists, varies depending on the item at hand. Some prostheses are still made by specialist makers, especially more complex products. However, even in this instance, users have more input to their products and generally receive products faster and cheaper than was the case in 2013. Prostheses are designed for disassembly so that users can upgrade or recycle them at will. Most prostheses connect via Wi-Fi to an ICT infrastructure. Users adopt both commercially available apps for their prostheses, while also creating their own. Smart prostheses are in far wider use than in 2013, thanks to generally lower costs and easier access. Consequently, the market has now diversified with exoskeletons becoming commonplace in sectors including construction and removals.

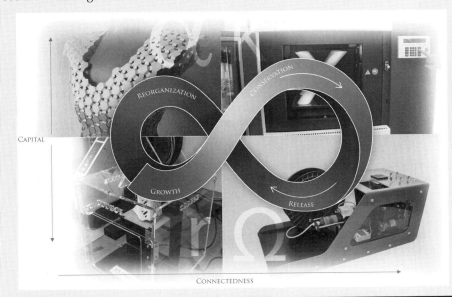

FIGURE 9-4 Panarchy as applied to 3D printing. Disruptive innovation at the edges drives the development of a commercial 3D printing market. Plastic recycling provides a continuous supply of materials, which are used to develop new product markets, such as 3D printed goods.

Plastics are just one of many material types that the 3D print and bionics community is experimenting with. Biological materials are another. Organovo's NovoGen MMX Bioprinter is the world's first commercial 3D bioprinter for printing human tissue. Inspired by the possibility of "being able to print human tissues on demand," the bioprinter is one of a growing number of technologies that explore either the replication or the manipulation of biological materials in fields as diverse as computing, lighting, furniture, and construction. While the process of replication or manipulation of biological materials is not biomimetic, the objectives are the same: to create materials as smart as those of nature that can self-organize to enable such things as self-repair, and to do so efficiently using elements in abundant availability, as do living things. Perhaps, by 2020, we may even be printing biological materials on our 3D printers at home.

A City for All Seasons: Biomimetics in the Built Environment

"This business of widening range has taken us through some weird territory; it means that most of the projects we make are hybrid in content as well as notion. They themselves are in a constant change of state, assembly, and value."

—Metamorphosis, Archigram 8; 1968

Codex Atlanticus documents da Vinci's inventions from the period 1478 to 1519. Within its atlas-like breadth, one will find a sketch of a "Living City" concept, of which the sewage system was modeled on da Vinci's observations of the hydrodynamics of natural water systems. While not biomimetic, the Living City nonetheless illustrated that da Vinci contemplated the application of systems thinking to city planning issues, and looked to nature for inspiration when doing so. However, as with so many other ideas explored by da Vinci, his Living City concept was hundreds of years ahead of its time. Step forward to the 1960s and another Living City by a radical London-based architectural group is attracting a lot of attention. Just like da Vinci, the members of Archigram were nonconformist, prolific in their visual output and revolutionary in their thinking. Archigram questioned the very premise by which cities and buildings were designed and in doing so identified new ways in which we could expect the built environment to perform better. The group pushed creative boundaries in all directions, applying an experimental approach to everything from the materials they used, to how they organized themselves, to the ideologies they embraced, to the technologies they applied, to ways in which they communicated their works. Archigram's Living City exhibition, which was held at the Institute of Contemporary Arts in London in 1963, was not born of biomimetics, yet recognized the city as a complex system, as an organism of which architecture is just one of many components.

Another iconic Archigram project—Walking City, created in 1964 by Ron Herron, explored a theme central to several of the group's projects—adaptability. Walking City's structures look like huge robotic insects migrating from one place to another. In Archigram's Instant City we saw an altogether different expression of portability. A "traveling metropolis," Instant City embedded concepts we are presently seeing expressed in "pop-up" cities, in which infrastructure migrates from community to community to bring a temporary taste of the metropolitan dynamic.

Archigram were not the only ones on the vibrant London 1960s architecture scene that were exploring the possibility of creating built environments with life-like properties. Cedric Price perceived four spatial dimensions to buildings and infrastructure: length, width, height, and time. A contemporary and a friend of Buckminster Fuller, Price believed a building should only last as long as it remained useful, and therefore ought to be designed for disassembly at the end of its lifespan. Whereas many architects would covet a listing for one of their buildings, when one of Price's works was cited for listing—his Inter-Action Centre—he successfully argued the building had outlasted its usefulness, and therefore ought to be demolished. A pioneer of anticipatory architecture, Price also recognized the advantages of creating adaptable structures comprised as a "kit of parts" that could reassemble as and when required, as seen in such projects as The Fun Palace (1960–1961). Together with the pioneering cybernetician Gordon Pask, in 1986 Price explored the potential of anticipatory architecture at the city scale. In their concept project, Japnet, we see a precursor to the smart city. Described by Price as a "community without propinquity," Japnet comprised complex feedback systems at the urban scale that migrated information to which dynamic infrastructure would respond. ICT concepts including the 24-hour postman (e-mail to you and I), enabled a city in continual transformation, constantly reorganizing and rearranging itself through processes of both expansion and retraction. Pask's "Conversation Theory" spoke of bottom-up decisions, hence was complementary to biomimetic systems thinking, because it explored how we might organize communications at an ecosystem level. John Frazer, a systems consultant that worked with Price and Pask posited that architecture should be a "living, evolving thing."

"The role of the architect here, I think, is not so much to design a building or city as to catalyze them; to act that they may evolve."

—Gordon Pask

Nicholas Negroponte was another early pioneer of built environment systems thinking. In March 1969 the *Journal of Architectural Education* published Negroponte's paper, "Toward a Theory of Architecture Machines," in which he discussed themes central to his book, *The Architecture Machine*. Negroponte envisaged "devices that can intelligently respond to the tiny, individual, constantly changing bits of information that reflect the identity of each urbanite as well as the coherence of the city." He spoke of structures that "evolve," that have "mutations" and "offspring." Given that Negroponte's citations list included the ilk of Warren McCulloh and Walter Pitts' paper "A Logical Calculus for Ideas Immanent in Nervous Activity," which was published in the Bulletin of Mathematical Biophysics in 1943, it is clear his terminological style was not incidental, not least when he states "communication is the discriminatory response of an organism to a stimulus." Negroponte, like Price and Pask, realized that to integrate systems thinking at the city scale would require smart technologies, stating, "It is possible to build an architectural seeing machine by developing a simple device that will observe simple models. Such a mechanism is the prelude to machines that someday will wander about the city seeing the city. In such a manner, architecture machines could acquire information beyond that which they are given and therefore would have the potential to challenge and to question."

Today, as discussed earlier, we do have machines that can do just that, endowed with sensors, apps, and Wi-Fi: Technologies including smartphones and drones can aggregate and transmit real-time information to any other Wi-Fi connected device, near or far. These technologies not only facilitate such concepts as were imagined by Price, Pask, Frazer, and Negroponte, but far more besides. By 2020, it is estimated that the machine-to-machine communication market will have expanded to 24 billion smart sensors and connected devices and be valued at US$1.2 trillion. A great many futurists have visualized future cities as high-rise concrete jungles with barely a tree in sight, a scene which was presented in feature films including Fritz Lang's *Metropolis* (1927), Ridley Scott's *Blade Runner* (1982), and Joseph Kosinski's *Tron* (2010). However, while such visions accounted for exponential growth in technological capacity (i.e., more processing power), what they did not account for was user behavior (i.e., human creativity). The legacy of innovation, be that technological or otherwise, is not always that which was initially intended. Case in point, when Alexander Graham Bell invented the telephone, he likely did not anticipate his invention would one day integrate with computing and photography to become a smartphone. Technologies are not an end unto themselves; each new technology presents new possibilities, which in some instances challenge that which came before. In his statement "unlearning is as important as learning," Negroponte summed up why da Vinci, Buckminster Fuller, Archigram, Price, and other visionaries were able to anticipate the future applications of new science and of new technology to cities with such uncanny precision. The future is not the exponential expansion of the present; it is a new reality unto itself.

"While technology solves problems largely by manipulating usage of energy, biology uses information and structure, two factors largely ignored by technology."

—Julian Vincent

Living PlanIT's Urban Operating System (UOS) provides cities with the capacity to access data from 100,000,000+ sensors across 4000 categories—data from land, sea, air, and space. The UOS could be considered akin to a city what a mycorrhizal network is to a forest and will provide a city's citizens and in turn its infrastructure, with access to more realtime data, the ever previously possible. PlanIT Valley, a cityscale living laboratory featuring the UOS, is in planning in the Municipality of Paredes in northern Portugal (see Fig. 95). Designated a Project of National Interest by the Portuguese Government, when complete, PlanIT Valley will serve as a testbed for smart technologies and as a global innovation hub for STEM research, startups, and intrapreneurship. Recognizing the farreaching potential of the vast quantities of realtime networked data the UOS will generate, Living PlanIT is building a global transdisciplinary collaborative community comprising organizations at all scales, a business ecosystem designed to diminish the traditional silos that undermine

FIGURE 9-5   PlanIT Valley under construction in Porto, Northern Portgual. Living PlanIT SA. (*Source*: PlanIT Valley, Porto, Portugal. ©Living PlanIT SA 2013—All Rights Reserved.)

innovation at the edges. While some facets of the UOS are relatively generic in smart city infrastructure, such as ICT that can improve energy efficiency, Living PlanIT is developing a platform with far broader applications: A platform that makes it easier for innovators to connect to the people that can enable and upscale their ideas, that recognizes the fact that real estate and construction are two of the least efficient, yet largest capital markets, and the potential for change therein, that applies the concept of "product lifecycle" to buildings and infrastructure, as did Price; that aspires to greater energy savings than that facilitated by smart metering alone; that calculates the optimum urban landscape topography for a city's renewable energy systems and optimizes city planning for citizen health—from the foundations upward a city is designed to enable citizens to walk, run, or cycle about, rather than be reliant on automated transportation.

While, as Julian Vincent stated, the technology of the past and the present largely solves problems "by manipulating usage of energy," the UOS is one indicator that the technology of the future will, like biology, put more emphasis on information and structure. London-based independent urbanist, designer, and futurist Liam Young of future's think tank Tomorrow's Thoughts Today, points out "one of the critical questions we are asking ourselves at the moment is what do we do as architects in a near future where the dominant building material exists outside the physical spectrum." In June 2012, Young, in collaboration with Eleanor Saitta, Oliviu Lugojan-Ghenciu, and Superflux, used components originally intended for aerial reconnaissance and police surveillance to create a flock of GPS-enabled quadcopter drones that broadcasted their own Wi-Fi network. Swarming into formation, the flock transmitted its pirate network, before dispersing undetected. The project, titled "Electronic Countermeasures," was one of several of Young's works which exploit the potential of cities that are planned around the speed of electrons, satellite sight lines, and big data, in which connection to Wi-Fi is more critical than connection to light and where the death of distance has created new forms of city based around ephemeral digital connections rather than physical geography. Wi-Fi–enabled flying quadcopter drones may not in themselves be biomimetic, but as with Living PlanIT's UOS, at a systems level they illustrate how emergent technology is enabling more life-like qualities to emerge within our built environments.

"The current approach to the production of architecture is ancient and yet the technology that could potentially revolutionize our approach to the construction of buildings is even older than the invention of concrete. This technology is life."

—Rachel Armstrong

Plectic systems architecture explores the possibility of developing environmentally responsive buildings that work with the natural energy flows within matter and that connect to Earth systems, not insulate from them. In development at the Advanced Virtual and Technological Architecture Research (AVATAR) group at the University of Greenwich in

FIGURE 9-6 Dr. Rachel Armstrong, Co-director of the Advanced Virtual and Technological Research group at University of Greenwich's, "wet architecture" in the form of protocell droplets.

London, the model and its methodology adopt a bottom-up design approach, in which wet architectures, such as protocells (Fig. 9-6) and slime molds, self-organize. Pioneered by AVATAR's codirectors, TEDGlobal Fellow Dr. Rachel Armstrong and Dean of the School of Architecture, Design, and Construction at the University of Greenwich Neil Spiller, plectic systems architecture presents possibilities such as creating "material computers," which are able to biologically calculate a range of environmental data, or, to put it another way, biological information communications technology (Fig. 9-7). Possibilities which Armstrong and Spiller's wet architectural system presents include, scrubbing the air of pollutants, carbon capture, resource production (energy and food), self-repair, and a range of novel aesthetic applications, such as bioluminescent algae that lights up in response to kinetic stimulus. What might emerge from this primordial architectural soup? Leroy Cronin, who leads a group of scientists that are also pioneering new bottom-up approaches to building materials, sums it up in the statement, "The key aspect of any living technology is its potential for autonomous adaptation, and its application to design and architecture could be profound in the extreme."

Another pioneering figure challenging the traditional built environment paradigm is Neri Oxman, founder and director of MIT Media Lab's Mediated Matter design research group. Oxman aims to enhance the relationship between built and natural environments and do so by applying biomimetic design principles. Named one of *ICON* magazine's top 20 most influential architects to shape our future (2009), Oxman speculates that in the future we will "grow living breathing chairs and construct buildings by hatching swarms of tiny robots." Experimenting with diverse 3D printable materials, including concrete and silk, Oxman reinterprets material applications and expands our understanding of the potential for composite materials within the building process. She sees potential for common building materials to perform a range of tasks, stating that "comprehending difference enables us to design repetitive systems, like bone tissue, that can vary their properties according to environmental constraints. As a consequence of this new approach, we will be able to design behavior rather than form." Design principles Oxman migrates from the natural world include growth over assembly, integration over segregation, heterogeneity over homogeneity, difference over repetition, and material as software. Using the metaphor of a spider, Oxman highlights how the creature is akin to a multimedia 3D printer that generates silk for a variety of uses. In February 2013, U.S. start-up Wobble Works LLC put its new invention, "3D Doodler," the world's first 3D printing pen, onto crowd funding platform Kickstarter. Having stated a target of $30,000, within weeks Wobble Works had raised $2,344,134. This begs the question: How long will it be before we see an upscaled version of the 3D Doodler

(a)

(c)

(b)

FIGURE 9-7   Hylozoic Flask: a closeup of Dr. Rachel Armstrong's protocell architecture within architect Philip Beesley's Hylozoic Ground installation at the Venice Architecture Biennale 2010.

pen, possibly mounted on a quadcopter drone that can sketch buildings in such fashion as Oxman observes in a spider building its web?

In his TED 2011 talk, Skylar Tibbits, head of the MIT Self-Assembly Lab asked, "Can we make things that make themselves?" Believing that self-assembly of large-scale structures is "not just for nature" and "there's something superinteresting about natural systems," Tibbits, together with his MIT colleagues, has demonstrated robotic parts encoded with blueprints of that which they intended to build, which self-organized into those structures. Working with 3D printer maker Stratasys and software firm Autodesk, Tibbits began developing a four-dimensional printing method to produce such self-assembling technologies. Cedric Price's vision for fourdimensional buildings is fast becoming a reality. The question therein is 'how might we use such assemblages within the built environment'? Might we use them to create

FIGURE 9-8   Banyan tree simulations by Biornametics, Dr. Barbara Imhof and Dr. Petra Gruber.

structures that mimic nature in other ways such as the banyan tree, which mediates a host structure—be that a tree, a building, or a bridge, for structural support, as explored by Dr. Barbara Imhof and Dr. Petra Gruber of research group Biornametics (Fig. 9-8). Or, might we create buildings that, like those of Archigram's Walking City, can migrate from place to place, an idea also explored by American architect Glenn Small. In 1965, citing nature as "the ultimate technology," Small began to envision a city in step with "the principles that nature sets up." A skilled illustrator, Small reimagined Detroit, then Los Angeles as "biomorphic biospheres"; huge, life-like ecosystems, which could both expand and retract as their populations required. Embedding sustainability to the core, Small described his visions as an "alternative to the mediocrity [that] traps pollutants and uses direct natural energy."

"As soon as a living creature ceases to absorb energy, it must submit to the law of entropy and disintegrate."

—Helmut Tributsch

Historically, one of the greatest challenges that any city can face is that of a major natural hazard. Given life, in all its many shapes and forms, has found novel ways to build resilience to extreme meteorological and geological events, might it be possible for a city to increase its general level of resilience by mimicking such strategies? Positing the potential of the city as a complex biomimetic adaptive system, my own research project, "Bionic City," underway since January 2010, explores this possibility. Asking the questions "how would nature design a city?" and "how might we apply leading-edge science and technology to that end?" has led me to scenarios that would not seem out place in a science fiction film set far in the future. However, developments such as those discussed in this chapter make these scenarios, extreme though they may be, increasingly plausible. Three such scenarios are below.

Seasonal City

A city in-step with the seasons, as does a deciduous forest biome, the Bionic City anticipates and adapts to each season. The city shape shifts from one month to another, as its buildings and structures adopt whichever assemblage is best suited to the environmental conditions of the moment. Chromatophoric (color-changing) surfaces enable the city to adjust its albedo effect at a city, district, and even building scale. In springtime, birds, bees, and drones fly about the city, the latter installing new features to buildings and infrastructure. Practical advantages of such features include increased resilience to extreme weather events, including heavy precipitation, high winds, and heat waves, as well as increased capacity to harvest natural resources, including solar, wind, hydro and kinetic energy, sunlight, and rainwater. Novel aspects of such seasonal adaptivity would include greater visual contrast throughout the year with the possibility of seasonal spectacles, such as buildings that bloom in springtime and that, like deciduous trees, form a riot of autumnal color. There is poetry and romance to this city, which like a forest enriches the senses—visual, audio, gustatory, olfactory, somesthetic—with sensations that enhance the human spirit.

Seeded City

DNA like information, including the material composition, structural design, and operational functionality, of items as small as pieces of jewellery and as large as buildings is stored in cloudbased data systems, which are backedup in multiple locations, both locally and remotely. In the event of a natural hazard, such as a wildfire, flood, landslide, earthquake, or even a volcanic eruption, a process of renewal initiates, wherein the Bionic City is 'sown' from the cloud. Dandelionlike seeds are released into the air to float until such time as onboard nanosensors and processors identify a suitable location in which to 'grow'. Like their biological counterparts, these seeds are embedded with the blueprints of the items they shall produce. Destroyed items become the raw materials, or to use another term 'the nutrients' that will make anew. Some seeds, such as those that would produce items of sentimental value, are encoded to reassemble 'as was'. Others, namely those of which the design could be improved, would 'evolve', such that the seeded item was in some way better than its predecessor. Interventions that would enable citizens to curate the regrowth process could include the necessity to add a substance, such as water, that enabled the seed to 'germinate'. Natural processes that might inspire a seeded city include serotiny, in which seeds are released in response to an environmental change, such as when the Banksia serrata opens its follicles to release seeds in response to a wildfire.

Successional City

A city organised around the principles of Panarchy, as defined by Lance Gunderson and C. S. Holling; it's a city with a lifecycle that revolves around growth, conservation, release and reorganization. In the aftermath of a natural hazard, 'pioneering' infrastructure and buildings lay the foundations for the introduction of complexity, which increases to the point of stability, until the next natural hazard unfolds. This ongoing cycle allows for evolution, wherein natural hazards act as catalysts for change. Just as natural hazards upend stable communities in ecosystems, such as those formed by phenomenon including allelopathy, so too do they upend stable communities in the Bionic City. This process allows for reorganization of the city, such that its infrastructure is continuously adapted to changing environmental and social conditions. Daytoday, further processes enable this city to constantly 'evolve', for example, buildings that shed materials and/or parts, using processes that mimic abscission in plants, thereon regrow those features at a later date.

The city of the past was, to quote my AVATAR colleague Rachel Armstrong, built of "inert materials that are belligerent to a changing environment." The city of the future will, I believe, be built of smart, biomimetic materials engineered from the nanoscale upward. These materials will facilitate urban fabrics with life-like qualities—with metabolism, the ability to maintain homeostasis, to respond to stimuli, to grow, to reproduce, and to adapt to their environment. Dubbed a "biomimetic imaginarium," the Bionic City is not so much a design blueprint, as a set of systems and processes migrated from the natural to man-made world—an "urbanature" hybrid.

Our Ever-Evolving Future: Possible Biomimetic Futures

"The only way of finding the limits of the possible is by going beyond them into the impossible."

—Arthur C. Clarke

While the future is a set of possibilities, not a set of certainties, for all the challenges humanity now faces, it appears there is a great deal to be excited about. The dots that are joining at this time are enabling humanity to see things on a bigger scale, but in more detail than ever before. We have the technology—the technology to make things happen that were impossible just a short while ago. Visionaries such as Hugh Herr, Miguel Nicolelis, and Max Ortiz Catalan are developing medical bionics with the potential to radically improve the quality of life for many millions of people. There is a very real possibility that in so little as 20 years, some of the most debilitating aspects of aging will be mitigated through the ilk of exoskeletons and brain-to-machine devices. Presently, an aging population is commonly framed as a growing societal problem, wherein those of retirement age and upward are labeled as an economic burden. Medical bionics is one of the technologies with potential to challenge that assumption. Despite the many warnings that by 2030, obesity could be at pandemic levels, as a consequence of factors including inactive lifestyles, we may well find that, to the contrary, within a decade or two thanks to factors including medical bionics, physical activity levels have increased and consequently population health is generally better, not worse than at present.

Biomimetic business models that adopt closed-loop systems, which reduce resource consumption yet enhance design performance, could potentially go a long way to addressing humanity's current resource issues. Information communications technologies within our cities, such as the Urban Operating System, have potential to improve resource efficiency yet further. Smart materials, such as those Skylar Tibbits, Neri Oxman, Rachel Armstrong, and Neil Spiller are developing, not only present the possibility of greater resource efficiency, but of materials that give back to the environment. Since the mid 1990s, interest in biomimetics has grown exponentially. A couple of decades ago, under 100 biomimetic papers were published a year; by 2005 that figure had grown to over 1000 and by 2012 was around 3000 a year. New science and new technologies, such as those discussed in this chapter, will doubtless present some downsides, too; the Catch 22 of innovation is that "you can't have it all," as technology comes with unintended consequences, not all of which are pleasant. However, had our ancestors abandoned fire because of the threat of arsonists, we would most surely be the poorer for it.

Bibliography

Alemseged, Z. http://www.ncbi.nlm.nih.gov/pubmed?term=Alemseged%20Z%5BAuthor%5D&cauthor=true&cauthor_uid=20703305.

Armstrong, A., Clear, N., and Spiller, N. Future Cities: Design as Research Conference, AVATAR, School of Architecture, Design and Construction, University of Greenwich, April 19–20, 2012.

Armstrong, R. "How Protocells Can Make Stuff Much More Interesting," *Architectural Design*, 81 (2), March/April, 2011.

Armstrong, R. "Living Buildings: Plectic Systems Architecture," *Technetic Arts: A Journal of Speculative Research*, 7 (2), Intellect, November 2009. DOI: 10.1386/tear.7.2.79/1.

Béarat, H.A. "Evidence for Stone-Tool-Assisted Consumption of Animal Tissues before 3.39 Million Years Ago at Dikika, Ethiopia." *Nature*, 466 (7308): 857, 2010. DOI: dx.doi.org/10.1038/nature0924810.1038/nature09248. http://www.ncbi.nlm.nih.gov/pubmed?term=B%C3%A9arat%20HA%5BAuthor%5D&cauthor=true&cauthor_uid=20703305.

Bejder, L. http://www.pnas.org/search?author1=Lars+Bejder&sortspec=date&submit=Submit.

Bell, E. ed. *Archigram*. New York, NY: Princeton Architectural Press, 212.995.9620, 1999.

Bentley-Condit, V.K. and Smith, E.O. "Animal Tool Use: Current Definitions and an Updated Comprehensive Catalog." *Behavior*, 147: 185–221, Koninklijke Brill NV, Leiden, 2010. DOI: 10.1163/000579509X12512865686555.

Bionic Brain Symbionts: The Next Page of Human Slurry, Future in Biotech, Episode 82, TWiT Netcast Network, June 24, 2011.

Birbaumer, N., Kubler, A., Ghanayim, N., Hinterberger, T., Perelmouter, J., Kaiser, J., Iversen, I., Kotchoubey, B., Neumann, N., and Flor, H. "The Thought Translation Device (TTD) for Completely Paralyzed Patients." *IEEE Transactions on Rehabilitation Engineering*, 8 (2), June 2000.

Bobe, R. http://www.ncbi.nlm.nih.gov/pubmed?term=Bobe%20R%5BAuthor%5D&cauthor=true&cauthor_uid=20703305.

Bradshaw Foundation. http://www.bradshawfoundation.com/chauvet.

Brown, M. "Samsung Cites *2001: A Space Odyssey* as Prior Art in iPad Patent Battle." *Wired*, August 24, 2011.

Cedric Price, Design Museum. http://designmuseum.org/design/cedric-price.

Chinese man spends eight years building himself bionic hands after DIY accident, Mail Online, August 14, 2012.

Cipriani, C., Controzzi, M., and Carrozza, M.C. "The SmartHand Transradial Prosthesis." *Journal of Neuroengineering and Rehabilitation*, 8: 29, May 22, 2011. DOI: 10.1186/1743-0003-8-29.

Clark, L. "Bionic Hand That Lets Amputees 'Feel' to Be Trialed This Year." *Wired*, February 18, 2013.

Connor, R.C. http://www.pnas.org/search?author1=Richard+C.+Connor&sortspec=date&submit=Submit.

Constrvct. http://constrvct.com.

Crisafulli, C.M. and Hawkins, C.P. Ecosystem Recovery Following a Catastrophic Disturbance: Lessons Learned from Mount St. Helens, Natural Processes, Status and Trends of the Nation's Biological Resources, 1, (n.d).

Cronin, L. "Defining New Architectural Design Principles with Living Inorganic Materials, Protocell Architecture." *Architectural Design*, 81 (2), March/April 2011.

Dale, V.H., Swanson, F.J., and Crisafulli, C.M. *Ecological Response to the 1980 Eruption of Mount St. Helens*. Springer Science + Business Media, 2005.

Darwin, C. *The Descent of Man, and Selection in Relation to Sex, Volume 2*. London, UK: John Murray, 1871.

Dickens, E. *The da Vinci Notebooks*. London, UK: Profile Books, 2005.

Digital Files and 3D Printing in the Renaissance? Smithsonian.com. blogs.smithsonianmag.com/design/2013/03/digital-files-and-3d-printing-in-the-renaissance/March 1, 2013.

Ekso Bionics. http://www.eksobionics.com/January 2013.

Electronic Countermeasures, Exhibited at Hack the City, Science Museum, London, June 22–September 8, 2012.

Extracts from *Living Arts Magazine*, 2, June 1963.

Fagan, W.F., Bishop, J.G., and Schade, J.D. "Spatially Structured Herbivory and Primary Succession at Mount St. Helens: Field Surveys and Experimental Growth Studies Suggest a Role for Nutrients." *Ecological Entomology*, 29: 398–409, 2004.

Filobot: Plastic Filament Maker. http://www.kickstarter.com/projects/rocknail/filabot-plastic-filament-maker.

Finn, J.K., Tregenza, T., and Norman, M.D. "Defensive Tool Use in a Coconut-Carrying Octopus." *Current Biology*, 19 (23): R1069–R1070, December 15, 2009. DOI: 10.1016/j.cub.2009.10.052.

Foerder, P., Galloway, M., Barthel, T., Moore, D.E. III., and Reiss, D. "Insightful Problem Solving in an Asian Elephant." *PLoS ONE*, 6(8): e23251, 2011. DOI: 10.1371/journal.pone.0023251.

Frazer, J. Electronic Version of an Evolutionary Architecture, January 1999. http://www.aaschool.ac.uk/publications/ea/intro.html.

Fuller, R. Buckminster, "Operating Manual for Spaceship Earth." Lars Muller Publishers. Zurich, Switzerland, 1969.

Garman, J. Geek Trivia: Strange (Water) Bedfellows, TechRepublic.com, August 1, 2006.

Gebeshuber, I.C., Gruber, P., and Drack, M. "A Gaze into the Crystal Ball: Biomimetics in the Year 2059." *Mechanical Engineering Science*, August 27, 2009. DOI: 10.1243/09544062JMES1563.

Geraads, D. http://www.ncbi.nlm.nih.gov/pubmed?term=Geraads%20D%5BAuthor%5D&cauthor=true&cauthor_uid=20703305.

Glenn Small Envisions the Biomorphic Biosphere Replacing LA, (n.d). http://www.artofthefuture.com.

Global Mobile Statistics. Part A: Mobile Subscribers; Handset Market Share; Mobile Operators, MobiThinking.com, December 2012.

Goyal, A. "Bioelectric Neural Circuitry, Computational Biology Group, Oxford University Computing Laboratory, Science and Spiritual Quest." *Proceedings of the 3rd All India Students' Conference*, Andhra Pradesh, India, December 2007.

Graham, S. Non-Plan, Citymovement.wordpress.com, April 11, 2012.

Gruber, G. and Imhof, B. "Science to Architecture: Design by Research." *International Bionic Engineering Conference*, Boston, September 18–20, 2011.

Gunderson, L. H, and Holling, C. S. Panarchy: Understanding Transformations in Human and Natural Systems, Island Press. 2001.

Heithaus, M.R. http://www.pnas.org/search?author1=Michael+R.+Heithaus&sortspec=date&submit=Submit.

Herr, H. "Exoskeletons and Orthoses: Classification, Design Challenges and Future Directions." *Journal of Neuroengineering and Rehabilitation*, 6: 21, June 18, 2009. DOI: 10.1186/1743-0003-6-21.

Herr, H. The World We Dream, Zeigeist Americas, 2012.

Herr, H. Extreme Bionics, Patterns That Connect, DLD (Digital Life Design), Munich, January 21, 2013.

Hickey, J.T. and Salas, J.D. Environmental Effects of Extreme Floods, U.S.–Italy Research Workshop on the Hydrometeorology, Impacts, and Management of Extreme Floods, Perugia, Italy, November 1995.

Holmes, K. Talking to the Future Humans—Rachel Armstrong, Vice Magazine, Vice.com (n.d).

ICT Facts and Figures, The World in 2011, ICT Data and Statistics Division, Telecommunication Development Bureau, International Telecommunication Union, 2011.

Information Resource Center. http://www.MountSt.Helens.com.

Inventech. http://www.invetech.com.au/portfolio/life-sciences/3d-bioprinter-world-first-print-human-tissue.

Krützen, M. http://www.pnas.org/search?author1=Michael+Kr%C3%BCtzen&sortspec=date&submit=Submit.

Kunicki, C. http://www.nature.com/srep/2013/130228/srep01319/full/srep01319.html \l auth-4.

Lakin, J.L., Jefferis, V.E., Cheng, C.M., and Chartrand, T.L. "The Chameleon Effect as Social Glue: Evidence for the Evolutionary Significance of Nonconscious Mimicry." *Journal of Nonverbal Behavior*, 27 (3), Fall 2003.

Lamont, T. "John Maeda: Innovation Is Born When Art Meets Science." *The Observer*, November 14, 2010.

Lebedev, M. http://www.nature.com/srep/2013/130228/srep01319/full/srep01319.html \l auth-3.

Lepora, N.F., Verschure, P., and Prescott, T.J. "The State of the Art in Biomimetics." *Biomimetics and Bioinspiration*, January 9, 2013. DOI: 10.1088/1748-3182/8/1/013001.

Lewis, S.J. Living PlanIT's City of Things Summit, M2M Explained, Greenwich, London, June 12, 2012.

Lewis, S.J. Living PlanIT/Deutsche Telekom M2M Partner Event, Budapest, September 26, 2012.

Lewis, S.J. Smart Infrastructure and the Digital Economy, ICT Summit, British Business Embassy, August 3, 2012.

Lillie, B. Writing a TEDTalk from the Future: Q&A with Damon Lindelof, TED Blog, blog.ted.com/2012/02/28/writing-a-tedtalk-from-the-future-q-a-with-damon-lindelof/, February 28, 2912.

Lindelof, D. We Are the Gods Now, TED Talks, http://www.imdb.com/video/imdb/vi2390728985, February 2012.

Living PlanIT. living-planit.com/what_is_living_planit.htm.

Living PlanIT. Urban Operating System (UOS™): Introduction to the Living PlanIT UOS™ Architecture, Open Standards and Protocols, Living PlanIT SA, 2012.

Mann, J. http://www.pnas.org/search?author1=Janet+Mann&sortspec=date&submit=Submit.

Marean, C.W. http://www.ncbi.nlm.nih.gov/pubmed?term=Marean%20CW%5BAuthor%5D&cauthor=true&cauthor_uid=20703305.

McPherron, S.P. http://www.ncbi.nlm.nih.gov/pubmed?term=McPherron%20SP%5BAuthor%5D&cauthor=true&cauthor_uid=20703305.

Miller, G. The Wildly Ambitious Quest to Build a Mind-Controlled Exoskeleton by 2014, Wired.com, February 2, 2013.

MIT Media Lab. http://www.media.mit.edu.

Negroponte, N. "Toward a Theory of Architecture Machines." *Journal of Architectural Education*, 23 (2), March 1969.

Neuroprosthetic Research Group. http://www.bme.miami.edu/nrg/.

Nicolelis, M.A.L. Brain-to-Brain Interface for Real-Time Sharing of Sensorimotor Information, Scientific Reports 3, Article number 1319, February 28, 2013. DOI:10.1038/srep01319.

Nicolelis, M. A Monkey That Controls a Robot with Its Thoughts. No, Really, TEDMED, http://www.ted.com/talks/miguel_nicolelis_a_monkey_that_controls_a_robot_with_its_thoughts_no_really.html, April 2012.

Oxman, N. "Architecture's Primordial Soup and the Quest for Units of Synthetic Life." *Architectural Design*, 81 (2), March/April 2011.

Pais-Vieira, M. http://www.nature.com/srep/2013/130228/srep01319/full/srep01319.html\l auth-2.

Peacock, H. "Cities in Competition: Branding the Smart City, Urban Environmentalism." *Urban Times*, July 14, 2011.

Phillips, A. Back to the Futurist: Liam Young, URBNFUTR. urbnfutr.theurbn.com, February 13, 2012.

Problem: Our "Take, Make, Waste" Economy (n.d). http://www.cradle2.org/2012/04/problem-our-unsustainable-economy/.

Raspberry Pi. http://www.raspberrypi.org/.

Reed, D. http://www.ncbi.nlm.nih.gov/pubmed?term=Reed%20D%5BAuthor%5D&cauthor=true&cauthor_uid=20703305.

Ridley, M. *When Ideas Have Sex*. Oxford, UK: TEDGlobal, 2010.

Ross, M.D., Menzler, S., Zimmermann, E., and Lett, B. "Rapid Facial Mimicry in Orangutan Play." 4 (1): 27–30, February 23, 2008;. Published online December 11, 2007. DOI: 10.1098/rsbl.2007.0535.

Sagan, C. Journeys in Space and Time, Cosmos: A Personal Voyage, Episode 8, November 16, 1980.

Sample, I. "Brain-to-Brain Interface Lets Rats Share Information via Internet." *The Guardian*, March 1, 2013. http://www.nature.com/srep/2013/130228/srep01319/full/srep01319.html\l auth-1.

Sherwin, W.B. "Cultural Transmission of Tool Use in Bottlenose Dolphins." *Proceedings of the National Academy of Sciences of the United States of America*, 102 (25), June 21, 2005. DOI: 10.1073/pnas.0500232102.

Silva, J. We Are the Gods Now, Festival of Dangerous Ideas, Sydney Opera House. http://www.youtube.com/watch?v=PjpC6GmeLGI, September 2012.

Sofge, E. Smart Bionic Limbs Are Reengineering the Human, Popular Mechanics. http://www.popularmechanics.com/May 28, 2012.

Sterry, M. *2020: A Brand Odyssey: The Brands of the Future*. Moscow, Russia: Aegis Media Autumn Session.

Sterry, M. Biomimicry in Design: Sourcing Sustainable Design Solutions for the 21st Century, Greengaged at the Design Council, London Design Week, September 23, 2009.

Sterry, M. "Building a Bionic City: Sci Fact or Sci Fi?" CIOB CPD Lecture, the THINKlab, University of Salford, February 9, 2012.

Sterry, M. "Building a Bionic City: The Place Where Science Fiction becomes Fact." MAA Spring Lecture, Institute for Advanced Architecture of Catalonia, Barcelona, April 23, 2012.

Sterry, M. Creating Resilient Cities in-Step with the Seasons, ThisBigCity.net, October 31, 2011.

Sterry, M. *Future City Scenarios*. Keynote Presentation, Forum for the Built Environment, Liverpool, UK, June 22, 2011.

Sterry, M. *Journey to the Center of Biomimicry*. Keynote World Congress on Sustainable Technologies, November 7, 2011.

Sterry, M. The Bionic City, EcoStruxure Solutions for Green Building, Schneider Electric Energy Efficiency Theatre, European Future Energy Forum, October 21, 2010.

Sterry, M. *The Bionic City: A Model for the Resilient Metropolis of the Future*? Keynote Presentation, International Bionic Engineering Conference, Boston, USA, September 19, 2011.

Super/Collider. http://www.super-collider.com.

Tate, R. 4D Printing Sees Materials Form Themselves into Anything." *Wired*, February 2013.

Technology Will Save Us. shop.technologywillsaveus.org.

The Infinite Variety, Life on Earth: A Natural History by David Attenborough, BBC in association with Warner Bros and Reiner Moritz Productions, transmitted January 16, 1979, UK.

The Next List: Neri Oxman, The Woman Who Wants to Print Buildings, CNN, whatsnext.blogs.cnn.com, December 4, 2012.

Thought-Controlled Prosthesis is Changing the Lives of Amputees, Press release, University of Chalmers website, November 28, 2012.

Tibbits, S. Can We Make Things That Make Themselves? TED2011. http://www.ted.com.

Torgovnick, K. The Interspecies Internet: Diana Reiss, Peter Gabriel, Neil Gershenfeld, and Vint Cert at TED 2013, TED Blog, February 28, 2013.

Tributsch, H. *How Life Learned to Live, Adaption in Nature*. Cambridge, MA: The MIT Press, 1982.

van der Jeijden, M.G.A. and Sanders, I. R. *Mycorrhizal Ecology, Ecological Studies*. Springer, p. 157, 2002.

Vasari, G. Le Vite de' più eccellenti pittori, scultori, e architettori, 1550, Torrentino, Italy.

Vincent, J.V., Bogatyreva, O.A., Bogatyrev, N.R., Bowyer, A., and Pahl, A-K. "Biomimetics: Its Practice and Theory." *Journal of the Royal Society Interface*, 3: 471–482, 2006. DOI: 10.1098/rsif.2006.0127, April 18, 2006.

Wang, J. http://www.nature.com/srep/2013/130228/srep01319/full/srep01319.html\l auth-5.

Wynn, J.G. http://www.ncbi.nlm.nih.gov/pubmed?term=Wynn%20JG%5BAuthor%5D&cauthor=true&cauthor_uid=20703305.

http://www.pnas.org/search?author1=William+B.+Sherwin&sortspec=date&submit=Submit.

CHAPTER **10**

Innovation Benchmarking

Klaus Solberg Søilen

Introduction

Innovation benchmarking is not just another benchmarking exercise, but primarily another "approach" to benchmarking. In innovation benchmarking we want to identify the factors behind the benchmark's success. These are factors that are not always measurable, so we need to go by some other variables first. Afterward we want to try to adapt and apply these factors to our own company.[1] In other words, what makes innovation benchmarking difficult is that there are no factors or *Key Performance Indicators* (KPIs) that specifically call out innovation and can be measured as such. Instead we need to identify what aspects belong to the innovation process from the indicators we can define and measure. Examples may include the number of new ideas, products, and services introduced in a company or organization, the size of research & development (R&D) staff, R&D budget, process improvements, or the amount of business investments. Or the KPIs may be more complex like leadership style and corporate culture, in which case they are seldom quantifiable, thus difficult to compare.

The goal of innovation benchmarking is the same as for other forms of benchmarking, to define and, in the end, achieve superior performance in the organization. But, to "be innovative" is not necessarily the same as "being profitable." Innovation and profitability may even be negatively correlated. At the start of the millennium the Danish construction toy manufacturer Lego did most things right when it comes to innovations. Still the company almost went bankrupt. After having started a number of highly acclaimed product lines, with popular themes such as Star Wars and Harry Potter, the company almost ran out of cash by 2003. Creative and cutting-edge products had been prioritized at the cost of financially sustainable growth.[2]

Innovation benchmarking is not just a concern for private companies, but has come to the attention of the public sector as well; and on all levels, from local government, to regional, and national governments. For the private company, innovation benchmarking is important to survive and prosper. For the state it is important to ensure future and long-term tax revenues, to finance its increasing economic obligations in society. Because of this shared concern, there is also an important ongoing discussion about who should contribute the funds necessary for innovation activities in a state, country, or company? When the United States built its space program after the World War II, most of the resources for R&D came from the government. By the late 1970s, the amount of funds coming from the private sector had surpassed that coming from the state. This trend has continued during the last two decades (see Fig. 10-1).

Total R&D spending in the United States is close to three percent of Gross domestic product (GDP), where more than two-thirds comes from the private sector. Different nations have chosen different directions and different degrees of involvement in this question. For example, in Sweden and the Nordic countries most research funds today come from state institutions and foundations. Applications for research funds are constructed solely on meritocratic principles. No wonder then that Nordic countries are placed among the top performers in *national innovation systems* (NISs), as in the ranking presented later in this chapter.

Innovation has become a special concern for states after the financial crisis. Many low-tech industries have outsourced to Asia and to developing countries, or have been out-competed by foreign companies. As a result the Western world has become more dependent on innovations in its high-tech industries to assure continued economic growth. As the effects of the financial crisis were felt in 2008 and unemployment rates rose to new highs,

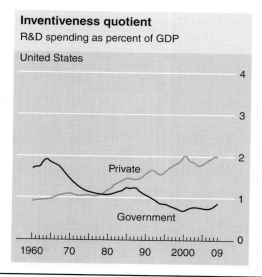

FIGURE 10-1   R&D spending as percent of GDP. (*Source*: National Science Foundation, OECD.[3])

Western governments have become even more preoccupied with the notion of innovation as a way—even as a strategy—out of the crisis. This has also been concluded in research. In its report from 2001, "The New Economy: Beyond the Hype—The OECD Growth Project," the Organisation for Economic Co-operation and Development (OECD) identified four related principal areas driving economic growth: innovation, entrepreneurship, Information and Communications Technology (ICT), and human resources.

It is recognized today, more than at any time in history, that innovation is a significant factor in a nation's economic growth. Thus, innovation has also become interlinked with Adam Smith's economic doctrine of the *competitive advantage of nations*. To gain a competitive advantage it is not simply enough to produce and export more goods; countries should aim for those categories of goods and services that can bring about sustainable growth. To succeed in this, they have to continuously monitor what other countries are producing and selling.

The problem for Western private companies and Western nation states is not that they have been investing less in R&D, but that competition has increased. We see this from statistics for the major Western economics over the past 20 years. R&D as a percentage of GDP has been relatively stable, but the competitive advantage of these countries has diminished (see Fig. 10-2).

According to an article in *The Economist* the companies that spend most on R&D belong to three industries: computing, cars, and drugs. According to Booz, a consultancy firm, the companies that invested most on R&D in 2012 were Microsoft, Toyota, and Roche. Microsoft spent $9 billion on R&D in 2011. They had 850 PhD scientists and 40,000 developers who are mainly preoccupied with projects on cloud computing. Toyota spent ¥730 billion ($9.2 billion)

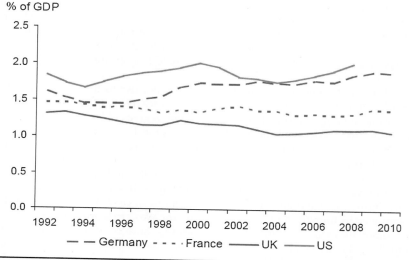

FIGURE 10-2   Research and development investment as a percentage of GDP. (*Source*: Eurostat.)

on R&D. They have 15 research centers, which investigate new materials and fuels for hybrid cars. Roche has over 330,000 patients enrolled in clinical trials and spent SFr8 billion ($9 billion) last year.*

If companies sometimes think they do not have competitors it is because they ignore the facts. Even those companies that do not have many *direct competitors* will still have *indirect competitors*. That is, all companies are, in the end fighting for the same customer money. Often they will be outcompeted by a company they hardly knew existed a few years ago. It happens all the time. In 2012, Ericsson had only one major competitor, Huawei. Ten years earlier it had five major competitors and none of them was Huawei. They found themselves in a major industry with major companies. One can only imagine how things are in the smaller companies, who have fewer resources available to pursue the competition and the market.

It's true that benchmarking takes time away from core activities. Perhaps this is why it is sometimes so unpopular. Alternatively, we can benchmark quickly by checking on a few of our competitors' products, to see how they compare to what we have. According to Munro[4] many engineers see benchmarking as a waste of time: "They're either pressed to get a project out of the door as quickly as possible or looking to create something new and exciting."

Definitions

It is common to differentiate between *innovation* and *invention*, where the latter is often seen as synonymous with a patented idea. Innovations take advantage of, and build on, already existing inventions. In its broader form an innovation may be defined as a new product, a new service, or simply a new way of working which makes an output better, cheaper, more sustainable, or more appealing aesthetically, like a new design for a smartphone. This broader definition is not unproblematic either. If innovation is defined as everything that is new and adds value for a customer, as some scholars argue, then the term *innovation* is synonymous with new value, as in products that create new value. Others think innovations have to be something unique, but views on what that are, or should, differ. Moreover, innovations seem to be linked to our perception of what is unique. For example most people when asked, think that the iPod was a great innovation, even though the product's technology existed long before Apple made their version of it.

Benchmarking is a form of comparison where we sometimes include all major players and define a "best practice," today usually on an international scale as markets have become truly global. As such we may say that benchmarking as a method is the opposite to a *case study*, were we to study one country or company at a time. In benchmarking we have many companies or industries or countries on the X axis and many variables to measure against on the Y axis. The amount of variables we choose to include depends much on the purpose we have for the benchmarking, our skills, and the amount of resources available.

The second part of benchmarking consists of defining and choosing the variables to compare, the so-called KPIs. Kearns[5] presented what later became one of the classical definitions of benchmarking:

> "Benchmarking is the continuous process of measuring products, services, and practices against the toughest competitors or those companies recognized as industry leaders."

Benchmarking has also been described as a practice that promotes *imitation*.[6,7] One can look outside of the organization to perform comparisons of both practices and performances enabling the process of acquiring external explicit/tacit knowledge.[8,9] A benchmark can also be said to be a point of reference from which measurements of any sort may be made (*Webster's Dictionary*). Camp[10] defines it as "the search for industry best practice that lead to superior performance." Other contributions to the discussion about the definition of benchmarking for innovation have been presented by Bessant et al.,[11] Cagliano et al.,[12] and Cox et al.[13] Bessant et al. point out that most definitions emphasize the value of learning from best practices. Cox et al. view benchmarking as a continuous learning activity.

Pages and Toft[14] provide a popular definition of innovation, as "doing things better, faster, cheaper, and greener." Benchmarking would then be how we measure this performance, between companies, industries, nation states, or regions.

There is overlapping research between benchmarking and related areas such as *knowledge management* (KM), asserting the notion that innovation is the knowledge-based perspective of the main source of competitive advantage. Stata[15] and Sinkula et al.[16] suggest

---

*The Economist, "Arrested development America and Europe are relying on private firms in the global R&D race", Aug 25th 2012

that *organizational learning* (OL), especially in *knowledge-intensive industries* (KIIs), not only leads to organizational innovation, but is the only sustainable competitive advantage in the long run.

A number of research articles are about benchmarking between different countries, where China, Korea, Japan, and Taiwan are often compared to Western countries. For example, benchmarking with China has been researched by Garg and Ma,[17] and Li-Hua and Simon.[18] Many of these articles show the speed by which Asian companies are gaining on their Western competitors.

Companies do not need research articles on innovation to know that the phenomenon is important. Innovation has been well understood for some time in the field of economics, but studied under different terms. Previously, companies and researchers have studied the problem from the perspective of R&D, entrepreneurship, product development, and market research, just to give a few examples. This is not to say that research on innovation and innovation benchmarking is of little importance. Much of the research presented during the past years has helped to create a broader understanding for the phenomenon at hand.

Even though there is some research on innovation benchmarking, today there are more public sector, think tank, and consulting reports on the topic. This public sector engagement appears to have been primarily motivated by the economic crisis. With higher unemployment figures there is an ultimate realization that the Western world may have lost much of its competitive advantage to Asian countries. This has been a strong trend which may continue if unabated.

Forms of Benchmarking

We can conclude that the private sector is more concerned with benchmarking than with benchmarking innovations, in part because the latter is more complicated and its results less certain. Most companies start by benchmarking inside of their own company (*internal benchmarking*) and then move on to their competitors (*external benchmarking*). The reason we may want to do this is because internal benchmarking is easier to perform and can still derive great benefits, especially in larger companies where it makes more sense to make comparisons. When we do external benchmarking it does not necessarily have to be against our closest competitors. It may also be with a company in another industry with a view of "best practices" or "world class" within an important function. For example, for more than a decade companies from a wide range of industries have gone to Japan to study just-in-time processes at Toyota.

Today, benchmarking has become a truly global phenomenon. There are even firms which start to benchmark themselves globally as soon as they are created, the so-called *"Born Global."* Examples of such companies are freelancers of IT and graphic designers, many of whom operate over websites like guru.com. These sites are themselves created as a *benchmarking system*, listing suppliers by geographical location, customer satisfaction, price, and competences. Companies, especially the larger ones, are realizing that benchmarking is not something we do a few days before Christmas, but that it is an ongoing process of measuring and making changes.

Benchmarking innovation has become an activity not only for private organizations but for nation states, regions, and municipalities as well. The state is building business incubators to youth entrepreneurship programs. It is also creating science parks. For the state, these benchmarking processes take a much longer time to complete, as we shall see later in the chapter.

For companies, benchmarking is a way to find out how they are doing, and to implement changes to the organization. If they only know how they need to change, much of the work is already done; thus, the idea of "best practices" as a standard to work toward. The emphasis is on studying practices and processes rather than gathering results and bottom-line data to try to surpass these.

There can be said to be three types of benchmarking:

1. *Process benchmarking*, which involves identification of best practices
2. *Strategic benchmarking*, which involves identifying emerging trends
3. *Comparative benchmarking*, which is more result-oriented

We will also notice that there is a difference between innovation benchmarking and *benchmarking innovations*. Benchmarking innovation can be seen as a form of contradiction. If we are doing something completely new—applying an invention in a new way—it means that others are not doing the same thing. Thus there is nothing to benchmark against. As it is new it cannot be compared. Thus benchmarking innovations make little sense as an idea.

	Apple iPhone 5	Samsung Galaxy 3	Nokia Lumia 920
Operating system	iOS 6	Android 4.0	Windows Phone 8
Display	4-in IPS LCD; 1136×640 pixels, 326 ppi	4.8-in HD Super AMOLED; 1280×720 pixels, 306 ppi	4.5-in AMOLED; 1280×768 pixels, 332 ppi
Price	$199.99, $299.99, $399.99	$199.99–$329.99, depending on carrier	Unannounced
Carrier	Sept. 21: AT&T, Sprint, Verizon	Now: AT&T, Sprint, T-Mobile, U.S. Cellular, Verizon	Unannounced
4G LTE	Yes	Yes	Yes
Camera	8-megapixel, 720p front-facing	8-megapixel, 1.9-megapixel front-facing	8.7-megapixel, 1.2-megapixel front-facing
Processor	Proprietary A6 CPU	1.5-GHz dual-core Qualcomm Snapdragon S4	1.5-GHz dual-core Qualcomm Snapdragon S4
Memory	16 GB, 32 GB, 64 GB	16 GB or 32 GB; 2 GB RAM	32 GB; 1 GB RAM
Expandable memory	No	Up to 64 GB	No
Battery	Capacity TBA (talk time up to 8 h on 3G); embedded	2100 mAh, removable	2000 mAh, embedded
NFC	No	Yes	Yes
Weight and thickness	3.95 oz, 0.3 in	4.7 oz, 0.34 in	6.5 oz, 0.42 in
Colors	Black, white	White, blue, red (AT&T); also, globally: black, brown, gray	Black, white, yellow, red, gray

TABLE 10-1   Benchmarking Smartphones (*Source:* CNET News At *http://news.cnet.com/8301 -17938_105-57510802-1/iphone-5-vs-galaxy-s3-vs-lumia-920-by-the -numbers/* [2012-10-05])

Innovation benchmarking, on the other hand, is different and can be understood as how to become or stay innovative.

As already mentioned earlier, there is also a great difference here between reality and perception. Sometimes people perceive something to be new, even though it is not. It may be that it is the packaging that is new, or some new service attached to the product. For more complex products, which offer a series of added values there may be one or two functions which can be described as an innovation. For example, if you are Apple Inc. ready to launch the first iPhone in 2007—a truly new product many would agree due to its touch screen and design—most of the components in the product were the same as used in other phones. The smartphones do basically the same thing, but new functions may persuade customers to see new models as more innovative. When the first iPhone went on sale it took much of the competition by surprise, sending three of them into turbulence: leading to the breakup of Sony-Ericsson, the near collapse of Motorola, and decline of Nokia. Ever since, Apple has benchmarked with the Android system used by Samsung, HTC, LG, and Motorola. Both companies check for variables such as general Central Processing Unit (CPU) performance, processor, battery life, amount of RAM, graphics, and apps. In fact, it's the software that is being benchmarked. Most of these phone companies will share the actual hardware or buy from the same suppliers. The reason BlackBerry lost market share and much of its customer base is primarily due to the company's software (or lack of), especially its lack of apps. When Apple launched its iPhone 5 in 2012, the company still faced stiff competition. Bench-marking figures are being scrutinized not only by producers but also by customers. Many customers complained that the iPhone 5 did not consist of the latest technology, and that companies like Samsung and HTC were more cutting-edge, and more innovative. In this extended perspective where both competitors and customers participate, benchmarking becomes a form of advertising.

According to Michael Stanleigh of the consulting company Business Improvement Architects, Apple's approach to innovation builds on the following key elements:

- Think differently
- Enable a product-oriented culture instead of one driven by technology or money
- Always make a profit to keep making good products
- Hire people who want to make the best things in the world
- Innovation comes from passionate, dedicated people
- Focus on where you think you can make a significant contribution
- Own and control the primary technology in your products

Google, who is also in the smartphone business, gives their employees 20 percent of their time to work on projects they are passionate about. The company also strives to get a prototype up and running as soon as they can, so customers can play around with it for a while and give their feedback before the final product is launched. This has shown to be incremental for the business success of their products. Having an early prototype then has become a KPI for Google, which we can measure in the number of months.

Finding the Variables

In the early days of the study of benchmarking, in the 1970s and 1980s, most people thought that it was all about new technology. This is no longer the case, even though new technology is still an important variable, and often the single most important. Today innovation can also be new services, design, new working processes, new ways of doing things, new markets to enter. If design was not a KPI, Apple would not score as high as it does and we would get the wrong picture of the company's potential performance. Still, going back 10 years there were few people in Sony-Ericsson who took the Cupertino company seriously in the phone industry. Apple was doing PCs and Sony-Ericsson had the better mobile phone technology. Superior technology was the most important thing, or so they thought. Today there is no Sony-Ericsson. Ericsson sold out its part in the mobile phone venture and Sony has moved most of its development back to Japan, where it is planning a comeback.

This raises a series of questions: How do we detect innovation in a competitor, what should we look for? According to Zairi,[19] the most important features reflecting strengths of innovation are

- Market position
- Competence
- Commercial desirability
- Technological feasibility
- Product application
- Creativity
- Product knowledge
- Investigative skills
- Company-wide commitment
- Wide involvement in innovation activities
- Ability to respond
- Continual investments
- Quality
- Training
- Pollution and environmental control
- Engineering expertise
- Energy of management
- Flexibility
- Speed of response

- Market knowledge
- Courage of our convictions
- Staying very close to the customer
- Operating in teams with freedom of expression
- Senior management commitment
- Removing fear in the organization
- TQM program
- Small size/ability to make rapid decisions
- Simple technology
- Idea generation
- Category focus
- Freedom to act
- Collaboration
- Senior management support
- Flair of development team
- Synergistic benefits with existing processes
- Customer consultation and support
- Corporate culture responsive to innovation
- Dedicated innovative resources

- Product base
- Technical strength
- Creative team strength
- International support
- Customer understanding/market knowledge
- Close working relationship between R&D and sales
- Well-trained technicians
- Sensitivity to changes in consumer needs
- Use of strong brands to build product ranges
- Sound R&D base

- Commitment to innovation
- Good marketing skills
- Networking throughout the company to iterate the optimum solution
- Speed
- Expertise and experience
- Company commitment to innovation
- Consumer/customer focus
- Teamwork
- Breakthroughs
- KAIZEN
- Limited priorities focused on strategy

This list may serve as a menu to choose from. No companies will select all the variables in these lists. Most important factors reflecting weaknesses in innovation are

- Right first-time delivery—linkage of brand development right arrow supply chain
- Strategic relevance
- Limited priorities
- External orientation
- International clarity of role
- Difficulty in defining innovation via technical edge
- Cash
- Resources
- Lead times too short
- Success depends on major retailer purchases not necessarily the consumer
- Confined areas of operation
- High cost of marketing required for new brands
- Complexity of relations with other companies in group
- Prioritization of projects
- Level of available resources
- Reacting to changing demands
- Overall company size
- Internal competitiveness
- Weak marketing ability
- Organizational structure
- New product failure rate
- Developments ahead of marketplace
- Replicating pilot processes on a full scale

- Identifying market opportunities
- Maintaining enthusiasm for constant change
- Inward-looking organization
- Scarcity of resources (financial, R&D, human)
- Poor financial incentives to support innovation
- Weak R&D base
- Lack of resources (personnel)
- Lack of knowledge
- Lack of innovation planning procedures
- Too much time spent on current quality problems
- Wide product range
- R&D spread too thinly
- Inadequate market research
- Development of test methods
- Limitations of plant capability
- Lack of systems
- Control of business not on manufacturing site
- Lack of trained personnel
- Manufacturing flexibility
- Forward planning
- Manufacturing response
- Technical resources
- Product development procedures
- Profit margins

In other words, there are numerous factors to consider. The complexity of all of these indicators demand nothing less than a corporate intelligence system. During the past decade, companies have built such information systems and capabilities under the terms *business intelligence* (hardware and software), *market intelligence*, and *competitive intelligence* (managerial practices).

Finding the Metrics and Choosing the Weight

The challenge in benchmarking is to find the right metrics. These metrics should fulfill the following criteria:

- They must be understood by the user
- They must be (easily) available
- They must be the best measures we can find for a given variable we want to measure, and
- They must be comparable, preferably quantifiable

According to Zairi[19] performance measures used in innovation include

- Product flown—definition and measurement
- Measurement of waste products to determine effective yields of manufactured products
- Profitability analysis of project (i.e., Can we do it? Can we afford it?)
- Performance—cost, ratio
- Plant utilization
- New products as a percentage of total turnover
- Number of development projects in key categories
- Measure wins in terms of new business
- Measure time and cost of development + value of each project
- Percentage of sales value by products launched within latest 3-year period

- Speed
- Commercial success (volume from new products)
- Tightly controlled budgets
- Number of innovations introduced to market per annum
- Customer service/cost of waste
- Percentage of NPS in innovation
- Time to market
- Brand share/category share
- Competitive portfolio strength
- Long-term big advertising ideas
- Mixes adopted internationally
- Percentage of on-time launches/relaunches

According to the consulting company *Innovation Point*, one-third of all Fortune 1000 companies have a set of formal innovation metrics in place. The most prevalent metrics include

- Annual R&D budget as a percentage of annual sales
- Number of patents filed in the past year
- Total R&D head count or budget as a percentage of sales
- Number of active projects
- Number of ideas submitted by employees
- Percentage of sales from products introduced in the past "X" year(s)

Companies choose a set or family of metrics for their specific purpose. This is necessary as companies and industries are not alike and there are limits to which metrics are available. It is also a question of time and budget constraints. A benchmarking exercise must be extensive enough to be meaningful, but it should not be so large that it takes more than a year to conduct. Innovation Point defines three categories for a metric portfolio (see Fig. 10-3).

Both input metrics and output metrics are essential for ensuring measures that drive resource allocation and capability building as well as return on investment assessment. The three categories contain the following metrics portfolio:

1. **Return on investment metrics (ROI)**

   ROI metrics address two measures: resource investments and financial returns. ROI metrics give innovation management fiscal discipline and help justify and recognize the value of strategic initiatives, programs, and the overall investment in innovation.

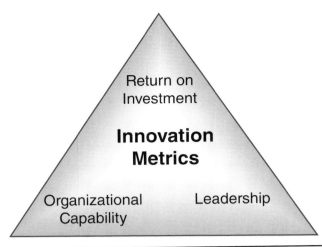

FIGURE 10-3   Three categories for metrics portfolio. (*Source*: Innovation Point.)

2. Organizational capability metrics

Organizational capability metrics focus on the infrastructure and process of innovation. Capability measures provide focus for initiatives geared toward building repeatable and sustainable approaches to invention and reinvention.

3. Leadership metrics

Leadership metrics address the behaviors that senior managers and leaders must exhibit to support a culture of innovation within the organization, including the support of specific growth initiatives.

Within each category there are input metrics and output metrics. Input metrics are the investments, resources, and behaviors that are necessary to drive results (see Fig. 10-4). Output metrics represent the desired results for the metric category.

Return on Investment Metrics	
Input Metrics	**Output Metrics**
• Percent of capital invested in innovation activities such as submitting and reviewing ideas for new products and services and developing ideas through an innovation pipeline • Percentage of "outside" vs. "inside" inputs to the innovation process (open innovation) • Number of new products, services, and businesses launched in new markets in the past year	• Actual vs. targeted breakeven time (BET) • Percent of revenue/profit from products or services introduced in the past X years • Royalty and licensing income from patents/intellectual property

Organizational Capability Metrics	
Input Metrics	**Output Metrics**
• Percent of employees who have received training and tools for innovation - e.g., instruction in estimating market potential of an idea • Existence of formal structures & processes that support innovation • Number of new competencies (distinctive skills and knowledge domains that spawn innovation)	• Number of innovations that significantly advance existing businesses • Number of new-to-company opportunities in new markets

Leadership Metrics	
Input Metrics	**Output Metrics**
• Percent of executives' time spent on strategic innovation versus day-to-day operations • Percent of managers with training in the concepts and tools of innovation • Percent of product/service or strategic innovation projects with assigned executive sponsors	• Number of managers that become leaders of new category businesses

FIGURE 10-4   Drivers of innovation metrics. (*Source*: Innovation Point.)

By dividing the metrics into three variables: return on investment, organizational capability, and leadership it becomes easier to see what needs to be done in essential functions of the organization. Using metrics to drive and assess growth is more than a one-time exercise. As an ongoing tool for innovation management, the approach involves

Planning: Involves key stakeholders in the identification of metrics to ensure the assumptions about the sources of value are explicit and clear, and metrics align to the firm's strategy.

Monitoring: A structured activity of monitoring metrics against goals as a means to gauge progress and define necessary adjustments to measures and strategies.

Learning: A continuous feedback loop that assesses progress, engages key stakeholders in identifying implications, and new opportunities to support the firm's metrics-driven goals.

When we have identified the metrics and want to use a quantitative method we need to allocate a certain weight to each variable. The weight represents the relative importance of each variable. This is not an easy task and we will probably have to go back and forth to alter certain values before we arrive at a correct or representative measure. The danger in this is that we alter the values to certain preconceived ideas or because we were not happy with the end results.

When the goal with the benchmarking exercise is to arrive at the organization's innovative capabilities, Innovation Point suggests the following steps:

1. Clarify enterprise strategic business objectives,
2. Define innovation goals to support growth objectives,
3. Identify required innovation capabilities for the future,
4. Identify desired innovation-related leadership behaviors,
5. Identify organizational processes and models required to drive incremental and strategic innovation,
6. Create a family of metrics that support the enterprise innovation strategy of the company,
7. Create cascading metrics that align business units, divisions, groups, and lateral process capabilities, and
8. Revisit and recalibrate strategies and metrics on an ongoing basis.

It is critical to engage key stakeholders in defining the metrics that will guide the organization into the future. The learning of lessons from these analyses must be captured and applied during the whole process. As such the benchmarking exercise should be seen as an ongoing process where we sow and harvest with certain intervals.

The Process of Benchmarking

Benchmarking is not simply about finding the entities, the variables, and doing the measurements. It is also about how we use the data we have collected. This means that it is about analysis, decision making, implementation, and follow-up.

According to Kaiser Associates there are three steps to a successful benchmarking: selecting key performance drivers or KPIs, selecting companies to benchmark, and allocating resources to the best value-added areas identified. This last stage is defined as a moment of truth. Many companies will know what to do, but fail to act. There may be many reasons for this: internal resistance in the organization, fear of making mistakes, or budget restraints.

Massa and Testa[20] found that benchmarking has an important effect on knowledge management and innovation. It is more than just a copying and imitation strategy. According to the study many firms perceived benefits in improving their innovative and creative capabilities even if these are not traditionally the targets for benchmarking studies. Their findings are suggested by the steps in following model (Fig. 10-5) for how to work with the benchmarking process.

We see how innovation is of importance for the preanalysis, the analysis, and the decision-making phase in the firm. In the preanalysis phase we start with socialization, when we identify the problems the firm is confronted with and draw up some potential performance drivers. During the externalization stage we define performance indicators and perform

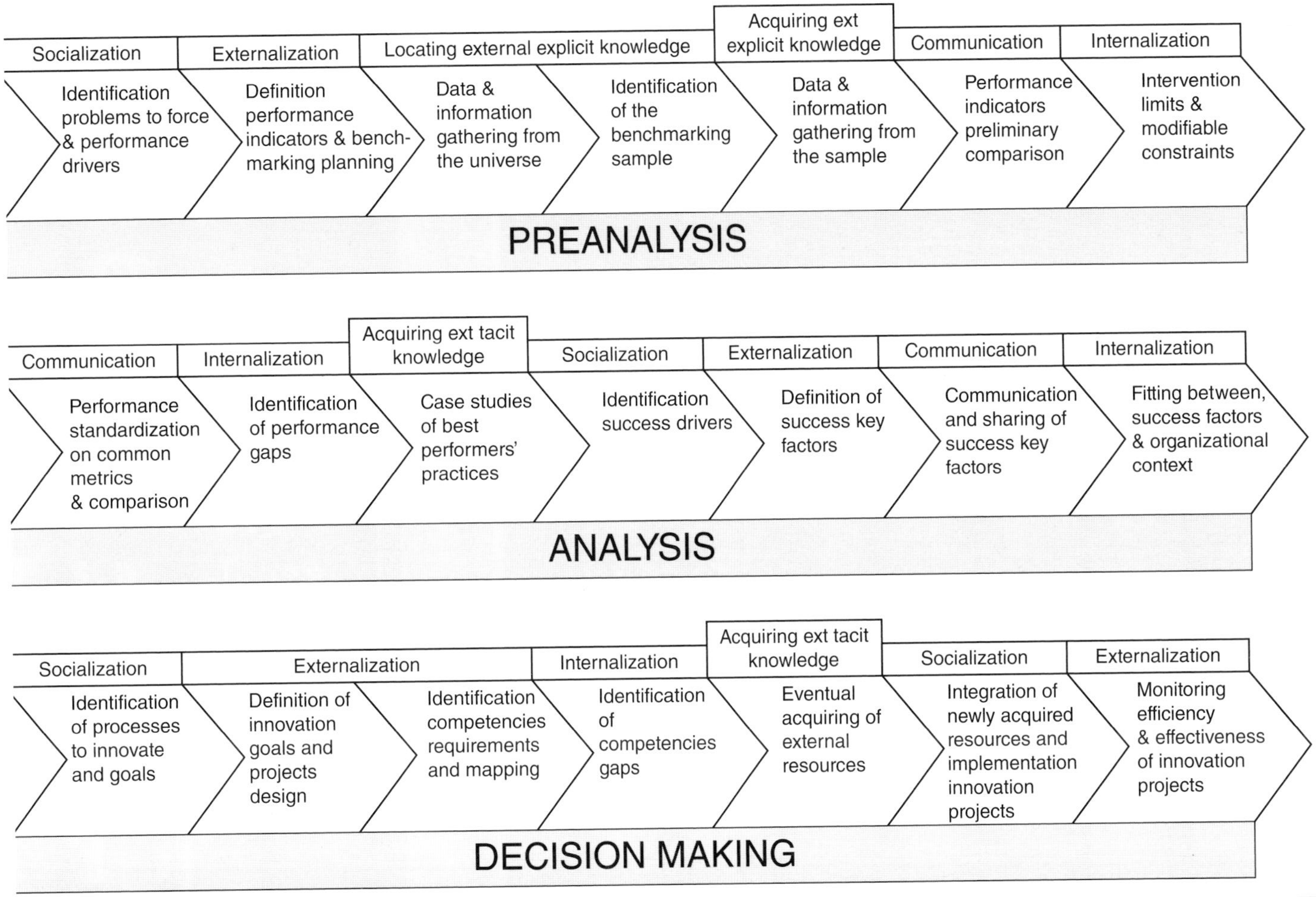

PREANALYSIS

Socialization	Externalization	Locating external explicit knowledge		Acquiring ext explicit knowledge	Communication	Internalization
Identification problems to force & performance drivers	Definition performance indicators & benchmarking planning	Data & information gathering from the universe	Identification of the benchmarking sample	Data & information gathering from the sample	Performance indicators preliminary comparison	Intervention limits & modifiable constraints

ANALYSIS

Communication	Internalization		Acquiring ext tacit knowledge	Socialization	Externalization	Communication	Internalization
Performance standardization on common metrics & comparison	Identification of performance gaps	Case studies of best performers' practices		Identification success drivers	Definition of success key factors	Communication and sharing of success key factors	Fitting between, success factors & organizational context

DECISION MAKING

Socialization	Externalization	Internalization		Acquiring ext tacit knowledge	Socialization	Externalization
Identification of processes to innovate and goals	Definition of innovation goals and projects design	Identification competencies requirements and mapping	Identification of competencies gaps	Eventual acquiring of external resources	Integration of newly acquired resources and implementation innovation projects	Monitoring efficiency & effectiveness of innovation projects

Figure 10-5 The RBP model. (*Source:* Massa, Silvia, p. 616, 2004.)

benchmarking planning. During the next stage we gather secondary data and information from the universe of data. We identify a possible sample size to work with and gather data from this group to compare it with the overall information we gathered first (universe). This should give us some performance indicators for preliminary comparison. During the last stage we fine-tune these indicators.

During the next phase, the analysis, we communicate the performance indicators we have found and are ready to use. In the internalization stage we identify possible performance gaps. Based on the external tacit knowledge gathered we can write a case study of best performance practices. Based on discussions we identify our KPIs or as they are also called, key success drivers (KSDs) or key success factors (KSFs). We communicate the KSFs to the organization and see how well they fit with what we are already doing in the firm.

In the last phase, the decision-making phase, we discuss the processes to actually innovate. We set innovation goals and define projects. We identify the competences required to solve the tasks, which again will reveal our *competences gaps*. Based on this we can start to acquire external resources and implement them in the organization. The last stage is to monitor the efficiency and effectiveness of the innovation project.

What we are looking for at the end is a list of *best practices* for our industry, an explanation of how the best performing companies in our industry succeed. Best practice is also a feature of accredited management standards such as ISO 9000 and ISO 14001. If a company does not want to perform the benchmarking themselves most major consulting companies will offer a form or template. There are also consulting companies who do nothing but benchmarking for one or several industries. They offer readymade reports. J. D. Power & Associates is an example of such a consulting company for the automotive industry.

Innovation Point defines the managed innovation process as a result of six key functions: (1) industry foresight, identifying emerging trends, opportunities; (2) identifying core technologies and competencies in terms of capabilities and assets; (3) drawing in consumer/customer insights; (4) identifying the organizational readiness, in terms of culture, structure, and processes; (5) identifying strategic alignment, readiness to change, and make rapid decisions; and (6) the support for the implementation (see Fig. 10-6).

These functions will then hopefully lead to a number of specific results, of various types. These results may be in the form of new growth strategies, new products and services, new ventures, new markets, new business models, new partnerships, new business practices, and sustainable innovations.

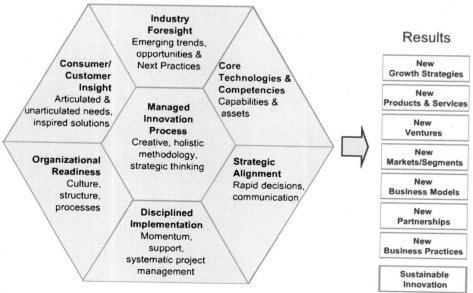

Figure 10-6 Working framework as defined by Innovation Point. (*Source:* Innovation Point.)

Finding the Data (or Benchmarking Intelligence)

How do we gather information and data for benchmarking? To treat this topic requires no less than a separate chapter. Some examples are given in Solberg Søilen,[21] Jenster and Solberg Søilen,[22] and Solberg Søilen.[23]

Most companies will want to start with secondary sources, searching patents, registers, the web, and articles. Examples include the Securities & Exchange Commission, Compustat, commercial websites like reportgallery.com, carol.co.uk, LexisNexis, and hoovers.com. In the next stage we will have to supplement our information set with interviews, both internally and externally, and possibly with market research. It may even be an idea to use a consultant here, as it is easy to be blind about one's own organization. We have worked there for too long, to the extent that we are so used to how things work that we no longer question it. It is therefore a good idea to have someone come in from the outside who can take a fresh look at the organization with a set of outside experiences as a reference point.

Patent statistics are an excellent way to start industry benchmarking[24] and can also be used for NIS. Patents tell us about who worked on what problem, what they found out, and where they tend to commercialize a complete product.*

National Innovation Systems

In the introduction we talked about the interest for innovation benchmarking among nation states. The interest for such studies have only increased in the Western world during the past years, much due to increased global competition, the state's dependence on its tax base for upholding public services, and the recent financial and economic crisis.

One example of an NIS is the Global Innovation Index (see Fig. 10-7), copublished and produced by INSEAD and the World Intellectual Property Organization (WIPO, a specialized agency of the United Nations). The index tries to say something about how innovative each country is.

From the ranking we see that Switzerland is number one. Other major contenders consist of Nordic countries, smaller states inhabited by a Chinese population, northern European states, and North American states. From this we can identify two leading *innovative cultures*, on one side White Caucasians and on the other Han Chinese. For a deeper analysis, see for example Solberg Søilen.[25]

The innovation metrics for nation states in the survey are presented in Fig. 10-8. For input we have institutions, human capital and research, infrastructure, market sophistication, and business sophistication. Just by choosing these variables we can already know which countries will place higher up in the ranking. For example, a country like India is producing a significant number of science graduates each year, but has systematic problems with their infrastructure. Thus they will score lower in the ranking.

The result of the analysis lies mostly in the selection of the metrics and the relative weight allocated to each. Many regions choose too few metrics, like education, infrastructure, and unemployment to draw conclusions on how innovative they are. An annual index of Silicon Valley places emphasis on local energy use, the cost of housing, and other quality-of-life measures. It is found that these are critical to the region's innovation capacity, because they affect the region's ability to attract and retain talent.

This raises a general question about the selection of metrics. For example, it may be difficult to get a representative picture of reality without including numerous macro factors; which may also include new laws, social trends, macroeconomic decisions, political elections, and ecological concerns. If we include these, we risk getting too many metrics, and we risk getting metrics that are difficult to measure like infrastructure or ecological awareness. It is easy to find a number of metrics to measure, but difficult to decide when this pool of metrics can be said to represent an organization's or a community's innovative ability. The answer can only come through thorough discussions and evaluations.

The problem when you think you have found the best metrics is to know how to weigh them. In other words, what is more important in our example earlier, institutions or human capital and research, and how do we weigh them? What percentage do we allocate to each metric and why? This is when benchmarking becomes a real methodological challenge. Very rarely do reports on innovation benchmarking show their methodology in detail and it is almost unheard of to see documentation of the process leading up to it.

---

*The OECD published a manual on how to analyze patents: OECD (ed.) (1994).

Country/Economy	Score (0–100)	Rank	Income	Rank	Region	Rank
Switzerland	68.2	1	HI	1	EUR	1
Sweden	64.8	2	HI	2	EUR	2
Singapore	63.5	3	HI	3	SEAO	1
Finland	61.8	4	HI	4	EUR	3
United Kingdom	61.2	5	HI	5	EUR	4
Netherlands	60.5	6	HI	6	EUR	5
Denmark	59.9	7	HI	7	EUR	6
Hong Kong (China)	58.7	8	HI	8	SEAO	2
Ireland	58.7	9	HI	9	EUR	7
United States of America	57.7	10	HI	10	NAC	1
Luxembourg	57.7	11	HI	11	EUR	8
Canada	56.9	12	HI	12	NAC	2
New Zealand	56.6	13	HI	13	SEAO	3
Norway	56.4	14	HI	14	EUR	9
Germany	56.2	15	HI	15	EUR	10
Malta	56.1	16	HI	16	EUR	11
Israel	56.0	17	HI	17	NAWA	1
Iceland	55.7	18	HI	18	EUR	12
Estonia	55.3	19	HI	19	EUR	13
Belgium	54.3	20	HI	20	EUR	14

FIGURE 10-7 Global Innovation Index ranking. (*Source:* INSEAD and the World Intellectual Property Organization.)

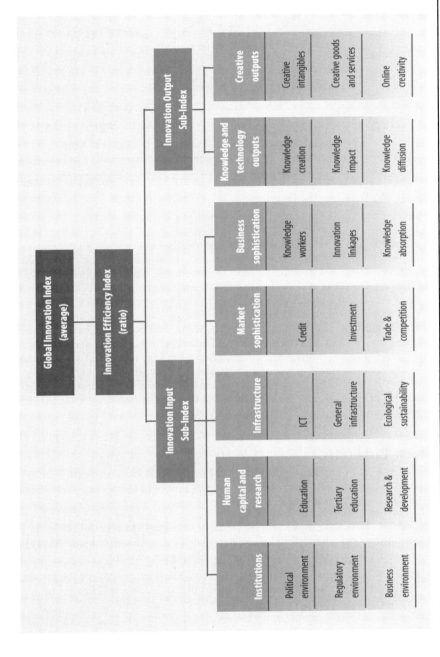

FIGURE 10-8 Framework of the Global Innovation Index 2012. (*Source:* INSEAD and the World Intellectual Property Organization.)

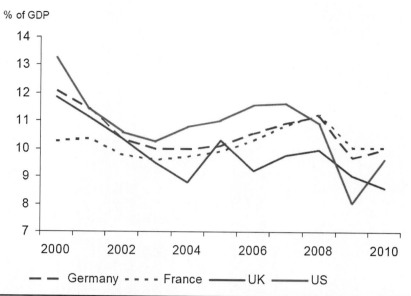

% of GDP

FIGURE 10-9   Business investment as a percentage of GDP.* (*Source*: Eurostat and US Bureau of Economic Analysis)

"The Kay Review of Equity Markets" is a report that attempts to give a measure of the degree of innovation among UK companies. There are no answers as to how much companies should invest, but it is possible to see how much they actually do invest. As a share of national income, business investment in the United Kingdom has declined over the past decade (see Fig. 10-1). Unlike in the United States, Germany, and France, investments have not rebounded in the United Kingdom after the crisis of 2008. Since the turn of the century a large part of England's industry has been shut down or sold out to foreign companies, much of which has gone to India. This trend of deindustrialization can be compared with the result of the Global Innovation Index, where the United Kingdom comes in fourth. If the United Kingdom is among the most innovative countries in the world, as is suggested by the first report presented earlier, this does not show clearly in terms of business investments, economic growth, or employment, as we can see in this report. Similar contradictions can be found for other countries as well (see Fig. 10-9).

Benchmarking Challenges

What the earlier-mentioned example shows is the difficulty of benchmarking innovations. To explain this, Jones and Kaluarachchi[26] point to a number of challenges with benchmarking. In their research the following critical points are identified:

- The benchmarking system must be flexible enough to accommodate the varying aspirations of different stakeholders while being robust enough to provide real insights into the underlying factors that affect performance.

- When benchmarking innovation one should always anticipate problems and ensure that feedback mechanisms are developed that do not appear to apportion blame to any party but seek to provide a balanced view of the issues.

- In circumstances where processes are in a state of rapid change benchmarking may not be the most appropriate tool to drive innovation forward. All too often the benchmarking process will be overtaken by events rendering the outputs irrelevant.

For example, a problem with NIS is that there is no one set of analytical framework applicable to all countries at any one time. The NIS approach is linked to the notion of the knowledge-based economy. A problem when measuring NIS is that nations are dependent on capital and expertise from abroad. For example, it is not possible to detect the innovation

*The Kay Review of UK equity markets and long-term decision making, 2012.

of Korea and Taiwan without considering knowledge sharing, foreign direct investments (FDI), and joint ventures with the United States and Japan.[27] U.S. patents were cited in 44 and 55 percent of the total patents—references of Korea and Taiwan, respectively, based on NBER patent citation data.[28] Japan's share is high in Korea but relatively low in Taiwan, with 34 and 14 percent, respectively. These factors are seldom taken into consideration when the innovation of any one country is measured.

Conclusion

Many companies do not bother with benchmarking because they are doing alright. That idea means we are forgetting that the competitive environment in which we operate is sending major companies into bankruptcy on a regular basis. The list of company failures is long. Just the past few years it includes large well-established firms like ESCADA, Blockbuster Inc., Circuit City Stores, Märklin, and many banks. Ignoring benchmarking also means that we are forsaking the opportunity to become better at what we are doing.

Innovation is not easy to benchmark. There is often also a discrepancy between what companies actually measure and compare and what the more successful companies say they do to stay innovative. Apple and Google are two good examples. On one side it seems to be all about new products and design, but on the other it is just as much, if not more about recruiting the right people and building a strong *corporate culture* based on a passion for the work which unites them.

References

1. Guimaraes, T. and Langley, K. "Developing Innovation Benchmarks: An Empirical Study." *Benchmarking for Quality Management & Technology*, 1 (3): 3–20, 1994.
2. Robertson, D. "Innovating a Turnaround at LEGO." *HBR*, Sept. 4, 2009.
3. OECD (ed.) "The Measurement of Scientific and Technological Activities: Using Patent Data at Science and Technology Indicators." *Patent Manual*. OECD/GD 94: 114, 1994.
4. Munro, A. "Benchmarking: Who Needs It?" *Machine Design*, p. 44, Nov. 3, 2011.
5. Kearns, D.T. "Quality Improvements Begin at the Top." In Bowles, ed. *World*, 20 (5): 21, 1986.
6. Maine, J. "How to Steal the Best Ideas Around." *Fortune*, 146 (8), 102–106, 1992.
7. Schnaars, S. *Managing Imitation Strategies*. New York, NY: The Free Press, 1994.
8. Drew, S. "From Knowledge to Action: The Impact of Benchmarking on Organizational Performance." *Long Range Planning*, 30 (3), 1997.
9. Lucertini, M., Nicolo, F., and Telmon, D. "Integration of Benchmarking and Benchmarking of Integration." *International Journal of Integration Economics*, 38, 1995.
10. Camp, R.C. *Benchmarking: The Search for Industry Best Practices That Leads to Superior Performance*. Milwaukee, MI: ASQC Quality Press, 1989.
11. Bessant, J., Kaplinsky, R., and Morris, M. "Developing Capability through Learning Networks." *International Journal of Technology Management & Sustainable Development*, 2 (1), 2003.
12. Cagliano, R., Voss, C., and Blackmon, K. "Small Firms under Microscope: International Differences in Production Operations Management Practices and Performance." *Proceedings of 6th EUROMA International Conference*, Venice, June 7–8, 1999.
13. Cox, J.R.W., Mann, L., and Samson, D. "Benchmarking as a Mixed Metaphor: Disentangling Assumptions of Competition and Collaboration." *Journal of Management Studies*. 34: 286–314, 1997.
14. Pages, E.R. and Toft, G.S. "Benchmarking Innovation." *Economic Development Journal*, Winter, 8 (1): 22–27, 2009.
15. Stata, R. "Organizational Learning: The Key to Management Innovation." *Sloan Management Review*, 30 (3): 63–74, 1989.
16. Sinkula, J.M., Baker, W.E., and Noordewier, T. "A Framework for Market-Based Organizational Learning: Linking Values, Knowledge and Behaviour." *Journal of Academy of Marketing Science*, 25 (4): 305–318, 1997.
17. Garg, R.K. and Ma, J. "Benchmarking Culture and Performance in Chinese Organizations." *Benckmarking*, 12 (3): 260–274, 2005.
18. Li-Hua, R. and Simon, D. "Benchmarking China Firm Competitiveness: A Strategic Framework." *Journal of Technology Management*, 2 (2): 105–118, 2007.
19. Zairi, M. "Innovation or Innovativeness? Results of a Benchmarking Study." *Total Quality Management*, 5 (3): 27–44, 1994.

20. Massa, S. and Testa, S. "Innovation or Imitation? Benchmarking: A Knowledge-Management Process to Innovate Services." *Benchmarking*, 11 (6): 610–620, 2004.

21. Solberg Søilen, K. *Introduction to Public and Private Intelligence*. Studentlitteratur, Lund, 2005.

22. Jenster, P. and Solberg Søilen, K. *Market Intelligence: Building Strategic Insight*. Denmark: Copenhagen Business School Press, 2009.

23. Solberg Søilen, K. *Exhibit Marketing & Trade Show Intelligence—Successful Boothmanship and Booth Design*. Berlin: Springer Verlag, 2013.

24. Grupp, H. and Maital, S. Innovation Benchmarking in the Telecom Industry. The International Center for Research on the Management of Technology, MIT Sloan, 1996.

25. Solberg Søilen, K. *Geoeconomics*. London, UK: Ventus Publishing ApS/Bookboon, 2012.

26. Jones, K. and Kaluarachchi, Y. "Performance Measurement and Benchmarking of a Major Innovation Programme." *Benchmarking*, 15 (2): 124–136, 2008.

27. Shin, J., Lee, W., and Park, Y. "On the Benchmarking Method of Patent-Based Knowledge Flow Structure: Comparison of Korea and Taiwan with USA." *Scientometrics*, 69 (3): 551–574, 2006.

28. Hall, B.H., Jaffe, A.B., and Trajtenberg, M. The NBER Patent Citations Data File: Lessons, Insights and Methodological Tools. National Bureau of Economic Research Working Paper, no. 8498: 3–53, 2001.

CHAPTER 11

Process, Practice, and Innovation

Michael Grieves

Introduction

As Einstein is reported to have said, "the definition of insanity is doing the same thing over and over again and expecting different results." Given that, if we take processes to their logical conclusion, we will never have any major or radical innovation. We can get incremental improvement, but we cannot get true innovation. Processes by their very definition are about doing the same thing over and over again.

That is not a bad thing. If we have a process that gives exactly the results we desire each and every time we perform it to obtain that result, then this is highly useful. In organizations, where we have tasks that are repeated over and over again, having processes with the above characteristic produces results efficiently and saves resources by never producing results that are erroneous.

In addition, we can take that process and divide its resource usage into two categories. The first category contains the resources that are absolutely needed to perform the task, and the second category contains the resources that add no value to the task. If we then eliminate the second category, we have a textbook definition of *lean*.

However, doing the same task over and over again is not all that we do in organizations, and, a good deal of the time, it is not the most important thing that we do. Often what we do is to try to accomplish things that we have not accomplished in the past. If business were simply about executing well-defined processes, there would be no innovation or new product creation. While we might get incremental improvements, there would be no new and innovative products.

One of the other things that we do in organizations besides processes is something I have labeled as "practices."* By definition, processes take one or more inputs, perform specified operations or routines on those inputs, and produce outputs. Processes are defined so that different people performing the same processes will produce equivalent results. While they may not be exact results, from the organization's standpoint, they will be effectively equivalent.

Practices are very different. With practices, we are looking for desired results. What we attempt to do is to look at all the inputs that we have available for selection and all the available operations or routines that we can perform on those inputs. We then attempt to select those inputs and operations that will give us our desired results. Oftentimes, the results that we are trying to obtain are contradictory or even mutually exclusive.

For example, we want vehicles that are both safe and fuel-efficient. Adding weight to a vehicle can make that vehicle much safer, but fuel efficiency declines because of the

---

*My concept of "practice" has its origins in the sociological research on practice. Practice in that respect is about constituting and reconstituting purposeful social relationships and transmitting intergenerational knowledge. (See Bourdieu, P. 1977. *Outline of a Theory of Practice* (R. Nice, Trans.). Cambridge: University Press and Bourdieu, P. 1990. *The Logic of Practice*. Stanford, CA: Stanford University Press.) While there are definitely some elements of that in the practices businesses engage in, I am more focused on the outcome orientation of practice, that is, producing a specific result. As I discuss later, rote practices have this element of social continuity.

additional weight. A process approach would be to create a formula that would balance safety and fuel efficiencies. A practice approach would be to find or create a new material that was both lightweight and strong.

I am sometimes accused of being antiprocess. This is not the case. I am very much in favor of trying to create and formalize processes. In fact, without processes, tasks are almost impossible to bring to conclusions. There is an old and not inaccurate saying that, "an engineering task is only completed when the time runs out."* What is implicit here is that when faced with a task that is result driven, that has innumerable input and routine combinations, and that has competing results that require trade-offs, engineers would never complete the task if they were not under some sort of deadline. Since there would always be room to improve the results, the engineering task would never be finished.

However, while I believe in the value of processes, I also believe that other things are going on in the organization. These things are practices and even art. While I will not discuss art, I would like to spend the bulk of this space discussing the process-practice distinction, with an emphasis on producing innovation.

There are a couple of reasons why process has and continues to get the bulk of attention. First, processes are a lot neater and well defined than practices. Second, there is a tendency for the technical disciplines to want to believe that what they do is well defined, orderly, and deterministic. Design, engineering, and information systems are all structured disciplines.

These technical professionals would like to believe that what they do is very "Spock" like. Like Mr. Spock, the unemotional Vulcan of Star Trek fame, technical professionals would like to believe that they are oh-so logical as they go about their tasks. The idea that they only do processes fits in very nicely with this perspective. In fact, if we ask these people how they go about their job, they will provide us with very neat, orderly, and well-defined flowcharts.

But as any ethnographer will tell us, we should not accept what people tell us about what they do, we need to look at what they do. If we look at what designers, engineers, and even support personnel do in performing their work, it rarely resembles the neat, orderly flowcharts that they would show us. In fact, if we look at the process maps of any group within an organization, we would find major gaps between what the process maps say that they do and what we would observe them to be actually doing.

One of the clearest ways to examine the difference between process and practice is to look at two examples. In Fig. 11-1, we have an example of two engineering processes. The top flowchart shows the process for defining a vehicle's weight. The bottom flowchart shows a process for approving a vehicle's design.

The top flowchart is a true process. Inputs are well defined and the routines to perform on those inputs are highly specified. Irrespective of the people that perform the process, the results that would be obtained would be equivalent. In fact, we could translate this flowchart into an algorithm or program that would provide a step-by-step definition of how to perform this process.

The bottom part of the flowchart, while a process, contains elements of practice. The box labeled as "define specifications to meet requirements," while in a neat little box, masks a very messy practice. Assuming that there are competing and even contradictory requirements that need to be accomplished, there may be a myriad of solutions that meet or at least are acceptable so that the specifications can pass the approval decision. In fact, the criteria used to define whether or not the specifications meets the requirements may be altered depending on what results are obtained.

If there are competing or contradictory requirements, trade-offs will have to be defined and some of the requirements might have to be relaxed in order to produce any solution at all. This is not something that lends itself to a process. There is no algorithm or code that could be produced in order to provide the solution. In point of fact, anyone who has sat in on one of these sessions would see heated arguments and intense negotiations taking place by the participants. This is hardly a well-defined and orderly process. It is disorderly and messy.

However, as a result of being messy, these practices can and do produce innovation. In attempting to produce a result, and, in many cases a result that has never been produced before, people will attempt to combine inputs and the operations on those inputs in new and unique ways.

Innovation is not unique to humans. Nature also produces innovation through what we call mutation or genetic programming. In nature's case, it tries to produce all possible combinations and permutations and then allows the environment to select out those designs that are best

---

*There is a more pointed and tougher saying favored by my friend Stephen Sharf, the former Vice Chairman and Vice President of Engineering at Chrysler, "There comes a time in every project where you need to kill the engineer."

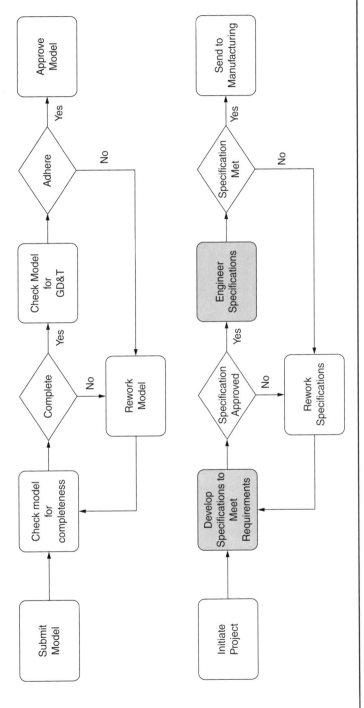

Figure 11-1 The two processes.

suited for survival. This works for nature because it has unlimited time and resources. Humans, without that luxury, need to do things differently. And we do.

What we do is to envision the results that we want and then use information that we have acquired about cause and effect in order to get those results. In doing this, we trade off information that we have in order to avoid wasting resources on attempted solutions that do not meet the required results. Nature has unlimited resources and no information. Humans have information and limited resources.

There are some notable exceptions where humans have used a genetic-like approach to obtain a dramatic innovation. The discovery of the electric light by Edison involved him trying thousands and thousands of experiments with different materials. However, even here, Edison used information about material properties to whittle down the universe of possible properties.

So while nature uses genetic programming in order to obtain innovation, we need to obtain innovation through practices. Process will not get us true innovation. This is my only issue with process. Unfortunately, process people fail to see this. In fact, there is almost a religious belief that process can get us innovation.

One of the more interesting discussions that I have had about this was with a NASA Six Sigma expert. At a meeting that we were in, she claimed that there was a Six Sigma process that produced innovation. My response to this was to ask her to code up this process so that it could be disseminated throughout the rest of NASA in order to create innovation. Her response was to inform me that I clearly didn't understand what she was talking about and then stomp out of the room.

While her reaction might have been an extreme one, people who only believe in processes have very difficult issues in addressing innovation. People armed only with a hammer believe everything is a nail. While I think that incremental improvements may be obtained through a process, true innovation requires practices.

Figure 11-2 shows the visual difference between process and practice. If we look at a spectrum, from ill defined to well defined, we can see the difference between process and practice. On the right side, the inputs and the routines on those inputs are well defined. The output is simply the result of running those routines against those inputs. At its most rigid and/or bureaucratic outcome, the output is accepted irrespective of whether it makes sense or not. With processes, we get the outputs that we get by running well-defined routines and operations on well-defined inputs. While that may not be the result that we really desire, it is what we get from processes.

One of the better examples of this that I have observed was going through security at the airport. A man had passed through the X-ray machine with a 10-oz bottle that had a couple of ounces of liquid in it. The TSA officer informed him that he could not take this bottle through security with him. When the man objected that the amount of liquid was less than the 3 oz that were allowed, the security officer, not without a sense of irony and amusement, informed the traveler, "Sir, I'm an official of the United States Government and not allowed to exercise common sense." The TSA official's process was that only bottles that were less than 3 oz in size could be passed through security. Even though the intent was to limit liquids to 3 oz, the TSA's process implemented a limitation of 3 oz in the physical size of bottles.

On the left side of Fig. 11-2, we have practices. As was mentioned previously, practices are result driven. We define the outputs that we want and then we select the inputs and the operations or routines on those inputs in order to produce the desired outputs. As indicated by the wavy lines of outputs, the outputs may not be as crisply defined as the outputs of processes. In fact, we may redefine the outputs as we deal with competing requirements and trade-offs.

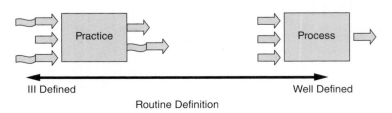

FIGURE 11-2   Process-practice continuum.

With practices, we start with the result that we desire and back solve to select the inputs and routines that give us those results. We can find out a number of things in doing this, especially if the results that we are looking for are competitive or contradictory. At worst, we may find out that what we're trying to do cannot be done, at least not with the inputs and routines that we currently have at our disposal. Unlike a process, where we will always get a result, a practice might or might not be successful, where success is defined as finding a set of inputs and operations on those inputs that will produce the results we desire.

When we are unable to succeed in our quest for a desired result, we can either give up, or we can adjust the results that we are looking to obtain. Once we adjust the results, we can then reselect our inputs and routines or operations in order to obtain these new, adjusted results. Except in the most trivial cases, practices are generally iterative in nature. If that were not the case, we could probably make a process out of a practice. By selecting our output and then trying all combinations of our inputs and routines in a systematic fashion, we would then produce our desired output. This is just what nature does.

As I will cover more in depth later, there are substantially different information requirements between processes and practices. With processes, we only want to give the minimal amount of information that will allow the process to be performed. Everything else is irrelevant. For our TSA official, the amount of liquid in the bottle was irrelevant. The only information that he wanted was the size of the physical bottle.

For practices, we need to provide as much information as possible. Since we do not know ahead of time what inputs and what routines or operations are relevant to obtain the desired result, we need to make available to the practitioners a vast pool of information from which they can select what they need. If we limit the pool of information, we most likely will limit the alternatives that can be considered.

Distinction without a Difference

While the idea of practices that is different from processes resonates well with most people, I occasionally get the objection that processes and practices are distinctions without any real differences. The objectors will say, "When I talk about processes, I'm really talking about practices, too." There is no question that we can be pretty fuzzy in our use of the terms processes and practices. For example, we almost exclusively refer to processes when we talk about what we do in our organizations. However, we then throw in the term of "best practices" that seems to refer to anything an organization does.

Huh? How is it that all this discussion is about processes and now we have this thing, best practices? Ignoring the fact that I take a very jaundiced view of best practices, the primary answer is that we clearly haven't thought through the differences between process and practice. As a result, the phrase, best practices, began being used and was unquestioningly picked up as a common phrase. Interestingly, I find that this phrase does bother some people, but they are unsure as to why.

The core issue is that there is a substantial difference between a process and a practice. In spite of some lexical sloppiness, we inherently know the difference between a process and the practice. In reality, we act on that difference all the time as we go about working in our organizations.

In the next section, I will outline the difference between process and practice. There are clear, characteristic differences between process and practice. This is not simply a difference in terminology. There is a difference in the way we do things.

Process and Practice—Different Characteristics

In Table 11-1, I have listed what I believe to be most of the characteristic differences that make processes distinct from practices.

Lean versus Innovative

Processes are intended to give us lean results in our organizations, while practices are intended to get us innovation. Processes, by definition, are structured activities that are intended to give us equivalently repeatable results. What we would like to happen with a process is to obtain the same results each and every time that we perform the process. We do not want processes to be idiosyncratic. That is we do not want processes to be dependent on

• Lean	• Innovative
• Predictive	• Unpredictable
• Input, Routine Driven	• Goal Seeking
• Precise	• Fuzzy
• Codeable/Calculable	• Arguable/Negotiable
• Efficiency	• Effectiveness
• Optimized	• Satisficed
• Frictionless	• Friction
• Training-based	• Education-based
• Minimal data	• Maximum Information

TABLE 11-1   Process versus Practice

the individuals who run them. We want the same results each and every time, no matter who performs the processes.

Processes by their very nature are definable. We define the inputs that we want for the process, and we define the operations or procedures that we will perform on those inputs. Because we have defined them, we have an understanding of the outputs that we will obtain from the operations that we are performing on these inputs. Our intent is to take variation out of the result. With processes, we would like to be able to be assured that we will get specific results if we have specific inputs. If we add up all the weights of all the components of our vehicle, then we will get the total weight of the vehicle. For this to be a process, each and every time we select the same components we will get the same weight.

How do we get to "lean" from this? The answer is that it does take some work. We cannot simply equate a process with being a lean process. To get to a lean process, we need to analyze all competing processes and pick the one that consumes the least amount of resources. Organizations can have and unfortunately often do have inefficient processes. While processes will give us defined results, we need to go the extra step to produce not only defined results, but results that are also lean.

With practice, we are looking for innovation. The focus of practice is to attempt to find the inputs and operations that will give us the result that we are looking for. With practice, we are not telling the participants what specific inputs to select or what specific operations to perform, we are asking them to obtain a result.

Again, we need to take an additional step to get to innovation. While we can define a result we know how to get, that really isn't the intent of practice. If we could do that, we ought to make it a process. After all, we have a result. We know what inputs and operations to perform in order to get that result, so it makes no sense to ask people to recreate this.

Innovation is another way of asking for results that we had not obtained before. With innovation in the product area, we are looking to create a product that performs certain functions better than other products or performs functions that we have not been able to perform before.[1] Before the computer chip, we had the ability to perform calculations, but the cost of those calculations was extremely expensive compared to performing those calculations on a computer chip. With the creation of cellular phones, we could do something that could not be done in the past, which was to talk to people within the phone system without being tethered to a copper wire.

In most organizations, however, we are not simply looking to be lean or innovative. We are looking to be both lean and innovative. For some tasks the objective is to be lean, as in repeatable manufacturing. For other tasks, the objective is to be innovative, as in new product development. What this requires is that we select either a process or practice, depending on what we want to accomplish. Processes alone will not suffice. If we're looking to reduce resources, that is, be lean, then we need to engage in the process. If we're looking to be innovative, then we need to engage in a practice.

Predictable versus Unpredictable

Processes are predictable, while practices are unpredictable. With a process, we are interested in repeatability. We would like to know that each and every time we have certain inputs, that we will get certain outputs. Even in the case of our not-so-humorous TSA example, there will be a very predictable output. Each time a container of greater than 3 oz passes a screening station, it will be rejected. This is very repeatable and predictable. It is one thing to be repeatable and predictable. The trick is to have the repeatable and predictable output to be the one that we really want. That is an issue that we will deal with later on.

With practices, we have unpredictability by its very nature. In one possible outcome, we may not ever be able to obtain the result that we desire. In our example earlier, where we want a vehicle that is perfectly safe but also has superior gas mileage, we may never be able to get to that result. The results that we desire may be unobtainable, or, in the case of opposing results, it may be impossible to obtain one result without precluding the competing result, without a major change in technology.

The other issue of unpredictability is that, even when the results are obtainable, we may not be able to predict how we are going to get those results. There may be multiple ways of selecting the inputs and operations that will produce the results we desire. Because the permutations and combinations of inputs and operations may be so large, there is no practical way of being able to predict which combination will be selected. This predictability is less of a problem than being able to predict whether or not the results can actually be obtained. But it does lead to the problem of evaluating how efficiently resources are being used in order to produce those results. I will deal with this issue a little later.

Not being able to predict whether results can be obtained or not is the problem that most companies face when introducing new products. If they do not produce the results that they set out to obtain, they fail in creating a new product. However, if they do not try to produce results that had never been obtained before, they will produce nothing that is innovative. The issue that organizations face is how to assess the potential success of their ability to actually produce something innovative.

While processes may help organizations in making sure that they do not spend too many resources in pursuing something that cannot be obtained, the actual activities creating innovation are practices. While we can define a product development process aimed at creating new and innovative products, within that process the actual innovation is a practice.

Input, Routine Driven versus Goal Seeking

As I described earlier, processes are about taking specific inputs and performing defined operations or routines on those inputs. For true processes, we have little or no say in which inputs that we take and which routines that we perform on those inputs. These have been defined by whoever defined the process. The outputs that we get are the outputs that we get. With processes, we will always get an output, even if it is not very useful.

In practices, we start out with a goal and attempt to determine which inputs and which routines or operations that, when combined, will produce our goal. With true practices, we have a lot of leeway in our selections. We may even have to create new operations and find new inputs in order to produce our desired outputs. With practices, we may never get the outputs that we seek.

Processes are sequential and the steps are fully specified. We finish one step and move onto the next step in a prescribed fashion. Practices are much messier. We coalesce around a solution, moving back and forth between ideas as we see how they work out.

Figure 11-3, my Practice Model Methodology, illustrates the type of holistic approach that humans engage in that differs greatly from the sequential approach that computers use and that process supports. The tendency would be to interpret the model as flowing sequentially from the bottom to the top. In that interpretation, the Definitional/Requirements Space would be fully completed before moving onto the Potential Solutions Space. Then all potential solutions would be enumerated before assessing their technical and environmental fit. The potential solutions that made it through would then populate the Feasible Solution Space. Those feasible solutions would be compared for trade-offs and cost or value. The solution or solutions that emerged from that analysis would populate the Final Solution Space.

However, while there certainly would be this general movement from bottom to top, humans are neither methodical nor single minded. We are intuitive and tend to approach this sort of activity holistically. If we watched this model in motion over time, we would see the different spaces being populated and pruned simultaneously. As soon as we saw the definitional/requirements space start to take shape, the other spaces, including the Final Solution Space would start to be populated. As humans, we cannot help it. We start to think about solutions as soon as we start to formulate problems.

While we need to guard against closing solution sets too early, and there are methodologies that help prevent that, working in this simultaneous way allows us to take the intuitive

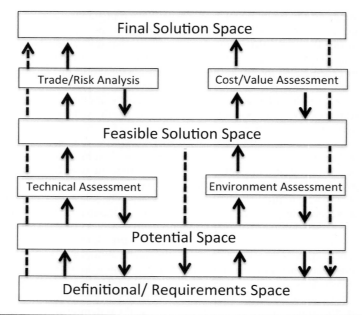

FIGURE 11-3 Practice model methodology.

leaps that innovation requires. Often this holistic approach presents us a solution that a sequential approach might have filtered out because we can tinker with the definitional/ requirements space to allow a highly innovative solution to take shape and provide functionality we did not know we needed until we invented it.

So while processes are forward-solving, practices are backward-solving, and sometimes backwards from a new, elegant innovation.

Precise versus Fuzzy

Processes are very precise. They have to be if we are to perform operations or routines on a selected group of inputs. Certainly we would need to be precise in our selection of both the inputs and the operations to be performed on those inputs. There can be some fuzziness in the inputs and operations if we're prepared to tolerate some variability in results. This may be why some people think that process and practice are interchangeable. However, even in these cases, we are still forward-solving from inputs, not back-solving from a result.

Practices, on the other hand, are very fuzzy. In some cases, we do not even know why we make the selections of inputs and operations that we do. We know we need a certain result, and somewhere in our mental processing faculties, we make the necessary connections and weightings in order to decide which inputs and operations we need to perform in order to get the results we desire.

Codeable/Calculable versus Arguable/Negotiable

In point of fact, humans are more comfortable with practices than with processes. Contrary to the model of the mind as a computer, our minds operate significantly different than computers do. We are not sequential computational machines. We operate by associating data within our brains and by weighting those associations in a particular fashion. In many situations, our brains do it automatically, without us even being aware we are doing it or being able to articulate why or how we did it.

Processes are codeable and calculable. If we are dealing with a true process, we must be able to define the inputs that we are going to deal with, and we must be able to define the routines or operations that we are going to perform on those selected inputs. We can create an algorithm that defines the selected inputs and the operations that we will perform on those inputs.

That being the case, we can also define the calculations that the operations are to perform on inputs. By calculations, I do not necessarily mean arithmetic ones. Calculations can and do include logical types of calculations of the type, "if this is the case, then do that." In fact, pure processes require us to be able to code them. As computer systems have become embedded in almost every facet of an organization, we have replaced humans with computers in many, many processes. Computers are ideal for processes, because they will in a slave-like fashion perform exactly the same operations on exactly the same selected inputs each and every time. Humans lack that sort of consistency.

Practices, on the other hand, are not codeable and calculable. While there may be codeable and calculable aspects to practice, as noted earlier, fuzziness prevents the precision necessary to perform calculations. Instead, practice exhibits the characteristics of argumentation and negotiation.

With a practice, there may be a very large universe of potential inputs and operations. There may be many different combinations and permutations that create the same result. Which inputs to select and which routines to run may be a matter of personal preference. Sometimes, we select one way of doing things over another simply because we are comfortable with that way or because we have been trained that way. Since we may feel strongly that only our way produces the best desired results, we may tend to engage in heated arguments, not because our way is the only correct way, but because we have so much invested in that way of doing things.

Arguments, stripped of their emotional overtones, are an attempt to provide rationale and evidence that our approach to obtaining a desired result is superior to an alternative one. If we were completely rational beings, which we are not, this might always be a successful method. However, oftentimes, the decision on which argument is better is often decided on whose rhetoric is better or, worse, who yells loudest. This is a pitfall of practices and is not to be underestimated.

In addition, practice involves negotiation. Sometimes a negotiation is around a trade-off of competitive or opposing results. This is useful when the goal of the negotiation is to produce the best result. It is not useful when the result of the negotiation is to preserve one of the party's vested interests at the expense of a superior result. Again, practices have a tendency to deteriorate from result-oriented negotiation into a vested interest negotiation.

An example of this process-practice distinction is a conversation I had with a human resource manager. They told me that they had a process for setting salaries. They had salary ranges set up for different positions. The employee would then come in, and the HR department would discuss the employee's performance. HR would then settle on a salary increase.

I explained to them that while part of this was a process, there was a major element of practice involved. Two identical employees could go through this process and come out with two different salaries. What actually was occurring with the employee was negotiation. While the employee was only negotiating within a certain range, he or she still was negotiating.

If arguing or negotiating is taking place, we are engaged in a practice not a process. This practice may exist within the confines of a process, but it is a practice nonetheless.

Efficiency versus Effectiveness

We come to a core issue regarding the distinction between process and practice. Processes are about efficiency. Practices are about effectiveness. Processes are about lean. Practices are about innovation. With processes, we want to strip out any unnecessary action in performing the process. We have already seen that we do not want to consider any data or information that is not actually needed to complete a process. We have selected the inputs and the operations, and we want to perform a process using the minimal amount of resources as possible.

Again, however, evaluating the efficiency of a practice is difficult. Two different people could have very different conclusions about how efficient a practice was. In fact, a practice that does not produce the desired results might also be very efficient. Determining that the results were unobtainable might have been very efficient, that is, it uses the minimal amount of resources to make that determination. Someone else might have used substantially more resources to reach the same conclusion. Practices are different. With a practice, we want to produce a result or a set of results. Going into a practice, we really don't know what combination of inputs in routines that we will need in order to produce our results. If the results that we are seeking are similar to results that we have obtained in the past, we already may have a good idea of the inputs and operations that will give us our desired results. However, if our results have never been obtained before, that is, they are highly innovative, it may be detrimental to have preconceived ideas about the inputs and operations that will be required.

With practices, constraining resources and methods may be highly detrimental. It does little good in a product development effort to create a new product quickly and using minimal resources if that new product does not produce the results that we require. If the time runs out too quickly on our product development efforts vis-à-vis the adage quoted above, we may have nothing to show for our efforts. If we have minimized resource usage but do not acquire our results, we have not minimized resources but wasted them.

The issue that we have to deal with is that we can evaluate effectiveness of a process, but it is difficult to determine efficiency of a practice. With processes, we have predetermined which inputs we are going to select and how we are going to select them. We have also determined what operations we are going to perform on those inputs. We therefore can calculate the time and resources we need to expend in performing a process. If we cannot, then we do not have a process.

With continuous improvement or kaizen, we are in essence creating successive processes, where each process uses fewer resources than the previous process. The issue here is that at the beginning we do not know which combination of inputs and operations minimizes our use of resources. As we try different combinations, we calculate the usage of resources by each of these processes and we select the process that uses the least amount of resources as the process we want to perform. While there may be some variability in the usage of resources, they're within acceptable range.

Since we know the results of our output that these processes will give us, we understand their effectiveness. If they did not give us the desired result, we would not be performing these processes.

With practice, we do not know whether we can obtain the result or not. Therefore, we may expend a great deal of resources without any results. Also, since we may be able to obtain the results in many different ways, it is difficult to create an *a priori* evaluation of the effectiveness of one practice over another. Evaluating the effectiveness of practice is very often a matter of experience and judgment rather than being able to use objective criteria.

I recently fired an attorney who was working on a legal matter for me. While this attorney had produced results, I felt that the amount of time that he worked on the case and, as a result, his fees were excessive. However, there were no objective criteria for me to prove to him that he was wasting resources. My criteria for evaluation was that I have spent many, many years dealing with attorneys and have spent millions of dollars on attorney fees, so I have a great deal of experience in knowing how long something should take. He was taking much longer than my experience told me he should.

However, practices are going to be inherently inefficient. Creating new knowledge takes resources. If information is a trade-off for wasted resources, conversely creating information requires wasting resources. The key is to continually test and experiment with our combination of inputs and routines to determine if we are headed in the right direction. Where there is real waste of resources in practice is to pursue an elaborate theory of how selected inputs and routines will produce our desired output without developing tests and experiments of parts of that theory along the way.

Frictionless versus Friction

We want processes to be frictionless. What this means is that we would like a process to be like a well-oiled machine. When we perform a process, we want it to go as smoothly as possible. We don't want to fumble around looking for the appropriate inputs. We don't want errors executing the operations. Ideally, we would like a process to use a minimal amount of resources each and every time no matter which people were involved in performing that process.

We do not want people asking questions about what they should do or worse questioning why they are doing it. We simply want the process to work fast and efficiently. To use our earlier example, computing the weight of a vehicle, we don't want people wasting time by not knowing either what all the components they need to add up or where to find the weights of these components. We don't want two different people performing this process and coming up with two different results. Instead, when we need to execute the process, we want the list of components with their weights readily available, and we want to perform the addition flawlessly. This is the hallmark of a process.

Practices are different. We want questions as to why we are doing things a certain way. In fact, we want to generate arguments or at least critical discussions about what is the best way to obtain a result. If we do not have that, we can easily fall into complacency whereby we produce nothing new or innovative.

Also using our earlier example, if we have no discussion about how best to meet our requirements in a design review, we most likely will wind up with pretty much the same product that we produced last time. If that is what we are trying to do, then we should have been engaged in a process not a practice. If, however, we are trying to produce something new and innovative, then by definition we need to do something differently.

This is why it's important to have nonhomogeneous groups when we are trying to be innovative. By homogeneous, I'm referring to people who think and act alike. If I want to produce new and innovative things, we need people that have differing views and differing perspectives.

I would like to make the distinction between friction and drag. By friction, I am referring to ideas that, even in opposition to each other, move the discussion forward and result in the synthesis of new ideas.[2] This is very different than people who are naysayers and attempt to find fault with any new idea. We can always prove that we cannot do something. That is pretty easy, and that is drag. The trick is to prove that we can do something new. The friction that I am encouraging in practice is about the synthesis of new ideas and not about negativity or drag.

Optimize versus Satisfice

Because processes can be codeable and calculable, we can attempt to optimize them. We can look at the operations for selecting inputs and we can look at the operations that we perform on those inputs. We can analyze and evaluate each step in the operation and determine whether it has any role in producing the result.

Clearly, the two steps in attempting to optimize the process are (1) eliminate all steps that do not enter into producing the result, and (2) evaluate alternative operations that may use less resources. The first step is really value stream mapping. Depending on the complexity of the operations and the potential number of alternative operations, the second step may or may not be feasible.

Practices are different because they are results driven, and there may be a universe of possible inputs and the ways to combine them in order to produce the desired result. When that is the case, it very well may be an impossible task to optimize a practice. Instead, what we should be looking to do is to satisfice. By satisfice what we mean to do is to get to a stage where we deem the result to be "good enough." This does not mean settling for mediocrity. But it does mean that we are evaluating the return on our efforts and determining that the future expenditure of resources will not produce an adequate return in the improvement of the desired result.

With true innovative products, the early version of the products may leave a great deal to be desired. Breaking into new territory often means that the functionality we are striving for is subpar or intermittent. Early automobiles often broke down. Many of the drivers of these machines often heard the derisive yell of "get a horse" as buggies passed them by. Automotive pioneers understood that they needed to get their product into the market in order to understand what to improve in their products. Again, this is not to endorse shoddy products but to point out that *perfect innovative products* is an oxymoron.

Autopilot versus Sense-Making

People engaged in processes are effectively on autopilot in more than a metaphoric way. As I defined processes, they are about people running well-defined routines on selected inputs. If they are engaged in normal, everyday processes, this is much like running on autopilot. People are comfortable with processes because, they are energy efficient. We use less energy performing a routine process than we do in facing a new, challenging problem.

This is not to demean people who engage in processes. Processes make for lean organizations if the processes are well crafted. Since the process-practice distinction is a continuum, processes can run from the truly mindless to those that require some thought in dealing with exceptions. However, autopilots work best when not faced with a lot of exceptions.

People engaged in practices are sense-making.[3] Practices are about finding the right inputs and routines to obtain the desired outputs. In essence, people work to make sense of the myriad of inputs and routines that they can select from in order to obtain those desired outputs. They are actively engaged in ferreting out those things that will allow them to reach their goal.

While processes define the inputs that need to be selected, practices need to select the inputs that will give the desired results. Often, those inputs have weak signals.* If they did not, it would be obvious that they were important. But being weak, they are not obvious.

---

*Weak signal inputs are inputs that do not have a prominence or an obvious role in creating an output. However, they do have a major role, and they are important. In what are called High Reliability Organizations (HROs), such as hospitals, nuclear reactor sites, or NASA, paying attention to weak signals may head off major problems. The increase in pressure readings on the Deepwater Horizon in April 2010 prior to the well blowout was a weak signal that should have been focused on. For more on HROs, see Weick, K.E., and Sutcliffe, K.M. 2001. *Managing the Unexpected: Assuring High Performance in an Age of Complexity*, 1st ed. San Francisco: Jossey-Bass.

People engage in sense-making in seeking them out to explain what is happening and how that will help them reach their goal.

Sense-making is more difficult and more resource consuming than running on autopilot. However, if we are looking to solve problems and not simply looking at getting through a process, then we need to continually make sense of the inputs we deal with. Changing the inputs and evaluating how the results change is a mechanism we constantly employ. Auto-pilot is fine for routine tasks. We need sense-making for problem solving and innovation.

A great deal of innovation is trying to make sense out of a technological advancement. Putting together different materials in different ways with different forces create phenomena that need to be understood in order to deploy. The innovation of the transistor was about putting different materials together that allowed current to flow. Making sense of this new phenomenon was critical to understanding how to deploy and use this radical new technology.

Sense-making creates the "Eureka" moment when all the pieces fall into place, and we understand how to produce our innovative result from the inputs and operations we perform on those inputs.

Training-Based versus Education-Based

We train people to perform processes, but we educate them to engage in practices. Since processes are well defined and codeable, we are interested in the people involved in process doing the same thing over and over in a repeatable fashion. We have already defined which inputs are to be selected. We have also defined which operations that we are to perform on these inputs. We are not looking for a personal interpretation of how to do this. We want the process to be performed in the same way no matter who performs it.

Given that is our criteria, we need to train people. It is not important that they understand why they do things. It is important that they perform the process in the same way each and every time. They do not need to know why they do it, but they do need to know what to do. For processes, we need to train people in what they do for as long as it takes to make them proficient.

Practices are different. For practice, we need to obtain a specific result. We have not defined the inputs, and we have not defined the operations to use. For true innovation, we need people to be free to think about different, and sometimes radically different, ways of doing things. For that, we need education.

We need people to understand and question what they need to do, because we cannot predefine the actions they need to take to get the desired results. But they do need to understand why we are attempting to obtain a result. It is only if they understand the "why" that they can create the "what" to do. With education, we teach people how to critically think about issues. Unlike processes, where we really want people to unquestioningly follow what we have determined that they need to do, with practice, we need people to question the "whys" in hopes of uncovering erroneous assumptions that can be disputed and changed.

As Fig. 11-2 shows, process and practice are on the spectrum. Certainly, there are processes that need to be followed in exact detail. Likewise, there are practices where the results have never been accomplished before, and any and all inputs and operations can and should be considered. In the middle of the spectrum, there are processes where there is some discretion in either selecting the inputs or executing the operations. As an example, the purchasing process may define that the lowest bidder is selected. However, purchasing may have some leeway in the selection if they do not think that the low bidder is really capable of performing. The judgment that they use is not defined by a formula but is guided by past experiences.

Likewise there are practices that only consider a small number of inputs and a small subset of operations. When we begin to develop a variation of a product, we do not start with a blank sheet of paper. Instead, we start with the last variation of the product and make modifications to that. Rarely is something that involves people either pure process or pure practice.

Processes Derive from Practices

To put it in a different way, processes were once practices. It may have occurred through performing ad-hoc practices and recognizing that one way of doing things was superior in terms of resource usage. Or it may have been through a planning process that looked at the desired output and determined that a particular set of inputs operated on in a certain way produced the desired output. Either of these methods are practices.

However, once the desired method is analyzed, formalized, defined, and implemented in a repeatable way, what was once a practice becomes a process. It is now a process, because it is no longer goal-seeking. The goal or output is now defined as what comes about by following the process. It is a good process if the output produced is always the one that is desired.

Mismatched Situations and Activities

There are some very common situations where processes should be used instead of practices and vice versa. The two that I will discuss are rote practices and bureaucratic processes.

Rote Practices

Rote practices are those activities where it looks like people are engaged in finding the right routines and inputs to obtain the desired result. However, for a number of reasons, the people that are engaged in this activity are just going through the motions. A common example of this is a meeting that is convened in order to solve a particular problem.

However, it is common knowledge by all participants that no matter what ideas are proposed and what discussions occur, the outcome of the meeting will be whatever the boss or decision maker has already decided. This occurs time and time again in numerous organizations. It may not occur all the time, but it occurs enough of the time that it is a common pathology.

The situation is not always a bad thing, and many times it becomes embedded in the organization, because it works. The boss or decision maker may be in the best position to decide how the organization should best proceed. However, regardless of whether this is an effective strategy or not, the organization is wasting its resources in pretending to engage in a practice. Instead, the organization would be far better off in recognizing that this indeed is a process. The process in this case is that the boss decides and conveys his decision to the rest of the participants. Suffice to say, not much innovation occurs here, unless the boss is someone like Steve Jobs. However, innovative people of Mr. Jobs' caliber are exceedingly rare.

The one valid use of rote practices is to reaffirm the social system for existing members and indoctrinate new participants. People have a need to understand the social system in which they operate and to understand their role and position in that social system. Even though rote practices may not produce a useful result in an efficient manner, rote practices may be necessary for the maintenance of the social system necessary for the smooth functioning of the organization.[4]

Bureaucratic Processes

The converse of the situation is the bureaucratic process. A bureaucratic process occurs where the inputs are defined and a specific routine is performed. However, it is only by random chance that the desired output is obtained. In fact, in numerous cases the output defies commonsense and appears to be absurd. The common cause of this is that the definers of this process have no real understanding of the environment in which this process is supposed to operate or simply do not care.

I recently encountered a process for a university where I am a professor. Last year, my first year teaching for the university, I had to submit an I-9 form and a letter that was signed by a third party who had inspected my passport that proved my U.S. citizenship. That certainly made sense. However, this year the university payroll department is requiring that I submit a new I-9 form and letter. If I was a U.S. citizen last year, it stands to reason that I am a U.S. citizen this year too. This process is silly and a waste of resources.

Process and Practice—Different Information Requirements

The process-practice distinction has the most impact on the role of information. A process orientation results in the perspective that participants should only be provided just enough information to do their job in the name of efficiency or leanness. Unfortunately, if what the participants are engaged in is really practice, this perspective leads to starving the participants of the raw material that is needed to really drive innovation. With practices, participants do not know what information they need until after they have successfully produced the output that they are searching for.

Minimal Data versus Maximum Information

With process, we have defined which inputs we are going to consider because we have also defined which operations we are going to perform. If that operation is not going to use an input, then there is no need for the participant to have access to that data or information. In a process, we only want to consider the minimal amount of information that allows us to perform the process and obtain a result.

It was irrelevant to the TSA official how much liquid was actually in the bottle. All the data he needed was the capacity of the physical bottle. The fact that that physical bottle held the allowable amount of liquid was irrelevant to his decision. Also irrelevant to his decision was the fact as to whether or not this passenger really needed that liquid or that this liquid was the most harmless substance in the world. The process that he was following specified that he was to select the physical capacity of a container and compare that capacity to 3 oz. If 3 oz or less, he was to pass the container. If greater than 3 oz, he was to reject the container. Other data or information was irrelevant.

Practices are considerably different. With practices, we would like to have the universe of knowledge at our disposal. We are attempting to get a desired result. What we need to do is to explore this universe of possibilities in order to produce that result. Because we are back-solving from a result, we do not know ahead of time which inputs or operations that we need to combine in order to produce the desired result.

This of course is the extreme position. We don't always need an entire universe of knowledge at our disposal. Most of the time, we have a pretty good idea of the potential inputs we need and operations that we can perform to produce the desired result. However, to produce real innovation, we need to make sure that we do not get locked into only considering a subset of available inputs and operations. We can get locked into a situation where what should be a practice degenerates into a process, because we have gotten so comfortable with only using a limited subset of inputs and operations.

It is somewhat amusing to see proponents of "everything is a process" to call for out-of-the-box thinking in order to solve a problem. Thinking of activities as a process when we should be thinking of these activities as a practice is a major reason for getting into the box in the first place. Out-of-the-box thinking should be second nature to us if we are thinking in terms of practices. However, our single-minded focus on process limits our ability to do so.

For companies that wish to be innovative, they will have to first realize that there is something other than processes that happen in their organizations. They will need to start discussing practice, educate their people on the role of practice, and identify where it occurs or should occur, and determine whether information and knowledge are being deployed appropriately.

Summary

This chapter explored what people in organizations really do. It proposed that contrary to the common wisdom that people only engage in processes, people also engage in practices. In fact, for true innovation other than incremental improvements, practices are required. While processes are input and routine driven, practices are results driven. Both processes and practices are needed for an organization to be successful.

The characteristics of processes versus practices include lean versus innovative, predictable versus unpredictable, input, routine driven versus goal seeking, precise versus fuzzy, codeable/calculable versus negotiable/arguable, efficient versus effective, frictionless versus friction, optimize versus satisfice, autopilot versus sense-making, and training-based versus education-based.

This chapter then defined the characteristics of process versus practices. The idea of processes and practices has implications for product lifecycle management, although it can apply to many organizational endeavors. The type, organization, and quantity of information needed for process differ markedly from the information needed for practice. The more innovation, the more information that is required and put together in different, unique ways that will not be apparent from the outset. Without a focus on practices, organizations will struggle to be innovative.

References

1. Christensen, C.M. *The Innovator's Dilemma: When New Technologies Cause Great Firms to Fail*. Boston, MA: Harvard Business School Press, 1997.
2. Leonard-Barton, D. *Wellsprings of Knowledge: Building and Sustaining the Sources of Innovation*. Boston, MA: Harvard Business School Press, 1995.
3. Weick, K.E. *Sensemaking in Organizations*. Thousand Oaks, CA: Sage Publications, Inc., 1995.
4. Berger, P.L. and Luckman, T. *The Social Construction of Reality: A Treatise in the Sociology of Knowledge*. Garden City, NJ: Doubleday & Company, 1967.

CHAPTER 12

Ethnography

Brett E. Trusko

Introduction

Ethnography is an element of anthropology dealing with the systematic study of cultures and human behavior. Conventionally, anthropologists would spend time studying or observing patterns of human behavior. Today, ethnography has become a consumer research method for uncovering unspoken customer needs or to gain insights into customer desires for specific positive experiences. Given the growing demand for innovation, the role of exciting user experiences in the success of new products, and the role of design innovations in making user-friendly products, ethnographic research has become the go-to tool for driving product innovation.

In his book *Ethnography*,[1] John Brewer states

> "Ethnography is the study of people in naturally occurring settings or 'fields' by methods of data collection that capture their social meanings and ordinary activities, involving the researcher participating directly in the setting, if not also the activities, in order to collect data in a systematic manner but without meaning being imposed on them externally." [sic]

Ethnography is more of a qualitative research methodology that has its own challenges surrounding accuracy, validity, and reliability due to the natural settings of the research, participants' activities and choices, design of the research, and selection of participants. Ethnographic research explores the beliefs, practices, artifacts, folk knowledge, and behaviors of a group of people.[2]

Need for Ethnography

The primary purpose of ethnography is to document or discover a faithful and accurate understanding of participants' ways of using certain methods or tools. Ethnography reveals subjective and subtle information about the behaviors of a target group of people that are difficult to identify in other scientific approaches to collecting data. As in any social science experiments, information gathered through ethnographic research is subject to known uncertainties surrounding individual behaviors. Also called field research, it is used to understand people's actions and their related and routine experiences. The methodology of such research is best summarized below:

> "Field research involves the study of real-life situations. Field researchers therefore observe people in the settings in which they live, and participate in their day-to-day activities. The methods that can be used in these studies are unstructured, flexible, and open-ended."[3]

This demonstrates that ethnographic researchers must become intimately familiar with the people being studied as well as their environments, activities, and situations. Ethnographers sometimes become a subject of the study itself to gain deeper insights of its relevance and meaning. Ordinary activities are studied without making it a planned experiment with unique data collection methods, and unstructured observations form the basis for analysis. At the end of the research, ethnographers may tell the story of their experience, present analysis of the information gathered, and make their own.

In the business environment, ethnography is used as a deliberate inquiry process[4] for the specific purpose of assessing the use of a certain product, or for creating a new product. It is not meant to be a formal, academic, and scientific research method to prove a hypothesis.

FIGURE 12-1   Ethnography.

Researchers may have a set of questions that they want answered about what, how, why (or not), when and where, and how well people will use new products. Observed user reactions and experiences are captured and examined to establish meaning, or the internalization of the user experience by the research subjects.

Meaning is commonly manifested in several ways: the removal of pain, connection to cultural practices, demographic preferences, relationships to other activities, changes in behavior, positive or negative impacts of interaction, and benefits of the experience. The more routine the behavior or benefit is to the user, the stronger the meaning of the experience will be. Ultimately, this information is used to create products that will provide enjoyable experiences and prevent painful challenges.

Ethnography and Innovation

Innovation can happen in form or function, or both. Form innovations are called design innovations and are based on more qualitative research including ethnographic research. Typically, research leads to new opportunities for form or design innovations, which at times lead to the development of new technologies. For example, when the mobile phone was initially designed by technologists with an old calculator type keypad, the marketing team rejected the idea because of its poor human interface and recommended instead the familiar phone type keypad for better acceptance by users.

Ethnographic research tends to be most effective for platform-type innovations—more so than for fundamental, derivative, or variation types where individual users would play a bigger role. For platform innovation, ethnographic research is a compelling form of cocreation.

We know that technology innovations increase a company's revenue, and we have learned that when design innovation is added to the mix, that revenue benefit is kicked up a notch or two. Enter ethnographic research!

Other uses in ethnography in innovation are utilizing the methods to find out what target groups might actually want. As discussed elsewhere in the book, managers, R&D professionals, marketers, and so forth, may not always understand the needs of the customer. A somewhat notable example of this process in action is in the 2000 film *What Women Want*, where a chauvinistic ad executive accidently "goes native" and discovers what it is that women really want in order to win an ad campaign from a female executive. Although his original motivation is purely about beating a woman out of the ad campaign, he begins to understand them better and soon becomes enlightened.

Imagine what a company could gain if they were to be able to really understand markets in which they compete.

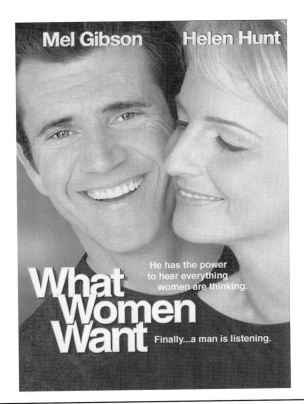

FIGURE 12-2   Thinking outside your paradigm.

Ethnography Skills and Tools

Ethnography is a holistic approach to studying a cultural system where the culture can change from place to place, or one market segment to another. Culture, a soft attribute, may include physical environment, historical behaviors, or traditions shared within a societal segment, and perceived priorities. All of these aspects evolved over time. Thus, ethnographic research is scoped by market segment and time in space. We begin the process by establishing a framework for study, outlining the purpose of the research, and intended target objectives.

Similar to other professions such as auditors or research scientists, ethnographers must have superior curiosity and the determination to dig deeper for information. They must be good at collecting available information, understanding what is not available, and thinking quickly of alternative sources. Strong communication skills are required to conduct thorough and probing interviews in a friendly and nonthreatening way, and they must be able to analyze information to recognize patterns. Moreover, an ethnographer must plan the research, understand the cultural aspects of the group, organization, or society under study, summarize findings in an easily understood form for use by the target audience, which generally includes product or service designers.

During the observation phase, the focus is on recognizing and recording obvious patterns giving attention to minute details of those activities that convey meaning or have a significant impact on the overall experience of the subjects—both implicit and explicit. Implicit expectations may be manifested through facial and verbal expressions (such as a frown, smile, curse, complaint, compliment, pain, and pleasure); a demonstrated need for extra effort (such as excessive time to perform, extra effort to push, extra movement, proneness to mistakes or getting hurt); unexpected or unintended ways of doing things; and requiring additional resources for certain activities. The deeper the understanding of human behavior and its relationship to the user experience, the more opportunities to discover new ways to achieve breakthrough innovations.

Using the right analytical tools is equally important to making insightful observations. Graphical tools for capturing patterns, descriptive statistical tools to summarize observations, and inferential statistical tools for comparative analysis, can be used to address sample size limitations and reduce the risk of drawing incorrect conclusions. Many ethnography-specific software applications are available for gathering and analyzing data. Throughout the research, ethnographers must remain vigilant to recognize what else needs be collected.

Interviewing is the next critical step for capturing and validating relevant information. A healthy respect and admiration for the culture of the group that is being studied underpins successful research by helping the ethnographer sort between the positives and negatives inherent in any group.

During the interviews, ethnographers must remain mindful of local customs, sensitivities, sensibilities, and conversation protocols. Leading with open-ended questions will prevent defensiveness and increase the quantity of information. Friendly and frank conversations lead to unbiased responses and more usable data. As appropriate, researchers may use audio and video recording tools so they can be analyzed thoroughly and repeatedly, if necessary.

All the data in the world would do no good if it is interpreted out of context. Interpretation must be bound by the purpose of the research, the culture of the group, and available data, but carefully interpreted and analyzed, they may lead to enjoyable user experiences. Ultimately, the innovation solution must work differently, look different, and offer different enjoyable experiences to create demand for the new product or service. The key being that a color, shape, or word that might be unappealing to an individual person, group, or culture may have the opposite effect on another group. An example of this in modern days can be seen by the attitudes toward sex as you move from one country to the next. In some countries television will show partial or full nudity and in other countries nudity is completely taboo. While your product may have nothing to do with sex, there are still norms that may inspire you to innovate specifically for that market.

Speeding Ethnographic Research

Ethnographic research can be expedited without detriment to its power of insights. Similar to rapid prototyping and other rapid product development methods, rapid ethnography has also been proposed.[5] A typical ethnographic research project that could take months to complete can be sped up and shortened by applying other techniques to the tasks of identifying the subjects of research and observation, capturing information, and generating insights.

One such technique is the Lead User concept[6] that has been used for identifying market opportunities by receiving feedback from early adopters of new solutions in the marketplace. Lead users are individuals who like to experiment with new technology or solutions, or they may have a direct use for the new solutions. It can also be used for ethnographic research to identify subtle requirements based on their qualified experiences.

Instead of identifying a random sample of the group under study, the ethnographer would identify representative lead users within the selected group for getting information. It is like working with sample averages versus the population. In doing so we may lose resolution or miss the finer inputs we could get from many individuals, but this can be mitigated by more rigorous preparation for the ethnographic research project.

Narrowing the scope of the project, too, will shorten the overall time of the project. Instead of studying a social group in a target market for a larger purpose, we could quickly collect the information about specific aspects of the intended innovative solution. Instead of one large ethnographic research, we may take multiple short trips to the group for specific information. Thus dividing and focusing on a smaller scope would accelerate the collection of input for new products.

As well, for greater efficiency, interviews and activities could be recorded in audio/video and physical objects could be captured in high-resolution photos during the data collection phase. The resulting findings from analysis derived in isolation can be validated separately with the so-called smart people in the target group.

Organizing ethnographic research such that various aspects could take place in parallel could also reduce time to completion. As parallel processing has been used in manufacturing to reduce cycle time for producing products, performing ethnographic research with a select sample of informants to perform interviewing, data collection, and recording could occur simultaneously with more researchers for field research. Today, webcams or similar technology could be used to gather data remotely over a longer period of time, or even continually.

Institutionalizing Ethnographic Research

Considering the success of a few products that utilized ethnographic research for capturing unmet requirements, it is becoming an integral and routine part of capturing and validating user requirements for new product development. To make it routine, it must become a standard process and be deployed economically and rapidly.

Organizations may take a stab at establishing a standard operating procedure for ethnographic research[7] using these steps:

1. Study people's behavior in natural settings rather than experimental ones.

2. Collect data specifically to identify unmet or hidden user expectations.

3. Be open minded in collecting data in any form; it is available rather than in a prescribed format.

4. Scope out the research based on the type of innovation, and target the platform for the greatest benefit.

5. Take detailed notes from many perspectives while in the field. Capture exact terms used or views expressed by participants.

6. Select lead users to gather more information faster, to shorten project duration.

7. Try to live the participant's experience in order to understand the emotional meaning of the experience.

8. While interviewing, let the participants express their responses and experiences in their own ways, terms, and expressions. Capture them all as is without questioning or formatting.

9. Identify multiple sources of information for a variety of data and cross-validate findings.

10. Anticipate and prepare probing questions ahead of time, but be prepared for impromptu discussion.

11. Be a courteous, caring listener and express appreciation for the participant's support.

12. Do not criticize any social group, its members, leaders, decisions, or activities.

13. Most importantly, have fun to bring out your own best.

14. Analyze, distill, and extract as many findings as possible, but pay more attention to unmet needs, subtle experiences, or insights.

15. Prepare a positive and supporting report.

Examples   There are many examples where ethnographic research was conducted during the product development phase that resulted in successful product launch. For one, a camera manufacturer sent its marketing staff and researchers to a department store to learn about consumer behaviors, questions, and expectations. As a result, a successful family of cameras was launched. In another, a car manufacturer sent its researchers to live in the state where its target customers reside. Some lived with likely customers to learn their experience of using the car. As a result, a successful luxury car was launched and gained a significant market share.

Many new products with progressive designs including kitchen appliances, tools, furniture, car interiors, and electronic gadgets have used ethnographic research to integrate them into a user's life style. The recent focus on design innovation brings new relevance and importance for ethnographic research methods.

Summary

Ethnography is a research method for social sciences. It is a relatively new process incorporated into the product development process for user-friendly innovation, to gain better acceptance and loyalty, and to capture greater market share. It involves the up close and personal participation of the researcher in a social group environment to capture hidden, unmet needs, and insights to make innovations more useful and to deliver a more enjoyable user experience. However, ethnographic research must make economic sense and be conducted proportionate to the type of innovation and market targeted. Literally, ethnography means a portrait of people.[8] Interestingly, today's technology can help capture the finer portrait faster and even virtually.

References

1. Brewer. *Ethnography*.
2. LeCompte, M. and Goetz, J. "Problems of Reliability and Validity in Ethnographic Research." *Review of Educational Research*, Spring, 52 (1): 31–60, 1982.
3. Burgess, R. *In the Field*. London, UK: Routledge, 1984.

4. Erickson, F. "What Makes School of Ethnography 'Ethnographic'?" *Anthropology and Education Quarterly*, 15: 51–66, 1984.

5. Millen, David R. *Rapid Ethnography: Time Deepening Strategies for HCI Field Research.* New York, NY: ACM, 2000.

6. von Hippel. 1988.

7. Venkatesh, A. The Home of the Future: An Ethnographic Study of New Information Technologies in the Home, Advances in Consumer Research Volume XXVII, Mary Gilly & Joan Meyers-Levy (eds.), Valdosta, GA, Advances in Consumer Research, pp. 88–96.

8. Harris, M. and Johnson, O. *Cultural Anthropology*, Needham Heights, MA: Allyn & Bacon, 2000.

Creativity Tools

Perhaps the best-known, but least understood part of innovation is creativity. Creativity is thought by many to be the essential ingredient, or the secret sauce of innovation. It is that ability that diminishes from the moment you are born until the day of your death. We all know the story; kids are born creative, but we lose the ability to be creative because of

- Age
- Education systems
- Responsibilities
- Marriage
- Realism
- Kids
- Something in the water

The authors in this section of the book would argue that all of these and the 100 other reasons you can come up with are all wrong. People never stop being creative.

Harrington and Voehl have compiled a list of more than 100 creativity tools. Of course, this book does not have the room to feature all of those tools. Instead they have prepared a chapter on the most common creativity tools. What they call the Sweet Sixteen are key creativity tools and techniques that you can use to develop creative and imaginative solutions to business problems. The rest of this section has chapters on the tools and techniques that we think are essential for innovation science.

Larry Thompson, president of Ringling College of Art and Design discusses his take on the Six Thinking Hats. This chapter highlights the need for whole-brain thinking and innovation. From the perspective of innovation science, this is critical because it demystifies the secret sauce myth of innovation. As Mr. Thompson discusses, anyone who is a left brain thinker, that section of the brain responsible for linear, sequential, analytical, logical, and detailed thinking, can learn and apply the right brain thinking of nonlinear, imaginative, emotional, conceptual, and holistic skills. This is important because too many linear thinkers do not believe that they can participate in innovation. In fact, the opposite is true. Releasing the creativity of someone who is disciplined in left brain thinking is perhaps one of our greatest opportunities for innovation.

In Richard Greene's chapter on mind mapping, you will become a converted mind mapper, if you are not already one. In fact, as much as I hate to admit it, I was converted after I read this chapter. I purchased and downloaded my mapping software, and started mind mapping everything I do now. The applicability to innovation is thoroughly discussed in the chapter, and the process for developing mind maps will be invaluable to anyone wishing to learn how to do so.

In the chapter on social networks, by Christopher Rollyson, the reader will find, much as in ethnography, that there is a whole new way of looking at the world. From the perspective of the working innovation professional, utilizing social networks for new and novel ideas seems natural. Monitoring, contributing, and evaluating the traffic in social media and in social networking harnesses the power of directly communicating with your customers or potential customers. Since the beginning of mass merchandising (since merchants were separated from their customers), businesses have been trying to understand their customers better. Social networking allows both one-on-one contact and facilitates the ability to monitor communications among customers.

In David Worrell Conley's chapter on combination methods, he examines the simultaneous use of many methodologies in an effort to find innovative solutions to challenging problems. Unlike some maps, innovation has many paths to a destination. Depending on the needs and goals of any situation, different methodologies can and should be used to achieve those goals. Combining the most effective methods into an ideal solution gets to the essence of creativity. For example, when you ask a kindergartner why the sky is blue, you may get a different answer from every child you asked. As adults we know why the sky is blue, which means that we may not entertain other solutions. Combination methods force us to solve problems in ways that we might not have tried before.

Fiona Schweitzer and Iris Gabriel explore the role of market research in innovation. Included in this section is a discussion of innovation and its role in the sustainability and lasting success and profitability of the business. They question the role of new product development and the fact that there are so many high failure rates. Additionally, in this world of increasingly shorter product life cycles, creativity and market research are more important than ever to integrate the customer's mind into the innovation process and force a new focus on market orientation. In many ways, this means that instead of researching trends, we are now moving into a period of researching minds. This research includes the detection of what is going on in the customer's mind, their latent needs and desires, and what they believe is important in a product.

Finally, we close this section with Nikolaus Franke's chapter on lead user analysis. Franke compares and contrasts Ford and von Hippel from the perspective of their view of the customer. For example, he points out that Ford's attitude (appropriate at the time) was that customer research would have pointed out that people actually wanted faster horses, not automobiles. When looking at innovative solutions from Apple, much of what Steve Jobs gave the world, we didn't even know we wanted. After all, many people could not imagine the Internet on their phone. On the other hand, von Hippel argues that we are shifting from manufacturer-centered innovation to a period where innovation is defined by the end user. These lead users are the essence of this chapter.

Creativity is one of the most fascinating subjects that we can study. So many of the innovations we take for granted in the world today such as the wheel, fire, and other essentials must have been thought of by creative minds at some point in history. Perhaps it was a kindergartner who thought of the concept of the wheel. But we think it was an adult simply trying to solve a problem, looked around the environment, and created that first wheel. Not much has changed in the modern world. We have modern-day da Vincis, not just in Silicon Valley, but everywhere in the world.

People never stop being creative. They go to their graves with ideas and concepts that either come to them at the end of life, or ideas that they have had for years but never knew how to develop. Maybe they never worked for a company that valued their ideas. One of the great tragedies of the world manifests itself when you talk to elderly people who tell you about all the world-changing ideas they thought of during their lives. But never did anything about. Imagine a world where all ideas are heard and acted on. Would the world be a better place? We believe so.

CHAPTER 13

Creativity Tools: Develop Creative Solutions to Problems and Opportunities

H. James Harrington and Frank Voehl

In the book *Problem Solving for Results*,[1] authors Bill Roth and Frank Voehl discussed the need for shaping the right attitude and perspectives for creativity to flourish in light of the need for improving creativity in problem solving in any business operation. Ten years later, on his seminal work on the subject, "Creativity," Mihaly Csikszentmihalyi said that an effective creative process in almost any type of problem-solving or opportunity-finding situation usually consists of five steps. Picking the right tools and techniques can be divided into four meta-categories[2]:

- Those used to improve the creativity and enhance the problem identification skills of the individual.
- General techniques used to improve the creativity of groups.
- Systems-oriented techniques used to work with problem networks or "messes" as systems scientists call them.
- Tools used to specifically measure an organization's productivity and innovation creativity IQ.

The Sweet Sixteen Creativity Tools and Techniques are organized within the following five-step method outlined in this chapter, and many of the tools can be found in two or three of the steps listed below. The five steps for improving creativity in any organization are

1. Preparation
2. Incubation
3. Insight
4. Evaluation
5. Elaboration

We've focused on these five steps and their associated tools, which are covered in this chapter, in order to provide a clear and practical way for you to think about creativity, and to use it in your everyday life at work. Over the years, we have seen as many as 20 steps involving a process for creativity often used with lesser impact and success than the five steps outlined in this chapter. We like the idea of five steps because—in the areas of creativity and directed problem-solving rules and steps—less is more.

Step 1: Finding Opportunities and Problems to Solve (Preparation)

Tool 1: The Quickscore Creativity Test

Start by taking the MindTools Quickscore 3-Minute Creativity Test,* which helps you assess and develop your business creativity skills. We must emphasize that creativity is all about finding fresh and innovative solutions to problems, and identifying opportunities to improve the way that we do things, along with finding and developing new and different ideas. As such, anyone of us can be creative, as long as we have the right mind-set and use the right tools. This test helps you to think about how creative you are right now. Take it, and then use the tools and discussions that follow to bring intense creativity to your everyday work.

We have found that there are at least three core alternative attitudes related to creativity found in many organizations.

1. Attitude 1: Research and development is the only creative area in the organization.

2. Attitude 2: All employees are creative. We just need to sit back and let things happen.

3. Attitude 3: We must proactively stimulate and encourage all of our employees to be creative and pursue new and different ideas.

In today's environment, the third attitude is the only one that is acceptable. Our employees are our most valuable resource and most organizations are making poor use of this resource. We are missing an important opportunity if we are not providing an environment that requires our employees to be creative. We can no longer afford to just tap the employees' physical abilities. We must also make effective use of their mental abilities and help expand this capability if the organization is going to survive. Not making effective use of both the physical and mental abilities of our employees is a lot like having a two-car garage and only parking one car inside.

Scientists have known for a long time that our brain's right- and left-hand sides perform very different functions (see Fig. 13-1). The left-hand side is the logical reasoning side and the right-hand side is the creative side. Felix, like the left-hand side of our brain, is the functional, logical, technical, and planning individual. He loves lists; he wants everything in

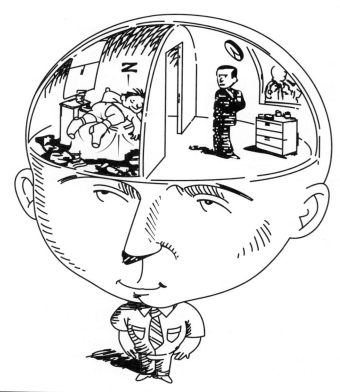

Figure 13-1   The Felix and Oscar sides of the brain.

*Based upon the MindTools Creativity Test. http://www.mindtools.com/community/pages/article/creativity-quiz.php.

order. He's upset in a confused environment. He has a strong desire to please other people. Oscar, on the other hand, is much like the right-hand side of our brain: very creative, very conceptual, has a tendency to get emotional very quickly, and thrives in confusion and ambiguities. Concepts that Oscar creates, Felix tries to put into a logical, structured order so that they can be implemented or he rejects the concept outright as being impractical.

The problem that we face in being creative is how we can get these two different parts of our brain to work together to accomplish previously unattainable results. But creativity alone is not enough. It has to be accompanied by innovation. Creativity and innovation are partners, but they are not the same thing. Creativity is developing new or different ideas. Innovation is converting ideas into tangible products, services, or processes. The challenge that every organization faces is how to convert good ideas into profit. That's what the creativity process is all about.

This means that the organization must have a system in place that will put an individual's ideas on the fast track of getting it approved and implemented. This requires that the creative or innovative process make a smooth transition from an individual, to a team, and to all of the impacted individuals. The creative thinking methodology uses the following process for accomplishing this type of organization:

- Embed creativity into the organization's culture and vision.
- Assess creativity status. An assessment of the creativity performance level of the organization should be made. Typical questions that would be answered are
  - Does the organization have a measurement of its return on its creativity investment?
  - What percentage of our effort is devoted to creative activities, and is that enough?
  - Does the organization have a chief creativity officer?
  - Do we have creativity goals and targets?
  - What percentage of our employees made a measurable creativity improvement in the past 12 months, and is that percentage high enough?
  - Are resources made available to support the refinement of new ideas?
  - What roadblocks are in the way of the organization becoming more creative?

Establish a creative thought process (see Fig. 13-2).

Example   The following are two flowcharts of different types of the thinking process. Figure 13-3 represents the conventional thought pattern. Figure 13-2 represents the creative thought pattern.

One factor that strongly affects an organization's creativity success rate is its attitude toward creativity and problem-opportunity-finding in the first place. On the creativity spectrum, it ranges from inactive to active, proactive, and hyperactive.* Regardless of the type of approach, creative people don't just sit around and wait for opportunities or problems to surface. Instead, they scan their environment for potential opportunities or issues, and they see this as exercising creativity time well spent, for in actuality they are often excited by the opportunity to change things. They aren't intimidated by change, rather, they embrace it.

Tool 2: Kano Analysis

Kano Analysis is one of the most useful creativity techniques for deciding which features you want to include in a product or service. It helps you break away from the common mind-set that says you've got to have as many features as possible in a product, and helps you think more subtly about the features that you do choose to include. The Kano model evolved based on the premise that a product or service can have three types of attributes or properties:

- Threshold attributes: Which attributes do customers expect to be present in a product.
- Performance attributes: Which attributes are not absolutely necessary, but which are known about and increase the customer's enjoyment of the product.
- Excitement attributes: Which attributes are the ones that customers don't even know they want, but are delighted when they find them.

---

*Russ, Ackoff discusses these four types and the resulting relationships that prevail in his outstanding book *Redesigning the Future*. The entire book has an explicit focus on Democracy as a System of Systems that cannot be "broken down" the way science strives to break down what it studies. In Democracy, as in Reflexivity, the engaged participants are wild cards, nothing can be predicted, agility and resilience are everything, and it is the relationships (the Yang) rather than the objects (the Ying) that really matter.

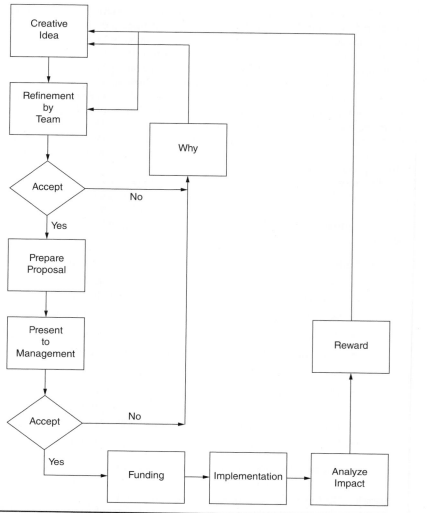

Figure 13-2 The creative thought process.

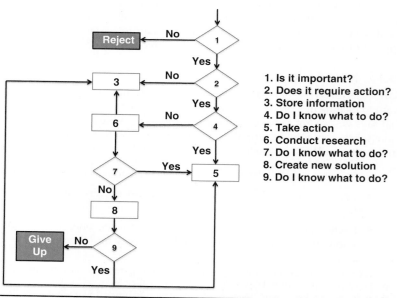

Figure 13-3 Conventional thought process.

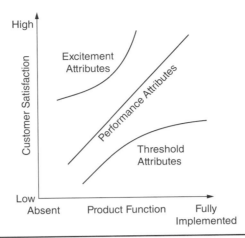

FIGURE 13-4 Kano analysis.

Threshold attributes affect customers' satisfaction with the product or service by their absence; if they're not present, customers are dissatisfied. And even if they're present, if no other attributes are present, customers aren't particularly happy. Using the example of a cell phone, the ability to store people's names and telephone numbers is a threshold attribute. While a cell phone without this function would work, it would be grossly inconvenient. According to Kano, it is usually on performance attributes that most products compete.

When we weigh up one product against another, and decide what price we're prepared to pay, we're comparing performance attributes. These are shown as the middle line on the graph in Fig. 13-4. For example, on a cell phone, performance attributes might be polyphonic ringtones or cameras, although to many teenagers, using polyphonic ringtones may be threshold attributes. Excitement attributes are things that people don't really expect, but which delight them. These are shown as the top curve on the graph in Fig.13-4. Even if only a few performance attributes are present, the presence of an excitement attribute will lead to high customer satisfaction.

To use Kano Model Analysis in a creative fashion, follow these steps:

1. Start by brainstorming all of the possible features and attributes of your product or service, and everything you can do to please and delight your customers.

2. Classify these as threshold, performance, excitement, or "not relevant."

3. Be sure that your program, product, or service has all appropriate threshold attributes. If necessary, cut out performance attributes so that you can get these included, as you're going nowhere fast if these aren't present.

4. Where possible, cut out attributes that are not relevant.

5. Next, look at the excitement attributes, and think how you can build some of these into your product or service. Again if necessary, cut some performance attributes, so that you can afford your excitement attribute.

6. Select appropriate performance attributes so that you can deliver a product or service at a price the customer is prepared to pay, while still maintaining a good profit margin.

Tool 3: Nominal Group Technique[3]

The Nominal Group Technique (NGT) is a powerful and time-tested group ideation and problem-solving technique involving the "triple crown" of problem identification, creative solution generation, and decision making. It can easily and consistently be used in groups of many types and sizes—groups or teams who want to make their decision quickly by voting—but who want at the same time everyone's input and opinions taken into account. This is opposed to traditional voting, where only the largest group is considered[4]—with the method of tallying being the essential difference and the main focus on establishing priorities rather than synthesizing.*

---

*The purpose is to encourage the useful and creative participation of all team members, as outlined in the Paper by Fran Voehl and Bill Roth called Picking the Right Tools and Techniques, 1996. Published as part of Chapter 3 in the book *Problem Solving for Results*, St Lucie Press, pp. 51–52.

First, every member of the group gives their view of the solution, with a short explanation. Then, duplicate solutions are eliminated from the list of all solutions, and the members proceed to rank the solutions, first, second, third, fourth, and so on. Some facilitators will encourage the sharing and discussion of reasons for the choices made by each group member, thereby identifying common ground, and a plurality of ideas and approaches. This diversity often allows the creation of a hybrid idea (combining parts of two or more ideas), often found to be even better than those ideas being initially considered.

In the basic method, the numbers each solution receives are totaled, and the solution with the highest (i.e., most favored) total ranking is selected as the final decision. There are variations on how this technique is used. For example, it can identify strengths versus areas in need of development, rather than be used as a decision-making voting alternative. Options may also be evaluated more subjectively.

The nominal group technique is particularly useful when any of the following conditions exist as inhibitors of creativity:

- When some group members are much more vocal than others.
- When some group members think better in silence.
- When there is concern about some members not participating.
- When the issue is controversial or there is heated conflict, or when there is a power imbalance between facilitator and participant or participants.
- When stakeholders like a (or some) quantitative output of the process.
- When the group does not easily generate quantities of ideas, or when all or some group members are new to the team.

Routinely, the technique involves five creativity-building stages:

1. Introduction and explanation: The facilitator welcomes the participants and explains to them the purpose and procedure of the meeting in terms of its core vision.

2. Silent generation of ideas: The facilitator provides each participant with a sheet of paper with the question to be addressed and asks them to write down all ideas that come to mind when considering the opportunity or question. During this period, the facilitator asks participants not to consult or discuss their ideas with others. This stage lasts approximately 10 minutes.

3. Sharing ideas: The facilitator invites participants to share the ideas they have generated. He records each idea on a flip chart using as close as possible the words spoken by the participant. The round-robin process continues until all ideas have been presented. There is no debate about items at this stage and participants are encouraged to write down any new ideas that may arise from what others share. This process ensures all participants get an opportunity to make an equal contribution and provides a written record of all ideas generated by the group. This stage may take 15 to 30 minutes.

4. Group discussion: Participants are invited to seek verbal explanation or further details about any of the ideas that colleagues have produced that may not be clear to them. The facilitator's task is to ensure that each person is allowed to contribute and that discussion of all ideas is thorough without spending too long on a single idea. It is important to ensure that the process is as neutral as possible, avoiding judgment and criticism. The group may suggest new items for discussion and combine items into categories, but no ideas should be eliminated. This stage lasts 30 to 45 minutes.

5. Voting and ranking: This involves prioritizing the recorded ideas in relation to the original question. Following the voting and ranking process, immediate results in response to the question are available to participants so the meeting concludes having reached a specific outcome.

Tool 4: Synectics

Synectics combines a structured approach to creativity with the freewheeling problem-solving approach used in techniques like brainstorming. It's a useful technique when simpler creativity techniques like *SCAMPER, brainstorming,* and *random input* (which are embedded within the synectics approach) have failed to generate useful ideas, as it uses many different triggers and stimuli to jolt people out of established mind-sets and into more creative ways of thinking. However, given the sheer range of different triggers and thinking approaches used within synectics, it can take much longer to solve a problem using it than with, say, traditional brainstorming—hence many experts classify its use as a "second-level tool" when other creativity techniques have failed. The problem is that no two experts view this tool in the same way, largely due to the inventor (William Gordon) purposefully incorporating a sense of vagueness in order to enhance its flexibility as a creativity-enhancing tool.

Generating ideas with synectics is a three- or four-approach/stage process:

1. Referring: Gathering information, and defining the opportunity in terms of direct analogies.

2. Reflecting: Using a wide range of techniques to generate ideas, including personal analogies.

3. Reconstructing: Bringing ideas back together to create a useful solution using a compressed conflict model.

4. Building fantastic energy: Users must let their imaginations ramble unrestrained, and to connect and concoct the most bizarre solutions imaginable, often described as the "fantastic analogy."

In the referring stage, you lay the foundations you'll use later for successful use of the tool.

Reflecting is where you creatively and imaginatively generate possible solutions to the problem you've defined. The emphasis here is on using a range of different "triggers" and "springboards" to generate associations and ideas. Just like brainstorming, reflecting is best done in a relaxed, spontaneous, and open-minded way with an emphasis on creative thinking rather than on critical assessment of suggestions. Where synectics differs from brainstorming and other creativity approaches is in the formal and systematic way it seeks to spark comparison with other approaches and situations, creating new ideas by making associations between these and the problem being solved. That said, a useful way of starting the synectics idea generation process is to brainstorm inside and around the opportunity or problem normally. This should generate a range of possible solutions to the problem.

If none of these solves the problem, the next step is to use some of the 22 possible triggers below to try to break free of existing thinking patterns. These triggers reflect things that you can do to transform your current product, service, or approach to try to solve the problem. They are

1. Subtract
2. Repeat
3. Combine
4. Add
5. Transfer
6. Empathize
7. Animate
8. Superimpose
9. Change scale
10. Substitute
11. Fragment
12. Isolate
13. Distort
14. Disguise
15. Contradict
16. Parody
17. Prevaricate
18. Analogize
19. Hybridize
20. Metamorphose
21. Symbolize
22. Mythologize

Tool Summary

Use triggers as starting points for brainstorming. Again, once you've done this, evaluate whether you have a satisfactory solution to the problem you're addressing. If you haven't, it's time to move to the next stage: using "synectics springboards" to stimulate new ideas. These are analogies between the current situation and other situations or things. They can be functional analogies (with other products, services, and approaches that do a similar job), analogies with other phenomena (for example, with an ocean storm, a rainforest or a mechanical digger), or stretched analogies (for example, comparisons with emotions or symbols). Reconstructing is where you collect all of the ideas you've created during the reflecting step, and evaluate them rationally, bringing them together to create practical and useful ideas.

Tool 5: Brainstorming or Operational Creativity

Brainstorming is often used in creativity because of its versatility. Brainstorming combines a relaxed, informal approach to problem solving with lateral thinking. It encourages people to come up with thoughts and ideas that can, at first, seem a bit crazy. Some of these ideas can be crafted into original, creative solutions to a problem, while others can spark even more ideas. This helps to get people unstuck by "jolting" them out of their normal ways of thinking.*

Therefore, during brainstorming sessions, people should avoid criticizing or rewarding ideas. You're trying to open up possibilities and break down incorrect assumptions about the problem's limits. Judgment and analysis at this stage stunts idea generation and limits creativity. Evaluate ideas at the end of the brainstorming session—this is the time to explore solutions further, using conventional approaches.

In most cases, brainstorming provides a free and open environment that encourages everyone to participate. Quirky ideas are welcomed and built upon, and all participants are encouraged to contribute fully, helping them develop a rich array of creative solutions. When used during ideation and problem solving, brainstorming brings team members' diverse experience into play. It increases the richness of ideas explored, which means that you can often find better solutions to the problems that you face. It can also help you get buy-in from team members for the solution chosen—after all, they're likely to be more committed to an approach if they were involved in developing it. What's more, because brainstorming is fun, it helps team members bond, as they solve problems in a positive, rewarding environment. Later, by using operational creativity, questions grow increasingly specific until the issue at hand is addressed.

While brainstorming can be effective, it's important to approach it with an open mind and a spirit of nonjudgment. If you don't do this, people "clam up," the number and quality of ideas plummets, and morale can suffer. While group brainstorming is often more effective at generating ideas than normal group problem solving, several studies have shown that individual brainstorming produces more—and often better—ideas than group brainstorming. This can occur because groups aren't always strict in following the rules of brainstorming, and bad behaviors creep in. Mostly, though, this happens because people pay so much attention to other people that they don't generate ideas of their own—or they forget these ideas while they wait for their turn to speak. This is called "blocking."

When you brainstorm on your own, you don't have to worry about other people's egos or opinions, and you can be freer and more creative. For example, you might find that an idea you'd hesitate to bring up in a group develops into something special when you explore it on your own. However, you may not develop ideas as fully when you brainstorm on your own, because you don't have the wider experience of other group members to draw on. Individual brainstorming is most effective when you need to solve a simple problem, generate a list of ideas, or focus on a broad issue. Group brainstorming is often more effective for solving complex problems.

With group brainstorming, you can take advantage of the full experience and creativity of all team members. When one member gets stuck with an idea, another member's creativity and experience can take the idea to the next stage. You can develop ideas in greater depth with group brainstorming than with individual brainstorming. Another advantage of group brainstorming is that it helps everyone feel that they've contributed to the solution, and it reminds people that others have creative ideas to offer. Brainstorming is also fun, so it can be great for team building!

---

*Madison Avenue advertising executive, Alex Osborn, developed the original approach to brainstorming and published it in his 1953 book, *Applied Imagination*. Since then, researchers have made many improvements to his original technique. The approach described here takes this research into account, so it's subtly different from Osborn's approach.

Group brainstorming can be risky for individuals. Unusual suggestions may appear to lack value at first sight—this is where you need to chair sessions tightly, so that the group doesn't crush these ideas and stifle creativity. Where possible, brainstorming participants should come from a wide range of disciplines. This cross-section of experience can make the session more creative. However, don't make the group too big: as with other types of teamwork, groups of five to seven people are usually most effective.

How to Use the Tool

You often get the best results by combining individual and group brainstorming, and by managing the process according to the rules below. By doing this, you can get people to focus on the issue without interruption, you maximize the number of ideas that you can generate, and you get that great feeling of team bonding that comes with a well-run brainstorming session.

Rule 1: Prepare the Group   First, set up a comfortable meeting environment for the session. Make sure that the room is well-lit and that you have the tools, resources, and refreshments that you need. How much information or preparation does your team need in order to brainstorm solutions to your problem? Remember that prep is important, but too much can limit—or even destroy—the freewheeling nature of a brainstorming session.

Consider who will attend the meeting. A room full of like-minded people won't generate as many creative ideas as a diverse group, so try to include people from a wide range of disciplines, and include people who have a variety of different thinking styles. When everyone is gathered, appoint one person to record the ideas that come from the session. This person shouldn't necessarily be the team manager—it's hard to record and contribute at the same time. Post notes where everyone can see them, such as on flip charts or whiteboards or use a computer with a data projector. If people aren't used to working together, consider using an appropriate warm-up exercise, or an icebreaker.

Rule 2: Present the Opportunity   Clearly define the opportunity that you want to solve, and lay out any criteria that you must meet. Make it clear that that the meeting's objective is to generate as many ideas as possible. Give people plenty of quiet time at the start of the session to write down as many of their own ideas as they can. Then, ask them to share their ideas, while giving everyone a fair opportunity to contribute.

Rule 3: Guide the Discussion   Once everyone has shared their ideas, start a group discussion to develop other people's ideas, and use them to create new ideas. Building on others' ideas is one of the most valuable aspects of group brainstorming. Encourage everyone to contribute and to develop ideas, including the quietest people, and discourage anyone from criticizing ideas. As the group facilitator, you should share ideas if you have them, but spend your time and energy supporting your team and guiding the discussion. Stick to one conversation at a time, and refocus the group if people become sidetracked. Although you're guiding the discussion, remember to let everyone have fun while brainstorming. Welcome creativity, and encourage your team to come up with as many ideas as possible, regardless of whether they're practical or impractical, or some unexpected ideas.

Don't follow one train of thought for too long. Make sure that you generate a good number of different ideas, and explore individual ideas in detail. If a team member needs to "tune out" to explore an idea alone, allow them the freedom to do this. Also, if the brainstorming session is lengthy, take plenty of breaks so that people can continue to concentrate. If you're not getting enough good quality ideas, try using an operational creativity technique to increase the number of ideas that you generate.

Tool Summary

After your individual or group brainstorming session, you'll have a lot of ideas. Although it might seem hard to sort through these ideas to find the best ones, analyzing these ideas is an important next step, and you can use several tools to do this. When managed well, brainstorming can help you generate radical solutions to problems. Brainstorming can also encourage people to commit to solutions, because they have provided input and played a role in developing them. The best approach to brainstorming combines individual and group brainstorming. During the brainstorming process, there should be no criticism of ideas, and creativity should be encouraged.

Tool 6: Six Thinking Hats

"Six Thinking Hats" is an important and powerful technique.[5] It is used to look at decisions from a number of important perspectives. This forces you to move outside your habitual thinking style, and helps you to get a more rounded view of a situation. Many people think from a very rational, positive viewpoint, which is part of the reason that they are successful. Often, though, they may fail to look at a problem from an emotional, intuitive, creative, or negative viewpoint. This can mean that they underestimate resistance to plans, fail to make creative leaps, and do not make essential contingency plans.

Similarly, pessimists may be excessively defensive, and more emotional people may fail to look at decisions calmly and rationally. The thinking is that if you look at a problem with the Six Thinking Hats technique, then you will solve it using any and all approaches. Your decisions and plans will mix ambition, skill in execution, public sensitivity, creativity, and good contingency planning. You can use Six Thinking Hats in meetings or on your own. In meetings it has the benefit of blocking the confrontations that happen when people with different thinking styles discuss the same problem. Since Chap. 14 discusses the Six Thinking Hats in greater detail, this is all we will discuss in this chapter.

Step 2: Gathering and Reflecting on Information (Incubation)

When you have a potential opportunity for idea creation or a problem to be solved, you must gather as much information as the time allows. As part of this process, investigate solutions that have been tried previously (both in your own organization, and in other areas), and identify ideas that might have surfaced, but were never acted on. At this stage it's also a good idea to step away from the problem for a while, and allow new thoughts and ideas to enter your mind.

When we allow ourselves to concentrate on one issue for too long, there is a tendency to latch onto one or two ideas at first, and this can block other good ideas. (One of the benefits of being proactive in your problem finding is that you have time to incubate ideas, rather than being pressured to find an immediate solution to a problem.)

There are three major tools in this area (in addition to synectics, brainstorming or operational creativity, and nominal group techniques outlined in step 1) as follows:

Tool 7: Attribute Listing, Morphological Analysis, and Matrix Analysis

Attribute listing, morphological analysis, and matrix analysis are good techniques for finding new combinations of products or services. They are sufficiently similar to be discussed together. We use attribute listing and morphological analysis to generate new products and services. To use these tools and techniques, first list the attributes of the product, service, or strategy you are examining. Attributes are parts, properties, qualities, or design elements of the thing being looked at. For example, attributes of a pencil would be shaft material, lead material, hardness of lead, width of lead, quality, color, weight, price, and so on.

Draw up a table using these attributes as column headings. Write down as many variations of the attribute as possible within these columns. This might be an exercise that benefits from brainstorming. The table should now show all possible variations of each attribute. Now select one entry from each column. Either do this randomly or select interesting combinations. By mixing one item from each column, you will create a new mixture of components. This is a new product, service, or strategy. Finally, evaluate and improve that mixture to see if you can imagine a profitable market for it.

Imagine that you want to create a new lamp. The starting point for this might be to carry out a morphological analysis. Properties of a lamp might be power supply, bulb type, light intensity, size, style, finish, material, shade, and so on. You can set these out as column headings on a table, and then brainstorm variations (see Table 13-1). This table is sometimes

Power Supply	Bulb Type	Light Intensity	Size	Style	Finish	Material
Battery	Halogen	Low	Very large	Modern	Black	Metal
Mains	Bulb	Medium	Large	Antique	White	Ceramic
Solar	Daylight	High	Medium	Roman	Metallic	Concrete
Generator	Colored	Variable	Small	Art Nouveau	Terracotta	Bone
Crank			Handheld	Industrial	Enamel	Glass
Gas				Ethnic	Natural	Wood
Oil/Petrol					Fabric	Stone
Flame						Plastic

TABLE 13-1   The Zwicky Box

known as a "Morphologial Box" or "Zwicky Box" after the scientist Fritz Zwicky, who developed the technique in the 1960s.

Interesting combinations might be

- Solar powered/battery, medium intensity, daylight bulb—possibly used in clothes shops to allow customers to see the true color of clothes.

- Large hand-cranked arc lights—used in developing countries, or far from a mains power supply.

- A ceramic oil lamp in Roman style—used in themed restaurants, resurrecting the olive oil lamps of 2000 years ago.

- A normal table lamp designed to be painted, wallpapered, or covered in fabric so that it matches the style of a room perfectly.

Some of these might be practical, novel ideas for the lighting manufacturer. Some might not. This is where the manufacturer's experience and market knowledge are important. Morphological analysis, matrix analysis, and attribute listing are useful techniques for making new combinations of products, services, and strategies.

Tool Summary

Attribute listing focuses on the attributes of an object, seeing how each attribute could be improved. Morphological analysis uses the same basic technique, but is used to create a new product by mixing components in a new way. Matrix analysis focuses on businesses. It is used to generate new approaches, using attributes such as market sectors, customer needs, products, promotional methods, and so on.

Tool 8: Generating New Ideas with Storyboarding

Standard idea generation techniques concentrate on combining or adapting existing ideas. This can certainly generate results. The storyboard approach is a creativity-enhancing tool in that it stretches your mind to forge new connections, think differently, and consider new perspectives. The storyboard technique combines brainstorming and lateral thinking with a studio-type system for developing film plots. The facilitator brings along a flip chart and corkboard, thumbtacks, and a good supply of 5×8 in blank index cards.[6]

The facilitator begins by describing the opportunity or issue to be resolved and the participants name the potential solution categories. For example, suppose the problem is—what should we do with the people who are part of an operation that is being shut down? The categories might include reorientation, relocation, release retraining, reduction to a part-time role, retirement, and so on. Each category is handprinted on a card, which are pinned along the top of the corkboard. A word of caution—while these techniques are extremely effective, they will only succeed if they are backed by rich knowledge of the area you're working on. This means that if you are not prepared with adequate information about the issue, you are unlikely to come up with a great idea even by using the techniques listed here.

The facilitator then asks a series of idea-generating questions such as *How can these employees be oriented in a way that will profit the company, the individual? The participants then write their ideas on the 5 × 8 in cards, one idea per card. The cards are collected, read aloud without naming the author and without criticism, grouped according to common themes, and pinned on the board under the proper category card. Depending on what he or she sees, the facilitator asks related questions to help generate more creativity and ideation.*

Remember, the following techniques can be applied to spark creativity in group settings and brainstorming sessions in order to break and/or create thought patterns. All of us can tend to get stuck in certain thinking patterns. Breaking these thought patterns can help you get your mind unstuck and generate new ideas. There are several techniques you can use to break established thought patterns:

- Challenge assumptions: For every situation, you have a set of key assumptions. Challenging these assumptions gives you a whole new spin on possibilities.

- Reword the problem: Stating the problem differently often leads to different ideas. To reword the problem, look at the issue from different angles. You might come up with new ideas to solve your new problem.

- Think in reverse: If you feel you cannot think of anything new, try turning things upside-down. Instead of focusing on how you could solve a problem, improve operations, or enhance a product, consider how could you create the problem, worsen operations, or downgrade the product. The reverse ideas will come flowing in. Consider these ideas—once you've reversed them again—as possible solutions for the original challenge.

- Express yourself through different media. We have multiple intelligences but somehow, when faced with workplace challenges, we just tend to use our verbal reasoning ability. How about expressing the challenge through different media? Clay, music, word association games, paint—there are several ways you can express the challenge. Don't bother about solving the challenge at this point. Just express it. Different expressions might spark off different thought patterns. And these new thought patterns may yield new ideas.

Some of the best ideas seem to occur just by chance. You see something or you hear someone, often totally unconnected to the situation you are trying to resolve, and the penny drops in place. Newton and the apple, Archimedes in the bathtub, examples abound. Why does this happen? The random element provides a new stimulus to your brain. You can capitalize on this knowledge by consciously trying to connect the unconnected.

Actively seek stimuli from unexpected places and then see if you can use these stimuli to build a connection with your situation. Some techniques you could use involve using random input—choose a word from the dictionary and look for novel connections between the word and your problem. Also, try mind mapping possible ideas: put a key word or phrase in the middle of the page. Write whatever else comes in your mind on the same page. See if you can make any connections. Over the years we all build a certain type of perspective and this perspective yields a certain type of idea. If you want different ideas, you will have to shift your perspective. To do so get someone else's perspective: Ask different people what they would do if faced with your challenge. You could approach friends engaged in different kinds of work, your spouse, a 9-year-old child, customers, suppliers, senior citizens, someone from a different culture, in essence anyone who might see things differently.

Next, play the "If I were" game: Ask yourself "If I were ..." how would I address this challenge? You could be anyone: a millionaire, LeBron James, anyone. The idea is the person you decide to be has certain identifiable traits. And you have to use these traits to address the challenge.

How to Employ Enablers

Enablers are activities and actions that assist with, rather than directly provoke, idea generation. They create a positive atmosphere. Some of the enablers that can help you get your creative juices flowing are

- Belief in yourself: Believe that you are creative, believe that ideas will come to you; positive reinforcement helps you perform better.

- Creative loafing time: Nap, go for a walk, listen to music, play with your child, take a break from formal idea generating. Your mind needs the rest, and will often come up with connections precisely when it isn't trying to make them.

- Change of environment: Sometimes changing the setting changes your thought process. Go to a nearby coffee shop instead of the conference room in your office, or hold your discussion while walking together around a local park.

- Shutting out distractions: Keep your thinking space both literally and mentally clutter-free. Shut off the BlackBerry, close the door, divert your phone calls, and then think.

- Fun and humor: These are essential ingredients, especially in team settings.

Tool Summary

When the process has been completed for each category, the groupings are discussed and ranked by voting. The ability to generate new ideas is an essential work skill today. You can acquire this skill by consciously practicing techniques that force your mind to forge new connections, break old thought patterns, and consider new perspectives.

At this point, several things can occur: (1) categories and groupings can be deleted if none of the ideas contained are popular enough, (2) new categories or groupings can be created with various idea formations and combinations, or (3) ideas can be shifted to an entirely new category or grouping. Note that ideas can also be duplicated and pinned under two or more groupings.* Along with practicing these techniques, you need to adopt enabling strategies too. These enabling strategies help in creating a positive atmosphere that boosts creativity.

Tool 9: Absence Thinking

Absence Thinking involves training your mind to think creatively about what you are "thinking" and "not thinking." When you are thinking about a specific something, you often notice what is not there, you watch what people are not doing, and you make lists of things which you normally forget. In other words, you try to deliberately think about what is absent and envision "what is not there."[7] Both individuals and groups can use it when you are stuck and unable to shift thinking to other modes. Also, they can use it when you want to do something that has not been done before.

How to Use It
1. Divide a piece of paper in half and consider what you are thinking about on one side, then do the same with what you are not thinking about on the other side.
2. Next, when you are looking at something (or otherwise sensing), notice what is *not* there.
3. You can also watch people and notice what they do not do.
4. Some use it to make lists of things to remember that you normally forget.

In summary, to be really useful, deliberately and carefully think about what is absent.

Tool Summary
Think about an artist who draws the spaces between things, or a manager for a furniture warehouse who wonders about product areas where customers have made no comment. He or she watches them using tables and notes that they leave the tables out when not using them. The creative act is that she invents a table that can be easily folded and stored. The psychology of the creative thinking process is such that while we are very good at seeing what is there, we need to do a better job at seeing what is not there. Absence Thinking compensates for this by deliberately forcing us to do what does not come to us naturally and easily.

Step 3: Opportunity Exploration (Insight)

Once you've identified and verified your opportunity, you can figure out what's really going on. Often, the initial issue that you identified will turn out to be a symptom of a deeper opportunity. Therefore, identifying the roots of the opportunity at issue is extremely important. There are six major tools in this area, of which three have already been covered and the remaining three are discussed below.

Tool 10: Breakdown (Drilldown) Tree Diagram

Breakdown or drilldown is a simple technique for breaking complex opportunities and/or problems down into progressively smaller parts. To use the technique, start by writing the opportunity statement or problem under investigation down on the left-hand side of a large sheet of paper. Next, write down the points that make up the next level of detail a little to the right of this. These may be factors contributing to the issue or opportunity, information relating to it, or questions raised by it. This process of breaking the issue under investigation into its component part is called drilling down.

---

*The notion of generating new ideas with Storyboarding was first developed by Frank Voehl and Bill Roth based upon the work at FPL and Wharton Business School, and the findings of the Tavistock Experiments, made popular by Eric and Beulah Trist at Wharton. Trist was heavily influenced by Kurt Lewin, whom he met first 1933 in Cambridge, England. Kurt Lewin had moved from studying behavior to engineering its change, particularly in relation to racial and religious conflicts, creating Sensitivity Training, a technique for making people more aware of the effect they have on others, which some claim as the beginning of the concept of Political Correctness. This would later influence the direction of much of work at the Tavistock Institute in the direction of management and, some would say, manipulation, rather than fundamental research into human behavior and the psyche. It was a partnership between Trist's group at the Tavistock, and Lewin's at MIT that launched the *Human Relations Journal* just before Lewin's death in 1947.

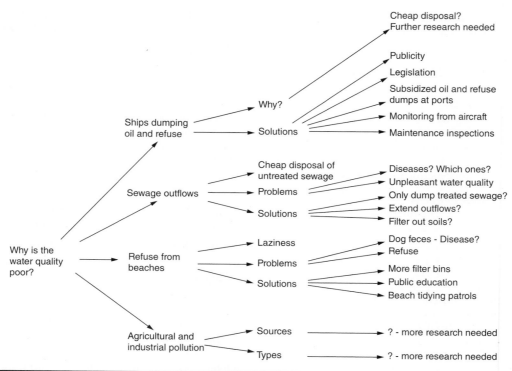

Figure 13-5 Drill down into problem of poor seawater quality.

For each of these points, repeat the process. Keep on drilling down into points until you fully understand the factors contributing to the situation or opportunity being explored. If you cannot break them down using the knowledge you have, then carry out whatever research is necessary to understand the point. Drilling into a question helps you to get a much deeper understanding of it. The process helps you to recognize and understand the factors that contribute to it. Drilldown prompts you to link in information that you had not initially associated with a problem. It also shows exactly where you need further information.

> **Example** The owner of a Florida windsurfing club is having complaints from its members about the unpleasant quality of the water close to the Biscayne Bay shoreline, which according to many members seems like it has been a huge problem for some years now. The owner brings in a Director of Creativity who carries out the drilldown analysis in the generic Fig. 13-5.

Tool Summary

Drilldown gives creative people a starting point in which to begin thinking about the situation and starts prompting their creativity and curiosity. It highlights where they do not fully understand the facts at hand, and shows where they need to carry out further research. Drilldown helps you to break a large and complex opportunity statement down into component parts, so that you can develop plans to deal with these parts. It also shows you which points you need to research in more detail.

Tool 11: Lotus Blossom

The Lotus Blossom technique was originally developed by Yasuo Matsumura, Director of Clover Management Research (Japan). This technique is based on the use of analytical capacities and helps to generate a great number of ideas that will possibly provide the best solution to the problem to be addressed by the management group.

It can be broken down into six major steps:

1. Draw up a lotus blossom diagram (see Fig. 13-6) made up of a square in the center of the diagram (the pistil) and eight circles (petals) surrounding the square.

2. Write the central idea or problem in the center of the diagram (light square).

3. Look for ideas or solutions for the central theme. Then write them in the flower petals (darker circles).

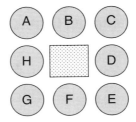

FIGURE 13-6 Lotus blossom.

Example The main idea of a company was to build a creative atmosphere within the Harrington Institute. The trainers, consultants, and staff wrote this core idea or statement in the center of a lotus blossom. Afterward, during a debate, they devised eight ideas about the main theme:

- Offer contexts for ideas
- Create a challenging atmosphere
- Organize meetings of creative thinking
- Generate paths to get out of the box
- Generate a positive attitude
- Make up a creativity board
- Make the job fun
- Expand the meaning of the job

4. These ideas were written in the circles around the main square. Each idea written in the circles becomes the central theme of a new lotus blossom (see Fig. 13-7).

5. Follow step 3 with all central ideas.

6. Continue the process until all ideas have been used.

Tool Summary

This technique has helped the Harrington Institute over the years to create many interesting ideas. Among the others, we helped to set up a special empowerment room for creative thinking which was furnished with multimedia creativity books, videos, educational toys, and so on. Moreover, the tool was used with drawings, and the room was decorated with these drawings made by the staff and family members to remind everyone that we are all born innocent and creative.

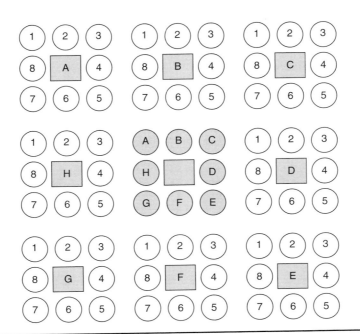

FIGURE 13-7 Lotus blossoms.

Step 4: Generating and Evaluating Ideas (Evaluation)

When you have clear insight into the essence of the issue, you can move onto generating ideas for a creative solution. Here you want to look for as many ways to inspire ideas as possible. Obviously not all of the ideas you have will be practical or possible. So, as part of this step in the creativity process, you need to decide which criteria you'll use to evaluate your ideas. Without a solid evaluation process, you'll be prone to choosing a solution that is perhaps too cautious.

The following are a few key creativity tools you can use for the generation and evaluation of new ideas.

Tool 12: TRIZ Analysis

TRIZ is a problem-solving methodology based on logic and data to solve problems creatively. As such, TRIZ brings repeatability, predictability, and reliability to the problem-solving process with its structured and algorithmic approach. More than 3 million patents have been analyzed to discover the patterns that predict breakthrough solutions to problems, and these have been codified within TRIZ.

The three primary findings of the last 65 years of research are as follows:

1. Problems and solutions are repeated across industries and sciences. By classifying the "contradictions" (see later) in each problem, you can predict good creative solutions to that problem.

2. Patterns of technical evolution tend to be repeated across industries and sciences.

3. Creative innovations often use scientific effects outside the field where they were developed.

Much of the practice of TRIZ consists of learning these repeating patterns of problems-solutions, patterns of technical evolution, and methods of using scientific effects, and then applying the general TRIZ patterns to the specific situation that confronts the developer. Figure 13-8 describes this process graphically.

The basic approach as explained by Sussman and Zlotin is to take the specific problem you face, and generalize it to one of the TRIZ general problems. From the TRIZ general problems, you identify the TRIZ solutions to those general problems, and then see how these can be applied to the specific problem you face.*

Since TRIZ is covered extensively in Chap. 24 of this book, we will not go into further detail. Just remember that it is number 12 on our list of tools.

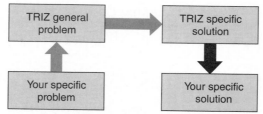

The arrows represent transformation from one formulation of the problem or solution to another. The gray arrows represent analysis of the problems and analytic use of the TRIZ databases. The black arrow represents thinking by analogy to develop the specific solution.

Figure 13-8   The TRIZ problem-solving method.

*TRIZ research began with the hypothesis that there are universal principles of creativity that are the basis for creative innovations, and that advance technology. The idea was that if these principles could be identified and codified, they could be taught to people to make the process of creativity more predictable. The short version of this is: Somebody someplace has already solved this problem or one very similar to it. Today, creativity involves finding that solution and adapting it to this particular problem.

Tool 13: SCAMPER

It can often be difficult to come up with new ideas when you're trying to develop or improve a product or service. This is where creative brainstorming techniques like SCAMPER can help. This tool helps you generate ideas for new products and services by encouraging you to think about how you could improve existing ones.*

SCAMPER is a mnemonic that stands for

- Substitute

- Combine

- Adapt

- Modify

- Put to another use

- Eliminate

- Reverse

You use the tool by asking questions about existing products, using each of the seven prompts above. These questions help you come up with creative ideas for developing new products, and for improving current ones. SCAMPER is really easy to use. First, take an existing product or service. This could be one that you want to improve, one that you're currently having problems with, or one that you think could be a good starting point for future development. Then, ask questions about the product you identified, using the SCAMPER mnemonic to guide you. Brainstorm as many questions and answers as you can. (We've included some example questions, later.) Finally, look at the answers that you came up with. Do any stand out as viable solutions? Could you use any of them to create a new product, or develop an existing one? If any of your ideas seem viable, then you can explore them further.

Let's look at some of the questions you could ask for each letter of the SCAMPER mnemonic:

Substitute

- What materials or resources can you substitute or swap to improve the product?

- What other product or process could you use?

- What rules could you substitute?

Combine

- What would happen if you combined this product with another, to create something new?

- What if you combined purposes or objectives?

- What could you combine to maximize the uses of this product?

Adapt

- How could you adapt or readjust this product to serve another purpose or use?

- What else is the product like?

- Who or what could you emulate to adapt this product?

Modify

- How could you change the shape, look, or feel of your product?

- What could you add to modify this product?

- What could you emphasize or highlight to create more value?

Put to another use

- Can you use this product somewhere else, perhaps in another industry?

- Who else could use this product?

- How would this product behave differently in another setting?

---

*Alex Osborn, credited by many as the originator of brainstorming, originally came up with many of the questions used in the SCAMPER technique. However, it was Bob Eberle, an education administrator and author, who organized these questions into the SCAMPER mnemonic.

Eliminate

- How could you streamline or simplify this product?
- What features, parts, or rules could you eliminate?
- What could you understate or tone down?

Reverse

- What would happen if you reversed this process or sequenced things differently?
- What if you try to do the exact opposite of what you're trying to do now?
- What components could you substitute to change the order of this product?

Some ideas that you generate using the tool may be impractical or may not suit your circumstances. Don't worry about this—the aim is to generate as many ideas as you can.*

Step 5: Implementation (Elaboration)

A common misconception is that creative people spend all their time thinking of new and interesting ideas. In fact, truly creative people recognize a good idea and run with it. A famous Thomas Edison quote supports this: "Creativity is one percent inspiration and 99 percent perspiration." For this final step, you need to be committed to taking your ideas and making them happen, and you need to be confident that you can, indeed, map out and propose innovative ideas which inspire change.

Tool 14: Mind Mapping

Mind maps are a powerful creative approach to note-taking *(also known as Mind Mapping, Concept Mapping, Spray Diagrams, and Spider Diagrams).†*

Have you ever studied a subject or brainstormed an idea, only to find yourself with pages of information, but no clear view of how it fits together? This is where mind mapping can help you. Mind mapping is a useful technique that helps you learn more effectively, improves the way that you record information, and supports and enhances creative problem solving. By using mind maps, you can quickly identify and understand the structure of a subject. You can see the way that pieces of information fit together, as well as recording the raw facts contained in normal notes.

More than this, mind maps help you remember information, as they hold it in a format that your mind finds easy to recall and quick to review. Mind maps were popularized by author and consultant, Tony Buzan. They use a two-dimensional structure, instead of the list format conventionally used to take notes. Mind maps are more compact than conventional notes, often taking up one side of paper. This helps you to make associations easily, and generate new ideas. If you find out more information after you have drawn a mind map, then you can easily integrate it with little disruption.

More than this, mind mapping helps you break large projects or topics down into manageable chunks, so that you can plan effectively without getting overwhelmed and without forgetting something important. A good mind map shows the shape of the subject, the relative importance of individual points, and the way in which facts relate to one another. This means that they're very quick to review, as you can often refresh information in your mind just by glancing at one.

This topic is covered in detail in Chap. 15, so a detailed discussion will not be done in this chapter. Once again, know that this is a very effective technique for creativity.

Tool 15: Affinity Diagram

The Affinity Diagram is a technique for organizing a variety of subjective data (such as options) into categories based on the intuitive relationships among individual pieces of information. It is often used to find commonalties among concerns and ideas. At its core, the affinity diagram is a creativity-building method of organizing ideas based on their natural relationships. It is particularly useful for organizing thoughts after a brainstorming session that has generated a large number of ideas. If used effectively, it is a creative process that lets new patterns and relationships between ideas be discovered, leading to more creative solutions. The technique

---

*To get the greatest benefit, use SCAMPER alongside other creative brainstorming and lateral thinking techniques such as Random Input, Provocation, Reversal, and Metaphorical Thinking.
†*"Mind Map" is a trademark of the Buzan Organization.*

is usually used in group problem solving, but the approach can also be used by individuals for the same objective.

Steps

1. State the problem that is to be considered in broad terms. Avoid detailed problem statements.

2. Brainstorm ideas on how to solve the problem. Record the responses on 3×5 in index cards.

3. Shuffle the cards and place them randomly in the middle of a table.

4. The group should silently sort the cards into piles of related ideas. Limit the number of piles to between five and ten. Team members should do this exercise quickly, relying more on their first impressions than logical thought.

5. For each pile of cards, pick a card that best represents the theme in the pile. Put that card on top. If no card clearly stands out as representative of the whole group, and if another idea comes to mind which does summarize the theme, then create a new card for that theme and put it on top of the pile.

6. Using the card groupings, record the grouped ideas on paper. Draw lines around each grouping.

Guidelines and Tips

- This technique works best with groups of six to eight people.

- When brainstorming ideas, short statements are best. Statements should be as concise as possible and still have a clear meaning. Use statements with a noun and a verb.

- The recommendation to sort cards quickly cannot be stressed enough. By using intuitive processes instead of logic in grouping the cards, new patterns and relationships can be found between elements of a problem. This can lead to more creative solutions.

- If using this technique as an individual, take some time off in between idea generation and using the affinity diagram to organize those ideas. Going from idea generation directly into organizing thoughts will not be as creative, since the associations that led from one idea to another in idea generation will result in established patterns of thought.

- One way to do it is to record the responses on 3×5 in post-it stickers and attach them onto a whiteboard as the ideas are generated. This makes it easy to move them around into groupings.

Examples   The Acme Fulfillment Company had the following issues involved in missing promised delivery dates:

- Table 13-2 is a list of ideas generated in a brainstorming meeting on this problem.
- Figure 13-9 is the same brainstorming ideas organized in an affinity diagram.

Tool Summary

The affinity diagram is a business tool used to organize ideas and data. The tool is commonly used within creativity sessions and allows large numbers of ideas stemming from brainstorming* to be sorted into groups, based on their natural relationships, for review and analysis. It is also frequently used in creativity contextual inquiry sessions as a way to organize notes and insights from field interviews. It can also be used for organizing other free-form comments, such as open-ended survey responses, support call logs, or other qualitative data. People have been grouping data into groups based on natural relationships for thousands of years; however, the term affinity diagram was devised by Jiro Kawakita in the 1960s and is sometimes referred to as the *JK method*.

Tool 16: Force Field Analysis Diagram

Force Field Analysis is a visual aid for pinpointing and analyzing elements that resist change (restraining forces) or push for change (driving forces). This technique helps drive improvement by developing plans to overcome the restrainers and make maximum use of the driving forces.

---

*For further info, see the Online AFFINITY tool at Discover 6 Sigma Lab.

• Hire freight carrier based on lowest rate available	• Wrong count by operators on production floor
• Dock is overcrowded	• Need classification by type of error
• Historical trends of errors	• Frequently change freight carriers
• What is a "shipping error"?	• Too slow getting replacement product or paperwork
• Time lag in order changes on computer	• Don't tell customer if shipping error is known but not detected by customer
• When big customer push, switch labels and ship	• Allow changes to orders over the phone
• Shipping errors only in certain product lines	• Sometimes substitute facsimile product is unavailable
• Data entry complexity	• Computer system too slow—use handwritten forms instead
• Difficult to measure true cost of errors	• Labels fall off boxes
• High turnover among shippers	• High turnover among data entry clerks
• New 11-digit code too long	• No place to segregate customer returns
• Bar codes damaged/unreadable	• Some new, reusable packaging has wrong bar codes
• Dock used to store material for return to vendors	• Lack of training for data entry clerks
• How many are paperwork errors—right product shipped?	• Newest employees go to shipping
• Shipping sometimes contracted directly by sales representative to rush orders	• Production bonus system encourages speed, not accuracy
• Old shipping boxes easily damaged, requiring replacement	• Dock is coldest place in winter, hottest place in summer
• Customer orders still initially handwritten	• How many customers lost that we don't know about?

TABLE 13-2 List of Ideas Generated from a Brainstorming Session

The force field analysis technique has been used in a number of settings to do the following:

- Analyze a problem situation into its basic components.
- Identify those key elements of the problem situation about which something can realistically be done.
- Develop a systematic and insightful strategy for problem solving which minimizes "boomerang" effects and irrelevant efforts.
- Create a guiding set of criteria for the evaluation of the action step.

The technique is an effective device for achieving each of these purposes when it is seriously employed. The level of the activity, to put it differently, is the starting point in the problem identification and analysis. In order to constitute a problem, the current level typically departs from some implicit norm or goal.* A particular activity level may be thought of as resulting from a number of pressures and influences acting upon the individual, group, or organization in question.

---

*Kurt Lewin, who developed force field analysis, has proposed that any problem situation be it the behavior of an individual or group, the current state or condition of an organization, a particular set of attitudes, or frame of mind—may be thought of as constituting a level of activity which is somehow different from that desired. For example, smoking, as an activity may become the basis for a problem when it occurs with greater intensity or at a higher level than one desires. Quality, as another example of an activity level, may become a problem when it is at a lower-than-desirable level. Depression or authoritarianism, as examples of attitudinal activity levels, become problems when they are too intense or at higher-than-desirable levels.

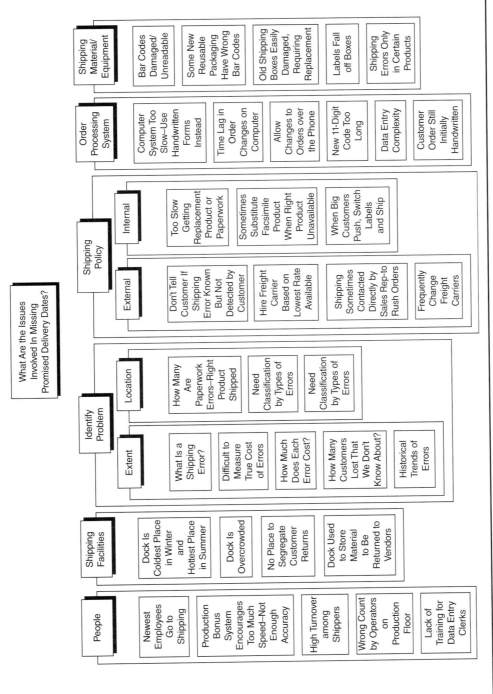

FIGURE 13-9 Affinity diagram of brainstorming ideas from Table 13-2.

These numerous influences Lewin calls "forces" and they may be both external to and internal to the person or situation in question. Lewin identifies two kinds of forces:

1. Driving or facilitating forces that promote the occurrence of the particular activity of concern, and

2. Restraining or inhibiting forces that inhibit or oppose the occurrence of the same activity

An activity level is the result of the simultaneous operation of both facilitating and inhibiting forces. The two force fields push in opposite directions and, while the stronger of the two will tend to characterize the problem situation, a point of balance is usually achieved which gives the appearance of habitual behavior or of a steady-state condition. Changes in the strength of either of the fields, however, can cause a change in the activity level of concern. Thus, apparently habitual ways of behaving, or frozen attitudes, can be changed (and related problems solved) by bringing about changes in the relative strengths of facilitating and inhibiting force fields.

In order to appreciate just what kinds of forces are operating in a given situation and which ones are susceptible to influence, a force field analysis must be made. As a first step to a fuller understanding of the problem, the forces (both facilitating and inhibiting) should be identified as fully as possible. Identified forces should be listed and, as much as possible, their relative contributions or strengths should be noted.

Once the problem has been recognized, and commitment is made by the appropriate stakeholders to change the problem situation, there are four basic steps used in the force field analysis activity to analyze the problem:

1. Define the problem and propose an ideal solution.

2. Identify and evaluate the forces acting on the problem situation.

3. Develop and implement a strategy for changing these forces.

4. Reexamine the situation to determine the effectiveness of the change and make further adjustments if necessary.

The first step includes defining the problem and proposing the ideal situation. In defining the problem, it is necessary to say exactly what it is.

1. Propose an ideal situation in a "goal statement." It can be prepared by answering the question "What will the situation be like when the problem is solved?" The answer must be tested to determine if it really gets to the heart of the problem.

2. Another possible question is "What would the situation be like if everything were operating ideally?" (See Fig. 13-10.)

Determining the precise goal statement is important because it guides the rest of the problem-solving steps. The second step is to identify and evaluate the forces that act on the problem, the goal, are restraining or inhibiting forces.

1. The facilitating forces tend to move the problem situation from reality toward the ideal. The restraining forces resist the movement toward the ideal state, and, in a state of equilibrium, counterbalance the facilitating forces. (See Fig. 13-11.)

2. You can visualize a problem situation by drawing a line down a sheet of paper and listing the facilitating forces on one side and restraining forces on the other side. Each of these forces has its own weight and taken together they keep the field in balance. (See Fig. 13-12.)

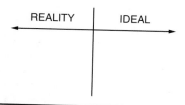

Figure 13-10   Reality versus ideal.

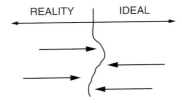

FIGURE **13-11** The "as-is" state in equilibrium.

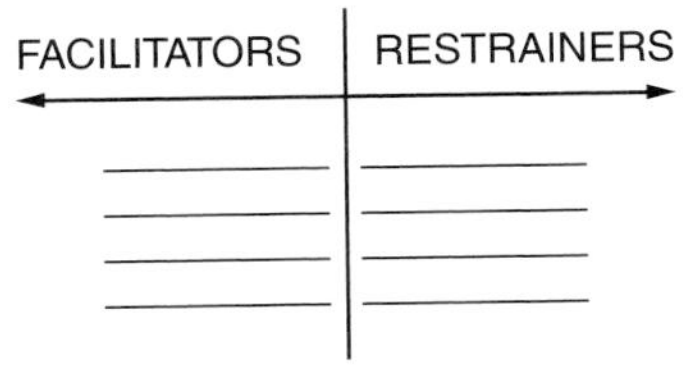

FACILITATORS │ RESTRAINERS

FIGURE **13-12** In balance.

3. In addition to helping make the problem situation visual, force field analysis provides a method for developing a solution. The most effective solution will involve reducing the restraining forces operating on the problem. (See Fig. 13-13.)

4. There are two reasons for reducing the restraining forces.
   - To move the problem toward solution.
   - To avoid the effect of having too many facilitating forces.
     Because the forces on each side of the situation are in balance, removing or reducing the restraining forces will cause movement of the problem toward solution.

5. On the other hand, adding facilitating forces without reducing restraining forces will likely lead to the appearance of new restrainers. Remember that although you may change the situation by changing a force, you may not have improved the situation.

6. An effective strategy cannot be planned without evaluating the restraining forces for two factors: first, whether and to what degree a restrainer is changeable, and second, to what degree will changing a restrainer affect the problem.

7. It is ineffective and a waste of energy to try to change an unchangeable force.

8. One way to begin planning a strategy is to evaluate each force to see how changeable it is. A simple three-point rating scale is sufficient:
   - A fixed, unchanging force
     - Example: A contractual item, a law, a fixed budget
   - A force changeable with moderate to extensive effort
     - Example: An item which involves the efforts and cooperation of many departments
   - A change that can be rather readily performed, perhaps by just revising a procedure and is probably within the control of the group

9. The change or removal of some restrainers may have little or no impact. You must consider the effect that changing the force will have. It is good, then, to also rate the restrainers for their effect on solving the problem.

FACILITATORS │ RESTRAINERS

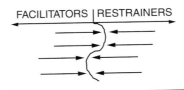

FIGURE **13-13** Reducing the restraining forces.

10. A three-point rating scale can be used to rate the effect a change will have on the problem:

    • No significant improvement will occur with the change.

    • Some minor improvement will occur with the change—perhaps up to 20 percent of the improvement needed to solve the problem.

    • A major improvement results from changing this force, that is, from 25 to 100 percent of the needed improvement.

11. After you have rated all of the forces operating on the problem situation, you can determine a priority for dealing with each force by adding together the numbers with which you rated each of the forces. The highest priority will be the restraining force; it will have the most effect and is the most changeable. After this will come those forces which you judge to have a large effect but are less changeable, and so on.

At this point in the force field analysis, you are ready to begin the third step: developing and implementing a strategy for changing the forces affecting a situation.

• In deciding the priority, strive for a balance of ease of change and the impact of the proposed change. Often the actions in dealing with any situation will require creative thinking. The balance between the facilitators and the restrainers is a clue for deciding which forces to change.

• A recommended tool is to remove the restraining forces to allow the point of equilibrium to shift. If the new point is not satisfactory, examine the driving forces and determine which ones you can successfully change.

The fourth step is to examine the situation. If you are still not satisfied with the new situation, determine which facilitating forces can be added.

Each time a change is planned, take the time to estimate and determine whether the change is worth it. Ask these questions:

• Will it produce the desired results?

• Which facilitating and restraining factors will be affected and by how much?

• How will the equilibrium point be affected?

• Is there a better way of getting the same results?

• Does the change have a negative impact on other parts of the process?

• What will be the return on investment?

Force field analysis is a straightforward tool. Using it with diligence and an ongoing evaluation of solutions will assure that it can work toward the achievement of your desired goal. Force field analysis is valuable because it goes beyond brainstorming by helping to develop plans and set priorities.

Summary

To implement your creative ideas successfully, develop a solid plan, using action plans for simple projects, and more formal project management techniques for larger, more complex projects. You'll also need to be able to sell your ideas within your organization. If your idea is likely to affect other people, you'll want to develop strong change management skills so that the people around you accept and use the products of your creativity. Once you bring one idea through to successful implementation, you'll be motivated and inspired to repeat the process again and again.

In the workplace where production occurs, some people are naturally more creative than others. However, that doesn't mean that we can't all learn to be more creative, and use creativity-enhancing tools and techniques in our daily lives. At its core, creativity is the ability to see familiar things in a new light, and the first step to being more creative is training yourself to look for opportunities to improve the products, systems, and processes around you. Then gather information, find the main cause of issues and problems, and generate and evaluate your ideas. And don't forget that the mundane work of implementing your ideas is key to being genuinely creative.

References

1. St. Lucie Press, and Delray, F.L., Chapter 2, Shaping the Right Attitude, Perspective and Vehicle, in *Problem Solving for Results*, pp. 13–27, 1996.
2. Ibid., pp. 29–32.
3. Delbecq, A. L. and VandeVen, A.H., "A Group Process Model for Problem Identification and Program Planning." *Journal of Applied Behavioral Science*, 7: 466–491, July/August, 1971.
4. Dunnette, M.D., Campbell, J.D., and Jaastad, K. "The Effect of Group Participation on Brainstorming Effectiveness for Two Industrial Samples." *Journal of Applied Psychology*, 47: 30–37, February 1963.
5. Bono, Edward de, "Six Thinking Hats." Little Brown, 1985.
6. Ibid., pp. 46–47.
7. Ibid. Problem Solving for Results, Chapter 1.

CHAPTER 14

Creativity Education: A Catalyst for Organizational Prosperity

Larry R. Thompson

Once upon a time, circa 2400 years ago, the greatest thinkers of Ancient Greece—Plato, Aristotle, Socrates—were so brilliant, charismatic, and persuasive that they left an indelible mark on the future of Western thought. Their intellectual prowess and influence was so powerful that they successfully established logical, linear, rational thought as the indelible standard of thinking for Western culture. Discovering "truth" required logic, analysis, sequential thinking, debate, judging, and adherence to what we have now come to understand as left brain–associated thinking.

Intuition, a "felt sense" involving the emotions, empathy, and creativity manifested themselves through the arts. Theater, music, art, poetry, sculpture, and writing were valued as a means to reflect about life and death, love and loss, and the anxieties commonly experienced in a struggling world. But while these pursuits (now identified as more right brain–associated) were appreciated as meaningful and certainly entertaining, they were not considered to be of the same import as the logical approach. They tended to take a back seat to left brain proclivities in the worlds of commerce, government, and education. Through the centuries, Western culture became so wedded to holding up logical thought processes as superior that such thinking became habitual and embedded.

Educational systems focused on this way of thinking and businesses, governments and social institutions followed suit. Logic, analytical thinking, numeracy, and sequential thinking became the norm in education, business, politics, and personal relationships. In short, Western culture became imbalanced and entrenched in a left brain "state of mind." Such thinking ultimately impacted its people's values, their wallets and, as time went on, their ability to compete. Now such "left brain only" thinking may be an obstacle to success in the future.

Of course, throughout history there have always been highly innovative and successful scientists, mathematicians, engineers, business and civic leaders who seemed to have a natural talent for combining what we now know to be right and left brain thinking in their work. Perhaps because the field of neuroscience didn't exist until fairly recently, society tended to attribute the abilities of these gifted individuals as more magical, accidental, and even mysterious than we now know to be the case. Rather than possessing some inherent "secret sauce" that made these people capable of being both highly logical *and* creative in tandem, today we understand that they had become adept at using their right and left brain capacities in combination to achieve very impressive results.

This little "Tale of Two Sides of the Brain" reflects the paradigm shift that is redefining how we live, work, play, and conduct business in the 21st century. Due to a number of reasons which we'll go into greater depth about later in this chapter, the time has come for us to take action to develop and utilize *both* sides of our brains. Since this is a chapter about creativity and innovation in business, we will focus on how bringing what author Daniel Pink calls "Whole Mind" thinking into the workplace can transform our work and ensure our competitive edge in the global marketplace.

FIGURE 14-1 Businessman or Artist.

FIGURE 14-2 Thinking Man.

Beginning with a brief overview of the contributions made by two key creativity in business thought leaders and an analysis of the impact of their thinking on business strategy and practice, we will make the case for *why* creativity at work works. We, in fact, view creativity as the fuel that will drive success in the global economy. The challenges and benefits of fostering creativity in the workplace, as well as the risks of *not* addressing this business imperative, will also be discussed.

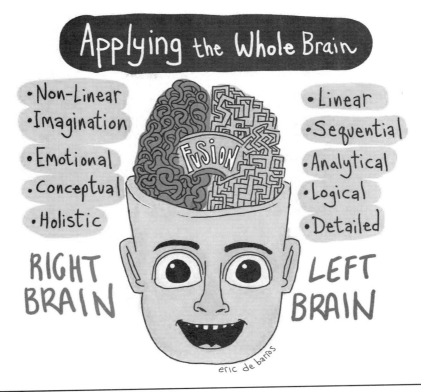

Applying the Whole Brain

RIGHT BRAIN
- Non-Linear
- Imagination
- Emotional
- Conceptual
- Holistic

FUSION

LEFT BRAIN
- Linear
- Sequential
- Analytical
- Logical
- Detailed

eric de barros

FIGURE 14-3   Applying the whole brain.

Our entire educational system, from pre-kindergarten through higher education, is the precursor to the workforce challenges our economy faces now, and even more so, in the future. Therefore, the current state of creativity education and the overemphasis on logic and analysis must be changed if our nation is to remain competitive. Using the Ringling College of Art and Design in Sarasota, Florida, as an example, we will demonstrate how the institution's whole brain educational approach has helped us develop into a college that graduates business-ready artistic professionals.

Finally, we'll describe how Ringling College developed the Ringling College Creativity Center (RC3) to offer creativity-based, immersive experiences with a business perspective. We view this as a prescription for helping business executives and their employees, as well as individuals, tap into their creativity. Real-world examples will illustrate how an immersive creative process can improve bottom-line results and breathe new life into business cultures and employee motivation. The chapter closes with a brief profile of what a creative leader looks like as well as the qualities, attitudes, and behaviors that are necessary for individuals to manifest creativity in their personal and professional lives.

Context: The Evolution of Creativity as a Business Imperative

There are numerous examples of brilliant and dedicated people who have played significant roles in bringing creativity to the workplace as a business imperative. Two such individuals—Edward De Bono and Daniel H. Pink—each in their own way, appeared at an inflection point in our post-industrial world and helped businesses and other societal institutions understand the importance of incorporating creativity into the workplace. Their work provides a context for understanding how the Ringling College Creativity Center stands on the evolutionary shoulders of these two influential thought leaders.

Edward De Bono's Impact

De Bono, himself the embodiment of a person who incorporated both right and left brain hemispheres in his work, came along at a time when business, government, and educational leaders were looking for new approaches. He was one of the first to recognize that in a rapidly changing

FIGURE 14-4   De Bono portrait.

and increasingly complex and shrinking world we needed to become far more creative and flexible in our thinking in order to remain competitive and to effectively tackle a multitude of challenges facing society. The traditional rational, argumentative (left brain–associated) methods for communicating and solving problems, while highly useful, just weren't enough.

In his work as a medical researcher, scientist, academic, and prolific author, De Bono gave great credence to intuition and creativity. He articulated the imbalance in our thinking and sounded the call for us to *think about how we think.* Since the 1970s his thinking methods have been applied throughout the world from classrooms to prisons, from juries to the highest echelons of business boardrooms. His ideas have contributed to a widespread recognition and respect for the importance of being more balanced in our thinking processes. He helped us become more cognizant of how the quality and nature of our thinking impacts business, education, international relations, and virtually all facets of our lives.

One of De Bono's greatest contributions was showing that creativity is a teachable skill. That was not the accepted perception when he began his work. His pronouncement, in combination with a variety of creative thinking tools that he developed, helped expand our view of creativity and our appreciation of its significance to all areas of life.

The Six Thinking Hats[1]®

De Bono's parallel thinking technique, known as the "Six Thinking Hats," has helped organizations move beyond a linear, one-track way of approaching problem solving and decision making. Six Thinking Hats is a method to teach the brain to look at a problem from multiple perspectives (hats) in order to promote open-mindedness, creative thinking, engage everyone in the conversation, and deter shutting down discussions due to defensive or offensive behavior.

In a group setting, each thinker is asked to put forward his or her thoughts in parallel with others—without criticizing or judging. By exploring different ways of thinking about the same issue, it encourages sharing ideas openly, viewing an issue from many different "hats," or perspectives, and avoiding the need to defend or attack what others say. The process facilitates more attentive listening and expands the individual and group perspective on issues. This leads to more creative approaches to problem solving, decision making, and communication.

Note: Analyzing usually white hat. Positivity usually yellow hat. Managing and Controlling usually blue hat. Intuition and Feelings usually red hat. Exploration and Creativity usually green hat. Danger Spotting usually black hat.

FIGURE **14-5**   The Six Thinking Hats.

The Six Thinking Hats Tool[2]

The leader guides each team member to separate his or her thinking into six functions and roles. Each role is identified with a colored symbolic "thinking hat." By mentally wearing and switching "hats" one can focus or redirect thoughts, conversations, or the meeting itself.

Below is a brief description of the function of each hat, along with illustrations inspired by the Six Hats tool that the De Bono Group created. Ringling College alumnus Van Jazmin has provided his own visual interpretation for each hat.

Lateral Thinking[3]

In 1967, Edward De Bono coined the term *Lateral Thinking*. Lateral thinking is a creative approach to solving problems using reasoning that is not immediately obvious. It incorporates ideas that may not be obtainable through step-by-step logic. Also referred to as *out-of-the-box* thinking, the goal of lateral thinking is to surface ideas that would not have arisen within the "box" of our usual way of thinking and perceiving ideas. Lateral thinking promotes openness to new possibilities and alternatives that might be overlooked when taking a purely linear approach.

De Bono developed a series of techniques to encourage out-of-the-box thinking to solve challenges and create new possibilities in all domains of life. Intrinsic to the lateral thinking technique is the understanding that creativity requires being free to make mistakes. This principle is fundamental to fostering creativity and innovation in business.

Figure 14-6 Dan Pink portrait.

Daniel Pink's Impact

The second major creativity thought leader is Daniel H. Pink. In his bestselling book, *A Whole New Mind, Why Right-Brainers Will Rule the Future,* he articulates the six senses, or aptitudes, that are necessary to succeed in what he calls the *Conceptual Age.* This work has redefined how businesses need to view and prepare for the future if they are to survive and thrive.[4]

Pink helps us recognize that our country is at an inflection point as we transition to a fully interconnected global economy. According to his thesis, preparing for this new world order and integrating key right brain skills into organizational cultures and practices will make the difference between whether we fail or succeed in the future.

The Conceptual Age of the 21st Century

Evolving from the Agricultural Age (farmers) through the Industrial Age (factory workers) to the Information Age (knowledge workers) has brought greater, more widespread affluence, technological advancement, and globalization. Now, he contends, we have transitioned into the Conceptual Age, a period in which right brain–associated skills such as design, storytelling, and empathy among others, are more crucial than the traditional, left brain–associated logical, linear-oriented analytical skills valued so much in the past. He states, "The future belongs to people who are creators and empathizers, pattern recognizers, and meaning makers such as artists, inventors, storytellers, caregivers, and big picture thinkers." In other words, to remain competitive this future requires putting creativity front and center!

Like De Bono, Pink points out that Western society has historically been dominated by a deeply analytical, narrowly "reductive" approach to life. According to Pink, a combination of three major forces, referred to as the "three As" (abundance, automation, and Asia) have projected us into a new age, an era that requires an emphasis on right brain aptitudes and skills, along with left brain thinking.

Abundance: Today we live in an age of prosperity and abundance with many choices about *what*, and *from whom*, to buy. For example, no longer does one have to settle for a pair of plain, functional walking shoes. Given the plethora of different types of walking shoes available to us at affordable prices (both through brick-and-mortar stores, and through the Web), we demand much more from our purchases. Why? Because, we can! So elements like style, beauty, uniqueness, feel, and meaningfulness factor into our buying decisions. In other words, today, design takes a "front row seat" in our choices. From wastebaskets to the cars we drive, to the health clinics we visit, or the appliances we use in the kitchen, design has become a differentiator when we're making purchases. In many cases it is *the critical factor* when making a choice.

Asia: Industrial Age and Knowledge Workers abound throughout Asia. They are well trained, highly motivated, and they work for *a lot* less money than their American and European counterparts. Manufacturing jobs, customer service positions, as well as information types of employment such as computer programming, accounting, and even radiology, are leaving our American shores due, in large part, to the lower costs of doing business in Asia.

Automation: Technology is taking over even some of our highest-level professional roles. Computers can now do medical diagnosis. Inexpensive Internet services are replacing the fees lawyers charge for legal work. Routine computer programming skills are being taken over by machines. There is even software that can write software, thus eliminating the need for certain kinds of software designers. If such left-brain thinking skills requiring logic, numeracy, and linear thinking can be written into an algorithm, then they can be done by machines. And, machines do not take vacations or coffee breaks. Nor do they unionize or complain!

The Case for Creativity or Why Right-Brainers Will Rule the World

Pink builds a persuasive and urgent case for organizations to deeply examine where they are headed in light of the three As. If work can be done cheaper overseas, faster by a computer, or with more appeal by a foreign competitor, we must recognize that we are a nation at risk. We must take concrete steps to ensure that we thrive in the Conceptual Age.

So, how can we remain competitive in this warp-speed era of change and globalization? Pink proposes that the answer lies in supplementing our logical, analytical abilities with attributes that are high concept and high touch.

According to Pink, "**High Concept** involves the capacity to detect patterns and opportunities, to create artistic and emotional beauty, to craft a satisfying narrative, and to combine seemingly unrelated ideas into something new. **High Touch** involves the ability to empathize with others, to understand the subtleties of human interaction, to find joy in one's self and to elicit it in others, and to stretch beyond the quotidian in pursuit of purpose and meaning."

One of his examples of high concept, high touch provides surprising, yet concrete evidence of how the world is changing and indeed, needs to change. He outlines how the curriculum of medical schools, long the bastion of pure left brain, logical and analytical thinking, is changing dramatically.

Medical students at some innovative universities are now being trained in "narrative medicine" to better understand the patient's story. This approach can provide diagnostic clues that nonhuman computer diagnostics cannot pick up. These students are studying painting to hone their powers of observation so they can improve their ability to notice subtle details about their patients' conditions. They are learning to "walk in their patients' shoes" and increase their empathy quotient by spending overnights in hospitals where they role-play being patients. They are learning mindfulness meditation practices to become more focused and less distracted by environmental stressors.

Pink's high concept, high touch premise serves as a wake-up call to examine how we must prepare for doing business in the future. In some ways it is frightening; in other ways it is exhilarating. It will take steadfast commitment, trial and error, and much work and growth. On the other hand, it is a positive, upbeat message of hope, provided we heed the call to do whatever it takes to prepare our workforce by helping them develop their right brain high concept, high touch abilities.

The Six Senses

Pink outlines the six senses, or aptitudes, required to succeed in the Conceptual Age. Like De Bono, he emphasizes that they are teachable and not limited to the domain of the artist, designer, poet, or therapist. Our own thinking is added to each sense.

Design: Design, aesthetic appeal, and positive emotional feel for customers and the products and services they buy are of paramount importance. In an age of abundance, design enables one's product or service to stand out from the competition. This has resulted in organizations now viewing themselves in new and exciting ways. In *A Whole New Mind*, Pink provides compelling quotes to drive home the importance of design:

"Businesspeople don't need to understand designers better. They need to be designers."
—Robert Martin, Dean, Rotman School of Management

"We don't make 'automobiles.' BMW makes 'moving works of art' that express the driver's love of quality."

—Chris Bangle, BMW

Story: The impact that stories have on creating a personal identification between customers and vendors can be significant. Context and narrative are more memorable, emotionally impactful, and make far deeper connections than facts. They build and strengthen relationships and hence, encourage customer loyalty to the brand.

Symphony: This is the ability to "connect the dots" and see the big picture by combining seemingly disparate elements into something new. Visual artists are exceptionally good at seeing how the various pieces come together. They learn proportion, relationships, perspective, light, shadow, negative space, and the space between spaces. Their training contributes to their being adept at viewing situations from multiple perspectives. Pink believes that as companies increasingly look for new employees with symphony attributes, those with MFA and design degrees will be more sought after than MBA graduates.

Empathy: Empathy is the capacity to walk in another's shoes—to accurately recognize what is being experienced by another person. Empathy is a powerful connector in client or business relationships. It builds trust and a sense of being understood. It even impacts product development and services. In the past the view was, "if you build it they will buy." Such thinking is going the way of the dinosaur. Now businesses must seek to understand customer needs and wants more deeply. They can then create products and services that meet those desires. In the Conceptual Age anthropologists work with research and development teams and designers to help companies better understand their customers and their behaviors so that they can anticipate future needs the customer may not even realize he or she will have.

FIGURE 14-7   The Six Senses.

Play: Humor and laughter have physiological and psychological benefits. Companies like Google and Zappos have Chief Happiness Officers who are responsible for creating a fun and energized work environment. Laughter clubs are sprouting up around the globe. Laughter and fun can lead to greater openness, creativity, productivity, and collaboration. Play encourages learning, taking risks, connecting with others, and "connecting the dots" thinking. These are all intrinsic skills that contribute to establishing an environment where creative ideas and innovation can flourish.

Meaning: There is a universal human desire to belong to something meaningful and larger than oneself. People want to work for something or someone they believe in. They will put in long hours and undertake herculean efforts if they believe their work will make an important difference.

The Compelling Case for Creativity at Work

According to a (2010) IBM study conducted among 1500 global CEOs,[5] the *most* important skill identified as necessary for current and future leaders to be successful in the world economy is *creativity*. Yes, creativity. Businesses are becoming increasingly global, productive, efficient, and entrepreneurial. Out of necessity they are also becoming much more competitive as new businesses are created to take away market share from existing companies. So how does a company maintain a competitive edge? How does it differentiate its products or services from those of competitors snipping at their heels?

They cannot remain competitive by doing things the same old ways they have always done them. They need leaders who are creative, who encourage and nurture creative ideas, and who can simultaneously organize, plan, and successfully lead and manage others. In short, organizations need gifted, creative leaders who are "brain bidextrous," or whole brain thinkers. They need individuals who are adept at using both right and left brain–associated skills and traits nimbly and flexibly in an increasingly complex and rapidly changing economy.

Figure 14-8   Brain bidextrous.

Reinvigorating Our Long History of Creativity and Innovation

The challenges faced today by businesses, governments, and institutions are part of a burning platform requiring us to shift our thinking. We must embrace right brain–associated thinking, along with left brain–associated thinking, by applying creativity and innovation to solve critical business issues. Traditional left brain thinking alone is still, of course, necessary. But it is no longer sufficient. We believe that creative and innovative thinking is one of the most—if not *the most*—critical asset needed for employees (at all levels) and businesses to succeed in the 21st century.

If businesses are to flourish—not just survive—in this global, economic landscape, they need the skills and talents of artists and designers like never before. While this may seem like a bit of an overstatement, a salient feature of the United States is that historically it has been extremely inventive and creative. Perhaps it hearkens back to the United States' Founding Fathers. Early on this group of individuals from diverse backgrounds creatively determined that the United States should be based on a democracy with individual freedoms guaranteed.

No one had done this before. This philosophy of freedom—the freedom of personal expression, ideas, and imagination, and working in collaboration with people from diverse backgrounds—is *exactly* what is needed now to again spawn the American spirit of creativity. As educators preparing artists to merge their creativity with practical business needs, we have, for a long time, been passionately expressing to business leaders the criticality of infusing art, design, creativity, and design thinking into their business strategies and processes. This is the road map for ensuring future growth and economic vitality. Many are taking notice.

What Does Creativity in Action Look Like?

But, what in the world are we talking about? How can business leaders make this happen? How can business leaders *themselves* become more creative? How can they balance left brain operational styles, strategies, and thinking with right brain creativity, design, and innovative thinking? Let's look at one classic example of a very mature and well-established manufacturer that, on the brink of survival, dared to reinvent itself as a design company that represents a whole brain approach to doing business.

General Motors Rises from the Ashes with an Identity Change

In 2009 General Motors (GM) learned the hard way that it was no longer an automobile manufacturer. What took GM off track and caused its eventual bankruptcy was its sole focus on its identity as a carmaker. When the financial crisis occurred they were forced to see themselves from a different perspective and to ultimately redefine themselves. Instead of continuing to view themselves as a manufacturer of cars, they realized that they were actually in the design business.

For GM, that meant coming to grips with the fact that as a design business, they needed to think like designers and pay intense attention to the look and feel of their automobiles. Cars today have become mobile pieces of art—mobile sculptures if you will—with interiors that are practically entertainment centers! So, GM reinvented itself as a customer-centric company that focuses on look and feel—the essence of art and design. Today, instead of recruiting only auto designers and engineers, they recruit sculptors, computer animators, and illustrators from places like Ringling College of Art and Design and other design schools.

GM also recognized that they needed to reexamine how automobiles were powered. Using creative thinking they developed the Chevy Volt, a car that switched to using electricity instead of gasoline as its energy source. It became an instant sensation. General Motors is now back with a vengeance.

The Fundamentals: Preparing Our Children for Success in the Conceptual Age

Before we further discuss infusing creativity and innovation into more traditional business settings like the one just described, let's take a step back and examine our educational system. After all, that is truly where our children and young adults' job preparation and expectations about work and employment begin. To reclaim America's "creativity differentiator," we must provide businesses with a workforce of imaginative and creative employees who will pave the way for a new future in American business. The preparation of such a workforce begins in the earliest grades and continues through high school and then into our institutions of higher learning. It is time to transform our educational system to embrace and nurture creativity and art and design as core values.

In many ways, our K–12 educational system is still rooted in the agricultural and industrial ages. It has not changed significantly in a century. What business could survive doing things the same way as a century ago?

Tragically, schools in the United States continue to beat creativity out of our children. They are being forced into learning the right answers, coloring *inside* the lines, following Aristotelian logic, and these days they are even outsourcing their brains to the Internet. Rather than being encouraged to engage in original thinking, they are being allowed to download information and regurgitate it without questioning its accuracy or its value. In order to succeed, they must get out their No. 2 pencils and fill in the blanks on multiple-choice standardized tests that measure left brain logical thinking but completely leave out creativity from the equation. We do not need *more* multiple-choice tests. What we really need are more tests with *multiple answers*. That is what creative thinking is all about.

Keeping Creativity Alive in Education

We, and our children, were *born* creative. Just spend some time observing young children and you will see that they exhibit the essence of creativity and imagination in their play. They're curious, they question freely, they see things with an open mind and they demonstrate an uncanny ability to connect the dots. Before being put into "mental straightjackets" by our left brain–dominated educational system, they do not really care what others think. They thrive on suspending judgment and engaging in "what if" thinking. In short, anything is possible for them before they learn otherwise as they struggle to conform to the culture of outdated educational training.

We believe strongly in the need for education in STEM (science, technology, engineering, and math) subjects. But, we will not survive economically if we focus solely on STEM. STEM *alone* moves us backwards, not forward. Remember the discussion about outsourcing to Asia? Asian countries are very accomplished in STEM subjects. We in America certainly need to catch up in those areas. Yet, instead of copying what Asia has done, or is doing, it makes more sense for us to take the lead in the area in which we have always excelled—and that is creativity.

FIGURE 14-9 Steam.

So, to fuel our economic engine for the future we should be focusing on teaching our young people STEAM. That means taking the STEM subjects and adding an "A" for the arts—the creative part of the equation. As Daniel Pink says, "The arts are no longer ornamental. They're fundamental. Today, as we try to find our footing on a new commercial landscape, they're fundamental to our economic future."

A sad example of our outdated educational system is the experience that the late Gordon MacKenzie, Creative Director of Hallmark, repeatedly had with young students. He would regularly talk to schoolchildren at various stages of their march through the educational system. He would ask kindergarten through first graders the question, "How many of you want to be artists?" and they would *all* raise their hands.

When he asked third graders the same question, approximately *half* the class would raise their hands. By the sixth grade he found that *not a single hand* would be raised. He watched the children look around to see if anybody had the "nerve" to raise his or her hand because now being an artist was viewed as *deviant behavior.*[6]

Reawakening Creativity in the Workplace

From schools to our businesses and institutions, the damage continues. By the time we are 25, studies have shown that only two percent of the population is able to think in "divergent or nonlinear ways," a key component of creativity. Contrast this with 98 percent of children ages 3 to 5; 32 percent of children ages 8 to 10; and 10 percent of 13 to 15 year olds.[7] Not only is this downward spiral in creative abilities tragic, it is devastating for the future of our country *and* our world.

Thankfully, neuroscience has given credibility to the need for left/right brain balance. We need to be "Brain Bidextrous." The reality is that we are not inventing something new. We are simply being called to make use of the potential that has *always* been there. We don't need to import creativity; we need to activate it—to give it nourishment and support, for it to take root!

We Are All Creative

The wonderful and highly encouraging reality is that we are, indeed, all creative. Creativity isn't magic. It can be taught. While many of us may never become professional artists, musicians, or poets, we can become creative and tap into our imaginations—just as we did as children.

Figure 14-10 We are all creative.

In many ways creativity flow is analogous to our body's circulatory system. We have clogged up the creative flow arteries. We need to clean them out and put stents in for the good of our economic future!

Creativity is "the ability to generate new ideas and new connections between ideas, and ways to solve problems in any field or realm of our lives."[8] When talking about young children, creativity involves curiosity, questioning, the ability to see things with an open mind, to connect the dots, and to develop novel solutions and approaches to problems. In adults it includes all of the above, plus the willingness to examine assumptions, explore options, suspend judgment, take risks, celebrate learning, and embrace failure. Creativity is linked to better job satisfaction, higher-quality leisure time, more positive emotions, and greater overall well-being and happiness.[9]

Bridging the Gap between Creativity and Commerce

In the Conceptual Age, creativity is often manifested in the workplace through what is known as design thinking. Design thinking is *the ability to come up with simple, elegant user-friendly solutions to complex problems and to deliver experiences that anticipate people's needs and delight them in unexpected ways.* This *is* the new competitive advantage. It is creativity, or design thinking that can't be downsized or outsourced. It can't be duplicated in assembly-line fashion. It can't be made cheaper or faster anywhere else. Design thinking is the missing key ingredient in our businesses, institutions, and governments. The good news is that it can, indeed, be taught, nurtured, and sustained.

The New Leadership—Blending Creativity with Business Acumen

The challenges many organizations face is that they, like our schools, have become entrenched in traditional approaches and strategies. Those approaches simply will not work in the Conceptual Age. Designers and artists can lead the way in challenging outdated modes of thinking, being and operating while also inspiring and helping to guide this much needed change.

They can do this by becoming fluent in the languages of business, art, and design. By attaining a cross-disciplinary mastery—and knowing how to communicate between these worlds—they can help to bridge the gap between traditional linear thinking and the creative thinking that leads to innovation. A person functioning at this level is not only extremely valuable in the workplace. She or he can facilitate and lead the transition to the workplace of the future.

Commerce in the global economy requires business leaders to completely redefine themselves. Artists and designers can help them to break out of their current perceptions of the world and patterns of operating. They can support leaders in approaching their businesses from novel and fresh perspectives. That translates into financial as well as personal rewards.

Forward-looking businesses are hungry for creative, imaginative, and innovative employees to create products and services that make them stand out from their competitors. They are looking for workers capable of balancing right and left brain–associated skills, talents, and approaches. This practical balance is attainable and those individuals who have this capacity will be the leaders who will drive the future economy.

Ringling College Example: Preparing Tomorrow's Business Leaders—A Design, Art, and Business (Whole Brain) Approach

At the Ringling College of Art and Design students matriculate toward a BFA in an art and design discipline. Unlike at traditional universities, our students are mostly creative, right brain thinkers. They come to us with great talent in the arts and we nurture and grow that talent throughout their education. However, in order to prepare them to be successful in the 21st century work world we believe we must help them to balance their creativity with critical left brain skills that are *also* necessary to business success.

With that goal the College created the Business of Art & Design (BOAD) program. It is the first undergraduate program of its kind in the nation to infuse the study of a business program with an art and design studio arts-based education. It teaches our primarily right-brain thinking students the left-brain thinking skills necessary for success in the business world. This academic major prepares students to bring design thinking into the field of commerce to meet critical business needs in the Conceptual Age.

Students gain a solid grounding in business skills, along with an in-depth understanding of the creative process. They receive training in the critical communication skills that will

FIGURE 14-11   Business of art and design education.

enable them to work effectively with both creative and business-oriented colleagues. They are expected to be competent in accounting, finance, marketing, organizational development, and leadership. They learn how businesses operate, the language of business, and how to channel design thinking into practical, real-world applications. Through their education they become living examples of the convergence of art and business. In some ways the BOAD degree is the quintessential business degree.

One example of how we teach art and design students to learn about, and contribute to, the practicalities of business is a new initiative known as the Ringling College Collaboratory (RC2).[10] The Collaboratory orchestrates opportunities for students to apply their design training or mind-set as they work on real-world business problems with real-world clients. This gives our students a business canvas on which to create, learn, experiment, and grow while infusing businesses and other organizations with creative talent and design thinking. It is clearly a win-win for everyone.

The Ringling College Creativity Center (RC3)—Sharing What We've Learned about Teaching Creativity

As a *learning community* the Ringling College of Art and Design is expected to continually evolve. Accordingly, we have added programs and expanded existing ones to meet the growing needs of the design-oriented future global economy. As an institution of higher learning we believe that we must reach beyond the boundaries of our campus to make contributions in areas that can benefit from our perspective, skills, and talents.

For several years we worked informally with organizations who asked us to help them solve problems creatively. As the economy transitioned into a global marketplace with increased urgency to meet challenges creatively, the need for our services expanded. Over time we realized that businesses needed much more.

Businesses began to understand that they needed to help their employees and leaders to change. They recognized that in order to continue to succeed and grow it would require operating in far more creative and innovative ways. However, there was no effective mechanism for organizations to learn how to reignite and sustain the intense level of creativity necessary to achieve those goals.

We realized that we had all the ingredients necessary to help businesses meet 21st century competitive demands. We could share our "know-how" for developing and nurturing creative talent through the College's arts-based programs. Plus, we could share what we have learned from preparing artists to apply creativity and design thinking to address fundamental business needs. Thus, the Ringling College Creativity Center was born.[11]

Creativity Immersion—An Alchemical Process for Changing Hearts and Minds around Work

The Center offers custom-designed retreats that expand participants' creative thinking and approaches to facing work-related challenges. By immersing senior executives, managers, and employees from both private and public sector organizations in creative, arts-based, process-oriented experiences, individuals and teams learn how to get back their "creative mojo."

We metaphorically perform "open heart surgery and put stents in" to get creativity flowing again for the benefit of both the organization and the individual. By learning to work together more imaginatively and practically, the retreat serves as a "jump start" for jolting individuals and teams out of habitual ways of doing business. It opens them up to new ways of thinking and solving problems.

The Center borrows from the whole brain approach so successfully integrated into the Ringling College of Art and Design's 4-year degree programs. It distills this educational process into a 2- to 5-day experience for organizations that want to become more skilled at creatively examining practical business processes and practices.

At the heart of the Center's mission is our conviction that creativity can be taught and that *all* people are creative. We are not attempting to create artists through the retreats. Rather our intent is to enable individuals and teams to experience the immense value that the creative process can bring to individuals and their organizations in terms of solving problems, working collaboratively, and increasing joy and meaning in the workplace. All of this is done to be better prepared to face the challenges of successfully competing in the global economy.

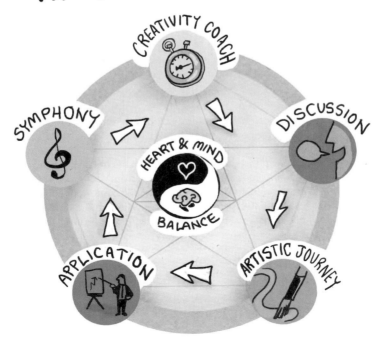

FIGURE 14-12   Alchemy.

Overview: The Whole Brain Approach to Learning

The retreat process provides attendees with an in-depth experience of blending right and left brain–oriented approaches. Through immersive, guided arts-based experiences, followed by business-relevant, open discussions, participants can view challenges and opportunities through a creative lens.

Instructors from both the creative and business worlds—including accomplished artists, professional educators, and business leaders—guide and teach participants to approach issues in innovative ways from multiple perspectives. This mélange of expertise is transformative. It leads to unexpected insights and groundbreaking ideas. Such surprises not only delight participants—they can have a profound impact on the future of the organization.

Learning Outcomes

So what are the outcomes of such an approach? Through immersion in the Center's retreat process, individuals and teams increase their capacity to bring greater creativity to their work back on the job. They learn to

- Question preconceptions and assumptions related to the organization's culture and see them from a fresh perspective
- Foster a culture and environment that encourages creative, flexible, and fluid thought and expression
- Create a climate of respect, honesty, trust, and open communication
- Actively seek new sources of inspiration and unexpected associations
- Embrace taking measured risks and accepting failures
- Suspend judgment and feel comfortable with "not knowing" the answers
- Design interdisciplinary encounters and remove boundaries
- Challenge the status quo
- Develop a consistent practice of creative development and idea generation
- Embrace and foster change and innovation

Both the organization and its individual workers benefit. Throughout the retreat process, creative facilitators (called C-Facilitators) point out how the creative process impacts *both* the individual and the team. To successfully develop an organizational culture that nourishes creativity, it must encourage and nurture creativity *person-by-person*. Just as the retreat content and design interlaces artistic experience with practical business-related discussions, we strive to maintain a balance between attending to individual and team-based creativity.

Retreat Overview

The retreat is designed to use immersion in the artistic process as the vehicle for fostering greater creativity and innovation back at work. A "symphony" of experiences from the artistic to the practical is introduced by the multidisciplinary retreat team previously described. Participants go from exploring new ways of thinking related to creativity and innovation in the workplace to immersive, artistic experiences that allow participants to understand how an artist thinks, sees, and responds to new challenges. This unleashes their creativity so that they can view things differently, look at issues and ideas from multiple perspectives, and engage in applying artistic, right brain thinking to solve organizational challenges. The whole brain approach is designed to gently, yet powerfully, open the doors to understanding the creative process itself and then to balance this experience with practicality and business acumen.

An Orchestrated Experience

Prior to the retreat, the Center engages in a discovery process to better understand the organization and attending team's strategic gaps. They explore what has been done and what has worked, or not worked, with regard to fostering creativity and innovation. The team chooses one to three projects to focus on during the retreat that may necessitate creativity and innovation. The retreat is then customized to enable them to work on their selected projects and address strategic gaps from new perspectives.

The Center's multidisciplinary retreat faculty includes

- A professional artist or educator who guides participants through the immersion process.
- A business leader experienced with applying cutting-edge creativity strategies in the workplace.
- A content expert who expands attendees' thinking about relevant creativity in business areas such as design thinking, storytelling, and the latest developments in neuroscience and creativity.
- Skilled group creativity facilitators (C-Facilitator/s) who manage the learning process and serve as "conductors" of the multifaceted process. They also later serve as C-Coaches (Creativity Coaches) in the 6 months following the retreat.
- Art illustrators who document the retreat visually and provide a wholly different perspective on the process.

A description of how the faculty interweaves multiple perspectives into the retreat program is provided later. Through this variety of different, yet related experiences, participants are able to loosen up their thinking and entertain different ways of perceiving their individual and group situations. Of course, this approach can be applied to any business wishing to enhance creativity.

The Guided Artistic Journey

Participants work with professional artists to learn how the artist thinks and to later apply the creative process to work-related problems and challenges. An artist serves as their guide. His purpose is to help participants better understand how the artist's approach can open the mind to new ways of viewing and working to meet a multitude of business-related goals. He provides a metaphorical road map to help them truly experience how they can access their latent creativity as they journey through the various stages of the creative process.

For example, as the participant is actually "playing in the clay" and creating a sculpture, the artist or guide describes their thought processes and the challenges involved in creating art. They share with the group how they

- Set the stage for creating a climate that leaves them open to inspiration
- Take an idea and explore it through various perspectives
- Look at an object they may want to create in a variety of ways—from the top, the bottom, the sides, close up, and far away
- Face challenges, obstacles, and "artist's block" and how they overcome barriers
- Use the creative process to nurture their artistry
- Are comfortable with letting go and not knowing exactly what will happen
- Cultivate an open, focused, and patient mind
- Welcome chaos, take risks, and accept failures as part of the process inherent to developing a significant work of art that meets their client's needs

The participants not only *hear* what the artist says. They *experience it*. The key is that participants are immersed in the guided sculpting experience and create their own works. They engage in the creative process first hand. The sculptor combines guidance, encouragement, challenge, and support with moments of reflection to help participants fully integrate the process on a personal level. The immersive experience also includes working collaboratively with the other participant artists on their team. They gain a deeper, more visceral understanding of how, through working together, "the whole becomes greater than any of the parts." They come to better understand how their egos must be left behind for collaboration to work.

From Art to Business—Translating the Immersive Experience

After an enriching creative or artistic experience that engages participants in the creative process and frees up their thinking, the facilitator helps the team apply this experience to relevant situations at work. They explore how working together collaboratively in new ways can create a strong and exuberant sense of both individual and team ownership. They learn how to challenge the ways they may have been doing things by looking at problems differently and from multiple perspectives.

For example, they may talk with their colleagues about how they can apply the creative process to uncover customer or market needs. They may explore how they might develop new or improved products or services to meet those needs. Or perhaps they may even begin thinking about creating a new product or service that will meet needs that are not yet known to their customers.

During the retreat they also begin applying the creative process to their one to three predetermined projects. By applying the lessons learned and experienced in the arts-based processes participants are encouraged to challenge their preconceptions and assumptions. They are asked to remain open to disparate ideas, to listen closely and respectfully to each other, to suspend judgment, take risks, practice patience, accept mistakes, and remain open to "not knowing." They are reminded to trust in the creative process and allow ideas and solutions to evolve in a natural manner. By practicing the artist's creative way of approaching issues, they come to better appreciate the mosaic of qualities and talents so necessary to fostering creative collaboration and innovation.

Cutting-Edge Creativity-Related Content

Examples and case studies about organizations that have successfully blended creativity and innovation with proven business practices are introduced and explored. For example, a design thinking expert is sometimes asked to assist a team in learning how to apply design thinking techniques to critical problems or needs of the organization. He helps them to frame, understand, and articulate exactly what the true need, or issue, is that concerns them. As a result, they often discover that the problem is different from what they thought it was at the outset. He elucidates the process of moving through a cycle of idea generation, priority setting, prototyping different solutions, and further testing and refining ideas and possible solutions.

By integrating a design thinking process with the multiple, and sometimes unanticipated, perspectives that arise out of the arts immersion process, potential outcomes are enhanced.

Participants are able to see the value of approaching the design thinking process from broader perspectives than they might have accessed otherwise. The essence of these experiences is captured in Albert Einstein's well-known quote

> "The world as we have created it is a process of our thinking. It cannot be changed without changing our thinking."

Depending on the specific needs of each retreat group, creativity-related content is selected to improve learning. For example, approaches such as storyboarding, storytelling, and/or the latest neuroscience findings about creativity might be presented.

Business-to-Business Learning

For each retreat the Center selects a creative, innovative business leader who shares and inspires participants with their organization's path to success. For example, in one retreat, the former Associate Vice President of Casting and Performance for Cirque du Soleil shared how the company applied creativity to its organizational practices. She presented this process to the product development group of a manufacturing company that sells wholesale and retail consumer products who were also looking to redefine their business processes and overall mission.

Bringing in business leaders to share their own company's stories of how they applied creativity to their business processes and products accelerates the participants' transfer of knowledge to real-world possibilities. This helps them gain a practical and experiential understanding of how the whole brain approach works. It demonstrates how powerful it can be when applied to all stages of an organization's growth.

Through the Eyes of Artists—A Running Documentary of the Retreat Process

Throughout the retreat, artist illustrators capture all aspects of the team's work visually—both during the art immersion exercises, as well as during discussions. This provides an in-depth record of the retreat from an artistic, nonverbal viewpoint. It is amazing how much more one gets out of a record of visual images than from a traditional, written record of retreat minutes.

Embedding Creativity: The Post-Retreat Process

The Center recognizes that the retreat is a "jump start" to jolt people out of their day-to-day ways of doing things in order to help their organizations transform. As stated, the experience is intended to open up employees and organizations to new ways of thinking and acting creatively both individually and collaboratively. But, so often the problem with attending conferences, institutes, or retreats is the reality that we have to go back to the (same old) office on Monday. All that work and the excitement generated from the retreat can get dissipated as we tend to slip right back into our old habits and familiar ways of doing things.

We wanted to find an effective way to keep the creative process alive and to ultimately help embed it into the daily operations of the team. So, following the retreat, the Center provides 6 months of creativity coaching services to attendees. They continue to work on the projects they began addressing during the retreat. Meeting with them once a month, their C-Facilitator now serves as a mentor, C-Coach, and resource. She or he helps the team integrate their retreat-gleaned learnings and insights and apply them to these selected projects. It is this on-the-job, practical application of the creative process that helps to embed creativity into the team's working culture and the long-term fiber of the organization.

Example: A Manufacturing Firm Redefines Itself

Let us revisit the manufacturer of wholesale and retail consumer products that was briefly described in the Business-to-Business leadership section. As mentioned earlier, their product development group attended the retreat with an expressed interest in redefining their business model, organizational processes, and perhaps even their overall mission.

Prior to the retreat the organization regarded itself as a manufacturing company charged with producing the same basic lines of "functional" products year after year. That changed dramatically over the course of the time they spent at the Center.

The combination of arts-based exercises, intermingled with cognitive business learning experienced over the 5-day session led the company's product development team to discover that it was *not* a manufacturing organization. Rather, it was a design organization. They continued exploring this idea over the next 6 months, utilizing their coaching sessions to gain greater clarity into how to infuse creativity and redefine their outlook of their business and its practices, plans, and communication processes.

Over the course of the next year they successfully repositioned themselves in the marketplace. They revised their overall vision, mission, goals, and objectives to incorporate a more design-centered approach to their business. Their product development group transformed its perception of their mission as well as their short- and long-term business strategies. Within a year they came to define themselves as a design organization committed to creating and delivering products to meet their customers' needs—needs that were not previously even known to these customers. To the great satisfaction of the company's owners and employees, their bottom line improved dramatically!

This "design organization's" approach is reminiscent of the groundbreaking restructuring of identity that Apple pioneered. It has resulted in more and increasingly diverse and creative product lines, much-improved business, and greater recognition as a leader in their industry.

What Does a Creative Leader Look Like?

The Center's goal is to help develop whole brain creative leaders who are both organizational and people-focused in ways that lead to meaning, creativity, joy, and productivity in the workplace. These leaders have the ability to develop and sustain a highly engaged workforce—one in which every employee is viewed as a potential innovator. They have the organizational, people-oriented, and business acumen to break down barriers to progress and innovation. They embrace failure as a necessary outcome of creative thinking. Plus, they provide the support and resources necessary to develop and bring groundbreaking products, technologies, or services to market quickly and efficiently.

Creative (Whole Brain) Leaders Are Organizational and People-Focused

Whole brain leaders demonstrate a critical set of qualities, traits, skills, and clusters of behaviors through their day-to-day leadership.

FIGURE 14-13 Attributes of a creative leader: open mind.

FIGURE 14-14 Attributes of a creative leader: reflection.

They seek to

- Inspire organizational and individual purpose
- Emphasize company-wide support and appreciation for creativity and innovation
- Eliminate barriers to creativity

FIGURE **14-15**   Attributes of a creative leader: taking risks.

- Willingly surrender unproductive, traditional mechanisms of control
- Foster a culture that includes and values time for reflection
- Promote and support autonomy
- Enable opportunities for mastery
- Yield to ambiguity and chaos when appropriate
- Suspend attachments to outcomes, at the right times
- Expand their self-awareness and self-reflective capacities
- Be openly vulnerable and evolve as needed
- Demonstrate empathy and are able to relate and connect to anyone
- Encourage collaboration
- Be open to new sources of ideas and/or approaches
- Embrace risk and accept failures as learning opportunities

The Creative Individuals' and Leaders' Toolkit of Qualities, Attitudes, and Behaviors

Creative individuals and leaders are committed to a process of ongoing experimentation, learning, and development. They seek to model the combination of qualities, attitudes, and behaviors that they also support in others. Many of the qualities, traits, and skills listed below may take a lifetime to master. The creativity path is characterized by excitement, challenge, satisfaction, and growth.

- Takes risks
- Comfort with chaos
- Ability to suspend judgment and "not know"
- Patience
- Respect

- Trust
- Honesty
- Curiosity
- Listening
- Empathy
- Collaboration
- Shares information
- Appreciation for diversity of thought, disciplines, cultures, races, generations
- Reflection
- Integrity
- Open-mindedness
- Support
- Willingness to give and receive constructive feedback
- Open to new learning opportunities
- Accepts mistakes or failures as part of the learning process
- Transparency of management
- Celebration of learning
- Shows appreciation

The Ongoing Journey

Awakening creativity and finding ways to activate and nurture the creative process in the workplace is challenging and perhaps even frustrating and painful at times. Ultimately, however, it is also highly rewarding. It will add tremendous joy and meaning to work. It will ensure the future financial success of the organization and, from a macro perspective, our nation.

We wish you well on your journey. Just as creative leaders must demonstrate patience, consistency, and gentle, yet powerful support toward others as they take risks and open themselves up more and more to the creative process, we encourage you to be kind, patient, and gentle with yourselves as you embark upon this meaningful path.

References

1. De Bono, Edward. *Six Thinking Hats.* Boston, New York, London: Little, Brown and Company, 1985.
2. Ibid.
3. Ibid.
4. Pink, Daniel H. *A Whole New Mind, Why Right-Brainers Will Rule the Future.* The Penguin Group, 2005. Daniel Pink has generously granted permission for us to present many of the core concepts and quotes from his book in this chapter. He holds an Honorary Doctorate of Arts from the Ringling College of Art and Design.
5. IBM Corporation. *Capitalizing on Complexity—Insights from the Global Chief Executive Officer Study.* 2009–2010.
6. MacKenzie, Gordon. *Orbiting the Giant Hairball: A Corporate Fool's Guide to Surviving with Grace.* Penguin Group (USA) Inc., 1998.
7. Robinson, Sir Ken. Former Chair, UK Government's report on creativity, education, and the economy—from a speech at the Scottish Book Trust in Glasgow, Scotland, March 2005.
8. *This Emotional Life Series*, PBS. Retrieved from http://www.pbs.org/thisemotionallife /topic/creativity/creativity.
9. *This Emotional Life Series*, PBS. Retrieved from http://www.pbs.org/thisemotionallife /topic/creativity/creativity.
10. The Collaboratory was initially funded by the Patterson Foundation, Sarasota, Florida.
11. The Gulfcoast Community Foundation, Venice, Florida, provided seed money for the Ringling College Creativity Center.

CHAPTER 15

Unlocking Your Creativity Using Mind Mapping

Richard B. Greene

Discussing creativity and its benefits both in the life of individuals as well as in business environments is important, but when it comes down to actually applying creative thinking many people run into difficulties. While it may sound odd at first, one of the best ways to think creatively is by taking advantage of tools that can help capture ideas and give them some structure. Mind mapping is the ideal choice for this because it allows people to quickly keep track of their creative ideas in a structured and useful manner.

What Is Mind Mapping?

Mind mapping is a concept in which an individual or group starts with a main idea or goal in the middle and diagrams ideas out from this one main subject. It has sometimes been referred to as concept mapping and augment mapping.[1] Adding words, brief descriptions, or even simple images to a mind map makes it easy to capture an idea quickly in a way that it can be reviewed again later. This can be done with a paper and pencil, but most people today use mind-mapping software because it is far faster and better organized. The creation of this chapter started with a simple mind map (see Fig. 15-1).

People have been using diagrams with words and images for centuries, but the specific name, layout, and concept of mind mapping as it is known today was developed and began by a British psychology author and television personality named Tony Buzan. Buzan hosted a TV series in 1974 on BBC called "Use Your Head," where he discussed this technique and the many benefits people can gain from it.[2,3] The concept continued to develop over time, and while there are many different specific types of mind maps, they all follow some general guidelines.

1. **Starting in the center**: Having the main idea to focus on listed in the center is important because it allows people to branch off in any direction allowing for sufficient room for multiple ideas on a single map.

2. **Use colors with meaning**: Most mind maps will have at least three different colors, each with a specific meaning. The meaning can represent anything from level of importance to type of task.

3. **Shapes and images**: Adding different shapes or images around an idea is a great way to make further categories on mind maps. It can be especially effective to have an image placed along with the center idea to invoke both the visual centers of the brain and those that are used for reading.

4. **Easy to follow lines**: There should be clear lines from one area to each subarea to ensure it is simple to follow the train of thought, as well as allowing for additional ideas to be added. Many people also suggest having thicker lines at first and slowly making them thinner as the map branches out. This is a simple visual representation of the natural flow of ideas from the center concept out toward those further along the map.

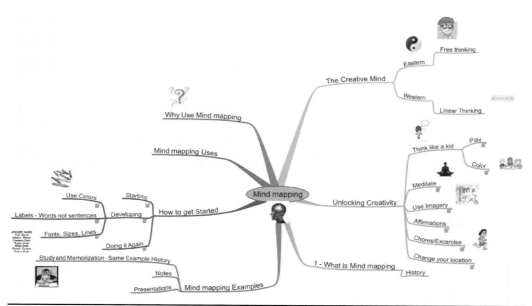

FIGURE 15-1   Mindmap of a Mindmap.

5. **Radial organization**: Ideas should be organized radially off of each other to keep them organized. In addition to allowing for the best use of space on the mind map it also forms an accurate picture of how the actual brain processes thought.

6. **Custom styles**: Each individual can come up with their own style for mind mapping that works best for their specific purposes, which makes this technique extremely flexible.

At first glance many people look at mind mapping as a tool to use for brainstorming ideas, and while it is great for that, it really goes much deeper. True mind mapping will take individuals or groups through every step of development for new ideas. Starting a new project by brainstorming a mind map is a good start, and then the different ideas can be organized further to put similar concepts together on the map. Having these items quickly placed in a visual way can actually help to stimulate additional ideas throughout this beginning phase.

Once a general outline of a particular idea or concept is completed the mind map can still be used to expand on ideas or get them to be organized in better ways. When using these maps with others or for a business it can be an invaluable tool for quickly introducing others to the project in a way which will allow them to gain an understanding of what the project is, and the steps that will be required to be completed. This is a great way to bring a project to management or finance teams since it is an easy to understand, high-level view which can be quickly displayed and explained to people whether they are involved with a project directly or indirectly.

Finally, when it is time to actually begin working on the project, mind maps allow individuals or groups to put each task in the proper order, so all work is done in the most efficient way possible. Whether it is organized by color, or it is numbered, this is the easiest way to divide work and get the most important things done.

After learning the basics of what a mind map is and how to use it, anyone can see how it would be effective for organizing and getting things done. Of course, it is also essential for promoting creative thinking and using any creative ideas in the best possible way.

Why Mind Mapping Promotes Creativity

People have been coming up with ways to improve their creativity for centuries, but mind mapping is different. It isn't just a clever idea or a set of rules to help people stay focused on coming up with ideas. When done properly, mind mapping can actually help to stimulate the brain's creative process and can cause people to think differently than they otherwise would.[4,5] Mind mapping organizes information in the same way in which the human mind works. People don't think in lists or sentences. For example, in Fig. 15-2, they think using

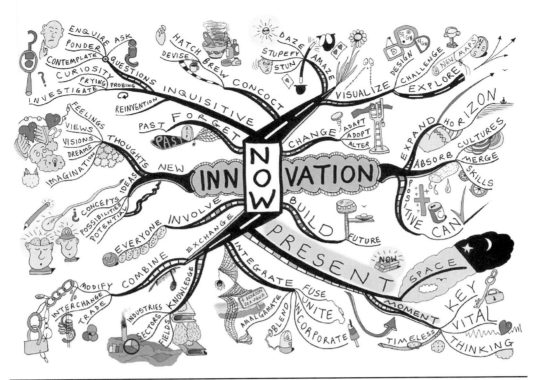

FIGURE 15-2   Innovation mind map.

images, themes, shapes, and patterns, all connected in some manner.[6] Understanding the reasons behind the guidelines of mind maps will help explain what makes mind mapping so effective for promoting creative thinking.

Working with the Brain

Science has proven that every brain is divided into two distinct hemispheres, which are responsible for different activities.[7] This is why so many left-handed people, for example, are also excellent artists. Many techniques for creative thinking attempt to harness the power of the eastern hemisphere of the brain, which is widely considered to be responsible for more creative and artistic thought processes.[8] The problem with focusing all the attention on this one side of the brain is that it can also cause individuals to become distracted or even forget the ideas they came up with before they are able to use them.

The "western hemisphere" of the brain is much better at linear thinking; it is where things like mathematics and most other complex thoughts take place.[9] While at first this may seem like it is not a very useful side of the brain for a creative thinking guide, it is actually essential for successful creative thinking.

Mind mapping combines the strengths of each hemisphere of the brain in a way that allows people to quickly come up with new creative ideas, and still document them in a logical fashion to ensure they are useful to the end goal of the project. The mind-mapping guidelines are specifically set up to help encourage both sides of the brain to take part in the process:

- **Radial organization**: The radial layout of mind maps allows the western hemisphere of the brain to concentrate on this organization and take advantage of the linear thinking. It allows the creative ideas that flow from the eastern hemisphere of the brain to be organized in a useful and productive manner. Since the radial organization is something that is not often used, however, the eastern side of the brain also appreciates this new and different way of focusing which helps to stimulate this creative powerhouse in your mind.

- **Colors**: It has long been known that when people see specific colors it can cause them to think or feel differently. Harnessing this concept, the mind map uses a variety of colors not only to organize the mind map further, but also to stimulate the creative centers in the eastern hemisphere of the brain. Once again this guideline is stimulating to the entire brain rather than just one area. Each individual or group can decide which colors they want to use for each mind map based on what they find to be most effective.

- **Images**: When words aren't quite enough to capture an idea people are encouraged to draw a quick picture, or find an image that more fully harnesses the concept they were thinking of. Adding images to a mind map forces the brain to adjust its thinking from the textual to images. Any time the brain has to adjust the way it is thinking it will stimulate different centers which can often trigger new and creative thought processes. In addition, when drawing or finding an image that properly captures a thought it can trigger additional concepts that would have never come up when working only with text.

- **Words over sentences**: While in some rare circumstances it is acceptable to write more than a few words on a particular part of a mind map, the bulk of the map should consist of one to four words. By avoiding complete sentences it keeps the brain from focusing on things like grammar and punctuation, which can distract from the creative process. Focusing on the most important words of an idea will also keep the mind map very focused on turning out ideas quickly rather than taking the time to make sure everything is perfect, which can smother the creative side of the brain.

- **Informational structure of the brain**: When most people picture an image of someone coming up with an idea it typically contains a cloud coming out of their head. This is actually a surprisingly accurate concept, because the brain often has ideas and thoughts which expand off of each other like a cloud. Mind mapping offers the perfect structure for capturing thoughts because the structure of a mind map is an accurate representation of how most people visualize their thoughts.

- **Better than a brainstorm**: While the first draft of a mind map will often be made in what could be considered a brainstorming session, it is actually much more than that. Rather than simply jotting down ideas as quickly as they come as with an unorganized brainstorming session, the mind map allows the brain to quickly come up with ideas but still has them organized in a logical and efficient manner. Rather than having every idea listed in the same area, mind mapping allows for instant grouping and subgrouping of ideas and concepts so they are not only more useful down the road, but the ideas can also build off each other in a much more effective way.

It is this combination of guidelines that helps to harness the full creative power of the brain that makes this method so effective for so many people. Most people find that this unique blending of organization and free thought allows their mind to think in a new and different way that they likely have not experienced in the past. Forcing the brain to use both hemispheres at the same time and with the same goal is not something that is done in other daily activities, which is why it is so effective.

The combined focus of both hemispheres of the brain helps to prevent distractions from interrupting the task at hand. When many people are working on an important project, for example, they are only using one side of their brain because it is typically either exclusively creative in nature, or exclusive linear, or logical, in nature. While this does cause one side of the brain to work hard, it leaves the other side free to wander which is why many people are easily distracted when they most need to focus on a specific task. Rather than fighting against this natural function of the brain, mind mapping works by keeping both sides of the brain stimulated throughout the activity.

By changing the way the brain is taking in and processing information it is presented with it just makes sense that it will begin producing different results than most people would come to expect. This is how the creative side of the brain and the logical side of the brain can truly work together to produce extremely powerful results. Regardless of whether the goal of the mind map is to produce exclusively creative ideas or only logical ideas, mind mapping helps ensure people get the results they are looking for much more quickly than could otherwise be possible.

The mind will be simultaneously taking in the information, which is on the mind map, while also producing new ideas to be placed into the map. This is, in effect, an incredible process, in which the mind is literally feeding itself new and creative ideas that it would not have otherwise come up with. It then processes these ideas in a way that it normally would not, which produces even more new and exciting ideas. This process can continue on as long as the individual wants to spend time working on the mind map.

Most people, even skeptics, are astounded at the amount of information that can be generated by an individual or a group when mind maps are used. Even people who have been able to successfully come up with ideas without these maps often find that when they start to use mind maps the number and quality of the ideas they are able to produce goes up significantly, and the time it takes them to come up with new ideas drops dramatically.

This is because of the fact that the individuals are able to harness both sides of the brain and cause them to work together on a single task. This rarely happens for most people in their normal lives so it can be quite shocking to see the incredible results. Everyone has heard the statistic that most people only use about 10 percent of their brain. Whether this is accurate or not, being able to think in this new way is sure to help individuals push their brain to new limits.

Getting Started with Mind Mapping

There are many different options when it comes to creating a mind map for the first time. Many people will pull out a large piece of paper and a pencil to physically write out their mind map. For groups of people it can be done on a chalk or whiteboard. Most people today, however, harness the power of computer software, which was designed specifically to perform mind-mapping tasks.

There are many different mind-mapping programs available for people to choose from, and they all work in much the same way. Most of these software options are web-based programs, which allow people to access their mind map from any computer that is connected to the Internet. They often also allow collaboration with multiple people on one mind map, which can be an extremely valuable feature. When working with multiple people on a specific project this allows everyone to access, update, and edit a single mind map even if they are not in the same room.

No matter what option for mind mapping is chosen it is important to follow the mind-mapping guidelines to ensure the best results. When choosing things like which colors represent different categories or concepts it is possible to either define this before starting, or let it develop naturally while creating the mind map, as long as it is clearly labeled by the time the first draft of the map is completed. In addition, if multiple people are working on a single map, having everyone be aware of what different colors or shapes mean is critical.

If everyone in a group is using a different color scheme or their shapes have different meanings it will cause everyone to have to stop and think about what the other people are doing, which will slow down, or interrupt the creative thinking of everyone involved. This is why it is important to have a "leader" or "facilitator," of any group who creates and manages the mind map itself. It is important to note that this leader does not have to be the highest ranked or most experienced individual. In fact, it is typically best if it is not someone in management, because they will then be tempted to attempt to direct more than just the general layout of the mind map.

The Main Word or Phrase

When beginning any mind map, for any project, the first thing to do is to write down the main overarching idea that will define the project. If, for example, an individual wants to write a fiction novel and is going to use a mind map to take down their thoughts, the center phrase might be the title of the book. If there is not yet a good title for the book in place then it can even be something as general as, "fiction book." Once a title is finalized the center area can be edited to reflect the changes.

This main word or phrase is key to the success of any mind map because it will stand out against the rest of the page. Many people use a much larger font than the rest of the map as well as making it bold or even in an extremely prominent color so it really stands out. The idea behind this is that this one word or phrase should always be drawing the attention of anyone who views the map, including the author who created it. This helps to keep the mind focused on this most important idea or concept. Everything else on the mind map should always come back to be supporting of this center item which is why it should always be on the mind of everyone involved in the project.

First Level Ideas

There are many different names used for the first level of ideas, which spring off of the main word in the center, and what each individual calls them is not important. These ideas should be directly tied to the main word in the center in a clear and obvious way. Going back to the example of a fiction author, the first level ideas would include things such as

- **Characters**: A list of all the characters would come off of this first level idea.
- **Settings**: Each different setting in the book would branch off of this.
- **Plot ideas**: Different plot lines could come off of this idea.
- **Conflict**: Specific conflicts can be listed here.
- **Possibilities**: Many people, especially authors and others who are working on creative projects, will add in an additional section specifically for ideas which they are not yet sure if they will be used or where they will go.

Of course, there will be several other first level ideas for most projects, and the exact number of them will depend on what the project is. There is no right or wrong amount as long as the number is sufficient to organize the entire project, but not too large to cause undue confusion.

Second Level Ideas

As might be expected the second level ideas will build off of the first, and will be specific instances of whatever the first level idea was. If the second level idea being worked on were "characters," it would have a list of all the main characters of the novel. It is easy to see how by even just the second level of a mind map it can quickly become extremely branched out. This is one of the great features of mind maps when used in creative thinking because it allows the mind to see the entire project at a glance.

While the conscious mind can't process every entry in the mind map at the same time, the subconscious mind can. It will be focused on all the different items within the map so new and interesting ideas will continue to be produced quickly and almost effortlessly.

Additional Levels

Since there is no limit to the depth a mind map can go there can be any number of levels. Going back to the characters example earlier, a third level idea would be the name of a character. Continuing down the fourth level might be a personality trait with the fifth level being a particular habit that the character has developed because of the personality trait.

Since there is no limit to the number of entries on each level it is easy to see how a very detailed character can be developed before the author even writes the first word of the actual book. This is an excellent way to be able to keep track of a wide range of characters, settings, conflicts, and more without getting confused and making errors in the plot line.

Of course, writing a fiction novel is just one of an endless number of things that mind maps can be used for. When people need to think creatively this is truly the ideal method to do it because it is such a powerful tool, which encourages creative thought and documents it in a very useful way.

Proven Creative Uses for Mind Mapping

While the benefits of using mind mapping are obvious, many people still have trouble coming up with ideas for what types of things they can use this technique for. The following are several creative ways to use mind mapping to come up with new and exciting ways to accomplish anything from an everyday task such as schoolwork and even projects at work. These uses of mind mapping are a great start, but just about any task which requires creative thought can be improved by using mind mapping.

Studying

When studying for a particular subject mind mapping can be a great way to engage both the eastern and western hemispheres of the brain. This will help most individuals retain more of the knowledge that they are studying as well as help them stay focused for longer and more successful study sessions.[10] Many people also make the mistake of thinking that studying is primarily a matter of memorizing specific facts or concepts so they can pass a test. The fact is, however, that when people approach virtually any type of studying from a more creative perspective, they end up doing much better on tests.[11,12]

The reason creativity and studying go together so well is because when people come up with new and interesting ways of remembering different things the mind is far less likely to discard that information. Many people perform poorly on tests because their mind processes the information which they are studying, but after a very short period of time it decides that this information is not useful and disregards it. This is sometimes considered a short-term memory problem, but in reality it is the brain functioning as it should.

If people work with the brain's natural process rather than attempting to force new information into it, they would find that they are much more successful at retaining the information they desire. Using mind mapping it is possible to put the topic of an upcoming test in the center and then create different subtopics all around it. When one gets to the third or fourth level they can start entering creative ideas to help them remember a specific idea.

If, for example, a student wants to study for a test on the state capitals for a geography class they could put state capitals in the center. Off of that would be each of the states followed by the capitals. When studying in a traditional fashion this would be all the information that is used to attempt to memorize them. With mind mapping, however, this is really just the beginning. Off of each capital the individual could come up with humorous or bold ideas to link the name of the state with the capital in their mind. Using these creative thoughts each individual can quickly come up with methods for memorizing facts which they would have undoubtedly struggled with had they not used creative thinking.

Presentations

Presenting information to a group is something that many people hate doing because it combines two of people's biggest fears: (1) public speaking and (2) teaching others. Similarly to studying, many people make the mistake of focusing exclusively on attempting to force a significant amount of information into their short-term memory so they can later present it to their audience.

Even people who can remember their entire presentation flawlessly during a rehearsal often stumble and forget what to say when they are in front of a group. This is because their brain is triggering a fear response due to the very common fear of public speaking, which is, at one level or another, a part of most people. When the brain is afraid it is going to disregard all nonessential information so it can quickly pull up survival information if needed. Of course, to the brain, which evolved over thousands of years ago, a presentation is hardly considered essential to survival.

If, however, the presentation is planned (see Fig. 15-3) and created using the power of mind mapping it is far more likely to retain the information and have it available for recall even in the most stressful situations. This is because the information is presented to the brain in a new and creative way, which it is not used to. Taking advantage of both sides of the

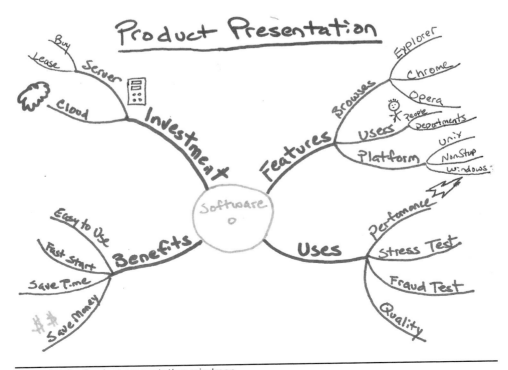

Figure 15-3 Product presentation mindmap.

brain, it is more likely to commit this information to the long-term memory where it can quickly and accurately recover it as it is needed.

In addition, mind maps are commonly used within presentations, to help describe specific concepts to the audience, which makes it even more essential for successful presenting. When done properly a mind map can bring to life the creative thought process which was used to come up with the presentation, this will help those in the audience focus on the presentation and remember what it was about making the concepts which are taught stick with them far better than most other presenting options.

Taking Notes

People are required to take notes for many different things in life, and one of the problems most people face is when they go back to review the notes they took it is very difficult to find the information they need. When taking notes most people make the mistake of attempting to write out detailed sentences of what the person who is teaching them says. If this were a good idea, it would be far better to simply record the entire lecture to listen to again at a later time. Notes are supposed to be a very quick way to review the topic which was taught on, and that's how mind mapping can help people document the information quickly and in a way which is extremely useful.

Mind mapping lends itself to note taking extremely well because it can quickly break up topics into related areas which makes it easier to review at a later time. It is also much faster to write or type out very brief words or phrases, which capture the concept of what is being taught. It can even be useful to sketch a quick picture at times to help document an idea in a more useful way.

Another mistake people make when taking notes is thinking that they are done taking the notes once the lecture is over and all the information has been given to them. Some of the most valuable parts of note taking should take place in the hours or, at a minimum, days after the lecture is over. As the mind is processing the information it took it will begin coming up with ideas or thoughts about it. This is part of the process of storing the memory, but it can be very beneficial for expanding the mind map with additional information.

People who add these thoughts to the mind map after the lecture is over often find that some of the most valuable points were actually thoughts that they had added on their own rather than simply quoting something the instructor actually said. This is because when the mind comes up with its own ideas about a subject it is far more likely to be able to remember it in the weeks and months following an event. Coming up with these creative reflections on the topics that were presented is much easier when using a mind-mapping tool to facilitate the process.

Brainstorming

One of the most obvious things which mind mapping is used for is brainstorming for creative ideas.[13,14] Most people can see that the mind map is set up in a way which really helps with taking down ideas quickly and in a well thought out manner. Unlike traditional brainstorming sessions, which simply write down ideas as quickly as possible with little or no organization at all. Once a normal brainstorming session is done it can be difficult and time consuming to go through and try to figure out which of the ideas are useful and which ones can be disregarded. In fact, it often happens that when reviewing the brainstorm list there are ideas which nobody can remember coming up with.

When using mind-mapping tools as a way to promote a creative brainstorm it is much easier to quickly sort out the ideas as they come. This is especially true when using software-based brainstorming tools rather than a piece of paper and a pen. It is possible to quickly click on different entries into the mind map and add additional layers. This will help ensure the results of the brainstorm are already in a logical format that is easy to use. Having each idea categorized where it belongs, also helps people remember what they were thinking when they first came up with it as well.

When people use mind maps for brainstorming they are also typically able to not only come up with more total ideas to list, but also better ideas. This goes back to the increased creativity, which is promoted by the design and guidelines used with a mind map. When attempting to come up with new and creative ideas during a brainstorming session there is no better way to get the desired results than by using a mind map.

Decision Making

When making difficult decisions many people are used to writing out lists of positives and negatives for each option they have available. This is a good start and it can help people to have a good understanding of all the factors involved with a specific decision, but it is not enough to make the best decision possible. This type of listing of options really only uses the western hemisphere of the brain, which thinks in a very linear way. While this is an effective way to get all of the logical or linear factors about a decision down on paper, most people will be missing the emotional or "free-thinking" impact of the decision.

Everyone knows that when making a decision there is more to the picture than just the facts of the scenario. People need to also think about how the decision will affect others and what type of impact it will have on virtually every aspect of their life. Depending on the decision that needs to be made, the list of pros and cons can be extremely long, but even then it is typically not enough.

Harnessing the power of mind mapping for decision making people can engage the entire brain. This will help come up with additional factors about both sides of every important decision, so individuals can make more informed decisions. The mind map will still have all of the typical pros and cons of any decision as the linear side of the brain is still extremely engaged in the process. Most people will find, however, that when using a mind map they will have many additional entries into their decision-making mind map.

When people look at the differences between a map they make using the traditional column method and one with a mind map they will almost always find that there are many more entries. They will also notice that most of the extra entries are much less concrete and more emotional or creative; based than the others. If, for example, someone had to decide whether or not to move to a new city to take a new job that they want, the normal list would be dominated by things such as the following:

- Pro—Higher income
- Pro—Opportunity for advancement
- Pro—Working for a good company
- Con—Have to sell the house and find a new one
- Con—Kids have to change schools
- Con—Starting over with seniority at a company

Of course, there would be many other items on the list, but it is easy to see that each of these pros and cons are all very concrete ideas. These are certainly important things to consider when making such an important decision, but they are not the only things that should be on the list. When people use mind mapping to help them make this important decision they will likely come up with some additional pros and cons such as the following:

- Pro—Career will be more fulfilling
- Pro—New city has better schools for children
- Pro—New city has great art and culture
- Con—It will be more difficult to attend family functions
- Con—New city has far longer winter months
- Con—Children will be sad to move away from friends

The types of things on this second list are typically going to be generated more from the creative side of the brain, which is also more concerned with the abstract or emotional things in life. By using the mind-mapping techniques for decision making people will get a fuller idea of the impact each decision will make. When relying only on the first type of list a lot of individuals will end up making a decision based only on those fact-based concepts. They may then find that it was the emotional impact on one's life, which was actually more important for that particular decision.

Gathering together all the information about important decisions like this is critical in order to make the best possible choice for the individual and others who will be affected. Without using a tool that engages both sides of the mind it is almost impossible to look at all the options in a way that will give a big picture of the overall impact it may have. With this type of tool, however, many people have been able to make decisions based on all the information and they are typically happy with the choices they have made. In addition, most people will find making the final decision is easier when using a mind map because they can be confident that they have properly thought through all their options and how they will affect everyone involved.

Overcoming Writer's Block

All too often people who are working on some sort of creative work like a script for a movie or a book will get to a point where they simply can't come up with the new ideas required to keep it going. This is so common that it has even been given a name, "writer's block." While there are many tips and tricks for getting through this often devastating condition, few can be as effective as using a mind map. Even if mind mapping was not used from the beginning of a project it is still possible to gain benefits from one.

Taking a fiction novel as an example, if a writer is unable to advance the plot any further because he can't come up with the right ideas for the story it can be very helpful to begin a new mind map. Placing the main point which he is having trouble getting past in the center the writer can then quickly list out as many different options as possible around the main problem. Everything from introducing new characters to going back and rewriting that specific plot line from the beginning can be placed in the map.

For the third levels and beyond the writer can write brief points which would result in the overall storyline if he were to choose that specific path for his characters. Some of the ideas will undoubtedly create some problems, so those can be dismissed quickly. Others might not be to the writer's liking so those too can be either discarded or set aside for possible solutions to use if nothing better comes up.

The fact is, however, that due to the increased levels of creative thinking which are brought about by the very act of using the mind map it is very likely that the writer will come up with the perfect solution to his problem. In many cases the mind map will then help serve as a guide for several more chapters or even the rest of the book so the writer can quickly get through it and produce a work that he can be truly proud of.

Unlocking Creativity with Mind Maps

While it is clear that mind mapping is a great way to encourage creative thoughts and ideas, it is not enough to just pull up a mapping program and let the creative juices flow. For projects that require a lot of new and innovative ideas, it is best to combine the mind-mapping techniques with some other proven triggers for creative thought. Most of these naturally fit with the guidelines of mind mapping, so they will be simple to do while filling out a map for just about any project. The point of using these tips, which are known for helping people think more creatively, along with the mind map is to further focus on the creative aspect of any project.

This will not eliminate the logical or linear thinking altogether, but it will become a secondary thought process. Allowing the creative portions of the brain to dominate the mind-mapping process will help create a wide range of new and interesting ideas, which may otherwise not have never come up. Using these techniques is typically only recommended when attempting to focus primarily on coming up with new or creative ideas. For projects which require a balance of the linear and nonlinear thinking, it is often better to simply use the mind-mapping tool on its own.

Each of the following techniques used for unlocking creativity can be used on its own or in conjunction with each other based on the desires of those participating in the project. Each individual will also find that some of these methods work better than the others for them.

- **Thinking like a child**: Children are well known for coming up with some of the most interesting and unique ideas imaginable. This isn't simply because they are younger, but also because of what they are often doing. While working on a mind map encourages everyone in the group to pull out a piece of paper and some crayons or markers to doodle with. This will keep the creative side of the mind flowing while distracting the linear portion of the brain. Having small toys available to play with can often accomplish the same goals.

- **Meditate**: When the ideas begin to slow, but more are still required, have everyone participating in the mind-mapping exercise take some time to meditate. Clearing one's mind for a short period of time before starting to focus on coming up with new creative ideas is a great way to get through roadblocks like this.

- **Imagery**: If a group of people is working on a mind map in the same room take some time to put up images about the project. Hanging posters or just snapshots of things related to the main subject will help force the brain to stay on task and give it additional imagery to use when coming up with the new ideas.

- **Affirmations**: Every idea that is produced should be considered a good one, and recognized as such by others in the room. Even if the idea is later discarded, the initial part of the mind mapping should always accept every possible idea. Receiving compliments or words of affirmation from others in the group releases pleasure-producing hormones in the brain which will help incentivize it for coming up with even more creative ideas.

- **Distractions**: If possible, avoid just sitting down at a desk the entire time while working on a mind map. Find something to do which will keep the conscious mind distracted and allow the subconscious part of the brain to really get running. Many people find things like doing the dishes or exercising to be perfect for this type of task.

- **Change of scenery**: If the mind-mapping process will take more than about an hour, make sure to move to new locations from time to time. Starting a map in a large conference room is great, but it will quickly get boring. After a while have everyone pick up and move to a new room, or better yet to an outdoor location if possible. Having regular changes of scenery prevents the brain from falling into a rut, which slows down the creative process. For projects that are expected to take a significant amount of time it is best to have four or five locations available to alternate between.

- **Play catch**: Similar to the games and distraction techniques, having one or more balls in a room full of people is a great way to keep everyone contributing creative ideas. Gently tossing the ball around the room keeps the muscles moving and the blood flowing so the mind will stay awake and alert. This method also ensures that people are distracted by the ball, allowing their mind to efficiently work in the background.

- **Music**: Having soft music playing in the background is another good way to stimulate the production of more creative thoughts. Typically classical music or sections of ancient chant are excellent for this because people won't be tempted to sing along or try to identify the artist, but simply allow the music to stimulate the mind.

- **Watching for new connections**: One of the most powerful benefits of a mind map is that it lays out an incredible amount of information in a fairly small area, and in a way that allows the brain to absorb it quickly and completely. Make sure everyone in the group is taking a good look at the map throughout the process to see if they can find new and innovative connections that people may have previously missed. Since the entire map is related to a single topic, people are often able to notice that items from one area of the map can also be used in another area, for example. Connecting different items together is simple on a mind map and will often lead to very interesting results.

There are many different things which people can use to help increase the flow of creative ideas whether they are in a group or on their own. Those who frequently need to think creatively will want to try each of the earlier techniques on their own, and also come up with some of their own ideas to try. Over time most people will narrow down these techniques to two or three of their favorites, which are most effective for them.

Each person is unique and it makes sense that some techniques that work great for one person will be extremely ineffective for the next. When it comes to creative thinking it should be no surprise that the best way to do it is by coming up with creative methods to help promote the creative thinking centers in the brain.

Beyond the Brainstorming—Benefits of Mind Mapping Throughout a Project

When it comes to using mind mapping for coming up with new and interesting ideas during the brainstorming phase of a project it is fairly simple to see how it works and understand the benefits mind maps bring. Some people, however, have trouble figuring out how to properly use a mind map for the rest of a project since the main concepts are already in place. This concept was touched on briefly earlier in this chapter, but it is important to go into some depth in this area to fully understand the power of the mind map.

Once the mind map is filled out with dozens or even hundreds of different ideas about a particular subject it is time to begin putting those ideas to use. The specifics of how this will work will vary greatly from project to project, but in general they all follow the same basic path. When working through a project, however, if a specific task listed below does not fit with the project it can be skipped without any concern of causing a problem.

First Revision of the Mind Map

With an initial draft of the mind map complete, the first thing to do will be to go through and clean it up. In most cases people will find that there are several ideas or concepts that are very similar or even identical. Eliminating duplicates will help avoid confusion or even conflict later on in the project so getting it done early is ideal. Go through each item on the mind map and locate anything that might need to be removed due to duplication, and mark it with a specific color or other item. After all the duplicates are marked, choose which ones will be kept and which ones will be eliminated. Make sure to get the input of the entire group if working with others on this project.

The next step in the initial revision is going through and eliminating ideas that don't fit well with the project. In most cases there will be a fair number of what can be considered "bad ideas" and that is actually a good thing. High numbers of bad ideas is an indication that everyone in the group was offering their ideas freely, which is a great accomplishment. Make sure that everyone has a say in which ideas are considered good and which ones are considered bad. Having open discussions about each idea before it is eliminated will help promote the full understanding of each idea.

Keep in mind that for this stage of the process everyone in the room should be given equal weight as far as evaluating ideas goes. Managers and employees and even executives should be comfortable sharing their opinions at this early stage in the process.

Organizing Logically

Up to this point one of the key benefits of the mind map was that it helped encourage creativity, but at this step in most projects it is important to be able to briefly put a hold on the creative thought and evaluate the organization of the map logically. Through the initial brainstorming it is very common for ideas to be placed with other ideas even though they would be better located in other groups. In addition there will likely be some ideas that were placed at the third or even fourth level which should actually have their own section directly off the main idea.

The way each item is organized will have a lot to do with the specific project that is being done and what exactly the project is. The key at this point is to put similar ideas together in the same groups in as logical a way as possible. Even after this stage there may be some significant changes made, but getting things into the areas where they seem to belong now is important. Getting things in the proper areas may take some time, but it will help ensure the project goes much more smoothly down the road.

Assigning Responsibility

Similar to identifying which parts of the project are done at which time, it is also important to assign each item to an individual who will be responsible for completing whatever actions are needed to finish it. If the project is being done by a single individual then this section can be skipped, but when it is a larger project it is important to make sure every thought on the mind map is assigned to an individual who will be responsible for bringing it to life.

During this stage individual tasks can either be assigned to an individual or a group of people, who are directly responsible for the action. Ideally, however, even those sections which have a group assigned to them will have a leader within the group who is the primary contact if anyone has any questions or concerns with how that step of the process is going.

Scheduling

Now that everything in the map is located where it belongs logically it is time to start scheduling which types of things need to be done first, and which ones can wait. For larger projects it is often best to use some sort of indicator to show the items that are being done in the first phase, the second phase, and so on. Many people will use color-coding or the shape of the box around an idea to make it quickly apparent to those looking at the map which ones are being done at which point in the project.

When in this stage it is necessary to have a good understanding of what resources are available to work on the project. In fact, having a section on the map labeled resources is a great way to help assign individuals to different projects. The more people that are working on a given project the larger the groups of things being done at any one particular time can be. For large or important projects it is often a good idea to have an experienced project manager take over the mind map for this phase, and most of the rest of the phases as well.

Tracking Tasks

Using the mind map it is simple to keep track of which tasks have been done and which ones haven't. Remember, since many of the items on the map are not set down specifically as a task but rather just an idea or a concept it is often possible to check off several of them together when a single action has been taken. The specific way this is done will vary greatly from project to project, but keeping track of the activities being done for this mind map is a great way to help ensure everyone is staying on target for getting things done.

This is even more important with projects that have different tasks that are dependent on each other, because one group may be waiting for another to finish their task before they can begin. Essentially, at this stage the mind map is functioning more as a scheduling tool than what it originally started as, but it is still extremely valuable to keep things on track and well organized.

Project Completion and Error Checking

When every item in the mind map has been certified as completed by the responsible parties the project is nearly done. The last thing that a mind map should be used for is going through the final result to review everything to make sure it was done properly. Checking for any problems is difficult during most projects, but when there is a detailed mind map which is well thought out and documented it is much simpler to go through and find any potential errors.

It is simple to see how using mind mapping can help at every stage of virtually every type of project. Whether it is being used for creative tasks like writing a book or music, or an important business project, mind mapping brings with it some incredible benefits which simply aren't available in other methods of organization. The mind-mapping tool is also extremely flexible, which makes it perfect for just about any type of project.

Throughout any project where mind mapping is used the participants will be much more productive because of the fact that it forces both sides of the brain to remain focused on the task at hand. The added creativity, which is encouraged through the use of this tool, will also help improve the results of a wide range of different topics. The bottom line is that whether an individual is a project manager at a large corporation or simply trying to write a book on their own, the mind-mapping concepts can help ensure the project goes smoothly from beginning to end. In fact, the end results of projects completed with this tool are often far superior to those that are done using any other type of project management method available.

Making Mind Mapping a Part of Everyday Life

It is easy to see how mind mapping can be used in a wide range of different projects in life, but that is really just the beginning. Once people realize that they can expand their creative horizons by using this extremely simple tool they can begin to apply it to virtually any situation in their life (see Fig. 15-4). Writing up a quick mind map can do everything from planning out tasks they need to complete that day, or creating a plan for your future.

The more people use mind maps the more they can apply new and creative solutions to their everyday problems. Before they know it, many of the things that used to cause them a significant amount of stress, are now handled quickly and easily by pulling out a sheet of

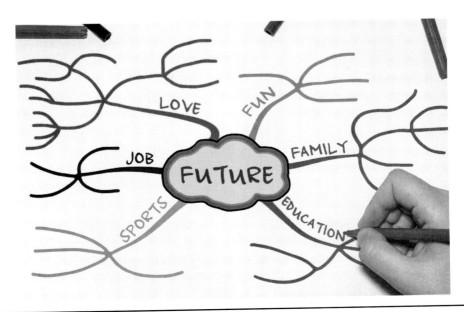

FIGURE 15-4   A simple mind map applied to "life."

paper and a pencil or pulling up their favorite web-based mind-mapping program. Some unorthodox ways in which mind maps are used in everyday situations are listed below, but always remember that every individual will have a different set of uses for this tool so take some time to think about how it can be applied in the day-to-day events of life.

- **Arguments with a spouse**: Every marriage has a litany of things in which the husband and wife constantly disagree with each other, and this can lead to arguments that can be a constant source of conflict. Rather than yelling at each other, write down the main topic in the center of a mind map and come up with some new and creative solutions to the problem. Both parties can write up their own mind map and compare them once they are complete. Many couples will be surprised at just how similar their solutions are.

- **Grocery shopping**: Virtually everyone has had those times when they are grocery shopping, spend a lot of money on food for the week but when they get home they feel like they still have nothing to make for dinner. This is because most people are very unorganized when they shop, and when they try to plan for shopping they can't think of meals they would like to make. Pull out a sheet of paper and write groceries in the middle and see where it goes. Jotting down some meals for the family throughout the week on the first level followed by ingredients on the other levels will quickly result in an easy to follow grocery list that guarantees a successful shopping trip.

- **Home improvement projects**: Planning every detail of a major home improvement project can be extremely difficult. Everything from the color of paint to the style of counter tops and much more need to be decided but unfortunately many people are thinking about these things using the linear or logical side of their brain since they are attempting to plan out the project. Pull up the mind-mapping software and most people will quickly find that their creative ideas begin to flow quickly and easily so they can decide exactly what they want their new kitchen or other room of the house to look like when they've finished.

- **Identifying allergies**: Even something medically related like allergies can be benefited from using mind maps. Many people have had to try to figure out what they or someone they love is allergic to after suffering a reaction. Making a list of everything they ate or interacted with can be difficult, but when it is done using a mind map people tend to remember far more than they otherwise could. They can also link some foods to each other when they are commonly eaten together to see if there is any link. This new and interesting approach is often able to help people narrow down the list of potential causes quickly and easily.

The list of everyday activities that can be done using mind mapping is virtually endless. If anyone is ever running out of ideas for what they can use a mind map for in their life all they have to do is create a mind map to help them think of additional uses for mind mapping. While it may sound comical the fact is that when used properly mind maps can improve virtually every aspect of one's life. They will quickly find that they are spending a lot less time on tasks that they dislike, because they were able to think of new and creative ways to get them done faster or better. Those things which they do enjoy doing in life, they can spend more time on, and get far better results as well because their mind was optimized using this great technique.

Even for people who are used to thinking creatively on a daily basis like artists will find that mind mapping truly takes their creativity to an entirely new depth. Adding the power of the linear thinking side of the brain to the creativity of the other side can help even the most artistic individual quickly come up with new, and possibly revolutionary, ideas for just about anything in their work. Perhaps the best thing about mind maps is that even someone who has never created one before can sit down and start working on one in just minutes. They are very flexible and easy to learn about so why not give it a try and see just how creative the mind can really be.

Tips for Mind Mapping

Anyone who is interested in giving mind mapping a try in their life can get started quickly by following these quick tips. Remember, however, that every mind map will be different and the key is to use mind maps in a way that works best for the people using it, and for the specific task it is being used for. The power of the mind map is in its ability to unlock creative thinking, so don't attempt to stifle that power by believing that these maps have a set of rigid rules that must be followed. Start the map and let the mind lead the map where it needs to go.

- **Use software**: Finding software to make mind maps is one of the best ways to get started fast. They have all the tools and guidelines built in to allow people to get up and running in just seconds. There are many great websites that provide basic mind-mapping software for free and they can create incredible mind maps on any topic.

- **Don't overthink**: One of the biggest issues many people have when first starting with mind mapping is overthinking the entire process. This will slow people down in their learning process and can cause them to become discouraged. If an individual step in the process is taking more than a couple of minutes then it is most likely being overthought.

- **Have fun**: Whenever attempting to increase creativity in any activity it is essential that everyone involved allows themselves to have fun with it. Making jokes or writing comical images that will help them with a particular area of the map is a good thing. If the mind map is feeling more like work than fun, it is most likely being done wrong.

- **Use them often**: As with just about everything in life, the more people do it the better they will get. Using mind maps for every possible task, especially at first, will quickly help people learn what works for them and what doesn't. Over time the mind map will become an incredible tool that they can pull out and use whenever necessary to get amazing results.

As more and more people are learning about how and why mind mapping works, and that it can turn just about anyone into a creative genius, it continues to get more and more popular. Many people believe that this method of thinking and taking down information will become the standard that is taught to people from a very young age in the near future, so don't be left behind. Join the creative thinking revolution and start using mind maps today!

References

1. Davies, M. "Concept Mapping, Mind-Mapping and Argument Mapping: What Are the Differences and Do They Matter?" *Higher Education*, 62 (3): 279–301, 2011.
2. http://www.mind-mapping.org/blog/mapping-history/roots-of-visual-mapping/
3. Tony, B. *The Mind Map Book*. BBC Books, 1993.
4. Tsinakos, A.A. and Balafoutis, T. "A Comparative Survey on Mind-Mapping Tools." *Turkish Online Journal Of Distance Education (TOJDE)*, 10 (3): 55–67, 2009.
5. Tucker, J.M, Armstrong, G.R., and Massad, V.J. "Profiling a Mind Map User: A Descriptive Appraisal." *Journal of Instructional Pedagogies*, 2:1–13, 2010.
6. Lewis, C. "Power Mapping: An Expression of Radiant Thinking." *Training Journal*, 4: 37–40, 2005.
7. Hugdahl, K. "Visual-Spatial Information Processing in the Two Hemispheres of the Brain Is Dependent on the Feature Characteristics of the Stimulus." *Frontiers in Neuroscience*, 7, 2013.
8. Cashford, J. "The Integration of the Cerebral Hemispheres in Poetry and Mystic Texts." *Gifted Child Quarterly*, 23 (1): 56–70, 1979.
9. Timofti, I.C. "The Complex Functioning of the Human Brain: The Two Hemispheres." *Broad Research in Artificial Intelligence and Neuroscience*, 1 (2): 177–179, 2010.
10. Edwards, S. and Cooper, N. "Mind-Mapping as a Teaching Resource." *The Clinical Teacher*, 7 (4): 236–239, 2010.
11. Abi-El-Mona, I. and Adb-El-Khalick, F. "The Influence of Mind-Mapping on Eighth Graders' Science Achievement." *School Science & Mathematics*, 108 (7): 298–312, 2008.
12. Dhindsa, H.S., Makarimi-Kasim, and Anderson, O. "Constructivist-Visual Mind Map Teaching Approach and the Quality of Students' Cognitive Structures." *Journal of Science Education & Technology*, 20 (2): 186–200, 2011.
13. Stuart, G. "Visual Brainstorm: Mind-Mapping Programs Help You Put Your Thoughts in Order." *Macworld*, March 2009.
14. Byrne, R. "Harness That Brainstorm." *School Library Journal*, 57:5, 2011.

CHAPTER 16

Social Networks

Christopher S. Rollyson

How Social Technologies Transform Innovation

Social technologies have profoundly changed the role of innovation in organizations and transformed its practice. Within the next 20 years, an increasing portion will regard innovation as currently practiced as an unimaginable relic because it was conceived and practiced in an era in which communication was not free, as it is now. They will ask, "What were they thinking?" This section and "Examples of Social Network–Powered Innovation" of this chapter will serve as a whirlwind tour to how social technologies have changed innovation and how leaders are already showing the way. They set the context for sections "Social Network Innovation Good Practices and Pitfalls" and "Getting Started: Using Social Networks for Innovation", which concern the practice of using social networks for innovation. They will suggest emerging social network–powered innovation models and close with how you can get started immediately.

In the spirit of social network–powered innovation, most of the illustrations have the words "social" and "innovation" hidden within them, so look carefully. See how many you can find!

The Social Enchilada

Digital "social" technologies and behaviors are changing the economics of communication and collaboration. Here is a primer on the unique characteristics of social technologies and practices, and how they transform innovation.

Technologies and Practices

Web 2.0 technologies form the technical infrastructure for social technologies. They are termed "many to many" because they run on networks that enable any member on the network to communicate with any other member. This makes them inherently social because everyone can talk with everyone in easily recognizable ways. Web 1.0, on the other hand, was "one to many" because relatively few websites on the network offered "many" users information and the ability to transact. Websites' features were not very interactive or natural for most people to use, although they were a major improvement of pre-web "enterprise" applications.

Social networks like Friendster, MySpace, Orkut, Facebook, LinkedIn, and, arguably, Google+ have played a pivotal role in the global adoption of social technologies. They are profile based and enable members to create networks of "friends" or "connections." By having profiles and friends networks, social networks encourage people to share digital content, to communicate, and to collaborate. Moreover, each member can control which friends see what kinds of things they share or say.

Social media refers to using social technologies as "media" in order to influence large audiences. Social media injects a semblance of familiarity and intimacy into its communications, but its key aim is using social technologies to distribute content. As of 2013, social media has been by far the first and most prevalent use of social technologies, so it has dominated peoples' and organizations' imaginations about what social technologies are and how they are to be used.

Social business is the practice of using social technologies to transform business, and it will eclipse social media as the prevalent use of technologies from 2017 to 2022. It takes the view that social technologies enable people to communicate and collaborate far more efficiently, and this is the root of social technologies' promise for innovation. Social technologies are the emerging knowledge economy's equivalent to the waning industrial economy's assembly line, which enabled mass production and led to unprecedented wealth in much of the world. Social networks offer an analogous quantum leap in productivity via communication and collaboration, but in order to tap it, organizations will have to dismantle barriers to communication and often change their cultures.

Digital

To appreciate how social networks work, it is useful to understand their technical traits, which enable users to change their behavior to become more productive. Most importantly, social networks are digital and channeled over the Internet, which is a global network of servers, routers, and other gear. This means that once a message or piece of content is "out there," it lives forever because parts of the network have records of it, even if they do not store it. Digital content is searchable by anyone that has access to that content and the majority of content is public.

Interactions in social networks are "transparent" and very different from conference calls or "In Real Life" (IRL) interactions because they are automatically recorded and can be accessed at any time in the future. Sharing is lightning fast and free. Although interactions may take place in private behind a firewall, because they are digital, anyone with access to the interaction can share it outside. Likewise, the social network in which the interaction has taken place may change its features or policies and make private interactions public. Hackers can gain access to any computer or network on the planet if they are determined, whether it is "social" or not.

Social networks' transparency inhibits some people from having certain kinds of interactions because they are uncomfortable about losing control over who may see or hear what they said. However, in a business or government context, relatively few elements of people's personal lives are the primary focus of interactions. Instances of collaboration on social networks can be spontaneously discovered by colleagues who may learn from and reuse the advice and ideas without even contacting the participants of that collaboration. Or they may join the collaboration some time after the fact, ask questions, and offer solutions. Therefore, transparency yields "discoverability" and "annuity." Conversations and collaborations are discoverable annuities because they may be found and reused, thereby paying more "dividends" over time. Contrast this with thousands of phone or e-mail interactions that occur every moment of the business day that are never reused; large portions of them are dealing with the same problems and solutions.

Asynchronous

People have long been accustomed to asynchronous communications like e-mail and postal mail, but social networks take them to a whole new level. Since they are instant and fast, social network communications have been growing quickly because they enable each person to participate when it is most convenient and the cost is lowest. This enables people to have communications that, when they happen in a networked environment, also harness the network effect[1] to dramatically compound their value. Just as powerful, social networks follow the "90-9-1 rule of participation," which holds that, for every 100 participants, 90 are passive "lurkers" that observe, nine are contributors, and one is a heavy contributor.[2] However, all 100 people are influenced by the interaction and feel some degree of commitment because they were "in the room" and may return to it at any time. When we consider that entire collaborations may be discovered, shared, and reused forever, we can see that communication is becoming essentially free.

Social

Social networks add yet another often-overlooked piece of the puzzle: they provide social context for information and communications. Many people get frustrated with digital "data" and "information" because, they say, it is so stupid. What they usually mean is that the information is not organized to serve the purpose at hand. As primates, humans are inherently social; Robin Dunbar even argues that sociality is our primary survival strategy.[3] Although very few people are aware of sociality, it is the primary determinant of "relevance" for most human activity. It turns out that when people interact in groups, they learn far more quickly about each other by observing how everybody interacts with everybody else. When you meet someone IRL, how long does it usually take you to learn what things make him or her mad, sad, or glad? If you become friends on Facebook, you can read back on their wall and learn an astounding amount about their personality, who their friends are, and how they interact with them. This type of digital information is unprecedented. Social networks, because they encourage "socially oriented" sharing, have made it a habit to share information and social context together, dramatically increasing the value of digital interactions.

Even more far-reaching, since they embody these digital, asynchronous, and social traits, social networks have changed the economics of relationships. It is now an order of magnitude less expensive to find people based on specific interests, to invite them to communicate and to maintain an interaction with them over time. That means people can have more relationships, and more diversity in their networks, than ever. Moreover, people use interaction to constantly monitor how much and under what conditions they may trust each other. Trust and common interests are the bedrock of relationships *and collaboration*.

Emergent

Social networks enable emergent collaboration, which means that the organization of people and groups emerges from spontaneous interactions. People can find each other based on their profiles and what they say during social interactions. They can invite each other to collaborate in social networks and form groups that communicate quickly and asynchronously for work or play. All of them may share all kinds of ideas, papers, inventions, software, etc. Groups can coalesce and get productive very quickly, and geography is no longer a limitation. Once the group agrees on the mission, its organization develops rapidly based on social interactions, which shows who has the most relevant knowledge of each subject, who is most motivated to lead, and how effective people are as leaders. People inherently prefer working this way because it is merit based: people listen to others who know what they're talking about within the context of the mission. In addition, people learn from each other very efficiently and quickly; they present, challenge, and synthesize ideas very quickly, educating the whole community.

This is why industrial economy organizations will change or disintegrate. Industries were built around large-scale mechanical production, which requires organizations that mirror its massive means of production. These organizations are too inflexible to enable emergent collaboration, so they will need to use social business to transform because people will be far more productive using social network–powered collaboration and innovation.

Most leaders have very limited experience with emergent collaboration, and they trust their organizations, which have made them successful. To many executives, it feels chaotic to let people determine organization spontaneously and very quickly.

Economics and Markets

We have outlined some of social networks' key characteristics and how they affect people, relationships, and collaboration, but that is only half the equation. As suggested earlier, social networks are a key driver that is transforming the global economy. The economy creates the context in which organizations succeed and fail, and it is the ultimate context for the organizations that pursue innovation.

Social technologies and social networks are examples of *knowledge economy means of production*.[4] To grasp the scale and scope of the social and economic transformation that all people are experiencing, it is useful to understand where we have been and where we are headed.

The knowledge economy is emerging due to three converging forces: (1) financial or political liberalization and (2) standardized web communications have led to true global peer-to-peer interactions, *en masse*, (3) social technologies have made communication and collaboration with social context free or "too cheap to meter."[5]

Companies and governments are large organizations that are disrupted and scrambling. To understand why, reflect on the profound shift in organization success factors, which drive organizational behavior at the deepest level (Table 16-1).

In the knowledge economy, people talk more about *everything*, including products and services they use or hear about. Moreover, when people talk, they consume novelty, one of the key fruits of innovation, which usually seeks to invent new products, services, or delivery. This has the perverse effect of commoditizing products and services because people identify product irrelevance, faults, mediocrity, and novelty overnight, which shortens product life cycles and usurps firms' ability to create and charge the vaunted "brand premium." In most product and service categories, markets are bifurcating; a small portion of the market is premium, and the rest is commodity. Customers using social networks consume innovation far more quickly than firms can produce it. In addition, competitors learn about innovation,

Industrial Economy Success Factors	Knowledge Economy Success Factors
Deliver vast quantity and low prices to relatively few broad demographics	Deliver unprecedented choice to many narrow customer segments; low prices table stakes
Tightly integrated command/control organization	Loosely coupled collaborative-networked organization
Failed new offering launches are the rule because it's a low-velocity, internally focused research/product/sell process	New offering launch processes are high velocity and feature customer input and collaboration with external specialists
Off-line consumers are isolated and passive, buying decisions usually made using company-influenced information, consumers can't easily find consumer-produced information	Online customers validate each other's interests and influence purchases, creating micromarkets with unique demands; this will be the rule in emerging markets, where online is the default for new middle classes
Focus on product/service features	Focus on customer experience

TABLE 16-1 Industrial versus Knowledge Economy Success Factors

copy it, and drive the price down. These developments have gutted the industrial economy's chief means of profitability. They have prolonged high unemployment and led to the global economic crisis of the early 2000s in heretofore "advanced" industrial economies, which has in turn dampened demand for emerging economies' products.

Even more profound, the Social Channel[6] argues that the concept of mass-produced products is a relic because product differentiation was focused on product features. Today, they rarely differentiate enough to recoup investments in innovation. Moreover, software-powered production is enabling mass customization for an increasing portion of products, and social networks enable prospective customers to form groups, collaborate, decide what products they like and "commission" products with the features they want.[7] The stage is set for firms to collaborate with customers to decide what products to make and to individualize production.

Another key point is that people deeply enjoy collaborating on things they like or find important. Collaboration is an order of magnitude less expensive, so many more people get involved, even if most are lurkers. The Social Channel proposes that *the act of social collaboration will be the engine of differentiation in the knowledge economy*. In many instances, whether the product customers designed is absolutely better is beside the point: they did it together and will enthusiastically promote it to their friends, neighbors, and family. This is how "differentiation" happens in the Social Channel.

Examples of Social Network–Powered Innovation

Now that we have covered social networks' key features, how they are changing behavior and how they are recreating the global economy from the ground up, let's learn from pioneering commercial and government organizations that are already using social networks for innovation.

3M: Creating a Global Research Lab Campus with a Social Network

• Used private social network to create a virtual R&D campus for 2000 researchers from 35 innovation labs in 65 countries • Boosted spontaneous and planned knowledge exchange with mobile connectivity and user-defined groups and forums • Increased sharing maintained rigorous I.P. security standards	**3M**

TABLE 16-2 3M Case Study

Challenge 3M operates 35 innovation labs in 65 countries, and they wanted to create a "digital campus" to encourage the flow of ideas and collaboration among R&D teams. They used a proprietary enterprise social network, Socialcast, to encourage sharing while maintaining security, as teams share very sensitive intellectual property.

Highlights

- Every company emphasizes its ability to "innovate," but 3M has made innovation an explicit part of its branding and business strategy. It wanted to explore social networks' potential for enhancing its existing innovation capabilities.

- Innovation is continuous and pervasive at 3M. The company operates R&D labs around the world. They wanted to increase the diversity and volume of communication among R&D teams, regardless of location. They sought to tap digital social networks' ability to enable people to connect based on common interests, without geographical limits.

- A key organizing principle of this initiative was addressing what John Wordworth, 3M's Tech Forum Chair-elect and Head of IT Lab Collaboration, called "knowledge at rest" and "knowledge in motion." Thirty-five labs have extensive digital artifacts organized somewhat differently, in multiple languages, in wikis, and databases. R&D recreates things they need but cannot find, so knowledge at rest has significant potential for reuse. The ability to make requests of fellow researchers increases reuse. Knowledge in motion is time-sensitive. The social network makes it possible for an engineer to make a request of a private, trusted global network of fellow engineers and scientists.

- The social network includes public and private groups and private messaging, too, which enables employees to self-organize. This flexibility is critical when the company wants employees to "make it their own."

- Given that R&D labs develop and share sensitive intellectual property, 3M wanted to maintain its rigorous security standards while it increased employees' ability to share, so John Wordworth led the effort to optimize sharing and security. He knew that the social network had to "feel natural," so people would really use it. "We wanted it to feel the same as walking down the hall to ask somebody a question."[8]

- The social network incorporates features and functions from popular consumer social platforms, which minimizes the need for employee training. This lowers the barrier to usage.

- The social network grew to 2000 global members during the first year. In addition, the social network software features mobile connectivity, which lets members maintain a continuous seamless connection with each other. This is critical in addressing "knowledge in motion." It encourages employees to use the social network in mission-critical situations.

Insights

- Relationships are built along two vectors: trust and common interest. In most cases, relationships begin when people have important interests in common, so they begin interacting, and each interaction increases or decreases trust. Moreover, trust is deeply accelerated by digital social networks because people observe each other interacting with others. Remember, this is the rationale for "communication is free." It will take one person time to have, say, ten relevant interactions with another person, to adjust the trust meter. However, by observing the other person's interactions with others, the first person can adjust trust based on fewer interactions between himself and the other person.

- Enterprise social networks, if the community is nurtured appropriately, become trusted spaces, so they encourage more sharing over time and become more valuable. Trust serves to reduce friction and increase the depth and breadth of sharing.

- Incorporating mobility is critical for almost every use case now. People in global markets increasingly expect mobile connectivity, and in many countries, the smartphone or tablet not the computer, is the primary gateway to the Internet. Even in computer-centric countries, mobile penetration is rapidly growing. Many employees won't take a social network seriously unless it has excellent mobile connectivity. Use Facebook as a benchmark for many users.

P&G: Using Social Networks to Scale Product Innovation

• Open innovation ecosystem sources 50% of P&G's innovation • Supported 7500 firm R&D staff with 50,000 external scientists • "Challenges" optimize innovation efficiency, quality, and cost • Social networks "required" to attract/retain the brightest people	**P&G**

TABLE 16-3   Case Study: P&G Consumer Products

Challenge   In 2000, P&G's new CEO, A.G. Lafley conducted a review of the company and found it to be in so-so shape. It had accelerating innovation costs, stagnating R&D productivity, and shrinking revenue growth. Only 35 percent of its products met company financial objectives. The new management team concluded that the company had to transform, and they launched the Connect + Develop strategy to drive transformation by sourcing half the company's innovations from outside the company.[9]

Highlights

- Connect + Develop represented a major culture and operational change and unfolded during the early 2000s. Its 7500 R&D engineers and scientists were threatened at first, but management operated under the principle that the purpose of Connect + Develop was to leverage their global R&D team better, and R&D got the message once they understood it was real.

- Connect + Develop was founded on two central ideas: using external resources was the right way to best leverage its R&D team. Second, the company cast a wide net in order to develop a diverse, robust network through which it intended to source one-half its total innovations. By 2006, its top 15 suppliers, among which InnoCentive was prominent, comprised about 50,000 researchers. They included entrepreneurs, retired scientists, universities, private labs, government labs, and more.

- One of Connect + Develop's most successful organizing principles was the concept of "the challenge" which was pioneered by InnoCentive. Challenges are a profound construct that's vital to crowdsourcing, one of social networks' most powerful tactics. Challenges are meticulously crafted units of work that are stimulating, actionable, and clear enough to be easily communicated to a "crowd" that is motivated and able to solve them. Along with this, the sponsor of the challenge only pays the solver if the challenge criteria are met. Moreover, other participants are on a level playing field; transparency is a key element. Challenges articulate the need, describe the problem, specify success criteria, and establish the inducements. The inducement is a critically important component because it telegraphs a tangible and measurable value to the world. The best challenges are universal and thus universally understood."[10]

- Like many companies, P&G discovered that they had to use social networking, blogging and micro-blogging tools if they wanted to attract and retain the smartest younger workers for whom these tools are critical to their productivity. In the mid-2000s, P&G innovation management noticed that social technologies were bubbling up everywhere, and they undertook a protracted effort to garner senior executive support.

- It is critical to set expectations and metrics to measure innovation and social networking initiatives, and P&G learned how to succeed. Innovation executive Stan Joosten explains, "P&G is very data driven, and ROI type measurements are very appropriate for stable processes. If your process isn't stable, ROI isn't relevant. You need results, like when your numbers go up. You can get away with this when you organize innovation into small chunks."[11]

Insights

- P&G's story epitomizes the top-down strategy-driven approach to tapping the transformational power of social networks to revitalize a global brand. The success of the Connect + Develop strategy is now legendary, and P&G is one of the most highly esteemed consumer products firms in the world.

- Organizations need to anticipate a stubborn vestige of industrial economy, command and control attitudes that feature authority and competition. In P&G's legacy culture, the "white coat" R&D engineers were trusted authorities that knew "what was best" for new products. Management needed to be clear and committed to its strategy, to make its R&D more productive. The external social networks were to be complementary and collaborative, not challenging. Collaboration, based on core competency, is the ethos of the knowledge economy. For example, P&G scientists undoubtedly knew the finer points of how to commercialize innovation. External scientists were probably as or more effective as internal staff for analyzing chemical compounds. Recognizing each party's core competency and structuring them to be complementary is a critical success factor.

- P&G's decision to create a diverse ecosystem is adaptive in the knowledge economy, which is far more volatile due to free communication and globalization. Diverse networks are more likely to notice and accommodate disruptive opportunities.

- Organizing work into "challenges" is vital to leveraging the power of networked environments. All networks work this way, including the human body, whose brain fires off requests that are only understood and executed by appropriate organs. Having well designed and encapsulated requests is critical to the function of open networks. A banal example is asking a question on a message board, which is making a (usually) simple request. The most successful questions provide enough context to provide potential respondents enough background to understand the question. Moreover, these questions are written to be actionable, given the network to which they are posed.

State Farm: Engaging Generation Y/Z with Digital/Physical Social Network

• Physical coworking space integrated with public social network • Focused on community, collaboration, and financial empowerment • Innovation research on how to be relevant to generation Y/Z • Physical/digital interconnection speeds learning and innovation	State Farm

TABLE 16-4   Case Study: State Farm Financial Services

Challenge   State Farm operates in the United States and Canada, and its major business is automotive insurance. Based on ongoing research, the company understood that generation Y/Z has a very different attitude vis-à-vis automobile ownership, which is a major threat to State Farm's main business. About 26 percent of generation Y do not have driver's licenses, and 46 percent of 18 to 24 year olds prefer access to the Internet to access to a car.[12] State Farm realized that it needed to "get inside the heads" of this demographic, so they partnered with IDEO to design Next Door Chicago, an integrated digital social network and coworking space on the city's North Side.

Highlights

- Next Door Chicago's physical presence is a free coworking space that features attractive yet functional furnishings and fixtures that invite Lakeview and Lincoln Park residents to "drop in and work" at open tables equipped with ample power outlets and free Wi-Fi. Generation Y is very comfortable with coworking.

- The space is open, literally, when weather permits. The entire front of the space is glass doors, which are open during the summer and inviting year-round since passersby see the activity inside. To become members of Next Door, people sign up on the digital social network. Membership enables them to reserve conference rooms and hold private meetings for free.

- Based on its research, State Farm realized that generation Y perceived insurance and financial services as needlessly complex and intimidating.

- Next Door holds financial education seminars that empower members to be more aware of the financial impact of life decisions. However, Next Door is not staffed by insurance agents, and no insurance is sold there. One of the most popular topics is starting up (a business) because this generation is not so excited about employment.

- Next Door conducts significant community outreach; it enables local organizations to sponsor local artist showings and local music during evenings.

- Six months after opening, Next Door has nearly 1500 community members. Its financial coaches have held more than 200 coaching sessions, and the center has hosted more than 50 classes and community events and has a 5-star rating on Yelp![13]

Insights

- State Farm is approaching disruption of its core business adaptively—by engaging with its stakeholders in an environment where they are most comfortable. One of the key goals of Next Door Chicago is engaging and learning about this cohort in a socially relevant context. The legacy approach would have organized a few focus groups and called it a day. Notably, Next Door is also focused on educating State Farm employees as they interact with clients and prospects.

- Integrating digital and physical social networks often creates powerful synergy because they are so complementary. Physical social networks offer rich, multisensory engagement, but at a price: participants can only be present in one physical space at a time. Online, people can be in several "places" simultaneously, so "presence" is lower cost. Digital social networks enable people to keep connected more easily.

- Next Door Chicago uses Facebook, Twitter, smartphone, and tablet apps to keep itself in front of members and to solicit their ideas. These channels are critical for proving relevance to this generation.

Fenland, Cambridgeshire: Social Network Integrates Fractured Community

• Public social network lets citizens to get involved in improvements • Citizens report on problems and follow resolutions online • Methodical approach used rich multichannel due diligence • Transparency boosted citizen engagement, reduced inefficiency • Got prelaunch agency commitment to interact/follow up online	**Fenland, UK**

TABLE 16-5   Case Study: Fenland, Cambridgeshire, UK Local Government

Challenge   Fenland is one of the most economically and socially deprived districts in its county; it has suffered the protracted economic decline of its food processing industry. Residents were very unsatisfied with their government, and the community had several disenfranchised populations. District, county, and related governments wanted to create an online social network to empower residents by giving them more ownership of events in their physical community.

Highlights

- The social network had a very practical focus. ShapeYourPlace.org was focused on connecting residents with various agencies of government, government partner agencies, and other residents.

- The project was very inclusive from the start, and it took a very methodical approach. First, the team mapped the online ecosystem of Facebook groups, blogs, online communities, and discussion forums to understand how citizens were already using online resources. Second, it surveyed residents and held focus groups to analyze residents' use of social networks, frustrations, and aspirations. Third, it held a stakeholder workshop to present findings and brainstorm ideas for designing ShapeYourPlace. These preparations enabled the team to generate requirements for the social network and include residents in usability testing of the beta site. Finally, ShapeYourPlace was designed to synergize with physical events called "community fairs."

- The project team recognized that residents felt helpless because they had no easy way to correct problems in their community: vandalism, trash, drunks hanging out in playgrounds, drug taking, poorly managed construction, broken infrastructure, etc. The social network enabled residents to report things they wanted fixed, but the critical piece was it also enabled them to be notified of the status of their reports, so they could see their impact. That became self-reinforcing and empowering. Not only that, residents could see the progress of each other's reports, which led to a sense of community in which residents worked together to address problems.

- Prior to ShapeYourPlace, residents had no idea of whom to contact among agencies to address specific problems, and this added to their sense of helplessness and alienation. ShapeYourPlace explicitly addressed this and got full agency commitment, including police, fire, and executive and legislative government, at district and county levels. The innovation team knew that consistent agency responsiveness was critical to success.

- Residents report community issues on the social network, and appropriate agencies respond; the social network's moderators follow up to make sure that resident issues get addressed within promised time frames.

- The social network offers blogs, microblog (Twitter) functions, ratings (thumbs up/down), commenting, integration with consumer social platforms (Facebook), and the county's separate community as well as e-mail integration. Moreover, it is moderated, although moderators take a minimalist approach and take care not to be perceived as controlling conversations.

- ShapeYourPlace addresses non-English speaking residents by offering foreign language functionality and support. The initiative conducts outreach to hard-to-reach residents like young parents, non-English speakers, the elderly, disabled, and youth.

- The county plans to roll out similar social networks in its other districts. In addition, the social network team is working with government agencies to import resident-reported issues into their "fault-reporting" systems.

Insights

- Off-line outreach was critical to success because residents live in a physical community, and the team sent a strong message about inclusion by holding the community fairs, which enabled all residents to participate, especially hard-to-reach residents.

- The social network dramatically increased transparency in the community, which increased residents' sense of empowerment and ownership. Residents can observe each other reporting problems and the government actions to resolve them. This counteracts residents' former feeling of helplessness because they can see resolutions. This is very self-reinforcing and leads to increased participation over time.

- Recalling the 90-9-1 rule, when the government serves residents by responding to issues they report, it achieves a 100× return on that service because *it vicariously serves all 100 people who are watching the interaction.*

- Transparency saves various government agencies a lot of time and money because they know the status of issues in real time, which reduces agency time by eliminating duplications of effort.

- Initiatives with very diverse users must expect that some stakeholders will require extensive support because they are not tech-savvy. Moreover, some people and departments cannot understand the concept of an "online social network" until they see it. "Colleagues in partner organizations only began to understand and see how they could use these tools—and how the site would work—once the website was up and running."[14]

Social Network Innovation Models

We have discussed several best practices and risks of using social networks for innovation, so we can now turn to specific models that you can use. The case studies highlighted private social networks (3M), a financial services firm that faced disruption head-on by collaborating with stakeholders (State Farm), and a consumer products company that mounted one of the most successful open innovation programs in the world (P&G). In addition, the two local governments used public social networks to engage their citizens to actively improve their communities by solving problems themselves, not paying for larger government.

These innovation models offer more leading edge examples of engaging stakeholders with social networks.

Community Company

- Members of (usually) public social networks ("community") manage key company business functions such as product selection.

- The community is self-reinforcing; members like or find meaning in what the "company" does, whether it's a start-up or established brand.

- "Community involvement" is a big part of the attraction and engagement.

The Challenge   Product and service companies spend billions a year launching and evolving their products and services based on sales, marketing research, and R&D investments. Many companies are even organized according to product and service businesses. As long as it has been measured, the average product launch failure rate is 95 percent. The process to develop a new product is long and ponderous. Meanwhile, the market is moving faster than ever due to all kinds of customers' constant interactions in social networks. They are exceedingly bored with products and services and refuse to pay premium prices with very rare exceptions.

As you may recognize, this describes an industrial economy product management process, which is built on assumptions that are no longer true. Most important, before social networks, there was no viable way to communicate with large groups of people, so firms created products in the absence of customer voices. Anyone in marketing knows that surveys and focus groups do not reflect reality very effectively *because the social context is wrong*. In most cases, when customers are talking among themselves about their challenges, in real life and without a product focus, the social context reflects the true usage of the product or service. And communication is free, so the stage is set to improve profit while slashing marketing costs. Note that this is true with business clients or consumers.

The Opportunity   Threadless is a Chicago start-up that was launched in 2000 for $1000 by two guys. It is an online community in which artists upload an average of 1000 designs per week, the community picks the top 10 designs, and the company prints on t-shirts and sells worldwide via the community. The company is private, but revenue is estimated at $30 million, with a 30 percent profit margin.[15] At the other end of the company size scale, Samuel Adams Longshot Home Brew Contest judges individuals' home brew recipes, and three winners have their beers brewed by Samuel Adams and nationally distributed for one year in special six-packs.[16]

Community Company (CommCo) gives customers access to the company's means of production. To understand its power, recall the 90-9-1 rule, which holds that 90 percent of participants watch, nine percent participate, and one percent actively participate. The nine percent might vote and the one percent might design. *All people feel ownership over the company and products because they are members of the community.*

Insights

- CommCo shows how most firms will differentiate in the knowledge economy. People care less about products, although they will need them and use them. They care about causes, connection, and achieving outcomes by using products and services. CommCo is about connection and helping people do things, not selling product. Note that a lot of product gets sold, but the focus is different.

- CommCo's DNA is creating community and collaboration around a cause or a fun activity, and everything can be fun. There will be Longshot contests for many categories of consumer products: mustard, apparel, tools, household products.

- In CommCo, members of the community do much of the marketing by getting excited about what the community is doing and telling their friends. In Threadless's case, they make product to order and carry no inventory. They reprint designs according to community feedback and have all kinds of contests. Since community members do so much of the work, CommCo can be far more profitable than the industrial economy brand, and brand employees are part of something bigger, so their satisfaction increases, too.

- Finally, do not assume that CommCo only applies to simple products: Local Motors designs and fabricates automobiles for customers and on commission from major brands.

Professional-Amateur

- Magnifies the impact of costly firm experts with the skills of energetic amateurs.
- Numerous success stories in astronomy, P&G's Connect + Develop was professional-amateur (Pro-Am).
- Amateurs and professionals can validate each other in different ways and build more support for their mission.
- Success is founded on using each party's core competency to create complements.

The Challenge   The Pro-Am model increases the business leverage of your organization's experts by collaborating with passionate amateurs. Industrial economy organizations and society were stratified, so people with common interests were largely prevented from collaborating. Producers were separate from consumers. Professionals were isolated from the biggest fans of their work. Brands created products in isolation with a 95 percent launch fail rate on average. Moreover, as discussed earlier, brands' profitability is falling in many categories because their innovation processes are too ponderous. Brands' costs for professionals are high, and a significant portion of their work is occupied by critical repetitive tasks.

The Opportunity   Pro-Am is only practical when using social platforms. In effect, digital social networks have created a huge volunteer work force that professionals can tap when they know how. In the industrial economy, amateurs were hobbyists and had relatively few opportunities to connect with specialists relevant to their passions. Social networks, through the dynamic discussed under Emergent earlier, enable passionate people to educate themselves very quickly, and collaboration among them is instant and free. Even more powerful is putting professionals in the digital room with many amateurs, and the professionals' expertise is highly leveraged. Amateurs are passionate and motivated and often have practical knowledge that is highly complementary to that of professionals.

Pro-Am works best when process owners identify tasks that are important, numerous, or repetitive and require a moderate skill level. Then they ask amateurs to help them with those tasks under their guidance. Astronomy was an early example; astronomers enlisted amateur astronomers to watch certain parts of the sky that they could not afford to monitor, and there have been many success stories in which the amateur will identify an astral event and the professionals train their telescopes on it to zoom in.[17] OhmyNews was another early success. Professional editors qualified and edited newspaper stories submitted by citizen journalists.[18]

Practically, the cornerstone of Pro-Am is understanding the capabilities and roles of each party and aligning them with tasks and projects. Professionals can set standards for work and educate amateurs about standards and how to use them, which dramatically increases professionals' value.

Professional crowdsourcing is a Pro-Am variant that is increasingly used by brands. Here, the brand makes a request of an online crowd of people who may be qualified or not. P&G is notable in its work with Innocentive, an online community of scientists and engineers that accepts challenges and offers cash prizes for the member that finds the solution. P&G realized that many retired scientists love challenges and were a powerful workforce that only needed to be paid when the problem was solved.[19] At the other end of the spectrum, Amazon Mechanical Turk enables brands to make requests of "Turkers," who complete often simple tasks for little recompense.[20]

Insights

- Pro-Am is different from general crowdsourcing or CommCo in that experts collaborate with qualified volunteers.

- Another valuable aspect of Pro-Am initiatives is that the community learns and improves its skills over time. Professionals can ask more of amateurs who although volunteer, self-select for a high interest in the topic.

- Classical Pro-Am unites professionals, who are passionate and studied in their fields, with amateurs who are passionate, so the model is usually volunteer, and the payoff for amateurs is being involved in important work.

- In certain cases, Pro-Am can benefit from having explicit recognition of amateurs who go beyond. Cultural nuances are very powerful, so make sure to have a firm grasp of the community's culture before trying this. For most amateurs, celebration of their efforts and successes happens automatically in the community, and recognition from professionals is another powerful motivator.

Service as Marketing

- Solves customer problems in public, and uses conversations to attract new business; prospects see other people getting problems solved, which reduces their objections to buying the product or service.

- The firm actively republishes conversations to support additional sales.

- Dramatically increases trust among the brand, customers, and prospects.

The Challenge Old-style business runs customer service as a cost center that is managed with efficiency and cost minimization as a key goal. Meanwhile, marketing is spending millions paying models to talk on TV about how thrilled they are with the company's products. Service as marketing (SAM) aims to redefine both of these functions and make them relevant in the social channel.

The Opportunity SAM recognizes that people who have specific common interests are compassionate and want to help each other vanquish problems and take advantage of opportunities. Social networks, tagging, and RSS enable the company to *help customers help themselves and make each other happier*. SAM has been around, but Web 2.0 technologies take it to a new level. In 1993, for example, Cisco launched the progenitor of Cisco Connect Online, a website for online technical support. Although by accident, they discovered that their clients loved helping each other, which improved service, boosted engagement, and slashed support costs.[21]

In phase 1, the company creates a public community in which customers can post questions and problems and company representatives serve them publicly, for all to see, and they validate all kinds of good-faith discussions, including those critical of products or the company. This sends a strong signal that the company puts customers first. The team builds the forum's features to enable people to find and discuss specific kinds of problems, and they nurture the growth of a trusted space by careful moderation; customers start helping each other, increasing the sense of community. The technical team creates social tags that catalog various aspects of questions and answers, and these make it very easy for people to locate very specific conversations. They also create a recognition system on the platform to recognize various levels of contribution. These are very mature features.

In phase 2, marketing and sales use service conversations to support presales, which can work for business clients or retail customers. Although it varies with the product or service, prospects are usually concerned about what happens when the product doesn't meet expectations. The tagging system and RSS[22] make it easy for them to filter and republish forum conversations on websites, blogs, Twitter feeds, etc. E-commerce sites can republish them alongside products, and direct sales forces can refer prospects to them.

Insights

- When prospects see other customers with problems getting served, they trust the information far more than any company source. Existing customers of the product or service are already using it and encountering problems or discovering new uses. They have same social and usage context as prospective customers, and these "service conversations" will lower barriers to buying because they address prospects' worst fears: "What if I don't like the product, what if…?"

- Old-style business puts happy stock images on websites and minimizes products' potential problems. In the Social Channel, publicizing customers' problems and *solving them in public aligns company and customer.*

- Recall the 90-9-1 rule: most customers will read and learn and not interact, but they will be influenced when they see the company serving people and putting people first, and they will refer their friends there. The nine percent and one percent will be raving fans and tell their friends; many will become advocates for the brand due to the trust and empowerment it has supported.

- The first company in the category that becomes the "Amazon.com of its category" will win the most customer and prospect attention for the entire category: the destination for answers.

Getting Started: Using Social Networks for Innovation

"Social network innovation good practices and pitfalls" outlined good practices and pitfalls and offered several models for using social networks for innovation. "**Getting Started: Using Social Networks for Innovation**" presents the nuts and bolts of planning and mounting your own initiative while significantly mitigating risks.

Strategy →	Execution →	Expansion
Focus on people	Use an agile approach	Lead with sociality
• Throw product thinking away	• Grow by pilots	• Build on trust
• External & internal diligence	• Learn and build teams	• Focus on community
• Minimize risk with strategy	• Scale gradually	• Practicing "social"

Focus on People

Throw Away Your Product-Service Thinking

All organizations say that they are people-focused, and they believe that they are. Very few are. The industrial economy was preoccupied with large-scale, mechanical tools like shop floor machines in huge factories, supply chains, and R&D efforts (i.e., pharmaceuticals). *The organization was focused on itself. It may have appreciated people, but it was rarely focused on people.* Knowledge economy communication tools are most powerful when people get involved emotionally in their work, which happens when they communicate and collaborate with other people. Making things is still important, but the *differentiation depends on what other people believe and say, not what your company says.*

As detailed in Social Channel One,[23] organizations now have the opportunity to shift their orientation away from their products and services to the people that buy and use their products or services for specific purposes. This is crucial because people with digital voices are more influential than any organization. The outmoded assumption is that firms produce gazillions of products very efficiently and subsequently market them. This model is a reptile, and temperatures are falling. People who buy products and services do not care about them. Rather, people are buying the outcomes they are able to achieve using products and services. Since communication is now free, for the first time, organizations can align with their customers by helping them reach better outcomes.

In practice, this means observing people with new eyes. Look for the outcomes people want. Here is how this works in stages.

Audit the Ecosystem (External Analysis)

The team begins by asking itself, "Who are the most important people to our business?" Most firms work with demographics, and some have detailed statistics and behavioral models about certain types of stakeholders, who might be customers, clients, employees, regulators, partners, government officials, etc. Based on the firm's business strategy, who is most important?

Social networks enable the team to *understand individual stakeholders for free*. As detailed in, stakeholders increasingly interact with others in public, giving firms who look for it valuable social information about their activities, which provides much better clarity about their motivations and goals. However, you need to get more granular than demographics, which were suitable for mass marketing. Now you will relate to individuals. Begin by developing personas or profiles of stakeholders, and rank them in importance to you. Make these descriptions as good as feasible, but realize they are just a starting point; you will iterate them by testing them online. For internal and hybrid innovation, stakeholders may be employees, alliance or supply chain partners, other business unit employees, etc.

Next, define stakeholder outcomes as best you can. Ask yourselves, "What actions are stakeholders taking when we become relevant to them?" Define these in as much detail as you can. Then define stakeholder workstreams, which are work processes that include the actions you just defined, but they also include other actions stakeholders take *before* the actions, and they include various outcomes after the actions. For example, a fitness equipment brand helps people build home gyms. One stakeholder's action might be "buy elliptical machine." The workstream might be: realize I don't like my body => consider gym membership or working out at home => buy elliptical machine => use elliptical machine to tone body or lose weight.

Most brands focus on selling their products, but the stakeholder wants a toned body and cares less about the product. Social networks enable you to engage at the outcome level.

Next, you need a specialist to design sophisticated filters to locate stakeholders and workstreams online. This is a detailed, iterative process whose output enables you to filter the entire web to isolate stakeholder and workstream conversations or interactions. By combining workstreams and stakeholders, you have a way to rank the social ecosystem's venues (sites) that are optimal places for you to engage stakeholders around their desired outcomes. The Social Network Roadmap(SM) Ecosystem Audit has more detail about auditing the ecosystem.[24]

Popular platforms are usually very inefficient venues in which to engage stakeholders, but most organizations invest in popular social networks like Facebook, LinkedIn, MySpace, and perhaps Google+. Most firms would achieve much better returns elsewhere, but their diligence is not rigorous enough. Thus, they expose themselves to the risk of better informed rivals who engage their stakeholders more deeply and efficiently.

Create Your Social Networking Strategy (Internal Analysis)

It is difficult to describe the insight that the team will have when the audit is complete—because information and insight that they now possess has never been possible before. Teams invariably get very excited and want to jump in and start relating. Most teams will get better results if they analyze their organizations and create a strategy for innovating with social networks.

The audit shows what stakeholders are doing and what they are trying to accomplish. The next question is, "Based on their goals, what unusual information or capabilities do we have that could help them achieve their outcomes?" Once you define what you have to give, evaluate it in terms of how easy it is for your organization to share. For example, if the information is only possessed by very few experts who are already highly leveraged, you will have limited access to it. You need to optimize the value of the help to stakeholders with the ease of sharing for you. Few organizations take this step and end up with inefficient initiatives that lose momentum over time.

The Social Network Roadmap(SM) uses a gated process in its organization audit.[25] You begin with a simple core competency analysis to determine what you have that is unique among your peers, followed by a two-step stakeholder survey process (online + telephone) to broaden your understanding of stakeholder workstreams. When this is complete, you synthesize your in-depth understanding of your uniqueness with stakeholder workstreams to create trial social networking pilots. Pilots are small, short, inexpensive projects for efficiently interacting with stakeholders. Then the gated process refines the trial pilots according to organization strengths and weaknesses. It reviews competitors and substitutes results with similar projects to learn from others' experience. It surveys employees for skills and knowledge relevant to pilots. Finally it reviews organization initiatives that might strengthen or weaken certain pilots.

Do as many of these steps as you deem feasible. All are short, simple processes that enable you to reprioritize your trial pilots as a function of organization. Auditing the organization mitigates risks of social networking initiatives by *assigning the highest priority to the optimal pilots*. In addition, in 2013, many organizations have experimented with social networking, but the majority have had mediocre results. Most organizations also have influential internal stakeholders who are uncomfortable with the empowerment social networking represents and would like to see it flounder. This is another reason why you want optimal pilots because success feeds on itself, and you can build momentum and garner support, even in the face of opposition.

Use an Agile Approach

Agile development has revolutionized enterprise software development, and people in myriad other disciplines like marketing are applying it to their functions because it mitigates numerous risks and is far more efficient. "Agile software development is a group of software development methods based on iterative and incremental development, where requirements and solutions evolve through collaboration between self-organizing, cross-functional teams. It promotes adaptive planning, evolutionary development and delivery, a time-boxed iterative approach, and encourages rapid and flexible response to change. It is a conceptual framework that promotes foreseen interactions throughout the development cycle."[26] Using agile provides a formidable advantage when managing social networking for innovation.

Make Small Investments to Test Strategy

Scope pilots to require small investments in cash and time, but design them to learn about stakeholders by interacting with them. Pilots should aim to achieve two goals: testing the strategy and building your team's skills. By interacting with stakeholders and *observing stakeholders interacting with each other*, your team will learn very quickly. Most pilots last 6 to 12 weeks. As Stan Joosten of P&G explained earlier, by having small investments, you don't get bogged down in resource negotiations. Social networking also gives you real data and results immediately, which makes it a unique kind of innovation initiative.

Measure, Measure, and Measure Outcomes

Social networking provides another advantage because it is grounded in digital interactions, which are far easier to capture and learn from than analog interactions were. However, take care to structure your metrics and measurements in terms of trust, commitment, and relationship. Most social media metrics and monitoring platforms are oriented around promotion, which is inappropriate for most social networking initiatives as discussed in "How Social Technologies Transform Innovation".

One example of relationship-oriented models is The Social Network Life Cycle Model,[27] which shows how to measure changing trust and relationship according to four phases of the life cycle. In addition, the Hierarchy of Social Actions reveals how each platform's social actions (defined interactions within that platform as governed by its functionality) have a hierarchy that indicates how effective your team's interactions are. When their interactions increase trust among members of the social network, others' interactions with your team will climb the hierarchy to higher-order trust or commitment interactions.

Adjust and Repeat in Tight Cycles

When you conduct rigorous due diligence, most of your pilots will show a considerable degree of success, but another reason for scaling them small is it becomes easier to kill those that don't work well enough. You should build expectations that you will unplug pilots to give priority to those whose results are most meaningful.

Pilots 6 to 12 weeks in length can easily be adjusted, renewed, and expanded according to continuous measurement and adjustment. Some will mature into ongoing initiatives that you will scale up considerably over time.

Lead with Sociality

Explicitly Develop and Test Intimacy Concepts

It is easy to overlook the significance of the current transformation in relationships between companies and customers (and other stakeholders) during the shift to the more personal knowledge economy. Executives unknowingly harbor the assumption that companies produce and market to demographics, not people. This is completely understandable since organizations have never related to individual customers in business. They don't know their customers or their other stakeholders personally, some business-to-business firms notwithstanding.

As detailed in "How Social Technologies Transform Innovation", social media uses social networks to repeat the old impersonal, promotional pattern. Don't make this mistake because you will miss the gold, which is to enable stakeholders to develop more personal connections to your company. When they are able to do that, they prefer you for personal reasons.

Your teams will probably find that the impersonal, promotional pattern is a tough one to break because it is so ingrained. The best way to evolve is to use trust-oriented metrics and to develop a hierarchy of social actions. The latter is a model for intimacy within that platform, as a function of its functionality. This example shows how it works for Facebook, a social network with which you are undoubtedly familiar.[28] Specifically, if your team is interacting effectively, "likes" start migrating to comments over time, as people feel more comfortable and committed. Then the organic posts grow, etc. Movement up the hierarchy should be a key measure in most social networking pilots.

Creating Community Is the Default

Most brands are focused on the relationship between themselves and customers. This is more industrial economy baggage. Far more powerful is focusing on community, which emphasizes members' relationships with each other within the community.

When you pilot with diligence, you attract highly desirable stakeholders, so investing in a social network community that helps them reach their goals reflects well on you and will keep them coming back. People are deeply attracted to relevant trusted places in which they can discuss and solve important problems and challenges. By focusing on community, you put your energy into the meat of the matter, creating a space that stakeholders take seriously, and to which they return.

Therefore, your team facilitates productive interactions among members of the social network by introducing them to each other or suggesting ways that could help each other, based on your team's knowledge of other people—and your team's attentiveness and consideration. When stakeholders see this behavior consistently, they will be impressed because most people find it inspiring when people serve other people and show considerate behavior. Your team's helpful interactions are immortalized and, if the social network in which they are interacting is public, anyone can find the interactions at any point in the future. *By relating to the few, you influence the many.*

The Importance of Mentoring and Modeling Behavior

Most team members whom you will enlist for pilots will have modest-at-best awareness of themselves and other people socially. Very few people at all have high social awareness. Anticipate this and plan for it because the higher awareness your teams have, the faster they can learn and scale your social networking initiatives.

Interacting in digital social networks makes social skills or lack thereof quite obvious because social interactions are transparent. Happily, stakeholders do not have much awareness of sociality, either, and people are accustomed to having low standards of social awareness. This is all unconscious, by the way.

Most important, by analyzing interactions and noting results, team members can learn quickly and mentor each other very purposefully in interacting to increase trust and building their social skills. In addition, many people learn more effectively by doing, so explain things to social network pilot contributors, but don't get caught up in theory; they will learn more quickly by doing and correcting.

Along with this, build teams on the premise that people respond to behavior, and their responses are largely unconscious but real. Modeling desired behavior is very effective because people recognize actions more than words. This holds true when interacting with stakeholders, too. If your team's interactions are admirable, they encourage stakeholders to interact that way, too. In groups, people mimic admirable behavior, so explicitly recognizing and defining that behavior will help your team get better results more quickly.

Conclusion

Social technologies and behaviors transform innovation, and people and organizations who realize it and practice it first will create defining value and remake industries. More sobering, people and firms that delay their transformation processes will become less relevant far more quickly that they imagine.

Social networks transform innovation at two levels

- They make communication free, so collaboration is an order of magnitude less costly, and it is much faster. Social technologies change the guts of innovation because collaboration is their core activity.

- In the knowledge economy, innovation carries most of an organization's value, so it is the tip of the spear. Today, most organizations are accustomed to the industrial economy's long product life cycles, which required intermittent innovation. Since innovation was a back room activity conducted by specialists, it was ponderous, slow, and largely unsuccessful.

The tools and techniques presented here can help you to innovate at an unprecedented level by leveraging social networks to the hilt. Carpe diem.

References

1. Network effect. *Wikipedia: The Free Encyclopedia*. Retrieved from http://en.wikipedia .org/wiki/Network_effect.
2. Nielsen, J. Participation inequality: Encouraging more users to contribute. *Nielsen Norman Group*, October 9, 2006. Retrieved from http://www.useit.com/alertbox /participation_inequality.html.
3. Rollyson, C. S. "Book Review/Grooming, Gossip, and the Evolution of Language." *The Global Human Capital Journal*, December 26, 2009. Retrieved from http://globalhumancapital .org/book-review-grooming-gossip-and-the-evolution-of-language/.
4. Rollyson, C. S. Knowledge economy products and the future of manufacturing. *Christopher S. Rollyson and Associates*, November 23, 2012. Retrieved from http://rollyson .net/knowledge-economy-products-and-the-future-of-manufacturing/.
5. Anderson, C. *Free, the Future of a Radical Price*. New York, NY: Hyperion Books, p. 75, 2009.
6. Rollyson, C. S. The social channel. *Christopher S. Rollyson and Associates*, August 7, 2012. Retrieved from http://rollyson.net/social-business-opportunity/the-social-channel -home-page/.
7. Rollyson, C. S. "Surprises in Emerging Chinese Consumer Market." *The Global Human Capital Journal*, March 9, 2006. Retrieved from http://globalhumancapital.org/surprises -in-emerging-chinese-consumer-market/.
8. Socialcast by VMware. 3M case study: Socialcast gives 3M's global labs a distinctly local feeling, 2012. Retrieved from http://www.socialcast.com/sites/default/files /Socialcast_CaseStudy_3M_11_28_2012.pdf.
9. Bingham, A., and Spradlin, D. The challenge driven enterprise. *The Open Innovation Marketplace*. Upper Saddle River, NJ: FT Press, p.20, 2011. Retrieved from http://www .innocentive.com/about-us/open-innovation-marketplace.
10. Ibid. p. 7.
11. Rollyson, C. S. "Always in Beta: How Big Business Can Benefit from 'Little' Innovation." *The Global Human Capital Journal*, October 17, 2007. Retrieved from http://globalhumancapital .org/always-in-beta-how-big-business-can-benefit-from-little-innovation/.
12. Rollyson, C. S. Federal reserve bank of Chicago economic forecast. *Christopher S. Rollyson and Associates*, October 17, 2007. Retrieved from http://rollyson.net/federal-reserve -bank-of-chicago-economic-forecast/.
13. IDEO. Next door for State Farm Insurance: A "financial learning space" for the next generation of customers, 2011. Retrieved from http://www.ideo.com/work/next-door/.
14. Local government improvement and development. Customer led transformation programme, case study—cambridgeshire, social media project in Fenland, p. 14. February 2011. Retrieved from http://www.local.gov.uk/c/document_library/get _file?uuid=2776159a-c529-4afa-8a7b-281c495bf4c3&groupId=10171.
15. Threadless. *Wikipedia: The Free Encyclopedia*. Retrieved from http://en.wikipedia.org /wiki/Threadless.
16. Samuel A. Longshot American homebrew contest, 2012. Retrieved from http://www .samueladams.com/promotions/LongShot2012/.
17. Hammel, H. (Photographer). Hubble view: Jupiter impact, 2009. [Web Photo]. Retrieved from http://apod.nasa.gov/apod/ap090731.html.
18. Oh my news. *Wikipedia: The Free Encyclopedia*. Retrieved from http://en.wikipedia.org /wiki/Oh_My_News.
19. InnoCentive. *Wikipedia: The Free Encyclopedia*. Retrieved from http://en.wikipedia.org /wiki/InnoCentive.
20. Amazon Mechanical Turk. *Wikipedia: The Free Encyclopedia*. Retrieved from http:// en.wikipedia.org/wiki/Amazon_Mechanical_Turk.
21. Seybold, P., and Marshak, R. *Customers.com: How to Create a Profitable Business Strategy for the Internet and Beyond*. New York, NY: Times Business, pp. 320–323, 1998.
22. RSS. *Wikipedia: The Free Encyclopedia*. Retrieved from http://en.wikipedia.org/wiki/Rss.

23. Rollyson, C. S. Post product customer relationships in the social channel. *Christopher S. Rollyson and Associates*, 2012. Retrieved from http://rollyson.net/post-product-customer-relationships-in-the-social-channel/.

24. Rollyson, C. S. The digital social ecosystem audit: The key to optimal interactions. *The Social Network Roadmap(SM)*, 2012. Retrieved from http://www.socialnetworkroadmap.com/index/the-digital-social-ecosystem-audit-the-key-to-optimal-interactions/.

25. Rollyson, C. S. The organization audit and social business strategy. *The Social Network Roadmap(SM)*, 2012. Retrieved from http://www.socialnetworkroadmap.com/index/the-social-business-organization-audit-and-social-business-strategy/.

26. Agile development. *Wikipedia: The Free Encyclopedia*. Retrieved from http://en.wikipedia.org/wiki/Agile_development.

27. Rollyson, C. S. "Realizing Value from Social Networks: A Life Cycle Model." *The Global Human Capital Journal*, 2009. Retrieved from http://globalhumancapital.org/realizing-value-from-social-networks-a-life-cycle-model/.

28. Rollyson, C. S. (2012). http://rollyson.net/using-facebooks-ladder-of-social-actions-to-build-community/.

CHAPTER 17

Innovation Combination Methods

David Conley

This chapter examines the simultaneous use of manifold methodologies for the purpose of producing innovative solutions to challenging problems. There are numerous paths through which problem solving and innovation can be pursued. Depending on the needs and goals of any particular analysis, different methodologies can be used to take advantage of differing system and process attributes and approach solution generation from different angles. A variety of methodologies will be discussed with a focus on the strengths and weakness of those methods and how combining them can accelerate the process of innovative problem solving. Most of the methods described are in and of themselves complex processes, which require study and practice in order to effectively apply them. Therefore, it is not my intention to focus on each of the individual methods discussed later, but rather to demonstrate how they can be combined into highly effective and targeted innovation processes. If you are interested in learning more about an individual methodology discussed in this chapter I urge you to seek additional information from other qualified sources.

Problem Solving versus Innovation

The term *innovation* has been used extensively in business and technical organizations to communicate the interests and needs of their operations. In fact, many organizations use the terms *problem solving* and innovation interchangeably as if the solving of a problem automatically results in an innovation. This is not the case, as innovation requires a solution type not created by most problem-solving efforts. Unfortunately, most organizations do not have a solid understanding of what the word innovation really suggests or what its pursuit entails. According to the *American Heritage Dictionary*, New College Edition, *innovate* is defined as "to begin or introduce something new, be creative." This definition is of course correct in the general usage of the word but it lacks the detail required to properly guide interested parties in its quest. Among experts in the field of innovation a more detailed and exact definition is utilized and provides general guidance to the innovation practitioner. From the engineering perspective, creation of a new invention (innovation) always manifests as the full or partial overcoming of a technical contradiction (limiting situation).[1] Rephrased, an innovation is an advancement that transcends a limiting situation within the system under analysis. Another way to describe these limiting situations is to refer to them as *contradictory requirements* within a system. Figure 17-1 provides several examples of system limiting situations, which can also be referred to as contradictory requirements. For example, the first example is in relation to the size of the engine in a car. The bigger the engine the more power it produces but also the more fuel it consumes. Therefore a car needs both a large and small engine simultaneously. This is a contradictory demand of the system called engine. Solving this engine problem with a system that produces both high power with low fuel consumption would be an innovative solution because it meets both contradictory demands simultaneously. It is important to note here that while all three examples in Fig. 17-1 involve contradictory requirements around the parameters of size or quantity, any system parameter could be at play including, but not limited to weight, speed, volume, density, and strength. Knowing that a contradictory situation must be solved in order to create an innovative solution, we now understand the importance of incorporating contradiction

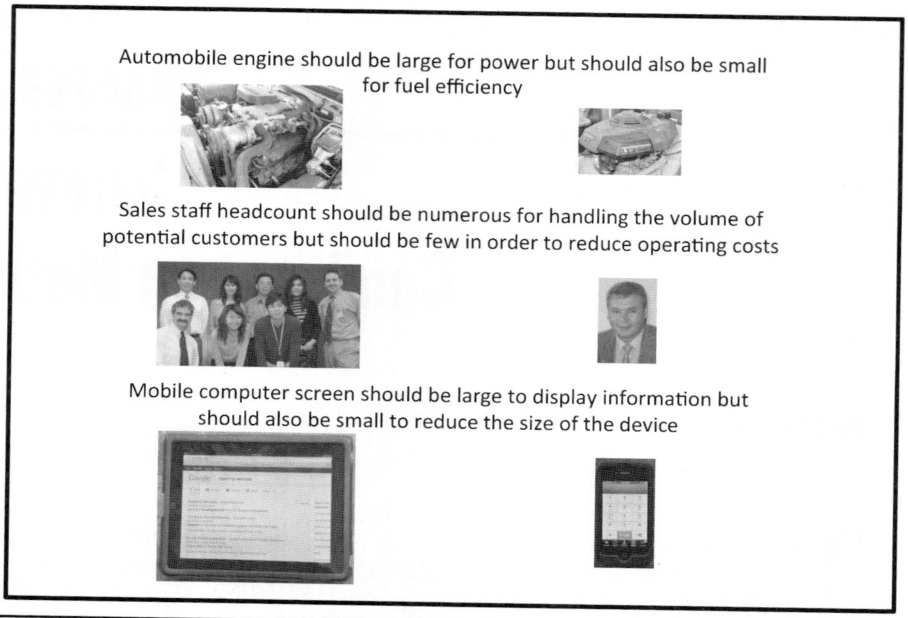

Automobile engine should be large for power but should also be small for fuel efficiency

Sales staff headcount should be numerous for handling the volume of potential customers but should be few in order to reduce operating costs

Mobile computer screen should be large to display information but should also be small to reduce the size of the device

FIGURE 17-1   Examples of contradictory requirements of a system.

resolution methods into our problem-solving processes if we indeed are seeking innovation. Therefore, innovative problem solving is a subset of problem solving in that a solution must resolve a limitation in the system under analysis in order to be an innovative solution. However, in many situations solutions are generated that indeed solve a problem but do not resolve any system limitations while doing so. Therefore, if a workable solution is generated then a problem has been solved. If that workable solution also happens to resolve contradictory requirements (see Fig. 17-1) of a system parameter, or attribute, then the solution can also be considered an innovation. Herein lies one of the weaknesses of most every problem-solving technique currently in use today; almost none of the popular methodologies address the issue of contradiction identification and resolution. Therefore, they do not regularly drive innovative solutions, at least insofar as the technical definition of innovation is concerned. This is not to say that an innovative solution can never result from one of the prevalent problem-solving processes but since the goal is systematic and repeatable innovation generation, occasional success is not acceptable. A second problem with the most popular problem-solving processes is that while they all have a step which calls for "creating a solution" to the problem at hand there are few, if any, algorithms or tools within the processes by which to generate those solutions. In practice most expert problem solvers would agree that the prevalent problem-solving methods offer varying combinations of strengths and weaknesses. If chosen wisely, different problem-solving methods, with complimentary sets of differently focused strengths, can be combined to create a solid, useful, and innovative process. Combining complimentary methodologies can result in fluid and well-rounded processes that provide effective tools and direction beginning with the initial problem identification step all the way through the verification of the implemented solutions. In the remainder of this chapter I will refer to methodologies as problem-solving processes if they at least support the solution generation process. Further, I will only refer to methodologies as innovation if they support the generation of solutions that are aimed at resolving contradictory system requirements.

More Detail on Defining Contradictions

Let us take a few minutes to better define what a *contradiction model*, or the modeling of a limiting system contradiction, looks like. As an example suppose that a wireless phone company needs a significant amount of equipment in order to improve the system's coverage area but the company also wants very little equipment because it is expensive. Generally speaking, there are two boundary conditions existing at opposite ends of a continuum that define the range of available solutions to this problem within the restrictions set forth by the existing system design. In other words, based on how our fictitious wireless phone service provider's technical systems are designed there are two extremes defined by the current

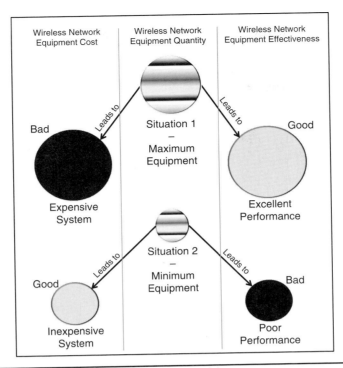

FIGURE 17-2 Conflicting situations graphic-problem models.

system limitations and therefore the solution can only exist somewhere between them. Figure 17-2 shows these two extremes. In situation 1 of Fig. 17-2 the amount of wireless network equipment is at a maximum (Striated circle in the center). This large amount of equipment drives two results: first is the cost of the system (Black circle on the left) which is very high and therefore represents a bad, or undesirable, situation; and second is the performance, or effectiveness, of the system (Grey circle on the right) which is excellent and therefore represents a good, or desirable, situation. In situation 2 of Fig. 17-2, the amount of wireless network equipment is at a minimum (circle in the center). This small amount of equipment drives two results: first is the cost of the system (circle on the left), which is very low and therefore represents a good, or desirable, situation; and second is the performance or effectiveness of the system (circle on the right), which is poor and therefore represents a bad, or undesirable, situation. Therefore the amount of equipment needed for services is in conflict with the need to minimize the amount of equipment for the purpose of controlling costs. On the one hand, the company wants a large amount of cell towers, repeaters, and switching circuits; and on the other hand, the company does not want to have to pay for any equipment at all. This represents a contradictory requirement that serves as a system limitation. So the problem to be solved is how the company can spend very little on network equipment but have the system perform as if there is a substantial amount of equipment in operation. In order to solve this problem with an innovative solution it is necessary to resolve the limiting system contradiction. The abstract model of such a solution is shown in Fig. 17-3. In this diagram the solution, which is not necessarily just based around equipment quantity, is shown as the "unknown" solution state shown as a white circle. The, as of yet unknown, solution must result in the best of both worlds as reflected in the two related predecessor contradiction models (see Fig. 17-2). First, the solution should be relatively inexpensive (circle at the left in Fig. 17-3) and therefore represents a good situation; and second, the solution should have a high level of performance (circle at the right in Fig. 17-3) and therefore also represents a good situation. Any solution that simultaneously meets the inexpensive and high performance requirements of the abstract solution model stated earlier will indeed be an innovative solution. If you recall from the previously listed definition, "an innovation is (one) that transcends a limiting situation within the system under analysis"; in order to create an innovative solution, it is necessary to resolve the contradictory requirements around the parameter of equipment quantity. How can the system simultaneously contain substantial equipment, in order to support the need for wireless services, and very little equipment, in order to control costs? The purpose of this example is not necessarily for driving to a solution but rather to demonstrate what contradictory requirements and abstract

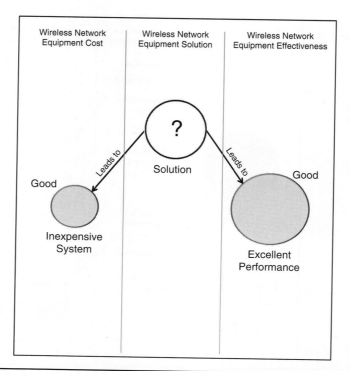

FIGURE **17-3**   Conflicting situations graphic-abstract solution model.

contradiction resolution models look like. However, quite simplistically, one solution could be to design mobile phones (i.e., smartphones) with repeater capabilities so that the collection of phones in the network are orchestrated to also act as the network itself.

Combining Methods

Unfortunately, it is not always obvious as to what the contradiction is in a system that needs resolving. Further, most organizations have preferred problem-solving methodologies in use within their operations. The methods that have been previously accepted by an organization have the obvious benefit of already being familiar to, and in use within, the organization. Additionally, the wide variety of problem-solving methods in use today can provide unique capabilities and insights to the problem solver. For example, Six Sigma incorporates rigorous statistical analysis methodologies, which can help define the problems, and measure their impact, plus support the evaluation as to whether the solution did indeed improve the situation. Lean (a.k.a.Toyota Production Process) provides an excellent framework by which to help us focus on the various types of waste that can be found in an operation or process. The Plan-Do-Study-Act (PDSA) cycle helps the user to consider important aspects of the problem-solving process. Root cause analysis (RCA) facilitates the discovery of the fundamental causes of a problem by allowing a look beyond the effects that often mask the true problem sources. However, none of the aforementioned methods provide any resources, or algorithms, designed for the generation of actual solutions, innovative or otherwise. So this leaves us with the question of which methodology is best suited to fulfill an organization's problem-solving and innovation needs? Actually, there are no individual methods that fulfill all problem-solving needs. In reality, whether problem solvers realize it or not, multiple methodologies are employed in conjunction whenever problem solvers analyze issues and create solutions. Observations during my career have revealed that the most common practice is the usage of a problem identification method (i.e., Lean, Six Sigma, Kempner-Trego, etc.) followed by the use of the standard fall back of brainstorming as the solution generation vehicle. This reality demonstrates that at the very least problem solvers utilize two methods when analyzing issues and generating solutions. I propose that the most effective advancements are indeed achieved by way of combining methods and the best combinations are dependent on the traits of the organization solving the problems and the type of problems that need to be solved. Dr. Craig S. Flesher, author, academic, and Chief Learning Officer of Aurora WDC, stated that "utilizing a purposefully sequenced combination of multiple analysis and problem-solving methods is typically the best way to create effective and actionable results in today's complex world of business and technology."

The Problem-Solving Path

How do humans generally solve problems? The complexity of our innate problem-solving processes might surprise you, especially since, for most problems we encounter, the process is executed somewhat unconsciously. The complexity lies in that when we solve problems we move through several problem modification steps in generating a solution. The problem-solving pathway shown in Fig. 17-4 demonstrates the generic problem-solving process we naturally use when generating solutions to somewhat easy problems. For example, if the slamming of the barn door scares the livestock we would quickly understand that it was necessary to keep the door from being blown shut by the wind (specific problem to simplified problem—Fig. 17-4). Next, we would determine that we need a way to keep the barn door open in the wind (simplified problem to general problem—Fig. 17-4). Then we would envision ways to keep the barn door open in the wind such as pushing against it or holding it back (general problem to general solution—Fig. 17-4). And finally, we would generate specific solution concepts in achieving the general solution, such as using a latch or propping the door open with a board (general solution to specific solution—Fig. 17-4). However, as mentioned, for a fairly easy problem such as the barn door blowing shut issue, your brain takes you through these steps so rapidly that you do not even know they are occurring. However, when the problem is more complex, and we cannot instinctively move through the solution generation process, we attempt to jump directly from step one (specific problem) to step five (specific solution) of the problem-solving pathway (see Fig. 17-4) creating unfocused or ineffective solutions (i.e., putting cotton in the livestock's ears during wind storms). Let's reexamine the problem-solving pathway by way of a more complex, yet still somewhat simple, problem. Let's assume that you need to determine how much carpet to buy for your home. Referring to Fig. 17-5 if you attempt to jump from

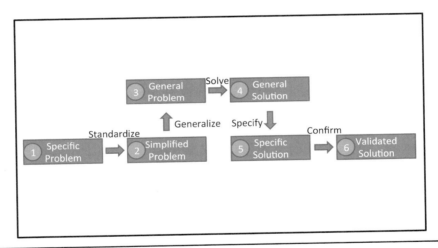

FIGURE 17-4 Problem-solving pathway.

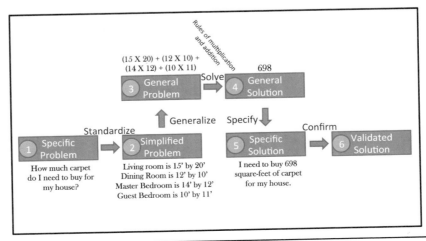

FIGURE 17-5 Problem-solving pathway—carpet purchase problem.

step 1 (specific problem—How much carpet to buy?) to step 5 (specific solution—uhmm.... 1000 sq ft?), you will most likely be wrong and either buy too much or too little carpeting. If however, we use all of the steps (see Fig. 17-5) it is quite easy to come up with an exact answer even though there is no way to arrive at that result without the intermediary steps.

What does the problem-solving pathway have to do with innovation? Clearly "I need to buy 698 sq ft of carpet for my house" is not an innovative solution. One reason why it is not an innovative solution is that no contradictory requirements were resolved in obtaining the answer. So once again, what does the problem-solving pathway have to do with innovation? It is the process by which innovative solutions can be created. However, as discussed previously, in order to create innovative solutions there are additional requirements of modeling (see step 3 of Fig. 17-4) and solving (see steps 4 and 5 of Fig. 17-4) contradictory system requirements within the problem-solving pathway. Further, we will use additional problem-solving methodologies at other steps of the problem-solving pathway thus creating combinations of methods for creating innovative solutions. The next several sections will analyze various problem-solving and innovation methodologies to understand what their contributions to an overall innovation process can be and where they will fit into the problem-solving pathway. Following the separate methodologies discussions I will then return to the options of combining those methods into an orchestrated processes.

Problem-Solving Methodologies

During the course of exploring and executing various problem-solving methodologies over the past couple of decades, I have come to define two distinct categories of methods: *administrative processes* and *technical processes*. An *administrative process* specifies what tasks need to be done and the order in which they should be accomplished but does not give any, or at least very little, insight as to how those tasks should be realized. An example of an administrative process might be a chore list left for your children. You could instruct them to straighten their rooms, dust the furniture, vacuum the carpet, and then take out the trash. The tasks, and their order, are specified but there is no technical detail as to how to execute any of the tasks. In comparison a *technical process* would specify not only what needs to be done, and in what order, but would also provide details of how to specifically to execute the various tasks (see Fig. 17-6). Therefore, problem-solving methodologies, such as those categorized as administrative, will often benefit from having technical methodologies inserted into their processes. The combining of administrative and technical methodologies can result in not only a comprehensive and well-ordered set of "what to do" instructions but also simultaneously provides the problem solver with detailed "how to" directions. Further, some technical methodologies are focused on individual steps of the problem-solving pathway (see Fig. 17-4). Combining multiple technical methodologies into the proper series can also result in a complete and detailed problem-solving road map. Additionally, rounding out a focused technical methodology by overlaying a decidedly broader but less detailed administrative process can also support the problem solver's goals. Looking in more detail there are generally five steps in problem solving (see Fig. 17-4). The first activity is to standardize the problem. The second activity is to generalize the problem. The third activity is to solve the

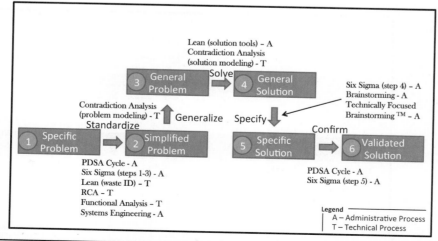

FIGURE 17-6 Problem-solving pathway—with transitional tool options.

problem. The fourth activity is to specify the solution for the situation at hand. And finally the fifth activity is to confirm that the applied solution is indeed valid. I am not aware of any single process (administrative or technical) that addresses each of the five steps thoroughly. Many problem-solving methodologies have good tools and processes for executing the first activity, standardization. Few methods have tools for executing the second and third activities, generalization and problem solving. Most problem solvers use free association or brainstorming to execute the fourth activity, specification. And finally, there are only a handful of methods that support the requirement of the fifth activity, solution evaluation. Popular problem-solving methodologies include Six Sigma, Lean principles, contradiction analysis, functional analysis, root cause analysis, systems engineering, and brainstorming methods.

Plan-Do-Study-Act

PDSA is generally an administrative process. The strength of the PDSA process is the guidance it provides to the planning in understanding how changes made to a system affect that system and the guided response to the measurement of the change effect. Most of the PDSA contributions fall into the standardization of the problems statement (moving from step 1 and step 2 of the problem-solving pathway) and the testing or validation of the specific solutions that would occur between step 5 (specific solution) and step 6 (validated solution) of the problem-solving pathway (see Fig. 17-4). The weakness of the PDSA process is that it has no tools in support of the development of solutions (changes), innovative or otherwise.

Six Sigma

Six Sigma is a process designed for the reduction of variation in processes. Therefore, it is a methodology intended to support the reduction of errors (misprocessing) in technical and business processes. Six Sigma is one of the most utilized, and therefore most developed, methodologies used in problem solving.

While the general steps used within the DMAIC (define, measure, analyze, improve, and control) and DMADV (define, measure, analyze, develop, and verify) methodologies are at first glance mostly administrative in nature the high level of integration with other tool sets pushes the Six Sigma process strongly toward the technical process end of the scale. Further, in relation to the Six Sigma process steps that are dependent on a high level of integration of the science of statistics, Six Sigma can definitely be considered a technical process, at least within those operations. The strengths of the Six Sigma process lie in its ability to capture analysis requirements for success, the quantification of system performance levels from before and after changes have been implemented, and the focus on follow-up and continuous monitoring. The first three steps of both DMAIC and DMADV fall into the standardization of the problem statement (moving from step 1 and step 2 of the problem-solving pathway—Fig. 17-4). The fourth steps of DMAIC (improve) and DMADV (design) align with the "specify" activity for the specific solution step but unfortunately gives almost no guidance as how to accomplish those tasks. The last step of both of the DMAIC and DMADV processes correlate with the confirm (follow-up or validate) activity supporting step 6 of the problem-solving pathway (see Fig. 17-4). The primary weakness of Six Sigma is that the fourth step of both DMAIC (improve) and DMADV (design) are poorly, if at all, supported by any technical processes within the mainstream usage of Six Sigma. In other words, the problem solver is instructed to improve and design at this step but left pretty much up to their own devices in how exactly to do so. Further, a *Fortune* article stated that "of 58 large companies that have announced Six Sigma programs, 91 percent have trailed the S&P 500 since."[2] The summary of the article is that Six Sigma is effective at what it is intended to do, but that it is "narrowly designed to fix an existing process" and does not help in "coming up with new products or disruptive technologies." Advocates of Six Sigma have argued that many of these claims are in error or ill informed.[3,4]

Lean (a.k.a. Toyota Production System)

Lean is a process for finding and eliminating waste within systems. Initially developed and utilized within manufacturing systems it now enjoys a broader application base including service, health care, and business processes in general. Lean is also a heavily utilized problem-solving tool within industry but having the tighter focus of waste elimination than Six Sigma's broader focus of variation reduction results in a smaller but more specifically associated set of tools.

Lean's tight focus on waste elimination along with fairly standardized solution generation tools pushes it somewhat toward the technical process end of the scale. The categorization of the seven wastes along with the standard solutions to responding to them does provide a somewhat specific "how to" directions to the problem solver. The strengths of Lean lie in its focus on waste reduction, opposed to a general focus on problem solving, and its somewhat well-defined operating definitions and guidance. The waste identification steps in the Lean process generally move us from step 2 to step 3 (generalize) of the problem-solving pathway (see Fig. 17-4) while the solution tools (i.e., workspace organization, error proofing, etc.) move the user from step 3 to step 4 (general solution) of the problem-solving pathway. The weaknesses of the waste elimination portion of Lean include a non-system level approach and the inability to thoroughly and technically direct the problem solver in waste elimination. Non-system level analyses can create micro changes that appear useful at the point of application but that ultimately harm the overall output of the entire system. The lack of technical direction in the "how to" of waste elimination can create situations when it is not obvious how to reduce waste and eliminate errors in the process due to Lean's lack of depth of analysis of the process itself.

Root Cause Analysis

Problems solvers often mistakenly focus on effects or symptoms of problems when looking for solutions because those effects and symptoms are what are most visible when analyzing an issue. Root cause analysis (RCA) is a graphical and textual technique used to understand complex systems and the dependent and independent fundamental contributors, or root causes, causing the issue, or problem, under analysis. It allows the problem solver to better understand and visualize the complex relationships between causes and helps point the way to the shortest solution path and the most effective solution options. While RCA is used extensively in engineering,[5] it is also effectively applied to other types of systems including computing, organizational, social, and political to name a few.

The RCA technique is a technical process in that it provides specific direction as to how to execute the method. One of the strengths of RCA is that it combines textual information (reasoning) with a graphical format (RCA mapping) to create a visual record of what situations are contributing to an issue. Further this visual record also reveals the relationships between the various RCA paths (chains) so that convergent solutions can be applied often creating more efficient solution paths. The RCA analysis process moves us from step 1 (specific problem) to step 2 (simplified problem) of the problem-solving pathway (see Fig. 17-4). The weakness of the process is that, like most other problem-solving methods, it has no solution generation tools to support elimination of the initial issue or the identified root causes. However, the resolution provided by RCA in capturing the various causes of an issue reveals exactly where solution development should be focused.

Functional Analysis

Functional analysis is a graphical and primarily qualitative methodology used to focus the problem solver on the functional relationships (good or bad) between system components. A function model or *functional model* is a structured representation of the functions (activities, actions, processes, operations) within the modeled system or subject area.[6] The process is effective because concentrating on system function is not only the correct concern (all systems exist solely to provide some function) but once functionality is the target then the mental inertia of how to address issues within the system is greatly reduced as there is no longer a focus on the system components. Functional analysis is particularly strong in supporting other analysis methodologies as it is an excellent problem-modeling tool that can be used for any and all system problems.

The *functional modeling (functional analysis)* technique is a technical process in that it provides specific direction as to how to execute the method. One of the strengths of functional modeling is that it combines textual information (components and functions) with a graphical format (functional mapping) to create a visual record of how a system operates and what its strengths and weaknesses are. The functional analysis process moves the problem solver from step 1 (specific problem) to step 2 (simplified problem) of the problem-solving pathway (see Fig. 17-4). The weakness of the process is that, like most other problem-solving methods, it has no solution generation tools to support elimination of the initial issue. However, the resolution provided by functional modeling in capturing the interrelationships between system components provides tremendous insight as to where system improvements can be made and therefore where solution development should be focused.

Systems Engineering

Systems engineering (system analysis) is a technique to ensure that full system effects, impacts, benefits, and responses are understood when looking at changes or problems within a system.

The various manifestations of the systems engineering methodology are more technical than administrative processes, as they are fairly specific as to how to create and utilize the various systems engineering models. However, the use of the systems engineering process is meant to guide the problem solver insofar as understanding that full system analysis is necessary in creating truly effective solutions. Therefore, I categorize systems engineering as more administrative in nature than technical. One of the strengths of system analysis (specifically system functional analysis) is that it combines textual information (components and functions) with a graphical format (functional mapping) to create a visual record of how a system operates and what the system's useful and harmful functions are. The system analysis process moves the problem solver from step 1 (specific problem) to step 2 (simplified problem) of the problem-solving pathway (see Fig. 17-4). The weaknesses of the modeling method are that it has no solution generation tools to support elimination of the identified issues. However, the resolution provided by systems analysis in capturing the interrelationships between system components provides tremendous insight into where system improvements can be made and therefore where solution development should be focused.

Contradiction Analysis

Contradiction analysis is the process of identifying and modeling contradictory requirements within a system, which, if unresolved, will limit the performance of the system in some manner. If a *limiting system contradiction* is resolved with the application of a solution then that solution can be considered an innovative solution. Contradiction analysis can be appended to most any problem identification process such as Six Sigma, Lean, functional analysis, and RCA.

Contradiction analysis is the process of identifying and resolving contradictory requirements within a system or process. The method is a technical process as the requirements for creating and solving contradiction models is very specific (see Chap. 24 on TRIZ: Theory of Solving Inventive Problems, for details of creating and solving contradiction models). Inclusion of contradiction analysis and resolution into a problem-solving process is required if the goal is to purposely drive toward innovative solutions. The strength of contradiction analysis is that it allows the problem solver to view a problem in its most elementary state in order to fully understand what within the system is making contradictory demands of the system. Contradiction analysis moves the problem solver from step 2 (simplified problem) to step 4 (general solution) of the problem-solving pathway (see Fig. 17-4). The weaknesses of the modeling method is that if the problem solver does not already have a good understanding of the system components, or root causes, that are responsible for the contradictory demands on the system then a method must first be employed to identify the contradictory system limitations. In other words, it is not a stand-alone methodology if the contradictions are not already identified.

Technically Focused Brainstorming

Technically focused brainstorming is a term I created to describe the use of standard brainstorming methods bounded by certain acceptable solution concept conditions and guided by the attainment of an ideal solution. More specifically, any solution concept is acceptable as long as that solution concept meets two important criteria. The first criterion is that the solution concept must plausibly resolve the contradiction identified by a previously accomplished contradiction analysis. Secondly, the solution concept must move the system under analysis in the direction of *ideal final result*. The concept of *ideal final result* is that in order to improve a system or process the output of that system must improve (i.e., volume, quantity, quality, etc.), the cost of the system must be reduced, or both. In other words, the system's or process's value must increase by the utilization of the solution concept being tested.

The methodology of technically focused brainstorming is a technique that guides the generation of solution concepts by ensuring that those solution concepts support the resolution of contradictory requirements of the system under analysis and renders that system to be of higher value than it was before the solution was applied. It is mainly an administrative process as it instructs the problem solver as to what the goals of the process are but generally leaves it up to the problem solver to devise how to accomplish the directives. The strength of technically focused brainstorming is that it bounds the less effective and more simplistic method of brainstorming by guiding the problem solver to create solution concepts that will resolve system contradictions and improve the value of the system under analysis. The technically focused brainstorming process moves us from step 4 (general solution) to step 5 (specific solution) of the problem-solving pathway (see Fig. 17-4). The weakness of the technically focused brainstorming methodology is that it can only be applied in problem-solving efforts where contradiction analysis and resolution have already been completed.

Methodology Merger

Now that several methods, including where they fit into the overall innovation process, have been discussed it is possible to better define how they can be tied together to create a seamless and powerful tool set. Each methodology brings with it certain strengths and serves to fulfill specific steps and activities represented on the problem-solving pathway (see Fig. 17-4). Therefore, when combined together and properly utilized, these methodologies create a very effective and useful outcome (see Fig. 17-6). This is one of those situations where the whole is more powerful than the sum of its parts. Studying Fig. 17-6 it can be seen that, except for the "generalize" transitional phase, there are multiple problem-solving methodologies associated with the transitional phases between the various problem-solving steps. There may very well be additional processes that could also be used at the generalize phase but contradiction analysis is the only one I propose using considering that the goal is to generate innovative solutions, not just solutions. There are many additional processes that could have also been included on other portions of the pathway but I have limited the inclusion to the most popular and most effective, at least from my experience. At least one methodology must be used for each transitional phase but it is also perfectly acceptable to use multiple methodologies at each transitional phase. In the section tilted "Problem Solving versus Innovation," I explained that in order for a solution to a system problem to be innovative it must solve a limiting system contradiction. Therefore, if it is an innovative solution being pursued then the transitions between steps 2 and 3 and steps 3 and 4 must use contradiction analysis as the transitional engine. Figure 17-7 shows some of the method combination options and the scenarios in which they could be used. For example, if the desire were to use an existing Six Sigma analysis as the basis for an innovation development project then referring to line item number 2 in Fig. 17-7 the problem solver would use the first three steps of the Six Sigma process to convert the specific problem into a simplified problem. Next, contradiction analysis would be used to convert the simplified problem into a general problem and then a general solution. Then, technically focused brainstorming would be used to convert the general solution into a specific solution. And finally, the problem solver would then once again return to the fifth step of the Six Sigma process to validate that the implemented solution concept had the intended effect. A real-world example of the use of this particular combination of methodologies can be seen in the section titled "Operating Room Utilization Improvement—Variation Reduction Innovation Study." As another example it is possible to innovate an existing technical system. For this process I would suggest using the methods of system innovation found on line item numbers 4 and 5 of Fig. 17-7. First, root cause or functional analysis (or both) would be used to convert the specific problem into a simplified problem (see Fig. 17-6). Next, contradiction analysis would be used to convert the simplified problem into a general problem and then a general solution. Then, technically focused brainstorming would be used to convert the general solution into a specific solution. Since there is no specific solution validation tool set associated with system innovation (see Fig. 17-7) the problem solver could execute the above process within the PDSA cycle process and use the PDSA cycle (steps 3 and 4—Fig. 17-7) as the solution validation engine. A real-world example of the use of this particular combination of methodologies can be seen in the section titled "Wireless Power System Improvement—System Innovation Case Study."

LI #	Problem Solving Focus	Transitional Methodologies for:				
		Standardize	Generalize	Solve	Specify	Confirm
1	Innovation with Change Control	PDSA	Contradiction Analysis	Contradiction Analysis	Technically Focused Brainstorming ™	PDSA
2	Innovation in Variation Reduction	Six Sigma	Contradiction Analysis	Contradiction Analysis	Technically Focused Brainstorming ™	Six Sigma
3	Innovation in Waste Reduction	Lean	Contradiction Analysis	Contradiction Analysis	Technically Focused Brainstorming ™	-
4	System Innovation	RCA	Contradiction Analysis	Contradiction Analysis	Technically Focused Brainstorming ™	-
5	System Innovation	Functional Analysis	Contradiction Analysis	Contradiction Analysis	Technically Focused Brainstorming ™	-
6	Variation Reduction	Six Sigma	-	-	Six Sigma	Six Sigma
7	Waste Reduction	Lean	-	Lean	Brainstorming	
8	Change Management	PDSA			Brainstorming	PDSA

FIGURE 17-7 Combination methods summary chart.

Case Studies

Let's look at some case studies to review how the various combinations of methodologies coordinate to produce accelerated results. It should be noted that since the goal is to create innovative solutions, both examples, regardless of what other methodologies are utilized within the process, will employ the use of contradiction analysis at steps 3 and 4 of the problem-solving pathway (see Figs. 17-4 and 17-5).

Operating Room Utilization Improvement—Six Sigma Case Study

I was hired by a hospital chain to help improve the utilization rate of the operating room (OR) suites. As expected a hospital's OR suites are very expensive resources and utilizing them at a high rate is crucial to effectively running hospital business. The hospital's staff of Six Sigma Black Belts had already conducted a statistical analysis of several parameters associated with the use of the ORs and one of those analyses involved understanding at what rate the ORs were being utilized. OR utilization was below 60 percent in general and below 50 percent for some of the more specialized OR suites such as the ones constructed and outfitted for brain surgery. Ultimately, over 35 contradictions were modeled and solved for this project but for the sake of brevity I will discuss only one of the identified issues. The main operational factor affecting OR utilization was turnover time between operations. This turnover time was even more impacted by utilizing a single OR back to back for two different types of surgery (i.e., heart surgery vs. knee surgery). Therefore, the transition from step 1 (specific problem) "low utilization of the ORs" to step 2 (simplified problem) "poor turnover time" (Fig. 17-8) was completed using the Six Sigma process. Next, the transition from step 2 (simplified problem) to step 3 (general problem) was completed by way of contradiction analysis. While there are usually several ways that any particular problem can be reflected as to its contradictory requirements I will discuss just one such contradiction. The simplified problem of "poor turnover time" can be analyzed insofar as whether or not the ORs are standardized in their layouts, equipment, and facilities. Highly standardized OR suites (situation 1 in Fig. 17-9A) with the flexibility to perform any type of operation in any of the ORs would be very expensive while also being very flexible. On the other hand, highly specialized ORs with low standardization (situation 2 in Fig. 17-9A), which could only host specific types of operations by OR configuration, would be much less expensive and, of course, not very flexible. High standardization would help OR utilization by making turn over time shorter. Once again referring to Fig. 17-8, the transition from step 3 (general problem) to step 4 (general solution) is represented by the merger of the two desirable results from the problem models (see Fig. 17-9a) created by some yet to be determined system design (Fig. 17-9b). Therefore, the abstract model that solves this contradictory requirement for the ORs to be both highly standardized and highly specialized simultaneously would look like the diagram in Fig. 17-9b. The "unknown solution" (represented by the large white circle) would result in less expensive facilities (represented by the small grey circle) while simultaneously providing a very flexible facility (represented by the large grey circle). The solution should result in the "good" effects represented by the two differing problem models (less expensive

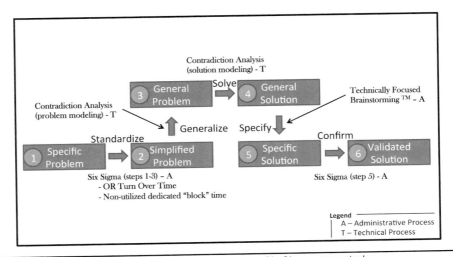

FIGURE 17-8 Problem-solving pathway—operating room Six Sigma case study.

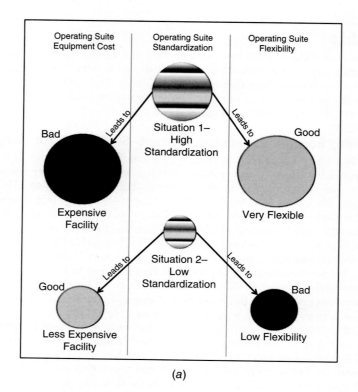

(a)

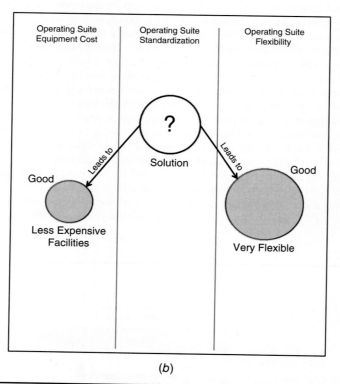

(b)

FIGURE 17-9 Conflicting situations graphic—OR utilization.

facilities and very flexible). Next, utilizing technically focused brainstorming (see Fig. 17-8) to transition the analysis from step 4 (general solution) to step 5 (specific solution), the team developed specific solutions that transcended the system's contradictory requirements, as reflected in the associated contradiction analysis, and simultaneously increased the system's value. While over 450 solutions were developed for the more than 35 analyzed contradictions, only a few will be shared. One of the more interesting solutions I developed, which is now in use at the organization's newest facility, is to completely eliminate the utilization issue of the expensive OR suite by operating directly in the inpatient hospital room. Clearly this did not completely solve the utilization issue as only a small subset of operations are suitable for executing in a standard hospital room, but it did greatly reduce the turnover requirements by removing some of the changeover demand and thus raising overall utilization. A second solution developed for the project was to incorporate a visual feedback system to the operational environment. It is understood by efficiency experts that providing real-time feedback to teams is necessary in order to drive the team's performance to highly effective levels. The conceptualized visual feedback system takes two primary forms: (1) display a large wall timer visible to all personnel responsible for OR turnaround that shows how long an OR has been out of service since the finish of the previous operation and (2) radio tag all equipment (including personnel) so that its location can be readily understood and it is easy to know when all necessary components are indeed where they are suppose to be. Following the analysis and solution phase of the project certain choice solutions were implemented and then Six Sigma was once again used to statistically measure the impact to OR utilization. Therefore, Six Sigma was used to transition the team from step 5 (specific solution) to step 6 (validated solution) of Fig. 17-8.

Wireless Power System Improvement—System Innovation Case Study

While working for the world's largest semiconductor manufacturing corporation I was asked to support the improvement of a new wireless power system for use on mobile computing and phone devices. Wireless power is a technology where power is provided to electrically powered systems without the use of the ubiquitous power cord. It is not hard to imagine the advantage of having battery-powered electronic devices that would never need to be physically plugged into an electrical outlet in order for them to be charged. The company had a working system but the range over which it would work effectively was quite short (several inches). The goal of the project was to extend the range, up to several feet, over which the wireless power system would operate. First it was necessary to establish the specific problem in step 1 of the problem-solving pathway (see Fig. 17-10). Many times the specific problem can be established by simply restating the problem as initially presented. Therefore, in this case the specific problem could be formulated as "the effective operating range of the wireless power system is very limited." To innovate solutions in improving the system I used the functional analysis system innovation combination of methodologies found on line item 5 of Fig. 17-7. It was first necessary to transition the specific problem (step 1) into a simplified problem (step 2) of Fig. 17-10. According to Fig. 17-7 the standardization process associated with line item number 5 calls for the use of functional analysis. In order to build a functional model some type of information (drawings, description, etc.) can be used to guide the development of the model. I used a schematic of a generic wireless power system as the basis for a functional model of the wireless power system (see Fig. 17-11). As can be seen it is necessary to understand how a system physically operates in order to understand how to functionally model that system. System experts are quite useful, and often necessary, in properly creating functional models. If the modeling is not correct there is a slim chance of creating any meaningful solutions to the initial problem, innovative or otherwise.

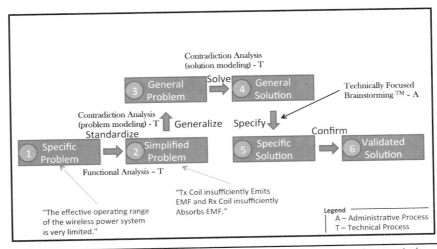

Figure 17-10   Problem-solving pathway—wireless power system improvement case study.

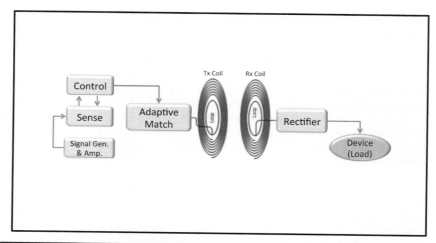

Figure 17-11 Wireless power system schematic.

In studying the resulting wireless power system functional model (see Fig. 17-12) it can be seen that there are four insufficient functions listed within the model. How did I know to make the functions of absorbs and emits EMF (electromagnetic field) insufficient within the functional model (see black oval highlights)? Quite simply, since the existing system is not capable of transmitting power over very long distances, the emission and absorption of the EMF must be insufficient. You will also notice that the production of EMF and current 3 are also listed as insufficient, but these issues will be ignored for this analysis as the first insufficiency can be rectified by simply increasing the power source and the second insufficiency will be rectified when the power transmission is improved. Returning to the power transmission issue, it was necessary to understand the reason for the insufficiencies in order to generate and solve a properly formulated contradiction model. Therefore, a little background on wireless power system deficiencies was useful. In summary, the primary system design parameters that control how well a wireless power system can transmit power across a significant distance are the shape, size, and orientation of the transmission (Tx) and receiver (Rx) coils (see Tx coil and Rx coil in Figs 17-11 and 17-12). Large round coils, around 30 cm

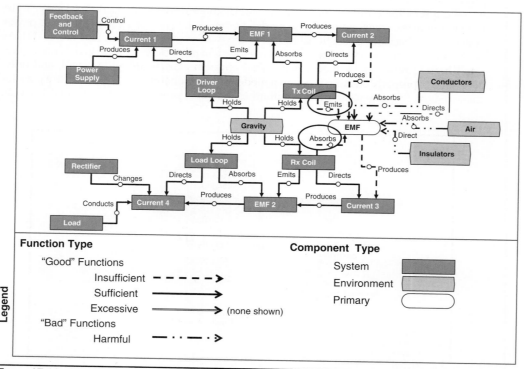

Figure 17-12 Wireless power system functional model.

in diameter, with their faces positioned parallel to each other can transmit and receive electromagnetic power up to several meters. Such a system can even work over greater distances if the coils are even larger. Because of the frequency, wavelength, and power levels used in these system a human can even walk into the path of the beam with no adverse effects while a modestly sized piece of equipment can be run by way of the invisible beam. The challenge of improving the system is the limitation in size of one or both of the coils based on the form factor (size and shape) of the device (i.e., cell phone) intended to take advantage of the wireless power source. Imagine a wirelessly power smartphone where the receiver coil is placed just inside the smartphone's case, limiting the coil to a diameter of approximately 5 cm. Such a system would most likely be limited to an effective charging range of no more than 15 cm.

Therefore, the transition from step 1 (specific problem—the effective operating range of the wireless power system is very limited) to step 2 (simplified problem—Tx coil insufficiently emits EMF and Rx coil insufficiently absorbs EMF) (see Fig. 17-10) was completed using functional analysis. Next, the transition from step 2 (simplified problem) to step 3 (general problem) was completed by way of contradiction analysis. The simplified problem of Tx coil insufficiently emits EMF and Rx coil insufficiently absorbs EMF can be analyzed insofar as the size and orientation of the transmission and receiver coils. Large coils (see large black circle represented in situation 1 of Fig. 17-13) would transmit over long distances (see large grey circle represented in situation 1 of Fig. 17-13) but would require the device transmitting the power and the device receiving the power to also be very large (see large striated circle in situation 1 of Fig. 17-13). On the other hand, small coils (see situation 2 in Fig. 17-13) would be much less effective at transmitting power (see small black circle in situation 2 of Fig. 17-13) but would allow the transmitting and receiving devices to have small form factors (see small grey circle in situation 2 of Fig. 17-13) since the coils would be small. Once again referring to Fig. 17-10, the transition from step 3 (general problem) to step 4 (general solution) is represented by the merger of the two desirable results from the problem models created by some yet to be determined system design.

Therefore, the abstract model that solves this contradictory requirement for the coils to be both large and small simultaneously would look like the diagram in Fig. 17-14. The "unknown solution" or general solution (represented by the large white circle in Fig. 17-14) would result in small device form factors (represented by small grey circle) while simultaneously providing long-distance power transmission (represented by large grey circle). The general solution (see step 4 of Fig. 17-10) should create the good effects represented within the two differing problem models (small form factors and long-range power). Next, utilizing technically focused brainstorming (Fig. 17-10) to transition the methodology from

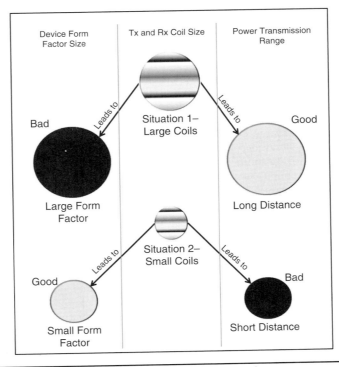

Figure 17-13 Conflicting situations graphic—wireless power system.

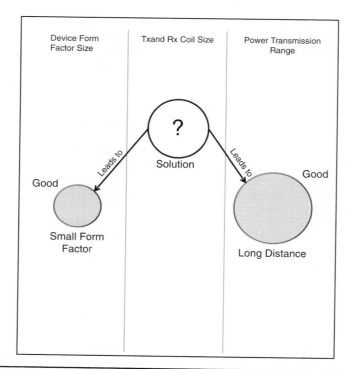

FIGURE 17-14   Conflicting situations graphic—wireless power system.

step 4 (general solution) to step 5 (specific solution), I developed specific solutions that transcended the system's contradictory requirements, as reflected in the associated contradiction analysis, and simultaneously increased the system's value. While more than 35 solutions were actually developed to address the contradictory system requirements, only a few are shared in this case study. First, and decidedly most simply, is to use flip-up and/or fold-out coils that can be expanded when power transmission is necessary. This keeps the system's form factor small during battery power (or mobile) operating modes but still allows for effective power transmission during charging states. Solutions 2, 3, and 4 are most complex and would require some technology development but completely solve the contradiction of needing both large and small coils for the system. Since the coil size is driven by the physics of electricity, it would be prudent to find another way to charge a remote device without the use of transmitted power. For example, instead of transmitting power and therefore needing large system coils, it would be possible to create power at the receiving device by the use of other transmissions. The first possibility is to transmit a different type of electromagnetic propagation that instead of being the actual power source, creates power at the receiving device by moving ferrous balls in an oscillatory fashion within a linear electric motor to create power where needed instead of transmitting that power. This eliminates the need for large transmission and receiver coils. A massively parallel array of these linear electric generators could be used to create the necessary device voltage for operation. The third and fourth solution concepts operate in a similar manner of creating power at the receiving device but use sound and heat as the triggering medium. In each case a nanomaterial (the first sound-sensitive and the second heat-sensitive) is used to convert transmitted sound or heat into electrical power at the receiving device. This once again meets the requirements of "transmitting power" but avoids the need for large coils in doing so.

Conclusion

There are dozens if not hundreds of problem-solving methods in use today. None of them are solely capable of providing all of the necessary steps for effectively solving problems. Further, only a couple of these methods are capable of creating innovative solutions to complex problems even though those methods still require other processes to round out their capabilities. By combining multiple methods (see Fig. 17-7), which have complimentary sets of strengths and weaknesses, complete and effective processes can be created that not only provide the resources necessary in effective problem solving but also allow for the generation of innovative solution concepts.

The problem-solving pathway contains six steps for effective problem solving:

1. Specify a problem
2. Simplify the problem
3. Generalize the problem
4. Generalize a solution
5. Specify a solution
6. Validate the solution

Different problem-solving and innovation methodologies are useful as process engines at different points of the problem-solving pathway (see Fig. 17-6). In order for a solution to be innovative it must solve a contradictory demand placed upon a system. To model and solve these system contradictions, contradiction analysis must be utilized when transitioning from step 2 to step 3 and from step 3 to step 4 of the problem-solving pathway. With a little planning and an understanding of how differing methods provide different capabilities, a very wide range of problems can be addressed with the generation of innovative solutions.

References

1. Altshuller, G. *The Innovation Algorithm—TRIZ, Systematic Innovation and Technical Creativity*. (L. Shulyak and S. Rodman, Trans.). Worcester, MA: Technical Innovation Center, Inc., p. 91, 2007. (Original work published 1973.)
2. Morris, B. "Tearing up the Jack Welch Playbook." *Fortune*. http://money.cnn.com/2006/07/10/magazines/fortune/rule4.fortune/index.htm. Retrieved November 26, 2006.
3. Richardson, K. "The 'Six Sigma' Factor for Home Depot." *Wall Street Journal Online*. http://online.wsj.com/article/SB116787666577566679.html. Retrieved October 15, 2007.
4. Ficalora, J. and Costello, J. *Wall Street Journal SBTI Rebuttal* (PDF). Sigma Breakthrough Technologies, Inc. http://www.sbtionline.com/files/Wall_Street_Journal_SBTI_Rebuttal.pdf. Retrieved October 15, 2007.
5. Reason, J. *Human Error*. New York, NY: Cambridge University Press, 1990.
6. FIPS Publication 183 released of IDEFØ December 1993 by the Computer Systems Laboratory of the National Institute of Standards and Technology (NIST).

CHAPTER 18

Market Research in the Process of New Product Development

Fiona Schweitzer

Introduction

Innovations represent a key factor for the sustainable, lasting success, and profitability of businesses. Despite the high importance of new product development (NPD), however, realization in practice also brings with it high failure rates.[1] In addition, increasingly shorter product life cycles require early identification of customer requirements and their consideration in product development. It is therefore all the more important for new product performance to integrate the customer's mind into the innovation process and to focus on market orientation. Customer focus is central to product development as the final decision about whether to purchase the product rests with the customer. Market research plays an important role in innovation management as it enables the company to look into the customers' minds, to detect and satisfy their latent needs, and to bring successful products onto the market.[2] A vast body of market research tools exists to support the whole innovation process, from ideation to market launch. This chapter covers the phases of the innovation process and focuses on the role of market research throughout the process.

Market Research

Definition of Market Research

Market research is defined as "the systematic and objective identification, collection, analysis, dissemination, and use of information that is undertaken to improve decision making related to identifying and solving problems (also known as opportunities) in marketing."[3] Thus, market research aims at facilitating decisions through a broad and reliable database and allows the successful development of new products which are matched to customers' needs, wants, and desires.[4]

The identification and satisfaction of various customer needs represent a major goal of marketing. For this purpose, marketing managers are reliant on information about customers, competitors, and external factors such as the general economic conditions, public policies and laws, the political environment, and social and cultural changes. In order to support marketing decision making, marketing research assesses information needs and makes relevant, reliable, valid, and current information available. Hence, in order to remain competitive and to avoid the high costs associated with making poor decisions based on unsound information, companies engage in market research. This has a positive effect on reducing uncertainty and improving the quality of decision making.[5] Further, market research enables companies to align their market-related activities to the realities of the market, which in further consequence can be judged as a precondition for the market success of companies.[6] The decision as to whether a company should invest in market research should be based on how the arising costs compare with the expected benefits in terms of information and if resources, especially time and money, are available to (1) conduct research and (2) implement the findings.[3]

Market Research in Innovation Management

Importance of Market Research for Successful Innovations

The understanding that innovations represent the key to the success of companies is not new. What is new is that due to an ever more complex and dynamic business environment, companies feel more and more under pressure to bring new products to the market at ever shorter time intervals in order to remain competitive.[7] This drastic shortening of product life cycles is partly traceable to increasingly more specific customer needs.[8] Thus it is more important for successful companies, not only to consider technical criteria by following the traditional technology push approach, but also to develop customer-oriented goods and services.[9]

Failure rates of new products and services are relatively high, so early preselection of promising ideas is becoming more and more important. Here, market research can make a significant contribution by not only eliciting the internal expert opinion (in the technical area) for new products, but also checking market acceptance.[10]

Types of Innovation

Innovation management deals with the collection of ideas for new and improved products and their development, implementation, and exploitation in the market. According to the degree of novelty, a key defining feature of innovations, radical and incremental innovations can be distinguished.[11] This degree of innovativeness rises with increasing severity of the following four dimensions[12]:

- **Technology:** Technical uncertainty of innovation projects
- **Market:** Targeting of innovations on new or not previously satisfied customer needs
- **Organization:** The extent of organizational change
- **Innovation environment:** Impact of innovations on the innovation environment

Radical innovations (high degree of novelty) are marked by a high level of activity in all four dimensions, while incremental innovations (low degree of novelty), are only weakly to moderately developed.[11] Figure 18-1 demonstrates these relationships.

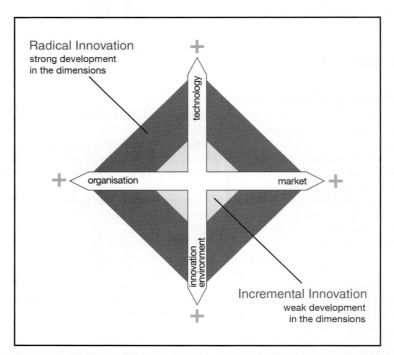

FIGURE 18-1 Dimensions of the degree of novelty. (*Source*: Illustration based on Salomo et al., 2003.)

Phase Model of the Innovation Process

Both in literature and in operational practice, there are a variety of concepts for the structuring of innovation processes which all strive to make the entire innovation process manageable and controllable.[11] For this, a subdivision of the innovation process in time segments is made, where each stage of the innovation process is associated with different tasks and specific methods and gates before each stage, where decisions on termination or continuation are taken.[8]

Generally, these process models include a front-end stage that includes tasks such as trends scoping, idea generation, idea evaluation, concept generation, and concept evaluation, a central stage (development, evaluation, and testing) and the launch phase which is the final back end of the process (see Fig. 18-2). In each case, a decision is taken on termination or continuation of the project. The funnel-shaped course of the phase model demonstrates that the ideas and product concepts are identified and sorted out step by step. By collecting, analyzing, and interpreting market data as well as providing information about the market and the customers within the company, market research helps in every phase of the innovation process to reduce market uncertainties.[13]

To ensure the efficient course of an innovation process, the definition of strategic search fields represents the first step in the front end of innovation.[7] It requires a thorough examination and consideration of current trends. Market research provides tools such as "netnography," outcome-driven innovation (ODI), Delphi, empathic design, or future workshops to explore and evaluate trends and to detect latent customer needs. This is followed by the generation of ideas. Here market research can contribute through idea competitions among users, creative focus groups, and other methods. In the subsequent evaluation of these ideas market research can provide early customer feedback, for example via virtual stock markets or focus groups. Based on in-depth analyses, in the next process steps the preferred ideas are put in concrete form step by step.[14] One or more concepts are developed and evaluated. Market research can assist and secure customer focus throughout these phases through techniques such as lead user workshops, focus group workshops, and customer codevelopment.

At the end of the early stages of the innovation process, these product concepts pass the first gate toward development. The development phase includes both the technical product development and the development of a detailed marketing concept. Before introducing the products onto the market, they have to be subjected to various tests. On the market side,

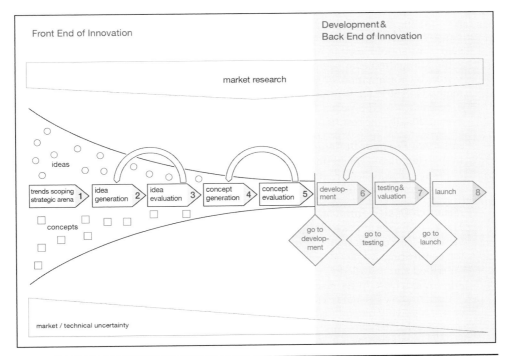

Figure 18-2 Phase model of the innovation process. (*Source*: Illustration based on Schweitzer et al., 2012.*)

---

*Schweitzer, F., Gaubinger, K., and Gassmann, O. "A Three-Track Process Model of the Front End of Innovation in the Machine Building Industry." Proceedings of the XXIII ISPIM Conference, Barcelona, Spain, June 17–20, 2012.

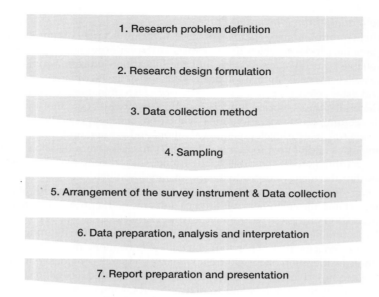

1. Research problem definition

2. Research design formulation

3. Data collection method

4. Sampling

5. Arrangement of the survey instrument & Data collection

6. Data preparation, analysis and interpretation

7. Report preparation and presentation

FIGURE 18-3   Market research process. (*Source*: Illustration based on Malhotra, 2009; Shukla, 2008.)

these tests check product acceptance by the potential users and help to segment the market, to assess market size for the new product in detail, and to position the product correctly. The tests can be prototype tests, product tests, or market tests. Prototype and product tests concentrate only on the reaction of customers to product features, whereas in market testing techniques, the whole marketing mix is tested.

Market Research Process

In order to be effective, marketing research should follow a systematic, predictable path and a planned process. This market research process defines the necessary tasks for conducting a marketing research study and consists of eight steps (see Fig. 18-3), where each step is equally important. Even though the market research process is presented as a sequence of steps, one should keep in mind that these steps have an interdependent and iterative character. Market research begins with identification or definition of the research problem, followed by formulation of the research design and determination of the data collection methods to be used. Sampling and questionnaire development represent the next phases of a market research process. Thereafter, the data are collected, prepared, analyzed, and interpreted. Finally, the results are presented to the decision makers.[5]

Research Problem Definition

In this step management defines the information needed via market research, the objectives of the research (research questions), and the basic population (e.g., for a survey of private persons general population aged 14 and older, households, etc.) for the research.[6]

Research Design Formulation

The research design, which may be fundamentally descriptive, exploratory, or causal (see Table 18-1 for details on the comparison of basic research designs), represents a framework for conducting a market research project.[5] In a descriptive study, the relevant facts are described and recorded as accurately as possible, but no correlations between the variables are studied. Descriptive research is often conducted via quantitative surveys or observation. An exploratory research design is superficially suited to the structuring and understanding of still relatively unexplored topics. Qualitative research tools, such as focus group sessions or in-depth interviews are most suitable for exploratory research. In the context of causal studies, relationships between variables based on preestablished hypotheses are the center of attention. Review of the hypotheses as well as explanation of which variables are the cause and which are the effect, form the main tasks of a causal study. Laboratory and field experiments as well as quantitative surveys are the methods used in causal research designs.[4]

Data Collection Method

This step consists of clarifying how the data are obtained from the respondents.[3] Various methods of data collection exist and it is important to select the one most suitable for a specific research problem.[6] Figure 18-4 gives an overview of potential data collection methods for market research.

	Exploratory	Descriptive	Causal
Objective:	Discovery of ideas and insights	Describe market characteristics or functions	Determine cause-and-effect relationships
Characteristics:	Flexible Versatile	Marked by the prior formulation of specific hypotheses	Manipulation of one or more independent variables
	Often the front end of total research design	Preplanned and structured design	Control of other mediating variables
Methods:	Expert/experience surveys	Secondary data: Quantitative	Experiments
	Pilot surveys	Surveys	
	Case studies	Panels	
	Secondary data: Qualitative	Observational and other data	
	Qualitative research		

TABLE 18-1   Comparison of Basic Research Designs (*Source*: Malhotra, 2009.)

Market research distinguishes two possibilities with regard to data collection: the collection of primary data (field research) and the collection of secondary data (desk research). These two ways of data collection are both associated with advantages and disadvantages (see Table 18-2).

Before starting primary data collection, a researcher should consider existing secondary information. Secondary data is data that has already been gathered for other reasons and exists either inside or outside a company. Only if this data does not fully satisfy the information needs, a primary research will be started. There are three different basic approaches to collecting primary data[3]: surveys, observations, and experiments. In the context of **surveys**, qualitative and quantitative survey methods can be distinguished. **Observation** means the recording of behavioral patterns of people, objects, and events in order to obtain information

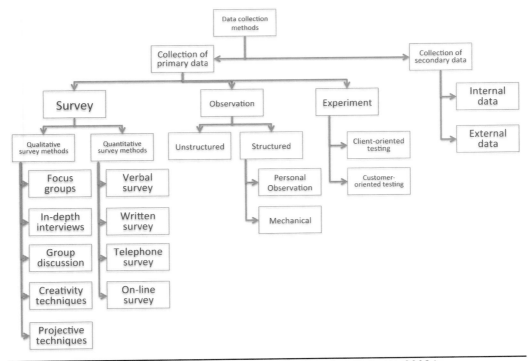

FIGURE 18-4   Data collection methods. (*Source*: Illustration based on Malhotra, 2009.)

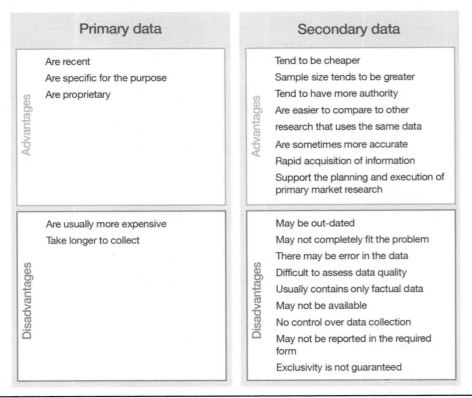

TABLE 18-2   Advantages and Disadvantages of Primary and Secondary Data (*Source*: Illustration based on Mooi and Sarstedt, 2011;[15] Malhotra, 2009.)

about the phenomena of interest. The observation can be conducted personally or mechanically. **Experiments** represent a mixture of surveys and observations in an artificial setting and can be summarized as test procedures.

Which of the three data collection methods—survey, observation, or experiment—ultimately proves to be the most appropriate, depends mainly on the five criteria listed (see Table 18-3). In general, it can be stated that high response rates are the greatest strength of observations. Reliability, simplicity, and ease of analyzing and interpreting the data represent the main advantages of survey methods. The strengths of the experiment lie in the fact that all influencing factors can be controlled in the artificial setting and therefore deep accurate insights can be gained for very specific issues.[16]

| | | criterion | | | |
data collection procedures		costs	time	response rate	quality	representation
	survey	-	+	-	+	+
	observation	o	-	+	o	o
	experiment	+	o	o	+	-

TABLE 18-3   Comparison of Data Collection Procedures (*Source*: Illustration based on Kamenz, 2001.)

	Qualitative Research	Quantitative Research
Objective	To gain a qualitative understanding of the underlying reasons and motivations	To quantify the data and generalize the results from the sample to the population of interest
Sample	Small number of nonrepresentative cases	Large number of representative cases
Date collection	Unstructured	Structured
Data analysis	Nonstatistical	Statistical
Outcome	Develop a richer understanding	Recommend a final course of action

Table 18-4   Differences between Qualitative and Quantitative Research (*Source:* From Malhotra, 2009.)

The collection of primary data in market research can be classified in qualitative and quantitative market research procedures. These two approaches differ not only in the method of information acquisition, but also in the way of selecting samples. **Qualitative research** represents an unstructured, exploratory research methodology, which makes use of psychological methods and relies on small samples, which are mostly not representative. Qualitative research aims at gaining insight into the problem and getting a qualitative understanding of the underlying reasons. In addition, unconscious motives, attitudes, and expectations can be identified through qualitative market research. Especially for finding new, previously unexplored areas, qualitative research delivers better results than quantitative market research, as intuition and creativity are activated. **Quantitative research** can be seen as a structured research methodology based on large samples. The main objective in quantitative research is to quantify the data and generalize the results from the sample, using statistical analysis methods.[5]

In contrast to quantitative research, the focus of qualitative market research is not on large data breadth, but on a great depth of data. Quantitative market research on the other hand is built more on numerical data. Qualitative research brings with it problems in identifying potential respondents, whereas the achievement of a high response rate represents a difficulty of quantitative methods. To sum up, quantitative research refers to describing and measuring, while qualitative research is more associated with explaining and understanding.[17] Table 18-4 emphasizes the main differences between qualitative and quantitaive market research.

Sampling
As time and cost constraints usually hinder researching all relevant objects of a population (census), only a few objects (partial census) are interviewed in corporate market research.[3] Figure 18-5 shows the difference between population and sample.

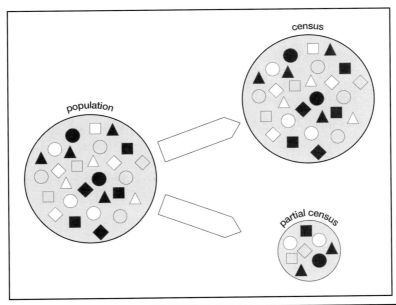

Figure 18-5   Census versus partial census. (*Source:* Illustration based on Kamenz, 2001.)

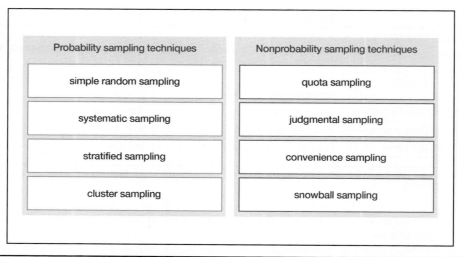

Figure 18-6 Sampling techniques. (*Source*: Illustration based on Malhotra, 2009; Shukla, 2008.)

Figure 18-6 provides an overview of existing sampling techniques, which can be classified into probability sampling techniques and nonprobability sampling techniques. The choice of an appropriate sampling technique depends on the chosen research method (quantitative or qualitative): probability sampling techniques and quota sampling are mainly used for quantitative research studies, whereas for qualitative research studies nonprobability sampling techniques are commonly used.[4,16]

In **probability sampling techniques**, a sample is randomly drawn from the whole population, for example by numbering each individual of a population and then randomly selecting numbers (e.g., via a technical computing software). Only individuals with the selected numbers will then be included in the sample. **Nonprobability sampling techniques** draw a sample according to specific and considered characteristics and are therefore based on the researcher's subjective judgment.[18] A detailed explanation of the different sampling techniques is provided by Malhotra.[5]

Arrangement of the Survey Instrument and Data Collection

In this phase the survey instrument is developed in detail (e.g., question content, question wording, question order, measurement scale, and the design of the survey). A pretest and a final revised version of the questionnaire complete this step. Then the respondents are approached and asked to participate in the study.[6]

The willingness of respondents to participate is essential for quality and relevance of the results of the study. In written surveys, a response rate of around 15 percent can usually be considered good. Significantly higher response rates can be achieved by a high involvement of respondents, clever design of the incentive structure, and a sophisticated questionnaire. Telephone and face-to-face interviews are usually accompanied by higher response rates than written surveys. In market research practice, a number of techniques to increase the response rate have already established[6]:

- **Incentives:** Financial or material reward for participating in the study
- **Personalization of the survey:** Consideration of personal information about the respondent
- **Facilitation of reply:** Stamped and addressed envelope for written surveys
- **Follow-up:** A second questionnaire for those who have not responded
- **Building confidence:** The data is evaluated anonymously and not shared with third parties

Data Preparation, Analysis, and Interpretation

In qualitative research, gathered data is transcribed, single cases are analyzed and compared in order to find similarities and differences to gain deeper insights into the subject of interest. For these analyses, nodes and in vivo coding via NVivo can be utilized[19] to detail defined themes, to identify emerging themes from the raw data, and to organize them into categories and meaningful themes.[20] A team of researchers should discuss the data in order to make data analysis more objective.

In quantitative research, the data preparation step contains the editing, coding, and transcribing of collected data. Here the completeness, readability, comprehensibility, comparability, consistency, and reasonableness of responses are checked and the responses are converted into numbers (coded) for standardized entry into a computer. In a further step, the coded data are entered and processed, usually via statistical software packages (e.g., SPSS, SAS).[3] This is followed by an analysis of the data using different statistical analysis techniques. Depending on whether one, two, or more variables are taken into account univariate, bivariate, and multivariate methods can be distinguished. In this context, a variety of different analytical methods are available (see Hair et al., 2009 for further details).[21] Finally, conclusions related to the marketing research problems are reached by an interpretation of the results. The objective of this phase is to generate meaningful information from complex and often unmanageable amounts of data.[4]

Report Preparation and Presentation

The resultant dissemination step involves creating a written report on the entire marketing research project, including all data, analysis, and applied methods, and presenting the results to the decision makers of the company. The level of detail in the interpretation of the results depends on the purpose-related information interests of the client. During the presentation of the results, answers to the original problem should be provided and recommendations should be made. Hence, these results serve as a basis for the final decision of the marketers.[3,6]

Use of Market Research Methods in the Innovation Process

Systematization of the Market Research Methods

As already mentioned, market research plays an important role in innovation management as it enables a company to look into the customers' minds, to detect their needs, and to bring successful products to market. Yet, there is an enormous amount of different market research tools that support the whole innovation process, from ideation to market launch. In the following, we provide a general classification of market research tools and their descriptions. In this systematization, the market research tools are classified in terms of five different clusters. The individual constructs of this matrix are briefly explained in Table 18-5.

- **Aim of research:** Market research has two major aspects, which however often go hand in hand: (1) research to identify problems and (2) research to solve problems. Problem-identification research aims at helping to identify problems that are not readily apparent or that are likely to arise in the future. Problem-solving research aims at solving already identified, specific marketing problems. This distinction serves as a basis for classifying the methods used in market research in (1) generating methods with opening character, which strive to increase the number of possible solutions, ideas, or product concepts; and (2) evaluation methods with closing character which enable marketers to assess the market acceptance and market potential of ideas and concepts.[3]

- **Phase of the innovation process:** Due to the different aims, not all market research techniques are suitable for all stages of the NPD process. For this reason, we indicate for which phase a certain technique is most suitable.

Aim of Research		Phase of the Innovation Process					Web Application		Nature of Research		Degree of Innovativeness	
Opening	Closing	Identification	Generation	Evaluation	Development	Testing	Traditional	Online	Qualitative	Quantitative	Radical	Incremental

TABLE 18-5 Systematization of Market Research Methods (*Source*: Schweitzer, 2013.)

- **Web application:** Carrying out online research is increasingly popular. Web 2.0 applications and other Internet-based tools have proven useful for identifying, collecting, analyzing, and disseminating market information. Online research has time and cost advantages compared with off-line research.[5] For this reason, it is important to differentiate between instruments that can only be carried out in traditional ways, tools where online and off-line versions exist, and new tools that can only be used online.

- **Nature of research:** As set out earlier in this chapter, market research methods can be classified as qualitative or quantitative.

- **Degree of innovativeness:** The degree of innovativeness is the extent of novelty of a product. Here it is possible to distinguish between radical and incremental new products. Systematization according to the degree of innovativeness seems to be interesting because some market research tools are more suitable for the generation of radical innovations, while other research methods are better for identifying incremental ideas.

Table 18-6 categorizes commonly used market research tools, according to aim of research, innovation phase, web application, nature of research, and degree of innovativeness.

Opening Methods

Analysis of Customer Complaints   Customer complaints represent a key source for new product ideas as well as improvements and modifications to existing products. Companies receive a broad and subtle insight into the world of consumers, their desires, needs, and requirements through the analysis of customer complaints.[22]

Brainstorming   At this point, brainstorming is explained as representative for all creativity techniques. In a market research context it is often used to gather new product ideas from customers, potential customers, or other important stakeholders. Brainstorming is an associative creativity technique that is mainly used to detect new approaches and concepts through mutual exchange and inspiration. Normally, brainstorming takes place in heterogeneous groups of four to seven people and prohibits any kind of criticism of the mentioned suggestions and ideas. A key advantage of this technique lies in the formation of associative chains, which means that ideas expressed by others will be assimilated consciously and developed further. Online or virtual brainstorming is popular for activating the wisdom of groups and allow several participants to share their ideas on an Internet platform and to collaborate on solutions with temporal and local flexibility.[4]

Contextual Inquiry/Empathic Design   Contextual inquiry is regarded as a structured, qualitative, user-centered design research method, which discovers customer needs indirectly by observing and interviewing users doing their normal activities in their own natural environment. A typical contextual inquiry is a one-on-one interaction of the user and the researcher. The empathic design method goes one step further by trying to uncover latent customer needs and ideas for new product concepts through observing consumers when using products in their own environment. The observed consumer behavior in routine situations and the associated problems with the use of the product make it possible to uncover needs which have not been articulated explicitly by the consumers. The data collection process within these methods is highly unstructured and requires further interpretation in order to use the results as a basis for the creation of new product concepts. The observation is usually combined with in-depth interviews and can include intensive contact between researchers and respondents throughout which they co-live certain situations in an ethnographic way.[23–25]

Cross-Industry Innovation   Cross-industry innovation is a method to achieve radical innovations at low risk. This approach attempts to identify promising findings and technologies from other industries and to transfer them to one's own industry. Through the integration of solution principles from other industries, yet undiscovered potential for radical innovations can be detected. Cross-industry innovation can be implemented in all phases of the innovation process, following a three-step process. In the abstraction phase, the solution space is opened. In the second phase, possible solutions for technological problems are developed through the formation of analogies. In the adaptation phase, the selected solutions are evaluated and finally adapted to one's own problem. From a methodological point of view, creativity workshops with the involvement of external experts are particularly suitable for the cross-industry innovation approach.[26,27]

Aim of Research	Market Research Tool	Phase of the Innovation Process					Web Application		Nature of Research		Degree of Innovativeness	
		Identification	Generation	Evaluation	Development	Testing	Traditional	Online	Qualitative	Quantitative	Radical	Incremental
Opening methods	Analysis of customer complaints		X				X					X
	Brainstorming	X	X				X	X			X	X
	Contextual inquiry/ empathic design	X	X	X		X	X				X	
	Cross-industry innovation	X	X	X	X	X	X				X	
	Crowdsourcing	X	X	X	X			X				X
	In-depth interview	X	X				X					X
	Lead user technique	X	X	X	X		X				X	
	Listening-in technique	X	X					X				X
	Netnography	X	X	X				X			X	X
	Outcome-driven innovation	X	X				X				X	X
	Quality function deployment	X	X	X	X		X					X
	SOPI/SIT		X				X					X
	Tracking/panel	X	X	X	X		X					X
Closing methods	Analytic hierarchy process			X			X	X		X	X	X
	Category appraisal	X				X	X			X	X	X
	Concept test/virtual concept test			X			X	X		X	X	X
	Conjoint analysis			X	X	X	X	X		X	X	X
	Store and market test					X	X			X		X
	Free elicitation		X				X		X			X
	Information acceleration			X				X		X	X	
	Information pump			X				X	X		X	
	Kelly repertory grid		X	X			X		X			X
	Laddering		X	X			X					
	Perceptual mapping		X	X	X		X			X		X
	Product test/product clinic			X		X	X	X	X			X
	Virtual stock market/ STOC	X	X	X				X		X	X	X
	Zaltman metaphor elicitation technique		X	X			X		X			X
Mixed methods	Customer idealized design	X	X	X			X				X	X
	Codevelopment	X	X	X	X	X	X				X	
	Expert Delphi discussion	X	X	X			X	X	X		X	X
	Focus group	X	X	X			X	X	X		X	X
	Future workshop	X	X	X			X		X		X	
	Toolkit		X	X	X		X		X			X

TABLE 18-6 Market Research Methods for the Collection of Primary Data (Sorted in alphabetical order.)

Crowdsourcing   Crowdsourcing, composed of the words "crowd" and "outsourcing," describes the outsourcing of business tasks to the intelligence and industry of a mass of free-time workers (professionals, amateur scientists, experts) on the Internet, who work together to generate content in order to solve certain tasks and problems. Crowdsourcing includes initiatives for seeking solutions for problems that companies do not manage to solve internally. For example, InnoCentive is a company that allows others to publish their problems, more that 250,000 registered users in more than 200 countries can potentially work on its solution, and the best solution is financially honored. Moreover, crowdsourcing can deliver new ideas and inspiration, for example via online idea competitions. Ideas can be made available online and others can be encouraged to contribute to an idea by amending or improving it. In the latter case, the crowd is not regarded as an indefinite mass of Internet users, but rather as a consolidated group of people who know and inspire each other.[28]

In-Depth Interview   An in-depth interview represents an unstructured, qualitative, personal interview by a highly skilled interviewer, conducted on a one-to-one basis. Its strength lies in its ability to uncover latent needs and underlying individual motivations, attitudes, and emotions toward sensitive issues. In-depth interviews are primarily used for qualitative, exploratory research. This market research tool is usually designed in a very dynamic and flexible way.[5,29]

Lead User Technique   The lead user technique represents a method for finding and assessing future needs for novel products and for developing new product ideas and concepts. Lead users are selected, sophisticated consumers whose present needs will become general in a marketplace months or years into the future, and who are highly motivated to find solutions and develop innovative ideas that meet their requirements. In addition, a high level of dissatisfaction with currently available products characterizes lead users. In applying this method, lead users are explicitly involved in an actual product development process and they interact with company engineering personnel in group problem-solving sessions. During this problem-solving process, lead user statements about their needs and problems generate possible solutions. A product concept which both meets the consumer's and the company's needs, represents the outcome of a lead user session.[30]

Listening-In Technique   The listening-in technique represents an indirect method for detecting previously unarticulated and unidentified customer needs by observing customer interactions. By using this Internet-based method, dialogues between customers and virtual advisors in a web-based sales recommendation system are observed, in order to finally gather in-depth information or ideas for new products or product features. For this purpose, a virtual advisor makes several product recommendations and raises a number of questions to the customers. Using Bayesian algorithm, utility functions and needs are derived from the responses and reactions of the customers.[31,32]

Netnography   The word Netnography is composed of the terms Internet and ethnography (observation of the behavior of groups and their members). By using Netnography, implicit and explicit needs, desires, and attitudes of consumers with regard to particular products and brands can be identified. Unveiled information can even include product prototypes that users designed and published on the web. Netnography is regarded as an exploratory method for subtle and unobtrusive observation of the (online) communication flow and social interaction of a group, without even having to join this group. The results of this analysis are condensed into so-called consumer insights, followed by a translation into customer-oriented products and services.[33,34]

Outcome-Driven Innovation   The outcome-driven innovation (ODI) method aims at revealing, assessing, and analyzing customer needs systematically. Using ODI, customers are not asked how a product could be improved; they are asked which activities and which results they want to have met with the help of a product. For example, customers do not want a cell phone but they want to communicate over distance. The qualitative interviews hence focus on gathering customer requirements. In a second step a quantitative survey is carried out in which respondents state the level of satisfaction and importance which they attribute to the elicited customer requirements leading to a clear ranking of the most important but unmet needs of customers.[35]

Quality Function Deployment   Quality function deployment (QFD) concentrates on translating the customer's language into the language of technology, and therefore combines tools and methods, which enables market-oriented product development. This methodology of consistently and systematically considering expectations, desires, and demands of customers and transforming them into a product, accompanies the entire product development process. Customer requirements are gathered through interviews or customer visits, clustered according to primary, secondary, and tertiary requirements and finally weighted.[36–38]

Sequence-Oriented Problem Identification/Sequential Incident Technique   Both the sequence-oriented problem identification (SOPI) method as well as the sequential incident technique (SIT) are guided by the basic idea that customers classify certain key experiences during a service process as particularly quality-relevant. By using the SOPI method, critical (positive or negative) events and problems of customers, which ultimately provide good ideas for new innovations and make customer requirements and expectations transparent, are identified in several steps. The SIT aims at generating ideas and improvements in processes. Additionally it enables collection of customer suggestions and ideas in order to solve previously identified problems. Based on analysis of the entire service process, customers are encouraged to describe step by step their thoughts and experiences (both positive and negative) in the different phases of the entire service process or at different points of contact with the company.[39]

Tracking/Panel   Tracking research is a more or less regularly repeated survey on structurally identical subjects and always on the same topic, where both questioning and observation methods can be applied. Panel surveys differ from tracking surveys to the extent that the same respondents (truly identical subjects) are always addressed. Both tracking and panel research help to determine changes over time. Despite the fact that panel and tracking studies usually involve considerable organizational costs and high continuous effort, their use in recent decades has been promoted as they enable the study of changes in behavior and preferences through longitudinal investigation.[3]

Closing Methods

Analytic Hierarchy Process   The analytic hierarchy process (AHP) is a highly structured decision-theoretical tool that is exclusively computer assisted for the early evaluation of concepts from a customer perspective and the exploration of customer needs. The AHP is a method for supporting decision-making processes by realizing the decision problem in its entirety, simplifying complex decision situations, breaking them down into their component parts, structuring them hierarchically, and solving them rationally. This idea-scoring method is based on the idea that the purchase decision of customers is determined by different criteria (price, value, preference, etc.), which are compared in pairs by the users. This decision-support tool is suitable for complex, multicriteria problems with both qualitative and quantitative aspects.[40,41]

Category Appraisal   Category appraisal is a method that shows the structure of markets and the position of products in a competitive framework as perceived or preferred by consumers. By using category appraisal, consumers evaluate a set of competitive products available on a certain market and determine how well products perform. Category appraisal identifies consumer needs indirectly and allows for identifying product opportunities and attributes that influence product choice. The data collection process is quantitative and highly structured, as multiple products are ranked and rated by the respondents on their perceived similarity, attributes, and preference in a questionnaire.[22]

Concept Test/Virtual Concept Test   By using concept tests, which are particularly suitable for the review of product ideas in the early stages of the innovation process, various alternative product concepts or variants are tested and assessed. A concept is a clear and precise statement of the essential characteristics of a potential product, for example in the form of product models, functional models, pictorial representations, or in written or verbally explained ideas. Without having already developed the new product, the respondent's reaction in the concept test provides information on his or her potential reaction to the new product and his or her buying intention. In virtual concept tests (VCTs), customers select their preferred concepts at varying prices from multiple virtual prototypes (holistic descriptions of products, digital depictions). Rather than having to wait for physical prototypes, this technique makes it possible to test market acceptance quickly and inexpensively.[42–44]

Conjoint Analysis   This quantitative market research technique aims at exploring consumer preferences for various (hypothetical) product concepts, which are composed of certain attributes and attribute levels. A conjoint analysis makes it possible to find out which attributes and which levels consumers prefer and how much they value the attributes. By rating holistic product concepts according to their preferences in a conjoint analysis, the respondents are forced to make trade-off decisions between a small number of different features or benefits. The amount of interest in the different product features is then calculated indirectly. Web-based conjoint analysis benefits from the opportunity to present products, features, and product use in streaming multimedia representations.[45,46]

Store and Market Test   In a store test, fully developed products are offered for purchase under controlled conditions in selected stores (usually around 30 shops) for several months just before the actual launch. In a market test, the area in which the product is tested is larger and can cover a whole city, region, or country. Both methods aim at testing the marketing mix of a new product and assessing the potential market success. Besides market potential (i.e., first and repeated purchase probability), different marketing measures such as price, placement, and promotional activities can also be assessed by this method.[4]

Free Elicitation   Free elicitation is a personal interviewing technique, which primarily aims at detecting consumers' existing knowledge of a particular product category. As respondents express the relevant attributes of a particular product (set) by using free elicitation, this technique puts the focus on intrinsically relevant attributes of products and less on extrinsic product differences. Various stimulus cues (usually words) are presented to the respondents, and they are asked to verbalize concepts that come to their mind. The respondents should know these stimuli or words in order to arouse relevant associations with products. The data collection process is unstructured and consumer needs are derived directly because the consumers themselves articulate their needs.[22,47]

Information Acceleration   Information acceleration is a concept testing method, particularly suited for testing radical products, which works with interactive multimedia stimuli and experimental setups. By using this method, the demand for new products can be estimated in a virtual buying environment by measuring consumers' attitudes and preferences. For this method future product scenarios or electronic representations of the product without prototyping are used to obtain consumer feedback. As a result, many variations of a product can be tested at a low cost at an early stage of the development process. This leads to an acceleration of product development learning and positively impacts time-to-market. Gathered data on purchase intentions in this virtual environment allow simulation of consumers' future response to this new product and prediction of sales potential.[22,48]

Information Pump   Information pump is a web-based question and answer game with a small group of participants (online focus group), which explores the customer evaluations of product concepts and prototypes and pumps information from customers on needs and perceptions. This method focuses on truth-telling and rewards truthfulness and creativity. Product concepts are defined and integrated into the game through a player with a special role, the detective. In this game situation, players make true or false statements about certain product concepts and thus allow the marketers to make a first assessment of how customers react to their new product ideas. This interactive method aims at generating a lot of unconventional ideas about a product. In addition, the fun factor plays an important role, which proves beneficial for the willingness to participate.[31,49]

Kelly Repertory Grid   The repertory grid methodology is a personal interviewing technique that is suitable for eliciting the characteristics or attributes of products by which consumers structure a product category or differentiate between products. When using this method, images of three randomly selected products from a large set are shown repeatedly to a customer, who is subsequently asked which two products are similar and thus differ from the third product. Hence, consumer needs are directly derived in this method and the data collection process is unstructured.[23,50]

Laddering   Laddering is a personal interviewing technique that helps marketers to understand the cognitive structure of consumers which links consumers' knowledge about product attributes with their knowledge about consequences and values. The interview focuses on a particular product (category) and employs a means-end chain approach, to elicit customers' needs directly and repeatedly.[22]

Perceptual Mapping   Perceptual mapping helps marketers to understand what customers think of current and future products as well as how products are perceived by consumers relative to competing products in the marketplace. This method results in visual representations of the perceived relative position in the mind of the customers. Using mathematical and statistical methods produces these visual representations. In these multidimensionally scaled perceptual maps, any product can be rated on a range of attributes in relation to other products as well as to the evaluative attributes. Hence, the products are described on the basis of selected attributes, which must ultimately be assessed by respondents according to their importance. The field of application of perceptual maps extends across the development and evaluation of concepts.[51]

Product Test/Product Clinic   In market-focused product testing, the respondents are asked—after the use or consumption of certain products (both prototypes and already existing real products)—about their subjective perception regarding product impact, product experience, and so forth. This testing method focuses on the subjectively perceived product characteristics. As a result, a product test provides information on customer satisfaction, potential market segments, and the degree of acceptance of a certain product. A product clinic is a special form of product testing, which can be used to assess durable consumer goods and identify consumer preferences. A well-known example is the "car clinic," where newly developed cars are tested by various observation, interviewing, questionnaire, and group discussion techniques.[52]

Virtual Stock Market/Securities Trading of Concepts   Virtual stock markets represent an alternative to traditional market forecasting tools for new products or product concepts, which usually make forecasts on the basis of historical data or by surveying experts or consumers. Virtual stock markets are regarded as artificial stock markets on which—as in real stock markets—product concepts and future market conditions can be bought and sold. As part of the securities trading of concepts (STOC) approach, new product concepts are traded as financial securities or virtual shares. The participants can express their desires, needs, preferences, and expectations about new product concepts and future events indirectly through the purchase and sale of virtual assets. While the participants often do not disclose their true prognosis due to lack of incentives in traditional surveys, material rewards and cash prizes in these stock market games increase the motivation of the participants.[43,53]

Zaltman Metaphor Elicitation Technique   The Zaltman metaphor elicitation technique (ZMET) is a projective technique that works with collages created by consumers about their feelings and experiences with a product or a topic. The consumers are required to provide at least eight pictures, photographs, or other visual images from magazines, catalogues, or photo albums that capture their feelings on the topic and afterwards explain their selection to the researchers in an interview situation. Each chosen picture represents a metaphor that expresses an important opinion about the product. In most cases, the participants are familiar with the product they have to elaborate on. The data collection process is unstructured and consumer needs at the level of benefits and values are elicited.[3,22]

Mixed Methods

Customer Idealized Design   By using customer idealized design (CID), current or potential customers develop their ideal product or service in small working groups in an unconstrained designing process. The participants initially consider all—even unrealistic—characteristics and requirements of the new product. The feasibility of designs plays no role; it is simply about the desirability of the solution. The discussion of the developed solutions within the group as well as the subsequent review of the solutions, taking into account the ideal product of the other group, is repeated until finally the conditions of all group members in terms of an ideal solution are met.[54]

Codevelopment   This method of joint product development represents the highest level of customer integration into the innovation process. For this purpose, selected customers with high expertise are integrated into the company in order to cooperate with the company's own researchers and jointly develop new innovations (mostly radical ones). Due to the intensive cooperation between the company's researchers and the customers, shared tacit and explicit knowledge can be established.[55]

Expert Delphi Discussion   By using this method, experts report in a more or less structured individual interview or group discussion about their experiences, insights, and opinions. This technique aims at forecasting the most probable outcome for some future state. In order to reduce the risk of mutual influence and the dominance of certain group members, the Delphi method, which is regarded as a structured survey method with an iterative process (at least three rounds) of consensus development across a group of experts with feedback loops, seems to be appropriate. During this process, the original problem or theory regarding the development of a trend is commented on in several rounds and the participants react to other expert opinions, until a consensual outcome is reached.[14,56]

Focus Group   A focus group session is a semistructured and interactive discussion among a small group of participants, usually 8 to 12, steered by one or two trained moderators. A focus group session aims at gaining insight—in a relatively short time—into a wide spectrum of opinions, views, and ideas of several people regarding a new product (concept) or a potential solution to a problem. This qualitative method can be used throughout the entire innovation process and requires further input for interpretation due to its flexible data collection process. The discussion can be complemented with various creativity techniques if it focuses on finding new ideas and solutions. If it focuses on preliminary assessment of the acceptance of concepts and prototypes different evaluation techniques can be integrated into the design of the focus group session.[5]

Future Workshop   Through the activation of people's creativity and problem-solving ability, a future workshop makes possible the development of visionary goals, solutions, long-term plans, and strategies that take into account existing constraints and innovative ideas. Participants can include a mix of suppliers (potential), customers, corporate employees, experts from other industries, and other stakeholders. In the first phase of a future workshop, the goals of the workshop, a clear statement of the problem, and a schedule are defined. Then the group conducts a critical analysis of the current situation using brainstorming techniques. In order to generate ideas about the future, participants have to be encouraged in the imagination phase to formulate creative wishes and fantasize about their own future in the form of utopias. Finally, the group members determine the feasibility of the generated ideas, extract the realistic aspects from the previous phase, and develop strategies to overcome potential constraints. Depending on the complexity of the problem and the makeup of the group, a wide variety of ideation and evaluation methods are applied in a future workshop.[26,57]

Toolkit   In the case of this method, key innovation tasks are outsourced to the users themselves. For the manufacturers, it is no longer just about understanding users' needs in full detail, but rather about letting the customer be creative and innovative. By using (virtual) toolkits, users are able to design and develop their own products. Via user toolkits, tacit knowledge of the consumers which otherwise would be unavailable or at least difficult to ascertain can be gathered. In order to guarantee the technical feasibility of the new product, toolkits also provide information regarding the capabilities and limitations of the manufacturing process.[32,58]

Summary

In this chapter, the role of market research in the different stages of the innovation process is presented to enable innovation managers to gain a concise overview of the possible applications of market research. The chapter highlights the need for market research throughout the innovation process along with the necessary steps to carry out market research and also provides a systematic overview of the vast field of different market research techniques.

List of Abbreviations

AHP	Analytic hierarchy process
CAPI	Computer-assisted personal interviewing
CATI	Computer-assisted telephone interviewing
CID	Customer idealized design
NPD	New product development
ODI	Outcome-driven innovation
QFD	Quality function deployment
SIT	Sequential incident technique
SOPI	Sequence-oriented problem identification
STOC	Securities trading of concepts
VCT	Virtual concept test
ZMET	Zaltman metaphor elicitation technique

References

1. Henard, David H. and Szymanski, David M. "Why Some New Products Are More Successful Than Others." *Journal of Marketing Research*, 38 (3): 362–375, 2001.
2. Nakata, C. and Subin, I. "Spurring Cross-Functional Integration for Higher New Product Performance: A Group Effectiveness Perspective." *Journal of Product Innovation Management*, 27 (4): 554–571, 2010.
3. Malhotra, Naresh K. *Basic Marketing Research: A Decision-Making Approach.* New Jersey, NJ: Pearson Education, p. 30, 2009.
4. Shukla, Paurav. *Essentials of Marketing Research.* Kopenhagen: PauravShukla & Ventus Publishing Aps., 2008.
5. Malhotra, Naresh K. *Basic Marketing Research: A Decision-Making Approach.* New Jersey, NJ: Pearson Education, 2009.
6. Homburg, C., Kuester, S., and Krohmer, H. *Marketing Management: A Contemporary Perspective.* New Jersey, NJ: McGraw-Hill Higher Education, 2008.
7. Rowles, K. *Market Research for New Products.* SMART Marketing Department of Agricultural, Resource, and Managerial Economics, Cornell University, 2000.
8. Cooper, Robert G. *Winning at New Products. Accelerating the Process from Idea to Launch.* Cambridge, MA: Perseus Publishing, 2001.
9. Evanschitzky, H., Eisend, M., Calantone, R.J., and Jiang, Y. "Success Factors of Product Innovation: An Updated Meta-Analysis." *Journal of Product Innovation Management*, 29 (supplement 1): 21–37, 2012. Doi: 10.1111/j.1540-5885.2012.00964.x.
10. Belz, C. "Fokus auf die Anforderungen des Kunden." *Marketing & Kommunikation*, 11: 19–20, 2004.
11. Tidd, J. and Bessant, J. *Managing Innovation—Integration Technological, Market and Organizational Change*, 4th ed. West Sussex: John Wiley & Sons, 2009.
12. Salomo, Sören, Gemünden, Hans Georg, and Billing, Fabian. "Interface Management for Radical Innovation Projects: Dynamics and Degree of Innovativeness." *Conference Proceedings to the Academy of Management 63th Annual Meeting*, Seattle, WA, 2003.
13. Böhler, Heymo. *Marktforschung.* Stuttgart: Kohlhammer, 2004.
14. Piller, Frank. Interactive Value Creation with Users and Customers. In: Huff, Anne S. (ed.), *Leading Open Innovation.* Munich: Peter-Pribilla-Foundation, pp. 16–24, 2008.
15. Mooi, Erik and Sarstedt, Marko. *A Concise Guide to Market Research: The Process, Data and Methods Using IBM SPSS Statistics.* Heidelberg: Springer Verlag Berlin.
16. Kamenz, Uwe. *Marktforschung. Einführung mit Fallbeispielen, Aufgaben und Lösungen.* Stuttgart: Schäffer-Poeschel, 2001.
17. Goodyear, M.J. Qualitative Research. In: Birn, R.J. (ed.), *The International Handbook of Market Research Techniques.* London, UK: Kogan Page Ltd., 2004.
18. Saunders, Mark, Lewis, Philip, and Thornhill, Adrian. *Research Methods for Business Students.* Edinburgh: Pearson Education Ltd., 2007.
19. Bazeley, Patricia. *Qualitative Data Analysis with NVivo.* Los Angeles, CA: Sage Publication, 2007.
20. Auerbach, Carl and Silverstein, Louise B. *Qualitative Data: An Introduction to Coding and Analysis.* New York, NY: NYU Press, 2003.
21. Hair, Joseph F. Jr., Black, William C., Babin, Barry J., and Anderson, Rolph E. *Multivariate Data Analysis.* Englewood Cliffs, NJ: Prentice Hall, 2009.
22. Tax, Stephen S., Brown Stephen W., and Chandrashekaran, Murali. "Customer Evaluations of Service Complaint Experiences: Implications for Relationship Marketing." *Journal of Marketing*, 62: 60–76, 1998.
23. Van Kleef, Ellen, Van Trijp, Hans C.M., and Luning, Pieternel. "Consumer Research in the Early Stages of New Product Development: A Critical Review of Methods and Techniques." *Food Quality and Preference*, 16: 181–201, 2005.
24. Beyer, Hugh and Holtzblatt, Karen. *Contextual Design: A Customer-Centered Approach to Systems Designs.* San Fransisco, CA: Morgan Kaufman Publishers, 1997.
25. Leonard, Dorothy and Rayport, Jeffrey F. "Spark Innovation through Empathic Design." *Harvard Business Review*, 75 (6): 102–114, 1997.
26. Fraunhofer Institute (n.d.). Cross Industry-Innovation. Retrieved from http://wiki.iao.fraunhofer.de/index.php/Cross_Industry-Innovation [01.02.2012].
27. Jungk, Robert and Müllert, Norbert R. *Future Workshops: How to Create Desirable Futures.* London, UK: Institute for Social Inventions, 1987.
28. Surowiecki, James. *The Wisdom of Crowds.* Anchor Books, 2005.

29. Willis, K. In-Depth Interviews. In: Birn, R.J. (eds.), *The International Handbook of Market Research Techniques.* London, UK: Kogan Page Ltd., pp. 283–298, 2004.

30. Von Hippel, Eric. "Lead Users: A Source of Novel Product Concepts." *Management Science*, 32 (7): 791–805, 1986.

31. Urban, Glen L. and Hauser, John R. "Listening-In to Find and Explore New Combinations of Customer Needs." *Journal of Marketing*, 68: 72–87, 2004.

32. Hemetsberger, Andrea and Godula, Georg. "Virtual Customer Integration in New Product Development in Industrial Markets: The QLL Framework." *Journal of Business-to-Business Marketing*, 14 (2): 1–37.

33. Kozinets, Robert V. "On Netnography: Initial Reflections on Consumer Research Investigations of Cyber Culture." *Advances in Consumer Research*, 25 (1): 366–371, 1998.

34. Füller, Johann, Bartl, Michael, Ernst, Holger, and Mühlbacher, Hans. "Community-Based Innovation: How to Integrate Members of Virtual Communities into New Product Development. *Electronic Commerce Research*, 6 (1): 57–73, 2006.

35. Ulwick, A.W. *What Customers Want.* New York, NY: McGraw-Hill, 2005.

36. Griffin, Abbie. "Evaluating QFD's Use in US Firms as a Process for Developing Products." *Journal of Product Innovation Management*, 9: 171–187, 1992.

37. Eversheim, Walter, Schmidt, Ralf, and Saretz, Bernd. "Systematische Ableitung von Produktmerkmalen aus Marktbedürfnissen." *io-Management Zeitschrift*, 63 (1): 66–70, 1994.

38. Kamiske, G.F., Hummel, T.G.C., Malorny, C., and Zoschke, M. "Quality Function Deployment—oder das systematische Überbringen der Kundenwünsche. Qualitätsplanungs—und Kommunikationsinstrument zwischen Marketer und Ingenieur." *Marketing ZFP*, 181–190, 1994.

39. Botschen, Günther, Bstieler, Ludwig, and Woodside, Arch G. "Sequence-Oriented Problem Identification within Service Encounters." *Journal of Euromarketing*, 5 (2): 19–52, 1996.

40. Coyle, Geoff. The Analytic Hierarchy Process (AHP). Pearson Education Limited, 1996. Retrieved from http://www.booksites.net/download/coyle/student_files/AHP_Technique.pdf [02.02.2012].

41. Saaty, Thomas L. *The Analytic Hierarchy Process.* New York, NY: McGraw-Hill, 1980.

42. Dahan, Ely, Kim, A., Lo, A., Poggio, T., and Chan, N. "Securities Trading of Concepts (STOC)." *Journal of Market Research*, 48 (3): 497–517, 2011.

43. Paustian, Chuck. "Better Products through Virtual Customers." *MIT Sloan Management Review*, 42 (3): 14–15, 2001.

44. Moore, W.L. "Concept Testing." *Journal of Business Research*, 10 (3): 279–294, 1982.

45. Green, Paul E. and Srinivaasan, V. "Conjoint Analysis in Consumer Research: Issues and Outlook." *Journal of Consumer Research*, 5 (2): 103–152, 1978.

46. Green, Paul E. and Srinivasan, V. "Conjoint Analysis in Marketing: New Developments with Implications for Research and Practice." *Journal of Marketing*, 54 (4): 3–19, 1990.

47. Bech-Larsen, Tino and Nielsen, Niels A. "A Comparison of Nice Elicitation Techniques for Elicitation of Attributes of Low Involvement Products." *Journal of Economic Psychology*, 20 (3): 315–341, 1999.

48. Deszca, G., Munro, H., and Noori, H. "Developing Breakthrough Products: Challenges and Options for Market Assessment." *Journal of Operations Management*, 17 (6): 613–630, 1999.

49. Klein, Bob. "Pump, Don't Pull." *MIT Sloan Management Review*, Spring, 15, 2001.

50. Steenkamp, J. and Van Trijp, H. "Attribute Elicitation in Marketing Research: A Comparison of Three Procedures." *Marketing Letters*, 8 (2): 153–165, 1997.

51. Chaturvedi, Anil D. and Carroll, J.D. "A Perceptual Mapping Procedure for the Analysis of Proximity Data to Determine Common and Unique Product-Market Structures." *European Journal of Operational Research*, 11 (1): 268–284, 1998.

52. Wildemann, Horst. *Produktklinik—Wertgestaltung von Produkten und Prozessen.* München: TCW, 1998.

53. Dahan, Ely and Hauser, John R. "The Virtual Customer." *Journal of Product Innovation Management*, 19 (5): 332–353, 2002.

54. Ciccantelli, Susan and Magidson, Jason. "From Experience: Consumer Idealized Design: Involving Consumers in the Product Development Process." *Journal of Product Innovation Management*, 10 (4): 341–347, 1993.

55. Brown, John S. and Duguid, Paul. "Knowledge and Organization. A Social Practice Perspective." *Organization Science*, 12 (2): 198–213, 2001.

56. Chakravarti, A.K., Vasanta, B., Krishnana, A.S.A., and Dubash, R.K. "Modified Delphi Methodology for Technology Forecasting: Case Study of Electronics and Information Technology." *Technological Forecasting & Social Change*, 58 (1–2): 155–166, 1998.

57. Kuhnt, B. and Müllert, Norbert R. *Moderationsfibel Zukunftswerkstätten verstehen anleiten einsetzen*. Neu-Ulm: AG SPAK, 2006.

58. Thomke, Stefan and Von Hippel, Eric. "Customers as Innovators: A New Way to Create Value." *Harvard Business Review*, 80 (4): 74–81, 2002.

CHAPTER 19

Lead User Analysis

Nikolaus Franke

User Involvement: Ford or von Hippel?

Many managers adopt what could be termed the "Ford attitude." The legendary founder of the automobile multinational once coined the aphorism, "If I had asked people what they wanted, they would have said faster horses." This position is often fueled by frustrating results of attempts to ask average customers about ideas for new products and services. Often, the outcome of such efforts is "I want what I have, only better and cheaper"—which is not exactly a precise specification sheet for a new product development project.

An opposite view is given by Eric von Hippel from MIT.[1] He claims that we are in the middle of a huge paradigm shift, from closed, manufacturer-centered innovation to innovations that are developed by users. He advises that "manufacturers have to learn how to adapt their business models to this."

What is right? And is there a new development? If we look back, we immediately see that innovations by users are not new. Rather, it is the primeval and archetypical mode of innovation: if I have a problem, I try to fix it. For a long time, user innovations might even have been the only mode. It is difficult to imagine prehistoric man inventing the use of fire, the handax, or pottery for commercial purposes. However, division of labor, industrialization, and increasing complexity of technology and production processes led to firms with specialized R&D departments and professional innovation functions, as we all know. Scholarly research on innovation management followed this and assumed since Schumpeter's models of innovation that the dominant mode is producer innovation—innovation projects are initialized and executed by those who expect benefits from selling the new product or service, not from using it oneself. The economic argument for this is scale effects. For an institution that is able to sell the newly developed standard product to many customers, the benefit is huge. It is the benefit for the individual user times the number of users. For the individual user, it is only the individual benefit. Given that markets usually do not consist of single individuals but of thousands or millions of customers, it follows that the incentive to innovate is much higher for a producer than for a user. Thus, theoretically, user innovations should hardly exist, at least not in industrialized and developed economies.

However, they do. A number of empirical studies have been demonstrating that user innovation is both a frequent and important phenomenon. The first type of study draws samples of users and analyzes what proportion reports having innovated.[2] Studies cover different industries, such as printed circuit CAD software, pipe hanger hardware, library information systems, surgical equipment, and several types of consumer products, and find that user innovation is a relatively frequent behavior. Recently, some studies have provided systematic evidence. In a new study, von Hippel, de Jong, and Flowers[3] draw a nationally representative sample of 1173 UK household residents aged 18 and over and find that 6.2 percent of them have engaged in creating or modifying consumer products they used during the previous 3 years. This represents 2.9 million people—about two orders of magnitude more than the number of product developers employed by all the consumer goods producing firms in the United Kingdom. Consumer product innovation spans a wide range of fields, from toys, to tools, to sporting equipment and to personal solutions for medical problems. Replications of this study in Japan and the United States show comparable numbers. There are also recent studies confirming that firms are frequent user innovators. De Jong and von Hippel[4] analyzed a nationally representative sample of 2416 small- and medium-sized companies (SME) in the Netherlands and found that 21 percent engage in user innovation; that is, they develop and/or significantly modify existing techniques, equipment, or software to satisfy their own process-related needs.

The finding that user innovation is a frequent behavior does not necessarily mean that it bears economic importance. Theoretically, user innovations could be "small potatoes"—trivial, irrelevant adaptations of limited value for other users. Thus, a second strand of studies investigated the economic value of user innovations. Here, researchers started by interviewing experts and, based on this, determined a list of the most important innovations in a given industry. Then they studied trade publications and conducted a series of interviews in order to find out who actually built the first prototype of each of these innovations. The stable finding in numerous industries such as petroleum processing, pultrusion, scientific instruments, windsurfing, skateboarding, and many others is that many of the most important innovations were originally developed by users.[2]

Of course, one immediately wonders why the significance of users as innovators has been underestimated for such a long time. Why is the idea of user innovators counterintuitive to most of us? Why is the first reaction of many students and managers straight disbelief? Why did older textbooks of innovation management often ignore this source of innovations completely or, in other cases, treat it as a somehow "exotic" exception? Typical quotes are "Customers should not be trusted to come up with solutions; they aren't experts or informed enough for that part of the innovation process," or "The truth is, customers don't know what they want. They never have. They never will. The wretches don't even know what they don't want." The reason is both a neglect of user innovation in most public statistics and innovation surveys and a general perceptual bias resulting from communication patterns. Producer innovations are being advertised and marketed to as many potential users as possible—after all, the producer wants to sell them. User innovations by contrast are being developed for the users' own use, thus often not very many other people get in contact with them. Users simply have a lower incentive of popularizing their achievements among the mass of users. Also in the frequent case of a producing firm picking up a user innovation and commercializing it, the producer has no interest in revealing the original source to the public. Claiming authorship might also appear subjectively justified as producers usually further develop and design user innovations before introducing them to the mass.

Another question is whether there is a trend, as Baldwin and von Hippel[5] suggest. Is the significance of user innovation increasing? There is no longitudinal study on this and any such analysis would be difficult to attain. However, a number of arguments make it plausible that the frequency and importance of user innovation has been increasing dramatically in the past years.

Foremost, the Internet and social network media have enabled individuals to exchange information much easier than in times where contacts were more or less restricted to friends, relatives, colleagues in work environments, and neighbors. Geographical and social impediments decreased, and users now have easy access to like-minded people around the globe. For individual innovators this means that it is quite easy to complement their creativity, their knowledge, and their technical capabilities, and thereby establish a critical mass often necessary for major innovative developments. Good examples of user innovations enabled by the Internet are open source software projects such as Linux, Apache, or Firefox, and digital products based on user-generated content such as Facebook, Wikipedia, or YouTube.

Developments in information technology also resulted in many tools that assist and support the individual user to actively convert an idea into a product. Personal computers and both general purpose and specialized software for writing texts, doing calculations, creating designs, and assembling machinery are cheap and easy to handle today and thus greatly decrease the costs of user innovation. For example, only a few years ago, creating a pop song recording required not only musical talent but also access to an extremely expensive studio, a professional producer, and much money for the physical manufacturing. Today it can be done with a PC at home and can be distributed via Internet.

A final question is whether any user is or can be an innovator or if there are specific patterns. The answer also bridges the apparently opposing positions by Ford and von Hippel mentioned earlier. Yes, many users are hardly able to provide information about future needs or even to develop their own ideas, concepts, and solutions to match those needs. But there is a specific subgroup of users who are indeed very creative and innovative. We term this subgroup *lead users*. Lead users are able to provide direct input in new product development tasks and have often prototyped new product solutions.

Lead Users

Who Are Lead Users?

Lead users are defined as members of a user population who display two key characteristics: First, they have strong needs that are not met by existing market offers. Second, they are at the leading edge of important trends in a given marketplace (see some well-known examples of lead users later).[6]

The first "high expected benefits" component of the lead user definition was derived from research on the economics of innovation.[7] Studies of industrial product and process innovations have shown that the greater the benefit an individual expects to obtain from an innovation, the greater that entity's investment in obtaining a solution will be. The benefits a user expects can be higher than those expected by a producer, for example, if the market is new and uncertain, if customer preferences are heterogeneous and change quickly in the market, or if the costs of innovation are lower for users than for manufacturers due to the "stickiness" of preference information. The first component of the lead user definition was therefore intended to serve as an indicator of innovation likelihood.

The second component of the lead user definition, namely being "ahead of an important marketplace trend," was included because of its expected impact on the commercial attractiveness of innovations developed by users residing at that location in a marketplace. Studies on the diffusion of innovations regularly show that some customers adopt innovations before others. Classic research on problem solving reveals that subjects are heavily constrained by their real-world experience through an effect known as "functional fixedness."[2] Those who use an object or see it used in a familiar way find it difficult to conceive of novel uses. Taken in combination, these findings led to the hypothesis that users at the leading edge would be best positioned to understand what many others will need later. After all, their present day reality represents aspects of the future from the viewpoint of those with mainstream market needs. The second component of the lead user definition therefore indicates the commercial attractiveness of an innovation created by such a user. If both components meet, there is a high likelihood that this individual will generate innovations of high commercial attractiveness.

Some Well-Known Examples of Lead Users

- *Jacques-Yves Cousteau* had a strong interest in sea exploration and made underwater films during WWII. At that time, divers could stay underwater only for a very limited time. Cousteau thus had a strong need for an innovation—for an aqualung, a product and a market that simply did not exist. When he met Émile Gagnan, they both devoted their creativity to solving this problem. In 1943 they came up with a device that was incredibly safe, reliable, and easy to use—and still is the basis for diving equipment today.

- *John Heysham Gibbon* was a vascular surgeon who could not stand seeing young patients suffering from rheumatic fever (that affected their heart valves) die during surgery. At that time, it was unsure whether the underlying technological problem could be solved and whether there would be a market at all. So Gibbon invented the first heart-lung machine himself. In 1953, he first used his machine on a human patient.

- *Tim Berners-Lee* is a British computer scientist who worked at CERN in Geneva. For his computer networking work, he dearly needed instant access to several sources of information and thus he joined hypertext with the Internet—and in 1991 the World Wide Web with websites and homepages as we know it today was born.

Are Lead Users a Distinct "Species"?

In many publications, lead users are treated as a specific population or "species." In essence, this implies a binary concept: individuals either are lead users or they are not, similar to individuals who are left-handed, firms that are listed in the Fortune 500, or scholars who have tenure. However, this is a simplification for two reasons.

First, the lead user construct is a characteristic that is distributed over a continuum, such as intelligence, creativity, or technical skill. There is no natural borderline that objectively distinguishes "lead users" from "non-lead users." Several empirical assessments reveal that the distribution of lead userness follows a normal bell-shaped distribution and is not even bimodal. For matters of stringency, it is useful to refer to lead users; however, it would be more precise to talk about "individuals who display high levels of lead user characteristics." It is important to keep in mind that the classification of a user as a lead user is necessarily subjective and arbitrary to some degree.

Second, an individual's lead userness is not a general characteristic. It always refers only to a specific need and trend. Thus an individual can be a lead user with regard to safety equipment, for example, and have an extremely low level of lead userness with regard to most other trends (such as gardening, high-speed processing of large data, heart surgery, etc.) at the same time. This means that lead users must be searched separately for every different trend and search field.

What Do Lead Users Do?

The most important and most prominent feature of lead users is, of course, that they may innovate; that is, they may develop totally new products and services. Recent studies also found that, beyond this, lead users are also more open to new products generated by others (be it firms or other users) and buy them earlier and more frequently than the bulk of the market.[8] They also influence many other potential buyers' purchase decisions. They are opinion leaders. This suggests that the lead user concept constitutes a valuable approach in other phases of the innovation process as well, such as new product forecasting, product and concept testing, product design, and the diffusion of innovations.

The Lead User Method

The lead user method is a managerial tool originally proposed by Urban and von Hippel[9] that allows companies to benefit from the creative potential of lead users. It is situated in the early stages of new product development. This so-called "fuzzy front end" is very important, as success and failure are largely determined here. A concept that is not responsive to (future) market demand will hardly be successful, no matter how meticulous the further process of development, market tests, and market introduction will be carried out. In the early stages, also, 75 to 85 percent of the total costs of the innovation project are already determined—which is another imperative of paying particular interest during this stage. Nevertheless, it is surprising how little attention the fuzzy front end receives in many firms. Among the methods used in these early stages, the lead user method is a particularly effective tool (see the box: A Performance Assessment of the Lead User Method). It is used by many innovation leaders worldwide (see Firms Using the Lead User Method in the box).[10]

A Performance Assessment of the Lead User Method

- In a natural experiment at 3M, the performance aspects of alternative methods of idea generation were compared.
- Researchers studied the outcome of 47 real innovation projects; 5 of them were based on the lead user method, and 42 of them used more traditional methods such as focus groups, etc.
- Doing so, they controlled for many alternative explanations.
- The impressive finding is that on average, the lead user method resulted in eight times higher commercial success (US$146 million vs. US$18 million projected sales) and provided the basis of a major new product line in all five cases—while this was only the case for one out of 42 non-lead user projects. Differences are highly significant.[11]

Firms Using the Lead User Method

- Many well-known international innovation leaders use the lead user method, such as 3M, Airbus, Hilti, Johnson & Johnson, Johnson Controls, Kapsch, Magna, OMV, Palfinger, Schindler, and Siemens.
- Also, among the most innovative SMEs, the lead user method is a standard method. An analysis of the companies that are granted the top 100 award of most innovative SMEs in Germany (www.top100.de) reveals that 82 percent of them regularly use the lead user method.

Of course, the lead user is not the only method that might help in this stage and it is not a "silver bullet" that will solve any innovation problem by any firm. There are other methods for harnessing user creativity and innovativeness, such as crowdsourcing and toolkits for user innovation and design as described later in this chapter. The investment in a lead user project is justified only if there is a probability of higher returns than by alternative methods (see the box below).

Checklist: Does a Lead User Project Make Sense?

- Is there a need for radically new product ideas and concepts in the company?
- Is it worth investing in this "fuzzy front end"—or does an improved input of ideas and concepts make no difference in your company or industry?
- Are user innovations possible? Is there "room for innovation" at all?
- Is the lead user method more effective and efficient than alternative methods, such as crowdsourcing or toolkits for user innovation and design?
- Does the company culture embrace necessary values such as openness to external ideas and the willingness to discard familiar and long-held beliefs and traditions? Or is there a strong "not invented here" attitude?
- Are there prohibitive problems with secrecy or social desirability?

The lead user method can be described in four phases.[12]

Phase 1: Getting Started

In the beginning, it is important to clarify the objectives and constraints of the project (see the box Checklist: How to Organize the Starting Phase of a Lead User Project). Typically, these objectives can be more focused or broad. Examples for the former are, for example, "finding an innovative solution to problem X" or "identifying an innovative product concept for the market segment Y." The advantage of narrowing the scope of the project is that this way, the project usually becomes much more effective. Of course this implies a decision against possible innovations. But with a narrow scope, subsequent phases can be managed much more easily and the danger of getting lost in a totally diluted project is at least reduced. One must keep in mind that most projects that start with a focused objective turn out to be much more complex and multifaceted than one expected in the beginning. The need to refocus is a constant issue in the later phases of the lead user method. A broad focus (identify a totally new product field) will most likely not lead to a concrete and usable product concept in a short time frame. On the other hand, it may provide fruitful insights into the further development of the market.

When staffing the project, it is of paramount importance to include company employees from different functions and departments. Of course the required competences will vary from project to project. In most applications, the functions of R&D, marketing, and production will matter most. Including a cross-functional team is important to ensure that solutions found are sufficient to fit with regard to (general and marketing) strategy, R&D, and production capabilities and objectives. Also, broad anchorage reduces the risk of "not invented here" problems arising from the fact that solutions external to the company are being sought. Usually, a team consists of three to five people who are able to devote enough time to the project.

Checklist: How to Organize the Starting Phase of a Lead User Project

- Define the search field clearly—complexity and vagueness will reduce efficiency considerably.
- Collect and prepare existing company knowledge: what do you already know about the problem? Which ideas, complaints, etc., are known? There is no use reinventing the wheel.
- Did you make a deliberate decision for either a focused problem definition (with the advantage of much more efficiency) or a wide problem definition (with the advantage of not excluding a potentially interesting scope from the outset)?
- Did you define constraints of the project (budget, time, production, corporate strategy, supply chain, etc.)?
- Did you include three to five company employees from all relevant functional areas (such as marketing, R&D, and production)?
- Did you ensure top management support (in case of some "not invented here" conflicts)?

Phase 2: Identification of Major Needs and Trends

In the second phase, the most important trends are identified. Trends are those dimensions on which lead users are far ahead of the mass market. The function of the trends in the lead user method is to narrow the problem and to allow a systematic search for lead users. Trends can be based on technology (e.g., a trend toward modularization or new materials), or on market information (e.g., a trend toward increasing security demands or wireless solutions). Basically, they answer the question in which direction the target market will move.

Naturally, most managers will have an immediate intuition about the trends in one's own domain. However, in most lead user projects, it pays to devote time and effort in a separate outside search for trends (see Best Practice Example Airbus). For example, users could be interviewed regarding needs not yet satisfied and problems they experience. Often, there is information regarding customer complaints already stored, and very much can be learned from scanning discussions among target market users in online forums. In most markets, there are also experts who have insights regarding the further development of customer preferences. Technology trends can be collected scanning information from academic and trade publications and again by interviews with experts. Often, it makes sense to go beyond the target market and look for developments in related or "analogous" markets. Sometimes, trends already visible in them foreshadow developments in the target field.

Best Practice Example Airbus

One might think that the needs and trends of a certain market are well known to the companies operating in this market for decades. However, familiarity precludes recognizing changes. Often, key assumptions have not been challenged for many years. Therefore, it is typical that including customers' or users' perspectives leads to the surprising detection of problems and aspects company experts have not realized up to that point.

The objective of the second phase in a lead user project by E&I Institute of WU Vienna and Airbus Operations GmbH on innovative hygiene concepts for aircraft lavatories was to uncover unmet needs of airline passengers for the predefined search field. Since airlines operate globally, trend research was conducted on an international level. Interview partners came from 32 countries worldwide. Based on systematic in-depth research including 342 interviews (233 target market, 109 analogous market), 242 secondary sources, and 37 online blogs, the project team identified 14 relevant trends (i.e., aggregated uncovered needs).

The users and experts were selected by different criteria: diverse end-consumer segments (e.g., flying purpose and frequency, age, cultural backgrounds); experts from target markets (e.g., airline personnel); experts from analogous fields (e.g., medical, interior design, shipping), and so forth. However, the most surprising insights came from users: A lot of passengers have developed their own strategies for the use of the lavatory—from behavioral adaption to bringing their own aids and appliances. Other findings were related, for example, to cultural or perceptual issues (e.g., auditive, visual). The data collected was coded and aggregated into 14 trends that were evaluated against criteria like "fit with the search field" and "relevance to target market." The team chose the three most relevant trends—(1) intuitive use of lavatory, (2) avoidance of contamination, and (3) minimizing of contact points—for the development of innovative solution concepts. The restriction to these trends proved essential to narrow down the search field. It enabled a focused search for lead users and largely contributed to the overall success of the project.

"It is worthwhile to invest considerable effort into complementing existing company knowledge regarding customer needs and trends. Company representatives regularly report that benefits go far beyond the lead user project. Some of the identified needs can be served right away with only little adaption of existing solutions—however huge effects for the users—and some insights offer promising starting points to continue working on in separate projects," says Ilse Klanner, project leader of the Airbus Operations GmbH lead user project.

Typically, the external trend search will also provide new and unexpected information beyond the trends. It is advisable to store valuable information such as existing ideas, problems, solutions, potential candidates for lead users, or starting points for their identification.

The selection of the trends to further pursue constitutes an important decision. Typically, many more important trends are identified than can be further pursued. Again the conflict between focus and breadth becomes visible. It is quite typical that managers are reluctant to drop most of the trends found. However, it is necessary to keep in mind that trends have a specific function in the lead user method: they allow for searching individuals lead them. If one would keep 30 trends and wanted to include 3 users per trend in order to ensure some cross-fertilization of ideas and knowledge, the result would be 90 lead users. It is clear from the outset that it is impossible to handle such a number in a single lead user workshop.

In most successful applications, three to five trends are followed, not more (see Checklist: Identifying Needs and Trends within a Lead User Project).

Checklist: Identifying Needs and Trends within a Lead User Project

- Narrowing the selection of trends will inevitably reduce the solution space. Make sure that you do not pick the wrong trends or save effort in a shortsighted way.
- Even if you think that you already know what the most important trends are, consider at least some additional external information.
- In most cases, market trends (aggregated customer needs) are most important. But do not forget technology trends.
- Make sure that you also analyze related or analogous markets. Only looking at the target market can be quite myopic.
- Ensure that other interesting information gathered during the interviews is stored.
- Usually, you will find many important trends. Ending with over 30 trends is not a rare figure. Select the most important three to five trends—not more.

Phase 3: Identification of Lead Users

The third phase involves a broad search for individuals who are far ahead with regard to the trends identified and have high personal benefits from innovations.

In earlier applications of the lead user method, usually a mass screening approach was employed—probably because getting large representative samples is a typical (and often valuable) goal in market research. Screening means that a large sample of users, often from customer databases, is systematically filtered in order to identify those users who score highest in both lead user dimensions. The advantage of this approach is that all members from the given population are analyzed. One does not miss a less connected individual within the population. As the search is done in a single wave, time is also saved. On the other hand, the search space is restricted from the outset; only those users who are contained in the original population have a chance of being selected. Individuals outside cannot be found. Also, there is no opportunity of learning from interview to interview. This is one of the three advantages of the alternative search method of "pyramiding."

Pyramiding is a search technique in which the searcher simply asks an individual (the starting point) to identify one or more others who he or she thinks has higher levels of expertise of the sought-after attribute—or better information regarding who such people might be. The researcher then asks the same question of the persons so identified, and continues the process until individuals with the desired high levels of the attribute have been found. Usually, such "search chains" are not very long. Most achieve their maximum after two or three steps. Starting with the right persons (probably identified during phase 2) shortens the length of the chains and increases the probability of success. An advantage of pyramiding (relative to screening) is that the interviews are opportunities to learn. Talking to knowledgeable users and experts offers many important insights. Sometimes this leads to further refining the trends selected, sometimes interesting ideas are found, or in the extreme, leading to additional projects or a refocus of the study. Recall that the interview partners are already trend leaders. A second and very major advantage is that it is an open search technique. Unlike screening it is not restricted to a prespecified population. Users might easily refer the researcher into totally different domains and analogous markets. A third advantage is that typically, pyramiding is more efficient in terms of monetary investments (see A Study on the Efficiency of Pyramiding versus Screening Search Techniques).

A Study on the Efficiency of Pyramiding versus Screening Search Techniques

- In four groups of individuals (total $n = 147$), 18 search topics (e.g., who in your group do you think has climbed the highest mountain?) were introduced. In each setting, people were asked about their own characteristics regarding the topics and to whom they would refer had they been asked this in a pyramiding study.
- On this basis, 1.8 million pyramiding searches were simulated in a Monte Carlo simulation and compared with the efficiency of screening.
- The impressive finding is that pyramiding offers an efficiency gain of up to 1000 percent compared to the efforts of screening.[13]

Typically, screening and pyramiding techniques are complemented by systematic searches within online communities. Lead users do not know everything. Generally, an individual may well generate an idea, but developing the idea into a functioning prototype often requires diverse and specific knowledge a single individual is unlikely to possess. As a result, lead users often join (or start) communities in order to complement their capabilities. Traditionally, these have been off-line communities of practice. In the era of the Internet and social software, online communities have gained much importance as they allow to exchange and access knowledge globally. Today, there are expert networks or other forms of virtual institutions on basically every topic. For a firm seeking lead users, the easy accessibility of these online communities is a major advantage. Given that the online community is closely related to the trend identified in the former phase, there are different possibilities of how to search.

For example, if the forum is particularly large, one possibility is to analyze the social network position of its members and focus on those who display particularly high levels of so-called "betweenness centrality." This network position has been shown to be an effective shortcut for the identification of lead users (see the box: A Study on the Social Network Position of Lead Users). A more elaborate alternative is to post questions regarding the searched solutions in the forums and follow up on those users who provide (helpful) answers. Of course, it is vital to study norms and behavior in the forum before posting questions—otherwise one might easily get "flamed." In smaller forums, content analyses have also proven helpful.

A Study on the Social Network Position of Lead Users

- In three studies (survey, network data, and field data from lead user studies) comprising over 2000 participants, the social network position of lead users was measured. The stable finding is that lead users have a particularly high so-called "betweenness centrality." This means that they act as boundary spanners. They bridge between groups of users that otherwise would be unconnected.

- In a fourth study, the instrumentality of this finding was tested. As information on betweenness centrality of users can be retrieved relatively easily with automated software, it might be helpful for the identification of lead users. For this purpose, 431,257 posts from all 13,287 members of a specific forum were analyzed. The lead user candidates identified this way were then compared with randomly drawn "average" users from the forum. While the latter have a lead userness of only 0.01, the individuals with particularly high betweenness centrality—the lead user candidates—exhibit an impressive mean score of 4.98, which equals a gain of over 40,000 percent. Obviously, analyzing the social network position can be a helpful shortcut for the identification of lead users within large online communities.[14]

Another question is where to search. Again, there, some development has to be noted. In earlier lead user studies, the focus of search was exclusively on users from the target market. Almost automatically, managers limited the search to those experts that have the most knowledge about the market in question. However, it was found that users from analogous markets are particularly valuable (see the box: A Study on the Efficiency of Sampling Users from Analogous Markets). Analogous markets are markets that are different from the target market but characterized by the same underlying trends. Consider the example of a lead user study which aims to find methods of preventing infections in clinical surgery. For this purpose, one important trend was "methods for increased air purity." But outside of leading hospitals, experts from the analogous field of chip production or CD production are also able to provide valuable creative input—their ideas might even be more creative. There are two reasons why it might make sense to ask such people: First, users from analogous fields might possess solution-related knowledge that is worth transferring from the analogous field to the target field. It might easily be that the analogous market is more advanced. In this case, the (less advanced) target market can, of course, profit from importing superior solutions, solution principles, and more experienced, qualified, or creative solution providers. Higher technological advancement can be assumed if the underlying problem is more extreme in the analogous market, if its consequences are more severe, or if firms and users have been looking for solutions with greater intensity for other reasons. Sheer volume (i.e., the number of firms and users) or the maturity of the market might also contribute to such advancement. For example, a firm in the inline skating market looking for ideas for protective gear might investigate the stunt industry or jackass-style shows (because the protection problem is more extreme there), protection gear for people with brittle bone disease (because the consequences are more grave for those suffering from this affliction), or the market for safety belts and other protection systems in cars (because the market is far larger).

The second factor is the greater creativity of people due to their analogous market origin. Research on problem solving clearly shows that individuals are often quite heavily constrained by their past experience. The "functional fixedness" effect implies that people unconsciously stick to mental schemes and problem-solving strategies that have proven helpful in the past, which impedes them from coming up with truly novel solutions. Lead users from analogous markets might have a fresh and unbiased perspective and therefore come up with novel ideas and original solutions to which the target market experts are "blind."

A Study on the Efficiency of Sampling Users from Analogous Markets

- In an experiment, 213 problem solvers from the three analogous markets of roofing, carpentry, and inline skating were sampled. The three markets share the analogous need to use safety gear to prevent serious injuries and to make improvements in the comfort of such gear; but the producers, products, and use contexts of these markets differ vastly.

- Each participant was asked about (1) solution ideas for his or her own problem field and (2) solution ideas for the two analogous problem fields (in random order). Thus, there was a total of 639 new product ideas and no bias between the markets was possible.

- The ideas were then evaluated by independent market experts blind to the origin of problem solvers and blind to the specific research questions underlying this project.

- The finding is that individuals from analogous markets yield clearly more innovative and original ideas than users from target markets. Calling in problem solvers from an analogous market (instead of the target market) increases the novelty of solutions provided for a given target market problem by 63 percent of the gains from asking lead users instead of average problem solvers.[15]

Of course it would not be advisable to only focus on lead users from analogous markets. Target market users also have some clear advantages. Primarily, they know the very problem quite well and might already have developed solutions to exactly that problem (which is usually not the case for users from analogous markets). Problem familiarity is quite valuable and difficult to achieve. Especially when problems are novel, complex, and ill structured, it is very challenging to fully shift problem information from one locus to another. The articulation and definition of a problem may not compensate for an insufficient understanding of its very nature, solution constraints, problematic side effects, important contextual factors, and so forth. Thus, it is a good idea to combine the strengths and compensate for the respective weaknesses of target and analogous market users.

How can one recognize a lead user? The bad news is that there are no objective criteria or clear decision rules. It is a matter of subjective assessment (see Best Practice Example Johnson Controls). The good news is that there are some robust indicators. The expected benefit component can be assessed along indicators such as the (estimated) degree of use benefit in the target market, the degree of use benefit when transferring ideas back to solve analogous market's problems, or (in some cases) by the benefit from selling the innovation. The trend position can be assessed by the activities of the person in question concerning the trend (relative to the state of the art in the field), the value of ideas of the person in question concerning the problem stated (again relative to the state of the art in the field), and of course by the number and quality of referrals to the person in question. Probably the best indicators are solutions they already have developed. It makes sense to include some users who lean more toward the trend position and have a lower personal benefit from an innovation (experts such as scholars) and also consider including some "tinkerers" who have a high need for innovations but do not necessarily lead the trend. Naturally, selected individuals must also be open, creative, willing to work jointly in a team with other lead users, possess sufficient verbal skills, and be motivated to join the workshop and willing to transfer all intellectual property generated during the lead user workshop to the firm (see Checklist: Identifying Lead Users).

Best Practice Example Johnson Controls

In order to find qualified problem solvers for your search field, you can apply certain identification approaches. The overall goal is to identify lead users who have the ability to solve the problem and also derive a benefit from solving it within a joint workshop.

In a particularly successful lead user project, the objective of the automotive supplier, Johnson Controls International, was to develop innovative solutions for the mechanical length-adjustment of car seats. After an excessive trend search, the project team had agreed upon the three trends: (1) need for efficient materials and alternative mechanisms, (2) standardization of adjustment mechanism, and (3) intuitive operability. In order to find lead users, more than 600 pyramiding interviews were conducted and 280 user group discussions were investigated. Finally, 43 well-suited lead users for solving the problem were identified.

A typical pattern is that lead users really stand out. For example, in two independent pyramiding interviews with professors from leading technology universities (TU Munich and TU Darmstadt), we were pointed to a specific engineer. In the parallel broadcast search in an online forum, he came up with innovative solutions to the problem himself. When we followed up on his suggestion, it was soon found that this single person had solved more than 1200 technical problems in this user group within just 1 year. He seemed to be a "prototypical" lead user.

The identification of lead users is the core part of the lead user method. "It is not only important to talk about possible solutions to the problem of a partner company. Beyond this, it is crucial to establish a positive climate in a number of personal or phone conversations. For example, we regularly update our interview partners about interesting solution concepts of other potential lead users. We then validate the solution quality of other people's ideas and show our interview partners that we appreciate their opinion," says Susanne Roiser, project leader, Johnson Controls International lead user project.

It is important to get a feeling for each identified problem solver and compile detailed lead user profiles to facilitate the final decision of who will make up the specific lead user workshop mix. When we made the decision of which of the 43 lead users we should invite into the workshop, we applied the criteria of innovativeness, practical experience, theoretical knowledge, and fitness for the workshop. These criteria had been rated by the interviewers on a numerical scale. We finally invited nine lead users to the lead user workshop. The lead user mix was quite diverse regarding the fields of expertise, ranging from aerospace, lightweight construction, special engineering or electrical metrology to industrial design, or hydraulic engineering.

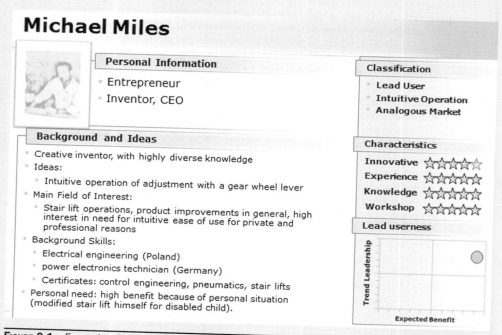

FIGURE 9-1   Example of a lead user profile.

Checklist: Identifying Lead Users

- Use pyramiding as the dominant search technique. Complement it with broadcasting. Screening is too inefficient.
- The choice of the right starting points shortens the pyramiding chain. Determine starting points by asking for whom is knowledge along the trend(s) critical?
- Standardized interview forms and interviewer trainings support the search process.
- Track whom you asked—people get upset if they are contacted twice.
- Ask for users from analogous markets.
- By all means, actively search for lead users from analogous markets. They are often less functionally fixed than target market lead users. Often, existing solutions in analogous markets are the basis for truly novel product concepts in your market.
- Keep in mind: Analogous markets need to be linked to the target market (via the trend identified in phase 2).
- Ideally, mix users from target markets and from analogous markets.
- Good lead users have technological knowledge, market knowledge, and user experience— ideally all at once.
- Prior ideas and developments are a very good indicator for the candidates' lead userness.
- Do not forget the social side: lead users selected must be communicative and work in a group.
- Ideally, the composition of the group balances extremes—heterogeneity works well.
- Try to find two to four lead users per trend identified.
- Make clear from the outset: IP will be transferred to the firm.

Phase 4: The Lead User Workshop

Once the lead users are identified, they are invited to a 2- or 3-day workshop in which company members from different functional areas also participate. Ideally, they are aware of the prior phases and their output.

It is important for the company to address the issue of intellectual property rights prior to the workshop to ensure that the ideas and concepts generated can be commercialized without the risk of legal infringements. Often, this is unproblematic as in many cases it is economically profitable for users to reveal their innovations freely to the firm.[16] One reason is that they expect to profit from the use of the resulting product. Also, it is often clear that there is a long way from an idea or a concept to a product—the users might simply not hold the resources to exploit the idea themselves. In many situations, the users might develop expectations regarding future cooperation or possible jobs. The most important set of motives for why lead users join lead user workshops, however, is psychologically rooted. It is great fun to join a team of other high-end lead users, get in contact with like-minded people who really push the envelope. Also getting recognition by an admired firm often plays an important role. Companies should be aware of these motives and behave fairly and transparently. If lead users sense that they are being exploited or treated unfairly, this will lead to considerable problems. Lead users might not attend the workshop or might refrain from actively contributing their insights and ideas (see the box: A Study on why Individuals Participate in Firm Innovation—Or Refuse to Do So).

A Study on Why Individuals Participate in Firm Innovation—Or Refuse to Do So

- In two experiments, 893 potential participants were asked to participate in an innovation project by a firm.
- Participation conditions were manipulated regarding (1) value distribution with regard to monetary profits, (2) reputation shared, (3) IP ownership, (4) transparency in the process, and (5) mode of profit sharing. Fairness perceptions and willingness to participate were measured afterward.
- The finding is that users not only wanted a *good* deal—they wanted a *fair* deal. Fairness perceptions with regard to the distribution of value between the firm and contributors (distributive fairness) and the fairness of the procedures leading to this distribution (procedural fairness) impacted the likelihood of participation even more than considerations of self-interest. It is important to note that they not only impacted the individuals' transaction-specific reactions, but also inform their future identification with the firm.[17]

The workshops usually take place in nice, remote hotels that facilitate undisturbed and focused work (see Checklist: Organizing a Lead User Workshop). As lead users might come from various countries worldwide, a close proximity to an airport is advisable. During these workshops, techniques such as brainstorming, group discussions, and others are used to capitalize on the creativity of the participants. Typically, it makes sense to alternate small group work with presentations and discussions with all participants. The small groups consist of three to five lead users within one trend. They should be accompanied by a firm employee in order to avoid "ivory tower" solutions. It is important that the employee (or someone else) takes over the role of a moderator in order to ensure that there is some structure and output orientation and not just "wheel spinning." On the other hand, too much conservatism and excessive constraints are also critical to avoid. After all, the objective is breakthrough—not incremental innovation. During the workshop, there should be some development from loose and totally open brainstorming of ideas to conceptualizing and refining the most promising ones. The great advantage of a lead user workshop is that there are individuals who are able to go beyond mere idea generation. They should be able to develop concrete concepts during the workshop (How does it function? How can it be developed, what are critical development issues? How could it be produced? Are there specific marketing issues? Who could be partners? What are next steps?). The output of a lead user workshop usually is a small set of three to five well-developed and particularly attractive new product concepts. In the next step, they are evaluated and further developed by the firm.

Checklist: Organizing a Lead User Workshop

- Choose a nice and remote hotel in close proximity to infrastructure. Do not use your firm facilities.

- Two to three days are neither too short nor too long.

- Ensure that IP is owned by the company.

- Consider participation motivations and fairness issues, and address them. Work on the sensation of a common mission. Establish norms of openness, sharing of ideas, creativity, and constructive critique.

- Alternate small group work with plenary presentations and discussions. Establish structure and time management.

- Progress from brainstorming of many ideas to elaborating the few most promising new product concepts.

- When the project is over, it is not yet over. Build a relationship with lead users. They might be beneficial in many ways.

The lead user project is not over after phase 4. Recall that for many lead users, recognition by the firm has been a major motivator. It is thus important to provide them with information about the further development of the concepts. Also, their contributions, knowledge, and creativity might be beneficial in other projects. Farsighted companies build a relationship with lead users, and do not treat the project as a temporary transaction only.

Best Practice Example Kapsch

Kapsch TrafficCom carried out a lead user project in order to generate new and innovative solutions to decarbonize transportation of goods and people in smart cities. After an extensive trend search and much effort to identify both target market and analogous market lead users, a series of three workshops was organized in which nine highly promising concepts were generated.

Out of those nine concepts, one concept stood out especially in terms of innovativeness. Investigating the development process of this concept, it became evident that it strongly profited from the knowledge of a participant from an analogous market. This lead user is a biologist and researches in the field of bionics, where he looks into the mechanisms behind complex systems in nature. In the workshop, he used that knowledge to inspire new approaches in the field of transportation. The other participants that were rooted in the target market helped and supported him to translate his highly original ideas into existing approaches and technologies in the transportation field. This "cocreation" process finally led to the generation of this highly innovative solution.

"From the experience in organizing lead user workshops it is very much about finding the right mix of people from analogous markets as well as from the target market. Of course a "common denominator" helps a lot. Also, using experience and facing similar problems help facilitate exchanges between participants. This was the case within the group concerning the lead user project with Kapsch that led to the concept with the highest innovative potential," says Stefan Perkmann Berger, project leader, Kapsch TrafficCom lead user project.

Variants of and Alternatives to the Lead User Method

A "full scale" lead user project as described earlier takes considerable time. Often, 4 to 8 months are being reported, sometimes even longer. It is self-evident, also, that considerable costs are involved. If external consulting during the project is being sought, typical costs are between €50 K to €150 K. Is this a sound investment? Of course there is no *a priori* answer—innovation projects are risky by definition. If we consider figures from the performance assessment study (see the box: A Performance Assessment of the Lead User Method), it appears that the returns possible are enormous, however.

On the other hand, SMEs often shrink back from such great investments. Their innovation budget is simply too small and needs to be spread over more projects and the whole development process. Thus, are lead user projects restricted only to the big players and multinationals? The answer is, no. There are many ways of conducting a "lean" and particularly cost-effective lead user study without forfeiting all its benefits. In all phases, there is room for savings. In all such attempts, it is important to keep the core principles of the lead user method in mind (see the box: The Core Principles of the Lead User Method). For example, a lead user study might still provide immensely helpful insights if a firm shortens the phase of trend search (probably because there is already good knowledge about the field from another study) and narrows the lead user search by a local instead of a global search. On the other hand, reducing the lead user method to just some workshops with arbitrarily chosen average customers (and thus eliminating phases 1 to 3) will likely not result in breakthrough ideas and concepts.

The Core Principles of the Lead User Method

- Do not assume the firm knows everything. Accessing external knowledge is paramount.
- Focus on the most important trends in your field.
- Look beyond your market and industry boundaries. Analogous markets are extremely valuable.
- Do not involve average users if you want outstanding ideas and concepts. Go for lead users.
- Listen and learn from them.

Of course one should also consider alternative methods. There are other methods that build on the insight that there is much valuable creativity outside the firm boundaries. Two of them are described in the following.

Crowdsourcing

Crowdsourcing means innovation-related tasks like new product ideation is outsourced to anonymous crowds outside the company. Firms issue open online calls for new ideas and solutions, and they offer rewards to those who submit the best ones. Participants are often quite large numbers of heterogeneous, self-selected, and voluntary individuals who engage in temporary, decentralized problem-solving activities for the firm. Well-known examples are Innocentive.com in the high-tech field and Threadless.com in consumer goods (in this case T-shirts).

The advantage of crowdsourcing as compared to a lead user study is that many more potential problem solvers can be active. There are examples of crowdsourcing tournaments where tens of thousands of ideas were submitted. Instead of a focused search process by the organizing firm, there is self-selection as a recruiting principle. Ideally, the many participants contain those few whose ideas really make the difference. Another advantage is that the online platforms necessary for crowdsourcing can be used across many tournaments, which makes crowdsourcing relatively cost-efficient.

Crowdsourcing is particularly helpful if and only if the underlying problem to which a solution is being sought can be formulated very clearly and briefly—after all, solution attempts are made in a decentralized way and without guidance by the manufacturer. Complex problems that need integrated solutions adapted to the specific situation of the firm and interaction during the solution process are difficult to externalize to crowds. In such situations, the lead user method appears more appropriate.

Crowdsourcing is also more efficient than a lead user study when creativity plays a more prominent role than specific knowledge. In such instances, the sheer number of users and the "wisdom of the crowds" will most likely outperform a few carefully selected lead users. Consider, as an example, the objective to find a new name for an innovative product. Brainstorming about names is a task that can be done basically by anyone. Granted, there are huge differences between individuals. But if 1000 motivated average users devote all of their creativity to this task and each comes up with a proposal, chances are quite high that there are some "lucky strikes" among them. Compared to them, 10 lead users will definitely be more creative on average, but it is unlikely that their best idea (out of the 10) is superior to the best idea of the crowd.

Toolkits for User Innovation and Design

A toolkit for user innovation and design is a coordinated set of design tools that allows individual users to self-design their own product according to their individual preferences and give visual and informational feedback on (virtual) interim solutions. If the customers like what they design, they can order their products. The toolkit provider in turn will produce them according to their individual design specifications.

They share two common principles: First, they all contain some form of design tools that enable the user to create and modify a design. Some come in a quite restricted form of lists to choose from, such as the preferred material for a watch strap (e.g., metal, ostrich leather, etc.) for self-designing a watch, seat types possible when self-designing a car, or predesigned doorknobs when self-designing a cupboard. There are also toolkits that allow users to combine product components modular-wise, similar to building an artifact with Lego bricks. Still others allow free design, like a graphic computer program. Some are very user-friendly and can be used intuitively and without further ado; others require some sort of specific training and learning. Depending on the product and the toolkit, the design tools allow users to determine different aspects of the product. They comprise functional aspects of the product (like material, size, shape, or functions included); the product's aesthetics (such as color, graphics incorporated, or other forms of style); and the possibility of personalization (e.g., by adding one's name or other symbols).

The second principle toolkits have in common is that they give some sort of feedback information during the design process. The most common form of feedback is a virtual, simulated visual representation of the current product design that is updated with every design change the user makes. It informs about how the product would look. Some toolkits provide additional context information; for example, the customer can upload a picture of himself in order to make it easier to evaluate if he actually would look good in his self-designed T-shirt or not. If the toolkit allows for functional product manipulation, feedback, of course, should also be functional. In sum, a good toolkit provides the user with information about the anticipated consequences of design activities, like a capable salesperson would do. This enables the user to conduct what is called trial-and-error learning.

Toolkits enable users who are less capable of the technical side and the "building" of the product to come up with individual products. If there is great heterogeneity of preferences among customers, it might be a better idea to outsource the task of specifying the individual product to the individual customer. Any standard product—be it generated based on the lead user method or another method—will be disadvantageous to them. In such instances, toolkits are superior to the lead user method. On the other hand, toolkits restrict the solution space. The typical outcomes are design variants, individualized products, new combinations, and sometimes incremental innovations. Breakthrough innovations can hardly be generated using toolkits. In some instances, the best individual products can be marketed as a new standard product. In a way, toolkits and crowdsourcing overlap in such a case.

Two Alternatives to the Lead User Method

Crowdsourcing: Open online calls for solutions to problems, rewards only for the best.

- Works better than the lead user method if the underlying problem can be formulated very clearly and briefly, and independent problem solving is possible.
- Works better if the wisdom of the crowds can be used effectively; for example, when creativity matters.[18]

Toolkits for user innovation and design: Design tools that allow individual users to obtain solutions to their specific problems—produced by the firm.

- Works better if customers have heterogeneous and quickly changing preferences.
- Works better if the objective is individualization and customization.[19]

References

1. Von Hippel, E. Address on the occasion of the awarding of his Honorary Doctoral Degree from the Technische Universität Hamburg-Harburg, April 2013.
2. Von Hippel, E. *Democratizing Innovation*. MIT Press, 2005.
3. Von Hippel, E., de Jong, J.P.J., and Flowers, S. Comparing Business and Household Sector Innovation in Consumer Products: Findings from a Representative Study in the UK, SSRN Working Paper, 2010.
4. De Jong, J.P.J. and von Hippel, E. Measuring User Innovation in Dutch High-Tech SMEs: Frequency, Nature and Transfer to Producers. MIT Sloan Working Papers. MIT Sloan School of Management, 2009.
5. Baldwin, C. and von Hippel, E. "Modeling a Paradigm Shift: From Producer Innovation to User and Open Collaborative Innovation." *Organization Science,* 22 (6): 1399–1417, 2012.
6. Franke, N., von Hippel, E., and Schreier, M. "Finding Commercially Attractive User Innovations: A Test of Lead User Theory." *Journal of Product Innovation Management*, 2006.
7. Von Hippel, E. "Lead Users: A Source of Novel Product Concepts." *Management Science*, 1986.
8. Schreier, M. and Prügl, R. "Extending Lead User Theory: Antecedents and Consequences of Consumers' Lead Userness," *Journal of Product Innovation Management*, 2008.
9. Urban, G. and von Hippel, E. "Lead User Analyses for the Development of New Industrial Products." *Management Science*, 34 (5): 569–582, 1988.
10. Von Hippel, E., Thomke, S., and Sonnack, M. "Creating Breakthroughs at 3M." *Harvard Business Review*, 1999.
11. Lilien, G.L., Morrison, P.D., Searls, K., Sonnack, M., and von Hippel, E. "Performance Assessment of the Lead User Idea-Generation Process for New Product Development." *Management Science*, 48 (8), 2002.
12. Lüthje, C. and Herstatt, C. "The Lead User Method: An Outline of Empirical Findings and Issues for Future Research." *R&D Management*, 2004.
13. Von Hippel, E., Franke, N., and Prügl, R. "Pyramiding: Efficient Search for Rare Subjects." *Research Policy*, 2009.
14. Kratzer, J., Lettl, C., and Franke, N. "The Social Network Position of Lead Users." Working Paper, 2013.
15. Franke, N., Pötz, M., and Schreier, M. "Integrating Problem Solvers from Analogous Markets in New Product Ideation," *Management Science*, 2013.
16. Jeppesen, L. and Laursen, K. "The Role of Lead Users in Knowledge Sharing." *Research Policy*, 2009.
17. Franke, N., Keinz, P., and Klausberger, K. "Does This Sound Like a Fair Deal?: Antecedents and Consequences of Fairness Expectations in the Individual's Decision to Participate in Firm Innovation." *Organization Science*, 2013.
18. Afuah, A. and Tucci, T.L. "Crowdsourcing as a Solution to Distant Search." *Academy of Management Review*, 37 (3), 2012.
19. Franke, N. and Piller, F. "Value Creation by Toolkits for User Innovation and Design: The Case of the Watch Market." *Journal of Product Innovation Management*, 21 (6), 2004.

Idea Management

I will quote from the first author (Cathie Currie) in the idea management section: "Innovation is not for the timid. Innovative thinkers boldly perceive opportunities in circumstances where conventional thinkers see few prospects. Innovators trawl among bountiful sources, whereas the less innovative find themselves fishing in an empty stream."

Idea management is a concept that has only recently become popular. Reasons for this include shortened product life cycles and the need to be an innovative organization. Innovation can't happen without ideas, and ideas must be plentiful. With a bounty of ideas, just as in anything else in life, there must be a way to organize and store these ideas so that they are not overlooked or left to rot in some forgotten place. Imagine a field of grain, with thousands of bushels of wheat. Early man discovered that this wheat could be harvested with the grain, separated from the stems, and be processed and stored in any number of ways. Ideas are like fields of wheat; we can leave the ideas to nature, in which case they will fall to the ground, likely rot, and be lost forever. Or we can utilize idea management systems to harvest the ideas, separate the good ideas from the bad ideas, process, store, and send them to market. It only makes sense therefore, that the ideas of the agricultural and industrial age be used in the management of ideas.

As mentioned earlier, Cathie Currie's chapter on managing the development of innovation ideas reflects the importance of bringing significant innovation ideas from concept to reality. Taking these unique innovation ideas and turning them into profitable ventures is the mark of a great company. Additionally, taking those great ideas and changing the world goes way beyond. This chapter discusses methodologies that allow and encourage organizations to move beyond simply coming up with ideas, and actually doing something about it.

In the quality of ideas chapter, we discuss the problem of separating out high-quality ideas from poor-quality ideas. Organizations at their very nature are expected to manage resources effectively. While many organizations wish to be innovative, lots of ideas do not necessarily equate to lots of good ideas. This chapter discusses how to generate ideas and how to evaluate their quality. Finding the few viable ideas among the thousands of potential ideas is a problem that highly innovative organizations had not anticipated.

Finally, Glassman and Walton in their chapter on idea management discuss the process and methods behind evaluating ideas. Many organizations venturing into innovation underestimate the number of ideas that will be submitted when they're successful. Managing thousands of average ideas that are generated when you are successful at creating them can be as difficult as having none at all. Additionally, if employees feel like their ideas are not being considered, they will no longer participate in innovation methods. Therefore, a successful idea management system will include the ability to robustly evaluate every idea, gather evidence that the idea is viable, perform analysis of the value proposition, and select the projects that should move forward.

Although this is a short chapter, we cannot stress enough that an idea management system is critical to an organization that is serious about innovation. Just as the production of wheat contains many procedures to assure that it eventually makes it into your bread; innovations, from idea conception to final product must be managed to ensure that the highest-quality ideas make it to market.

CHAPTER 20

Managing Development of Innovation Ideas

Cathie M. Currie

Innovation is not for the timid. Innovative thinkers boldly perceive opportunities in circumstances where conventional thinkers see few prospects. Innovators trawl among bountiful sources, whereas the less innovative find themselves fishing in an empty stream.

Entrepreneurs and fast-track companies see breakthrough innovation ideas that offer opportunities to disrupt markets.[1,2] But discovering a significant innovation idea is far from innovation. Unique business ideas only become innovation when they are put into play, either as an effective internal process improvement or as a new product that benefits the public sphere.[3]

Boldly innovative companies muscle their innovative ideas into market reality. These forward-thinking companies connect perceptive employees with potential sources of innovation ideas or information, then drive innovation development projects that introduce well-designed entries into receptive markets. Skilled innovative companies are fast trackers. They appreciate how efficiency and effectiveness in their innovation development can accelerate their chance of producing a stable of winning products, services, or technology. They also are keenly aware that innovation missteps confer a high risk of producing laggard or failed products.

Yet, according to Booz & Company's 2012 annual R&D spending study, only one-quarter of the Global Innovation 1000 companies believe they were effective in generating ideas and then converting the ideas into development efforts.[4] Looking further into the data, less than half of these innovative companies rated themselves as highly effective at generating innovation ideas. Even less, barely more than a third, consider themselves effective in converting ideas into product development projects. These companies are the best of the best in innovation. How effective are lesser companies in transforming innovation ideas into valued innovations? We can generally assume that many companies operating below the Global Innovation 1000 have much less success in pushing innovation ideas along their development pathways. Notable exceptions are the small superstar entrepreneurial companies.

What Factors Limit the Acquisition and Development of Viable Innovation Ideas?

Scarcity of innovation opportunities: Some markets have matured into commoditized exchanges. However, creative or dissatisfied thinkers inside and outside of companies frequently envisage novel improvements to most products and processes, or outright devise new useful goods and services. Whenever company strategists assume the low-lying fruit have been harvested, they are likely to become less vigilant for new design opportunities. With less active scanning, they are at risk for a more astute company to assail the market with a breakthrough product.

Permeability to innovation idea sources: Information and idea seeking differs greatly among companies. Some companies resist external ideas as "not built here" with little consideration of value. Many companies forego the wide variety of potential market information by relying on traditional market research or supplier advice. Other companies seek diverse information from external sources, some of whom follow explicit policy to circumvent patents held by competing firms. Companies that ply multiple sources and methods for potential innovation opportunities gain a step up in most product, technology, and service markets.

Communication of innovation information: Employees who are tasked with perceiving market and product information vary greatly in their ability to evaluate potentially significant information, and to convey qualified information to pertinent receivers in the product development stream. Determining requirements to screen and filter the innovation value of information is often difficult to establish.

Organization of internal boundaries: Within companies, employee silos often isolate chains of command and communication which can impede the progress of a valuable idea through product development. Even differences in vocabularies and cognitive styles between workgroups can modify descriptions or change emphasis on design features in ways that undermine effective product development.

These are but a few of the inadvertent disconnects that can derail a company's innovation development, even in organizations that function competently in most other capacities. The objective of this chapter is to identify methods for avoiding disconnects and for guiding the ideas that are derived from the various internal and external sources (patents, customers, problem-solving ideation, etc.) successfully through the full cycle of development into market products.

As a first step, we need to focus on a substantial, but underappreciated, premarket effect that is often engendered by innovation efforts within an organization.

Innovation Activity Disrupts at All Levels: From Employees to Market

Successful innovation is celebrated for its disruptive effects on far-flung global markets. Business and public policy leaders hail disruptive innovation as the surefire means to accelerate national economic growth. Disruptive innovation is eagerly sought in the marketplace by investors and financial analysts due to the substantial rewards garnered by savvy investors. Yet, a major innovation creates a cascade of market reorganizations that often jeopardize less savvy investors, companies, managers, employees and their families, sometimes with earthquake intensity.

A less recognized effect is that innovation activity often disrupts and realigns business organizations, from business leaders all the way down to workgroups and individual employees. Disruptive destabilizations of markets may be a boon to those well positioned for change, but the changes can be a bust for many businesses and people in direct and related fields.[5]

Disruptive effects in the innovation developing organization are valid decisions rarely related to idea content, though some innovation ideas can introduce controversy or have other response-inducing qualities. Within-organization innovation disruption is caused by realignments of employee roles, responsibilities, and relationships that are brought about by the processes necessary to developing innovative ideas into marketing successes.

The process of innovation development requires employees to interact with idea sources, to critically perceive unique features and qualities in proposed ideas, to judge value and validity of business opportunities, to acquire rights to use ideas, and to communicate ideas in compelling ways. Teamwork is required to develop the idea into a product, and then to guide the product through marketing and sales, into customer hands. Many of these activities and abilities require autonomy and decision making that is less permissible in other tasks. Individual employees, and their coworkers, may be less comfortable with the responsibility required in self-directed skills. Those who enjoy autonomous work may be unwilling to relinquish autonomy in other areas of their work.

Innovation development generally has a more urgent schedule compared to routine corporate responsibilities. Innovation strategists often want to exceed Wexelblat's Scheduling Algorithm: they want innovation developed faster, cheaper, and better.[6] To achieve their innovation goals, routine tasks may be pushed onto employees who are less involved in the innovation development project.

In many companies, innovation's biggest barriers are the internal boundaries surrounding business function channels, which are not aligned for converting innovation ideas into market efforts. Companies, managers, and their employees may urgently want the products of innovation, but they may not perceive or welcome needed structural changes that could improve fast-track throughput for innovation development.

Thrashing or Formal Reliable Creative Process

Several years ago, I encountered Steve McConnell's very useful application of a computer term, thrashing, to describe ineffective human workgroup activity.[7] Computers are said to "thrash" when two or more operations run headlong into each other, with each preventing or degrading the other's ability to accomplish its task. Thrashing puts computers on hold until a decision rule or human operator resolves the conflict in favor of one process over the other. Most of us who do our work in workgroups can readily appreciate McConnell's analogy.

Many of us participate our first formal workgroup experiences in education settings, where mixtures of socializing, work, resistance to work, and power competition greatly slow down productive accomplishment, but usually not without some pleasant by-product while we waste our time. Eventually, students part ways to work on their own or suddenly develop a surprising synergy that gets the tasks done when deadlines loom. Thrashing through company work time produces a vastly different set of by-products, few of which have positive valences.

McConnell suggests interventions to thrashing, but before we look at them, we need to understand why thrashing occurs. And why thrashing may occur more often in creative and technical endeavors than in more mundane business work.

My concern about thrashing began in research and clinical team settings, though I did not have the benefit of McConnell's observations then. Research team meetings were more reminiscent of grade school schoolyard posturing than technical exchanges of information and fact-based decision making. Most ineffective was the paltry exchange of useful information, and the high level of emotional responses. Astonishingly, I was months into postdoctoral clinical rotations before I heard a basic fact-filled clinical rundown, the sort that you see in every episode of television's "ER." I complimented the presenter, and asked where she was trained. "Egypt," she replied. She commiserated with me about the lack of routine information exchange, and the subsequent efforts we invested to clarify information and arrive at valid decisions.

Later in my research career, I quietly waited out these initial team thrashing forays until I could gain task leadership, marshal the necessary information, and mentor the team into valid decisions.

Eventually, I became a research project leader. Then the fun began. Kurt Lewin, the father of social psychology during the early half of the 20th century, had a famous saying: "If you want to learn about something, try to change it." I tried to prevent thrashing by providing structure and methods, to little avail. I learned how deeply thrashing was embedded into teamwork interactions for most researchers. Thrashing, they believed, was what researchers did in meetings.

Digging into the underlying motives, I discovered that most people believe nonroutine interactions are creative thinking. They believed that a "loosey-goosey" style stimulates creativity. Much of their belief seems to be hinged on the phenomenal public displays engineered by artists and rock musicians. Most people do not realize that successful musicians have well-practiced technical and teamwork skills, and they adhere to the formal structures within their fields. Much of their publicized avant-garde chaos is pure show, though post-performance behavior is sometimes real.

I learned to address the issue of thrashing head on, before we began project work. I strove to demonstrate how we could be more creative by following a reliable stepwise project management process. Unproductive thrashing abated, and several beneficial aspects of thrashing surfaced. Cooperative expertise-based thrashing is often constructive.

McConnell describes thrashing as effort lost in unproductive work. Software developers expect a limited amount of thrashing will occur throughout a project, that mistakes will be made, wrong turns taken, but the team will get back on track quickly with little loss of productivity. This tolerant view, McConnell has found, can provoke resistance toward "process" methods that are designed to circumvent thrashing. Those with an antiprocess orientation, want to hire talented people, offer them access to unlimited resources, and let them work in their own way. They view process as an unacceptable add-on cost, and an imposition of "rigid, restrictive, and inefficient" methods that will hamper the team, waste time, and drive up costs.[8]

McConnell agrees that if thrashing remained at tolerable levels throughout a project, time and resources expended on using process methods would be senseless. However, in his experience, unmitigated thrashing often builds over time, accumulating as complexity and unforeseen problems thwart progress. Process is added too late, after significant problems have occurred. For example, defects in tracking systems are set up after defects have gone unnoticed. Component features are nonuniform across vendors. Requirements are revised. Schedules are revised. Each change adds its own thrashing component. And each belated addition of process begins to consume productive work time, especially when tasks need to be redone or personnel require additional supervision or retraining to bring performance to standards. The outcome is not good. As McConnell notes,

"The lucky projects release their products while they are still eking out a small amount of productive work. The unlucky products can't complete their products before reaching a point at which 100 percent of their time is spent on process and thrashing. After spending several weeks or months in this condition, these projects are typically canceled when management or the customer realizes that they are no longer moving forward. If you think that attention to process is needless overhead, consider that a canceled project has an overhead of 100 percent."

Anyone who has worked on an innovation development project will resonate with the dismay of a good project gone down an ineffective path. McConnell suggests,

1. Investing in the introduction of process methods, for example ground rules, standards, technical vocabulary, etc., at the initiation of the project will produce a large return on investment over the course of the project.

2. By midpoint in a project, time and effort focused on introducing and maintaining process methods will diminish as the team acquires common methods, coordinates efforts, and systematizes their work.

3. The final stages of project work will experience limited amounts of thrashing within a generally aligned framework, with most of the team effort resulting in productive work. Residual thrashing is constructive: team members alerting others of an unforeseen problem, naturally occurring errors, or newly appreciated detail.

McConnell acknowledges that inept process methods can "create an oppressive environment in which. . . creativity and management goals are placed at odds. . . but it is just as possible to set up an environment in which those goals are in harmony and can be achieved simultaneously." Finding a balance between establishing systematic process and turning over the reins is key.

Innovation Process Must Be Creative and Reliable

Innovation development is approached in some organizations as brute force projects. Internal and external resources are churned for opportunity information, task groups form for high-energy brainstorming and strategy meetings, leaders delegate assignments, and everyone fervently hopes for innovation to percolate to the surface. Many of these projects discover a viable business idea that succeeds in the marketplace, while others fizzle out at various junctures along ad hoc development paths. At some point, even the most enthusiastic brute strength advocates perceive a need to establish reliable, repeatable methods to enhance their creative work.

Peter Drucker placed great emphasis on using reliable systematic methods to develop innovative business ideas. Nowhere in his writing is the currently popular breathless admiration of "Eureka" discoveries and superstar inventors. Drucker reported that few of the successful innovators he had worked with in his career had an obvious "entrepreneurial personality." Instead, he said, their only commonality was "a commitment to the systematic practice of innovation."

For Drucker, purposeful efforts to do good work supported and stimulated creative effort. Drucker advocated a deep, thoughtful, committed approach to achieving an enterprise's economic potential. In his advice on maximizing innovation development, he underscored the interplays of human ingenuity and knowledge, and inspiration for "hard, focused, purposeful work."[9] Drucker went on to say "If diligence, persistence, and commitment are lacking, talent, ingenuity, and knowledge are of no avail."

Chris Murphy, editor of *Information Week*, noted that while many employees view innovation project work as exciting, they generally prefer their usual work routines over continual rounds of innovation campaigns.[10] The name of his article, "The Innovation Hamster Wheel," invokes visions of sincere, competent employees beset with—and perhaps baffled by—innovation fatigue. Those of us who are keen on innovation work, need to realize that innovation projects are not everyone's cup of tea. Employee resistance to continual upheavals of innovation projects, he says, creates a major push for developing reliable processes that "not only keep disruptive ideas coming, but to guide them safely through the gauntlet of ways companies can kill even great ideas." Murphy tells of one pharmaceutical services CIO who ingeniously gives away his innovative information technologies ideas to outside companies, gaining technology process innovations without incurring development burdens. Murphy concludes that companies need to develop an innovation system that is granted company recognition, and fits the company culture. He says, "Having such an effort acknowledges that continuous, disruptive innovation, while an unnatural act, is nonetheless critical to keeping companies and products from stagnating."

Not every innovation consultant values fully reliable methods. During my first year in Columbia's graduate school, I met an alumna who specialized in organizational development. I asked her about the process she used to create change in companies. "Oh," she said, laughing, "I just go in and stir things up. Then I see how differently they sort it all out." We were on a deck linking her Greenwich, Connecticut, house and pool. I began calculating how many companies had paid how much for her theory-free "stir up." As I became more knowledgeable about organizational development, however, I realized she used Gestalt techniques. Gestalt psychology proposes that our thinking becomes "frozen," and change is only possible when we are "unfrozen" by some startling or obvious environmental change, which then allows us to think freely enough to form new ideas. After which, we "refreeze" to retain the new change in our thinking or behavior. Releasing frozen thinking is a good first step, if it is followed by reliable process methods.

Afraid of Failure?

Failure is an inevitable concern in innovation development. A senior IT executive, writing under a pseudonym, says consultants claim "the failure rate for technology projects is anywhere from 37 to 75 percent" but he favors the lower estimate.[11] B.C. Forbes observed, "History has demonstrated that the most notable winner usually encountered heartbreaking obstacles before they triumphed. They won because they refused to become discouraged by their defeats."

Some project failures are temporary delays in product design, others are due to market changes, and still others are only detected after deployment. For some innovations, the jury is still out:

- WD-40, selling strongly 60 years after its market entry, after 39 previous versions developed by the inventor, Norm Larsen, failed as a water deterrent. Hence the name WD-40.

- Tesla invented a working electrical alternating current automobile engine in 1888.

- Daniel Cowen worked at Redstone Arsenal in Huntsville, Alabama, when they decided to do away with a beaver colony that continually flooded the Army base. Their dubious solution was to import alligators from Louisiana, based on an assumption the reptiles would die in the first winter, after they had eradicated the beavers. Cowen reports that the beavers are gone.

- Hydrogenating food oil extends shelf life, but it also reduces metabolization for humans.

- Michael O'Leary, CEO of Ryanair, requested the aviation authorities to allow him to eliminate copilots from flight crews, estimating that his innovation would save the airline industry a fortune. His plan: flight attendants would fill in for pilots in emergencies.

Scott Painter, CEO for TrueCar, noted at a recent The Economist innovation event that all innovation ideas have a fatal flaw. Brandon Kelley, writing in his Innovation Excellence blog, added that we need to find our fatal flaws in innovation ideas early enough to cull them so as to not waste investment. He noted that fatal flaws are often less detectable than benefits, and therefore difficult to discover before a product is widely tested in the market.[12]

What is the outcome if your innovation project fails? If the response is "Everyone packs up and goes home" your project may be doomed before you start. Magellan circumvented the globe to find a shortcut to the Spice Islands. What most of us did not learn, or forgot, is that Magellan died in the Pacific, and only one of his five ships and 18 of his 270 crew returned to Spain. The voyage attained a very small profit, but the crew was never paid in full. Magellan's voyage is a model for discovery and global trade, but a cautionary tale for innovation teams.

Insecurity or fear is a deterrent, not a motivator. Walking on a gymnast's 4-in wide balance beam would make most of us hesitant or anxious, though we would fearlessly walk a 4-in wide stripe that was painted on the floor. Anxiety and performance quality have a U-shaped correlation: very high and very low states of anxiety are linked to performance degradation, whereas experiencing medium levels of anxiety motivates us to high-performance levels.

Incurring personal risk while working on innovation projects may reduce chances of success. Regina Dugan, Director of the U.S. Defense Advance Research Projects Agency (DARPA) says that fear puts limits on researchers. Her response, "Don't be afraid of failure. Really go for it." But how does that play out in DARPA reality? Suggestions for failure plans, including a 3-strike rule, are floating in innovation discussions. Chris Murphy, editor of *Information Week*, says the best companies "lay their mistakes bare so that their teams can improve." Employees will be reluctant to expose flaws if they know punishment or termination is a likely outcome.[13] Innovators who openly discuss risks will be better prepared to prevent project failures.

Fast-Track Innovation Development Pathways

Implementing systematic innovation management processes improves innovation development, but many experts now advocate more radical systemic change to ensure a steady pipeline of successfully developed innovation projects. To maximize innovation success, organizations institutionalize innovation process into company leadership, the organizational structure and culture, management processes, metrics, and training. In addition, benefits, recognition and rewards, and all organizational planning incorporate innovation development priorities in ways that increase business idea conversion efficiency and effectiveness.[14]

What does a company need to do to increase their innovation capabilities? Gary Hamel, visiting professor at the London Business School, says innovation-striving companies need to implement innovation training, funding opportunities, and metrics at all organization levels. In addition, Hamel says innovation should be the largest factor in the executives' long-term bonus plans. Hamel says that less than one percent of companies have established these key innovation resources. Hamel offers questions to assess a company's innovation capability:

1. Do you have clear innovation accountabilities that impact your compensation, that are part of your key performance indicators, with things that are clearly measurable?

2. How many innovations has your unit generated? And what is the average quality of those ideas and how innovative are they? And how fast are they moving through the pipeline?

3. How many people have been trained as innovators?

4. What proportion of your time are you spending mentoring those innovation teams?[15]

Mechanistic companies with strong hierarchical structures and rigidly defined work roles, while suitable for stable industries, tend to impede innovation development, according to Tom Burns and G.M. Stalker. They note a growing trend among innovation-focused companies to adopt a more organic structure with fluid work roles that encourage both lateral and vertical interactions across job functions.[16]

Innovation-intensive companies are building intraorganization pathways for innovation development to circumvent function and information silos in traditional hierarchical organization structures. The innovation pathways offer greater access to innovation development teams to collect valid information to generate high-potential ideas, and fast-track articulation with resources to efficiently and effectively speed up their innovation pipeline. These pathways allow the innovation team to leverage market information, reduce development costs, and to gain an early jump into promising markets without interruptions from regular business functions.

Many of these foreword-thinking companies develop separate cross-functional teams that focus their efforts on innovation projects. Rob Preston, VP and editor in chief for *Information Week*, observes that these separate teams "seek ideas from different stakeholders—technologists, salespeople, marketers, financiers, business developers—and they're careful to not build ivory towers that are dismissed, even loathed, by the rest of the enterprise."[17] These cross-function teams are similar in purpose to entrepreneurial roles discussed by Peter Drucker and other earlier innovation experts. The new innovation teams use reliable innovation processes, while the earlier efforts were more reliant on employee personality characteristics.

Samsung innovation projects are developed by a specialized innovation group built around TRIZ systematic methods. TRIZ is the core process for Samsung development and creativity efforts.

Seeking Innovation Ideas

We use a story to illustrate the importance of systematic search methods in research. The story goes like this: Late one evening, a man walking his dog comes upon a person who is searching the ground under a street lamp. The dog walker asks the searcher what he is doing.

"I'm looking for my lost keys," he replies. The dog walker helps search for the keys, but they have no luck. "Are you sure you dropped them here?" asks the dog walker. "I have no idea where I dropped them," says the searcher. "Then why are you looking under this street lamp?" "Well. . . " replies the searcher, "If I lost them here, I will see them. If I lost them where it is dark, I cannot see them."

The lesson is that by not devising a way to search in both areas, where the lamp lights the ground and in the dark, we've reduced our ability to find the lost keys.

Using a systematic and repeatable process to find what we are looking for is much more effective than nonsystematic search strategies. Do you remember the joke that we usually find what we are looking for in the last place we looked? Psychologically, the joke has a more complex truth than we first appreciate. If a search or problem-solving strategy is effective once, we tend to repeat that strategy even when circumstances are substantially different. We've learned the successful outcome, but failed to calculate the strategy's likelihood of success in a new scenario. So the general rule for searching, if the results are important, is to use a standardized method that prevents us from using ineffective personal rules. Scuba divers use a grid for marine salvage searches. Researchers use maps to sample population areas.

Areas of Potential Opportunity

Drucker suggests seven areas of potential opportunity for innovation ideas. By scanning these areas, information with a high potential for suggesting innovations can be collected. Unexpected occurrences such as higher gasoline prices encouraged Americans to add patios and decks to their homes. Incongruities, such as the abundance of mass-produced food and scarcity of local produce developed a resurgence of small local farms. Process needs sparked Twitter and other personal broadcast media. Industry and market changes upended the music business. Other spark-producing needs might include demographic changes such as an aging multidiversity population, changes in perception such as an increased acceptance of snacking or new knowledge such as renewable energy.

Perspectives

When gathering information from sources, do not merely read or observe. Participatory experience is vital to establishing a full familiarity. If you are designing washing machines, wash loads of clothes. This lesson was brought home to me when a colleague came to my office, obviously in psychological distress. He denied anything was wrong, but I knew a serious problem was troubling him. Finally, he admitted he was having surgery the next morning. This physician had sent thousands of patients to surgery, and he had done surgery, but his first personal experience of being a surgical patient was as novel as it was for any non-medically trained patient. Fortunately, the surgery was elective and minor. He later said his experience gave him a perspective to improve his patient care. The experience differs vastly depending on the role you find yourself in.

On the other hand, an outsider can provide information that experts in the field cannot perceive. From time to time, ask someone who has not been part of the design or revision team to sit in, and ask questions. Astonishingly, outsiders' naïve questions can sometimes point out an error or an area that has been ignored.

Decisions

People seem never to be happy with making decisions. "Whatever you do, there was always a better choice," says veteran actor, John Hurt. Most people choose to satisfice, which is to choose a "good enough" option rather than making the most optimal choice. But the real problem to avoid in decision making is to be aware that we naturally tend to look for and remember information that supports a preferred choice instead of trying to find information that may warn us against making that selection. This error is called conformation bias. We see this bias in police procedures, when a detective says he "likes suspect X for the crime, and we know the story will continue with the detective willfully ignoring all evidence to the contrary. It makes for a good detective story, but not a good product development decision.

How Do We Form Innovative Ideas?

Brainstorming, though a popular participatory activity, is not effective in solving larger complex problems. Larger problems need to be approached with systematic methods. I personally like TRIZ, and think that it will become a standard method for solving problems in many fields.

But the really intriguing question is how we form innovative ideas in our minds, that is, at the cognitive level. Brain scans have recently tossed out the idea that we have left brain or right brain functioning, at least in the geographic sense of left and right. Brain activity occurs on both sides during intense thinking or problem solving. Essentially, we gather information over time, and analyze the object or concept into its component parts at various levels of detail. Eventually, at some point, several of these pieces of information—whether in visual or semantic form—unite in a new way in our mind. Sometimes the juxtaposition and its resolution are jarring, and we are startled with the discovery. When the solution is valid, it can be very difficult to not see the new idea because it obliterates the previous state of thought. Innovation is an interplay of our fund of knowledge and process.

Does Everyone Innovate?

Some people are discomforted with the suggestion that everyone should be an innovator. The pushback seems to be based on the idea that if everyone were innovative, everyone would want to lead, and no one would do any following. This assumption has got hold because many, if not most, work environments interact in an authoritarian style.

I use an example of a working sailboat to explain the authoritative style of leadership and interaction. Every crew member must know as much as possible about the boat, its operations and maintenance, and safety. Part of the captain's responsibility is to make sure that all members of the crew are competent to run the boat, if need be, in an emergency. But when the captain gives an order, the crew "falls in" to accomplish the designated task. Authoritative leadership is based on expertise and civic regard, and does not use dominance.

Power struggles, which are continually evident in authoritarian groups, are not part of authoritative groups. People have defined roles and responsibilities in authoritative groups, which offers substantiation and reduces resentment of the top-down power that is characteristic of authoritarian interrelationships.

In an authoritative environment, everyone could be innovative without vying for leadership. Authoritative groups are, unfortunately, not prevalent.

Conclusion

Innovators have a propensity for eagerly searching for problems and flaws, and ardently exploiting opportunities to create change. They look for flaws that other people do not see in existing products or processes. The innovators' opportunistic perspective can go against the grain of general corporate perspective to communicate product value and convince potential consumers to focus on the available product features while overlooking any shortcomings in design or function. In short, innovators sometimes are viewed as trouble makers unless they are company leaders. But innovative minds bring value to their companies, and an environment that is conducive to their thinking style is most probably supportive of high-quality thinking in other realms of endeavor in the organization.

References

1. Abernathy, William J. and Kim B. Clark. "Innovation: Mapping the Winds of Creative Destruction." *Research Policy*, 14 (1): 3–22, 1985.
2. Christensen, Clayton. *The Innovator's Dilemma: When New Technologies Cause Great Firms to Fail*. Harvard Business School Press, 1997.
3. Coates, Donald. Interview by author. Phone interview. Naples, FL, January 23, 2013.
4. Jaruzelski, B., J. Loehr, and R. Holman. "The Global Innovation 1000. Making Ideas Work." *Strategy + Business Magazine*, 69, 2012.
5. Preston, Rob. "Our Economic Glass Is More Than Half Full." *Information Week*, December 17, 2012.
6. Richard Wexelblat. *Wikipedia, the Free Encyclopedia*. http://en.wikipedia.org/wiki/Richard_Wexelblat. Accessed May 12, 2013.
7. McConnell, Steve. "The Power of Process [software processes]." *Computer*, 31 (5): 100–102, 1998.
8. McConnell, Steve. "Article by Steve McConnell." Steve McConnell's Home Page. http://www.stevemcconnell.com/articles/art09.htm. Accessed May 12, 2013.
9. Drucker, Peter F. "The Discipline of Innovation." *Harvard Business Review*, 80: 95–104, 2002.
10. Murphy, Chris. "The Innovation Hamster Wheel." *Information Week*, 1350, 2012.
11. Meshing, Coverlet. "Why Tech Projects Fail: 5 Unspoken Reasons." *Information Week*, April 22, 2013.

12. Kelley, Branden. "Innovation Excellence/Innovation and Entrepreneurship Fatal Flaws." *Innovation Excellence.* http://www.innovationexcellence.com/blog/2012/09/23/innovation-and-entrepreneurship-fatal-flaws/. Accessed May 13, 2013.

13. Murphy, Chris. "Mistakes." *Information Week*, October 8, 2012.

14. Gupta, Praveen. "Institutionalizing Innovation for Growth and Profitability." *The Journal of Private Equity*, 9 (2): 57–62, 2006.

15. Denning, Steve. "Gary Hamel on Innovating Innovation—Forbes." Forbes.com. http://www.forbes.com/sites/stevedenning/2012/12/04/gary-hamel-on-innovating-innovation/. Accessed May 14, 2013.

16. Burns, Tom and G.M. Stalker. *The Management of Innovation*, Rev. ed. Oxford: Oxford University Press, 1994.

17. Preston, Rob. "CIOs: Innovate or Go Home." *Information Week*, April 22, 2013.

CHAPTER 21
Quality of Ideas

Brett E. Trusko and Praveen Gupta

Introduction

The 21st century has become a century of an experience economy, mass customization, "real-time" marketing, shortened product life cycles, and co-creation with customers. This requires better and more frequent innovations that are profitable and enjoyable. Performance of Apple, Google, and Facebook shows that customer-friendly innovations are profitable innovations. Social media and the Internet are making concept-to-commercialization cycle times shorter, approaching innovations in real time for the service sector. Quality seeds, germinated and nurtured appropriately, will produce a much better yield than seeds casually thrown to the ground. Similarly, quality ideas, inspired and nurtured will enable more and better innovative products. Ideas are seeds of innovation; however, knowing quality of ideas in order to nurture and invest in them is critical to improving the success of new products. This chapter presents tools to evaluate ideas for prioritization and reducing the risk of failure.

The Concept of Ideas

If someone asks us to come up with a creative idea, many of us would not really know or understand how to develop a creative idea. Unless people know how to generate new ideas, they would approach it randomly and the results would probably be of little value or practical use. This is how most of us feel, especially when new ideas start to pop into our minds. This is especially true when the idea seemed to be particularly crazy. But ideas are the seeds of innovation, the foundation for innovation, the place where innovations begin. Developing quality ideas (practical, affordable, and viable) is an essential ingredient in the development of new products. It is well understood that after filtering, trial and error, and prototyping, very few ideas actually turn in to profitable products. We also know that the vast majority of new ideas are not actually high quality. Many of those ideas are developed with her friends over a cocktail or shouted out and paste when we are frustrated or tired of a product that doesn't work or a service that proves the inadequate. We are also aware that thousands of ideas result in hundreds of activities, tens of processes, and a few innovative products. Managing too many low-quality ideas, takes away from the time and resources needed to manage high-quality innovations and ideas. To organize and manage thousands of ideas rapidly or continually, an organization requires everyone available to think of and manage new, high-quality ideas. In order to meet the demand for mass customization and massive innovation, we need people beyond R&D to generate and develop ideas for innovative new products or solutions.

What is an idea? An idea is a fuzzy thought experiment to create something new. It begins with curiosity, often in the form of a question. The question might take the form of "what if we," or "I wish someone would," for the ultimate, "you know what makes me mad?" For most of us, if an idea sounds interesting, we pursue it further. If an idea receives positive feedback from our friends, coworkers, managers, and others, we get excited and build on it either by ourselves or as a team. A creative idea is a thought experiment of combining two or more things, ingredients, processes, products, business models, or organizations, uniquely. A creative idea might also come from frustration, playfulness, or a moment of euphoria. Different moods and situations can create higher- or lower-quality ideas. Sitting around coming up with ideas while under the influence of chemicals might lead to a hit song, but probably not a unique process in the manufacturing of a widget. Uniqueness depends upon its distinction from its predecessors.

Conventional Suggestion Systems

Corporations have been experimenting with ways to capture and exploit employee suggestions for decades. Brainstorming has been a commonly used method to develop ideas. Since the late 1980s, many companies have tried implementing "suggestion box" systems to solicit employee ideas. In general, these suggestion boxes collected a few suggestions every month, a committee of a few people reviewed those suggestions, acknowledged ideas, and eventually these ideas died because of poor response. Perhaps the most telling thing about suggestion boxes, and the seriousness to which management takes them, is the monthly or annual award for the best idea. A $5 gift certificate at the company cafeteria for the best idea of the month is hardly an incentive for employees to save the company millions of dollars. Many of these suggestions are actually complaints about supervisors, peers, working conditions, or money or budgets. By the time initial ranting is over, the suggestion boxes are removed or ignored. The committee is dismantled due to the cost of running a program that produced little or no benefit to the company.

Alternatively, many companies today offer employee suggestion programs or employee idea initiatives that utilize IT infrastructure to collect employee input, primarily to improve business processes. Some large companies have run idea jam sessions for identifying new frontiers or improving business performance. However, participation has remained stubbornly low in most companies. Once again, misaligned incentives for employees feeling that their ideas have little value, stifles creativity in the organization. Companies like Google actually allow their employees to work on innovation on company time and allow them to share in any new ideas or innovations that translate to money. People ask, "How many ideas should we expect from employees?" Or, "Who decides what is and isn't a good idea?" Many are simply apathetic because, at the end of the day, they perceive that the ideas that are pursued are those of the management staff anyway.

Need for Ideas

Because of higher expectations for innovation, coupled with shortened product life cycles, there is a constant demand for new products or services, which is challenging most organizations' existing infrastructure for new product or service development. These expectations demand new products that are highly innovative and easily reproducible. The production process must be leaner than ever and able to quickly change over from one generation of a product to the next. Moreover, just developing an innovative product is not sufficient. New products must offer customers a total and thoroughly enjoyable experience. For many organizations, this might mean that R&D becomes a department of long-term innovation and the "innovation group" becomes the department of short-term innovation. Of course, this creates a whole new dynamic in the organization that must be managed and coordinated, which is outside the scope of this chapter but addressed elsewhere in the book.

Innovation begins with ideas. Sometimes a single idea, such as changing the size of the iPad or iPod Touch or iPhone, can lead to a breakthrough innovation.* Many times, ideas must be built in layers to create sufficiently attractive innovative experiences. Ideas generated for a known opportunity is one approach, but continual generation of ideas that intersect with continually arriving opportunities can lead to breakthrough or dramatically innovative solutions. Creating a culture where people are inspired to constantly look for new opportunities and then match those opportunities with relevant and novel ideas can lead to rapid innovation. It's often heard in many organizations that ideas are "a dime a dozen"—plentiful, but not that valuable. People forget that without ideas, one cannot create innovations. Ideas are the concept car, or the artist's rendering. Organizations are challenged to inspire employees for generating numerous ideas, any one of which can lead to major innovations. However, all ideas are not equal. Some ideas stand on their own while others need to be better thought out in order to become qualified ideas. Thus, there is tremendous opportunity for improving methods for generating, evaluating, and managing ideas especially ideas that could be considered "of high quality."

---

*It is often discussed, that at Apple, the iPad was invented before the iPhone. When Steve Jobs was evaluating the iPad, he realized that if you added a phone chip and made it smaller, you would have a very cool product.

Sources of Ideas

Ideas are a little like money; no matter how much you have, it always seems like you need more. Even organizations with plenty of innovative ideas need more innovative ideas. This is especially true today. Even the most innovative companies do not hang onto their lead for long if they fail to continue to innovate. Ideas require an owner who makes ideas usable and valuable. Knowing that conventional brainstorming methods are fractionally effective, researchers have studied methods to improve the ideation process. Ideas can come from anyone in the supply chain. Employees, customers, users, suppliers, or stakeholders—all could generate ideas for growth. Employees certainly have more at stake as their welfare depends on their intellectual contributions to their organization, thus being inspired to come up with new ideas. Progressive organizations will create or purchase a platform where anyone can easily contribute ideas. This means that it is not only the employee responsible for contributing new ideas, but it is also incumbent on the organization to harness employees' creativity and then manage those ideas effectively. After all, what we learned from the suggestion box is that if we do not effectively manage employees' ideas, contributions will taper off. To enlist employee ideas, organizations must create an environment or culture of innovation and a system for processing those ideas.

An Environment for Engagement

The environment for engaging employees intellectually includes an element of inspiration. We know that employees react positively to the perception that they are cared for, or when there is a relationship that is manifested by listening to their stories and not simply collecting ideas. This means employees receive feedback for their idea and trust that it is fairly considered for next steps. Information technology can automate idea capture, evaluation, feedback, and post-processing without consuming significant human resources.

One of Dr. Deming's 14 principles of quality management recommends driving out fear in order for employees to participate in quality. Similarly, when it comes to innovation and creativity, it has been observed that when employees experience an environment of fun or freedom, they are more engaged. Studies have found that when people are in a more positive mood they tend to be more engaged in generating ideas,[1] and it has been understood that quantity leads to quality. Classroom exercises have shown that when people participate in generating Good, Crazy, Stupid, and Funny[2] ideas to address a specific opportunity, the number of ideas on average decreases as the fun of the exercise diminishes. However, when the exercise is repeated during training, the number of fun ideas increases since it is not as restrictive as the first time.

The positive mood of employees contributes to an environment that is safe, unproblematic, relaxed, loose, explorative, and risk-friendly. The fun environment represents freedom from constraints and encourages ideas without any consideration of implementation. To create such environment, playful physical and mental activities have been used. Physiological and psychological relaxation has been found to help employees relax and experience an environment without fear of judgment. We have observed that people tend to hold their ideas if they believe the ideas might be ridiculed or considered "dumb." Thus, the methodology of allowing and encouraging good, crazy, stupid, and funny ideas has been found to create a relaxed environment for generating ideas.

Experience and Energy Levels

It has been seen (or at least it is believed) that younger people tend to have more ideas, while experienced people generate fewer ideas. Experience teaches constraint, which isn't always a bad idea when working as a team. Team members with institutional knowledge can offer valuable insight so long as it is not criticized or mocked. We have learned that less experienced people tend to be bold, risk taking, and impetuous, while the more experienced people tend to think of more rational and knowledge-based ideas. In other words, less experienced people generate more ideas while experienced people contribute a higher quality of ideas. Engaging people with diverse levels of experience can offer the best opportunity for generating useful creative ideas.

Methods for Generating Ideas

Most common methods used for generating ideas are the "coming together" of two or more people. Meetings in social settings have been reported to be even more conducive for generating wild ideas. In the Internet age, people are deploying virtual meetings and platforms to collaborate toward idea generation. We have learned that understanding and applying concepts of psychology and the human brain can improve conventional methods of generating ideas. Below are some known methods for generating ideas that we have found to be useful.

Brainstorming

Brainstorming has been considered as any discussion about a topic where people build on each others' ideas. It is the most commonly used method for generating ideas. Typically, an individual who is interested in soliciting ideas or discussing solutions to a problem calls a meeting. The typical meeting includes five to nine people, where one or two people dominate the meeting while others participate passively or wait for their turn to share their ideas. Many times, before most people have an opportunity to share their ideas, the meeting time is over. Ever wonder why you feel so much better leaving a meeting than entering it?

The problem is that these meetings are only partially effective, but they can be made more effective if properly managed. This means giving everyone the opportunity to share their thoughts and ideas. Sometimes, a designated meeting facilitator can help in making brainstorming meetings more effective.

PARC, a Xerox company that offers breakthroughs in providing primarily custom R&D services, has recommended ways to brainstorm for quality ideas.[3] The following are recommended do's and don'ts for brainstorming:

1. Don't call it a "brainstorm." Instead call it something unique.
2. Know what you plan to do with the outcomes.
3. Go somewhere—anywhere that encourages idea generation.
4. Adopt a "yes, and" protocol during the discussion.
5. Provide advance reading and questions.
6. Brainstorm constraints first.
7. Develop ideas iteratively.
8. Avoid "sticky idea" syndrome.
9. Design small experiments and check back.
10. Embrace failure.

The above recommendations, coupled with good working guidelines or ground rules, can help with facilitating productive brainstorming sessions.

Thinking Innovatively

The conventional brainstorming approach does not encourage the intellectual engagement of many participants in a group. Therefore, it is recommended that everyone be given a chance to be intellectually engaged or contribute his or her ideas. One of the easiest ways to do this is to give everyone a 3×5 card to jot down their ideas and submit them to the team leader. While this may seem simple, many times people just want to be asked. Once asked, they become more likely to contribute to meeting. We have discovered that many people do not have any ideas or are unable to create or express their ideas. Therefore, soliciting ideas from everyone is a challenge. There is a need for training people in asking questions, thinking of ideas, writing and articulating their ideas.

We have developed an approach to engage people intellectually and generate ideas. This idea generation process consists of four stages along the lines of four "qualities" of creative ideas. These approaches include fluency, flexibility, originality, and usefulness.[1] As mentioned earlier, the four types of idea generation include good, crazy, stupid, and funny. Each participant spends 3 minutes for each type of idea and writes them down on a card. We use green, yellow, blue, and purple colors for good, crazy, stupid, and funny ideas, respectively. Exercises conducted in a group have resulted in as many as 60 ideas in 12 minutes. Figure 21-1 shows the number of ideas generated by various groups over a period of time. It has been observed that good ideas represent a memory dump, and thinking begins with crazy, stupid, and funny ideas. Also, the crazy ideas are more about extending an existing performance parameter while stupid ideas are about expanding or creating altogether diverse ideas. For example, driving 75 mi an hour in a 35-mi/h zone would be considered crazy, while crossing over the lane markers could be considered "stupid." However, stupid ideas to address issues force people to think about something in a totally different way. The funny ideas bring purpose to the creative thinking process, thus creating an enjoyable experience for the user. Using this or a similar approach forces people to think differently and ensures that each member has an opportunity to think of or express creative ideas.

	Good	Crazy	Stupid	Funny	Max	Min	Avg	Stdev	N
Portugal	5.6	3.0	2.7	2.5	21.0	8.0	13.8	4.5	13.0
Malaysia	3.5	3.8	2.5	1.8	20.0	3.0	11.6	5.7	12.0
India	7.3	3.6	3.7	3.0	33.0	6.0	17.6	8.1	7.0
Singapore	4.7	5.0	4.1	3.4	23.0	13.0	17.3	4.3	7.0
Mexico	9.3	7.8	6.7	8.3	57.0	11.0	32.2	18.8	6.0
Brazil	3.8	3.1	2.6	2.3	29.0	7.0	11.8	5.7	13.0
USA	6.0	3.9	3.0	3.6	33.0	9.0	16.4	8.3	9.0
Average	5.7	4.3	3.6	3.6	30.9	8.1	17.2	7.0	9.6
Std. Dev.	2.0	1.7	1.5	2.2	12.7	3.3			

FIGURE 21-1   Idea generation across cultures.

Studies have shown that pure brainstorming is less effective than exercises that allow people to think first before brainstorming. Before any brainstorming or group meeting, people should go through individual idea generation exercises to ponder their creative ideas. Then, get together in a group to discuss them and build on those original thoughts to generate bigger and better ideas.

Online Collaboration

In the Internet age, interaction and collaboration with others around the world have become routine. Instead of organizing a group in a room together, one can convene an online brainstorming or idea generation session within minutes so members can participate remotely. Because of this, collaboration among peers or with subject matter experts has become more practical. Online collaboration tends to mitigate emotionally negative interaction and allows more objective and positive interaction for generating creative ideas and even solutions.

Evaluating Ideas

Our minds are continually assaulted with new information and ideas to the point that we ignore most ideas, since we don't know which ones to pursue. One of the burning questions of our time (the innovation age) is to sense or know as early as possible, which innovations have the best chance of success and make a positive return on investment. The question of which ideas and innovations to pursue with limited time or information requires that we understand the idea of idea *quality*. Traditionally, product or service quality refers to the ability to perform or execute as intended or expected. Similarly, defining idea quality refers to how well it serves its intended purpose. If an idea can be applied to multiple solutions, then its quality would vary for different solutions. The attributes of an idea are identified, and then the idea is evaluated against those dimensions in an effort to determine the quality of the idea. One of the challenges associated with evaluating an idea is the great deal of uncertainty associated with its ultimate success. It is never one idea alone that succeeds; however, any idea can be the seed that can lead to success.

Emotional Evaluation

The first evaluation of an idea is the human response. If, after listening to an idea, the person becomes disinterested, the conversation is over. If enough people who hear the idea show no interest, then the idea needs to be rethought or reformulated—or discarded. If enough people were excited about the idea, then you may want to consider it for further evaluation. Design innovation is defined as humanizing innovations, or making the product human-friendly. Similar criteria could be used to quickly judge the quality of an idea. Figure 21-2, emotional evaluation of an idea, shows four aspects, usability, usefulness, experience, and expression, and their corresponding reactions. If the reactions to an idea tend to be at "2" for each aspect of emotional evaluation, it would certainly be worth pursuing for further development. Otherwise, it would need more rigorous value evaluation.

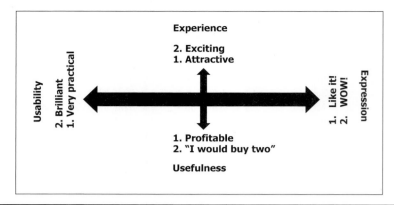

Figure 21-2 Emotional evaluation of an idea.

Value of an Idea

It has been understood that quantity feeds the quality of ideas. Thus, idea generation and the quality of an idea go together. It's not the "quality" of the generic idea but the quality of the *best* idea that matters.[4] Isolating the best idea and its quality makes it difficult to evaluate the overall quality of ideas. Accordingly, ideas are generated and evaluated for their quality, then the distribution of quality of ideas is examined (or the variance in quality) evaluated in order to select the best ideas based on probabilities. The challenge remains in meaningfully determining the best idea.

Similar to experiences about product quality, there is a learning curve in improving the quality of ideas. It is not so much in the quantity of ideas produced, but the *experience* of repetitively producing ideas that contribute to ideas' quality. Studies show that quality ideas occur during the middle of an ideation session where quality improves as the group builds upon other early ideas.

Academic literature has dealt with the quality of ideas without relating to its ultimate value. Real-life experience shows that identifying the best idea depends on the awareness of judges, identification of the opportunity, and relevance of the idea to the opportunity. The quality of the idea can be assessed for selection in terms of its usability, usefulness, and novelty. Experience has shown that idea generators must be aware of the opportunities that exist or have the domain expertise for seeing trends and anticipating future opportunities.

Assuming that the quality of an idea translates to the likelihood of its being used for creating value, utilizing a process model for determining the quality of ideas will help assure the best quality ideas to rise to the top. The decision to use the idea would depend on the opportunity or market potential, the idea champion, the novelty of the idea, the emotional impact, utility, usability, affordability, or value proposition. Once the idea is considered for development, it can be prioritized based on an idea priority index, a tool similar to that used for prioritizing projects. Quantifying the idea requires domain expertise in market segment, its performance, and its trends.

Figure 21-3 captures an empirical model for evaluating ideas. Key attributes to evaluate an idea include impact, sponsor, value, trade-off, usability, and cost. Each attribute is evaluated at three levels: low, medium, and high, and corresponding values of 1, 3, and 9, respectively. This matrix gives a relative indicator of the value or the quality of the idea to pursue at initial stages.

Attributes	Description		Low (1)	Medium (3)	High (9)
Impact	Market Opportunity (Total available market and trends)				
Sponsor	Leadership (Passion, Network, Resourceful)				
Value	Unique Value Proposition (Extent of innovation)				
Trade off	Adverse Impact on current performance levels				
Usability	Ease of use (Reproducibility)				
Cost	Affordability (Availability)				

Figure 21-3 Idea evaluation matrix.

Idea	Impact	Probability of Acceptance	Cost of Idea Development	Time to Develop	IPI
1	40,000	0.7	25,000	1.5 Year	0.75
2	100,000	0.9	10,000	1 Year	9
3	200 M	0.7	10M	0.75 Year	18.7

Idea 1. - Sailing Security device
Idea 2. - An App to find perfect shoes
Idea 3. - A major SEO business

FIGURE 21-4   Idea prioritization matrix.

Prioritization

In an "innovative organization," there will be many ideas that contend for resources in time and material. It is not uncommon for many projects to be funded without proper evaluation. These are at a high risk of being discontinued at the earliest difficulty or challenge, or pursued with no economic benefit to the organization. Prioritization tools exist to rank projects based on cost-benefit analysis for improving the success rate of selected projects. One such method is called the project priority matrix, adapted herein as the idea priority index (IPI). The IPI prioritizes ideas based on the potential cost-benefit analysis, associated risks, and likely time to commercialize the idea, using the following relationship:

$$IPI = \frac{\text{Annualized potential impact of the idea (\$)} \times \text{probability of acceptance}}{\text{Annualized cost of idea development (\$)} \times \text{time to commercialize (year)}}$$

Figure 21-4 shows three classroom project ideas considered for prioritization using the above relationship. The figure shows calculated IPI of 75, 9, and 18.7, respectively, for the three sample project ideas. IPI of around 10 or better is recommended for new products or services, and around 5 or better for internal process improvements. The SEO business idea has the highest IPI, thus it is prioritized for development over other ideas.

Excellence in Idea Management

Behind all human progress there must be an idea to trigger action or exploration. Generating an idea is an "exploratory" activity. In the ideation process, almost any idea is a combination of two or more "ingredients," or inputs, which creates an output of creativity from the brain. How do we come up with ideas? It is a thought experiment resulting from curiosity caused by questioning the status quo. Ideation is a natural activity. Thus there are two aspects of idea generation, thinking of new ideas when needed, and capturing the ideas when they occur unexpectedly or randomly. Excellence in management is about inspiring people to create ideas, followed by capturing, evaluating, prioritizing, and evaluating ideas for further development. Finally, it's about being able to organize those ideas for future retrieval or reuse.

Excellence represents the achievement of a performance level equal to or better than a designated target, without waste, and creating value. In idea management, the target is the intellectual engagement of people in an organization approaching 100 percent. To prevent waste in idea management is to minimize the leaking of good ideas from the system, which misses opportunities, or pursuing a questionable or irrelevant idea, which leads to inefficiency. Measures for excellence in idea management include employee engagement, success rate of idea implementation, and value creation, all while maintaining a culture of curiosity, fun, celebration, and continual participation.

Figure 21-5 shows the "key idea management process" beginning with browsing existing ideas preserved in a database system. With future needs in mind, anyone can submit one or more ideas into a database that has company-specific filtering criteria. This provides the submitter with instant feedback and facilitates improvement of the idea consistent with the desired measures of excellence in idea management. Therefore, idea generation is an interactive activity. Once the system is satisfactory, the idea can be processed for its value

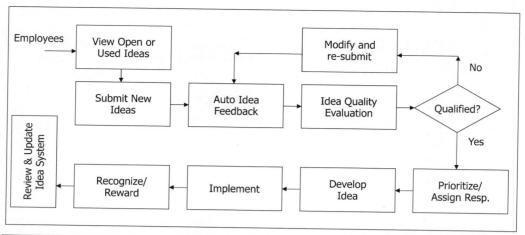

Figure 21-5 Key idea management process.

consistent with the attributes identified in the idea evaluation matrix. Further refinement can be made to the idea or put on hold for further development of the idea until the idea achieves sufficient value proposition. The criterion for evaluating an idea is generally company specific or limited to a proprietary body of knowledge.

Having multiple ideas, even hundreds of ideas to pursue is a possibility in a well-run idea management system. The next step is to continually prioritize ideas for return on investment (ROI). At this stage, any estimation of cost and benefits or other factors are very preliminary and inaccurate. Thus the criteria to pursue an idea are refined continually based on an analysis of its idea management system. Once an idea or a group of ideas are selected for development, passionate idea sponsors and necessary resources are assigned for exploiting or serving the market opportunity. Once the solution is developed and implemented successfully, the entire team is recognized and rewarded fairly and openly. It is a critical step in sustaining the intellectual engagement of employees. Any compromise or biased recognition is demoralizing for employees. They can shut down the process very quickly.

Periodically, the idea management system is reviewed against the achievement of its intended targets, the results are shared with participants, and improvement actions are taken to make the idea management system more efficient and effective.

Summary

Over the years, employee suggestions or ideas have been received sporadically at best, and have been deficient in the numbers needed for sustaining the profitable growth of an organization. In a rapidly changing and globally competitive economy, the need for ideas has become greater than ever. Distance measured in time has decreased to 6 to 10 seconds between customers and suppliers, demand for new products or services is continually growing, product life cycles are shrinking, and commoditization is accelerating. Management and R&D employees are not enough to keep up with the demand for innovative new products and services. Utilizing all available intellectual resources offers the greatest competitive advantage to an organization. Inefficiency and failures in idea management and subsequent product development processes can be expensive and lead to the demise of an organization. Success of an organization begins with excellence in utilizing its intellectual resources in a timely and economic fashion.

References

1. Vosburg, Suzanne K. "Mood and the Quantity and Quality of Ideas." *Creativity Research Journal*, 11 (4): 315–324, 1998.
2. Gupta, Praveen. *Business Innovation in the 21st Century*. BookSurge, 2007.
3. Ernst, Jennifer. Brainstorming for Quality Ideas: 10 Tips. http://blogs.parc.com/blog/2009/10/brainstorming-for-quality-ideas-10-tips/. Accessed October 6, 2009.
4. Girotra, Karan, Terwiesch, Christian, and Ulrich, Karl T. "Idea Generation and the Quality of the Best Idea." *Management Science*, 56 (4): 591–605, 2010.

CHAPTER 22

Idea Evaluation and Management

Brian Glassman and Abram Walton

Introduction

This chapter reviews the process and methods behind evaluating ideas. It further places a specific emphasis on managing this complex activity and discusses management approaches and common issues. Unlike others who may think evaluating ideas is all about selecting final ideas, this chapter describes a robust evaluation process that requires preliminary screening of ideas, gathering evidence, performing analyses, and then a final screening of ideas resulting in the selected ideas proceed into the product development process, or converting those ideas into formal business plans. Theory and practice are well balanced in this chapter so that the readers understand the basic premise behind the process and can adjust it properly to their organizations.

Overview of Idea Evaluation and Management

One of humanity's greatest strengths is our innate ability to be creative; nowhere else in the animal world can you find a species even 1/1000ths as creative as us. We marvel when animals exhibit creative behaviors like when a chimpanzee uses a stick to fish out ants, when a dolphin uses bubbles to scare fish, or when an otter uses a rock to crack open clams. Yet, all of these simple inventive tasks can be deduced by a 3-year-old child. Humanity's ability to collaboratively create can easily be our greatest strength, helping us evolve from hunter-gathers, to modern farmers and mass producers, to builders of large cities and advanced technologies. Another key difference between animals and humans is our ability to purposefully identify problems, evaluate solution alternatives, and create value through problem-solution optimization.

Ideas are vital links in the innovation process and ideas that are properly assessed and selected can create value and build revenues in the form of new products and services, or reduce costs and improve operations resulting in higher profits and increased financial stability. Yet, resource limitations, the cost of ineffective product launches, ineffective cost-saving projects, and missed market opportunities underpin the importance of creating and effectively evaluating ideas. As an early task in the innovation process, mistakes in the idea evaluation phases are magnified as selected ideas are turned into less than ideal projects, which expend resources as they pass through the product development pipeline. For entrepreneurs who have very limited time and resources, properly evaluating and selecting the core idea for their next venture can make the difference between success and wealth or failure and bankruptcy.

These authors assert that the idea evaluation process and management tasks are more frequently performed incorrectly and are most often mismanaged. Part of that is due to a number of factors, namely, lack of training on proper idea selection and evaluation methods, an overreliance on market evidence (like market studies), an underuse of direct customer evidence, poor selection of evaluation tools, a miscategorization of ideas, typical psychological decision biases (like looking only for confirming evidence), and a lack or total absence of an idea evaluation process.

Because of these downsides, this chapter details an idea evaluation and management process. The major activities in this process are in idea evaluation: preliminary screening of ideas, gathering evidence for the ideas, analyzing the evidence, and the final screening of ideas. The preliminary screening of ideas involves a very rough screening of an idea to determine if it should even be considered for further evaluation. For example, should General Mills consider ideas to sell consumer electronics related to in-box cereal products and toys? Next, entire batches of ideas are categorized, where similar ideas are evaluated by first gathering evidence for and against those ideas. Once evidence is gathered, the ideas are analyzed using the appropriate tools and methods with the goal of assessing against key decision criteria. Finally, the evidence and analysis results are used in the idea selection activities, where a single or multiple selection tools are used to compare and ultimately select ideas to continue on with development. Ideas based on emerging customer needs or emerging technologies are difficult to gather evidence and analyze, hence the process may repeat several times in an iterative cycle.

First and Final Screen

First Screening of Ideas

The first idea-screen should occur upon receiving the idea, which is usually conducted by an employee formally tasked with overseeing the ideation process. The first screen is useful in that it eliminates ideas having low market value or that would not fit the business' core competencies. Yet, the first screening process can also be extremely harmful if setup is wrong, in that it might weed out potentially valuable ideas, creating a self-reinforcing view which closes the organization to new ideas. The following section provides an in-depth discussion of the theory and methods for conducting the first screen, the tools and resources for such tasks, and common problems with the first screen.

Unfortunately, there are oftentimes more reasons to filter out new ideas than to keep them, and for the idea evaluation process to function properly, ideas must be submitted on a regular basis. The truly successful companies and visionary entrepreneurs only place barriers to ideas where they actually exist, not those based on biases in the minds of strict managers trying to keep their organization on a predictable track. Unfortunately, innovation does not follow a predictable path. Accordingly, successful entrepreneurs have expanded their companies in nontraditional paths, such as from a music record company into an airline company, from simple airplane manufacturers into satellite launch and management companies, from simple retail stores into the largest supply chains on the planet (Richard Branson, Howard Hughes, and Sam Walton, respectively). These entrepreneurs are examples of those who espouse the saying "think outside the box," which is further delineated later in real and strategic terms (see section Theories).

The methods for conducting the first screen are described in the next section, including a discussion of the individuals who should be tasked with capturing and screening ideas. When screening ideas, an organization can choose from over a dozen screening methods; however, it is critical that at least nine of them be strictly avoided for the first screen in order to avoid unnecessarily filtering otherwise potentially valuable ideas. The table in this section provides a comprehensive review of screening methods and reasons why a particular method should or should not be used for a first screen.

As with all other vital processes inside an organization that receive supporting tools and resources, capturing and screening ideas should also be equally supported. The tools and resources section discusses the provision of instructional handouts and guides, having meetings to explain the idea submission process and, more importantly, explains the connecting roles that innovators have inside the organization.

The final section discusses common problems with the first screening of ideas, namely an organization where employees are unaware of how to submit ideas, when managers fail to apply the first screen correctly, and the selection and implementation of poor screening criteria. Readers should expect a comprehensive view of the first screen helping them to conduct it properly and avoid easy but nonobvious mistakes.

Final Screen

After an idea passes the first screen, it then goes into a repository where it may be selected to move forward in the idea management process. Managers with a need for a certain type of idea may scan through the repository (which acts as an idea vault, idea bank, or idea database) and then the manager selects a category or group of ideas to evaluate. An evaluation plan is created and evidence is gathered on this group of ideas and then is analyzed. With this newly gathered information, the uncertainty and ambiguity associated with the ideas is reduced and a more robust final screening decision method can be used to select which ideas pass onto product development or into a rigorous business plan creation process.

The final screen of ideas is much more rigorous in nature and uses the gathered evidence and analysis results to make an informed decision with respect to the ideas; whereas, the first screen is almost a totally uninformed decision with very little or no information. The goal of any screening or decision point is to select the best or most valuable ideas for the organization's or department's particular and changing needs. Given the tremendous ambiguity and uncertainty associated with new ideas, especially disruptive ideas, an evaluation period is vitally needed to gather evidence and perform analysis to have some solid information and facts to make a decision.

Methods for Conducting Screening

When performing the initial idea-screen, an organization must carefully choose those individuals who will perform the evaluation and have transparent methods to hold them accountable for the results. Especially in the case of software-based screening, innovation administrators should monitor the ideas rejected by the first screen via a monthly meeting. For entrepreneurs, they must become accustomed to being accountable to perform the first screening of ideas and not rely on the biases of their friends, family, or advisers, because the idea is often in a nascent stage and only loosely defined and thus not ready to be judged by these individuals.

In assigning individuals to the screening process, consideration must also be given to their personality types, and those receiving the ideas should be open, affable, friendly, and have limited, if any, biases. Organizations can internally choose managers, line employees, specialists, or experts to serve on the review team, understanding, however, that each should be purposefully chosen due to their unique backgrounds. Managers are usually the most aware of a corporation's needs and challenges, and thus are more likely to let known limitations on resources cloud their judgment; yet because of their responsibility, they are often more available to meet with individuals or answer the phone and return e-mails. Line employees can be chosen when their work allows them to depart frequently to meet with people. Specialists or experts are extremely knowledgeable in a particular business area or technology space and can be used to enrich the idea as it is being screened; however, if they are rigid in their approach to technology, they may not be able to grasp (or want to imagine) the next generation of disruptive technologies.

The screening of ideas can occur on an ongoing basis, during assigned hours, or as a team or batch process; but each should be linked with a strategic initiative, either at the corporate, functional, or departmental levels. The screening processes should be designed to allow for a template approach for submitting ideas while keeping process formality to a minimum. Due diligence must be given in developing the criteria and process for idea submissions so as to ensure idea-submitters do not self-select out of submitting simply due to the process or criteria. In other words, a poorly designed submission process is in itself a form of filter, but is the most wasteful of all. During a batch screening process, a meeting is held and ideas that have been submitted pass through the first screen, and if accepted, are recorded into the idea bank. The benefit of this format is that when ideas are accepted, individuals can be collectively asked "who will help gather or submit evidence for the idea(s)?" If the evaluation process is team based, a diversity of individuals would be recruited to take on the idea evaluation process. As a culture of innovation grows, the social pressure to submit a quality idea and also to participate in the evaluation and evidence-gathering processes would strengthen, especially once the core group of innovators grows to between 8 and 20 people.

As early as 2000, robust software has been widely available to enable organizations to capture and screen ideas; however, this practice does create additional problems that must be monitored. Typically, ideas can be submitted through a website or web portal but often the submission page has limitations that hurt an employee's ability to submit additional documents, excel files, or even show physical prototypes, thus further increasing the chance an idea-submitter would self-select out of the process. More importantly, if the idea is complicated and requires some depth of thought, the web portal–based software would strip the submitter of the instant feedback usually provided from submitting an idea to an individual. Given that feedback is how people learn and ideate, this is a major downside to technology solutions for idea submission processes. This is why there should always be a secondary mechanism for submitting ideas directly to an individual. Yet, if done properly the software systems can provide a way of capturing, screening, and evaluating a large number of ideas. For instance, Glassman and Walton describe an idea stock market that allows users to submit ideas and screen them for appropriate alignment with a corporate or departmental initiative.[1] Such software has been known for increasing engagement rates in innovation activities from an average of 30 percent to nearly 80 percent.

Beyond software, there are many methods for screening ideas; however, the methods chosen for the first screen should be simplistic because one wants the employees to (1) see the first screen as fair and unbiased, (2) eventually predict whether their ideas will pass the first screen, and (3) focus on finding or creating ideas that will pass the first screen as it helps the organization focus on achieving its goals. Table 22-1 outlines screening methods for the first screen of ideas and it also delineates why certain screens should be used while others should be avoided.

In stark contrast to the first screen, the methods used for the final screen can be much more involved and should look at the evidence and analysis conducted for each idea. The final screening is a highly organized and carefully selected set of screening methods, partially chosen to meet the needs of the company and the department requesting ideas, and partially chosen to ensure a solid value proposition and return on investment. It is common for companies to rank order ideas so that the most valuable are identified, but a carefully designed final screen will achieve several things. First, it will align the group of ideas to the company's strategy and goals. Second, it will consider multiple criteria and develop a discussion among decision makers over which criteria are (a) critical, (b) important, and (c) that are merely niceties. Finally, it will look objectively and subjectively at the idea's pros and cons, carefully weighting them while avoiding biases and common decision-making mistakes.

A company's strategy is meant to unify action and drive individuals to achieve a major vision or objective. Ideas that present major deviations from the standard departmental business unit, or corporate strategy will meet major resistance throughout the development process. It is best to screen ideas for their match against these different levels of strategy using scoring, ranking, or checklist methods listed in Table 22-1. For example, new microwave concepts would be a major shift away from Samsung's shift toward higher-margin appliances and thus receive a −5 score for disruptive innovativeness, or not meet the checklist criteria for further consideration. Be sure to understand the departments, business unit, or corporate strategy before using the strategy-based screen methodology.

In some instances your company may want to develop new strategic options; only in these cases would unconventional ideas be rated regardless of the deviation from the existing strategy. Most corporations can handle minor deviations, but significant ones (e.g., Singer Sewing Machines venturing into personal computers) are rarely successful and thus rarely tolerated. Hence, the ranking, rating, or point scoring of ideas that present new strategic options must have a range of comfort built-in based on the core capabilities of the company to produce products or services in that area, or in their capabilities to develop strategic alliances or joint ventures for contract manufacturing. For example, Intel may score new LED-monitor ideas as a 9/10, but they would stay away from developing new manned rocket ships by scoring it a cumulative 1/10 based solely on their capabilities to produce that item. However, Intel may innovate with regard to chips used on new rockets or zero gravity applications, thus still enabling them to enter a new market while leveraging their core competencies.

Developing and establishing the final screening criteria is the next task for the innovation administrators, who should involve key stakeholders from marketing, sales, research and development, operations, and manufacturing to provide their input on the decision criteria. For example, a marketing company may use multiple screening methods to determine the marketability of an idea, while the VP of manufacturing would just want to ensure that the selected ideas can be either produced by a private labeler or made in-house and thus settle anything outside these two options as an exclusion criterion. Internal deliberations regarding final screening criteria should be transparent and include high-level decision makers to ensure the company's strategic goals are being met. Any limitations with a company's business model become apparent through these conversations, and criteria such as minimum profit margins, cost structures, and other financial considerations often become key exclusion criteria for the final screen. Unfortunately, less tangible, but more important criteria often become more subjectively scored and lesser ranked when compared to common financial criteria. Such subjective measures might include "innovativeness of the idea(s)," anticipated total accessible market size, development times and cost, product-market leadership, branding and positioning opportunities, and possible market penetrations. Finally, personal preference criteria, such as product attributes or the finer details of customer attributes (e.g., look and feel, manufacturing cost, customer type) lose their weight since the discussion may not be in relation to a specific idea and rightly fall out of the final

	Definition	Acceptability for a First Screen?	
Theme criteria	• Multiple themes are created; if an idea fits into those predetermined themes, then it is accepted (e.g., operational improvements, branding improvements, revolutionary products). • Used when an organization wants to focus on a few core areas. • Downside is if the themes are too broad or too narrow then all ideas or too few ideas will be accepted. Themes have to be updated every few months to align with strategic goals.	Yes	• This is the most acceptable first screen because employees can self-check if their idea fits a predetermined theme. Further, it forces employees to find ideas that fit those themes. Finally it is fair; if an idea matches a theme then it is accepted into evaluation.
Exclusion or inclusion criteria	• Multiple inclusion criteria (10 or more) should be set. These are general criteria, such as, does it help expand the brand, improve customer relationships, etc. • Exclusion criteria should be set carefully and only really be around the values of the company. For example, does the idea violate "being honest with the customer?"	Yes	• Inclusion criteria for a screen are preferable; does the idea help increase operational efficiency? If yes, then keeping inclusion criteria broad around certain themes is helpful. • Exclusion criteria must be set carefully as it is too easy to create a situation where few ideas are eligible to be included.
Grouping or tiers	• Much like themes that categorize ideas based on similarities, groups can be helpful in evaluation as the methods used to evaluate them are similar. • Tiers like top ideas, or worst ideas related to operational improvements can be used very effectively as a first screen. • However, both grouping and tiers can only be used in a batch evaluation process, not a continuous process.	Yes	• Evaluating a group of similar ideas is much easier than dissimilar ideas. • Evaluating top tier ideas in a particular theme helps maximize and focus scarce resources.
Idea sponsor	• A member of the selection committee can decide to sponsor an idea. The number of ideas they can sponsor can be limited based on fairness or resource limitation. This allows executives to push ideas they see valuable through the corporation.	Yes	• If the idea has a department head that is sponsoring it, then it will receive the needed resource in the following phases of gathering evidence and analysis. • Sponsored ideas are an old method of bringing ideas to fruition.
Checklist or threshold	• An individual idea's list of attributes must match the checklist or threshold in order to pass (e.g., be implemented in 6 months, profit at least $500,000 and require no more than two employees).	No	• It is too easy to poorly choose an attribute and create a barrier to early-stage ideas, especially where solutions to problems with the idea have not been derived.
Personal preference	• A manager, director, line employee, or even expert is used to screen an idea based on his or her own preferences.	No	• This should not be used as it is highly biased and viewed as unfair for any type of screen. • Experts are often called in to perform a first screen. They can be helpful if the idea is accepted, but due to their depth of knowledge they almost always tend to evaluate the idea more fully, finding immediate holes and issues which should only be addressed in the later stages of evaluation.

TABLE 22-1 Screening Methods and Their Acceptability for Use during the First Screen

	Definition	Acceptability for a First Screen?	
Voting	• An individual can vote openly or in a closed ballot, i.e., blind/peer review. • Voting can be weighted; an individual can give multiple votes to a given idea; for example, an expert in a technology area could be given 20 votes compared to a nonexpert's one vote.	No	• Voting to pass an idea into evaluation will seem unfair to the idea-submitter. • Personal preference and politics play a large role, see reasons for rejecting that screening method earlier. • More importantly, it is difficult to predict if an idea will pass a first screen and thus it may prevent employees from submitting ideas.
Point scoring	• An individual uses a scoring sheet to rate a particular idea on its attributes (e.g., an idea that can be implemented in 6 months gets +5 points, and one that can make more than "X" dollars gets +10 points). • The points are then added together and the top ideas are ranked ordered by highest point scores.	No	• For the same reasons as described in voting.
Rating scales	• An individual rates an idea on a number of preset scales (e.g., an idea can be rated on a 1 to 10 on the implementation time, any idea that reaches a 9 or 10 is automatically accepted).	No	• For the same reasons as described in voting.
Ranking or forced ranked	• An individual must rank ideas (#1, 2, 3, etc.)—this makes the group consider minor differences in ideas and their characteristics. For forced ranking there can only be a #1 idea, #2 idea, and so on.	No	• For the same reasons as described in voting.
Resources and capability	• Determine the resources or capabilities required to develop and implement the idea into the final product, service, or operation.	No	• As the idea has not been evaluated, it is not possible to accurately estimate the resources and capabilities required to develop it; thus one must use educated guesses, which is very much open to personal bias. • The process for evaluating ideas finds solutions for acquiring resources and capabilities, and thus this screen can only be used as a final selection method.
Delphi method	• Creates social agreement on idea selection via an iterative process whereby a variety of evaluators not only assess the idea, but are allowed to see other evaluators' opinions during their review.	No	• For the same reasons as described in voting.

TABLE 22-1   Screening Methods and Their Acceptability for Use during the First Screen (*Continued*)

screening process; whereas in an informal screening process they may have unreasonably affected the selection of the final ideas.

After upper management of a business unit or corporation completes this screening exercise, they should have a set of criteria and process for conducting a final screen that meets their goals and can be disseminated to all stakeholders for transparency, inclusion, and engagement purposes. From here on, errors in idea selection will no longer be assigned to individuals, but can be traced back to a final screening process or method that underlaid the decision and be quickly adjusted as to avoid further errors. This is critical in further developing a culture of openness and innovation, where the system is accountable and adjusted as needed versus individuals being unnecessarily scrutinized.

From Table 22-1, one can see that most screening methods are not appropriate for a first screen. Simply put, the first screen is a guide for those submitting ideas; it must be fair, clear, and allow a reasonable number of ideas into the evaluation processes. Only 3 out of the 11 methods are acceptable for a first screen and they are theme criteria, exclusion or inclusion criteria, grouping and tiers methods.

The opinion of these authors is that the theme criterion is the most simple and valuable method for conducting a first screen. In this method, a number of themes are proposed that align with the organization's strategic aims. So if the organization's aim is to increase productivity, reduce defects in manufacturing, and come up with new products, then these should be the themes. Any ideas that fit into these themes would be automatically accepted; those that do not might be placed on hold, but not eliminated. Six to nine themes are acceptable; however, one should avoid themes that are overly broad, like "ideas that create revenue," or "ideas that build brand." The benefit of themes is they are easy to communicate to employees, easy to change, and force employees to find or create ideas that fit the themes.

The exclusion or inclusion criterion is a very valuable screening method, but requires a careful application. Inclusion criteria can relate to a special project that the company has undertaken, or other obvious acceptance rules (e.g., any idea that pays for itself in less than 4 months and improves efficiency, production, or reduces manufacturing defects). Exclusion criteria must be set very carefully as they can negatively impact the creation and submission of ideas. An example of a poor exclusion criteria would be "any idea that increases brand risk" because (1) one will not know the risk until after the idea is evaluated and (2) risky ideas, especially branding ideas, can have significant payoffs. These authors only recommend using exclusion criteria that relate to company values, so if the corporate value is being "honest with the customers" and a submitted idea is based on deceiving pricing practices, then it would be excluded by the criteria. When using exclusion criteria, good and bad examples should be provided in writing so that employees do not misinterpret or misuse this portion of the screening process.

Finally, the grouping and tier methods are useful, especially when the idea evaluation process is conducted in batches. Groups can be formed by any attribute, but generally grouping is organized after the ideas are submitted. This is different from themes, in that themes are predetermined and ideas are then categorized into them. Grouping only occurs at the end of the idea-submission batch process; examples of idea-groups include those that would improve operations, branding, increase efficiency, new product suggestions, and so forth. Tiers are slightly different because they break idea-groups into layers of predetermined qualities; for example, best, average, acceptable, or top, middle, bottom, as measured against some preset criteria. The authors caution that tiers should be purposefully labeled, and words such as "worst" should be avoided when describing idea-tiers because one does not want to ever discourage the submitter. Tiers are incredibly useful in cases when a large number of ideas are captured but only a few can be accepted into further evaluation. Further, placing ideas into tiers, one must be careful to do so in a way that seems fair but that is also measurable. Please note that when placing ideas into tiers, there is ample opportunity for error and bias due to normalization; that is, a great idea may be labeled an average idea simply because that round of ideation created more great ideas than previous rounds. Moreover, a poor idea may be placed in a top tier merely due to an overall deficiency in idea quality among a batch. Thus, the laws of averages apply, and organizations should attempt to preset criteria for what constitutes a great idea. Moreover, as more and more ideas are stored in a database, the problem of single-round normalization becomes less of an issue, as reviewers can access prior ideas to help further contextualize the tier standards or criteria.

First Screening Tools and Resources

The first screen can be improved by ensuring proper tools and resources are provided to those individuals submitting ideas and conducting the screening. To prepare the submitter for the first screen, it is wise to have an instructional handout or webpage describing the first screen methods. This generally increases the willingness of individuals to submit ideas (because they know what to expect) and they can prepare an idea in advance that would meet the first screen criteria. Further, proper dissemination and promulgation of corporate idea-needs increases the likelihood employees will attempt to identify and submit ideas.

Many software-based idea management systems (IMSs) have web portals where ideas can be submitted. However, rarely do these web portals have the first screening criteria listed on the submission page, which is a major oversight. Currently, many of these IMSs have screening criteria integrated to the webpage via predefined drop-down boxes. Unfortunately, if an idea is eliminated via the first idea-screen, the submitter is usually not notified, which creates a feedback gap and thus deprives the submitter of learning from their experience; worse yet, the lack of feedback often results in lower ongoing engagement levels.

Both of these problems can be eliminated via an instructional handout and live meetings. Webinars are sufficient for ongoing introductory sessions; but frequent, live, open sessions are important since there will be and should be questions about idea submissions, screening, gathering evidence, analysis, and the final selection of ideas. In other words, explaining the idea evaluation and management process is a very valuable way to disseminate innovation practices, encourage their use, and create a culture of innovation.

To further assist in submitting ideas and to perform the first screen, the innovation director, VP of innovation, or an assigned innovation manager should share their contact information with the organization and be available to answer questions. To conduct an internal audit of the transparency of an innovation program, there is a simple test: "ask a line employee to get in contact with the head of innovation at your company so they can ask a question about submitting an idea." If they cannot reach the innovation leader even with the assistance of their functional director, then your innovation leader has seriously failed; being connected with and available to the entire organization is one of their primary responsibilities. Innovation personnel are a bridge between organizational functions and provide the framework for interdisciplinary solutions to propagate beyond a single product development or marketing division. Active innovation cultures require a large number of people with a "connector" personality, or as Hardagon and Sutton termed them, "knowledge brokers."[4] Without these personality types, the innovation functions must rely on more formal communication processes, which often present barriers to the organic growth necessary for developing a culture of innovation. Without an individual pushing for a constant and valuable communication or idea, new ideas often fail to gain traction. The head of innovation should therefore embody these key characteristics, and have both a depth and breadth of background and experiences in order to better make connections between knowledge domains, as well as communicate with a myriad of other colleagues across business functions in order to build an innovation ecosystem.

Finally, not every idea is easy to explain in written words. Design concepts usually require visual aids and video-based methods to explicate or delineate their essence. Thus, submission of ideas should include methods by which documents, pictures, videos, and even physical prototypes can be captured, stored, and evaluated. For example, a fishing lure company could take prototyped lures as submitted ideas, and subsequently photograph them and store them so they can be properly evaluated (which, in this case may require reproducing the original and testing them via simulated fishing in a large stocked fish tank).

Common Problems

Common Problems with the First Screen

There are a number of common problems that occur with the first screening of ideas, usually around the submission of an idea, the individuals performing the first screen, the first screen methods, and methods for storing of ideas for future evaluation.[2] If you ask an employee how to submit an idea and they answer "I don't know," or "I am unsure," then there clearly is a problem. If they know where to submit an idea but are unaware of the first screening criteria, there is a problem. All idea management systems (software or person based) are useless unless employees are aware of their existence and functionality. Meetings should be held, memos should be spread, instructional guides should be easily accessible to all to promote and instruct employees how to submit ideas, how to pass the first screen, and what to expect from the process.

Another common problem occurs with the "idea manager" who may be developing improper criteria or misapplying the screening methodology. Simply asking employees how long it takes for an idea or innovation manager to respond to a given idea is the first step in uncovering inefficiencies and barriers in an innovation process. It is vital that idea managers be as accessible as possible to provide clarification. It is important they be affable and willing to apply the screens correctly. It is very easy to offend an individual submitting an idea; judging their idea, making negative comments about it, or not giving it the proper consideration, which permanently dissuades that individual from submitting subsistent ideas. Considering that creative individuals are 100 times more likely to submit an idea, it becomes very easy for one bad or poor idea manager to alienate the creative population of an organization and harm the entire idea submission and evaluation process. In some cases, managers become overzealous and choose to apply their own screens and disregard preset screening criteria in order to push a personal agenda. Idea managers who use personal preferences in lieu of a predetermined idea-screen should be replaced. Failing to ensure consistent, transparent, and efficient personnel policies creates a culture of distrust, which stymies innovation.

It is obvious when an organization does not have a first screen in place; however, it is less obvious when it is in place but the screening criteria are ineffective (e.g., dissuading employees from submitting ideas, or screening out valuable ideas). It is highly recommended that the first screening methods be limited to the three on Table 22-1, reviewed every 4 months, and updated as necessary to reflect alignment with the organization's strategic priorities. Further, after a major change to the screening criteria, it is useful to interview employees as to their experience in submitting ideas. Even organizations that are experts in gathering ideas could make a small change that inadvertently eliminates ideas of potentially significant value (e.g., consider the Innovative Nokia's failure to enter the touch screen phone market only to be almost put into bankruptcy).

Common Problems with the Final Screen

There are also several potential problems with the final screen, the first of which is getting all the key decision makers together to agree on reasonable screening criteria. One meeting may not be enough to conclude on criteria, and since there are no real ideas under consideration, most decision makers are hesitant to waste time on such an activity. To avoid this, have one individual tasked with putting together a final screening processes consisting of multiple screening methods. Present them to the group, get their feedback, then as a group, pick and choose methods to create the best model.

Another common problem is not being satisfied with the final screening criteria; this may be acceptable, however, as long as the group is regularly reviewing and appropriately updating the final screening methodology. However, the case where individuals bypass the final screen can be troublesome, such as in the case of an executive's pet project and other individuals' special interests. Requiring them to enter the idea evaluation process helps validate them, gain necessary support and resources, but, critically, avoids creating the perception that only special interests are approved based on politics or bias. Protecting the process' credibility and that of those who manage it is of paramount concern.

Another reason idea evaluation decisions are postponed stems from business leaders anticipating that the idea will morph and grow through the business plan creation process into a better and more feasible product or service. This is correct in the sense that creating a business plan does force more creativity and development around the concept, but this does not change the value of the underlying or fundamental concept. Despite how much development is done on a concept, it is still fundamentally constrained by the market; such development does not address industry competition and other uncontrollable factors associated with that product or industry. These higher-level attributes can and should be evaluated in the final screen to avoid clouding the decision-making process with extraneous inputs and wasting time and money during business case development.

Gathering Evidence

A wise man once said "go with your gut in blackjack, but check and evaluate carefully in business." This is also true about evaluating an idea. All too often, managers, directors, and even VPs at organizations evaluate an idea with nothing more than their own market knowledge, a feeling in their gut, and their years of experience. Although their many years of experience may help them in the evaluation of ideas that are merely incremental improvements (e.g., a better mousetrap), experience alone is insufficient when the idea represents an opportunity that is even a slight departure from the norm. But one cannot blame them for taking a shortcut and skipping the step of gathering evidence. Why? Because gathering evidence for an idea takes time and money, and delays selection of ideas for weeks, months, and even years in the case of fundamental new technologies; hence, people's proclivities tend toward minimizing efforts and making the process more efficient. Efficiency, however, is sometimes the worst enemy of effectiveness, and thus a process is needed to ensure consistent, reliable results.

The next section on gathering evidence is of significant use when formulating an argument to upper management for the vital activity of idea development. It also dives into a number of topics, such as what is considered evidence (is a single customer complaint evidence?), the different sources of evidence (can suppliers provide evidence?), what type of evidence is needed when evaluating a particular idea (what evidence is needed to evaluate a new business model?), and the amount of evidence that needs to be gathered (this helps avoid gathering too little or too much evidence). The first section also discusses the difficulties in gathering evidence, namely dealing with limitations due to cost, time, shortage of people, and potential legal issues (note that these limitations are always cited by managers and directors as arguments against gathering evidence for a formal evaluation of an idea).

The second section dives into the practices behind gathering evidence, namely the different methods and tools for gathering evidence (both primary and secondary research methods). Sadly, many corporations have lost touch with their customers because they rely so much on secondary evidence gathered through purchased market reports. Primary evidence obtained by the company may be directly gathered from the customers, and makes a significant difference in evaluating and justifying a new idea or concept (cite same voice of the customer articles); the methods of and tools for which are delineated below.

Types of Organizational Needs and Decision Criteria

Ideas can range from simple manufacturing process improvements to company-changing disruptive products. For the former, an average line manager may have the authority to approve the idea, and his primary concern may be cost versus effectiveness; while for the latter, a full business case may be needed that accounts for all the evidence and analysis related to aspects of the business from marketing to manufacturing. Unfortunately, a manager's functional specialty often skews his or her view of what constitutes a key performance metric (KPI); for instance, directors of marketing and sales will view customer support, customer retention, customer acquisition costs, and overall sales as important, whereas directors of manufacturing view ramp-up time or switchover costs as more important (see Table 22-2 for further examples). Hence, having an interdisciplinary team will yield the most thorough and comprehensive list of KPIs. Therefore, gather evidence and make a list of key decision makers and their core decision criteria; once this list is made it should be shared with anyone developing business cases, gathering evidence, or conducting an analysis on ideas so that all parties are aligned with the KPIs.

During the evidence-gathering phase, additional considerations should be given to how much of the evidence and subsequent analysis might be needed during future phases. Oftentimes data created via first screen processes provide vital insights during later phases. However, if organizations are not purposeful in capturing and storing these additional data points, they may be unavailable for future phases, thus precluding other decision makers from making fully informed decisions.

Again, every organization has its own criteria for evaluation and selecting ideas, but it is highly biased toward its current business model. A company that requires high profit margin product due to their high overhead rates will almost always ignore low profit margin products; this is often true even if the overall revenue or profit generated could exceed that of their current product lines due to the ability to capture more market share. This class of organizations is driven by "business model criteria," which can affect anything from the required profit margins, product development times, to market sizes, or minimum orders, which are independent of strategy, company culture, or personal preferences. Generally,

Title	Most Important Decision Criteria
VP of marketing	Market size, possible market share penetration/acquisition, potential revenues and gross profit, promotional and advertising costs, channels for advertising
VP of manufacturing	Cost to manufacture per unit, volume, cost and time to modify the manufacturing processes, product quality issues due to manufacturing error
Line manufacturing manager	Number of units to be manufactured per month/week, changes in manufacturing process, training requirement for employees
VP of sales	Sales potential, profit margin, price compared to competitors, sales channels, product attractiveness, cost to close each sale
Venture capitalists	Accurate estimates of company value and competitive attractiveness for acquisitions in 3 to 6 years. Uniqueness and technological advancements
CEOs	The strength of the internal support for the idea, risk versus reward, required capabilities to compete, competitiveness of the market

TABLE 22-2 Examples of Decision-Makers' Decision Criteria

ideas with KPIs that fall outside of these criteria require a different business model to be created, and one must gather evidence on how such a business would be set up and operated.[3] Any proof of similar businesses and their success is very much required.

The next type of organizational need would involve criteria driven by a company's culture and core values. Examples include Johnson & Johnson for those who put their families first and need to keep a safe and clean home, Whole Foods for those who need to focus on healthy eating, or Starbucks for those who have an appreciation for coffee, customer service, and sustainability. Ideas that are counter to the culture and core value are rejected. However, innovative companies recognize that while an idea may be counter to its values, there may be a possibility to spin out a new company and still create value for shareholders; and thus gathering evidence will still enable decision makers to evaluate and possibly launch the new product.

Other companies espouse industry-specific preferences and products that introduce biases in their evidence-gathering activities, such as those that tend toward technologies, retail products, specific market segments, personal values, synergies between industries, or specific customer segments. Specifically, organizations can dissect their markets via a host of geographic, demographic, psychographic, utility, benefit, and usage or volume segmentations. They also include the risk tolerance and general capabilities that are required. The list can be very long and it is important to write them down and see which criteria are very firm and which are flexible. Then collect evidence on the firm criteria first (showing that an idea is low risk) and, if time exists or the information is found in passing, note the evidence that supports the flexible general preferences (e.g., we like fast-moving products and these doggy treats are high-volume sellers).

Methods of Gathering Evidence

> *"A delusion is something that people believe despite a total lack of evidence."*
>
> —Richard Dawkins[4]

Various types of evidence are fairly easy to gather, while others are much more difficult; some may be openly available and written in books or magazines, or secretly held and orally kept. When venturing to gather evidence, the first distinction needing to be understood is that between primary and secondary sources of evidence. Primary sources are gathered directly from the source; for instance, if new customer opinions were required to justify a new product, then customer interviews, focus groups, or surveys would suffice. However, primary sources are unfiltered, thus extraneous data must be removed, and useful meaning must be made of the remaining data. Moreover, minor insights and other valuable information may also be obtained that would normally be omitted in a formal market research report. Although organizations can glean some value from the raw primary evidence, much analysis is required to extract accurate conclusions.

The major benefits of primary data sources include the ability to choose what information and data should be gathered, which gives an interviewer the ability to dive down into the granular detail of the customer's needs, problems, or product features, pulling out critical insights. Yet another benefit is the ability to gather follow-up information that pre-canned market reports do not provide. The major drawbacks of collecting primary data are the time and expense required to identify, locate, and acquire the customer data; interviews, focus groups, customer visits, and on-site observations all take time to set up and conduct, trained moderators and often budgets.

Secondary data sources involve evidence gathered from someone other than the primary source of the information. Most media outlets, magazines, books, articles, trade journals, market research reports, or publisher-based information are considered as a secondary source of evidence. These individuals or companies take evidence and compile it, screen it, analyze it, format it, and therefore effectively present the evidence and conclusions in a way they see fit. This treatment of the evidence can be very useful and explicate vitally important conclusions; however, it could also eliminate or obfuscate the very evidence that organizations are looking for to evaluate an idea. Further, because the secondary data was prepared with a specific audience in mind, there may be significant amounts of information that were redacted or overlooked, and yet other data that is not relevant to your particular situation. Finally, because publishers and other media outlets put out this evidence, there is still a cost associated with accessing it, a cost that sometimes exceeds that which would have been required to obtain new, primary data.

Since there are limits and barriers to gathering evidence, managers may attempt to cite these barriers, and thus avoid conducting a formal evaluation of an idea. While this tactic may be purposeful or just ignorance, it further emphasizes the need for a formal process with preset criteria and evidence-gathering requirements. Common barriers to gathering evidence may include limited money, limited time, limited people, and legal limitations; fortunately, there is a way around each to ensure organizations can gather the required evidence. Methods of overcoming barriers may include partnering with other, noncompetitor organizations, local universities, government agencies, departments of labor, commerce, or technology transfer, and other nonprofits whose mission is to enhance their economic regions.

In the case of limited money, one can always turn to Internet sources of information and low-cost popular publications. Governments around the world provide census and industry information for free, and a number of public and private institutions, especially those listed on stock exchanges, provide a wealth of information about the markets and customers they serve. Further, primary evidence can be very low cost; given a commission-based model, it may cost very little to make a large number of phone calls, or to send large e-mail blasts.

When time is a major limitation, one can switch to secondary sources of information such as market research reports, or subject matter experts who have detailed knowledge and current information on market trends and customer preferences. Similarly, forming a team to gather evidence may shorten lead times. When access to personnel is the limitation, organizations should use secondary sources. Certain primary sources, such as online surveys, are excellent methods to gather large amounts of information with little required resources. Finally, when legal limitations stand in the way of gathering evidence, one may have to switch to simulations, alternative yet similar markets, or find ways to gather the evidence through observation.

There are many methods of gathering evidence and most are discussed in marketing research articles and books. Table 22-3 lists a few methods to gather evidence under each class of primary or secondary sources.

Analysis

"If I would have judged my research's success by the number of mice that died, I would have missed the vital breakthrough that I found from analyzing the one mouse that lived."

—Dr. Roger Cunningham

Evidence analysis is vital and required for drawing out insights, key statistics, and conclusions from the gathered evidence. Alone, each piece of evidence may offer limited insights, but when combined and appropriately reviewed, the resulting key conclusions and insights can significantly aid decision makers. When analyzing the evidence, innovation managers must select and apply the appropriate analytical methods, of which there are a great number; so care must be taken in preselecting and deploying the methods.

Theory on Analysis

Raw unanalyzed evidence cannot be provided to the decision maker; they would balk at the sight of it, and thus the evidence must be analyzed. An analysis is performed to extract a focused insight, data points, or conclusion, such as potential total market, competitor market share, or expected profit margins. Most analysis methods are "fixed outcome analyses," which attempt to produce a given set of metrics or results, while fewer are "exploratory analyses," which are open to revealing any number of results. The goal of performing any analysis is to provide information regarding the key decision criteria required by the decision makers (see Table 22-2). Due to the straightforwardness of objective data gathering and analysis, most managers will tend to use quantitative methods; however, qualitative and mixed method analyses also provide significant insights when evaluating data.

When analyzing the evidence, it is easy to spend too much time analyzing data with the goal of performing an all-encompassing review. At the other extreme, reviewers can expend limited efforts on analyzing the evidence, thus resulting in limited or useless insights. Analysts must attempt to obtain a minimum level of findings that provide insights near the minimum required threshold set forth by decision makers. It is key to remember that while decision makers need information to aid in their process, they do not require 100 percent in predictive results; to obtain such data would assuredly require significantly more resources, with every additional data point ultimately providing diminishing returns. For example, if Manager Jones, the VP of finance, requires a profit margin above 5 percent at a 60 percent confidence level, then an analyst might reasonably stop gathering and analyzing evidence when a 50 to 55 percent confidence level is reached, and extrapolate to infer the difference. The next question may be "how does one determine the confidence level of the decision makers?" Test the manager by asking hypothetical questions such as "If I surveyed only 10 percent of the potential customer market and found this idea was highly valuable, would you consider that sufficient to move to the next step?"

Methods of Gathering Evidence from Primary Sources			
	Largest Benefit	**Largest Limitation**	**Largest Cost Dependency**
Interviews	Deep questioning	Not easily scaled	Hourly costs
Focus groups	Deep questioning	Not easily scaled	Participant compensation
Online surveys	Large sample size	Expertise to create	Cost to create
Handout surveys	Large sample size	Limited distribution	Cost to distribute
Online polls	Large sample size	Limited information	Cost to distribute
In-person polls	Broad sample	Limited to locations	Cost to distribute
Focus groups	Deep questioning	Limited by number of moderators	Participant compensation
Group conversations	Deep questioning	Limited by number of moderators	Participant compensation
E-mail questioning	Large sample size	Limited information	Hourly cost required to respond and analyze
Phone interviews	Deep questioning	Not scalable	Interviewer compensation
Web conferencing	Deep questioning	Limited by number of moderators	Moderator compensation
Online polls	Large sample size	Limited information	Cost to distribute
Interactive tours	Deep questioning	Not easily scaled	Participant compensation
In-person competitions	Deep questioning	Not easily scaled	Participant compensation
Online competitions	Large sample size	Building awareness	Participant compensation
Methods of Gathering Evidence from Secondary Sources			
	Largest Benefit	**Largest Limitation**	**Largest Cost Dependency**
Employee web access	Free sources of information	Screening out less valuable information	Employee hourly costs
Library access	Massive stores of knowledge	Their archive and online access	Subscription costs
Paid consultant/ contractor	Experts know information or has access to key sources	Their expertise areas	Consulting hourly costs
Automatic alerts (Google alerts)	Up-to-date news and information	Unscreened information	Employee hourly costs
Paid databases (Scopus. com, Jigsaw.com)	Accurate up-to-date statistics	Range of statistics and information	Subscription costs
Academic journals	Expert accurate knowledge	Range of topics	Subscription costs
Trade journals	Expert accurate knowledge	Range of topics	Subscription costs
Published magazines	Quickly digestible information	Limited depth of information	Subscription costs
Search engines	Large range of information	Screening out low-quality information	Employee hourly costs
Online dictionaries	Large range of topics	Out-of-date information	Employee hourly costs
Professional organizations	Focused topics	Limited focus	Membership costs
Patent databases	Easy-to-find information	Limited to patented	Employee hourly costs
Marketing/industry research reports	Quickly digestible information	Limitation of the study	Costs per report
Paid marketing research	Quickly digestible information	Research budget	Consulting costs
Published books	Depth of information	Focused topic	Employee cost to read
Government (census)	Depth of information	Unanalyzed information	Employee cost to read
Social media sites	Large sample size	Lots of false information	Employee cost to read
Online websites (Amazon, competitor home pages)	Large range of topics	Lots of false information	Employee cost to read

TABLE 22-3 Methods of Gathering Evidence from Primary and Secondary Sources

Functional Area	Example Analysis 1	Example Analysis 2	Example Analysis 3
Financial	Return on investment	Cash flow	Revenue stability
Market	Market segmentation	Niche market growth	General market metrics
Customer	Customer needs	Customer behavior	Market segmentation
Technology	Technology adoption	Technology road mapping	Technology forecasting
Products	Feature gap	Substitution	Usage trend
Manufacturing	Value chain analysis	Supply chain stability	Machinery analysis
Patents	Obviousness	Mapping	Citation
Sales	Product velocity	Sales growth	Profit margin
Strategy	SWOT	Porter's five forces	Pest
Engineering	Feasibility	Engineering costs	Capabilities matching

TABLE 22-4   Analysis Methods

Gathering evidence and conducting analysis can often be performed concurrently. Online surveys can be analyzed as the data is gathered; interviews can be analyzed to extract insights and trends. In a few cases, all the evidence must be collected before it is analyzed (e.g., manufacturing assessments). Plan to perform a preliminary analysis on the evidence with any idea evaluation projects, because in some cases the arguments against the idea may be so strong that the evaluation is stopped prior to the final screening of the idea (this is a good outcome as it saves resources).

The key decision criteria and performance indicators dictate which evidence should be gathered, as well as the methods used to analyze that evidence. There is a large number of analysis methodologies, so many in fact they would fill several books; however, each roughly falls into a particular functional area. In law, for instance, a patent attorney may perform an obviousness test to determine the patentability of a new invention over the prior art (inventions). Table 22-4 provides a list of functional areas; however, one should determine a list of analysis methods relevant to their own specific needs when creating their evaluation plan.

After the analyses are completed, one must present the finding in a way that is useful for the final screening. Because most final screenings comprise a panel of people, it is very useful to create a one-to-three page summary of the idea and results of the key decision criteria. Therefore, should a certain part of the final screening process require analysis of a potential profit margin, those figures and corresponding confidence levels are then placed prominently on the first page. A detailed report does not have to be created for each idea if the one who performed the analysis (analyst) can be called upon at the final screening meeting (available by phone or in person). If access to the analyst is limited, then the key decision criteria and additional information can be presented in a mini report.

Summary

Evaluating an idea properly takes work and is a rigorous process. Those who skip the process steps of conducting the first screen, gathering evidence, and conducting an analysis, and instead jump straight to the final screening of the idea have very little information to aid in making an educated decision. Poorly developed innovation processes result in major bottlenecks in the innovation process and result in the selection of ideas that are thusly based on hunches, personal preferences, and political biases. Those organizations that follow a proper idea evaluation process have the opportunity to increase profits and success rates of new product launches by evaluating ideas prior to investing a large amount of time and effort in research and development. These saved resources can be redeployed to further sift through a larger number of ideas to find the most likely winners.

The proper, effective, and profitable evaluation of ideas begins with a first screen, where ideas are evaluated against high-level screening criteria to eliminate those that are not aligned with strategic level goals for the company. If ideas are valuable but not immediately deployable, those ideas are stored in an idea bank or repository. Next, several ideas or batches of ideas are chosen for evaluation. In order to evaluate ideas against the key decision criteria set forth by the decision makers, a plan is created that spells out what evidence is required and what analysis methods need to be performed. Because gathering evidence can be done in myriad ways, the plan can be adapted to deal with limitations in terms of money, time, available labor, or legal constraints. The evidence is then put through the chosen or most appropriate analysis methods in order to extract the conclusions relevant to the key decision criteria. After the vast majority of the key decision criteria are analyzed to a level of confidence required by the decision makers, the idea is summarized along with the results of key decision criteria in a report format or brief slide deck. The final screening of the ideas requires the committee members to have predecided on a general, organized, and rigorous process. The final screen uses different screening methods to evaluate if the ideas meet the requirements of the company, firm preferences, or individual requirements. The final screen provides a highly objective screening of ideas where fervent debate can take place. After the ideas are put through the final screening process, one, or a batch of ideas are selected that best meet the company's current and future needs, strategic goals, and other operating requirements. These ideas can be put into the product development process, or if a clear plan is required to proceed, they can be developed into a formal business plan. Because the idea evaluation process quickly gathers the vital evidence and enables the critical analyses needed, the resultant ideas are more likely to proceed with employee backing to market launch with the support and resources of key internal managers.

References

1. Walton, A., Glassman, B., Sparer, T., and Sandall, D. Managing Ideation, Sonae.com Research. Paper presented at the 2010 Business Innovation Conference and Exhibition, Wheaton, IL.
2. Glassman, B. Improving Idea Generation and Idea Management in Order to Better Manage the Fuzzy Front End of Innovation. PhD Dissertation, Purdue University, West Lafayette, IN, 2009.
3. Osterwalder, A. and Pigneur, Y. *Business Model Generation*. Hoboken, NJ: John Wiley & Sons, 2010.
4. Brainyquotes.com. Quote reference, Brainy Quotes 5-9-2013. http://www.brainyquote .com/quotes/quotes/r/richarddaw447499.html. Accessed September, 2012.

Methodologies

The methodology section of this book is, to put it mildly, the mother of all sections. Many readers will have purchased this book just for the methodology section. For the 12 chapters we've included in this book, there are probably 24 chapters that our readers will tell us we should have included. As we developed the book, we understood that we were building a body of knowledge in a rapidly changing and involving field. Therefore, we tried to include what was most appropriate at the time it was written. We have included a chapter on types of innovation in an effort to capture some of the less common methodologies. We've also included some chapters that people may not consider innovation. This section offers a perspective on innovation that will allow you to evaluate all other methodologies in a whole new light.

As mentioned earlier, the general types of innovation are both wide and deep. From the beginning of time, or at least since commerce began, humans in competition with other humans, specifically in business, have found a constant assault on their product or service. Generally, this assault comes from competitors who have newer, better, faster, or just different offerings. Vern Burkhardt points out that in the early 1800s, ice was supplied from frozen lakes typically in New England. These were the days of ice houses as a means to store ice through the summer. The industry employed lots of people throughout the world. Global competition for ice was intense and spread throughout the world as a profitable venture. Into the early 1900s, refrigeration methods were invented. This led to giant refrigeration plants, which meant cutting ice from lakes was no longer necessary. Eventually, we all made our own ice in our freezers and the industrial ice business has almost disappeared.

The point of this story is that the world changes. More recent changes have included the displacement of the typing pool by the personal computer. Accordingly, we must recognize that the world is constantly changing and that without innovation methodologies, we will find ourselves going the way of industrialized production or worse, the way of the New England lake ice producer.

The rest of this section focuses on those innovation methodologies that are most common. For example, Ellen Domb's chapter on TRIZ outlines the theory of solving inventive problems. Developed post-World War II between 1946 in 1985 in the former Soviet Union by Altshuller, TRIZ has become the best-known method for systematic, logical, left brain creativity and innovation. Although this sounds like an oxymoron (logical left brain creativity), anyone who has experienced TRIZ done well will understand the value of this methodology.

Another interesting methodology is addressed in Gupta's chapter on Brinnovation. This chapter discusses the trial and error process that has been undertaken by humans since the beginning of our existence. One might also call this the cycle of experimentation, knowledge, and innovation that continually repeats itself. One needs to only look at a baby learning to walk. As most of us know, walking on two legs is a highly complex interaction between brain, nerves, and muscles. Humans, unlike many animals, are not born knowing how to walk. Somewhere in the second year of life, humans start to walk. Learning to walk becomes an experiment in balance, in strength and, more important, in endurance. Eventually, however, the baby has experimented with all the wrong ways to walk (fallen down), and has learned how to walk.

Another modern innovation methodology is discussed by Juan Vincente Garcia Manjón in his chapter on crowdsourcing. Crowdsourcing is a fascinating activity that presumes that the wisdom of the crowd is greater than that of any one individual. Gone are the days when the chief executive officer of the company dictates all of the new products and services and expects to be successful. Instead, companies now employ methods of crowdsourcing for anything from writing to product or service design. Is the crowd always right? Perhaps not. If Henry Ford had listened to the crowd, he would have been breeding faster horses. Instead, he gave people what they didn't even know they wanted. But even in those cases where the crowd doesn't appear to know what they want, there are always a few dreamers among them. Even if the public didn't know they wanted a Model T, there were already members of the public who were, in fact, dreaming of the automobile.

Another relatively new approach to innovation is that of open innovation. Scholten and Temel outline in their chapter the Chesbrough model of open innovation as a tool for the transfer of knowledge in and out of the organization. The theory being that this would increase the effectiveness of innovation management and eventually generate revenue and profit for the organization. Much of this is based on the idea that research and development has long been considered an internal process, kept secret from the rest of the world. Utilizing business partners, customers, and others, open innovation allows borders to perish and innovation to thrive.

Systematic innovation, on the other hand, brings invention into the discussion. Coates, in his chapter, tackles the issue of systematic innovation. He defines innovation as:

- The solution has potential utility for users.
- There is novelty, where novelty means the state or quality of being new, exciting, unusual, or unique.
- The solution to a problem is not obvious to those skilled in the art; that is, there could be some psychological inertia that prevents others from seeing the solution.

Accordingly, this means that innovation takes something new and puts it into place for popular or mutual good. Just as in TRIZ, systematic innovation attempts to organize the innovation process. Ways of doing this include the general scientific method and all its offshoots. Six Sigma and lean tools can be included as well many other tools from the total quality management movement.

Of course, some of the methods aren't as methodical as others. Meyer discusses in her chapter on Eureka that sometimes innovation is just about an "aha" moment. We have all heard and experienced the "aha" moment. Mine tend to come to me in the middle of the night or while I'm driving. This is perhaps because the mind is occupied with other thoughts. Of course while you're sleeping or driving is the worst time to capture those ideas. Meyer discusses the roots of this "aha" moment and how you can encourage these insights to come when you need them.

Stage-gate is another popular methodology that has recently come to the attention of the innovation community. Di Benedetto recognizes that controlling innovation is a critical piece of information management. Everything from idea inception to product launch must have some degree of order. Even in the most innovative shops in the world, methodology prevails. When you look at an organization like IDEO or Apple, you'll find stage-gates throughout the whole process. This is due to a number of factors, most specifically the fact that investment and uncertainty tend to move in opposite directions. While this may be a remnant of the past, it is still prudent not to undertake large investments without first understanding the risks involved. In innovation, risk is often not easy to evaluate. Therefore, establishing stage-gates allows you to reevaluate the project on a regular basis.

A popular methodology in innovation comes from the design industry. In his chapter, Alexis explains that for too long, companies have ignored design as a feature that isn't necessarily important. Even automobile manufacturers, who thought design was all about how cars looked, did not understand that design goes to the core of products such as an automobile. It includes eliminating vibrations, improving comfort, and numerous other factors that allowed competitors to move ahead of U.S. manufacturers late in the 20th century. To be precise, design matters, and more and more companies are starting to understand it.

Similar to design innovation, service innovations have become popular in recent history. Vidyarthi and Tyagi in their chapter on service innovations discuss the criticality of paying attention to innovation in service. They explain the characteristics and challenges of service innovation, its methodology and measures of service.

Brocato and Schmidt in their chapter on Social Innovations touch on one of my favorite questions I ask of students and crowds at conferences. Do you think Captain Kirk or Captain Picard is the better manager? Obviously, Kirk is more exciting and always wins the day; but Picard has trained a team of highly skilled lieutenants to handle any problem that comes up, even if he is incapacitated. This is the central question addressed in the social innovation chapters. In a world of changing organizational dynamics, the Picard model has to win out to encourage innovation.

Innovation in the social sector is the topic of Weichert's chapter. "How is this a methodology," one might ask? You only need to look at the way the Gates Foundation has changed the face of nonprofit innovation. The term venture philanthropy has snuck into the lexicon of the community as donors and funders have begun to ask for accountability in how their money is spent. The innovations discussed in this chapter will help all innovators consider the way they do their work.

Cross Industry is another new innovation methodology that illustrates the point that everything old is new again. Old and wizened innovators have long known that the best new ideas for their business can come from other industries. The most popular story is how hospitals have learned customer service from hotels. With a good imagination, one can think of a hundred different ways that cross-industry innovation can be a highly effective method. Ferreira, Raposo, and Fernandez discuss cross-industry cooperation as a key innovation methodology utilized by many innovative organizations and innovators.

Once again, there are probably dozens of tools that could be considered innovation methodologies. How these tools are used varies from organization to organization. The most innovative organizations understand these tools and how to apply them to maximize their innovation return. As an innovator, you should always strive to use the right tool for the right job.

CHAPTER 23

Types of Innovation

Vern Burkhardt

Introduction

Most businesses operate in an environment of ever growing uncertainty and are subject to being blind sided by the entry of new competitors, price competition, technology advancements, the "next best thing" being introduced into the marketplace, societal trends, global competition, and a myriad of other possibilities that disrupt their current strategies. In the face of uncertainty the natural tendency is to retreat, seek to increase efficiencies, and reduce costs. For a period of time this approach may serve shareholders' and owners' emphasis on enhancing profits, but inevitably time will run out.

The natural ice trade (cut from frozen lakes) began in New England in the early 1800s (Fig. 23-1), and businesses in various parts of Europe entered the fray to profit from the sale of ice to countries near and far—even as far as India. Capital investment and expansion of operations continued into the early 1900s despite early signs of other innovations, ice produced in cooling plants and even more disruptive refrigeration cooling systems. Most of the natural ice traders responded by increasing efficiencies in their operations, but were soon overwhelmed by the new technologies and went out of business. The ice traders had reveled in the comfort that natural ice was a superior product and failed to anticipate how innovation would result in the inferior cooling plant ice becoming cheaper and of higher quality. Ultimately refrigeration led to the elimination of the need for ice for cooling, and therefore the need for ice chests in private homes and ice in commercial applications.

The response to increasing uncertainty and competition cannot be to keep doing the same things but only bigger, better, faster, and cheaper. This is true unless long-term survival is not the goal. The best response is the development of an internal culture that reflects a "sense of urgency" for change. John P. Kotter describes such a culture as a huge asset in our "turbulent, fast-moving world"[1] and Scott D. Anthony advises, ". . . the new normal is perpetual change."[2]

Symptoms

Clinging to a business-as-usual operating model is not a recipe for survival much less higher profitability. Trying to squeeze more juice out of the same lemon eventually reaches the point of diminishing returns. Jeffrey Phillips describes how mature firms tend to be drawn to a business-as-usual operating model, which can result in the acceptance of "poor decisions, unwieldy processes, poor communication, contradictory reward structures, and the absence of clear goals *without question*."[3]

Organizations, large or small, that have a business-as-usual model built into their cultures, business practices, approaches to risk, and human resources practices tend to have inherent impediments to innovation. Disruptive new products, services, and business models are direct threats to business-as-usual. Failure to respond may result in cutting ice blocks in winter with plans to sell them next summer. Meanwhile the refrigerator has eliminated virtually all demand (Fig. 23-2).

Every reader will be able to name well-established businesses that have failed. Being in business longer does not guarantee survival. As can be seen in the following chart the survival rates of businesses follow the same downward curve irrespective of when they commenced business (Fig. 23-3).[4] Sobering indeed. One question business leaders must ask

Figure 23-1   New England ice harvesters in early 1902. (Library of Congress.)

themselves after their previously successful business has failed is why they didn't allocate a portion of their resources to innovation while they were profitable. "Creative destruction" is an apt term to describe the births and deaths of businesses. Since 1975, when Joseph Schumpeter[5] coined this term, the pace of change has exponentially escalated in the never-ending search for new and imaginative methods, processes, technologies, products, services, and business models.

The conclusion: innovate or die. Firms that are able to innovate successfully and continually will differentiate themselves from their competition. This is especially true as organizations lead, or are impacted by, technology change at an ever increasing pace.

It has been observed that most organizations rely on their traditional attempts at "innovation"—improvements to business-as-usual—to provide security, and hope they might stumble upon a disruptive innovation, if they are lucky. This is not a formula for longevity. Many people think innovation is random and/or requires creative genius. We believe

Figure 23-2   Business options: ice or refrigerator?

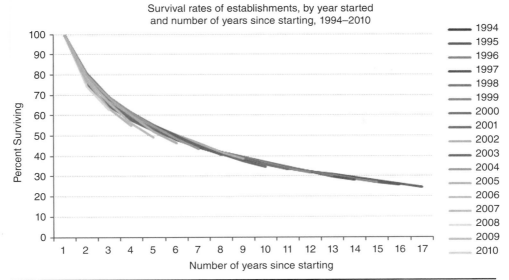

FIGURE 23-3   Handbook of innovation: U.S. survival rates of businesses.

that by putting in place the right structures and systematic steps most organizations could improve their odds of creating profitable growth and a sustainable competitive advantage.

Much attention has been placed on game-changing, radical, breakthrough innovation. A lot of hype has been generated around examples such as Apple's iPod and iTunes, IBM's rapid transition from hardware supplier to business consulting services and from mainframes to grid computing, W.L. Gore's flat organizational structure, 3M's goal of achieving 20 percent of its revenue from products created in the last 3 years, P&G's target of 50 percent of ideas coming from outside its R&D organization, University of Texas' development of 3D printing, and Better Place's interchangeable battery system for electric cars. There are many more.

Successful companies incorporate processes for generating both sustaining and disruptive initiatives for their innovation pipeline. The majority of organizations may never achieve a radical innovation, yet they are able to remain viable and profitable. This is because they have adopted change as a fact of life and are engaged in more than one type of innovation.

Based on a survey of 1130 CEOs worldwide in 2008, IBM described the enterprise of the future as aiming "beyond articulated needs and wants, creating first-of-a-kind products, services, and experiences that were never asked for—but are precisely what customers desire."[6] The enterprise of the future will not result from innovation chaos. While not as tight as a manufacturing process with all its process controls, innovation will be done in an orderly way, and integrated into the product development process.

In the rest of this chapter we will discuss some of the prevalent types of innovation (Fig. 23-4), though it is sometimes difficult to separate them because organizations are

Process Innovation

Functional Innovation

Design Innovation

Product Innovation

Service Innovation

Business Model Innovation

Co-creation Innovation

Open Innovation

FIGURE 23-4   Types of innovation.

often engaged in more than one type at the same time. We will start with process innovation that often, but certainly not always, uses primarily internal resources. We will then proceed to discuss functional, design, product, service, business model, cocreation, and open innovation. As will be seen, these types are not pure because multiple types are usually involved in any innovation.

Process Innovation

Process innovation is the innovation of production processes. To varying degrees all organizations are involved in the implementation of new or improved production or delivery methods. Improved production may involve the use of new machines, new ways of using existing machines, new methods of production, new ways of serving customers, or even the use of different software to enable more efficient processes. New or improved delivery methods may occur in all aspects of the supply chain in an organization. The cornerstone to process innovation's dramatic results potential is information technology—a key enabler of process innovation.

Management initiatives such as Six Sigma, *kaizen*, Lean manufacturing, and JIT inventory levels derived from the Toyota Production System (TPS), Theory of Constraints, TRIZ, lateral thinking, design of experiments, and many others have been employed to improve profit margins by removing nonvalue added steps from workflow. The dichotomy is that the customers' cost-reduction objectives must be met, while also making the production process more capital-efficient in order to maintain vendor margins. Michael Dalton has developed a process titled Speed-Pass that allows innovation teams to continue with their innovation process, provided the project's goals and objectives are being met—the "hinge assumptions"—rather than going through a stop and gate review which is typical of most stage-gate product development processes.[7]

David Magee describes Toyota's TPS as follows: "As a company, Toyota understands the pursuit of greatness as a long-term activity, resulting from the continual improvement of its processes. Toyota works toward tomorrow by getting better today."[8] Toyota was by no means the first to focus on process innovation. Continuous process improvement was the message behind Walter Shewhart's "Plan-Do-Check-Act" cycle, first published in 1939, and in the 1950s W. Edward Deming changed the "Shewhart Cycle" to "Plan-Do-Study-Act" in order to emphasize process analysis over inspection (Fig. 23-5). Plan a step or process. Follow your plan. Study the results and analyze what you achieved. Based on what you learn, improve that step, or a similar next step.

The Theory of Constraints (TOC), which was first described in 1984 by Eliyahu M. Goldratt,[9] is based on the principle that the output of any system is constrained by its lowest performing element—the "bottleneck." Thus, rather than wasting resources to improve those parts of

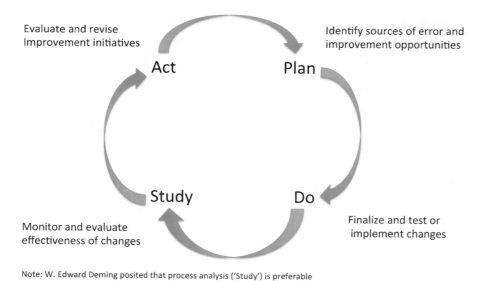

Evaluate and revise
Improvement initiatives

Identify sources of error and
improvement opportunities

Act   Plan

Study   Do

Monitor and evaluate
effectiveness of changes

Finalize and test or
implement changes

Note: W. Edward Deming posited that process analysis ('Study') is preferable
over quality inspection for achieving continuous improvement

Figure 23-5   Deming's continuous improvement model.

the process that are already over-performing compared to the bottleneck and creating process problems such as storage costs and even inventory obsolescence, efforts should be devoted to finding the bottleneck, setting a goal for improvement, and making the small (or large) changes which will have a big effect. It's about focusing your innovation on achieving the ideal capacity at the leverage point, the one step that constrains the overall process. Once that bottleneck has been addressed attention should then be focused on the second ranking bottleneck, and so on. Some bottlenecks cannot be significantly improved so the innovation may be to reduce the surplus output of adjacent processes (e.g., a high cost drug-testing process). This applies even to complex systems with many processes. "While TOC grew out of manufacturing, this approach applies to innovation for any market—whether consumer or B2B. TOC has even been used in service businesses."[10]

During my college years, one of my summer jobs was at a factory that spun fiberglass insulation out of melted sand mixed with catalysts—like candy floss at community fairs. The plant had been closed for several months for a major upgrade to increase throughput and product quality. When it reopened operations there was a major process problem. The manufacturing process would infrequently generate large unspun globs of the raw materials scattered throughout the insulation, rendering it unsaleable. The engineers and chemists scrambled to come up with solutions to the bottleneck process problem while under intense pressure from the production manager due to the daily loss of tens of thousands of U.S. dollars. The bottleneck was in the jets, which spun the insulation, but during this time attention and resources were diverted to improvement of other lower priority processes, such as improving warehousing and the hydraulics in the packaging part of the assembly line process. They even shut down for preventative maintenance, which delayed testing of adjustments being made to the bottleneck process. The company wasn't aware of TOC. It was before Eliyahu Goldratt had his insight!

Process Innovation is often associated with incremental innovation but it need not be. In manufacturing major process innovations may be realized by reducing the number of parts in a product and therefore the complexity of assembly. Ford reduced the number of parts in the Ford Taurus by 28 percent compared to the Ford LTD. Examples of process innovations which were fundamentally disruptive include the Henry Ford team's assembly line which brought components to the worker during fabrication of the Model T, fabricating microchips using the lithography method, Dell Corporation's customer-driven manufacturing of computers, radio frequency identification in logistics operations, use of scanners and bar codes, use of the Internet to enable customers to track and trace their parcels being shipped through couriers, and the list goes on . . . and on.

Dramatic changes in processes can enable an organization to realize a breakthrough innovation. When manufacturing an existing product there is low uncertainty, so innovation of the product itself often tends to be incremental in nature, with increasing focus placed on process innovation. New product innovations have a higher level of uncertainty and their production process is more uncertain and idiosyncratic, thus there is more reliance on trial and error learning.

Functional Innovation

It could be said that the etiology of "functional innovation" is "systems thinking," in which a problem is seen as being part of a total system. This is in contrast to addressing a problem in isolation with the potential for wasted effort or unintended consequences. Not only are the components viewed in the context of their interrelationships within a whole system, but also in their relationships with other systems.

Functional innovation involves identifying the functional components of a problem or challenge and then addressing the processes underlying those functions which are in need of improvement. Through this process overlaps, gaps, discontinuities, and other inefficiencies can be identified.

The key to functional innovation is breaking a challenge into its functional components so it can be fully analyzed and understood. It is then possible to identify the functional problems that need to be addressed. This approach enables even those entrenched within the problem, or those in one of the specialized disciplines in the system, to disaggregate a complex situation. It should also assist in breaking down barriers as it becomes more evident, if not obvious, where there are opportunities to change the whole system by making changes in the functional components.

For example, let's assume our organization has a legacy accounting system which is negatively impacting the business. Rather than simply replacing the system with a more modern system, following a functional innovation approach would have us identify the functions involved in supporting our financial processes, their interrelationships, and

Design
Engineering
Head of Innovation
Sales and Marketing
Information Technology
Management
Accounting
Early Adopter Customers
Ex customers
Zero gravity thinkers
Suppliers

Disorder in the System (The Challenge)

Note: Cross-functional teams are effective for product development, designing services initiatives, and solving specific problems.

FIGURE 23-6 Cross-functional teams.

opportunities for streamlining and enhancing the user experience for our accounting, line and management staff, customers, suppliers, and others involved. Accounting would be broken into various functions in order to identify opportunities for innovation. Accounts payable would have a number of subfunctions, including the decision to purchase, procure, commit, receive, inspect (if goods), verify against contract (if services), resolve deficiencies, request payment, process payment, print check/electronically transfer funds, archive files, and audit. Then each subfunction would be analyzed with the aim of streamlining processes, minimizing duplication of effort, reducing end-to-end processing time, and reducing errors. This would be done in light of the constraints or opportunities afforded by a possible new accounting system.

Functional innovation can be applied to challenges such as engineering problems, customer account management, and improving relationships with customers. This is often where someone with an outside, clean-slate perspective can be useful, and why one of the values of consulting firms is their ability to use well-honed templates as well as questioning and thinking frameworks, to clearly identify functions in need of improvement and provide recommendations for that improvement within the context of the whole system.

A common approach is to establish cross-functional innovation teams, which come together to work on a specific problem, product development, or service initiative (Fig. 23-6). Members of the team would come from each of the affected business units in different areas of a company. Selected outsiders, such as known early adopters, noncustomers, supplier partners, and others may be included during some, if not all, of the innovation process. The teams may also include what Cynthia Rabe called "Zero-Gravity Thinkers."[11] They are outsiders who ask questions from a "naïve" point of view about the technology being addressed. As Rabe explained of her role as a Zero-Gravity Thinker at Intel ". . . it became clear I added value precisely because I wasn't an expert and because my naïve questions, my out-of-left-field questions that nobody who had a technical degree would think to ask, caused the engineers and technologists I worked with to look at problems from a very different angle. So at times we came up with solutions to problems that led to technology patents, to marketing or business-related solutions that they were not on a path to come to otherwise."[12]

Cross-functional teams lead to a diversity of perspectives and insights, which would likely not result if the entire team came from one business unit or work area. Involving participants from wide and deep within the organization avoids the blinders of "silos," "groupthink," and "expertthink."

Design Innovation

Let's start by touching on what design innovation is not. It is not focus groups, quantitative market research, traditional market-testing techniques, asking customers what they need or want, though some products and services are designed in this manner, or even asking customers or potential customers what they think about a product. It is not restricted to focusing on quantum leaps in product performance that are enabled by breakthrough technologies. If at the beginning of the 20th century an entrepreneur had asked the providers of pigeon post what they wished for by way of invention, it is unlikely they would have identified telegraph, telephone, or wireless communications. If operators of stagecoaches had been asked what innovation would disrupt their business, it is unlikely they would have identified the steam locomotive or automobile. "Breeding homing pigeons that could cover a given space with ever increasing rapidity did not give us the laws of telegraphy, nor did breeding faster horses bring us the steam locomotive."[13]

Over 55 years ago Peter Drucker made a prophetic statement when he said, "The customer rarely buys what the company thinks it sells him. One reason for this is, of course, that nobody pays for a 'product.' What is paid for is satisfaction(s). But nobody can make or supply satisfactions as such—at best, only the means of attaining them can be sold and delivered."[14] This wisdom still applies today.

In 1974, Theodore Levitt introduced the compelling concept that customers acquire products to accomplish a particular task. They don't set out to buy products, such as a drill, as ends in themselves. As a problem arises in their lives, such as how to make a ½ in hole in a piece of wood, they look around for the product best suited to solving it. The drill they buy is a way to get that job done.[15] This means that when designing products (and services) the aim should be discovering what jobs customers need to get done, which are not getting done very well, or at all, with existing solutions.

Understanding the customers' job-to-be-done is the beginning of designing an innovative customer value proposition. As in the example of the carrier pigeons and horses, customers generally cannot imagine the convenient, easy-to-use, reliable, and affordable instrument they need in order to accomplish a specific job. There are opportunities for innovation when there is no adequate solution to an important job-to-be-done, where people are resorting to workarounds, or tolerating unsatisfactory products or services. Opportunities also emerge when one realizes that with the exception of some dedicated brand followers, most customers don't care what products or services meet the fundamental problems they need solved, or what will fulfill their social and emotional needs.[16]

Design innovation goes further than focusing only on the functional dimension of the job-to-be-done. It also involves the social and emotional dimensions, which are sometimes more important than the functional. Great architects understand, when designing a structure, that all of its components should have a purpose or they will appear out of place and perhaps even irritate its inhabitants.

Roberto Verganti refers to products and services as having a twofold nature, the utilitarian dimension of function and performance, and the more abstract dimension of "meanings."[17] Meanings are involved with symbols, identity, and emotions, and both incremental and radical innovation of meanings can occur in any industry. This leads to a number of possibilities, such as predicting what customers will want in terms of meanings before they are aware of it. Apple introduces new products and tells potential customers how they will use them, how they will impact their lives—this is a type of magic and it consistently led to much customer anticipation of Steve Jobs' product launches, not to mention the more important follow through of product launch sell outs. Wherever there are customers with an unsatisfied job-to-be-done there is an opportunity, and if the solution satisfies the functional need and also provides high-value meanings it may be a disruptive innovation.

The challenge when designing new products for a global market is that different cultures give different meanings to the same products, because different cultures have different values, beliefs, norms, and traditions. This is a challenge, but understanding it can also lend a competitive edge when designing what are hoped to be breakthrough products globally. For example, if an auto manufacturer introduced a luxury car with a lot of space in the front seat and less in the back into an emerging economy where the custom is to have a chauffeur, this vehicle would have a different meaning than if introduced into North America or Europe. The owner would have less room than the chauffeur! (Fig. 23-7).

Even high-tech companies need to include design innovation as an integral part of their product development. Verganti calls this a "technology epiphany" which can help make the company a market leader. He identifies Nintendo Wii, and Apple iPod integrated with iTunes, as examples. The Kindle and Kobo eBook readers are other good examples. He also provides additional useful advice: "If you only compete with technology to improve performance you will realize only incremental change in the meanings of your products. This will limit your potential for profitable growth. This is the case when you are responding to what customers in the marketplace are saying, and even when there are radical changes in technology. Radical changes in the meanings of products occur only through design-driven innovation, even when there are radical changes in technology."[18]

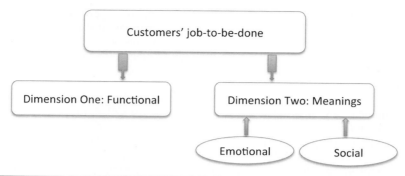

Figure 23-7   Innovative customer value proposition.

A design innovation perspective suggests that we shouldn't design a product with so much functionality or so many options that it confuses or causes negative feelings about the product—"featuritis." We are all aware of electronic products that are overburdened by functions that add unnecessary complexity.

When you implement a new feature, within a month or two another company will almost definitely provide the same feature but with improved performance. Aspects of products or services that connect emotionally or symbolically with people are not as easily replaced or overcome. Their competitive advantage is in the innovation of the meaning of the product. Other auto manufacturers may replicate the functionality of a BMW, but they have difficulty competing with its meaning—"I'm special."

How does one go about design innovation? The fundamentals of design are observation, brainstorming, and prototyping. Observing involves watching people to understand what job they want to get done, or observing people, processes, companies, or technologies to discover an idea that could be used effectively in a different context, perhaps with modifications.[19] Tom Kelley uses the term "spontaneous team combustion" to describe brainstorming sessions at IDEO. Building even simple prototypes is a great way to communicate the essence of an idea and to enable others to build on it. Tom Kelley says, ". . . a good prototype is worth a thousand pictures."[20] He might also have said it is worth more than ten thousand words.

Designers are good at not starting to design until they've understood the real job that needs to get done. Rather than starting from the idea of the product or service, good designers start from the point of view of the customer, the user.

Product Innovation

A lot of product innovation is occurring worldwide, but judging by the end result, not all of it is useful or successful. The numbers of new patents can measure the pace of innovation. The U.S. Patent and Trademark Office granted 2,622,917 utility patents prior to 1998 and since 1998 it has issued 2,369,275.[21] In 2011 there were 224,505 utility patents granted in the United States, 238,323 patents, utility models, and microorganisms granted by Japan, 172,113 by China, and 62,112 by Europe, as well as many in other countries.[22]

Product innovation is a multidisciplinary process usually involving many different functions within an organization and, in large organizations, often coordination across continents. In the product development process one of the most important interfaces is between marketing and R&D (Fig. 23-8). Marketing will identify jobs-to-be-done opportunities and, without making the typical error of over estimating potential sales, estimate the extent of the market potential. Marketing, or in some cases the Chief Innovation Officer, will be responsible for drawing conclusions from the observations of, and discussions with, early adopters in a market. They will also obtain input from existing customers, and attempt to understand what would attract noncustomers.

Gaining a competitive advantage through product innovation also requires manufacturing and logistics that are capable of providing an uninterrupted supply of the product.

Product innovation occurs in the complex milieu of the marketplace, where people you know about, as well as some you don't know about, are working at an ever-faster pace to compete with you. They are working on incremental innovations to increase their market share, and product and technology breakthroughs aimed at dominating your industry.

Almost 200,000 new products are introduced into the marketplace each year. Many fail. There are many reasons for failure, besides that of a poor showing in the categories of

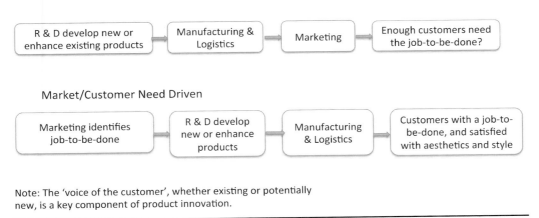

Note: The 'voice of the customer', whether existing or potentially new, is a key component of product innovation.

FIGURE 23-8   Approaches to product development.

function and meaning. Many physical and online products fail because they didn't have a sense of purpose from the customers' perspective. They didn't fulfill a job-to-be-done for enough potential customers. Doubtless some disappointments and failures could have been avoided by carefully target testing in a small market area before a full launch of the product nationwide or beyond.

Other reasons for product failures include undercapitalization, a flawed business plan, poor leadership, product quality problems, brand loyalty, commoditization of the product line, stiff competition, manufacturing problems, poor distribution channels, marketing failures, and the price being higher than potential customers' perceived value of using the product. Some might also say luck or the power of surprise plays a role.

One thing to keep in mind when innovating consumer products is that aesthetics and style matter. If the product is beautiful and looks functional, customers are more likely to be patient if they have to struggle to learn how to use it. "Even if you're trying to design something profound and noble, it still benefits from being sexy."[23] You will know this is true if you have ever purchased an Apple product—even the packaging is part of the emotional experience.

The scope and complexity of product innovation has also changed because informed and demanding customers are able to impact the success of a product through social media—both positively and negatively. "More demanding customers who place a higher premium on the experiential qualities of using a product—ease of use, how it makes them feel, how it fits into their lives, what it communicates to others—that go above and beyond familiar objective criteria like performance and price."[24]

Whirlpool has integrated its stage-gate project management process into its innovation processes ("I-pipe") and product development process. This reengineering was motivated by the CEO's dictum "Innovation will come from everywhere and everyone. . . ." The company established a set of tools to ensure that the "voice of the customer" is included in the product innovation process. The goal is to focus on the best innovative ideas which could both delight consumers and keep the company's brand promise.[25]

In many industries there can be an initial flood of new competitors into the market with comparable products, but using different technologies. Each may be promoting a different standard—for example, different voltages of electricity during Edison's era or Sony's Beta recorders versus VHS. Eventually one design or standard becomes dominant industrywide, either through the success of one company or as established by an industry association. Today there is one standard size and shape of Universal Serial Bus (USB). Europe and other countries have 220-volt electrical plugs, North America 110 volt, and all appliance and electrical companies comply. Once an industry-dominant design or standard is established products tend to become similar, with function and features relatively undifferentiated other than for look and feel. The smartphone is another example with the leaders striving to compete on feature and function with each new release.

Commoditization can lead to stiff competition and almost zero margins for all participants, resulting in some competitors going out of business or abandoning that part of their market. Over time the tendency is for the competitive environment to reach relative stability with some large companies producing relatively undifferentiated products within a relatively stable market share, and some smaller companies serving market niches.[26] This will continue until a new entrant hits the market with a disruptive product.

The goal of product innovation in a "volume-operations" market, such as consumer products, is to attract a large number of customers over a short period of time. This necessitates taking a breakthrough idea and executing it hard and fast—what Vijay Govindarajan and Chris Trimble call "the other side of innovation."[27] Complex product offerings may be targeted at smaller groups, perhaps only hundreds or thousands of customers. In these cases, the capital risks are high, manufacturing and logistics are complex, the product usually includes partner-supplied products and services, and marketing needs to be much more intensive. Geoffrey A. Moore and others call these "tornado markets."[28]

Much has been written about the advantage of being the first to market with a disruptive, radical product innovation. Being a pioneer or first is not always a sure bet for dominating the market, or even for survival. Often there will be a considerable number of niche players in the market all vying for growth and prominence. The firm that provides the product that becomes the dominant design often realizes long-lived first mover benefits. If this doesn't happen first, movers who devoted extensive resources in the rush to be first can end up being marginalized by a dominant player. This dominant player can be a "fast-second"[29] player who has avoided the high cost of the disruptive innovation, but is able to enter, scale up, and dominate the market. Often this has to do with being well capitalized, having existing distribution channels, partnering with key players in the industry, acquiring technology, and having a good business strategy. The early example of the IBM compatible personal computer applies.

Service Innovation

As we move increasingly to an innovation economy, where knowledge, technology, entrepreneurship, and innovation are paramount, service innovation will become ever more significant. Even today services represent an estimated 80 percent of the U.S. gross domestic product and 60 to 80 percent in the other advanced economies.

Service innovation is not substantially different than product innovation in that the goal is to satisfy customers' jobs-to-be-done, wow and retain customers, and ultimately optimize profit. Focusing innovation on your predetermined paradigm rather than on what customers need tends to lead to the same pitfalls, which often trouble product innovation—failure to meet primary functional and emotional needs. Customers are often able to describe their needs for service innovation but there will always be opportunities to innovate solutions that will positively surprise customers and earn their brand loyalty.

Lance Bettencourt argues that service innovation should focus on the customers' job-to-be-done. "So long as you're focused on service, whether on improving an existing service or creating a new one, your understanding of customer needs will be less than optimal to guide innovation. If you make the job [to-be-done] the focal point, if you focus on what customers are trying to accomplish, then you're free to talk to customers about their needs in a way that isn't constrained by the services you offer today or even those you might envision."[30]

A focus on jobs-to-be-done better enables a company to deconstruct the various steps in its process and look for ways to streamline service delivery, eliminate or improve irritating steps, and even eliminate nonvalue-added components. Each step that impacts the customer should be taken into account and measured by the customers' experience of its outcomes. This systematic process should identify many opportunities for innovation. Missing value-adding steps can be identified and added to the service process. Steps resulting in confusion and frustration can be amended or eliminated. New processes or steps that would increase customer loyalty can be added.[31]

Often the difference between a highly positive and a neutral or negative customer experience is in the process necessary to obtain the service. A very delicious restaurant meal is less satisfying if the server is not attentive, friendly, or helpful. The customer need is not just for a great meal; it is for a great dining experience.

A focus on job-to-be-done and outcomes can help uncover other services which would satisfy a need, resulting in new services, or even new types of services. Suppliers of mountain climbing equipment might decide to help customers learn the job-to-be-done of safe mountain climbing. A canoe or kayak retailer may find there are a sufficient number of customers wanting to enjoy the sport without having to store the equipment—a rental opportunity.

Previously we counseled against customer surveys as a method of obtaining input into product development. Customer satisfaction surveys for service innovation is a different matter, provided the focus is on how customers measure success (outcomes) when working with a particular service provider, rather than on their opinions about features of the current service. ". . . [A] traditional customer satisfaction survey in a retail store might ask for

Note: Manufacturers and distributors would also examine service opportunities in its consumption chain jobs-to-be-done

FIGURE 23-9   Focus on jobs-to-be-done.

customers' level of satisfaction with the presence of knowledgeable sales associates. The problem with this approach is that knowledgeable sales associates are simply a solution. 'Aha,' company managers think, 'we can improve our service offering by making sure our sales associates are knowledgeable.'"[32] Unless managers learn why customers value a knowledgeable sales associate—what it is that customers expect a knowledgeable sales associate to help them with—they will continue to make less-than-optimal decisions regarding how to innovate their service offerings.

Another useful approach to service, especially for manufacturers or distributors, is to look for "consumption chain" jobs-to-be-done when a product is purchased (Fig. 23-9).[31] The consumption chain begins with product selection and continues with purchase, installation, learning how to use, actual use, warrantee claim, moving, storage, maintenance, upgrade, disposal, and replacement. Service opportunities exist, for example, for helping customers with the job of environment-friendly recycling. Another lens might be "asset investment recovery" where the customers' job-to-be-done is obtaining optimum benefit from the product, including the ability to sell it when they are done with it. The example of eBay comes to mind.

One difference between service and product innovation is that customers are more frequently involved in determining the type and level of a service they will receive. But this line is becoming blurred as more and more customers participate online in designing a product they will receive (more on this later under Cocreation Innovation). Of course, many service offerings do not permit customer involvement in finding solutions. In these cases great care has to be taken to meet the target customers' jobs-to-be-done. Warrantees and extended warrantee offerings are examples of solutions that can miss the job-to-be-done. This occurs when the vendor does not offer the warrantee service or if offered, does not expedite the claim process but rather refers customers to service centers in other locations. Dealing with a warrantee claim thus becomes the customers' responsibility, thereby negatively impacting the emotional aspect of the total experience.

There is a service element involved with the provision of any product to a consumer or B2B customer, even when the product is bought online. Adam Richardson talks about a new type of innovation challenge—X-problems—which arise from the collision of several factors, one of which is: "The need to create integrated systems of physical products, software, online experiences, and services that work as a single whole. Often these integrated systems are the keys to expansion beyond core areas, as well as to meeting customer needs in ways impossible from a more isolated offering."[33]

Services are often bundled with a product and are at least as important in ensuring a positive outcome for customers. Services can be the differentiator for products in a highly competitive environment, and they don't always have to have a high cost. But they do have to be right from the customers' perspective. "Say what you are going to do, and do what you said you were going to do" is good advice. The key point is that both functional and emotional jobs are critical, and you can systematically uncover and prioritize both from a customer perspective, which can then lead to incremental improvements to existing services, and the creation of breakthrough new services.

Business Model Innovation

A business model is the method by which an organization creates and delivers value to its customers and how, in turn, it will generate revenue (capture value). If there are potential customers who have a job-to-be-done and you can satisfy that need, even if they don't consciously know they have this need, there is an opportunity to make money through an effective business model.

The first step in developing a business model is to identify target customers and the value that will be created for them. The value proposition can be framed in terms of what job the target customers need to have done. What problem is being solved in the marketplace? The next step is determining how this value will be delivered to these customers. This involves identifying what capabilities the enterprise must have or acquire in order to deliver this value to customers. It can be useful to think in terms of how to deliver value to one customer and subsequently rapidly scale to many customers. The third step is identifying how value will be captured in order to generate the resources needed to sustain and grow the business.

With the rapid pace of change in technology, the marketplace, and the global socio-economic environment, stability in business models has become a dream of the past. But the jobs-to-be-done lens enables people in existing businesses to see the world through customers' eyes and therefore see and think beyond the blinders inherent in their existing organizational perspectives and business models.

The most successful enterprises sometimes face the largest challenges. Their culture, organizational structure, silo structured business units, human resource practices, marketing strategies, and production resources are targeted at scaling and executing their core business models to compete for market share. As long as the current business model is viable the employees in the organization need to continually improve performance in order to contribute toward creating, delivering, and capturing value.

Problems arise if everyone including the leadership of the enterprise is focused on improving the performance of the existing business model, and no one is assigned the task of exploring entirely new business models. Kodak, a leader in its field, failed to change its business model despite being one of the first to invent a digital image capture technology, and resulted in its need to seek bankruptcy protection. While any business failure is a complicated affair, Kodak waited too long, hoping revenues from its photographic and motion picture film lines of business would not continue to decline.

There are many other examples. Saul Kaplan tells companies, "Don't get Netflixed" like Blockbuster did.[34] Blockbuster's business model of enabling customers to enjoy movies of their choice in the comfort of their homes was sound, albeit their unforgiving late return charges irritated customers. They failed to respond to the possibilities inherent in DVD technology, which enabled Netflix to mail movies to customers' homes using first class postage. Protecting their bricks and mortar structure blinded Blockbuster, or, perhaps it would be fairer to say, prevented them from developing a new and radically different business model. Waiting too long to innovate new products and services can be a terminal business strategy. Another way of thinking about it is problems arise when everyone is focused on achieving "best practice" and no one is specifically tasked with developing "next practice."

A start-up that tries to be a disruptive entrant into the market can have great difficulty going head to head with the established competition in the predominant markets. On the other hand, established competitors will often yield the low-margin ends of the market without fanfare when, as a business strategy, a start-up goes down market and innovates a simpler, cheaper, and more accessible offering. From this low-cost business model it may be possible to generate adequate revenue and gradually work up from the foothold market by adding features and functions. Many highly successful businesses started in this manner.

A disruptive business model either meets the lower product or service need of customers who have been overshot, or targets potential new customers who are not having their needs met in an affordable, accessible, or simple manner. A variation on this approach is when a large enterprise spins off a start-up or identifies a business unit to operate using a different business model than the core business. Once the fundamentals of the job-to-be-done for identified potential customers are understood, the new venture should start small with a low-cost business model and a basic offering that can be tested in a representative market. It may be found that the targeted customers are uninterested, or some aspects of the offering or business model may have missed the mark.[35] The new venture should be given only enough resources to allow it to develop the offering, market test key assumptions, and prove its viability. Providing too many resources removes the sense of urgency to fine-tune and prove the new model with the result deadlines are continually missed.

The 2010 IBM face-to-face survey of more than 1500 CEOs worldwide found that most of them understood the need to question "obvious" industry practices, to ask what a new entrant to their industry, having no legacy burden, might do. ". . . [T]he CEOs we met with talked of adopting a new approach to planning—iterating their business strategies with greater frequency and instituting continuous change through business model innovation. This calls for operating models designed for extreme flexibility and the surety to act with speed."[36]

This contrasts with the past, when most CEOs (and employees) could assume business model stability and focus on improving organizational performance. Now the challenge is to continually improve the current business model, while designing, prototyping, testing, and scaling up entirely new business models, even if they fundamentally challenge the existing core business model. Change is the new normal.

A business model serves two purposes. Leaders, managers, and employees will be on the same page if they are able to visualize and describe their business model. They will have a common mental model. It tells the story of the enterprise's value proposition, which can be an "inspiration accelerator" for everyone in the organization.[37] Second, when creating new offerings it is useful to clearly understand the existing business model before generating ideas for possible alternative business models for each product, service, or technology. Options range from offer it for free (Facebook), to turning fixed costs into variable costs (Better Place), to premium pricing for premium service, and this is only one lens. By asking "what if" questions, ideas for transforming business models become easier to envision. For example, what if rural Kenyans working in Nairobi or diaspora abroad could send money home to their families even though they don't have bank accounts? The business model that emerged was M-PESA.

Emerging markets have about 85 percent of the world's consumers and projected population growth rates at least twice that of developed countries. Merely modifying products and prices designed for developed countries is insufficient. To compete in developing countries completely new solutions with new business models must be innovated. Often these new solutions must be developed within those developing countries in order to ensure they fit in with the culture and affordability of the local people who have a job-to-be-done.

Alexander Osterwalder and Yves Pigneur have designed a useful online template to guide the preparation of a business model. The Business Model Canvas is composed of nine "building blocks": customer segments served; value propositions; communication, distribution, and sales channels; customer relationships with each customer segment; revenue streams derived from value propositions offered to customers; key resources needed to deliver the value proposition; key activities; key partnerships with outside enterprises; and cost structure.[38] A great business model needs to be implemented well. Think big but start small. Test early and validate assumptions, adjust the plan's building blocks accordingly, then, once proven, scale up with speed.

Few companies outperform the revenue expectations of their employees. Business model innovation (Fig. 23-10) focuses on continually adjusting a company's value proposition in order to increase both outgoing and incoming value—perhaps this is the best way to promote higher stretch targets for revenues because it points the way.

Steps in developing a business model

1. Identify target customers & value to be created for them (value proposition)

2. Identify how this value will be delivered to these customers

3. Identify how value will be captured to generate resources to sustain and grow the business

Jobs-to-be-done lens enables seeing and thinking beyond blinders inherent in existing business model

An enterprise must have some people exploring entirely new business models while most focus on improving the performance of the existing business model

A sense of urgency must replace complacency given the pace of change in technology, the marketplace, and the global socioeconomic environment

Act despite uncertainty. Take risks that disrupt legacy business models

Figure 23-10   Business model innovation: key points.

Co-creation Innovation

Co-creation is a way to introduce external catalysts, unfamiliar partners, and disruptive thinking into an organization in order to ignite innovation. The term co-creation innovation can be used in two ways: co-development and the delivery of products and services by two or more enterprises; and co-creation of products and services with customers.

The first, co-development by two or more enterprises goes beyond partnerships and consortia to a model where the participants agree to jointly develop products and share certain aspects of their knowledge and technology under well-defined agreements. Participants are able to reduce development costs, access external capacity, and share risks. Innovation through global collaboration of a network of firms may be the new source of competitive advantage in the 21st century.

For example, one party may have capital resources while others have valuable technology and R&D capability. Others may find it is no longer possible to generate a respectable return on capital due to increasing costs and reduced product life cycles (e.g., biopharma industry). It may be that none of the parties has the capacity (technology, resources, and the other side of innovation—execution) to develop the offering on their own.

The end result may be that all participants are able to market and distribute the products under their own brand or one of the participants may be responsible for delivering value to customers. This approach requires trust among the participants, shared governance and risk, allocation of deliverables among the various R&D organizations, peer interaction across enterprise boundaries, clearly defined intellectual property sharing agreements, a complex project management structure, and well-established deliverables for all parties. Divorce processes need to be considered in advance should the innovation process fail or trust break down.

Co-development between Procter & Gamble and Clorox Company resulted in Glad food wraps, containers, and the ForceFlex trash bag technology.[39] In the services area, SpineConnect is a collaborative knowledge network where spine surgeons can share knowledge, develop new treatments, discuss top challenges, and create technological solutions. Medical device firms can also benefit from the shared knowledge of this network by acquiring a better understanding the jobs-to-be-done for spinal surgeons and their patients. Raymond Miles et al. refer to "collaborative entrepreneurship" as being the creation of value based on jointly generated ideas arising from businesses sharing information and knowledge.[40]

Self-directed communities of practice involving participants from different parts of the organization, selected customers, vendors, and partners are also collaborative knowledge networks which accelerate innovation.

The second, and more common model for cocreation innovation, involves collaboration with customers to create value. No longer should customers be viewed solely as partaking in consumer interaction. At the early stages of a product launch lead users will most often have useful insights about the design of existing or new products, and their insights can also be useful for developing marketing messages.

Social media websites and weblogs are a great source of information about what customers are saying about products (Fig. 23-11). Initiatives to establish co-creation through social media needs to be done in a manner which is engaging, listens to customers' creativity, and requires transparency and considerable access to the business' decision makers and R&D staff. Many successful businesses embrace this valuable information rather than being defensive even when comments are not highly complementary. The information can be used to make changes to a product or related service in order to enhance value to customers. Through the social media customers may be given a voice in the selection of the final design of a product.

Eliciting comments from customers through various types of feedback processes, such as focus groups, surveys, and interviews, most often result in relatively unhelpful information and few useful insights. Projective techniques, which collect data related to customers' unconscious impressions, emotions and feelings, are more likely to generate insights which will drive innovation.

Video game producers use cocreation to build customer loyalty and sustained value. Apple's App Store is an example of cocreation with developers. There are many other examples such as Adidas' soccer shoes, Danone's yogurt "Fruchtzwerge," Macdonald's "Just Stevinho Burger," Fiat's "Mio" car concept, and Nivea's Black & White Deodorant.

Many industries provide opportunities for customers to personalize a product or service to meet their unique requirements. Online templates that allow customers to create the total experience of how their job will be done often facilitate this. Simple examples include online composition of multicolor publications, personalized banking, and the design of products from a menu of options with the product displayed as it is being created. The transportation device of the future will be co-creation with the customer, replacing the obsolete business

Early adopters provide feedback on new product offerings, and new applications/opportunities

Social media can benefit product adoption and market share OR slag the product's shortcomings

Blogosphere can build on the ideas of others, and may be a form of peer review

Open Innovation Portals invite customer input, but must be well managed or becomes a marketing liability

Customers' unconscious impressions, emotions, and feelings are a key source of insights

If enabled, personalization provides a wealth of information about customers' jobs-to-be-done

FIGURE 23-11 Sources of cocollaboration innovation.

model of vehicle dealers having stock on hand, hoping a buyer will chance along who happens to match one of the choices presented. Why are almost all light trucks in North America manufactured with uncomfortable back seats making them unnecessarily long and heavy for the urban market? In the future it will be standard practice for customers to design their perfect truck online.

It is the difference between standardization and customization. Perhaps nowhere is co-creation as exciting as in the healthcare industry. Increasingly "patients" will be what Don Tapscott et al. call "prosumers of wellness" rather than passive consumers. Collaborative health care will be enabled by advancements in technology that give patients easy access to medical information about themselves. Patients will be involved in very timely, precise, and personalized diagnosis, treatment, and lifestyle preventative measures. Geneticist and Cardiologist Eric Topol has developed smart phone apps that allow patients to monitor their own vital signs and enable doctors to provide individualized treatment with continuously updated vital signs data transmitted through the Internet.[41]

IBM's 2010 survey of CEOs worldwide found that co-creation with customers tops in their minds. "The most successful organizations cocreate products and services with customers, and integrate customers into core processes. They are adopting new channels to engage and stay in tune with customers. By drawing more insight from the available data, successful CEOs make customer intimacy their number-one priority."[42]

From a business model standpoint, large enterprises can divide their operation into two components, a customer-facing front end responsible for developing, packaging, and delivering customized solutions for each customer, and a back end where processes are standardized in order to optimize efficiencies.

Co-creation with customers can excite them and generate their long-term loyalty. It will be a key differentiator for businesses which provide this capability in a responsive, easy to use manner. Some approaches to co-creation of products and services could be called Open Innovation, the last type of innovation discussed in this chapter.

Open Innovation

For every brilliant person in your R&D group there are at least 1000 equally brilliant people worldwide with expertise in the same area. Many of these experts will be more brilliant. If you involve only your 25 scientists in your R&D effort you are missing out on the brainpower and creativity of 25,000 other bright minds. This example is scalable. If you have 4000 scientists, engineers, and other researchers in R&D you are missing out on the smarts of 4 million others who might have the ability to solve your innovation problems. Tapping into this incredibly large outside resource of knowledge and skills is the essence of open innovation (Fig. 23-12).

In the traditional closed innovation approach, R&D is vertically integrated and highly secret with tight controls over all aspects of the innovation process. Product development moves from the enterprise's science and technology base, through internal R&D, and on to

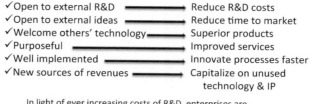

In light of ever increasing costs of R&D, enterprises are increasingly turning to Open Innovation for survival. It's the wave of the future.

Figure 23-12   Open innovation.

market. The goal is to develop new offerings in secret and get first-mover advantage in the marketplace. Acquiring external technologies are seldom, if ever, entertained.

"Not Invented Here" versus "The Wheel Not Reinvented"

Open innovation makes use of external ideas and technologies to enhance the enterprise's internal technology base, reduce the cost of R&D and time to market, and achieve superior product, service, or process innovations. At the same time, unused intellectual property and technology—latent internal intellectual capital—is made available for other firms to license and use. Other possibilities for these unused valuable assets include forming joint ventures or spin-offs to enter new markets or serve existing markets with different business models.

There are many benefits to taking an open innovation approach enterprise-wide. There can be significant cost savings at a time of ever decreasing margins and increasing competition. Superior solutions can be developed to meet the needs of customers because ideas for the development of new, or enhancements of existing products, services, and processes can come from experts anywhere in the world. Solutions can be found quickly with offerings getting to market faster. Exceptional solutions can be brought to market faster with methodologies to rapidly access global talent.

The rising cost of technology development is becoming almost prohibitive for many but the largest enterprises, and even they are challenged with the cost pressures of long product development time frames. On the other hand, the ever-shorter life cycle of new products reduces the length of time a product has traction in the marketplace before being upstaged by the "next best thing." The electronics industry is a good example—smartphones, laptops, MP3 players, and apps are only a few examples. The pharmaceutical industry is equally under pressure with long R&D time frames and ever increasing costs to develop new drugs, with the possibility that a competitor might beat them to it, and the certainty that the availability of generic drugs is around the corner. Examples abound in every industry.

It is easy to understand how open innovation can benefit product development. Henry Chesbrough describes how it can also benefit the development of services, which are composed of a series of process steps. Openness can be a service enhancer by providing customers with the value of a one-stop shopping location, while sharing infrastructure costs among participating companies. A good example would be an online shopping site for multiple brands of a given commodity group, like mountain climbing equipment.[43] Chesbrough calls this "leveraging outside-in openness for economies of scope." Rather counterintuitive is "exploiting inside-out openness for economies of scale." This is where a core service is shared with another firm, even with a competitor, and revenues are generated with every transaction. An example would be where a manufacturer permits the selling of some or all of its products under a competitor's brand, as well as its own, with the core manufacturing and distribution channels supporting both.[44] Would licensing its technology to Samsung rather than resorting to litigation have been a better business and public relations strategy for Apple?

An ever-growing number of companies are technology scouting worldwide to help them with their innovation strategies. Many have set up portals where they invite submissions from owners of technologies that may meet their business needs. Some examples are Syngenta, Unilever, GE, General Motors, Ford, IBM, 3M, Kellogg's, LEGO, Clorox, Intuit, Medtronic, GASF, Beiersdorf Pearlfinder, General Mills, DSM, and P&G. Most simply set up a portal website and wait. "Build it and they will come" does not generate positive results. Also, some do not have a preestablished process for reviewing submissions in a timely way.

Syngenta, a large multinational in agribusiness, has taken open innovation seriously with its "ThoughtSeeders" portal. Joe Byrum says this about their portal, "People want to know their technology will be evaluated. They want truthful acknowledgment not only that it has been received but also that it is being adequately and thoughtfully evaluated. This is what ThoughtSeeders is intended to do."[45] This company also does outreach to inform universities and others about the portal and to encourage submissions.

Another interesting case is Psion Teklogix's portal called "Ingenuity Working." Psion designs and manufactures rugged, handheld computers primarily for data capture in supply chain logistics operations. Rather than modifying its base handheld product to meet the needs of customers, it has opened the underlying software in the product to allow distributors who are closest to their customers to add features and specialized keys. This is a form of open innovation. "We want to be the experts in creating a modular chassis or platform so that our partners can add onto it, and take it into niche markets based on their expertise. . . . Rather than us trying to be the experts in everything, we've come up with an Open Platform. We support our partners through software and hardware development kits to enable them to take the base chassis, build onto it, and create a derivative device that's particular to their market. That's where we're going to focus our approach."[46]

A number of global "innovation marketplaces" have been established to act as intermediaries between organizations that need ideas and technologies, and those that believe they either have or can develop them (Fig. 23-13). These intermediaries facilitate creative crowdsourcing by providing a platform for their own crowd, and ensure experts in various fields actively participate by stimulating them with a constant flow of potential opportunities to work on challenges. They are, in effect, aggregations of the innovation portals of multiple enterprises—one-stop shops. They also provide project management in the relatively new field of crowdsourcing. They help frame the problem to be solved, ensure clarity in the description of the challenge, advise on how to incentivize the crowd, moderate and prescreen submissions from the crowd, and manage the transfer of intellectual property from the developers to the organization seeking the technology. Usually the crowd works on a fee contingency basis, being paid only if their solution is chosen.

IdeaConnection Ltd., founded in 2007, prequalifies "experts" from all over the world to work on R&D challenges. In response to challenges from organizations it establishes a select number of teams with a diversity of knowledge and expertise, each led by a facilitator, which work collaboratively online. The teams are motivated by the chance to win a full or partial award if their solution meets preestablished evaluation criteria and the business needs of the organization seeking the solution. Successful teams assign the intellectual property to the client upon receiving the award. IdeaConnection Ltd. also searches for existing and breakthrough technologies through its network of experts who have their own networks—a "network of networks." Idea Rallies of world-class thought leaders are incentivized to participate in strategic research questions and provide peer reviews. It also enhances the likelihood of university researchers, labs, and others with technologies for sale or license to submit

IdeaConnection Ltd	R&D Solutions by facilitated virtual teams of experts worldwide Technology Scouting through international networks Idea Rallies for peer review of strategic questions Submissions to firms' open innovation portals
NineSigma	Assessment of technology areas Technology search via RFP or targeted partner search Grand Challenge for positive societal outcome
Yet2	Matches buyers and sellers of technologies Online marketplace for technologies for sale Supports companies' open innovation portals
YourEncore	Connects companies to retired scientists & engineers
TopCoder	Access to digital designers, algorithmists, and software developers for online digital products competitions
Innocentive	Challenge-based crowdsourcing Enterprise Open Innovation platform
Choosa	Crowdsourcing of creative designs in graphics & websites

FIGURE 23-13   Open innovation intermediaries.

proposals to firms' Open Innovation portals. A free online weekly newsletter is provided on a subscription basis with a wide range of topics about Open Innovation.

NineSigma, founded in 2000, provides an assessment of selected technology areas to identify development partners, potential competitors, and new white space opportunities. It also facilitates technology searches through requests for proposal and targeted partner searches. Grand Challenges are designed to integrate prize-based challenges for development of transformative technologies with a company's strategic public relations and marketing. The goal is to focus on an umbrella topic that has a positive societal outcome.

Yet2, founded in 1999, focuses on later-stage technologies, rather than ideas, bringing together buyers and sellers of technologies. Searches for technology are done through a worldwide network. It also offers companies an online marketplace for technologies they have for sale. Yet2 also supports clients' open innovation portals by filtering submitted technology prospects.

YourEncore, founded in 2003, connects organizations with retired scientists and engineers who have expertise in aerospace and defense, specialty materials, food sciences, consumer sciences, and life sciences.

TopCoder administers contests in computer programming through a standards-based methodology. It operates a platform which digital designers, algorithmists, and software developers use when working on digital project competitions.

InnoCentive, founded by Eli Lilly in 2005, operates a prize-based crowdsourcing open innovation model. Challenges are posted and anyone registered to be a problem solver may submit their solution. If awarded the prize the recipient transfers the intellectual property to InnoCentive's client. It also offers an enterprise open innovation platform to harness ideas from employees, customers, partners, and suppliers.

Choosa, founded in Argentina in 2009 and operating simultaneously in English, Spanish, and Portuguese, crowdsources designs in graphics, printed products, websites, logos, and other designs from freelance creative designers. The designs are posted as they are submitted and designers may improve on others' designs by submitting what they believe is an enhancement (each designer is restricted to 10 submissions per contest).

With shrinking budgets for R&D and increasing costs of developing technologies, open innovation can provide a competitive advantage. Likely the future will see open innovation as an integral part of every successful company's business strategy. It is best summed up thusly: "The optimum 'balance' of open and closed innovation for a large corporation will be found through fostering a culture and attitude where 'Open Innovation' is *always* considered as an option for new knowledge, and the onus is on those who wish to remain closed to make their case."[47]

Conclusion

The cost of innovation through traditional R&D is rising in many industries outpacing owners' and shareholders' expectations for a return on capital. This is due to a number of factors including increased demands from customers, increased global competition, shorter product life cycles, increased technological complexity, and stricter environmental and safety regulations. This necessitates new and different approaches to innovation. It necessitates a portfolio of approaches.

Getting a handle on the types of innovation discussed in this chapter is highly desirable. Including them in an enterprise's business strategy is a must do for long-term survival. This also applies to not-for-profits and other kinds of organizations—even to governments. Besides, learning to welcome business change through innovation and making it an integral part of the organization's culture will help us with another frontier—learning to welcome personal change.

Innovate and live!

References

1. Kotter, John P. *A Sense of Urgency.* Boston: Harvard Business Press, p. 187, 2008.
2. Anthony, Scott D. *The Little Black Book of Innovation: How It Works, How to Do It.* Boston: Harvard School Publishing Corporation, p. 27, 2012.
3. Phillips, Jeffrey. *Relentless Innovation: What Works, What Doesn't—And What That Means for Your Business.* New York, NY: McGraw-Hill, p. 19, 2012).
4. United States Department of Labor, Bureau of Labor Statistics. "Entrepreneurship and the U.S. Economy."

5. Schumpeter, Joseph. *Capitalism, Socialism, and Democracy*. New York, NY: Harper, pp. 82–85, 1975.

6. IBM Corporation. "The Enterprise of the Future, IBM Global CEO Study," 2008, p. 28.

7. IdeaConnection interview with Michael Dalton, "Theory of Constraints, Part I." http://www.ideaconnection.com/open-innovation-articles/00165-Theory-of-Constraints-Part-I.html, February 1, 2010.

8. Magee, David. *How Toyota Became #1: Leadership Lessons from the World's Greatest Car Company*. New York, NY: Penguin Group, p. 40, 2007.

9. Goldratt, Eliyahu M. and Cox, Jeffrey (eds.). *The Goal*. Great Barrington: North River Press, 1984.

10. IdeaConnection interview with Michael Dalton, "Theory of Constraints, Part I." http://www.ideaconnection.com/open-innovation-articles/00165-Theory-of-Constraints-Part-I.html, February 1, 2010.

11. Rabe, Cynthia Barton. *The Innovation Killer: How What We Know Limits What We Can Imagine – And What Smart Companies Are Doing about It*. New York, NY: AMACOM: American Management Association, 2006.

12. IdeaConnection interview with Cynthia Barton Rabe, "Zero-Gravity Thinkers." http://www.ideaconnection.com/open-innovation-articles/00028-Zero-Gravity-Thinkers.html, May 12, 2008.

13. Edward John von and KomorowskiMenge. *The Quarterly Review of Biology*. Chicago: University of Chicago Press, 1930.

14. Drucker, Peter F. *Managing for Results*. New York, NY: Butterworth-Heinemann, 1955.

15. Levitt, Theodore "Marketing Myopia." *Harvard Business Review*, July 2004.

16. Christensen, Clayton and Raynor, Michael E. *The Innovator's Solution: Creating and Sustaining Successful Growth*. Boston: Harvard Business Review Press, 2003.

17. Verganti, Roberto. *Design-Driven Innovation: Changing the Rules of Competition by Radically Innovating What Things Mean*. Boston, Harvard Business Press, p. 28, 2009).

18. IdeaConnection interview with Roberto Verganti, "Innovation of Meanings, Part 1." http://www.ideaconnection.com/open-innovation-articles/00175-Innovation-of-Meanings-Part-1.html, March 6, 2010.

19. Dyer, Jeff, Gregersen, Hal, and Christensen, Clayton M. *The Innovator's DNA: Mastering the Five Skills of Disruptive Innovators*. Boston: Harvard Business Review Press, p. 96, 2011.

20. Kelley, Tom. *The Art of Innovation: Lessons in Creativity from IDEO, America's Leading Design Firm*. London: Profile Books Ltd, p. 112, 2004.

21. U.S. Patent and Trademark Office. "Patents Granted by Country, State, and Year—Utility Patents," December 2011.

22. World Intellectual Property Organization. "World Intellectual Property Indicators, 2012 Edition."

23. Berger, Warren. *Glimmer: How Design Can Transform Your Life, Your Business, and Maybe Even the World*. Random House Digital, Inc., 2009.

24. Richardson, Adam. *Innovation X: Why a Company's Toughest Problems Are Its Greatest Advantage*. San Francisco: Jossey-Bass, p. 3, 2010.

25. Tennant Snyder, Nancy Tennant and Durarte, Deborah L. *Unleashing Innovation: How Whirlpool Transformed an Industry*. San Francisco: Jossey-Bass, pp. XV, 91, 2008.

26. Utterback, James M. *Mastering the Dynamics of Innovation*. Boston: Harvard Business School Press, 1994.

27. Govindarajan, Vijay and Trimble, Chris. *The Other Side of Innovation: Solving the Execution Challenge*. Boston: Harvard Business Review Press, 2010.

28. Moore, Geoffrey A. *Dealing with Darwin: How Great Companies Innovate at Every Phase of Their Evolution*. London: Penguin Group, 2005.

29. Markides, Constantinos C. and Geroski, Paul A. *Fast Second: How Smart Companies Bypass Radical Innovation to Enter and Dominate New Markets*. San Francisco: Jossey-Bass, 2005.

30. IdeaConnection interview with Lance A. Bettencourt, "Service Innovation—Getting the Job Done." http://www.ideaconnection.com/open-innovation-articles/00227-Service-Innovation-Getting-the-Job-Done.html, November 21, 2010.

31. Bettencourt, Lance A. *Service Innovation: How to Go from Customer Needs to Breakthrough Services*. New York: McGraw-Hill, pp. 109–133, 2010.

32. IdeaConnection interview with Lance A. Bettencourt, "Service Innovation—Getting the Job Done." http://www.ideaconnection.com/open-innovation-articles/00227-Service-Innovation-Getting-the-Job-Done.html, November 21, 2010.

33. Richardson, Adam. *Innovation X: Why a Company's Toughest Problems Are Its Greatest Advantage*. San Francisco: Jossey-Bass, p. 3, 2010.

34. Kaplan, Saul. *The Business Model Innovation Factory*. Hoboken, NJ: John Wiley & Sons, Inc., 2012.

35. IdeaConnection interview with Mark Johnson, "Innovate to Grow." http://www.ideaconnection.com/open-innovation-articles/00195-Innovate-to-Grow.html, May 22, 2010.

36. IBM Corporation. "Capitalizing on Complexity: Insights from the Global Chief Executive Officer Study." 2010, p. 52.

37. IdeaConnection interview with Saul Kaplan, "Staying Relevant." http://www.ideaconnection.com/open-innovation-articles/00327-Staying-Relevant.html, September 17, 2012.

38. Osterwalder, Alexander and Pigneur, Yves. *Business Model Generation: A Handbook for Visionaries, Game Changers, and Challengers*. Hoboken, NJ: John Wiley & Sons, Inc., pp. 12–44, 2010.

39. IdeaConnection interview with Jeff Weedman, "Connect + Develop with Proctor & Gamble." http://www.ideaconnection.com/interviews/00070-Connect-Develop-with-Procter-Gamble.html, November 20, 2008.

40. Miles, Raymond E., Miles, Grant, and Snow, Charles C. *Collaborative Entrepreneurship: How Communities of Networked Firms Use Continuous Innovation to Create Economic Wealth*. Stanford: Stanford University Press, 2005.

41. Topol, Eric. *The Creative Destruction of Medicine: How the Digital Revolution Will Create Better Health Care*. New York, NY: Basic Books, 2013.

42. IBM Corporation. "Capitalizing on Complexity: Insights from the Global Chief Executive Officer Study," 2010, p. 9.

43. Chesbrough, Henry. *Open Business Models: How to Thrive in the New Innovation Landscape*. Boston: Harvard Business School Press, 2006.

44. Chesbrough, Henry. *Open Services Innovation: Rethinking Your Business to Grow and Compete in a New Era*. San Francisco: Jossey-Bass, pp. 75–88, 2011.

45. IdeaConnection interview with Joe Byrum, "Global Head of Soybean Seeds and Traits R&D, Syngenta" http://www.ideaconnection.com/interviews/00304-Syngenta-Thoughtseeders.html, May 15, 2012.

46. IdeaConnection interview with John Conoley (CEO), Mike Doyle (Chief Technology Officer), and Todd Boone (Director of Market Development), Psion Teklogix Inc. (http://www.ideaconnection.com/interviews/00235-An-Open-Source-Business-Model.html, January 21, 2011.

47. Golightly, John, et al. "Realising the Value of Open Innovation." (Big Innovation Centre—an initiative of The Work Foundation and Lancaster University) p. 9, 2012.

CHAPTER 24

TRIZ: Theory of Solving Inventive Problems

Ellen R. Domb

TRIZ is the anglicized acronym for the Russian phrase, "theory of solving inventive problems." TRIZ was developed in the post-WW II period of 1946 to 1985 in the former Soviet Union by Genrich Saulovich Altshuller (1915–1998) and his research colleagues. (*Creativity as an Exact Science, Innovation Algorithm*) TRIZ started as an analysis of wartime patents, looking for ideas that could help restore the Soviet economy, and has become the best-known method for systematic, logical, "left brain" creativity and innovation. Altshuller's insight that the global patents demonstrated repeated patterns of problems and solutions whereby categories of problems were solved by groups of solutions and historical patterns of problems were matched by historical patterns of solutions led to the development of the TRIZ methodology for analyzing problems and developing solutions. The work of Altshuller and his colleagues has now spread worldwide, and TRIZ continues to grow in scope and application among global researchers (Fig. 24-1).

TRIZ is a dynamic methodology. Here's a sampling of the meetings held just this year (2011–2012):

- Malaysian TRIZ Association
- India TRIZ Association
- Iberoamerican Innovation Congress (11 countries, organized by AMETRIZ, the Mexican TRIZ Association)
- European TRIZ Association TRIZ Futures Conference (11th Dublin, 12th Lisbon; 27 countries were in Lisbon. 13th is scheduled for Paris in October 2013.)
- MATRIZ (Russia-based International TRIZ Association)
- Altshuller Institute (U.S.)
- TRIZ France
- Apeiron (Italian TRIZ Association)
- Iran Institute for Creatology
- Japan TRIZ Symposium
- Korea Global TRIZCON
- Israel TRIZ Association
- Systematic Innovation Society, Taiwan and TRIZ Association of Taiwan
- China TRIZ Association
- TRIZ Centrum (German-speaking countries)
- UK TRIZ Group

Not to mention training courses for groups that range from 600 people in Saudi Arabia to 15 people in Zacatecas, Mexico, to individuals working with books, online courses, and instructors all over the world.

FIGURE 24-1 Global scope of TRIZ in 2010. Solid dot: European TRIZ Association survey of 2009. Open dot: Author's personal knowledge of TRIZ practitioners. Dark gray: Industry only. Light gray: Academia and industry.

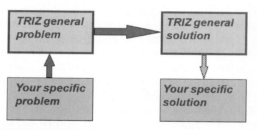

FIGURE 24-2 The logic of TRIZ problem analysis and solution.

TRIZ is very different from brainstorming-type creativity tools. It is an analytical and data-based system in which the practitioner analyzes the situation, matches the type of innovation called for with standard patterns, and then applies the solution techniques that have a strong history of creative solutions to similar situations (Fig. 24-2). To some people this sounds almost antithetical to innovation—following formulas and procedures to find innovations! But it works. It lets people who have limited intuitive creative ability solve problems in unique ways. It lets people with no personal knowledge of the history of system development or the patterns of change of technology and business become skilled strategic planners and product developers. Boris Zlotin developed this diagram, which is frequently used to demonstrate the method.

In Fig. 24-2, the light gray arrow indicates that the problem solver's personal experience is added to TRIZ methods and databases when interpreting the solution for a specific problem.

The goal of this handbook chapter is to help the reader learn enough about TRIZ to know what kinds of innovation problems will benefit from TRIZ, and to learn enough of the TRIZ vocabulary to be able to learn what he/she needs to benefit from TRIZ. The unique structure of this handbook makes the essential database references of TRIZ available to handbook users. (Contradiction Matrix, 40 Principles for Problem Solving, 76 Standard Solutions.) The chapters on systematic innovation (Chap. 28) and combination methods (Chap. 17) give details on how TRIZ integrates with many other innovation, organizational improvement, and product development methodologies, including Six Sigma, Lean, Theory of Constraints, and others.

For example, Fig. 24-3 shows the many different ways that TRIZ is helpful during difficult transitions in a stage-gate process:

In Fig. 24-3, you can see that TRIZ is not used at just one point in the development of a product/service. The patterns of evolution are used in the strategic plan stage, and the problem-solving methods are used throughout development, delivery, and after-market support by engineering, manufacturing, marketing, supply chain management, and service professionals. Note that despite TRIZ's origin in patent analysis, it has proven useful in generating creative solutions to problems in business, management, government, and science, as well as in engineering.

When To Use TRIZ?

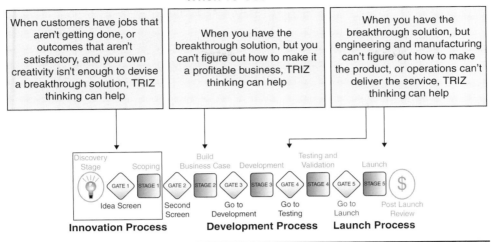

FIGURE 24-3   When to use TRIZ?

The balance of this chapter is organized as follows:

1. Problem solving

2. Patterns of evolution

3. TRIZ research—how TRIZ is changing

Problem Solving with TRIZ

TRIZ research began with the hypothesis that there are universal principles of creativity that are the basis for creative innovations that advance technology. If these principles could be identified and codified, they could be taught to people to make the process of creativity more predictable. The short version of this is:

- *Somebody someplace has already solved this problem (or one very similar to it).*

- *Creativity is finding that solution and adapting it to this particular problem.*

The research has proceeded in several stages during the last 60 years. The three primary findings of this research are as follows:

1. Problems and solutions are repeated across industries and sciences. The classification of the contradictions in each problem predicts the creative solutions to that problem.

2. Patterns of technical evolution are repeated across industries and sciences.

3. Creative innovations use scientific effects outside the field where they were developed.

Much of the practice of TRIZ consists of learning these repeating patterns of problems-solutions—patterns of technical evolution and methods of using scientific effects—and then applying the general TRIZ patterns to the specific situation that confronts the developer.

Before applying the detailed solution methods to the problem, it is necessary to analyze the problem and understand the details of the situation that gives rise to the problem. TRIZ tools help the developer generate insights into the problem situation that make innovative problem solutions possible. Each TRIZ teacher and practitioner develops her/his own method of using the tools. In general, the term "ARIZ" is used for "Algorithm for Inventive Problem Solving." The last version that Altshuller worked on personally was ARIZ 85-C. A general flowchart version of a simplified ARIZ is shown in Fig. 24-4, and an expanded version in Fig. 24-5.

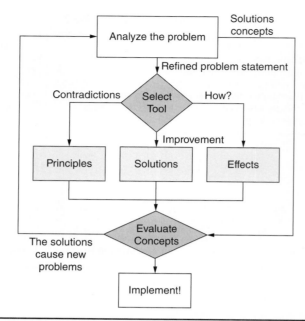

FIGURE 24-4 Simple flowchart for TRIZ tools. (Based on "Practical Innovation: Applied TRIZ" by Ellen Domb.)

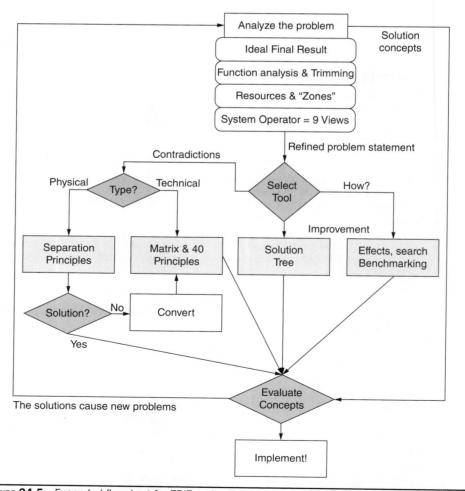

FIGURE 24-5 Expanded flowchart for TRIZ tools. (Based on "Practical Innovation: Applied TRIZ" by Ellen Domb.)

TRIZ Tools for Problem Analysis

The ideal final result (IFR) is an implementation-free description of the situation after the problem has been solved. It focuses on customer needs or functions needed—not the current process or equipment. The goal of formulating the IFR is to eliminate rework (solve the right problem the first time!) by addressing the root cause of the problem or customer need. The IFR helps you reach breakthrough solutions by thinking about the solution, not the intervening problems.

A basic principle of TRIZ is that systems evolve toward increased ideality, where ideality is defined by the value equation (borrowed by TRIZ from value engineering in the early 1950s):

$$\text{Ideality} = \Sigma\text{Benefits}/(\Sigma\text{Costs} + \Sigma\text{Harm})$$

Evolution is in the direction of:

- Increasing benefits
- Decreasing costs
- Decreasing harm

The extreme result of this evolution is the IFR: it has all the benefits, none of the harm, and none of the costs of the original problem. The IFR describes the solution to a technical problem, independent of the mechanism or constraints of the original problem. The ideal system occupies no space, has no weight, requires no labor, and requires no maintenance. It delivers benefit without harm. For this reason, beginners usually regard the IFR as the scariest tool of TRIZ—if you are the person who maintains the system, the IFR tells you that you are out of a job! For this reason, some of the ARIZ methods put the IFR later in the sequence of tools, to avoid scaring the problem solvers at the beginning of the process.

Loosely interpreting the equation, the IFR has the following four characteristics:

1. Eliminates the deficiencies of the original system
2. Preserves the advantages of the original system
3. Does not make the system more complicated (uses free or available resources)
4. Does not introduce new disadvantages

Example: Consider the power lawnmower as a tool, and the lawn as the object to be cut. The lawnmower is noisy, uses fuel, requires human time and energy, produces air pollution, throws out debris that can endanger children or pets (or the legs of the person pushing it), and is difficult to maintain. If our job is "improve the lawnmower," we could immediately set up and prioritize solutions for a number of TRIZ problems to improve fuel usage, reduce noise, improve safety, and so forth. But if we define the IFR, we can get a much better perspective on the future of the lawnmower, and the lawn-care industry.

What does the customer want? Whenever I ask this question, I get the same answer—the customer wants nice looking grass with no problems. The machine itself is not part of the desired solution. It should come as no surprise to find out that at least two companies that make lawnmowers are experimenting with "smart grass seed"—grass that is genetically engineered to grow to an attractive length, then stop growing.

Suppose your assignment is not quite so global as planning the future of the whole lawnmower industry. Can you still benefit from the IFR? Yes! To continue with the lawnmower example, if your assignment is to reduce noise, what is the IFR? It is a quiet lawnmower, or a quiet lawn-care system.

What is the difference between "less noisy" and "quiet?" To reduce noise, most engineers add baffling, add dampers, muffle the noise, or in other ways add parts, thereby adding complexity and reducing reliability. To make the lawnmower quiet, the designer has to look at the sources of noise, and remove them. This will make the lawnmower more efficient as well as achieving the original objective of less noise, since noisy engines are inefficient, noise from vibration wastes energy, and so on.

The IFR is a psychological tool that orients you to the use of the technical tools. Formulating the IFR helps you look at the constraints of the problem, and consider which constraints are required by the laws of nature, and which are self-imposed ("But we've always done it that way!"). You may choose to accept the constraints in solving your problem, but at least you are then conscious of the choices. For example, in the "quiet lawnmower" case, we can choose to continue using metal cutting blades, accepting the maintenance and safety problems; but we replace the gasoline engine with an electric motor to eliminate the most significant source of noise.

Start your problem solving by formulating the IFR. It will help you

- Encourage breakthrough thinking
- Inhibit moves to less ideal solutions (reject compromises)
- Lead the discussions that will clearly establish the boundaries of the project

The IFR will position you to use the technical tools of TRIZ effectively in solving the right problem.

In terms of the value equation, the IFR has all the benefits that the customer requires, and none of the harm caused by the original system. In many cases, developing a clear statement of the IFR will lead directly to a solution to the problem, and frequently leads to a solution at a very high level, since the technology-independent definition of the IFR will lead the problem solver away from traditional means of solving the problem.

For example, in Altshuller's classic problem about the candy factory, small chocolate bottles are filled with a gooey liqueur/sugar syrup mixture. To increase the factory's productivity, the filling is heated to make it run faster; but the hot filling melts the chocolate. (This is also an excellent problem for practicing identifying technical contradictions and physical contradictions.)

TRIZ students frequently struggle to identify the IFR. As long as they use the word "fill" in their statement, they will concentrate on finding ways to pour the liquid into the bottle; but if they choose a technology-independent formulation, such as,

"The goo is on the inside and the chocolate is on the outside."
  "The chocolate encloses the goo."
  …they then see solutions that are not dependent on pouring, such as the classic one of freezing the goo in the final shape, then dipping it in melted chocolate; and not-so-classic ones, like blown injection molding, where the pressurized goo provides the propulsive force to shape the chocolate in a mold, or inverting the mold so that the goo is poured into the large diameter end of the bottle.

Other cases are more resistant to solution at the stage of formulation of the IFR. In these cases, a procedure is required to guide the TRIZ student from the statement of the IFR to a redefinition of the problem to be solved and then to the solutions to the problem. This piece of ARIZ (Algorithm for Creative Problem Solving) is outlined in the following steps:

1. What is the final aim?
2. What is the IFR?
3. What is the obstacle to this?
4. Why does this interfere?
5. Under what conditions would the interference disappear? What resources are available to create these conditions?

Gasanov et al. use a charming story to illustrate this methodology. Consider the problem of raising rabbits. The rabbits need fresh food constantly, but they cannot be allowed to roam free, because they will pursue fresh food and not be where the farmer can find them. The farmer does not want to spend all his or her time bringing fresh food to the rabbits. Using the five steps above, the TRIZ student formulates the problem as follows:

1. What is the final aim? The rabbits can feed on fresh grass.
2. What is the IFR? The rabbits feed themselves fresh grass.
3. What is the obstacle to this? The walls of the cage are immobile.
4. Why does this interfere? Since the walls don't move, the area of grass available to the rabbits doesn't change.
5. Under what conditions would the interference disappear? When the enclosure moves to fresh grass whenever the rabbits have eaten the grass inside it. What resources are available to create these conditions?

The solution is frequently obvious from step 5. (Put the enclosure on wheels, so that the rabbits themselves can push it to a fresh grazing area.)

If the solution is not obvious, reexamine all the resources available in the problem. Elegant, high-level solutions to technical problems are frequently found by using resources in multiple ways. To continue the problem of the rabbits, the solution is to move the enclosure to fresh grass, but it might not be obvious how to move it. The only energy resources in the problem are those of the rabbits and the farmer. Since the objective of the problem was to find a way that the farmer could avoid using his own time and energy, we should look at the rabbits as the source of energy for moving the enclosure. The general list of possible energy sources is:

- Use "harmful" energy, force.
- Use free energy, force.
- Look for an engine standing idle nearby.
- Lessen the loss of energy, force.
- Put together a very simple machine.

In this case, the rabbits are a source of free energy already present in the problem.

Darrell Mann popularized the "Itself" method of formulating the IFR by asking "how can the problem solve itself?" In these examples, the focus is on how can the grass "mow" itself (later rephrased as "the grass keeps itself short") or how can the rabbits feed themselves? See his cases in the February, March, and April issues of www.triz-journal.com and in *Hands-On Systematic Innovation*.

Function Analysis and Trimming

Function Analysis is a standard method of systems engineering that has been adapted and adopted into TRIZ. The "subject-action-object" method is most frequently used now, although the "substance-field" method was historically important in the development of TRIZ. Figure 24-6 shows an example of subject-action-object diagramming.

In the diagram method, the relationships are plotted first, then the possibility of removing elements of the system (trimming) is explored. Notice how the diagram makes it very obvious that the harm done by C to D should be removed; but if C itself is removed, the good that is done to A must come from somewhere else. Function analysis and trimming are done early in the TRIZ analysis because, if trimming is possible, the resulting system is significantly simplified.

Frequently asked questions

- Can an element do both beneficial and harmful things? YES—look at C in the example. This is the source of contradictions—we want C and we don't want C.

- Can two elements have multiple arrows between them? YES—it is not shown, because of the size of the picture, but it is a frequent real situation. For example, see Fig. 24-7.

This is the most frequently used set. Another common version has a straight arrow for beneficial, squiggly arrow for harmful (∿∿∿►).

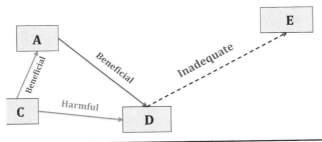

FIGURE 24-6   Illustration of the diagramming method of function analysis. Object C does beneficial action to object A, harmful action to D. D does an inadequate action (good but not good enough) to E.

The bottle contains water (useful, beneficial) and the water leaches bad-tasting chemicals out of the bottle (harmful) or the bottle contaminates the water (harmful).

FIGURE 24-7   Symbols.

Material Resources	Energy Resources
• System elements	• Energy in system
• Inexpensive materials	• Energy in environment
• Modified materials	**Space Resources**
• Waste	• Under-used space – inside, outside, under, above, around . . .
• Raw meterials	**Function Resources**
Time Resources	• Harmful functions: "Make your enemy be your friend"
• Parallel operations	• Primary and secondary effects of *features or attributes*
• Pre/post work	
Information Resources	
People Resources	

TABLE 24-1   Resources

Another method called the Problem Formulator starts with the relationships between functions, rather than the relationships between objects.[*]

Zones of Conflict and Resources[†]

Classical TRIZ emphasizes that the problem situation must be well understood before one attempts to solve the problem. The term "zones of conflict" refers to the temporal zone and the operating zone of the problem—loosely the time and space in which the problem occurs. In more modern English terminology, it is to understand the root cause of the problem and to couple this step with the function analysis so that cause and effect chains can be analyzed. Since root cause analysis is the subject of numerous quality improvement texts, Six Sigma manuals, and so on, it will not be addressed here. But the TRIZ technique of examining the resources that are available in the problem is most easily done when the researchers are immersed in the problem, attempting to understand the situation.

TRIZ defines the resources of a problem in a very comprehensive way, including materials, sources of energy, information, time, and so on as shown in Table 24-1.

The identification of resources is most easily taught by a progression of examples. The simplest everyday environment is the grocery store. Four recent examples have been very stimulating to students, helping them find many more examples in the grocery, then generalizing those examples to their business environment. See Fig. 24-8.

1. Honey, catsup, and other viscous liquid products are packaged with the dispensing orifice at the bottom, so that gravity can help with easy dispensing, instead of hindering it.

[*]See Terninko, J., Zusman, A., and Zlotin, B. *Systematic Innovation.* Boca Raton, FL: CRC Press, 1997. http://www.taylorandfrancis.com. For more in-depth information on the subject-action-object method, see Z. Royzen, "Tool, Object, Product Function Analysis," http://www.triz-journal.com/archives/1999/09/d/index.htm, or tutorials and examples in "TRIZ Power Tools" Skills 2, 3, and 4, http://www.opensourcetriz.com/2012. For more information on the substance-field model, see section on Belski, Iouri, *Improve Your Thinking: Substance-Field Analysis.* TRIZ4U, Melbourne Australia. 2007. www.triz4u.com or from www.aitriz.org.

[†]Based in part on "Learner-Focused Teaching Applied to the Use of Resources in TRIZ Problem Solving" by Ellen Domb, Miller, Joe A., and Czerepinski, Ralph, published in the proceedings of the European TRIZ Association TRIZ Futures Conference, 2011, Dublin, Ireland.

Chenopodium album, Lambs-Quarters

Amaranthus retroflexus, Redroot Pigweed

FIGURE 24-8   Use of the resources of gravity (far left, honey), unsaleable food products (pretzel and carrot pieces), and wild food (right side of picture) as examples of resources easily found in or near a grocery.

2. Pretzel bits are a new "product" made by flavoring fragments of pretzels left in the production facility after whole pretzels are produced.

3. "Baby carrots" are not a new species. They are carved from large carrots that are considered too ugly to be sold whole.

4. Edible plants grow wild in many urban areas. Lambs-quarters and pigweed are frequently found in parking lots, including those at grocery stores. They help make the point that knowledge is necessary to recognize the utility of a resource.

A complex case involves physical resources (gravity, water), governmental resources (federal, state, and local government agencies, including the state university), knowledge resources (particularly methods of changing governmental policies), and financial resources (private investment companies). The case demonstrates resource identification and utilization at system, subsystem, and supersystem levels and demonstrates that good resource utilization frequently enables further opportunities.[*]

Water in Arizona is vital for survival and growth. It is scarce, tightly regulated, and competed for by businesses, municipalities, and even other states. Higher education in Arizona is also vital for economic survival and growth. It is expensive, rapidly becoming more so, generally restricted to larger cities, and subject to constantly increasing demand.

The town government of Payson, a small (16K) largely residential community in the mountain area of central Arizona and totally surrounded by United States National Forest Preserve, has found unique ways to obtain water for current needs and future growth, and

---

[*]An expanded version of this case appears in the material at http://dld.bz/tuEm.

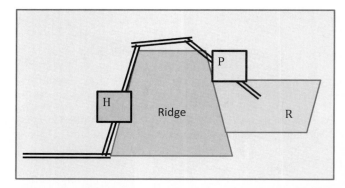

Figure 24-9   Schematic diagram of the water supply for Payson, AZ. R = reservoir, P = pumping station, H = hydraulic generating station that provides power to the pumping station.

thereby created a unique opportunity to help provide high-quality education at costs much lower than available in large urban centers.

Payson's traditional water supply has been some 50+ deepwater wells. To enable expansion, Payson has acquired (through a complex political process) an existing reservoir, originally named Blue Ridge and built for the purpose of providing replacement water for very distant copper mining. That reservoir is annually replenished by snowmelt from the Ponderosa pine forest. The difficulty: Blue Ridge is on the other side of a mountain area, many miles from Payson. The solution: use the resource of gravity, pump the water uphill some 200 yards, drop it down the face of the Mogollon Rim (famous in Wild West literature) some 500 yards, and then transport it to the town. The roughly 300-yard difference in the pumped height and the drop height allows the water to be used to generate sufficient electricity in a dedicated hydroelectric station to power the initial pumping operation. This required the cooperation of a large multinational mining corporation, the U.S. Federal government, the National Forest Service, the State of Arizona, and a quasi-governmental water resources management organization, as well as some US$35 million to US$40 million. The result: Payson has water for 40,000 people, and is able to realistically plan for a university branch with a capacity of 6000 on-campus students as depicted in Fig. 24-9.

The Payson town mayor and government have realized the opportunity to provide for the demands for higher education by using nongovernmental resources, partnerships with private investors and multinational companies to build and operate a campus with all university facilities plus a research park, business center incubator, convention center, hotel, performing arts center, and regional high-speed network, and by using advanced methods for construction with low environmental impact technologies on available national forest land. This is a radical break with past methods of developing a university. Using private funding means that the university can focus on developing the curriculum, faculty, and student systems, while investors focus on the land, building, and infrastructure. Tuition cost to the students is estimated at 50 to 70 percent of that at urban universities. The town will realize many benefits from the presence of the university and related facilities.

Historically, TRIZ has had structured and more informal approaches to the use of resources.

Structured: Many of the specific steps within ARIZ are expressly for the purpose of identifying resources, that is, steps:

1. Formulating the mini problem
2. Defining conflicting elements
3. Defining substance and field resources

Mobilizing and Utilizing Substance-Field Resources, is directed toward systematic procedures for modifying existing resources to achieve problem-solutions. Two relationships between the ARIZ view of resources and basic TRIZ concepts become obvious when resources are catalogued in detail:

- If there is a contradiction, it's because we are misusing a resource.
- Breaking "psychological inertia" is challenging assumptions about how resources are used.

In ARIZ, the "mini" problem is one that is solved without introducing new elements. It is very clear when working on a "mini" problem why we have to understand resources, since the emphasis in the "mini" problem is on solving the problem without introducing anything new to the system. It is equally important, if not more so, for working on the "maxi" problem, since it will reduce the time and complexity of the solutions.

Informal: Contemporary TRIZ training manuals typically show these steps:

1. List all resources in the problem and environment (sometimes with a list of things, information, energy or the acronym TIE as guidance)

2. List attributes of the resources

3. List what the problem needs to be solved (without guidance about how to do this, which can require in-depth knowledge of the effects that can be accomplished with the attributes of the resources)

4. Look for matches between lists

Understanding the resources in the system—the functions and attributes of the resources—is a preliminary step for easy application of TRIZ problem-solving methods.

System Operator, or the 9 Windows

The system operator (also called "9 windows" or "multiscreen" method) is a visual technique that is used frequently in the initial stages of TRIZ as part of problem definition. The method begins with the construction of a 3 × 3 matrix, with the rows labeled as the system, subsystem, supersystem; and the columns labeled past, present, and future. Since time travel is not yet an available technology, the "past" column can also be labeled as "preventive," as a short form of, "If we could have done something in the past to prevent the problem from occurring, what would it have been?"

As with many matrix tools, documentation of the contents of each cell is necessary apart from the creation of the template. For example, Fig. 24-10 shows step 1 of the case study of a soggy piece of pizza. Separate documentation is necessary to define the problem in detail: "the pizza was baked at a pizza shop, delivered to a home, the delivery takes typically 25–35 minutes (with statistics if available); what fraction of customers complain about sogginess; are the people who are trying to solve the problem the home purchaser or the pizza shop or the suppliers to the pizza shop," and so on. In a rigorous example, a separate set of system operators might be constructed for each of the possible problem solvers; in this example, opportunities for all of them are considered together.

In step 2, the subsystems, or components, of the system are listed and the supersystem is identified (Fig. 24-11). Sometimes there are multiple possible supersystems. In this case, the pizza is part of the family dinner supersystem, and it is also part of the pizza production and delivery supersystem. In this informal use of the system operator, just listing these elements of the problem (and the accompanying discussions if a team is working on the problem) may suggest solutions to the problem or pathways to be investigated to generate solutions. For example, listing "mushrooms" as a component leads to a discussion of how mushrooms emit water when cooked. Maybe if the mushrooms were kept in a separate

	Past (Preventive)	Present	Future (Corrective)
Subsystem			
System		Soggy Pizza 	
Supersystem			

FIGURE 24-10   Step 1 of the simplified system operator: what is the problem at the system level?

	Past (Preventive)	Present	Future (Corrective)
Subsystem		Crust, Cheese, Sauce, Pepperoni, Mushrooms	
System		Soggy Pizza 	
Supersystem		Pizza, Box, Carrier Pouch, Delivery Car, Driver	

Figure 24-11 Step 2 of the simplified system operator: what are the subsystems, components, and supersystems?

	Past (Preventive)	Present	Future (Corrective)
Subsystem	Can we change a component to prevent going soggy?	Crust, Cheese, Sauce, Pepperoni, Mushrooms	Can we do anything to a component to re-crisp the pizza?
System	How can we prevent the pizza from going soggy?	Soggy Pizza 	How can we make a soggy pizza fresh and crisp again?
Supersystem	Can we prevent waiting by changing the packaging and delivery system?	Pizza, Box, Carrier Pouch, Delivery Car, Driver	Can we use the package and delivery system to re-crisp the pizza?

Figure 24-12 Step 3 of the simplified system operator: ask how to remove the problem for each box.

container, the pizza would be crisp? This is a path to investigate (and an example of inventive principles two, take out and three, local quality, too).

In step 3, the problem solvers explicitly consider the questions (Fig. 24-12):

- "How could we prevent the problem, operating at this level?" for the "preventive" column.

- "How could we fix this problem, operating at this level?" for the "corrective" column.

Examples are shown in Fig. 24-13 for all three levels. It is easy to expand the system operator in many directions. In the pizza example, the box collects moisture that evaporates from the hot pizza and condenses on the cold surface of the box. Considering separately what to do about the box, the moisture, the heat transfer, and so forth, could produce new thinking about crisp pizza. Figure 24-14 illustrates how "What to do?" can lead to solution expansion, by asking the question for each of the five elements of the complete system in each of the 9 windows of the system operator.

Solving Problems by Eliminating Contradictions

A basic principle of TRIZ is that a problem is defined by contradictions. That is, if there are no contradictions, there are no problems. The benefit of analyzing a particular innovative problem to find the contradictions is that the TRIZ patent-based research directly links the

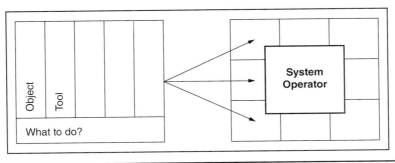

FIGURE 24-13 Ask "What to Do?" questions.

type of contradiction to the most probable principles for solution of that problem. In other words, the general TRIZ model shown in Fig. 24-2 is particularly easy to apply for contradictions.

Contradictions

TRIZ defines two kinds of contradictions, "physical" and "technical." These labels are artifacts of the early translations of TRIZ works, and should be thought of as reference labels—neither is more or less "physical" than the other! Classical TRIZ had a third category of "administrative" contradictions (you can't solve the problem because the boss won't let you), which is now solved as one of the other two types.

Technical contradictions are the classical engineering and management "trade-offs." The desired state can't be reached because something else in the system prevents it. In other words, when something gets better, something else gets worse. Classical examples include

1. The product gets stronger (good) but the weight increases (bad).
2. The bandwidth of a communications system increases (good) but requires more power (bad).
3. Employees get in-depth training (good) but they are away from their regular assignments for a long time (bad).

Before starting to solve the problem, it is useful to examine the trade-off to find out if assumptions about the solution have already been made. Removing the assumptions can remove the problem, which may be more ideal. For example, assumptions were made about all three of the cases above:

1. **Assumption**: The way to make something stronger is to add more material. Suppose we made it stronger by changing the shape? Or changing to a composite material (as advanced as nano-structures or as ordinary as fiberglass and resin over foam)?
2. **Assumption**: The information coding doesn't change, so the only way to push more data through the "pipe" is a bigger pipe, which requires more power.
3. **Assumption**: The only way to do good training is in a classroom, during business hours.

Physical Contradictions are situations where one requirement has contradictory, opposite values. Everyday examples abound:

1. When pouring hot filling into chocolate candy shells, the filling should be hot to pour fast, but it should be cold to prevent melting the chocolate. This is one of Altshuller's teaching examples.
2. Aircraft should be streamlined to fly fast, but they should have protrusions (landing gear) to maneuver on the ground. Better way to say it: The wheels should hang down; the wheels should not hang down.
3. Highways should be wide for easy traffic flow, but narrow for low impact on communities.
4. I want to know everything my 17-year-old child is doing; I don't want to know everything my 17-year-old child is doing.

Worsening Feature → Improving Feature ↓		Volume of moving object	Speed	Force (Intensity)	Stress of pressure	Shape	Reliability	Object-generated harmful factors	Ease of operation	Ease of repair	Device complexity	Difficulty of detecting and measuring
		7	9	10	11	12	27	31	33	34	36	37
9	Speed	7, 29, 34	+	13, 28, 15, 19	6, 18, 38, 40	35, 15, 18, 34	11, 35, 27, 28	2, 24, 35, 21	32, 28, 13, 12	34, 2, 28, 27	10, 28, 4, 34	3, 34, 27, 16
10	Force (intensity)	15, 9, 12, 37	13, 28, 15, 12	+	18, 21, 11	10, 35, 40, 34	3, 35, 13, 21	13, 3, 36, 24	1, 28, 3, 25	15, 1, 11	26, 35, 10, 18	36, 37, 10, 19
11	Stress or pressure	6, 35, 10	6, 35, 36	36, 35, 21	+	35, 4, 15, 10	10, 13, 19, 35	2, 33, 27, 18	11	2	19, 1, 35	2, 36, 37
12	Shape	14, 4, 15, 22	35, 15, 34, 18	35, 10, 37, 40	34, 15, 10, 14	+	10, 40, 16	35, 1	32, 15, 26	2, 13, 1	16, 29, 1, 28	15, 13, 39
15	Duration of action of moving object	10, 2, 19, 30	3, 35, 5	19, 2, 16	19, 3, 27	14, 26, 28, 25	11, 2, 13	21, 39, 16, 22	12, 27	29, 10, 27	10, 4, 29, 15	19, 29, 39, 35
33	Ease of operation	1, 16, 35, 15	18, 13, 34	28, 13, 35	2, 32, 12	15, 34, 29, 28	17, 27, 8, 40		+		12, 26, 1, 32	32, 26, 12, 17

FIGURE 24-14   Selected rows and columns from the contradiction matrix.* The numbers in the cell refer to the principles that have the highest probability of resolving the contradiction. See Appendix 1 for the 40 principles. The circled cell is discussed in the example in the text.

Stating the problem in the physical contradiction formalism can help you separate the issues that contribute to the problem. These may be assumptions, as in the trade-offs, or they can be multiple issues that have been combined so often that you have forgotten that they can be separate. For example:

No. 1: I want to make the candy quickly. Making the filling flow fast by heating it is only one way to make it flow fast, and it is a really bad choice for the survival of the chocolate.

No. 4: I want to be able to trust my child so that I don't have to watch him or know what he's doing in detail.

When using the TRIZ research findings, in general the most comprehensive solutions come from using the physical contradiction formulation, and the most prescriptive solutions come from using the technical contradiction. In terms of learning, people usually learn to solve technical contradictions first, since the method is very concrete; then they learn to solve physical contradictions; then they learn to use both methods interchangeably, depending on the problem.

Resolving Technical Contradictions

The TRIZ patent research classified 39 features for technical contradictions. Locate the features in the contradiction matrix. See Fig. 24-14, above, for an extract.

Find the row that most closely matches the feature or parameter you are improving in your "trade-off" and the column that most closely matches the feature or parameter that degrades. The cell at the intersection of that row and column will have several numbers. These are the identifying numbers for the principles of invention that have most frequently led to a **breakthrough** solution instead of a trade-off.[†]

For example, consider the proposal to change the speed of inflation of the airbag to reduce injuries to small occupants. The trade-off is that injuries in high-speed accidents increase. Translating this into the TRIZ matrix terms, the parameter that improves is "duration of action of a moving object" (row 15) and the parameter that worsens is "object-generated harmful effects" (column 31). The cell at the intersection has the notation "21,39,16,22" which are the identifiers for four of the principles of invention.

---

*The classical (Altshuller) Contradiction Matrix in Excel worksheet format is available at http://www.triz-journal.com/archives/1997/07/ and the explanation of the features in the matrix is at http://www.triz-journal.com/archives/1998/11/d/index.htm. An automated version of the matrix is available at www.triz40.com. For more recent research, see Darrell Mann, Matrix 2010. www.systematic-innovation.com
†See Ref (Darrell Mann Matrix 2010 IFR Press, available from www.systematic-innovation.com) for a newer version.

The 40 principles of invention are listed in Appendix 1, and versions with examples from a variety of fields (software, healthcare, electronics, mechanics, ergonomics, finance, marketing, etc.) are available. Some TRIZ practitioners use the contradiction matrix to select which principles to apply to a specific problem. Others try each of the principles for every problem, rather than depend on the historical frequency of use for guidance.

Regardless of how you select the principles that you will use, read each, consider the examples, construct analogies between the examples and your situation, and then create solutions to your problem that build directly on the concept of the principle and the analogies to the examples. For example, in the case of the automobile airbag:

Principle 39: Inert atmosphere

A. Replace a normal environment with an inert one.

- *Prevent degradation of a hot metal filament by using an argon atmosphere.*

B. Add neutral parts, or inert additives to an object.

- *Increase the volume of powdered detergent by adding inert ingredients. This makes it easier to measure with conventional tools.*

What does the damage is the encounter between the person and the airbag, before it is fully inflated. The outer surface of the bag grabs the child's face, then twists her head. So something that would "soften" the surface would be the equivalent of an "inert" material—it does not prevent the original purpose (inflate the bag and protect the person from hitting solid objects); but it prevents the bag from grabbing when it first contacts the skin. How can this be implemented? Change the structure of the bag—make it corrugated, or make it of filaments, or use multiple crushable layers, or make the outer layer slippery, or lubricated, or...

If the problem is better expressed as a *physical contradiction* (where one parameter has opposite requirements) rather than a technical contradiction, then the contradiction matrix won't work—it has no entries on the diagonal, so you can't look for "X gets better but X gets worse."

TRIZ has four classical ways to resolve physical contradictions:

1. Separation in time. For the airplane, have the wheels hanging down a few seconds before reaching the runway.

2. Separation in space.

3. Coexistence of the contradictory properties in different subsystems or different regions of phase space. For physical situations, this could include phase transitions such as liquid/solid/gas, paramagnetic/ferromagnetic, superconductor/normal conductor, etc. For human situations, consider separating someone's authority from his personality, or her decision making from her musical talent, etc.

4. Solve the problem in either the supersystem or in a subsystem. For the airbag, prevent all accidents (solve the problem in the supersystem of the car and the highway) or make the airbag textile very slippery (change a subsystem).

Examination of the 40 principles shows extensive overlap with these four methods, since they are based on the same research on the same collection of solutions to a wide variety of problems. There *is* a fifth way to resolve a physical contradiction: convert it to a technical contradiction. The conversion may be obvious or subtle: the most useful technique is to separate the elements of the contradiction and ask, "WHY?" Returning to the candy problem:

I want the filling to be hot. WHY? To make it flow faster.
I want the filling to be cold. WHY? So it doesn't melt the chocolate.

The trade-off here is that the speed of the filling gets better, but the shape of the chocolate gets worse. You can return to the matrix and find that principles 35, 15, 18, and 34 are the most frequent creative solutions to this problem, or you can use all 40 principles and find a wide spectrum of solutions.

Since people use TRIZ to solve problems that have been unsolvable by other means, it is no surprise that they become fascinated by the 40 principles and by having large numbers of possible solutions. It usually takes 2 to 3 years for practitioners to learn to appreciate that having a small number of high ideality solutions is more efficient, since they don't have to spend time rejecting unworkable solutions. Repeated iterations of physical-technical-physical

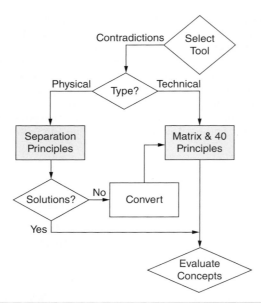

Figure 24-15 Beginner's approach to physical and technical contradictions. More advanced practitioners make several iterations between the two formalisms.

approaches will refine the definition and the solutions. (In classical TRIZ this is called "peeling the onion.") See Fig. 24-15 above.

Solving Problems Using the 76 Standard Solutions[*]

The "76 standard solutions" of TRIZ were compiled by G.S. Altshuller and his associates between 1975 and 1985. They are grouped into five large categories as follows:

1. Improving the system with no or little change — 13 standard solutions

2. Improving the system by changing the system — 23 standard solutions

3. System transitions — 6 standard solutions

4. Detection and measurement — 17 standard solutions

5. Strategies for simplification and improvement — 17 standard solutions

Total: 76 standard solutions

This list was developed from referenced works and published in a comparison with the 40 principles to show that those who are familiar with the 40 principles will be able to expand their problem-solving capability by learning Su-field (the common abbreviation for "Substance-field") analysis and the 76 standard solutions.

The 76 standard solutions are useful for Level 3 inventive problems. Level 3 inventions significantly improve existing systems and represent 18 percent of the patents included in the initial study. An inventive contradiction is resolved within the existing system, often through the introduction of some entirely new element. This type of solution may require

---

[*]This section is extracted from the review of substance-field modeling and the 76 standard solutions written by John Terninko, Joe Miller, and Ellen Domb. See
http://www.triz-journal.com/archives/2000/02/d/article4_02-2000.PDF Substance-Field Analysis
http://www.triz-journal.com/archives/2000/02/g/article7_02-2000.PDF 76 Standards, Section 1
http://www.triz-journal.com/archives/2000/03/d/index.htm 76 Standards, Section 2
http://www.triz-journal.com/archives/2000/05/b/index.htm 76 Standards, Section 3
http://www.triz-journal.com/archives/2000/06/e/index.htm 76 Standards, Section 4
http://www.triz-journal.com/archives/2000/07/b/index.htm 76 Standards, Section 5

several hundred ideas, tested by trial and error. Examples include replacing the standard transmission of a car with an automatic transmission, or placing a clutch drive on an electric drill. These inventions usually involve technology integral to other industries but not widely known within the industry of the inventive problem. The resulting solution causes a paradigm shift within the industry. A Level 3 innovation lies outside an industry's range of accepted ideas and principles.

Typically, the 76 standard solutions are used as a step in ARIZ, after the Su-field model has been developed and any constraints on the solution have been identified. The model and the constraints are used to identify the class and specific solution.

In these examples, the language of Su-field is used. A typical function is described by "S2 acts on S1 by means of a field (or force) F."

A few examples from above will illustrate the use of the model and the constraints in developing solutions.

Complete an incomplete model

If there is only an object S1, add a second object S2 and an interaction (field) F. *Example*:

- If the system is just a hammer, nothing happens. If the system is a hammer and a nail, nothing happens. There must be a complete system—hammer, nail, mechanical energy for the hammer to act on the nail and a guidance/feedback system to know if the hammer and energy have successfully moved the nail.

- If a truck has no fuel, it won't move. The complete system is the truck, the fuel, and the conversion from the chemical energy of the fuel to mechanical energy of the truck.

- In many organizations, a single individual cannot get anything done. The system must be complete with the original individual (S2) acting on others (S1) by means of persuasive arguments (F).

The system cannot be changed, but a permanent or temporary additive is acceptable

Incorporate an internal additive in either S1 or S2. *Examples*:

- Adding aerated slag particles can reduce the density of concrete.

- Doping of silicon wafers changes properties.

- Use a carrying agent in spray paint.

- In baking, add ammonia bicarbonate to increase the loft of bread (larger cells).

- Add vitamin C (to a person) to strengthen the immune system.

- In a heart-lung machine, blood clotting is a problem that is reduced by the addition of heparin. Later, the heparin metabolizes.

A pattern of large/strong and small/weak effects is required

The locations requiring the smaller effects can be protected by a substance S3. *Examples*:

- Small glass ampoules of medicine are sealed by flame, but the heat from the flame can degrade the medicine. Immerse the ampoules in water to keep the medicine at a safe temperature.

- Use heat sinks during soldering to protect elements that could be damaged by high heat.

- Use masks during the fabrication of silicon wafers to allow dopants to penetrate certain regions and to prevent them from penetrating other regions. Likewise, masking tape or stencils can be used to keep paint in the region where it is desired.

Useful and harmful effects exist in the current design

It is not necessary for S1 and S2 to be in direct contact. Remove the harmful effect by introducing S3. *Examples*:

- The hands of the doctor S2 are used to perform surgery on a patient S1. Wearing sterile gloves (S3) eliminates germs.

- A house jack (S2) will damage the main carrying timber (S1), but a steel plate (S3) between the jack and the timber will distribute the load.

The flowchart in Fig. 24-16 shows the flow of information and the decisions that are made to use the 76 standard solutions.

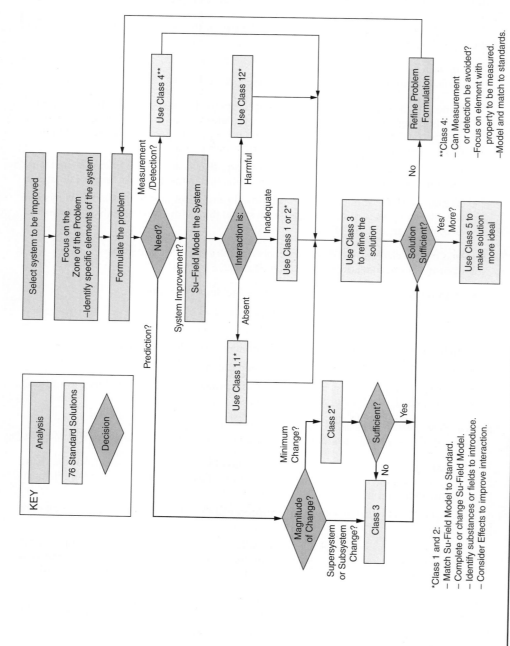

FIGURE 24-16 The flow of information for the use of the 76 standard solutions.

Solving Problems Using "Effects" from Other Technologies

In TRIZ, the term "effects" refers to methods, usually scientific methods. The term was adopted because of the common usage such as the Bernoulli effect in fluid dynamics, the Seebeck and Peltier effects in semiconductors, and so on, but many other terms besides "effect" are used:

- Archimedes' principle in fluid transfer
- Ohm's law in electric power
- Newton's laws of dynamics (for baseball as well as for space travel)
- Snell's law of refraction of light in optics
- Maslow's hierarchy in human psychology
- Mendel's laws of inheritance of characteristics in genetics

In TRIZ problem solving, the use of "effects" is closely linked to function analysis and resource analysis. The steps are as follows:

- Identify the resources present in your problem.
- Identify beneficial and harmful functions, using general vocabulary, not industry-specific jargon.
- Use effects to obtain the functions needed, or to make changes in the functions that are in the system, using the resources in the system if possible.
  - Look for science outside your own field.
  - Use many search methods and resources.

The reason for looking outside your own field is very practical—if the solution lies inside your field, you have probably already tried it. The conclusion that a problem is hopeless or impossible frequently means that the people trying to solve it don't know of a solution.

There are specialized TRIZ resources as well as general resources for looking for effects, such as patent search tools, general web search tools, best practices reports from professional societies, to name a few. TRIZ resources are indexed with specific reference to functions and the changes to be made to those functions, and include

- The Invention Machine Corporation, http://www.invention-machine.com. Effects database in the Goldfire software
- CREAX, http://www.creax.com Function Database online
- Oxford Creativity "effects database" at http://www.triz.co.uk

Examples:
- When the cost of electricity quadrupled in California in 2003, farmers who had been drying cow manure using electric ovens needed to find a new way. A patent search quickly found a method for concentrating fruit juice that used hydrophilic gas to remove the water, instead of using heat.
- When the medical community of South Africa needed a way to treat AIDS/HIV without the expense of pharmaceutical "cocktails," they turned to ozone treatment, which had been used for many years to clean public water supplies (and which has been more successful!).
- Theory of Constraints has been successful in manufacturing process improvement for many years, but it is having an extraordinary impact in education, as a method for teaching people to deal with conflict.

Many TRIZ practitioners use the "effects" method informally. For a structured method, especially oriented toward nonengineering applications, see Hongyul Yoon's "Pointers to Effects for Non-Technical Problem Solving."[*]

---

[*]http://www.triz-journal.com/archives/2009/03/03/.

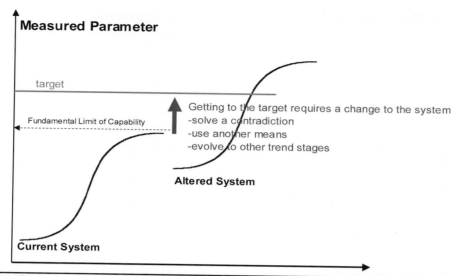

Measured Parameter

target

Fundamental Limit of Capability

Getting to the target requires a change to the system
-solve a contradiction
-use another means
-evolve to other trend stages

Altered System

Current System

FIGURE 24-17 S-curves of improved functionality (measured parameter) versus time. The version shown demonstrates that a new system may have lower performance when first introduced, but has higher potential for future development.

Trends of Evolution[*]

The TRIZ trends of evolution are easy to use in problem-solving and strategic decision making. They are based on the same data as the 40 principles and the 76 standard solutions, analyzed by time sequence rather than by type of problem solved. The basic premise, promulgated by Altshuller and used by TRIZ practitioners for more than 50 years, is that systems change in predictable patterns. The patterns exist because the needs of the users change, environments change, and people who develop those changes use the methods of problem solving, consciously or not, that fall into predictable patterns.

The S-curves shown in Fig. 24-17 demonstrate step-change system evolution. The trends of evolution can be used any of these three ways:

1. Find new S-curves as one of the ways of overcoming contradictions.

2. Improve any part of the system that has been determined "insufficient" or "excessive."

3. Make smaller improvements to components or processes.

The TRIZ trends are also increasingly used in innovation management. When used as a strategic decision-making tool the trends can be used to

1. Identify gaps in the innovation effort and therefore opportunities for innovation.

2. Help to foresee future evolution possibilities of a product.

3. Create innovation strategies and help users commit to step-changes for their product or system.

Classical TRIZ describes eight generic patterns of evolution for technical systems.[†]

1. Increased ideality

2. Stages of evolution

3. Nonuniform development of system elements

4. Increased dynamism and controllability

---

[*]This section is based on the research and applications of Darrell Mann, www.systematic-innovation. com. For comprehensive tutorials and databases see *Mann's Hands-On Systematic Innovation*, 2nd ed. Published by IFR, Clevedon, UK, 2007.

[†]*The Tools of Classical TRIZ*, Ideation International, 1999, www.ideationtriz.com and Fey, V. and Rivin, E., "Guided Technology Evolution." http://www.triz-journal.com/archives/1999/01/c/index.htm.

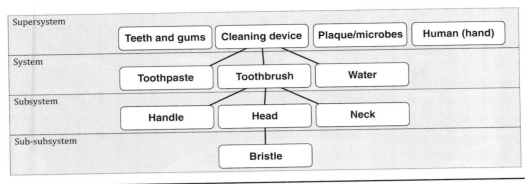

FIGURE 24-18   System hierarchy for the toothbrush.

5. Increasing complexity, then simplicity (popularly known as mono-bi-poly)

6. Matching and mismatching of parts

7. Transition to micro level and use of fields

8. Decreased human interaction (increased automation)

Researchers expanded the original set of trends to provide a less abstract set of trends that are easy to apply in problem solving and strategic decision making. Mann's approach to expanded trends, presented in a sequence of stages, is used here for ease of interpretation in many situations.

The toothbrush case study demonstrates the trends of evolution methods (Fig. 24-18).

The toothbrush has evolved from its early beginnings in China (in the 15th century) and Europe (in the 19th century) as a simple small brush made from animal hair and a bone/wood handle. Before the first toothbrush, the most common way to clean teeth was to chew a stick or use a toothpick. Toothbrush development really took off in the 1950s with the arrival of nylon bristles and injection-molded handles.

The toothbrush makes an interesting case study because, although the device is fairly simple, the task of tooth cleaning is still far from having reached its IFR. Some of the unresolved problems in tooth cleaning exist because teeth are complex shapes (convex, concave) with deep creases and slits between the teeth:

- Teeth and gums have completely different properties.
- Parts of the mouth are difficult to reach and clean.
- Many people do not enjoy cleaning their teeth.
- Toothbrushes accumulate dirt.
- Hands are often wet and slippery.

Each figure below shows the evolutionary trends of part of the system hierarchy.

Looking at the handle, it is easy to see how the surface segmentation trend has improved the ergonomics of the toothbrush (Fig. 24-19). These changes in surface and shape help provide the grip needed when hands are wet and slippery. The recent improvements due to the use of a two-shot molding technique are also being used to make the brushes look and feel more desirable. The examples shown of the head profiles are only a small sample of the huge number of different designs available. Most manufacturers have tried to innovate over the last five years to both improve reach inside the mouth and deal with the complex shapes of teeth.

Looking at this dynamization trend (Fig. 24-20), it is possible to see that no matter how sophisticated the manual toothbrush gets, it is unlikely to evolve much further than multiple joints. The brush has evolved to improve the reach and cleaning in the difficult parts of the mouth. To move to the next step, a completely new way of cleaning teeth has to be developed. In the second set of examples it is possible to see a range of different technologies that have been adapted and introduced into tooth cleaning. These new ways of cleaning all start to overcome some of the problems with the complex surfaces to be cleaned.

The trimming trend shows that it is possible to reduce part count (material efficiency) in a very concrete way (Fig. 24-21). In the first example these changes do constitute an improvement in overall efficiency, but they take the original product as the starting point

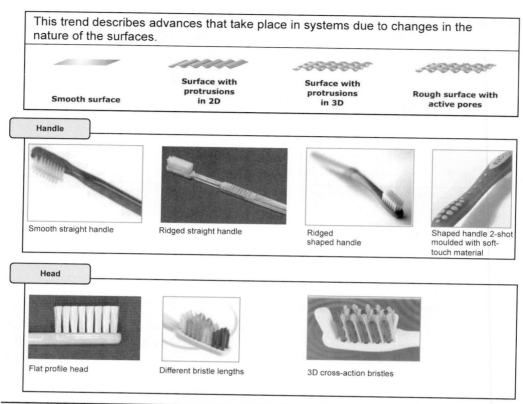

FIGURE 24-19   Surface segmentation trends.

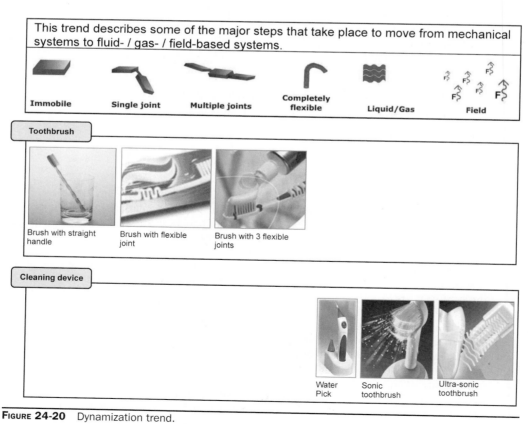

FIGURE 24-20   Dynamization trend.

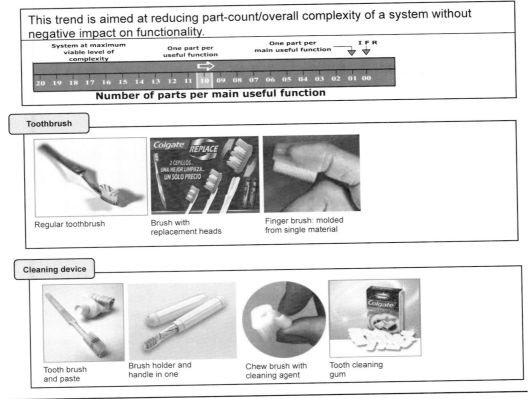

This trend is aimed at reducing part-count/overall complexity of a system without negative impact on functionality.

FIGURE 24-21   Toothbrush and cleaning device.

and do not really address any of the unresolved problems in tooth cleaning. Applying this trend at a higher systems level "cleaning device" results in more fundamental changes. When trimming at the systems level like this, looking at the prompts "one part per useful function" and "one part per main useful function" will likely result in significant changes to the system.

Examples from the trends above have shown us that, even in a very simple system, we can identify plenty of relevant trends and identify where current solutions can be placed on those trends (Fig. 24-22). Taking the point of view of evolution to make products more sustainable shows other trends in action, such as this example:

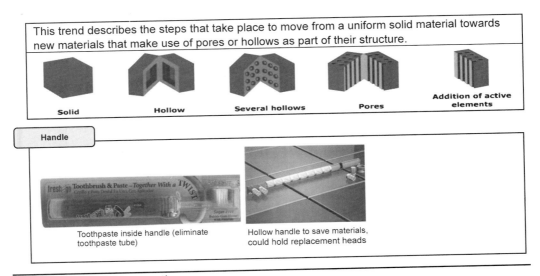

This trend describes the steps that take place to move from a uniform solid material towards new materials that make use of pores or hollows as part of their structure.

FIGURE 24-22   Space segmention.

The second idea for a hollow handle is one where the handle has five sets of replacements heads. This design concept is the equivalent of five normal brushes and is therefore likely to be more sustainable than current systems. It may even be possible to develop a sales model where the heads are sold in refill packs.

The trends can be organized into families that are applicable to sustainable design, to manufacturing cost-effectiveness, to customer behavior, and so on but the same basic patterns show up repeatedly.

Strategic Use of the Patterns of Evolution

The evolution trends are finite and therefore they inherently indicate that a system will evolve up to that point. This point is sometimes called the "evolutionary limit" of a design. This limit can always be overcome by working at a higher level of the system hierarchy. However, normally the evolutionary limit is used in innovation strategy to identify the unexploited potential.

The evolutionary potential diagram is a "spider chart" or "radar chart" that makes it easy to discern the difference between the current system and the evolutionary limit of the trend as shown in Fig. 24-23.

To evaluate a system, select the trends of evolution that characterize that system. These trends are displayed as the "legs" of the "spider." Then decide on the level of evolution of the current system for each trend (the dark-colored central area of the graph). The evolutionary potential is the light-colored outer area. It is tempting to conclude that all trends should evolve to their highest limit, but factors including the customers' readiness for new systems, the suppliers' ability to support production of the system, and the impact on the environment of the changes must also be considered. The general steps for the use of the trends of evolution are as follows:

1. Formulate the IFR. Decide if it is appropriate to move directly to the IFR, or if intermediate stages of development are needed.

2. Analyze the history of the system. Construct the S-curves for all important functions.

3. Identify the trends of evolution that are demonstrated in the system. Construct the evolutionary potential graph and identify the high-potential changes for the system.

4. Formulate the problems that must be solved to achieve the changes. (Include failure prevention, reliability, robustness, etc.)

5. Solve the problems using TRIZ problem-solving methods.

6. Select the development to be implemented.

The discipline of the steps and the detailed guidance of the trends of evolution take much of the guesswork out of strategic decision making, and have been a key factor in the adoption of TRIZ.

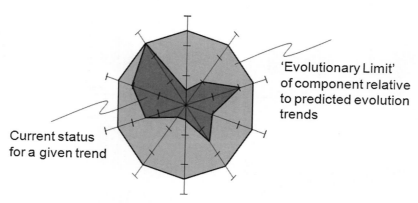

'Evolutionary Limit' of component relative to predicted evolution trends

Current status for a given trend

Each spoke in the evolutionary potential radar plot represents one of the known trends identified by TRIZ researchers for a particular system

Figure 24-23   The evolutionary potential diagram.

Research and Changes in TRIZ

The TRIZ patterns of evolution tell us that any system that is designed to solve a problem will change—partly because the environment of the problem will continue to change, and partly because the people who need the solution will continue to evolve their needs. TRIZ itself is no different from other systems—it has undergone continuous change since the research began. This results in a classical TRIZ contradiction:

- Good news: TRIZ continues to evolve and be relevant to the needs of people and society in an evolving world.

- Bad news: The multiple alternate methods loosely classified as "TRIZ" can be confusing to newcomers to the field.

Another contradiction is that during Altshuller's life, he judged and endorsed some variants and rejected others. Now there is no one authority; researchers publish their TRIZ concepts, discussion takes place at the meetings and in online groups, and the most useful of the variants develop communities of users. Table 24-2 is a very brief summary of some

Name	Description	Primary Researcher	Reference
Systematic Innovation	Combines TRIZ for problem solving and patterns of evolution with studies of social trends and changing market dynamics for product development, and studies of organizational dynamics for process improvement.	Darrel Mann	Y1
SIT, ASIT (Structured Innovative Thinking)	Simpler, fewer rules. More general, not tied exclusively to engineering. Only two main principles and five problem-solving tactics. Best for improving existing systems, not creating new ones.	Roni Horowitz	Y2
USIT (Unified Structured Inventive Thinking)	Spun off from the early version of ASIT, but took a deeper look at the mental processes underlying all structured, problem-solving methodologies. It develops thinking paths based on the role of metaphors in stimulating both brain hemispheres in generating solution concepts for a given problem.	Ed Sickafus	Y3
OTSM (General Theory of Powerful Thinking)	The core of OTSM is the challenge, "How is it possible to work on complex nonstandard problems, which may be represented as networks of interdisciplinary nonstandard problems, which develop and change over time? At that, the speed of these changes is commensurable with the time necessary for solving the problem." A family of axioms are used to guide the problem-solving methods; classical TRIZ has been absorbed into OTSM.	Nikolai Khomenko	Y4
GTI (General Theory of Innovation)	GTI is concerned with all the artificial (man-made) systems and was derived from TRIZ to enable "invention on demand." GTI focuses on the evolutionary (market) success/failure (i.e., on the Value Proposition) and the relationships between a system and its environment that required introduction of such concepts, while TRIZ focuses on a system itself.	Gregory Yezerski	Y5
MUST (Multilevel Universal System Thinking)	MUST users analyze, evaluate, and define change level of any artificial systems (including methods and techniques) by means of "customization" levels instead of hierarchy ones: result (satisfied need), method, technology, means, and parameters. Systems thinking is used at each level.	Gregory Frenklach	Y6

TABLE 24-2   Research and Changes in TRIZ

current TRIZ derivatives, with references for learning more about each of them. There are many TRIZ instructors and consultants who have trademarks or copyrights for their various teaching and consulting methods (including the author) and this list is not meant to exclude any of those; rather, it is intended to identify for the reader some of the alternate terminology that is emerging.

- TRIZ research in English was reported in the TRIZ Journal, www.triz-journal.com from 1996 to 2010.
- TRIZ research in Russian is reported in the proceedings of MATRIZ, www.matriz.ru and the publications of individual conferences.
- The proceedings of the European TRIZ Association are archived at www.etria.net.
- The proceedings of AMETRIZ (the Mexican TRIZ Association, including the proceedings of the Iberoamerican Innovation Congress) are at www.ametriz.com.
- The TRIZ homepage in Japan has extensive translations of papers from the Japan TRIZ symposium, reports on other conferences, and original papers, at http://www.osaka-gu.ac.jp/php/nakagawa/TRIZ/eTRIZ/.

This handbook chapter will be obsolete as soon as it is published, since TRIZ will continue to evolve and change, to enable people to innovate in the changing world. I encourage the readers of this handbook to contribute to those changes, and to the evolution of TRIZ.

Bibliography

Altov, H. (Altshuller pseudonym). *And Suddenly the Inventor Appeared*. Translated by Lev Shulyak. Technical Innovation Center, Waltham, MA, 1994.

Altshuller, G. *Creativity as an Exact Science*. Translated by Anthony Williams. NY. Gordon & Breach Science Publishers, 1988. Available from Taylor and Francis, 1-800-326-8917 or www.crcpress.com

Altshuller. G. *The Innovation Algorithm*. Translated by Lev Shulyak and Steven Rodman. Technical Innovation Center, 1999. http://www.aitriz.org

Gasanov, A.M., Gochman, B.M., Yefimochkin, A.P., Kokin, S.M., and Sopelnyak, A.G. *Birth of an Invention: A Strategy and Tactic for Solving Inventive Problems*. Interpraks, Moscow, 1995.

Zlotin, B. *The Tools of Classical TRIZ*. Ideation International, 1999, and Belski, I. *Improve Your Thinking: Substance-Field Analysis*. TRIZ4U, Melbourne, Australia, 2007.

"Case Study: Pizza and the System Operator for Teaching" by Domb, Ellen, Miller, Joe A., and Czerepinski, Ralph G. www.triz-journal.com, March 2008.

Mann, Darrell. "System Operator Tutorials." *The TRIZ Journal*, September, November, and December 2001 and January 2002.

Serediski, Avraam. "System Operator and the Methodology of Prediction." *The TRIZ Journal*, January 2002.

Miller, Joe and Domb, Ellen. "Applying the Law of the Completeness of a Technological System to Formulate a Problem." *The TRIZ Journal*, January and December 2007.

Golden Classics of TRIZ. Ideation International, Inc., 1996, and *Tools of Classical TRIZ*, Ideation International, Inc., Southfield, MI, 1999.

"Invention Machine Laboratory," version 1.4, 1993. Invention Machine Corporation. (In Russian) Chapter 6 and Appendix 9 and Appendix 5.

Terninko, J., Zussman, A., and Zlotin, B. *Step-by-Step TRIZ*. Responsible Management, Nottingham, NH, 1997.

Terninko, J., Domb, E., Miller, J., and MacGran, E. http://www.trizjournal.com/archives/1999/05/e/index.htm.

www.systematic-innovation.com. Books by Darrell Mann: *Hands-On Systematic Innovation* (both technical and business versions), *Systematic (information) Innovation, TrenDNA* (versions for UK/US, China, Germany), Matrix, 2010.

www.sitsite.com, www.start2think.com.

Sickafus, E.N. *Unified Structured Inventive Thinking—How to Invent*. www.u-sit.net.

http://otsm-triz.org/en, www.tetrisproject.org.

www.strategicinnovation.com.

http://www.bmgi.com/sites/bmgi.com/files/Multi-level%20Problem%20Solving.pdf.

http://www.linkedin.com/groups/MUST-Innovation-3825363?goback=%2Egde_3825363_member_181884111%2Egmp_3825363%2Egde_3825363_member_181846459.

Appendix 1. The 40 Principles for Inventive Problem Solving

Every TRIZ textbook has a listing of the 40 principles, with examples. Learning the principles by identifying examples, and learning to apply them to the specific issues of the new TRIZ practitioner's business by identifying examples already in use are well-established methods, which have led to the creation of a large number of sets of examples.

The TRIZ Journal has sets of examples of the 40 principles in a wide variety of fields, including the following (http://www.triz-journal.com/archives/contradiction_matrix/for links)

- **Technical Examples and Tutorial**, The TRIZ Journal, July 1997
- **Business Examples**: The TRIZ Journal, September 1999
- **Social Examples**: The TRIZ Journal, June 2001
- **Architecture Examples**: The TRIZ Journal, July 2001
- **Food Technology Examples**: The TRIZ Journal, October 2001
- **Software Development Examples**: The TRIZ Journal, September, November 2001 and (PDF) August 2004
- **Microelectronics Examples**: The TRIZ Journal, August 2002
- **Quality Management Examples**: The TRIZ Journal, March 2003
- **Public Health (fighting SARS)**: The TRIZ Journal, June 2003
- **Chemistry Examples**: The TRIZ Journal, July 2003
- **Ecological Design Examples**: The TRIZ Journal, August 2003
- **40 Inventive Principles with Applications in Service Operations Management** By: Jun Zhang, Kah-Hin Chai, Kay-Chuan Tan: The TRIZ Journal, December 2003
- **40 Inventive Principles with Applications in Education** By: Dana G. Marsh, Faith H. Waters, Tabor D. Marsh, The TRIZ Journal, April 2004
- **The 40 Inventive Principles of TRIZ Applied to Finance** By: Stephen Dourson, The TRIZ Journal, October 2004
- **40 Inventive Principles in Marketing, Sales and Advertising** By: Gennady Retseptor, The TRIZ Journal, April 2005
- **40 Inventive Principles with Examples for Chemical Engineering** By: Jack Hipple, The TRIZ Journal, June 2005
- **Application of 40 Inventive Principles in Construction** By: Abram Teplitskiy, The TRIZ Journal, March 2005
- **40 Inventive Principles in Customer Satisfaction Enhancement** By Gennady Retseptor, The TRIZ Journal, January 2007
- **40 Inventive Principles in Latin Phrases** By Gennady Retseptor, The TRIZ Journal, January 2008

CHAPTER 25

Brinnovation (Breakthrough Innovation)

Praveen Gupta

Innovation has always been a part of mankind. Since the discovery of fire by rubbing two stones together, humans have been innovating. Innovation is probably the oldest known process; in other words, innovation is an extension of a person's creativity. Imagine when the human evolved and discovered fire. What was the knowledge level then based on what we know today? What was the level of excitement at the discovery of fire? As people gain new understanding by trial and error, they transform it into new knowledge and then use that knowledge to gain new understanding, discover more unknowns, and become even more curious. Thus the cycle of experimentation, knowledge, and innovation continually repeats. The outcome of the knowledge-experience cycle has led to continual creativity and innovation.

Evolution of Innovation

Before discovering fire, humans discovered simple rocks that could be used as tools. Getting ideas from human or animal teeth, a thought of a dagger could have arisen, and so daggers of stone were made. A dagger provided protection from animals and probably was used as a tool to prepare cold food, which was then warmed by the sun's heat. Daggers could easily have evolved into knives and spears. These tools could be used to tame animals or even for hand-to-hand fighting.

Humans discovered fire more than 50,000 years ago. Fire, which could be very destructive if not controlled, could be a great friend when controlled. The discovery of fire led to further human knowledge, as the fire could be used for making tools, keeping humans warm, keeping animals away, cooking meals, lighting dark caves, or even melting ice. Therefore, the discovery of fire could be considered a great breakthrough in human evolution because it was critical for survival.

How could the early humans or hominids get an idea about fire? They must have observed fire caused by lightning, or sun heat, or volcanic eruptions. They could have even observed fire while throwing rocks, which produced sparks when they hit other rocks. The discovery of fire led to humans thinking about how to use fire and how to protect themselves from it.

Thousands of years later, humans did invent the bow. The idea of a bow could have come from tree branches loaded with fruit. In thunderstorms or high winds, tree branches often throw their fruit far away. The tree branches may have been the catalyst for the invention of slings for throwing rocks, and slings led to bows to launch arrow-like spears. The arrow could be considered an evolution of spears adapted to work with bows for throwing longer distances.

The discovery of daggers, knives, fire, and bows and arrows may have led to the preparation of warm meals. Warm meals resulted in warmer bodies and may have led to the need for clothes to satisfy the demand for warmth. Clothes made out of grass and roots evolved to clothes made of animal skins with the help of a needle. Therefore, the discovery of the needle was a breakthrough. The early needles were like a hook to stitch two pieces of skin or fabric to replace the series of knots previously used to put two pieces together. The knots could have been discovered from natural entanglements of long string-like objects, or even tree branches or bushes.

Early civilization appears to be based on the seven metals, as the remaining known metals were discovered since the 13th century. The seven metals are gold, copper, silver, lead, tin, iron, and mercury. Early tools and weapons were made of copper, which was discovered around 4000 BC, and tin and iron were discovered around 1500 BC. The discovery of copper was more significant, as the first set of tools, implements, and weapons were made of copper. Early applications of copper were made with hammer and chisel.

Copper smelting was probably learned while throwing copper waste into fire. The first copper-smelted artifacts were found in the form of rings, bracelets, chisels, and weapons about 500 years after the discovery of copper. By the 17th century, an additional five metals were isolated, which are platinum, antimony, bismuth, zinc, and arsenic. By this time, metallurgy was a well-developed discipline. Post 17th century discovery of metals accelerated as 12 new metals were discovered in the 18th century.

Great Innovators

Understanding the roles of, and methods deployed by, Einstein and Edison requires looking at their accomplishments in a "big-picture" context. Figures 25-1 and 25-2 enable us to

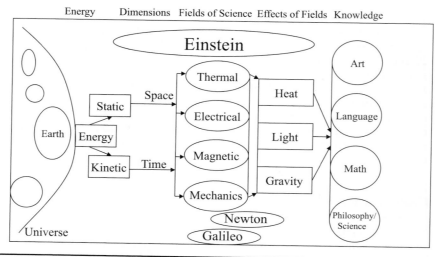

FIGURE 25-1   Fundamental innovations.

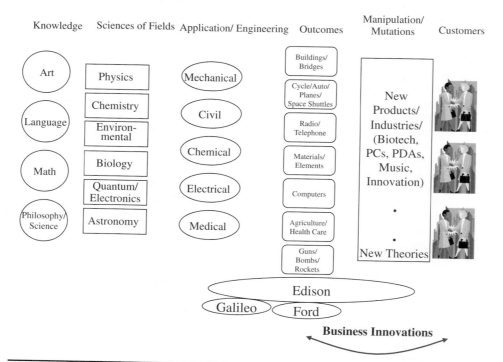

FIGURE 25-2   Business innovations.

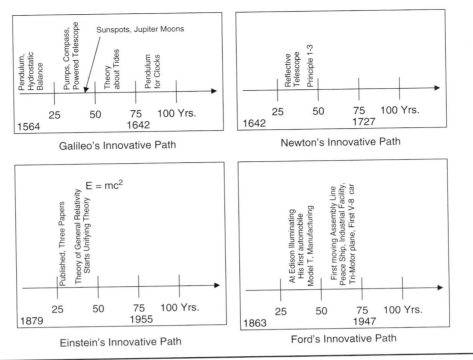

FIGURE 25-3 Great innovators.

review scientific evolutions at a higher level from the beginning. These two figures display the role of each great innovator in the context of the big picture, and they also provide a comparative analysis.

The work of Galileo, Newton, Edison, Ford, and Einstein represents a period from the 16th century to the 20th century, which was an era of super scientific discoveries. These innovators either made significant contributions or recognized major natural phenomena that helped humans understand the universe better, thus leading to extensive further discoveries.

To establish a standard process of innovation, observation of the details of innovations from great innovators should occur. Figure 25-3 depicts a lifetime engagement in discovery and innovation from some of the greatest individuals who took a unique approach to their accomplishments. Galileo, Newton, Einstein, Ford, and Edison all excelled in scientific, technical, industrial, and business innovation.

Galileo was an Italian astronomer and mathematician. He studied the works of Euclid and Archimedes, developed a pump for raising water and a high quality refractory telescope to study stars, invented a geometric compass and a thermoscope (i.e., thermometer), developed hydrostatic balances, designed pendulums, and applied the concept of pendulums to clocks. He observed that objects of different densities achieved the same rate over an inclined plane. Galileo utilized his own telescope to make celestial observations about sunspots and star formations, and he discovered new stars and their movement.

Galileo was a mathematician as well as a craftsman. He made great observations and posited bold theories when the war between religion and science was at its peak. He was convicted by religious leadership and imprisoned in his house for his belief that Earth rotates around the sun. Risk, part of an innovator's life, was experienced firsthand by Galileo. Galileo explored and innovated for about 55 years in a variety of fields, even during his imprisonment.

Galileo died on January 8, 1642, and Isaac Newton was born on December 25th of the same year. Newton studied mathematics and started to make his mathematical entries in 1664, completing his legendry publication *Anni Mirabliles* in 1666. In 1669, Newton developed a reflecting telescope and studied light and colors. Newton then studied planetary motion and observed Halley's Comet in 1682.

Newton's best published work is *The Mathematical Principles of Natural Philosophy*, or *The Principia*. Newton proved Kepler's Third Law about the motion of elliptical bodies in orbits. The Principia, completed in 1686, became the most comprehensive book on forces of gravity and motion between objects. *The Principia* includes definitions, rules, and laws of momentum (mass and motion) and forces (inertial, impressed and centripetal), and definitions of time, space, and motion.

Newton then applied his laws of motion to the motion of planets, moons, and comets as well as to the behavior of Earth's tides. Newton built on the work of Kepler, Galileo, and others. *Principia* was the most scientific approach to studying physics at that time. In Newton's own words, he researched his topics of interest and advanced the prior work. He also recognized his own possible "defects" in such a difficult subject and encouraged further investigation by his readers. Newton was a trained scientist who studied natural phenomena in a systemic fashion and who knew how to innovate methodically.

At the age of 5 years, the movement of the needle of his personal magnetic compass mesmerized Einstein. The continual orientation of the needle northward pointed to the existence of some invisible forces somewhere. At the age of 12, when he studied Euclidean plane geometry, he concluded that certain truths can be proven without doubt and with a sense of certainty. Einstein developed an uncanny ability to concentrate on topics leading to fundamentals and to clear a multitude of distractions out of his mind.

Einstein, generally a good student, was outstanding in mathematics, and hated memorization. He enjoyed studying mathematics, physics, and philosophy. He was even considered a distraction in his class at times.

In 1901, Einstein had a temporary teaching assignment and worked in the patent office in Bern from 1902 to 1909. While in the patent office, he wrote on theoretical physics on his own without being associated with a science community. During this time, he also earned his Ph.D. from the University of Zurich in 1905, which happened to be the year he made history. He revealed his breakthrough research in March, May, June, and September of 1905. In March, he published his theory of light quanta (i.e., the particles of energy versus the conventionally accepted theory of light as oscillating electromagnetic waves). In May 1905, Einstein submitted his theory of kinetic energy explaining the so-called Brownian motion, which reinforced the kinetic theory and helped in the study of atom movement. Actually, Einstein's light quanta theory was based on his experiments on particles.

In June 1905, Einstein published his work unifying the application of the relativity principle between electromagnetic waves and motion. Earlier Galileo and Newton had studied relativity for mechanical objects, while Maxwell and Lorentz studied the effects of relativity on electromagnetic effects. Their electromagnetic theory predicted that the velocity of light would show the effects of motion, but they could not prove it in the lab.

Einstein theorized that both mechanical and electromagnetic effects would be affected by the principle of relativity. In September 1905, continuing his work on the principle of relativity, Einstein reported that if a body emits certain energy, the mass of the body must be reduced proportionately. The relationship between mass and energy was defined by the famous equation, $E = mc^2$. Einstein unified interactions among particle motion, optics, and electromagnetic waves based on the prior work of Galileo, Newton, Maxwell, Lorentz, and many more.

Henry Ford, born in 1863, had an interest in mechanical activities. He worked with steam engines, farm equipment, and factory equipment. He had a sawmill business in early adulthood. Later he joined the Edison Illuminating Company where he became chief engineer in 2 years. During that time he experimented in internal combustion engines, which led to the development of his own self-propelled quadricycle and the formation of the Ford Motor Company in 1903.

Ford became a social entrepreneur and improved his manufacturing processes to produce cars at a reasonable price. He combined precision machining, standardized processes, interchangeable parts, division of labor, and assembly-line manufacturing, where the product passes by the worker for assembly. On the assembly line, Ford used conveyor belts, conducted time studies, and accelerated the Industrial Revolution. He significantly reduced the cost of manufacturing for his famous Model T cars.

Around 1920, Ford built the world's largest industrial complex which included a steel mill, glass factory, automobile assembly line, rolling mills, forges, assembly shops, and foundries. Basically all processes from refining raw material to the finished automobile were now performed at one plant. Ford created the concept of mass production and all the components associated with it.

As a social entrepreneur, Ford cared about his employees' lives at home and at work, built cottage industries in rural areas, established schools in several areas countrywide, and created the Henry Ford museum to preserve past innovations for future generations. At the Ford museum, the evolutionary nature of the innovation process is seen in living color when viewing old methods of washing clothes compared to automated washing/drying using the latest washers and dryers. Clearly innovations are built on what came before or on observations of an existing natural fact.

Thomas Alva Edison was born in 1847, a few years ahead of Ford. Edison had only about 3 months of formal education and became the greatest innovator of all time. He established a chemical laboratory at the age of 10. Edison's exemplary discovery was the light bulb, and hence the entire lighting industry was born. Edison's greatest contribution was the first practical electric lighting.

Edison invented the phonograph, telegraph, and telephony components, such as the carbon microphone, motion picture camera, electronic vote recorder, and the universal stock ticker. Edison also assisted on the production of the typewriter, electric pen, paraffin paper for wrapping candies, wireless telegraphy, dictating machines, shaving machines, improved electric railways, roller machines to break large masses of rocks, the fluoroscope, storage battery, Portland cement, electric motor, phonograph, kinetophone (sound and motion), and carbolic acid for explosives in World War I.

Edison figured out the innovation process in his early childhood. Edison epitomized the innovation process by combining scientific, industrial, and business innovation through his desire for continual innovation and growth. Edison continually grew professionally, personally, and financially through his endeavors. He was a gifted innovator who believed in working hard, learning from everything he did, and improving and innovating on everything he did. As a result, he expanded experience more, learned more, and innovated continually. He was so fascinated by creating new products that he set up the first modern-age research laboratory where he facilitated and accelerated innovation.

Edison really mastered the innovation process. He had over 1000 patents—the most issued to an individual. His famous quote: "Genius is one percent inspiration and 99 percent perspiration" still reverberates throughout the world. Edison loved physical and intellectual work. His heart, head, and body must have been busy all the time. He believed that innovating new things was a good task to undertake in order to gain fame and fortune while also benefiting society.

Knowledge Innovation

With the advent of the Internet, information is becoming available quickly. The Internet has already provided tools to collaborate among people globally; in other words, we can have clusters of innovation without clustering geographically. Moreover, the rate new information is being added on the Internet is itself exploding exponentially.

This information explosion looks like it will continue, and the protection of intellectual property could become trivial in many cases. The future need for innovation on demand in real time will mandate that corporations create new solutions to meet customer needs and then quickly move on to create other innovative new solutions to meet the next wave of customer needs. In other words, the laws to protect intellectual property will have to be reexamined as the rate of innovation increases. Current bureaucratic systems will not be able to keep up with the explosion of innovative solutions and related intellectual property.

In the upcoming and exploding knowledge age, customer-supplier relationships will appear to be very close, interdependent and insistent on innovative solutions. Current application of lean thinking demonstrates that business systems will be designed to produce to order, rather than produce to create demand. In such a scenario, if each item shipped by a company is unique, the innovation process must be institutionalized throughout corporations.

The expected extent of innovation goes far beyond development of products and services. Instead future customer demands will mandate innovation at every level in an organization in order to be able to serve customers and grow profitably. In other words, this type of innovation outlook is how a company can grow making millions of unique widgets.

Utterback, in *Dynamics of Innovation*, explores the relationship between process and product innovation. Product innovation leads to process innovations and vice versa. However, the success of a business is not ensured by just product or process innovation. Sometimes, very innovative products do not live up to their potential, and simple innovations exceed their expectations.

Figure 25-4 shows paths to innovation beyond process and products. A process is composed of activities which are outcomes of ideas. The resulting products, when sold to customers, bring more business; that is what the corporations are all about. In other words, if there are no ideas, there will be no new activities or experiments, and if no new activities are performed, new processes will not evolve. Fewer new products or solutions, therefore, will be available for fueling business growth. Lots of ideas are needed on a continual basis to make the idea-to-business-cycle work. Dauphinais, in *Straight from the CEO*, confirms this

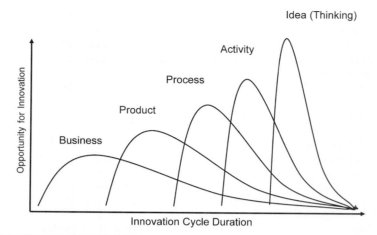

FIGURE 25-4   Dynamics of innovations.

and likewise mentions a lot of ideas are needed to arrive at one new innovative product. Dauphinais also asserts that businesses need many new products to hit a homerun.

If a business is internally driven to create demand, a continual stream of ideas leading to new products must be flowing. In other words, innovation begins with ideas. To generate a stream of ideas, a corporation must utilize all its intellectual resources (i.e., its entire workforce). When a corporation is engaging in *customer-driven* innovation on demand, speed of innovation counts. In some cases, the current sequential innovation process tends to be ineffective; instead the knowledge age innovation process is the place where many minds collaborate to create an effective solution on demand. In either case, many minds in a corporation must be involved to generate new ideas for innovative products and solutions. Excellence in idea management will become a corporate imperative in order to grow profitably.

Institutionalizing Innovation

Involving all employees requires that we understand how to utilize their intellectual potential in creating the new intellectual property. The process must be standardized to a great extent, with some exceptions. Therefore, the innovation process must be well understood. One cannot accelerate innovation as an art; it must become science in order to accelerate. The paradigm of innovation must be that it is a science as much as it is an art. In order to innovate with a higher probability of success, corporations must look into various elements of the innovation process and practice them just like any other existing standard process.

In the early years after the discovery of electricity, businesses used to have a Chief Electricity Officer. As electricity matured and commoditized, it became a utility, and no longer was there a need for the electricity officer. Similarly in the information age, we have a Chief Information Officer to glean tons of information. To utilize the information, businesses now are trying to extract intelligence, so business intelligence is becoming an important issue that people are addressing through dashboards and scorecards. However, the application of business intelligence is to create new knowledge and new solutions. Therefore, the position of Chief Innovation Officer is going to become a natural evolution of the changing business model.

Studies show that innovation is built on the past. In other words, all innovative solutions are based on past knowledge, continual experimentation, and extension of this past knowledge and experimentation. The *process* of innovation appears to be evolutionary in nature as well. People must understand this evolutionary nature of the innovation process, open new doors to new insights in the world around them (or even the universe), and search for new solutions. Einstein implied that all innovations are merely discoveries. Therefore, people must continually strive to discover new aspects of business and the world. Once people accept that innovation is a result of the discovery process, not a subconscious effort, the process of innovation can be easily understood and established as a predictable system.

A New Framework for Innovation

Striving for immortality or comfortable living drives the human appetite to innovate better ways of living. The innovation can be in drugs, foods, tools, communication, or even astronomy. The drive to innovate originates from a fear of extinction at the extreme, or out of the need to make life more comfortable or free from suffering. The focus of successful innovation is to fulfill human needs, which may be health, food, work, communication, security, and knowledge. Any drug for longevity, delicious food for better health, instruments for earning a salary, devices to communicate faster, weapons for protection, and methods of knowledge acquisition to perpetuate an appetite for more will lead to successful innovations, if they are affordable.

Peter Drucker (2002) observed that the innovation must be purposeful and begin with an analysis of the opportunities. Accordingly, the innovation must also be simple and capable of performing at least one specific task. Drucker identifies seven sources of innovation that include a flash of genius, exploiting incongruity or contradiction, growth in demand, changes in demographics or perceptions, and creating new knowledge. Recent knowledge innovations include history-making innovations such as the personal computer, cellular phone, iPod, the hybrid, and the Internet; older innovations include the airplane, wireless technology, electricity, cement, penicillin, and many more. Finally, innovations include the discoveries in space and materials, such as finding new planets like Neptune or discovering new elements such as Uranium. These innovations conventionally require commitment, hard work, and perseverance.

As mentioned earlier that at the early stages of the discovery of electricity, corporations used to have Chief Electricity Officers; in the information age, businesses appointed Chief Information Officers; now in the knowledge age, businesses are appointing innovation officers more frequently. Innovation is moving toward becoming a standard process similar to other functions in business, such as purchasing, sales or quality control. Becoming a standard process implies a standard task must be performed; the outcome is predictable to some extent; personnel for the process are designated; a box on the organization chart exists; and a room for standard processes is labeled in the facility.

Some companies already have appointed innovation chiefs, including Coca-Cola of the United States, DSM of the Netherlands, the Health Science Centre in Canada, Publicis Groupe Media of France, and Mitsubishi and Hitachi of Japan, implying their focus and resource commitment to innovation in sustaining profitable growth.

Companies like 3M, Procter & Gamble, AT&T, IBM, Siemens, Sony, Toshiba, Airbus, Unilever, Ford, GM, Tata, and Birla have been innovating for many years. Even large corporations, however, are realizing that the innovation process utilized thus far may not be as effective competitively, as it was in the past. IBM offers innovation-on-demand consulting services to other businesses, yet even experts at IBM are realizing that their current understanding of innovation must improve.

IBM's first Global Innovation Outlook (GIO) studies concluded that business must have mistaken invention for innovation. One of the obvious changes is that innovation is occurring faster and more frequently.

Participants of the GIO consequently recognized the need to redefine innovation. Accordingly, the consensus opinion was that "We must define 21st century innovation as beginning at the intersection of invention and insight: We innovate when a new thought, business model, or service actually changes society." This redefinition of innovation demonstrates that businesses must adjust their understanding of innovation in the knowledge age.

When comparing innovation in the 20th century with the 21st century, innovation in the 20th century was a forte of large corporations with tremendous resources for research and development. The smaller corporations followed their lead by developing some derivative products. New knowledge had protection from reuse without paying royalty for it. Large corporations grew and employed thousands of people. The standard of living improved and families saved money. At some point when risk was manageable, entrepreneurs tried something new related to their work at larger corporations, and the idea spun off. New companies formed and became larger corporations. The new, large corporations did not fund the basic research and development as their predecessors did, however, because they could not afford to continue to do the research.

In the 21st century, however, knowledge acquisition is now a decentralized process due to the invention of the Internet. People have access to knowledge everywhere there is access to the Internet. With the Internet, the control of knowledge has fragmented all the way down to individuals. As a result, the rate of innovation is changing, and large corporations cannot keep up with it. Many new companies start up with new ideas funded by the resources of venture capital firms. Therefore, larger corporations must recognize the new model of innovation. Some of the large firms, like Procter & Gamble, have set goals to seek a certain percentage of innovation from outside the corporation's boundaries. Such outsourcing of innovation is called "open innovation."

Key Resources	Time and Material/ Physical Age	Information/ Knowledge Age
Material	Raw material	Information
Tools	Machines and tools	*Brain (to be understood)*
Methods	Repeatable Processes for well understood machines	*Repeatable process to be developed*
People	Workers for physical effort	Workers with thinking effort
Environment	Comfortable for producing goods	Learning and creative
Expectation	High volume reproducible products	High volume customized solutions
Measurements	Productivity and Performance	Performance and productivity

FIGURE 25-5   Ages of innovation.

Examining the last century highlights the changes that have occurred from horse cart to space shuttle, labor to automation, material flow to information flow, and physical resources to intellectual resources. For example, the physical resources included time and material, while the intellectual resources imply knowledge. Figure 25-5 shows a comparison of innovation in the age of time and material with the knowledge age in terms of material, machines, methods, people skills, testing instruments, and the environment.

For the innovation process to be repeatable and available for any innovator, from the individual to larger corporations, some standard process must be established. In order for the innovation process to become repeatable, it must first be understood. To understand the innovation process, a basic framework must be discovered that can then be used to fill in the blanks for producing a repeatable innovation process. The process must work with any brain, be easy to understand, and be logical enough for many people to use repeatedly. As one of the professors working in a major university said, "We do not understand the innovation process well enough for it to be standardized."

In the knowledge age, with access to the Internet, a networked individual is the building block of innovation. As shown in Fig. 25-6, now the individual has access way beyond the network of a few individuals; instead, practically the whole world is the network. Once the network really becomes ubiquitous, the individual becomes the building block of innovation with access to laboratories, universities, experts, corporations, or trademark and patent offices. This phenomenon has already begun; however, its full impact is yet to be realized.

Once the building block of innovation is established, defining the innovation is equally important. Creativity and innovation are interchangeably used; however, experts define innovation as the process where a creative idea is applied to produce value to society. The current process of innovation is from idea to commercialization. Until the idea becomes a success, it remains a creative idea. Once the creative idea becomes successful, however, it is a major breakthrough—an innovation.

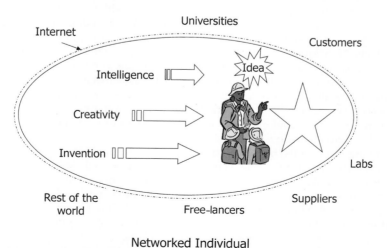

Networked Individual

FIGURE 25-6   Building block of innovation.

In the knowledge age, utilizing the processes of both Einstein and Edison to produce knowledge solutions is needed. Einstein helped in learning about the theory of innovation, and Edison helped in understanding the methodology of solution development. Interestingly, Einstein did no physical experiments, while Edison had laboratories and the innovation room for developmental experiments. Einstein believed that every innovation is discovered, and Edison believed innovation could be produced on demand. Innovation thus becomes the discovery of an innovative solution on demand. With this understanding, a framework for innovation can be developed, and a methodology for innovation can be established.

An Innovation Model

Einstein's exhibits in the Boston Museum of Science show that most of his innovative work was published through four papers in 1905. Einstein tried to put the puzzle together with various pieces of nature, which he then tried to include in his unsuccessful effort in developing a unifying theory. Even with the availability of various innovation methodologies, tools, and practices, a framework for innovative thinking is yet to be developed. Without such a framework, the predictability of methodologies and the repeatability of the innovation process cannot be established with confidence. The author has developed a model called Gupta's Einsteinian Theory of Innovation (GETI) to provide this needed framework for innovation.

GETI is based on Einstein's famous equation $E = mc^2$, where "E" represents energy, "m" represents mass, and the "c" represents the speed of light. Einstein's equation delineates the relationship inherent in the conversion between mass and energy. Every activity in nature is a conversion process.

Human beings are also energy converters. People consume resources and convert energy through actions or rest. Energy conversion can be physical or intellectual. The intellectual burning of energy occurs through thinking. When thinking, the ability to focus or channel energy into a direction, start associating various experiences with one another, and generate thoughts based upon those associations, is critical. Thinking is continual, purposeful, or inadvertent. When inadvertently thinking people think randomly, while when purposefully thinking people channel their thinking into some direction to get an answer.

Thus, innovation begins with an idea, which is an outcome of the thinking process; it must have some energy associated with it. Sometimes a lot of energy is required to think of an idea for some purpose. People even literally scratch their heads to stir up and stimulate thinking. Thus, every idea must have some energy associated with it that is an outcome of effort and the speed of the thought.

All discoveries occur in the human brain. As shown in Chap. 6, certain parts of the brain contribute to the innovation process. The brain has a cortex that consists of billions of neurons (cells) and trillions of axons (connectors). Neurons and axons form connections called synapses. With billions of neurons and trillions of axons, the number of possible synapses is practically infinite. The way the brain continually processes information is by comparing stored and received information. The speed of thoughts relates to how fast the brain processes the information stored and received, thus generating an idea.

For example, a patch of cortex consisting of 75×30 neurons is shown in Fig. 25-7. In order to match object A with an object A', the brain could compare one cell at a time to see if the objects match. Such a process may take thousands of comparisons. However, an experienced mind that has built anchors can quickly hop from object A to object A' and make the match.

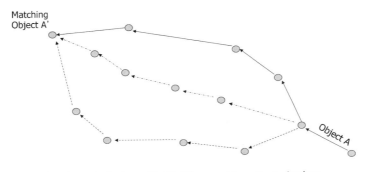

Example: Sample Cortex with 75x30 neurons   ◉ => Anchor

Figure 25-7   Speed of thinking.

One such comparison can take about 5 ms. If the brain makes thousands of comparisons, it can take hours, but if the brain hops through the anchors, comparisons can happen quickly.

The numbers or matching time could be astronomical—practically forever—if the person makes comparisons one cell at a time. The speed of thought, however, can accelerate by installing anchors based on multidisciplinary experiences. The actual brain speed can be one synapse per 5 ms. However, since the brain has billions of neurons and trillions of axons, billions or trillions of synapses can form in parallel within 5 ms. Moreover, if the anchors already exist, the speed can be faster.

As to the comparison with the speed of light, the speed of thought can be much faster if the anchor is already established in the brain for two points irrespective of their distance. If a person has been on the moon or to a location light years away, the brain can hop to that location right away in the mind on demand. Interestingly, the capacity of the brain is practically infinite with respect to visualizing the universe. Richard Restak, M.D. (1995), in his book *Brainscapes*, mentioned that if the number of synapses is about 10^{80}, that number is considered to be the same as the number of atoms in the universe. The brain can handle many objects, perform associations, discover new things, and innovate.

Thus, an innovation is a transformation of one set of ideas into another set of productive ideas. Therefore, the speed at which a person can process these thoughts becomes an important factor in accelerating or creating innovation on demand. Applying Einstein's equation to the process of innovation, one can equate "E" to the energy (value) associated with innovation, "m" to the physical effort or resources allocated to innovation, and "c" to the speed of thought, which can be faster than the speed of light. Restating Einstein's equation with proper substitutions, GETI delineates the following relationship:

$$\text{Innovation Value} = \text{Resources} \times (\text{Speed of Thought})^2$$

where the speed of thought can be described by the following relationship:

$$\text{Speed of Thought} \equiv \text{Function (Knowledge, Play, Imagination)}$$

The units of the innovation value can be represented in terms of resources and ideas over the unit of time, which can be equated to a new unit, Einstein (E), with the maximum value of "1." Thus, the innovation value can be increased with more resources or faster generation and processing of ideas. The innovation value accelerates with better utilization of intellectual resources rather than merely allocating more physical resources to innovation.

The following matrix defines various terms and demonstrates an example of the quantification of innovation.

Resources (R)	Knowledge (K)	Play (P)	Imagination (I)	Innovation Value (IV)	Comments
Degree of resources or time committed	Extent of knowledge based on research and experience	Percentage (%) of possible combinations of various variables explored	Dimension extrapolated as a percentage of ideal solution for breakthrough improvement	Estimated Innovation Level	This is an initial estimation of the proposed model. Further work is required.
50% (Limited time and insufficient resources)	75% (Significant knowledge and experience gained, some latest work is to be explored)	40% (Percentage of combination of variables explored mentally, experimentally or through simulation. Work is in progress)	66% (Selected dimension is extrapolated such that improvement is expected to be about 30%, which is about 66% of the breakthrough improvement)	0.182 (Long way to find an innovative solution due to lack of effort and play. To accelerate, one needs to improve all elements of innovation)	Innovation value = 0.5 × ((0.75 + 0.4 + 0.66)/3)² **= 0.182 Einstein**

Note: In the absence of a fully developed relationship among the variables, an additive relationship has been used to determine the innovation value (Gupta, 2005).

TABLE 25-1   Matrix for Using GETI for Assessing Personal Innovation

In other words, the innovation value is equal to the resources (commitment) times a function of knowledge, play, and imagination (KPI) squared. More than its numerical value, the equation identifies elements of innovation in order to maximize the innovation value. Most innovations are based on research, current experiments, and innovative thinking. Measuring knowledge and quantifying combinatorial play are possible, but measuring imagination is difficult due to the complexity of mental processes. Therefore, imagination is transformed in quantifiable terms by understanding that pure imagination is a random extrapolation. Thus, imagination becomes a measurable component by nature of extrapolation.

Innovation Categories

Reviewing contributions of great innovators, specifically Einstein, Galileo, and Edison; Einstein engaged in mostly theoretical innovation, Edison innovated practical or business solutions, and Galileo did a combination of both. Einstein's work was fundamental in nature, while Edison's work was more tangible. Einstein conducted mostly thought experiments, like riding the light wave, while Edison conducted his real experiments in his laboratory. Looking at various innovations, they can be classified into four categories based on the amount of effort and the speed-of-thought component. The four categories of innovations are the following:

1. Fundamental
2. Platform
3. Derivative
4. Variation

The **fundamental** innovation is a creative idea that leads to revolution in thinking. Such innovations are based on extensive research and are extremely knowledge driven, are theoretically proven, and lead to follow-up research and development. Such innovations occur with the collaborations of academia, commercial laboratories, and even corporations. The fundamental innovations may lead to changes in thinking, extend an existing theory, or be a breakthrough concept with enormous impact, perhaps leading to the evolution of a new industry.

Actually, such innovations contribute to human evolution. Examples of such innovations could be Einstein's theory of relativity, light quanta or photon, electricity, penicillin, the telephone, Xerox, wireless communication, the transistor, computer software, UNIX, the Internet, the Fractal, the Edison effect, and planes. The fundamental innovation has a significant academic component of science, which makes it available for the common good and thus less commercially protected.

The **platform** innovation is defined as one that leads to the practical application of fundamental innovations. Such innovations normally are launching pads for a new industry. Examples of platform innovations include personal computers, silicon chips, cell phones, digital printers, web technology, Microsoft Windows, databases, CDMA, Linux, drug delivery devices, satellites, and the space shuttle. The platform component increases the portion of the laboratory or development component more than do fundamental innovations. The platform innovations launch industries, change people's way of living, and fulfill the basic purpose of innovation, which is to live longer and more comfortably.

The **derivative** innovation is a secondary product or service derived from the platform innovation. Derivative innovations include new server-client configurations based on the new network architecture or operating system for a cell phone, for example. Derivative innovations are slight modifications of the main product. In the case of Microsoft-like software, the platform is Windows, and derivatives are a new office suite; for CDMA-like platforms, derivative innovations are various features available to service providers; for a major satellite system, the derivative innovations are various launching options or capabilities offered to users.

The **variation** innovation is the tertiary level of innovation that requires much less time and is a slight variation of the next-level products or services based on the derivative innovations. For example, variation innovations in cell phones are various color covers, ring tones, camera features, and more software-based optional features. In the case of Microsoft software, variation innovations are various applications developed and based on the Microsoft platform and derivative innovations. Typically, the variation innovation occurs close to the customer and may be the candidate for reaching the ultimate in speed of innovation or innovation on demand in real time.

Types of Innovation	Primary Drivers	Key Aspects	Deliverables	Frequency	Time to Innovate	Ownership
Fundamental	University/ Laboratories	Science/ Knowledge	Concepts/ Revelations	Rare	Years – Months	Govt. (s)
Platform	Corporate R&D	Technology/ Large Sys.	Equipment/ Capability	Sporadic	Months – Weeks	Govt./ Business
Derivatives	In-house/ Outsourced	Application/ Small Sys.	Product/ Service	Regular	Weeks – Days	Business/ Individuals
Variations	Networks/ Individuals	Disposables/ Ideas	Packaging/ Integration	Continuous	Days – On-demand	Individuals

FIGURE 25-8   Attributes of innovation.

Understanding types of innovations and their relevance to a business helps in establishing appropriate goals for innovations and devising correct measures of innovation. Figure 25-8 lists various aspects of innovation. Innovation on demand can mean different things to different levels of innovation.

Over time, responsibilities regarding where to put resources and who can reposition such resources must be understood. For example, switching systems, chip manufacturing facilities, and basic material or technology research have gone beyond the affordability of businesses; their collaboration with one another, or the government's support, will come into the picture in order to further fundamental or platform innovation. Based on the commercial success of an innovation, the innovation can move to the next level up or higher. For example, a cell phone like Razor (Motorola) becoming so successful can become a platform in itself (rather than a derivative innovation of a larger strategy). Microsoft Office is a platform innovation based on its success, and many additional, diverse, next-tier products or applications may be developed.

Various types of innovations are achieved by differing degrees of the speed of thought. For example, a fundamental innovation may require a more meditated process to think of theories, concepts, or solutions without major experimentation. In fundamental innovation, knowledge and imagination are key components. Interestingly, most of Einstein's work was completed in his mind rather than in a laboratory. He typically conducted "thought experiments."

The platform innovation involves relatively less knowledge and imagination and more play or experimentation. Figure 25-9 shows that variation innovation requires more play or more development effort than research and reflection. The chart helps in understanding

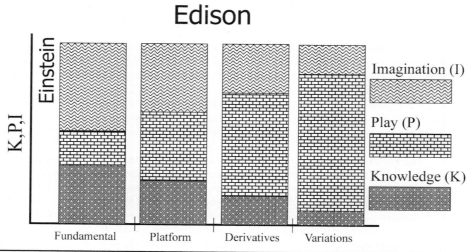

FIGURE 25-9   Type of innovation.

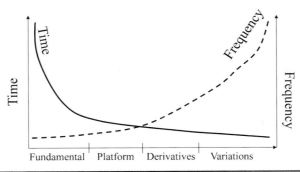

FIGURE 25-10 Extent of innovation.

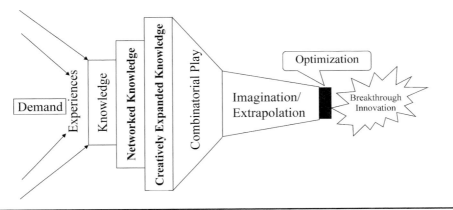

FIGURE 25-11 Breakthrough innovation (Brinnovation) process.

how various innovators focus on a particular area for their work, and how various innovations are achievable by focusing on the right component of speed of thought.

Figure 25-10 shows that fundamental innovations can take a much longer time than do the variation innovations. More variation innovations will result than will fundamental innovations. The fundamental innovation is a rarity, while the variation innovations are continuously occurring.

Figure 25-11 graphically depicts the innovation process, which appears to be linear at first glance. However, any step within the linear process has nested loops or divergence. As an overall process, the innovation process must be streamlined and appear linear in order to show progress. The innovation process is based upon the innovation framework and is designed to produce breakthrough innovation (Brinnovation) on demand. In other words, the innovation begins with a demand, and it must be purposeful.

The Breakthrough Innovation Process

The first step of the innovation process is to listen to the requirements for innovation and then gather knowledge about the topic in order to identify necessary inputs to the innovation. The networked individual, or the innovator, gathers more knowledge to achieve a certain level of competency in the field quickly. At this stage, process thinking helps in identifying input for the intended innovative solution. This step is a critical one and differs from the current methods of innovation, where an innovator searches for a solution or the outcome.

The following steps summarize the basic innovation process to realize breakthrough solutions through innovative thinking:

1. Understand the need for innovation and its purpose.

2. Research a topic individually, collectively, or through the networked resources, and gain a deeper understanding of the subject. Do not immediately solve the problem without proper research and knowledge.

3. Identify the potential variables affecting the problem. Make the list as long as possible and expand it using creativity tools, such as benchmarking, brainstorming, mind mapping, and TRIZ.

4. Test "what if" scenarios to isolate unlikely combinations of variables and identify likely combinations of variables. The objective is to remove obviously unrelated variables and retain related innovative solutions.

5. Establish the dimension of improvement or the performance characteristic(s).

6. Investigate likely combinations that could improve the performance characteristic(s).

7. Extrapolate the dimensions of interest and validate potential outcomes.

8. Expand your thinking by applying appropriate TRIZ-like principles to explore potential innovative solutions for generating significant change, thus making innovation obvious or disruptive.

9. Continue to explore and formulate alternative solutions. Select a solution that produces expected breakthrough improvement for further validation, optimization, and implementation.

The TEDOC Methodology

The above breakthrough innovation process has been organized in five phases. The phases are target, explore, develop, optimize, and commercialize, abbreviated as TEDOC. The target phase highlights a clear need for innovation based on opportunity analysis. The explore phase requires research, benchmarking, and analyzing the opportunity, and developing domain expertise. The develop phase mandates alternative innovative breakthrough solutions to maximize innovative value. The optimize phase requires optimizing the final solution for affordability and profit margin. The commercialize phase necessitates early access to the marketplace and customers to ensure premium margins and above market return on investment.

Target

Defining an opportunity for innovation is critical. In order to develop breakthrough innovations, a business needs to know what to innovate. To determine what to innovate, they must look at existing needs. These needs are found in complaints, nagging or chronic problems, indecision, frustrations, technical limitations, circumstances, and competitors' organizational limitations. A business should also look at the maturity of its industry, trends in supplier performance, SWOT (strengths, weaknesses, opportunities, threats) analyses, industry performance, and the available market. Once potential innovation opportunities are identified, the innovation team must document the key benefits of the solution to be innovated and determine the key measures of its success.

Explore

A company needs to fully and quickly research its opportunities to beef up its necessary competencies. The innovation team should identify and research keywords associated with the opportunity for innovation, generate new ideas, answer questions, discover new questions, and produce more new ideas. These ideas then need to be combined, filtered, analyzed, and prioritized. They are analyzed as input to the solution to be developed. Then, the team experiments with them to find solutions. Tools in this phase may include creativity, research, brainstorming, affinity diagrams, failure mode and effects analysis (FMEA), and process thinking.

Develop

Innovators need to develop alternate solutions that are significantly innovative. Experience shows that following the "rule of two" (discussed in Chap. 1) helps stretch imaginations as people experiment. According to the rule of two, in order for a solution to be a breakthrough innovation, it must affect the performance of the desired features by a factor of two (dividing or multiplying). In other words, if less is better, halve (divide by two) it, and if more is better, then double (multiply by two) it. The change is expected to force a different approach to the current position.

The extent of innovation depends upon the innovation team's efforts (the amount of available time committed to the desired innovation), knowledge (domain expertise), ability to play (experimentation), and overall imagination (extrapolation to achieve breakthrough innovation). In order to create a unique selling proposition and overcome barriers or competition, a company must try to maximize innovation rather than just create a minimal innovation. Tools used in this phase include the competency necessary to create new knowledge, creativity for proposing alternative solutions, evaluation and analytical methods, and the facility to conduct experiments.

Optimize

Many great innovations remain marginally successful and have limited shelf life because they are not effectively and economically reproducible. A great design alone does not provide a good return on innovation. The optimize phase focuses on maximizing the economic benefit of the innovation. In the current R&D-driven product development environment, the optimize phase is the most significant step missing for ensuring a product's success. Due to a lack of optimization in the design or preproduction stage(s), manufacturing operations suffer from design constraints. Today, most designs are quickly verified for their functionality and performance, but only on a limited sample size of potential process conditions during a product's life cycle. The prototype or pilot run that appears acceptable may actually result in continual rework and field failures leading to a significant adverse impact on profit margins. The tools typically used in this stage are process management, optimization software programs, and the facility to conduct the necessary experiments.

Commercialize

Many entrepreneurs and innovators fail in this phase—an innovative solution exists, but not enough people who would value it know about it. Without development, there is no creativity; without optimization, there is no profit; and without commercialization, there is no innovation. The commercialization of a solution converts creativity into innovation. Every innovator, therefore, must learn the process of commercialization and develop the knowledge necessary to create value. In the commercialization phase, an innovation team must practice strategic thinking about methods of pricing a solution, messages of value proposition, viral marketing, business planning, and making deals for licensing or selling the breakthrough solutions.

Leadership guru Steven Covey says to begin a task with the end in mind. In the case of innovation, begin innovating with commercialization in mind. Often, commercializing is tougher than discovering the innovative product. The full cycle of innovation starts with the identification of the need for an innovative solution and ends with the commercialization of the innovative solution.

Developing the ability to innovate on demand makes the task of commercialization easier, as the innovative solution has already been sold. However, improvement in the success rate of demand-driven innovation depends on the speed of the innovation. Once a company masters the process of innovation through practice and commitment, it can innovate quickly.

However, after a company has invested in deploying innovation through cultural transformation, it must take steps to sustain the culture of innovation. Every company should begin its innovation journey with the end in mind; in this case, an effort to sustain innovation must be carefully planned and practiced to perpetuate the culture of accelerated introduction of new products or solutions.

The idea generation process in the current environment focuses on ideas about the potential solutions and then picks the one that justifies the use of resources for a tryout or its novelty. The TEDOC methodology process incorporates a system for creativity or divergence, and systematic approach to convergence. The planned convergence process, or an algorithm associated with it, can speed up the innovation process to identify ingredients or sources of innovation. Once the purpose and sources of innovation are identified, then the extrapolation is utilized to achieve the desired extent of innovation.

The current process of imagination focuses more on subtle aspects, such as visualization, dreaming or using the subconscious mind. Actually if a person is introspective, the current process of imagination can be described as an ability to imagine various possibilities. In order to understand the imagination process better, looking at its boundary condition, which is pure imagination, is necessary.

Pure imagination appears to be conceiving very random thoughts or possible solutions and then playing with them in the mind by stretching them to their limits. Typically, when people imagine and stretch, they tend to go to the ultimate limits, which are beyond business needs. People imagining to that extreme often get lost and forget the purpose of innovation as well as their chain of thoughts. Thus, innovation on demand requires purposeful imagination, which is the identification of practical solutions and the extrapolation of the best solution in the direction of innovation.

Innovative Idea Generation

One of the challenges in developing innovative solutions is the practice of innovative thinking. In many brainstorming sessions, most of the ideas appear to be on the line of "been there, done that, nothing new, and same old, same old." Many ideas or suggestion programs fail because of triviality or the purposelessness of ideas. Many people do not even consider themselves innovative individuals. Even the "perceived dumbest" mind has enough neurons and axons for truly innovative thought. In order to stimulate human thinking, a process, which is simple and powerful but initially perceived to be trivial, has been developed. However, in the many sessions the author has conducted this simple process works.

The first step is to **clear the mind**. This step is achieved by asking people to write down good ideas they have about a topic without talking to other people. People love this step, as they already have so many good ideas that are clogging their minds. Once these ideas are written down, the mind is open, biases are out, and resistance is down. Reviewing these ideas shows that most of the ideas are the "same old" ideas everyone has already thought of and found to be useless.

Having cleared the mind, people are now asked to write crazy ideas about the topic. "Crazy" here is defined as **logically stretching** the mind by thinking about what can be done to the subject of innovation at its extremities. The left hemisphere of the brain usually drives the crazy thinking. These ideas stretch current performance levels. Some people continue the "good idea" process; thus, many of the "crazy" ideas still look like "good" old ideas. Some people really struggle as they try to conceive crazy ideas.

The next step is to involve the right hemisphere of the brain by asking participants to write down stupid ideas. "Stupid" ideas here represent unintelligent ideas, which really are **unrelated to the subject**. Participants see the difficulty in conceiving stupid ideas, and they learn to appreciate stupid ideas (as they really are well-thought-out innovative ideas).

The right hemisphere of the brain usually drives spatial thinking, which broadens the space of innovation. This thinking represents the creative aspects of innovation that people are afraid of thinking about for the fear of being called "stupid." They must recognize that stupid ideas are innovative ideas as well as some of the possible combinations of variables. At this stage, practically everyone tries to avoid looking "stupid." With enough prodding; however, participants do generate some ideas. The objective is to learn to think innovatively by utilizing all available mental resources and gaining speed of thought by practicing thinking. Some people are good at thinking of stupid ideas. Such individuals have a sense of uniqueness and differentiation.

At this point, people have learned to apply thinking on demand (i.e., flexibility of thinking). Mental agility is fundamental to developing **innovative thinking** quickly. The final step is to write down funny ideas about the subject of interest for innovation. Here practically everyone stumbles, with a few exceptions. People have to think hard to come up with funny ideas. The innovation process looks like an improvisation. At this stage, people are now practicing combinatorial play, or freely trying to associate various things they know about the subject of innovation.

Reviewing several sessions on applying the above process indicated that idea generation does take time. Actually, generating innovative ideas takes even more time than does generating good ideas. Figure 25-12 shows that funny ideas take the longest and are more innovative than are good ideas.

Many leaders like to communicate to their employees that they want them to have fun at work. Measuring whether employees have fun at work or not is difficult. The objective is not just to have fun, but also to have fun productively. Learning to give funny ideas demonstrates what can be a measure of having fun at work. Having fun means employees are free to give funny ideas without any fear. A measure of purposeful funny ideas is a great measure of the innovative thinking of an organization's human capital.

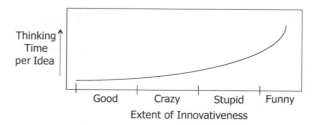

FIGURE 25-12 Innovative thinking.

Encouraging Innovation in an Organization

To achieve higher performance continually, business leadership must be cognizant of the intellectual potential of employees. Committing to innovation throughout the organization will accelerate the performance, as more employees are collaborating with the leadership team rather than resisting it. Such interaction in the organization reduces friction among employees and managers, identifies new opportunities for innovation, reduces cost, and creates a sense of ownership among employees.

Commitment to continual innovation requires a good understanding of theory, practice, and the results of the innovation methods. Scattered and successful application of innovation methods demonstrates that the innovation process is more than random creativity. Just like any other processes, one can benchmark the best innovators and organizations to learn and innovate. Einstein happened to be the best thinker, and Edison was the best innovator. Einstein's discoveries are more fundamental, but Edison's work was product-based innovation.

How did Edison innovate so frequently? He understood the innovation process, built his laboratory in Menlo Park, New Jersey, and guided his researchers to produce solutions on demand. He built factories and products based on his innovations and accelerated that growth in wartime. Edison was not acting as an innovative person; instead, he was seeking opportunities and producing innovative solutions. He expanded his knowledge, changed his expertise from one field to the other, and innovated on demand.

To institutionalize innovation on demand, benchmarking against Einstein and Edison is important. Einstein mastered thought experiments and saw something in nothing, and Edison perfected product innovation to produce innovative solutions on demand, as evidenced by the number of patents (over 1000) assigned to him. The corporate process for breakthrough innovation must consider the following aspects.

Defining Innovation

Clayton M. Christensen and Michael R. Raynor in their book *The Innovator's Solution* emphasize sustained innovation in achieving corporate business growth (Christensen and Raynor, 2003). A successful era of superior performance in the life of a corporation occurs due to some innovative disruption. Sustaining innovation requires not just the ideas, but also the packaging of ideas for growth opportunities. Even Six Sigma emphasizes breakthrough improvement; however, methods do not currently exist to produce breakthrough solutions.

Everyone has a different understanding of innovation and creativity. The extent of innovation can be incremental (implying a little change), radical (representing a disruptive change), or of a general purpose (implying a new discovery). One way to define innovation is "doing differently." The question then is how much change should the innovation create? Considering the process of evaluating change, a change is statistically significant when the change exceeds at least 47.5 percent in the desired characteristics. The 47.5 percent change corresponds to two standard deviations from the current process' typical performance, as shown in Fig. 25-13. When the change in a parameter is statistically significant, the probability of occurrence is small, and the change is a breakthrough innovation. Thus, innovation happens when an activity occurs differently in order to create value through products, services, or solutions.

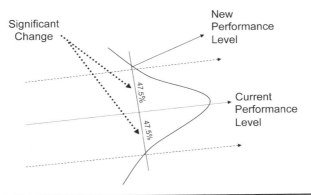

FIGURE 25-13 Breakthrough solutions.

Innovation happens daily, whether it is a simple container by Rubbermaid, an Apple iPod, Panasonic's robust laptop, a cell phone from Motorola, Sony, LG, or Nokia, a grill from Weber, new drugs from a pharmaceutical company, or a printer from HP, Canon, or Xerox. Many companies are serially innovative, while others are sporadically innovative. Thus, innovation is not a new thing, but it is often sporadic. In order to accelerate and sustain innovation for meeting continual demand for new products, services, or solutions, corporations must deliberately recognize organizational needs and address them. This recognition requires an understanding of strengths and weaknesses, company leadership for innovation, and a clear link between innovation and the corporate values and organizational strategy.

In order to understand an organization's strengths and weaknesses, an assessment must occur using simple tools like a checklist, survey, or performance analysis. A thorough analysis must include social, operational, financial, customer, and leadership aspects of an organization. The social aspects may include corporate values, teamwork, and employee participation. The operational assessment includes an emphasis on creativity in process management and daily activities, the ability to take risks, and a general decision-making approach. The financial aspects include resources committed to innovation-related activities, training, rewards, and revenue generated from innovative solutions. The leadership aspects include creativity at the leadership level, inclination for risks, recognition for success, understanding for genuine innovation failures, and keen participation in innovation. The assessment's objective is to understand how an organization can transform into an agile and thinking organization. A thinking organization is one that promotes learning new skills, experiencing new domains and productively applying lessons learned for developing innovative solutions.

In a training session, people are always looking for a trick, a case study, software, or a formula to apply quickly so they can reject the tool for its differences and application difficulty. Quickly applied methods, however, are not adequately developed and will not produce desired results. This rote application of a technique is called reproductive thinking. Accordingly, if everyone learns a technique to design a product and applies it the same way, then the expected result will also be predictably the same.

Every problem and every company are different. Thus, the application of a technique must be adapted creatively to the opportunity in consideration. Figure 25-14 illustrates different thinking types. A problem can be solved just by doing something, or many impractical creative ideas can help to solve the problem. The case of "just doing something" to solve a problem often leads to new problems, though, while in the case of using "imaginative ideas," nothing really happens. Thus, the solution needed for the problem often lies in the application of creative ideas.

One of the challenges in a corporation is to allow time for thinking. Companies often hire smart people, keep them busy fighting fires, and give them no time to think. The 3M Company allows employees to spend 15 percent of their work time as they wish thinking of something new, learning something new, or doing whatever they like. In order to justify time for thinking, or the investment in innovation, following the approach of systems thinking, as shown in Figure 25-15, must take place.

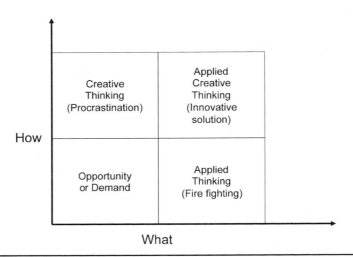

FIGURE 25-14   Thinking types.

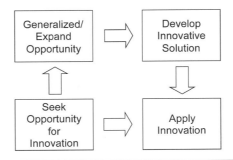

FIGURE 25-15 Systems thinking.

As the figure shows, innovating a solution for an opportunity at a level higher than the level of its symptoms is essential. For example, if a department is having a problem at one network node, the innovative solution should be implemented above the node level (i.e., at the server), so all nodes can benefit from it. Similarly, if an opportunity is identified for an innovative solution, the opportunity must be defined at a higher level. This elevation of the opportunity creates more value, due to its expanded scope of application, and justifies resources for developing the innovative solution. Once the solution is developed, then it is applied to specific situations.

Systems thinking requires that the corporate leadership practices process thinking, establishes measurements for monitoring performance, and promotes risk-taking in developing innovative solutions. In today's economy, diffusion of opportunities globally mandates that every society continually innovate in its domain of expertise in order to create value. Otherwise, by the law of diffusion, migration of opportunities from higher-cost to lower-cost locations will cause societal frustration.

Summary

To maintain market leadership, a company must launch a multipronged approach to grow the top line as well as the bottom line. Many improvement efforts to perfect the bottom line eventually lead to a business downsizing due to lack of sales. Businesses must develop new products, services, and solutions. The demand for innovative products or solutions has become a norm. To commit to continual innovation requires a good understanding of theory, practice, and the results of the innovation methods. Scattered and successful application of innovation methods demonstrates that the innovation process is more than random creativity.

An organization attempting to institutionalize innovation must determine its methodology of choice. As many innovation methodologies are out there as there are innovation consultants. Therefore, the reasons for selection must be based on a better understanding of the innovation process. Some methodologies are measurement heavy, which leads to a number-driven innovation approach without a predictable outcome. A successful innovation methodology must incorporate inspiration from leadership, involvement of employees, and outcomes for higher value. Such methodology will include planning, organization, a process, tools, measurements, collaboration, and celebration.

Innovation begins with the intellectual involvement of employees through their ideas. The process of getting employee suggestions, ideas, or recommendations has been in existence for a long time. However, its effective implementation and success have been far from satisfactory for many reasons, including the lack of understanding of its value and importance for improving corporate growth and profitability, and the lack of an established process of idea management.

Bibliography

Altshuller, G. *And Suddenly the Inventor Appeared: TRIZ, the Theory of Inventive Problem Solving*. Worcester, MA: Technical Innovation Center, 1996.

Chasan, Emily. CEOs Find Innovation Hard to Achieve—Survey. Innovathttp://prelaunch .reuters.com/sponsoredby/AMEX/article.asp, 2006.

Christensen, Clayton M. and Raynor, Michael E. *The Innovator's Solution*. Boston, MA: HBS Press, 2003.

Drucker, Peter F. *Innovation and Entrepreneurship: Practice and Principles*. New York, NY: Harper & Row, 1985.

Drucker, Peter F. "The Discipline of Innovation." *Harvard Business Review*, August 1985.

Edison Effect. http://www.ieee-virtual-museum.org/collection/tech.php?id=2345876&lid=1.

Edison Effect. http://www.bookrags.com/sciences/physics/electronics-wop.html.

Gupta, Praveen. "Beyond PDCA: A New Process Management Model." *Quality Progress*, April 2006.

Gupta, Praveen. "Innovation and Six Sigma, Six Sigma Columns." www.qualitydigest.com, December 2004.

Gupta, Praveen. "Innovation: The Key to a Successful Project." *Six Sigma Forum Magazine*, August 2005.

Gupta, Praveen. *Six Sigma Business Scorecard: A Comprehensive Corporate Performance Scorecard*. McGraw-Hill, NY, 2003.

Gupta, Praveen. *The Six Sigma Performance Handbook*. New York, NY: McGraw-Hill, 2004.

Gupta, Praveen. *Business Innovation in the 21st Century*. Charleston, SC: BookSurge, 2007.

Gupta, Praveen. *The Innovation Solution*. Charleston, SC: CreateSpace, 2012.

http://college3.nytimes.com/guests/articles/2003/07/15/1100994.xml.

http://galileo.rice.edu/chron/galileo.html.

http://neon.mems.cmu.edu/cramb/Processing/history.html.

http://topics.developmentgateway.org/knowledge/rc/BrowseContent.do~source =RCContentUser~folderId=3212.

http://trendchart.cordis.lu/scoreboards/scoreboard2003/index.cfm.

http://userpage.fu-berlin.de/~rober/linguistics/origins.html.

http://web.class.ufl.edu/users/rhatch/pages/13-NDFE/newton/05-newton-timeline-m .html.

http://www-groups.dcs.st-and.ac.uk/~history/Mathematicians/Galileo.html.

http://www.aip.org/history/einstein/index.html.

http://www.c3.hu/scca/butterfly/Kunzel/synopsis.html.

http://www.eurescom.de/message/messageMar2003/International_Symposium_on _Innovation_Methodologies.asp.

http://www.hfmgv.org/exhibits/hf/default.asp.

http://www.idrc.ca/en/ev-33214-201-1-DO_TOPIC.html.

http://www.indianexpress.com/full_story.php?content_id=69486.

http://www.lucidcafe.com/library/96feb/edison.html.

http://www.oecd.org/document/28/0,2340,en_2649_34273_34243548_1_1_1_1,00.html.

http://www.thomasedison.com/biog.htm.

http://www.usembassy-china.org.cn/sandt/stconfaug99.html.

http://www.virtualclassroom.net/tvc/internet/fire.htm.

McCarty, Thomas, Daniels, Lorraine, Bremer, Michael, and Gupta, Praveen. *The Six Sigma Black Belt Handbook*. New York, NY: McGraw-Hill, 2004.

McCarty, Tom, Daniels, Lorraine, Bremer, Michael, and Gupta, Praveen. *The Six Sigma Black Belt Handbook*. New York, NY: McGraw-Hill, 2004.

Porter, Michael E. Clusters of Innovation Initiative, Final Report, Council of Competitiveness. http://www.compete.org/pdf/pitts_final.pdf.

Utterback, James M. *Mastering the Dynamics of Innovation*. Boston, MA: HBS Press, 1996.

www.compete.org.

CHAPTER 26

Crowdsourcing: Tapping into the Talent of the Crowd

Juan Vicente García Manjón

Introduction

A meaningful first impression of crowdsourcing is offered by Jeff Howe,[1] who stated the following:

> Crowdsourcing isn't a single strategy. It's an umbrella term for a highly varied group of approaches that share one obvious attribute in common: they all depend on some contribution from the crowd.

It is no wonder, really, that crowdsourcing is becoming increasingly common in evolving social and economic environments. A robust discussion of the main drivers that have facilitated its rise in popularity helps us understand why. Jeff Howe in his book *Crowdsourcing: Why the Power of Crowd is Driving the Future of Business* refers to four fundamental developments to explain the phenomena.

The Renaissance of Amateurism

According to Wikipedia,* "an amateur is generally considered a person attached to a particular pursuit, study, or science without formal training, also referred to as an autodidact."

More and more people are engaged in their own interests and passions, from photography to software development or star gazing. In many cases, such as photography, the availability of cheaper technical equipment and the emergence of new digital technologies have overcome the barriers of manipulating and distributing their works, creating a huge community of individuals ready to create and share new content. That is the reason why services like iStockphoto (iStockphoto.com) have been created and have been competing with traditional and professional photo services. iStockphoto was founded by Bruce Livingstone in May 2000, with the intention of being a free stock imaging website. However, over time, it has evolved into its current micropayment model. On February 9, 2006, iStockphoto was acquired by Getty Images for US$50 million. Applicants must send three original samples of art for the filtering process. Contributors are paid a base royalty rate of 15 percent for each file downloaded and up to 45 percent if exclusive contributors (a minimum of 250 downloads).

Crowdsourcing operates on two fundamental bases: the assumption that there is a global pool of talent that can be effectively tapped and the acknowledgment of the community allowing genuine meritocracy. As referred in *Crowdsourcing Translation*,[2] "The idea behind crowdsourcing is that 'the many' are smarter and make better choices than 'the few,' and that the 'crowd' has a huge potential for which they often find no outlet." The same document adds, "There are more and more people who have knowledge and competences but do not have the chance to use them in their professional lives. Now, crowdsourcing offers them the opportunity to pursue their interests at an amateur level."

The Open Source Software Movement

It is best to understand crowdsourcing by referring to the open source movement. In 1983, MIT computer scientist Richard Stallman founded the GNU project in an effort to create

---

*http://en.wikipedia.org/wiki/Amateur.

computer-operating system software that would be completely open to the community. He subsequently established the Free Software Foundation in 1985, demonstrating how eager people are to dedicate their efforts and to freely share their results. To guarantee that no one could copyright this free software, he promoted the GNU General Public License (GPL), which specifies the free distribution and incorporation to other works of any piece of software under this license.

By 1991, Linus Torvalds had created his own operating system, which was open to collaboration and today has become arguably the most ubiquitous piece of software in the R&D domain. The development of Linux is one of the most relevant examples of free and open source software collaboration, since the underlying source code may be applied, modified, and commercially or freely distributed by anyone under GNU licenses.

As well, the Wikipedia initiative founded by Larry Sanger and Jimmi Wales started as an attempt to create a free encyclopedia available to everyone with Internet access. In 2001, Wikipedia went live for anyone to collaborate, and today it is competing face to face with Encyclopedia Britannica, which scrapped its printed edition in 2012.

Wikipedia offers more than 4 million articles in English, all contributed and updated by the community. Wikipedia explains that "anyone with Web access can edit Wikipedia, and this openness encourages inclusion of a tremendous amount of content." There are about 77,000 editors who regularly edit Wikipedia. The contributors have a wide variety of profiles, from expert scholars to technical enthusiasts or just individuals who want to collaborate.

Availability of New Tools of Production

Jeff Howe drew our attention to the fact that what we know as a consumer is becoming an antiquated concept, since amateurs are fueling the crowdsourcing engine and they are profiting from the open source movement as a way of production to take part in the economic processes long dominated by commercial enterprises. If there is an industry in which this case is genuinely paradigmatic, that is without a doubt, the media market.

New and evolving technology has been a primary driver in the media market, making everything faster and cheaper to create through reduced hardware costs and the emergence of incredibly powerful but user-friendly software. Moreover, free how-to-use information is becoming more familiar with the use of these new technologies. The Internet has been the primary trigger for the evolution of massive and collaborative methods to facilitate the rise of crowdsourcing in the media.

Self-Organized Communities

Another factor to be aware of is the rise of vibrant self-organized communities focused around peoples' shared interests. In the past, communities of interest were formed primarily based on geographic criteria due to the communication and interaction barriers of physical distance. However, expanding Internet access became a tipping point for the emergence of a vast pool of enthusiastic amateurs creating and sharing their knowledge within and among communities.

Those communities are no longer governed by traditional rules and policies; rather, they are self-policing, they prioritize the best ideas to rise to the top, and merit has become the most valuable asset of the participants. Hierarchies can be characterized as fuzzy and contributors within these communities are self-motivated.

Community efforts can now be defined in global terms. For example, the spare-time contribution of individuals now aggregate into thousands of working hours across the community. Even though money is not a primary motivator for contributors, financial incentives are also important in a community environment. People also respond well to competition where one can show their own skills and performance.

Above all, communities are a new and important stakeholder in the global marketplace. Large online communities, with the use of social networks, have the power to influence individuals, businesses, industries, and society as a whole. The self-organizing character of the community is relevant and available to anyone who wants to exploit the power of its services.

The Crowdsourcing Definition

The term "crowdsourcing" encompasses in a contraction two terms: "crowd" and "outsourcing." The outsourcing process is defined by Wikipedia* as a way to "transfer or delegate to an external service provider the operation and day-to-day management of a

---

*http://en.wikipedia.org/wiki/Outsourcing.

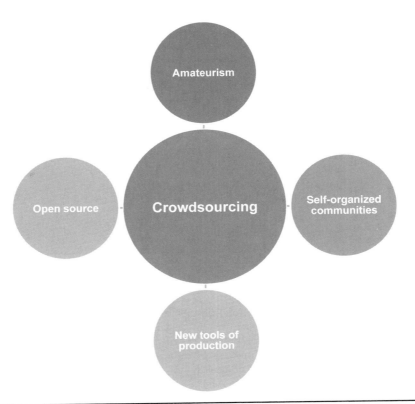

FIGURE 26-1   Crowdsourcing framework.

business process." Taking this into account, we could define crowdsourcing as the process of outsourcing some task or work to the *crowd* itself. The term was disseminated after the publication by Jeff Howe on the June, 14, 2006, in the article "The Rise of Crowdsourcing" in *Wired Magazine*.

Howe proposed the following definition: "Crowdsourcing represents the act of a company or institution taking a function once performed by employees and outsourcing it to an undefined (and generally large) network of people in the form of an **open call**. This can take the form of peer-production (when the job is performed collaboratively), but is also often undertaken by sole individuals. The crucial prerequisite is the use of the **open call format** and the wide network of potential laborers."[1,3]

So, the key point is the existence of an open call and the main difference between outsourcing and crowdsourcing is that while in an outsourcing process the firm defines and plans needs and seeks a supplier whose relation is formalized in a contract; crowdsourcing is based on an open call to the crowd who respond to it providing services to the firm on a voluntary basis or for some incentive.

Howe further offered in his blog[*] another two explanations of the term. The first one, cited as "The White Paper Version," defines crowdsourcing as "the act of taking a job traditionally performed by a designated agent (usually an employee) and outsourcing it to an undefined, generally large group of people in the form of an open call"; while the second one referred to as "The Soundbyte Version" offers, "the application of Open Source principles to fields outside of software."

Additionally, the *Financial Times Lexicon*[†] defines crowdsourcing as "a business model or function that relies on a large group of users as third parties for outsourcing certain tasks." This definition stresses that "popular use of the Internet makes communication and coordination progressively cheap: tasks that would have been impossible to communicate and coordinate before have become extremely easy to set up and coordinate. Crowdsourcing can add significant value to a product or service, and can also generate valuable connections between the users and the company."

---

[*]http://www.crowdsourcing.com/.
[†]http://lexicon.ft.com/Term?term=crowdsourcing.

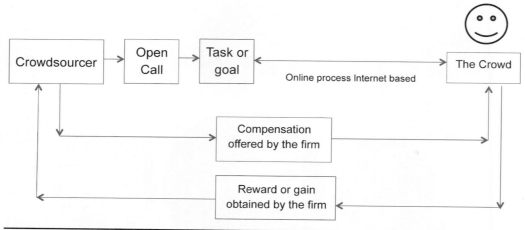

FIGURE 26-2   The crowdsourcing process.

Wikipedia[*] offers another definition: "Crowdsourcing is a process that involves outsourcing tasks to a distributed group of people. This process can occur both online and offline. Crowdsourcing is different from an ordinary outsourcing since it is a task or problem that is outsourced to an undefined public rather than a specific body. An example of specific body is paid employees from a company. Crowdsourcing is related to, but not the same as, human-based computation, which refers to the ways in which humans and computers can work together to solve problems. These two methods can be used together to accomplish tasks."

Moreover, the work of Estellés-Arolas and Gónzalez-Ladrón-de-Guevara[4] in which they gathered the different definitions of crowdsourcing from scientific literature offers an integrated definition of the term:

Crowdsourcing is a type of participative online activity in which an individual, an institution, a non-profit organization, or company proposes to a group of individuals of varying knowledge, heterogeneity, and number, via a flexible open call, the voluntary undertaking of a task. The undertaking of the task, of variable complexity and modularity, and in which the crowd should participate bringing their work, money, knowledge and/or experience, always entails mutual benefit. The user will receive the satisfaction of a given type of need, be it economic, social recognition, self-esteem, or the development of individual skills, while the crowdsourcer will obtain and utilize to their advantage that what the user has brought to the venture, whose form will depend on the type of activity undertaken.

In this context, the authors identify the following topics to be considered:

- A clearly defined crowd
- The existence of a task with a clear goal
- A way of compensating the crowd for its input
- The crowdsourcer or the firm who initiates the crowdsourcing process
- A clearly defined gain for the crowdsourcer
- An online process, using the Internet
- Based on an open call

These conclusions are presented in Fig. 26-2, including the crowdsourcing process and all its elements, participants, and relationships.

Taking into account that crowdsourcing is still emerging, the limits and approach of the concept remain unclear, leading us to explore related concepts such as open innovation, user innovation, and free and open source software. According to Schenk and Guittard,[5] there are many common points as well as dissimilarities among these concepts.

The relationship between open innovation and crowdsourcing is tight. The term "open innovation" was coined by Chesbrough,[6,7] who defined open innovation as "a paradigm

---

[*]http://en.wikipedia.org/wiki/Crowdsourcing.

that assumes that firms can and should use external ideas as well as internal ideas, and internal and external paths to market, as the firms look to advance their technology." According to the author, the open innovation paradigm is based on several factors such as the increasing availability and mobility of skilled workers, the growth of venture capital markets, the external options for companies' ideas, and the increasing capability of external suppliers.

Open innovation and crowdsourcing share a common paradigm[8]: knowledge is distributed and the opening of a firm's R&D processes can be a source of competitive advantage. However, there are also dissimilarities between both concepts, since

- Open innovation is focused on the innovation process itself while, crowdsourcing is targeted to many other fields and applications.

- Open innovation is more related to knowledge flows between businesses, while crowdsourcing has a wider approach including the crowd as an undetermined group of individuals. Following the crowdsourcing approach, the crowd itself becomes a knowledge supplier for the firm.

Likewise, we consider the related concept of "user innovation." This term was first used by Eric von Hippel[9] when he observed that users are actively participating in the development or fine tuning of many products and services, usually at the implementation and usage stages. All those ideas from users are moved back into the supply network for the producers. End users actively participate and contribute to the innovation process sharing their ideas freely with producers ("free revealing") hoping to get a product fulfilling their specific needs. User innovation is a "nonlinear" dimension of the innovation process where user and market feedback is a source of novelty for the innovating firm.

A more interesting concept, perhaps, is FOSS (free and open source software) that is also a similar concept to crowdsourcing. FOSS is a twofold term that encompasses both open source software and free software. Both approaches are based on similar development models, but they have differing cultures and philosophies. While the free software movement focuses its efforts on the fundamental freedoms it gives to users, open source software targets its peer-to-peer development model.

According to Raymond, crowdsourcing and FOSS rely on the idea that knowledge and competencies are distributed and that "given enough eyeballs, all bugs are shallow."[10] However, as it is stated by Schenk and Guittard,[5] "there are significant differences between these concepts. In crowdsourcing, firms usually make use of traditional IPR (e.g., by patenting their output), while FOSS makes (at least partly) use of Copyleft* licensing."

Characterization of Crowdsourcing Activities

Given these definitions and characterizations of crowdsourcing, it is clear that the concept embraces different approaches and applications. According to Jeff Howe's vision, crowdsourcing manifests itself in four primary and distinctly different commercial settings.

We look specifically at the **use and application of collective intelligence**. Bill Joy, cofounder of Sun Microsystems, stated, "No matter who you are, most of the smartest people work for someone else," assuming that there is a lot of talent and intelligence outside the boundaries of the firm. The term *collective intelligence* was first coined by Wheeler, an entomologist who, by observing ants has concluded that "seemingly independent individuals can cooperate so closely as to become indistinguishable from a single organism." More particularly and according to Tapscott,[11] collective intelligence is "mass collaboration under the principles of openness, peering, sharing and acting globally."

The **Diversity Trumps Ability Theorem** states, "a randomly selected collection of problem solvers outperforms a collection of the best individual problem solvers." Based on that assumption, the Internet has eased the collective intelligence process, which can take on different forms in the way in which it is applied to the crowdsourcing phenomena, varying from making predictions to tackling problems and brainstorming. The following chart includes examples of these three applications (Table 26-1).

The second approach to crowdsourcing is to produce **mass creative work**. There are a large number of examples in which the crowd is producing reliable, creative, and valuable content. This content is being used to feed a business model, which in turn adds value to this content by categorizing, peering, evaluating, and distributing it back to the crowd.

*http://en.wikipedia.org/wiki/Copyleft.

Collective Intelligence	Example of Application
Making predictions	**Trendwatching.com** is an independent opinion-tracking firm, scanning the globe for the most promising consumer trends, insights, and related hands-on business ideas. It bases its activity on a network of spotters in more than 90 countries.
Tackling problems	**Globalservicejam.org** comprises people from different countries facing common challenges and solving global problems together.
Brainstorming	**Ideascale.com** provides tools to organize collective brainstorming online.

TABLE 26-1 Collective Intelligence Crowdsourcing Examples (*Source*: Author's elaboration.)

Company or Service	Explanation
YouTube.com	Users provide videos which are uploaded and freely available to the public. The crowd members themselves are the producers.
Istockphoto.com	A vast collection of images contributed by thousands of photographers who distribute their work online for compensation.
Threadless.com	The users submit T-shirt design ideas to be ranked by the community and produced. The community includes 2.3 million members, 254,000 designs submitted, and total compensation to contributors of US$7 million.
Businessmodelgeneration.com	Cocreated by 470 practitioners from 45 countries, this book becomes a bestseller. The authors used collective intelligence to produce an innovative approach to business model innovation.

TABLE 26-2 Crowdsourcing Examples of Mass Creative Work Production (*Source*: Author´s elaboration.)

Above are some examples of crowdsourcing for the production of creative work (Table 26-2).

The **organization of large amounts of information** is another key use of crowdsourcing. Bradly Horowitz, Vice President of Yahoo's advanced development division, stated the "1:10:89 rule," which means that only one percent of people visiting a site will contribute with new content, another 10 percent will vote or rank content and the remaining 89 percent will simply view the contents. The work of this 10 percent of people is crucial for many crowdsourcing services to function. In fact, the majority of services such as Threadless and YouTube are profiting from these users' work, and are aware that without it, none of their services would be feasible.

A good example of this is the customer's review on Amazon. The company solicits consumers' opinions about books or music, adding value to the selection and buying process. Consumers freely give their opinions to support a new release or, on the contrary, to reject it.

Finally, crowdsourcing is also about **financing**. One of the most interesting applications of crowdsourcing is using the financial power of the crowd to fund projects that otherwise could not be released. The concept is known as **crowdfunding**, which Wikipedia describes as "the collective effort of individuals who network and pool their money, usually via the Internet, to support efforts initiated by other people or organizations. Crowdfunding is used to support a wide variety of objectives including disaster relief, citizen journalism, fans support for artists, political campaigns, Startup Company funding, movie or free software development, invention development and scientific research."

In October 2012, *Time* published an article entitled "How to crowdfund your creative project." The publication refers to services such as Kickstarter, Indiegogo, Rock the Post, Gambitious, and other lesser known funding sources.

But crowdfunding also has social applications such as Kiva.org, where it is possible to make a loan to an individual beneficiary. Contributors are updated about the evolution of their loan when borrowers repay the money and it becomes available in their account.

Additional well-known examples of crowdfunding are Sellaband, Zazzle, ChipIn, RocketHub, Crowdtilt, and GoFundMe, among others.

Even though this first approach to a classification of crowdsourcing services is highly relevant, several other classifications of the term have been proposed in different academic fields such as information society, economics, or management. According to Geiger et al.,[12] the existing work of crowdsourcing classifications "often focuses on specific applications of crowdsourcing (e.g., open innovation or human computation) and does not consider crowdsourcing as a generic method."

The authors propose an academic classification of crowdsourcing approaches based on a literature review which is included in Table 26-3 below. The table explains the field of application, the motivation or purpose, the dimensions involved, and which area the crowdsourcing service is related to.

Reference	Field	Motivation or Purpose	Dimensions	Relates To
Doan et al. (2011)[19]	IS, Computer Science	Global picture of crowdsourcing systems on the Web	Nature of collaboration	Process
			Type of target problem	Task
			How to recruit and retain users	Stakeholders
			What users can do	Task
			How to combine inputs	Process
			How to evaluate inputs	Process
			Degree of manual effort	Task
			Role of human users	Task/stakeholders
			Stand-alone vs. piggyback	Process
Corney et al. (2009)[13]	IS, Outsourcing	Foundation for identifying methodologies or analysis methods	Nature of the task	Task
			Nature of the crowd	Stakeholders/task
			Nature of the payment	Process/stakeholders
Schenk and Guittard (2011)[5]	Management	Understanding crowdsourcing from a management science perspective	Integrative/selective nature of the process	Process
			Type of tasks	Task
Rouse (2010)[14]	IS, Outsourcing	Clarifying the notion of crowdsourcing	Nature of the task/ supplier capabilities	Task/stakeholders
			Distribution of benefits	Stakeholders/process
			Forms of motivation	Stakeholders/process
Zwass (2010)[15]	IS, Cocreation	Taxonomic framework as prerequisite for theory building in cocreation research	Autonomous vs. sponsored	Stakeholders/task
			Performers	Stakeholders
			Motivation	Stakeholders
			Process governance	Process
			Task characteristics	Task
			Principal mode of product aggregation	Process
			Economic beneficiary	Stakeholders
Malone et al. (2010)[16]	Collective Intelligence	Identifying the building blocks of collective intelligence approaches	What (goal)	Task
			Who (staffing)	Stakeholders
			Why (incentives)	Stakeholders
			How (structure/process)	Process

TABLE 26-3   Classifications of Crowdsourcing Approaches

Reference	Field	Motivation or Purpose	Dimensions	Relates To
Piller et al. (2010)[17]	Open innovation	Analyzing strategies for customer participation in open innovation	Stage in innovation process	Task
			Degree of collaboration	Process
			Degrees of freedom	Task
Quinn and Bederson (2011)[18]	Human computation	A common understanding of human computation systems	Motivation	Stakeholders/process
			Quality control	Process
			Aggregation	Process
			Human skill	Stakeholders/task
			Process order	Process
			Task-request cardinality	Process

TABLE 26-3   Classifications of Crowdsourcing Approaches (*Source*: Geiger et al 2011)[12]. (*Continued*)

But perhaps the most valuable classification of crowdsourcing practices is depicted by Doan, Ramakrishnan, and Halevy[19] who discussed crowdsourcing systems on the Web from a variety of perspectives. The authors define crowdsourcing as a system that "enlists a crowd of humans to help solve a problem defined by the system owners." The authors consider several parameters to make their classification: characteristics of tasks and stakeholders, the nature of the collaboration, the architecture where the system operates, the necessity of recruiting users, and the problems targeted by the system.

The first distinction made by the authors is the explicit or implicit nature of the crowdsourcing system. The explicit nature refers to a system's enlisting a crowd in order to perform a task for its own benefit, while the implicit nature of the system means that users collaborate when they are doing another task or activity, such as in ESP* games. The authors also explain differences between crowdsourcing systems working on a stand-alone basis or those whose architecture piggybacks on another system.

We find another interesting list of tasks or activities users can do within *explicit* crowdsourcing systems, such as: evaluating (reviewing, voting, and tagging), sharing (structured or textual knowledge items), networking, building artifacts (software, textual, or structured knowledge bases and systems), and task execution. The tasks normally performed in *implicit* systems are, among others: playing games with a purpose, betting on prediction markets, using private accounts, solving CAPTCHAs,† playing massive multiplayer games, keyword searching, buying products, or browsing websites.

An overview of the classification posed by Doan et al.[19] is included in Table 26-4 below.

To conclude, we cite Dawson and Bynghall[20] who make a taxonomy of crowdsourcing based on the different business models supporting the process.

Even though there are many simple ways that crowds are helping businesses to achieve their objectives (design, software development, advertising campaigns, or product design), this approach does not fundamentally change the nature of the company, since this simply replaces sourcing suppliers or gets some of the organization's supporting functions performed in a different way.

This approach characterizes the entire orientation of a business using the crowd as a primary resource to get and monetize value. Citing Dawson and Bynghall,[20] we depict a framework of crowdsourcing business models in Table 26-5.

---

*According to the Wikipedia, the ESP Game is an idea in computer science for addressing the problem of creating difficult metadata. The idea behind the game is to use the computational power of humans to perform a task that computers cannot yet do (originally, image recognition) by packaging the task as a game.

†A CAPTCHA is a test that is used to separate humans and machines. CAPTCHA stands for "Completely Automated Public Turing test to tell Computers and Humans Apart."

			What Users Do?	Examples	Target	Problems/Comments
Nature of collaboration: Explicit	**Architecture:** Stand-alone	**Must recruit users?** Yes	Evaluating • Review, vote, tag	• Reviewing and voting at Amazon, tagging web pages at del.ici.ous.com and Google Co-op	Evaluating a collection of items (e.g., products, users)	Humans as perspective providers. No or loose combination of inputs.
			Sharing • Items • Textual knowledge • Structured knowledge	• Napster, YouTube, Flickr, CPAN, programmableweb.com • Mailing lists, Yahoo! Answers, QUIQ, ehow.com, Quora • Swivel, Many Eyes, Google FusionTables, Google Base, bmrb.wisc.edu, galaxyzoo, Piazza, Orchestra	Building a (distributed or central) collection of items that can be shared among users	Humans as content providers. No or loose combination of inputs.
			Networking	• LinkedIn, MySpace, Facebook	Building social networks	Humans as component providers. Loose combination of inputs.
			Building artifacts • Software • Textual knowledge bases • Structured knowledge bases • Systems • Others	• Linux, Apache, Hadoop • Wikipedia, open mind, Intellipedia, ecolicommunity • Wikipedia infoboxes/DBpedia, IWP, Google FusionTables, YAGO-NAGA, Cimple/DBLife • Wikia Search, mahalo, Freebase, Eurekster • newspaper at Digg.com, Second Life	Building physical artifacts	Humans can play all roles. Typically tight combination of inputs. Some systems ask both humans and machines to contribute.
			Task execution	• Finding extraterrestrials, elections, finding people, content creation (e.g., Demand Media, Associated Content)	Possibly any problem	
Nature of collaboration: Implicit	**Arch.:** Stand-alone	**Yes**	• Play games with a purpose • Bet on prediction markets • Use private accounts • Solve CAPTCHAs • Buy/sell/auction, play massive multiplayer games	• ESP • intrade.com, Iowa Electronic Markets • IMDB private accounts • recaptcha.net • eBay, *World of Warcraft*	• Labeling images • Predicting events • Rating movies • Digitizing written text • Building a user community (for purposes such as charging fees, advertising)	Humans can play all roles. Input combination can be loose or tight.
	Arch.: Piggyback	**No**	• Keyword search • Buy products • Browse websites	• Google, Microsoft, Yahoo • Recommendation feature of Amazon • Adaptive websites (e.g., Yahoo! front page)	• Spelling correction, epidemic prediction • Recommending products • Reorganizing a website for better access	Humans can play all roles. Input combination can be loose or tight.

TABLE 26-4   Basic Crowdsourcing Types on the Web (*Source*: Doan et al. 2011).[19]

Business Model	Definition	Deliverables
Media and data	Creation of media, content, and data by crowds	Knowledge sharing Data Content
Crowd services	Services that are delivered fully or partially by crowds	Labor pools Managed crowds
Marketplaces	Matching buyers and sellers of services and financing through mechanisms including bidding and competitions	Service marketplaces Competition markets Crowdfunding Equity crowdfunding Microtasks Innovation prizes Innovation markets
Crowd ventures	Ventures that are predominantly driven by crowds, including idea selection, development, and commercialization	Crowd ventures
Crowd processes	Services that provide value-added processes or aggregation to existing crowds or marketplaces	Crowd process
Platform	Software and processes to run crowd works and crowd projects, for use with internal or external crowds	Crowd platforms Idea management Prediction markets
Content and product market	Sale of content or products that are created, developed, or selected by crowds	Content markets Crowd design
Nonprofit	Use of crowd to support nonprofit projects	Citizen engagement Contribution Science

TABLE 26-5　Crowdsourcing Business Models (*Source*: Dawson and Bynghall 2012).[20]

Steps to Organize a Crowdsourcing Process

When developing a crowdsourcing strategy, critical success factors include the careful identification of the objective as well as the following aspects:

- Target participants
- Methodology and required resources
- Timeline for completion
- Incentives and compensation for participants

Belsky[21] suggests "Crowdsourcing should spawn long-term relationships between clients and creative rather than be a one-off experiment." In this statement the author draws our attention to the fact that there is a clear distinction between going into the development of a crowdsourcing strategy and the organization of some brand engagement competitions.

Taking the above into account, it is also important to be aware of the different objectives, applications, and orientations of crowdsourcing services, and therefore, how a crowdsourcing strategy can be applied. One key requirement is to identify the dimensions of the crowdsourcing processes to be implemented. We are informed by the work of Geiger et al.[12] who identifies four main dimensions:

1. **Preselection of contributors**: The crowdsourcing organization filters the crowd of potential contributors to the process. The authors state that most of crowdsourcing processes try to benefit from as much diversity and number of participants as possible, so they do not limit the contributors even though many contributors may be excluded *ex post* if they do not fulfill the requirements or reach the quality

standards required. However, many other crowdsourcing processes require certain knowledge or skills to contribute (qualification-based preselection), and others limit it to their own employees or to their customers as suitable contributors.

2. **Accessibility of peer contributors**: The organization needs to decide to what extent individuals can view others' contributions. Geiger et al.[12] suggest that the accessibility to other's contributions can vary from none, view, assess, or modify. When no one can see others' contributions it is impossible to reuse, complement, or react to other content, perhaps for privacy reasons or as a way to ensure diversity. On the contrary, the highest level of accessibility is modify, where contributors can rework others' contributions in order to check, update, or otherwise improve them.

3. **Aggregation of contributors**: The organization must decide how to aggregate the results provided by contributors. According to Schenk and Guittard[5] cited in Geiger et al.[12] the "nature of the process" can be designed around integrative (the issue is to pool complementary input from the crowd) or selective contributions (the crowdsourcing process is designed to choose an input from among a set of options that the crowd has provided). In the first case, all contributions are reused for the final outcome unless they fail to meet certain quality requirements, while the second follows a more competitive approach.

4. **Remuneration for contributors**: The organization needs to establish compensation mechanisms for the contributors. Thus, the authors distinguish between fixed, success-based, and no remuneration models. Fixed remuneration means "that all contributions that adhere to the respective terms and conditions to generate a fixed payment regardless of their value to the final outcome." Second success-based remuneration means "that contributions will be paid depending on their individual value to the crowdsourcing goal." Finally, there are many crowdsourcing services that do not offer remuneration at all, where contributors participate for recognition only.

Examples, Uses, and Applications of Crowdsourcing

In the former section we talk about the characterization of crowdsourcing practices and describing the different approaches to implement crowdsourcing services. However, there is no more practical way to illustrate the uses and application of crowdsourcing than giving real examples of crowdsourcing services that are already functioning. The following case studies are organized taking into account the scope of the crowdsourcing service, more than other factors.

Creation of Content

One of the most compelling reasons for implementing a crowdsourcing service is to create robust content. To fulfill this objective, the crowdsourcing service may count on the participation of a suitable cohort of individuals wishing to participate collectively in the provision of content. These individuals may include but are not limited to: writers, musicians, software developers, photographers, designers, and others. This can be broken into three main areas: creative, multimedia, and knowledge sharing.

Creative: A good example for this area is Behance (behance.net), a crowdsourcing platform for creative works such as graphic design, photography, industrial design, branding, fashion, and so on. The functionality is quite simple: an interested recruiter can post a job to the community who can then apply for it. The recruiter can consult the professional portfolio of the candidates in order to make a selection. Selected candidates get the job and receive compensation for their work. Recruiters pay around US$200 for posting a job during a period of 60 days. Other interesting examples in this area are crowdSPRING (crowdspring.com: marketplace for logos, graphic design, and naming) and 99designs (99designs.com: logo, web design, and other features). It is important to note that these services have dramatically changed the business model in their markets, since they are able to make an offer worldwide with no fixed cost.

Even large companies are promoting crowdsourcing initiatives to enhance creativity, as in the case of Pepsi, which crowdsourced Beyonce's Superbowl intro. Pepsi asked fans to snap and upload images of themselves in a series of specific poses to be used in the on-air intro video as the star takes to center field. "Pepsi is bringing to life its 'Live for Now' mindset which places fans at the center of this experience," said Angelique Krembs, vice president of marketing for Pepsi. "Pepsi is looking for new ways to involve fans in the halftime show experience, and 'live' it like never before."

Multimedia: Current TV, or Current (current.com), is a media company promoted by former U.S. Vice President Al Gore and Joel Hyatt. The Current cable television network went on the air in the United States on August 1, 2005. Current TV's offering is quite similar to other cable TV channels; however that's where the similarities end. Current TV is supported by a social network where individuals can upload their own content, post valuable comments, and vote which segments should go on air. Community members can also post news items to Current.

But initiatives in this field are popping up everywhere. One notable example is Amazon Studios, who announced plans in December 2012 to produce pilots for six original comedy series, following an open call. "Since launching our original series development effort, we have received more than 2000 series ideas from creators around the world with all different backgrounds, and we are extremely excited to begin production on our very first set of pilots," said Roy Price, Director of Amazon Studios.

Knowledge sharing: One paradigmatic example in this area is Wikipedia (Wikipedia.org), the enormous online encyclopedia written collaboratively by largely anonymous Internet volunteers, who developed themselves the policies and guidelines to improve the encyclopedia. Contributors do not receive any remuneration for their work. Since its creation in 2001, Wikipedia has grown rapidly into one of the largest reference websites in the world, attracting 470 million unique visitors monthly (February 2012). It currently has 77,000 active contributors working on over 22,000,000 articles in 285 languages. As of early 2013, there are 4,134,838 articles in English.

Distributed Work

Frequently, crowdsourcing is used to match supply and demand for a distributed online labor pool. This work may cover simple or complex tasks and it is available on demand. Virtual workers can perform a variety of activities and may receive compensation for their work. Some examples for this area follow below.

Microtask (microtask.com) is a crowdsourcing-based company whose aim is to help businesses digitize handwritten forms. The client can upload scanned forms by mail or "FTP" to Microtask and they distribute the work to their community who simply type the data into databases. Each form is broken down into separate pieces and all data is typed, guaranteeing anonymity. Once the work is performed, the data is returned to the client duly digitized, helping businesses achieve a paper-free environment. Microtask launched a joint project called "digitalkoot" with the National Library of Finland, using a voluntary workforce to correct Optical character recognition (OCR) errors in old newspaper archives.

Another good example in this area is Clickworker (clickworker.com). This is a specialized community with more than 300,000 "clickworkers" (Internet users registered with clickworker.com) to help clients to perform data processing that computers cannot do. Based on the belief that online jobbing fits into today's networked and globalized world of work, this service specializes in the processing of large quantities of unstructured data, such as text, photos, or videos: providing text creation, writing or editing of simple texts, translation and keyword assignment, image capturing and categorization, product reviews and opinion polls, and web research.

Collective Intelligence and Distributed Knowledge

As stated above, talent and intelligence can be elsewhere. The challenge of crowdsourcing is tapping into this collective intelligence for the creation of new knowledge that can be applied to many fields, such as the development, aggregation, and building of new knowledge, sharing knowledge and information through open questions and answers, user-generated knowledge systems, news, citizen journalism, and forecasting.

One phenomenon that has been possible thanks to crowdsourcing is citizen journalism. Individuals become journalists and participate in the information process reporting on whatever happens and wherever they are at that moment. The use of social networks, high-speed communications and smartphones has fostered this trend. There are many examples of citizen journalism, both in relation to established traditional media and new media based entirely on crowdsourcing practices.

For example, CNNiReport (ireport.cnn.com) is a crowdsourcing-based news service powered by CNN. As it is stated on the site, "iReport is an invitation for you to be a part of CNN's coverage of the stories you care about and an opportunity to be a part of a global community of men and women who are as passionate about the news as you are." CNN provides contributors with tools and support to encourage their work. The service is also supported by social networks.

On the other hand, we have examples of stand-alone crowdsourced news services. This is the case for Global Voices (globalvoicesonline.org), an internationally based community of more than 700 authors and 600 translators who feed the site. Global Voices was founded in 2005 by Rebecca MacKinnon and Ethan Zuckerman. The seminal idea for the project grew out of an international bloggers' meeting held at Harvard in December 2004.

But collective intelligence could also be applied to science. There are many scientific projects that extend beyond the collective and distributed knowledge of the crowd. Traditionally, scientific communities have shared their findings through specialized networks. There have been an increasing number of projects based internationally, though, that have been sharing knowledge to increase the pace of scientific progress. However, it boils down to already established scientific communities, while the crowdsourcing-based approach opens the scientific process to the crowd. One good example in this field is Open Source Drug Discovery (osdd.net). "OSDD is an Indian Consortium with global partnership who aims to provide affordable healthcare to the developing world by providing a global platform where the best minds can collaborate and collectively endeavor to solve the complex problems associated with discovering novel therapies for neglected tropical diseases like Malaria, Tuberculosis, Leshmaniasis, etc."

Another simple and valuable example of how crowdsourcing can be used to make progress in scientific fields is the CrowdHydrology initiative. CrowdHydrology is an experiment currently run by Dr. Chris Lowry at the University of Buffalo Department of Geology with the goal of developing innovative methods to collect spatially distributed hydrologic data. Their objective is to create freely available data on stream stage (water levels) in a simple and inexpensive way. They do this through the use of crowdsourcing, which means they gather information on stream stage from anyone willing to send them a text message of the water levels at their local stream. This data is then available for anyone to use, ranging from universities to elementary schools.

We can also find some examples in the field of ecology, such as iNaturalist (inaturalist.com). This is a site where you can record what you see in nature, meet other nature lovers, and learn about the natural world. iNaturalist began as the final project of a Master's Degree at UC Berkeley's School of Information in 2008. Individuals can contribute by adding observations to the service.

Crowdfunding

According to the site crowdsourcing.org, "Crowdfunding is an approach to raising capital for new projects and businesses by soliciting contributions from a large number of stakeholders following any of three types of crowdfunding models: 1 donations, philanthropy, and sponsorship where there is no expected financial return, 2 lending, and 3 investment in exchange for equity, profit, or revenue sharing."

In the field of philanthropy we refer to Razoo (razoo.com), a platform for fundraising. The platform enables donors to contribute money to an existing cause, or to promote their own cause for fundraising. The site reports a total fundraising of US$135 million and thousands of fundraising initiatives. Razoo takes a 2.9 percent processing fee for all transactions.

In the field of lending funds, in addition to the example of Kiva identified above, we suggest Prosper (propsper.com), who defines itself as "the market leader in peer-to-peer lending, being a popular alternative to traditional loans and investing options." The idea is to cut out the middleman to connect people who need money with those who have money to invest, so everyone prospers. The functionality is simple: on the one hand, borrowers choose a loan amount, purpose, and post a loan listing. On the other hand, investors review loan listings and invest in projects that meet their criteria.

Crowdfunding is also about investing and within this field there are many examples to consider. Perhaps one of the best known is Kickstarter (kickstarter.com), which, as reported by its own site, "is a funding platform for creative projects. Everything from films, games, music, art, design, and technology. Kickstarter is full of ambitious, innovative, and imaginative projects that are brought to life through the direct support of others." Since April 28, 2009, over US$350 million has been pledged by more than 2.5 million people, funding more than 30,000 creative projects, the company reports.

Open Innovation

One of the best known applications of crowdsourcing is supporting open innovation processes, where the importance of the open call cannot be overstated. In this area, it is obligatory to refer to one of the best known crowdsourcing platforms in the field of innovation—InnoCentive. InnoCentive defines itself as "the open innovation and crowdsourcing pioneer that enables organizations to solve their key problems by connecting them to diverse sources of innovation including employees, customers, partners, and the world's largest problem solving marketplace." InnoCentive relies on a network of more than 270,000 solvers from nearly 200 countries.

Also, large companies such as Nokia are putting their names on crowdsourcing and open innovation. This is the case of InventwithNokia (inventwithnokia.nokia.com), where the company makes a quest to collaborate in developing new mobile products and solutions for their consumers. Nokia makes an open call to those who want to submit an invention for consideration by following a few easy steps. If Nokia considers that the invention is of interest, the contributor makes an agreement with the company and is compensated for it.

Another interesting example is ScienceExchange (scienceexchange.com) which is an online marketplace for science experiments. Their goal is to make it easier for researchers to access core resources across institutions. ScienceExchange brings together research scientists seeking to outsource experiments with other scientists at core facilities of major research universities who have the capacity to conduct the experiments.

Conclusion

Crowdsourcing is a rapidly growing method of organizing relationships among organizations and individuals to create new synergies. The main challenge of a crowdsourcing initiative is finding and optimizing the dispersed talent of the crowd to establish new ways to organize production, enhance creativity, and create value.

Conversely, crowdsourcing benefits from modern social and economic conditions, such as the renaissance of amateurism, the availability of new tools of production, the appearance of self-organized communities, and the impact of the open source software movement.

Among related concepts such as outsourcing, open innovation, user innovation, or free and open source software, crowdsourcing is distinguished by the existence of an open call to the crowd in order to accomplish a specific task using Internet-based social network technologies.

Crowdsourcing is used for different purposes, the most important being the use and application of collective intelligence, the production of mass creative work, the organization of large amounts of information, and collaborative project funding. In fact, the first three of these purposes converged to produce this very handbook!

References

1. Howe, J. *Crowdsourcing: Why the Power of the Crowd Is Driving the Future of Business.* New York, NY: Crown Business, 2009.
2. European Commission. *Crowdsourcing Translation.* Brussels, 2012.
3. Howe, J. "The Rise of Crowdsourcing." *Wired Magazine*, 14: 6, 2006.
4. Estellés-Arolas, E., González-Ladrón-de-Guevara, F. "Towards An Integrated Crowdsourcing Definition." *Journal of Information Science*, 38 (2): 189–200, 2012.
5. Schenk, E. and Guittard, C. "Towards a Characterization of Crowdsourcing Practices." *Journal of Innovation Economics*, 7 (1): 93, 2011.
6. Chesbrough, H.W. "Why Companies Should Have Open Business Models." *MIT Sloan Management Review*, 48 (2): 21–28, 2007.
7. Chesbrough, H.W. *Open Innovation: The New Imperative for Creating and Profiting from Technology.* Boston: Harvard Business School Press, 2003.
8. Albors, J., Ramos, J.C., and Hervas, J.L. "New Learning Network Paradigms: Communities of Objectives, Crowdsourcing, Wikis and Open Source." *International Journal of Information Management*, 28: 194–202, 2008.
9. Von Hippel, Eric. "Lead Users: A Source of Novel Product Concepts." *Management Science*, 32, (7): 791–805, 1986.
10. Raymond E. *The Cathedral and the Bazaar: Musings on Linux and Open Source* O'Reilly Sebastopol, 1999.
11. Tapscott, D. and Williams, A.D. *Wikinomics: How Mass Collaboration Changes Everything.* New York, NY: Portfolio, 2006.
12. Geiger, D., Schulze, T., Seedorf, S., Nickerson, R., and Shader, M. Managing the Crowd: Towards a Taxonomy of Crowdsourcing Processes. *Proceedings of the Seventeenth Americas Conference on Information Systems.* Detroit, MI, August 4–7, 2011.
13. Corney, J.R., Torres-Sanchez, C., Jagadeesan, A.P., and Regli, W.C. "Outsourcing Labour to the Cloud." *International Journal of Innovation and Sustainable Development*, 4 (4): 294–313, 2009.

14. Rouse, A.C. A Preliminary Taxonomy of Crowdsourcing, In *ACIS 2010 Proceedings*, 2010.

15. Zwass, V. "Co-Creation: Toward a Taxonomy and an Integrated Research Perspective." *International Journal of Electronic Commerce*, 15 (1): 11–48, 2010.

16. Malone, T.W., Laubacher, R., and Dellarocas, C.N. "The collective intelligence genome." *MIT Sloan Management Review*, Spring, 51 (3): 21–31, 2010.

17. Piller, F.T., Ihl, C., and Vossen, A. *A Typology of Customer Co-Creation in the Innovation Process*. SSRN eLibrary, 2010.

18. Quinn, A.J. and Bederson, B.B. Human Computation: A Survey and Taxonomy of a Growing Field, in *Proceedings of CHI*, 2011.

19. Doan, A., Ramakrishnan, R., and Halevy, A.Y. Crowdsourcing Systems on the World Wide Web, *Communications of the ACM*, 54 (4): 86, 2011.

20. Dawson, R. and Bynghall, S. *Getting Results from Crowds: The Definitive Guide to Using Crowdsourcing to Grow Your Business*. New York, NY: Advanced Human Technologies, 2012.

21. Belsky, Scott. "Crowdsourcing Is Broken: How to Fix It." *BusinessWeek—Business News, Stock Market & Financial Advice*, January 27, 2010. http://www.businessweek.com /innovate/content/jan2010/id20100122_047502.htm.

CHAPTER 27

Open Innovation

Victor Scholten and Serdal Temel

Introduction

Open innovation is a relatively new approach to developing a business model for managing innovation within companies and public organizations. It was introduced by Chesbrough[1] as a business model concept with a more flexible approach to innovation management that would stimulate the flow of external knowledge into the company as well as the transfer of internal know-how and technology to external parties. This in turn would increase the effectiveness of innovation management and thus the innovation outcome and eventual revenue and profits.

This open innovation approach has added a new perspective on conducting innovation activities. Before, innovation was mainly an internal, in-house R&D activity for most firms. In particular, for large firms in the 1980s and early 1990s, the main strategy was to grow larger through diversifying into related markets. The internal R&D department then contributes through developing platform technologies that are applied in a variety of products in each of the lines of business. Following this strategy, R&D departments could learn from one line of business and apply that to another line of business. In this approach the company would benefit from R&D department in terms of economies of scope. The returns of R&D are then expected to be high, since the findings from single R&D activities could benefit various kinds of products and it's multiple lines of business.

Despite these positive contributions of internal R&D and closed innovation, the open innovation approach has gained much interest among firms, especially among large firms. Similarly, smaller firms such as SMEs (small and medium-sized enterprises) and start-ups and research organizations such as universities and academic institutes do benefit as well from the trend of more open innovation. The search for more flexibility in their R&D activities has turned the interest of large established firms toward small and innovative firms that are developing new technologies, as well as research organizations that are conducting basic research and new technology development. These considerations result in a number of interesting questions to investigate, such as, what are the reasons that large companies changed their innovation toward a more flexible and open innovation approach? What actually is an "open innovation" approach and how does it differ from traditional in-house R&D and innovation activities? What are the (dis)advantages and the associated challenges of open innovation?

Furthermore, it is an approach that emerged in developed countries. The question at hand is also how it manifests in emerging countries. Is it relevant as well and how do SMEs in emerging countries engage in open innovation activities? In order to answer these questions, we discuss the principles of open innovation and how it becomes more beneficial for mostly small firms within emerging economies. We then discuss the case of Turkey and the effects of innovation strategy and university collaboration on the performance of Turkish SMEs.

Networks and Innovation

Innovation is vital for a firm to capture new positions of competitive advantage. Rather than accepting current structures, successful firms change industry boundaries; and by constructing new product/service configurations generate demand and create new markets.[2,3] Under these conditions, strategy must be seen, in and of itself, as a process of innovation. Due to fierce global competition it is more difficult to hold onto an advantage long enough to pay for significant internal R&D and innovation costs and for the activities to generate new goods and services with sufficient speed. Since strategy must be constantly revisited and reformulated, an exclusive reliance on internal R&D and closed innovation processes no longer makes strategic sense.[4] In order to successfully pursue innovation as strategy, firms are increasingly developing external linkages, and relying on opportunities they developed through information outside the firm's boundaries.

This open innovation approach allows access to the aggregation of knowledge among other players in the process; the innovating organization "discovers locations in the landscape that it may never have reached had it been in charge of all choices."[5] The external partners that provide the firm with new information could include organizations such as universities, public sector organizations, competitors, suppliers, customers in their own or in related industries.[6-8] Following these reasons, Tidd and Bessant[9] formulated four main arguments pushing for a higher level of networking for innovation:

1. **Collective effectiveness:** In a complex and highly competitive business environment, it is hard to sustain support R&D and innovation expenses, not just for SMEs but also big companies. Networking allows firms access to different external resources like expertise, equipment, and overall know-how that has already been proven with less cost and in a shorter period.

2. **Collective learning:** Networking not only helps firms to access expensive resources like machinery, laboratory equipment, and technology, but it also facilitates shared learning via experience and good practice sharing events. This brings new insight and ideas for firms' current and future innovation projects.

3. **Joint risk taking:** Since innovation is a highly risky activity, it is very difficult for a single firm to undertake it by itself and this impedes the development of new technologies. Joint firm collaboration minimizes the risk for each firm and encourages them to engage in new activities. This is the logic behind many precompetitive consortia collaborations for risky R&D.*

4. **Intersection of different sets of knowledge:** Networking creates different relationships to be built across knowledge frontiers and opens up the participating organizations to new stimuli and experience.

Following these arguments we witnessed more and more firms collaborating in joint innovation projects.

Open Innovation: When a Firm's Borders Perish

Various studies have provided evidence that open innovation allows companies to be more competitive during the early stages of the product life cycle. Open innovation significantly differs from the traditional innovation approach. The traditional innovation approach is considered as the in-house innovation model. In-house innovation has long been considered the source of competitive advantage of a firm. The specific knowledge and know-how being generated through in-house R&D and internal innovation processes can be unique to the firm if not shared with other firms. Having the unique knowledge provides a competitive advantage if the knowledge is able to provide value to customers and the firm is able to appropriate the value itself.

From Traditional Innovation to Open Innovation

Traditional innovation is often regarded as the in-house innovation processes that large firms adopt. Their main argument is that for successful innovation you need to fully control it in order to bring about innovative products that contribute to the strategic goals of the firms. Innovation is often regarded as something difficult to predict and difficult to govern. It is about searching into technological regimes that are more distant to current ones. This requires creativity and experimenting and developing knowledge, skills, and capabilities that are new to the firm. If these activities are not well governed, they may drift away from the strategic

---

*Joint risk taking is the logic behind many precompetitive collaborations for R&D activities. As it stated R&D is very risky business and it is very difficult for a company to face that risk sometimes. Therefore, sometimes companies unify to collaborate (put money and know-how) on precompetitive (not on final goods and services) areas to minimize and share the risk.

goals of the firm and thereby the organization may develop the wrong knowledge, skills, and capabilities that may not contribute to the firm and thus result in a low return on investment.

In order to fully control the innovation process, the firm needs to own the laboratories and personnel and equipment. The decision to invest in new technology, market or business model depends on the likelihood it will contribute to the strategic goal of the firm. This requires large investments in laboratories, personnel, and equipment (the best equipment to conduct the research) in order to attract the best scientists and marketers. But at the same time, when fully controlled, it can be protected more easily from imitation by competitors. Thereby, the investments become a strategic asset and a barrier to competitors. The full control of the R&D investments, new technology, market, and business model development allows the firm to fully benefit from its investment. Since no other organization is involved, the firm doesn't have to share the revenues from the new products, markets, and business models with other organizations. It can fully appropriate the value it has created from its investment and thereby make new investments in its own R&D, further strengthening the firm's competitive advantage.

In the early 20th century, however, large multinational companies with their established laboratories tried to develop high numbers of new products for each of their market segments. This resulted in many new technologies of which many had no clear commercial application. This was mainly due to lack of input from other departments of the company, and because most large firms allocated resources only to those innovation projects that were supported by at least two different departments, resulting in significant reduction of newness of innovation. When the interests of multiple departments needed to be satisfied, the outcome of innovation was often a compromise where the radical ideas were often excluded. To include more radical new ideas based on external knowledge became more difficult. Firms needed experts in various fields that could track external developments and evaluate whether to further develop the technology in-house.

Soon, these firms recognized that it is hardly impossible to attract the best researchers in all relevant disciplines and not all the smart people can work for a single firm. Through collaboration in R&D, development firms can shorten the time to market because they do not have to learn the new skills or acquire the new knowledge themselves. Also, some of the investments and risks associated with innovation can be shared with partners with whom they collaborate. At the same time, firms, especially larger firms, have recognized that licensing strategies and spinning off or selling good ideas that lie beyond the firm's strategy are new sources for raising the level of income. Chesbrough[1] provides a relatively simple illustration to clarify the argument, as seen in Fig. 27-1.

In this figure, we can see the closed business innovation model on the left. In this model, a firm carries out all its own R&D activities internal to the firm. These developments may lead to new products that generate market revenues. In mature markets, internal development is often related to small, incremental changes of existing products and is mostly aimed at process innovation to make production more efficient. The likelihood of success is more predictable and investments relatively small, which often results in positive cash flows between firm revenues and the internal costs of development.

However, when firms move forward in the product life cycle, as presented by the next closed business model, the internal cost of development will increase due to the fact that

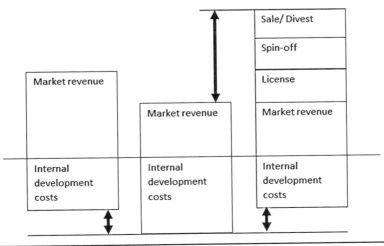

FIGURE 27-1 Comparative innovation models.

new technologies and new market developments are more difficult to assess in this stage. The likelihood of failure increases and more investments are needed to learn and experiment with new technologies, markets, and business models. At the same time, the competition in this stage is more about acquiring market growth and customers will select certain products that will become the dominant ones. As a result, the product lifetime will be shorter and regular updates are needed to extend its lifetime and grow into mature markets. New investments in innovation and internal development will increase the costs of innovation and, since the firm is making all the investments alone, it will also bear all the risks associated with the development of the new technology, markets, and business models.

The model on the right hand of Fig. 27-1 is the open innovation model. Here the firm acknowledges that it is impossible to have all smart people working for the firm. Many experts in various technological fields and market segments may work outside of the firm; and to tap into their expertise, firms need to engage in collaborative relationships with them. This will reduce the costs of internal development as well as the risks associated with shared new product development. Besides bringing ideas from outside the firm into their mainstream business, firms can also consider bringing ideas they developed themselves but are not considered strategically relevant outside of the firm. This can be done through spinning-off the ideas or selling the venture unit to outside firms. Firms increasingly consider the benefits from licensing their IP to other firms, thereby increasing the stretch of their ideas while avoiding the need to make investments themselves. This is often to expand to other markets and to allow partner or competing firms to use the technology under a license agreement. The licensing firm can generate additional revenues when the partner firm is selling the products in a different market.

As a result, open innovation brings about three types of benefits to the firm. First, the spin-off, licensing strategies, and the selling of business units or new ideas that do not fit the strategic interest of the firm will increase the revenues of the firm. The second benefit is that firms can reduce the internal cost and risks of development. But more important, the third benefit, they are open to a broader scope of opportunities and speed up their learning processes about new technologies and thereby make significant progress in the learning curve. Staying ahead of the learning curve and being more adept to recognizing opportunities provides the firm with the advantage of selecting the more promising opportunities and establishing a competitive advantage. These advantages are more prominent in the early stage of the product life cycle when firms need to position themselves in the market in order to acquire a large market share before market growth takes off.

SMEs and Open Innovation

Open innovation is considered to be a major driver for firms wanting to enhance their business performance.[10,11] Albeit the increasing interest in open innovation research exists in all sizes of firms, SMEs have been excluded from the mainstream.[12] Various explanations can be given as to why small and medium-sized enterprises have mostly chosen to forego the open innovation discussion.

First, open innovation is more easily studied in larger firms, as small firms have less ability to access external resources and fewer technological assets that they can exchange than big firms.[13] Second, small firms use noninternal means of innovation more than large firms, as they consider networking as a way to extend their technological capability,[14,15] which means that innovation in small firms already has an external focus and the concept is not new to them. However, their collaborations tend to be limited to strategic alliances with larger firms[16] and outsourcing, mainly via other small firms. Considering the fact that firms that have multiple types of ties are more innovative than those that have one type of tie,[17,18] it is therefore very important to seek the potential benefit of different types of ties in the context of open innovation for SMEs.

Compared to large firms, SMEs often lack a sophisticated resource base, including financial capital, to invest in research and development and recruit talented engineers who are crucial to successful innovation.[19] They can compensate for this lack of resources by initiating and exploiting connections with external sources of knowledge.[1] Research has shown that a firm's search strategy for new technologies can influence its innovation performance considerably.[20] Universities and research institutions especially are accumulators of specific knowledge, and firms that work together with universities can improve their knowledge base and thereby increase their innovation performance. More specifically, various studies have shown that, when they work together with universities, SMEs can benefit from increasing their access to useful knowledge and skilled graduates, and increase their technological problem-solving capacity[21–24] and innovative capability.[25,26]

Although these studies suggest that having an innovation-based strategy and being linked to universities will improve firm performance, other researchers have found inconclusive evidence with regard to the existence of such a causal connection.[27–29] This may be due to the complex, systemic, context-related, tacit, and noncodified nature of innovation[30] and of the knowledge that is transferred from universities to SMEs,[31] which often requires more detail than can be obtained through traditional publications like conferences, journals, and patents.[32–34]

Another problem is the fact that academics and businesses have divergent goals and scopes.[35] Academics are often evaluated on their number of publications and teaching activities. Technology commercialization and contract research with individual companies are often aimed at understanding the technological principles, but adapting it to customer needs is often not a core task of an academic. Business practitioners are often less interested in the technological principles and want to apply the technology quickly to be able to gain a competitive advantage. Too much focus on technology and understanding the principles may delay market introduction and increase the costs, neither of which contribute to the business goals of profit making. As a result, different mind-sets, systems, and processes are required for both academics and practitioners that can enable the conversion of scientific knowledge into products[36] and overcome the diverging goals and scopes that exist between scientists and engineers in industry.[35] Zahra et al.[37] emphasize the importance of having a knowledge conversion capability when university-based start-ups try to exploit scientific knowledge in the market. Similarly, we believe that such a knowledge conversion capability is essential to the transfer of scientific knowledge to SMEs.

Open Innovation among SMEs in Emerging Countries

Small and medium-sized enterprises account for a large part of employment in most countries. In emerging economies, they are the dominant type of firm. Within these emerging economies, SMEs are increasingly changing their orientation from a product-based focus toward a knowledge-based focus. Similar to companies in developed countries, they know that to keep up with new developments in technology and changing demands in customer markets, they need to broaden their scope and learn quickly about new technological, market, and business model developments.

In particular for SMEs, it is important to consider each investment carefully. The investments in one technology or product development may provide the benefit that, when it becomes successful, the company has a competitive advantage. In that case the innovating firm does not need to share the revenues generated from a new technology or product with other companies. Yet, choosing the path to innovate is one where risks and uncertainty will increase. Following the open innovation argument, it is better to share the risk and uncertainty through collaboration with partners.

In emerging economies, SMEs were mainly product-oriented and focused on low-cost and efficient production for mature markets based on existing and mature technologies and products. Moving forward in the product life cycle increases the importance of innovation and firms need to build on a unique knowledge-based competitive advantage rather than a cost-based competitive advantage. In these early stages of the product life cycle, new technology is far from mature. It is surrounded by various unknowns, such as technical feasibility, market size, and the share the company is able to obtain. These unknowns bring about significant issues concerning the financial projection. Because it is unclear whether technological problems may occur and how long it will take to solve these problems, the moment of market introduction becomes unclear as well. When introduced into the market, the question of potential market size and achievable market share may pose additional problems when calculating the returns that are likely to be obtained. In sum, the return on investment is more difficult to assess.

This is even more difficult for SMEs that are not familiar with knowledge-based and innovation-driven strategies. They are facing different problems and need different solutions compared to cost-based and efficiency-driven strategies in mature stages of the product life cycle. This poses additional problems to the SMEs as they need to change their mind-set and apply various skills and techniques to address the challenges in the early stages of the life cycle.

Currently, the impact of innovation strategy and university collaboration on the performance of SMEs is a major concern of technology and innovation policies in emerging economies. Governments of emerging economies build on the common understanding that a greater focus on an innovation-based strategy and more university collaboration will contribute to the knowledge base of SMEs and their innovation output. This in turn will make them more competitive in a global economy. As a result, emerging economies like Turkey have invested in policy programs to nurture the innovative competitiveness of SMEs and provide incentives for them to collaborate with university institutes.

In particular, collaboration with universities is believed to be beneficial to SMEs because it provides them with access to new knowledge and technologies[38,39] and can increase their legitimacy and prestige.[40,41] Although policy makers increasingly encourage this kind of collaboration to increase local economic development,[42] research has yielded ambiguous results concerning the effect of collaboration with universities. Lee et al.[28] found no direct connection between working together with universities and sales growth, although their findings indicate that SMEs with high levels of technological capabilities can benefit from such collaboration. Lawton Smith and Bagchi-Sen[43] and Pickernell et al.[44] argue that the impact of universities on the development of industries needs to be inter-preted more carefully. Similarly, at the moment, the effects of having an innovation-based strategy on firm performance are largely unclear, with some scholars finding a positive relationship between the two,[45] while others argue that the effects are negative[46] or that there is no effect at all.[47]

In an extensive review, Capon et al.[48] and Song et al.[29] found inconclusive evidence for either positive and negative effects concerning the relationship between product innovation and firm performance, and they proposed including more interaction terms, which prompted Li and Atuahene-Gima[49] to examine the moderating effect of environmental factors on the relationship between having an innovation-based strategy and business performance. In their research on Chinese new ventures, they found negative effects of strategic alliances and dysfunctional competition on the relationship between having an innovation-based strategy and firm performance, and a positive effect with regard to environmental turbulence and institutional support. They suggest that environment-based and relationship-based factors moderate the effect of having an innovation-based strategy on firm performance. These mixed results indicate that the relationship between having an innovation-based strategy, university collaboration, and SME performance warrant further research.

Innovation Programs in Turkey

In emerging countries, SMEs increasingly change their orientation from a product-based and efficiency-driven organization toward a knowledge-based and innovation-driven orien-tation. The governments of emerging countries support this transfer because it will make the companies more innovative and competitive in a global market. In Turkey we have wit-nessed similar developments and the innovation capacity of SMEs is a major target of tech-nology and innovation policies. The underlying assumption is that a greater focus on an innovation-based strategy and university collaboration will contribute to the knowledge assets of SMEs, which will in turn make them more competitive in a global economy. Emerg-ing economies like Turkey have invested in policy programs to nurture the innovative com-petitiveness of SMEs and provide incentives for them to collaborate with university institutes.[50]

In the early 1990s, after the liberalization of the Turkish economy, Turkish firms have faced increasing international competition, which made innovation and university–industry collaboration more important.[50] In response to that, several public institutions, including the Directorship for Small and Medium-Sized Enterprises (KOSGEB), Directorship for Technology and Innovation Assessment (TEYDEB), and Technology Development Foundation of Turkey (TTGV), were established in the mid-1990s to facilitate innovation and high-tech knowledge transfer between academia and industry.[51,52]

After 1994, consecutive Turkish governments launched programs, introduced incentives, and founded organizations to support and encourage firms (mostly SMEs) to become more innovative.[53] These institutions are designed to help and guide firms in developing their own innovation projects, providing financial support through various programs. They also encourage firms to collaborate with universities and research centers in order to be eligible for further subsidies. The ultimate aim of these support programs is to enhance the firms' innovative capacity and thus increase their business performance. Although compared to most Western countries, Turkey implemented support programs relatively late, the develop-ment of its innovation infrastructure has been remarkable. The recently introduced financial innovation support schemes encourage companies to collaborate with universities in their innovation and this cooperation model is almost the only way firms can gain access to most of the financial grants.

For instance, the "Industrial Theses-SanTez" innovation support scheme requires the development of joint research projects leading to postgraduate degrees. Since universities are among the major organizations in the Turkish National Innovation System,[54,55] more firms are looking for opportunities to create sustainable links with academia to gain access to various support schemes and other innovation-related financial incentives. This is reflected by the rate of R&D-oriented companies receiving funding from the government, which gradually grew from around 1 to 1.4 percent in 1995, and public R&D support funds increased substantially to 2.1 percent in 1997 and 2.5 percent of governmental funding in 2000.[56]

Although most Turkish SMEs are still labor-intensive and produce low value-added products, their focus on innovation is increasing.[57] Turkey is one of the fastest growing economies when it comes to R&D and innovation, and the number of SMEs that have the potential to collaborate with academia is increasing continually. As such, the Turkish situation provides us with an opportunity to examine the early effects of an open innovation-based strategy among SMEs, how they collaborate with academia, and the effect on firm performance. Although previous international studies have identified a positive effect of university collaboration on the innovative capabilities and performance of SMEs,[58–60] as yet no study has examined Turkish SMEs. Very little is known about the impact of universities on the innovative focus and performance of SMEs, which is why we examine, among other things, whether or not university collaboration has an impact on the profit growth of SMEs.

Evidence from the Aegean Region

In order to get a better understanding of an open innovation strategy among SMEs in an emerging country, we examined the innovation strategy of SMEs in the Turkish Aegean region. The SMEs were selected through samplings from the regional technology transfer database at Ege University Science and Technology Centre (EBILTEM). This organization has been working as an interface between academia and industry since 1994 and aims at enhancing the competitiveness of SMEs via technology transfer, innovation, and university–industry partnerships in the region.

SMEs operating in different sectors in the Aegean region were included to examine the role of universities in R&D and innovation in different sectors. The sample represents the agricultural, plastics, chemicals, machinery, and electronics sectors—the five largest sectors in the region. From each sector, we selected SMEs with an R&D department. A final sample of 100 SMEs was asked if they would be willing to take part in the research, and 86 SMEs filled out the questionnaire, representing an initial response rate of 86 percent. Of the 86 firms, 7 were unwilling to share their financial data, bringing the total sample to 79.

To assess the university–industry collaboration in Turkish SMEs, we used the Wageningen Innovation Assessment Tool (WIAT) that was developed by Fortuin et al.[61] The design of WIAT is based on the well-established NewProd model,[62] which has been used extensively to measure the success and failure of product development projects. We translated and adapted the questionnaire to the needs of Turkish SMEs and extended it with questions on university collaboration, following the work by Hanel and St-Pierre.[63]

First, we conducted a pilot study among seven firms to adapt the WIAT questions to the context of Turkish firms.

Second, the data were collected through a survey between May and August 2009 and interviews to validate the answers and ask follow-up questions. Meetings were held with at least two representatives of each company. Among the representatives were directors (45 percent), managers (37 percent), and staff members (18 percent) with a good insight into the firms' practices, including their university–industry network and innovation. The measurement items were based on existing 7-point Likert scales (1 [strongly disagree] to 7 [strongly agree]). A complete list of the questions is provided in the appendix to this chapter. We measured the performance of a firm, its innovation strategy, the focus on open innovation, and the collaboration with university and public organizations. Firm performance was measured on the basis of profit margin growth, a commonly used indicator to understand the impact of an innovation-based strategy on firm performance.[49] Profit margin growth was measured as relative profit margin growth; for example, Profit Margin (2008)—Profit Margin (2007)/Profit Margin (2007). Three questions measured the focus on an open innovation strategy. These questions were based on the literature discussions by Chesbrough[1] and Van de Vrande.[10] The questions included the collaboration for R&D through joint ventures and alliances; through joint innovation projects with external partners such as knowledge institutions, suppliers, or buyers; and the extent that the firm is open to license out inventions that do not fit with the mainstream business model of the company.

Another indicator of open innovation was more focused on the public institutional environment. These questions investigated the extent that the firm used public support in terms of networking, to promote an innovation-favorable atmosphere, to increase and contribute to technology-transfer, and to provide guidance when needed in other aspects of innovation activities.

We measured the innovation strategy of the firm using three items on how much emphasis firms put on keeping track of their innovations, capture what they learn during the process, and provide clear incentives to stimulate innovation. These three types of innovation activities are considered important to the success of technology-oriented firms[64] and their performance. University–industry collaboration was measured by looking at the extent to which the firms collaborated with universities and research centers in their innovation process. Besides these innovation-related questions, we also asked several questions that are indirectly related to the innovation strategy of the firm. Therefore we asked about the firm's awareness of intellectual property rights (IPR). Here we asked the extent to which the firm distinguishes itself positively from its competitors by the protection of the products and processes through patents, licenses, and so forth. Furthermore, we investigated the competition in the market using three items that reflected threats in the business environment and opportunities for growth.

For all the independent variables in our model, we examined the "unidimensionality" and convergent validity of the constructs with principal components factor analysis. All items loaded on their respective constructs, and each loading was large (>50). We included several control variables that are commonly used in studies on the connection between innovation-based strategy and firm performance, as well as control variables that are more specific to Turkish firms. Firm size is a control variable that measures the number of employees, which we subsequently log transformed. We also included the relative share of export sales of the total sales, since the firms in the Aegean region are located close to Izmir's harbor and provide a better export infrastructure than other regions in Turkey.[65] We included these control variables because there is some theoretical basis for expecting the variables to have a systematic relationship with the independent variables or with the dependent variable. For instance, larger firms may have stronger relationships with universities.[66] Furthermore, we analyzed the models including sector-specific dummy variables for the four main sectors involved in this study. These sector-specific control variables were excluded from further analyses because we found no significant differences with regard to the main variables between the four main sectors.

Results

Before we investigate the effect of open innovation strategies, university collaboration, and competition in the environment on the performance of SMEs, we conduct some sector analyses. We have distinguished four sectors: the chemical sector (3 firms), the machinery and electronics sector (45 firms), the plastics sector (12 firms), and the agro industry sector (15 firms). For each of the sectors we rated the extent that firms focused on an innovation strategy, used public institutional support, adopted an open innovation strategy, collaborated with a university, and used IPR to protect their innovation assets. Table 27-1 shows the results and although none of the values significantly differ across the sectors, they do provide us with some feeling with the orientation toward open innovation and innovation management. The focus on innovation strategy is almost equal among the sectors, but the machinery and electronics sector is making relatively more use of public support for innovation than the other sectors. This is mainly because the government has addressed this sector as the leading one for Turkey. Furthermore, it seems that the agro industry has the strongest orientation toward open innovation, while the plastics industry has the lowest focus on looking beyond firm boundaries to exploit their IP through licensing or making use of joint R&D. However, both the plastics and agro sectors have a strong focus on protecting their R&D efforts through IPR. It seems that plastics sector firms use their IP to protect their R&D and keep it more in-house whereas the agro industry is more open and uses its IP to license and collaborate. The low focus on IPR in machinery and electronics sector can be explained

	Chemicals	Machinery & Electronics	Plastics	Agro Industry	Total
	3	45	12	15	75
Innovation strategy	3.08	3.27	3.15	3.12	3.21
Public inst. supp.	1.80	2.63	2.44	2.31	2.50
Open_innovation	3.11	3.07	2.69	3.69	3.14
IPR_awareness	3.67	3.51	4.5	4.53	3.88

TABLE 27-1 Sector Differences for Innovation Indicators

for two reasons. First, various companies in emerging economies act as subcontractors, producing on behalf of international companies that also hold the IPR rights; and second, the lifetime of electronic products is generally short and patenting becomes too expensive.

We furthermore analyzed how the open innovation variables varied across different levels of industry interaction. We grouped the companies that collaborate with a university into three groups. The first group is that of companies that have no or very little collaboration with a university. The second group contains the companies that have a moderate collaboration with a university and the third group has an intense collaboration with a university. Table 27-2 gives the distribution of the number of companies in each group and the level of each variable for each group.

We found various significant differences between the levels of university collaboration for each of the innovation indicators. Innovation strategy differed significantly between group 1 and group 2 (Delta = 0.48; p = .022) and between group 1 and group 3 (Delta = 0.64; p = .01) indicating that firms with a stronger focus on innovation strategy are also more actively collaborating with a university. It seems that firms that collaborate with a university also place more emphasis on innovation strategy. Also, the level of open innovation differed significantly between group 1 and group 2 (Delta = 1.09; p = .006) and between group 1 and group 3 (Delta = 1.63; p = .001), which provides us with the same conclusion that firms that emphasize an innovation strategy focus considerably more on an open approach to their innovation activities. Moreover, for IPR awareness, we found only a significant difference between group 1 and group 2 (Delta = 1.10; p = .072) but not between groups 1 and 3 or groups 2 and 3. These findings may indicate that firms that moderately collaborate with a university are most focused on protecting their research findings.

Maybe the group that has little or no collaboration does not have research findings to protect, while the group that has a strong and intense collaboration with a university has research findings for which it may be too early yet to apply for a patent because the eventual application is unclear yet. It might be as well that in these strong and intense collaborations with a university, the patents are owned by the university and the firm has a license to use the patent and therefore IPR awareness is of less concern.

For public institutional support, we found significant differences across all the groups, which is in line with the intensity of collaboration with a university. The stronger the intention of collaboration with a university, the more public institutional support the firm is receiving. This is due in part to the fact that if SMEs collaborate with a university they are more likely to receive additional public research funding.

The indicators of innovation across the sectors trigger the interest to further investigate how it affects firm performance. First we conducted some descriptive analyses, and second we examined a subset of variables using regression analyses. The regression analyses included the variables of innovation-based strategy, university collaboration, market competition, and the control variables, firm size and export sales. The role of public institutional support, open innovation, and IPR awareness did not have significant effect on firm performance and because of reasons of parsimony, we left out these variables from the regression analysis. Table 27-3 provides the means, standard deviations, and correlations for the variables used in the regression analysis.

The average relative profit margin growth between 2007 and 2008 was 0.07, with a standard deviation of 0.35. In 2007, profit margin growth ranged between 0 and 78 percent, with a mean of 21 percent; while in 2008, it was between 0 and 50 percent, with a mean of 18.5 percent. The average firm's focus on innovation-based strategy was 3.17 (SD = 0.83) on a 7-point Likert scale, indicating that, in our sample, the firms have a relatively moderate

	Group 1. Little or No Collaboration	Group 2. Moderate Collaboration	Group 4. Extensive Collaboration	Total
	32	27	16	75
Innovation performance	3.96	4.51	4.69	4.31
Innovation strategy	2.90	3.38	3.54	3.21
Open innovation	2.40	3.49	4.02	3.14
IPR awareness	3.38	4.48	3.88	3.88
Publ. inst. supp.	1.67	2.88	3.52	2.50

TABLE 27-2 University Collaboration and Innovation Indicators

		Mean	**SD**	**1**	**2**	**3**	**4**	**5**
1	Profit growth	−0.07	.35					
2	Innovation-based strategy	3.17	.83	−.312**	(.68)			
3	University collaboration	3.19	2.05	−.019	.202*			
4	Market competition	3.30	1.14	−.305**	−.035	−.225*	(.63)	
5	Firm size	60.00	107.3	−.004	.143	.198*	−.084	
6	Export sales	13.40	24.3	.070	.144	−.004	−.180†	.297**

$N = 79$; $^{**}p <.01$; $^{*}p <.05$; $^{†}p <.10$; Cronbach alpha between brackets.

TABLE 27-3   Descriptive and Correlations

focus on innovation. Average market competition is 3.30, which is moderate as well, although the standard deviation is 1.14, indicating that firms perceive different levels of competition in their markets.

At the same time, we found that, on average, university collaboration scored 3.19, with a relatively large standard deviation of 2.05, which suggests that some firms are engaged in minor collaboration with universities, while others have a more active collaboration relationship. With regard to export sales, the data show that, on average, 13.7 percent of total sales is export-related. Furthermore, we found that firms on average employed 60 people. With regard to the correlations, we see that having an innovation-based strategy and competitive strength are both negatively associated with profit margin growth. Among the independent variables, we observe low levels of correlation, indicating that there are no problems with multicollinearity. Between brackets, we included the Cronbach alphas for the multi-item constructs.

The estimated OLS regression models are presented in Table 27-4. Model 1 includes the control variables and the main effects. In Model 2 we were interested in whether the influence of an innovation-based strategy on firm profit differed under the condition of fierce market competition or when they are more actively collaborating with a university. Therefore, in Model 2 we included the interaction effects of innovation-based strategy with university

	Model 1	**Model 2**	**Model 3**
	Profit Growth	**Profit Growth**	**Profit Growth**
Innovation-based strategy (H1)	−.327**	−.245*	−.226*
University collaboration (H2)	−.023	−.065	−.828*
Market competition (H3)	−.311**	−.248*	−.230*
Firm size	.003	.008	−.005
Export sales	.059	−.004	−.002
Innovation-based strategy × Market competition (H4)		.263*	.240*
Innovation-based strategy × University collaboration (H5)		.134	−.125*
University collaboration square (H6)			.810*
R^2	.201	.251	.316
Adj R^2	.146	.176	.236
Delta R		.060	.059
F	3.633**	3.349*	3.981**
Delta F		2.307†	6.549*

$N = 79$; $^{**}p <.01$; $^{*}p <.05$; $^{†}p <.10$.

TABLE 27-4   OLS Regression Analyses

collaboration and with market competition. In Model 3, we were more interested on whether there is a nonlinear relationship between university interaction and profit growth and consequently we examined the inverse U-shaped relationship by adding the squared term of university collaboration.

In all models, the relative profit margin growth between 2007 and 2008 is the dependent variable. We centered the innovation-based strategy, the market competition, and the university collaboration variables prior to multiplication and creation of the interaction terms.[67] For each of the predictor variables, we calculated the maximum variation inflation factor (VIF), which was below 1.40, suggesting no serious multicollinearity problems.[68] For each model, Table 27-4 presents the standardized coefficients of the independent variables as well as the R^2, ΔR^2, the adjusted R^2, the F and ΔF.

Model 1 assesses the contribution of the main effects and the control variables on the relative profit margin growth between 2007 and 2008 and shows that a focus on innovation-based strategy is statistically significant but negatively related ($b = -.327$, $p < .01$) with profit margin growth. This may seem to contradict the common understanding that an innovation-based strategy has an important impact on a firm's competitive position. By engaging in an innovation-based strategy, firms can develop new products and services or introduce new features to existing products and services that add value for customers. This may strengthen the loyalty of existing customers and help recruit new customers, which may help prevent firms from having to compete on prices and declining sales. However, Turkey is an emerging economy with significant economic growth.[50] Emerging economies provide SMEs with significant scale or first-mover advantages where experience effects and network externalities are important to building a dominant position for themselves.[69] In emerging markets, an initial strategy focusing on growth rather than on profitability is more important in gaining additional market share.[70,71] Hence, large investments in innovation will reduce the firms' immediate profitability.[71]

For university collaboration we found no immediate effect on profit margin growth. SMEs have relatively large investments in innovation and face generally more risk because technological developments require well-equipped labs and experienced researchers while assumptions concerning future market demands are uncertain. Hence, one would expect that SMEs are more willing to engage in collaboration with a university to share the costs and risks associated with innovation. The role of competition in the industry seems to have a negative effect ($b = -.311$, $p < .01$) on profit margin growth. In an emerging economy like Turkey, firms that focus more on an innovation-based strategy, generally speaking, have lower profit margin growth. This can be due to the labor-intensive and relative low value-added character of most firms in emerging countries. Their competition is often based on price and speed of production[72] rather than on innovation, and by focusing on efficiency and the exploitation of existing production capital, they may reach higher levels of profit margin and can gain a larger market share.[71] In industries where price competition is dominant, buyers often have more access to a greater variety of firms making it easier to switch to other firms, and give them a stronger position to negotiate price.[73]

Model 2 provides information related to the interaction effects and depicts a statistically significant and positive effect of market competition ($b = .263$, $p < .05$) on the relationship between having an innovation-based strategy and profit margin growth. Several researchers argue that the relationship between firm level capabilities, such as innovation, and firm performance depends on environmental factors.[49,74] The effects of technological capabilities on firm performance are nested in national framework conditions,[74] such as the level of competition, the availability of a labor force, and institutional support. In particular, in industries with fierce competition, buyers can choose among a larger variety of products and services, allowing them to switch to other firms more easily.[75] To reduce switching opportunities for buyers, SMEs need to make buyers more dependent on their products and services, by distinguishing themselves from the competition by, for instance, adopting an innovation-based strategy. Therefore, we conclude that firms with a stronger focus on innovation can better bind buyers to their products and services and, as a result, charge higher prices and increase their profit margins.

With regard to the role of university interaction, we did not find a significant effect that it is positively moderating the relationship between innovation-based strategy and profit margin growth. Although university interaction may help SMEs in developing innovations based on specific knowledge and at lower costs, it would appear that, in an emerging economy like Turkey, where labor-intensive and low-cost production is more common, firms benefit more from innovation when competition is fierce. It may also be that the relatively recent focus on university–industry collaboration in Turkey[50] explains why the effect of collaboration with a university on the relationship between an innovation-based strategy and profit margin is not significant. It is only recently that the Turkish government, universities, and SMEs have started to explore the benefits and best practices of university–industry collaboration and, as a result, their knowledge conversion capability[37] is not yet well developed and the potential gains of collaboration may not be fully realized.

Following the argument of the knowledge conversion capability, several other researchers have argued that the benefit of interaction with universities in part depends on how well firms are able to utilize and commercialize their research findings.[66] Collaboration with universities is characterized by the transfer of noncodified information and experience,[31] which requires a common language and often involves face-to-face interaction between university researchers and industry researchers.[26] In particular in Turkey, where strong growth in public funds for R&D support is more recent,[56] the learning experience for university and industry researchers alike with regard to their collaboration needs to mature and SMEs need to learn best practices to benefit fully from the support available. This requires time and may have a negative effect on the immediate profit ratio.

To investigate the effect of this learning experience in the relationship between university collaboration and firm profit growth, we included in Model 3 the squared term of university collaboration, which is statistically significant and positively related to profit margin growth ($b = .810$, $p < .05$). This relationship is further illustrated in Fig. 27-2, which clearly shows a U-curve type of relationship between university collaboration and firm profit growth. At low levels of university collaboration, profit margin growth is high, and decreases when there is more collaboration; but at a certain point, profit margin growth increases again, which indicates that firms need to exceed a certain level of university collaboration to reap the rewards. It seems that SMEs that emphasize innovation as a strategy can benefit from direct collaboration with a university,[17] but the benefits are more present when they exceed a certain threshold in the collaboration with a university[34] and hence are able to convert scientific knowledge into commercial applications.[37] This can be explained by the argument that collaboration needs a certain minimum intensity to better absorb the technologies and knowledge that they develop together with universities.[21,76]

Collaboration with universities can range from short-term direct contract research and services, and temporal contract research, to long-term in-depth collaboration and knowledge exchange. In the case of short-term direct contract research, we argue that the costs are likely to outweigh the immediate benefits. Under such circumstances, firms may not take enough time to understand the noncodified elements of the knowledge that is produced in collaboration with universities.[31] In addition, they may experience difficulties in bridging the diverging goals and scopes between academics and practitioners[35] and therefore find it difficult to reap the rewards of commercial application. In other words, it is only beyond a certain level of effort and commitment that scientists and practitioners will be able to better understand each other[34] and improve their knowledge conversion capability.[37]

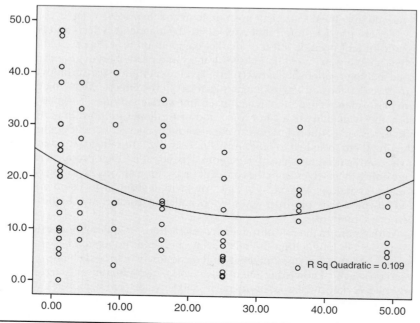

FIGURE 27-2   Relationship between university collaboration and firm profit growth.

Conclusion

Open innovation brings considerable benefits for SMEs. We discussed the concept of open innovation within SMEs and how SMEs in an emergent country such as Turkey can benefit from open innovation. Using a sample of 79 small and medium-sized enterprises located in the Aegean region of Turkey, we found that Turkish firms have lower profit margin growth when they focus more on innovation or operate in more competitive markets. However, when SMEs operate in more competitive industries they benefit from a focus on innovation strategy. Moreover, the findings reveal that collaborating with universities provides considerable benefits for SMEs in terms of profit margin growth, but they need to exceed a certain level of on intensity in the interaction. The implication and management recommendation from these findings is that the scientific information that is transferred from a university to the SME often requires more detail than is available in a publication.[31,34] Direct interaction between scientists and practitioners can make the transfer smoother, and SMEs enjoy the benefits more if both scientists and practitioners understand each other better and can bridge their diverging goals and scopes.[36]

Therefore, we argue that low levels of collaboration are not enough for scientists and practitioners in SMEs to fully understand each other and benefit from their collaboration. It takes a certain threshold of university collaboration intensity. When both scientists and practitioners put more effort and commitment into their collaboration, they are better able to understand each others' goals and scopes, which improves their knowledge conversion capability.[37] A more intense collaboration between SMEs and universities increases the scientists' perception of what the SMEs need and how they create value for their customers. Similarly, the SMEs can better translate the scientific knowledge from the universities and apply and develop that knowledge in products and service improvements. Finally, the findings show that for open innovation to work well, a minimum level of interaction between parties involved is required to get to know each other and benefit fully from the potential that open innovation has to offer.

References

1. Chesbrough, H. *Open Innovation: The New Imperative for Creating and Profiting from Technology.* Boston, MA: Harvard Business School Press, 2003.
2. Kim, W.C. and Mauborgne, R. "Value Innovation: A Leap into the Blue Ocean." *Journal of Business Strategy*, 26 (4): 22–28, 2005.
3. Kim, W.C., Mauborgne, R., Beaver, J., Marks, B., and Mortensen, W. *Crafting Winning Strategies in a Mature Market: The U.S. Wine Industry in 2001.* Harvard Business Publishing.
4. Gould, R.W. "Open Innovation and Stakeholder Engagement." *Journal of Technology Management & Innovation*, 7 (3): 1–11, 2012.
5. Almirall, E. and Casadesus-Masanell, R. "Open versus Closed Innovation: A Model of Discovery and Divergence." *Academy of Management Review*, 35 (1): 27–47, 2010.
6. Bogers, M. "The Open Innovation Paradox: Knowledge Sharing and Protection in R&D Collaborations." *European Journal of Innovation Management*, 14 (1): 93–117, 2011.
7. Jarvenpaa, S.L. and Wernick, A. "Paradoxical Tensions in Open Innovation Networks." *European Journal of Innovation Management*, 14 (4): 521–548, 2011.
8. Maehler, A.E., Curado, C.M.M., Pedrozo, E.Á., and Pires, J.P. "Knowledge Transfer and Innovation in Razilian Multinational Companies." *Journal of Technology Management & Innovation*, 6 (4): 1–13, 2011.
9. Tidd, J. and Bessant, J. *Managing Innovation.* John Wiley & Sons Ltd, 2009.
10. Van de Vrande, V., de Jong, J.P.J., Vanhaverbeke, W., and de Rochemont, M. "Open Innovation in SMEs: Trends, Motives and Management Challenges." *Technovation*, 29 (6–7): 423–437, 2009.
11. Elmquist, M., Fredberg, T., and Ollila, S. "Exploring the Field of Open Innovation." *European Journal of Innovation Management*, 12 (3): 326–345, 2009.
12. West, J., Vanhaverbeke, W., and Chesbrough, H. "Open Innovation: A Research Agenda." In: Chesbrough, H., Vanhaverbeke, W., and West, J. (eds.), *Open Innovation: Researching a New Paradigm.* New York, NY: Oxford University Press, 2006.
13. Narula, R. "R&D Collaboration by SMEs: New Opportunities and Limitations in the Face of Globalisation." *Technovation*, 25: 153–161, 2004.
14. Edwards, T., Delbridge, R., and Munday, M. "Understanding Innovation in Small and Medium-Sized Enterprises: A Process Manifest." *Technovation*, 25: 1119–1120, 2005.
15. Rothwell, R. "External Networking and Innovation in Small and Medium-Sized Manufacturing Firms." *Technovation*, 11: 93–112, 1991.

16. Rothwell, R. and Dodgson, M. Innovation and Size of Firm. In: Dodgson, M. (ed.), *Handbook of Industrial Innovation*. Aldershot: Edward Elgar Publishing Limited, pp. 310–324, 1994.

17. Baum, J.A.C., Calabrese, T., and Silverman, B.S. "Don't Go It Alone: Alliance Network Composition and Startups' Performance in Canadian Biotechnology." *Strategic Management Journal*, 21 (3): 267–294, 2000.

18. Powell, W.W. and Owen-Smith, J. "Network Position and Firm Performance: Organizational Returns to Collaboration in the Biotechnology Industry." *Research in the Sociology of Organizations*, 16: 129–159, 1999.

19. Caputo, A.C., Cucchiella, F., Fratocchi, L., Pelagagge, P.M., and Scacchia, F. "A Methodological Framework for Innovation Transfer to SMEs." *Industrial Management & Data System*, 102 (5): 271–283, 2002.

20. Katila, R. and Ahuja, G. "Something Old, Something New: A Longitudinal Study of Search Behaviour and New Product Introduction." *Academy of Management Journal*, 45 (6): 1183–1194, 2002.

21. Cohen, W. and Levinthal, D. "Absorptive Capacity: A New Perspective on Learning and Innovation." *Administrative Science Quarterly*, 35 (1): 128–152, 1990.

22. Salter, A. and Martin, B.R. "The Economic Benefits of Publicly Funded Basic Research: A Critical Review." *Research Policy*, 30 (3): 509–532, 2001.

23. Azagra-Caro, J.M., Archontakis, F., Gutiérrez-Gracia, A., and Fernández-de-Lucio, I. "Faculty Support for the Objectives of University-Industry Relations versus Degree of R&D Cooperation: The Importance of Regional Absorptive Capacity." *Research Policy*, 35 (1): 37–55, 2006.

24. Kodama, T. "The Role of Intermediation and Absorptive Capacity in Facilitating University-Industry Linkages—An Empirical Study of TAMA in Japan." *Research Policy*, 37 (8): 1224–1240, 2008.

25. Kaufmann, A. and Tödtling, F. "Science-Industry Interaction in the Process of Innovation: The Importance of Boundary-Crossing between Systems." *Research Policy*, 30 (5): 791–804, 2001.

26. Balconi, M. and Laboranti, A. "University-Industry Interactions in Applied Research: The Case of Microelectronics." *Research Policy*, 35 (10): 1616–1630, 2006.

27. Capon, N., Farly, J.U., and Hoenig, S.M. "A Meta-Analysis of Financial Performance." *Management Science*, 16: 1143–1159, 1990.

28. Lee, C., Lee, K., and Pennings, J.M. "Internal Capabilities, External Networks, and Performance: A Study on Technology-Based Ventures." *Strategic Management Journal*, 22: 615–640, 2001.

29. Song, M., Podoynitsyna, K., Van Der Bij, H., and Halman, J.I.M. "Success Factors in New Ventures: A Meta-Analysis." *Journal of Product Innovation Management*, 25(1): 7–27, 2008.

30. Autio, E. "New Technology-Based Firms in Innovation Networks Symplectic and Generative Impacts." *Research Policy*, 26 (3): 263–281, 1997.

31. Agrawal, A. "Engaging the Inventor: Exploring Licensing Strategies for University Inventions and the Role of Latent Knowledge." *Strategic Management Journal*, 27 (1): 63–79, 2006.

32. Mowery D., Oxley J., and Silverman B. "Strategic Alliances and Interfirm Knowledge Transfer." *Strategic Management Journal*, Winter Special Issue (17): 77–91, 1996.

33. Almeida, P. and Kogut, B. "Localization of Knowledge and the Mobility of Engineers in Regional Networks." *Management Science*, 45 (7): 905–917, 1999.

34. Owen-Smith, J. and Powell, W. "The Expanding Role of University Patenting in the Life Sciences: Assessing the Importance of Experience and Connectivity." *Research Policy*, 32 (9): 1695–1711, 2003.

35. Dasgupta, P. and David, P.A. "Toward a New Economics of Science." *Research Policy*, 23 (5): 487–521, 1994.

36. Zahra, S.A. and George, G. "Absorptive Capacity: A Review, Reconceptualization and Extension." *Academy of Management Review*, 27 (2): 185–203, 2002.

37. Zahra, S.A., Van de Velde, E., and Larrañeta, B. "Knowledge Conversion Capability and the Performance of Corporate and University Spin-Offs." *Industrial Corporate Change*, 21 (3): 569–608, 2007.

38. Adams, J.D. "Comparative Localization of Academic and Industrial Spillovers." *Journal of Economic Geography*, 2 (3): 253–278, 2002.

39. Lee, J. and Win, H.N. "Technology Transfer between University Research Centres and Industry in Singapore." *Technovation*, 24 (5): 433–442, 2004.

40. Baum, J.A.C. and Oliver, C. "Institutional Linkages and Organizational Mortality." *Administrative Science Quarterly*, 36 (2): 187–218, 1991.

41. Podolny, J.M. "Market Uncertainty and the Social Character of Economic Exchange." *Administrative Science Quarterly*, 39 (3): 458–483, 1994.

42. Packham, G., Pickernell, D., and Brooksbank, D. "The Changing Role of Universities in Knowledge Generation, Dissemination and Commercialization." *International Journal of Entrepreneurship and Innovation*, 11 (4): 261–263, 2010.

43. Lawton Smith, H. and Bagchi-Sen, S. "University Industry Interactions: The Case of the UK Biotech Industry." *Industry and Innovation*, 13: 371–392, 2006.

44. Pickernell, D., Packham, G., Brooksbank, D., and Jones, P. "A Recipe for What? UK Universities, Enterprise and Knowledge Transfer." *International Journal of Entrepreneurship and Innovation*, 11 (4): 265–272, 2010.

45. Dowling, M.J. and McGee, J.E. "Business and Technology Strategies and New Venture Performance: A Study of the Telecommunications Equipment Industry." *Management Science*, 40 (2): 1663–1677, 1994.

46. Bloodgood, J., Sapienza, H.J., and Almeida, J.G. "The Internationalization of New High-Potential U.S. Ventures: Antecedents and Outcomes." *Entrepreneurship Theory and Practice*, 20 (4): 61–76, 1996.

47. Zahra, S.A. and Bogner, W.C. "Technology Strategy and Software New Ventures' Performance: Exploring the Moderating Effect of the Competitive Environment." *Journal of Business Venturing*, 15 (2): 135–173, 2000.

48. Capon, N., Farly, J.U., and Hoenig, S.M. "A Meta-Analysis of Financial Performance." *Management Science*, 16: 1143–1159, 1990.

49. Li, H. and Atuahene-Gima, K. "Product Innovation Strategy and the Performance of New Technology Ventures in China." *Academy of Management Journal*, 44 (6): 1123–1134, 2001.

50. Pamukcu, T. "Trade Liberalization and Innovation Decisions of Firms: Lessons from Post-1980 Turkey." *World Development*, 31 (8): 1443–1458, 2003.

51. Beba, A. and Saatcioglu, K. "Financing R&D Projects of Innovative SMEs: National and International Funds." *International Entrepreneurship Congress: SME's and Entrepreneurship* (Ed: Katrinli), ISBN: 978-975-8789-32-0, pp. 70–80.

52. Turkoglu, M. and Celikkaya, S. "R&D Support for Small and Medium Size Enterprises." *International Journal of Alanya Faculty of Business*, 3 (2): 56–71, 2011.

53. Yaniktepe, B. and Cavus, M.F. "Investigation of Policy and Incentives on the Industrial Research and Development in Turkey." *African Journal of Business Management*, 5 (22): 9214–9223, 2011.

54. Chung, S. "Building a National Innovation System through Regional Innovation Systems." *Technovation*, 22 (8): 485–491, 2002.

55. Arikan, C., Akyos, M., Durgut, M., and Goker, A. *National Innovation System*. TUSIAD, TUSIAD-T/2003/10/362, 2003.

56. Taymaz, E. Development Strategy and Evolution of Turkey's Innovation System. In Suh, J.H. (ed.), *Models for National Technology and Innovation Capacity Development in Turkey*. Seoul: Korea Development Institute, pp 63–104, 2009.

57. Cetindamar, D. and Ulusoy, G. "Innovation Performance and Partnership in Manufacturing Firms in Turkey." *Journal of Manufacturing Technology Management*, 19 (3): 332–348, 2008.

58. Bleaney, M., Binks, M., Greenaway, D., Reed, G., and Whynes, D. "What Does a University Add to Its Local Economy?" *Applied Economics*, 24: 305–311, 1992.

59. Love, J. and McNicoll, L. "The Regional Economic Impact of Overseas Students in the UK: A Case Study of Three Scottish Universities." *Regional Studies*, 22 (1): 11–18, 1998.

60. Wright, M., Clarysee, B., Lockett, A., and Knockaert, M. "Mid-Range Universities' Linkages with Industry: Knowledge Types and Role of Intermediaries." *Research Policy*, 37 (8): 1205–1223, 2008.

61. Fortuin, F., Batterink, M., and Omta, O. "Key Success Factors of Innovation in Multinational Agrifood Prospector Companies." *International Food and Agribusiness Management Review*, 10 (4): 1–24, 2007.

62. Cooper, R.G., Edgett, S., and Kleinschmidt, E.J. "Portfolio Management for New Product Development: Results of an Industry Practices Study." *R&D Management*, 31 (4): 361–380, 2001.

63. Hanel, P. and St-Pierre, M. "Industry–University Collaboration by Canadian Manufacturing Firms." *The Journal of Technology Transfer*, 31 (4): 485–499, 2006.

64. Boer, H. and During, W.E. "Innovation, What Innovation? A Comparison between Product, Process and Organizational Innovation." *International Journal of Technology Management*, 22 (1/2/3): 83–107, 2001.

65. TURKSTAT. *Statistical Database*. Available at: http://www.turkstat.gov.tr, 2010.

66. Mansfield, E. "Academic Research and Industrial Innovation: An Update of Empirical Findings." *Research Policy*, 26 (7/8): 773–776, 1998.

67. Aiken, L.S. and West, S.G. *Multiple Regression: Testing and Interpreting Interactions*. Newbury Park, CA: Sage, 1991.

68. Hair, J.F. Jr., Black, W.C., Babin, B.J., and Anderson, E.E. *Multivariate Data Analysis*, 7th ed., Pearson, 1998.

69. Lieberman, M.B., Montgomery, D.B. "First-Mover Advantages." *Strategic Management Journal*, 9 (Special Issue: Strategy Content Research): 41–58, 1988.

70. Katz, R. and Alen, T. "Investigating the Not Invented Here (NIH) Syndrome: A Look at the Performance, Tenure, and Communication Patterns of 50 R&D Project." *R&D Management*, 12 (1): 7–19, 1982.

71. Steffens, P., Davidsson, P., and Fitzsimmons, J. "Performance Configurations over Time: Implications for Growth and Profit-Oriented Strategies." *Entrepreneurship, Theory and Practice*, 33 (1): 125–148, 2009.

72. IZKA. *Izmir Regional Innovation Strategy Report*. Izmir Development Agency (IZKA) Publication, ISBN:978-605-5826-07-9, 2012.

73. Porter, M. E. *Competitive Strategy*. New York, NY, Free Press: 1980.

74. Goedhuys, M. and Srholec, M. "Understanding Multilevel Interactions in Economic Development." *UNU-MERIT Working Paper Series* 003, 2010.

75. Williamson, O.E. "Strategizing, Economizing and Economic Organization." *Strategic Management Journal*, Winter Special Issue, 12: 75–94, 1991.

76. Hitt, M.A., Dacin, M.T., Levitas, E., Arregle, J.L., and Borza, A. "Partner Selection in Emerging and Development Market Contexts: Resource-Based and Organizational Learning Perspectives." *Academy of Management Journal*, 43 (3): 449–467, 2000.

CHAPTER 28

Systematic Innovation

Donald A. Coates

Introduction

What is systematic innovation?

I cannot count the number of times the term innovation has been misused in advertisements or discourse. Its definition is very important for the understanding of systematic innovation and this chapter.

First, invention is part of innovation. But invention alone without bringing it into the marketplace or in widespread use will not create innovation. If a hermit invents something in a shack in the woods, it does little to no good unless he exposes it to more people who value its use.

Invention is defined as

1. The solution has the potential utility for the users.

2. There is novelty where novelty means: "… the state or quality of being new, exciting, unusual or unique."[1]

3. The solution to a problem is not obvious to those skilled in the art, that is, there could be some psychological inertia that prevents others from seeing the solution.

Second, the definition of innovation is very important to understanding systematic innovation. Innovation means that something new is put in *popular* use for the mutual good. This is close to Chesbrough's definition of innovation[2] that is "…Innovation is invention implemented and taken to market." What constitutes newness is debatable and does not always mean a patent is required. Many new concepts are purposefully not patented. To me, something new means a *significant* problem was solved through an inventive concept that was derived. A problem is defined as a gap from a current situation (real or perceived) to where you *want* to be. Therefore, when an inventive concept is brought into popular use, it becomes innovation.

To have innovation, invention is necessary but not inclusive. Examples of innovation were given by Joseph Schumpeter in the 1930s in a paper by Rogers[3]:

1. The introduction of new goods or qualitative changes in an existing product

2. Process innovation new to an industry

3. The opening of a new market

4. Development of new sources of supply for raw materials or other inputs

5. Changes in industrial organization

Third, with the foregoing definitions, systematic innovation is an inventive concept that is brought to popular use by a *method*. In an insightful review of Peter Drucker's classic work *Innovation and Entrepreneurship*,[4] Professor Goldstein of University of North Carolina paraphrases Drucker's belief in systematic innovation and its benefits: "Like the discipline of management, which he helped found, Drucker believes that innovation can be undertaken in a systematic way and, when it is, the results are consistently positive."[5]

Many people have tried to measure innovation via different indicators (number of patents, research dollars, number of new products, investment in machines, investment in marketing, etc.) and these are discussed thoroughly in a paper by Rogers.[3] Unfortunately the various measures of innovation are customer and company subjective and cannot be easily correlated with innovation. To simplify this situation a limited number of systematic methods will be discussed with emphasis on one method in particular. The purpose of these methods and this chapter will be to consistently and profusely produce financial benefit from innovation while other specific indicators of innovation are left for further study.

Brief History of Systematic Innovation

For the earliest evidence of systematic innovation we go back to prehistoric times. Anthropologists seem to be divided on when man "began to demonstrate an ability to use complex symbolic thought and express cultural creativity."[6] One camp believes this occurred about 50,000 years ago when man "…began to exhibit full behavioral modernity."[6] I believe this is indicative of man's ability to use systematic innovation although some may argue that this ability occurred much earlier and may even have been as early as 5 million years ago when we split from the chimpanzee.[7] Regardless of when, the results clearly indicate the ability of humans over other animals to solve problems to advance and adapt their way of life while other animals have remained virtually static.

Coupled with problem-solving methods, is the process for dissemination of the results of problem solving for public use. More will be said about this as part of the business process but the process is discipline specific (for instance scientific research where one would publish the results versus product design, where the product is sold). In this chapter, one method applicable to product and process design will be explained.

Trial and error problem solving without regard for previous successes could be argued as the zero form of systematic innovation. Surely animals and archaic man exhibited this method and that did not separate man from his animal ancestors.

The next higher form of systematic innovation is believed to be reproductive thinking[8] or, as I term it, analogical thinking for an individual. This is reproducing experiences of previous successes (possibly original trial and error successes) to solve similar problems. Again some animals exhibit limited ability to do this, but over time this did not improve their way of life beyond basic behavior and therefore they have or will become extinct with major changes to the environment.

The scientific method I regard as one of the keystones systematic innovation. It is based on testing and observations. It is reported to originate over 1000 years ago and basically involves the steps of[9]

1. Formulating a question that needs answering based on experience and available data

2. Formulating a hypothesis that can explain the behavior and that can be shown to be false

3. Prediction of the consequences of the hypothesis

4. Testing that can show the predictions are true or not false and therefore that the hypothesis is not false

5. Analysis of the results to determine if the hypothesis is verified and determine a modification of the hypothesis if it was not verified

There are different versions of the scientific method and they often depend on the field of study but they are similar to this process.

More recently, especially in the 1900s, there has been a flurry of methods for initiating systematic innovation inventive concepts but they do not take it all the way to public use. So the following list focuses on the problem-solving aspect of the systematic innovation process. More will be said regarding the rest of the process later in the chapter. These methods are useful to the process of defining and finding solution concepts but they are woefully incomplete with respect to a business model that accompanies them for public use. They represent starting points. I have identified ones that I believe are helpful mainly to engineers and business people. Each will be described briefly. The reader should study the references for a fuller appreciation of the method so please forgive the brevity. Most methods are covered in detail in the book *The Inventor's Toolkit*.[10]

Someone once presented a slide at a conference with a thousand problem-solving methods, but only a limited number are needed for any innovator. It is like your tool box or fishing tackle box. Only a limited number of tools are needed for most jobs. I created a list of problem-solving methods that I find helpful.

1. **Trial and error:** Attempts at successful solutions to a problem with little benefit from failed attempts. This is not a good method.

2. **Analogical thinking and mental simulations:** Using past successes applied to similar problems by mental simulations and testing.

3. **Theory of inventive problem solving (TRIZ):** TRIZ will be covered in subsequent chapters and is felt to be the most powerful problem-solving tool although it takes training. It originally started with analogical thinking but has additional tools that are very powerful.

4. **Scientific method[11]:** A classical method that uses hypothesis based on initial observations and validation through testing and revision if needed.

5. **Edison method:** The beauty of the Edison method[12] is that it consists of five strategies that cover the full spectrum of innovation necessary for success. Not only did he learn from his many experiments, he appreciated user's needs, was a great collaborator both in research and business, was a practical experimental scientist not a theoretical one, appreciated intellectual property, developed the popular concept of a research laboratory that prevailed after World War II, and most of all had great perseverance.

6. **Brainstorming[13]:** Recording many ideas without initial criticism that could solve a problem followed by organization and evaluation. This is one of the most used methods and several versions have been developed. I criticize it since it must be done carefully and tends to create the same ideas from the same group of participants after several tries.

    6.1. Osborn method: Original brainstorming method developed by Alex F. Osborn[13] by primarily requiring solicitation of unevaluated ideas (divergent thinking) followed by convergent organization and evaluation.

    6.2. Six Hats: Structured method of brainstorming through different steps to control thinking and emotions that can speed up the process of brainstorming.[14]

7. **Problem detection and affinity diagrams:** Focus groups, mall intercepts, mail and phone surveys that ask customers what are the problems are forms of problem detection.[15] Essentially the responses are grouped according to commonality (affinity diagrams) to strengthen the validity of the response. Developing the correct queries and interpreting the responses are critical to the usefulness of the method.

    7.1. Five whys[16]: An extremely simple but effective method of asking why a problem occurred five times. After each answer ask why again using the previous response. It is surprising how this may lead to a root cause of the problem but it does not solve the problem.

    7.2. Explore unusual results: Unusual results can be investigated for how they occurred and what problems they could solve. Many great inventions occurred this way (Nylon, Post-Its, etc).

    7.3. Fishbone diagrams a.k.a. Ishikawa diagrams[17]: A mnemonic diagram that looks like the skeleton of a fish and has words for the major spurs that prompt causes for the problem.

    7.4. Ethnography[18]: Observing and recording what people do to solve a problem and not what they say the problems are. It is based on anthropology but for use on current human activities. It is based on the belief that what people do can be more reliable than what they say.

    7.5. Function analysis and fast diagrams[19]: Analyzing a system for the different functions by which it operates is believed to generate more ideas than focusing on the physical part. It was developed during World War II when materials were scarce and alternate materials that could perform the same function might be available.

    7.6. Kano method[20]: Based on the idea that features can be plotted using axes of fulfillment and delighters. This defines areas of: must haves, more is better, and delighters. The latter is used to excite the customer and close a sale. Cup holders were such a delighter although now they have become must haves.

    7.7. Medici effect: The book by this name[21] describes the intersection of significantly different ideas that can produce cross-pollination of fields and create more breakthroughs. The book starts in Horta, a city in the Azores, and makes an interesting story especially for sailors. There are several other methods below that have allegiance to this approach.

7.8. Technology mapping and recombination: I created this method but have subsequently discovered others may have developed a similar method (see Morphological Design[9]). Mine is a matrix-based method that lists the various technologies that can perform a function and then examines combinations that have not been tried to see if there is enhanced performance or features. Figure 28-1 shows a partially completed example for a vacuum cleaner. It is similar to morphological design except the marketing benefits are shown and ranked at the bottom of the matrix and solicits augmentations that may resolve problems from implementation. There are other subfunctions that can

			Hoover Bagless	Dyson	Hoover Cross Flow cyclone	Hoover Bag	E	F	OPPOSING FORCES/Limitations			
			NOVEL COMBINATIONS									
	1	Interception includes VanderWaals & rainbow filt										
	2	Electrostatics Dust agglomeration			Could cone be electrostatic under ePTFE. Mesh is charged opp to particles	X?			Hoover BL still requires cleaning			
	3	Gravity	X Both cups		X Cyclone cup & Cycl cone	X			Volume too small to use by itself			
	4	Sieving (interception, inertia, diffusion)	X cartridge & Screen	X Between cyclones & Post	X in cyclone surface	X			Is double sieving needed for large particles??			
	5	Inertia: Cyclone, etc	X Screen	Cyclone	X cross flow cyclone & use of inertia & self cleaning				Press. Drop is high			
	6	Diffusion incl VanderWaals							Random motion req too low a velocity and too much time			
	7	Evaporation							Phase change to separate types of materials may not apply here.			
	8	Thermal							Particles move from hot to cold surface			
		New way to empty										
		Self cleaning: Vibration Wiping										
		Surface Filter	ePTFE		could cycl cone be ePTFE HEPA over spun B and mesh							
		Parallel actions		8 cyclones reduce Dp								
		Auto wiping for self cleaning										
Overt Benefit	1			less cleaning & const suc	Self Cleaning ePTFE							
Reason TB	2											
Dramatic Difference	3											
3) Total Benefit (Incremental- Breakthrough):												

Vertical labels on left: "Basic ways to solve problem; Study-data-collection-discovery; Combinations may resolve limitations" and "Augmentation,May solve limitations"

FIGURE 28-1   Technology mapping.

help the primary technologies. The various options may lead to problems that need further solution.

7.9. Abundance and redundance: This is an old saw but so true that if you want a good invention that solves a problem you need lots of ideas. Most people stop too soon. Sir James Dyson says in his vacuum cleaner advertisements that he made 5000 prototypes before he found the right one.

7.10. Hitchhiking (author named this method): When a breakthrough occurs, it is a fertile area for innovators. They should hitchhike on the breakthrough to create new applications and improvements that can be inventions. Many engineering advancements come from companies vigilant and with alacrity that use these breakthroughs to create new and cost-effective products.

7.11. Kepner Trego[22]: This method is very useful for processes that have were performing well and then developed a problem. It is a good step by step method that is based on finding the cause of the problem by asking what changed. Since the process was working fine originally, something must have changed.

7.12. Quality function deployment (QFD) *aka* the house of quality: This creates a matrix that looks like a house that can mediate the specifications of a product or process.[23] There are subsequent houses that can follow that further mediate downstream implementation issues. The *difficulty* of bringing together marketing, sales, and engineering to a set of optimized customer and competition specifications cannot be understated.

7.13. Design of experiments: This method is a statistically based method that can reduce the number of experiments needed to establish a mathematical relationship between a dependent variable and independent variables in a system. This will resolve a control issue in a manufacturing or business process. A simple explanation is given in *The Inventor's Toolkit*.[10] It requires computer software and an expert to help guide the group but it can take a lot of guesswork out of running a process or system especially when there is variability in the data.

7.14. Failure mode effects analysis[24]: A matrix-based method developed by NASA to investigate potential serious problems in a proposed system prior to final design (such as a moon landing). It creates a risk priority number that can be used to create a ranking of the biggest risks and then ranks the proposed solutions. Using the method an overall reliability can be estimated based on certain assumptions.

I feel that problem-solving methods are the most important element of the fuzzy front-end process. To me this is the heart of systematic innovation. If a systematic method for problem solving is not outstanding, the rest of the systematic innovation breaks down.

I have taken the liberty to give a recommended list of the various problem-solving methods based on my experience and these are summarized in Table 28-1. They are divided into groups by double lines. Within each group they are considered equally useful. Some people will undoubtedly argue with the selection and rankings. It is intended to help innovators with a short list of relative information and ranking. It is not intended to prevent the exploration of these or other methods.

Why We Need Systematic Innovation

You probably have heard it said that "you cannot invent on a schedule." The same can be said for innovation. I believe this chapter will convince you this is becoming less true and that the opposite is needed in today's environment. In the book *Innovate Like Edison*[12] the authors quoted Professor Vijay Govindarajan of the Dartmouth Amos Tuck School of Management regarding innovation:

> We've got to get every member of the organization, from the top to the bottom, literate in innovation just like we make them literate in finance, or literate in marketing, or literate in any other management discipline. Innovation is not about ideas and creativity, it's a whole discipline about how you turn an idea into reality. Innovation literacy has to be across the board. It's got to be done.

The beauty of this quote is that it is one of the key points of this chapter. Innovation involves systematic methods and many skills and is not just inventing.

Many of the systematic innovation methods have been around for some time, but we did not recognize them for what they are. As these methods become better and new ones are developed (case in point, The Theory of Inventive Problem Solving together with the business model counterpart), the probability of conducting innovation on schedule is higher.

Method **Ranked from *Most* Favored to *Least* Favored within Groups of Equal Favor (Double Line Indicates Groupings)**	**Comments**
TRIZ: Theory of inventive problem solving	A multitooled method that defines conflicts well, IDs an ideal system, inventive principles using analogical solution, function analysis, trends of evolution, analyzes resources, and standard solutions. See later chapters.
General scientific method (general approach and contained in many other methods) and Edison method	Generally accepted approach. The Edison method expands on the scientific approach with skills and processes.
Redundance and abundance, analogical thinking, mental simulations	Data bank of many solutions, mixed and matched with problem is more efficient than trial and error.
Kano, surveys, focus groups, mall intercepts, ethnography, and problem identification	These are good at identifying features and problems but do not solve problems in general.
Function analysis	Breaks down main problem into functions and allows solving of simpler problems.
FMEA: Failure mode and effects analysis	Matrix method that utilizes parts and their functions, IDs failures and prioritizes and identifies risks but does not solve problems.
Kepner Trego	Defined steps and useful for manufacturing process problems.
Brainstorming, Osborn method, and Six Hats	Somewhat easy and has been refined by Osborn and de Bono. It does not solve problem unless obvious and is dependent on audience.
Hitchhiking	Utilize recent advancements to create new inventions.
Utilize unusual results	This is the reverse of problem solving since it recognizes an unusual result and then finds a problem that it solves. Similar to Analogical Thinking except for vigilance for unusual results. Solution is already available but the problem has to be found.
QFD: Quality function deployment aka house of quality	A matrix method that creates compromise definition of problem (customer, marketing, engineering).
Design of experiments	Statistical technique that can solve control problems with many variables.
Fishbone diagrams and affinity diagrams	Organizes and suggests categories for cause of problem but is dependent on audience knowledge and does not solve the problem directly.
Technology mapping and morphological design, and Medici effect	Combining different technologies and ideas to discover if a new combination can introduce a new feature but does not solve all the problems that may arise from the combination.
5 Whys	Easy to use and hope solution becomes obvious.
Trial and error	Still used but inefficient and time-consuming.

TABLE 28-1   Recommended List of Problem-Solving Methods

One cause for systematic innovation evolution is that growing population produces significantly more competition. The social brain hypothesis has been proffered by British anthropologist Robin Dunbar[25] that larger population has caused man to evolve for survival and interaction. We certainly see that today in global trade.

Bicycle Innovation Timeline

* 1813 Running Machine
* 1840 Pedals
* 1845 Brakes
* 1884 Transmission
* 1890 Tires
* 1897 Freewheeling clutch
* 1905 Derailleur shifting system
* 2007 Continuously Variable Transmission

FIGURE 28-2   Long timelines for past products.

The need for systematic innovation and better versions is also shown by comparing the timeline for past and present innovations. Figure 28-2 shows the time to develop a modern bicycle that took almost 200 years. Contrast that with today's product where the life cycle may only be a few years as demonstrated by Moore's law that says the number of integrated circuits on a chip doubles approximately every 18 months.[26]

There are several post-World War II factors that caused the demise of what is called closed innovation, where companies developed their innovation internally and did not rely on outside sources. When there was more outside competition for the same ideas,[2] called open innovation; I believe it shortened the timeline of innovation. Some of these factors that he lists are:

1. The increasing availability and mobility of skilled workers partly due to the GI bill
2. The availability of venture capital
3. Options external to the originating company for ideas that sat on the shelf
4. The increasing capability of external suppliers
5. Great minds go down the same path. When the technology is ready, multiple inventors develop the same solution almost simultaneously and the time between patent applications may be days, hours, or minutes.[27] So companies do not want to wait to introduce them for fear of independent introduction.

These factors ushered in the era of open innovation that significantly shortened the timeline available for innovation. This also put pressure on companies to create fast cycle times for innovation.[28]

Summary

This Introduction has defined systematic innovation, a brief chronology of systematic innovation with a summary of popular methods and why systematic innovation is important. We will now drill deeper into a generic systematic innovation method for products with application to processes as well.

Anatomy of Systematic Innovation for Product Development

As mentioned, a general process for systematic innovation for products will be the focus of the chapter. With slight modification this also fits process development. Figure 28-3 shows the stages for the systematic innovation process. The first two stages are what, I call the fuzzy front end. This name fits because the fuzzy front end was always thought to be mysterious and left to the inventor whose process was far from systematic. If it is poorly done and the results are "thrown over the wall" (to the development stage to meet a schedule, pressure by executive management, paralysis by analysis, procrastination, etc.), the results can mean project delays, finger pointing, and even worse significant quality problems. Many early stage-gate methods did not have a systematic method for the fuzzy front end.[28] The fuzzy front-end stages are extremely important because if well done they are an important start to successful innovation. In today's fast-paced and high-pressure environment, even the best final development stages can release products with serious problems. Pilot runs with test market studies can help reduce these problems but their corrections are usually expensive.

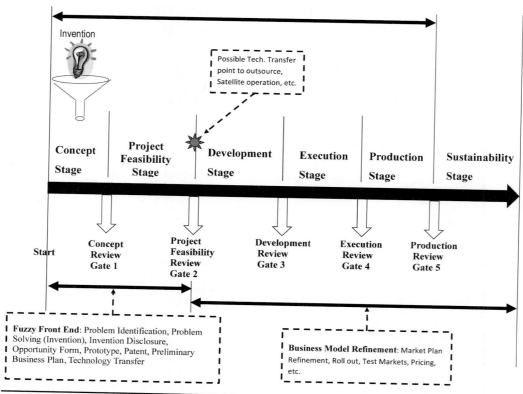

FIGURE 28-3   Systematic innovation stages.

The concept stage is the first stage that has most of the following activities:

1. It includes problem identification, problem dissection into smaller problems, ideation for potential solution of the smaller problems, and combinations of these potential solutions into concepts that could solve the larger problem. The problem-solving methods previously described can be used to solve these problems and to refine the need.

2. Also documentation in the form of a witnessed lab notebook, information disclosure sheet (IDS), and opportunity form (similar to a suggestion sheet) as shown in Figs. 28-4 and 28-5, respectively, are recommended for this stage.

The concept review gate at the end of this stage is concluded with an interdisciplinary mid-management team reviewing the results to make a decision as to whether the opportunity warrants proceeding with the feasibility stage. This is shown in the top half of Fig. 28-6. Their decision will give direction to write up a preliminary business plan for review by an executive review committee, or they can recommend reworking the idea, or shelve it for later use in planning, or kill it.

The second stage is the feasibility stage where concept feasibility is established by

1. Prototypes of key subsystems or theoretical validation of a solution to the main problem. They should be adequate to establish a preliminary basis for technical feasibility.

2. Assessment of patent potential and patent infringement. The results may indicate the need to submit a provisional patent to protect the date of invention or submit a patent application.

3. The feasibility stage-gate is concluded with an analysis of a preliminary business plan as shown in Fig. 28-7 (which includes preliminary versions of market size estimation, capital expenditure plan, resource plan, project schedule, risk abatement recommendations, and financial plan). This plan is based on preliminary design specifications and industrial design renderings. It is based on conservative back-of-the-envelope estimates of the innovation potential. It is not intended to be an

NOTEBOOK: ABC09112003   PAGE: ___/____ (attach additional sheets if needed)

Title of invention:
_____
_____

Background of conception:
_____
_____

Purpose or problem solved:
_____
_____

Description (structural) and operation (functional) description:
_____
_____

Drawings or Pictures:

All possible applications of invention:
_____
_____

Novel features of invention:

Invented by:_____Date:_____
Invented by:_____Date:_____

The above confidential information is witnessed and understood by:
_____Date:_____
_____Date:_____

Figure 28-4   Information disclosure sheet.

Opportunity Form

Date: __/__/__   Name:_____

- **PROPOSAL:**

- **TYPE OF PROGRAM (e.g. new product development, sourcing, productivity, capacity, etc):**

- **BENEFIT TO Organization (estimated):**

    Picture if available

Figure 28-5   Opportunity form.

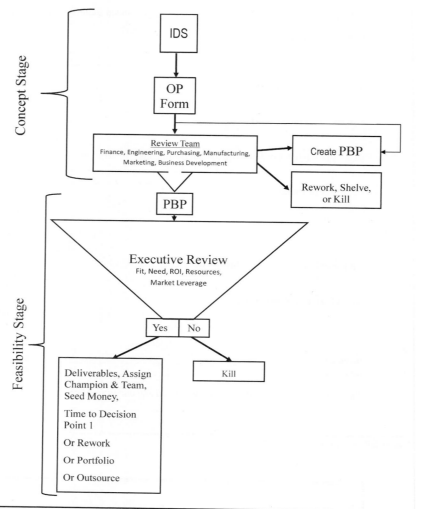

FIGURE 28-6 Preliminary business plan review process.

in-depth financial analysis. It couples the technical part of the opportunity to the business side, both of which are needed to create innovation. Chesbrough[2] identifies the elements of the business model as

3.1. Value proposition to the target audience
3.2. Target audience who will purchase the product
3.3. Value chain that describes where the company resides and what value it brings in the chain that delivers the product to the customer
3.4. How will the company collect money
3.5. Cost/margins that are required to make the product or process
3.6. Value network of ancillary suppliers that enhance the product but may not be in the direct chain to the customer
3.7. Competitive strategy that will give the company longevity

Defining and reviewing a preliminary version of the business plan including the associated business model is believed to be one of the cornerstones of successful innovation.

The gate review for this stage, as shown in the lower half of Fig. 28-6, is very critical since at this point, Decision Point 1 (DP1), the executive team makes a decision to allocate resources and move the project to the next decision point, shelve it for long-term planning, send it back for revision, pursue patents, sell or license the concept, pursue outside venture capital funding, or outsource the manufacturing. An executive team is needed that has the authority to make the resource decisions. A project must be a good opportunity to deviate from an existing strategic plan; however, a company cannot wait if the opportunity is great. More will be said about the strategic planning later.

Preliminary Business Plan

Proposal	*Sales / Profit*		
Description:		Year 1	Year 2	Year 3

	Year 1	Year 2	Year 3
Estimate Units			
Unit Cost			
Total Sales $			
Total Gross Profit $			
Gross Profit%			
Market Share %			
Simple ROI			

Market Overview and "Business Model"
(add additional sheets if needed.)
Timing:

Issues / Risks	*Investment / Costs*			
		DP1	DP2	DP3	Total

	DP1	DP2	DP3	Total
Total Investment				
Development & Tool Expense				
Marketing Expense				
Man-years				

Commercial:

Technical:

Key Assumptions:

FIGURE 28-7   Preliminary business plan.

If the product proceeds to the development stage the financial commitment is usually large compared to what has been spent to that point. The business plan will be reviewed continually and refined as the project proceeds to production.

As can be seen from the preliminary business plan review process, the fuzzy front end is iterative and is represented in Fig. 28-8, as an iterative systematic partial innovation process. I call this the fuzzy front-end process. The reason for representing the process this way is to focus on a front-end systematic system and the key elements that make it more productive, simple, and sustainable. The two elements, problem solving and business analysis, have been described; the new ones are environment and strategic planning.

The environment is defined to mean the physical surroundings, the organizational benefit policies and the human personal interactions (motivation, vision, leadership, trust, friendship, etc.). The environment here is meant to be what an inventor incurs while conducting innovation in an enterprise or while working alone. The environment acts as a catalyst or amplifier in the operation of the fuzzy front end. A better environment produces higher productivity for innovation. Some stress is helpful for invention and too much stress is harmful, but even under the most stressful situations some innovation can occur. Much more needs to be said about this factor. Some suggested readings are *Jump Start Your Business Brain*[29] and *Skills of an Effective Administrator*.[30]

The last element of the fuzzy front-end process is strategic planning. It is meant here to be the time-related sequencing of future innovation actions to maintain the business's continued viability and growth if desired. These actions must relate to the whole business mission, goals (e.g., become a $1 billion business in 2019), current channels, and product offerings, and so on. It is a very integrative interdisciplinary activity.

Figure 28-9 shows two strategic plans in cooperation with each other. The technology platforms must coordinate with a marketing plan and vice versa. A short list of deliverables for a strategic plan is given here:

1. Validate or change mission and major goals.

2. Review plans and progress against objectives.

3. Validate review team plans to meet goals and priorities.

4. Complete any opportunity or preliminary business plans review in follow-up meetings.

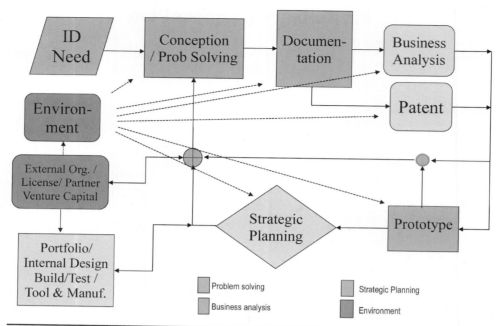

FIGURE **28-8** Fuzzy front end process.

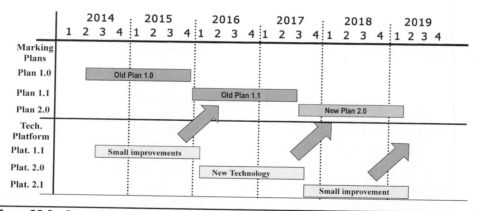

FIGURE **28-9** Strategic planning: merger of technology platforms with marketing plans.

5. Utilize the data bank of old "opportunity forms," PBP's, portfolio data bank, brainstorm new ideas, and recommend top ones for consideration in strategic planning.

6. Rank recommended projects, balance the risk of projects, as shown in Fig. 28-10, and put them into the strategic plan for a new long-term plan. Remember it is "not a plan if it is needed all at once!" This sounds silly, but given great projects,

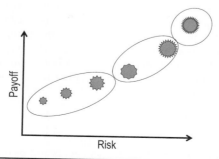

FIGURE **28-10** Project balance.

marketing will always want it all as soon as possible. For a company to survive in the long term there must be short-term, medium-term, and long-term innovation. The shorter-term innovations have less risk but less reward. The longer-term projects can be a new platform and/or a disruptive breakthrough.

During the strategic planning it is important to consider how much research is needed to be best in class. Too much can drain the company of rainy day reserves and too little can let it fall behind the competition. Some items to consider are

1. The market is a major factor for what is adequate. What do your customer base and competition demand? This is tempered by the balance needed in the project portfolio.

2. Most projects should be equal to slightly better than competition.

3. The plan must consider the limitation of resources.

4. The plan should create a technological forecast and establish a norm, for example, every 3 years major introduction; with upgrades in intervening years.

5. The plan should reflect management goals.

6. One or two projects should be considered that represent breakthrough (disruptive) opportunities.

Once completed the strategic plan must be reviewed periodically. Monthly reviews are good periods. I would be negligent to develop such a plan and miss the first-year goals, so several reviews are warranted during the year to make adjustments as needed.

Returning to the systematic innovation process, the third stage in Fig. 28-3 is the development stage. Design manpower and other resources are allocated in the form of a project team to create the deliverables in this stage. Most important in this stage are functions related to design (manufacturability, usability, performance, serviceability, sustainability for maintainability, and recycling), full prototype fabrication, design specification refinement (such as by refinement of quality function deployment from the fuzzy front end), validation testing, reliability testing, industrial design, and detailed mechanical design for tooling. There are many other deliverables that are needed such as: marketing plan update, bills of materials, manufacturing/procurement plan, capital plan update, capital and expenditure requests, risk abatement plan update, aesthetic approvals, patent prosecution, and risk reduction analyses such as failure mode effects analysis (which may be carried over from the fuzzy front end). These can be tailored for a particular industry. Once the above items are deemed acceptable, the design must then pass the development review gate. Subsequently, the capital and expenditure requests are reviewed and, if appropriate, approved by management. The project then passes onto the executions stage.

There are many project management stage-gate processes that treat the above items in more detail than space will allow. The stage-gate process was first introduced by R.G. Cooper in 1986. His landmark work as published in his book *Winning at New Products*[31] is a recommended reference. A collage of stage-gate diagrams is given by a Google webpage,[32] which demonstrates the plethora of stage-gate systematic systems.

In the execution stage the approved capital and expenditures are spent on tooling and manpower to carry out the design in preparation for production. The marketing plan is further developed with a launch strategy, facilities are readied for tooling, reviews are conducted and sought (aesthetics, internal safety, agency approvals, quality in the form of overall design conformance, vendor part approvals, and reliability), preproduction trials run, final bills of materials released, and a production plan developed. Upon the successful completion of these items the execution stage-gate is passed and the production stage begins.

The production stage is the first production of units for sale. There may be preproduction runs in the execution stage for field-testing but these units are usually given to company employees and reclaimed or not given warranty support. The development engineering, manufacturing engineering, process engineering, part quality, and customer assurance need to issue close-out reports with nonconforming items identified and then turn the project over to a support engineering function. Successfully done these define the production stage-gate.

The sustainability stage is the maintenance of the product in use via service personnel, customer support mail, web and telephone lines, warranty, customer assurance monitoring of field performance, and recycling. The ultimate goals are satisfied customers and environmental consciousness and if something has been missed it needs to be corrected via the above functions. This could involve telephone support, onsite service, unit replacement, and even recalls in the event of safety-related items.

Conclusion

Innovation involves more than just inventing. The business model is part of the innovation process. It is critically important to have both technical development of your product or system as well as business development. The groundwork for this can be laid during the fuzzy front end of the development process.

The chapter presents a systematic process that can be used to produce innovation. It is general and can be used for most product- or process-related businesses. This process should provide a consistent flow of innovation but it is not without effort. If it were easy, everyone would be doing it. It requires hitting on all cylinders of the process. Many books, experts, and systems stress one or two methods and ignore the other aspects of creating innovation.

One of the key cylinders is the fuzzy front-end process that can continuously feed good inventive solutions for the rest of the innovation process. The fuzzy front-end process has four parts that are most important for inventive solutions: problem solving or problem identification, business analysis, strategic planning, and environment.

In the fuzzy front-end process, problem solving or problem identification has to be regarded as the most important of the four elements. The reasoning is that without a good solution the rest of the process and for that matter the rest of the development process is mediocre at best. One of the following chapters will present one of the most powerful problem-solving methods namely the theory of inventive problem solving. I recommend it; especially for the inventors in fuzzy front end. They should study this method carefully and even become TRIZ masters. The other methods mentioned in this chapter nicely complement TRIZ but they cannot compare to its systematic approach to problem solving/problem identification which is critically important to systematic innovation.

It can be argued that the rest of the development process is labor-intensive but with a greater probability of completion given a good fuzzy front-end process. Some may say that all the problems cannot be foreseen. To answer that question, it is paramount to systematic innovation that powerful problem-solving methods are available both at the beginning and latter stages of innovation.

References

1. LoveToKnow Corp. *Your Dictionary.* http://www.yourdictionary.com/novelty (accessed November 27, 2012).
2. Chesbrough, Henry W. *Open Innovation.* Boston: Harvard Business School Publishing Corporation, 2003.
3. Rogers, Mark. "The Definition and Measurement of Innovation." *Melbourne Institute Working Paper 10/98,* 6, 1998.
4. Drucker, Peter. *Innovation and Entrepreneurship.* New York: Harper Collins Publishers, 2006.
5. Goldstein, Burton. http://www.entrepreneurship.org/en/resource-center/a-guide-to-druckers-systematic-innovation.aspx (accessed November 28, 2012).
6. Wikipedia. "Behavioral modernity." http://en.wikipedia.org/wiki/Behavioral_modernity (accessed November 29, 2012).
7. "Human." *Wikipedia.* November 28, 2012. http://en.wikipedia.org/wiki/Human (accessed November 29, 2012).
8. Michalko, Michael. "How Geniuses Think." *The Creativity Post.* April 28, 2012. http://www.creativitypost.com/create/how_geniuses_think (accessed November 29, 2012).
9. Wikipedia. "Scientific method." http://en.wikipedia.org/wiki/Scientific_method (accessed November 29, 2012).
10. Silverstein, David, Samuel, Philip, and DeCarlo, Neil. *The Inventor's Toolkit.* Hoboken, NJ: John Wiley & Sons, Inc., 2009.
11. Wikipedia. "Scientific method." http://en.wikipedia.org/wiki/Scientific_method (accessed November 29, 2012).
12. Gelb, M. and Caldicott, S. *Innovate Like Edison.* New York: Dutton, 2007.
13. Wikipedia. "Brainstorming." http://en.wikipedia.org/wiki/Brainstorming#Osborn.27s_method (accessed December 2, 2012).
14. de Bono, Edward. *Six Thinking Hats.* Boston: Little Brown and Company, 1985.
15. Burkhard, Peter. *Burkhard Works.* n.d. http://www.burkhardworks.com/pdr.htm (accessed December 2, 2012).
16. Brainstorming. November 27, 2012. http://en.wikipedia.org/wiki/Brainstorming#Osborn.27s_method (accessed December 2, 2012).

17. Ibid.

18. de Bono, Edward. *Six Thinking Hats*. Boston: Little Brown and Company, 1985.

19. Wixson, James. *Function Analysis and Decomposition Using Function Analysis Systems Technique*. September 10, 2008. http://www.srv.net/~wix/622_p113.pdf (accessed March 11, 2007).

20. Wikipedia. "Kano Model." http://en.wikipedia.org/wiki/Kano_model (accessed December 11, 2012).

21. Johansson, Frans. *The Medici Effect*. Boston: Harvard Business School Press, 2006.

22. Kepner, Charles H. and Trego, Benjamin B. *The Rational Manager: A Systematic Approach to Problem Solving and Decision Making*. New York, NY: McGraw-Hill, 1965.

23. Hauser, J. and Clausing, D. *The House of Quality*. May–June 1988. http://www.csuchico.edu/~jtrailer/HOQ.pdf (accessed December 11, 2012).

24. Google. "Failure Mode Effects Analysis." http://en.wikipedia.org/wiki/Failure_mode _and_effects_analysis (accessed December 11, 2012).

25. Wikipedia. "Evolution of human intelligence." http://en.wikipedia.org/wiki/Evolution _of_human_intelligence (accessed November 29, 2012).

26. Wikipedia. "Moore's law." http://en.wikipedia.org/wiki/Moore's_law (accessed December 1, 2012).

27. Most people know Alexander Graham Bell invented the telephone but few know that Elisha Gray who cofounded Western Electric in Chicago actually filed what is today a provisional patent for the telephone hours before Bell filed his patent. However, after legal wrangling Bell won out. Many still believe that Gray actually invented the telephone first. (Interesting note: Gray was the great grandfather of Elisha (Bud) Gray II who was an MIT graduate that later worked at the Whirlpool Corporation from 1938 to 1977. He became the company president and later chairman for 13 years. Bud was a wonderful person, whom I had the honor to know. He contributed much to create the company's success.)

28. Meyer, C. *Fast Cycle Time*. New York: Simon & Schuster, 1993.

29. Hall, Doug. *Jump Start Your Business Brain*. Brainbrew Books, n.d.

30. Katz, Robert L. *Skills of an Effective Administrator*. Boston: Harvard Business Review, 1974.

31. Cooper, R.G. *Winning at New Products*, 4th ed. New York, NY: Basic Books, 2011.

32. Google. "Stage-gate process." http://www.google.com/search?q=stage+gate+process &hl=en&tbo=u&rlz=1C1AFAB_enUS453US453&tbm=isch&source=univ&sa=X&ei= GRDGULzOHIaW8gTGy4HADQ&ved=0CEkQsAQ&biw=1366&bih=643 (accessed December 10, 2012).

CHAPTER 29

Eureka! What Insight Is and How to Achieve It

Andrea Meyer

Introduction

The flash of lightning, the shout of "Eureka!," the *aha* moment—we've all heard descriptions of creative achievements and new idea breakthroughs that feature these magic moments. No doubt you've had such an experience yourself, when an answer just pops into your head. But what precisely occurs when we get this flash of insight? What happens that brings it about? What *is* this *aha* moment and how can we encourage these insights to come when we need them, like to solve a problem or come up with an innovative idea?

The purpose of this book is to explain insight. It is written for people who don't want to simply be awed by the achievements of others, but who want to know how to produce such achievements themselves. By the end, you'll understand not only *what* insight is, but *how* to get insights to happen to you. You'll be able to answer:

- What happens when you get an insight?
- How do insights form?
- How can I get insights when I need them?

Many people assume that it's impossible to take steps to get insights. Insight has been characterized as a mysterious, inaccessible event—something that just happens to you or that just struck you, but not something that you could actively bring about.

Now, however, research in cognitive science, psychology, and neuroscience is letting us piece together what insight really is. This book integrates the latest research in these fields and presents a theory of what goes on in your brain when you get an insight. Then it provides specific steps that you can use to get insights when you need them.

Myths about Insight

To understand insight, let's first debunk some long-held myths. Popular stories of insights by luminaries like Newton, Darwin, Dostoyevsky, and Mozart recount how an amazing insight arrives seemingly out of nothing—the proverbial bolt from the blue—or is inspired by a single trigger event that, once seen, creates the whole and complete insight.

The danger of these myths arises when we compare these stories of insights to our own insights. For most of us, moments of such complete and whole insight are rare. We'd probably characterize our own insights as being partial or half-baked. Sometimes they're no more than fleeting feelings. So, the notion that some people have brilliant insights makes it seem that they have something innate that we ourselves lack. But the truth is that the stories are misleading—upon deeper inspection, you see that the truth of the insight wasn't a flash. And, more importantly, because it's not something innate, we can learn how to do it, too. In this first section, we'll debunk the myths:

- The *complete and whole* myth that an insight arrives complete and whole, out of nothing

- The *single trigger* myth that a single event instantly crystallizes the creative breakthrough

In each case, we'll explore the mythical version of the insight, the origins of the myth, and the reality behind the myth.

Dostoyevsky and the Complete and Whole Myth

The first myth is that insights arrive complete and whole out of nothing. Debunking this myth requires only looking deeper into some of those myth perpetuators. For example, Fyodor Dostoyevsky—Russian author of highly-acclaimed books like *Crime and Punishment* and *The Brothers Karamazov*—seems at first glance to support that insights arrive in a finished state. In a letter Dostoyevsky wrote to a friend, he said that an author or poet "is not a creator, because certainly a creative work comes suddenly, complete and whole, finished and ready, out of the soul of the poet." In this view, an insight comes to the creator full-blown, as a coherent whole. It is not preceded by anything in particular but just arrives, and the creator is almost just a transcriber.

But did Dostoyevsky understand his insight process?

Instead of accepting what Dostoyevsky said, let's look at what he did. In particular, we can look at Dostoyevsky's notebooks, his work in progress. For example, Dostoyevsky kept notebooks as he wrote his novel, *The Idiot*. The novel dramatizes the influence of one man on the members of several noble families in 19th century Russia. Dostoyevsky divided the novel into parts, and his notebooks show that he began by listing numerous economic and emotional relationships among the characters, trying to weave those relationships into a novel. Rather than arriving complete and whole, the novel actually involved working on a lot of pieces and parts.

Nor was Dostoyevsky's novel finished and ready once he had the parts. From his notebooks, we see that Dostoyevsky worked on a total of eight different plans for just the first part of the novel. And each of these plans was full of sidesteps and alternative paths. Dostoyevsky was not sure where the correct direction of his novel lay, and he tried many different options. For example, we see him label one passage in the plan as "the main point" but then say "or else…." A passage in the fourth plan is called "the chief idea of the novel" but the next entry is "well now, this opens up a new path. What is to come now?"

In contrast to Dostoyevsky's comments about writing, his notebook entries show us a much different view of his creative process. We see that the ideas and relationships in *The Idiot* did not spring to Dostoyevsky complete and whole, finished and ready. Rather, we get a more accurate and realistic picture of Dostoyevsky's hard work and meandering progress while developing the novel.

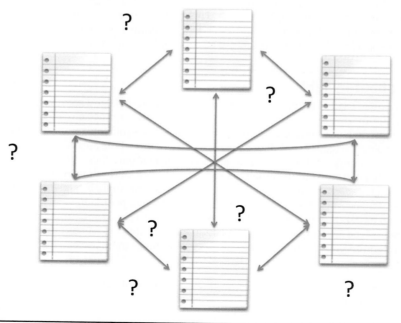

FIGURE 29-1   Weaving several stories into one.

Dostoyevsky didn't simply receive a vivid insight one day that let him sit down and crank out a novel. Rather, he had to carefully work out the details, test different paths, see the implications, overcome challenges, and refine which alternatives worked the best. In the end, he had a coherent whole—but the novel did not begin that way.

Newton, Darwin, and the Single Trigger Myth

The second myth about insight is that a single trigger event magically spawns the creative accomplishment. In this myth, a single event plays the culminating role because it causes all the other pieces to fall into place. The most often-cited example of this myth is that of Newton's apple. Some popular creativity books describe Newton's serendipitous sighting of a falling apple as the key that unlocked his theory of gravitation. As one book puts it, "Suddenly everything falls into place when a sudden new insight pops into the mind. After long years of study, one day in 1685 [sic] Isaac Newton, seeing an apple fall, produced his law of gravitation."

How did the Newton myth come about? It has some basis in truth—Newton did see an apple fall—but what role did this event play in Newton's thoughts? The apple, it turns out, garners undue attention in stories and biographies because of its story-worthiness, not its accuracy.

Biographers, such as Richard Westfall, counter the apple myth by pointing out that Newton didn't produce his theory on his first try. Westfall says that the story of the apple "vulgarizes universal gravitation by treating it as a bright idea....Universal gravitation did not yield to Newton at his first effort. He hesitated and floundered, baffled for the moment by overwhelming complexities." Newton didn't just happen upon the laws of universal gravitation in the orchard. He pondered them for a surprisingly long time after the apple dropped. Newton saw the apple fall in 1666, but he didn't publish the law of gravitation until 1687, more than 20 years after the apple event. During those intervening years, Newton wrote: "I keep the subject constantly before me, and wait till the first dawnings open slowly, little by little, into a clear and full light." Newton's insight and achievement was made possible by an active, ongoing process, not a single fortuitous event. We'll look in more detail later at the role the apple played, and how you can use the same principles to get your own insights.

The final example is that of Charles Darwin, who, struggling to explain how evolution occurred, purportedly got his key insight about survival of the fittest while reading an essay by Malthus on the inevitability of human population booms and disastrous crashes. The originator of the myth is none other than Darwin himself, who wrote in his 1887 autobiography:

> In October 1838, that is, 15 months after I had begun my systematic enquiry, I happened to read for amusement Malthus on population and being well prepared to appreciate the struggle for existence which everywhere goes on from long-continued observations of the habits of animals and plants, it at once struck me that under these circumstances favorable variations would tend to be preserved, and unfavorable ones to be destroyed.

In reading Malthus' discussion of human population pressures and the struggle for existence, Darwin saw that competition among living things would lead to a selection process. The fittest would survive. But did Malthus' essay actually produce a sudden insight that struck Darwin at once? When Darwin wrote his autobiography in 1887 (almost 50 years after his discovery), he certainly suggests that the insight was sudden. But Darwin also wrote near-daily entries in his notebooks while he was in the throes of developing his ideas. Darwin's notebooks from 1838 show us a much different picture of his insight process.

When we look at Darwin's notebook entries about his reading of Malthus, the first thing that we are struck with is that the entry looks no different from his other entries. There's no boldface "eureka!" entry on the day he read Malthus. Second, Malthus' essay didn't even fully capture Darwin's imagination at the time. Instead, Darwin continues to write about other lines of thoughts that he was pursuing at the time. The third noteworthy observation is that Darwin took more than a month before he wrote the passage that succinctly summarized his theory. This delay shows that he had to do more thinking and working even after he read Malthus— his insight did not immediately fall into place upon the reading. The Malthus essay certainly influenced Darwin's thoughts over time, but it didn't produce the theory in a flash.

Technique #1: Emulating the Insightful

The insidious side of these two myths is that they characterize the creator as a passive recipient of the insight. It's not the creator doing the work, but some muse gifting the idea to the creator. Whereas the flippant or misattributed statements of the famous make insight seem mysterious and unattainable by mere mortals, the notebooks of these same great inventors, scientists, artists, and writers reveal the true sources of insight in the methods they used. Rather than thinking we can't do as they say, we can learn to do as they do. Dostoyevsky and Darwin kept notebooks of their works in progress that we can examine. Newton likewise described the nature of his thinking as keeping the topic continuously in mind until he worked out the theory.

Like these creative stars, you can increase your insight by taking steps such as:

- Keep a notebook (on paper or online) of your ideas, observations, and readings. Make daily entries just like Darwin did.

- Sketch or list the relationships between the players, parts, or components just like Dostoyevsky did.

- List alternative solutions, play with them, and experiment.

- Comment on your own comments, work through the implications, and reflect on your work.

- In short, immerse yourself in the work rather than assume the insight will drop in your lap in a eureka moment.

The Process of Insight

Having described what insight is *not*, let's turn to what insight is. Three sources of evidence—research using chimpanzees, analysis of how people solve insight problems, and further study of history-making insights—give us carefully controlled data about when insight occurs and what produces it. This lets us define insight and start to understand the process of insight.

The Role of Experience in Insight

In a famous set of experiments on cognition, German psychologist Wolfgang Kohler placed a banana just out of reach of a chimpanzee in a cage. He also placed within view a stick long enough to reach the banana. Typically, the chimps would first try to reach the banana with their hands, fail, and then would sit back for a while. Then, in an "insightful" moment, they would use the stick to retrieve the banana. Kohler proposed that the chimps saw the stick and the banana and put the two together in a moment of insight. But was Kohler right?

Experimenter Herbert Birch saw a flaw in Kohler's experiments. Kohler had used chimps captured from the wild, not chimps raised from birth in captivity. This made a difference, Birch reasoned, because Kohler didn't know what role experience played in the insightful behavior. For example, had the chimps used sticks in the forest to reach things or had they seen other chimps using sticks?

FIGURE 29-2   The monkey and the banana.

So Birch repeated Kohler's experiments, but this time using only chimps raised in captivity. Like Kohler, he put the banana and the stick arranged within sight, but this time the chimps showed no insight concerning how the stick could be used as an extension of their arms. What explains the captive chimps' lack of insight? Birch explained that Kohler's chimps' insight was actually based on specific past experience with sticks that they had had. They had probably used sticks to reach bananas or extend their reach in the past. In contrast, Birch's chimps had no such prior experience, and they were unable to reach the insight. A further experiment added more evidence: once Birch gave the captive chimps opportunities to handle sticks in daily life, they, too, were able to solve the insight problem.

The implications of these experiments are that **experience, involvement, and interaction with the problem's elements help us get insights**. In his book, *Theories of Creativity*, Robert Epstein prescribes "nonalgorithmic interactions with the cognitive and physical materials of the projects" as necessary antecedents of insights. In other words, Epstein recommends playing around with the ideas and objects. *Nonalgorithmic* means that you try actions whose results you can't predict for certain, whose results you don't know. You test things out to see what will happen. You let yourself be surprised. This could mean physically manipulating objects if that's possible, such as in a lab experiment or with artistic media. Or, if you're working with ideas, you can create combinations of words or test putting various facts together to see to which conclusions they lead you. Or you can juxtapose words or images in your mind. The main goal of these interactions is to gain an understanding of the elements from different perspectives and viewpoints.

The Role of Recall and Adaptation

A laboratory study from experimental psychology illustrates the process of insight more clearly. Robert W. Weisberg, professor of psychology at Temple University, has been studying the creative process for over four decades and countering the notion that leaps of genius are required. His books *Creativity, Genius and Other Myths* and *Creativity* are excellent further reading on this topic. Here, we'll examine a specific laboratory study that Weisberg conducted to investigate insight. Weisberg refers to it as the "candle problem." In this experiment, subjects are given a candle, some thumbtacks, and a box of matches. They are told to attach the candle to the wall. See Fig. 29-3.

Weisberg also asked the subjects to think out loud as they were solving the problem. That is, asked them to voice what they were thinking. He didn't ask them specific questions, but just had them verbalize what they were thinking as they worked on solving the problem. Before reading further, you may want to take a moment to think how you would solve this problem.

Your Solution

Weisberg's results: Weisberg found that everyone first thought of attaching the candle directly to the wall with the tacks, and then discovered that this solution didn't work. (The tacks were not long enough.) Next, the subjects considered variations on these attaching solutions that involved platforms of tacks. Finally, they incorporated the box itself into the solution. This approach led to a workable solution: using the box as a platform on which to

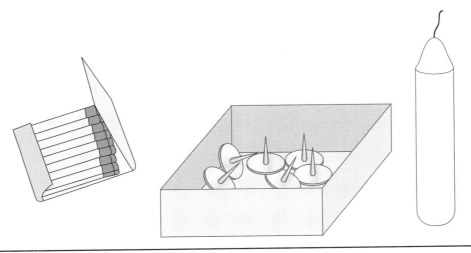

FIGURE 29-3   (a) Candle problem.

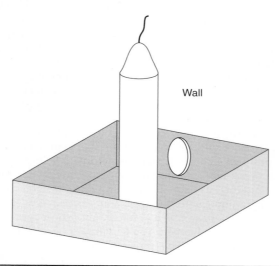

Wall

FIGURE 29-3   (*b*) Candle problem.

put the candle and attaching the box via tacks to the wall. Once the subjects had decided that what they needed was some sort of platform for the candle, they noticed the box and realized that it could serve as a platform.

This experiment shows that if you see only the initial problem situation and the final solution, it's hard to see where the solution came from. It appears "insightful." Yet when you hear the problem solvers' thoughts as they search for a solution, you learn the intermediate steps they took to reach the solution, and you see that each step was directly based on what preceded it. Specifically, Weisberg found that all subjects, whether in the end they found a solution or not, started at basically the same place: that of attaching the candle directly to the wall. When they discovered that one solution didn't work, they tested other solutions. Those who arrived at defining the problem as "I need a platform on which to put the candle" eventually saw that the box itself could be used. People didn't consider the box as a tool at first—they just saw it as something the matches came in. Once they had the goal of finding a platform for the candle, they realized that the box could be used as a platform.

This experiment illustrates the role of **recalling information from memory** in the insight process. The candle problem was based on recalling what you knew about sticking things to walls and adapting that knowledge to the current situation. Because this was a new problem, simply reusing a previous solution wasn't possible. Rather, insight required recalling information that could be useful to a solution (e.g., a shelf or platform can hold something on a wall) and adapting it to fit the new situation (e.g., a matchbox can be used as a platform).

The process probably went something like this: You've likely never had to affix candles to the wall, but you've stuck others things on walls, and you've probably used tacks. So when given the task of attaching the candle to the wall, your first step was to search your memory for ways of attaching things to walls. You matched the tacks available now with the instances in memory of using tacks to attach things to the wall. Or perhaps you were reminded of the notion of gluing, that gluing things makes them stick to something, and you connected that knowledge with your knowledge that candles melt into a sticky wax substance. You may have tried melting part of the candle and using the wax as a glue. Again, it was a process of retrieving some information from your memory and matching it to the problem at hand.

If you tried these solutions and they didn't work, you may have used the new information to modify what the solution might be instead. That is, you had to go beyond what you knew and modify or adapt partial solutions. You may have tried a row of tacks as a platform and thought of the attribute of platforms that would be useful: the fact that platforms are flat. This attribute of "flatness" led you to notice the matchbox, which had a flat surface, and to realize it was thin enough to attach to the wall with tacks. Or maybe you bypassed these steps because you recalled an image of candle holders in old-time castles and sought something to function as a holder immediately. Whatever specific memories you used, you probably did recall something you knew and adapted it to the problem. How quickly the process went depended on your familiarity with the task or task elements and their properties.

The Role of Curiosity, Questions, and Goals

Curiosity is a trait often found among creative achievers. Curious people spend time exploring and asking. A curious entomologist observes the behavior of ants and explores what happens when he drops a bit of food into the ant colony or blows on it. He's usually posing a "what happens when I do this?" question and observing the results.

These types of explorations are called *play* by some and *nonalgorithmic interactions* by others. As defined earlier, *nonalgorithmic interaction* means that you can't predict exactly what the result of your action will be. Adding row after row of *1 + 1 =* won't give you much insight because the result is always the same and predictable. But what would you learn if you added successively increasing numbers? That type of interaction can bring you new information.

Successful scientists, artists, and other creators spend a lot of time in nonalgorithmic interaction. They explore the attributes, properties, and characteristics of what they work with. Metallurgic scientists study the behavior of metals under different conditions; artists mix colors or try new actions on media; poets play with words. Moreover, these creators not only explore individual properties, but they explore combinations and relationships as well. Painters can explore not just colors and blends of colors, but their combination with different glazes, surfaces, placements in relation to each other, gradations of luminescence, and so forth. Programmers test what a line of code produces, what combinations of lines do, how changing one aspect of an equation changes the results, and so on. Poets play with sounds, syllables, cadence, double meanings, and combinations of words. New product developers explore attributes of a product and combine them, or they think of the next higher level of abstraction in order to find other ways to accomplish the goal that the product satisfies.

When you engage in nonalgorithmic interactions, you're asking yourself questions all the time. Questions like, "What happens if I do this?" "What happens if I add that?" "How strong is this material?" "How far can I change this factor before it all collapses?" and so on. Your questions are both attribute-based (to discover the properties of the objects or ideas) and goal-based. Goal-based questions include, "Where else can I use that?" and "What problems could this solve?"

Douglas Hofstadter, in *Metamagical Themas*, describes the benefits of this prolonged contact and interaction with your topic: "The more different manifestations you observe of one phenomenon, the more deeply you understand that phenomenon, and therefore the more clearly you can see the vein of sameness running through all those different things. Or put another way, experience with a wide variety of things refines your category system and allows you to make incisive, abstract connections based on deep shared qualities."

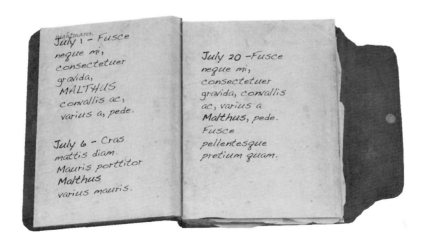

Darwin's Competitor: Parallels and Differences

We can learn even more about the process of insight if we look at how two different people—Charles Darwin and Alfred Russel Wallace—created the same great insight. Wallace also developed the theory of evolution based on natural selection, and he, too, read Malthus and ascribed a significant portion of his solution to this reading. At first glance, it would seem that again Malthus was a final click that brought the insight.

But let's read Wallace's account:

One day something brought to my recollection Malthus' "Principles on Population," which I had read about twelve years before. I thought of his clear exposition of the 'positive checks to increase'—disease, accidents, war, famine—which keep down the population of savage races to so much lower an average than that of more civilized peoples. It then occurred to me that these causes or their equivalents are continually acting in the case of animals also; and as animals usually breed much more rapidly than does mankind, the destruction every year from these causes must be enormous in order to keep down the numbers of each species, since they evidently do not increase regularly from year to year, as otherwise the world would long ago have been densely crowded with those that breed most quickly. Vaguely thinking over the enormous and constant destruction which is implied, it *occurred to me to ask the question, why do some die and some live? And the answer was clearly, that on the whole the best fitted live.* [italics added]

Did you notice that Wallace did not get his insight at the time he read Malthus? Indeed that he read Malthus some 12 years before he formed his theory? If the Malthus' essay was so powerful, why hadn't Wallace gotten the insight the moment he read it? The answer is that single pieces don't trigger insights.

At the time that Wallace read Malthus, he had not yet been on a naturalist voyage and had not seen the tremendous variation of species like Darwin had. Even more crucial, Wallace didn't just go on a naturalist's voyage—he visited an archipelago. Archipelagos are key in this case because an archipelago's small scale makes evolution's action more clearly visible. Archipelagos provide a unique opportunity to see the variation in species. Thus, exploration of archipelagos was crucial for both Wallace and Darwin.

Another noteworthy fact is that Wallace finalized his insight while on his voyage. Seeing the archipelagos stimulated him to think about evolution, and it was then that he recalled the Malthus' essay. He recalled Malthus only after his thinking about evolution had matured to the point that he understood how the Malthusian principle fit into his theory. This example illustrates the way in which novel ideas may be forgotten until the structure of which they are to become a part is sufficiently complete.

Comparing Darwin and Wallace, we see that Darwin had made his voyage early in life, pondered about evolution, later read Malthus, recalled his voyage, and crystallized his theory. In contrast, Wallace read Malthus first, and then went on a voyage, pondered evolution, recalled Malthus, and developed his version of the same theory. Both men made a synthesis between something immediate and something remembered.

Insight Defined

What do all these studies and historical examples tell us about the nature of insight? They show that **insight is a linking or connection between ideas in the mind. The connections matter more than the pieces.** Both Darwin and Wallace needed all the pieces and needed to spend time thinking about them to work out the implications.

Insight is a phenomenon of connection in memory. That is, during the process of insight, we make a connection between something we know and the current situation. **Insight depends on recalling a piece of information from memory and adapting it to the current situation.** Both Darwin and Wallace formed their insights on evolution from the connection between the natural forces that create variations among creatures like those seen in archipelagos and the Malthusian forces that winnow down those populations. They both sought a connection between the idea of evolution and the need for some mechanism to explain it.

Now that you've got the basic idea of insight, let's make the theory more succinct through cognitive science terminology. In these terms, an insight occurs when you get a

Darwin	Wallace
Going on voyage to archipelago	Reading Malthus
Reading Malthus	Going on voyage to archipelago
Coming up with theory of evolution	Coming up with theory of evolution

TABLE 29-1   Parallels between Darwin's and Wallace's Paths to a Theory of Evolution

reminding (you recall something) from memory and match it with the current situation. You may retrieve a partial match that you can then adapt to work as a solution. (More discussion of these terms and the details behind them can be found in books by Roger Schank, especially the one entitled *The Creative Attitude*.)

Technique #2: Making the Creative Connection

Insight usually arises from a connection of ideas gathered over time and adapted to achieve some goal. To create insights, you need to

1. Have a need, goal, or curiosity to look for some explanation that's not readily apparent

2. Accumulate knowledge and experiences to recall and work with (whether it's sticks to reach a banana or math to compute an orbit) so that you have elements to connect

3. Reinterpret or adapt things you know to fit the new situation or purpose

How to Get More Insights

Making the creative connection deserves more explanation, so we'll delve into that next. The three steps in technique #2 can be expanded into more steps. Technique #3, for example teaches you the types of questions you can ask to help remind you of ideas, solutions, and experiences that might lead to insight into a problem. Technique #4 teaches you how to learn and remember more effectively by indexing ideas in multiple, more remindable ways. Technique #5 teaches you think about partial matches and how to adapt them.

With each technique, we'll have some practice exercises at the end. The exercises will include a hypothetical scenario and some answers—that is, you'll see some possible topics for questions. The answers provide a reference and comparison point, but they are not the only answers or necessarily even the best answers. Rather, they're intended to give you a flavor of different possibilities.

Technique #3: Using Questions to Find Insightful Answers

The cognitive description of insight now leads us directly to techniques we can use to get insights. **Questions** help us gain insight in three ways. First, questions play a crucial role in curiosity. Curiosity manifests as a set of questions: why did X happen? How could we improve it, where else can it be applied? Questions help us frame the problem to focus the insight process on particular attributes or goals. Finally, questions also create remindings—remembering some idea or experience from the past, like Wallace remembering Malthus' essay 12 years later. We usually let remindings occur to us naturally and unconsciously, but you don't have to rely on the whims and imprecision of an unconscious process. By taking some conscious control, you can look for remindings directly.

What Does Newton's Apple Really Mean?

Let's look back at Newton and explore what role the apple really played. Instead of accepting flowery descriptions and miraculous productions, we can learn more about insight by examining what went on in Newton's mind when he saw the apple fall. We have some clue of Newton's thoughts from *Memoirs of Sir Isaac Newton's Life*, written by his personal friend Dr. William Stukeley:

> After dinner, the weather being warm, we went into the garden and drank thea [sic], under the shade of some apple trees, only he and myself. Amidst other discourse, he told me he was in just the same situation as when formerly the notion of gravitation came into his mind. It was occasion'd by the fall of an apple, as he sat in a contemplative mood. Why should that apple always descend perpendicularly to the ground, thought he to himself. Why should it not go sideways or upwards, but constantly to the earth's centre? Assuredly, the reason is that the earth draws it.

From this example we see that the falling apple didn't so much spark an answer as spark a question. The event became valuable because Newton followed it up with questions. He saw the apple fall and began to wonder why apples always fall perpendicularly to the ground, why not sideways or upwards.

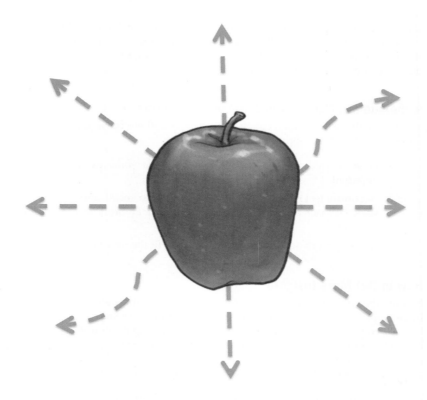

What were the attributes of the apple, the earth, and other heavenly bodies that caused them to move as they did? It was these questions, and Newton's persistence in finding some explanation for them, that led to his creative achievement. Newton was curious about why apples fall perpendicularly to the ground, and he began speculating about possible reasons for it and testing out those explanations and their implications for other phenomena. Newton's process was that of questioning and of forming explanations. Stukeley writes of Newton's progress in pursuing the question of the apple's fall to the insight of gravity:

> Therefore the apple draws the earth as well as the earth draws the apple. That there is a power, like that we here call gravity, which extends its self thro' the universe. And thus by degrees he began to apply this property of gravitation to the motion of the earth and of the heavenly bodys [sic], to consider their distances, their magnitude and their periodical revolutions; to find out, that this property conjointly with a progressive motion impressed on them at the beginning perfectly solv'd their circular courses; kept the planets from falling upon one another, or dropping altogether into one centre; and thus he unfolded the Universe.

How to Do It
You can improve your insights by using questions to explore a problem and to remind you of other related key ideas or memories. Because insight depends on recalling a piece of information from memory, the first task is how to recall or get reminded of something that can help you when you're faced with a problem or task. The best way to get reminded is to pose a question to yourself that will retrieve the information you seek. For example, Newton asked "why does the apple fall straight down?" and Wallace asked, "why do some die and some live?"

There are three types of questions that you can pose to yourself: questions based on attributes, questions based on abstract goals, and questions based on time or place—"eras" for short.

Ask Attribute-Based Questions
Questions based on attributes are ones in which you look for a specific attribute of an object or idea. For example, when trying to mount a candle to the wall, the flatness of a platform was the attribute that reminded us of the flatness of the matchbox. Similarly, if you needed to pound a nail into the wall and didn't have a hammer, you might look for objects that have the attributes *graspable, heavy, and hard.*

Solid objects are not the only ones that have attributes—ideas have attributes as well. For example, when Darwin was looking for a mechanism that explained species variation, he was looking for an idea that would explain another idea. But he had definite attributes in mind that a potential solution had to possess. One attribute was *incorporates adaptive variation, reproduction, life-span, and changing conditions of life*. Newton looked for something that would explain the motions of objects like apples and planets. Other scientists might look for theories that have the attribute *parsimonious* or equations that are simple and elegant. A poet may look for the attribute *rhymes with nevermore*. Having these attributes in mind helps you to recognize a potentially good idea because if you encounter an idea that possesses these qualities, it will get your attention and you will take a closer look at it.

Another advantage of looking at the attributes of an object independently of the object itself is that it enables you to see a quality of the object that's usually embedded in the object and thus not recognized. For example, you may not consider that your door is also a piece of wood, flat, taller than a person, water resistant, and opaque.

Ask Goal-Based Questions

On the other hand, there are times when you want to go beyond known attributes to consider how else you could solve the problem. In that case, goal-based questions work best because they pose the end goal without specifying the means or locking you into particular attributes. "Design something to keep liquid hot and handy" gives you more options than "design a coffee cup" because the latter conjures up specific images of the shapes and materials used in coffee cups that may limit you to exploring only coffee cup designs. A more general goal might help you think about insulated soup bowls, thermos bottles, and even a thick padded bag with a spigot. Goal-based questions are also useful when you don't know what type of attributes would fit the solution and thus can't specify the attribute.

To ask goal-based questions, phrase your question in more general terms, based on the goal you are trying to accomplish: "something to pound with" or "something to set the candle on." Goal questions can lead to attribute questions. The time to use a goal question is when you want to expand the number of options you have. Goal questions are broader. "Staying warm" has more ways of achieving that goal than "lighting a fire." In this way, goal questions are also a good way of getting unstuck.

The advantage of attribute questions over goal questions is that attribute questions are more specific and concrete—attribute questions make good search images because you're searching for one or two key things. In the candle problem, the question, "what can I use that's flat?" gives you a specific search image that helps you identify a matching solution. You look for something that's flat. This can lead you to a faster solution because of the specificity of the search image.

FIGURE 29-4   Hot and handy design.

Figure 29-5 Chairs.

Ask Era-Based Questions

If you encounter a situation in which neither attribute questions nor goal questions work because the area you're thinking about is just too new to you, then there is a third type of question format you could try: era-based questions. To do an era-based reminding, you put yourself in the position of thinking about a question in a different time or place from the one you are currently in. For example, instead of asking yourself the general question, "have I ever seen a situation like X before?" ask yourself if you have ever experienced anything like X when you were in a different town or when you were in school. Specifying a question in this way changes your point of view considerably, and thus lets you recall items that are related, but are related in non-obvious ways. The era-based aspect of the question aids you in recall because it provides an additional hook to search by.

Let's work through an example: chairs. It's easy to imagine a chair and then say some things about what kinds of chairs you like or don't like. Now, let's add in an era-based question: what kinds of chairs did you experience in college? At the airport? When you were a child? In trying to answer these time or place-oriented questions, you're recalling more and more examples of chairs, probably examples that didn't come to mind initially when you simply listed features of chairs randomly. These questions function mainly as triggers to get you thinking about the problem in different ways. They're not specific and they're not solution-based, so they won't lead you directly to an answer; but they will get you started and they will help you recall non-obvious relations.

To summarize: posing the right question to yourself brings to mind ideas you know, which you can then apply to new situations.

Practice Exercises

Let's work through some exercises to practice asking questions that can remind you of other ideas, experiences, or solutions to a problem. Imagine you are tasked with innovating air travel. To start, list some attribute-based questions.

Your Solution

Some possible answers: Your answers might take many forms and lead you in different directions. For instance, you might consider utilitarian attributes for air travel such as speed, schedule, cost, and reliability. You might ask, "what else has speed?" Asking questions about these attributes and other objects, people, or companies that have these attributes might remind you of other ideas, experiences or companies that also have attributes of speed, schedule, cost, and reliability. From that you might create insights into airline operations that reduce waiting times, eliminate unneeded costs, or avoid potential problems that create unscheduled delays or canceled flights. Or you might consider questions about attributes of the ideal flying experience, such as restful silence, entertainment, adventure, or luxury. Asking questions about these attributes to remind you of related ideas or experiences might help you create insights into the personalities of the flight attendants you hire, the interior of the

Figure 29-6   Air travel.

plane, the in-flight entertainment, or other amenities. Or you might think of the attributes of air travel besides the flight itself that negatively impact people's total experience of air travel, such as long concourses, security hassles, crowded gates, or uninviting baggage claim areas. These questions might lead to insights in selecting airports for your airline, lobbying the government to change security procedures, or supporting airport improvement projects.

Next, think of some goal-based questions. Of course, these questions could relate to the attributes such as goal of low cost or high reliability, but try to think of some higher-level goals for why people travel.

Your Solution

Some possible answers: The core goal of air travel seems to be to arrive at a particular place by a particular time. That is, many travelers have a specific place-and-time goal such as an 8 am business meeting at a client's office or being with your family for Thanksgiving dinner. Perhaps this goal question might help you see how the goal of air travel is much like the goal of a FedEx delivery (a package at a particular place and time) or even the goal of a live theater performance (all the actors in their particular places and times). Or one might realize that being early is fine but being late is unacceptable. Even the appearance of a plane being potentially late creates stress in travelers, so the goal is to never seem to be running behind schedule.

You might even take a step back from this goal and realize that the specific time-and-place goal doesn't apply to all travelers. Some people just want a week on the beach in January and don't care which airport or exactly what day they fly. Or, someone might be planning a European holiday and not care whether they fly first to Amsterdam, Paris, Frankfurt, or any of a range of cities on their intended list of cities to visit. You might realize that these people hate the standard flight reservation process that first asks for the specific destination airport and travel days.

Finally, think of some era-based questions around the topic. Era-based questions are a rich source of insights for the air-travel innovation problem because air travel is an experience that many of us have had, read about, or heard about from family and friends.

Your Solution

Some possible answers: Everybody's answers will be different and will include both horror stories (surly flight attendants, capricious cancellations, idiotic delays, uncommunicative airline staff, etc.) and delightful stories (surprise upgrades, staff that went the extra mile, great seat mates, generous compensation for a problem, etc.). To get even better questions yielding better insights, you might expand the era questioning process. Think about the difference in your own travels for pleasure versus business. And what about the differences in travel when you were a child, young adult, older adult, or a parent? Or you could expand the experience questions to travel by other modes. What was it like to travel by car, bus, train, or boat?

Technique #4: Indexing Ideas for Better Reminding

You may be wondering how to optimize your ability to get reminded and recall experiences, facts, and ideas in flexible ways. The way to do this most effectively is to index the information you know and as you learn. We file things into memory all the time, but we usually do it unconsciously.

In this section, you will learn how to index. The value of consciously controlling how you index is that you'll increase the amount of information and experiences you can recall and connect to create more insights. Moreover, you'll be able to recall it flexibly, in many different ways. This is important because the more of these remindings that you have, the more possible solutions you have available to use in a problem situation. Creativity thrives on such possibilities.

When you consider what insight really is, you realize that the heart of insight is the ability to recall partial solutions and adapt them to a new situation. Indexing helps you recall more potential solutions. (In the next section, we'll explore how to adapt those partial solutions.) Most people aren't as insightful as they could be because they can't access all the partial solutions they might have hiding somewhere in their brain. Indexing will provide you with that access.

So, what is indexing? Indexing is providing a tag for a fact, piece of information, or experience, so that you can retrieve it when you want it. It's like an index in the back of a book or on a blog which gets you to the page that has the information you seek. What types of things should you index? You'll want to index what you want to access, which means, first of all, indexing attributes and goals.

Index Your Experiences by Attributes

The first step in indexing attributes is to identify what the attributes of an object, situation, or idea are. For example, we can return to the candle problem and think about the attributes of a candle. A traditional candle is made of wax, which has the property of melting and also the property of coming in different shapes or being able to change shape. Wax also floats and can leave a water-resistant mark on paper (think crayons). Another attribute of the candle besides the wax is the wick. The wick is like a piece of string, and it burns. If you needed a string, would you think of removing the wax of a candle and using the wick? Another attribute of a candle could be its color. You could think of the candle as a vessel for its color if you needed color (in decoration, for example).

Those are some characteristics of traditional candles. Perhaps other types of candles came to your mind, if you've encountered them, such as floating candles, which are wicks floating in oil, or ear candles, which are intended to help soften earwax. That brings us to what a candle is and what purposes (goals) candles could fulfill. One of the goals that a candle can fulfill is to provide light. Candles also provide some heat, and the flame of the candle could be used to set fire to other things.

Consider what's absent: candles provide light but they don't require electricity. But they do require some external means of lighting them. The attribute of soft, gently flickering light that candles give off can satisfy the goal of imbuing a romantic atmosphere, for example, in a restaurant.

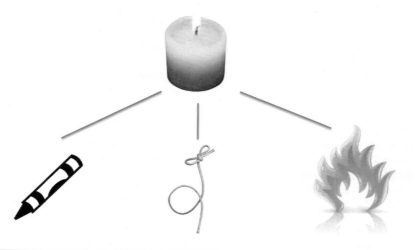

Figure 29-7   Candle attributes.

As you can see, there's interplay between attributes and goals that naturally occurs while you think. Don't worry about whether something should be put exactly in an attribute box or a goal box, because overlap does occur, and our purpose is to get reminded of something useful, whether it is by way of an attribute or by way of a goal.

Index Your Experiences by Goals

To index by goals, ask yourself, "What goal could this example accomplish? When could it be useful?"

Indexing by goals can be tricky, because there are two opposite problems to be aware of. On the one hand, experiences labeled too specifically and narrowly limit their recall for other situations. For example, maybe you heard of a good retail sales technique (such as putting little impulse-buy items near the cash registers) but hadn't considered that it could apply to other sales or negotiation situations as well. On the other hand, solutions that are too general don't include enough detail to indicate when to apply them. For example, the advice "look before you leap" warns you to pause before rushing into something, but it doesn't contain enough detail about when to apply the advice—and when "he who hesitates is lost" would be the more appropriate attitude.

The way to overcome this double-bind of too specific or too general is to keep the details of the experience intact; that is, note that the sales technique was used in a retail store, but then abstract out some of the goals (such as that it was a successful sales technique) so that it can be accessed as a successful sales technique for potentially other situations as well.

Let's look closer at the nature of abstraction by goals. The purpose of indexing on a more abstract level is so that you can recall the information even if surface features don't match at all. Less insightful people tend to recall information that is superficially similar rather than usefully similar because they haven't thought about the information in different contexts or at different levels of specificity. For example, in the retail sales example, they would file it only under retail stores as opposed to considering it on a higher level of abstraction to discover that it could be applied to other sales or negotiation situations (such as a last-minute offer of some extra vacation days if a job seeker agrees to take the current offer of less salary). Nor have they explored the attributes or features of the technique that contributed to its success and mulled over how to incorporate those features into other situations.

You'll notice that the key aspect of indexing is the process of thinking and exploring an object, situation, or idea. Perhaps in the examples described above you have thought the indexing process felt forced or artificial. Who'd want to do such a cumbersome task as indexing? Frankly, there are probably hundreds of times that you've indexed without knowing it: you've done it whenever you were curious about something. Being curious about something means that you wonder about it and want to understand it. You naturally explore it from different sides and angles. For example, why are more people buying from your competitor than from you? Is it because of a price difference? Quality? Service? Does your competitor offer more features or have easier financing terms? Do each of you have a different type of customer? In asking these questions, you've been thinking about attributes and goals simply because you've been curious about a situation. It turns out that a big part of indexing is being curious about things.

Mulling over Your Experiences for More Indexing

Another facet of nonalgorithmic interaction is that of mulling. That is, indexing occurs when we mull over a situation, when we consider what happened and why. Mulling is important because the less you think about something, the less likely you are to remember it. Put another way, if you haven't thought sufficiently about something, you're less likely to remember it. Mulling improves your recall because you remember something better the more completely you have understood it, and that usually involves having considered it from different vantage points. Mulling is a process whereby you turn things over in your mind and explore them from different sides.

We rarely give much thought to events and actions that happen every day. They get lumped into a general, nondescript category. "Got up, went to work, came home, had dinner, watched TV" is often the recital you get when you ask people what they did yesterday. And if that's how they've stored the events of the day, that's the only level of detail they can recall. Greater specificity or meaningfulness requires more attention. The mulling attitude is one of exploration of details (like something unusual happened today), not dismissal of the generalities (today was just another day). For example, some people will dismiss someone's different behavior (such as texting instead of meeting face to face) as "crazy," whereas others will wonder and ask questions about what might lead that person to behave that way. Do they know something that I don't? Have they found a better way? Are they an example of a new type of customer? By wondering, speculating, and investigating, you create more indexes and hooks that can be called up later.

One convenient feature of mulling is that it can be done at any time, and it can be done after the fact as well. You can call up a memory and examine it from some new angle or relate it to other things you know. Mulling after the fact does require, however, that you paid attention to a lot of details, **because you can't recall what you didn't notice in the first place**.

Another time to mull is when you notice a spontaneous reminding—that is, when you spontaneously get reminded of something, ask yourself, "why did that come to mind?" This not only gives you clues to how you've indexed in the past, but it also lets you explore the similarities between the two memories.

Tagging Your Index

Let's return now to consciously directed indexing. Besides attributes and goals, what other types of things would you want to index? What other access points would be useful? Some of the other categories for indexing might include

- Backgrounds or preconditions (what caused a situation)
- Side effects (both positive and negative)
- Results (what happened)
- Trade-offs
- Roles people play
- Problems encountered en route to a goal
- Examples that illustrate a situation (like examples of a good customer mix or a good interview)
- Alternatives
- Warnings or indicators (what to watch out for)

What makes a good index tag? The best tag is distinctive but not unique. That is, to tag a solution "how to settle a conflict between Mary and Tom who both want the same office" shouldn't be the only tag you use because it will restrict its recall to the unique situation of a conflict between Mary and Tom. But the lessons learned from settling the conflict between Mary and Tom may apply to other conflicts as well, so you want a tag that will allow for overlap to other situations as well. The best indices, therefore, are those that are general enough to be applicable in a variety of situations but distinct enough to be recognized. A useful tag to describe the Mary and Tom conflict could be "how to settle a conflict when both parties want the same thing."

Choosing good indices can be difficult because it's not easy to determine beforehand which features will be useful in the future. That's why mulling over various features is worthwhile—you don't have to predecide which feature is the index point because you will find several as you mull.

Exercise: Indexing by Attributes of an Object

What are the attributes of wristwatches? List as many attributes as you can.

Your Solution

Some possible answers: Some of the categories of attributes include:

- color
- analog/digital display
- arrangement of features
- material
- where it is sold
- where it is made
- breakdown/repair
- what you do with it when you want a new watch
- shape
- size
- types of features
- price
- instructions
- battery
- who designed it

Exercise: Indexing by Goals of an Object

What goals does a wristwatch satisfy? Think of the purposes why people buy watches and list as many as you can.

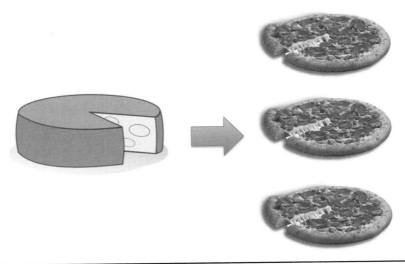

FIGURE 29-8   Cheese pizza.

Your Solution

Some possible answers:

- to know what time it is
- to express your personality
- to mark a special day (anniversary, birthday)
- to fulfill additional functions (date, calculator, worldwide time, alarm, stopwatch, pedometer, altitude meter, GPS, etc.)
- for a specific type of person: child, elderly, blind, scuba diver
- as a piece of jewelry
- to show your wealth

Exercise: Indexing by Attributes of an Idea

You can also index a situation or a clever idea that you encounter. Say you read an article about Ed the entrepreneur who buys surplus cheese from the government and uses it to make cheese pizzas that he sells to school lunch programs. Why is this a clever idea? List the attributes of this idea.

Your Solution

Some possible answers: One attribute is that the source material (the cheese) is cheaper than usual because it is surplus. This means that Ed can offer his product at a lower cost without sacrificing quality. Another attribute is that schools are on tight budgets, so cost is an issue—so Ed has a potential competitive advantage on price over his competitors. Furthermore, school lunch programs are awarded by contract, so Ed knows that he will sell a certain amount of product. (Notice that the attributes of an idea might be more than one-word descriptors.)

Exercise: Indexing by Goals of an Idea

What goals does Ed's idea exemplify? That is, what lessons can be learned from this story? What types of problems could it be a solution to? State some broader goals that this story is an example of. List some ways to index an idea like Ed's.

Your Solution

Some possible answers:

- Run a business without a lot of capital
- Create a source of competitive advantage
- Add value to a commodity product
- Have a low-cost strategy
- Find steady customers

When thinking about the goals and when to apply them, you may want to mull over the other index tags (such as preconditions, warnings, etc.) to help you in understanding when else the story can be applied. For example, one precondition that must be met in the Ed story is that you need a low-cost source.

You could also apply the lessons of this story to other situations. For example, what other kinds of surplus materials are there? How could you sell them or add value to them as a basis for a new business?

Conclusion

The purpose of this section on conscious indexing has been to increase the number of memory access points you have to your experiences. You can now access a much greater number of facts in greater depth, and in many different ways. You have depth as well as flexibility. What indexing enables you to do is to retrieve partial matches. Retrieving and adapting partial matches is the key to insight. Most people can't do it well because they haven't indexed to the depth and breadth needed to make partial matches effectively. Less insightful people may not see the connection between some current problem and some past experience or idea because they only remember the past idea in one simple way. That is, they haven't thought about experiences and ideas in multiple ways in order to recall the information in different contexts or at different levels of specificity. This section explained how to index so that you can retrieve more potential matches from more directions. The next section explains what to do once you have a potential match: how to adapt partial matches into a solution.

Technique #5: Adapting Partial Matches

The preceding four techniques help us work with ideas, make connections, ask insight-provoking questions, and index what we see and learn. Now let's turn to the final step of the insight process: adapting the old to create the new. To do this, let's return to the example of Darwin and Wallace and their independent insights about evolution.

Recall the two experiences or ideas that aided Darwin and Wallace. Naturalist voyages gave Darwin and Wallace (and others) exposure to the diversity of life. And Malthus' essay helped both Darwin and Wallace see how something like natural selection would create fit creatures. Malthus' work was first published several decades before Darwin or Wallace read it. Obviously, many other intelligent people had read Malthus, and many naturalists had visited archipelagos, but only two of them produced a theory of evolution.

This was because the Malthus essay, on its surface, does not provide the answer in and of itself. That is, the ideas in Malthus' essay were only a partial match to the explanations sought by Darwin and Wallace. In fact, Darwin and Wallace had to reinterpret Malthus in a completely opposite way. Whereas Malthus saw disease, famine, and war as the inevitable, dismal consequence of over-reproduction, Darwin and Wallace saw how this destructive process yielded the constructive process of new and ever-more adapted creatures. That is, Malthus' essay was a partial solution that needed adapting.

Darwin and Wallace saw the parallels between the ups and downs of human populations described in Malthus' essay and the plant and animal populations they observed as naturalists. It was then a matter of seeing the implications of Malthus' population dynamics. Similarly, we don't think of apples and planets having much in common. Yet Newton spent years working on the mathematics to explain both the straight, downward fall of the apple and the circular orbits of the heavenly bodies. These insightful scientists thought about how one thing was like another even if those things weren't identical. That led them to a line of thought that, after some work, produced an insight. People and plants or apples and planets may be very different things, but they can also follow similar laws when it comes to survival or the influence of gravity, respectively.

Thus, the goal of getting reminded is often to recall a partial solution that you can adapt to your situation or problem. This is where the heart of insight really lies—it's more than simply recalling a previous solution, because no prior solution fits perfectly if you're trying something new. Rather, you have to adapt or modify potential solutions and partial matches. People have difficulty with this because they tend to reject a partial match as improbable before exploring it. They don't know how to adapt it effectively. They too quickly say, "that won't work." There's an art to partial matching, because you have to decide which features or attributes are relevant and how they can be altered.

Substitution in Partial Matches

So, how can you make connections between elements that don't seem obviously connected? You can do this by identifying and labeling more attributes and goals of the object, idea, or situation.

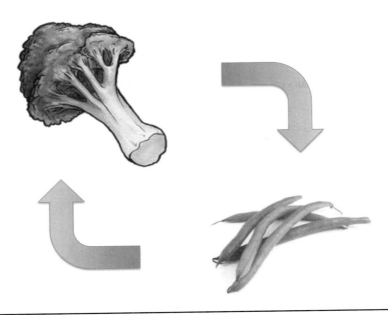

<small>**Figure 29-9**</small>   Substitution.

The more attributes and goals you can identify, the more potential connections you can make. To get a feel for partial matching, let's use the attributes and goals we listed for wristwatches and use them as a basis for designing a new wristwatch.

One manipulation you can do with partial matches is a substitution. Substitution is most effective when you are substituting attributes that fulfill the same function. For example, when cooking a beef and broccoli dinner, you can substitute green beans for broccoli. Both broccoli and green beans are vegetable side dishes—they have the same function.

In our wristwatch example, we could try substituting one shape for another. For example, you could ask if there is a pleasant shape that you like and if it could be made into a watch. The city of Boulder, Colorado, for instance, is very proud of its distinctively shaped flatiron rock formations, and many tourist items incorporate this shape. A watch could be made into this shape. Or, one can make partial match on something a watch has that other things have too, like a face. Other things like people, coins, animals, and planets have faces. Perhaps one can make a custom watch that has a person's face on it.

Likewise, one color can be substituted for another. You can also vary attributes such as:

- Magnitude (the strength of a color; strength of a material: finer and finer—how thin can it get?)
- Size (how small or large can you make it?)
- Placement (can it be worn elsewhere like on a ring or as a pendant)
- Motion (the second hand, other moving dials, the pulse of the digital colon, maybe it can change color with each pulse)

Non-Obvious Matches
Beyond substitution, you can think about a whole other range of possible changes that are much less obvious. Changes that are based on goals are deeper and less immediately apparent. In the indexing section, we identified various goals that a wristwatch can satisfy. The next step is to choose a goal and then ask what else is like that goal and what attributes it has, and then transfer those attributes back to the wristwatch. For example, we identified that one goal of a wristwatch was that it was like a piece of jewelry. So in this case, you could think about other kinds of jewelry and what attributes they have. For example, jewelry is sometimes sold in matching sets—like earrings and a necklace or a set of cufflinks. Could you sell your watch in a matching combination with something else? See Fig. 29-10.

Another feature of jewelry is that some people make it themselves. Could you add any do-it-yourself features to your watch? For example, one could sell a variety of shiny metal, sturdy ceramic, or colorful plastic links in bead shops for do-it-yourself watchbands. Another goal that wristwatches can satisfy is to express one's personality. Think about what kinds of things people like to express about themselves. It could be their values, their heritage, their

FIGURE 29-10   Watch and cufflinks.

religion, or their favorite hobby. How could you incorporate these into your watch? Perhaps your watch could be in the shape of an endangered animal, or display the logo of a cause. Could you make your watch distinctive by giving it a flavor of ethnicity in style? Perhaps you could appeal to people's values beyond the watch itself, such as by having a trade-in offer in which you recycle old watches or donate them. Maybe you even offer trade-in credit, as is offered on cars, to appeal to the thrifty. Some people like a sense of humor—what if the second hand on your watch runs counterclockwise? Or the numbers are backwards? Or a digital watch that has two human-hand shapes so that it, too, has a second hand?

As you can see, matching on the level of goals can bring entirely new perspectives that even an exhaustive list of attributes wouldn't reveal.

Practice Exercises in Partial Matching

Let's work through an exercise to practice partial matching. Imagine you've been tasked with creating a portable coffee mug with a lid that can't be lost. Think about adapting other things for attaching things together or avoiding loss.

Your Solution

Some possible answers: One possible answer is that a lid is like a door of a house or car. Adapting the door idea to the coffee mug might include creating a hinged lid, just like a door. That adapted idea could lead to other adaptations, such as adding a little locking knob to keep the lid sealed and spill-proof, or adding a little peephole to see how much coffee is left. You might even add a little "doggie door" to your lid door to make it easy to gulp the drink or to add sugar and cream.

For a second possible adapted answer, you might think about strong metal chains that can securely link two objects together. Although you could have a literal chain connecting the lid and cup, you might notice that the handle of a coffee mug is shaped like a link of chain. Adding a corresponding loop to the lid would make a simple chain with only two links molded directly into the lid and mug. The chain motif might make your mug look rugged, too.

A third answer might recall the loss-prevention devices in retail stores—an alarm that goes off if someone tries to walk off with the merchandise. You might adapt that to the lid-and-mug, so that a little sensor in the mug lets out a beep if you leave the lid on the counter and walk off with the mug. Adapting a high-tech solution might lead to other products, such as camera lens caps that can't be lost or sunglasses that can't be lost.

Insights in Action: Some Final Examples

Sydney Opera House

Now that we've explored reminding and adapting, let's look at them in action. We'll take the example of creative insight in architecture: the design for the Sydney Opera House in Australia by J. Utzon. This example is based on matching and adapting at a very abstract level.

Figure 29-11   Sydney Opera House design.

The Sydney Opera House has dramatic visual appeal—whether viewed from any side or from above. What makes it particularly pleasing is that it is expressive of its location on Bennelong Point, jutting out into Sydney harbor. It combines a strong platform of the lower parts of the construction with a white billowing roof line that's reminiscent of the sails of the ubiquitous small boats seen in the surrounding waters. See Fig. 29-11.

How did Utzon develop this design? We can trace the development of Utzon's ideas through his article, "Platforms and Plateaus: Ideas of a Danish Architect." For Utzon, the key idea was that of the interplay between roof and platform. This idea developed from his studies and personal experiences of many diverse architectures. In his article, Utzon suggests how his "notion of platform and its capacity to evoke a sense of strength and calmness was gleaned from sources as diverse as Mayan architecture and Chinese house and temples." He explicitly acknowledges a relationship between precedents such as these and the strong platform element in the Sydney Opera House. But even more important, Utzon felt that a flat roof was insufficient to express the flatness of a platform. So he instead combined a strong platform element with curved concrete shells to create a form that "provides adequate contrast to the flatness of the platform and creates a feeling of a floating roof and a whole exterior [that] radiates lightness and festivity."

In short, the starting point for Utzon's design was a notion of platform and the interplay between roof and platform that he abstracted from previous existing structures he had seen and which he adapted to fit the requirements of the design site. In the language of insight, Mayan architecture, Chinese houses, and temples provided a partial match to what Utzon wanted. Sailboat sails offered another–much more abstracted–partial match. Bringing the pieces together created the balance of strength with festivity.

This ability to move to a higher level of abstractness allows you to pull in more elements that at first glance don't seem related (such as sailboats and opera), and thus their combination is more surprising and is seen as more insightful and creative.

How can you characterize something more abstractly? One way is to think of it in terms of its goals or the results it achieves. Utzon noticed in his observations of different architectural structures the results produced by different roofs and platforms. He abstracted out the quality of stability and permanence in a flat platform and the possibilities of a dynamic festivity in a contrasting roof structure. He applied these abstract ideas to the Sydney's Bennelong Point site to design an opera house that combined these ideas in a highly effective and original way.

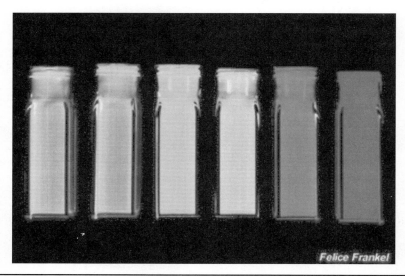

FIGURE 29-12   Nanocrystals.

Felice Frankel's Photographs

Another example shows the use of images and data visualizations to spark remindings that lead to innovative results. For example, Felice Frankel is a scientific photographer and researcher at the Massachusetts Institute of Technology. She works with scientists to help display their data in authentic, yet intriguing ways. That is, all her images are scientifically accurate data visualization, but she strives to make them captivating and beautiful as well.

In one project, for example, Frankel approached Moungi Bawendi with the idea of photographing his work. Bawendi, an MIT scientist who'd been working with nanocrystals for years, agreed. While exploring how to photograph Bawendi's work, Frankel zeroed in on the idea of photographing the cuvettes that hold the nanocrystals. The nanocrystals take on different colors under different circumstances. For her photograph, Frankel decided on an abstraction that zoomed in on the cuvettes, cropping the top and bottom of each cuvette. The resulting image removed all reference points to the nanocrystals' containers.

Seeing Frankel's photograph, Bawendi was reminded of bar codes, because the photograph looked like a colored bar code. The reminding gave Bawendi an idea for a potentially new application for nanocrystals. The image reminded Bawendi of a colored bar code, and bar codes are used as a form of labeling. Seeing the nanocrystals arranged like a bar code gave Bawendi the idea to use them as an alternative to fluorescent organic dyes that are currently used for labeling, imaging, and monitoring biological systems, particularly in their response to cancer.

To apply this technique in your own situation, look at images from different disciplines to see new ways to present your data or visualize the problems and solutions that you work on. Or, create images with different arrangements, even abstractions, of your data or system to reveal new patterns in your data or ideas.

Dostoyevsky Revisited: *Crime and Punishment*

Remember Dostoyevsky's quote, in which he said that ideas spring to authors full-blown? We saw from his notebooks that this was not the case, and we can get an even better idea of Dostoyevsky's process when we learn about his life, not just his work. What this expanded view reveals is that his life and work were intertwined and that the situations that he wrote about in his novels were adaptations based on his own life experiences.

Let's take the example of his most well-known book, *Crime and Punishment*. In the novel, the main character murders a pawnbroker (which he rationalizes because she was old, mean, and dishonest) but then his guilt feelings make him want to confess his crime. How did Dostoyevsky get the idea for his novel? Looking at Dostoyevsky's life, we see that he was arrested for reading a letter aloud in public that the police considered treasonous. He was sentenced to death, but reprieved at the last moment and sent to Siberia. While he was in Siberia, he spent much of his time considering a book based on a criminal's need to confess. A second parallel between Dostoyevsky's life and his *Crime and Punishment* novel are that while he was writing the novel, Dostoyevsky himself was in constant financial trouble (as the main character is) and borrowed a large sum of money from an elderly woman who was similar to the novel's pawnbroker in many ways.

Thus we see the interplay between Dostoyevsky's life and work. Dostoyevsky adapted experiences from his own life to the task of his work. His insight was based on experiences gleaned from life, which he explored and examined from different angles and wove together to fit the broader goals of his novel.

Putting It All Together: Hunt's Insight and Beyond

J.B. Hunt, founder of the J.B. Hunt Transport, had an insight that he turned into a business success. J.B. Hunt Transport is now a $4.53 billion company, but back in the 1940s, Hunt was a truck driver. While on his rounds one day, Hunt noticed that rice mills in Arkansas were disposing of their rice hulls (a waste product generated in the processing of white rice) by burning them. That gave Hunt an idea. He got a contract with the mills to haul away their rice hulls, and then he sold the hulls to poultry farmers as chicken-house litter.

Hunt's idea was a clever and profitable insight. How was he able to make it? What enabled Hunt to see rice hulls—a waste product—as something that could be useful to poultry farmers? The first question would be if any reminding took place. When you look at the life of J.B. Hunt, you discover that before he was a truck driver, he worked at a saw mill. The saw mill sold wood shavings to poultry growers for chicken-house litter. It was Hunt's knowledge of this fact, and his ability to recall it and adapt it to a new situation, that led him to his insight. Let's map out the connection:

Saw Mill (Fact in Memory)	Rice Mill (New Situation)
• Wood-shaving attribute: small absorbent pieces	• Rice hull attribute: small absorbent pieces
• Wood-shaving goal: dispose of economically	• Rice hull goal: dispose of economically
• Action: shavings sold to poultry farmers	• Action: burn as waste (insight action): sell to poultry farmers

Hunt saw the partial match on the attributes and goals and made a substitution on the action. The solution to the problem of wood shaving waste could be applied to rice hull waste. The abstraction that Hunt had to make was to see both rice hulls and wood shavings as examples of a waste product with similar material properties. Then he had to remember that wood shavings were not a waste product for poultry farmers, but were valuable. Then he had to consider that perhaps rice hulls could be useful to poultry farmers as well. It was a multistep process, based on Hunt's knowledge, a reminding, a partial match, and an adaptation to a new situation. Although Hunt's insight may seem easy when explicitly mapped out as we have done, nobody gave Hunt such a map—he had to make it on his own. But by seeing his process in terms of this map, we're in a better position to use the process to make insightful connections of our own.

Exercise: Mulling over Hulls

Let's take the Hunt story one step further. It's fine to recognize insight in this one story—but don't stop there. How else could you use this story to get other insightful ideas? Try to abstract out some additional goals and purposes from the Hunt story. What other partial solutions does it hint at? How else could you use them? List situations similar to the hull story.

Your Solution

Some possible answers: A key feature we identified was that of "valuable waste"—an unneeded by-product of one process that is actually useful to someone else. To make new connections, then, you could list other products that are by-products or waste products. Next, you could list who (what other type of person or organization) could consider that waste product a useful product. Sticking close to the Hunt example, we can say that wood shavings are similar to sawdust. Both are by-products produced at paper mills. Thus wood shavings and sawdust match on attributes and goals. Who else could use sawdust? That's a broad goal—probably too broad to work with, so go back to the fact that poultry farmers use wood shavings. What do they do with the shavings? They use them to spread on the chicken house ground. Who else spreads things around on the ground? Plant nurseries use mulch to spread around. Plant nurseries could use sawdust. Check how that match works by checking attributes: mulch is fine and biodegradable, and so is sawdust.

Now let's think more about waste products. Old tires are a waste product. Old tires aren't small like sawdust is. But what about making them smaller by grinding them up? How could they be used then? For example, one could use them in highway material.

As you can see, the process of insight involves connections and substitutions. You go up the ladder of abstraction to broader goals and then come back down it to instantiate them in new ways.

Conclusion

At the beginning of this chapter, we set out to answer three questions:

1. What happens when you get an insight?
2. How do insights form?
3. How can I get insights when I need them?

To answer the first question, we delved into historical accounts of insight to counter the myths that keep insight a mysterious process. We discovered that insight does not arrive complete and whole, out of nothing; nor does a single piece make an insight. Rather, we discovered **that insight occurs when you're reminded of something you know which you can shape into a new solution.** This answered the second question and was substantiated by experiments in the lab, which demonstrated that the mental processes that lead to insights are knowable and learnable.

To answer the third question and be able to get insights when we need them, we need to recall a useful fact or experience that could be applied to the new problem. One way to "get reminded" is to ask yourself a cuing question. Another method, indexing, provides deeper and more diverse recall through attributes and goals.

The most pleasing and exciting insights come from a matching on the goal level: when you can see two objects or ideas as instances of the same, broader goal. Insights at this highest level are the most satisfying (and the most valued) because they capture a similarity that's not readily apparent. Who would have thought that wood shavings and rice hulls had much in common? No one until Hunt saw that they both fit a common goal.

Some insights take a long time to work out, because the attributes must be explored and structured to fit the broader goal. The less you know about the particular attributes or goals, the longer it takes. Other insights, however, like Hunt's, can be mapped more directly.

Now you know what to look for in your experiences and knowledge and how to index by attributes and abstract out broader goals. So now you, too, can get insights in a directed, more reliable way.

CHAPTER 30

Stage-Gate

C. Anthony Di Benedetto

The Phased Review Process

In order to deal with the risks and uncertainties inherent in new product development (NPD), many firms have implemented formal procedures to identify new product opportunities, and to take these from the idea stage through launch and postlaunch. Usually, this procedure is depicted as a phased review process, alternating between phases (activities in the development process) and evaluative tasks (go/no go decisions as to whether the process should be continued to the next phase). The term *Stage-Gate*, coined by business professor and consultant Robert Cooper of the Product Development Institute Inc., is sometimes used to describe this phased review process. We will use the more generic term *phased review process* which is intended to include Stage-Gate as well as other variants such as that found in Crawford and Di Benedetto[1] and elsewhere.

Very briefly, a new product starts as an idea. There may be dozens, or hundreds, of ideas generated for even one successful new product.[2,3] Most of the ideas may not be very good and will be screened out, but some are promising and are further developed into concepts. The best concepts are selected, then early prototypes are produced, tested, and refined. A market test may be conducted at this point. Finally, if the market test is positive, the new product is launched into the marketplace. This process will be presented in greater detail later, but note in Fig. 30-1 that as the large number of ideas is whittled down to a single launched product; the uncertainty is being managed out. At the same time, the phases are increasingly expensive, in terms of financial commitment as well as time and people commitment. So the underlying principle in a phased review NPD process is to keep the amount at stake low while uncertainty is high, and to screen out ideas and concepts periodically at the evaluative task points to gradually reduce uncertainty and justify further investment.[4]

The evidence suggests that the firms that are most successful in NPD have implemented phased review processes. According to studies commissioned by the Product Development and Management Association (PDMA), over 65 percent of the best firms have some kind of phased review process, while under 45 percent of the "rest" of the firms have one; additionally, almost half of the firms studied had some kind of clear evaluative tasks after each phase.[5-9] About 40 percent of firms in the PDMA study assigned a process manager to supervise the implementation of the phased review process.[8] High-profile firms that have reported major new product success with a phased process include 3M, Lego, Guinness, Corning, and Exxon[4] (see Fig. 30-2).

Abbie Griffin, one of the authors of the PDMA study, notes that the phased process is vital to firms involved in product development.[5] She also reports that without senior management support, the process will not work as well. For example, senior management has the perspective to make decisions on how many financial resources should be allocated to each project.

Objectives of a Phased Review Process

According to Robert Cooper, Stage-Gate and similar phased review processes accomplish several objectives.[4] These are briefly summarized in Fig. 30-3 and detailed below.

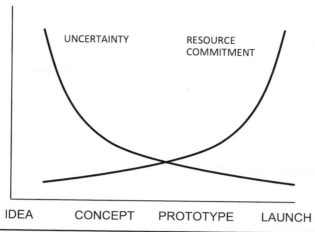

FIGURE **30-1**   Investment and uncertainty through the phased process.

Use of Phased Process in Top Firms

3M: Has long used a phased new product process to blend creativity and discipline in new product development. As a result, has an enviable new product track record.

Corning: Established its own version of a phased process in early 1990s, now all new product projects or new processes are developed according to this process.

Guinness: Even though the beer it brews is unchanged, Guinness innovated the widget (a capsule that releases nitrogen when the beer can is opened, guaranteeing a nice head of foam). Guinness has its own phased process known as NaviGate.

Exxon: All chemical new product development at Exxon is managed through a phased process, after initial success within its polymers business unit.

Lego: manages its new product development through a phased process that plans around a two-year cycle from opportunity identification (start of Year 1) to launch (holiday season of Year 2).

Hewlett-Packard: has a Phased Review process in place since the mid-1960s which accounts for its many successful new products through the years.

FIGURE **30-2**   Use of phased process in top firms.

Focus the new product team on quality new product development efforts.
Prioritize the new product team's efforts.
Encourage parallel processing and accelerate time to market.
Ensure cross-functional interaction on the new product team.
Ensure that the voice of the customer drives new product development.
Require that technical and commercial assessments are done early, and well.
Assess the product's competitive advantage throughout the development process.

FIGURE **30-3**   Objectives of a phased review process.

First, a phased review process focuses the new product team on quality NPD. It ensures that the new product team completes all activities important in the NPD process, follows quality control checks, and does not ignore the typically difficult up-front activities (such as concept development and screening).

A second objective is to provide a means for prioritization of the firm's efforts. Financial and human resources are always scarce, and good, strong evaluative tasks ensure that the new product team remains focused on the most promising concepts. The new products process can be thought of as a funnel; weaker ideas are dropped and resources are allocated to the highest-value ones. Gathering sufficient information about the market throughout this process is obviously important in making sound go/no-go evaluative decisions.

Third, a phased review process helps accelerate time to market, since it encourages activities to be done in parallel. The NPD process is cross-functional, involving members from marketing, R&D, operations, and manufacturing from the beginning and throughout all the phases. The phased review process should minimize or eliminate silo thinking—engineering has to be 100 percent done, then the product goes into the production silo, then finally when it is ready to be launched, the marketing silo takes over. Instead of this time-consuming process, all functions are working jointly from the beginning, which shortens time to market.

Fourth, the phased review process guarantees that the teams really are working cross-functionally. It is easy to pay lip service to a new product team that really has no power or authority, or whose members do not feel any real commitment to the team. In a real cross-functional team, participants will receive release time to work across functions (and, in a heavyweight, very powerful team, may be fully devoted to the project team). The team has authority and responsibility, and results are rewarded in terms of bonuses or salary increases.

A very important fifth objective of the phased review process is that it guarantees that the voice of the customer drives NPD. Marketing is a member of the team from the earliest phases, and is responsible for bringing information about customers and the market to the team. The cross-functionality of the team ensures that the team is market-oriented and at all times concerned that the products it develops meet customer needs.

Sixth, the phased review process requires that early technical and commercial assessments are done well before any real product development work occurs. That is, ideas are rapidly eliminated if they do not fit with the firm's product strategy, would not be feasible technically, or do not pass early assessments of market attractiveness.

Seventh, and finally, the phased review process requires assessments of product competitive advantage throughout. If effectively implemented, each evaluative task will ask questions about whether the product would offer some point of superiority over competitive products. This focus on competitive advantage results in truly superior products, rather than me-too, run-of-the-mill products.

Overview of the Phased Review Process

A firm employing a phased review process views NPD as a series of phases. The number of phases will vary; in its very simplest form, one could consider a simple three-phase process consisting of predevelopment (or fuzzy front end), development, and launch activities.[10] Most textbook references include five or more phases,[1,11,12] and in actual practice there can be any number of phases or subphases, depending on the needs of the firm. Note that the launch stage only begins at the time of market introduction; it is understood to continue until a key objective has been reached and success has been achieved.[1] Thus, the launch stage consists of continuous monitoring and learning about customer and competitive responses, as well as management and control to stay on track with objectives. For many firms, this launch stage is the riskiest and most expensive stage in the NPD process.[13,14] The actual number of phases is not so important; more relevant is the idea that each phase moves the idea closer to launch, and at the end of each phase is an evaluative task, where a decision is made on whether to go to the next phase or not. The phased review process, therefore, systematically reduces uncertainty from phase to phase, as the amount invested in the product increases through time. A useful starting point appears in Fig. 30-4, which shows not only the phases in the NPD process, but also typical evaluative tasks at the end of each stage.

One way to envision the phased review process is to track the development of the product, from idea, to concept, to prototype, to the ultimately launched product. An idea can be thought of as a starting point: a new technology, an unmet customer need, or a product form that would be new to the market.[1] Many of these ideas simply do not fit with the firm's strategic plan, its technological abilities, or its desired target market, and are dropped early. The surviving ideas are developed into concepts, which can be considered as more fully formed ideas. If the idea sprang from an unmet customer need, the new product team would have to provide the technology and/or the product form in sufficient detail to elicit meaningful customer purchase intention data. An idea might be, for example, simply to make educational

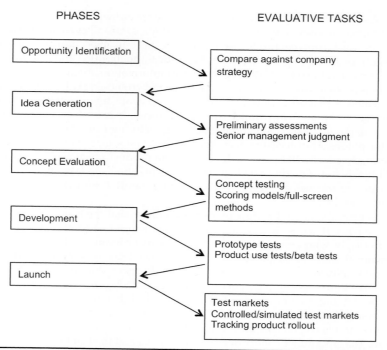

PHASES   EVALUATIVE TASKS

FIGURE 30-4 A phased process showing evaluative tasks. (*Source*: Adapted from Crowford and Di Benedetto (2011).)

DVDs. It is just an idea at this stage as many of the details still need to be filled in (The DVDs teach you what? What educational level? Are they expensive? Are they any good?). A fully formed concept might be a series of popular movies on DVD, including audio tracks in eight popular foreign languages, to be used in language learning. The package would include online instruction and testing, and completing the program would be equivalent to 2 years of college education in the target language. The package would retail for $500. At this stage, of course, no DVDs have been made yet; but there is enough information at this point to test the concept with potential consumers to assess their likelihood of purchase, and reasons they liked or did not like the concept. Once the concept is approved, it goes into development and early prototypes are made, which can be tested with customers. These do not need to be fully functioning (a prototype phone may be made out of wood, for example, but its proposed dimensions, size of buttons and screen, and so forth, can be tested with customers before expensive product development is initiated). After feedback, later and possibly more complete prototypes are developed and tested. Finally, preparations are made for the launch of the actual product in the marketplace; once launched, the product's progress is tracked and corrective action is taken should it fall behind sales or profit objectives.

The Phases in Detail

As practice varies considerably, we will stay close to the structure for a phased review process as presented by Crawford and Di Benedetto[1] and which has been previously shown in Fig. 30-4. Figure 30-5 provides additional details on each of the phases in the process, which will be further expanded below.

Phase 1: Opportunity Identification

The first phase occurs even before the product planners begin to search for ideas, as it is essentially a step that defines where they should begin looking. This is the opportunity identification phase. As noted by Crawford and Di Benedetto,[1] there are four categories of opportunities that may present themselves to the firm.

1. A firm may seek to expand its use of an underutilized resource, such as a manufacturing plant or a solid franchise with distributors.

2. A firm may have discovered a new resource with market potential. A chemical firm may have isolated a new compound with unusual properties, or a pharmaceutical firm may have found a cure for a disease.

3. There may be an external mandate for innovation, such as a weak market, strong competition, or changing customer needs. Kellogg's shifted its NPD efforts in the early part of the 2000s to add more snack products to their line, seeing weaknesses and lack of growth potential in the cereal business.[15]

4. Finally, there may be an internal mandate. The firm may have established an aggressive sales or profit growth objective that could not be met simply by selling the current line harder: NPD is seen as a way to close the gap between current and targeted sales.

The reason the firm goes through this opportunity identification stage is to help direct the activity of the new product team, once it starts looking for ideas. New product researcher Peter Koen talks about the *sandbox*: even before starting to think of specific product ideas, the best managers ask, "What sandbox should I be playing in?"[16] The perimeter of the sandbox defines areas that the new product team should look—and also should not look—for new product ideas, in consideration of the firm's strategy. Without consideration of the opportunities in the marketplace, and the firm's corporate strategic plan, there is no perimeter to the sandbox, no delineation between good and not-so-good product ideas, and ultimately, any idea would be acceptable.

Phase 2: Idea Generation

The second phase shown in Fig. 30-5 is idea generation. As noted, ideas can come from just about anywhere: customers in focus groups and interview settings, marketing and sales employees, engineers working in R&D, senior management, designers, suppliers, distributors,

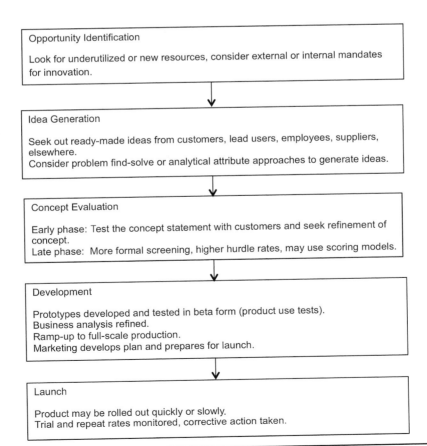

FIGURE 30-5 A detailed phased process. (*Source*: Adapted from Crowford and Di Benedetto (2011).)

government or university laboratories, competitors, and many others. Some ideas are referred to as *ready-made*, meaning that the firm receives them in relatively complete form, and some estimates are that at least 40 percent of new product ideas fit this category.[1] A particularly useful source of ready-made product ideas is lead users: customers who are ahead of the curve, and have product needs today that might become mainstream in the future.[17] A research hospital working on very advanced blood analysis for a new treatment may not find diagnostic equipment that is accurate enough, and may even make the machine itself to suit its needs. An equipment manufacturer can benefit by partnering with this research hospital and develop a commercial version of this diagnostic machine. If that treatment becomes widespread, there will be wide demand for this machine in the future.

Many firms are now turning to partner firms for ideas to complement their own, home-grown ideas, in an open innovation framework.[18] Open innovation implies an external search for innovative ideas and products. In the pharmaceutical industry, a small biogenetics company may do all the discovery work for a new treatment, then work in an open innovation partnership with a big pharma firm, which can afford to do regulatory testing and full-scale manufacturing, and can provide adequate marketing support.

Now, many firms are complementing traditional customer research with crowdsourcing to generate ready-made product ideas.[19] Customers can be encouraged to send in product ideas online, and the firm can follow blogs and online communities to get further ideas. Apple reportedly relied heavily on crowdsourcing when getting ideas for the iPad.[20]

There are many creative ways to generate ideas, to complement the ready-made ones. Two broad categories of approaches exist. In the find-solve approach, the goal is to find problems with the current product, then to identify solutions to these problems: these would be potential new product ideas. James Dyson famously found problems with conventional vacuum cleaners, and later with hand dryers and fans, and in each case developed products that effectively eliminated those problems. Regular fans are hard to clean, tend to tip over due to a high center of gravity, and the spinning blades can pose a safety risk. Dyson's air multiplier fan has no blades, thereby making cleanup simple and removing the safety risk altogether; the revolutionary design is bottom-heavy and will not tip over.[21]

In the second category, the analytical attribute approach, one starts with the attributes of the product, and makes changes to these to develop new product ideas. For example, a flashlight can be defined in terms of its length, weight, buoyancy, flexibility, even its explosiveness.[1] A change in any of these constitutes a new flashlight concept. Allowing a flashlight to be flexible results in Black & Decker's Snake Light. And although an exploding flashlight is probably not a good idea, a flashlight that shoots flares might be useful to hikers who could get lost. Familiar perceptual maps commonly used in marketing similarly suggest directions in which products can stretch into underserved market areas.

Phase 3: Concept Evaluation

At this phase, concepts are assessed, weaker ones are thrown out, and good ones are improved. Keep in mind that the concept evaluation phase ideally contains two parts. At early concept evaluation, the hurdle rates are kept reasonably low. One does not want to risk killing a concept with high potential. Rather, at this point, improvements to a promising concept might be sought out. Later in this phase, however, it is critical that the hurdle rates are set much higher. The reasons for this will be explained in the next section on evaluative tasks.

To begin the concept evaluation stage, the firm develops a concept statement (recall the language learning DVDs discussed earlier). This statement may be verbal only, or accompanied by a picture or sketch. While several different measures may be taken here (perception of how novel the concept is; how frequently one would buy the product if it were available), a very important measure at this point is purchase intention, often recorded on a five-point scale (with extremely likely to purchase and extremely unlikely to purchase as the anchor points). The top-two-boxes score is recorded (that is, the percentage of respondents who said they were extremely likely, or likely, to buy the product) and compared to similar products, to get an assessment of customer reaction to the concept. In earlier days, a concept test would have been conducted by mail. Now, participants are likely to be in a focus-group style room, within reach of a laptop. They can be shown concepts in images or text form and asked to respond quickly with their purchase intention, on a 1-to-5 or 1-to-10 scale. Then, the researcher can get immediate feedback on what was liked or disliked about the concept, make adjustments, and then measure again. This way, improved versions of the concept can be identified.

Concept screening later in this phase is much more formal, and examples of the techniques used here are presented in the next section. Realistic forecasts and financial projections may be attempted here, based on customer information. For example, based on concept screening data, a firm may be able to predict what percentage of the target audience will try the new product, and what percentage of these will become regular customers. These predictions of likely trial and repeat purchase can be used to generate a rough forecast (if half the market tries the product, and 20 percent of this group become regular customers, then long-term market share is forecasted at 10 percent). This forecast can be further adjusted to take into account likely awareness and availability of the product, which are related to the firm's spending on promotion and distribution, respectively.

Phase 4: Development

This is the phase where the product concept takes form. The marketing people are still working hard at this time, however, as this is generally when the marketing plan for the new product is fleshed out. It is important for the firm to consider its level of resource preparation at this phase. If this will be a close-to-home new product, this may not be an issue. But for a new-to-the-world or new-to-the-firm product, the firm may be treading into unknown territory. The firm may be lacking key resources or manufacturing capabilities, or may even need to rethink its project review system.[1]

The R&D and engineering personnel may be developing early prototype versions of the product, even nonworking ones, to gauge customer interest and guide further development. The newer to the world the product is, the more likely the firm will need to rely on prototype testing to get this important customer feedback. The development personnel may be working here to match product specifications, which are supported by market research. For example, a development team working on a new printer knows that customers want fast but clear printing. Ideally, they want a balance of both (superfast but sloppy printing will be unacceptable, and so will crystal-clear but superslow printing). Techniques such as the House of Quality can be used to provide guidance on this kind of trade-off.[22] Finally, beta testing with customers may be attempted in order to ensure that customer specifications are being adequately met.

Another characteristic of this phase is a further business analysis review. The financial review and sales forecasts attempted at the concept evaluation stage are further refined, now that the firm has more recent, and more realistic, information from customers on likely purchase behavior.

Finally, as the launch date approaches, the manufacturing and production personnel prepare the ramp-up to full-scale production. At the same time, marketing personnel do a kind of marketing ramp-up to prepare for the launch: the distribution channel is primed, the last few packaging, branding, and labeling decisions are made, advertising and promotional copy is prepared, and so forth. This is done to ensure that the launch is not delayed by inadequate marketing preparation.[1]

Phase 5: Launch

The launch is, not unreasonably, usually considered to be a single event—it occurs the day the product is commercialized. New product personnel, however, view launch differently, as a phase which starts with that day, and continues until the product, hopefully, reaches its sales and profit targets. That means that the launch date must be followed up with a launch management program. A trial-repeat model as described above can be used here, except the data are no longer projections, but now are real sales figures from the marketplace. If a product is underperforming, the firm can determine if the underlying reason is poor trial, poor repeat, or both. Poor trial may point to inadequate promotion or distribution, which may be adjusted as part of launch management; good trial but poor repeat generally means that the culprit is the product itself. Good new product firms will track all of the expected problem areas (not enough sales support, low cooperation at the distributor level, etc.), and will take corrective action if any of these measures fall below critical levels for a significant amount of time.

Evaluative Tasks in the Phased Review Process

In addition to the phases, Fig. 30-4 also indicates typical evaluative tasks. As the product moves from idea to launch, and its form becomes clearer, the evaluative tasks change as well. However, note that all the way through the process, the same basic questions are asked: questions of technical viability ("can we make this?") and of commercial viability ("will they buy this?"). These questions are answered more accurately and in more detail the further through the process the product gets, and the more information is available about the technology and the markets. That is, the evaluative tasks comprise a kind of rolling evaluation.[1] The product is evaluated continuously, and early, weaker information is replaced by later, stronger information when it becomes available. The closer one gets to the launch date, the more accurate and detailed the information is, and therefore the more accurate the financial projections and sales forecasts drawn from this information.

Incidentally, the PDMA studies mentioned above also examined the use of common evaluative tasks by the sampled firms.[9] It is probably not surprising to learn that the firms that did the best with NPD also were significantly more likely to have used several evaluative tasks than those that fared worse with their new products.

At the earliest phase of the NPD process is the idea screen. At this point, there is not much to judge except the initial idea, which may be quite lacking in detail. Hence, simple immediate judgments will be made by firm executives, on whether the idea fits the firm's product strategy, or whether it appears to be technically and commercially viable. Details are built in, and one or more concepts can be developed from the same basic idea. Concepts can be tested with customers. A concept statement is presented (this is a description of what is new and unique about the product, how it can benefit the customer, and expected price). This is much more detailed than the idea screen described above, as customers may be expected to provide their likely level of adoption of the product at that price point. If customers do not like certain features, they can be adjusted and the concept test can be redone; this procedure can be repeated several times until a concept is at its highest level of acceptance.

Up until now, the hurdle has been set relatively low; in the early phases, one does not want to kill off potentially promising ideas and concepts. But, in late stage concept testing, the hurdle is set at a much higher level. By this point, weaker ideas are already weeded out, and good concepts have been refined into better ones. What remain, then, all are rather good concepts; but the firm only has the human and financial resources to pursue a small number of these. Other promising ideas must be dropped. That means, for the first time in the process, there is the possibility of incurring opportunity costs (on rejected product ideas that had market potential) in addition to significantly higher real costs. New product managers often have difficulty at this point, since all the remaining concepts look good. Lacking guidance in this decision, they either (seemingly) randomly pick one or two and hope for the best, or they approve all the concepts, guaranteeing that each concept will be underfunded.

There are several evaluative tasks available after the concept development phase. Foremost among these are full-screen scoring models. Figure 30-6 shows a set of questions that can be generally useful in a full-screen scoring model, while Fig. 30-7 shows a scoring model used by the Industrial Research Institute. Notice that both of these scoring models ask basically the same questions about the new product—can we make it and will they buy it—as seen earlier, but now there are more detailed data available, so one can obtain a richer understanding of the technical and market characteristics pertaining to each concept. For example, instead of just asking about technical accomplishments, at the time of the full screen, one asks about the difficulty of the task, the amount of research and development needed, the rate

Technical Accomplishment	Commercial Accomplishment
Technical task difficulty	Market volatility
Research/development skills required	Probable market share
Technical equipment/processes	Product life
Rate of technological change	Sales force requirements
Design superiority	Promotion requirements
Patent/security of design	Target customer
Manufacturing equipment/processes	Distribution channel
Competitive cost	Unmet customer need
Vendor cooperation	Competition
Speed to market	Field service requirements

FIGURE 30-6   Commonly-used criteria for full-screen scoring models.

Technical Success Factors	Commercial Success Factors
Proprietary Position	Customer/Market Need
Competencies and Skills	Market/Brand Recognition
Technical Complexity	Channels to Market
Access to/Use of External Technology	Customer Strength
Manufacturing Capability	Supply of Raw Materials/Components
	Safety/Health/Environmental Risks

FIGURE 30-7   Full-screen scoring model used by the Industrial Research Institute.

of technological change, the likelihood of getting the product to market quickly, and so on. Similarly, instead of simply asking if the product is commercially viable, one asks about the size, volatility, and growth rate of the market, who the target market is, whether there is a significant unfilled need, how strong the competition is, and so forth. Note that it is still technical and commercial viability that is being assessed: weaker estimates based on judgment are replaced by stronger, more detailed data as part of the "rolling investigation."

When the product is in development, early, nonworking prototypes may be tested with customers for a quick assessment; later, almost-ready versions may be beta-tested. Beta tests, or product use tests, are undertaken to make sure that the product performs at levels satisfactory to the customer. Anyone who has downloaded and tried out a beta version of a new software product is actually taking part in a kind of product use test. As the product moves through the phased process, a formal test market may be conducted (the product may be made in sufficient quantities and actually sold in several cities, perhaps at different price points or with different levels of promotional support). Increasingly, firms are turning away from big-scale test markets like this, and using a quicker and more controlled version of the market test, especially if the perceived risk in the product does not warrant expensive and time-consuming test marketing. An old-fashioned test market may still be conducted, however, if perceived risks justify it. Starbucks, for example, extensively test marketed Via instant coffee packets to make sure they would be accepted by Starbucks customers and would not tarnish the Starbucks image.[23]

Recall that this phased process is not meant to put constraints on new product development, or to limit the creativity of the team members. Robert Cooper provided several thoughts to keep in mind, so that the phased review process is implemented effectively.[4]

1. The phased process is not a relay race, where each phase is managed by different individuals who throw their contributions over the wall to the people assigned to the next phase. A real problem with this old-fashioned phased process is that it is very time-consuming, and therefore will not work in today's competitive market where new products must go through the process speedily and efficiently.

2. The phased process is not rigid. Phases or tasks could be skipped or done out of order. Flexibility is built into the process to take advantage of market opportunities.

3. The phased process is not bureaucratic. It is not conducive to endless meetings, reports, and committee work. The process may be systematic, but that does not mean it is top-heavy with bureaucracy. This also slows down the process and should be avoided.

4. The phased process is not project management. Management of individual product projects should be viewed as taking place within the phases of the phased process. Project management, for instance, could be employed to manage a difficult new product project through a lengthy development or launch process.

Radical Products and the Phased Review Process

As briefly mentioned above, the phased review process may be somewhat different for new-to-the-world products. Certainly, the firm's expectations will be different: new-to-the-world products may have a lower survival rate than incrementally new products, but if successful, can be more profitable for the innovating firm.[24] Nevertheless, the first phase is probably the same: opportunity identification and strategic planning. Firms that are very committed to new-to-the-world products may make significant organizational changes. Corning, for example, created a senior position in the R&D department, vice president for Strategic Growth and New Business Development, who reports directly to the chief technology officer. In addition, project teams assigned to radical innovation projects receive the support of a new business development team, which finds applications and customers for the new technology.[25]

Prototype may play a very important role for radical new products. Firms might employ a probe-and-learn strategy.[26] Here, nonworking prototypes may be developed in rapid succession and tested with potential customers, and feedback is sought on each prototype. After a dozen (or a few dozen) nonworking prototypes have been built, each incrementally better than the previous one, the firm has a good idea of the product features desired by the target market and specifications are detailed. Thus, in radical product development, there may be more prototypes, and they may be developed and tested at earlier phases in the development process.

Third-Generation Processes

The phased review process depicted in Fig. 30-4 seems rather orderly and regimented, but the discussion of radical NPD suggests that it in fact should be applied with a certain amount of flexibility. Indeed, in his more recent work, Cooper notes that the better firms employ a more flexible process, which he refers to as "third-generation Stage-Gate."[1, 27–29] Some of the hallmarks of this flexible process are as follows.

First, strict go/no-go decisions are not required at each evaluative task, since the quest for sufficient information may slow down the process. Instead, *on* or *conditional go* decisions are permitted. Evaluative tasks that allow this intermediate outcome are called fuzzy gates. As an example, a new cola might do well in a concept test, but not so well that the firm is really convinced of its viability. They may make an *on* decision, and commit to the next step, which might be a product use test (an on-site or at-home beta test with real consumers). But the implication is that the fuzzy gate must have teeth: that is, if the cola fails the product use test, a no-go decision must be made. A fuzzy gate simply means that the product is allowed to proceed to the next step, and an appropriate amount is invested in this next step, and a firm go/no-go decision must be taken at that time. It is not an excuse for the CEO's pet project to keep moving through the process regardless of how negative the test results are!

While parallel activities were mentioned above, they are very much stressed in the flexible process. The term used is *overlapping phases*, the term referring to the fact that one phase need not be totally completed before work on the next phase begins. For example, a new sports car may be 80 percent engineered, while the front-end appearance is not yet finalized. The industrial designer on the new product team may work on alternative designs that provide both aesthetic beauty and ergonomics, while the marketing team member obtains data on market preferences and desires. Bringing in the designer and marketer before the blueprint is finished reduces the chance of costly and time-consuming do-over work later in the process. A strong team leader can deal with any power struggles among team members that arise, as well as any technical complexities that may be confronted.

And, as noted before, the phases may sometimes be done out of order, for example, in the case of a totally new-to-the-world product.[1] This is done to increase flexibility in the development process, a requirement in radical product development. But, regardless of the radicalness of innovation, the NPD process always must start with the first phase in the phased review process: a consideration of marketplace opportunity and the firm's strategic plan for new products.

Best Practices in Phased Review Processes

According to the originator of the Stage-Gate system, Robert Cooper, firms should aspire to follow several best practices in order to get the most out of their phased review processes. These include the following:

1. Gates need strong teeth, so that projects that should be phased out are phased out. Otherwise, too few resources get spread across too many development projects.

2. Focus on projects that will result in products with competitive advantage and customer value.

3. Maintain a focus on market orientation and listen to the voice of the customer. Measure the economic value of the product to the customer.

4. Do not neglect the early phases: aim for sharp, early, and stable product definition and don't allow scope creep.

5. Use the phased process to speed the product to market and not get bogged down with an over-the-wall approach. Be sure the team members work together effectively.

6. Cross-functional teamwork means: there is a defined team leader, a fluid team structure (key members can be brought in or dropped as needed), and the team is accountable for its own success.

7. The process remains lean and efficient throughout.

8. There are effective metrics used throughout the process to assess performance and results.

References

1. Crawford, Merle and Di Benedetto, Anthony. *New Products Management*, 10th ed. New York: McGraw-Hill/Irwin, 2011.
2. Booz, Allen and Hamilton. *New Products Management for the 1980s*. New York: Booz, Allen and Hamilton, 1982.
3. Stevens, G.A. and Burley, J. "3,000 Raw Ideas = 1 Commercial Success!" *Research Technology Management*, 40 (3): 16–27, 1997.
4. Cooper, Robert G. *Winning at New Products: Accelerating the Process from Idea to Launch*. New York: Perseus, 2001.
5. Griffin, Abbie. *Drivers of NPD Success: The 1997 PDMA Report*. Chicago: Product Development & Management Association, 1997.
6. Cooper, Robert G., Edgett, Scott, and Kleinschmidt, Elko J. "Optimizing the Stage-Gate Process: What Best-Practice Companies Do—I." *Research-Technology Management*, September–October: 21–27, 2001.
7. Adams, Marjorie. *Competitive Performance Assessment (CPAS) Study Results*. Chicago: PDMA Foundation, 2004.
8. Kahn, K.B., Barczak, G., and Moss, R. "Perspective: Establishing a NPD Best Practices Framework." *Journal of Product Innovation Management*, 23 (2): 106–116, 2006.
9. Barczak, Gloria, Griffin, Abbie, and Kahn. Kenneth B. "Perspective: Trends and Drivers of Success in NPD Practices: Results of the 2003 PDMA Best Practices Study." *Journal of Product Innovation Management*, 26 (1): 3–23, 2009.
10. Calantone, R.J., Schmidt, J.B., and Di Benedetto, C.A. "New Product Activities and Performance: The Moderating Role of Environmental Hostility." *Journal of Product Innovation Management*, 14 (3): 179–189, 1997.
11. Urban, G. and Hauser, J.R. *Design and Marketing of New Products*, 2nd ed. Englewood Cliffs, NJ: Prentice Hall, 1993.
12. Cooper, Robert G. and Kleinschmidt, Elko J. *New Products: The Key Factors in Success*. Chicago, IL: American Marketing Association, 1990.
13. Di Benedetto, C. Anthony. "Identifying the Key Success Factors in New Product Launch." *Journal of Product Innovation Management*, 16 (6): 530–544, 1999.
14. Guiltinan, J.P. "Launch Strategy, Launch Tactics, and Demand Outcomes." *Journal of Product Innovation Management*, 16 (5): 509–529, 1999.
15. Naughton, Keith. "Crunch Time at Kellogg." *Newsweek*, February 14: 52–53, 2000.
16. Koen, Peter. "Tools and Techniques for Managing the Front End of Innovation: Highlights from the May 2003 Cambridge Conference." *Visions*, October, 2003.
17. Von Hippel, Eric. *The Sources of Innovation*. New York: Oxford University Press, 1988.
18. Chesbrough, H. *Open Innovation*. Cambridge: Harvard University Press, 2003.
19. Pisano, Gary P. and Verganti, Roberto. "Which Kind of Collaboration Is Right for You?" *Harvard Business Review*, 86 (12): 78–86, 2008.
20. Jana, Reena. "Apple iPad Product Development Approach." The Conversation blog, *Harvard Business Review*, January 27, 2010.
21. Smithers, Rebecca. "Latest for the Dyson Touch: The Fan without Blades." *The Guardian*, October 13, 2009.
22. Hauser, John R. and Clausing, Don. "The House of Quality." *Harvard Business Review*, May–June: 3–13, 1988.
23. Jargon, Julie. "Starbucks Takes New Road with Instant Coffee: Company Launches Marketing Campaign and Taste Challenge to Tout Its Portable, Less Expensive Product Via." *The Wall Street Journal*, September 29: A29, 2009.
24. Min, S., Kalwani, M.U., and Robinson, W.T. "Market Pioneer and Early Follower Survival Risks: A Contingency Analysis of Really New versus Incrementally New Product Markets." *Journal of Marketing*, 70 (1): 15–33, 2006.
25. O'Connor, G.C., Liefer, R., Paulson, A.S., and Peters, L.S. *Grabbing Lightning: Building a Capability for Breakthrough Innovation*. San Francisco: Jossey-Bass, 2008.
26. Lynn, Gary S. and Reilly, Richard R. *Blockbusters: The Five Keys to Developing Great New Products*. New York: HarperCollins, 2002.
27. Cooper, Robert G. "Perspective: Third-Generation New Product Processes." *Journal of Product Innovation Management*, 11 (1): 3–14, 1994.
28. Cooper, Robert G. "What Leading Companies Are Doing to Reinvent Their NPD Processes." *Visions*, 32 (3): 6–10, 2008.
29. Cooper, Robert G. "The Stage-Gate Idea to Launch System." In B. Bayus (ed.), *Wiley International Encyclopedia of Marketing*, Vol. 5, Product Innovation and Management. Chichester, West Sussex: John Wiley & Sons, 248–257, 2011.

CHAPTER 31

Design Innovation

Jeremy Alexis

Designers tend to think about and act on problems differently compared to engineers, scientists, and people with traditional business training. The results of design-driven projects tend to be more creative (providing a large set of nonobvious elegant solutions) and empathetic (aligned with the needs and behaviors of the target user group) compared to other approaches. This does not mean that the design approach is in any way superior to the other processes.

For example, the design approach would be useless when trying to synthesize a new compound for pharmaceutical development and less than ideal when trying to optimize a manufacturing facility. However, when an organization needs to address issues that seem impervious to traditional methods or find ways to grow organically, the design process can provide unique perspective and a distinctive process.

Over the last 15 years, companies such as Proctor & Gamble, Boeing, S.C. Johnson, Target, Intuit, McDonald's, Hyatt, and AB/Inbev (just to name a few) have used design innovation to create new products, services, and business models. Design innovation has been a critical part of successful new offerings such as the McCafé, the interior of the 787, the Swiffer, and Target's store design. Although there were variations in process at each organization, they all applied a form of the human-centered design approach. In this chapter, we will discuss why design innovation is particularly useful for global companies looking for organic growth; outline the approach and process; and then highlight specific tools that have been proven effective for implementing this process in organizations.

It is important to distinguish that design can be both a *state* ("look at my design, it is a chair") or a *process* ("we need to design a new chair"). This chapter will focus on the latter, discussing design as a process. The following pages include various design frameworks, tools, and methods to help illustrate the design innovation process.

Why Design Innovation Matters: The Innovation Gap

Successful companies often start with an insight about an unmet need. To grow the company, management focuses on leveraging this insight. Over time, as the organization pursues efficiency through optimization of organizational functions, the organization becomes disconnected from the customers and interactions that created the initial insight. Companies naturally get better at knowing *how to make things*, but, due to the focus on operations, actually get worse at knowing *what things to make* (Fig. 31-1). This phenomenon was observed and named the *Innovation Gap* by Patrick Whitney, the dean of IIT's Institute of Design.[1]

For example, Facebook's early success was based on an insight about exclusivity[2]: social networks become more attractive and sticky when a user has control over their connections and groups (as opposed to Friendster and MySpace which allowed your profile to be seen by anyone). Facebook grew quickly, and added departments like marketing, human resources, and accounting. With a focus on growth, profits, and a forthcoming public offering, the company naturally focused on optimization and taking advantage of "low hanging fruit" or obvious revenue opportunities made possible by their large user base.

Facebook hired platoons of exceptional programmers and coders. The company became a powerhouse of digital development—they could *make* anything. This led to some unintended consequences: Facebook was frequently in the news as customers reacted negatively to changes in their privacy policies. So much of their focus was on *how* to make things that they lost focus of the original insight that paved the way for their success. Starting in early

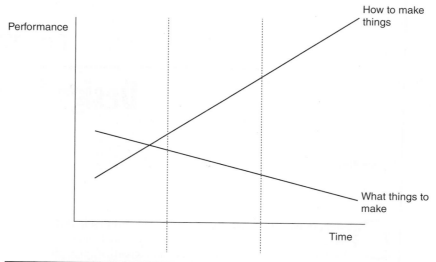

FIGURE 31-1   The innovation gap.

2012, the company has been hiring some of the best design talent available.[3] This suggests that they have recognized the importance of understanding and designing for their users—in other words, refocusing they refocused on *what* to make.

The Facebook story is unusual in the speed at which the innovation gap was created, recognized, and is currently started to close. For most companies, the gap has developed over decades, and requires a significant shift in culture and process to close. A first step for closing the gap is to adopt the balanced breakthrough model for decision making.

The Balanced Breakthrough Model

One of the foundational, conceptual frameworks of design innovation is the balanced breakthrough model. The model suggests that successful new products and services are desirable for users, viable from a business perspective, and technologically feasible. The model is helpful both in synthesis (create ideas that balance all three criteria) and analysis (prioritize and evaluate using the three criteria). The balanced breakthrough model is sometimes called the DVF model (desirable, viable, possible) (Fig. 31-2).

McCafé is a great example of a balanced breakthrough. It provides a desirable option for users: a less expensive, crafted coffee drink available anywhere there is a McDonald's restaurant. It is viable: McDonald's can charge more for the drink than it costs to make it. It is also feasible: sophisticated supply chain management, drink machine technology, and service blueprints and training allow this concept to be delivered on a global scale.

Napster is an example of a service that was desirable to users, technically feasible, but did not have a sustainable business model. It ultimately failed. iTunes, which also provides

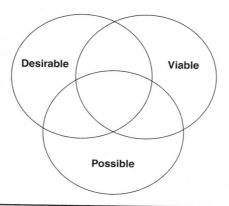

FIGURE 31-2   The balanced breakthrough model.

music, has enjoyed longer-term success since, as a concept, it better balances the three elements. So, even though it is not as desirable as Napster (free music is hard to beat!), iTunes includes a viable business model and is based on a reliable technology platform. A balanced breakthrough does not mean equal weight for each criterion, rather that the concept addresses each of the criteria at the appropriate level for the demands of the market.

Great innovators see the world through the lenses of DVF. They use the framework to both develop and evaluate ideas. The model becomes more powerful when paired with a process; in other words, a reliable way to create ideas that balance DVF.

The Four-Square Model for Design Innovation

Jeff Semenchuk, who is currently the chief innovation officer at Hyatt, noted, "all innovative companies that I have worked at use the same process, they just call the steps different things...."[4] As practiced, most design innovation processes are some variation of the four-square model.

This model is composed of two sets of polar extremes: understand/make and abstract/concrete (Fig. 31-3). When training designers, we want to develop both *understanding* and *making* skill sets. *Understanding* includes research and critical thinking. *Making* includes finding nonobvious connections, abductive thinking (which is a fancy way of saying guessing), and actually making things like models, drawings, prototypes, and communications. We also want designers to be able to think abstractly (focus on ideas and concepts, not just real objects) and concretely (in order to make something real and feasible). These polar extremes form the model and create a design process with five steps:

1. **Problem framing:** In this stage, we identify what problem we intend to solve and outline a general approach for how we will solve it.
2. **Research:** In this stage, we will gather qualitative and quantitative data related to the problem frame.
3. **Analysis:** In this stage, we will unpack and interpret the data, building conceptual models that help explain what we found.
4. **Synthesis:** In this stage, we will generate ideas and recommendations using the conceptual model as a guide.
5. **Decision making:** In this stage, we will conduct evaluative research to determine which concepts or recommendations best fit the desirable, viable, and feasible criteria.

Like all process models, the reality of practice often includes iteration, skipping of steps, and subprocesses. Also, there is an array of methods and tools employed during the process steps. The remainder of this chapter will highlight four of the five steps (research is covered in other parts of the book):

Reframe the problem: We will dedicate a large portion of this chapter to this process step, because it is can be misunderstood and is often skipped.

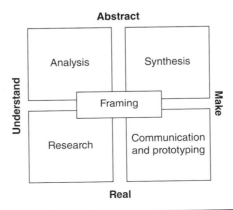

FIGURE 31-3 The four-square design process model.

Analysis methods: We will discuss four approaches for analyzing the type of data you might collect during a design project.

- 5E's
- Customer journey
- Three levels of why
- Empathy map

Synthesis methods: Although brainstorming is a key component of design innovation, it is covered elsewhere in this book. We will focus on How Might We (HMW) statements. HMW statements distinguish design innovation from other approaches.

Decision making: Here we will focus on using prototypes to reduce uncertainty related to new products and services.

Reframe the Problem

The first step in the process sounds simple but is actually the most complex—reframing the problem. Our goal is to restate the problem at the right level of focus and abstraction in order to ensure our process is effective and the results are aligned with the needs of the business.

Imagine that you are a store manager for Starbucks. Your shift manager from the afternoon shift gives you a call—noting that a new employee is making the drinks very slowly and should be fired. What would you do?

When faced with this situation, most people would first investigate the situation before determining a course of action. You might talk to the new employee, observe them working, or even talk to the other baristas to get their point of view. In short, you would make sure that the problem as framed is actually the real problem. You certainly would not just fire the employee.

At a conceptual level, the problem was narrowly framed (the new employee is slow) and there is a solution in mind (fire them). This is the same problem structure that is often provided to designers (for example, "We are losing market share so make a new website"). In the store manager scenario, we feel compelled to reframe. Too often in the design scenario we proceed working on the proposed solution.

There are powerful and hidden biases that lead to projects being framed improperly. Table 31-1 outlines some of the most common human biases that drive this condition.

Experienced design leaders are always looking for these biases. It is not uncommon for a problem frame to include several biases. Once they are detected, the team can work on actively addressing them. This chapter is not focused on the social dynamics of problem framing.

When working on a design innovation project, it is important to frame the problem in such a way as to allow for exploration of issues and solutions. This will ensure that you develop concepts that are new and novel, not just incremental improvements or a rehash of

Bias	Narrow Framing	Anchoring	Attribution
Description	People tend to frame problems with short-term, tactile focus. People overvalue what is happening now, undervalue the impact of the future state.	People often anchor to a piece of information (or an idea) early in the project, even if it's not a solution idea.	People often identify the wrong problem drivers. People rarely think they can be the problem.
Implications	Design often requires a long timeline. Most problems are not framed as true design challenges.	People may not accept new ideas / information if they have already been anchored.	If we solve for the wrong drivers, the problem will not actually be resolved (and could get worse).

TABLE 31-1   Framing Biases

an existing idea. If the frame presented is not relevant, and you move quickly to the proposed solution, you will find yourself in *repair service behavior*.

In repair service behavior, the designer acts like a repair person—you are constantly in reactive mode, fixing small problems. Moreover, if the problem is not solved by your simple fix, you are blamed for the failure (even though the reason for the failure is more likely the incorrect frame).

To frame projects correctly, the designer should

- Work at the right level of abstraction.
- Focus on activities, not problems and technologies.

Work at the Right Level of Abstraction

Consider two projects:

1. Design the next version of the iPod Nano that can be released next year.
2. Design a new way to experience music.

Which project is likely to produce more innovative, creative results? The second project (a new way to experience music) has been framed in a very abstract way. The project has no clear solution and contains a wide variety of potential activities. Of course, there is a downside to this level of abstraction. The results of this project will be highly variable—meaning that you do not know if the concepts will even be implementable or usable.

On the other hand, the first project is more tightly framed—you *will* get a new version of the Nano. As long as the team followed the charter, the results will not be variable—you will get minor improvements to the existing music player.

Neither framing is necessarily correct or better—until you include organizational context. If an organization is trying to create a breakthrough innovation, they should frame the project as abstractly as possible. If they need a new product that can be released the following year, the frame should be more concrete.

The *abstraction ladder* is a powerful tool for problem framing. Designers can use the ladder to show how making a problem more or less abstract will result in different outcomes. This helps the organization match the problem frame (and degree of abstraction) to the business need. The example in Fig. 31-4 shows how the abstraction ladder can be used to frame a project on air travel.

In this example, a very concrete project might be, "add a button to the airport kiosk that allows the passenger to enter their frequent flier number." A slightly more abstract project would be, "re-imagine the check-in experience." If your airline was receiving thousands of calls per day asking for a button to add a frequent flier number, project one is likely correct and spending time thinking about the "check in experience" will only further aggravate customers.

However, experience tells us that most projects are framed at too narrow and concrete of a level to fully leverage the design process. A more abstract frame results in more varied and unique ideas. The power of design often relies on the power of abstraction—it gives us a larger canvas and allows for a more deep and thorough exploration of an opportunity space.

Good design leaders gauge the tolerance for abstraction of their clients, and then challenge them just enough (by making the problem slightly more abstract) to ensure that the solutions feel different and new. As part of our process, the designer should work with the client to identify their ambition as well as the overall willingness to change. Although all client situations are different, the following chart can help guide your conversations (Table 31-2).

Note: Most design/innovation projects leverage the last three levels of abstraction. The most concrete level of abstraction (functions and features) often does not require all of the steps of the following process in Table 31-2.

To help illustrate this step in the process, imagine that an office furniture company has asked your design team to develop a new podium or lectern. During your initial conversations with them, you have determined that they do not want simple improvements to their existing line of podiums, but instead want a more breakthrough product aligned with new types of classroom design. The following example shows different levels of abstraction and the resulting projects (Table 31-3).

In this case, the team eventually picked *lecturing* as the right level of abstraction for the project. This decision was supported by the fact that you did not know enough about how lectures are currently delivered to immediately narrow down the frame to *delivering a lecture to 10 to 30 people* (or any other more narrow problem frame, the team had no facts to support

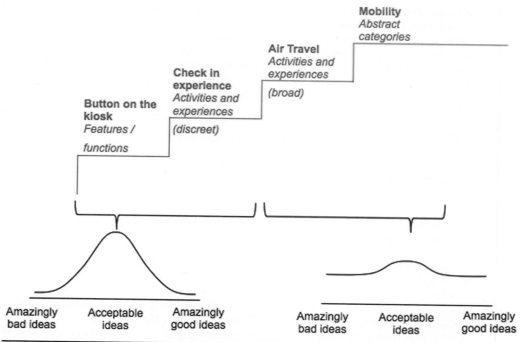

FIGURE 31-4   The abstraction ladder.

Frame As	When the Client Needs
Functions/features	Near-term, reliable improvements
Discreet experiences/activities	To improve an existing business unit or specific product offering
Broad experiences/activities	To invent a new business unit or create a product portfolio
Abstract categories	To reinvent the company or industry

Note: Most design/innovation projects leverage the last three levels of abstraction. The most concrete level of abstraction (functions and features) often does not require all of the steps of the process above.

TABLE 31-2   Levels of Framing Abstraction Matched to Business Needs

Level of Abstraction	Podium Example
Functions/features	Design a more flexible (in terms of technology) podium
Discreet experiences/activities	Delivering a lecture to 10–30 people
Broad experiences/activities	Lecturing
Abstract categories	Learning

TABLE 31-3   Example of Abstraction Ladder for Framing a Project

such a narrow frame). Tackling *learning* seemed too broad—this was such an abstract category that the team may not even create a product for the classroom. The most concrete frame *design a more flexible podium*, seemed too narrow, plus there was a team already working on those types of improvements. The frame you selected is fairly broad, and will give the team plenty of space for exploration and concept generation.

Focus on Activities Not Problems (or Technology)

Perhaps the most powerful quality of user-centered design is that we focus on designing for user activities, not just improving existing components. This notion is highly aligned with the concept of *growth platforms*[5] being advanced by leading business schools. The underlying premise of the growth platforms theory is that successful new businesses are built on underserved customer activities. This is not news to most designers, but it does lend credibility to the approach of user-centered design.

Successful innovation projects will focus on designing for user activities and then later thinking about organizational capabilities and business models. This does not mean that we ignore the operational and economic issues of new ideas, only that we start the process by identifying what people are really doing and what they need.

It can be a challenge to frame projects in terms of activities, since projects are often presented as problems (we are losing market share) or technologies (a faster way to open garage doors). The goal is to identify the underlying activities related to the problem or the technology (we are losing market share in the Health and Beauty section, so we will focus on the activity of *aging gracefully*; and instead of garage door technology we might focus on *quickly accessing my car*).

To further this discussion, we should revisit two overused but still relevant quotes:

- "We can't solve problems by using the same kind of thinking we used when we created them." (attributed to Albert Einstein)
- "The customer wants a hole in the wall, not a drill." (Harvard Business School marketing professor Theodore Levitt[6])

If a project is focused around a problem (increasing market share), we have already limited the project to a limited range of solutions. Chances are that people at the organization have already thought about and tried to change all of the obvious levers related to market share. So, focusing the project on this area will ensure that your work feels derivative and repetitive. This is related to the Einstein quote—market share is likely still a problem since the company is trying to solve it as a market share problem. The actual drivers of the underlying issue likely live outside of the company-centered problem definition.

The quote about "hole in the wall, not a drill" helps us understand that organizations are often built around technologies (manufacturing better drills). So, when they want to innovate, they see the world through the lens of new drills. As designers, we can help them understand new opportunity spaces by refocusing them on the customer activities that originally made drills relevant, but may have changed or become more nuanced since the original technology was developed.

Since the next step of the process is to create a framework for analysis based on a high level activity, it is important to spend some time making sure you have selected the right activity. You may try identifying several candidate activities, and then outlining the implications of selecting each activity. For example, in the health and beauty example, the activity of aging gracefully was selected since the activity was aligned with the client organization's target customer.

To help select an activity, you may employ the distinctiveness/relevance map. We will use this map later in the process as well—it is a key tool for decision making since we should always be looking to work on activities or ideas that are distinctive (meaning that our organization or team is unique in pursuing this activity, it is an open opportunity space) and relevant (meaning this activity is important to an attractive group of users). Below is a map from our health and beauty example (Fig. 31-5).

Activity Models: The 5E's Framework

The 5E's framework may be one of the most used frameworks in all of design innovation. The framework helps you break down and analyze existing experiences—it will also be helpful in synthesis later on. It is best used on experiences where there is a clear beginning and a clear end. For example, you can use the framework to break down the experience of a single lecture. However, if you wanted to show how, over a longer period of time, a lecturer developed their content, you would likely use a different framework—the customer journey, specifically.

An experience model outlines and identifies the different phases a customer goes through during an end-to-end service experience. Breaking an experience into discrete segments allows for more in-depth analysis (i.e., sequencing, pacing, time). Pleasure and pain moments can be overlaid onto the model to provide more subjective analysis around the overall experience, and highlight what is working well and what needs to be addressed.

There are five stages to the framework, outlined below (Fig. 31-6).

We will use a table format to fill out the framework. The rows have been laid out in order—try to complete the top row first before continuing to the next row (Table 31-4). You should be

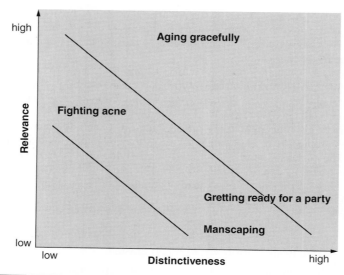

FIGURE 31-5 Distinctiveness/relevance map.

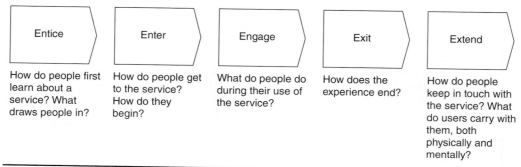

FIGURE 31-6 The 5E's model.

	Entice	**Enter**	**Engage**	**Exit**	**Extend**
User activities at this stage					
What users want at this stage					
What is provided					
Resulting gaps (underserved activities and unmet needs)					

TABLE 31-4 5E's Model Template

using data from your spreadsheet to fill out the first three rows. Some interpretation (or filling in the gaps) is acceptable, but if you are missing large chunks of data you should review your research and potentially conduct additional, targeted research to fill in any holes.

This activity is best completed in stages. Have one of the team members take a first pass at the table, filling in the obvious cells. When complete, have the team spend time working through the table, adding and deleting where appropriate.

The goal of this exercise is to generate a set of insights across the user's experience.

The result of the exercise may be useful in a top-line report to the client or manager. Again, it should not take the place of a custom framework, but it can help frame your initial insights.

You may work on design problems where you are required to design for more than one user. In the case of lecturing, it is feasible to design for the instructor, students, and even the instructor's support staff. When it is appropriate to consider multiple users, use the following refinement to the 5E's framework that includes the key elements of service design.

This refinement results in the development of a service blueprint—you will document the experience from both a customer and employee perspective. According to Denis Weil, one of

	Entice	Enter	Engage	Exit	Extend
Customer activities					
Onstage employee activities					
Backstage employee activities					
Support processes					
What users and employees want at this stage					
Physical evidence (key objects, environments, interactions)					
Pleasure and pain moments					

TABLE 31-5   5E's Model Service Design Template

the inventors of service design: "Juxtaposing the customer and employee experience should highlight issues and, in turn, potential opportunity areas for mutualistic innovation."[7]

The components of this framework include

- *Customer activities*: The activities of the customer during the service experience—what are they are doing.
- *Onstage employee activities:* The activities of the service staff that interact with the customer.
- *Backstage employee activities:* The activities of the service staff that do not have physical contact with the customer.
- *Support processes*: The systems and tools that support service delivery.
- *Physical evidence*: The tangible elements of the service that the customer interacts with.
- *Wants at this stage*: Both the expressed and latent needs for both customers and employees in each stage.
- *Pleasure + pain moments:* Moments in the service that delight and frustrate both the customer and the staff.

Here is the table laid out (Table 31-5) to include the elements of service design.

Activity Models: Customer Journey

Sometimes we are studying an activity that is not contained in a single experience, but rather grows and changes over a lifetime (or other extended period of time). When this is the case, we can use the customer journey framework. As the title suggests, this framework outlines the main stages of a typical journey, starting with discovery through trial, achieving mastery, and eventually settling into a period of maintenance.

The value of this framework comes from understanding how and why users move from one stage to another, as well as why they may drop out at each stage. So, if you are trying to understand golf, you will derive insight into why so many golfers leave the sport before they can achieve mastery.

The phases of the journey are as follows (Fig. 31-7):

- *Discovery:* The period of time where the customer initially finds out about the activity. We need to know what draws them in and keeps their interest. The period of discovery is critical—this is often called the *first moment of truth*,[8] when a user discovers a new product or service. Although often the shortest of the phases, it is important to know the range of entry points for any activity.
- *Trial:* Once discovered, the user begins to "try on" the new activity. The trial phase is often where they experience the activity for the first time (considered the *second moment of truth*) and determine if it fits into their life and lifestyle (or, it may positively shape their life and lifestyle). It is important to note that in most cases in the trial phase, the connection between the user and the activity is tenuous—they may leave it at any time.
- *Mastery:* During trial, the customer begins to learn and improve. They leave the trial phase once they achieve mastery—this usually requires an investment of time from

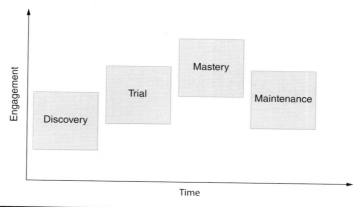

FIGURE 31-7   The customer journey framework.

	Discovery	Trial	Mastery	Maintenance
User activities (what are they are doing)				
Objectives (what are they trying to achieve)				
Defining characteristics				
Primary needs				
Triggers to next step				
Barriers to next step (why they drop out)				

TABLE 31-6   Customer Journey Template

the customer. In mastery, the user has a strong connection to the activity—it may define them and their lifestyle.

- *Maintenance:* It can be hard to remain a "master" of any activity over an extended period of time. Often, customers slip into a phase of maintenance where they are no longer engaged at the same level of mastery.

The following map shows how these stages unfold over time:
Table 31-6 will help you develop insights.

Activity Models: Three Levels of Why

This tool is less a framework and more a process. The goal is to isolate an activity that seemed newsworthy and/or unexplained and try to determine *why* it is happening. The *five whys* were originally parts of root cause analysis for manufacturing quality improvement. We have borrowed this approach in design innovation.

For example, while studying recycling in the City of Chicago, a design team observed that many residents stored their recycling garbage cans in their garage. This is counter intuitive, since garbage cans often attract rodents and bugs, and putting an empty food container in your garage may bring unwanted visitors inside your home. So, we started with the observation, and kept asking *why* to try and develop a deeper level of explanation (Fig. 31-8).

There are limitations to this approach—it is easy to overthink or overanalyze an activity, giving it more meaning than actually exists.

This tool is best used when your insights feel superficial and not deep. Often this approach will suggest additional research. You can ask why more than three times. At some point the answer will be "that is obvious, silly, or too pedantic." Back up one level and you should have a deep insight.

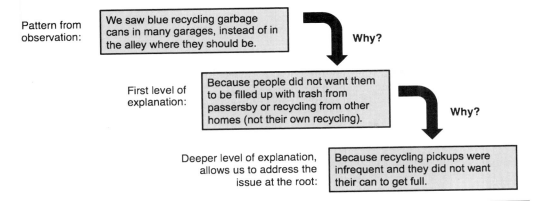

Pattern from observation:

We saw blue recycling garbage cans in many garages, instead of in the alley where they should be.

Why?

First level of explanation:

Because people did not want them to be filled up with trash from passersby or recycling from other homes (not their own recycling).

Why?

Deeper level of explanation, allows us to address the issue at the root:

Because recycling pickups were infrequent and they did not want their can to get full.

FIGURE 31-8   Three why examples.

User Models: The Empathy Map

The empathy map is a simple but powerful way to capture what a user is thinking, feeling, and doing related to the activity you are studying (Fig. 31-9). These maps are particularly useful when working with clients—they provide a comprehensive but digestible snapshot of the users and their needs. This approach was popularized by "d.School" at Stanford.[9] There are many variations of this method, but it remains one of the easiest to implement and is very useful for design innovation.

When filling out the framework, the more specific the user or customer the better. So, "busy moms with more than two kids" is better than "people in North America."

You can fill out the template using a composite user (single template for all of our busy moms noted above). This will help you generate broad insight about the target group. You might also try filling out one template for each of the individuals you observed. You can then generate insight comparing the different results.

User Name

Enter the name of the target customer or user (the individual or group for whom you are filling out this template).

What Are They Thinking?

Enter three to five statements that summarize what people are thinking about an activity. These statements require you to do some interpretation. It can be hard to determine exactly what people are thinking. The best way to write these statements is to imagine their internal dialogue—what they are saying to themselves as they prepare for and engage in this activity. These statements are often more personal and less confident than what they are saying out loud.

User name

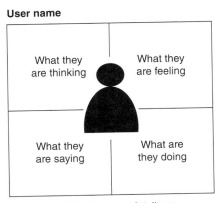

What they are thinking

What they are feeling

What they are saying

What are they doing

+ ...What they won't tell you

FIGURE 31-9   The empathy map.

What Are They Feeling?

Enter three to five statements that summarize how people are feeling during the activity. Again, these statements require some interpretation. You can ask people how they feel, but you should also base these statements on your observations. Write these statements as effect or cause; for example, they are tired because the commute to the office was long. So, state how the user is feeling, and then if possible detail the reason they are feeling that way. Feelings can be written as adjectives like tired, confused, excited, and so on.

What Are They Saying?

Prompt: Enter three to five statements the user is actually saying about the activity. These statements should be quotes from the users or customers. The other sections of this template require some interpretation. This section, however, should be primarily verbatim. It is fine to use quotes that summarize what people are saying, but be sure it is actually something they would say.

What Are People Doing?

Write down at least five activities that people are doing as part of the larger identified target activity. This requires no interpretation: you should note the patterns of activities you observed. Use the gerund phrase structure again. This would include activities like *picking vegetables* or *updating the shopping list* and *browsing granola bars*.

What Won't They Tell You?

Write down one to five things the user would not want anyone to know (related to the activity). This is often called the *dirty little secret*. This might be something like "I don't want people to know how little I planned." When looking for insight, these statements can help unearth the real conflicts and needs of your users.

When the map is complete, use the four questions below to help generate insights:

- What was new? Was there anything that surprised you?
- What was different? Was anything different than you thought it would be?
- What was frequent? Did you see a pattern?
- What do you want to act on? Was there a problem or need you feel compelled to solve?

"How Might We" Statements

Engineering design is focused on creating detailed, measurable user requirements. Any legitimate design project will include a requirements definition. However, requirements definition can stifle innovation at the early stage of a project. Teams often start to generate requirements (what the solution should do) before they have exhaustively explored what the solution could be.

Requirement definition is better saved for the decision-making phase of an innovation project. When we are in synthesis, we are trying to explore all of the possible solution directions, even if some of those directions are not feasible. To ensure that you generate a large set of creative ideas, it is helpful to frame your brainstorming sessions with "how might we" (or HMW) questions.

Good HMW statements are derivative of your insights, and include a constraint. Below is an example development of an HMW statement (Table 31-7). In this case, the design team was working with the Judd Goldman Adaptive Sailing program. The team discovered that parents of disabled youth preferred sports where they could watch their kids play and interact with other parents. The insight was phrased as "there are no sidelines in sailing."

When developing HMW statements, consider:

- You may need to write two or three HMW statements for each insight—this is why it is helpful to distill all of your research into a single driving insight.
- A typical project may include up to nine HMW statements. If your HMW statements are good, the ideas should generate themselves. If you can't see good ideas coming from your statements, try rewriting them to be more specific, less high level, and add more detail to the constraint and context.
- HMW statements should have teeth: this means that the statement will send you in a specific direction and is not overly vague.

Process Step	Example	Tips
Start with your driving insight	"There are no sidelines in sailing"	If you have several insights, you need to create an equal number of HMW statements.
Rewrite your insight as a problem that needs to be solved, starting with the phrase "how might we"	"How might we better integrate parents and caregivers into the sailing experience?"	Your insight may require several problem statements; it is better to have several statements and know the problem is covered compared to trying to make one single statement.
When appropriate, add a key constraint to the statement	"How might we better integrate parents and caregivers into the sailing experience without compromising the freedom and self-direction that are so critical to sailing?"	Adding the constraints makes your HMW statements naturally more challenging to answer. Great ideas are often the answer to a tension or a paradox. Adding the constraints will help you create a tension or paradox.

TABLE 31-7 Developing "How Might We" Statements

- HMW statements should present a clear challenge or paradox; this is usually because there is a good constraint or element of context.

- HMW statements should be controversial or have a counterargument—you should be able to make an alternate argument (no, we should be designing this way...); this means that the HMW statement is clear and thoughtful.

- HMW statements should not have a single solution in mind or include a solution in the text.

- A set of HMW statements should not be variations on one theme.

Prototyping

Much like design, the word *prototype* can be both a noun and a verb. We will not discuss how to build prototypes in this chapter. Instead, we will focus on the verb form of prototype. Specifically, we will address how prototyping can help reduce uncertainty about new concepts.

Proof, during innovation projects, can be elusive. Even with surveys and focus groups, it can be hard to prove that you are pursuing the right concept direction, or even that your concept is valid. Since ideas are often abstract and vague for a good portion of the project, it can be challenging to derive a definitive yes or no from potential customers. If your concept is truly new and breakthrough, there is no way to gauge likely purchase using a concept board.

Instead, we want to isolate the key assumptions required to be valid in order for our concept to succeed. Once we have identified these assumptions, we can design simple prototype experiments to reduce uncertainty. Again, the only way to prove that the concept will work is to launch it into the marketplace. But, with iterative prototyping, we can answer key questions and gather data about possible outcomes. This will help us make more informed implement or don't implement decisions.

In the spring of 2012, Hyatt approached the Institute of Design in Chicago with the challenge of transforming the experience of the hotel's meeting attendees.

The student team started with hours of interviews and observations of the behaviors of meeting participants during breaks outside of the meeting rooms.[10] It became clear that there were no accessible quiet or private places where people could make phone calls or catch up on e-mail. Many of the guests ended up on floors or in corners to make their calls.

The team developed an early concept called the *chatter box* with the goal to accommodate private phone conversations in a small, soundproof booth—think telephone booth without the phone. The chatter box could coexist within the noisy and crowded breakout spaces outside of meeting rooms.

The students built a very simple mock-up of the chatter box—it was full size but made from acrylic glass, acoustic foam, and foam board. Like all prototypes, the first iteration answered a question: If we offered a private booth for phone and e-mail, would guests know to use it?

Guests immediately started using the chatter box, but it was not quite the glowing success the team initially hoped it would be. Many of the first guests to try the chatter box agreed it was addressing an unmet need, but felt the design was missing key elements. They expressed a need for more space, a comfortable place to sit, and a counter for resting their things.

Since guests were so explicit with their feedback (one of the benefits of a full-scale prototype), the team redesigned the concept to provide greater space and comfort. To save money and time, the next prototype was built using some of the materials from the first prototype. The team also used existing stools and shelves at the hotel.

The second round of testing yielded two important lessons. First, the guests wanted increased privacy, but also the ability to see out of the box (the old social science concept of refuge and vista). The second insight was that making calls wasn't the only way guests stayed connected. Guests needed amenities that catered to their entire suite of work devices.

In the next version of the prototype, the team reframed the idea from a box to make phone calls to the *privacy pod*. They observed a range of activities (talking on Skype, charging devices, shopping, and making travel arrangements) that needed to be supported by the concept. Instead of accommodating a single user, it needed to cater to multiple users at the same time. Furthermore, it needed to serve as a robust and welcoming workspace that guests could use for extended periods of time.

Again the team used inexpensive materials to build the prototype. For the final version, the teams worked to better integrate the privacy pod into the overall hotel environment. The decision was validated almost immediately when guests began using the pods with no encouragement from the team. The pods and guests needs turned out to be a natural fit. The pods enabled guests to breakaway from large meetings and noisy public spaces for productive work time. One guest went as far as to host an online lecture from the pod. It was not uncommon for guests to spend up to 1 hour in the pod making a conference call.

This cycle of iterations took less than 2 weeks and cost under $500 in materials. The process did significantly reduce the uncertainty of launching the concept system wide for the hotel. The feedback helped make the concept stronger, and was based on real needs (not the needs one might perceive while reviewing a rendering of the chatter box).

In order to engage your users for more detailed feedback, particularly related to the main assumptions of the concept (our assumptions about why it will be adopted, how it will be used, and so on), you will need to design an experiment and build rapid prototypes. Remember, you are prototyping to answer questions, not to create a representation of what the concept will look like for implementation. Below is a simple template to help you plan an experiment like the Chatter Box case study above.

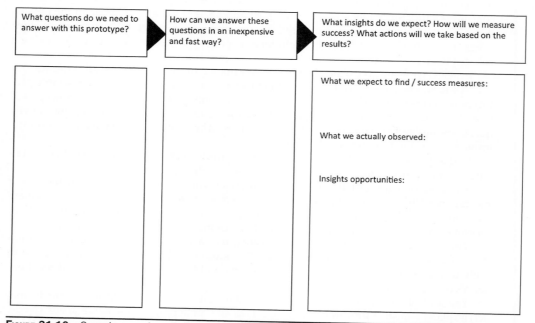

FIGURE 31-10 Sample experiment plan.

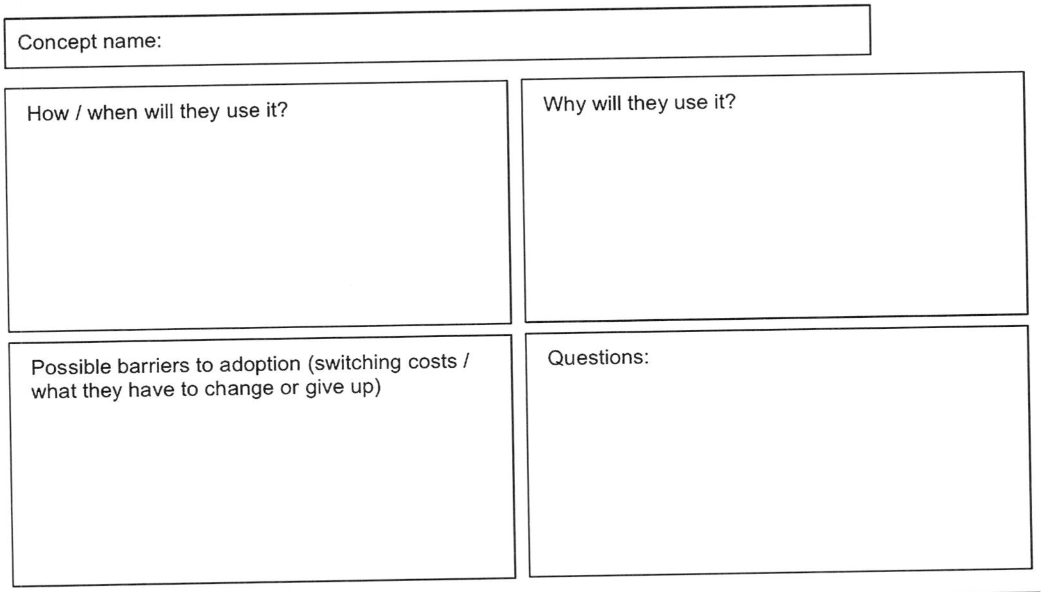

Concept name:

How / when will they use it?	Why will they use it?
Possible barriers to adoption (switching costs / what they have to change or give up)	Questions:

FIGURE 31-11 Prototyping feedback form.

Here are some tips for planning effective experiment designs:

- Look for creative ways of prototyping, like skits or full-scale mock-ups.
- Always plan what to build before you build it.
- Prototypes should be full scale and interactive: the user should be involved in the prototypes. This should not be a passive experience.
- Never try to sell your prototype; do not lead the user to give you the feedback you want to hear.
- Set context: Explain when and how they will encounter the concept in real life. Make this as real as possible. Also make expectations for feedback clear. You want to know what worked and what did not, and how closely you came to meeting their needs.
- Ask them to narrate: When appropriate, ask users to narrate the experience—what they are thinking and feeling. Depending on the cycle time of your prototype, you may ask them to move through the interaction once silently, and then go through it again while narrating.
- Seek feedback: Use the form to get feedback on the concept (Fig. 31-11). Be sure you are addressing the key areas of risk/concern/assumption. The form above has proven effective for gathering feedback from a prototype experience. Note that it does not ask people if they would buy or use the prototype, but rather how they would use it. The more clear and concrete the user is when describing how they will use the prototype, the more likely they will actually use it.

References

1. Berner, Robert. "Design Visionary." Bloomberg Business Week online, June 2006.
2. Piskorski, Mikołaj Jan, Chen, David, and Knoop, Carin-Isabel. MySpace: Harvard Business School Case Study, July 2, 2008.
3. Boyd, E.B. "How Facebook Finds the Best Design Talent, and Keeps Them." www.fastcodesign.com, April 4, 2012.
4. Semenchuk, Jeff. ID Strategy Conference. https://vimeo.com/43850000, 2012.
5. Laurie, Donald L., Doz, Yves L., and Sheer, Claude P. "Creating New Growth Platforms." *Harvard Business Review*, May 2006.
6. Levitt, Theodore. "Marketing Myopia." *Harvard Business Review*, 1960.

7. Paynter, Ben. "Making over McDonald's." *Fast Company*, October 1, 2010.
8. Berner, Robert. "P&G: New and Improved." *Business Week*, July 6, 2003.
9. http://dschool.stanford.edu/wp-content/themes/dschool/method-cards/empathy-map.pdf.
10. Espinosa, Angelica, Strawn, Brian, Skelton, Will, and Duke, Kyle. "Hyatt Chatter Box Case Study." IIT Institute of Design White Paper, Spring 2012.

CHAPTER 32

Service Innovation: Introduction, Methodologies, and Key Findings

Navneet Vidyarthi and Rajesh Tyagi

Introduction

Service innovations are a firm's complete management activities focused on developing and marketing new services to meet customers' needs. Literature review shows that innovations are critical for business development and survival. Research on product development is very large and varied; however, research in service innovation is still very limited. Unlike manufacturing companies that take advantage of product features and characteristics, knowledge-intensive business firms need to find new methods to ensure success of new services. Services and service innovations are key drivers of economic growth and knowledge-driven services, but the research is very limited. Some literature review suggests that the models from product innovation can be adapted for service innovation by incorporating service characteristics; however, services are known to be intangible, inseparable, and perishable in nature, thus more research is required in this field. Service innovation extends to service quality, service encounters/experiences, service design, relationship marketing, customer retention, and internal marketing. Factors that greatly affect service innovation are organizational culture, information systems, operation processes, and human resources.

In one form or another, service innovations are being adopted by various industries like hotels, telecommunications, electronics, media, airlines, banks, and health care, among others. Various studies have developed models and processes to define, measure, and implement innovation. Service innovation focuses on innovations beyond technological innovations and focuses on organizational or operational innovations that firms can adopt based on their resources and competencies. The various kinds of service innovations that have been developed are discussed below.

Types of Service Innovations

Innovation typologies have been defined either with respect to their outcome or initial focus. Most of the service innovations can be categorized into one of the following groups:

- Incremental or radical based on the degree of new knowledge
- Continuous or discontinuous innovation depending on its degree of price-performance improvements over existing technologies
- Sustaining or disruptive innovation relative to the performance of the existing products
- Exploitative or evolutionary innovation in terms of pursuing new knowledge and developing new services for emerging markets

Research indicates that if a firm uses customer orientation, it is more likely to adopt incremental service innovation; whereas if it uses competitor orientation, it is more likely to adopt radical service innovation. Also, incremental and radical service innovations lead to greater market performance and in turn, better financial performance. Yet, another classification of types of innovations comprises four broad aspects: product or service, production or process, organizational, and people innovation. Studies have also classified innovation as an intentional activity, a semi-intentional activity, and a bricolage. For management, innovation is intentional and imposed; whereas for employees, it is bricolage and intrinsically motivated. Another classification includes an innovation radar covering 12 types of innovations that affect business performance based on (1) product, (2) platform, (3) solution, (4) customer need, (5) customer experience, (6) brand, (7) process, (8) value capture, (9) organization, (10) ecosystem, (11) channel, and (12) supply chain. The innovation radar helps streamline innovation strategies across functional areas as well as organizational structure. It also helps in benchmarking competition and identifying opportunities to grow and improve efficiencies.

Importance of Service Innovations

New service development (NSD) is an important aspect for business growth and survival. Firms in experiential services study service innovation in order to better understand customer perspectives and experiences. Through service innovations, firms can develop on a number of dimensions like customer solutions, customer experiences, supply chains, and operations. As business evolves, companies can change one or more dimensions of the innovation radar and create substantial new value for customers. Firms need to constantly innovate not just for better performances and results, but also to maintain a competitive advantage. New service development relies on the expertise and cooperation of individuals working in teams during and after development. NSD and its basic equivalent of new product development (NPD) is increasingly seen by managers as embracing more than the improvement of core performance attributes. Factors affecting innovation include innovation orientation, external partner collaboration, and information technology capability. Innovation orientation and IT capability were found to be the key drivers that lead to service delivery innovation that further leads to improved financial and nonfinancial performance.

How Companies Get Involved in Service Innovation

Many organizations adopt various models and frameworks that best engage their resources and competencies. For example, health firms have transformed complex multiagent frameworks involving policy makers, public service organizations, and consumers into a model for health services innovation. Various studies have helped frame the strategies for innovation in health services by dividing the processes into front-end or back-end competencies, depending on interaction with clients. Five dimensions of a service innovation model have been identified: organizational, product, market, process, and input innovation. Each of these may further undergo radical or incremental innovation. Also, to successfully innovate, firms must communicate those changes internally so that employees perceive the change as needed and also to successfully adopt the change. Literature review suggests the involvement of frontline employees in the NSD process specifically in firms where the final product is a result of the interaction between the customer and employee. Engaging employees in the process also helps in better implementation and adoption of the new service. The healthcare system faces huge barriers with its retiring workforce, increasing number of elderly patients, cost-efficiency demands, and expectations of high-quality care that utilizes all the latest advances in technology and knowledge. Involvement, motivation, and support in the form of training consultants and technical support when provided by key stakeholders in the organizational context have a positive effect on the adoption of innovation practices. In another example, an innovation and development team was created at the Bank of America to carry out rigorous service and product innovations along with service delivery techniques to strengthen customer relationships. They developed a test market in one of the branches to accurately measure and compare results with other control branches for higher customer satisfaction and revenues. A five-stage process was developed and it led to high customer satisfaction and enhanced creative and innovative thinking. Similar experiments were adopted by other branches.

Characteristics of Service Innovation

Classical characteristics of services are intangibility, and simultaneous production and consumption. The intangibility aspect creates a challenge for managers as service innovation needs to go through prototyping and testing phases. Even small aspects of customer experience could be overlooked and could cause issues during the implementation phase.

Simultaneous production and consumption adds complexity in terms of errors and strategies of quality management. Customers might be involved at each, some, or none of the stages of service innovation. These stages could be classified as initial stages, execution or delivery stage, and later stages (commercialization). Within delivery, we can think of use technology, experience, and channels. User involvement is usually more intense at initial stages (idea generation, idea screening) and at later stages (commercialization). Service components are intangible and are in the form of processes, human skills, knowledge, competence, and material.

Luteberget[1] identified research models based on service characteristics, innovation process, form of innovation, and service performance. The inputs include (1) service characteristics: intangible, inseparable, heterogeneous, perishable, and information; (2) innovation process: involvement and customer characteristics; (3) form of innovation: innovativeness and innovation type (service or process); and (4) service performance: process quality, customer value, market performance, and profitability. Results based on interviews and surveys indicate that innovation process and form of innovation affect service performance while service characteristics have a moderating effect on innovation process, service performance, and form of innovation.

Challenges in Service Innovation

Management of service innovation is a daunting task from multiple perspectives falling into the categories of *What* and *How*. The *what* category includes the focus area of service innovation. The first important aspect is the understanding of the service innovation itself. Innovation can happen along any dimension, as indicated in 12 dimensions of the radar.

The *how* aspect provides ways to implement the service innovation strategies and ways to protect the competitive advantage. Service firms do not have traditional research and development departments and often cannot protect their innovation using patents. A lower percentage of revenue is invested in innovation at service firms. It could be argued that this lower percentage is due to the fact that innovation in services is not confined to just new service innovation. Incremental innovation, an essential part of the continuous improvement and innovation strategy, can easily be copied by other service firms.

Another challenge in services is due to added variability caused by customer presence in the system. Customer introduced variability is caused by variability of customer arrival, customer effort, customer perception, and capability. Any innovation in services needs to take this variability into account and needs to identify strategies either to minimize (or to accommodate) this variability.

Service firms work in partnership with other firms to offer an experience to customers. For example, an airline may provide check-in services via a hotel desk, or a financial institution could use partners to offer solutions to customers. In such a fragmented world, a service firm needs to work with partners to identify service innovation opportunities. Service delivery is often harder to scale and distribution economies often may not work favorably.

Methodologies in Service Innovation

There are number of innovation theories and frameworks proposed so far. In this section, we discuss some of them. For example, Hunter[2] conducted an empirical study proving the difference in the adoption of innovation practices between manufacturing and service firms. It highlights which factors are most common and also which factors can further be implemented for better innovation practices. The inputs to the study include adoption of innovative practices that depends on strategic human resource management, technology, HR, organizational culture, and factors in both manufacturing and services: TQM, teams, job rotation, job sharing, and flexitime. Research revealed that manufacturing and services are very different in their adoption of innovation practices.

In a study conducted by Hagedoorn and Cloodt,[3] the authors offer a multi-indicator approach to measure the effect of all the indicators on innovative performance. The inputs include innovation indicators such as R&D, patent counts, patent citations, new product announcements, and survey-based measures. Factors such as research input, size of invention activities, quality of inventive output, and level of new product development; all are measured through innovative performance. The composite approach suggested by the authors is crucial for the pharmaceutical sector.

Tellis et al.[4] attempt to fill the gap in past studies by emphasizing that all inputs do not necessarily lead to innovation. The authors suggest the importance of radical innovation and introduce metrics to measure the effects and returns. The paper developed a hypothesis to study the effect of various selected variables on radical innovation and assess the metric of radical innovation against the traditional patent and R&D metrics. The factors affecting innovation (variables): religion, distance from equator, country culture, intellectual property protection, scientific input, and company culture. The controls include citation-weighted patents, R&D expenditure as a percentage of sales, GDP, and population; whereas dependent variables include innovation output and firm level financial performance. The study found that radical innovation has a greater impact on financial performance than patents, R&D, and other control variables. It is also a powerful metric to assess the firm's performance.

Tseng et al.[5] proposed an empirical research model based on identifying the factors affecting the innovation process specifically for the hotel industry. Inputs include innovation sources: technological, organizational, and human capital. The inputs were collected through questionnaires and interviews with hotel managers. Based on the responses, they identified 10 innovation factors (mix of three sources) specifically for the hotel industry. The sample size of 116 hotels was then divided into four clusters based on the innovation factors. The study found that innovativeness is considered very important for the hotel industry and is increasing its value. Technological, organizational, and human-capital sources, if well managed, increase the firm's innovativeness. Other factors like culture, information systems, operations, and human resources also increase firm value.

Cheng and Krumwiede[6] developed a model that tests market orientation, service innovation, and new service performance in multiple sectors. The inputs to the model include customer orientation, competitor orientation, and interfunctional coordination (market and financial performance). These three main concepts were adopted from the different studies and measured. The study found a positive relationship between customer orientation and incremental innovation which leads to improved new service performance. Also, competitor orientation is associated with radical innovation.

Chen[7] extended the original study on innovation radar and identified measures to test the 12 dimensions of an innovation radar: product, platform, solution, customer need, customer experience, brand, process, value capture, organization, ecosystem, channel, and supply chain. Using in-depth field interviews and web-based questionnaires, the authors found measures to test various dimensions: reflective measures to obtain an overall metric for the actual level of innovativeness at each dimension, and formative measures to gain insight into activities or factors that contribute to the observed level of innovativeness.

Lee et al.[8] present a business model innovation that links the 3V (industry level) and 3D (firm level) models for better understanding the importance of service innovations. Their model comprises the 3V innovation model—value creation, value deployment, value appropriation—and the 3D innovation model—new service design, development (NSD), delivery. The paper presents a link between the 3V and 3D models and formulates strategies for the service sector both at the industry and firm levels.

Measures of Service Innovation

Service innovation could take place along multiple dimensions, for example, by offering new services, by providing existing using new channels, or by using a new service delivery model. Managers at service firms face additional challenges in terms of management of service innovation since patents cannot be granted for most service innovations. As discussed in literature, innovation in services could be viewed as a traditional, demarcation approach (service innovation being inherently different from innovation in the manufacturing setting) to an assimilation approach (process of innovation is considered similar in both service and manufacturing settings). *Synthesis approach* combines the demarcation approach while applying existing tools and concepts to the service setting.

Considering various dimensions of service innovation, we discuss measures of innovation under the following three categories:

- Employee engagement-related measures: Employee engagement is critical in service firms and employee engagement measures are proxy measures for service innovation.

- Idea management-related measures: Considering service innovation as a process, we could think about measures at the front end (number of ideas), execution level (length of time taken from ideation to commercialization), and back-end levels (revenue from new services).

- Other measures: Investment in innovation.

Dimensions	Measures
Employee engagement	Recruitment (employee profile/capability experience, knowledge, competencies, educational level, demographics, diversity)
	Human-capital investment (hours of training/employee, training cost)
	Productivity (satisfaction, engagement)
	Employee relations (absenteeism, work/life balance, turnover, length of employment, employee benefit, compensation)
Innovation measure	Number of suggestions (or ideas) per (full-time) employee
	New services commercialized per employee
	Percentage of total revenue obtained from new services (e.g., in the past 3 years)
Technological knowledge	Business R&D expenditures
	Acquisition of other external knowledge
Nontechnological change	Changes in strategy, management, marketing, organization
Sources of knowledge/ diffusion	Share of innovative firms cooperating with others
	High use of suppliers
	High use of customers
	High use of competitors
	High use of research institutes
	High use of universities
	ICT expenditures (ICT office machines, data processing equipment, data communication equipment, and telecommunications equipment and related software and telecom services)
Commercialization	Sales of new-to-market goods/services
	Sales of new-to-firm not new-to-market goods/services
Intellectual property	Share of firms that use patents, trademarks, designs

TABLE 32-1   Common Measures of Service Innovation*

We suggest using three types of metrics for service innovation: One measure for engagement and two others measures related to the process of innovation (for example, revenue share from new services, and velocity and amount of flow of innovative ideas through the system, such as launch rate, commercial success rate, and time to market).

Case Studies

Edvardsson and Enquist[9] conducted a study through a case analysis of the furniture company IKEA to understand the importance of corporate social responsibility and sustainability in achieving service excellence and how service innovation is affected by the firm's ethical values. The researchers concluded that innovation management is illustrated by business platforms, service brand, marketing communication, and sharing of corporate values. The researchers developed the Service Excellence and Innovation (challenges, competencies, opportunities) model that focuses on creating and managing resources to get customer support in innovation activities.

- Literature review demonstrates the problems associated with the purchasing of services since services cannot be easily measured and specified like products and also illustrates how the supply chain management of services can be improved to create better opportunities. The study showed that managers feel purchasing of services is much more difficult than goods because of their intangible and perishable nature, and because the measure of quality depends on each customer's perception. The study identified various drawbacks in mismanagement of services purchasing from lack of resources to IT to cost structures that lead to poor firm performance.[10]

*Adapted from *Investing in Your Company's Human Capital: Strategies to Avoid Spending Too Little—or Too Much*, by J. Phillips, AMACON, 2005; Tyagi, R.K. and Gupta, P. "A Complete and Balanced Service Scorecard: Creating Value through Sustained Performance Improvement." FT Press, 2008; TrendChart Report: Can We Measure and Compare Innovation in Services? by M. Kanerva, H. Hollanders, and A. Arundel, 2006.

Reference and Objective	Innovation Process
Storey and Easingwood[18]: To identify the key factors necessary to introduce new products/services in the market and consequently to study the factors affecting the performance of new products based on the type of product, type of service sector, and marketing efforts.	A conceptual model has been developed that explains the relationship between the success of the new product in the marketplace based on the product offering, marketing support, development process, and the organizational culture. 153 responses were collected from a sample of 299 with a response rate of 51%.
Safizadeh, Field, and Ritzman[19]: To study the orientations of various service organizations and to understand which orientation (front office or back office) is adopted by the best performers based on its characteristics. To study the organizational orientation it is important to understand the delivery processes of a company.	Identified 13 measures of new service performance covering sales performance, enhanced opportunities, and profitability. The empirical results showed that only 6 out of 13 were not related to the performance dimensions. The final seven performance measures are: company-product fit, staff, importance, distribution strength, communications, product quality, and service quality.
Thomke[20]: To demonstrate the importance of innovation and development for effective service innovation techniques that would strengthen the bank's relationships with its customers and increase efficiency.	An Innovation and Development Team was formed at Bank of America to carry out rigorous service and product innovations along with service delivery techniques to strengthen customer relationships. They developed a test market in one of the branches to accurately measure and compare results with other control branches for higher customer satisfaction and revenues. A five-stage process was developed: evaluate ideas, plan and design, implement, test, and recommend. Outputs included a 10% failure rate, high customer satisfaction, and enhanced creative and innovative thinking. Experiments were adopted by other branches.
Alam[21]: To document the study of New Service Development in business-to-financial service firms in India (both multinational firms and local and endogenous) and highlight their services, strategies, and operations, as well as how the global firms modify their NSD processes to better suit an emerging economy like India.	• Processes with a front-office orientation have more interactions with customers whether face-to-face or telephonic as compared to back-office orientations (customization). • Front-office orientations associated with high costs and flexibility (more use of IT). • Back-office orientations tend to have more standardized processes and thus have more practical service delivery. • Front-office orientations are capital-extensive while back-office orientations are labor-intensive.

TABLE 32-2 Innovation in the Financial Sector

- An analysis of financial services organizations was performed to study the orientations and to understand which orientation (front office or back office) is adopted by the best performers based on its characteristics. The study concluded that processes with a front-office orientation have more interactions with customers whether face-to-face or by telephone as compared to back-office orientations (customization). Front-office orientations are associated with high costs and flexibility (more use of IT) and back-office orientations tend to have more standardized processes and thus have more practical service delivery.[11]

- Researchers also explored various opportunities in the retail sector by improving transaction services or increasing the power of customers and identifying new markets or new services in existing markets. The authors developed a model to identify the key elements affecting interactive retail services and the possible opportunities for innovation. The five key areas were power of consumers, channel synergies, pre/posttransaction service, resource usage, and customer heterogeneity in interactive retail services.[12]

- Another study concluded that most of the firms do not innovate in an integrated manner and innovations are mostly driven by factors initiated outside the industry like assortment and technology. Financing, regulation, and time constraints were the main barriers hindering innovation in the hospitality industry. On the other side, firms reported that innovation activities lead to higher revenues, low costs, and improved quality.[13]

- The construction industry works in tandem with the manufacturing industry, thus more innovations are adopted rather than produced. The adoption and generation of innovations in the construction industry lead to a lot of changes in systems, procedures, and skills. The key success factors identified were business strategies (building and enhancing relations, attracting new clients, and awareness of business issues), human resources, and technological capabilities for better innovation practices and strategies.[14]

- Researchers studied the complex factors affecting the health sector and identified how these interactions determine the success of innovations. Service characteristics do not always depend on product characteristics, especially in the health sector for services like surgery and pharmaceutical drugs. The key difference between service and product characteristics is based on whether the industry is knowledge-intensive or capital-intensive. Policy makers' competences and preferences are affected by service providers and, on the other hand, policy makers influence the innovation of service providers and users.[15]

- Service innovation greatly affects customer experience and behavioral intention, and experiential marketing is an important criterion for service innovation in ethnic restaurants and the hospitality industry. The researchers introduced the idea of servicescape for innovation in various dimensions like equipment, music, and color of the restaurant.[16]

- Researchers have emphasized that service innovations move beyond physical products, and high technology and innovation activities in investment banks are incremental rather than radical. Service innovation is distributed in organizational culture and it should be a continuous process. Service innovation is directly linked to hiring/promotions and is greatly affected by senior management's leadership qualities.[17]

References and Objective	Innovation Process
Windrum and García-Goñi[22]: To demonstrate the complex interactions between service providers, patients, and policy makers and how these interactions with the different types of innovations determine the success of innovation in the health sector.	The authors extended previous models on innovation by encompassing five types of innovations: organizational, product, market, process and input innovation, as well as the economic, social, and political spheres that affect the selection of an innovation in both services and manufacturing. Process/product characteristics as identified: tangible assets, intangible assets, human resources, and organizational resources. Based on literature review and empirical analysis, various innovation models have been modified and key characteristics of service innovation have been identified. The models provide a distinction between radical and incremental innovation.
Fuglsang[23]: To demonstrate that repetition and impact are key characteristics for service innovation to build skills and routine. Service innovation process is seen as implementation, diffusion, and replication rather than radical or intentional to demonstrate the importance of a process approach over practice-based approach.	By taking "elderly care" as a case study, researchers identified three types of innovations and distinguished among them as either process-based or practice-based innovation to demonstrate various key factors involved in innovations.

TABLE 32-3 Innovation in the Healthcare Sector

Reference and Objective	Innovation Process
Edvardsson and Enquist[9]: To understand the importance of corporate social responsibility and sustainability in achieving service excellence and how service innovation is affected by the ethical values.	The study is conducted through a case analysis of the furniture company IKEA and empirical evidence was collected from four firms based on their long-term commitment to value-based services and success factors. The researchers extensively studied documents, conducted interviews and narratives, and recorded observations.
Djellal[24]: To analyze the changes in the cleaning industry with the increasing use of information technology like robotics, tools, and small machines, and how these changes affect employment opportunities. The authors have highlighted innovation trajectories applicable to the cleaning industry.	The study has been conducted through qualitative interviews with managers from different firms to review the areas in which innovation has taken place to form innovation trajectories and then link those trajectories to employment opportunities.
Rothkopf and Wald[25]: To emphasize innovation activities in commoditized service industries like the passenger airline industry and to determine various characteristics to develop a framework for innovation activities in these industries.	The paper focuses on a qualitative analysis of 30 airline firms selected by a sampling method based on different criterions. The selected firms were analyzed in-depth to understand their innovation processes. The paper also uses case studies of two firms to stress that innovation practices differ between companies based on their market position.
Voss and Zomerdijk[26]: To study organizations delivering experiential services to the customer and focuses on experiences rather than functional benefits. Study includes empirical data from 17 companies along with interviews with senior managers.	The study targeted two types of organizations: experiential services and design agencies, and consultancies. Various dimensions of NSD were identified: process, market research, tools and techniques, metrics and performance measurement, and organization.
Tseng, Kuo, and Chou[5]: To study and evaluate the importance of innovation in the hotel industry, identifying various innovation configurations that affect hotel performance and to identify various indicators to measure performance in terms of quality and customer satisfaction.	Depending on the organizational configurations, firms may select an innovation configuration (technological, organizational, or human capital). The study identifies 68 innovation activities grouped into 10 innovation factors ranging from technological, human-capital, and organizational innovation.
Hertog, Gallouj, and Segers[13]: To demonstrate that innovations in "low-tech"' hospitality industries are more varied than researched, and to study technological and nontechnological innovation and organizational innovation.	The paper presents an empirical analysis through a survey of 613 firms in the Dutch Hospitality Industry followed by a comparison with other service industries to measure innovation and identify key factors associated with innovation.
Su[16]: To explore the role of service innovation in ethnic restaurants and the relationship with customers. The authors studied the concept of experiential marketing and how service innovation can add value to the firm.	This paper conducted a literature review followed by an empirical analysis of a sample of 360 (response rate 64.4%) restaurants to understand the effects of customer experience and behavioral intention. The restaurants were asked to fill out a questionnaire to gather perceptions about customer service and service innovation.

TABLE 32-4 Innovation in Other Sectors

Discussion

Service innovation goes beyond providing new services to customers. Service innovation is affected by employee engagement and other performance systems in place at a particular service organization. At a typical service firm, the cost of human capital is usually more than 50 percent of revenue, and can be even higher for a professional services firm. Employee engagement and involvement directly affects employee turnover, employee satisfaction, and innovation at a service firm. The methodologies to measure the impact of human engagement on corporate performance fall into three broad categories: the Accenture Human Capital Development Framework, the Human Economic Value Added (HEVA) approach, and the Return on Investment (ROI) approach. A recent Watson Wyatt survey report (February 2002) showed the relationship between its human-capital index and shareholder value. Specifically, five key human capital practices are associated with a 30 percent increase in shareholder value: recruiting excellence, clear rewards and accountability, collegial and flexible workplace, communications integrity, and prudent use of resources. The survey represented 60 percent of firms in service settings.

Idea management is a critical element of a service innovation management system. As discussed in *Ideas Are Free* by Allan Robinson and Dean Schroeder*, a firm named Wainwright found that idea-generating employees helped the culture of the company as well as the company's performance. Service innovation aspects and human resource aspects are intricately related to corporate performance. High-performing firms have a deeply rooted innovation culture as well as happy employees.

In this chapter, we propose an adapted version of the Radar incorporating specificity of service business (the Radar included the 12 different ways for companies to innovate and the Top 10 Lessons on the New Business of Innovation). The adapted version of the Radar includes multiple stakeholders of innovation such as customer, partners, multiple hierarchy of service organization (process, service, service value chain), and other ways to innovate in services. Unique types of service innovation are on-demand innovation, employee-driven innovation, and open innovation. Various dimensions of the Radar and examples are shown in Table 32-5.

On-demand innovation appears to be very similar to our customer need innovation in the product setting. Customer coproduction is similar to the customer experience, which could be called interface innovation. Employees (human asset) are an integral part of service organizations and deserve their rightful place (Southwest Airlines saying that employees are their number one customers). Management or business model innovation is highly relevant for service businesses (more than for products). In manufacturing settings, innovation can be broadly classified as product-related or process-related. Service firm innovation places emphasis on human skills and collaboration with service chain partners. However, services can innovate along a third dimension (and along more dimensions)—namely organizational change. This dimension of innovation is nonexistent in the manufacturing context. Similarly, open innovation is another concept more pertinent to services, which is highly related to partnerships or ecosystems.

Dimension	Examples of Company/Service
Offerings (WHAT)	
Service concept	Tele-Medicine
Solutions	IBM providing solutions to customers
Customers (WHO)	
Customer experience	Disney, Cirque de Soleil (moving away from dance and pure circus)
Customer coproduction	Airlines (Southwest), Delta (check-in)
Employee (human asset)	Southwest Airlines, Ritz Carlton
Process (HOW)	FedEx (online tracking), Wal-Mart
Business model (organization)	eBay, NetFlix, Priceline.com, Amazon.com, Dell
Service chain partnership (service delivery)	Hotels (with airline check-in)
Presence (WHERE)	Enterprise rent-a-car (we will pick you up)
Presence	Multiplex cinema, mortgage brokers
Brand	Numerous examples
Open innovation	Dell (IdeaStorm), Netflix, Linux
On-demand innovation (convenience)	Progressive Insurance (24 × 7 adjustment), DirecTV University of Phoenix

TABLE 32-5 Radar for Service Innovation

*Robinson, A.G. and Schroeder, D.M. *Ideas Are Free: How the Idea Revolution Is Liberating People and Transforming Organizations*, 2 reprint ed. Berrett-Koehler Publishers, 2006.

References

1. Luteberget, A. "Involvements Enhance New Service Success?" *Engineering and Science*, May 2005.
2. Hunter, L.W. "The Adoption of Innovative Work Practices in Service Establishments." *The International Journal of Human Resource Management*, 11 (3): 477–496, 2000.
3. Hagedoorn, J. and Cloodt, M. "Measuring Innovative Performance: Is There an Advantage in Using Multiple Indicators?" *Research Policy*, 32 (8): 1365–1379, 2003.
4. Tellis, G.J., Prabhu, J.C., and Chandy, R.K. "USC Marshall School of Business Innovation in Firms across Nations: New Metrics and Drivers of Radical Innovation." *Social Science Research*, February 2007.
5. Tseng, C.-Y., Kuo, H.-Y., and Chou, S.-S. "Configuration of Innovation and Performance in the Service Industry: Evidence from the Taiwanese Hotel Industry." *The Service Industries Journal*, 28 (7): 1015–1028, 2008.
6. Cheng, C.C., and Krumwiede, D. "The Effects of Market Orientation and Service Innovation on Service Industry Performance: An Empirical Study." *Operations Management Research*, 3 (3-4): 161–171, 2010.
7. Chen, Jiyao. "Defining and Measuring Business Innovation: The Innovation Radar." *Innovation*, May 2010.
8. Lee, C.-S., Chen, Y-G., Ho, J.-C., and Hsieh, P.F. "An Integrated Framework for Managing Knowledge-Intensive Service Innovation." *Technology*, 13: 20–39, 2010.
9. Edvardsson, B. and Enquist, B. "The Service Excellence and Innovation Model: Lessons from IKEA and Other Service Frontiers." *Total Quality Management & Business Excellence*, 22 (5): 535–551, 2011.
10. Ellram, L.M., Tate, W.L., and Billington, C. "Services Supply Management: The Next Frontier for Improved Organizational Performance." *California Management Review*, 49 (4): 2007.
11. Safizadeh, M.H., Field, J.M., and Ritzman, L.P. "An Empirical Analysis of Financial Services Processes with a Front-Office or Back-Office Orientation." *Journal of Operations Management*, 21 (5): 557–576, 2003.
12. Berry, L.L., Bolton, R.N., Bridges, C.H., Meyer, J., Parasuraman, A., and Seiders, K. "Opportunities for Innovation in the Delivery of Interactive Retail Services." *Journal of Interactive Marketing*, 24 (2): 155–167, 2010.
13. Hertog, P.D., Gallouj, F., and Segers, J. "Measuring Innovation in a "Low-Tech" Service Industry: The Case of the Dutch Hospitality Industry." *The Service Industries Journal*, 31 (9): 1429–1449, 2011.
14. Anderson, F. and Schaan, S. Innovation, Advanced Technologies and Practices in the Construction and Related Industries: National Estimates Innovation, Advanced Technologies and Practices in the Construction and Related Industries: National Estimates. Statistics Canada. (Working Paper), 2001.
15. Windrum, P. and García-Goñi, M. "A Neo-Schumpeterian Model of Health Services Innovation." *Research Policy*, 37 (4): 649–672, 2008.
16. Su, C.-S. "The Role of Service Innovation and Customer Experience in Ethnic Restaurants." *The Service Industries Journal*, 31 (3): 425–440, 2011.
17. Lyons, R.K., Chatman, J.A., and Joyce, C.K. "Innovation in Services." *California Management Review*, 50 (1): 174–192, 2007.
18. Storey, C.D. and Easingwood, C.J. "Determinants of New Product Performance: A Study in the Financial Services Sector." *International Journal of Service Industry Management*, 7 (1): 32–55, 1996.
19. Safizadeh, M.H., Field, J.M., and Ritzman, L.P. "An Empirical Analysis of Financial Services Processes with a Front-Office or Back-Office Orientation." *Journal of Operations Management*, 21 (5): 557–576, 2003.
20. Thomke, S. "R&D Comes to Services: Bank of America's Pathbreaking Experiment." *Harvard Business Review*.
21. Alam, I. (Ian). "New Service Development in India's Business-to-Business Financial Services Sector." *Journal of Business & Industrial Marketing*, 27 (3): 228–241, 2012.
22. Windrum, P. and García-Goñi, M. "A Neo-Schumpeterian Model of Health Services Innovation." *Research Policy*, 37 (4): 649–672, 2008.
23. Fuglsang, L. "Bricolage and Invisible Innovation in Public Service Innovation." *Journal of Innovation Economics*, 5 (1): 67, 2010.

24. Djellal, F. "Innovation Trajectories and Employment in the Cleaning Industry." *New Technology, Work and Employment*, 17 (2): 119–131, 2002.
25. Rothkopf, M. and Wald, A. "Innovation in Commoditized Services: A Study in the Passenger Airline Industry." *International Journal of Innovation Management*, 15 (4): 731, 2011.
26. Voss, C. and Zomerdijk, L. Innovation in Experiential Services—An Empirical View. Advanced Institute of Management Research, London Business School, 44 (0): 0–39, 2007.

CHAPTER 33

Social Innovation: Post-Fordist Globalization and New Horizons

Billy R. Brocato and Thomas Schmidt

Introduction

In an earlier research project, we examined the ambiguities surrounding leadership in a postmodern, global world.[1] We investigated the rapid changes that placed added emphasis on companies' capabilities to thrive in multicultural settings. We wanted to understand better the contrasts from the managers of the Fordist era who wielded inordinate responsibility for profit and loss and the new postmodern leaders of the global economy, who are responsible for developing talented teams.

What we found was surprising. Although the world as we know it had ever so quietly moved into rapid production and real-time communication technologies, in general, employees in vertical or horizontal positions across company structures were becoming less appreciative of the mythic, lone hero. The Captain Kirks of the famed *Star Trek* television series were an anachronism alongside the mentor and collaborator workers favor in the guise of Captain Jean Luc Picard of the *Star Trek: The Next Generation* movies and television programs.

Furthermore, in a knowledge-based economy, disembedded teams of employees scattered amid modern and not-so-modern nations were adopting new standards of individual and group conduct. Although their primary tools and resources are fixed in real-time information communication technologies, their actions are no longer tied or linked to their physical locations. What happens in Japan is immediately reflected in Europe and the United States, and vice versa. This disembedding—the lifting out from physical and temporal localized contexts—has reshaped the social practices that once informed cooperative behaviors[2] in the workplace. The notion of rationally driven communication, as detailed in *The Theory of Communicative Action*, by Jürgen Habermas, remains applicable. Nevertheless, the concept of rationality has taken on new dimensions, elevating truth statements from nature's observable laws[3] to include the decentering and critique of all knowledge claims. This is important because social innovation relies on expanded communication structures—synchronous and asynchronous.

Nowhere has this decentering occurred more frequently than in virtual and brick-and-mortar workplaces. Business journals like the *Harvard Business Review* are filled with buzzwords linked constantly to concepts of social innovation, social networking, social capital, and human capital. A web query for social innovation returns nearly 750 million hits within a second, social networking about half that many at 350 million hits. There are social innovation journals, and a plethora of scholarly and business articles written to reveal the truths of these new communication schemas. However, can we define social innovation? Can we describe what social innovation looks like?

Social Innovation

In recent history, we can point to the success-failure of the Apollo 13 mission to the moon as an exemplar. Memorialized in film by Ron Howard with film stars Tom Hanks, Kevin Bacon, Bill Paxton, Gary Sinise, Kathleen Quinlan, and Ed Harris, we recall the anxiety-provoking scene where carbon dioxide is building up in the astronauts' spacecraft. A team of engineers are in a room and shown a conglomeration of basic materials at the astronauts' disposal in space. They must develop a workable air filter—fit a round peg in a square hole and save the astronauts' lives. Through miraculous-seeming collaboration, they accomplish their goal. However, it is not simply a miracle of science over nature that wins out that fateful day; instead, it is the dominant force of social innovation, or cooperative communication that brought men together who knew little of each other or of one another's expertise that wins out.

Unfortunately, defining social innovation and its social networking tools is more cumbersome as more experts ponder the nuances of innovation. Simply, the literature is overwhelmed with descriptions that link social innovation to creative ideas designed to address societal challenges—cultural, economic, and environmental issues—that are no longer simply a local or national problem but affect the well-being of the planet's inhabitants[4] and ultimately, corporate profits and sustainability. Some researchers are investigating social innovation as a process that is not value-driven as much as it is a consequence of creative problem solving. For example, Jérôme Auriac has dubbed social innovation at the multinational corporate level as a "cocreative" endeavor, where knowledge workers use their intellectual capital to better mitigate "economic, social, and environmental inequalities between the so-called developed countries and the rest of the world"[5] to meet stakeholders' requirements.

Similarly, Arjun Appadurai's *Modernity at Large: Cultural Dimensions of Globalization*, goes to great lengths to explain the disjuncture across global communities and the birth of ethnoscapes—where the traditions of perception and perspective affect processes of representation in the social world[6] as new entities of creative life (biographies, imaginative exercises) that help inform a people's connectedness to their communities and the world. Ethnoscapes are the creations of disembedded individuals who are seeking ways to reestablish communicative frames across shifting social landscapes.

Maggie Brenneke and Heiko Spitzeck offer a definition of social innovation derived from "social intrapreneurs," a new idiom to describe workers from within a company that effectively allocate their social skills to ease the structures of bureaucracy that so often culminate in stifling creativity.[7] Linking their construct to social entrepreneurship is an important theoretical development. By describing *social intrapreneurs* as agents of change, they are able to demonstrate that it is in the communicative realm, within schemas of rational and irrational discourse that purposive driven social innovation emerges.

Social Capital and Social Innovation

The globalized economy is dependent on knowledge workers who must rely on electronic communication media to interact, conduct basic research, and develop foundational approaches to market demands. However, companies that rely on Fordist concepts of knowledge acquisition remain cemented to traditional standards of innovation and social networking, and may miss the importance of interaction and propinquity. The *Boston Globe* reports: "(a)n analysis of a decade of Harvard biomedical research collaborations—including Selkoe and Walsh's—found that the closer the offices of key research partners, the more influential their joint papers were likely to be. It mattered whether collaborators were riding the same elevators in a building in Longwood, or working in labs on opposite banks of the Charles."[8]

We are not arguing against the necessity or reality of workers increasing their innovation skills in proximate physical settings. We would like to point out that in a globalized world filled with ever-changing or emerging ethnoscapes, the structures or patterns of social networking and social capital accumulation emerge as a process of discourse. Moreover, it could be that academics are actually poor social innovators: the nature of academic inquiry being deep rather than broad, building social connections is not first nature in the field.

However, post-Fordist companies rely on a qualitatively different use of social innovation. One method highlighted by Jane Jacobs is the idea of externalities, as overviewed in Ellerman.[9] Information does not just flow within one industry but it flows across industries. A diverse work environment and a diverse manufacturing economy will have many people who can transfer knowledge between fields. Jacobs takes up Adam Smith's example of the pin-making factory. She points out that the growth and innovation in pin making did not originate with Adam Smith's division of labor. The division of labor enabled much greater production of pins by individual hand workers in the factory. However, she points out that a new, innovative process—the later invention of the pin-making machine—destroyed individual hand making of pins through a new division of production. Surprisingly, a person working in the pin manufacturing industry did not develop the pin-making machine; it was an innovation in another industry: printing.

Certainly, the social and management sciences are replete with research that traces social innovation and social networking skills as emergent cultural tools from the dawning of the industrial era.[10] However, more recently, the French social scientist Pierre Bourdieu made an important discovery while studying primitive tribes and the French elite academicians in the 1970s and 1980s.[11] He found that theories that attempted to explain social structures and the consequent social stratification within and across groups from a modern capitalist market paradigm were too simplistic or one-dimensional.

Perhaps social behaviors in the past were shaped by existing economic stratification across a society. As mentioned earlier, Jacobs had already destroyed Adam Smith's *invisible hand* metaphor[12] as a determinate structure that alone led to economic growth; innovation requires not the specialization that division of labor requires, but diversity of people and skills to supply new insights and generate new products. Bourdieu likewise questioned the efficacy of singular determinate structures institutionally embedded in contrast to the varying cultural behaviors that constantly adapted and introduced novel social interactions. Bourdieu points out that economic capital in the form of accumulated wealth was in itself not enough to reveal the social advantages that different groups gained across time and space. Simply stated, Bourdieu recognized as American pragmatists had long discussed, that the dialectical processes, or dynamic interactions of social structure, culture, and socially constructed identity, can and do foster novel changes in individual and community behaviors. As Martin Kilduff and David Krackhardt demonstrate in their book, *Interpersonal Networks in Organizations*,[13] social interaction is predicated on existing social patterns or structures, but cannot be explained by these structures alone. They reveal that the complexity of relationships emerges because of the social networks that people find themselves embedded within. The authors assert that social networks provide the insight to understand how actors develop relationships that bolster creativity or undermine it, promote social innovation or weaken it.

Social Bonding, Social Bridging, Diversity, and Social Innovation

Contemporary research into social capital defines the term in relation to the use of human capital resources emerging from social relationships.[14] As Agnitsch et al. explain, social relationships based on reciprocity result in the development of accumulated social relations, that is, tangible and intangible resources for use at some future date. The idea hardly seems novel when applied at the most local or personal level. If I borrow your lawn mower, I can anticipate later that you will ask a favor of me.

What is it about social capital that informs social innovation? We contend that it is the rational communicative schema or pattern of social conduct that sets the stage for efficient resources exchange, and thus, the emergence of social innovation. For example, a community or neighborhood of people who share communicative patterns of conduct based on established forms of social capital would be better equipped to mobilize their resources in times of emergency. Nevertheless, the problem that persists is how to describe the relations that emerge from social capital that result in cooperative conduct leading to social innovation. The social science literature is quite transparent on this phenomenon. Two schemas comprise social capital: social bonding and social bridging, where bonding refers to those communicatively dynamic interactions and bridging is the communication schemas across groups that foster cooperative behaviors.[15]

The key factor here is that the different forms of social capital—bonding and bridging—are recognized as qualitatively different social relations, but provide structure or patterns of conduct that increase or diminish network connections. Robert Putnam,[16] who argued that the sustainability of any community remains ineluctably linked to the local social networks, popularized the importance of bonding and bridging capital; we would expand this and state that communication schemas are not simply composed of bonding or bridging social capital. We believe that it is the qualitative expansion of the two network relations—one internal and localized and the other external and globalized that fosters the emergence of social innovation across industries.

Quality in this instance refers to the heterogeneity or homogeneity of the social network relations. Putnam's work demonstrates that bonding social capital can become too exclusive, inward looking, and protective and undermine the emergence of socially dynamic processes. Conversely, bridging social capital can be a catalyst for the development of new innovative approaches to pressing challenges, but risks alienating different network members. Similar to Appadurai's ethnoscapes of creative life biographies that provide an embedded framework of initial communicative interaction, bridging social capital informs participants within a social bonding network to the externalities that are available to them.

These social exchanges lead to shifts in actors' perceptions regarding the world at large. Although some might view such social exchanges as a disruptive influence, undermining long-established patterns of conduct, others might become better equipped to promote the group's survival or sustainability. Hence, social innovation emerges as bridging social capital expands and increases diversity within actors' networks. In this sense, social innovation becomes a collective resource or asset shared across the communal network or workplace. Everyone has access to it, everyone can incorporate some aspect or dimensionality in his or her resource toolkit, but no one person can claim ownership of a diverse process that fosters innovation across temporal locations.

The diversity of social networks is shown as critical in the work of Jane Jacobs. Its importance to her comes about because of the need for different sources of information and ideas to add into a final product. One example of this comes from studies that show that the language that a person speaks literally makes certain concepts inexpressible in their minds.[17] Some studies show that Spanish-language speakers were unable to conceive of certain topics that English or German speakers were. The reason has nothing to do with intelligence or social connections, but rather with the fact that the Spanish language is organized in a logical structure that makes it unique to that language's semiotic structure.

As we have argued throughout, the emergence of social innovation—in the workplace, the community, in the home—relies on rational communicative structures that promote or foster social networks across time and space to address varying concerns. Whether in an office sharing the same proximate space, or communicating through advanced communication technologies in a virtual setting, social innovation is less about a particular value-oriented approach and more about developing cooperative or competitive behaviors. For example, researchers point out that in the last several decades some major social innovations have occurred that foster acceptable community values, but have also furthered economic growth: charter schools, emissions trading, Fair Trade, International Labor Relations Standards, Microfinance, Supported Employment of the Disabled and Individual Development Accounts to help the poor assimilate as their societies industrialize.[18]

Social Media and the Innovator's Dilemma

It is not hard to imagine walking into a Monday morning briefing and listening as a product manager says, "Innovate or die." Certainly, the business trade press is awash with the constant cry of introducing new or improved products, services, methods, and processes. With the introduction of Tom Hayes' *Jump Point: How Network Culture Is Revolutionizing Business*[19] and Byron Reeves and J. Leighton Read's *Total Engagement: Using Games and Virtual Worlds to Change the Way People Work and Businesses Compete*,[20] social networking has matured.

Nevertheless, what is the engine that drives the success of social media? Is it simply a catchy phrase, an interactive website, a dynamic design? The authors mentioned above would say yes to all of these and provide an additional caveat: remove the shackles from your employees and clients, allow them to develop social networks across platforms, industries, and communities.

In a globalized business environment tightly wound about the use of analytics[21] to establish metrics of sustainability and advantage,[22] social innovation has become the outcome of countless shared networked alliances. Just as the point-of-sale system critically reshaped retail stores' inventory-control standards, social innovation has provided substantial new methods of successful business practices. We would argue that corporate boardrooms today do not need to be sold on the value of social responsibility while maintaining fiduciary commitments. Further, we would argue that the more successful global competitors (Wal-Mart, Apple, Microsoft, Samsung, Google, Starbucks, Proctor & Gamble) are demonstrating that using social media to enhance product innovation is a solid investment. Moreover, the fearless use of social media also implies that workers and consumers are free to offer up their perspectives or ethnoscapes within their various social milieus to the marketplace.

For example, international business researcher Ronald Jean Degen demonstrated that developing socially innovative products through social media bolstered Starbucks image at a time when the business community began wondering if the retail coffee company had peaked. What he especially praised was the company's insight to solicit product ideas from their consumers over the website "My Starbucks Idea."[23] "They also ask that consumers only post truly unique ideas (ideas that don't already exist on the website). Both Starbucks and P&G track these networks and provide incentives to people posting on their websites, thereby creating specific virtual communities to suit their purpose."[24]

Similarly, authors Timothy Galpin, J. Lee Whittington, and R. George Bell in *Leading the Sustainable Organization: Development, Implementation, and Assessment,*[25] argue that the human resources department's onetime role as administrative liaison with other departments should transition so that HR will become a more integral player in employee development as core contributors to the company's sustainability. They point out that the leaders of the global economy cannot assume their employees will automatically buy in to increased workloads or responsibilities without having a basic trust relationship in place. As we have argued elsewhere, we believe the development of trust begins with the recognition that each individual is a valuable contributor to the company's success. We further argue that employees left to languish without a social network linked to their individual biographies and company lives, stifles social innovation. Again, we are not advocating that employees can surf the Internet and ignore their daily duties. However, we do believe that corporate leaders, who extract maximum conformity from employees, are returning to the Fordist management principles in the last century that fostered the U.S. manufacturing sector's competitive loss and stand as relics of a long-since vanished past. The acquisition of social networking tools and their effective implementation derives from the emergence of socially innovative communication structures. Jürgen Habermas states "Participants find the relations between the objective, social, and subjective words already preinterpreted. When they go beyond the horizon of a given situation, they cannot step into a void; they find themselves right away in another, now actualized, yet preinterpreted domain of what is culturally taken for granted."[26] Habermas is most concerned with elucidating the communicative structures that foster innovative public policies and economic well-being. He points out that novel situations arise, but responses can never occur outside actors' basic moral or social norms.

Social innovation occurs because actors develop rational methods to evaluate their social exchanges in changing circumstances. Social networks in and of themselves are of little value unless the participants are acting on goal-directed behaviors. Habermas is right to point out that the domains of our existence remain linked to culturally specific modes of conduct. Increasing the productivity of work groups and thus social innovation requires a freeing of interaction among players who can discover among themselves the relevancy of their contexts.

For example, let's attempt a *Gedanken experiment.* If I work for a city government department and there is a need to develop a risk analysis regarding the city's public works projects, it would do little to enhance my study if I only examined known risks documented previously. Additionally, if I were to consider only the likelihood of some unknown event at some unknown time in the future, there would be less likelihood that my risk analysis would seem rational to the uninitiated. However, relying on the bridging capital I have accumulated or the bridging capital of persons I am close-knit with, I could develop a plan that speaks to current and future events that would appear rational to participants in the communication exchanges. I could point out that climate change is producing ever-larger storm systems that result in increased property damages and potential for lives lost. Essentially, I have enlarged my resource toolkit and adapted a localized problem to a number of different circumstances because of my expanded understanding of the problem. This socially innovative technique of local and global problem solving would become available to my social network cohorts who could then expand or modify, producing another innovative iteration.

This is what James Coleman refers to as a significant "potential for information" exchanges that once acquired can be used to expand "social relations…for other purposes."[27] Social capital and human capital (intellectual and tacit knowledge) combined result in the dynamic or dialectical process that promotes social innovation and expanded social networks. Susie Weller acknowledges that social capital formation and the subsequent implementation of accumulated resources however dynamic are consistently tied to socially specific contexts.[28] Although her studies focused on youth and ethnic groups, she found that social capital formation is an observable positive consequence of communicative exchanges, regardless the age group.

Why Social Innovation

Finding new markets for innovative products is not a simple matter. When Honda first came out with a motorized bicycle, the bike would not sell through regular dealer channels to regular buyers of motorcycles.[29] It was undersized and underpowered for that use. Regular motorcycle drivers would not desire the smaller lighter less powerful bike. Therefore, instead of undermining its potential social sales network, the company examined the marketplace and found there were weekend enthusiasts who liked to dirt bike in the hills around Los Angeles. These new users in a different customer group became the target audience for the innovative new bike from Honda. The point to remember is Honda did not discover a new market; the users in discussion and experimentation with each other created it. Where Honda showed social innovation was in allowing this socially created market to become the major vector for their motorcycle sales in the United States. Additionally, the company understood that motorcycle showrooms were notoriously small for any inventory display. The decision to target another audience is not an easy process—but a process nevertheless that relies on the accumulated bonding and bridging capital that made up the motorcycle market in Los Angeles. Without input from dealers, mechanics, salespeople, and consumers, Honda's first attempt to develop a socially innovative product would have failed. Having leveraged the social capital and innovative strategies that made the motorized bicycle a hit in the crowded streets of Tokyo, Honda expanded its social network and looked for new opportunities.

Finding new markets for what Jane Jacobs calls bifurcations, or splits off the main evolutionary tree,[30] is a consequence of a diverse environment. When Honda tried targeting its smaller, lighter bikes within the area of motorcyclists, it failed. Through communicative structures comprised of various social networks, Honda used social innovative capital and discovered a new area of applicability.

The Long Tail, Social Media, and Innovation

Chris Anderson's concept of the long tail[31] explains why some current businesses in the United States are facing extinction. Rereading the article today—it was written in 2004—allows us to glimpse the time when there was still a company known as Blockbuster that rented videos in the physical market and a company known as Tower Records that sold records in a music store. Moreover, Borders Books sold its merchandise through physical outlets. What caused the death of these companies was not a failure to address the needs of the customers. Instead, we argue that the companies succeeded and ultimately failed because they had a *hits*-based model for sales. For example, a bookstore might be limited to stocking 80,000 copies or 80,000 different types of books. However, that limits book buyers' choices and ignores their diverse wants.

Contrast Amazon.com. Amazon.com can sell the greatest hits books and sell them competitively to any market segment. More importantly, Amazon.com has the flexibility to stock titles that are less likely to be available to the general consumer since the company has developed a social network of cooperating businesses. It's not about competing with one another; it's about decreasing transaction costs. This is the concept of the long tail. By aggregating, the lots of one- and two-selling books, Amazon can achieve the same profitability as a bookstore by aggregating demand across a continent-wide market of more than 300 million customers (Fig. 33-1).

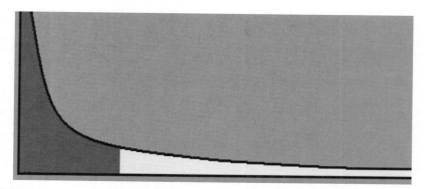

Figure 33-1   The hits threshold below which it is unprofitable for a physical store to stock an item; example of a power-law graph. (*Source:* https://upload.wikimedia.org/wikipedia/commons /1/13/Long_tail.PNG.)

Social media are critical in this process. Amazon's recommendation engine can help people travel from one of the greatest hits books that they might buy from Amazon or a regular bookstore and go down the chain to find related items that were also bought by customers of that book; this recommendation engine increases sales at the less popular end of the long-tail distribution. Social media compound the effect. When combining recommendations for social media with search engines in a much more efficient and effective communicative structure, Amazon is fostering new kinds of social innovation in the realm of consumer purchasing.

Social Media and the Wisdom of Crowds

The book *The Wisdom of Crowds* by James Surowiecki[32] is an explication of the process of Bayesian knowledge generation from distributed diverse crowds of people. The classic first example is a county fair in England where participants were asked to guess the weight of a slaughtered bull. The person who guessed the weight exactly would win all the meat from the slaughtered bull, a considerable prize at the time. Francis Galton, first cousin to Charles Darwin, and noted statistician, noted that no one guessed the weight exactly. However when he added up all the guesses that were decipherable, and divided by the total number of guesses, he discovered that the crowd had in fact as a group guessed the weight of the bull exactly.

The book discusses the phenomenon in other areas. For example, in 1986 when the space shuttle Challenger exploded, four companies listed on the stock market were involved with the disaster. During the day, those stocks plummeted. By the end of the trading that day, however, three of the four stocks that had been involved in building the space shuttle had recovered. Only one, Morton Thiokol, remained down on the day.

A presidential commission investigated the space shuttle Challenger accident. Led by physicist Richard Feynman, the commission studying the cause of the disaster eventually decided after 2 years that the space shuttle had exploded because of a problem with the rocket supplied by Morton Thiokol. In other words, using the wisdom of crowds, focused through the mechanism of the stock market, the source of the disaster was discovered nearly instantaneously.

Social media are an enabling technology for accessing the wisdom of crowds. Social media allow us to discover knowledge that we hold collectively but each of us cannot know in total. This is a spur to innovation because the knowledge is tacit, not explicit. Social media make the tacit knowledge explicit and allow us to access the ideas that we know collectively but cannot know individually.

Conclusion

The purpose of this chapter was to explicate the meaning of social innovation from the myriad instances of its use in the social and management sciences literature. We were content to define social innovation as an emergent group phenomenon structured or patterned after goal-directed behaviors.

We insist that social innovation is the consequence of communicative structures, and rational and irrational discourses that lead to novel actions. The patterns that foster social innovation are socially constructed, but are not deterministic of social innovation. A dynamic process, a dialectical interplay of social networks and actors' purposes, social innovation fosters survivability and sustainability. In the community, actors can rely on social networks—internal and external—to nurture relations that broaden communication schemas that lead to effective social innovation. In the marketplace, social innovation arises as knowledge workers leverage human capital resources to meet challenges or develop opportunities. Diversity of perspectives and diversity of challenges reveals an underlying foundation from which social networks develop social innovation. Just as social capital is composed of bonding and bridging network activities, social innovation represents the successful alchemy of purpose-driven communicative structures.

Fundamentally, social innovation is an intangible, tacit experience temporally located and embedded in those spaces—physical or virtual—where social actors meet. Paradoxically, social innovation is not necessarily present because actors find themselves in a communicative setting. Social innovation is a potentiality—it is an emergent phenomenon linked to rational concerns. Conversely, conspiracy theorists develop effective social networks and believe their concerns are rational, and are able to develop socially innovative proofs. However, the outcome is not a new, rational communicative structure that emerges to solve existing problems.

It is instead, the adapting of an irrational method to resolve simply untenable logical fallacies. Social innovation occurs in those moments when communication is pulled to more abstract levels of reasoning that promote imaginative but empirically observable or testable conclusions. Unlike the Fordist era, where a mythic lone hero is responsible for legitimizing communicative structures of action, the postmodern-era knowledge worker relies on the facilitator who marshals actors' trust and commitment to promote the organization's survival, sustainability, and growth.

References

1. Brocato, B.R., Jelen, J., Schmidt, T., and Gold, S. "Leadership Conceptual Ambiguities: A Post-Positivistic Critique." *Journal of Leadership Studies*, 5 (1): 35–50, 2011.
2. Stones, R. "Disembedding." New York, NY: The Wiley-Blackwell Encyclopedia of Globalization, 2012.
3. Habermas, J. *The Theory of Communicative Action, Volume 1: Reason and the Rationalization of Society*. T. McCarthy (trans.). Boston, MA: Beacon Press, 1981.
4. Phills Jr., J.A., Deiglmeier, K., and Miller, D.T. "Rediscovering Social Innovation." *Stanford Social Innovation Review*, 6 (4): 34.
5. Auriac, J. "Corporate Social Innovation." Organisation for Economic Cooperation and Development. *The OECD Observer*, 279: 32–33, May 2010.
6. Appadurai, A. *Modernity at Large: Cultural Dimensions of Globalization*. Public Worlds, Volume 1. Minneapolis, MN: University of Minnesota, 1996.
7. Brenneke, M. and Spitzeck, H. "Social Intrapreneurs—Bottom-Up Social Innovation." *International Review of Entrepreneurship*, 8 (2): 157–176, 2010.
8. Johnson, C. "Collaboration: The Mother of Invention." *The Boston Globe*, May 8, 2011.
9. Ellerman, David. "Jane Jacobs as a Development Thinker." *Challenge*, 48 (3): 50–83, May/June 2005.
10. Wren, D.A. *The History of Management Thought*, 5th ed. New Jersey: John Wiley & Sons, 2005.
11. Bourdieu, P. "The Forms of Capital." In *Handbook of Theory and Research for the Sociology of Education*, J.E. Richardson (ed.). New York, NY: Greenwood Press, pp. 241–258, 1986.
12. Smith, A. *The Wealth of Nations*. C.J. Bullock (ed.). New York, NY: Barnes & Noble, Inc, 2004 (1776).
13. Kilduff, M. and Krackhardt, D. *Interpersonal Networks in Organizations: Cognition, Personality, Dynmaics, and Culture*. New York, NY: Cambridge University Press, 2008.
14. Agnitsch, K., Flora, J., and Ryan, V. "Bonding and Bridging Social Capital: The Interactive Effects on Community Action." *Community Development: Journal of the Community Development Society*, 37: 36–51, 2006.
15. Bridger, Jeffrey C. and Alter, Theodore R. "Place, Community Development, and Social Capital." *Community Development: Journal of the Community Development Society*, 37: 5–18, 2006.
16. Putnam, R.D. *Bowling Alone: The Collapse and Revival of American Community*. New York, NY: Simon & Schuster Publishers, 2000.
17. Pinker, S. *The Stuff of Thought: Language as a Window into Human Nature*. New York, NY: Viking, 2007.
18. Phills Jr., J.A., Deiglmeier, K., and Miller, D.T. "Rediscovering Social Innovation." *Stanford Social Innovation Review*, 6 (4): 34.
19. Hayes, T. *Jump Point: How Network Culture Is Revolutionizing Business*. New York, NY: McGraw-Hill, 2008.
20. Reeves, B. and Read, J. Leighton. *Total Engagement: Using Games and Virtual Worlds to Change the Way People Work and Businesses Compete*. Boston, MA: The Harvard Business Press, 2009.
21. Davenport, T.H. and Harris, J.G. *Competing on Analytics: The New Science of Winning*. Boston, MA: Harvard Business School Press, 2007.
22. Barney, Jay B. and Clark, D.N. *Resource-Based Theory: Creating and Sustaining Competitive Advantage*. New York, NY: Oxford University Press, 2007.
23. Degen, Ronald J. "Social Network Driven Innovation." *The ISM Journal of International Business*, 1 (1): 1–29, 2010.
24. Ibid, p. 5.
25. Galpin, T., Whittington, J. Lee, and Bell, R. George. *Leading the Sustainable Organization: Development, Implementation, and Assessment*. New York, NY: Routledge, 2012.
26. Habermas, J. *The Theory of Communicative Action, Volume 2: Lifeworld and System: A Critique of Functionalist Reason*. T. McCarthy (trans.). Boston, MA: Beacon Press, 1981.

27. Coleman, James S. "Social Capital in the Creation of Human Capital."*American Journal of Sociology,* 94 (suppl): S95–S120, 1988.

28. Weller, S. "Young People's Social Capital: Complex Identities, Dynamic Networks." *Ethnic & Racial Studies,* 33 (5): 872–888, 2010. Academic Search Complete, EBSCOhost (accessed April 5, 2011).

29. Christensen, C. *The Innovator's Dilemma: When New Technologies Cause Great Firms to Fail.* Boston, MA: Harvard Business School Press, 1997.

30. Jacobs, J. *The Economy of Cities.* New York, NY: Random House, 1969.

31. Anderson, C. "The Long Tail." *Wired,* 12:10, October 2004.

32. Surowiecki, J. *The Wisdom of Crowds.* New York, NY: Anchor Books, 2005.

CHAPTER 34

Nonprofit Innovation: Rethinking Value Creation for the Social Sector

Margaret Morgan Weichert

Introduction

Philanthropic organizations evolve in response to diverse cultural, religious, social, and political realities. The needs, priorities, and motivations of not-for-profit organizations (nonprofits) are similarly diverse. By their very definition, nonprofits focus on goals and objectives that transcend the profitability drive. Leaders of most nonprofit organizations are committed to a chosen mission, which is often a lofty, intangible objective, rather than a concrete profitability goal. Nevertheless, most nonprofits stand to benefit from the application of relevant business practices to the business of charity. Innovation is a critical area for nonprofits will benefit by leveraging business best practices, enabling them to be even more effective in pursuing the underlying mission. In the 21st century, many innovative nonprofits are blurring the lines between philanthropy, investment, and business to drive greater efficiency and effectiveness.

The nonprofit sector is a large and growing component of most economies around the world. In the United States alone, more than 1.43 million nonprofits accounted for $1.87 trillion in revenues and $4.3 trillion in assets in 2009.[1] Charitable giving and volunteering account for a material share of gross domestic product (GDP), with estimates of charitable giving as high as 1–2 percent of GDP in developed countries like the United States, Israel, and Canada.[2] Even in developing countries like Brazil, India, and China, charitable contributions are growing at impressive rates, as professional nonprofit organizations are established to address the unique needs of those countries.

Despite growth in the sector, however, nonprofits face increased challenges as they compete for funding. Global economic uncertainty has depressed private and corporate philanthropic giving. At the same time, government funding of social programs has been squeezed by global economic challenges, putting even more pressure on nonprofits. To offset these challenges, nonprofits have become more innovative, identifying new ways to maximize the power of corporate donors, foundations, and smaller, individual contributors. In this competitive funding environment, corporate donors are looking for business savvy nonprofits that are data-driven and transparent. Foundations are using innovative measurement schemes as they search for sustainable solutions to bigger problems. Smaller donors are being offered more innovative-giving opportunities, as technology, social media, and Internet marketing combine to unlock new sources of funding.

Context for Not-for-Profit Innovation

By any measure, the business of philanthropy is a large and growing global phenomenon. In the United States alone the sector accounts for nearly $2 trillion in revenues.[3] No comparable global statistics are available globally, but work done in the 1990s at the Johns Hopkins Center for Civil Society Studies identified over 20 million nonprofit organizations

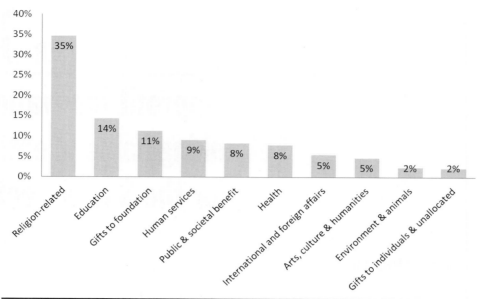

FIGURE 34-1 Percent of U.S. charitable contributions by type of organization in 2010. (*Source*: Giving USA Foundation, Giving USA 2011.)

in the 35 countries studied, excluding religious congregations.[4] Nonprofits are large employers and major contributors to the economy generating gross domestic product (GDP) and in-bound financial resources for developing economies.

Most of the nonprofits in the United States are quite small, with 56 percent of nonprofits collecting less than $25,000 in gross receipts.[5] The remaining 628,700 nonprofits drive the majority of the nonprofit activity and account for most nonprofit growth, with growth rates of 50 percent in the period 1999–2009.[6] Nonprofits address a wide range of activities, many of which face big, intractable problems that are also the subject of significant government and independent corporate investment (Fig. 34-1).

Nonprofits rely on a range of revenue sources to sustain their activities, with nearly 75 percent of nonprofit revenue derived from fees or payment for services (Fig. 34-2). This

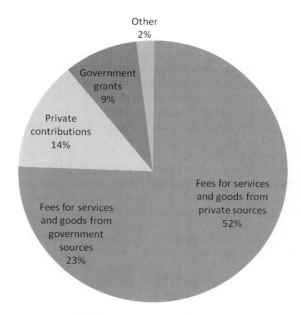

Figures for US Public 501 c3 charities that report financials to the IRS, covering over 1 million nonprofits and accounting for 75% of nonprofit revenues
Note: Figures do not sum to 100 percent because overall investment income was negative 0.2% for the sector

FIGURE 34-2 Sources of revenue for U.S. charities. (*Source*: Urban Institute, National Center for Charitable Statistics, 2009.)

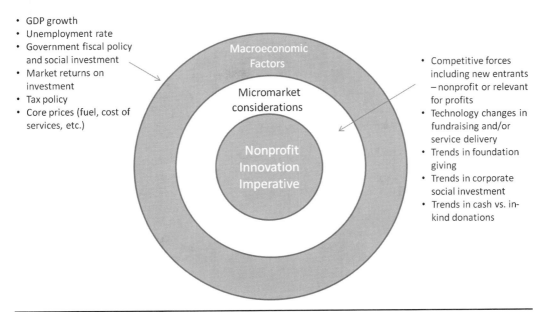

- GDP growth
- Unemployment rate
- Government fiscal policy and social investment
- Market returns on investment
- Tax policy
- Core prices (fuel, cost of services, etc.)

- Competitive forces including new entrants – nonprofit or relevant for profits
- Technology changes in fundraising and/or service delivery
- Trends in foundation giving
- Trends in corporate social investment
- Trends in cash vs. in-kind donations

FIGURE 34-3   Market forces affecting nonprofits.

statistic makes it clear that revenue-oriented innovation is as critical to nonprofits as for their for-profit companies.

Driven by greater transparency around not-for-profit overall expense ratios and fundraising efficiency, the sector is also using innovation to improve the expense equation. In 2011 the average program expense among the largest 200 U.S. charities was 86 percent.[7] Average fundraising ratios for the same group were 90 percent, meaning it cost the nonprofit $0.10 to raise $1 in funding.[8] Innovation has played a key role in helping many nonprofits improve their expense position, with some nonprofits creating models that actually deliver 100 percent, of benefits to the targeted cause, typically in partnership with other organizations.

Pressure for innovation in revenue and expense management is heightened by challenges in the political and macroeconomic environment. The global economic downturn of 2007–2009 had a chilling effect on most nonprofit revenue sources, while also driving increased demand for services. In this period nonprofit revenue was affected in multiple ways: (1) individual donations dropped; (2) market returns on foundation investment portfolios plummeted; (3) government contributions declined in economies facing fiscal crisis; and (4) corporate donors were also squeezed. Increases in commodity prices on food to fuel also put pressure on nonprofit expenses. Nonprofits continue to face revenue and expense challenges associated with tax policy changes (including charitable contribution deductions), and fiscal spending controls.

Nonprofits are impacted by the same market forces that affect financial performance of for-profit companies. Thus, nonprofits benefit from surveying how the broader market has responded using innovation. Figure 34-3 highlights areas of consideration.

This market context can help nonprofits improve planning and catalyze innovation in areas facing adverse market impacts. Analysis of competition in each nonprofit micromarket can further enhance innovation. Since players both in the nonprofit sector and the for-profit world face similar challenges, there is much to be gained through observation of the marketplace. The innovations of other nonprofits, socially responsible businesses, corporate social responsibility programs, and relevant for-profit initiatives may provide a springboard for nonprofit innovation.

The End Game: Profit versus Mission

Nonprofits and profit-driven enterprises have much in common. For-profit companies focus on shareholder value creation by providing products and/or services to its customers. Similarly, not-for-profits create value and drive results related to a specific mission, serving a set of beneficiaries and/or society in general. Although the social benefits driven by nonprofits are often difficult to quantify, their purpose is still to create value.

What sets nonprofits apart from their profit-driven counterparts is the difficulty in measuring and quantifying success. Profit-oriented companies have a clear benchmark for value

"P&L" Line	Key Innovation Activities	Examples
Revenue	• Business model innovation	• Microfinance • Mobile benefits delivery • Socially conscious spending • Socially responsible investing • Philanthro-capitalism
	• Innovative use of technology and social media in fundraising	• Mobile giving/fundraising • Crowdfunding • Online games as fundraisers • Social media fundraising
Expense	• Innovative cost reductions	• Financial inclusion • Online education • Open platforms/digital commons • Equal access to technology
Bottom line	• Innovation in measurement	• Reporting: charting impact • Expected return calculations • Nonprofit outcome measurement • Impact investing metrics

TABLE 34-1   The Bottom Line: Innovating to Enhance Value Creation

creation in the form of a profit and loss (P&L) statement. Companies that create sufficient market value are rewarded by customers and shareholders in tangible, measurable ways. By contrast, nonprofits have a greater challenge to demonstrate value creation. As a result, innovative leaders in the nonprofit realm are beginning to use the language and tools of traditional businesses to align not-for-profit objectives with outcomes and results. Use of a P&L framework provides an objective perspective on the bottom line of value creation even in a nonprofit context.

A disciplined, business-oriented innovation approach offers concrete benefits to nonprofits, with useful frameworks, tools, and metrics to improve the charitable bottom line. Nonprofit leaders can work closely with corporations, entrepreneurs, and academics on innovative partnerships that use the metrics and discipline of traditional business to achieve nonprofit goals. These collaborative efforts are tackling and making measurable progress on some of the most challenging global social issues in health care, education, human services, and the environment (Table 34-1).

There are a host of areas where nonprofits are demonstrating the results of innovation in revenue generation, operations, overhead management, and organization effectiveness. However, four areas stand out as the most promising opportunities for nonprofit innovation:

- First, *business model innovation* is transforming the connections between donors and recipients, using direct, tangible economic incentives to drive symbiotic relationships between donors and recipients. Social entrepreneurs, philanthro-capitalists, microlending organizations, and peer-to-peer socially responsible lending networks, are examples of innovative business models that blend social responsibility and business-oriented financial objectives. These new business models also leverage technology, data, and collaboration to empower individuals and organizations to impact the lives of others in tangible and direct ways.

- Second, *innovative uses of technology and social media in fundraising* are revolutionizing the ability of nonprofits to energize and supercharge the global funding power of small donors. Moving beyond e-mail marketing campaigns, nonprofits have pioneered a range of innovative fundraising solutions that leverage mobile phone technology, social networks, and innovative payment solution. These innovations make it easier to facilitate instant donations by individuals, while also reducing the friction for nonprofits to accept electronic donations.

- Third, *innovative cost reductions* that expand nonprofit access and reach are proliferating as innovators use technology to remove cost and geographic barriers. Innovators have transformed entire sectors with compelling, low-cost solutions for underserved segments. Remote healthcare delivery, mobile banking, online encyclopedias, and free educational institutions are examples of nonprofit innovations that have used innovative cost models to serve populations that had previously been out of economic and/or geographic reach.

- Finally, *innovation in measurement* has brought greater quantitative accountability to nonprofits, improving overall results and ROI. Significant progress in this area has been driven by strong leadership from several large, innovative foundations. In particular, newer foundations run by entrepreneurs and business leaders are introducing consistent, metric-driven criteria into the grant-making process, encouraging new generations of nonprofit professionals to focus on measurement and effectiveness. Building on past nonprofit measurement activities focused on efficiency, recent innovations have helped crystallize methods for introducing quantitative metrics focused on nonprofit results.

Innovation Best Practices for Nonprofits: Insights from Business Innovators

The examples above are a sample of the innovations taking place in the nonprofit segment today. Literally thousands of other case studies could be included here. Far from being a conservative space, the nonprofit universe shares much in common with the denizens of entrepreneurial hubs like Silicon Valley, London, Hyderabad, and Shanghai. It is extremely useful and relevant to analyze nonprofit innovation in the context of recent best practices from business innovators.

Nonprofit leaders are motivated by a strong sense of purpose, a drive to achieve big changes, and are undeterred by their lack of resources. Entrepreneurs and nonprofit leaders are passionate about what they do, but often must spend more time than they like on finding the financial backing to pursue their ultimate mission.

Given these similarities, it is critical for nonprofit innovators to learn from their for-profit counterparts, and leverage relevant innovation experiences and best practices from the for-profit world. Although not a comprehensive list, the following are critical innovation best practices (Fig. 34-4) that will benefit nonprofit innovators across disciplines:

- **Drive to make meaningful change**: A bold, audacious goal provides the critical rallying point for innovation.

- **Cultivate market knowledge**: Deep understanding of customers, market realities, and other critical stakeholders provides critical context for innovation, identifying unmet needs, pain points, and opportunities for massive change.

- **Harness technology**: Technology, mobile telecommunications networks, and social media are powerful enablers of connectivity allowing information exchange, communication, and engagement at a low cost, enabling profound shifts in fundraising, service delivery, and collaboration.

- **Use data as an innovation tool**: Innovative use of data and analytics can be used to drill down on critical problems, examine process improvement opportunities, and measure and communicate results, improving outcomes via better transparency.

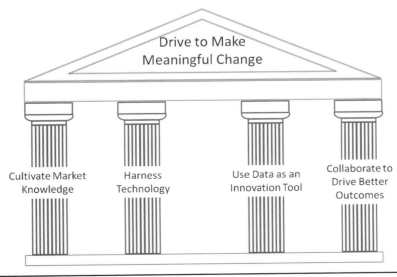

FIGURE 34-4   Drive to make meaningful change.

- **Collaborate to drive better outcomes**: Collaboration and *"coopetition"* are fundamental to driving breakthrough innovation, allowing each of the innovators to leverage the work of others and accelerate the pace of change.

Key Innovation Best Practices for Nonprofits

Drive to Make Meaningful Change

One of the common elements identified as the source of great companies is well known to most nonprofits. Each nonprofit has a **mission** that articulates the nonprofit's reason for being its purpose. In the for-profit world, however, the mission or purpose is often lost in the drive for profitability. By contrast, nonprofits usually have a laser-like focus on their mission, which provides a clear benchmark for action and motivation for its people.

Research on successful, innovative companies has proven that a transcendent mission is a critical part of innovation and business success. James Collins and Jerry Porras outlined the importance of having a **big, hairy, audacious goal** (BHAG) as a rallying point for innovation and business action in their book, *Built to Last: Successful Habits of Visionary Companies*.[9] Collins and Porras suggest that a 10- to 30-year goal that describes an "envisioned future" as critical to driving innovative action. "A true BHAG is clear and compelling, serves as a unifying focal point of effort, and acts as a clear catalyst for team spirit. It has a clear finish line, so the organization can know when it has achieved the goal."[10] Whether the organization is Amazon.com, Ford Motors, or AIESEC, a BHAG articulates the underlying mission in a clear and typically quantifiable way.

Another innovation thought leader, Nikos Mourkogiannis, articulates a similar concept called **purpose**. In his book *Purpose: The Starting Point of Great Companies* argues that a set of values and transcendent ideas are critical to motivating employees and driving meaningful success.[11] Although Mourkogiannis focuses on for-profit companies, and makes a clear distinction between corporate philanthropy and purpose, his insights about using purpose as a rallying point also apply to nonprofits.

Another popular business writer, Daniel H. Pink, examined what motivates people to achieve success in business and other disciplines, and concluded that purpose was critical to success in most fields of endeavor. In his book, *Drive: The Surprising Truth about What Motivates Us*, Pink identified purpose as a key trait present in successful nonprofit and for-profit organizations alike.[12] Pink examines companies like TOMS, which combine for-profit objectives with philanthropy, is motivated by a purpose. TOMS itself describes its purpose as being, "a for-profit company with giving at its core." Similarly, Pink found that worldwide membership in cooperatives has been increasing due to the combined pursuit of business objectives and an underlying, meaningful purpose.[13]

What this means for nonprofit innovation is that a focused, clear purpose or mission is critical. Mission creep, unclear goals, and lack of quantifiable objectives can all hinder innovation.

Cultivate Market Knowledge

Most innovation textbooks and thought leaders believe that successful innovations are rooted in a strong understanding of customer and market dynamics. Whether an organization is building cars, creating financial service innovations, or promoting new strategies for education, there are some common best practices for cultivating innovation. Best practices in the for-profit world show that innovation ideas can be systematically generated by observing the marketplace to understand critical unmet needs, market disruptions, and major challenges that face market players. This systematic approach can be easily leveraged in a nonprofit context.

One classic approach to innovation is outlined in Robert Cooper's book, *Winning at New Products: Accelerating the Process from Idea to Launch*.[14] Cooper recommends using customers, suppliers, competitors, private innovators, and universities as a source of innovation, emphasizing an approach that looks to use every market resource available to identify innovation that will provide solutions to the target challenge or purpose.

Similarly, a new innovation framework, outlined in *Innovation X*, by Adam Richardson, has significant applicability for nonprofits. Richardson's premise is that large, intractable problems and challenges lend themselves well to breakthrough innovation.[15] One of the keys to unlocking innovation in these challenging areas is through the concept of **immersion**, requiring a deep, 360-degree understanding of the nature of the problem from the perspective of the customer (and other key stakeholders). The notion of immersion suggests that a central component of innovation comes from deep understanding of a range of factors,[16] including:

- Competitors (direct and adjacent)
- Comparative solutions: from companies or products that may provide useful lessons but aren't competitors in the same arena
- Objective view of the organization's capabilities, brand, and values
- Broad cultural and economic trends
- Available technology enablers

IDEO, a leader in driving breakthrough innovation in the for-profit world, takes the idea of finding innovation by understanding customers and the market to an even deeper level. IDEO uses the notion of *design thinking* to drive innovation by linking human needs and desires with technology in functional and feasible ways. IDEO uses analytical tools and techniques including business model prototyping, data visualization, quantitative and qualitative research, and other tools to create an integrated view of human needs.[17] Recognizing the benefits of using this approach in the context of nonprofits and social enterprises, IDEO, with funding from the Gates Foundation, has created a *human-centered design (HCD) toolkit* for NGOs that was developed in collaboration with ICRW and Heifer International.[18] Use of this process has led to breakthrough nonprofit innovations including the HeartStart defibrillator, CleanWell natural antibacterial products, and the Blood Donor System for the Red Cross, solutions which now benefit millions of people.[19]

Each of these approaches emphasizes how important it is that organizations systematically survey the market landscape to cultivate the knowledge and insight to drive innovation. Quantitative and qualitative market research, field observation, competitive benchmarking, and understanding economic and technology trends are all important. And given the resource constraints typically faced by nonprofits, creativity and innovation are important to find ways to glean this type of market knowledge with finite resources. Use of third-party information and toolkits like the IDEO HCD toolkit, data sets from multilateral organizations, and even publicly available information from for-profit enterprises can all be used to further the market knowledge required for innovation.

Harness Technology, Networks, and Social Media

Although technology has long been viewed as a critical component in many business innovations, its power to transform all aspects of human endeavor has become self-evident as the Internet, mobile technology, and low-cost global telecommunications have transformed how all organizations are able to interact and engage in their communities. The ubiquity of global technology networks and the social media that are enabled by these networks have engaged players from every discipline, geography, and socioeconomic sphere. More importantly, a profound shift to open platforms, cloud-based services, and networked business models has democratized the relationships between companies and their stakeholders, giving more power and influence to groups of individuals than ever before. Consequently, nonprofits can learn a great deal about how to leverage the power of new technologies, networks, and social media, by observing innovative for-profit pioneers.

Economists David Evans, Andrei Hagiu, and Richard Schmalensee describe how technology platforms drive transformational innovation by creating a space where solution providers and end users can interact directly.[20] This powerful characteristic of today's technology platforms is evident in platforms like iTunes, Facebook, and PayPal. Each of these platforms enhances accessibility and interaction between players in a complex ecosystem, and has driven major changes in the cost of access and accessibility.

Technology networks have unique characteristics that connect people, creating platforms for entirely new ways of interacting. Social media phenomena like Facebook, Twitter, YouTube, Instagram, and Pinterest are perfect examples of the types of technologies that can be leveraged successfully in support of commercial and nonprofit enterprises. Forrester researchers Charlene Li and Josh Bernoff outline the powerful potential of this type of technology in their book *Groundswell*. Groundswell technologies have a high impact because (1) they enable people to connect to each other in new ways; (2) they are effortless to sign up for; (3) they shift power from institutions to people; (4) the community helps generate content and value; and (5) they are open platforms that invite partnerships.[21]

These networks can enable nonprofits to create strong, direct linkages between donors and beneficiaries at a low cost. Use of these networks can enable new business models, lower the cost of solution delivery, and act as a new fundraising tool. More importantly, these tools can help involve the beneficiaries more directly in the solution, regardless of distance, allowing them to help identify priorities, influence how money is spent, and provide data and feedback on the results of nonprofit work.

Many of the technology players that are at the epicenter of the groundswell are making it easier for nonprofits to tap into the power of those technologies. Google.org provides access to technology, data, in-kind resources, and financial support to help address global challenges from health care to energy use. PayPal offers low-payment rates for nonprofit donations. YouTube, Facebook, and Twitter all have similar specialized solutions for nonprofit organizations, making it easier than ever for nonprofits to leverage innovative networked technology from third parties.

Use Data as an Innovation Tool

Since the dawn of the age of the information superhighway, pundits have been heralding the importance of data and analytics in business. However, dramatic decreases in the price of data storage hardware have finally made this prediction a reality, and the era of *big data* has truly arrived. Innovative businesses are using data to enhance their understanding of customers, refine products, target sales, and track results in more ways than ever. This technology evolution has dovetailed with a demand for greater financial transparency driven by regulation like Sarbanes Oxley. Together, these factors have created a true breakthrough in innovative use of data and analytics in business.

As mentioned earlier in this chapter, one critical component in the use of data is to enhance transparency and focus through better measurement. One of the most obvious uses of this type of data is to improve communications with donors about results. Moreover, better data mining techniques also play an increasingly important role in targeting potential donors. With techniques from the for-profit world, nonprofits, political campaigns, and social enterprises have become sophisticated in the practice of data mining to enhance outcomes and response rates for direct mail and online campaigns. There are a host of studies and practical guides to using data in this manner. One book, *Fundraising Analytics: Using Data to Guide Strategy* by Joshua Birkholz showcases how data can be used to target potential donors, hone nonprofit strategies, and better showcase results by appropriately managing and disseminating data.[22]

As the focal point for one of the world's largest data sets, Google is in a unique position to leverage data for social purposes. In the healthcare realm, Google leverages its search engine data in partnership with national health organizations to provide a Flu Tracker that has been remarkably accurate in describing the path of flu epidemics. Innovative nonprofits can tap into and use tools like Google's data sets to identify solutions to other problems.

An emerging group of data scientists and innovators are also working together to promote opportunities to use big data for the common good.[23] Organizations like Where and Data Kind are bringing professional data scientists together with governments and aid organizations to use data to coordinate aid activity and improve outcomes. By linking weather-related data, economic data, and environmental and public health information, these organizations have been able to bring new insights for tackling major health, social welfare, and environmental problems like malaria, malnutrition, and environmental degradation.[24]

Collaborate to Drive Better Outcomes

Observations of virtually all recent innovation success stories have a common theme: much of their success has been attributable to the power of open networks. As far back as the triumph of the Microsoft business model in the personal computer market, organizations that leverage the power of a broad range of stakeholders have created tremendous market value via innovation. The concepts of **collaboration** and **coopetition** are hallmarks of an open innovation approach, acknowledging that customers, competitors, and a range of other stakeholders may have an important role to play in moving the innovation ball forward.

Much of the academic and business literature that has examined the recent success stories in the new networked economy, have identified an important new trend that is as important for nonprofits as it is to for-profit entrepreneurs. The old business paradigm of the "competitor-as-enem" mind-set does not lead to the breakthrough innovations required by customers and key stakeholders in today's dynamic markets. Rather, a new, more powerful innovation mind-set is driven by collaboration between customers, complementors, and other key stakeholders.[25]

Leading business thinkers have begun studying this new collaborative paradigm, highlighting important elements of **user-centered innovation** that incorporate direct input from customers and stakeholders in the development of products and services. Eric von Hippel's research into this subject, outlined in the book, *Democratizing Innovation*, highlights the critical role that users play in the innovation process, leading to more impactful and ultimately more profitable innovations.[26] Peter Gloor expands upon this concept with the concept of **collaborative innovation networks (COINs),** suggesting that in today's highly networked economy, networks of collaborators are even more powerful drivers of business success.[27]

The lessons of collaborative innovation have clear applicability to innovation in the nonprofit sector. Nonprofits historically have, by necessity, a strong track record of collaborating closely with other nonprofits, universities, and government agencies. However, the new collaborative imperative is far broader. Collaborative innovation lowers the costs and provides a more direct link between innovations and the end populations that benefit from those innovations, creating a strong and tangible link between effort and outcome.

A host of literature draws the connections directly between the importance of collaboration and innovation in the nonprofit realm. Increasingly, the importance of collaboration between business and nonprofits has become an important topic in innovation circles. Leading researchers and nonprofit experts began focusing on the benefits of this type of collaboration in the late 1990s as books like Shirley Sagawa and Eli Segal's *Common Interest, Common Good* articulated a new case for business and nonprofit collaboration.[28] In the new millennium, as these types of partnerships have expanded, the theory and research on the benefits of these partnerships have grown. James Austin provides concrete insights on the common requirements and best practices for effective nonprofit organizations in his book *The Collaboration Challenge.*[29] More recently, the Wikipedia model, highlights another even more open model of collaboration. The Wikipedia model is examined in detail in the book, *Good Faith Collaboration*, which outlines how a nonprofit can actually have collaborative as a foundation, relying on its community for content, quality assurance, and usage.[30]

Ultimately, the lesson for nonprofits is straightforward. The more collaborative nonprofits become the more effective and efficient they will be at achieving their mission. Aligning their activities with end users, donors, and other stakeholders interested in similar outcomes will help leverage resources, improve coordination, and speed implementation. Moreover, whenever nonprofits think of ways in which they can strategically align with large and powerful business interests, they have the potential ability to drive a much larger agenda and shepherd greater resources toward the end objectives of the nonprofit.

Conclusion

The spirit of nonprofit innovation is alive and well in the 21st century. Nonprofit leaders, driven by a powerful sense of mission and purpose, are using new business models, technology tools, and collaborative approaches to solve problems and create a lasting impact. Consistent and systematic application of innovation lessons from the for-profit realm can create even more leverage and impact for these nonprofits.

Like their for-profit counterparts, nonprofit innovators must take a holistic bottom-line-oriented approach to innovation. And although the nonprofit bottom line is measured in terms of impact and mission, many of the key innovation components are the same. Innovative nonprofits can dramatically improve their revenue situation through innovative business model changes and effective use of technology and social networks to enhance fundraising activities. Similarly, nonprofit innovators can leverage technology and partnerships to achieve dramatic cost savings that enhance accessibility and breadth of impact. Finally, nonprofits benefit tremendously from taking a quantitative and measurable view of their end-to-end innovation efforts, to ensure optimal focus on the areas of highest impact, while creating a more compelling story of the effectiveness of nonprofit programs in accomplishing the desired mission.

In taking a holistic end-to-end approach, nonprofit innovators can be guided by a set of best practices that are well established in the current for-profit innovation landscape. Most notably, entrepreneurial nonprofits must reinforce the importance and focus of their mission and purpose, using it to rally employees, partners, and key stakeholders to drive for meaningful impact. In pursuing that transcendent mission, nonprofits will benefit from using a range of innovation tools used by for-profit innovators, including deep understanding of their market environment, use of technology and data to support the innovation agenda, and a broad approach to collaboration to ensure the most efficient, effective, and well-implemented execution of innovative ideas.

Together these tools and a business-like approach enable nonprofit innovators to link the power of the mission with the focus and execution-orientation of the business world. This combination and the ability of nonprofits to adapt and learn from the successes and innovations of others will escalate the ability of nonprofits to have a meaningful, measurable impact in fields as diverse as health care, the environment, education, the arts, and social services.

References

1. Roeger, Katie L., Blackwood, Amy, and Pettijohn, Sarah L. *The Nonprofit Sector in Brief: Public Charities, Giving and Volunteering, 2011.* The Urban Institute, Center on Nonprofits and Philanthropy, 2011.

2. Salamon, Lester M., WojciechSokolowski S. et al. *Global Civil Society Dimensions of the Nonprofit Sector, Volume Two* Published in "Private Philanthropy across the World." The Johns Hopkins Center for Civil Society Studies, 2004.

3. Ibid.

4. Salamon, Lester, WojciechSokolowski, S. and List, Regina. *Global Civil Society: An Overview.* Johns Hopkins Center for Civil Society Studies, p. 37, 2003.

5. Ibid.

6. Ibid.

7. "The 200 Largest U.S. Charities for 2011." *Forbes*, November 30, 2011. Program expense is the measure of how much of the charity's total revenues went to the charitable cause.

8. "The 200 Largest U.S. Charities for 2011." *Forbes*, November 30, 2011.

9. Collins, James and Porras, Jerry. *Built to Last: Successful Habits of Visionary Companies.* Harper Collins, 1994.

10. Ibid.

11. Mourkogiannis, Nikos. *Purpose: The Starting Point of Great Companies.* Palgrave MacMillan, 2006.

12. Pink, Daniel H. *Drive: The Surprising Truth about What Motivates Us.* Riverhead Books, 2009.

13. Ibid.

14. Cooper, Robert G. *Winning at New Products: Accelerating the Process from Idea to Launch.* Addison Wesley, 1993.

15. Richardson, Adam. *Innovation X.* Jossey-Bass, 2010.

16. Ibid.

17. www.ideo.com/about.

18. http://www.ideo.com/work/human-centered-design-toolkit/.

19. Ibid.

20. Evans, David, Hagiu, Andrei, and Schmalensee, Richard. *Invisible Engines: How Software Platforms Drive Innovation and Transform Industries.* Massachussetts Institute of Technology, 2006.

21. Li, Charlene and Bernoff, Josh. *Groundswell: Winning in a World Transformed by Social Technologies.* Harvard Business Press, 2008.

22. Birkholz, Joshua M. *Fundraising Analytics: Using Data to Guide Strategy.* Wiley, 2008.

23. Carlson, Virgina (Metro Chicago Info Center) and Porway, Jake. (Data Without Borders). "Big Data for the Common Good." PowerPoint presentation given at the O'Reilly Strata Conference on March 1, 2012.

24. Rooney, Ben. "How Big Data Can Help Developing Nations." *Wall Street Journal* Tech Europe, October 5, 2012.

25. Hax, Arnoldo C. and Wilde II, Dean L. *The Delta Project: Discovering New Sources of Profitability in a Networked Economy.* Palgrave MacMillan, 2001.

26. Von Hippel, Eric. *Democratizing Innovation.* MIT Press, February 2006.

27. Gloor, Peter A. *Swarm Creativity: Competitive Advantage through Collaborative Innovation Networks.* Oxford University Press, 2006.

28. Sagawa, Shirley and Segal, Eli. *Common Interest, Common Good: Creating Value through Business and Social Sector Partnerships.* Harvard Business Review Press, December 1999.

29. Austin, James E. *The Collaboration Challenge: How Nonprofits and Businesses Succeed through Strategic Alliances.* Jossey-Bass, April 2000.

30. Reagle Jr., Joseph Michael. *Good Faith Collaboration: The Culture of Wikipedia.* MIT Press, September 2012.

CHAPTER 35

Cross-Industry Cooperation as a Key Factor for Innovation

João J. M. Ferreira, Cristina I. Fernandes, and Mário L. Raposo

Introduction

To a greater or lesser extent, innovation is referred to worldwide as a driver of growth and profitability even while there is growing acceptance that incremental improvements do not enable entities to attain genuine competitive advantages. Hence, radical innovation is deemed necessary. However, the leading and broadest reaching innovations are increasingly difficult to generate because the majority of industries have reached fairly mature stages in their development. Products, services, and business models are nevertheless significantly shaped by the mentalities prevailing in their respective industries. Sourcing innovation from beyond the borders of an industry may therefore prove to be one means of opening up new and interesting perspectives and represent a significant generator of innovation.

The term cross-industry innovation refers to innovations stemming from cross-industry affinities and approaches and involving transfers from one industry to another. Such affinities and applications may be drawn from across a variety of levels ranging from technological applications through to business models. In itself, the search for cross-industry innovation is also hardly new. Schumpeter[1] has long since identified how the majority of innovations do not represent anything more than the recombination of already existing knowledge. Furthermore, in the majority of cases, this recombination is closely restricted to the knowledge and technology developed within the company or the chain of value built up by partners throughout the industry.

Nolf et al.[2] exemplify this cross-industry innovation concept through the cases of Toyota and Coca-Cola and how these corporations looked beyond the confines of their own industry as a means of driving innovation. Many years have passed since Toyota successfully studied American supermarkets and adapted the techniques in effect to establish the just-in-time production system. The distributors of Coca-Cola's Freestyle soda deploy micro-dosage technology that the pharmaceutical industry first developed for the dispensation of dialysis and anticancer medical treatments. Installed at points of sale, Coca-Cola distributors enable their clients to tailor the flavor or combination of flavors that they themselves individually desire. This innovation furthermore provided Coca-Cola with a lower cost and lower risk model for the introduction of new products while also evaluating their level of take-up and market performance. Each one of these companies deliberately adopted the mentality of seeking out innovation beyond the confines of their own industry. Their managers understand the potential value this approach contains.

David Levin[3] puts forward another three examples of cross-industry innovation. The new BMW iDrive controller is one example based on the tried and tested Joystick technology. Another example is a leading player in the plumbing sector, Geberit, which recently generated a six-digit reduction in costs. The company began planning its plumbing installations with a tool Zühlke had originally developed for constructing power plants that was then swiftly adapted to meet the special needs in effect at Geberit. A third example is a sensor contained in the presser foot of Bernina sewing machines. This controls the needle to such an extent that stitch lengths are always constant regardless of the direction and speed of the fabric being fed in. After carrying out a research survey of other industries, Zühlke engineers also opted in favor of technologies already existing in other industries.

Studies on cross-industry innovation related phenomena have focused on similar thinking patterns as a source of competitive advantage[4] and increased company performance.[5] Nonobvious affinities may require highly innovative solutions because the appropriate combination of individual pieces of more distant knowledge is associated with a greater level of innovation potential.[4–7]

According to Holton,[8] cross-industry cooperation does work, while there are many differences between companies serving different industrial sectors, synergies and potentials susceptible to leveraging are also present. The process of examining the technologies and applications within the context of other configurations strips them of some of their traditional assumptions and limitations. This opens up possibilities that those directly involved in the operations are not able to grasp in dealing with the same processes on a daily basis. Beyond product development innovation, problem resolution is another field of cross-industry cooperation that may prove highly fruitful. Cooperation between industries is a means of leveraging and aggregating knowledge and generating direct benefits in terms of productivity.

In fact, some of the best sources of both incremental and disruptive innovation may derive from beyond the regular spheres of company contact and even from beyond its own industry. Cross-industry innovation, beyond incorporating external knowledge into the company itself, may also be deployed as a tool for transferring company owned technology and patents to industries on an international scale. While the outside-inside process requires a greater capacity for innovation, it also generates additional momentum at the cost of relatively little effort.[4]

Nevertheless, any successful search for similar solutions and their subsequent retranslation and multiplication inherently demands new or at least adapted processes and innovation management tools, skills, competences, and technologies. Herstatt and Kalogerakis,[9] along with Herstatt and Engel,[10] discuss possible means of structuring cross-industry processes and tools based on an analogous perspective. Gassmann et al.[11] conceptualize cross-industry innovation phenomena in the automobile sector and specifically target inter-industrial cooperation between small start-up companies and established corporations. Gassmann and Zeschky[4] also demonstrate how cross-industry distances encourage similar thinking and may be incorporated into product innovation at the company level.

This chapter approaches cross-industry cooperation as a key factor in innovation and evaluates in what ways cooperation between companies from different sectors may contribute to cross-industry innovation.

The chapter is structured as follows: following this introduction, we move onto reviewing the current literature on this theme. In the third section, we set out our sample and the statistical methodology underpinning the empirical study. In the fourth section, we discuss the results obtained before finally closing with our conclusions.

Literature Review

Cooperation for Innovation

Various empirical research projects have confirmed that face-to-face contacts and geographic proximity are important factors in spreading innovation[12] and in specific forms of exchanging knowledge[13,14] and in providing better access to information and technology.[15]

In one recent study, Bell[16] finds that Canadian companies locating in industrial clusters and in central positions in networks experience greater incremental innovation. Taking the biotechnology sector into consideration, Aharonson et al.[17] similarly consider companies grouped into clusters attain higher rates of innovation than those in dispersed geographic locations.

Sonn and Storper[18] also confirm the positive effect of geographic proximity on innovation. In another study, Almeida and Kogut[19] approach the differences existing between regions in terms of both the location of different knowledge types and the mobility of locally generated patents. Other researchers have also defended how small companies explore new technological fields and tend to prosper in locations with dense networks.[19]

In their study of German companies, Audretsch et al.[20] verify that a location within the vicinity of universities represents a factor of major importance to company performance.[21] Despite a generalized consensus around the role played by geographic proximity in achieving higher performance levels in terms of innovation, there is another branch of the literature that is critical of the theory. For example, despite Boschma[22] highlighting the importance of proximity between institutions and organizations, in another study Boschma and terWal[23] demonstrate how what matters are not only locally based connections but also the global level of connections able to boost company innovation performance. Other studies portray how social bonds and connections may be more important than simple geographic proximity.[24–26]

This growing emphasis on social factors implies they bear direct implications to the existence of networks and social relationships and hence it is not only geographic proximity influencing innovation related issues.[27] In a study of Canadian companies, Doloreux[28] identifies how they also maintained connections with clients and suppliers from beyond their region with these connections and relationships enabling their own innovations. This result is also backed up by Oerlemans et al.[29] in a study of Dutch companies displaying strong external connections with these relationships bearing a strong impact on company innovation performances.

Therefore, in accordance with these studies, the presence of trust in social relationships ongoing in industrial districts is one of the most important mechanisms in fostering innovation. However, it is important to verify whether or not the innovation production mechanisms are confined only to these relationships. Furthermore, some findings in the literature call for the reconfiguration of intersectoral mechanisms within the framework of the economic and geographic space.[30]

Innovation in Industries

Innovation is fundamental to every sector of the economy in order to ensure businesses are able to thrive and survive within the context of an increasingly globalized context. Innovation enables companies to generate the responses necessary to the increasingly diversified demand patterns undergoing constant evolution and bring about improvements across the fields of security, health, communications, the environment, and quality of life in general. Innovation therefore plays the role of a motor driving progress, competitiveness, and economic development.[31,32] Muller,[33] after criticizing the traditional dichotomy between goods and services, posits innovation as an association of processes. Alternatively expressed: Is the distinction made between product-based innovation and process-based innovation relevant to the analysis of innovative interactions between the transformative and service industrial sectors?

Should we adopt the classification of company (transformative) innovation proposed by Pavitt,[34] we find this rests essentially on four innovation drivers: (1) dominant suppliers; (2) intensive-scale companies; (3) science-based companies; and (4) specialist equipment suppliers. This author also details how the company's respective area of activity conditions whether innovation processes tend to be faster or slower paced. The Pavitt-developed classification was transposed to the service sector by Soete and Miozzo[35] as follows: (1) dominant service suppliers; (2) service networks; (3) intensive product and scale services; and (4) specialist technology and science-based services.

According to Miles et al.,[36] when seeking to identify the ways in which service innovations take place, the following types should be taken into consideration: (1) product innovations: deriving from innovation processes and very often corresponding to demand side requests; (2) process innovations: especially those stimulated through new technologies; and (3) delivering innovation: through the application of new resources and methods, such as support structures for interactions between service companies and their clients. Within the same framework, Gallouj[37] proposes the formalization of innovation activities as follows: (1) preemptive innovations, described as the most authentic form of innovation and correspondingly the least frequent and the most difficult to implement and essentially consisting of producing something entirely new; (2) objective innovations, as the most frequent and least risky, and which primarily consist of exploiting new methodologies or recycling those already existing; and (3) valuing innovation, which essentially involves adopting already existing experiences and in the specialization of capacities and knowledge leading to the emergence of new ideas and solutions.

From Muller's perspective,[33] the distinction between product innovation and process innovation makes little sense to the extent that the different forms of innovation are inherently interrelated and, in the majority of cases, the innovation emerging is characterized by the existence of strongly varying and different facets. Nevertheless, this distinction remains common to the literature on transformative industrial companies. Meanwhile, den Hertog[38] considers what might be important for the introduction of a specific product into one market might prove totally irrelevant to other products. In practice, most innovations would seem to be made up of a mixture of large and small modifications and adaptations to already existing products or services.

Studies on innovation have thus contributed to a better understanding on the evolution of industries as well as the connections between knowledge and the better performances of certain sectors to the detriment of others.[39–42] The central point of these research findings is exactly that economic growth across different sectors derives out of the specialization of companies in their predominant core activities, the utilization of different types of knowledge, and the identification of all latent opportunities.[42]

Another innovation-related field of study considers how financial performance depends on diverse contingencies and specific circumstances.[43] Taking these perceptions into account, Malerba[44] interrelated the various modes of innovation with the specific characteristics of different sectors. This author maintains that sectors are mutually unified by certain groups of products as well as by shared knowledge. Companies within a particular sector turn out to be simultaneously similar and heterogeneous. Within this framework, the growth of knowledge and development in the sector mutually interweave as processes self-reinforcing the capacities in effect across the sector.[39]

The processes of professionalization, systematization, and specialization of activities are important mechanisms to strengthening and deepening the knowledge present across different economic sectors.[45,46]

Professionalization and systematization nurture significant developmental trajectories in the knowledge structures of sectors. In conjunction with the accumulation of experience, they reconfigure the core structural tasks in sectors where they merge with the experiences of employees and the codified knowledge of senior management.[47]

Methodology

Sample and Methods

This study incorporated a questionnaire survey of a sample of 61 companies engaged in different sectors of activities in two neighboring countries (Portugal and Spain) exploring their cooperation and innovation activities. Respondents were requested to answer different questions based upon a five-point Likert scale. The operational processing of the variables is described in Table 35-1.

The data obtained was analyzed through recourse to SPSS Software version 18.0. The numerical variables are summarized by their average, median, minimum, maximum, and standard deviation while their qualitative dimensions are expressed according to their absolute and relative frequencies. For comparative bivariate analysis between Portuguese and Spanish companies, we applied the Mann-Whitney test and the t-test to the continuous variables and the Chi-squared test for the various categorical variables. In multivariate terms, we adopted linear regression as the statistical methodology for analyzing company innovation and cooperation and evaluating any possible differences between the countries involved (Portugal and Spain). We furthermore deployed probit regression models for cases when the dependent variables are binary (low or average turnover volumes) and ordinal regression models (with probit link) for the ordinal turnover volumes (average, average/above average, and high). We finally applied the Nagelkerke Pseudo R^2 for calculating the determining coefficient.

Results

Company Profile

This point describes the profiles of companies by their location (Fig. 35-1) with the results displaying statistically significant differences ($p < 0.05$) between Portuguese and Spanish companies with regard to their core business activities. Of the Portuguese companies, 46.2 percent belong to the transport sector while, across the border in Spain, production and distribution accounts for 54.3 percent of firms. There were also statistically significant differences ($p < 0.05$) with regard to the number of company employees with the majority of Spanish companies providing employment to less than 10 members of staff (60 percent) while their Portuguese counterparts, in the main, employ between 10 and 49 workers (61.5 percent). According to the European Commission (1996),[48] these firms are classified as micro- and small-sized companies, respectively.

Commercial relations with partners in the Norte, Centro, and Lisboa e Vale do Tejo regions of Portugal have risen substantially ($p < 0.05$) more in Portuguese companies than in Spanish companies while Spanish firms have seen a significant increase ($p < 0.05$) in their commercial relations with firms located in the Castela e Leão, Centro, Sul, and Other regions of Spain (Fig. 35-2).

With regard to informal relations, our results show that Portuguese companies have significantly greater levels of interaction in the Portuguese regions of Centro, Norte, Lisboa Vale do Tejo, and Algarve while Spanish entities experience that boost in interaction in the national regions of Castela y Leão and Norte (Fig. 35-3).

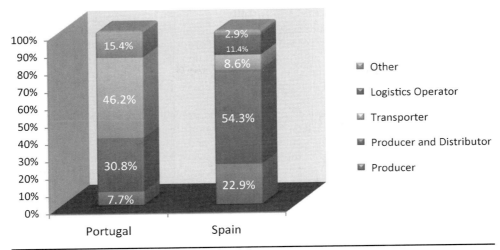

Figure 35-1 Company profile by sector of activity.

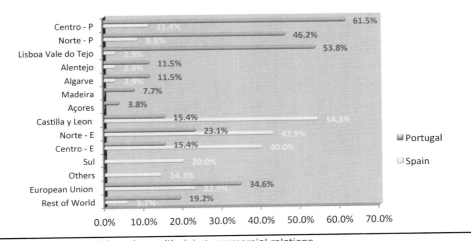

Figure 35-2 Geographic regions with rising commercial relations.

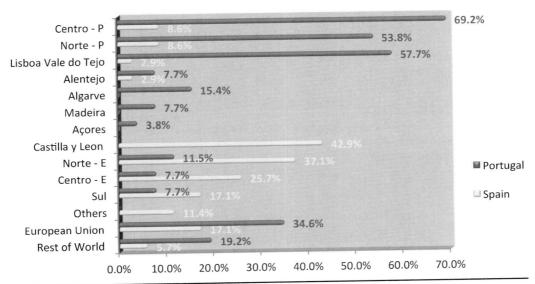

Figure 35-3 Geographic regions with rising informal partner relationships.

Business Cooperation and Innovation

The level of cooperation between the diverse business sectors does not return any statistically significant differences ($p > 0.05$) between Portuguese and Spanish companies. Furthermore, our results for the configuration and duration of cooperation, the level of importance attributed to clients and the level of importance of the different innovation types all report no statistically significant differences ($p > 0.05$) between Portuguese and Spanish companies (Fig. 35-4).

In relation to the intensity of the relations prevailing and the location of companies engaging in innovation-related cooperation, the results return statistically significant differences ($p < 0.05$) between Portuguese (Fig. 35-5a) and Spanish (Fig. 35-5b) companies in terms of having partners located in the regions of Castela y Leão, Norte, Centro, and Sul of Spain as well as those located in the European Union and in the Rest of the World with a significantly higher level of intensity in Spanish companies.

There are statistically significant differences ($p < 0.05$) between Portuguese and Spanish companies with regard to the scale of the companies that they cooperate with for purposes of innovation. Spanish companies most commonly cooperate (45 percent) with firms employing over 250 members of staff (large companies) while Portuguese businesses predominantly (75 percent) cooperate with companies employing between 10 and 49 employees (small) (Fig. 35-6).

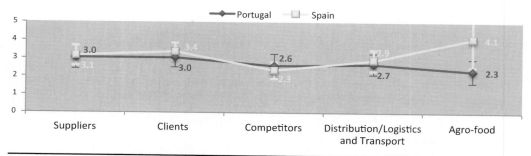

FIGURE 35-4   Types of cooperation.

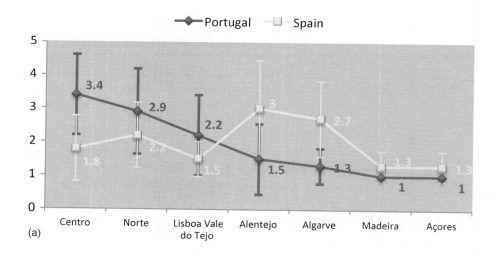

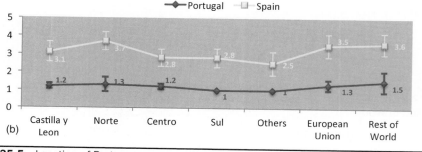

FIGURE 35-5   Location of Portuguese companies engaged in cooperation.

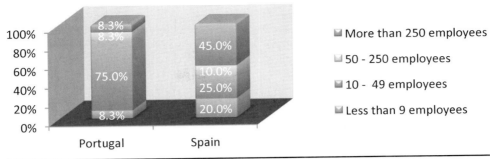

FIGURE 35-6   Size of innovation partner companies.

In relation to the level of concurrence over the advantages deriving from such coopera-tion, Portuguese companies register significantly higher levels of agreement ($p < 0.05$) in terms of strengthened technical knowledge, improved management, higher turnover, and marketplace flexibility, as well as an overall increase in competitiveness stemming from the knowledge obtained from partners furthermore leveraging better client satisfaction levels and the marketplace positioning of the company (Figs. 35-7a, 35-7b, and 35-7c).

With regard to the level of agreement about cooperation, Portuguese firms return sig-nificantly higher levels of such concurrence ($p < 0.05$) with regard to the cultural differences between partners, the nonrevelation of all information to partners, a greater postcooperation capacity to satisfy clients, the noninterference of cooperation with company independence, and avoiding unnecessary expenditure (Figs. 35-8a, 35-8b, and 35-8c).

With regard to the level of difficulty encountered in generating innovation, Spanish companies perceive significantly stronger obstacles ($p < 0.05$) due to a lack of both company investment capital and external financing, very high wage costs, difficulties both in forecast-ing demand, and organizing innovations in addition to a shortage of qualified staff in the fields of R&D, production, and marketing/sales (Figs. 35-9a and 35-9b).

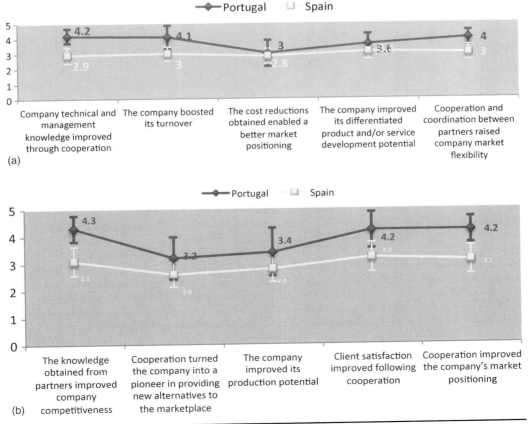

FIGURE 35-7   Level of agreement with the advantages obtained from cooperation.

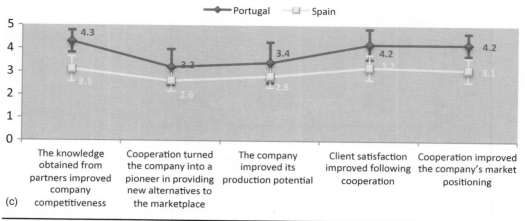

FIGURE 35-7 (Continued)

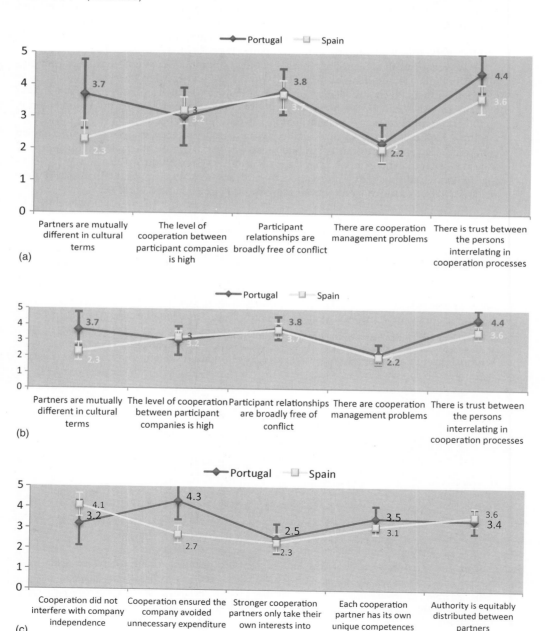

FIGURE 35-8 Level of agreement about cooperation.

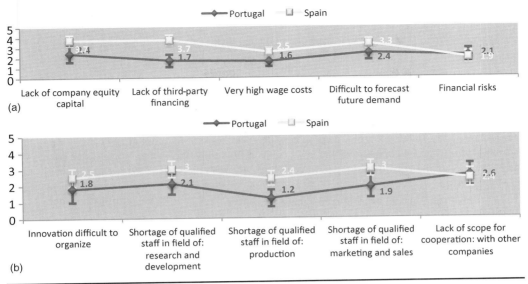

FIGURE 35-9   Innovation-related difficulties.

In terms of the regional factors and their level of importance to company innovation capacities, our results show Spanish companies report significantly higher ($p < 0.05$) levels of risk capital.

Research laboratories, universities, study institutes, and the specialist press are the main sources of information on product innovation cooperation of greatest significance ($p < 0.05$) to Spanish firms.

With regard to process innovation, Portuguese companies attribute statistically significant ($p < 0.05$) importance to their suppliers while Spanish entities attribute the highest levels of importance ($p < 0.05$) to competitors, research laboratories, universities, and study institutes.

Clients, suppliers, competitors, and research laboratories take on statistically significant ($p < 0.05$) greater importance to Spanish companies in terms of organizational innovation processes.

Multivariate Analysis: Modes of Cooperation

Table 35-1 presents the characteristics of the variables adopted for analyzing their influence over the various types of cooperation ongoing at Portuguese and Spanish companies.

Furthermore, we also calculated separate linear regression models for Portuguese and Spanish companies in order to infer just which variables they significantly associate with the level of importance attributed to cooperation with: (1) suppliers; (2) clients; (3) competitors; (4) distribution/logistics and transport sectors; and (5) agro-food sector companies.

The level of importance attributed by Portuguese companies to cooperation with suppliers (Table 35-2) is significantly linked to that attributed to the suppliers ($B = 0.64$, $p < 0.05$) factor.

The greater the importance attributed to suppliers, the greater the level of importance paid to this cooperation type. The level of importance attributed to cooperation with suppliers by Spanish companies is significantly associated with the importance awarded to both the qualified human resources ($B = 0.77$, $p < 0.05$) and state support ($B = -0.73$, $p < 0.05$) items. Additionally, the greater the importance attributed to qualified human resources, the greater that conferred on cooperating with suppliers with the inverse observed in the case of state support and susceptible to being perceived as an obstacle to this type of cooperation.

The level of attention attributed to cooperation with clients by Portuguese companies significantly correlates with that paid to suppliers ($B = 0.77$, $p < 0.01$) and consultants ($B = -0.48$, $p < 0.05$). The greater the importance attributed to suppliers, the greater the emphasis placed upon this type of cooperation and with the inverse observed with the consultants item and again able to be approached as an obstacle within the framework of this cooperation type. At Spanish companies, the level of importance awarded to cooperation with clients does not attain statistical significance for any variable.

With regard to cooperation with competitors, no variable achieves any statistically significant level in terms of the importance attributed to this factor by Portuguese companies while

Dimension	Variables	Quantitative Measurement	Qualitative Measurement
Modes of cooperation	1. Suppliers 2. Clients 3. Competitors 4. Distribution/logistics and transports 5. Agro-food sector company	Likert scale from 1 (not at all important) to 5 (very important)	
Factors of cooperation	1. State support for development 2. Consultants 3. Research 4. Suppliers 5. Qualified local labor resources 6. Clients 7. The company undertakes productive activities 8. Qualified human resources	Likert scale from 1 (not at all important) to 5 (very important)	
Cooperation configuration	1. Improvements to business processes 2. Exports 3. Distribution agreements 4. Outsourcing 5. R&D agreements	Likert scale from 1 (not at all important) to 5 (very important)	
Financial performance	1. Business turnover	Quantitative ranges <under 2 million 2–5 million 5.1–10 million >10 million	Low Medium Medium high High
Company profile	1. Company age	Quantitative ranges <5 years 5–15 years 16–36 years 36–70 years	Young In growth In Maturity Aged
	2. Number of employees	Quantitative ranges <10 employees 10–49 employees 50–249 employees	Micro company Small company Average company

TABLE 35-1 Description of Analytical Variables

Spanish companies attribute a statistically significant level of importance between cooperating with competitors with that paid to the capital ($B = 0.49$, $p < 0.01$) factor. The greater the importance attributed to this factor, the greater the significance endowed to cooperating with competitors.

The level of importance paid to cooperation with distribution/logistics/transport companies by Spanish companies is significantly linked to company age ($B = 1.39$, $p < 0.05$) and the importance attributed to the regional risk capital ($B = −0.45$, $p < 0.05$) factor. Companies under the age of 15 award significantly greater importance to cooperation with such companies and the greater the importance attributed to risk capital, the lower the importance attributed to cooperation and hence representing a potential obstacle to this mode of cooperation. At Portuguese companies, the level of importance of cooperating with distribution/logistics/transport companies bore no statistically significant association.

The level of importance attributed by Spanish companies to cooperation with agro-food sector companies statistically significantly correlates with both the company age ($B = −1.22$, $p < 0.1$) and the importance awarded to qualified regional human resources ($B = 0.38$, $p < 0.1$) factors. Companies in business for less than 15 years attribute significantly less importance to cooperating with agro-food sector companies and the greater the importance paid to qualified human resources, the greater the priority paid to cooperation. With regard to Portuguese companies, cooperating with agro-food sector companies did not attain any statistically significant correlation with any of the variables.

Dependent	Country		B	Std. Error	Beta	t	P	R²
Suppliers	PT	(Constant)	1.01	0.84		1.20	0.253	0.382
		Suppliers	0.64	0.24	0.62	2.73	0.018**	
	SP	(Constant)	4.06	1.27		3.20	0.008**	0.506
		Qualified human resources	0.56	0.20	0.61	2.82	0.017**	
		State support for economic and technological development	−0.73	0.31	−0.51	−2.37	0.037**	
Clients	PT	(Constant)	2.56	0.72		3.57	0.004**	0.618
		Suppliers	0.77	0.18	0.96	4.22	0.001***	
		Consultants	−0.48	0.20	−0.55	−2.41	0.034**	
Competitors	SP	(Constant)	4.12	0.75		5.51	0.000***	0.328
		Risk capital	−0.49	0.20	−0.57	−2.42	0.032**	
Distribution and transport sector	SP	(Constant)	4.17	0.63		6.63	0.000***	0.541
		Company age ≤15 years	1.39	0.53	0.54	2.64	0.023**	
		Risk capital	−0.45	0.17	−0.54	−2.61	0.024**	
Agro-food sector	SP	(Constant)	2.61	0.63		4.13	0.002**	0.510
		Qualified human resources	0.38	0.18	0.48	2.20	0.050**	
		Company age ≤15 years	−1.22	0.62	−0.43	−1.97	0.074*	

$*p <0.1, **p <0.05, ***p <0.001.$

TABLE 35-2   Modes of Cooperation: Multiple Linear Regression Model

We calculated the linear regression models to determine just which factors statistically influence the intensity of cooperation across diverse configurations with statistically significant associations obtained only in the case of Portuguese companies (Table 35-3).

With regard to the intensity of export-focused cooperation, we find the factors bearing greatest influence to be whether companies are engaged in productive activities ($B = 1.72$, $p <0.01$) and the importance of the climate of innovation ($B = 1.15$, $p <0.001$), and state support ($B = −0.72$, $p <0.01$). Product-based companies report a greater intensity of engagement in this cooperation type, with the greater the importance attributed to the regional innovation climate, the greater the intensity of cooperation with the inverse in effect with regard to the state support factor.

The intensity of distribution agreement cooperation is significantly influenced by the importance wielded by the suppliers ($B = 0.80$, $p <0.05$) and state support ($B = −0.60$, $p <0.1$) factors. The greater the importance attributed to suppliers, the denser the intensity of distribution agreement related cooperation with the inverse observed in the case of the regional public support factor.

The intensity of cooperation within the framework of representation agreements is significantly influenced by the importance played by the regional clients ($B = 0.71$, $p <0.05$) factor and where the greater the importance attributed to clients, the greater the corresponding intensity of cooperation.

Regarding the intensity of cooperation with commercial associations, the findings show that only companies engaged in productive activities return a statistically significant influence ($B = 1.17$, $p <0.1$) for this variable and where these companies report a greater intensity of cooperation.

The intensity of training/technical assistance related cooperation is significantly influenced by company age ($B = 5.58$, $p <0.01$) and with companies in business for less than 15 year returning a greater degree of intensity in this cooperation type.

Dependent		B	Std. Error	Beta	t	p	R²
Export cooperation	(Constant)	−0.47	0.64		−0.74	0.478	0.802
	Production with or without distribution	1.72	0.37	0.67	4.60	0.001***	
	Importance played by the following regional factor: Innovation climate	1.15	0.21	1.11	5.35	0.000***	
	Importance played by the following regional factor: State support for economic and technological development	−0.72	0.19	−0.75	−3.69	0.004**	
Distribution agreement	(Constant)	2.60	1.20		2.18	0.052*	0.465
	Importance played by the following regional factor: Suppliers	0.80	0.27	0.71	2.96	0.013**	
	Importance played by the following regional factor: State support for economic and technological development	−0.60	0.31	−0.47	−1.96	0.076*	
Representation agreement	(Constant)	−0.49	0.97		−0.51	0.620	0.393
	Importance played by the following regional factor: Clients	0.71	0.26	0.63	2.79	0.016**	
Commercial association	(Constant)	1.33	0.49		2.72	0.019**	0.212
	Production with or without distribution	1.17	0.65	0.46	1.80	0.097*	
Technical training/ assistance	(Constant)	1.92	0.52		3.66	0.003**	0.576
	Company in business less than 15 years	5.58	1.38	0.76	4.03	0.002**	
Outsourcing	(Constant)	0.08	0.60		0.13	0.896	0.597
	Importance played by the following regional factor: Suppliers	0.71	0.17	0.77	4.22	0.001***	
Joint production agreement	(Constant)	−1.76	0.98		−1.80	0.099*	0.575
	Importance played by the following regional factor: Clients	0.79	0.21	0.87	3.84	0.003**	
	Production with or without distribution	1.36	0.60	0.51	2.25	0.046**	
R&D agreement	(Constant)	1.48	0.91		1.63	0.142	0.936
	Importance played by the following regional factor: Consultants	−1.99	0.34	−1.06	−5.79	0.000***	
	Importance played by the following regional factor: Qualified human resources	1.79	0.28	0.82	6.34	0.000***	
	Importance played by the following regional factor: Capital	0.57	0.23	0.33	2.51	0.036**	
	Importance played by the following regional factor: Transport infrastructures	0.87	0.27	0.41	3.22	0.012**	
	Importance played by the following regional factor: Innovation climate	−0.95	0.39	−0.45	−2.43	0.041**	

*$p < 0.1$, **$p < 0.05$, ***$p < 0.001$.

TABLE 35-3 Linear Regression: Intensity of Cooperation by Configuration (Portugal)

The intensity of outsourcing cooperation is significantly influenced by the suppliers ($B = 0.71$, $p <0.01$) factor. The greater the importance attributed to suppliers, the greater the intensity of outsourcing based cooperation.

The intensity of production agreement related cooperation is significantly influenced by the clients ($B = 0.79$, $p <0.01$) factor and whether the company itself engages in productive activities ($B = 1.36$, $p <0.05$). The greater the importance attributed to clients, then the greater the intensity of cooperation and where product companies demonstrate a greater prevalence toward joint production agreements.

The intensity of R&D agreement related cooperation returns a statistically significant influence for the regional consultants ($B = -1.99$, $p <0.01$), qualified human resources ($B = 1.79$, $p <0.001$), capital ($B = 0.57$, $p <0.05$), transport infrastructures ($B = 0.87$, $p <0.05$), and the climate of innovation ($B = -0.95$, $p <0.05$) factors. The greater the relevance attributed to qualified human resources, capital, and transport infrastructures, the greater the intensity of cooperation and the greater the importance awarded to consultants and the regional climate of innovation, the lesser the intensity of R&D based cooperation. Figures 35-10a, 35-10b, and 35-10c display the differences between the different cooperation agreement configurations between the two countries.

Multivariate Analysis: Cooperation for Innovation

In Table 35-4, we calculate linear regression models in which the dependent variable is the intensity of cooperation by configuration and the independent variables are innovation capacity and company age. Significant associations were reported only in the case of Portuguese companies.

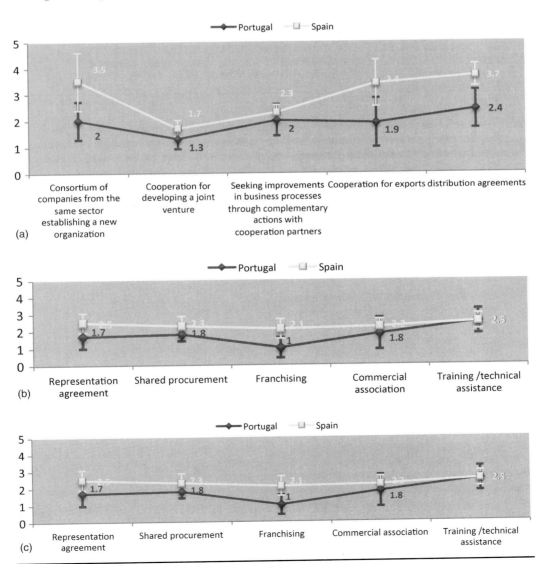

FIGURE 35-10   Configuration of cooperation.

		B	Std. Error	Beta	t	p	R²
Seeking improvements to business process through complementary actions	(Constant)	0.54	0.57		0.94	0.375	0.492
	Process innovations	0.55	0.20	0.70	2.79	0.024**	
Export cooperation	(Constant)	0.27	0.45		0.61	0.560	0.750
	Organizational innovations	0.76	0.16	0.87	4.90	0.001***	
Distribution agreement	(Constant)	0.36	0.74		0.49	0.639	0.623
	Product innovations	0.81	0.22	0.79	3.64	0.007*	
Outsourcing	(Constant)	0.05	0.64		0.07	0.946	0.665
	Organizational innovations	0.77	0.21	0.83	3.61	0.009**	
	Company operational <15 years	2.19	1.10	0.45	1.98	0.088*	
Joint production agreement	(Constant)	0.35	0.54		0.63	0.543	0.554
	Process innovations	0.59	0.19	0.74	3.15	0.014**	
R&D agreement	(Constant)	0.83	0.16		5.09	0.001***	0.922
	Company operational <15 years	2.83	0.33	0.92	8.70	0.000***	
	Process innovations	0.17	0.05	0.32	3.06	0.018**	

$*p < 0.1, **p < 0.05, ***p < 0.001.$

TABLE 35-4   Linear Regression: The Innovation Capacity Influence over Cooperation Intensity (Portugal)

Regarding the intensity of cooperation undertaken with the objective of seeking out improvements to business processes, this statistically correlates only with the level of importance paid to process innovations ($B = 0.55$, $p < 0.05$). The intensity of export related cooperation is significantly associated with the emphasis placed on organizational innovations ($B = 0.76$, $p < 0.01$). In the case of distribution agreement cooperation intensity, there is a statistically significant relationship with the importance attributed to product innovations ($B = 0.81$, $p < 0.01$). The intensity of outsourcing based cooperation statistically correlates with company age ($B = 2.19$, $p < 0.1$) and the respective importance paid to organizational innovations ($B = 0.77$, $p < 0.01$). The intensity of joint production cooperation agreements is statistically significantly linked with the level of importance attributed to process innovations ($B = 0.59$, $p < 0.05$). Finally, the intensity of R&D agreements is significantly associated with company age ($B = 2.83$, $p < 0.001$) and with the level of importance attributed to process innovations ($B = 0.17$, $p < 0.05$). In all regressions, the greater the importance awarded to the respective aforementioned innovations, the greater the intensity of the cooperation types with companies in business for less than 15 years reporting the greatest levels of cooperation.

Multivariate Analysis: Factors and Obstacles to Innovation

In Table 35-5, we set out the description of the variables deployed in our analysis of the factors and obstacles to innovation.

With the objective of analyzing the importance of different types of innovation (and any differences between Portugal and Spain), we applied linear regression to analyze which factors proved determinant to the level of innovations in each country in terms of: (1) process, (2) product, (3) organization, and (4) the introduction of already existing products in new markets.

In relation to process innovations, no variable generates a statistically significant effect ($p > 0.05$) on the level of importance attributed, whether by Portuguese or by Spanish companies.

The level of importance attributed by Portuguese companies to product innovation (Table 35-6) is significantly associated with the importance attributed to the following factors of innovation: (1) state support ($B = -1.81$, $p < 0.01$), (2) suppliers ($B = 1.31$, $p < 0.01$), (3) clients ($B = 1.40$, $p < 0.01$), and (4) the climate of innovation ($B = -1.25$, $p < 0.05$).

Dimension	Variables	Quantitative Measures	Qualitative Measures
Modes of innovation	1. Product innovation 2. Process innovation 3. Organizational innovation 4. Introduction of existing products into new markets	Likert scale from 1 (not at all important) to 5 (very important)	
Factors of innovation	1. Local labor 2. Innovation climate 3. Regional consultants 4. Risk capital 5. Qualified human resources (HR) 6. Transport infrastructures 7. Research	Likert scale from 1 (not at all important) to 5 (very important)	
Company profile	1. Company age	Quantitative ranges <5 years 5–15 years 16–36 years 36–70 years	Young In growth In maturity Aged
	2. Number of employees	Quantitative ranges <10 employees 10–49 employees 50–249 employees	Micro company Small company Average company

TABLE 35-5   Description of Variables

Dependent		B	Std. Error	Beta	t	p	R^2
PT	(Constant)	5.37	0.49		10.99	0.000***	0.962
	Importance of the state support for economic and technological development factor	−1.81	0.21	−1.49	−8.52	0.001***	
	Importance of the suppliers factor	1.31	0.23	1.09	5.76	0.005**	
	Importance of the clients factor	1.40	0.26	1.15	5.30	0.006**	
	Importance of the innovation climate factor	−1.25	0.32	−1.05	−3.96	0.017**	
SP	(Constant)	4.12	0.89		4.65	0.001***	0.716
	Importance of the innovation climate factor	−0.69	0.23	−0.55	−3.01	0.013**	
	Company operational <15 years	−2.34	0.61	−0.66	−3.85	0.003**	
	Importance of the local labor factor	0.73	0.24	0.56	3.06	0.012**	

*p <0.1, **p <0.05, ***p <0.001.

TABLE 35-6   Linear Regression: The Importance of Product Innovations

The greater the importance attributed to suppliers and clients, the greater the level of product innovation with the inverse observed for the state support and climate of innovation factors. Thus, the greater the importance attributed to these factors, the lower the importance attributed to product innovations and may be perceived as obstacles to fostering this innovation type.

For Spanish companies, the company description variable "company age" holds influence over product innovations ($B = -2.34$, $p < 0.01$) and the importance attributed to factors of innovation: (1) innovation climate ($B = -0.69$, $p < 0.05$) and (2) local labor ($B = 0.73$, $p < 0.05$) significantly influence the importance attributed to product innovation. Therefore, companies in business for a period of less than 15 years (young companies) confer greater importance on the climate of innovation factor and attributing significantly less importance to product-based innovations, which may therefore be perceived as an obstacle to innovation. Hence, that Spanish companies tend to fall into this category implies a lower level of overall product innovation. In the case of the local labor factor, the greater the importance awarded then the greater the profile attributed to the product innovation factor.

The level of importance attributed by Portuguese companies to organizational innovations (Table 35-7) is significantly associated with the importance attributed to the following factors of innovation: (1) clients ($B = 1.20$, $p < 0.05$), (2) state support ($B = -1.25$, $p < 0.05$), and (3) research ($B = -0.61$, $p < 0.1$). The greater the importance attributed to clients and research, the greater the perceived priority of organizational innovations and with the inverse reported by the state support factor in which the greater the importance attributed corresponded to a lower importance going to organizational innovations, and thus a factor that may represent an obstacle to innovation.

Meanwhile, Spanish companies consider the importance of organizational innovations as associated to the following regional factors: (1) research ($B = 0.31$, $p < 0.1$), (2) consultants ($B = 0.84$, $p < 0.05$), and (3) innovation climate ($B = -0.53$, $p < 0.05$), as well as the variable describing company "number of employees" ($B = 3.36$, $p < 0.01$). Companies currently employing between 50 and 249 (medium-sized companies) members of staff are those awarding the highest level of relevance to organizational innovations and the greater the level of importance attributed to consultants and research, the greater the level of importance attached to organizational innovations. Furthermore, the inverse is observed with the innovation climate factor in which the greater the importance received corresponds to a lesser importance being attributed to organizational innovations, and may therefore be considered an obstacle to innovation.

Dependent		B	Std. Error	Beta	t	p	R^2
PT	(Constant)	1.43	0.99		1.45	0.207	0.799
	Importance of the clients factor	1.20	0.32	1.01	3.71	0.014**	
	Importance of the state support for economic and technological development factor	−1.25	0.37	−1.04	−3.33	0.021**	
	Importance of the research factor	0.61	0.28	0.57	2.13	0.086*	
SP	(Constant)	−0.31	0.79		−0.39	0.707	0.803
	Importance of the research factor	0.31	0.18	0.30	1.71	0.099*	
	Between 50 and 249 employees	3.36	0.70	0.75	4.78	0.001***	
	Importance of the consultants factor	0.84	0.26	0.66	3.26	0.01**	
	Importance of the innovation climate factor	−0.53	0.21	−0.46	−2.47	0.036**	

$*p < .1$, $**p < .05$, $***p < .001$.

TABLE 35-7 Linear Regression: The Importance of Organizational Innovations

	B	Std. Error	Beta	t	p	R²
(Constant)	2.19	0.42		5.24	0.003**	0.988
Importance of the clients factor	−0.73	0.13	−0.59	−5.46	0.003**	
Importance of the qualified human resources factor	1.43	0.09	1.62	16.15	0.000***	
Importance of the local labor factor	−1.85	0.17	−1.67	−10.90	0.000***	
Company operational <15 years	2.40	0.29	0.79	8.35	0.000***	
Importance of the transport infrastructure factor	0.81	0.14	0.85	5.68	0.002**	
Importance of the research factor	−0.55	0.12	−0.57	−4.47	0.007*	
Importance of the capital factor	0.51	0.14	0.49	3.62	0.015*	
Between 50 and 249 employees	0.69	0.33	0.17	2.11	0.089*	

$^*p < 0.1, ^{**}p < 0.05, ^{***}p < 0.001.$

TABLE 35-8 Linear Regression: Dependent Variable: Level of Importance Attributed to the Introduction of Already Existing Products into New Markets (Spain)

The importance endowed by Portuguese companies to the introduction of already existing products into new markets is not statistically associated with any variable. However, at Spanish companies, the level of importance attributed to this factor (Table 35-8) is statistically linked to the importance attributed to the following factors of innovation: (1) clients ($B = -0.73$, $p < 0.01$), (2) qualified human resources ($B = 1.43$, $p < 0.001$), (3) local labor ($B = -1.85$, $p < 0.001$), (4) transport infrastructures ($B = 0.81$, $p < 0.01$), (5) research ($B = -0.55$, $p < 0.01$), and (6) risk capital ($B = 0.51$, $p < 0.05$), as well as the company description items: (1) company age ($B = 2.40$, $p < 0.001$) and (2) number of employees ($B = 0.69$, $p < 0.1$).

Companies with less than 15 years of operational experience (young) and with between 50 and 249 employees (medium-scale companies) attribute significantly greater importance to the introduction of already existing products into new markets and the greater the importance attached to qualified human resources, transport infrastructures, and capital, the greater the emphasis on the introduction of such new products into new markets while observing an inverse association with clients, local labor, and research and may be seen as an obstacle to this innovation type.

Conclusion

This chapter strives to contribute toward a better conceptual understanding and enhance the capacity to shape the interweaving factors inherent to innovation. The study suggests that innovation between industrial sectors, through cooperation, may be deemed an effective method for the dissemination of innovation related efforts throughout more traditional companies within the framework of cutting response times to newly emerging market needs and requirements and thereby restoring competitiveness to companies. In fact, cross-innovation in other industries introduces already existing and tested solutions from these different sectors and thus generating the competitive leveraging effect necessary to responding to ever more sophisticated and demanding markets. Furthermore, these solutions may result from technologies, patents, specific knowledge, or new business processes.

The literature analyzed suggests that the innovation type itself influences the way in which the spread of the innovation is approached, whether through direct cooperation or through "coopetition." Thus, in considering how understanding the competitive realities and markets depends on the companies in the field and in how grasping the workings of the innovation systems present in companies located in peripheral regions is fundamental to developing the public policies able to foster the transfer of innovation between industries, we undertook this empirical study involving companies from two neighboring countries in the expectation of encountering different intensities for innovation and providing due explanation for such variations.

The results demonstrate the influence of cooperation on cross-industry innovation promoting the introduction of product and organizational process innovations.

Global competition increasingly impacts on the way companies operate, compete, and innovate leading them to open up their innovation processes through integrating into cooperation networks with external partners (suppliers, clients, universities, etc.) to bring about the interactions able to broaden and deepen their own innovation capacities. From a business strategy perspective, this study demonstrates that a higher level of cooperation, in particular with competitors, results in a higher level of innovations being introduced into companies and helping to counter the trend for traditional companies to return only low levels of innovation. In addition, at a time of market crisis, the agility of companies to engage in fundamental programs of development, through means of cross-industry innovation, is crucial not only to business competitiveness but also to the very survival of companies.

This study thus contributes to knowledge on cross-industry innovation processes and how cooperation may prove a key approach in successful strategies. We correspondingly believe there are all the grounds for continuing to advance knowledge in this field in order to ascertain the extent to which cross-industry innovation is susceptible to leveraging company development and performance and especially in more competitive, international contexts.

References

1. Schumpeter, J. *Business Cycles: A Theoretical Historial and Statistical Analysis of the Capitalist Process*. New York, NY: McGraw-Hill, 1939.
2. Nolf, B., PanagiotisTsiakis, P., and Sambukumar, R. Market Scan: Why Cross-Industry Innovation Is Important in Building Supply Chains. Wipro Consulting Services. www .wipro.com/consulting, accessed on November 19, 2012.
3. Levin, David. Managing Director of Zuhlke Engineering Ltd. http://www.zuehlke .com/, accessed on November 19, 2012.
4. Gassmann, O. and Zeschky, M. "Opening up the Solution Space: The Role of Analogical Thinking for Breakthrough Product Innovation," *Creativity and Innovation Management*, 17 (2): 97–106, 2008. DOI: 10.1111/j.1467-8691.2008.00475.x.
5. Gavetti, G., Rivkin, J., and Levinthal, D. "Strategy Making in Novel and Complex Worlds: The Power of Analogy." *Strategic Management Journal*, 26 (8): 691–712, 2005.
6. Holyoak, K.J. and Thagard, *Mental Leaps: Analogy in Creative Thought*. Cambridge, MA: MIT Press, 1995.
7. Hargadon, A. and Sutton, R.I. "Technology Brokering and Innovation in a Product Development Firm." *Administrative Science Quarterly*, 42 (4): 716–749, 1997.
8. Holton, D. "Cross-Industry Collaboration is Key to Innovation," *Manufacturing Engineering*, Guest Editorial (April): 12–14 , 2012.
9. Herstatt, C. and Kalogerakis, K. "How to Use Analogies to Generate Concepts for Breakthrough Innovations." *International Journal of Innovation and Technology Management*, 2 (3): 331–347, 2005.
10. Herstatt, C. and Engel, D. "Mit Analogien neue produkte entwickeln." *Harvard Business Manager*, 8: 2–8, 2006.
11. Gassmann, O., Stahl, M., and Wolff, T. The Cross-Industry Process: Opening up R&D in the Automotive Industry. In *R&D Management Conference*. J. Butler (ed.). Lisbon: Blackell, 2004.
12. Jaffe, A.B., Trajtenberg, M., and Henderson, R. "Geographic Localization of Knowledge Spillovers as Evidenced by Patent Citations." *Quarterly Journal of Economics*, 108 (3): 577–598, 1993.
13. Morgan, K. "The Exaggerated Death of Geography: Learning, Proximity and Territorial Innovation Systems." *Journal of Economic Geography*, 4 (1): 3–21, 2004.
14. Gomes-Casseres, B., Hagedoorn, J., and Jaffe, A. "Do Alliances Promote Knowledge Flows? *Journal of Financial Economics*, 80 (1): 5–33, 2006.
15. Porter, M. *The Competitive Advantage of Nations*. New York, NY: Free Press, 1990.
16. Bell, G.G. "Clusters, Networks, and Firm Innovativeness." *Strategic Management Journal*, 26 (3): 287–295, 2005.
17. Aharonson, B., Baum, J., and Felman, M. Industrial Clustering and the Returns to Inventive Activity: Canadian Biotechnology Firms 1991–2000. DRUID (Danish Research Unit for Industrial Dynamics). Working Papers, Nos. 04–03, 2004.
18. Sonn, W. and Storper, M. The Increasing Importance of Geographical Proximity in Technological Innovation: An Analysis of U.S. Patent Citations 1975–1997. Paper presented in the *Conference What Do We Know about Innovation?* In honor of Keith Pavitt, University of Sussex, Brighton, November 2003.

19. Almeida, P. and Kogut, B. "The Exploration of Technological Diversity and Geographic Localization in Innovation: Start-Up Firms in the Semiconductor Industry." *Small Business Economics*, 9 (1): 21–31, 1997.

20. Audretsch, D., Lehmann, E., and Warning, S. "University Spillovers and New Firm Location." *Research Policy*, 34: 1113–1122, 2005.

21. Audretsch, D.B., Lehmann, E., and Warning, S. "University Spillovers and New Firm Location." *Research Policy*, 34: 1113–1122, 2006.

22. Boschma, R. "Proximity and Innovation: A Critical Assessment." *Regional Studies*, 39 (1): 61–74, 2005.

23. Boschma, R. and terWal, A.L.J. "Knowledge Networks and Innovative Performance in an Industrial District: The Case of a Footwear District in the South of Italy." *Industry and Innovation*, 14 (2): 177–199, 2007.

24. Agrawal, A., Cockburn, I., and McHale, J. "Gone But Not Forgotten: Labor Flows, Knowledge Spillovers, and Enduring Social Relationships. *Journal of Economic Geography,* 6: 571–591, 2006.

25. Sorenson, O. Social Networks, Informational Complexity and Industrial Geography. In *The Role of Labor Mobility and Informal Networks for Knowledge Transfer*, D. Formahl and C. Zellner (eds.), pp. 1–19, 2003.

26. Balconi, M., Breschi, S., and Lissoni, F. "Networks of Inventors and Role of Academia: An Exploration of Italian Patent Data." *Research Policy*, 33 (1): 127–145, 2004.

27. Breschi, S. and Malerba, F. Sectoral Innovation Systems: Technological Regimes, Schumpeterian Dynamics And Spatial Boundaries. In *Systems of Innovation: Technologies, Institutions and Organisation*s, Edquist, C. (ed). London: Pinter, pp. 130–156, 1997.

28. Doloreux, D. "Regional Innovation Systems in Canada: A Comparative Study." *Regional Studies*, 38 (5): 479–492, 2004.

29. Oerlemans, L.A.G., Marius, T.H., Meeus, F., and Boekema, W.M. "Firm Clustering and Innovation: Determinants and Effects." *Papers in Regional Science,* 80 (3): 337–356, 2001.

30. Amin, A. and Cohendet, P. "Geographies of Knowledge Formation in Firms." *Industry and Innovation,* 12 (4): 465–486, 2005.

31. Romer, P.M. "The Origins of Endogenous Growth." *Journal of Economic Perspectives*, 8 (1): 2–22, 1994.

32. Johansson, B., Karlsson, C., and Stough, R. *Theories of Endogenous Regional Growth, Lessons for Regional Policies*. Berlin: Springer–Verlag, 2001.

33. Muller, E. *Innovation Interactions between Knowledge Intensive Business and Small and Medium-Sized Enterprises*. Heidelberg, New York: Physica-Velarg, 2001.

34. Pavitt, K. "Sectoral Patterns of Technical Change: Towards a Taxonomy and a Theory." *Research Policy*, 13 (6): 343–373, 1984.

35. Soete, L. and Miozzo, M. Trade and Development in Services: A Technological Perspective. Working Paper No. 89-031, Maastricht: MERIT, 1990.

36. Miles, I., Kastrinon, N., Flanagan, K., Bilderbeek, R., den Hertog, P., Huntink, W., and Bouman, M. *Knowledge Intensive Business Services. Users and Sources of Innovation*. Brussels: European Commission, 1995.

37. Gallouj, F. *Economie de l'innovationdans les services (Economics of Innovation in Services)*. Paris: Editions L'Harmattam, 1994.

38. Den Hertog, P. "Knowledge Intensive Business Services as Co-Producers of Innovation." *International Journal of Innovation Management*, 4: 491–528, 2000.

39. Metcalfe, J. "On the Optimality of Competitive Process: Kimura's Theorem and Market Dynamics." *Journal of Bioeconomics*, 4 (2): 109–133, 2002.

40. Nelson, R. *National Systems of Innovation: A Comparative Analysis*. Oxford: Oxford University Press, pp. 3–21, 1993.

41. Malerba, F. and Torrisi, S. "Internal Capabilities and External Networks in Innovative Activities. Evidence From the Software Industry." *Economics of Innovation and New Technology*, 2 (1): 49–71, 1992.

42. Antonelli, C. "Collective Knowledge Communication and Innovation: The Evidence of Technological Districts." *Regional Studies,* 34 (6): 535–547, 2000.

43. Rosenberg, N. 1976. *Perspectives on Technology*. Cambridge: Cambridge University Press.

44. Malerba, F. "Innovation and the Evolution of Industries." *Journal of Evolutionary Economics*, 16 (1): 3–23, 2006.

45. Feynman, R.P. [as told to Leighton, R.]. *What Do You Care What Other People Think?* New York, NY: Norton, 1988.

46. Savage, D.A. "The Professions in Theory and History: The Case of Pharmacy." *Business and Economic History*, 23 (2): 130–160, 1994.

47. Langlois, R.N. and Savage, D.A. Standards, Modularity and Innovation, the Case of Medical Practice. In *Path Dependency and Path Creation*, Garud, R. and Karnoe, P. (eds.). 149–168, 2001.

48. European Commission. The New SME Definition User Guide and Model Declaration, Enterprise and Industry Publications. European Commission, 1996. http://ec.europa .eu/enterprise/policies/sme/files/sme_definition/sme_user_guide_en.pdf, accessed October 5, 2012.

Measurements

Measurement of innovation seems like another one of those topics that needn't be addressed. After all, how would one know whether they're successful in innovation if they don't measure where they are compared to where they were? Unfortunately, this is exactly what many organizations do. Unfortunately, the term fuzzy, as in fuzzy front end of innovation, has permeated its way throughout the innovation community. The stereotype of the Silicon Valley start-up with wild parties, free food, and good times reflects an attitude that doesn't exist in the real world. Sure, innovative companies seem to start in garages, but eventually some degree of discipline is necessary. This is where measurement becomes important.

In Andersen and Walcott's chapter on sites, the Nordic region of the world is presented as an example of Innovation Radar. Utilizing the measured and managed innovation (MMI) model, they demonstrate how the Nordic region developed Innovation Radar on 12 dimensions to evaluate companies based upon innovation capabilities.

Innovation as an intuitive and creative process is a difficult one to measure. Conventionally, innovation is measured in terms of finance or quantity. Whenever someone attempts to create an index of innovation, they fall short. For example, should the number of patents be the standard measure for an innovative organization? If this is the case, then what does this mean when you compare IBM to Pandora? How does one measure the motivation of an organization, how much freedom employees feel, or how much trust they have in management? What about the organization's attitude toward idea time, playtime, conflict resolution? Even more difficult to measure is support, debate, or the amount of risk that is allowed? In our chapter on business innovation indices, we attempt to clarify some of these measurements; if not for your organization, at least to stimulate your thoughts.

CHAPTER 36

The Innovation Radar and Enterprise Business System: Innovation in Five Nordic Countries and Beyond

Jørn Bang Andersen and Robert C. Wolcott

Introduction

In this chapter we will present how both companies and countries can develop an innovation strategy and new innovation policy by applying new tools and innovation frameworks such as the Innovation Radar in a systematic way and over time. The chapter is structured in three main parts with the first part outlining the argument and case for piloting and using the Innovation Radar in 100 Nordic companies from five countries. Second, the main results from the Nordic programs "InnoTools" and "Measured and Managed Innovation" (MMI) are presented. This is followed by a discussion of key questions raised by 800 company managers and policy makers. The questions were raised from 2008 to 2013, which covered the InnoTools and MMI programs from start to finish. Finally, we summarize the findings into lessons learned from working with business model innovation at both the company and country levels, and how to take stock of these lessons with regard to future actions.

Innovation as a Discipline

Innovation is a relatively new faculty compared to disciplines like accountancy, economics, or organizational theory. This is also reflected in the hierarchy within companies and other organizations where it is rare to see the title "Innovation Manager" among board members or at C-level. But innovation matters and it does so increasingly not only at the company level but also in relation to nations' prospects for economic growth. William Baumol for example wrote in 2002 that "virtually all of the economic growth that has occurred since the 18th century is ultimately attributable to innovation."[1] Innovation policies have also become central to national economic policy in almost all industrialized countries and in many cases have overtaken what formerly constituted national industrial strategies for economic development. Michael Porter's seminal work *The Competitive Advantage among Nations*[2] falls within this category of focusing on "innovation as imperative for growth and welfare for companies, regions and entire nations." All of this belongs to a tradition dating back to the German economist Friedrich List (1878), Joseph A. Schumpeter,[3] Eric Dahmén (1950s), and so forth. In short, a long line of evolutionary economists have continually emphasized the importance of innovation, entrepreneurship, technology, and skills as the key ingredients for growth, prosperity, and competition among companies as well as nations.

In opposition to this, many neoclassical economists advising governments in the United States, the Nordic countries, and elsewhere argue that nations, for example the United States, are not in global economic and innovation competition with other nations such as Sweden, Denmark, or South Korea. The most famous economist making this claim is the Nobel

Laureate economist Paul Krugman. The main argument proposed by the neoclassical economic camp against national competitiveness is in short that only companies compete with each other because, for example, the Nordic, European, American, Asian, or Latin American markets and consumers are also each other's export markets and consumers.

In this chapter we will show how we combined these two opposing views on innovation by applying the business model innovation tool "Innovation Radar" developed by professors at the Kellogg School of Management to a large number of companies in the five Nordic countries and testing and documenting how innovation can be relevant at both the enterprise and national levels if applied systematically to a large number of companies and over time.

We therefore looked for innovation management tools that could address the issues of measured and managed innovation and be applied in companies in Denmark, Sweden, Norway, Finland, and Iceland. We also wanted to find tools that could give us a better understanding of how future public innovation programs can strengthen the innovation capabilities within companies. The two goals of the exercise were to strengthen Nordic companies' competitiveness and make radical changes in terms of thinking about policy innovation support programs.

Business Model Innovation Applied to Companies and Nations

Businessweek reported a few years ago that most company innovation programs don't work. Among the various studies they cited, one from the Doblin Group claimed that nearly 96 percent of all innovation attempts fail to beat targets for return on investment. Another, from Booz Allen Hamilton, concluded: *"There is no relationship between R&D spending and the primary measures of economic or corporate success, such as growth, enterprise profitability and shareholder return."*[4] Given these dismal results, should we stop investing in innovation? As will become apparent, we don't think this a real option for companies that want to develop and survive.

It seems, however, that innovation is still in its infancy as a management discipline, and it seems that if companies start approaching innovation in a more systematic way—for example, through the application of measured and managed innovation—they could increase their return on innovation at no or small additional cost. That is another way of saying that around 90 percent of all innovation efforts are never commercialized or used in general. If Nordic companies could raise the ROI by just 10 to 20 percent, this would give them a significant competitive advantage in global competition.

Second, the Nordic nations top the global league in terms of public and private investment in innovation (government, business expenditure on R&D combined—GBERD). Productivity is the standard measure for a nation's well-being and productivity is largely driven by innovation—especially the adoption of new technologies including business models and organizational changes in the individual enterprise.

As a discipline, innovation management is, as mentioned, young compared to, for example, accountancy's more than 500 years of history or the science lab's 75 to 100 years. It is therefore not surprising that in spite of the massive focus on innovation management in business, at business schools, and in economics, it is still not executed and understood very well in many organizations. Another factor complicating our understanding of innovation management is that the term "innovation" has no shared meaning in the way that science or marketing have. In fact innovation has, for decades, been seen in the context of the research; just think about how often the two are presented as research and development or innovation (R&D) departments by organizations. Gary Hamel argues in *Innovation to the Core* that if you ask 100 company CEOs if they have an innovation strategy and if so, to please explain it, 99 percent of the time you get a blank stare. They have none, according to Hamel.

On the other hand, as early as 1925 the Russian economist Nikolai Kondratiev wrote about "major economic cycles," how generic technology breakthroughs changed the landscape for economic activities, and how such technologies tended to come in waves of 30- to 50-year intervals. In the 1930s the Austrian economist Joseph A. Schumpeter introduced the notion of "creative gales of destruction," by which he meant that entrepreneurs and companies engage in innovation to capture new markets and create temporary monopoly rents from their investment. This lasts until others jump on the bandwagon and competition takes its course. Peter Drucker wrote in 1955 in *The Practice of Management*[5] that entrepreneurial activity has only two functions: "to create a customer which is marketing and to keep the customer which is innovation." Theodore Levitt's groundbreaking paper "Marketing Myopia"[6] is a more recent decisive contribution to our understanding of innovation and underscores the catastrophic results of being "product-oriented" rather than "customer-oriented."[7]

Another contribution that influenced our project was Theodore Levitt's argument that

"There is no such thing as a commodity. All goods and services are differentiable."[8]

During the 1980s and 1990s, a number of European scholars and managers made contributions to the field of innovation policy and innovation management. The contributions were and continue to be made within the European Association of Evolutionary Economists, where people like Christopher F. Freeman, Giovanni Dosi, Cantwell, and Carlotta Perez, to name but a few, have been influential in the discussions at both policy and company levels.[9]

Of more recent origin, our current understanding of innovation and how to look at it has come from: W. Chan Kim and Renee Mauborgne's *Blue Ocean Strategy*[10]; Procter & Gamble's change from in-house to open innovation as described by A.G. Lafley and Ram Charan[11] in *The Game Changer*, where they describe how the introduction of open innovation management made P&G regain its strength and competitive edge; and Henry Chesbrough, who coined the term open innovation as described in *Open Business Models*.[12]

Clayton Christensen introduced the notion of disruptive innovation—a process where a product or service takes root initially in simple applications at the bottom of a market and then relentlessly moves "upmarket," eventually displacing established competitors. C.K. Prahalad's[13] demonstration of the market potential and unrealized frugal innovation potential from focusing on the markets of the billions of people living at the bottom of the pyramid in Asia, Africa, and elsewhere should also be included in the current discussion on understanding innovation management. The Nordic InnoTools and MMI[14] projects dealt with in this chapter should be seen as a contribution to the ongoing search for a better understanding of how to practice innovation management, and hopefully the results can add new insights to the theoretical discussions in management institutes, academia, and within innovation policy circles.

Hypothesis and Research Question

Consider two companies, operating under the same business framework conditions, within the same industry, of similar size, and within the same country. How do we explain that company A outperforms company B? Business framework conditions are important, but more recent studies indicate that a company's innovation performance is primarily a result of the company's internal values, capabilities, and culture for risk and innovation. In addition, research on the subject has shown that companies that take a holistic view on innovation and work from a mind-set of business model innovation outperformed those that didn't. Hence, we decided to examine, select, and use a particular framework for business model innovation, in order to have a structured focused comparison methodology in our research. That is, we used the same innovation tool and asked the same questions across all 100 Nordic companies in the five Nordic countries: Denmark, Sweden, Norway, Finland, and Iceland.

Research Design

We shall not go into detail with regard to the initial phase of research for available innovation management tools that seemed to fit the purpose of the program, but we ended up doing a pilot phase with two innovation management tools. In pursuit of launching a pilot project, dealing with the issues of raising companies' ROI through measured and managed innovation, we scanned the global market for articles and references and for tools dealing with innovation in a structured systematic way.

From this desk research we singled out two tools, namely the Innovation Radar developed by Mohanbir Sawhney, Robert C. Wolcott, and Inigo Arroniz, known for *The 12 Different Ways for Companies to Innovate*[15] at the Kellogg School of Management, and Five Disciplines of Innovation developed by former CEO Curtis Carlson, Stanford Research International (SRI).[16]

Tool: The Innovation Radar

The Innovation Radar (Fig. 36-1) was partly chosen because it had already been tested and statistically documented among companies in the United States. More importantly, the Innovation Radar was chosen because it gives a holistic view of innovation and supported the empirical findings from IBM's global 2006 CEO study. The study shows that companies that take a business model perspective on innovation outperform companies that focus more narrowly on just product or process innovation.[17] Finally, the Innovation Radar provides a tool for measured and managed innovation and it is easy for a company to answer the online questionnaire on which the company's innovation profiling is based.

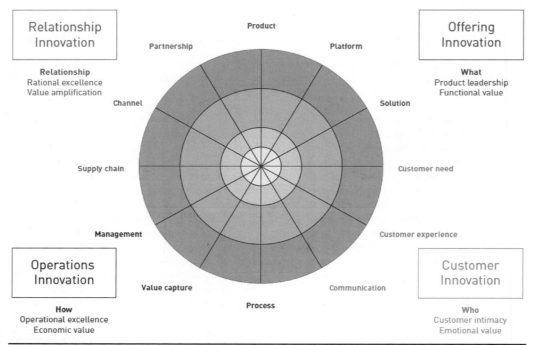

FIGURE 36-1   The Innovation Radar illustrated.

Within the Innovation Radar framework, innovation is defined as an initiative in any dimension(s) of the business system, to create substantial new value for customers and the firm. This innovation definition emphasizes three points: originality (an initiative to create new value); a holistic view (an initiative in any dimension(s) of the business system); and customer outcomes (the value generated by the initiative for customers and the firm). The Innovation Radar creates a visually compelling profile of the firm's current innovation strategy through our measurement of all 12 vectors. The Innovation Radar profile helps to create better alignment on innovation strategy across functional areas and seniority.[18]

The point to be considered in the context is that a company *can innovate along* any of 12 different dimensions shown in Fig. 36-1 and explained in Table 36-1 below. The key message is that business model innovation is far broader than just product or technological innovation, as demonstrated by successful companies in a number of industries. Starbucks is the classic case of a company who got consumers to pay $4 for a cup of coffee-latte, which until then typically cost 50 cents. This was not necessarily the result of better tasting coffee from Starbucks, but was instead because the company created a customer experience referred to as "the third place." Together, the 12 dimensions of innovation can be displayed in a new framework called the "Innovation Radar," which companies can use to manage the increasingly complex business systems through which they add value (Table 36-1).

To give an example of how to interpret and use the Innovation Radar tool, a dynamic and stylized company Innovation Radar profile for Apple is shown in Figs. 36-2 to 36-4.

Following the thinking behind the Innovation Radar, a company might strengthen its innovation strategy by focusing efforts on a select number—one to five—of the 12 innovation dimensions instead of putting innovation effort into every one of the 12. Moreover, as many companies tend to focus on the WHAT of product innovation and the HOW of process innovation macro dimensions, there is a considerable potential for gaining competitive advantages by differentiating along other dimensions. The stylized version shown here is the company Apple in a dynamic perspective over time. When Steve Jobs came back to Apple in the 1990s, the company was in serious trouble and had been written off by many industry experts. The first thing Apple did back in 1996 was slash and reduce its product line, which at the time contained many different products ranging from printers to personal computers. Moreover, the company was losing money. What Steve Jobs did was focus on a product line with only a couple of computers that were very well designed but much more expensive than other PCs or laptops. In view of the Innovation Radar, we think of Apple's turnaround from 1996 to 2000 as innovation strategy, focus and effort based on offering (a few newly designed Macs), value capture (slashing product lines and cutting costs while creating revenues from fewer offerings), and management turnaround within the company.

1. Offering	Development of new products or services.
2. Platform	Use of common new components or building blocks to create derivative offerings.
3. Solution	Creation of integrated and customized offerings that solve end-to-end consumer problems.
4. Customer Needs	Discover unmet customer needs or identify underserved customer segments.
5. Customer Experience	Redesign of interactions that customers have with the company in order to build customer loyalty based on positive emotional response.
6. Marketing Communication	Implementation of creative marketing communications to position, promote, or brand products and services.
7. Process	Redesign of core operating processes to improve efficiency and effectiveness.
8. Value Capture	Creation of new ways to get paid for products and services.
9. Management	Invention and implementation of a significant change in organization structure or management methods to further organizational goals.
10. Supply Chain	Thinking differently about sourcing and fulfilment.
11. Channel	New routes to the marketplace or innovative points of presence for customers to find and buy products and services.
12. Ecosystem/ Partnership	Creation of innovative partnerships and collaborative relationships with suppliers, partners, vendors, resellers, and so on, to create joint offerings.

TABLE 36-1 The innovation radar 2.0: 12 dimensions explained[19]

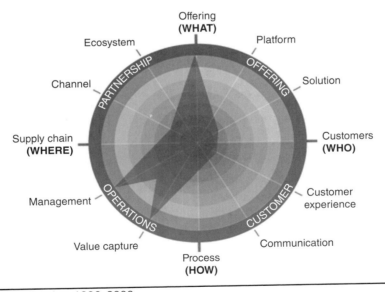

FIGURE 36-2 Apple Inc. 1996–2000.

In the second phase of Apple's turnaround from around 2000 to 2006, it is apparent that once the basic product line, management, and revenue streams were safely back in line, Apple started to focus on the offering dimension, and other new innovation dimensions came into play.

The introduction of Apple stores was groundbreaking for the industry in terms of the customer experience of the purchase of consumer electronics. Second, what is less well known is that Apple developed a world-class supply chain that was probably more efficient even than Dell's, the leader at that time.

Finally, when Apple opened its iPhone and iPad platform to developers, it rapidly developed into a global ecosystem of more than 200,000 "entrepreneurs" building profitable

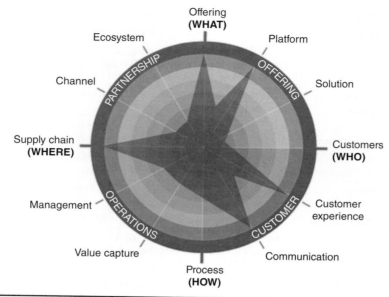

FIGURE 36-3   Apple Inc. 2000–2006.

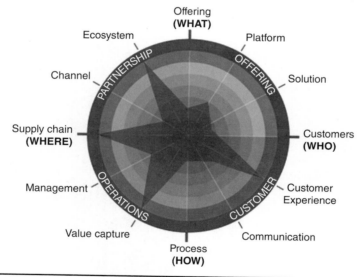

FIGURE 36-4   Apple Inc. 2006–2012.

"apps" for the iPad, and so forth. Moreover, looking back some years it would appear that Apple's primary innovation focus was on building the iTunes platform, raising the bar in terms of value capture and cutting down on offerings. In other words, the case of Apple shows how a company can use the Innovation Radar framework in a dynamic way to reassess its innovation strategy and focus, thereby moving into "blue oceans" or "white spaces"[20] of a given industry business innovation model.

The Pilot Project: InnoTools

The pilot study InnoTools was in essence very simple and comprised the following steps. A number of companies did the Innovation Radar questionnaire and consequently had their company Innovation Radar profile mapped. Each Innovation Radar profile was followed up with a deep-dive workshop where around five to eight managers from the various company departments had their profile presented. In the workshops each company went through the following exercises:

- Presentation of the 12 dimensions of the Innovation Radar, each dimension illustrated with examples from their company and industry.

- Presentation of the company's Innovation Radar profile and a discussion of the profile in respect to their stated innovation strategies.

- Presentation of the approach to each dimension in view of the company's internal divergence, with a discussion of possible reasons for the variation between departments.

- Each company session was concluded with the companies doing a mapping of their "preferred" future profile, selecting a maximum of two to five radar dimensions, where they wanted to differentiate from the competition within their industry. Additionally, the companies identified an innovation project to work with in the future.

- SRI did a selected workshop based on the selected innovation project with each company. In this workshop SRI explained and applied the methodology of Five Disciplines of Innovation to the companies' innovation project.

- The participating companies and a steering committee finalized the pilot study in a workshop where all the companies' innovation radar profiles were presented and discussed. At this workshop, the specific projects were also developed within the framework of the Five Disciplines of Innovation.

Given that both of the innovation tools for measured and managed innovation were developed in the United States and generally tested there, we decided to test the "measure" and "manage" tools' robustness in a Nordic setting. The test of applicability in a Nordic setting meant that eight Nordic-Baltic companies of different sizes from seven different business cultures and from eight different industries were selected for the pilot study.[21]

Both the Innovation Radar and the Five Disciplines of Innovation proved to be relevant and valuable tools, according to the companies' own evaluation. Based on the evaluation among 56 CEOs and senior managers from six countries (Denmark, Norway, Iceland, Finland, Estonia, and Lithuania), consisting of eight companies ranging from 10 to 15,000+ employees and from different industry sectors, both the Innovation Radar and the Five Disciplines of Innovation were accepted as a basis for the study. A steering group consisting of representatives from the five Nordic national innovation agencies followed the project and agreed with the companies that they had gained important knowledge from this method of working with innovation, at a company and policy level.

In addition, a number of success criteria were stipulated for the InnoTools pilot study. The main success criterion was that:

- At least 75 percent of the participating company managers will evaluate the tools at 3.5 or above on a scale of one to five (five being the highest score) as an evaluation of the innovation tools' usefulness in comparison with other innovation tools of which the company managers were aware. The average score for the Innovation Radar was above four for all participating managers.

- The InnoTools project addressed an important and relatively new aspect of public innovation programs, namely how to develop innovation capabilities within companies through public support mechanisms. InnoTools also opened the black box of innovation management in a way that was both new and challenging.

Scaling the Pilot: Measured and Managed Innovation in 100 Nordic Companies

Following the positive results from the InnoTools pilot project, there was a case to be made for changing the focus and mind-set regarding innovation policy and innovation focus within Nordic companies. In establishing the case for this change, a fairly simple overview of the evolution of dominating paradigms for innovation policies and their consequences at company levels was presented.

Figure 36-5 demonstrates that innovation policy has to adopt and change in line with changes taking place in other parts of the economic landscape of society and industry. Globalization, the Internet, climate issues, and the like can drive such changes. In short, it was

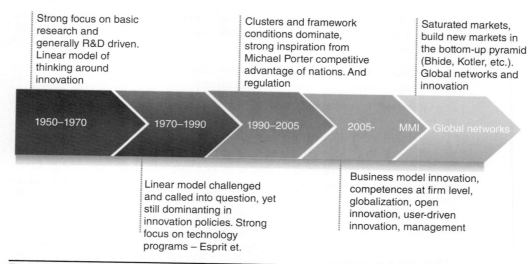

FIGURE 36-5   Time view for dominant paradigm for innovation policy and thinking.

argued that the MMI program should be viewed in this evolutionary context, where innovation policy within companies had been dominated by various paradigms. Each paradigm has, moreover, been linked to a certain set of tools and competencies based on certain assumptions regarding what constitutes the most important drivers for economic growth and competitiveness. For example, how did research dominate views on innovation and economic development from around 1950 to 1970? This paradigm began to be challenged during the 1970s because of economic crises, yet it continued to play an important role in innovation policy and within companies. From the beginning of the 1990s, the notion of clusters and economic framework conditions seized the agenda and until around 2005 had a dominant influence in funding decisions from OECD, Nordic, and EU innovation programs.

The MMI program[22] was presented in 2009–2010 as a new way of looking at innovation, both for organizations and policy makers. Focus was placed on business model innovation, open innovation, and the need to "open up the black-box" of management within companies in order to support competence-building measures. Nordic Innovation and the five national Nordic innovation agencies decided to launch a full-scale program involving 100 Nordic companies of various sizes and from different industries. Roughly 20[23] companies participated from each Nordic country. This program was labeled Measured and Managed Innovation (MMI)[24] and it was decided that the Innovation Radar tool would be used. This was decided because the Innovation Radar tool allowed for both data collection and the qualitative follow up deep-dive workshops in each participating organization.

Based on the follow-up evaluation of the program, the overall setup of the MMI program with activities, outputs, results, impacts, and purpose is summarized in Fig. 36-6 below.

Findings from the Nordic MMI Program: Phase I 2010–2011

From the outset of both the InnoTools pilot and MMI program, we had a guesstimate that most Nordic companies would have a strong orientation toward product and process innovation. Moreover, looking at public sector innovation programs in the Nordics and European Union there seemed to be a strong bias toward favoring "technology innovation" and very few innovation programs within marketing, customer experience, or building global supply chains. Thus, it can be argued that public innovation support programs were reinforcing the limited innovation focus of product and process.

The key findings from the MMI program's first round of surveying the 100 Nordic companies and 800 managers can be summarized as follows:

- Only 41 percent of the 100 Nordic MMI companies had an innovation strategy.

- Companies with an innovation strategy had higher innovation effort, higher innovation focus, and higher internal alignment from the perspective of "direction" and "future focus." In other words, there seemed to be a positive correlation between "just having an innovation strategy" and "having a focused effort in innovation."

- The majority of Nordic companies primarily innovate within the WHAT and OFFERING dimension. In fact 64 percent had offering (offering, platform, and

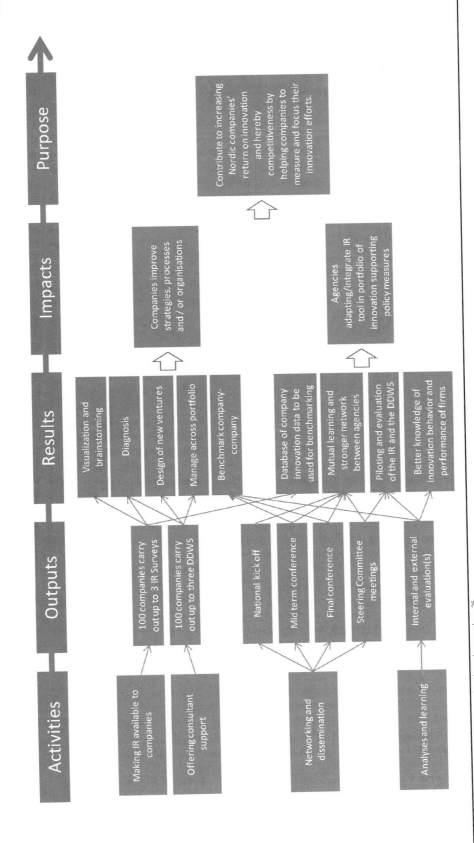

Figure 36-6 Chart 1: MMI organizational overview.[25]

589

solution dimensions) as the primary macro innovation focus. Moreover, offering innovation increases to 72 percent when Nordic MMI companies had to choose the most important macro radar dimension for the future.

- A majority of Nordic MMI companies had an unfocused innovation approach or low innovation effort levels. More precisely, 62 percent had a low innovation focus.

- Only 20 percent of the MMI companies had a "high" effort on innovation, leaving the remaining 80 percent with a "low" effort on innovation.

- The MMI also revealed that companies operating in a low to medium competitive market had a higher innovation focus than companies in markets with high or significant competitive situations.

- A somewhat puzzling result was the response to the question of primary future business challenges versus the most important future Innovation Radar dimension. The MMI companies singled out "Operations" as the biggest future business challenge, with 44 percent. However, 72 percent singled out Offering as the most important future Innovation Radar dimension.

- Only four percent of the MMI companies chose the macro dimension "Partnership" innovation (supply chain, channel, ecosystem) as the preferred future dimension.

- We also learned that more than 75 percent of all companies wanted to change the way they work with innovation as a result of their shown Innovation Radar profiles and deep-dive workshops.

- A majority of MMI companies rated the Innovation Radar tool as very useful.

- Finally, we did a benchmark with the U.S. Innovation Radar findings among 314 U.S.-based companies. The benchmark is shown below.

The overall observation is that the 100 Nordic companies have a much higher focus on "Offering" (that is to say on product innovation) than the U.S. companies. Second, the U.S. companies show a much higher focus toward innovation around partnerships than the Nordic companies. These are aggregate findings from the Innovation Radar surveys in the Nordics and the United States and industry samples and size of companies may explain some of the differences. Nevertheless, with new organizational forms such as open innovation, global supply chains, and the new channels offered by the Internet (dimensions within partnership), one can speculate that Nordic companies are somewhat still operating in the "industrial production era of the 20th century" and therefore, at least so far, have been unable to seize the opportunities to withstand future challenges.

Some more general observations from the InnoTools and MMI phase 1 are as follows:

- The Innovation Radar is very easy to communicate and understand within an organization, which makes it easy to use without a lot of prior information and investment.

- In the Nordics, the MMI program and the Innovation Radar framework has established a common language for discussing and understanding "business model innovation" among companies and across Nordic public innovation agencies. This alone is an accomplishment.

- At the time, the InnoTools pilot project was a serious challenge to most innovation policy makers and the business model innovation concepts introduced were a

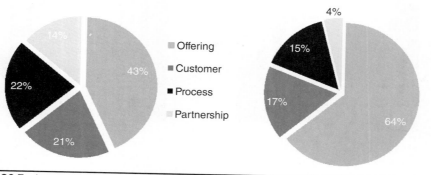

FIGURE 36-7   Innovation Radar: United States-Nordic benchmark.

novelty to most companies. Hence, starting with a pilot was an important first step to change mind-sets and entrenched views on innovation before embarking on the larger MMI program.

- The InnoTools and MMI results have created interest outside the Nordic region and been presented at the EU Commission's evaluation of their Improve3 tool and received positive recognition internationally.

Findings from the MMI Program: Phase II 2012

Phase II of the MMI program was with a slight reduction in the sample set of the original 100 companies and thus 70 companies went on to complete phase II. The results presented here are, however, taking this into consideration and hence the data set and comparison between the 2010 and 2012 time series, and so on, is based on the 70 companies that participated in both phases.

At least two things are important in understanding the Innovation Radar in relation to companies' innovation strategies. The first is whether a company has a focused innovation strategy. Second, is whether the organization puts low or high effort into innovation. With these two parameters, we can construct generic Innovation Radar profiles and use them to visualize and better understand what it means to for a company to have a "high focus" and "high effort" in its Innovation Radar profile and strategy. The four generic Innovation Radar profiles are shown in Fig. 36-8 and are useful in picturing the results presented in the sections below.

The OECD writes in "The OECD Innovation strategy: Getting a head start on tomorrow" that appropriate measurement of innovation is critical for policy making. Firm level analysis such as Innovation Radar in the Nordic MMI program represents important analytical insight and new methods for innovation at company and country levels. In the following section we will describe the main results from the MMI program's phase II.

Figure 36-9 shows that in 2010, when we carried out the first phase of the MMI program, 29 percent of the sample set of companies had a focused innovation strategy. Focus means here that they had a clear innovation profile according to the Innovation Radar framework shown in Fig. 36-8 below. During the 2 years the MMI program was in place, this went up to become 47 percent or almost half of all participating companies. Testimonials from the

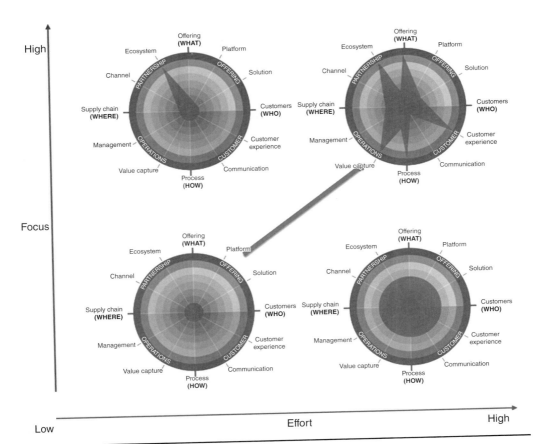

FIGURE 36-8 Generic Radar profiles.

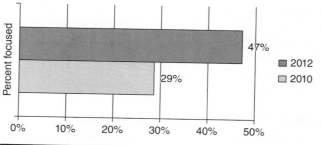

Figure 36-9 Nordic level change of innovation focus 2010–2012.

companies indicate that this correlated with "working with the Innovation Radar tool" and "through the deep-dive workshops."

"The MMI process has made us prioritize innovative work to a greater extent and we have put innovation more on the agenda in our strategic work. In the beginning we were struggling to use the process for the whole organization but we made it clear for the whole management team who participated. The project has assisted with more systematic ways to work with innovation and we plan to use the Innovation Radar on yearly basis."

—Norwegian MMI company[26]

"As an outcome of the MMI-participation, we are focusing on results, not on luck."[27]
"It took some time to get used to the tool, but then it opened our eyes for a new mind-set on innovation."[28]

Perhaps the change of mind-set is the most important but also the most difficult element for companies when moving from doing unfocused business development, often based on incremental changes on existing products and processes, to a focused and holistic business model innovation strategy.

Another insight from the MMI program is that the time series shows medium and large companies improving with regard to focus. The explanations for the initial variation of focus and company size might be that in smaller companies it is easier for the management to communicate and align any given innovation strategy. Hence, the relatively high score on innovation focus. On the other hand, large companies didn't score very highly to begin with but improved tremendously over 2 years. The explanation here could be that large companies face a bigger challenge in communicating to everyone in the organization, however, once large companies put their superior organizational, financial, and human resources behind change they actually can make significant changes and improvements.

The data from the MMI project also revealed that the mere existence of a defined innovation strategy correlated positively with the level of innovation effort as well as with having a focused Innovation Radar profile. This by itself is interesting as it indicates that a large majority of companies could easily improve their innovation performance, and possibly their return on innovation, just by formulating a clear innovation strategy.

An important aspect of using the Innovation Radar as a tool for business model development is to discuss and assess the company's range of possibilities among all 12 innovation dimensions. In the MMI program, efforts to change mind-set from a narrow product-process

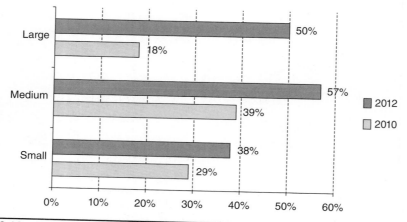

Figure 36-10 Company and change of innovation focus.

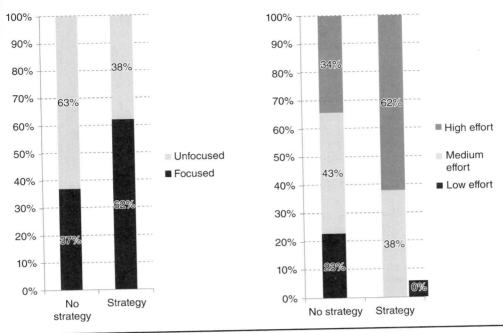

FIGURE 36-11   Strategy equals higher focus and effort.

innovation mind-set to a broader business model mind-set were intended. The four most prevailing focus dimensions were therefore tracked over time in order to determine the impact of the program and Innovation Radar tool on this. Figure 36-12 shows that in 2010, the three dimensions—offering, platform, and solution—were totally dominant across all five Nordic countries. All three dimensions within the WHAT macro vector of the Innovation Radar indicate a somewhat narrow product focus on innovation. In 2012 the dimensions offering, platform, and solution were still among the four preferred focus dimensions, but in Finland and Denmark the partnership dimension had become one of the four preferred innovation focus dimensions of the Innovation Radar. The time series shows that existing patterns of innovation are not easily changed. Both company managers and policy makers should acknowledge this when trying to facilitate changes in strategy and mind-set for innovation.

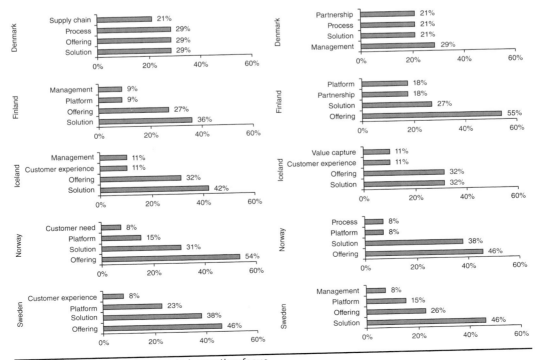

FIGURE 36-12   Country changes in innovation focus.

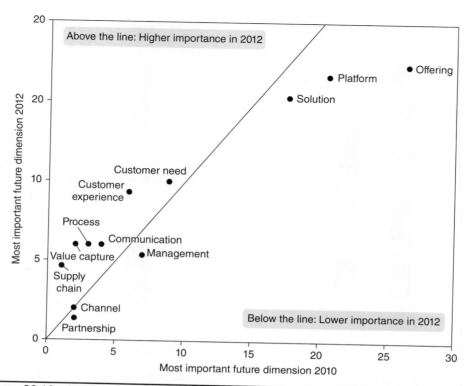

FIGURE 36-13   Changes of future most important Innovation Radar dimensions.

In 2010 the Nordic companies in the MMI sample had a strong preference for the Innovation Radar's dimensions of "solution," "platform," and "offering" when asked about their future preferred innovation focus. When asked the same question in 2012 the companies had changed significantly and mentioned "value capture," "supply chain," and "process" as the three most preferred future dimensions. We don't know if this will lead to actual changes within the companies, but it shows that you can change and broaden the innovation mindset of company managers by introducing business model innovation tools.

The overall conclusion from the aggregate data of the MMI program is that at the outset in 2010, only four percent of the companies had "high effort" and "high focus" with regard to their innovation profiles. Two years later at the end of 2012, 24 percent of all MMI companies had "high effort" and "high focus" for the Innovation Radar profiles. This arguably shows that within a reasonable span of time—that is, 2 years—it is possible to facilitate serious changes by systematically working with strategic innovation management tools like the Innovation Radar.

Throughout the MMI program and the company deep-dive workshops, a number of issues and questions were raised on several occasions. These particularly included the following Innovation Radar issues:

- Innovation focus and business life cycle
- Incremental versus radical innovation
- Innovation ecosystems and partnership innovation

Developments 2010 → 2012		
	Low effort	High effort
High Focus	24% → 23%	4% → 24%
Low Focus	56% → 33%	16% → 20%

FIGURE 36-14   Innovation Radar profile change 2010–2012.

We will elaborate on each issue in order to provide a guide for discussing and dealing with these questions.

Innovation Radar Focus and Business Life Cycle

Based on the insight gained from doing innovation strategy deep-dive workshops with companies from the Nordics, Baltics, the United States, Portugal, Austria, Russia, South Korea, and Mexico, there is a tendency for companies and people that work with innovation to have a perception that a new company or a new business will always need to have its focus on the WHAT or "offering" dimensions of the Innovation Radar. This is the case because you need to have something to sell. Although this argument has some credence, it nevertheless misses the essential point of the Innovation Radar framework and business model innovation in general. The reasoning stems from a relatively narrow perspective where innovation is a product or process.

The track of business model innovation is to apply a more holistic perspective on innovation. Consider the classic case of the photocopier. Xerox invented the copy machine, and dominated the global market for more than a decade. When Canon and other Japanese companies entered the copy machine market in the 1980s they didn't "invent a new product, that is, copy machine." What Canon and others did was to take an existing product and focus its innovation around the customer. Xerox had for years focused on the business-to-business market and almost completely neglected the business-to-consumer and consumer-to-consumer market. Canon spotted that niche market and used its core competence of miniaturizing electronics to make a copy machine for the private consumer at a lower price and available at retail stores.

The history of innovation also suggests that most radical and commercially successful innovations share at least one characteristic. They are often fast followers. They pick up on what someone else has invented and develop a new market.

Container Shipping

A case in point is the shipping industry's container. The first shipping container was invented and patented in 1956 by an American named Malcolm McLean, who later established the company Sea-Land Shipping.[29] Some years ago the American company Sea-Land Shipping was bought by the Danish company A.P. Moller Maersk and became Maersk-Sealand. Today it is named Maersk Line and the world's leading container shipping company. Maersk Line didn't become the world's leading container shipping company by focusing on the container as a new product offering.

Maersk Line became the container shipping industry leader by developing excellent logistics (i.e., supply chain in relation to the Innovation Radar) built around company-owned satellites that could monitor ocean shipments. In addition, for years the company invested and built its own harbors throughout the world (i.e., platform innovation) and became the most profitable, (i.e., value-capture innovation) in the industry. And Maersk Line has been one of the most punctual container shipping companies (i.e., process innovation) for several years.[30]

After the financial crunch in 2008, Maersk Line had decreasing revenues from its container shipping. In response to this the company carried out a global customer survey about how Maersk Line could become better at adding value for its customers. The result was three new focus areas for business excellence and innovation. The first was reliability and Drewry, an independent shipping analytic firm, has several times voted Maersk Line to be 15 percent or more reliable and punctual than the competition. Second, Maersk Line has built a new generation of container ships called the "Triple-E-Ships" (i.e., efficiency, economy of scale, and environment), which will emit 50 percent less CO_2 than the industry average. Finally, Maersk Line has introduced new Internet-based customer solutions that are among the easiest to use in the container shipping industry.[31] Maersk Line has increased its revenues by 30 percent since it made these changes. The example of Maersk Line demonstrates how a fast follower can become an industry leader. And it shows how a company can maintain its leading position through constantly monitoring and working with three to five Innovation Radar dimensions in a dynamic way.

Fanatical Support and Security

A third example is the American company, Rack Space, which offers hosting, cloud computing, and ICT services. When the founder of Rack Space announced that the company would be going into this business, industry experts deemed the venture would be a failure because the hosting industry was competing on "price, price, and price" and the only way to drive down prices was through gigantic server parks, which again was a barrier to entry for a company like Rack Space. However, the company did its own survey among companies and asked them "what was most important to them when outsourcing their ICT routines, e-mail handling, and so forth." The two most important issues turned out to be "security of data" and "speed of service in case of problems or blackouts." This insight led Rack Space to launch the company with a focus on fanatical support and high security, provided at a premium price. In other words, Rack Space turned the industry standard of competing on price on its head and raised the price tag by developing a company culture of fanatical support.[32] In relation to the Innovation Radar framework, what Rack Space did was to set a new gold standard for customer support and experience.

In summing up on this discussion we will refer to Theodore Levitt's seminal article "Marketing Myopia," where he argued that

"Every major industry was once a growth industry. But some that are now riding a wave of growth enthusiasm are very much in the shadow of decline. Others which are thought of as seasoned growth industries have actually stopped growing. In every case the reason it is threatened, slowed or stopped is not because the market is saturated. It is because there has been a failure of management. The railroads did not stop growing because the need for passenger and freight transportation declined. That grew. The railroads came in trouble not because the need was filled by others (cars, trucks, airplanes, even telephones) but because it was not filled by the railroads themselves."[33]

We think the three examples given and Theodore Levitt's argument are important insofar as they serve to challenge the still widespread view that innovation starts from a product and moves toward how to make and process that product. Business model innovation is by definition much broader, and if it is discussed from cases challenging the product and process view and made meaningful by using, for example, the dimensions of the Innovation Radar framework and its dimensions, the product-process dogma may be broken. In this way business model innovation can become a discipline and activity that can be discussed, communicated, managed, and measured and above all companies can continuously be on the lookout for ways of differentiation.

Incremental versus Radical Innovation

The second discussion often encountered when carrying out deep-dive workshops on strategic innovation in companies or when presenting the Innovation Radar has been the discussion as to whether innovations are radical or incremental. This discussion is usually connected to the Innovation Radar functionality, where an Innovation Radar profile hovering in the center or middle of the Innovation Radar signifies a company focusing its innovation effort close to the industry average and showing a low level of differentiation. Equally, a company with an innovation profile close to some of the outer edges of the Innovation Radar dimensions indicates a strategy challenging industry norms and possibly radical or even game-changing innovation strategies.

The incremental versus radical innovation is an old but still relevant discussion. When Isaac Newton reportedly told Robert Hooke that if he had seen far it was "by standing on the shoulders of giants" he coined in essence what this debate is still about, namely that most scientific and technological progress and therefore innovation is based on work, data, and insight already in existence.[34]

From this follows the argument that industrial progress and innovation is almost by definition incremental in nature. In economics, the discussion dates back at least to the British neoclassic economist Alfred Marshall, who argued that nature took no leaps and stressed continual incremental progress by managers and workers that would accumulate over time.[35]

The first to challenge the incremental view and introduce the notion and importance of radical innovations in economics was the Austrian born economist Joseph A. Schumpeter who in the 1930s stressed the importance of innovative leaps that were disruptive, dramatic, and discontinuous in nature. Schumpeter stated:

"Add successively as many mail coaches as you please, you will never get a railway thereby."

He argued that the core of economic development is about deploying existing labor and land services differently.[36] This line of argument is also advanced by Theodore Levitt in the *Harvard Business Review* article "Marketing Success through Differentiation—Of Anything." Levitt goes on to argue that there is no such thing as a commodity, all goods and services are differentiable and as such prone to different levels of innovation.[37]

From the 1990s and onward the notion of radical innovation became increasingly accepted to be as relevant as incremental innovation. In particular, the Harvard professor Clayton Christensen has advanced the debate. Though he uses slightly different terms—sustaining instead of incremental and disruptive instead of radical—the meaning attached is very similar. Clayton Christensen's main point is that most technological or methodological improvements in a given industry are sustaining in character whereas disruptive technologies or methods bring to market a very different value proposition than that available up to a given point in time. Moreover, radical innovations tend initially to be targeted toward niche markets that are overlooked or considered unattractive by incumbent industry leaders. Another feature of radical innovations is that they tend to be able to solve a problem and serve its niche customer market at a fraction of the price of what is considered to be the industry standard.[38] Take the example of Canon printers for retail consumers versus Xerox's much more expensive business-to-business solution. In other cases, disruptive innovations come about as a result of technological breakthroughs that make existing solutions obsolete.

Consider the company Johnson & Johnson's longtime success with selling Turtle Wax to car owners for polishing their cars. Turtle Wax is likely to be disrupted by the possibilities stemming from using nano-tech solutions on all kinds of surfaces including glass and steel on cars. This doesn't mean Johnson & Johnson cannot continue serving the existing market for cleaning car surfaces, for example by adopting nano-tech solutions. But it does mean that if the company doesn't abandon its production of Turtle Wax it is likely to become extinct in that market.

In summing up the discussion of incremental versus radical innovation, it should not be forgotten that an estimated 85 to 95 percent of corporate research and innovation portfolios are made up of incremental projects.[39] As such, most of what is happening within a company will be dedicated toward incremental improvements to existing products and services and in maintaining and building stable cash flows and profits from those existing products and services. However, if companies and their managers don't spend any resources on tomorrow's next and as yet unknown business, they will have few weapons in their armory when the next entrepreneur or start-up challenges their market position. Companies should probably not spend more than 5 to 10 percent of their resources on more radical innovations. This may not seem like much, but that 5 to 10 percent will often mean the difference between long-term business survival as opposed to leaving the company's fate to chance or, at best, hope. The five percent spent on strategic innovation is in this respect not the single biggest budget item, but arguably the most important from a strategic business point of view. To quote the French existentialist philosopher Albert Camus, "not to choose is also to choose." This also counts for having a clear and defined innovation strategy.

Ecosystems and Partnership Innovation

We learned from both the Nordic MMI program and the American sample of companies having an Innovation Radar profile that only four percent of Nordic companies had their innovation focus on the ecosystem innovation macro-dimension and the three dimensions for partnership innovation. The American companies had higher focus (14 percent), than the Nordic companies, but of the Innovation Radar macro-dimensions, ecosystem or partnership innovation was still the lowest of the four possible focus areas of the Innovation Radar.

To be clear on this discussion, there is nothing new about joint ventures and alliances among companies and all innovations require some form of collaborative arrangements for development, testing, and commercialization. But one can reasonably ask: why ecosystems and partnerships in relation to company innovation?

In a major research project on "where do innovation ideas come from," Cooper and Edgett looked at 160 firms and 18 possible sources for product innovation. Among the 18 sources of innovation the companies ranked the most effective: (1) ethnography, (2) customer visit teams, and (3) customer focus groups were scored as the three most effective, in that order. Perhaps more importantly, among the 18 possible sources for innovation, partnership, and vendor agreements were only ranked 12 out of the 18 sources.[40] In the following section we will show why we think that partners and vendors are extremely relevant and important to innovation and competitiveness and likely to increase in importance.

Distribution: BP Castrol and Hindustan Unilever

The following three examples will illustrate the importance of partnership innovation and show why the idea is by no means new. For several decades BP Castrol, the producer of lubricants for vehicles and machinery, has managed to keep a dominant position in the Indian market. The reason for BP Castrol's grip on the Indian market is not that its lubricants are particularly better or different from that of competitors such as Exxon or Royal Dutch Shell. BP Castrol's stronghold in India has to do with the fact that the British-Dutch company Unilever has been operating in India since 1888. In 1933 it became Hindustan Unilever Limited, with distribution covering over 2 million retail outlets and products available in over 6 million outlets across the country.[41] BP Castrol has for decades had a partnership with Hindustan Unilever for the Indian market and this is a major reason for its competitive advantage there. The parallel story is that in Latin America, Exxon has the upper hand due to its decade-long partnership with Procter & Gamble, who possess a historically similar distribution network to that of Unilever in India but in Latin America. This gives Exxon equivalent market advantages in the same market but in a different geographical region.

The Panama Canal II: Winning Because of Its Spare Parts Network

A similar story can be told about the building of the Panama Canal II, which is one of the largest engineering projects over the past 100 years and will be finished by 2014. In the work to dig out the Panama Canal II, the American company Caterpillar won a large majority of the contracts for trucks and other equipment over the Swedish company Volvo Trucks and the Japanese Tomatsu Trucks, the other two major competitors in this industry. Caterpillar won its contracts not because its technology or trucks are better than Volvo's or Tomatsu's, but because Caterpillar has a much more finely knit and decentralized network of agreements in Latin and Central America with vendors who can supply spare parts in hours as opposed to their competitors (Volvo who has a much more centralized distribution system and takes much longer to supply spare parts). In operations such as the building of the Panama Canal II or mining, machinery standing idle for hours means millions of dollars lost. Caterpillar's partnership arrangements beat Volvo and Tomatsu on the risk of losing money because machinery is idle in an engineering project on a scale such as that of the Panama Canal II.

B&O and Apple: Design for Aesthetics or Market Lead

Our next example is more speculative and could be labeled "what if?" Take the Danish company Bang & Olufsen. From the 1970s to the 1990s, the company became a global leader in the luxury market for TV sets, Hi-Fi, and other consumer electronics. Its appeal lay in its iconic design and the company had already built specialized Bang & Olufsen stores in the 1980s, where the purchase of TV or Hi-Fi equipment was handled by specially trained personnel and integrated into the environment of a designer living room equipped with furniture of the type to be found at the New York Museum of Modern Art. B&O didn't manufacture the tubes for its TV sets, they were sourced from Philips and the Hi-Fi equipment components were also sourced externally. What B&O did successfully for decades before anyone else was two things, creating iconic design and making the buying process a special experience. In the early 2000s, B&O also collaborated briefly with Samsung on making a luxury mobile phone. In 2010 the computer maker Asus and B&O began cooperating to make a sleek computer design.

From around 2000, Steve Jobs and Apple turned B&O from being a design icon into a company competing for survival. How did Apple do this? They made a personal computer with a sleek design, introduced the iPod and integrated it with the iTunes music system, the iPhone was introduced and later on Apple TV was added. Moreover, Apple sources all its products from the Taiwanese company Foxconn. Apple introduced Apple stores to give its customers a special buying experience for products targeted toward people willing to pay a premium for design and customer experience. Finally, it is estimated that in 2012, more than 200,000 entrepreneurs worldwide were busying themselves with developing new applications for Apple's iPhone and iPad. Every time an app is paid for and downloaded on an iPhone the successful entrepreneur pays Apple 33 percent in commission. In 2012, Apple had revenues of about $4 to $5 billion per quarter from this global ecosystem of app developers, yet Apple has invested nothing in any of these app developers. As has been observed, this is probably the best business model created by any company in the last 100 years.

A central question here is what did Apple do that B&O couldn't have done? To begin with, the only thing that B&O didn't make was the personal computer. But the real question B&O should have asked itself was: What will it take for us to go from high-tech low-volume markets to high-tech high-volume markets? We think that there is a case to be made that if B&O had applied the concept of partnership and ecosystem innovation in 2000, they might have realized their opportunities for global leadership by entering into strategic long-term partnerships with companies such as Samsung and even Sony and shared the spoils of being the world's most valuable company in 2012. However, things turned out differently and today B&O must content itself with being able to announce it has entered into a contract with Apple to produce a music system for the iPhone, iPad, and iTunes systems.

Every day there are firms that start up, survive, succeed, and fail. This is the nature of a free market and an enterprising system. One should not forget that not so long ago Apple was close to bankruptcy and its phenomenal stock market rise in 2012 took a dramatic downturn the first quarter of 2013. The point to be made here is that partnership innovation is probably the dimension most overlooked by company managers and policy makers. However, partnership innovation is bound to become increasingly decisive for survival and success. This is so because the phenomena of globalization and the Internet offer more and more opportunities for new constellations, supply chains, and access to talented people. Changes are already being realized by leading companies. One example is how a small company like the small British semiconductor design company ARM[42] is able to challenge a global industry leader like Intel by deploying a networked organizational model, which is not just about designing the chips but also orchestrates the rest of the value chain for making new and more energy-efficient chips.[43]

Lessons for Innovation Managers

Managers learn from experience and often ask for "practical tools" or "hands-on" approaches if they are to participate in activities like the InnoTools and MMI. On the other hand, experience alone is not enough. Managers and others may learn little from their experience unless they have a framework for classifying and analyzing it.

Moreover, as one company manager put it: "We would never have taken the time to research available innovation tools or frameworks and systematically test them out as was the case in this project. But we have learned a lot from it and from meeting other companies within the same project." Another important insight is the educational aspect of introducing a new concept such as the Innovation Radar and business model innovation to company managers.

Yet, in the words of Henry Mintzberg: "Education is hands-off; otherwise it is not education. It has to provide something different, conceptual ideas that are quite literally unrealistic and impractical, at least seemingly so in conventional terms. People learn when they suspend their disbeliefs, to entertain provocative ideas that can reshape their thinking."[44]

State of the art research and business concepts from business schools such as the Kellogg School of Management, INSEAD, and the like can be turned into a valuable managerial experience and tools if designed in a way that is meaningful for all parties involved.

Some of the stakeholders in the program argued that the MMI should only focus on making the companies take the online Innovation Radar survey and save the money spent on qualitative follow-up with deep-dive workshops. However, the deep-dive workshops proved crucial for discussions with managers and in challenging current thinking on innovation among managers within most of the companies. Hence, technology and the Internet seemingly cannot replace the value of human interaction, especially if the goal is to challenge perceptions and "how things are done" within companies.

Lessons for Policy Makers

The Nordic countries usually score high on international rankings and indexes for innovation such as the EU's Community Innovation Surveys, World Competitiveness Report, and INSEAD's Innovation Index, and so forth. In spite of their usefulness, it should be remembered that such rankings represent aggregated data at a national level and say little about innovation performance within particular industries let alone within individual companies. Hence the rigorous testing of innovation focus and effort within 100 Nordic companies with 800 managers across five countries has revealed that, for both managers and innovation policy makers, all is perhaps not perfect, at least not if the innovation focus is broadened from product and process innovation to business model innovation.

As mentioned in the introduction to this chapter, there is still disagreement among economists about the notion of competition between nations. In spite of these theoretical arguments, we think that the systematic application of the Innovation Radar in 100 companies across the five Nordic and Baltic countries supports the view that company-level focused programs can change mind-sets for both innovation policies as well as outcomes.

In the "OECD Innovation Strategy—Getting a Head Start on Tomorrow" it is stated that

"Sound evidence on the sources of technological and non-technological innovation, is found through firm-level analysis, which provides more detailed insight than country level analysis."[45]

Moreover, research teams from 21 OECD and non-OECD countries conducted econometric analyses attempting to calculate the link between innovation and productivity. It was found that firms introducing both product and process innovation derive, on average, 30 percent more "innovation sales" per employee than firms introducing only one of the innovations.

The OECD research provides some interesting facts in relation to the Nordic MMI and the Innovation Radar business model innovation program, namely:

"Firms active in international markets are 40 to 70 percent more likely to innovate than other firms and after correcting for the fact that not all firms are innovative, firms involved in collaboration spend 20 to 50 percent more on innovation than non-collaborating firms... In addition, firms further away from the technology frontier invest less in innovation per employee and have lower returns from innovation (i.e. lower sales per employee) than those closer to the frontier."[46]

"Moreover, a firm's productive capability is proxied by how far its productivity level is from the top productive firms worldwide. Firms with a large productivity gap were considered far from the technology frontier. While those with a small productivity gap compared to the top productive firms are close to the technology frontier."

A final result stemming from introducing the Innovation Radar in the Nordic countries is that Sweden's national innovation agency Vinnova has established a special unit for business model innovation. It should, in this context, be noted that the EU, OECD, and World Economic Forum, and so on, have ranked Sweden for years as a world leader in innovation policy and investment in innovation.

Conclusion

Companies are usually established to do specific tasks and solve specific problems for their customers. As companies grow they will develop capabilities and routines that reflect their mission and business focus and, over time, they will try to become better and better at what they do.[47]

In becoming more and more efficient at delivering a product or service without significant change, companies also risk becoming irrelevant. This is because the market is always in flux, and customers' tastes and preferences change as they are introduced to competing offerings. Of course regulation may dictate changes, demographics, entrepreneurs, and many other variables may also contribute to changes in market demand.

Over time, a robust innovation strategy can seemingly no longer just be based on a product or process alone. Evidence from the marketplace increasingly tells a story of companies that take a business model innovation perspective outperforming those companies that don't.

The MMI program also revealed that merely having an innovation strategy had positive effects on companies' innovation effort and focus. In view of the fact that many companies still don't have a clear innovation strategy, this seems like an obvious and easy place to start working with innovation.

We still have relatively little knowledge about how different types of organizations can influence types of innovation. For example, do hierarchical organizations favor incremental product and process innovation and do these correlate with each other?

Likewise, how do companies go from a technology-focused culture to a customer-oriented one, or embark on open innovation and partnership innovation if they have a strong tradition of in-house development?

Concerning policy makers and innovation programs, some of the lessons from the Inno-Tools and MMI projects include (1) if the goal is to change the mind-set for what can be done and push boundaries, it makes sense to start with thorough research on "what is out there and what has already been done" and (2) start with a pilot project. Make sure that it is designed to be scaled up and make sure that the crucial questions are evaluated for later documentation.

Build a brand around the project. If a project has no name or identity it is less likely to gather interest or support. And build a coalition with inside and especially with outside stakeholders.

References

1. Baumol, William. *The Free-Market Innovation Machine: Analyzing the Growth Miracle of Capitalism.* Princeton, NJ: Princeton University Press, 2002.
2. Porter, Michael. *The Competitive Advantage of Nations.* New York, NY: The Free Press, 1990.
3. Schumpeter, Joseph A. *Capitalism, Socialism and Democracy.* New York, NY: Harper & Row, 1950.

4. Jones, Chris. *Why Most Innovation Programs Fail*. Chief Innovation Officer—CEO. http://www.the-chiefexecutive.com/features/feature85954/, June 1, 2010.

5. Drucker, Peter. *The Practice of Management—The Classic Collection*. United Kingdom: Elsevier Ltd., p. 29, 2007.

6. Levitt, Theodore. *The Marketing Imagination: Marketing Myopia*. New York, NY: The Free Press, pp. 141–173, 1986.

7. Levitt, Theodore. *The Marketing Imagination: Marketing Myopia*. New York, NY: The Free Press, pp. 141–173, 1986.

8. Levitt, Theodore. "Marketing Success through Differentiation of Anything." *Harvard Business Review*, January-February 1980.

9. Archibugi, Daniele, Howells, Jeremy, and Michie, Jonathan (eds.). *Innovation Policy in a Global Economy*. United Kingdom: Cambridge University Press, 1999.

10. Kim, W. Chan and Mauborgne, Renee. *Blue Ocean Strategy*. Boston: Harvard Business School, 2005.

11. Lafley, A.G. and Charan, Ram. *The Game Changer*. New York, NY: Crown Business, 2008.

12. Chesbrough, Henry. *Open Business Models: How to Thrive in the New Innovation Landscape*. Boston: Harvard Business School Press, 2006.

13. Prahalad, C.K. *The Fortune at the Bottom of the Pyramid*. Philadelphia, PA: Wharton School Publishing, 2005.

14. http://jornbangandersen.com/wp-content/uploads/2012/09/IspimPaperFinal1.pdf.

15. Wolcott, Robert C. and Lippitz, Michael J. *Grow from Within: Mastering Corporate Entrepreneurship and Innovation*. New York, NY: McGraw-Hill, 2010.

16. Carlson, Curtis R. and Wilmot, William W. *Innovation: The Five Disciplines for Creating What Customers Want*. New York, NY: Random House, 2006.

17. Pedersen, Kris. Expanding the Innovation Horizon, IBM, Powerpoint, May 16, 2006.

18. Sawhney, Mohan, et al., 2006, op cit.

19. Chen, Jiyao and Sawhney, Mohanbir. *Defining and Measuring Business Innovation: The Value-Oriented View of Innovation*. Kellogg School of Management, 2010.

20. Johnson, Mark W. *Seizing the White Space: Business Model Innovation for Growth and Renewal*. Boston: Harvard Business Press, 2010.

21. Andersen, Jørn Bang. "InnoTools: Global Initiative for Measured and Managed Innovation." *Nordic Innovation*, September 2009.

22. The Innovation Radar for the MMI programme was version 2.0 a revised survey design developed by Mohan Sawhney and Jiyao Chen, Kellogg School of Management, 2008–2010.

23. MMI sample: 800 business managers from the Nordic countries. Company distribution was: 49 small companies (less than 50 employees), 30 medium sized (250–499), and 23 large companies (+499).

24. Andersen, Jørn Bang, et al., "Measured and Managed Innovation." *Nordic Innovation*. www.nordicinnovation.org/mmi. Oslo, 2010–2012.

25. MMI: Presentation of the External Evaluation: Final Steering Group Meeting, Copenhagen. Nils Gabrielsson-Inno Scandinavia AB, Stockholm Sweden, February 6, 2013.

26. "Nordic Innovation." *Measured and Managed Innovation 2010–2012: Final Report*, Norden, 56, 2012.

27. Inno Scandinavia. "Final Evaluation of the MMI Programme." *Nordic Innovation*, 30, 2013.

28. Op cit, p. 29.

29. http://www.isbuinfo.org/all_about_shipping_containers.html.

30. Authors' own interpretation of Maersk-Sealand applied to the Innovation Radar, 2012.

31. Arvid, Steen. "Containerrederriet som snudde skuten." *AGENDA, MediaMagnet*. Norway, 13, March 2013.

32. Based on presentation by Rack Space at KIN dialogue event in Las Vegas, NV, 2010.

33. Levitt, Theodore. "Marketing Myopia." *Harvard Business Review*, September–October 1975.

34. Dutfield, Graham M. and Suthersanen, Uma. *The Innovation Dilemma: Intellectual Property and the Historical Legacy of Cumulative Creativity*. Sweet & Maxwell Ltd. and Contributors, 2004.

35. Nasar, Sylvia. *Grand Pursuit: The Story of the People Who Made Modern Economics*. United Kingdom: HarperCollins Publishers, p. 190, 2011.

36. Nasar, Sylvia. op cit.

37. Levitt, Theodore. "Marketing Success through Differentiation: Of Anything." *Harvard Business Review*, January 1980.

38. The Innovators Toolkit, *Types of Innovation: Incremental and Radical Innovation*. Boston: Harvard Business Press, 2009.

39. Leifer et al., *Radical Innovation: How Mature Companies Can Outsmart Upstarts.* Boston: Harvard Business School Press, 2000.

40. Cooper, R. and Edgett, S. Ideation for Product Innovation: What Are the Best Methods? In *PDMS Visions Product Development Management Association.* United Kingdom, March 12–16, 2008.

41. http://en.wikipedia.org/wiki/Hindustan_Unilever and information from BP Castro'sl global marketing management.

42. http://www.forbes.com/sites/haydnshaughnessy/2012/02/24/intel-vs-arm-battle-of-the-business-model/.

43. http://www.innovationmanagement.se/2012/02/20/entrepreneurs-of-the-world-unite-in-eco-systems/.

44. Mintzberg, Henry. *Managers Not MBAs: A Hard Look at the Soft Practice of Managing and Management Development.* New York, NY: Prentice Hall, p. 249, 2004.

45. OECD. *The OECD Innovation Strategy: Getting a Head Start on Tomorrow: Innovation Trends.* OECD, Paris, France, 2010.

46. OECD, op cit., p. 31.

47. Prahalad, C.K. and Hamel, Gary. "The Core Competence of the Organisation." *Harvard Business Review*, 2003.

CHAPTER 37

Innovation Measures and Indices

Brett E. Trusko and Praveen Gupta

Innovation as an intuitive and creative process is a difficult process to measure. Conventionally, innovation is measured in terms of financials or counts. Innovation, being a complex and unknown process, proves to be a challenge when defining clear and correlating measurements. The financial and count-type measurements include product- or service-specific sales or revenue growth, and count-type measurements include items like the number of patents, trademarks, articles, and product or service versions produced. However, experience shows these measurements do not correlate to the innovation activity; therefore they should not be used as a business measure of performance.

In order to establish measures of innovation, understanding the innovation process first is a must. Corporations implement innovation through the network-centric, pipeline-fed, and opportunity-driven approaches. The network-centric approach, which is taught in colleges, is based on collaborative brainstorming. The concept is that more minds are better than one at a given time (without understanding the "why"). Inventors who work in research drive the pipeline model and development environment on a specific topic, explore new ideas, and develop new products and services. The pipeline model, which is driven by chance or innate genius, is a somewhat common perception of the innovation process.

The opportunity-driven model is more representative of street-smart individuals who take an idea at the right time and the right place, devise a solution, know how to market it, and capitalize on their breakthrough. They also appear to be lucky, which is defined as an intersection of continual preparation and opportunity. Their success represents a once-in-a-lifetime windfall out of the blue sky (i.e., fortuitous occurrences). Another innovation process, which is a combination of collaboration and opportunity, is called the "open innovation" process and leads to products such as Linux and the Internet.

Besides measuring innovation successes in a business, innovation needs be measured at personal, national, and global levels as well. Innovation has become a local, national, and global initiative for competitiveness, and also to maintain standards of living. Most innovation initiatives start with a set of measurements. There has been a major effort in establishing a scorecard for measuring innovation at national, European Union, and global levels.

Difficulty with Current Innovation Measures

Such variants of the innovation process and its outcome are difficult to measure. Peter Drucker's process, detailed in his book, *Innovation and Entrepreneurship*, identifies various phases of innovation, including the phases of opportunity identification, analysis, acceptability, focusing on a core idea, and leadership. The act of innovation, though, is still not clearly explained. Measuring innovation effectively is contingent on understanding details of the innovation process, its inputs and outputs, and its controls.

Measuring innovation is an important issue, as business growth and profitability in the knowledge age depend on innovation. Continual acceleration in innovation will sustain revenue growth, which will then fuel more innovation. Therefore, sustainable growth requires sustainable innovation, which requires that innovation be institutionalized and its output made predictable.

Today, corporations are skeptical of adding new measurements to the existing portfolio of measurements. The current measurements do not, however, get fully utilized through analysis for extracting business intelligence or continually creating new opportunities. Instead, too many companies end up measuring too many things for too little value. Currently, a variety of dashboards and scorecards focus more on display instead of extracting intelligence out of the data. Such tools are more appropriate for data mining. The waste continues to pile up without extracting a proper understanding of the related processes as well as a planned application of the lessons learned from the measurements.

A similar approach is taken regarding the measures of innovation. Several institutions, corporations, and consultants are developing measurements of innovation. They are interested in developing innovation scorecards, indices, radar, or dashboards. However, most of the measures lack a consistent definition of innovation and its elements.

As a result, innovation surveys are the most commonly deployed tools to determine an organization's readiness for innovation, its innovation capability, and innovation performance. Several players are trying to bite the innovation apple from different sides. Eventually, all these measures will converge when we have a better understanding of the innovation process. Because the understanding of innovation is currently fragmented, however, so are the measurements.

Research shows that the correlation between various measures of innovation and its impact on the output is somewhat limited. Measuring innovation is therefore challenging, because current measures do not provide statistical analysis or relate the impact of innovation with any degree of confidence. Measures of innovation are not available for strategic planning because of the uncertainty associated with measures of the financial impact of innovation. Ultimately, a lack of financial, organizational, and cultural structure around innovation exists. Research, however, has been done on the effects of innovation on workplace organization, corporate performance, organizational success, and organizational climate.[1]

State of Measures of Innovation

The climate of creativity and its effect on innovation was analyzed by The Creative Problem Solving Group to evaluate the link between climate and organizational innovation. The study included the following nine criteria:

- Challenge/motivation
- Freedom
- Trust
- Idea time
- Play/humor
- Conflicts
- Idea support
- Debates
- Risk taking

The normal gap between an average company and an innovative company is about 25 percent, and an additional 40 percent gap exists for the above categories between the worst company and an innovative company. Analysis of the data shows that innovative companies outperform in the areas of risk taking, play/humor, challenge or motivation, and idea support. The most significant factor that differentiated an organization for innovation is risk taking. Innovative companies encourage risk taking by their employees. They create a culture of risk and reward in order to intellectually engage employees.

The PA Consulting Group[a] identified nine dimensions, which the consulting group uses, to measure an organization's innovativeness:

- Committed leadership
- Clear strategy
- Market insights
- Creative people

---

[a]http://www.paconsulting.com/

- Innovative culture
- Competitive technologies
- Effective processes
- Supportive infrastructure
- Managed projects

Recognizing measures of innovation is a challenge because of the mismatch between the financial cycles and the long innovation cycle of concept to commercialization. Determining a real measurable output, in the time frame when the cost is incurred, is difficult. Moreover, the output of the innovation process is sometimes dependent on many direct and indirect contributors in the organization. Arriving at predictable measures of innovation appears to be almost impossible. Some examples of innovation measures cited on the PA Consulting Group website include speed of development processes, competency metrics, and number of patents.

Tim Studt, in an editorial in *R&D Magazine,* explored the "measures of innovation" topic.[2] He suggests that measuring innovation is a do-able task with complexities of its own. The role of innovation has changed in the last 30 to 50 years, from the research and development (R&D) efforts to broader efforts in an organization. The measures of innovation, therefore, are shifting from primarily R&D spending to measures on tangible processes, product enhancements, and intangible investments. Because of the variation from product-to-product and organization-to-organization, as well as the subjectivity of an organization's innovation capabilities, any comparison of organizations using a set of measurements will be subjective. According to this editorial, key components of successful innovation include:

- Funding for innovation
- Trained and educated staff
- Collaborative environment
- Key individuals
- Corporate infrastructure
- Strategic planning

These factors consist of subjective and objective measures or components of innovation.

Innovation scorecards are another way to measure innovation. Innovation scorecards are a graphical display of measures and the status of an organization. However, the challenges with this method are the measures to include in the scorecard, and how to collect data for those measures, both internally within an organization and externally for benchmarking purposes.

In 1986, a group of industrial, academic, and labor leaders organized to form the Council of Competitiveness (www.compete.org) to address the trends in U.S. competitiveness and to act as a catalyst to launch national initiatives for improving U.S. competitiveness. The Council developed a national innovation index for assessing U.S. innovation capacity. The index consists of the following four types of measurements[3]:

- The quality of the common innovation structure
- The cluster-specific innovation environment
- The quality of linkages
- Other measures

The variables that constitute the innovation index are patents, R&D personnel and expenditure, trade regulations, protection of intellectual property, investment in education, and university and private industry research participation. The index is utilized to assess the relative competitiveness of various developed and developing economies and to project emerging centers for innovation.

The Council launched the national initiative, similar to several European national initiatives, for improving competitiveness and raising the need for accelerating innovation. The Council's national agenda highlights three key components—talent, investment, and infrastructure. Accordingly, elements of these key components include improving education, developing innovators, opening new frontiers, promoting entrepreneurship, rewarding risk-taking, protecting intellectual property, strengthening manufacturing capacity, and developing growth strategies.

The European Business School and Arthur D. Little jointly developed the Innovation Scorecard for creating value through innovation management. The scorecard measuring innovation performance is driven by the following elements:

- Innovation strategy
- Organization favoring innovation
- Innovation process
- Innovation culture
- Resource deployment

The European Union Regional Innovation Policy also developed a scorecard for ranking the strengths and weaknesses of a region against a number of criteria. The main elements of this innovation scorecard include:

- Employees in medium- and high-tech employment
- Internet users per 100 inhabitants
- Business enterprise R&D
- Research infrastructure
- University research income
- University research strength
- Knowledge workforce
- Qualification level

The Kellogg School of Management[4] presented "Innovation Radar" that is designed to create a holistic framework to visualize, diagnose, benchmark, and improve the innovation process. An initial set of measurements incorporated in the innovation radar include:

- What (offerings), brand, networking
- Where (presence), supply chain, organization
- How (process), value capture, customer experience
- Who (customers), solution, and platform

Benefits of such a representation of the corporation's innovation performance are to identify strengths and weaknesses in the innovation process in order to develop a sound innovation strategy. However, similar to the other measures of innovation, the following questions remain to be answered:

- What data to collect?
- How to collect data?
- How to analyze the data?
- How to interpret the data?
- How to drive improvement?

Extensive research conducted at the University of Melbourne addresses issues with measuring innovation. Accordingly, the accuracy of innovation measurements is critical to assess the extent of innovation and its impact on economic and social well-being. A fundamental challenge in economic analysis is that static indicators are the basis for measuring dynamic processes. Paul H. Jensen and Elizabeth Webster of the Melbourne Institute of Applied Economic and Social Research (MIAESR) identified four specific dimensions to the problem of measuring innovation.[5] They are as follows:

1. The innovation process may take years from concept to commercialization.
2. In the narrow sense of innovation, the novelty of products or services is difficult to benchmark, and the process measurements are difficult to adjust.
3. Time carries an important economic value for the innovation process. Therefore, innovation measures must have some way to adjust in value over time.
4. Much of the innovation activity is categorized as unobservable and is not reported in conventional methods.

The authors, Jensen and Webster, identified three main characteristics of innovation measures: type of innovation, stage of pathway, and firm characteristics. The measures of innovation in their reported research included patent applications, trademark application, design application, expert assessment, journal counts, and survey of managers. A challenge exists in identifying a complete list of innovation measurements.

An analysis of various innovation measures identified the main industries active in innovation, including manufacturing, wholesale trade, finance, and insurance. This analysis involved examining several hundred firms and found no significant correlation between measures of innovation and the firm's size. However, data were collected using surveys, which did not appear to be a good measure of innovation.

Measures of Innovation for Developing an Innovation Index

Mark Rogers of the Melbourne Institute at the University of Melbourne has attempted to establish measurements of innovation at the corporate level. Rogers also identified input and output measures of innovation, along with their descriptions and (more importantly) the source of data collection. The recent effort in collecting data for measures of innovation has led to difficulties in determining what data to collect. Should the data collected be data that are already available in public records (annual reports), data that can be requested (through surveys or interviews), or data that are yet to be collected? Some of the input and output measures are listed in Table 37-1.

Each measure of innovation has some validity, but none can be used as a stand-alone measure of innovation. However, combining various measures to develop an index of indicators must consider tangibles and intangibles, economic and noneconomic measures of innovation-related resources, processes, deliverables, and value.

Based upon the current research, the innovation process has been a fuzzy one at best. In order to establish a set of working measures of innovation, one must identify common characteristics of the innovation process, their interrelationships, and well-defined deliverables. Figure 37-1 shows the process of innovation using the 4P (Prepare, Perform, Perfect, Progress) model. The model illustrates that for an innovation process to be standardized, identifying inputs, in-process activities, and outputs must occur.

Input Measures	Output Measures
R&D	Introduction of new or improved products or processes
Acquisition of technology	Percentage of sales from innovative products or processes
Expenditures associated with innovative products or processes	Intellectual property
Marketing and training expenditures	Firm's financial performance

Table 37-1   Sample of Input and Output Measures of Innovation (*Source:* Rogers.[6])

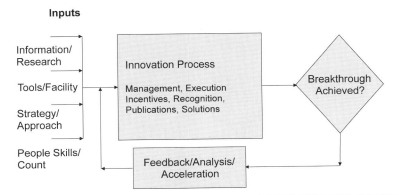

Figure 37-1   Understanding measures of innovation.

Supplier (Source)	Input	Process	Output	Customers
Customer, fundamental research, supply-chain	Demand from customers, demand defined based on market research, partner expectations	Establish targets for innovation	Solution for variety of selected and valuable demands	End users, marketplaces, businesses
Customer requirements-driven strategic plan, collaboration tools	Identified domain expertise or competence, field of solution, collaboration process	Teamwork with necessary knowledge base	Collaborative work, superior output than that of individuals	Organization
Internet-based access to research databases, publications	Variety of information, benchmarking information	Research the topic	Expanded understanding of the domain and related domains, topic of interest, exploration capability, applicable alternate sources of solution	Management, innovators, organization, Knowledge repository, publications or patents
Knowledge repository, innovators, management	Alternate solutions, internal capability	Make, "acquire," or "innovate" decision	Commitment to innovate or acquire	Management, innovators
Team members, management, suppliers, collaborators	Knowledge, resources, environment, methodology, tools	Play to innovate	Good ideas, crazy ideas, funny ideas, innovative ideas	Innovators, team members
Management, innovators	Culture for creativity, good ideas, crazy ideas, funny ideas, innovative ideas	Develop alternate solutions	Evaluation, performance classification, patents, publications	Organization, patents and trademark office
Management, organizational expectations	Alternative solutions, organizational expectations	Select a solution	Product, service, process, platform	Organization, marketing, and sales
Management, organization	Performance measures, utilizing the solution	Verify the solution for economic value	Alternate applications, customer review	Marketing, sales, and customer
Management	Market leadership intent, market demand, resources ($), target market, need identification	Develop marketing plans	Market plan to achieve necessary market recognition, customer's interest	Marketplace, customers, end users
Management, resources	Strategic sales plan, supply chain relationships	Commercialize	Growing product sales, new customer relationships, sales and distribution channels	Organization, society
Management, process owners	Implementation of Six Sigma business scorecard or equivalent measurement system, effective data collection	Monitor impact on business performance	Improving measures of innovation in business performance index, higher customer satisfaction	Shareholders, stakeholders
Inside or outside data sources	Market research, data collection	Assess impact on market capitalization	Improved shareholders' equity, innovative corporate image	Shareholders, end users
Process owners, industry sources, management	Model of innovation index, data collection	Measure Innovation	Innovation index	Management, marketplace

TABLE 37-2 SIPOC Analysis of the Innovation Process

The inputs include elements such as "information," "tools used in the innovation process," "the approach to innovation," and "targeted innovation output." The target may be a service, a product, or a certain change in product or service characteristics. The resources may be headcounts, equipment, or the acquired knowledge itself. The process includes execution, incentives, recognition, collaboration, and research.

The SIPOC (Supplier, Input, Process, Output, and Customer) model can be used for analyzing the innovation process. Table 37-2 shows various elements of the innovation process. Depending upon an organization's needs, these elements can become the measures of innovation.

The analysis of the innovation process shows many process steps and dozens of measures that can be used for monitoring innovation. The challenge is that people want to devise some magical measures of innovation that can tell the whole story and serve as predictors of innovation. Most management people would like to identify some measures, set targets, provide incentives, and start monitoring them. Even with a better understanding of the innovation process, a lot more thinking still needs to occur before selecting appropriate measures of innovation for an organization.

Given the current understanding of the innovation process, establishing an adequate and accurate measurement system right away is unlikely; instead starting an initial set of measures is a more appropriate way to begin measuring innovation. Figure 37-2 shows how sets of hierarchical measures are established to measure the effectiveness of the innovation process. Figure 37-2 also shows how people can get confused in devising a measure of innovation without fully understanding its components.

For example, if one is developing a process measure to ensure its effectiveness, one needs to look into its inputs, activities, and outputs. If one is interested in developing an innovation index for an organization, one must consider factors such as variation between entities and key selected processes of measures. When developing a process measure, considering its effectiveness in producing the desired result and its relationship with the inputs and activities is essential. In other words, when establishing measures of innovation, establishing a clear objective and purpose for doing so is a must. Once the purpose is defined, and the scope of measures is established, then critical inputs, activities, and outputs are identified. Based on the feasibility of data collection already existing or yet to be created, the aggregation of measures and their interpretation, communication, and resultant process adjustments must be thought through in order to select meaningful measures of innovation.

The Business Performance Index[7] (BPIn) consists of seven elements and 10 measures for monitoring business performance. The seven elements and 10 measures are shown in Table 37-3, BPIn Measures, and include measures for idea management, sales growth, and employee recognition for exceptional improvement. All three of these measures are measures of an organization's innovation performance at various stages. The most critical

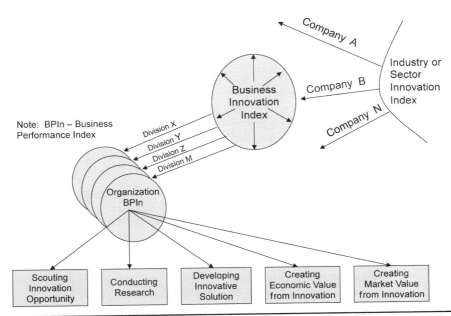

Figure 37-2 Sample processes for measures of innovation.

Six Sigma Business Scorecard Elements	BPIn Measures
Leadership and profitability	CEO recognition of employees for exceptional value creation
	Corporate profitability
Management and improvement	Managing rate of improvement
Employees and innovation	**Employee ideas for improvement and innovation**
Purchasing and supplier performance	Suppliers performance (Sigma level)
	Cost of purchase goods/supplies
Operational excellence	Aggregate process performance (Sigma level)
	Cycle time variance
Sales and distribution	**Sales for new products, services, or solutions**
Service and growth	Customer satisfaction

TABLE 37-3 BPIn Measures

resource with minimal financial impact is the effective intellectual participation of employees in developing innovative solutions as a way of doing work. This requires a robust employee idea management system, which can enlist employee ideas daily, filter criteria continually, and escalate the value-driven ideas for improvement or innovation for further implementation.

CEO recognition of employees and incentives acts as a catalyst for innovation, idea management is the process, and sales of new products, services, or solutions is an output of innovation processes at a corporation. Combining these three (or similar) measures to develop a corporate innovation index is simple and allows for the assessment of key aspects. The objective of innovation index is to provide trends in performance and identify areas for adjustment to accelerate innovation.

A Process for Developing Measures of Innovation

The question, then, is how to identify good measures of innovation. Victor Basili established a Goal-Question-Metrics (GQM) paradigm to identify process and product measurements in the software engineering environment.[8] Accordingly, GQM utilizes a top-down approach to define the goals behind measuring the software process, and then utilize these goals to determine precisely what to measure. Using the GQM approach, measurements are a means to realizing the end but are not the ultimate end. Instead, measurements must be focused on specific goals, applied to all life-cycle products, processes, and resources, and interpreted based on the organizational context, environment, and goals. GQM helps in identifying organizational, need-driven, dynamic measurements in achieving business objectives. The GQM approach consists of the following conceptual, operational, and quantitative level understandings of processes:

- Goal is defined as an intent or conceptual understanding in terms of products or outputs, processes or activities, or resources or inputs.

- Questions provide an operational understanding of measurements that can be used to assess, realization of goals and objectives.

- Metrics represent the data that provide a quantitative understanding of the answers to the questions in assessing performance against goals. The data can be objective or subjective, or the object itself along with the viewpoint from which the data are taken.

Why does one measure innovation? What are the objectives behind measuring innovation? The objectives may be to establish a relationship between innovation and market capitalization, predictability of the innovation process (or the rate of innovative products), to determine performance of the innovation process, to assess availability of necessary resources for the innovation process, or an overall assessment of the innovation process for an organization. Considering each question one by one requires a different approach for answering each question appropriately. Data needed to quantitatively answer each question is different, and thus the measurements for innovation are different.

Accordingly, in order to identify innovation measures, understanding the purpose of innovation, its environment, and the input, in-process, and output parameters is essential.

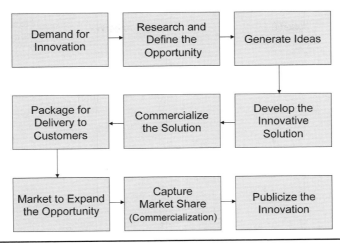

Figure 37-3   General cycle of innovation.

Furthermore, the relationships between input and output innovation variables must be implicit. To determine measures of innovation, understanding the role of each process in creating the desired innovation is essential.

Figure 37-3 shows steps in the innovation cycle, starting with the demand for innovation to publicizing innovation. The demand for innovation may be internal as well as external; therefore, measuring the location of innovation demand is important, because demand will drive the innovation activities within an organization. If developing measures for the entire innovation cycle is desirable, then an aggregate measure or a set of measures for innovation is critical.

With this understanding of measurements, an organization attempting to develop measures of innovation must clearly state its objectives before establishing the measures of innovation. Given the presence of a glut of measurements with no use in most organizations, an addition of nice-to-know measures is often perceived as "additional" work and not received well within the organization. Therefore, the following is a list of steps used to establish measures for a process or an activity:

1. Define the **purpose** of innovation in the organization.
2. Establish expected **deliverables** (basic and specific) and their contribution to business performance, including growth and profitability.
3. Determine the **measures** of success of key deliverables.
4. Identify challenging **opportunities** for improvement in the innovation process.
5. List **activities** that must be performed to accelerate innovation.
6. Identify **input, in-process, and output variables** that are critical to the success of innovation in the organization. If these variables are not monitored and managed effectively, the innovation outcomes will be adversely affected.
7. Establish an Innovation Index and determine the data collection capability for selected measures of innovation.
8. Establish reporting and communication methods, and monitor (levels and trends) critical and practical measures of innovation to drive business growth and profitability.

The above steps can facilitate the development of meaningful measures of innovation. Appropriately, the author has attempted to define innovation first, before identifying its measures. Innovation is the application and commercialization of new ideas to products, processes, or any other activities. Innovation means significant improvement can be realized through creation or collaboration or diffusion. Innovation can mean different things to different entities. Therefore, the measures will also vary from one organization to another organization.

Factors to Consider

Having understood the innovation process, measures of innovation, and elements of innovation indices, all that remains is actually implementing an innovation measurement system at one's organization. Simply copying another organization's measures of innovation is not sufficient, as they may reflect different priorities in performance and resource commitment.

Industry Innovation Indicators	Innovation Index Measures	Process Innovation Measures
Innovation funding, including R&D	Resources: funding, culture of risk-taking, rewards, tools	Excellence in research, innovation management, time allocation (%)
New products, services, or solutions	Activities: targets for innovation, process of innovation, extent of institutionalization, idea management, internal and external publications, knowledge management, internal and external collaboration, recognition	New idea deployment; extent of improvement or change; degree of differentiation, disruption, or innovativeness; time to innovate
Market capitalization	Outputs: patents; new products, services, or solutions; sales growth; market position or ranking; customer perceptions	Rate of innovation, savings, opportunities

TABLE 37-4   Measures to Consider for the Innovation Index

Table 37-4 lists a variety of measures that can guide thinking in the right direction and facilitate development of appropriate measures of innovation. Good measures of innovation, being specific, measurable, and actionable, catapult the innovation process and produce significantly more innovative outcomes.

Current corporate measures do not include most of the measures of innovation. Being a new process evolving toward standardization, difficulties are expected in collecting data for various measures of innovation and benchmarking. However, institutionalizing and measuring the innovation process must start in order for innovation acceleration to occur in a measurable way.

Return on Investment in Innovation

Most studies show that establishing a correlation between innovation and corporate performance is a challenge. Some CEOs may assume an adverse relationship between investment in innovation and corporate performance. Such incorrect perceptions of innovation by executives may be contributing to the confusion concerning the topic of innovation, as well as for the lack of commitment to systematic innovation. The best way to sustain innovation is to ensure there is a return on it.

Many companies consider growth in revenue as a return on innovation; many times, growth in revenue, however, does not translate into more money for the organization so there is no return on innovation. Though the revenue growth will somewhat reflect the innovation in a company, it does not say anything about the effectiveness of the innovation. To ensure a return on innovation, profit growth must also be guaranteed. Innovative products not only provide more opportunities for revenue growth, they also enable better margins on sales.

The return on investment (ROI) in innovation can be calculated as a ratio of profit increase divided by the investment in R&D and Innovation (RDI). In other words, if a company invests $4 billion in R&D and Innovation, and increases profit by $400 million, one can say that ROI in innovation is $400 million/$4 billion, or 10 cents on a dollar. It would take many years to recoup the original investment. Interestingly, such a return was acceptable in a conventional business where one could wait 6 to 7 years to recover the investment. However, in the 21st century with reducing product life cycle, corporations do not have 6 to 7 years available to recover all their investment in R&D and innovation. One approach could be to calculate ROI on various types of innovation separately. In other words, ROI on the short term and on the long-term innovations would have different goals as well as performance.

Corporations have an objective to be profitable on a quarterly basis. Managing profit by quarters leads to decisions for a quarter that require mostly actions in a short term that means most likely cutting costs. While taking actions to cut costs, leaders often "cut out" innovation investments. Such an approach is counterproductive to creating a culture of innovation, and creating new opportunities for growth for stakeholders. Organizations prioritize research and development projects based on their ability to provide returns in the short term. This strategy haunts these organizations in the long term, if they survive.

Organizations must apportion resources both for research on the long-term fundamental and platform innovations, and the short-term development of derivative and variation innovations. Large organizations that sacrifice the long-term technological research and development in favor of the short-term product development activities step into sudden crashes. Especially, technology firms in the interest of speed of innovation cutting the basic research for future products will face a void of new products after a few years.

Linking the corporate strategy to profitable growth will lead to planning for innovation at all levels. Successful companies continually look at their innovation portfolio with annual 10- or 20-year outlooks, in order to sustain profitable growth by perpetuating a culture of innovation.

Most, Best, and Managed Innovative Firms

Many magazines and organizations publish their list of most innovative companies. We see that their ranking can change quickly. However, if one wants to learn the difference between many innovative companies it is difficult to tell what to differentiate then except, maybe, the random luck of the draw. We do not know what one can learn from the listed most innovative companies, except the obvious growth in sales or their stock performance. There are many very innovative companies that do not make the list. Can innovation ranking really change annually? How long does it take for a company to become innovative? What makes a company innovative? Does a product, culture of company, processes, sales of new products, an innovative marketing campaign make a company innovative?

Typically, the criteria for most innovative companies include stock performance, revenue growth, profit growth, and patents. Research and experience shows that patent and stock performance are not very effective measures given that a small percent of (about five percent) of patents are gainfully deployed. Using the knowledge of intent, process, and tools of the innovation leads to the three types of innovation companies—"Most," "Best," and "Managed" innovative companies.

If the innovation is defined in terms of reproducibility of the product, process or a solution, the sales volume is a measure of reproducibility. Thus, the "most" innovative company is the one that sells the most number of new units and increases its revenue, or has the highest total revenue. In other words, the company with the most revenue can be considered the most innovative company.

A company is "best" innovative when its increase in the profit is the most from increase in sales or the revenue increase of new or innovative products. If a company increases its revenue but does not generate net income then the company may be an innovative but not the best. Losing innovations cannot last long, thus innovation would not be sustainable. For example, if company A develops new products with a revenue of $10 billion, and a profit of $300 million; and company B develops new products with a revenue of $8 billion, and a profit of $400 million. Then company A may be more innovative than B, but company B is better innovative than company A.

The "managed" innovative company is the one that has created a culture of innovation, implemented innovation processes and tools, routinely enlists employees for new ideas, maintains a portfolio of innovations for the short-term and long-term profitable growth, and establishes measures of innovation or innovation index to assess its innovation successes. If this company invests in R&D and innovation, it can predict its performance in terms of profit and revenue growth, and achieves its targets. In other words, this company has understood the causative relationship between innovation and business objectives, and it can deliberately sustain profitable growth innovatively.

National and Global Measures and Indices of Innovation

Process-Based National Measures

Following the Six Sigma Business Scorecard framework, the following measurements can be aggregated into a national innovation index that would facilitate our national objectives of sustaining economic, social, and political leadership in the world.

M1.   Research, innovation, and development funding (RID)

M2.   Extent of innovation education in schools and colleges (ISC)

M3.   Number of start-ups per year (SPY)

M4.   Employee fun index (EFI)

M5.   National awareness to innovate at all levels (NAI)

M6.   Creative resources for children (CRC)

M7.   New fundamental and platform innovations (NFP)

M8.   Government innovation award reach (GIA)

M9.   Business innovation index (BII)

M10.   Total new/lost job wages ratio (NLR)

M11.   Per capita GDP growth (PCG)

M12.   Export/import ratio (EIR)

It has been observed that developed, developing, and underdeveloped economies will have different needs for innovation. For example, underdeveloped economies will be interested in innovating basic necessities such as for health, safety, and food. The developing economies will be interested in innovating infrastructure such as roads, transportation, preservation, communication, and construction. The developed economies will be interested in productivity, comfort, entertainment, and defense. Accordingly, measures of innovation would change.

Measures of Innovation: USA

In April 2007, the "Advisory Committee on Measuring Innovation in the 21st Century" issued a call, seeking proposals for ideas for new or improved innovation measurements, to identify firm-specific data items that could enable benchmarking and aggregation.[9] These measures could include company culture, incentive, structures, and change management. It was also to identify what data would be needed to differentiate innovative firms from non-innovative ones. Holes in the data collection systems were also to be identified. The proposals were supposed to address how the new or improved measures would provide necessary signals of changes in business performance and policy decisions. About 30 proposals were submitted for consideration by the "Advisory Committee on Measuring Innovation in the 21st Century," reporting to the Secretary of Commerce of the United States. The Committee has about 10 industry CEOs or presidents, and five academicians. The final report of recommendations was issued in January 2008. As an attempt to establish a common understanding of the innovation itself, the committee proposed the following definition of innovation.

> "The design, invention, development, and/or implementation of new or altered products, services, processes, systems, organizational structure, or business models for the purpose of creating new value for customers and financial returns for the firm."

The Committee recommended the following guidelines for establishing meaningful measures of innovation for the U.S. government and businesses:

- Innovation data collection efforts should build on the way firms assess the effectiveness of their innovative activities.

- There needs to be tolerance for qualitative and subjective measures.

- Instead of being static, innovation measurement needs to be like an ongoing dialogue.

- Innovation measures should allow for analysis at firm, industry, national, international, and regional levels.

- Institute firm-level measures of innovation to establish causative relationship between innovation activities in the firm and impact on the firm performance.

- Develop and implement best practices in innovation management.

The "Institute for Defense Analysis Science and Technology Policy Institute" conducted research on measuring innovation and prepared a comprehensive report[10] based on review of the business and financial literature about measurement of innovative activities, examination of methods used to measure innovation process in many countries, and interviewed business leaders.

The IDA report reviews multiple definitions of innovation by various agencies in the United States and Europe. It identifies attributes of innovation in terms of inputs that are assets or ingredients, and involves activities for the purpose of creating economic value. Knowledge is key to innovation, and the process of innovation is complex. Complexity implies nonlinearity, iteration, and uncertainties. Inputs and outputs to innovation deal with knowledge. The report references to evolution of innovation metrics by generation, moving from R&D expenditures, discrete outputs such as products and patents, benchmarking and surveys, and finally to intangibles and management techniques. It is interesting that as we move from first generation to next generation of measures complexity increases, causative relationship to innovation outcomes weakens, and thus increases uncertainties with the measures of innovation. Two approaches to measures of innovation include aggregation and monetization, that is, developing scorecard, radar, or dashboard representation of measures of innovation, and economic value of innovation including return on investment in innovation.

Measures of Innovation: China

The main purpose of innovation is growth. Economic growth in China is an evidence of significant innovations happening there. China has innovated manufacturing capacity so rapidly at such a large scale that the economy is growing at about 10 percent annually. Manufacturing innovations have made trillions of dollars for China. China has now committed to individual and corporate innovations in the area of new products and solutions. Factors positively affecting China's growth include the following[11]:

- The scale and wide-ranging capabilities of its manufacturing sector enabling reverse engineering of existing products and leading to new product innovations with major manufacturing capabilities
- Focus on expanding education system producing a large supply of science and engineering skills for greater innovation capacity
- Necessary supply of capital by state-owned and private banks to support innovative firms
- Opportunities created by successful penetration of the global market supported by the growth of its domestic market
- Pro-business and entrepreneurial culture in many provinces supporting small firms and start-ups
- Potential for further innovation in underdeveloped service sector
- An urban development strategy to build efficient, green, and innovative cities creating innovation opportunities in the infrastructure domain

In addition to the above factors fueling innovation in China, there are constraints that if resolved,[11] will accelerate innovation for sustaining economic growth:

- Macroeconomic policies need to encourage growth of the domestic market
- Underutilized potential for innovation of state-owned enterprises' (SOE) huge assets including human talent
- Improvement in recruitment of faculty in leading universities for driving research and innovation
- Inexperience, and thus maturity of venture capital industry
- More strategic and mutually beneficial collaboration and interdependence between multinational and Chinese firms can advance innovation capabilities
- Current inefficiency, thus improvement in smart and sustainable urbanization and long-term fiscal planning to sustain the infrastructure will be needed for sustaining a culture of innovation
- Improvement in interaction between the market forces and its educational institutions will guide and facilitate experimentation and direction of innovation

Measuring innovation in China has been more focused on assessing intellectual capital in terms of patents, literature citations, and growth in research and development (R&D) in terms of addition of R&D functions in corporations, R&D expenditures, and R&D personnel as a percentage of total employment. The Chinese National Bureau of Statistics (NBS) conducted a nationwide innovation survey of about 70,000 enterprises in 2007. The survey included innovation behavior, and product and process innovation in three industrial sectors; mining, manufacturing, and electricity.

Measures of Innovation: Russia

In 1945 being considered equal to the United States, and thriving on inventions in a state-controlled industrial environment, Russia has been a superpower. A country of large natural resources it has been considered between developing and developed economies.

With changing political scene, resizing, and state of economy, Russia is now considered equal to emerging economies such as China, India, and Brazil. Due to Russia's long history of being a major player in the world, it has far greater aspirations. Today, Russia is integrating into the world economy and wants to play a major role again. However, its legacy infrastructure, policies, lifestyle, and mind-set have to change in order for transformation to become successful.[12]

Being a major industrial economy with many inventions, and per capita income much greater than Brazil, China, and India, Russia has made changes in innovation policy.[13] Innovation has become a driving force at the top level in Russia, development institutes and resources were set up to accelerate innovation. However, these innovation catalysts have not yielded in desired results yet due to lack of coordination between research and commercialization. Russia's transition from the Soviet era to a modern Russia is going through challenges of a large organization.

Measures of Innovation: India

Being a free and resource-limited society with a diverse culture and increasing population, creativity (as they call it "Juggad") and innovation (making something from nothing or little available resources) have been core aspects of Indian society. Immediately after independence, India committed to technology through formation of the Indian Institute of Technology in collaboration with Western technology institutes. Innovation and entrepreneurship have been prevalent throughout India. In 2005, the National Knowledge Commission was formed to formalize and accelerate the use of India's intellectual resources. India's Prime Minister declared, "The time has come to create a second wave of institution building of excellence in the field of education, research and capability building so that we are better prepared for the 21st century."

Innovation is becoming a key driver in India's economic growth. Its leadership in software and business services and globalization of Indian firms has led to global collaboration and increase in innovation intensity, or percentage of revenue derived from new products or services. The following findings from the Innovation in India report published by the National Knowledge Commission highlight the extent of innovation in Indian businesses.[14]

- "Internal processes for Innovation such as maintaining a specific Innovation department, allocating funds, rewarding innovative employees, forecasting probabilities of success, formalizing processes and systematic attempts, maintaining physical locations for Innovation and constituting cross-functional teams all lead to firms being more innovative. Further, firms with greater R&D spending, Innovation spending and strategic prioritization for Innovation are also more likely to be more innovative.

- Firms with their primary market in India have higher Innovation Intensity than those with primary markets abroad. On the other hand, a greater proportion of firms with their primary market abroad are Highly Innovative (i.e., have introduced more "new to world" innovations) as compared with firms with their primary market in India.

Barriers to innovation in India are considered to be a shortage of skills in industrial innovations, and systematic problem solving, in other words, formalizing its "jugaad" methods. Collaboration between educational institutions and industry, and government regulations are also major deterrents to accelerating innovation. Similar to other developing countries, there has been hesitation in investing too heavily due to lack of intellectual property protection that discourages investment in innovation. Due to global competition for domestic markets as well as exports, Indian firms are developing strategies and processes to remain competitive innovatively.

India has launched its Innovation Portal to network people, ideas, experiences, and resources to mobilize the innovation community in India. The sectors targeted for innovation include Health, Governance, Science and Technology, Food and Agriculture, Environment and Nature, Resources, Energy, Education, Infrastructure, and ICT. The India Innovation Portal was launched by Prime Minister Dr. Manmohan Singh in November 2011 on the occasion of releasing the National Innovation Council's Report to the People 2011.

The Indian innovation system is complex in terms of users, verticals, income disparities, differentiated markets, distributed resources, state autonomies, and political system measures of innovation are more specific to implementation of a specific policy initiative in a certain region or state. Overall indicators of economic activity such as exports, financial reserves, and funds allocation are well captured representing progress in the innovation initiative in India.

Measures of Innovation: European Union

The European Union established its goal to become the most competitive and dynamic knowledge-based economy with sustainable growth, and more and better jobs to maintain its quality of life or the standards of living. Innovation was the most critical aspect of this directive. As a result, a European Innovation Scoreboard was established. The first scoreboard was designed for 17 countries with 16 indicators based on the results from innovation surveys. By 2008, the number of countries increased to about 30 and the number of indicators increased to 29. These 29 indicators are grouped in three blocks covering enablers, activities, and outputs. Enablers are the main drivers of innovation such as funding, available skilled people, or incentives by governments. Activities include an organization's activities such as its own investments, internal entrepreneurship, collaboration, and intellectual property or assets. Outputs capture internal or external innovative solutions or new processes or products.

The European Innovation Scoreboard later modified into the Innovation Union Scorecard that can be used to help monitor implementation of innovation in Europe by benchmarking among 27 member states, and to identify opportunities for strengthening research and innovation.

The framework of the Innovation Union Scorecard[15] is shown below. The framework consists of three blocks, eight dimensions, and 25 indicators. Based on the average innovation performance, the member states are grouped in four performance groups. These performance groups are Innovation Leaders, Innovation Followers, Moderate Innovators, and Modest Innovators. Innovation leaders demonstrate that their research and innovation system is balanced across all eight innovation dimensions.[15]

European Union Innovation Scorecard Framework: Blocks (3), Dimensions (8), and Indicators (25)

Summary Innovation Index (SII)

> Enablers
>> Human resources
>>> New doctorate graduates,
>>> Population aged 30–34 when tertiary education
>>> Youth with at least upper secondary education
>> Open, excellent, attractive research systems
>>> International scientific copublications
>>> Top 10 percent most cited scientific publications
>>> Non-EU doctorate students
>> Finance and support
>>> R&D expenditure in the public sector
>>> Venture capital
> Firm activities
>> Firm investments
>>> R&D expenditures in the business sector
>>> Non-R&D innovation expenditure
>> Linkages and entrepreneurship
>>> SMEs innovating in-house
>>> Innovating SMEs collaborating with others
>>> Public-private copublications
>> Intellectual assets
>>> PCT patent applications
>>> PCT patent applications in societal challenges
>>> Community trademarks
>>> Community designs
> Outputs
>> Innovators
>>> SMEs with product and process innovations
>>> SMEs with marketing or organizational innovations
>>> High-growth innovative firms
>> Economic effects
>>> Employment in knowledge-intensive activities
>>> Medium and high-tech product exports
>>> Knowledge-intensive services exports
>>> Sales of new-to-market and new-to-firm innovations
>>> License and patent revenues from abroad

Global Innovation Index						
Innovation Input Subindex					Innovation Output Subindex	
Institutions	Human Capital Resources	Infrastructure	Market Sophistication	Business Sophistication	Knowledge and Technology Outputs	Creative Outputs
Political environment	Education	ICT	Credit	Knowledge workers	Knowledge creation	Creative intangibles
Regulatory environment	Tertiary education	General Infrastructure	Investment	Innovation linkages	Knowledge impact	Creative goods and services
Business environment	Research & Development	Ecological sustainability	Trade & competition	Knowledge absorption	Knowledge diffusion	Online creativity

FIGURE 37-4 Global Innovation Index framework. (*Source*: GlobalInnovationIndex.org)

Measures of Innovation: Global Innovation Index

In collaboration with the business school INSEAD, World Intellectual Property Organization, an entity of United Nations (WIPO) published the Global Innovation Index (GII). The GII measures the extent of a country's innovation and how it is integrated into its political, business, and social aspects. GII looks into capability of economy to innovate, and its innovation performance. GII is a dynamic innovation index that seeks to improve and update the way innovation is measured.[16]

The GII consists of two subindices, Innovation Input Index and Innovation Output Index.[16] Each subindex is made of pillars. The input subindex pillars represent institutions, human capital resources, infrastructure, market sophistication, and business sophistication. The output subindex is built on two pillars: knowledge and technology outputs, and creative outputs. Each pillar is divided into three subpillars and each subpillar is a collection of individual indicators (Fig. 37-4). GII includes 84 indicators in its 2012 edition.[17]

The GII report publishes national performance for various pillars and subpillars, and analyzes the extent of innovation by regions, income, and population. General finding is that average innovation ranking increases with income levels of a country. Reviewing the report by regions, North America leads in innovation followed by Europe, Southeast Asia, Northern Africa and Western Asia, Latin America and the Caribbean, Central and Southern Asia, and sub-Sahara Africa.

The GII ration between the best to worst region is 2.2, but there does not appear to be a significant difference among regions for innovation efficiency. Implying that it is not the effort of regions transforming innovation inputs to innovation outputs, instead it is a function of the inputs, that is, resources. In other words, GII is a function of a region's resources not its ability to transform its resources into output value. Efficiency for most regions is less than one highlighting the need for improving innovation process and tools, or advancing the innovation science.

References

1. "The Climate for Creativity, Innovation and Change." The Creative Problem Solving Group, Inc. www.cspb.com.
2. Studt, T. "Measuring Innovation… Gauging Your Organization's Success." *R&D Magazine*. www.rdmag.com.
3. *National Innovation Initiative, Innovate America Report*, 2nd ed., Washington, DC., 2005. www.Compete.org.
4. Walcott, Robert P. "The Innovation Radar." Perspectives to Date, 2003. http://www.technologymanagementchicago.org/meetings/presentations/03-10.pdf.
5. Jensen, Paul H. and Webster, E. "Firm Size and the Use of Intellectual Property Rights." *Economic Record*, 82 (256): 44–55, March 2006.

6. Rogers, M. *The Definition and Measurements of Innovation*. http://ecom.unimelb.edu.au /iaesrwww.home.html. The University of Melbourne, Australia, 1998.

7. Gupta, P. "Six Sigma Business Scorecard." New York, NY: McGraw-Hill, 2003.

8. Basili, V.R. "Software Modeling and Measurements: The Goal Question Metric Paradigm." Computer Science Technical Report Series, CS-TR-2956 (UMIACS-TR-92-96). College Park: University of Maryland, 1992.

9. The Advisory Committee on Measuring Innovation in the 21st Century, Innovation Measurement, Tracking the State of Innovation in the American Economy, January 2008.

10. Stone, A., Rose, S., Lal, B., and Shipp, S. Measuring Innovation and Intangibles: A Business Perspective. Conducted under contract for the Bureau of Economic Analysis, IDA D-3704, Institute for Defence Analyses, 2009.

11. Schaaper, M. Measuring China's Innovation System National Specificities and International Comparisons. Working Paper, OECD Directorate of Science, Technology and Industry, 2009.

12. Goh, C., Deichmann, U., and Fitzgerald, B. *Russia: Reshaping Economic Geography*. Report No. 62905_RU, World Bank, 2011.

13. Gokhberg, L. and Roud, V. The Russian Federation: A New Innovation Policy for Sustainable Growth. Higher School of Economics, Russian Federation, Global Innovation Index, 2012.

14. http://knowledgecommission.gov.in/downloads/documents/NKC_Innovation.pdf. Innovation in India Report, 5, 2007.

15. Hollanders, H. "Measuring Innovation: The European Innovation Scorecard." UNU-MERIT, Maastricht University, 2011.

16. Jewell, C. "Global Innovation Index 2012." *WIPO Magazine*, April 2012.

17. Benavente, D. and Dutta, S. The Global Innovation Index 2012: Stronger Innovation Linkages for Global Growth. Innovation Union Scorecard, European Union, 2012.

Deployment

If method was the mother of all chapters in innovation, then deployment is surely the father. The steps involved in the deployment of an innovation program are numerous. While not all are necessary to invoke or renew in every situation, looking into the history of the organization will generally demonstrate that each step happened somewhere along the way. This section is organized logically for any type of life cycle approach. Starting with inspiration we move to strategy, followed by organizing, developing excellence, establishing an innovation culture, a measurement system that protects intellectual property, and finally, effectively launching an innovative product.

Like all great quests, the illustration for your innovation process might come from a prophet or muse. It's more likely, however, that it came from an employee inside your organization. In McGowan and Di Resta's chapter on the inspiration for innovation, they discuss the creation of an inspirational programmer company. Innovation is a process of inclusion, allowing people to express their ideas, inspirations, and aspirations. Creating the context in your organization with clear expectations for both the employee and the employer establishes the basis for inspiration. There's a reason why so many organizations in Silicon Valley are so playful. This is a direct effort to create an organization that frees the mind and inspires the soul. This chapter reviews the process of innovation and discusses motivational frameworks that you can apply to bring innovation to life in your organization.

In Dubey's chapter on strategy for innovation, he discusses a deployment strategy for innovation. His chapter delves into changes in mission and vision required of an innovative company. Understanding the capabilities of your organization from the perspective of competencies allows you to critically explore whether or not you really have the ability to be the next Apple. Samsung, for example, might be considered more innovative than Apple at the time of this writing; however, they use a different strategy to get there.

Armando and Gondim in their chapter on organization for innovation talk about organizational models of both past and future. They have an extensive section on the main characteristics of an organic, or innovative, organization versus one that is more mechanistic, or traditional. While organizations are all different, the lessons to be learned from this chapter can be applied broadly to almost any enterprise.

In her chapter, journey to innovation excellence, Andrea Long discusses her very personal journey through several consumer packaged goods companies and their successes and failures in creating long-term sustainable innovation. She points out the importance of building the right foundation for innovation as well as her work with innovation strategy. She discusses the importance of the right portfolio mix as well as product mix (or, as she calls it, strategic focus areas). The remainder of the chapter reviews the process for building an organization built for excellence in innovation.

Not to be overlooked is the value of the culture for organization. Maria Thompson and Nina Fazio write in their chapter about the importance of managing the change in culture required to support a company trying to innovate. These steps include understanding where you are, where you would like to be, and how to get there culturally. While similar in scope to change management, there are differences in moving to an innovation organization, because the transition is so much more dependent on the definition. Since so many organizations struggle with innovation in the first place, it can be akin to looking for a white zebra with black stripes in a herd of black zebra with white stripes. Poorly defined change merely gets you to another place and not necessarily the place you would like to be.

In Forsman and Temel's chapter, we again look at the question of measurement, but from the perspective of deployment. How do we tactically measure innovation? How do we define success? What are the dimensions of the measurement?

In the intellectual property chapter, Nabor and Parmelee discuss the challenges of protecting intellectual property. Anyone who has been paying attention to the news in the last several years understands that nearly as important as the creation of an innovation is the protection of that innovation. This is an international problem that takes on greater significance as the stakes get higher. Understanding your legal rights, responsibilities, and options is a must for all innovators.

We have chosen to divide product launch into two chapters. The first is written by Di Benedetto and covers the strategic aspects of a product launch, while the second is written by Almeida, Monteiro, and Santos, and specifically deals with tactical launch details.

Finally, we complete the section with a discussion of the Business Innovation Maturity Model (Carlson and Gupta). Essentially, this chapter is about the sustainability of your innovation organization. This chapter has roots in several parts of this book including program sustainability and several other maturity models. Of course, developing the measures of success for your innovation program is not easy, but working through the model helps mitigate the uncertainty in the process.

CHAPTER 38

Inspiration for Innovation

Heather McGowan and Ellen Di Resta

Introduction

Perspective: There Is a Place for Everyone

We view opportunities for innovation along a spectrum that encompasses the optimization of your current business, the development of new products and services for known and existing problems, and the discovery and formulation of new offerings for new needs that may require new business models. In this chapter we will explore the inspiration and motivation of innovation focused primarily on the formulation of new offerings. However, within an organization, we emphasize the need to balance the focus on inspiring and motivating innovation across the entire spectrum. The formulation end of innovation is not for everyone and it doesn't need to be.

Inspiration, Motivation, and Assessment

In this chapter we focus on inspiration in innovation. We firmly believe that it is nearly impossible to inspire a team to innovate without first assessing both where you are and who you are as an entity to set realistic strategic direction for the innovation team. It is also necessary to develop an understanding of the theories and drivers of motivation to identify, structure, and inspire the team members best suited for innovation explorations.

What to Expect from This Chapter

Why is there this intense focus on innovation? Today companies can no longer keep up with the competition by slowly evolving their current businesses. The life cycle of products is too short, customers have too many options, and global forces allow competition to come out of nowhere. Companies need to simultaneously optimize their current strengths to delight the customers today while striving to reinvent their leadership position with new products and services that meet the changing customer needs tomorrow.

This is a scary proposition for leadership and management. It means that if a company embraces innovation, then what is good today may no longer be good tomorrow. Rather than only offering incentives to employees to motivate them to do more of what they do today, companies also need to identify employees who are skilled and inspired to figure out what's next for tomorrow.

Where do you begin? We believe it is essential to understand the following: 1 where you are as a company, your expectations, the composition of your current team, and future talent needs, and 2 the motivational frameworks along with tools for inspiring and structures for rewarding innovation.

Once the context for inspiration is established, we believe that a well-understood problem is the catalyst for innovation.

In this chapter we present a series of models to illustrate our ideas that were informed by experts from academia, consulting, and corporate. All of the models are intended to emphasize similar points about the need to understand your core capabilities, your team talents, and the type of innovation the opportunity requires. While the points are similar, the focus and representation differ so that they will resonate with different levels of experience and viewpoint of the audience.

Define Context and Expectations for Innovation

What Is Innovation?

Simply stated, innovation is creating new value. That's a broad definition, and we view it on a continuum from optimizing the existing business, to developing new products and services for known needs, to the discovery and formulation of entirely new models to serve currently unknown customer and user needs.

Most companies were formed at one time around a need they could solve better than their competition. The company's unique abilities to solve this need are referred to as their core capabilities. Figure 38-1 depicts the core capabilities of development and optimization. Most companies continue to develop new products and/or services and optimize the financing, sales, manufacturing, etc. of those solutions to best meet their customers' needs and beat their competition. To illustrate our model, we use the example of the mousetrap. In the develop and optimize framework, there is always ample opportunity to innovate as you seek new ways to build a better mousetrap.

In Fig. 38-2, we offer the full spectrum of innovation, which includes development and optimization but adds two additional areas called discovery and formulation. Continuing with the mousetrap analogy, we add a front-end phase of discovery around the needs of our user and a deeper understanding of the problem itself. In this discovery phase we understand that we don't actually want to catch mice at all. From this key discovery or problem identification we move into the formulation phase where we begin to further frame the problem as to answer the right question: "How do we achieve mouse-free premises?" The difference between these two cluster types of innovation is the difference between building a better mousetrap and achieving a mouse-free premise. In this chapter we will focus on methods of structuring and inspiring your team in the discovery and formulation types of innovation.

Innovation: Optimization to Development to Formulation

In this chapter we speak to the need to clearly distinguish among the optimization team whose focus is to optimize core capabilities to out-execute the competition, the product development team that develops new products and services for known consumer needs, and the discovery and formulation team that seeks entirely new needs and related business models.

Key Consideration: Know Thyself

Organizations need to clearly define their goal for innovation. At the early stage of the innovation program this must include an honest assessment of the current context of their business, as well as their expectations for what they would like to achieve from an innovation program.

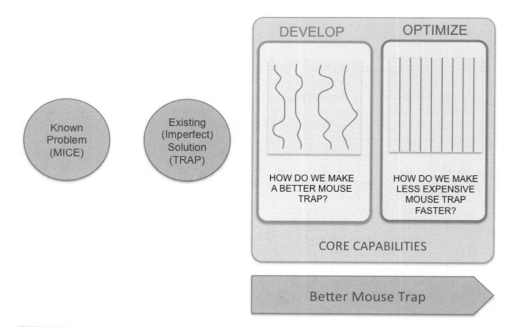

FIGURE 38-1 Core capabilities: develop + optimize.

DISCOVER
FORMULATE
DEVELOP
OPTIMIZE
HOW DO WE ACHIEVE MOUSE-FREE PREMISES?
WE DON'T WANT TO CATCH MICE. HOW DO WE DETER THEM?
HOW DO WE MAKE A BETTER MOUSETRAP?
HOW DO WE MAKE LESS EXPENSIVE MOUSETRAPS FASTER
Mouse-Free Premises
Better Mouse Trap

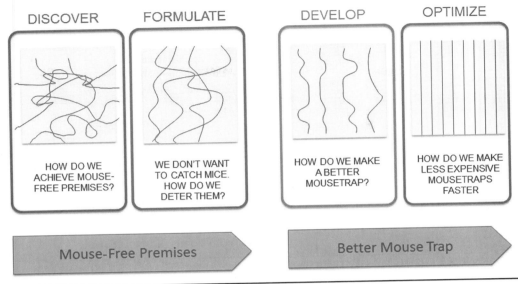

FIGURE 38-2 Spectrum of innovation.

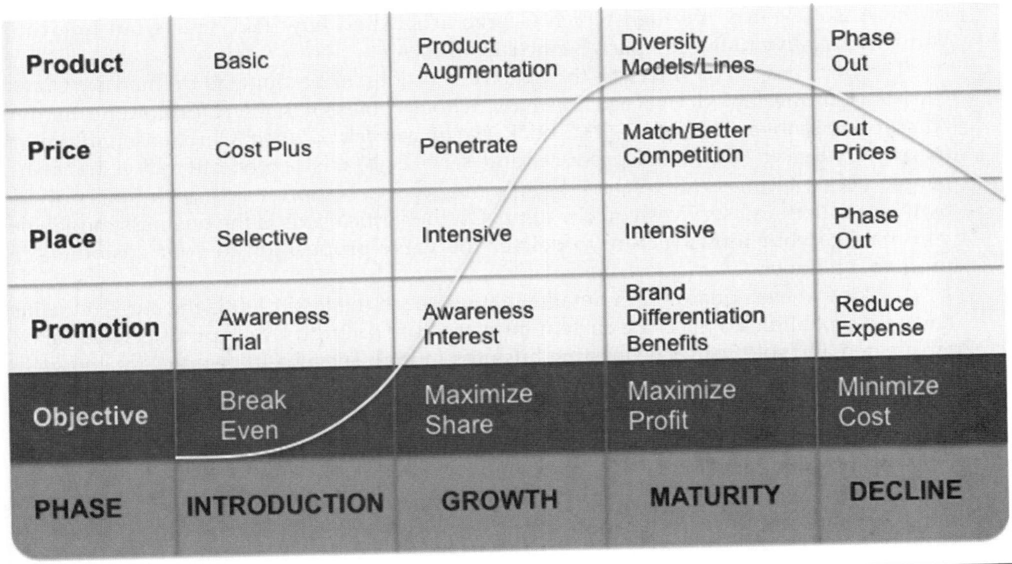

Product
Basic
Product Augmentation
Diversity Models/Lines
Phase Out
Price
Cost Plus
Penetrate
Match/Better Competition
Cut Prices
Place
Selective
Intensive
Intensive
Phase Out
Promotion
Awareness Trial
Awareness Interest
Brand Differentiation Benefits
Reduce Expense
Objective
Break Even
Maximize Share
Maximize Profit
Minimize Cost
PHASE
INTRODUCTION
GROWTH
MATURITY
DECLINE

FIGURE 38-3 Product life cycle. (This chart was offered by Dr. Sarah Beckman at UC Berkeley.)*

Are you in start-up, turnaround, or realignment mode or simply striving to sustain your success and leadership position? Is your industry mature? Is it in growth mode or in decline? Understanding the status of your company and the state of your industry is the key in determining where to begin.

Expectation: What Result Do You Want from your Innovation Efforts?

Given the status of your company, what type of innovation effort do you need at this time? We will illustrate this with a look at how Procter and Gamble (P&G) frames its expectations for innovation.

Story from the Field: Innovation at Procter and Gamble Procter and Gamble is an example of a company that fully understands the need to simultaneously build internal capabilities and external networks with their connect and develop model. P&G is very clear on what they are

*Dr. Beckman should be listed as the Chief Design Officer of the Institute of Design at UC Berkeley.

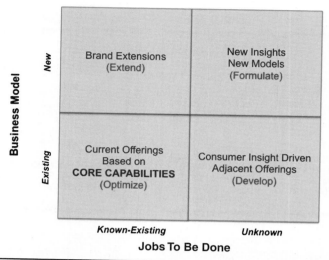

FIGURE 38-4   P&G innovation pathways.

good at and different ways to grow. George Glackin is an intrapreneur at P&G, and is responsible for building innovation capability within the company. Figure 38-4 shows an evolution of the Ansoff Matrix, through which George articulated how P&G maps out innovation pathways relative to the company's core capabilities.[1]

The horizontal axis refers to the spectrum of jobs to be done as defined by Clayton Christensen, the famed Harvard Business School Professor known for coining the term "disruptive innovation." In his 2003 book *The Innovator's Solution*, Christensen offered the following advice: "Don't sell products and services to customers, but rather try to help people *address their jobs-to-be-done*."[2] Simply put, jobs-to-be-done are the solutions provided by the products and services you consume. On the vertical axis is the business model, how you manage your infrastructure to deliver the value proposition to your customers in a profitable manner.

Looking at each quadrant in detail, what we see in the lower left is the current business and core capabilities. This is the current infrastructure through which products and services are created and sold using the existing business model. The jobs-to-be done are known and the business model exists. Innovation in this quadrant consists of new positioning and marketing messages, product improvements, or technological and manufacturing advances that evolve existing offerings. This quadrant is about optimizing your core capabilities to execute your business model in a way that beats your competition.

Moving to the lower right we are working within a known business model, but the jobs-to-be-done are unknown. Innovation in this quadrant is about harnessing consumer insights to develop or invent new products or services in your existing or adjacent domains.

The upper left quadrant is similar to the lower right quadrant in that it is also evolutionary. However, instead of using consumer insights to expand your portfolio of existing offerings, innovation in this quadrant is about extending your brands and/or expertise in the creation of new business models.

The upper right quadrant is where things get scary. Innovation in this quadrant requires the formulation of new consumer needs and insights as well as new business models to deliver. Companies that serve as examples of delivering this type of disruptive innovation include Netflix, Zappos, Southwest Airlines, and Apple.

An innovation process can be managed such that it delivers any or all of the types of efforts listed earlier. At this phase, once the entity status is defined and the innovation effort is understood, the leadership team needs to assess their internal capabilities to determine whether to *build, bridge,* or *buy* the capabilities necessary to deliver on the innovation.

Types of Innovation

Although often thought of in terms of tangible products as described in the mousetrap example, opportunities for innovation can have much broader reach. The P&G model below offers both new products (Swiffer, Fabreze) and new business models (Tide Dry Cleaners, Mr. Clean Car Wash). In 1998 the Doblin group identified ten types of innovation that cluster along the discover, formulate, develop, and optimize spectrum that they call business

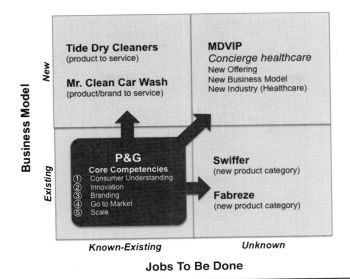

FIGURE 38-5 P&G innovation pathways.

formulation (profit model, network, structure, process); the more commonly thought of offerings (product performance and product system) and the experience (service, brand, channel, and customer engagement). In this chapter we encourage you to consider innovation as Doblin does, along this much broader spectrum.

Figure 38-5 shows how Procter and Gamble's recent efforts exemplify such reach.

Innovation at Procter and Gamble   P&G has spent the better part of the last decade honing their internal innovation capabilities, and building their external network of collaborators. They have built core capabilities in consumer understanding, innovation, branding, go to market, and scale, in which they can out-execute their competition. These core capabilities also provide a solid foundation from which they can build new pathways to innovation with the help of strategically selected partners.[3] In the lower right quadrant they have partnered with consulting firms such as IDEO, Continuum, and others to gather new consumer insights to create new product and service categories such as Fabreze and Swiffer. In the upper left quadrant, they have extended their brand equity in new business models, partnering with consulting firms such as Innosight for programs like Mr. Clean Car Wash and Tide Dry Cleaners. The upper right remains the Wild West with the example of their recent acquisition of MDVIP, a concierge healthcare offering in which the customer pays a premium for first class primary care attention from their doctor. This represents a new industry (health care), a new value proposition (premium-priced wellness care), and a new business model (customer funded vs. insurance funded).

While this new line of business is radically different from existing or adjacent offerings, it is clear that P&G's core competencies will provide the necessary executional excellence to ensure that the new proposition succeeds. The fundamental question here is how many of your core competencies can you leverage in these new opportunities, and not select them at random.[4]

We will now look at how to structure a process to achieve these types of innovation.

Process of Innovation

What Type of Innovation Program Should Be Established?

A successful innovation process needs to be managed with the same rigor, accountability, and attention to detail as any other work being done in the company. As mentioned earlier, however, innovation spans a broad spectrum, which can make it difficult to develop a process for it.

In Fig. 38-6, we expand on the mousetrap example from Fig. 38-1 and Fig. 38-2 to illustrate the types of knowledge each type of innovation requires. Below the dotted line is the known, where the activity is focused on creating incremental, evolutionary improvements to the product or service or means of delivering the known value. Above the dotted line you are crossing into the unknown, seeking new ways of meeting a customer's needs perhaps with solutions, products, or services that are entirely new.

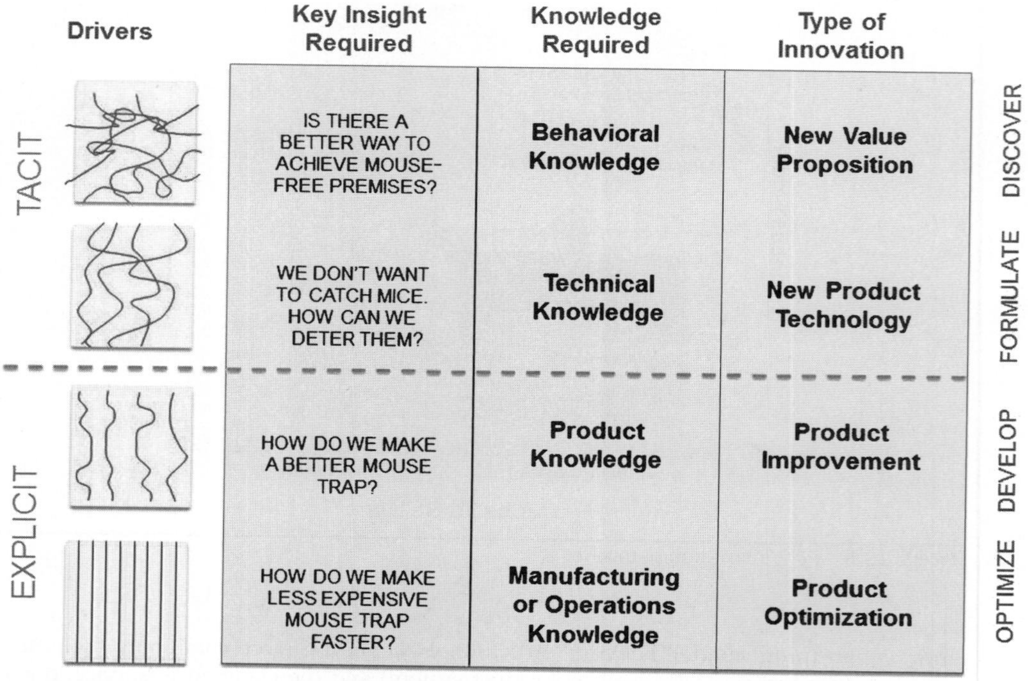

Drivers	Key Insight Required	Knowledge Required	Type of Innovation	
TACIT	IS THERE A BETTER WAY TO ACHIEVE MOUSE-FREE PREMISES?	**Behavioral Knowledge**	**New Value Proposition**	**DISCOVER**
	WE DON'T WANT TO CATCH MICE. HOW CAN WE DETER THEM?	**Technical Knowledge**	**New Product Technology**	**FORMULATE**
EXPLICIT	HOW DO WE MAKE A BETTER MOUSE TRAP?	**Product Knowledge**	**Product Improvement**	**DEVELOP**
	HOW DO WE MAKE LESS EXPENSIVE MOUSE TRAP FASTER?	**Manufacturing or Operations Knowledge**	**Product Optimization**	**OPTIMIZE**

FIGURE 38-6   Four dimensions of innovation.

Explicit Drivers for Evolutionary Improvement

The lower left box in Fig. 38-6 presents a challenge that is very familiar to people who work in the manufacturing process of most organizations. The product being produced doesn't change; the benchmarks for success are clear and measurable.

The box above it presents a challenge that may be familiar to people who work in the new product development process of most organizations. The solution is not the same as what was previously produced, although the type of product is the same. The benchmarks for success may not be initially evident, but can usually be discovered through direct feedback from or observational research with current consumers.

The type of innovation represented by both of these boxes can typically be managed from within an organization's current manufacturing or product development processes. Where it gets interesting is in the two boxes at the top. The dotted line is offered to delineate between the known and unknown or between evolutionary product developments to revolutionary innovation explorations.

Tacit Drivers, Insights, and Revolutionary Innovation

Above the dotted line, the end solution may leave both the mouse and the mousetrap behind. We are looking to create an entirely new value proposition that may require new customer behavioral insights or new technology development. Methods to acquire the knowledge required for these types of innovation may or may not exist. The innovation process referenced in the rest of this chapter has been developed to provide a framework for the acquisition and application of new technical and behavioral knowledge to achieve our innovation goals.

Process of Innovation

There are many models out there for design thinking and innovation process. For the purposes of this chapter, we prefer the simple four-quadrant framework as articulated later by Sara Beckman, PhD and Michael Barry (see Fig. 38-7) in their seminal work on *Innovation as a Learning Process: Embedding Design Thinking*.[5] The process is comprised of four main phases:

1. **Understand:** observing to understand the market and the user needs
2. **Frame:** extracting insights and framing the challenge
3. **Create:** developing ideas that satisfy the frameworks
4. **Deliver:** building the ideas into experiences to test and then deliver value

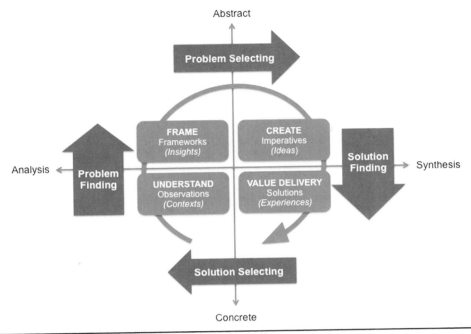

FIGURE 38-7 Beckman/Barry/Owen model for innovation process.

The planned progression is one from problem framing, to problem selecting, to problem solving, and finally to solution selecting. It is an iterative process with continual assessment often necessitating returning to prior phases, such as when new insights inspire questions that were not initially considered or when testing a solution with the users yield negative responses.

Understanding Your Team: Problem-Solving Preferences and Innovation Roles In thinking about how to assemble the innovation team, Beckman and Barry have correlated Kolb's Learning Styles Inventory to the process as shown in Fig. 38-8. In applying the Kolb assessment some interesting insights emerged.

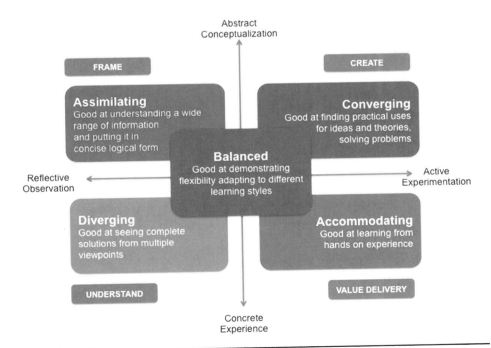

FIGURE 38-8 Beckman/Barry/Kolb model for learning styles and innovation.

Beckman and Barry found that MBA students have a higher tendency for directing their thinking toward the converging quadrant.[6] When we worked with Dr. Beckman at Philadelphia University in the creation of the Kanbar College of Design, Engineering, and Commerce, we screened hundreds of freshman design, engineering, and business students in the Kolb's Learning Styles Inventory and found that the designers and engineers tended to dwell in the lower half of the model depicting concrete thinking. The business students tended to gravitate toward convergence, similar to the MBA students. In speaking with Dr. Beckman about her work, she notes, *"It's helpful to think about this as it relates to your process. Often you don't have the opportunity to perfectly craft your teams but it is good to know what you are working with. For example, in our research on new product development teams, we've learned that when you have more than one converger on your team there is a statistically lower team satisfaction rating. We believe that this is because the convergers rush to close, argue about their different solution ideas, and can't bring themselves back into the framing space to determine whether or not they addressing the right problem."*[7]

An innovation team needs to strive for a balance between these styles to ensure that new insights are fully assimilated, and new solutions are fully explored and vetted in the consumer's context.

Motivational Framework

What Is Inspiration?

The word inspiration is from the Latin word *inspirare,* meaning "to blow into," as in to instill something into the heart and mind of someone. When people talk of being inspired this essence seems to come from within them.

Although related, motivation is different from inspiration, as its source is external and it is often used as a means to use external influence to tap into personal ambitions and values. It is often expressed as an exchange. Some simple examples include: if you work extra hard, I'll pay you more; if you get good grades, you can go to a better college, etc. It's clearly good to have a motivated work force.

You know how to structure your teams, motivate, and reward them to deliver on the business you are leading today but how do you structure teams, motivate them, and inspire them to discover the leading innovations for tomorrow? This is where we need to delve more deeply into the correlation between motivation and inspiration.

For this we look to the work of Nitin Nohria, PhD and Boris Groysberg, PhD of Harvard Business School, and Linda-Eling Lee, PhD of the Center for Research on Corporate Performance from their article on employee motivation.[8] We also look to the work of psychologists Timothy Butler and James Waldroop of Harvard Business School from their article on life interests and job sculpting for employee retention.[9] In the next section we will summarize key points from both sources.

Motivational Drivers

The first source by N. Nohria, B. Groysberg, and L-E Lee presents a new model for employee motivation based on cross-disciplinary research from the fields of neuroscience, biology, and evolutionary psychology. It defines four basic emotional needs or drives that underlie everything we do.

The Drive to Comprehend

Some people are driven to make sense of the world around them through creation of frameworks and theories that make events meaningful and valuable. In Daniel Pink's book *Drive,* he concludes that the drive to comprehend is the most resilient motivator. These people want to be challenged by their jobs, to grow and learn.

The Drive to Bond

Some people extend common kinship bonding to larger collectives such as organizations, associations, and nations. We find that many experiences in innovation are interdisciplinary, cross-functional, and as a result collaborative. These collaborative experiences often result in a renewed sense of teamwork and bonding in the organization.

The Drive to Acquire (aka the carrot)

Some people are driven to acquire tangible goods such as food, clothing, housing, and money, but also intangible goods such as experiences or events that improve social status. The drive to acquire is the most commonly used lever in many organizations.

The Drive to Defend (aka the stick)

This drive is rooted in the basic fight or flight response to real or perceived threats that is common to most animals. The drive to defend includes defending your role and accomplishments. Fulfilling the drive to defend leads to feelings of security and confidence. For the purposes of our discussion and in our framework the defend drive is included, but we believe to create an environment for innovation, the drive to defend must be well-met.

The authors suggest that these four drives must all be addressed, as there is no hierarchy among them. For example, you can't just pay people more money and expect them to be happy if the organization is not perceived to treat people fairly. These are levers that organizations can manage, and so implementing a strong, holistic balance is important to establish a healthy culture in which employees will be motivated to work.[8]

DELI: Leveraging Life Interests in Vocational Performance

The second source we will reference by T. Butler and J. Waldroop,[9] describes the impact of DELIs or deeply embedded life interests. The authors describe that DELIs are not the same as hobbies, nor are they the objects of topics of enthusiasm. Rather, they are defined as long-held, emotionally driven passions, and they impact the type of work people seek, much more strongly than the topic of the work. We see them as the source of internal inspirational drivers. To quote Butler and Waldroop

> *Life interests start showing themselves in childhood and remain relatively stable throughout our lives, even though they may manifest themselves in different ways at different times.*

They define eight DELIs as listed below:

Theory Development and Conceptual Thinking

These people love thinking and talking about abstract ideas. They love the "why" of strategy more than the "how." They may enjoy business models that explain the reasons behind the competitive position of a business. In an innovation process, these people are likely to gravitate to the front end thriving on uncovering insights, finding problems, and framing the innovation pursuit.

Creative Production

These people love beginning projects, making something original, and making something out of nothing. This can include processes or services as well as tangible objects. They are most engaged when inventing unconventional solutions. In an innovation process, these people may thrive on the ideation phase, creating multiple solutions to the identified problems.

Application of Technology

These people are intrigued by the inner workings of things. They may be engineers, but even if not, they like to analyze processes, "get under the hood," and like to use technology to solve problems (business or technical). Like those engaged by creative production pursuits, those intrigued by the application of technology are often also found in the problem-solving, ideation, and solution-selecting end of the process.

Quantitative Analysis

These people love to use data and numbers to figure out business solutions. They may be in classic quantitative data jobs, but may also like building computer models to solve other types of business problems. These people can fall into two camps, descriptive and prescriptive as data analysis can both uncover new opportunities as well as optimize the existing operations. Google is a great example of a highly innovative company that relies extensively on quantitative analysis to uncover new forms of value such as the Google Flu project, which tracks pharmacy purchases and Google searches for flu systems to map and predict cold and flu outbreaks globally.

Influence Through Language and Ideas

These people love expressing ideas for the enjoyment of storytelling, negotiating, or persuading. This can be in written or verbal form, or both. They also enjoy thinking about their audience and the best way to address them. In an innovation process these people can be very valuable in engaging consumers with stories and depictions of what is possible to gather new insights. On the optimization end of the existing business operations, those engaged in influencing through language and ideas can be found in all forms of deal making from sales, negotiations to even legal.

Counseling and Mentoring

These people love teaching, coaching, and mentoring. They like to guide employees, peers, and even their clients to better performance. It is ideal to have the innovation leaders have some passion for this interest as it will help them develop the team through the ambiguous process of innovation garnering the best performance out of the collective raw talent of the team.

Managing People and Relationships

As opposed to counseling and mentoring people, these people live to manage others on a day-to-day basis. They are less interested in seeing people grow, and more interested in working with them to achieve the goals of a business.

Enterprise Control

These people love to run projects or teams and control the assets. They enjoy "owning" a transaction or sale, and tend to ask for as much responsibility as possible in work situations.

Butler and Waldroop suggest that DELIs drive job satisfaction far more strongly than whether or not the employee has strong skills in their job. They cite several examples of people who were excelling at their jobs, but were ready to leave their companies to seek out more fulfilling jobs that more closely matched their DELIs,[9] which underscores the importance of alignment among them.

Finding the Motivational and Inspirational Levers for Formulation, Development, and Optimization

We found a strong correlation among the life interests and motivational drivers. We believe that to create an environment that fosters innovation, the fundamental needs must be met before others can be considered.

We will now look at these linkages as they relate to the spectrum of innovation as shown in Fig. 38-9.

Revolutionary Innovation: Discovery and Formulation

We believe those charged with discovering the unknown of today are motivated by life interests based on inquiry and discovery. To that end, on the formulation end of the spectrum it is best to engage team members with life interests in theory, conceptual development, creative production, and applications of technology with the drive to comprehend. Those with interests in quantitative analysis can be very helpful if the analysis uncovers new opportunities. Similarly those with interests in language and persuasion can be instrumental

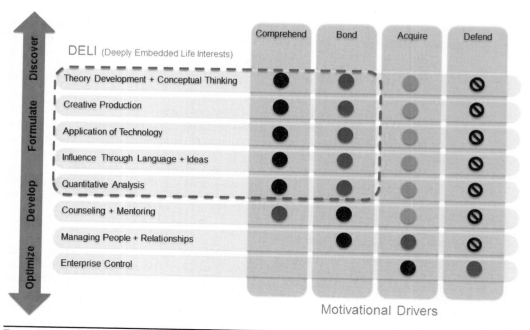

Figure 38-9   Motivational framework for innovation.

in their use of storytelling to help customers, users, and partners discover their unmet need or pain points. The drive to bond often comes up as a strong motivator in innovation projects as they are inherently collaborative, bringing together diverse user groups or entity functions to find new opportunity. This process of discovery often results in new feelings of belonging and pride about the entity.

Evolutionary Growth: Development

While creating new products is innovative, we consider it evolutionary innovation because you are often solving known problems with new solutions. Those with life interests in creative production are often happy in product or service development as are those with interest in applications for technology and persuasion with language and ideas. The drive to comprehend, bond, and often acquire are found here as well.

Incremental Improvements: Optimization

We refer to optimization as the pursuit of improving your existing business offering with a faster, cheaper, stronger, better form of your existing product or service solution. It is often easier to understand success and failure in this realm. As a result it is often easier to tangibly reward performance around management, enterprise control, mentoring and counseling, and the aspects of quantitative analysis that improve existing performance, as well as the use of language and persuasion in the known such as in selling existing products and services.

When planning the innovation process and building the team, the key is to understand if you are seeking revolutionary innovation, evolutionary product development, or optimization to know how to best compose, lead, and motivate your team. These factors should be considered in conjunction with the assessment of the current context and setting the expectations for innovation for your company.

Bringing Innovation to Life

"It is not the strongest of species that survives, nor the most intelligent that survives." [Rather] "It is the one that is the most adaptable to change."

Charles Darwin (1809–1882)

Mapping Your Innovation Space

As consultants, we find that many embark on innovation and design thinking workshops looking simply at consumer insights without, first, an assessment of the resources their team has to deliver and sustain innovation. If outsourcing innovation, consulting firms can deliver the needed innovation for the moment. However, without an assessment of how they will fit into the current context, or fill the current capability gaps, their ability to help to achieve innovation goals will fall short of expectations.

For this reason, it is necessary that a company defines the innovation space in which they will play. The innovation space is depicted in Fig. 38-10. This model shows how all of the factors we have discussed come together to define an innovation space that will take current context and expectations into account, as well as consider team motivation and alignment.

Innovation Process Implementation

When we talked about the four dimensions of innovation (see Fig. 38-7), we mentioned that the bottom two boxes could most likely be accomplished with current processes. However, the top two boxes will require a process that is more loosely managed than that of the typical development process. A good innovation implementation should accomplish three goals.

First, it should provide points of integration with the broader organization. Innovation teams exist to meet the future goals of the organization. Their work will ultimately need to feed the market and/or supply pipelines at a future date, and therefore must be relevant toward that end.

Second, it should also provide necessary separation from the influences of the broader organization. They need to be able to embrace the positive constraints that will define their project, unencumbered by existing constraints that would otherwise be imposed upon them by default. It's important that the process facilitates the former and discourage the latter.[10]

Third, a successful implementation should provide an infrastructure that will support the innovation process. We will look at the space and structures, rewards, and culture that will support innovation and inspire the people who work on innovation projects. But first,

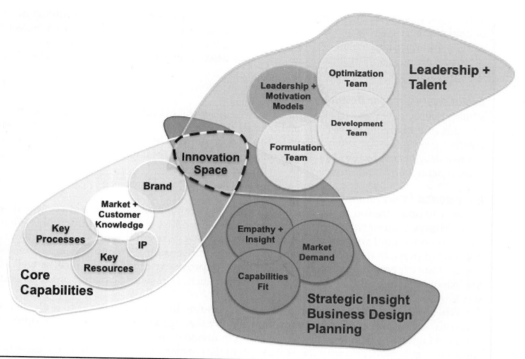

Leadership + Talent

Optimization Team

Leadership + Motivation Models

Development Team

Formulation Team

Innovation Space

Brand

Market + Customer Knowledge

Key Processes

IP

Key Resources

Core Capabilities

Empathy + Insight

Market Demand

Capabilities Fit

Strategic Insight Business Design Planning

FIGURE 38-10 Finding your innovation sweet spot.

we will walk through Beckman's process by using the development of Procter and Gamble's Swiffer as an example.

In addition to Beckman and Barry's four-phase framework, we have added two steps that provide important points of integration with the broader organization. The first step is called strategic direction, which aligns the work of the innovation team with the needs of the organization, and a last step is called support, in which the larger organization assimilates the innovation team's work into the broader infrastructure.

Note on the Iterative Process

The process depicted below is not linear and it is not predictable. It may require several iterations to get to a new offering from the initial assessment of a broad opportunity space. For example, you may go through the process once to get to the level where the external sources also inform the strategic direction. You may then go through the process again with a focus on specific opportunity spaces, and then again within a selected opportunity space with a focus on identifying the right offering. Famously, Steve Jobs spoke of his time away from Apple after he was fired in 1985 and he created NeXT computers. By most measures, NeXT was a massive failure, but Jobs repeatedly stated that he learned everything that made the second coming of Apple possible. Your iterations may not be comprised of numerous failed companies, but it should be kept in mind that iteration and taking several passes through the process to hone an idea is time well-spent.

In addition, the intent of the process is to guide the thinking and movement of the team's progress from a completely blank page to a refined opportunity for the organization to develop. However, it does not define specific tasks or tools that should be used along the way.

Strategic Direction

Align the innovation team with the business challenge to be met. This is a point of alignment with the broader organization, and the business goal must be articulated so that the challenge is clear. The team needs to understand the organization's assessment of who they are, where they are relative to the market, and the types of innovation challenges they want to pursue. Known hurdles in achieving the goal must be articulated, as well as the boundaries and scope of the work to be done.

> **Story from the Field: Procter and Gamble: Swiffer**
> *In the mid-1990s, Procter and Gamble CEO, John Pepper, famously announced that 50 percent of the company's revenue in the next 5 years would come from new product innovations. Given the company's resources and capabilities, it quickly became clear that the only way to achieve this goal was to align with external partners with complementary capabilities. One of the topic areas assigned was to develop new discoveries in the cleaning solution business unit. This was the initial direction for the project.*

Note on the Innovation Activities

The next three phases work iteratively together, and are a segment of work that needs to be unencumbered by the paradigms of the existing organization. The points of interaction may need to be less integrated than at the beginning and end of the project, and as visually articulated earlier, the innovation space integrates input and influencers from both the internal operations of the entity as well as external factors of the market and partners.

Understand the Market   This is achieved through primary and secondary research. Develop a deep understanding of consumer and stakeholder decision-making processes, and create criteria for successful solutions. This understanding becomes the constraints that must be embraced for the opportunity to be successful in the market.

> **Understand**
> The first step was to understand cleaning in its entirety. The team researched cleaning solutions, materials, and methods. They built sample flooring materials to try prototype solutions, and engaged in deep ethnographic research to understand how people cleaned floors.

Frame the Insights   Next step is to sort through the observations to try to find out what is important. This is often one of the most difficult parts of the process. Often the wrong product or service is created because the insights were not framed properly and the team sought to solve the wrong problem.

> **Frame**
> To summarize a very exhaustive research process, the outcome was twofold. First, it was noted that although cleaning the floor was a mundane chore, people stated that having a clean floor was a reflection on them as good or bad people—a much stronger emotional component than was originally anticipated. Second, after performing a task analysis the team realized that people were spending more time cleaning the thing they used to clean the floor than they were to clean the floor itself. Although consumers stated that they wished the process was faster, they were willing to spend time to ensure that it was as clean as it could be. Therefore, although speed was important, it was more important to ensure that the process was clean. This became the criteria through which new solutions were framed.

Create Opportunities   Identify business opportunities that satisfy the criteria and constraints and define specific attributes that deliver market value. Evaluate against desired experiences, and not necessarily against metrics applied to current products. This is one method for avoiding tacitly imposed, negative constraints.

> **Create**
> With the earlier hypothesis in mind, the team created many concepts to aid in floor cleaning. As suspected, those that were faster, but not perceived to be as clean were not valued as highly as those that were perceived to be at least as clean as conventional methods. In verbal concept tests, consumers also stated that they did not like the concept behind the Swiffer, since it used a disposable component. The team did not give up on the idea, because the analogy to a paper towel highlighted the fact that this may not be entirely true. Consumers were buying lots of paper towels, and no one was saying

that they wished they weren't disposable. This led the team to try more experiential evaluation methods. When using a prototype, consumers liked the fact that the disposable piece eliminated the mess. It made the process cleaner, in a way that was also faster. The dislike of disposables seemed to disappear—just like with the paper towel analogy. This highlights the importance of getting the context of the evaluation right.

Deliver Ensure the business model is aligned with the attributes of the offering that are delivering the value. Define the levers of value that will inform how the market will access the solution that satisfies the opportunity. Again, this needs to be defined specifically for the project, avoiding current paradigms and constraints that may be relevant to current business but irrelevant when considered in the context of a new value proposition.

Value Delivery

Procter and Gamble was primarily a consumables company, and did not typically make products with a durable component. It would be understandable if they decided to pass on the opportunity. However as leaders in supermarket distribution channels, they were able to secure shelf space for the slower-moving component in order to sell the entire system, and ensure that the disposable component would move quickly enough to make up for the durable space.

Support This is another point of integration with the organization. The innovation team needs to help the organization to understand and internalize the opportunity. The innovation team must guide technology, acquisition, development direction, and socialization within the organization. The organization needs to fully understand the opportunity, and begin to plan which existing capabilities and resources can be used to carry the innovation to market. Additionally, the support and reward structure must balance the need to reinforce relationships with channels and partners necessary to deliver the value to the market.

Support

Procter and Gamble supports their external teams by providing internal resources to be fully focused on the innovation project from beginning to end. During a project, while the external team is delving into the unknown, the internal team is instrumental at helping to navigate internal politics, and to socialize new ideas appropriately. They have welcomed the development of new metrics that were more appropriate to the needs of innovation, giving new ideas a greater chance to be developed into something of value. This happened with Swiffer, when in verbal concept tests consumers stated that they didn't want a disposable product component. However, P&G was receptive to developing experiential evaluations, which resulted in very positive reception of the idea. Once experiencing the cleaner experience, the negative perception of the disposable became irrelevant to consumers.

Innovation Process and Drivers

How does this process inspire innovation? First, it provides structure for the work that allows the team to see progress toward an end-goal, reinforcing the sense of accomplishment among the team and throughout the company. Without that, the team could become disassociated from the broader organization in ways that could be detrimental to their focus.

Directly relating the process for innovation to the motivational drives and DELIs of the team members for each phase of the process would be too prescriptive. However, it becomes apparent that the applicability of the different drives and DELIs can vary at different process stages. While it's clear that the first and the last phases, which require integration with the broader company, will favor those with strong influencing through language and ideas of DELIs.

Because the three middle phases which comprise the heart of the actual innovation project work will iterate, multiple DELIs will have the opportunity to be expressed throughout the course of the project. The team will need to bond and provide support during times where there is a high degree of ambiguity, which can be stressful. As the team will be making sense of highly ambiguous situations, they will need to comprehend the current world around them, and develop theories that will support new ideas and test their potential for success. There will be a wide open field of opportunity ahead of them, and the innovation team will often need someone who can quantify the most financially fruitful areas to play in the market.

Returning to our conversation with George, we learned more about how P&G thinks about adding talent to the innovation capability.

On Building Innovation Capability—George Glackin

We do a lot in the screening process. We ask applicants "Do you want to execute the status quo with excellence or would you prefer to navigate new space?" That is scary for someone who wants to win on the known spaces and we find many people self-select out when this is in an uncomfortable fit. At the same time, we know we need to find the right people and support them. We find and most have found if you look at companies like P&G, that it is essential to separate the innovation activity from the core business. The core team needs to feel supported and they can't feel that the new experiments are in any way taking away from them. At the same time, the innovation teams need their own space, their own leadership, and their own budget, so nothing as seen as taking away from the core business that generates the revenue and feeds everyone today.[11]

George reminds us that people who are well-suited to work in the discovery and formulation end of the innovation spectrum should be like-minded in that they work well with ambiguity, and can thrive without preexisting, clear-cut goals. They embrace the challenge of creating new goals as part of the project. However, within the general like-minded characteristics, there should still be a healthy balance of the different learning styles as Beckman suggests.

We will now look at the additional infrastructure elements that need to be implemented. They should be put into place to enable the innovation team to work effectively, without over constraining the process.

Space and Structures that Enable Innovation

Often, organizations talk about establishing an innovation program yet they fail to follow-through with tangible expressions of their intent. They may have a process that they follow intermittently, or they may fail to dedicate people fully to the task. One visible expression of the seriousness with which they intend to innovate is the establishment of a space for innovation. Another is the creation of a leadership role dedicated to innovation such as a vice president for innovation.

Space

A good space for innovation work does two things, and it is important to achieve a good balance of both. First, it enables the team to gather and share ideas, experiences, goals, artifacts of their research, etc. This space needs to be located such that the team can have lively discussions without being disruptive to others in the company. The group time is important, since this is the time the team needs to share experiences, and form connections that will feed into new ideas. This space needs to be a "project home" space that the team uses to immerse in the current thinking of the project. It needs to be dedicated to the team specifically for this purpose without needing to be cleaned out for other company uses.

Second, the space in which people are working ensures that people have quiet spaces to which they retreat. The retreat time is just as important, since people need quiet time to reflect on what they are learning and experiencing so they can bring fresh perspectives to the team. This can be regular desk space if that's how the office is currently set up, but it should be clear that people need both.

The emotional environment is also important. Many people want their companies to be innovative, but don't tie their reward structures to that goal. This point is illustrated in the story below:

Story from the Field-Enabling Innovation

Janet is a senior executive in an education company. She manages hundreds of millions in revenue, a very large team of product developers, and her executive bonus is based on a metric that calculates against top-line and bottom-line targets, her team's performance with a very small percentage at the sole discretion of her direct supervisor for additional contributions. Janet noticed that a large multimillion dollar content provider outside the education space could be an ideal strategic partner creating new lines of revenue for both the content provider and the educational partner. Janet independently navigated into the content provider, found her way to the decision makers, and brokered a multiyear, multimillion-dollar deal with long-term value to both entities. That year, her boss had little discretion to reward her for this game-changing partnership. Although it created innovations across multiple product lines in and out of her division for years to come, and was spoken about in the industry, she received little tangible reward because the structure could not reward this activity.

Reward Structures Both Encourage and Enable Innovation

In order to encourage this work, people need to know that they will be emotionally supported as they work on innovation projects. In the example above, it would have been great if Janet were rewarded for her contributions to the bottom line even though they didn't fit into the current work structure. At the very least, people need to know that they won't be fired for taking such a risk. In Janet's case, it worked out well, but what if the project had not turned out as it did? Creating new ideas that will be ultimately relevant to the organization from within a highly ambiguous context requires courage. As the process requires that the innovation team figure out new solutions, they need to figure out what the problem will be. They must be comfortable saying things or contributing ideas that turn out to be wrong. In many companies, the paradigm is such that people don't present ideas until they know they are right. In the innovation process this cannot happen, as the right idea cannot be known ahead of time.

If we think back to the motivational drives and DELIs, we can see that Janet's drives to acquire and comprehend were met. Although she didn't receive monetary rewards equal to the magnitude of the impact of the project, she did have a solid understanding of the value of this project to the organization. Due to her level in the company, she was able to satisfy her drive by pursuing the project and being able to claim the victory in terms other than monetary. However, one could see that with respect to the innovation team, this type of work could threaten those with the drive to defend. If Janet had expectations of monetary rewards, she could have felt resentment toward the perceived unfairness of the reward relative to others in the organization. On the other hand, if people in the company do not understand the work the innovation team is doing, they may perceive unfairness if they see that a team is rewarded for what may at times look like "failing" when that would not be tolerated in other areas of the organization.

Risk and Failure

Rewarding failure can also be a touchy subject that companies who want to take on innovation challenges must manage. The team needs to be able to feel comfortable exploring many different paths before the right path can be identified as such. Alternatively, the rest of the organization cannot work that way, as the day-to-day process needs to reward accuracy and efficiency.

One way to achieve this mind-set is for organizations to reframe their definition of failure. For innovation teams, true failure would be the lack of trying new ideas whose outcome is not yet known to be positive. Without these attempts, teams would only develop ideas that are incrementally better than what exist today. Instead, they should see this work as an ongoing series of experiments, from which a completely new, yet relevant, direction will be born. This mind-set should be incredibly motivating to people with DELIs that include creative production or the drive to acquire, as many new starts are necessary for this work to be successful.

Again, our conversation with George from P&G sheds light on the application of this idea.

On Risk and Failure—George Glackin

The challenge is that the refined models we use to test known offerings in existing markets don't apply in innovation. So there is a leap of faith. Despite the lack of testing models, you have to be clear about the difference between failure and mistakes. Failure is when you have a theory and you test it and mistakes are errors unconnected to a plan. Forgive failure and be careful of mistakes. Make sure there is a correlation between your failure rate and your learning rate. Are you able to understand the failure and build upon it? I am a fan of learn cheap, learn fast.[12]

Making Space for Innovation Activity

Establishing the right framework to support the environment and reward structures is also important, and can have many unanticipated side benefits. Work structures that encourage employees to pursue their passions at work are becoming more popular, with Google famously encouraging people to spend 20 percent of their time doing whatever it is that interests them. This structure has led to 50 percent of the company's new products and services, most notably Gmail[13]. It has come under fire in the last year or so with the closing of Google labs, as some outsiders perceive that 20 percent of each employee's time is a huge investment in random innovation that may not pay off as well as structured processes.

We would argue that these criticisms are missing the point. It is true that the Google 20 percent is not a structured innovation program. However, what it does is provide a way for people to feel empowered to pursue their passions without criticism. This enables them to feel comfortable enough to expose potential vulnerabilities at work, which is what is needed when people are assigned to an innovation project. It also enables employees to satisfy their drive to bond with each other, and physically express who they are to their companies. Further, the 50 percent of new products and services launched as a result of the 20 percent time may not all be big revenue drivers such as Gmail but there are undoubtedly essential key learnings in all those beta tests. While this program may not be a structured innovation program, it is an important element in establishing an innovative culture.

Tools and Resources to Inspire Innovation

Tools do not provide solutions. They provide a framework to structure thinking so that a new solution can be found. Most high-functioning teams use established tools as starting points, and they tailor them to the specific needs of their project as they go along. Below is a partial list of tools that we find to be useful in meeting these goals, a full inventory can be found at both the *IDEO Human Center Design* website here as well as at *The Design Exchange* built by Berkeley Expert Systems and Technology (BEST) and the Berkeley Institute of Design.

Descriptive Tools for Mapping What Exists

Business Model Generation Canvas

We find most people confuse business models, financial models, and business plans. We think the car analogy works best: the business model is the engine, the financial model is the fuel, and the business plan is the roadmap or where you plan to go. We find it is most helpful to first understand your engine. The business model canvas, created by Alex Osterwarlder and 470 collaborators including the authors of this chapter, is a strategic management and entrepreneurial tool comprised of the building blocks of a business model. The business is expressed visually on a canvas with the articulation of the nine interlocking building blocks in four cluster areas: *offering*—value proposition, *customer*—customer segments, customer relationships, *infrastructure*—distribution channels, key resources, key partnerships, key activities, *value*—cost structure and revenue model. Essentially, we think these nine blocks represent four clusters of consideration answering the fundamental questions every business should thoroughly understand, notably, (1) What problem are you solving (*offering*)? (2) For whom are you solving it (*customers*)? (3) How are you solving it? (*infrastructure*)? (4) Why are you solving it (*value/return*)? More information on the business model canvas can be found here.

Prescriptive Tools for Need Finding: Mining for Insights

Customer Profile—Empathy Map

The technique of creating a profile of your customer beyond the simple demographics of age, gender, income has been in use for some time. We in particular like the model created by the consulting firm Xplane that has been used in the book *Business Model Generation*; a link to the customer empathy poster can be found here.

Journey Map and Experience Evaluations

A customer journey map is a diagram that illustrates the steps your customer(s) go through in engaging with your company, whether it is a product, an online experience, a retail experience, a service, or any combination. Sometimes customer journey maps are "cradle to grave," looking at the entire arc of engagement. At other times, journey maps are used to look at very specific customer-company interactions.[14]

There are many resources available for information on Journey Maps, but it is difficult to determine an original creator of the concept. A good collection of examples and templates can be found at Pete Abilla's website www.shmula.com[15]

Additionally, here is a blog from *Harvard Business Review* which shows the work of Frog Design in creating a framework that explores the customers questions, barriers, motivations, and activities as they move through phases of research, purchase, use, and evaluation. We find this full spectrum, including post-use assessment, truly captures the full range of opportunities for innovation.

Time of Day

Time of day was developed by Professor Tod Corlett at Philadelphia University as a unique twist on the "day in the life" tool of journey or experience mapping. This tool focuses the participants not on a task or an experience but rather what happens or doesn't in 2- to 4-hour chunks in a person's day and what opportunities may appear. The exercise requires the division of the 24-hour day into 2- to 4-hour chunks such as midnight to 2 am and assigns specific focus to record what happens during those hours from which natural discussions of what needs might exist that are not currently met by products or services. A twist on this model might be to first consider the times of day then to map when users may use, need, or even think of your product, service, or brand and why.

Emotional Rollercoaster

Jump Associates bills itself as a strategy and innovation consulting firm that helps entities discover "What is next?" They specifically stay away from product or service consulting labels and constraints as their process is based in ethnographic and other forms of social research to uncover opportunities that may be products, services, processes, or entirely new business models. In their experience they noticed that the highly ambiguous nature of their work often left their clients on a wild ride of an emotional rollercoaster and as such they presented to their clients the likely stages they may go through in their process of discovery. We find that this same notion, similar to journey mapping, is a great opportunity to identify areas of high anxiety in a process and as such opportunities for new solutions.

7-14-28 Processes

The 7-14-28 process is a task-analysis assessment developed by Dean Michael Leonard at Philadelphia University. It involves breaking a process down into 7 tasks, then breaking it further into 14 tasks, and then another level further into 28 tasks. This exercise forces the participant to break down the components of a task until it becomes possible to see which tasks could be decoupled to form an improved process or where clear opportunities exist for new products or services.

Liveins, Shadowing, and Immersion Labs

Large companies such as Procter and Gamble and Nike have the resources to develop extensive labs to test their products and services in real-world environments. Some of the labs resemble the retail or home environment and gather extensive information about the product purchase or use. These labs are used to both test the known, new product launches, as well as to observe user behavior to find new opportunities for products, services, or systems.

Moccasins

Short of these resources at your disposal to create full immersion labs, you can gather a lot of information from assuming the role of your user walking in their shoes. "Moccasins," a term created by Jump Associates, requires the team to assume the role of the user whether they are a truck driver, ER nurse, or Target retail employee and actually experience their pains and joys in real time to uncover opportunities for improvement or new products, services, or systems.

Brains, Behavior, and Innovation (IIT)

The Brains, Behavior and Design group is dedicated to exploring how insights from the fields of cognitive psychology and behavioral economics can be used to design better products, services, experiences, and business strategies.

"The Brains, Behavior & Design toolkit began our foray into bringing behavioral economics out of the lab and into the design practice. The value of behavioral economic research was clear to us—designing for people means designing for the quirks of their brain, the challenge was how to bring this knowledge into field that is rich in methodology but short on time. We created the BB&D toolkit to help designers quickly understand and apply research on people's irrational decision making." This project was undertaken by six Master's of Design candidates as part of our year-long capstone project in 2009–2010, at the IIT Institute of Design.[16]

Ideating Tools

Associative or Mind Mapping

Mind mapping is a simple process of visually articulating the relationship between bits of information, namely concepts. Often this is used to take insights uncovered in early phases and unpack them to fully explore what those words or concepts mean and draw analogies to other solutions or more understood contextual references.

Image Board, Storyboarding, Role Playing

Image boards are collections of physical manifestations (image collages or product libraries) of the desirable (or undesirable if you are using that as a motivator) to help generate ideas, or to facilitate conversation with users about what they want. Storyboarding is a way of creating a narrative about the product use or the experience of consuming the service of the new proposed value proposition. Role playing is often used to uncover unmet needs kinesthetically.

Evaluation Tools: Assessing the Opportunity

AEIOU

The AEIOU framework for observations was developed by ethnographer Dr. Rick Robinson at Doblin in 1991. The acronym stands for activities, environments, interactions, objects, and users. It serves as a series of prompts to remind the observer to record the multiple dimensions of a situation with textured focus on the user and their interactions with their environment.

PESTEL Frameworks

The PESTEL framework is similar to a SWOT analysis. Whereas a SWOT analysis focuses on the strengths, weaknesses, opportunities, and threats that impact a specific business or market, the PESTEL framework focuses on the macroeconomic factors that influence a business. These factors are

1. **Political factors:** These refer to government policy such as the degree of intervention in the economy.

2. **Economic factors:** These include interest rates, taxation changes, economic growth, inflation, and exchange rates.

3. **Social factors:** Changes in social trends can impact on the demand for a firm's products and the availability and willingness of individuals to work.

4. **Technological factors:** New technologies create new products and new processes.

5. **Environmental factors:** Environmental factors include the weather and climate change. Changes in temperature can impact on many industries including farming, tourism, and insurance.

6. **Legal factors:** These are related to the legal environment in which firms operate.

Summary

To inspire innovation requires much more than simply allowing people to have fun, get crazy, and do anything they want. In this chapter we have shown the importance of understanding the current context within which your company operates, articulating your core capabilities, and defining your expectations for an innovation program. However, this is not enough. The discovery and formulation end of the innovation spectrum is very ambiguous, and can be an unsettling place for people who are not comfortable working under such uncertainty.

Understanding what the work entails is the first step in ensuring that the right people are aligned with the process. To that end, we have also offered a motivational framework to help you to understand how to inspire your development and formulation teams. This understanding in conjunction with a full understanding of the organization's expectations and capabilities for innovation will ensure that people are aligned to do the type of work best suited to their skills and interests.

We suggest that in order to get started, it is often best to do one of two things. On the one hand, your company can pull together a small, cross-functional team to begin the assessments, and possibly start on a small innovation project. On the other hand, it may be best to bring in outside help in facilitating these assessments and identifying skill and motivational gaps that may exist within the company. With either approach, the first thing to do is to be sure that you have a good understanding of where you are, and where you want to go before you start. A room full of koosh balls and post-it notes in which people get crazy for a day may be fun—for a day. But to inspire a sustainable innovation program that will yield results well into the future will require focused, hard work, and then the fun can begin.

References

1. Glackin, G. Interview with the author, November 29, 2012.
2. Christensen, C. *Innovator's Solution*. Harvard Business School Press, 2003.
3. futureworks.pg.com/about.asp.
4. Glackin, G. Interview with the author, November 29, 2012.
5. Beckman, S. and Barry, M. "Innovation as a Learning Process: Embedding Design Thinking." *California Management Review*, 50 (1), Fall 2007.
6. Beckman, S. and Barry, M. "Teaching Students Problem Framing Skills with a Storytelling Metaphor." *International Journal of Engineering Education*, 28 (2): 368, 2012.
7. Beckman, S. Interview with the author, November 28, 2012.
8. Nohria, N., Groysberg, B., and Lee, L-E. "Employee Motivation: A Powerful New Model." *Harvard Business Review*, July–August, 2008.
9. Butler, T. and Waldroop, J. "Job Sculpting: The Art of Retaining Your Best People." *Harvard Business Review*, September–October, 1999.
10. Di Resta, E. http://www.synapticsgroup.com, 2008.
11. Glackin, G. Interview with the author, November 29, 2012.
12. Ibid.
13. royal.pingdom.com/2010/02/24/google-facts-and-figures-massive-infographic/.
14. Richardson, A. HBR Blog, November 15, 2010.
15. http://www.shmula.com/customer-journey-map-continuous-improvement/10494/.
16. Toolkit authors. http://www.brainsbehavioranddesign.com/aboutus.html.

CHAPTER 39

Developing an Innovation Strategy for a Growing Firm

Rameshwar Dubey

"A great deal of business success depends on generating new knowledge and on having the capabilities to react quickly and intelligently to this new knowledge."

Richard Rumelt[1]

The above statement of a renowned professor of business strategy clearly implies that those firms who want to remain competitive and continue their business dominance need to be innovative, reactive, and agile.

What is innovation? This is perhaps one of the most deluding questions to me. There is a lot of confusion among practitioners regarding innovation. However, experts define innovation in one simple word, "strategy." It provides a cutting edge in the market and kills competition. How innovation can be a source of competitive advantage, the steps a firm needs to adopt to sustain competitive advantage, and how one can use contextual tools to formulate innovation strategy for their firm are the central propositions of this chapter, including

Strategy

- Definition
- Sources of firm competencies
- Sustainable competitive advantage

Systemic capability

- DNA of systemic capability
- Capability of competitors
- Industry dynamism
- Exploitation of firm resources
- Building customer responsiveness
- Innovation as a source of competitive advantage
- Antecedents of innovation
- Building an innovation framework
- Recommendations

Despite the complexity, continuous change, and uncertainty in today's market, there is a rational approach to innovation strategy available to companies today. In this chapter, our focus is to understand how to formulate an innovation framework that is most suitable for a given situation and how to exploit it for gaining competitive advantage. Before we drill down to the framework, let us first understand the fundamentals of strategy, competitive advantage, sustainable competitive advantage, sources of competitive advantage, systematic capability, how innovation can be a source of competitive advantage, and how to build a contextual framework using the innovation strategy model (ISM).

What Is Strategy?

Though there are several theories on the origin of the word strategy, we consider one theory that connects the origin of strategy with the Greek word "*strategia*," meaning generalship."[2] The amount of literature on strategy development is vast and growing with the passage of time. Strategy has been used loosely in present business literature, *which sometimes creates confusion in the minds of readers because, if it is a strategy, then why does it fail?* Despite the large amount of research on this subject, there is no single proven approach for strategy development. In one study, strategy is defined as a winning statement. As a result, a wide range of conceptual frameworks exist for the formulation and implementation of strategies.[3]

Peter Drucker appears to be one of the first to address strategy in a business context. In 1954, he spoke about it only in terms of answering the questions, "What is our business? And what should it be?" In 1962, Chandler was one of the few who made an attempt to define strategy as the determination of the basic long-term goals and objectives of an enterprise, and the adoption of a course of action and allocation of resources necessary for carrying out these goals. The first writers to focus on the concept of strategy in terms of its development and implementation were Andrews and Ansoff. Andrews defined strategy as the pattern of objectives, purposes, goals, and major policies and plans for achieving the goals.[4]

In a seminal work, Ansoff[5] viewed strategy as a composition of five integral components including market scope, growth direction, competitive advantage (unique opportunities in terms of product or market attributes), synergy generated internally by a combination of capabilities or competencies, and sourcing decisions. Henderson[6] proposed that the fundamental rule of strategy is to induce competitors not to invest in those products, markets, and services where firms expect to invest most. To achieve strategic victories, firms must use corporate resources to substantially outperform a competitor with superior strength.

Strategy definition is a deliberate search for a plan of action that will develop a business's competitive advantage in a way that compounds it. Strategy is the direction and scope of an organization over the long term. It ideally matches its resources to its changing environment.[7] In one seminal article in 1996, Porter has critically argued, "What is strategy?" In that article, he clearly states that in a highly competitive era, the sustainability of competitive advantage is more important than a simple or singular competitive advantage. *Under pressure to improve productivity, quality, and speed, managers have embraced tools such as TQM, benchmarking and reengineering, which resulted in operational improvements, but rarely have these gains translated into sustainable profitability,*[8] and gradually, these tools have taken the place of strategy.

In brief, there is no single definition of strategy. According to the author(s), strategy is the combination of

Strategic Intent

A firm must be clear about the nature of their existence and why it will continue to exist in coming years. A respected work by Theodore Levitt in the early 1960s titled "*Marketing Myopia*," reveals that many companies failed due to lack of *strategic intent*. This is a very important consideration for those firms who want to continue to be respected companies in years to come.

Mission Statement

The mission statement of a firm must be clearly defined. If we read the mission statement of any leading steel company, it is clearly written as, "*to be world's largest steel producer*." But if we critically analyze this statement, it clearly shows that even a steel company, after reaching to the top in its field, fails to sustain their competitive edge in an intensely competitive era where steel is itself being challenged by other materials that are replacing the demand for steel in the automotive and construction sectors.

If firm wants to continue to exist, then their mission statement has to be more generic and free from myopic views.

The mission statement must be

1. Rational, feasible, and implementable.

2. Clear, logical, and actionable.

3. Motivating for the firm's stakeholders including employers, employees, society, government, and investors.

4. Precise—neither too broad nor too narrow.

5. Unique and distinctive, to leave an impact in everyone's mind.

6. Analytical and results oriented; it should analyze the key components of the strategy.

7. Credible, all stakeholders should be able to believe it.

Vision Statement

A strategic vision is a description of a firm's future direction and business makeup. The focus of the firm's mission tends to be on the present, whereas the focus of a vision is on the company's future. A vision statement answers the question, "What will our business look like in 10 to 15 years from now?" A strategic vision is a blueprint of a firm's future, sometimes called future vector, highlighting market focus, product offerings, price decisions, and levels of service to be offered. Forming a strategic vision of what the firm's future business makeup will be and where the organization is headed is needed to provide long-term direction, delineate what kind of enterprise the company is trying to become, and infuse the organization with a sense of purposeful action. Strategic vision charts the course for the organization to pursue and create organizational purpose and identity. It spells out a direction and describes the destination.[9]

Sources of Firm Competencies

A firm's competencies arise from two complementary sources: its resources and capabilities. A distinctive competency is a unique strength that allows one company to achieve superior efficiency, quality, innovation, or customer responsiveness thereby to attaining a competitive advantage—the primary objective of a company's strategy. Consequently, the company will earn a profit rate substantially above the industry average. Details of these concepts and relationships are presented in the following sections.

What Is Competitive Advantage?

There is hardly any commonality in the available definitions for competitive advantage. In the early 1960s, Alderson[10] was one of the first to recognize that firms should strive for unique characteristics in order to distinguish themselves from competitors in the eyes of customers. Anderson critically argued that differential advantage can be achieved through lowering prices, selective advertising appeals, and/or product improvements and innovations.[11] Hall[12] asserted that, for a business to succeed in a hostile environment, it ought to either achieve the lowest cost or most differentiated position. Henderson emphasized that organizations that are able to adapt best or fastest will gain an advantage relative to their competitors. The message to managers is to respond to changes in the business environment by providing the best option in terms of product or service or be the quickest to respond to the needs of the market.[13]

Caves[14] was among the first to introduce competitive advantage, but without an explicit definition. Rather, the focus was on the commitment of resources to establish entry barriers that would enhance the performance of a firm.[15] In contrast to the competitive advantage terminology used by Caves, this conception of competitive advantage appears to be linked to a firm's being more competent in the market than its competitors.[15] Porter[16] asserted that competitive advantage comes from the value that firms create for their customers that exceeds the cost of producing that value. The key concern for a business is to capture that value which is greater than its cost. He also identified two types of competitive advantage: cost leadership and differentiation.

Competitive advantage is self-evident; there is no apparent need to define its exact meaning.[17] However, the differentiation based on key selling attributes of a product is the foundation of an advantage. This difference must be due to some resource capability that the firm possesses that competitors do not. Three conditions must be met for competitive advantage to have meaning: (1) that customers perceive differences between one firm's product or service attributes and those of its competitors; (2) the difference is the result of a capability gap between the firm and its competitors; and (3) that the aforementioned difference in attributes and the capability gap are expected to endure over time. Competitive advantage can even be defined as a firm that has gained an above-average return as compared to its competitors in its industry.

Barney[18] tried to define competitive advantage with strategy in mind. He stated, "A firm is said to have a competitive advantage when it is implementing a value-creating strategy not simultaneously being implemented by any current or potential competitors." Barney's definition is useful because it incorporates the idea that the creation of value, competition among firms, and the durability of that value are all fundamental to the conceptualization of sustainable competitive advantages.[15] However, it does not explicitly link competitive advantages to the resulting financial performance of a firm. In order to achieve competitive advantage, a company must implement a value-creating strategy.[18] Value creation is measured by the difference between value to the consumer and the cost of production.[16,18]

Sources of Competitive Advantage

According to Hill and Jones,[19] there are three factors that build competitive advantage: efficiency, quality, innovation, and customer responsiveness. These factors are highly interrelated.

Efficiency

A business is simply a device for transforming inputs into outputs. Inputs are basic factors of production such as labor, land, capital, management, and technological know-how. Outputs are the goods and services that the business produces. Efficiency can be measured as the ratio between outputs and inputs (efficiency = inputs/outputs). The more efficient a company is, the fewer the inputs required to produce a given output. Thus efficiency helps a company attain a low-cost competitive advantage. The most important component of efficiency for many companies is employee productivity, which is usually measured by output per employee. Holding all else constant, the company with the highest employee productivity in an industry will typically have the lowest costs of production. In other words, that company will have a cost-based competitive advantage.

Quality

Quality products are goods and services that are reliable in the sense that they do the job they were designed for and do it well. The impact of high product quality on competitive advantage is twofold. First, providing high-quality products increases the value of those products in the eyes of consumers. In turn, this enhanced perception of value allows the company to charge a higher price for its products. The second impact of high quality on competitive advantage comes from greater efficiency and lower unit costs. Less employee time is wasted making defective products or providing substandard services and less time has to be spent fixing mistakes, which translates into higher employee productivity and lower unit costs. Thus, high product quality not only lets a company charge higher prices for its products, but also lower costs.

Innovation

Innovation is the single most important building block of competitive advantage. Innovation can be defined as anything new or novel about the way a company operates or the products it produces. Innovation includes advances in the kinds of products, production processes, management systems, organizational structures, and strategies. Successful innovation is about developing new products and/or managing the enterprise in a novel way that creates value for customers. Successful innovation of products or processes gives a company something unique that its competitors lack. By the time competitors succeed in imitating the innovator, the innovating company has built up such strong brand loyalty and supporting management processes that its position proves difficult for imitators to challenge. Uniqueness lets a company differentiate itself from its rivals and charge a premium price for its products, or reduce its unit costs far below those of competitors.

Translating Competitive Advantage into Sustainable Competitive Advantage

The idea of a sustainable competitive advantage (SCA) surfaced in 1984, when Day suggested types of strategies that may help to sustain the competitive advantage. The actual term *SCA* emerged in 1985, when Porter discussed the basic types of competitive strategies that a firm can possess in order to achieve a long-run SCA.[11] Recall that, according to Barney,[18] a firm is said to have a sustained competitive advantage when it is implementing a value-creating strategy not simultaneously being implemented by any current or potential competitors and when these other firms are unable to duplicate the benefits of this strategy." He further asserted that "a competitive advantage is sustained only if it continues to exist after efforts to duplicate that advantage have ceased."

Flint[15] suggests short-term and long-term competitive advantages: short-term competitive advantages that would last through a business cycle and long-term competitive advantages that would last over more than one business cycle. If there were a competitive advantage that was or had the potential of being over the entire length of the foreseeable future, then one could label that as an "unthreatened competitive advantage."[15] Hoffman[11] states, "SCA is the prolonged benefit of implementing some unique value-creating strategy not simultaneously being implemented by any current or potential competitors along with the inability to duplicate the benefits of this strategy."

In brief, SCA is simply a competitive advantage that has been maintained for a period of time. The durability of competitive advantage depends on the maintenance of the advantage, the ability of competitors to duplicate the advantage, or the ability of competitors to somehow obtain the benefits of the advantage.

What Firms Need to Do for Innovation?

One thing is for sure, an innovation capability will not develop on its own; it needs to be consciously formulated, resourced, and driven into place. Some key questions are

- How can systematic innovation capability be built into the firm?
- Are staffs recognized and rewarded for their contribution to innovations?
- Do you have the right resources, skills, and systems in place to achieve systematic innovation?
- Does your firm want to grow through innovation?
- Do you have a strategy in place for innovation?
- Is there any performance measure mechanism available that can capture the impact of innovative measures on firm profitability or market share?

How to Build Systematic Innovation Capability?

Before we discuss the DNA of systematic innovation capability, let's understand what is meant by *systematic innovation capability.*

Systematic Innovation: The Ultimate in Adding Value

These are some important open-ended questions to ask:

- What business strategies do systematically innovative firms formulate and implement, and how do they go about this?
- How do these firms resource their innovation capabilities and activities?
- What measures of innovation and innovativeness are being used by innovation leaders?
- How do innovation leaders reward, recognize, and promote staffs?
- How do these companies drive culture and behaviors toward innovation?
- What barriers exist to doing even better in terms of innovation?
- How much and how is sustainable development being applied and used in innovation-oriented companies?
- Who in these firms are the critical contributors and catalysts of innovation?

DNA of Systematic Innovation Capability

Business strategy must be aligned with findings of innovative solutions to customers' problems. This orientation also allows firms to win the *war for talent*. Innovative companies offer infinite challenges and job satisfaction to superior performers.

Systematic innovation needs to be properly resourced and processes must allow for some experimentation, thinking outside the square, and taking carefully judged and calculated risks when needed. This includes stimulating creativity in all staffs, which is a training and skilling-up opportunity. Knowledge management is an opportunity here too, requiring systems capabilities and forums for exchanging ideas between staffs.

If a firm is serious about systematic innovation capability—more than just paying "lip service"—then innovation must be measured and be a central part of the business key performance indicator (KPI) system of the organization. Remember the saying that is, indeed, a truism: "What gets measured gets done!"

The business innovation measures are even more powerful when they are translated into personal incentives for all staffs. This means that staffs are recognized, rewarded, and promoted at least partly on their contribution to innovation capability and innovations. Without this, the staff can get away with not "buying in," which can defeat the purpose; whereas, with this factor in place, the staff achieves personal gains while doing great innovative things in the business and for clients. When business measures are strongly aligned with personal and team success drivers and incentives, a huge amount of energy is unleashed in the workforce!

Emphatic leadership of behaviors and culture works wonders. When we see our senior executives demonstrating some thinking outside the square, trying new initiatives, demonstrating and encouraging some sensible appetite for risk and tolerating the occasional failure as a learning opportunity, then fear is removed, people get on board with innovation, and it can become a reality.

Capability of Competitors

According to work done by Ghemawat,[20] a major determinant of the capability of competitors to rapidly imitate a company's competitive advantage is the nature of the competitor's prior strategic commitments. When competitors already have long-established commitments to a particular way of doing business, they may be slow to imitate an innovating company's competitive advantage. Another determinant of the ability of competitors to respond to a company's competitive advantage is their absorptive capacity, that is, the ability of an enterprise to identify, value, assimilate, and utilize new knowledge. Taken together, factors such as existing strategic commitments and low absorptive capacity limit the ability of established competitors to imitate the competitive advantage of a rival, particularly when the competitive advantage derives from innovative products or processes.

A company with a competitive advantage will earn higher average profits. These profits send a signal to rivals that the company is in possession of some valuable distinctive competency that allows it to create superior value. How quickly will rivals imitate a company's distinctive competencies? This is an important question, because the speed of imitation has a bearing on the durability of a company's competitive advantage. The critical issue is time. The longer it takes competitors to imitate a distinctive competency, the greater the opportunity the company has to build a strong market position and reputation with consumers, which is then more difficult for competitors to challenge.

Barriers to imitation are a primary determinant of the speed of imitation. They are factors that make it difficult for a competitor to copy a company's distinctive competencies. The greater such barrier to imitation, the more sustainable is the company's competitive advantage. Imitating resources is the easiest distinctive competencies for prospective rivals to imitate. They tend to be those based on possession of unique and valuable tangible resources because these resources are visible to competitors and can often be purchased on the open market. Intangible resources can be more difficult to imitate. Brand names are important because they symbolize a company's reputation. Marketing and technological know-how are also important intangible resources. Technological know-how is protected from imitation by the patent system. Imitating a company's capabilities tends to be more difficult than imitating its tangible and intangible resources. Since capabilities are based on the way decisions are made and processes managed deep within a company, it is hard for outsiders to discern them. To sum up, since resources are easier to imitate than capabilities, a distinctive competency based on a company's unique capabilities is probably more durable than one based on its resources.

Industry Dynamism

A dynamic industry environment is one that is changing rapidly. The most dynamic industries tend to be those with a very high rate of production innovation. In dynamic industries, the rapid rate of innovation means that product life cycles are shortening and the competitive advantage can be very transitory.[21] In summary, the durability of a company's competitive advantage depends on three factors: the height of barriers to imitation, the capability of competitors to imitate its innovation, and the general level of dynamism in the industry environment. When barriers to imitation are low, capable competitors abound; and when the environment is very dynamic, with innovations being developed all the time, then competitive advantage is likely to be transitory. On the other hand, even within such industries, companies can achieve a more enduring competitive advantage if they are able to make investments that build barriers to imitation.[19]

Exploiting Resources of Firm

The notion of firm resources was introduced into the strategic management field in the early 1970s when Ansoff[22] categorized skills and resources based on major functional area, that is, research and development (R&D), operations, marketing, general management, and finance.

But it wasn't until the mid-1980s that the concept of resources as a source of sustainable competitive advantage became dominant in the field of strategy. There has been a resurgence of interest in the role of the firm's resources as the foundation of firm strategy. A firm's resources can be defined as stocks of available factors that are owned or controlled by the firm. The final products or services are produced by using a wide range of other firms' assets and bonding mechanisms such as technology, management information systems, incentive systems, trust between management and labor, and more.[23] Similarly, in a similar work, resources are defined as the inputs into the production process, which are used as the basis of analysis.[24]

Porter[25] confirmed that resources are not valuable in and of themselves; but they have value because they allow firms to perform activities that create advantages in particular markets. Resources are only meaningful in the context of performing certain activities to achieve certain competitive advantages. Several resource level categorizations have been presented in the literature. One of the most famous classifications of resources is that of tangible and intangible resources. Physical or tangible resources are normally obvious to firms, competitors, and customers. Intangible resources are less apparent to competitors and customers, or even to the firm itself. Intangible resources include brand names, technological know-how, organizational capabilities embedded in a company's routines, process, and culture, reputation, tacit design, production know-how, customer relationships, and organizational culture. Briefly, resources can be defined as a firm's financial, physical, human, technological, and organizational capital. They can be divided into tangible resources (land, buildings, plant, and equipment) and intangible resources (brand names, reputation, patents, and technological, or marketing know-how).

Customer Responsiveness

To achieve superior customer responsiveness, a company must be able to do a better job than competitors of identifying and satisfying the needs of its customers. Consumers will then place more value on its products, creating a differentiation based on competitive advantage. Improving the quality of a company's product offering is consistent with achieving responsiveness, as is developing new products with features that existing products lack. In other words, achieving superior quality and innovation is an integral part of achieving superior customer responsiveness. Customization of goods and services to the unique demands of individual customers or customer groups is another aspect of customer responsiveness. One aspect of customer responsiveness that has drawn increasing attention is customer response time, which is the time that it takes for a good to be delivered or a service to be performed. Besides superior quality, customization and response time, other sources of enhanced customer responsiveness include superior design, superior service, and superior after sales service and support. All these factors enhance customer responsiveness and allow a company to differentiate itself from its less responsive competitors. In turn, differentiation enables a company to build brand loyalty and to charge a premium price for its products.

Briefly, efficiency, quality, innovation, and customer responsiveness are all important elements in gaining a competitive advantage.[26] Superior efficiency enables a company to lower its cost; superior quality lets it both charge a higher price and lower its cost; superior customer responsiveness allows it to charge a higher price; and superior innovation can lead to higher price or lower unit costs. Together, these four factors help a company create more value by lowering costs or differentiating its products from those of competitors, which enables the company to outperform its competitors.

We can conclude on the basis of the above discussion that to give rise to a distinctive competency, a company's resources must be both unique and valuable. A unique resource is one that no other company has. A resource is valuable if it in some way helps create strong demand for the company's products. A company may have unique and valuable resources, but unless it has the capability to use those resources effectively, it may not be able to create or sustain a distinctive competency. It is also important to recognize that a company may not need unique and valuable resources to establish a distinctive competency so long as it has capabilities that no competitor possesses. In summary, for a company to have a distinctive competency, it must at a minimum have either (1) a unique and valuable resource and the capabilities (skills) necessary to exploit that resource, or (2) a unique capability to manage common resources as they allow organizations to achieve superior efficiency, quality, innovation, or customer responsiveness, thereby creating superior value and attaining a competitive advantage. A company's distinctive competency is strongest when it possesses both unique and valuable resources and unique capabilities to manage those resources.

How Innovation Can Be a Source of Competitive Advantage and What Are the Antecedents of Innovation

According to Professor Danny Samson, "Innovations can be in the form of new products or services, cost-reducing process improvements, or innovative business models and methods."* The benefits of innovation are reflected on a company's profit and loss (P&L) statement: innovators drive additional sales volume, achieve price premiums, and reduce costs through process improvements.[27] In addition to the financial benefits, innovation goes hand in hand with sustainable development initiatives, as both require progressive leadership and an appetite for change, combined with a tolerance of experimentation and some risk.[28]

Innovative firms have developed a *systematic innovation capability*, which assures them of a series of innovations that deliver business value. Innovation success starts with strategy and leadership, in which innovation is prioritized as important to the business. Guided by this strategic direction, these firms support innovativeness in their operations, and invest in their workforce's creativity. They measure innovation and recognize it as important in their workforce, and some reward their staffs for contributions to innovation. Through strong senior executive leadership of innovation, staffs are encouraged to contribute to innovation and the behaviors and culture lead to a deep embedding of the innovation mind-set and culture. This innovativeness is attractive in labor markets and allows these firms to attract and retain talented people.

The aim of this article is to develop the contextual relationship between antecedents of innovation management, using interpretive structural modeling and classifying these antecedents on their dependence and impact. The opinions of several cited experts form the basis of the relationship matrix that is used for building the ISM model.[29–32]

Theoretical Discussion on Antecedents of Innovation

The objective of this section is to identify antecedents of innovation and its impact on firm performance. Journals like the *European Journal of Innovation Management, International Journal of Innovation Science, Managing Innovation: Integrating Technological, Market and Organizational Change, Technovation, International Journal of Entrepreneurship and Innovation Management, Creativity and Innovation Management, IJOPM,* and books in related fields were used to identify the antecedents of innovation management. Keeping the research objectives in mind, the focus is on antecedents of innovation. Table 39-1 identifies the reference sources for each of the antecedents discussed.

Antecedents of Innovation Management	References
Qualified staff	33–36
Leadership	33,37
Financial support	33,34
Education	33,38
Management expertise	33,35
Competence	33
Time	33
R&D	34,35
Culture	33,39
Bureaucracy	34,35
Organizational structure	36
Knowledge of the new product	33
Competitors' copying products	33
Access to technology providers	33
Scarce resources	33

TABLE 39-1   Antecedents of Innovation Management

*www.innovation.gov.au/.../IndustryInnovationCouncils/.../.

Theoretical Discussion on Innovation

The antecedents to the innovation process are discussed widely in innovation literature in different dimensions. Larsen and Lewis[39] categorized antecedents of innovation as financial support, marketing support, management and personal characteristics support, and other support. Similarly, Segarre-Blasco et al. categorized antecedents of innovation as cost, knowledge, and market.[40] Almus and Czarnitzki identified antecedents of innovation as financial support, risk, competence, organizational structure, and legal.[41] Marketing skills such as customer focus,[42–45] face-to-face contact with customers,[46] and marketing intelligence[47,48] have been cited as the most important antecedents for new product success.

Resolving Issues Related to Model Adoption

Some of the issues relevant for discussion are given below.

Which Antecedents

Previous work was seriously lacking consensus with regard to which antecedents of innovation should be included. This becomes apparent with the review of numerous articles in highly recognized scholarly journals published from as early as 1985 to as recently as 2012. The variables included in these studies and their impacts on firm performance vary significantly. The variables included in the present research are financial support, marketing support, organizational structure, leadership, competence, R&D, innovation culture, legal structure, and knowledge of the new product.

Theory on Model Classification

Despite the ascendancy of the bundle approach, there still is room for differences among researchers in selecting how to bundle the practices in each variable. In the main, two approaches are widely seen in the literature, even though there remains limited theory specifying how practices in each innovation variable should be grouped together. One approach examines the total items in each function by integrating all practices in that function into a single index to measure the extent a company utilizes these innovation practices,[38,48,49,50] while the others develop and empirically verify key dimensions of each function[38,51] through methods such as factor analysis.

From various publications, it can be seen that researchers have established the linkage between independent variables and dependent variables in two categories. One is a direct linkage between independent variables and dependent variables. The other, namely moderating linkage, added a moderating construct between independent variables and dependent variables in hopes of throwing more explanatory light on the direct linkage.

Methodology

This study has been conducted to closely examine the factors that are associated with successful innovation at the enterprise level. First, we wanted to find companies that have consciously adopted a competitive strategy that is at least partly if not significantly based on their innovation capability. Second, we wanted to find 10 companies that have different approaches to innovation; hence we considered and examined firms in a range of industries and of various company types and sizes. The interpretive structural modeling method was used for the study, a method that is used to identify and clarify the factors leading to innovation and their interrelations in terms of their impact value. It is an interactive planning methodology whereby a set of directly and indirectly related factors are structured into a comprehensive systematic model. For complex problems like the one under consideration, a number of antecedents may be affecting innovation management. However, the direct and indirect relations between the barriers describe the situation far more accurately than do the individual barriers in isolation. Therefore, ISM develops insights into a collective understanding of these relationships.

ISM-Based Model

From the final reachability matrix (RM) and level partitions, the structural model is generated by means of vertices or nodes and lines of edges (see Fig. 39-1). The result is called a graph or digraph, and is described in more detail later. It is ultimately converted into the ISM model and is subject to review to check for conceptual inconsistencies and make the necessary modifications.

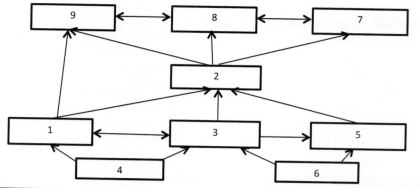

FIGURE 39-1   Digraph.

Findings of the Antecedents from Literature Review

In this step, the innovation-related literature was thoroughly reviewed. As a result of the review, 20 antecedents were found to be effective in innovation processes. A problem-solving group was formed comprising four academicians who have studied in innovation and seven experts from R&D. The group was asked to analyze the listed antecedents of innovation, take out the repeated ones, and rearrange the list including the ones thought to be valid in relation to conditions in India. The first tour resulted in reducing the list of antecedents to 15. Through the second round, 12, and after the third round, nine total antecedents were determined, as follows:

1. Financial support

2. R&D

3. Competence

4. Culture

5. Leadership

6. Scarce resources

7. Legal structure

8. Organizational structure

9. Marketing support

Structural Self-Interaction Matrix

Establishing a contextual relationship between elements with respect to the pairs of elements is examined below.

A questionnaire was formed to investigate the relationships among nine antecedents. The questionnaire was sent to 10 experts who were responsible for innovation projects in their companies. It was tested for content validity, which primarily depends on an appeal to the propriety of content and the way it is presented.[52] The instrument developed in this study demonstrates the content validity, as the selection of measurement items was based on both an exhaustive review of the literature and detailed evaluations by academicians and executing managers during pretesting. Developing a structural self-interaction matrix (SSIM) of elements demonstrates pair-wise relationships between elements of the system (see Table 39-2). To analyze the antecedents, a contextual relation of "achieve" was chosen. This means that one antecedent will achieve another antecedent; the latter will be achieved by another antecedent; the two antecedents will help achieve each other or the antecedents will be unrelated. For analyzing the barriers in developing SSIM, the following four symbols have been used to denote the direction of relationships between barriers (i and j):

V = Antecedent i will help achieve antecedent j

A = Antecedent j will be achieved by antecedent i

X = Antecedent i and j will help achieve each other

O = Antecedent I and j are unrelated to each other

	9	8	7	6	5	4	3	2	1
1	V	X	O	A	A	A	V	V	
2	V	O	X	A	A	A	V		
3	V	A	O	A	A	A			
4	V	V	O	A	V				
5	V	X	O	A					
6	O	V	X						
7	V	X							
8	X								
9									

TABLE 39-2   Structural Self-Interaction Matrix

Reachability Matrix

The SSIM has been converted into a binary matrix, called the RM (see Table 39-3) by substituting X, A, V, and O by 1 and 0. The substitution of 1s and 0s are as per the following rules:

- If the (i,j) entry in the SSIM is V, the (i,j) entry in the RM becomes 1 and the (j,i) entry becomes 0.
- If the (i,j) entry in the SSIM is A, the (i,j) entry in the RM becomes 0 and the (j,i) entry becomes 1.
- If the (i,j) entry in the SSIM is X, the (i,j) entry in the RM becomes 1 and the (j,i) entry also becomes 1.
- If the (i,j) entry in the SSIM is O, the (i,j) entry in the RM becomes 0 and the (j,i) entry also becomes 0.

After incorporating the transitivity that states that if A leads to B, B leads to C, then A definitely leads to C, the final RM is shown in Table 39-4, in which the driving power and dependence power of each barrier are also shown. Driving power of each barrier is the total number of barriers (including itself) that it may help achieve. On the other hand, dependence is the total number of barriers (including itself) that may help achieve it. These driving powers and dependencies will be used later in the classification of barriers into the four groups of autonomous, dependent, linkage, and independent (driver) barriers.

Level Partitioning

From the final RM, the reachability and antecedent set for each barrier are found. The RM consists of the element itself and other elements that it may help achieve; whereas the antecedent set consists of the element itself and the other elements that may help achieve it.

	1	2	3	4	5	6	7	8	9	Driving Power
1	1	1	1	0	0	0	0	1	1	5
2	0	1	1	0	0	0	1	0	1	4
3	0	0	1	1	1	1	0	1	1	6
4	1	1	0	1	1	0	0	1	0	5
5	1	1	0	0	1	0	0	0	1	4
6	1	1	0	1	1	1	0	1	0	6
7	0	1	0	0	0	1	1	0	1	4
8	1	0	0	0	1	0	1	1	0	4
9	0	0	0	0	0	0	0	1	1	2
Dependence Power	5	6	3	2	5	3	3	6	6	

TABLE 39-3   Initial Reachability Matrix

	1	2	3	4	5	6	7	8	9	Driving Power
1	1	1	1	0	1*	0	0	1	1	6
2	0	1	1	0	0	0	1	0	1	4
3	1*	1*	1	1	1	1	0	1	1	8
4	1	1	0	1	1	0	0	1	0	5
5	1	1	0	0	1	0	0	1*	1	5
6	1	1	0	1	1	1	1*	1	0	7
7	0	1	0	0	0	1	1	1*	1	5
8	1	0	0	0	1	0	1	1	1*	5
9	0	0	0	0	0	0	1*	1	1	3
Dependence Power	5	7	3	3	5	2	5	8	7	

*Transitivity property checked

TABLE 39-4 Final Reachability Matrix

1	Level 3
2	Level 2
3	Level 3
4	Level 4
5	Level 3
6	Level 4
7	Level 1
8	Level 1
9	Level 1

TABLE 39-5 Level Matrix

Then the intersection of these sets is derived for all elements. The element for which the reachability and intersection sets are the same is the top-level element in the ISM hierarchy. The top-level element of the hierarchy would not help any other element above their own level. Once the top-level element is identified, it is separated from the other elements. Then, the same process finds the next level of the element. This process continues till the levels of each element are found (see Table 39-5). These identified levels help in building the digraph and final model.

Digraph

From the final RM and level partitions, the structural model is generated by means of vertices or nodes and lines of edges. If there is a relationship between the barriers j and i, this is shown by an arrow which points to from i to j. This graph is called graph or digraph.

Here, 1 to 9 represent the antecedents as

1. Financial support
2. R&D
3. Competence
4. Culture
5. Leadership
6. Scarce resources
7. Legal structure
8. Organizational structure
9. Marketing support

Conclusions and Recommendations

Conclusions

The digraph indicates that scarce resources and innovation culture leads to legal structure, organizational support, and marketing support. It shows that scarce resources and innovation culture play key roles in defining leadership, creating competence and financial support, which together leads to R&D competence. Due to the scope covered as well as the method used, this research is thought to provide significant contribution to defining the right antecedents and determining their interaction, and provides an agenda in particular for the unique aspects and conditions in India. Determining the actual variables would lead to reaching the desired conclusions; therefore, future research, taking into consideration the antecedents and their interactions mentioned in the model, would provide considerable contributions to the analysis of the perceptions of the antecedents as well as to scrutinizing the performance of the process.

Findings from the Study

For those firms trying to achieve and sustain competitive advantage, there are some important realities to face, in terms of sustainable competitive advantage (SCA). This comes from true differentiation, cost leadership, and market focus through innovation. And while no single innovation lasts forever, what can last for a long time is the sustained advantage of a superior *systematic innovation capability*![27] To achieve this advantage, innovation needs to be the key focus of all the fundamental blocks of *your organization*, as is the case in leading innovation-oriented firms. Internationally well-known examples of this phenomenon include Apple, Google, Samsung, Sony, 3M, Nokia, and HTC. It must be acknowledged that no firm is always successful with its innovations, including the global "innovation masters" such as Apple, Samsung, and Sony, due to the challenging and uncertain factors in introducing new products, services, processes, etc. However, it is reasonably assumed that those with a robust and systematic innovation capability are more likely to have higher probabilities of success in any single new innovation activity and significantly more such success overall. A precursor to achieving a systematic innovation capability is to consciously and purposefully engage in such a strategy. Then and only then, will it get sufficient resources, priorities, and company-wide attention to succeed. Strategy is usually best made plain and explicit so that all stakeholders can understand and align with it; hence we would expect to see systematically innovative companies "talking the talk" of innovation at all levels of the organization, on the way to "walking the walk" of innovation. This is certainly the case at 3M in terms of product and service innovation that leads to revenue growth, and similarly in Toyota in terms of process innovation, which leads to increased productivity and quality. In this aspect of strategy and "mind-set", some companies have the dynamism of innovation "in their DNA," and some simply don't, with all shades of gray existing in between the extremes. In summary, it is possible to recognize the extent to which innovation is central to a business' competitive strategy. If it is not a key part of its stated competitive strategy, then that is the first building block to work on, assuming that systematic innovation is a desired outcome.

Recommendations

On the basis of expert opinions, we found that although they differed very much in size and structure, industry and product or service offerings, there were, deep down, some common success factors.

There Is No Single Formula for Innovation

Every firm must chart its own innovation course. It is different in detail, every time. By this we mean that the practices of innovation, the subject of innovation, and even who is centrally involved in innovation can vary a lot from firm to firm. Companies all have quite unique circumstances and pressures on them, different opportunities for innovation, different technologies, different resources and different leaders available, each with their own ideas, who do innovation uniquely. Therefore, innovation strategies are different in their details across firms. Some focus on new products and services, some on continuously improving internal processes, others on their business models and yet others on technology and process-based cost reduction. Different circumstances refer to the history of the company, the external forces of competition, the nature of their product life cycles, and the pace of technological change in their products and process technologies. There are also differences in structure, size, and market opportunities for innovation across companies, which means the details of innovation practices are different every time.

Strong Leadership

Innovation success is driven by dynamic leaders. In some cases, mostly the smaller companies, innovation is driven by these leaders in a very "hands-on" manner. In others, such as Tata Motors or Godrej & Boyce, it has been matured and become embedded in the broader culture.

Innovation Helps Firms to Attract and Retain Talents

Talented people at all levels are attracted to companies doing interesting things, not just the same old, same old; hence innovation supports reputational advantages that help in winning the "war for talent." These innovators have low turnover and have achieved a "switching on" of staff to the company's mission, and a quest for progress. Innovation is associated with high staff motivation, low turnover, and, ultimately, labor market success. Tata Motors and Godrej & Boyce achieve what can only be described as differentiation through its "people capability," meaning that the skills, knowledge of process improvement, and motivation of its staff keep them at the edge of cost reduction and product quality that is hard to match for their competitors.

Innovation in Supply Chain Network

Innovative firms drive innovation in their partners in the supply chain network. They are changing fast in terms of their products, services, processes, etc., so they need suppliers who can match such capability.

Innovation Can Become Fully Embedded in a Firm's DNA

This means that it is central to the company's market positioning, resource-base, measures and values (whether formal or informal), and rewards system. This leads to the development of an innovative culture and set of behaviors, until ultimately, systematic innovation is expected as a part of every activity. Hence it is debatable as to whether these firms have all achieved innovation capability in the firm's DNA. For those which are not yet at a mature point in this regard, that should indeed be a goal, and until it is maturely in place, it constitutes a risk factor.

Innovation Mind-Set

It is a way of working. In addition to doing the daily work of production, sales, and support functions, these innovative companies involve most or all staffs having a second job in mind, which is finding new ways to create value. This is not disconnected from daily work, but rather is applied to all aspects of daily work. Innovation can be a way of working, a part of and manner of doing the job. Problems are solved creatively, rather than routinely and in a similar manner to the past. Staff looks at products and services, even as they produce them, with a constructively critical eye, and with the skills to seek out improvement in them. While doing processes "by the book" in these firms, there is a continuous desire to improve the book's standards.

To Become Systematically Innovative, A Firm Must Be Competent at Quality Management

Quality management was a very large movement in the 1990s and 2000s in India, with most companies trying it, and most being unsuccessful in getting lasting improvements from it. Quality management requires a strong customer focus, commitment to process control and improvement, and a company-wide skill-set on the tools that support that philosophy and methods. Companies that have achieved quality maturity have a strong sense of discipline, of doing things "right the first time" and of focusing on customer value creation.

Innovation Is Not Free. It Requires Investment in Capability Building, Training, and Development

Firms that aspire to be innovative must be prepared to take risks, and hence have an appetite for some risk and a tolerance for efforts that result in failure. This means putting some capital at risk, and having a culture that understands risk-taking and tolerates rather than blames individuals when things do not always succeed. Fundamentally, if a company only is prepared to do things that succeed predictably, then only well-known technologies, products, and processes will be undertaken and innovation is banished.

Innovative Firms Attract Customers

As consumers, we are naturally curious and inquisitive about new things, and indeed that applies generally to the human condition. If there is a new solution to a problem, a clever one, that creates net positive value for customers, then customers will want it. Customers want low risk in their purchased services and products. Customers do not even like variability in specifications, much less risk of product or service failure. However, they do want more value, meaning they do want better solutions.

Further Scope

The findings of the study indicate that innovation culture and scarce resources drive innovation in India. This clearly establishes that frugal innovation (Jugaad) is deeply rooted in the Indian mind-set and has tremendous potential, but it requires an organized structure. Most experts in India and abroad have criticized this attitude, but it must be appreciated that frugal innovation cannot be denied and it should be considered a specialized division of innovation. In the future, it is recommended that an empirical study can be used to test the relationship among antecedents and extended to developing countries where there are scant resources. It provides a strategic framework for managing innovation in those countries where resources are constrained.

References

1. Rumelt, R.P. "The many faces of Honda." *California Management Review,* 38, no. 4 (1996): 103-111.
2. Long, C. and Vickers-Koch, M. "Using Core Capabilities to Create Competitive Advantage." *Organizational Dynamics,* 24 (1): 7–22, 1995.
3. Feurer, R. and Chaharbaghi, K. "Defining Competitiveness: A Holistic Approach." *Management Decision,* 32 (2): 49–58, 1994.
4. Andrews, K.R., Christensen, C.R., and Bower, J.L. *Business Policy: Text and Cases.* Homewood, IL: Richard Irwin, Inc., 1965.
5. Ansoff, H.I. *Corporate Strategy.* New York, NY: McGraw-Hill, 1965.
6. Henderson, B.D. *Henderson on Corporate Strategy.* Cambridge, MA: Abt Associates Inc., 1979.
7. Johnson, G. and Scholes, K. *Exploring Corporate Strategy—Text and Cases.* London, UK: Prentice-Hall, 1993.
8. Porter, M.E. "What Is a Strategy?" *Harvard Business Review,* 74 (6): 61–78, 1996.
9. Kouzes, J. and Posner, B. "Envisioning Your Future: Imagining Ideal Scenarios." *Futurist,* 3: 14–19, May 1996.
10. Alderson, W. *Dynamic Marketing Behavior: A Functionalist Theory of Marketing.* Homewood, IL: Richard D. Irwin, Inc., 1965.
11. Hoffman, N.P. "An Examination of the Sustainable Competitive Advantage Concept: Past, Present, and Future." *Academy of Marketing Science,* 2000 (4): 1, 2000.
12. Hall, R. "The Strategic Analysis of Intangible Resources." *Strategic Management Journal,* 13 (2): 135–144.
13. Yamoah, F.A. "Source of Competitive Advantage: Differential Catalytic Dimensions." *Journal of American Academy of Business,* 4 (1/2): 223–227, 2004.
14. Caves, R.R "Economic Analysis and the Quest for Competitive Advantage." *American Economic Review,* 74 (2): 127–132, 1984.
15. Flint, G.D. "What Is the Meaning of Competitive Advantage?" *Advances in Competitiveness Research,* 8 (1): 121–129, 2000.
16. Porter, M.E. *Competitive Advantage: Creating and Sustaining Superior Performance.* New York, NY: The Free Press, 1985.
17. Coyne, K.P. "Sustainable Competitive Advantage: What It Is, What It Isn't." *Business Horizons,* pp. 54–61, January–February, 1986.
18. Barney, J.B. "Firm Resources and Sustained Competitive Advantage." *Journal of Management,* 17 (1): 99–120, 1991.
19. Hill, C.W.L. and Jones, G.R. *Strategic Management: An Integrated Approach.* 5th ed. Houghton Mifflin Company, 2001.
20. Ghemawat, P. "Sustainable Advantage." *Harvard Business Review,* 64 (5): 53–58, 1986.
21. Tidd, J. A Review of Innovation Models. Discussion paper 1/1, Imperial College London, 2006. Available at ict.udlap.mx/projects/cudi/sipi/files/Innovation%20 models%20 Imperial%20 College%20 London.pdf (accessed 1st March, 2013.
22. Ansoff, H.I. *Corporate Strategy.* New York, NY: McGraw-Hill, 1965.
23. Amit, R. and Schoemaker, P.J.H. "Strategic Assets and Organizational Rent." *Strategic Management Journal,* 14 (1): 34–46, 1993.
24. Grant, R.M. "The Resource-Based Theory of Competitive Advantage: Implication for Strategy Formulation." *California Management Review,* 33 (3): 114–135, 1991.
25. Porter, M.E. "Toward a Dynamic Theory of Strategy." *Strategic Management Journal,* 12: 95–117, Winter 1991.
26. The New Zeland Tourism Research Institute, AUT University, 2007. Available at http://www.dol.govt.nz/er/bestpractice/productivity/researchreports/foodbeverage/food-and-beverage.pdf.

27. Teece, D.J. "Business Models, Business Strategy and Innovation." *Long Range Planning*, 43(2–3): 172–194, 2010.

28. Piatier, A. *Barriers to Innovation*. London, UK: Frances Printer, 1984.

29. Mandal, A. and Deskmukh, S.G. "Vendor Selection Using Interpretive Structural Modelling (ism)." *International Journal of Operations and Production Management*, 14 (6): 52–59, 1994.

30. Singh, M.D, Shankar, R., Narain, R., and Agarwal, A. "An Interpretive Structural Modeling of Knowledge Management in Engineering Industries." *Journal of Advances in Management Research*, 1 (1): 28–40, 2003.

31. Sharma, S.K, Panda, B.N., Mahapatra, S.S, and Sahu, S. "Analysis of Barriers of Reverse Logistics: An Indian Perspective." *International Journal of Model Optimization*, 1 (2): 101–106, 2011.

32. Tiwari, R. and Buses, S. Barriers to innovation. In: "SMEs: Can Internationalization of R&D Mitigate Their Effects?" Working Paper, No 50, *Proceedings of the First European Conference on Knowledge for Growth: A Role and Dynamics of Corporate R&D*, Spain, 2007.

33. Uzun, A. "Technological Innovation Activities in Turkey: The Case of the Manufacturing Industry." *Technovation*, 21 (3): 189–196, 1997.

34. Drucker, P.F. *Innovation and Entrepreneurship*. New York, NY: Harper & Row Publication, 1985.

35. Segerra-Blasco, A., Garcia-Guevedo, J., and Teruel-Carrizosa, M. "Barriers to Innovation and Public Policy in Catalonia." En*trepreneurship and Management Journal, 4 (4)*: 431–451, 2008.

36. Tonge, R., Larsen, P., and Ito, M. "Strategic Leadership in Super-Growth Companies—A Reappraisal." *Long Range Planning*, 31: 835–844, 1998.

37. Sund, K.J. "Innovation in the Postal Sector: Strategies, Barriers and Enablers." Ecole Polytechnique Federale de Lausanne, Sap, 2008.

38. Hadjimanolis, A. "Barriers to Innovation for SMEs in a Small Less Developed Country (Cyprus)." *Technovation*, 19: 561–570, 1999.

39. Larsen, P. and Lewis A. "How Award-Winning SMEs Manage the Barriers to Innovation." *Creativity and Innovation Management*, 16 (2): 142–151, 2007.

40. Segarra-Blasco, A. et al. "Barriers to Innovation and Public Policy in Catalonia." *International Entrepreneurship Journal*, 4 (4): 431-451, 2008.

41. Almus, M. and Czarnitzki, D. "The Effects of Public R&D Subsidiaries on Firms' Innovation Activities: The Case of Eastern Germany." *Journal of Business and Economics Statistics*, 21 (2): 226–236, 2003.

42. Clifford, D.K. and Cavanagh, C. *The Winning Performance—How America's High-Growth Midsize Companies Succeed*. London, UK: Sidgewick and Jackson, 1985.

43. Mondiano, P. and Ni-Chionna, O. "Breaking into the Big Time." *Management Today*, 11: 82–84, 1986.

44. Tidd, J., Bessant J., and Pavitt, K. *Managing Innovation: Integrating Technological, Market, and Organizational Change*. Chichester, UK: John Wiley and Sons, 2001.

45. Foley, P. and Gren, H. A successful high-technology company. In: Foley, P. and Gren, H. (eds). *Small Business Success*. London, UK: The Small Business Research Trust, Paul Chapman Publishing, 72–80, 1995.

46. Freel, M.S. "Barriers to Product Innovation in Small Manufacturing Firms." *International Small Business Journal*, 18: 60–80, 2000.

47. Wren, B.M., Souder, Wm. E., and Berkowitz, D. "Market Orientation and New Product Development in Global Industrial Firms." *Industrial Marketing Management*, 29: 601–611, 2000.

48. King, N. Innovation at work: the research literature. In: West, and Farr, J. (eds.). *Innovation and Creativity at Work*. Chichester, UK: John Wiley and Sons, pp.15–59, 1990.

49. Wolfe, R. "Organizational Innovation: Review, Critique and Suggested Research Directions." *Journal of Management Studies*, 31 (3): 405–431, 1994.

50. Avlonitis, et al. "Assessing the Innovativeness of the Organizations and Its Antecedents: Project Innovstrat." *European Journal of Marketing*, 28 (11): 5–28, 1994.

51. Lipparini, A. and Sobrero, M. "The Glue and the Pieces-Entrepreneurship and Innovation in Small Firm Networks." *Journal of Business Venturing*, 9 (2): 125–140, 1994.

52. Nunally, J.O. *Psychometric Theory*. New York, NY: McGraw-Hill, p. 701, 1978.

CHAPTER 40

Organization for Innovation

Eduardo Armando and Eduardo Vasconcellos

Introduction—Reasons for Organizational Innovation Boosters

The high speed of recent changes represents a challenge to man's ability to organize. The traditional structures have demonstrated their failure to deal with the high rate of change in the world in which we live in.[1] This chapter deals with forms of organizational structure that facilitates innovation. New products and processes as well as new business models require integration of multiple specialties and traditional forms of structure that tend to group the human and material resources into separate units.

Even without considering the innovation, the complexity of the modern world now demands structural adjustments. A list of influencing factors is presented below:

- More turbulence and uncertainty require increasingly agile structures able to integrate the various business units to detect threats and opportunities as well as to implement appropriate changes.

- Technological changes that continue to occur intensively lead to the creation of technological intelligence units.

- Increasing internationalization of business requires coordination units of the innovation process in the various subsidiaries to avoid duplication and to use synergies.

- Increasing government role in the economy.

- Increasing complexity of organizations, reaching limits not matched by similar organizations in the past, demanding effective processes for periodic evaluation of structure and redesign.

- Evolution of unionism, concerns about the ecology, with the substitution of energy sources, and the emergence of consumer protection organizations make the network of constraints to be managed even more complex.

- New technologies of information and communication require structure adjustments.

In many productive sectors, the increasing degree of turbulence in the environment and the lack of ideal conditions for the operation of traditional structures have disappeared, which made many structures inappropriate to the new conditions.

The discussion is not recent. Decades ago authors[2] indicated that high-performing organizations are those that are best suited to dynamic environmental conditions. Also, it is pointed out that any organizational form is problematic and that, even in these high-performing organizations, managers are constantly looking for new ways to improve its operations.

The functional structure, widely used by businesses, groups together human and material resources using the criterion of similar functions performance. An organization structured by departments such as finance, marketing, and production is an example of a functional structure. The departmentalization can also be used in technological areas such as, in geology, steel structures, and foundations, as a criterion. These technology areas are also considered as functional. The big problem is that innovation requires integration among specialists in all these areas, making this type of departmentalization complex to be implemented. Some factors contributed to the following:

- Tendency to bureaucratization and stagnation reduces the efficiency and effectiveness as the size of the organization increases.

- The organization's growth and geographic decentralization make it more difficult to respond effectively to market needs, requiring new coordination mechanisms.

- Increasing the level of workers' education makes them less satisfied with routine activities.

Main Characteristics of the Structures for Innovation Compared with Traditional Structures

Organizational structure is the result of a process through which the activities and authority are distributed and certain communication flows are established. The structure has two critical components: defining roles and departmentalization. The departmentalization exists because, for example, a president cannot handle a couple of thousand people. A chief executive of an organization of this size is bound to create positions of directors and distribute them among the people. In the same example, a possible situation is each director is responsible for approximately 400 employees and, in turn, is bound to create positions of managers. The criteria for grouping employees will form the structure. Examples of criteria traditionally used are functional, by products, by geographic area, by customers, by business area, etc. Also in this chapter it is shown that these criteria are not adequate to promote innovations.

The departmentalization result is a figure called an organizational chart. The rectangles represent management positions and the lines indicate the hierarchic relations (who is subordinate to whom). Examples of charts will be presented throughout this chapter.

The attributions definition explains activities and decisions and certain communication flow.[3]

In the literature,[4] there are explanations about the differences between mechanistic organizations (more focused on routine) and organic (more focused on innovation) and argue that these models require different organizational structures. Burns and Stalker (1966) studied two successful and respected organizations. One of them was more focused on routine activities and the other on innovation. It has been verified that both profiles were very different.

Table 40-1, which is based on the propositions of the authors, compares mechanistic organizations with organic organizations:

The two management systems, mechanistic and organic, represent the polar extremities of the forms that systems can take when they are adapted to a specific rate of technical and commercial change.

Both types represent a rational form of organization and may be explicitly deliberated, created, and maintained to exploit the human resources of an organization in the most efficient manner feasible in the circumstances.

While organic systems are not hierarchic in the same sense as they are mechanistic, they remain stratified. Positions are differentiated according to seniority, that is, greater expertise. Seniors frequently take the lead in joint decisions, but it is an essential presumption of the organic system that the lead, that is, authority, is taken by whoever shows himself as most informed and capable. The location of authority is settled by consensus.

A second observation is that the area of commitment to the organization—the extent to which the individual yields himself as a resource to be used by the working organization—is far more extensive in organic than in mechanistic systems. One further consequence is that it becomes far less feasible to distinguish between informal and formal organization.

Thirdly, the emptying out of significance from the hierarchic command system, by which cooperation is ensured and which serves to monitor the working organization under a mechanistic view, is countered by the development of shared beliefs about the values and goals of the organization.

Finally, the two forms represent a polarity, not a dichotomy. There are intermediate stages between the extremities labeled as mechanistic and organic. An organization may (and frequently does) operate with a management system that includes both types.

Organic	Mechanistic
The contributive nature of special knowledge and experience to the common task of the organization	Specialized differentiation of functional tasks
Realistic nature of the individual task, which is seen as a set by the total situation of the organization	Employees tend to pursue the technical improvement of means, rather than accomplishment of the ends
The adjustment and continual redefinition of individual tasks through interaction with others	The reconciliation of performances for each level of the hierarchy by the immediate superiors
The heading of responsibility as a limited field of rights, obligations, and methods	Precise definition of rights and obligations and technical methods attached to each functional role
The spread of commitment beyond any technical definition	Translation of rights and obligations and methods into the responsibilities of a functional position
A network structure of control, authority, and communication	Reinforcement of hierarchic structure by the location of knowledge exclusively at the top of the hierarchy
Omniscience no longer imputed to the head of the organization; knowledge about the technical or commercial nature of the here and now task may be located anywhere in the network; this location becoming the ad hoc center of control authority and communication	Tendency for interaction between member to be vertical, between superior and subordinate
A lateral, rather than vertical direction of communication through the organization, communication between people of different rank, also resembling consultation rather than command	A tendency for operations and working behavior to be governed by instructions and decisions issued by superiors
A content of communication which consists of information and advice rather than instructions and decisions	Insistence on loyalty and obedience to superiors as a condition for membership
Importance and prestige attached to affiliation and expertise valid in the industrial, technical, and commercial milieux external to the firm	Greater importance and prestige attached to internal (local) knowledge, experience, and skill

TABLE 40-1   Comparison between Mechanistic and Organic Management Systems

Forms of Departmentalization That Stimulate Innovation

As already mentioned, the traditional ways of structuring do not favor innovation because they tend to separate people in different departments, avoiding collaboration. Creating a new product requires the cooperation of marketing people, who know what the market wants, with production people, who know what can be manufactured, research development and engineering people who master the technological knowledge, etc. The following are exemplified forms of structure that allow integration between specialists to produce innovations.

Structure by Projects

Often, innovations in products and/or processes are carried out through projects. Projects are teams looking to get a result within a given time and resources. The departmentalization by projects occurs when people are grouped according to the project in which they are allocated. In this type of departmentalization, people are grouped using the project they are involved in that particular moment as criteria. Each project is like a "temporary department" whose head is the project manager, and the team is the project team, until the project is concluded.

In the structure by projects, people tend to develop more diverse skills. This type of structure shows great flexibility and high performance as the environment changes. A new requirement is immediately transformed into a project through the rapid formation of a team. If a contract is terminated abruptly, experts can be distributed to other projects reducing the impact on the organization.[5]

The functional structure was not able to handle projects involving several functional areas. When the projects were executed within each specific area, the functional structure was successful. However, modern organizations are complex and very often the projects involve many areas of knowledge. In these circumstances, serious problems arose, impeding the functional structure.

Briefly, the advantages of the structure by projects are

- Greater diversification of technicians, because on a project an expert should ideally know more than one area within their specialty to avoid too great team and idle capacity.
- Greater technicians' satisfaction by having a broad overview of the project and a better understanding of the relationship between their specialty and the others.
- Greater technicians' satisfaction for having opportunity to interact with wider variety of people and situations.
- Greater integration among technical areas of the project.
- Better accomplishment of deadlines.
- Improved customer service by the existence of a manager (of project) that integrates the resources aiming at the project objectives.
- Relief to senior management, because they would make the integration.
- Existence of a single person with authority and responsibility over the project as a whole.
- Easier and more efficient management of integrated projects.

Table 40-2 compares the advantages of pure functional structure and pure project.[6]

Matrix Structure

It is the simultaneous use of two or more types of departmentalization on the same group of people. It is usually a combination of functional and project types or by-products, though other combinations are possible.[7]

In the literature,[8] it's found that organizations use matrix structures because the environment obliges them to have high performance on two or three activities. In situations where the environment and the company's strategy demand technical excellence and speed in products launching to market, the result, as the organizational structure, is the matrix. Another reason for adopting a matrix structure is the possibility of sharing specialized resources, usually costly.

The matrix structure has striking features that distinguish it from other types of structures. The matrix form has emerged as a solution because of inadequate functional structure for integrated activities, that is, those to be performed, require interaction among functional areas. The matrix is one way of maintaining the functional units, creating horizontal relationships among them.

When two or more types of structures are used simultaneously on the same members of an organization, the resulting structure is called a matrix. A particular aspect of the matrix structure is its dual or multiple subordinating. An example of matrix structure is one in which a certain expert responds simultaneously to the functional manager of the technical area to which he is allocated and the project manager for which he is providing services. This situation, of dual subordination, tends to increase the level of conflict, especially if the expert provides services for more than one project simultaneously.

In situations where the matrix structure works well, the two leaders communicate to detect early problems and to prevent unnecessary conflicts. In those cases, the experts who will be responsible for a project are selected together, and so their goals are discussed.[9]

The matrices vary in complexity. The simplest may have one or two dimensions. There are also matrix structures of greater complexity, including those that contain one or more matrices within the main structure. The more complex matrix structures have several dimensions—three, four, or even more. And there are those matrix structures that include various dimensions with hybrid forms, that is, arranging the structure not only in matrix, but also using other forms.

Figure 40-1[10] provides an example of matrix structure.

A division or business unit specific to e-business, called Consumer Connect, was created in 1999 to leverage the power of e-business initiatives, in the case of a major innovation in the sales process, involving technological and management aspects, which had a president

Advantages of extreme types Comparison of structure factors	Advantages of Pure Functional Structure	Advantages of Pure Project Structure
1. Breadth	More specialized technicians	More diversified technicians
2. Institution technical prowess	Experience exchange among the area technicians avoids endeavor duplication and improves the institution technical prowess. More focus on technical improvements makes the technical prowess even better. Easier to organize "technical files."	Continuous interaction with technicians from other areas makes clear to each individual the relations among his/her own specialty and others, making the institution more prepared to develop integrated projects
3. Project quality	Superior technical quality	
4. Projects being on time		Timely delivery of projects
5. Technician satisfaction	Higher because of being with the same specialty technicians all the time Work group stability Feeling the higher focus of leadership on peoples' development Being assessed by a competent technician of his/her own specialty Better defined career	Higher because of having the opportunity to interact with other areas' employees and learn from the interrelation of other specialties Being able to interact with a wider variety of people and circumstances
6. Customer interaction		Better, because there is someone clearly accountable for the project as a whole
7. Use of resources	More efficient use of resources and materials	
8. Having a specific employee accountable	There is someone in charge of each group of employees having the same specialization: the functional manager.	There is someone leading one and only one integrated project.
9. Supervision	Easier and more efficient regarding the functional areas	Easier and more efficient regarding the integrated projects

TABLE 40-2   Organizational Structure Comparison—Pure Functional and Pure Project

himself who is also a vice president (VP) within the corporation as a whole. The appointment of that executive was made by that company's global CEO. That is, the organizational structure of e-business is fully integrated with the company, in the form of a business unit.

This structure shows a matrix mode of operating with respect to units worldwide (Asia Pacific, South America, etc.) with a block responsible for the creation and incubation of joint ventures between the automaker B and Internet technology companies (Microsoft, Yahoo!, Web Methods) among others, to create companies that provide the products required by the automaker around the e-business.

Once created, the joint ventures, those responsible for contact with the enterprises, are in the "operate" group. This operate group consists of several project teams, as shown in Fig. 40-1, for example, the project team Covisint, the call center project team, the direct sales team etc. Each team had people from Ford and partner companies.

The "accelerate" group is responsible for identifying the highest priority initiative for the automaker and makes it happen faster, integrating them to the activities of Ford (to inject automaker activities). These are people who are in the automaker B and are part of the e-business unit Consumer Connect. Its function is then to bridge those initiatives with the corporate areas of Ford, prioritizing projects and ensuring their incorporation in the corporate areas that are clients of these projects, in the sense that they will benefit from the resulting products. For example, the purchasing area of the company will use the e-business tools generated by the project Covisint, and so on.

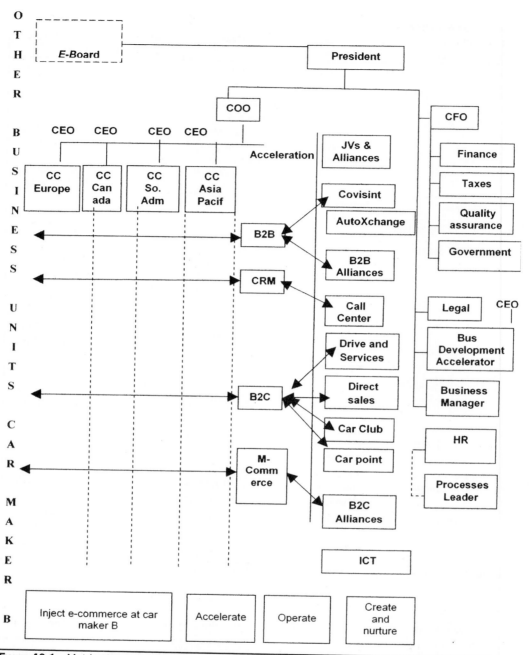

Figure 40-1   Matrix structure example—e-business organizational structure at Ford headquarters.

To inject the automaker e-business initiatives, the group that makes this link is organized in groups called B2B, B2C, B2M (mobile or telematics) within the framework Consumer Connect, liaising with corporate areas of the company. This structure involves representatives from South America, Asia, and Europe and reached the number of 350 employees. It is also interesting to note that this structure has its own finance, HR, and legal areas, among others, thus becoming an autonomous unit, but with links and sharing of Ford infrastructure. Therefore, it was not constituted in a completely unrelated unit of the company, as its vice president had status of Ford vice president.

Since the matrix is a combination of structures, there are a great variety of possible matrix structures, depending on the dosage that each type of structure participates in this combination. The balanced matrix is the matrix structure, having the following characteristics:

- Project managers and functional managers have the same hierarchical level and similar degrees of authority, although in different areas.

- All interdisciplinary project managers only manage projects, without simultaneous functional roles.
- Communication between the project manager and the project technical team is always direct, without involving the functional managers.

Cellular Structure

It is an organizational form whose characteristics are almost completely absent structure and high flexibility. It is a structural form in which informality is very high. Its existence is only viable in small organizations with human favorable climate.[11]

There is a successful case of a research institute division, of a 60-people cellular structure with no heads of departments and sectors. The case showed that there was only a small structured group, which was divided into subgroups to perform the tasks. The size of the subgroups varied and the team leader was determined by the characteristics of the task. Anyone could talk directly to the division director. However, there was a ranking of researchers, according to their experience (assistant researcher, senior A, senior B, and so on). Thus, when an assistant researcher had a problem, they reported to a senior researcher, although they had access to the director of the institute. This division became one of the most successful of the institute. Its director became president of the organization.

However, over the years, organizations that develop and commercialize products of high complexity, such as EMBRAER,[12] which has a cell (see Fig. 40-2), also began to use this cellular structure coupled to a matrix structure with three axes. In these larger and more complex structures, either by product or by size, the degree of formalization is greater. More details about EMBRAER structure for development of new products are discussed later in this chapter. Therefore, the cell concept allows multiple configurations.

The cell concept was first implemented in manufacturing. Volvo in Kalmar, Sweden, in the 1960s, had been facing problems of lack of motivation and high turnover of workers due to the monotony on the assembly lines. The solution was to split the line in big steps. Each step has to be done by about 20 workers, who learned the operation and worked together to build that part of the vehicle. Then, the vehicle was shipped to another location where another cell performed other operations. The workers themselves divided the tasks among them. In situations where the goals were met or exceeded, the workers could rest and decide when and how long they could enjoy the rest.

In the following decades to that experience reported in Sweden, other companies like Ford and General Motors began adopting cells in their manufacturing areas, increasing productivity and reducing costs. An additional attraction of the cells, depending on the model used, is to reduce the hierarchical level of supervisors. This occurs when leaders chosen by the group in a rotation procedure manage cells.

The next stage was to use the cells in management activities, beyond manufacturing. The example shown in Fig. 40-2 fits in this context. The program of the EMB 170/190 described later in this chapter is administered by a cell of 10 people, led by the program director. This cellular structure is replicated in other levels, incorporating concerns such as production, marketing, finance, etc., as shown in Fig. 40-2.

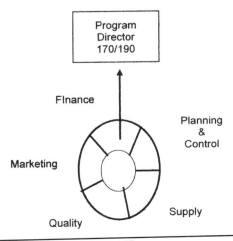

FIGURE 40-2 Cellular structure example—EMBRAER.

Specific Structure for New Ventures

The main concept of this type of structure is that innovation activities should be separated from routine activities. Thus, new ventures must have a proper structure.

When innovation is accepted and the new venture is operating, goes to the operational area.

The group of new ventures starts working on the next innovation.[13]

In the literature, there are examples of companies organized in this way. An example is a steel company, which created an Expansion Group, responsible for planning the expansion of facilities. This group performs all stages, from the initial planning of new plants to the first steel race. If the system is approved, the area of operations takes the control.

Telecommunications companies and others which provide continued service as, for example, industries as varied as electricity and credit cards, are often structured, separating the new facilities development group or new customer acquisition from the operations group, which is responsible for operating and maintaining networks and existing customers.

Figure 40-3[14] shows an example of a specific structure for new ventures.

To understand better the real case of the company Cia de Concessões Rodoviárias (CCR), we need a brief explanation of the context in which this company was founded and prospered. Since the mid-20th century, road transport has prevailed in Brazil, both for people and goods transportation. Costs of maintenance of such an extensive road system proved to be too high for the state. During the 1990s, with the deterioration of the main connecting routes, it was evident. Thus, in 1993, the Brazilian program of highway concessions began, establishing bidding rules for contracting public and private companies to manage the highways.

This initiative of the Brazilian government to privatize highways brought business opportunities for the private sector, to the extent that the concessions would allow the collection of highway tolls, for managing them. Several Brazilian business groups with interests in civil construction, particularly large works, decided in 1998 to create the CCR to jointly administer federal and state concessions.

In 2001, the CCR became the first company to join the so-called New Market of BM & FBOVESPA, the Brazilian market segment that congregates companies that are committed to the highest levels of governance. CCR became the largest road concession company in Latin America and one of the largest in the world.

The existence of a centralized area to address new business sets the strategy for expansion of activities to related areas. A specific board provides the necessary focus on the task for which it is designated, better use of resources, and the possibility of greater integration with corporate strategy. Obviously, there is the risk of duplication of work and idleness, to the extent that such businesses were negligible in the initial phase of the company.

In structures with the CCR configuration, when the new business typically starts operation—can be a new customer connected to the base of the company—it is transferred to the operation. In the example of Fig. 40-3, as indicated by the dashed line, the new business is transferred to a business unit or, as the case may be, a new unit may be created.

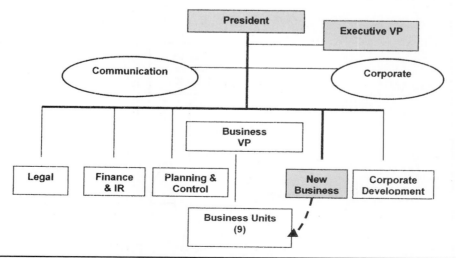

Figure 40-3   Structure for new business development of Cia de Concessões Rodoviárias do Brasil (CCR).

Cases of Structures Organized for Innovation

ALPARGATAS

History

ALPARGATAS was founded in 1907 in São Paulo by investors of British origin to produce and sell ALPARGATAS (plotted in jute rope soles) and other products such as tarps and covers. Its shares have been listed on the stock exchange since 1913.[15]

In the 1930s and 40s, the company began to diversify its product line, from traditional ALPARGATAS to other types of shoes such as leather shoes, sandals, and sneakers. In 1940 the name was changed to SãO PAULO ALPARGATAS S/A (recently the term "São Paulo" was removed), expanding its lines of fabrics and manufacturing of tarps and coverings for the garment industry. In the 1950s, it was responsible for launching the first Brazilian jeans and in the 1960s launched its flagship product, the Havaianas sandals. In the 1970s, the company created the brand Topper and acquired the brand Rainha.

The 1990s were marked by an extensive restructuring process in order to transform the culture focused on production into a culture focused on the market. The company abandoned its various unprofitable businesses and others that were not part of the new strategy of the company. At that time, the company closed its clothing activities and its partnership with Nike.

In return, the company started to license the brand Mizuno and Timberland, aiming to expand international exposure, to adopt new production technologies and ensure the presence of international brands in its portfolio. The intensification of footwear operations also included the expansion of Havaianas brand, expanding the product range and reaching a record 100 million pairs sold.

In January 2003, Camargo Correa SA, a leading Brazilian business group, increased its share holdings to 61.3 percent of the voting capital, becoming the controlling shareholder. In June of the same year, ALPARGATAS enrolled in Level 1 of Corporate Governance of BM & FBOVESPA, expanding the commitment to transparency and the adoption of higher standards of disclosure of information to investors.

The company keeps factories in various Brazilian states and regions abroad, has opened retail stores, and employs 17,600 people, 5000 in operations outside Brazil. The company receives 28 percent of its revenue in hard currency and its products are sold in 82 countries.[16]

In 2011, sales revenue reached R$2.575 million, with more than 249.6 million pairs of shoes sold. Table 40-3 presents some of the key financial indicators from 2005 to 2011.

Organizational Structure

According to ALPARGATAS, definition, it is a company focused on brands and anchored by three major pillars: profitable growth, operational excellence, and people. There is a strong concern in the defense of design and quality, which translates into strong focus on product innovation and the ability to add value to the brand.[17]

The new strategic plan guided the actions of ALPARGATAS from 2005 on, when a new CEO took over management of the company. The new strategy aims to support the growth of the organization and includes major actions that aim to strengthen their competitive advantages as a company of global brands.

The company revealed in its strategic planning the importance of internationalization. The decentralization of foreign trade area to each business unit, as well as setting objectives and targets of foreign sales per unit, has demonstrated the adaptation of the organizational structure to achieve this goal.

Indicator	2005	2006	2007	2008	2009	2010	2011
Sales	1.369	1.565	1.616	1.659	1.927	2.232	2.575
EBITDA	205	293	215	245	290	400	405
Net profit	165	198	210	173	123	303	307
EBITDA margin (%)	15.0	18.7	13.3	14.8	15.0	17.9	15.7
Net margin (%)	12.1	12.7	13.0	10.4	6.4	13.6	11.9

TABLE 40-3 Financial Performance from 2005 to 2011 (in million R$).[18] (*Source*: ALPARGATAS, 2007 e 2012.)

Innovation has become an even more important priority with the strategic decision to increase the degree of internationalization. It has become a necessity, and a prerequisite for success was to increase investments in R&D to develop new products and processes. Innovation has become one of the company's values, represented by the word "entrepreneurship."[19]

To support this strategic plan emphasizing internationalization, organizational structure has been redefined by grouping business units' brands with synergies.

The new organizational structure played an important role in innovation due to the following reasons:

- Facilitated the integration between R&D and business units.
- Facilitated the integration between business units and partners' distributors during the internationalization process, mainly to the United States.
- Facilitated focus on brands seeking innovation and functional support areas.

The manufacturing operation is now coordinated by the new industrial management, with the aim of adopting and disseminating best practices of production and increase productivity rates. The production capacity of the sandals factory and optimization process of the manufacture of sports shoes started to obtain gains in scale and increased productivity.

By making the industrial area a "corporate area," the company withdrew from the business units the control of factories. The units began to acquire the production of the industrial area, which has intensified efforts to optimize the manufacturing plants and identify production opportunities outside Brazil, hitherto unheard in the history of the company.

The area of products' research and development was formed on another board of directors as a way to strengthen the capacity for innovation, research, and development, considered vital by the company to work in the fashion segment. This area aims at the continuous improvement of products and processes and assists in the development and application of new technologies to increase ALPARGATAS competitiveness in national and international markets. Currently, the research and development department, focused on the innovation aspect, is located in the city of São Leopoldo, State of Rio Grande do Sul, in southern Brazil (the ALPARGATAS headquarters is located in the city of Sao Paulo, State of São Paulo, located in the southeastern region of the country).

The new research and development and industrial boards, integrate the corporate area, which operates in support of business, such as finance and investor relations, human resources, communications, and media. The audit board responds to the board of directors.

Figure 40-4 shows the ALPARGATAS organizational structure in the late 2000s.

Investments in innovation in ALPARGATAS are concentrated in the Havaianas product line, flagship of the company until today. The company commercializes this product line in franchised stores in 80 countries, with approximately 150 outlets.

In general, until early 2013, the organization chart remained the same. It included an area of support, new business strategy. The activity of supply chain returned as an area, and now has a board. As for innovation, the marketing and business area calls and R&D develops, in order to provide what was asked.

Although the matrix is not very strong, the sample to be studied is interesting because it shows how such a structure can be used for innovation. In this case, the department of research and innovation (R&D) supports all the company's business. Thus, R&D can support all business and create synergies, as an area of innovation can be reused in another, in the creation of other products. Another aspect of interest is the evidence that the matrix structure may be used in activities that are not technologically complex. The structure contributed to the development of hundreds of new products, disseminating the company internationally.

Construtora Norberto ODEBRECHT

History

The Construtora Norberto Odebrecht (ODEBRECHT) is the leader company of Organização ODEBRECHT in the engineering and construction business. Considered as the largest company in Latin America in its sector, it is among the 25 largest international construction companies, whose classification criterion takes into account the revenues obtained outside the home country of the company.[21]

The company's focus is to provide integrated engineering services, procurement, construction, installation, and management of civil, industrial, and special technology works.

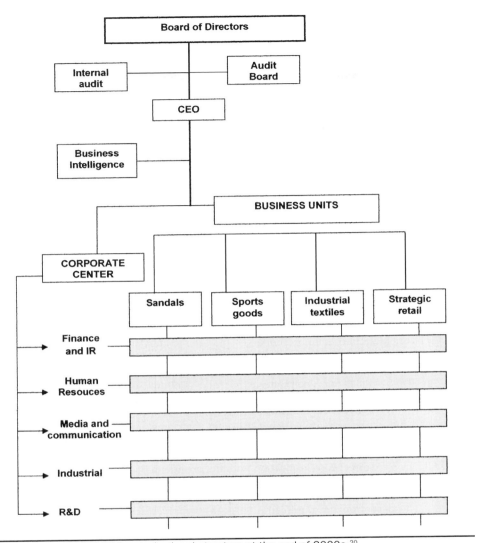

FIGURE 40-4   ALPARGATAS organizational structure at the end of 2000s.[20]

The company offers more than the construction of a hydroelectric plant to the customer, for example, because it proposes to deliver business solutions to put it in operation. Furthermore, we must deliver value to the community according to their needs. ODEBRECHT also develops real estate projects and participates in special projects in the energy, infrastructure, mining, and utilities.

Throughout its history of nearly 70 years of operation, the dedication included the construction of buildings, dams, thermal power plants, steel mills, hydroelectric, nuclear, and petrochemical plants, tourism and real estate complexes, subways, railways, highways, bridges, ports and airports, equipment for the oil and gas and mining projects, sanitation and irrigation, totaling more than 1800 works.

The company was founded in 1944 by engineer Norberto Odebrecht in Salvador, Bahia, as an individual company, which was later transformed into Construtora Norberto ODEBRECHT. Norberto ODEBRECHT provides services to Petrobras since the year of its foundation in 1953. In the 1960s, ODEBRECHT expanded its operations to other states in the Northeast of Brazil, following the development of industrial infrastructure in the region, which increased due to government incentives. In the late 1960s, ODEBRECHT was contracted to build the headquarters of Petrobras in the state of Rio de Janeiro, a work that symbolized the company's expansion into the southeast region of Brazil.

In the late 1970s, ODEBRECHT started internationalization, with the construction of hydroelectric Charcani V, Peru. The reason why it chose this country for the beginning of its internationalization was the characteristics of the neighbor country, similar to those in Brazil, for a team that had worked in the Amazon for years.

In the 1980s, ODEBRECHT acquired competitors, increasing the business segment of heavy construction and industrial assembly, including international business. In 1986, it started working in Argentina and in the following year, in Ecuador. The expansion in Europe began in 1988, with the acquisition of a local company. In 1991, the company entered the world's most competitive market, the United States of America. In the late 2000s, the company operated in 18 South American countries, Central America and Caribbean, North America, Africa, Middle East, and Europe.

The engineering and construction segment, plus the real estate area, had revenues of R$23.992 billion in 2011 out of a total of R$71 billion revenue of ODEBRECHT group. The engineering and construction features grandiose figures: Exports of goods and services to the U.S. were $1.28 billion in 2011, with almost 107,000 employees including 2102 Brazilian expatriates, besides another 751 expatriates from other countries.[22]

Organizational Structure

ODEBRECHT, even before the internationalization, in the last 25 years, operated with an organizational structure whose main features were administrative decentralization and delegation of authority.[23] These features make sense, since the geographic spread is large and the business is specific.[24]

Figure 40-5 shows the macrostructure of ODEBRECHT, which is arranged horizontally.

ODEBRECHT is organized horizontally from the shareholders to the customer, but it is a dynamic organization. The area of business support (RAE) supports all DSs (executive directors) and other directors, including the business leader and vice versa. Business areas are divided geographically and more specifically in Brazil by areas of interest. These large areas are identified as large companies and their leaders are the DSs. Each DS is responsible for and has absolute power of decision, being accountable for his/her goals, agreed with the business leader, their results and losses. Small businesses are close to the customer and each one also has its leader. The example that follows allows us to understand more clearly how the organization works: in Peru, the big company is ODEBRECHT Peru, and the small company, and the contracts administered by it, is a construction. Thus, there may be several small companies serving the same customer, Petrobras or a government.

The vice president of international business is focused only on businesses outside South America and this position is due to the great expansion of the company in the global market. All DSs have direct communication with the president, but treat basically of the macrostrategy of the company, since there is delegation of powers and each director is responsible for his/her area, accounting for all contracts with the customer.

The decentralized administration enables rapid response of businesses that are geographically dispersed, adapting more efficiently to their needs. The disadvantages of a decentralized structure, however, are that many of them are administered competently and do not directly affect the company's performance, being reversed in advantages for their own. One of the disadvantages, the idle capacity of resources and equipment, due to the wide geographical dispersion and the nature of the business, is often reversed, because experts and equipment are transferred from one project to another, as projects are completed. Other work can be at the same customer, on the same large company, or other projects. It is common that the expert is moved to other country to accomplish a new project, or to run a small business. This phenomenon occurs throughout the construction sector, as projects begin and end. This internal mobility offered to employees is a fundamental feature of ODEBRECHT.

EMBRAER

History

For decades, large companies seek management tools that make it possible to gather the competitive advantages of its size, with the agility of small businesses. This implies maintaining specialized areas and at the same time, high levels of integration between them. In the last decade, the integrated product development is increasingly being considered a model, almost a paradigm to address this challenge. Globalization has increased the complexity of this process requiring search and management of suppliers and customers in several countries while demanding flexible structures to accommodate different cultures and geographical separation, but at the same time allowing design, manufacturing, and sales with speed and quality in several global locations.[25]

The organizational structure of the program EMBRAER EMB 170/190 allows integrating the efforts of thousands of people, many of them in different countries and companies, to create, design, and build new aircraft, with deadlines, prices, and competitive performance.

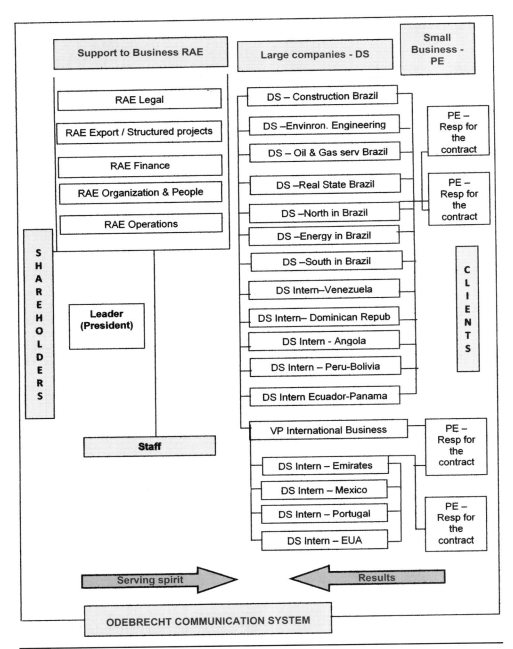

FIGURE 40-5 ODEBRECHT macrostructure.

This program is referred to in the literature as an example of the use of advanced structure, leading to rapid commercialization of a new complex product. The result meant that only 30 months after the beginning of the project, the new aircraft was flying.[26]

EMBRAER is a producer of capital goods of high unit value and long life cycle, (decades) for technically sophisticated organizations, both civilian and military. Such aerospace systems are complex and technologically sophisticated, which entails complex development process.

The company conducts military programs, such as the ground attack jet, "AMX," in partnership with Italian companies. It also produces the military trainer "Tucano." Formed as a state organization in 1969, the company was privatized in 1995. When privatization occurred, EMBRAER brought strong technological expertise, but was weak in terms of management and finance. At the time of privatization, the company, in the regional aviation market, sold the successful turboprop EMB 120 Brasilia. It also developed its first regional jet and found commercial success in the ERJ 145. In 2011, sales revenue reached R$9.9 billion. For the year, EMBRAER exports totaled U$4.2 billion, placing the company among the five largest Brazilian exporters.

Indicator	2005	2006	2007	2008	2009	2010	2011
Sales	9.046	8.265	9.983	11.747	10.813	9.381	9.858
EBITDA	1.077	952	1.322	1.500	1.157	1.069	923
Net profit	709	622	657	429	895	574	156
EBITDA margin (%)	11.9	11.5	13.2	12.8	10.7	11.4	9.4
Net margin (%)	7.8	7.5	6.6	3.7	8.3	6.1	1.6

TABLE 40-4 Financial Performance from 2005 to 2011 (in million R$)[27]

Also in 2011, a total of 216 aircraft were delivered. Of these, 105 were for the commercial aviation market, 103 E-Jets, and two ERJ 145s. Ninety-nine jets were delivered to the business aviation market, including 13 Legacy 600/650, three Lineage 1000, 42 Phenom 300, and 41 Phenom 100. Besides these, four F-5s were delivered to the defense and security market, brought up to date by the F-5M FAB program. Finally, eight Super Tucanos were produced: two for Brazil, three for Ecuador, and three for an undisclosed customer.

EMBRAER has more than 17,000 employees, excluding nonintegral subsidiaries. Table 40-4 presents some of the key financial indicators from 2005 to 2011. Information on investment in R&D reached approximately 2.6 percent of annual sales revenue.

EMBRAER has also built manufacturing facilities abroad, including China, the United States, and Portugal. In China, EMBRAER holds 51 percent of the capital stock of Harbin EMBRAER Aircraft Industry (HEAI) with a factory in Harbin, in association with the Chinese state-owned AVIC, which is currently pending approval by the Chinese government to eventually manufacture the executive jet Legacy 650.

In February 2011, the company opened a factory in Melbourne, Florida, its first Phenom jets factory outside Brazil that, in December, was complemented by the opening of the newest global Executive Aviation Customer Service Center. This new unit will be initially dedicated to the assembly of Phenom jets and to support the executive jet business in the U.S. market. In addition, EMBRAER announced the establishment of a center in this location for engineering and technology as part of the development of its aviation operations in the United States. The center will carry out research and development activities of products and technology for all EMBRAER lines of business and will be the second research and development center located outside the company's headquarters in São José dos Campos. The first one, located in the state of Minas Gerais, was announced in November 2011.

In Portugal, EMBRAER is about to conclude the construction of two new centers of excellence for the production of metallic structures and composite materials in Évora, in the Alentejo region, where it is planned to install an aviation hub. The new facility will absorb an investment of around €170 million and its opening is expected shortly. After the assembly lines in Harbin (China) and Melbourne (Florida, USA) plants, Évora will be the first total production unit of EMBRAER outside Brazil, as the company's global presence develops and its business lines diversify.

In early 2012, EMBRAER Defense and Security increased its stock holdings in the Aeronautical Industry of Portugal (OGMA), specializing in aircraft maintenance and aeronautical production. Until then, control was divided with the European Aeronautic Defense and Space Company (EADS). EMBRAER Defense and Security holds 65 percent of the capital stock in OGMA. The remaining 35 percent belong to the Portuguese government through the Portuguese Defense Company (EMPORDEF), which is also EMBRAER's partner in the development of KC-390 military cargo.

To support after sale operations, EMBRAER owns service centers and spare parts sales groups in São José dos Campos (SP), Fort Lauderdale (Florida), Mesa (Arizona), Nashville (Tennessee), and Windsor Locks (Connecticut) in the United States, in Villepinte (near Roissy Airport—Charles de Gaulle), France, and in Singapore, in addition to a specialized network which already has about 60 service centers owned and authorized in the world. To provide support to customers, EMBRAER also maintains distribution centers of spare parts and specialized teams in Louisville (USA), Beijing (China), and Dubai (UAE).[28]

The main customers of the company are regional air transport operators or military forces. In the case of regional jets, customers are regional operators of air transport worldwide, with a concentration in the United States and Europe. In this regional segment, operation costs and customer services are of exceptional importance in the defining demand, as well as funding for essential products that cost about U$20 million. The customer places their order in two ways. A firm order means an order committed to the delivery date. One option means paying a fee to reserve a place at the company's production line, reserving the right to decide later the purchase of the aircraft. The option has deadline constraint to be accomplished.

Sales of aircraft require certification of each new aircraft design with the national agencies that are responsible for airworthiness. These agencies establish requirements that the project must meet to receive permission to fly in that country. In Brazil, it is the *Instituto de Fomento e Coordenação Industrial (IFI*, an institute that aims at fostering and coordinating industrial endeavors*)*, and an agency of the *Centro Técnico Aeroespacial (CTA*, technical aerospace center*)*. In the United States it is the Federal Aviation Administration (FAA). Typically, FAA certification is critical as many countries follow their lead.

In terms of production, the production is limited and generally customized—from a few hundred to a few thousand planes—and the complexity of the product requires the coordination of many suppliers. In the case of EMBRAER, many of these suppliers are overseas and are major global companies. For example, the supply of jet engines, which represent 30 percent of the planes cost, is a market created by few companies: General Electric, Pratt & Whitney, and Rolls Royce. Therefore, EMBRAER must deal with a global network of suppliers, which in turn, have a global perspective. So the company opted to assume a role of solutions integrator in a product intended for the global market, the regional jets. The ERJ 145 involved 350 suppliers, 95 percent of them located outside Brazil. Four of them were partners at risk.

The ERJ 145, with 50 seats, was a commercial success and has become the company's sales champion. The plane made its first flight in August 1995 and was certified by the FAA in December 1996. Aiming to meet the customer's requirements, EMBRAER developed other jets using what it had learned in the ERJ 145 process; the ERJ 137, with 37 seats, and ERJ 140, with 44 places, completed the line of airplanes.

With this success, the strategic gap, obviously, turned to the lack of products for other types of regional aviation between 50 and 110 seats. The company has filled this gap with a full range of jets: the 145 family, with 37 to 50 places, and the 170/190 family, with 70 to 118 places. The introduction of the 170/190 line began with the first commercial flight in the new EMB 170 jet in January 2004.

The design of a new aircraft line with 70 to 118 passengers—the EMB 170/190 Program—proved to be the most technically complex. It required an investment of U$900 million and included a new level of commitment between partners and customers. Combined with a more aggressive competitive environment, EMBRAER faced new organizational challenges.

Organizational Structure

Given the complex combination of technologies and the diversity of resources involved in the development of aerospace system, the organization felt the need to create a matrix structure to accommodate the program and present a strategy to deal with other competitors. Such organizational structures needed to share authority over resources and management responsibilities between the two hierarchies: the traditional company hierarchy, involving functional areas such as manufacturing, engineering, procurement, and trading and the hierarchy that is responsible for the progress of the development program at the aerospace system. Depending on the circumstances, the functional hierarchy or program can dominate, posing challenges to the matrix structure.[29]

The development of matrix structure in EMBRAER consisted of two main steps. The first occurred with the development of ERJ-145, essentially the ERJ program. The ERJ 145 program allowed the deployment of a concurrent engineering concept in EMBRAER, receiving the name of integrated multiprojects management. As an instrument of implementation of integrated multiprojects management, the program board of industrial vice presidency was created. Initially this matrix approach was restricted to engineering, then within the technical board. It was a weak matrix, that is, the functions and power of the program coordinators were limited. However, in the late 1990s, the matrix view, which served the deployment of concurrent engineering was extended to all the industrial vice presidents with the creation of the program board that is subordinate to the industrial VP. To get an idea of the importance of this structure for EMBRAER, in the early 2000s, the company had nearly 8000 employees, of whom 6000 were under the industrial vice presidency, and 700 were engineers. The new program board began coordinating all projects and programs under the industrial VP, covering the engineering board, the production board, and the procurement board.

The second stage of the matrix structure's use occurred in the development of series 170/190. With the program EMB 170/190, the company identified a need for a dedicated board. At any given moment, there was a board for the EMB 170/190 program and another for the other programs. Figure 40-6 illustrates that during the first half of the 2000s,

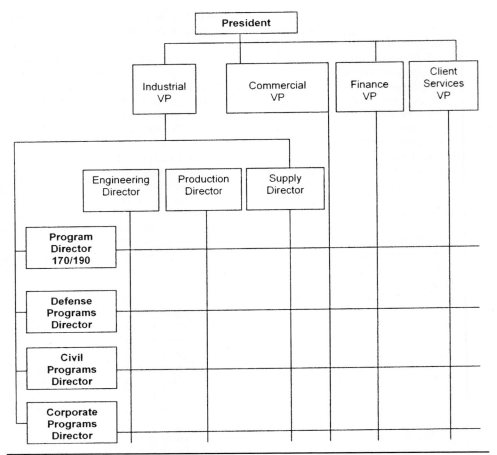

Figure 40-6 EMBRAER organizational structure.

EMBRAER appealed to program directors, each responsible for various programs targeting the company's markets: defense, civil, and corporate. The program 170/190, as well as others, operates in matrix structure with the various units of the vice presidencies: industrial, commercial, and customer services.

Figure 40-6 shows the EMBRAER organizational structure.

Airplanes are complex and technologically sophisticated products, which lead to a complex development process. This process follows how the resources in the development of aerospace systems are structured. In the design phase, work on a project of this type is multi-disciplinary, but restricted to a small group of high-level specialists and uses relatively few material resources. This step creates the product tree, without precision, in the definition of interfaces. As aerospace systems are complex, combining technologies and resources of diverse nature, the product tree has many levels, passing into subsystems and down to equipment and components. At each level, there are specifications and interface documents, absolutely essential to the intellectual division of labor that characterizes the development efforts of such equipment. In a natural way, product tree introduces modularity in the project and construction of aerospace systems. Thus, the necessary management element is introduced to sustain the division of intellectual and operational labor required to such systems. Once defined as the product tree, a natural structure for the program emerges. The natural is to assign development of each subsystem to a group that will set technical goals in the interface documents and costs and deadlines defined in the management documents.

Thus, in a large product development project, the time comes that great teams are required to develop the project. This great team does not function as an undifferentiated whole. Rather, it is placed in a hierarchy and divided into smaller teams, who take care of the various parts of the project. To define the work and the labor and responsibilities division, a product tree is created and based on it is a structure of division of project activities (work breakdown structure). In the past, this hierarchical structure had always been thought as defining the responsibilities of leaders, subprograms, projects, subprojects, equipment, components, and others.

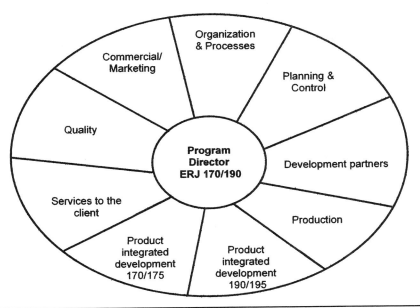

FIGURE 40-7 EMBRAER organizational process for the ERJ 170/190.

In recent decades, however, the need for a cross-functional team to manage large projects has been emphasized. But why not imagine that instead of just a leader team at the top of the program, you reproduce this multifunctional team structure in the various levels of activities breakdown structure? That is, why not a team responsible for each level of the structure divisions? Here we find the idea of creating management cells for all those subprograms, projects, systems, subsystems, equipment and components, and other activities. Why not organize these teams in a similar way? That is, why not include in each management cell the main functions necessary for the integration and coordination of each part of the project, as well as the project as a whole?

The EMB 170/190 program is administered by a cell (program manager center) of 10 people led by a program director, as can be seen in Fig. 40-7. For each pair of aircraft (170/175 and 190/195) there is a chief engineer and all participating in the program manager cell. In addition, representatives from the areas of quality, planning, and control, etc. are also part of the cell. One characteristic of the cellular structure is that it may quickly vary. Above this cell is the industrial VP. For purposes of investment decisions, however, the decision center is the EMBRAER board of trustees. This board has endorsed the major investment decisions to be made. The authorization to initiate studies of the new concept, form a group to design the new aircraft and the search for partners, and form a group to create a new program, have been the main decisions of the board in the development process, as well as the strategic monitoring of the program. This arrangement lasted for the first half of the 2000s. Later, among other changes, it evolved into two chief engineers, one for the 170/175 and another for the 190/195 plane.

Each chief engineer, in turn, activates a managing cell for their aircraft, called the technical center, as shown in Fig 40-3. This cell is formed by a number of product development managers (PDMs). There is a PDM for aeronautics, another for structures, another for electro electronic systems, etc. Some of these PDMs are responsible for design teams (DBTs, design build teams) and others for integrated systems, such as aircraft or structures (IPTs, integrated product teams). The DBTs have the function of ensuring they can physically build the airplane. That is, the functional subsystems are spread throughout the aircraft and someone must ensure that, when ready, all fit and function as expected when assembled on the plane. Each DBT is responsible for a piece of the physical plane (nose, tail, etc.). IPTs take care of the functional vision. That is, their role is to ensure a good project of the aircraft subsystems: propulsion subsystem, hydraulic subsystem, etc. Each PDM, in turn, leads a cell related to his unit. For example, the aircraft PDM leads a cell formed by elements connected to IPTs as aerodynamic, flight quality, performance, etc. As can be seen in Fig. 40-8.

All these cells operate in matrix, as can be seen in Fig. 40-8.

It is important to point out that, Fig. 40-8, in particular, is a simplification of the real structure, since the program involves partners in different countries. In all teams, EMBRAER is the leader and the partner is present in basically all teams, that is, system or segment teams.

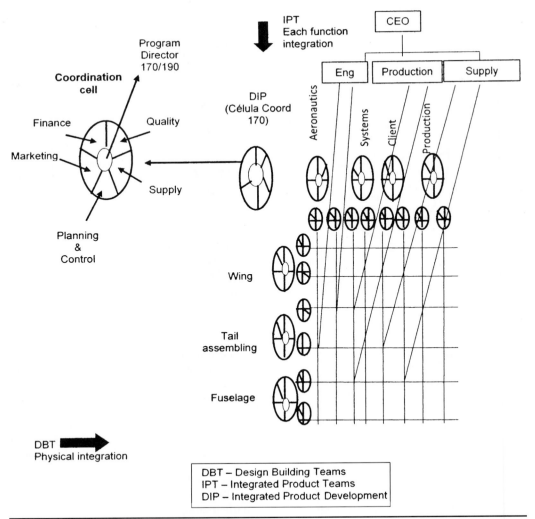

IPT
Each function
integration

DIP
(Célula Coord
170)

Program
Director
170/190

**Coordination
cell**

Finance   Quality

Marketing

Supply

Planning
&
Control

Wing

Tail
assembling

Fuselage

CEO

Eng   Production   Supply

Aeronautics

Systems

Client

Production

DBT
Physical integration

DBT – Design Building Teams
IPT – Integrated Product Teams
DIP – Integrated Product Development

Figure 40-8   EMBRAER structure to the program ERJ 170/190.

EMBRAER matrix combines a three-axis structure with cells at junctions between airplane parts and the integrating systems of the various parts. This facilitates innovation because it integrates better teams that develop cockpit, landing gear, rudder, with these components integrating systems. This form of structure is an innovation in itself because, instead of a manager at matrix intersections, we have a manager with a team formed by elements with competencies in several areas. Without the cell, the manager would have to call several meetings with experts. The cell has previously scheduled meetings while, at the same time, the cell members have a greater responsibility before the company when comparing the call for a meeting.

This structure is particularly important in this case because the engineering of an aircraft involves 1000 engineers and the work has strong safety requirements and a great interrelationship between specialties.

Thus, the structure formed by cells is beyond the limits of EMBRAER to interact with other companies in different countries, each with different languages and cultures. This structural form was essential to meet the challenges of the program. As a result, the structure of the aircraft parts development must be similar to its counterparts in other levels of the product tree. This is what the circles in Fig. 40-8 represent. Considering the levels of the product tree a dimension, one can speak of an organization that, with minor modifications, is maintained at various levels. In the case of an aircraft, this division is still necessary to overlay the inclusion of suppliers and the division between physical integration of the aircraft parts and the functional integration of its subsystems. This is the meaning of the circles representing the teams in the various levels of product trees. Each one would represent the need of multifunctional teams in each level and element of the tree products.

Conclusion

The importance of organizational structure for innovation is evident in the text, according to the various authors dealing with the theme and the real examples of organizations that use the design of the structure to stimulate and manage the innovation, the creation of new products and services, as well as the implementation of new ideas of management processes in general.

The option by large and complex real organizations in the chapter is due to the fact that it is easier to emphasize the relevance of organizational structure for the occurrence of successful innovation. The majority of organizations chosen were Brazilian, since the authors are Brazilian, but consult with global companies. The global presence is an additional fact that highlights the importance of the organizational model for the company success that, in the contemporary world, requires the continued existence of innovation.

It should be emphasized that the organizational complexity is not only in technology, that is, the same activities with less sophistication in this respect, can benefit from structures that facilitate the occurrence and innovation management. A useful example in this regard is ALPARGATAS, Brazilian multinational—although with roots in Argentina—which operates in nearly 90 countries.

The contribution of organizational structures presented in this article, while fostering innovation and enabling business success, benefits not only its shareholders, but also an entire network of stakeholders, with direct and indirect relationship with the studied companies. In addition to shareholders, it can be included in this group of beneficiaries, employees, customers, suppliers, and so on. Among the clients, we can include some governments, as ODEBRECHT, for example, has built hydroelectric power plants and other large works in many countries worldwide. In fact, the communities where these works are developed are the big beneficiaries of ODEBRECHT. Therefore, by allowing a country government to run a work in more favorable conditions of cost and time, the innovative management of the organization ends up benefiting the population of a country and highlights the amplitude of the benefit you can receive from an innovative structure. The same occurs in circumstances in which ODEBRECHT provides services to Petrobras, the largest Brazilian company, since its founding 60 years ago.

The case study of EMBRAER, a company that belonged to the Brazilian government since its foundation in 1969 until its privatization in 1995 also demonstrates the breadth of innovation activities. When it was privatized, EMBRAER, from the standpoint of management and finance, was fragile. It can be stated that the organizational structure contributed to raising EMBRAER to a leading global industrial firm. The aircraft manufacturing sector is of high technological complexity. The organizational structure contributed to the enablement of the product development process including integration of a diverse group of suppliers from different countries. From the moment a new product is conceived, it's necessary to consider the most suitable organizational model for the task.

In summary, in the three companies studied in this chapter—ALPARGATAS, ODEBRECHT, and EMBRAER—the organizational structures contributed significantly to innovation.

This text aims to bring the major forms of organizational structure that enable and encourage innovation, to give some examples and cases in which this phenomenon can be verified. Of course, there is no intention of examining these companies from anything other than the organizational aspects that are necessary for innovation to occur. Also, since the dynamics of the environment and organizations imply constant structure changes and adaptations, some information may be out of date by the time this chapter is published.

Therefore, the reader may want to examine these companies more thoroughly, investigating other aspects that may be contributing to the innovation success. Also, information about organizational structures can be updated and its evolution analyzed and compared over time. Finally, it is noteworthy that the organizational structure is a critical component for successful innovation, but is not isolated. It is necessary that the company's strategy is appropriate; it is necessary that processes are appropriate to enable the structure and people's behavior and company culture to complement this scenario.

References

1. Vasconcellos, E. and Hemsley, James R. *Organization Structure*. São Paulo: Editora Guazelli, 1997. (Available in Portuguese).
2. Lawrence, Paul R. and Lorsch, Jay R. *Organization and Environment: Managing Differentiation and Integration*. Homewood, IL: Richard Irwin, Inc., 1969.

3. Vasconcellos, E. and Hemsley, James R. *Organization Structure*. São Paulo: Editora Guazelli, 1997. (Available in Portuguese).

4. Burns, T. and Stalker, G.M. *The Management of Innovation*. London, UK: Tavistock Publications, 1996.

5. Vasconcellos, E. and Hemsley, James R. *Organization Structure*. São Paulo: Editora Guazelli, 1997. (Available in Portuguese).

6. Vasconcellos, E. Matrix Structures in Research and Development Institutions in the State of São Paulo. Habilitation (Post-Doctoral) Thesis. School of Economics, Business and Accountancy at the University of São Paulo—FEA/USP, 1977. (Available in Portuguese).

7. Vasconcellos, E. "A Model for a Better Understanding of the Matrix Structure." *IEEE Transactions in Engineering Management*, 26: 58–64, 1979.

8. Galbraith, Jay R. *Designing Matrix Organizations that Actually Work: How IBM, Procter & Gamble, and Others Design for Success*. San Francisco, CA: Jossey-Bass, 2009.

9. Galbraith, Jay R. *Designing Matrix Organizations that Actually Work: How IBM, Procter & Gamble, and Others Design for Success*. San Francisco, CA: Jossey-Bass, 2009.

10. Zilber, S.N., Vasconcellos, E., and Stelmach, J. How to structure the e-Business activity: a success case at Ford. Vasconcellos, E. (ed.). In *Competitiveness and e-Business: Brazilian Companies Experiences*. São Paulo: Editora Atlas, 2005. (Available in Portuguese).

11. Vasconcellos, E. and Hemsley, James R. *Organization Structure*. São Paulo: Editora Guazelli, 1997. (Available in Portuguese).

12. Vasconcellos, E., Nascimento, Paulo Tromboni de Souza, Lucas, Paulo Cesar de Souza, and Nelson, Reed E. Matrix—cellular structure to develop high complexity products: the EMBRAER case. Vasconcellos, E. (ed.). In *Competitive Internationalization*. São Paulo: Editora Atlas, 2008. (Available in Portuguese).

13. Vasconcellos, E. and Hemsley, James R. *Organization Structure*. São Paulo: Editora Guazelli, 1997. (Available in Portuguese).

14. Pimentel, João Eduardo Albino and Vasconcellos, E. Internationalization and the binomial strategy—structure: the case of Cia de Concessões Rodoviárias. Vasconcellos, E. (ed.). In *Competitive Internationalization*. São Paulo: Editora Atlas, 2008. (Available in Portuguese).

15. Franco, Arnaldo de Mello, Queiroz, Maurício Jucá de, and Vasconcellos, E. Internationalization strategy and impacts on the organizational structure of São ALPARGATAS. Vasconcellos, E. (ed.). In *Internationalization, Strategy and Structure*. São Paulo: Editora Atlas, 2008. (Available in Portuguese).

16. ALPARGATAS. Annual Report 2011. Available at ri.ALPARGATAS.com.br/informacoes_financeiras/relatorio/2011/index.html. Accessed February 23, 2013.

17. Franco, Arnaldo de Mello, Queiroz, Maurício Jucá de, and Vasconcellos, E. Internationalization strategy and impacts on the organizational structure of São ALPARGATAS. Vasconcellos, E. (ed.). In *Internationalization, Strategy and Structure*. São Paulo: Editora Atlas, 2008. (Available in Portuguese).

18. ALPARGATAS. Annual Report 2011. Available at ri.ALPARGATAS.com.br/informacoes_financeiras/relatorio/2011/index.html. Accessed February 23, 2013.

19. Franco, A. Phone conversation with executive, 2013.

20. Franco, Arnaldo de Mello, Queiroz, Maurício Jucá de, and Vasconcellos, E. Internationalization strategy and impacts on the organizational structure of São ALPARGATAS. Vasconcellos, E. (ed.). In *Internationalization, Strategy and Structure*. São Paulo: Editora Atlas, 2008. (Available in Portuguese).

21. Almeida, Alda Rosana Duarte de and Vasconcellos, E. Knowledge management, organizational structure and internationalization: the case of Norberto ODEBRECHT Construction. Vasconcellos, E. (ed.). In *Internationalization, Strategy and Structure*. São Paulo: Editora Atlas, 2008. (Available in Portuguese).

22. ODEBRECHT. Annual Report 2011. Available at http://www.ODEBRECHTonline.com.br/relatorioanual/2012/. Accessed February 25, 2013.

23. Almeida, Alda Rosana Duarte de and Vasconcellos, E. Knowledge management, organizational structure and internationalization: the case of Norberto ODEBRECHT Construction. Vasconcellos, E. (ed.). In *Internationalization, Strategy and Structure*. São Paulo: Editora Atlas, 2008. (Available in Portuguese).

24. Vasconcellos, E. "A Model for a Better Understanding of the Matrix Structure." *IEEE Transactions in Engineering Management*, 26: 58–64, 1979.

25. Vasconcellos, E., Nascimento, Paulo Tromboni de Souza, Lucas, Paulo Cesar de Souza, and Nelson, Reed E. Matrix—cellular structure to develop high complexity products: the EMBRAER case. Vasconcellos, E. (ed.). In *Competitive Internationalization*. São Paulo: Editora Atlas, 2008. (Available in Portuguese).

26. Muller, Amy, Hutchins, Nate, and Pinto, Miguel Cardoso. "Applying Open Innovation Where Your Company Needs It Most." *Strategy & Leadership*, 40 (2): 35–40, 2012.

27. EMBRAER. Annual Report 2011. Available at ri.EMBRAER.com.br. Accessed February 23, 2013.

28. EMBRAER. Annual Report 2011. Available at ri.EMBRAER.com.br. Accessed February 23, 2013.

29. Vasconcellos, E., Nascimento, Paulo Tromboni de Souza, Lucas, Paulo Cesar de Souza, and Nelson, Reed E. Matrix—cellular structure to develop high complexity products: the EMBRAER case. Vasconcellos, E. (ed.). In *Competitive Internationalization*. São Paulo: Editora Atlas, 2008. (Available in Portuguese).

CHAPTER 41

Journey to
Innovation Excellence

Andria Long

In over a decade of leading innovation at several consumer packaged goods (CPG) companies, one thing I have learned is that achieving innovation excellence is a journey. In this chapter, I will bring this journey to life and share the key factors to success in creating long-term sustainable innovation.

There are several key attributes to be best in class in innovation. In Fig. 41-1 you can assess where you are in your innovation journey based on those key attributes.

I have found there are seven key enablers for successful innovation:

1. Discipline in building the right foundation
2. Innovation strategy development
3. Portfolio strategy and management
4. Prioritization and focus
5. Process and executional excellence
6. People and capabilities
7. Culture of innovation and commitment

Discipline in Building the Right Foundation

One of the biggest areas of opportunity is failing to perform due diligence up front for innovation. A common mistake is thinking innovation starts with ideas. Innovation must be managed with the same diligence and discipline as other business processes whose success depends on a strong foundation. Building the right foundation is a key capability and differentiator for achieving bigger innovation versus closer in commercialization.

To build the right foundation for innovation there are several things you need to do:

- Define what innovation is to your organization
- Develop an innovation strategy
- Establish innovation metrics

Define What Innovation Is to Your Organization

The way I define innovation is solving problems (*needs*) in a new or different way. There are two important components to this definition. One is it must solve a problem or a need. Some will argue, especially in the case of technology like the iPhone, that consumers can't articulate what they don't know. But what consumers *can* do is tell you what they are dissatisfied with and what can be better, which leads into the second important part of my definition, solving problems in a new or different way. If there is no differentiation, there is no innovation. Any product or service can always be made better.

FIGURE 41-1   Journey to innovation excellence.

Types of Innovation

There are three main types of innovation, starting with the most difficult and infrequent type of innovation.

1. **Breakthrough/disruptive/new to the world (Fig. 41-2):** Paradigm shifts that reframe existing categories. Disruptive innovation drives significant, sustainable growth by creating new consumption occasions and transforming or obsolescing markets.

FIGURE 41-2   Examples of breakthrough innovation.

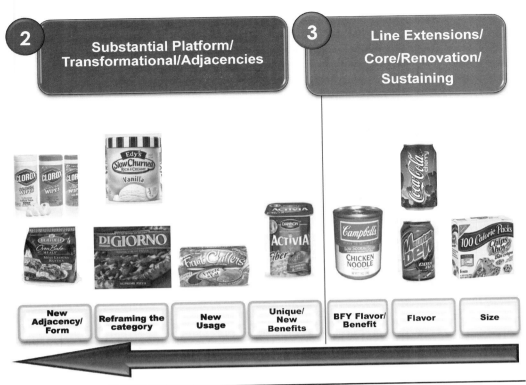

FIGURE 41-3   Examples of substantial innovation and line extensions.

2. **Substantial platform/transformational/adjacencies (Fig. 41-3)**: Innovations that deliver a unique or new benefit or usage occasion within an existing or adjacent category.
3. **Core/line extensions/renovation/sustaining close-in (Fig. 41-3)**: Innovation that extends and adds value to existing line or platform of products via size, flavor, format. It is incremental improvement to existing products.

Innovation Strategy Development

In order to achieve excellence in innovation you need to set the strategic direction. The innovation strategy should be

- Company-wide
- Clear, cascaded, and energizing
- Tied to corporate, brand, and portfolio strategy with technology enablers
- Based on insights

An innovation strategy should answer

- How much innovation do you need? Metrics
- What are the desired type, mix, and size of innovation? Portfolio mix
- Where will you innovate? The strategic focus areas

How Much Innovation You Need

Here you define specific goals and financial objectives that will be delivered by the innovation. There are two approaches for determining how much innovation you need:

- Top-down
- Bottom-up

Top-Down

Top-down is simply a revenue goal that is usually set from the top by the senior leadership team. It is usually a dollar revenue goal or a percentage of revenue target from innovation.

Bottom-Up

Bottom-up is a buildup of portfolio requirements to meet business objectives. Several key questions must be answered to build a bottom-up innovation number.

- What is the overall strategy to grow revenue?
- What is the base business expected to do (without product changes)?
- How far can you stretch the existing offerings? For example, refresh or improve taste of existing SKUs?
- What is the potential of existing innovation? This includes line extensions.
- What will new innovation add?
- What is the total innovation effort based on the above answers?

What Is the Desired Type and Mix of Innovation? Portfolio Mix

The purpose is to set direction on the specific type and mix of innovation you will develop to reach your goals. An example is shown below.

Innovation Types	Mix
Breakthrough/disruptive/new to the world	5%
Substantial platform/transformational/adjacencies	70%
Core/line extensions/renovation/sustaining	25%

Size

Setting the right metrics is crucial. Too many companies arbitrarily set the size of prize at $100 million. This can have a negative impact in multiple ways. One, you're immediately setting yourself up for failure by setting a goal that is unrealistic and unattainable. What happens is some really great ideas do not move forward because they do not hit that hurdle or, on the flip side, the numbers get rounded up just to get something into the pipeline. This is also very demotivating to your teams since it is unrealistic. According to Symphony IRI Group 2011 New Product Pacesetter Report from April 2012, a vast majority of products fail to meet $7.5 million in first year sales, with only 2.1% reaching $50 million and only point as noted above three percent achieving over $100 million.[1]

In addition, one of the biggest issues in a large company is that economies of scale and efficiencies can work against you. Initially, innovation is often not going to achieve the same margins as something that you have perfected and have been making for 20 years. So using the same criteria as an existing business can weed out a lot of great ideas.

As brands grow, it is increasingly difficult to sustain compounded annual growth rate expectations. Most mature companies have to create organic growth of four to six

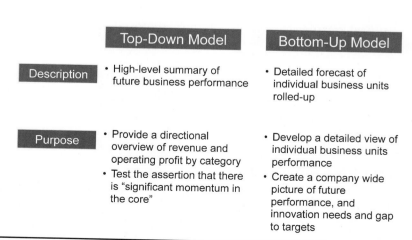

	Top-Down Model	Bottom-Up Model
Description	• High-level summary of future business performance	• Detailed forecast of individual business units rolled-up
Purpose	• Provide a directional overview of revenue and operating profit by category • Test the assertion that there is "significant momentum in the core"	• Develop a detailed view of individual business units performance • Create a company wide picture of future performance, and innovation needs and gap to targets

Figure 41-4 Top-down and bottom-up approaches to determining how much innovation you need.

percent year in, year out. For some large CPG companies, that can be the equivalent of building a $4 billion business every year. This is a growth mandate that requires change and acting differently.

There is not a one-size-fits-all with $100 million as the size of prize. This is why it is critical to benchmark and set metrics with both an internal and external view. From an internal perspective you should be looking to understand what is needed for growth based on historical contributions from innovation as well as forward projections to meet your growth objectives. From an external perspective you should seek to understand what is needed to stay relevant in your category versus the competition. This can be achieved by understanding the right innovation frequency for your category and the right dollar size for customer sustainability.

Where Will You Innovate? The Strategic Focus Areas

Platforms or Opportunity Areas

You must clearly define the platforms the organization will focus on. These choices will further focus the work of the organization and enable important decisions. The importance of developing platforms is that it provides multiple ideas and opportunity areas versus one-offs.

A platform is a consumer need–based opportunity that inspires multiple innovation ideas with a sustainable competitive advantage to drive growth. A strategic opportunity assessment will help determine what potential platform areas to pursue. You should begin by determining and valuing key insights by leveraging consumer and category segmentations, and translating insights into actionable opportunity areas.

A critical element of each platform or opportunity should be the consumer need. Consumer-driven innovation leverages the proper use of consumer research to drive the strategy. It is important to validate the platform's potential early on by conducting a business case with the appropriate rationale to support it.

Business Case

The business case is the insight-driven marketing opportunity that leverages sound business logic. It is a "living" document that is updated with new information and learnings throughout the innovation process. The business case properly vets each of the platform opportunity areas. It will increase your chances of success in the concept phase because it is based on consumer needs and insights and the appropriate up front due diligence. It also provides the foundation for scalable product, packaging, and technology platforms.

Your business case should answer several key questions.

What Are the Key Elements of the Platform?
- What is your hypothesized sustainable point of difference and how will you win?
- What is the need or unmet need?
- What is the market situation and competitive assessment?

Size of Prize
- What is the quantified size of opportunity?

Risk Assessment and Mitigation
- What will be the fact-based approach to identify challenges and uncertainties?
- What are the risk mitigations and probabilities for:
  - *Timing*—Risk of not meeting, in particular a launch date
  - *Project cost*—Risk of exceeding costs in terms of investment and resources required
  - *Technical execution*—Risk of the product not meeting the defined concept and specs
  - *Cost*—Risk in product costs, higher cost of goods sold (COGS), risk to P&L

Fit to Strategy
  How does the project fit in the strategy?
- From a business unit (BU)/brand strategy?
- From a portfolio strategy?

Consumer Insights
- What specific needs and trends can be identified in the market, relevant to the platform?

- Which targets do you have for consumer attributes (including price)?
- What are the results so far on consumer feedback/tests? Example metrics are:
  - Purchase interest
  - Frequency
  - Value
  - Meets a need
  - Is seen as unique or different/better
  - Incrementality/cannibalization

Competition
- What is the market landscape, what competition do you face?

Go to Market Strategy
- How will the product be marketed?

Capabilities
- What are the manufacturing platforms required to produce the product concept?
- What is the manufacturing strategy? Define the strategy you have including internal versus external manufacturing, both short and long term.
- What are the technical hurdles and what is the approach to overcome these?
- What are the capital requirements?
- What is the buildup of COGS, including ongoing assumptions like line capabilities and commodity prices?

Investment

Capital Investment
- Manufacturing equipment
- Packaging equipment
- Change parts

Go to Market Investment
- **Trade/Customer Investment**
  - Slotting fees for distribution
  - Trade spending
- **Consumer Investment**
  - Advertising dollars (creative development, media spend)
  - Promotional (coupons like FSIs, on-pack, Catalina)
  - Digital/social media

Technology Road Map

Technology road mapping is aligning consumer platforms to technology and manufacturing capabilities. Developing a technology road map is important to ensure enabling capabilities are in place to support your innovation pipeline. This will map out what innovation capabilities are needed to support each of the platforms. Figure 41-5 links the technology platform with the consumer needs and platform initiatives.

Platform Management

For each platform, you should have several initiatives defined to meet your innovation objectives. If there are several validated consumer concepts in each platform opportunity area, it's a good sign that you have a thoroughly vetted business case. Once you have validated concepts, develop a road map and unified go-to-market approach.

Portfolio Strategy and Management

A good portfolio strategy should enable prioritization and focus, be able to proactively and realistically manage risk, and have the right measures in place to ensure overall objectives are being met.

Portfolio strategy is about making choices for prioritization and focus. There are several ways you can look at a portfolio strategy:

- By business unit: Is there a particular business unit you want to focus your innovation efforts on?

- By category/segments: Are there particular categories or segments that you want to focus your innovation on where there is high growth potential?

- By brand: Are there brands in your portfolio that have more opportunities than others?

How it all comes together:
An Innovation Strategy Grounded in Insights & Consumer Needs

The clearer, the more precise, and the more accurate the
definition of consumer needs—the higher the ratio of success.

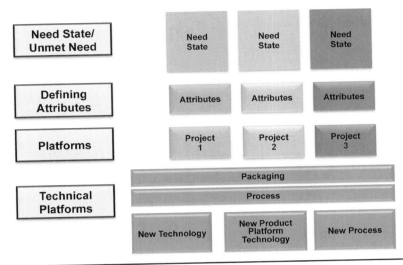

FIGURE 41-5 Consumer needs.

Portfolio Management

Portfolio management is the ongoing management of innovation to make sure you deliver against your goals and innovation strategy. Portfolio management is all about choice and making sure you are working on the right projects to achieve your goals.

A balanced portfolio of innovation will have a mix of type and time horizons, and be prioritized based on strategic importance. It will help manage risk, reward, investment, and different time-to-market horizons.

- Type: Spanning smaller incremental innovations to bigger more disruptive innovations
- Time horizons: A long range view and staging across multiple time horizons (short, medium, and longer term)
- Prioritization: Based on strategic importance as set earlier in your portfolio strategy

Role of Portfolio Management

The role of portfolio management is to maximize value while managing risk to achieve your innovation goals. You can maximize your portfolio value by selecting the portfolio of projects that help you achieve the best revenue and profit growth. You can manage risk by creating a balanced mix of time, type, and investment levels. Balancing projects across stages helps maximize organizational capacity utilization for better time to market. You should also be looking at pipeline value from both a forward- and backward-looking perspective (projected and historical) to ensure you will achieve your goals.

Criteria for Evaluating and Prioritizing Projects

Effective portfolio management should enable you to objectively screen ideas and projects, and enable formal measuring and monitoring of performance. It should answer several key questions:

- How will you evaluate innovation? This includes go or no-go criteria
- How will you measure the innovation program effectiveness? What are the key performance indicators (KPIs)?
- How will you resource the work?

One of the most common and, quite frankly, overused terms is "bigger, better, faster" innovation. Sometimes you will also hear "fewer, bigger, better." Given the industry emphasis and usage of these terms, I have shown the different metrics for ensuring work is focused on bigger, better, and faster innovation in Fig. 41-6.

Decision Making

Now that you have established criteria and have visibility across the portfolio of initiatives, it requires disciplined decision making and governance to make it work. There are trade-offs that have to be made so resources are not dragged down with initiatives that should not be worked on. This will only fragment and suboptimize the really great projects. Once clear

Critical Success Factors	Key Measures	Why
Maximize Value = **Pipeline Value**	• **Projected 3 Year <u>NSV</u>** (by year) • **Projected 3 Year <u>Profit</u>** (by year) $ Profit & %	• Focus on <u>**sustainable**</u> innovation • 3 year profit view provides **realistic view/ongoing run rate** without launch year costs
<u>**Sufficiency**</u> **Do we have enough projects in process and planned to meet growth targets?**	Above plus….. • **Stage** each initiative is in (feasibility, development, commercialization or launch) • **Target launch date** • $ Target for Pipeline by BU	• Activities can be coordinated, resource needs anticipated and investment decisions made. • Ongoing **visibility to fill the gaps** against strategic targets • Indicates how well the innovation needs are covered • Set a $ target rather than a % of sales target to provide further clarity and definition. *(absolute numbers vs. a percentage given continued base growth)*

(a)

Critical Success Factors	Key Measures	Why
Balance Make sure there is a healthy mix of projects to increase odds of success.	• Mix by <u>**Type**</u> of Innovation • Mix of projects by phase of development and target launch dates	• Helps <u>**manage risk**</u> • Help ensure you have a <u>**continuous pipeline**</u> • Assist with planning (capacity + next generation process/ equipment needs) • Balance across stage gate phases – <u>**capacity planning**</u>
Definition What types of projects are tracked as part of the portfolio process? Which projects are counted in the metrics?	• Includes line extensions, platforms • Does <u>**not**</u> include change management – reformulation, net weight changes or regulatory, etc.	• What's included in line with <u>**industry standards & true picture of growth**</u>

(b)

FIGURE 41-6   Key metrics for bigger innovation.

Critical Success Factors	Key Measures	Why
Capability **Pipeline Performance- =** **In Market Results** Organization capable of delivering innovation	In Market Performance • **Historical 3 year NSV** from recent launches • **Historical 3 year profit** from recent launches • Benchmarks against peers	• Indicates how **effective** the innovation process & organization have been over recent years (how we did vs. targets set) • Post Launch Assessments/ Scorecard • Indicates how **realistic assumption**s are around potential gains from Innovation projects • Indicates whether existing **capacity** for projects is large enough
Sufficiency How much and what do we need for Innovation by category and for overall pipeline?	• Built bottoms up from BU • Target $ NSV • Target $ Profit & %	• Need to understand what's needed to stay relevant in the category - what Type of innovation, Number – how many, Frequency how often , Size by Category (to be sustainable on shelf)

(c)

FIGURE 41-6   Key metrics for better innovation. (*Continued*)

Critical Success Factors	Key Measures	Why
Efficiency **Speed to market** **Speed to shelf**	• Track **weeks in each phase** by project (stage gate data) Should track by type (LE vs. Innovation) • ACV build time to shelf, & Vs. ACV target	• Helps identify **capacity constraints** (certain phases are much more labor intensive) • Helps identify **bottle necks** • Sets realistic expectations on timing • Benchmark vs. your past performance • Benchmarks vs. competitors
Prioritization & Focus (FEWER, Bigger, Better)	• All Above portfolio Metrics combined with Stage Gate Metrics	• Best utilization of resources • Ensure you are working on the right things

FIGURE 41-7   Key metrics for faster innovation.

hurdles on size of prize expectations are established, it is critical that the leadership team has the willingness to kill small and bad ideas so the right projects are being worked on.

A simple company-wide scorecard with a snapshot of how you are doing will help ensure you stay on track and deliver against your goals. The scorecard should have a long-term focus—minimum of 3 years—and include the appropriate financials metrics as outlined earlier.

"And it comes from saying no to 1000 things to make sure we don't get on the wrong track or try to do too much. We're always thinking about new markets we could enter, but it's only by saying no that you can concentrate on the things that are really important."

Steve Jobs

Prioritization and Focus

Ruthless prioritization of projects is critical to ensure you appropriately focus your resources, both people and financial, on the right things. With the right focus your speed to market will increase because you are not spreading your resources too thin and suboptimizing opportunities.

Type

You need an effective way to sort and prioritize projects to ensure bad ideas are killed. To do this you should establish a defined process and set of criteria for evaluating and prioritizing projects. As mentioned earlier, you will need to classify your projects by innovation type to ensure you are managing your risk and portfolio of initiatives to get the desired results.

Another Type, But It's Not Innovation

The following maintenance or change management activities are *not* types of innovation:

- Cost savings
- Ingredient or product change
- Regulatory change
- Label change

Given you are using the same set of resources, these are often lumped in and considered with innovation, but they are *not* innovation. While these types of initiatives are important to running a business, they should be managed with their own set of criteria and separated from the new product mix.

Hurdles by Type

While you will measure the same things across the different types of innovation, the hurdles for success will not be the same. This is why it's important from the beginning to establish your goals as an organization and to define what innovation is and what mix and type you want in order to achieve the sustainable growth results needed.

Why is it important to look at prioritizing within type?

Given the nature of uncertainty, risk, probability of success, and the investment level necessary for bigger, more substantial innovation, you need to look at things by type. Otherwise the short-term, less risky, smaller initiatives like line extensions and closer-in initiatives will potentially look more favorable. It is human nature to be risk averse and gravitate toward short-term, more defined, closer-in initiatives, and the here and now because it is inevitably more comfortable. If you are truly going to master bigger, better innovation it is crucial to have the culture and commitment as an organization to enable it to truly happen.

Organizational Capacity

It is important to understand what your organizational capacity is, that is, the number of projects that can be handled based on current resources. There are a number of ways to approach this depending on where you are on your innovation journey. If you track R&D hours you can leverage that data to understand capacity. Another simpler way in the beginning stages is to understand project capacity by function. Obviously, bandwidth will differ based on the type and complexity of initiatives as well as the stage the initiatives are in. The key will be to monitor it and keep a dialogue going with your cross-functional teams to understand if there are pinch points. As you manage the process you will see where projects get caught up or slowed down and be able to diagnose and address those issues.

People and Capabilities

When it comes to innovation there is a definite need for improved and distinct innovation competencies. Not all great brand managers make great innovators—you need to be comfortable with risk, ambiguity, and a lot of change.

Characteristics and Capabilities for Success in Innovation

Capabilities	Key Characteristics
Results driven	- Sets and achieves ambitious goals - Able to differentiate between critical/noncritical activities to focus on what really adds value - Ability to solve problems/generate ideas—leveraging both internal and external resources

Patience and Acceptance of Risk and Failure

Perhaps one of the biggest barriers to achieving breakthrough innovation is the lack of a culture that promotes patience and "successful failure." Patience to let ideas properly incubate and to take the time to do the up front due diligence will increase your chance of success and save your time in iterating. Risk aversion can manifest itself in several ways: delaying decisions, always asking for more data, more testing, and more levels of certainty. Leadership needs to encourage and model risk-taking, knowing that the innovation process will include some trial and error and recognize that some failures are prerequisite to success.

"Failure is success if we learn from it."

Malcolm Forbes

Risk Aversion

Innovation is risky. Let's face it: failure rates are high. Most people do not want to take risks on innovation, especially if it is going to impact their career potential. Fear of failure impacts the ability to innovate in several ways. One is watering down a great idea to make it less risky. An example would be making the idea fit to current capabilities instead of investing in new capital, which suboptimizes the original proposition and sabotages what made it a big idea in the first place. Another way is requiring more information or proof to make a decision, as a way to avoid taking risks. This can mean more meetings and more research, which impact speed to market.

"Only those who dare to fail greatly can ever achieve greatly."

Robert F. Kennedy

Accountability

A performance culture effectively reinforced with compensation, recognition and rewards, and accountability embedded in objectives is the best way to achieve sustainable innovation. In order to build a dedicated innovation team that is committed to innovation revenue growth, you will want to incent them on several measures. The two major ones will be pipeline value and pipeline sufficiency. The senior leadership team should also have innovation metrics as part of their incentives as well; usually both pipeline value and pipeline performance at a minimum of a 3-year horizon.

Time to Think versus Do

So much time is spent in meetings that the only time anyone has left to think is at the end of the day or when they are commuting. This does not help foster innovative thinking. This ends up eventually frustrating the more creative innovation thinkers who end up leaving and going out on their own to make it happen. And now technology has mitigated many of the huge barriers for launching your own product. So why not?

Some companies have already recognized this and are building in time for their people to innovate. They are allowing and incentivizing their employees to spend 10 to 15 percent of their time on innovation ideas they are passionate about.

Commitment

- Leadership support and champion or owner
- Long-term mind-set

"There is no radical innovation without inspiring leaders."

Roberto Verganti

Leadership Support and Champion or Owner

Senior management support is mandatory for successful innovation. Senior management must ensure everyone in the organization knows the importance of innovation but they cannot micromanage the process. Leadership must recognize that innovation is critical to success and communicate its importance throughout the organization. They can do this by delivering a consistent message around the innovation goals, vision, and focus on long term and having metrics as part of management annual objectives. It is important that leadership provides strong support and empowerment to team members by leaving day-to-day decisions to the team. They should also be strongly committed and involved in go or no-go decisions.

Another key component is commitment to investment. This can be achieved by ensuring the appropriate resources are allocated to innovation (time, people, dollars, etc.) and are not sacrificed for short-term needs. Commitment to multiyear support is necessary for impact and sustainability. This means you do not launch and leave or shift support to the next launch. Leadership also needs to be open to a new set of metrics to allow for lower margins or different rates of return versus existing, well-established businesses.

Long-Term Mind-Set

Short-term versus long-term mind-set, now versus later Another issue is short-term versus long-term thinking. The now always tends to win over the later. Today's business fire drills are always going to get the attention before next year's plans and unproven ideas. This is another great reason to have an executive leader and team who is focused on innovation.

The time to make these changes is now. You do not want to procrastinate, because if you do not do it, someone else will.

Reference

1. http://www.symphonyiri.com/Portals/0/ArticlePdfs/T_T%20April%202012%20NPP.pdf.

CHAPTER 42

Culture of Innovation

Maria B. Thompson and Nina Fazio

Introduction

Innovation is just a sexy name for change management or management of change. This chapter is about management of culture change in support of innovation. Our premise is that if an organization nurtures a capability for management of change, it will create and reinforce a culture of innovation.

Innovation is different than invention. For our purposes in deploying a culture of innovation, invention is the conversion of cash to ideas. Innovation is the conversion of ideas to cash. We define culture as the attitudes, feelings, values, and behaviors that characterize and inform society as a whole or any social group (organization) within it."[*] Organizational culture, for our purposes, includes items such as beliefs, values, goals, work processes, procedures, environment, techniques, and capabilities.

We will discuss how to successfully align the accountable organizations' attitudes, feelings, values, and behaviors with the needs of an innovation project. Whenever you have an idea that you want to convert to a successfully commercialized innovation, you need to evaluate the alignment of your people (skills), processes, tools, metrics, stakeholders and sponsors, management model, organizational structure, and reward and recognition mechanisms.

Assess Impact of the Change

The first culture change step for each innovation project is baselining the amount of change your current culture will require for you to achieve your innovation project's goals. In order to organize for managing successful change, any innovation project must be baselined against the existing organizations' business model and capabilities. It is best to identify the unique aspects of an innovation project versus those projects successfully managed in the past. Then manage the innovation project resourced with a dedicated team with targeted skills for those unique aspects, leverage shared organizational resources, and use the ongoing operations team, sometimes referred to as the "performance engine," sparingly to do *more* of what they already do well.[1]

There are several organizational innovation capability assessments that one can use. Among these are the North Carolina State University Innovation Management Maturity Assessment (IMMA)[2] and the Business Innovation Maturity Model (BIMM)[†] which offers a road map to deploy innovation in stages, and TEDOC (target, explore, develop, optimize, and commercialize).[3]

Motorola Technology Transition Framework Overview

A disciplined process for influencing and managing change in organizations is essential for true adoption of new technologies. The Motorola technology transition framework (c. 1993) was developed and enhanced based on 20+ years at both Bell Laboratories and Motorola, introducing new tools and technologies built or acquired in research to product development. No one likes to change, and when faced with customer deadlines, product developers and engineers do not want to accept even temporary losses in productivity. Change programs

---

[*]Per dictionary.com at http://www.dictionary.com.
[†]See Chap. 47.

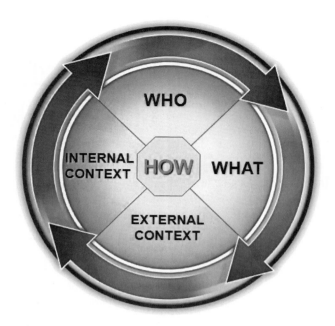

FIGURE 42-1 Elements of change.

must be marketed, managed, and monitored closely in order to truly "take" and become part of "business as usual" within the target organization.

The Motorola technology transition framework provides a series of phases through which organizations must proceed to successfully implement a change program.

The technology transition phases we will reference for culture change in support of innovation include

- Initiation
- Customization
- Fan out
- Institutionalization

During the initiation phase activities, you document the need for the change, document the change, and then prepare a change plan. This phase will end when there is a change plan in place for instituting the change and the decision to institute the change program has been made.

During the customization phase activities, you evaluate the change against the specific capabilities of the organization. The change and the organization are tailored to provide a best-fit match before the change is put into practice. The customization phase activities address inconsistencies between the change and the organization's existing culture. Culture includes items such as beliefs, values, goals, work processes, procedure, environment, techniques, and capabilities. Tailoring may need to be done to the change, the organization, or a combination of both. In addition, during this phase, the change support and measurement environment is defined and instantiated.

During the fan out phase activities, you deploy the change across the entire organization. Fan out projects can be run in parallel to deploy the change simultaneously across several different subgroups of the organization. This phase ends when the change has been deployed throughout the organization.

During the institutionalization phase activities, you integrate the change into the "everyday culture" or ongoing operations within the organization's performance engine. The change program or innovation project is no longer thought of as new or different. This phase ends when the infrastructure and maintenance resources are in place to sustain the change as business-as-usual in the organization.

Although the framework phases are presented in a sequential manner, it is important to understand that *you may return to reexecute activities in earlier phases when issues arise*. This may be the result of changes in the project requirements or environment. This may also occur as you gain a better understanding of the innovation project and its impact on the organization.

Figure 42-2 shows the activity flow between the technology transition framework phases.

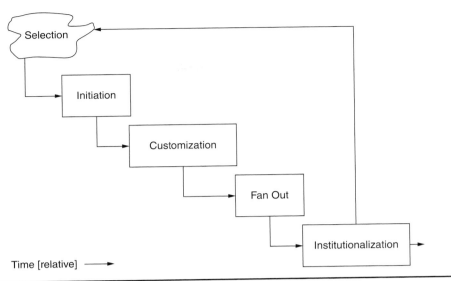

FIGURE 42-2 Technology transition workflow.

For purposes of this chapter we will include the steps for effectively moving a change program or innovation project from initiation through customization, fan out, and institutionalization, once an assessment has been done on the candidate project and target organization. The ensuing discussion maps to the intersection points of each technology transition phase (row) and activities (column) in Table 42-1.

Technology Transition→ Phase/Activity	Vision	Motivation	Skills Development	Implementation	
Initiate	• Develop vision & goals • Link to strategy • Develop business case for change • Develop communication plan	• Understand current organizational behaviors • Analyze the change impact • Perform maturity assessment • Develop motivational strategy	• Identify required skills • Define training strategy • Plan the training • Train the change team	• Document key change characteristics • Define current state, desired future state, gaps, and actions	• Define plan to manage personal transitions • Evaluate current reward and recognition systems for change project potential
Customize	• Reinforce vision and goals	• Create motivational infrastructure • Execute motivational plan	• Develop training program • Deliver training to early adopters	• Prepare support materials • Create support structure • Collect data	• Recognize training achievements • Reward early adopters
Fan out	• Reinforce vision and goals	• Reemphasize motivational plan	• Deliver training to broader audience	• Implement the change • Provide "war room" support • Collect & analyze data • Develop maintenance plan • Identify early wins	• Recognize training completions • Reward early adopters • Recognize improvements

TABLE 42-1 Technology Transition Phases and Activities

Technology Transition→ Phase/Activity	Vision	Motivation	Skills Development	Implementation	
Institutionalize	• Integrate vision and goals of the change into standard (management) operations	• Integrate motivational activities into standard (HR) operations	• Define refresher training for the change • Formalize training and integrate into standard (HR) operations	• Execute maintenance plan • Mainstream support network • Collect & analyze data • Identify improvements	• Recognize improvements • Sustain ongoing rewards and recognition for high performers
	Sponsorship	• Communication	• Project Management		

TABLE 42-1   Technology Transition Phases and Activities (*Continued*)

In the following sections, we present technology transition activities necessary to execute each of the aforementioned phases.

Initiate

Initiate/Vision

Here we are at square one. Be not afraid. You are here because you've already come up with a truly compelling innovation of some kind, and you've been sanctioned by executive sponsorship to get this done. Dig out the original business case—the one with the monetized return on innovation (ROI)—that resulted in the formation of a core team. You're already in good company. These are the passionate ones. And you just can't fail with all that passion behind you as long as you keep your eye on the ball. That's the point of the technology transfer phase or activity matrix. Always start on row 1, "initiate," and step through the activity columns, left to right.

Back to that ROI for a moment, Rowan Gibson postulates, "Why is innovation ROI hard to measure, and how may it be done better?" Asking "What's the expected ROI?" or "How profitable is this opportunity?" can often end up killing great ideas prematurely. Instead, consider this approach: ask first about the size of the idea, then about the feasibility, and only then about the profitability.[4]

Job one in square one is all about communicating the vision and goals to create an awareness of the need for the innovation project. You may remember this as describing "burning platform" from which you're asking people to jump. But remember that every innovation is not for everyone. Across the board enthusiasm is not necessarily the goal. You need some passionate early adopters to get on board and everyone else who matters to this project will follow.

From this point forward, it is important to distinguish this effort as uniquely innovative; it's not another standard "performance engine" project. Customize your company's standard

Technology Transition→ Phase/Activity	Vision	Motivation	Skills Development	Implementation	
Initiate	• Develop vision & goals • Link to strategy • Adapt business case for change • Develop communication plan	• Understand current organizational behaviors • Analyze the change impact • Perform maturity assessment • Develop motivational strategy	• Identify required skills • Define training strategy • Plan the training • Train the change team	• Document key change characteristics • Define current state, desired future state, gaps, and actions	• Define plan to manage personal transitions • Evaluate current reward and recognition systems for change project potential

TABLE 42-2   Checklist: Initiating the Innovation Project (Change Program)

70% of Change Initiatives Fail Due to	90% of Innovation Initiatives Fail Due to
Lack of alignment with strategic priority Insufficient reasons for change No financial estimate Can't be completed on predicted schedule	Lack of alignment between business and innovation strategy No budget allocation to future problems
No clear & measurable goals Not staffed with the right people or enough time	No concise & shared problem statements Not staffed with the right people or enough time
Key stakeholders unwilling to try new solutions Key stakeholders and managers not committed	Key stakeholders risk averse Key stakeholders invent themselves
Starting projects with no understanding of risk Ignoring early red flags	Starting projects with no understanding of competitive landscape Ignoring early red flags—litigiousness of competitors
Lack of understanding of customer experience and needs	Lack of understanding of variety of customer perspectives & issues, functional perspective lacking

TABLE 42-3   The Define Phase Is Critical

business case format for your innovation. Use compelling language to describe the vision and goals. Show credible linkage to the company's strategy. Apply these concepts throughout the project and keep these critical caveats (see Table 42-3) in mind.

Check out the Communication section below and use it to saturate the affected organization with everything they need to know about your vision and specific goals. The best delivered message is communicated five times in five different ways. Set the bar here.

Questions to ask

- Who is the audience and what is their perspective?
- Are the vision and goals communicated in the audience's terminology?
- Do you need to communicate the vision and goals differently for the different audiences (i.e., senior management, middle management, individual contributors)?
- Do those impacted by the innovation project understand the vision and goals?
- Are all stakeholders committed to the vision and goals?

Initiate/Motivation

Once the organization begins to understand and accept the need for the innovation, it's your opportunity to make them want to support it and actively participate. Remember that motivation is very personal. Make each and every individual feel that they are being personally invited to join the core team in this project. Demonstrate that the team has already gotten permission to challenge old assumptions. This applies to sponsors as well as stakeholders.

Use the output from your assessment to get a clear picture of current organizational behaviors and map this to your impact of change analysis. This will reveal the motivation "sweet spot(s)."

Season the communication plan heavily with these insights, and make creative use of promotional tools to capture those hearts and minds. Mix proven techniques common to the performance engine organization with creative methods tailored for your project. It's not another "program *du jour*"—it's innovation! Make it look and feel differently. Use whatever feedback mechanisms you can muster to confirm (or not) the desired surge in motivation.

Questions to ask

- What are the forces (e.g., attitudes, measurements, recognition programs, initiatives, management style, values, or philosophies) in the current organization which are reinforcing or supporting the current behaviors?
- Who supports the existing behavior?
- What is an appropriate motivation strategy for the different levels of the organization which must modify their behavior?
- Can you use existing motivation structures?

- Do you have an innovation advocate at every global site?
- Do you have an existing new product development funnel that you can leverage to support innovation projects?

Initiate/Skills Development

Skills development ensures that all participants have the skills they need to perform their roles and responsibilities, as well as information about the unique behaviors, processes, tools, systems, skills, job roles, and techniques necessary for successful implementation of the specific innovation project. There are two kinds of training that must be addressed. First is the training which will enable people to effectively implement the change (e.g., training for innovation team members and sponsors). Second is the training needed to allow people to perform their roles and responsibilities once the innovation project and deliverables are part of ongoing business operations and the companies' product line—it has been instituted in the organization (e.g., training on the innovation itself).

During the initiation phase, identify the unique skills necessary for the dedicated innovation project team members and organizations impacted (e.g., finance, program management, quality assurance, supply chain, etc.), and define the specific skills required to execute the change. This includes skills required to make the change happen as well as for the change itself.

Questions to ask

- What roles are going to be executed?
- What skills are needed to perform these roles?
- What are the specific knowledge and skills that are required?
- Are there prerequisite skills required (e.g., business model training before tool-use training)?

Define the training strategy and the specific skills required to implement the innovation project. This includes skills required to make the culture change happen as well as for executing on the innovation project.

Questions to ask

- Are there classes or resources available to provide the required training?
- Are there different training or orientation requirements for different parts or levels of the change or innovation project participants (implementors, middle management, senior management)?
- What are the objectives of each type of training?

Prepare the high-level training plan that will be implemented throughout the life cycle of the innovation project. This training plan should define who should be trained, in what areas, and using what training methods—centered on the unique aspects of culture change required for successful implementation of the innovation project.

Questions to ask

- How will we monitor training compliance and completion?
- How will we measure training effectiveness?

Train the innovation project team. Ensure that the team is staffed with a good mix of people with the innovator's DNA skills: observing, networking, experimenting, questioning, and associating[5] and the team has a process that ensures successful infusion and execution of these five skills in their daily work.

Questions to ask

- What is the status of training completion for individual team members?
- Have team members demonstrated they have learned and are able to apply the course materials?

Initiate/Implementation

Implementation activities consist of the core work that is associated with the initiation phase. It can be considered the "just do it" set of activities. It is all about turning the knowledge gained in the previous phases into action. The innovation project team demonstrates their capability to implement the change program successfully.

Document the key change characteristics and specifically the needs of the organization with respect to the innovation project to be implemented. This is the requirement statement for the culture change to ensure successful implementation of the specific innovation project.

Questions to ask

- What are the technical details associated with implementing the innovation project?
- What are the behavioral impacts of supporting and motivating the changes required for successful implementation of the innovation project?

Define the current state of the organization, the desired future state, gaps, and actions. Evaluate the current systems and environment to determine the forces that are acting to maintain current behaviors and actions. Perform a preliminary gap analysis to understand the scope of the modifications required.

Questions to ask

- What behaviors, specifically, will need to change when the innovation project is completed and part of ongoing operations?
- How are we going to know when we are successful in deploying the necessary changes?
- What are all the critical success factors and how might we measure them?
- What are all the objective criteria we might assess to perform gap analysis before and after implementation of the innovation project?

Perform a detailed gap analysis between the innovation project's environmental requirements and the current organization's environment. Then describe how each needs to evolve to facilitate successful introduction of the change.

Questions to ask

- What is the best way (or combination of ways) to characterize the organization? Might we sit in on meetings, give structured interviews, read documentation, perform formal appraisals?
- What are the current behavioral norms of the organization?
- What are the desired behaviors of the organization?
- What are the differences (gap analysis) between the current and desired states?
- What steps need to occur to move the organization from the current to the desired state?

Now is the time, in the initiate phase, to understand and document your approach to reward and recognition for the project. Keep those personalization principles and the behaviors you want to reward at the forefront, because individual members of your audience will experience a variety of emotions as the project progresses through implementation.

You might use the organization's existing reward and recognition systems as a starting point; but you'll want to eliminate the trite elements (e.g., people may not get inspired by another free corporate cafeteria meal) and spice it up with something unique. Be innovative! In this age of "gamification" and social networking, dig deep—or ask a college student— and find a golden nugget or two that will differentiate your projects' recognition mechanisms from all the others. Greater risks should reap greater rewards, and *greater* doesn't need to be measured monetarily; it needs to be *meaningful*.

Questions to ask

- Does everyone in the organization understand where to submit their ideas and are they confident their ideas will be considered and acted upon?
- Are there some measurements which would indicate progress relative to the change or the effectiveness of the change?
- How will you be able to demonstrate that the change has actually helped the organization?
- How will you demonstrate appreciation for key contributors and early adopters?

How people experience and react to change and its transitions can have a significant impact on the success of the initiative. The illustration below provides a good overview of how people experience change.

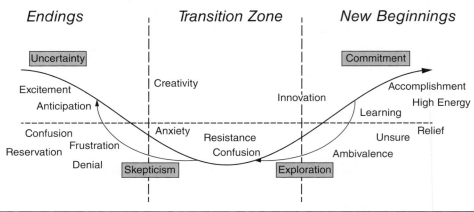

FIGURE **42-3**   Managing transitions. (*Source:* adapted from Managing Transitions, William Bridges)

Customize

Group Selection Activities

One of the keys to a successful customization phase is the selection of an appropriate group to perform the customization activities. It is important to select a group that is willing to take some risks. However, it is useful if they are viewed by the rest of the organization as being "representative" of the organization as a whole or leaders in the organization. It is best not to select a group that is working on the critical path of a major project unless the project plan is on track and includes leeway for such a trial. There may also be a critical path project that wants to trial a whole new approach, due to unique market window opportunities, for example, cycle time reduction initiatives.

Customize/Vision

As you plunge into the *customization* phase, hew to your communication framework and get as much feedback as you can. Is there evidence that people are truly aware of the project's goals and vision? Stop at nothing to check the pulse of the organization. Ask your team members and early adopters to ask at least one person everyday if they know about the project. Ask them to state the project's (compelling, succinct) vision and goals. If they can recite them on the fly, you've got something going.

Questions to ask

- Are the vision and goals still valid for the organization?
- Have organization environmental changes impacted the vision and goals?
- Are the change activities still focused on the vision and goals?

Technology Transition→ Phase/Activity	Vision	Motivation	Skills Development	Implementation	
Customize	• Reinforce vision and goals	• Create motivational infrastructure • Execute motivational plan	• Develop training program • Deliver training to early adopters	• Prepare support materials • Create support structure • Collect data	• Recognize training achievements • Reward early adopters

TABLE 42-4   Checklist: Customizing the Innovation Project (Change Program)

Customize/Motivation

The core team should be immersed in the communication strategy by now. Continuously reevaluate and execute each of the communications and activities contained in your communication framework. Keep a 30-day look-ahead on the project's key review cycles so you have plenty of time to adjust or refresh your plan and stay ahead of resource requirements for next steps. Check, too, for any additional approvals or budgetary considerations coming up in the current or next phase.

Questions to ask

- Have you presented the problems with the status quo and the necessity of the change?
- Have you provided a description of the benefits of the change?
- Is there evidence that your message is being heard?
- Have there been any environmental changes that would drive change to the project or its communication strategy?

Customize/Skills Development

The customize phase is where the innovation and the organizational culture are tailored to provide a "best-fit" match before the innovation is put into practice. Tailoring may need to be done to the innovation project, the organization or a combination of both. In addition, during this phase, the innovation project supports and measurement environment is designed and deployed. During the customize phase, training materials must be developed which support the change. In addition, training must be deployed to the innovation project team.

Design and implement the training program. The details of the training program will be influenced by the key characteristics and consideration of the change program and the organizational culture to be implemented.

Questions to ask

- Who is to be involved in the customize phase activities?
- What change characteristics or considerations specific to the innovation project require training?
- What new skills or knowledge are needed for the innovation to become institutionalized?
- What new skills or knowledge are needed to sustain the innovation?
- How might you train others in creative problem-solving techniques?[6]
- What custom training is needed for management on their unique sponsorship and management responsibilities for this specific innovation project?

Deliver training to early adopters. Provide training to the people involved in customization on the changes necessary to tailor the organizational culture to support the innovation project. In addition, provide full training to the innovation project team, sponsors, and supporting organizations on the skills needed to perform the innovation project support functions.

Questions to ask

- Does the customization team have the required training and skills for this phase? If not, who needs what training?
- Is the customization team adequately trained or skilled to understand the impact of the innovation project?
- Does the training exist that will help the customization team effectively perform tailoring of the organization or the innovation project successfully, or do we need to identify external training resources or consultants?

Customize/Implementation

Prepare support materials, including preparation for the build or install procedures for the change. It also includes understanding and defining the type of data that needs to be collected, and configuring the data collection mechanisms.

- What procedures need to be written to build and install the innovation?
- What level of detail is needed for the change or innovation build and installation procedures? What medium should you use (online help, hard copy manuals, software developer kits (SDKs), application interfaces (APIs), etc.)?
- What data need to be collected to ensure the build and installation were performed correctly and completely?

Create the support structure, tailoring the specific details of the innovation project structure to better fit the organization.

Questions to ask

- What parts of the changes necessary are easily tailored?
- Are the tailored changes consistent with the initial innovation project vision and goals?

Tailor specific parts of the organization or environment to accept the change better.

Questions to ask

- How might the organization be better configured to accept the change associated with the innovation project?
- What are the minimum prerequisite activities or conditions that must be in place before the changes can be made?
- How many of the prerequisite activities or behaviors is the organization already doing?
- Are there cultural changes that need to be addressed?
- Are there reporting structure changes that are required?
- Is there a technical information structure required?
- Would it be useful to create an advisory board for the innovation project?

Training and tools are only a few of the prerequisites for introducing a change. Some changes may have wide-ranging impacts on the way people do their jobs (their formal and informal work process). The change program must take this into account and ensure that the appropriate infrastructure is in place to support the innovation project. It is important to consider both the organizational maturity and the maturity of the proposed changes associated with the innovation project during this process.

For example, consider the introduction of a mandated metrics collection project. The real issue may not be centered on the act of collecting data and generating the charts for the metrics, rather the performance of the activities necessary to generate the data (which is then collected and shown on charts). If the prerequisite data generation resources, capabilities, or activities are missing, then no amount of metrics collection will make the project a success. In cases such as this, the customization activities will prepare and implement a project that will move the organization to the point where it is ready to begin the production and interpretation of metrics charts.

Collect information on the progress being made and feed it back into the project management activities.

Questions to ask

- Can you use the organization's existing project tracking and data collection procedures?
- Are you collecting data that will allow you to anticipate future change activities more effectively?

Assure that you are doing what you said you were going to do.

Questions to ask

- Are you following the culture change deployment process?
- If not, are you capturing the differences and making conscious, agreed-upon decisions on how to modify the change process for success?

- What techniques will effectively monitor the change progress (audits, peer reviews, management reviews, etc.)?
- How often should each technique be activated?
- How will the team assure that the appropriate activities are being done, and done correctly?

Make a go or no-go decision to move forward to the next phase based on the following considerations:

- The inconsistencies between organizational culture and the proposed change have been identified and addressed.
- The change support and measurement infrastructure is ready to support further deployment of the change.

Iterate the Customize Phase Based on Information Learned from Other Phases   In some cases the inconsistencies between the proposed change and the organizational culture may not be understood during the initiate or customize phases. In these cases, the customize activities may be repeated. The experience and information gained from the fan out phase activities are fed back into the process and then the appropriate changes are made before moving forward into the fan out phase again.

Customize/Implementation

With project implementation well under way at this point, it's time to show you care—a lot. Broadcast widely the team's training achievements. ("More than 50 percent of the organization has completed training!") Find those most passionate early adopters (outside the core team) and put their face on the next round of communications—literally! Ask *them* to ask at least one person everyday if they understand why this innovation is important. Send letters of acknowledgement and appreciation to their managers. Crank up the cash bonus machine. No one's expecting real reward this early in the project, so it will be memorable and it will once again distinguish the project as innovative.

Questions to ask

- Have you communicated positive feedback from training?
- Have the early adopters emerged yet?

Fan Out

Fan Out/Vision

Now in the fan out phase, continue to hew to your communication framework and get as much feedback as you can.

Questions to ask

- Are the vision and goals still valid for the innovation project?
- Have organization environmental changes impacted the vision and goals?
- Are the change activities still focused on the original vision and goals?

Technology Transition→ Phase/Activity	Vision	Motivation	Skills Development	Implementation	
Fan out	• Reinforce vision and goals	• Reemphasize motivational plan	• Deliver training to broader audience	• Implement the change • Provide "war room" support • Collect & analyze data • Develop maintenance plan • Identify early wins	• Recognize training completions • Reward early adopters • Recognize improvements

TABLE 42-5 Checklist: Fan out the Innovation Project

Fan Out/Motivation

Repeat, continuously reevaluate, and execute each of the communications and activities contained in your communication framework. Keep a 30-day look-ahead on the project's key review cycles so you have plenty of time to adjust or refresh your plan and stay ahead of resource requirements for next steps. Check, too, for any additional approvals or budgetary considerations coming up in the current or next phase.

Questions to ask

- Have you presented the problems with the status quo and the necessity of the change?

- Have you provided a description of the benefits of the change?

- Is there evidence that your message is being heard?

- Have there been any environmental changes that would motivate changes to the project or its communication strategy?

Fan Out/Skills Development

Now you're ready to deliver training to the broader audience. Extend appropriate training to those who will provide support for any required culture changes. In addition, all change participants are trained on the actual change being made and the skills needed to perform the change process. Also, provide training to the change support team on the change itself and the skills needed to perform the change support function going forward.

Questions to ask

- How might we provide just-in-time or on-the-job training?

- Does the innovation project team or support team need consulting, facilitation, or coaching skills training?

- Are there any new innovation project team members or supporting participants that need training?

- Do all the change participants have the appropriate skills to make the innovation project successful?

- Do all the change support teams have the appropriate skills to make the innovation project successful?

- Are there any new participants or sponsors recently added to the project or supporting organizations who need training?

Fan Out/Implementation

Implementing the change means converting knowledge gained in training to execution ability.

Questions to ask

- Does everyone clearly understand their roles and responsibilities?

- Has everyone taken the necessary training?

Establish a war room for the innovation project team, sponsors, and support staff to congregate, strategize, and plan on a regular basis. The Project Management Knowledge website[7] offers a detailed description of an effective war room, though it tends to focus on a physical space. If you don't have the luxury of having your project core team in the same physical location, you can replicate the physical space as a virtual water cooler on a project-specific intranet page.

Collect information and analyze data on the progress being made during this phase and feed it back into the innovation project management activities.

Questions to ask

- Do the data collection mechanisms cause significant extra work for the group?

- How might we automate and facilitate data collection?

- Do the data collection systems need to be scaled up to allow use by the entire organization?

- Analyze the impact data as it is collected to assure this phase is on target. This analysis will allow you to determine whether modifications should be made in the current phase.

Questions to ask

- Are there any early indicators that there are problems with the change program?
- Is the data consistent with earlier projections of the impact?

Analyze the impact data once the change has been implemented in this phase. This analysis will be used in a go or no-go decision for moving forward from fan out to institutionalize.

Questions to ask

- Does the impact analysis indicate requirements modifications are necessary before the change can be fully implemented across the organization?

Develop a maintenance plan and ensure the change management plan is revised and updated to meet changing organizational requirements or needs.

Questions to ask

- What are the plan, schedule, and milestones for the fan out activities?
- Who, from the change and innovation support team, will provide coaching for the fan out phase participants?
- What is an appropriate work breakdown structure for maintaining and sustaining the changes in the organization?
- What are the risks associated with the culture changes if they are not maintained?
- Do you have sufficient resources to ensure institutionalization of the culture changes in the next phase?
- Are any resources or support required from other organizations to institute and support the innovation longer term?
- Who will perform continuous analysis of the data collected?
- What level of user support will be required?

Monitor and update the change plans as a result of the work completed during this phase.

Questions to ask

- Does the experience so far result in lessons learned necessitating any modifications to the change plan?

Identify early wins or incremental successes and advertise them (see Communication Framework).

As you complete the fan out phase, build on the reward and recognition efforts defined in the customization phase. More than ever, show that you care by recognizing training completions and adding more people to your "hero's circle." Those early adopters will become the lubricant for smooth transitions for those that follow.

Now is also the time to communicate specific improvements. Haul out the project metrics defined during the customize and fan out implementation phases and give them a prominent place in your "information radiator" mechanism. Use collected data to interpret performance to goals and broadcast results early and often. Show how results related to specific strategy and the degree of progress achieved in making targeted improvements.

If you've done the required milestone evaluations and resets throughout the project, and the outcome of the fan out phase warrants it, declare victory and go directly to institutionalization!

Questions to ask

- Are you measuring the level of engagement across the innovation project?
- How closely are results tracking to stated goals?
- Are there specific individuals or groups making notable contributions?
- Is your sponsor communicating positive results upward, and broadly?
- Are you tracking your innovation project's success criteria relative to industry peers?

Technology Transition→ Phase/Activity	Vision	Motivation	Skills Development	Implementation	
Institutionalize	• Integrate vision and goals of the change into standard (management) operations	• Integrate motivational activities into standard (HR) operations	• Define refresher training for the change • Formalize training and integrate into standard (HR) operations	• Execute maintenance plan • Mainstream support network • Collect & analyze data • Identify improvements	• Recognize improvements • Sustain ongoing rewards and recognition for high performers

TABLE 42-6   Checklist: Institutionalization of the Innovation Project

Institutionalize

Institutionalize/Vision

In this phase, a successful fan out triggers integration of the innovation into standard operations. The first step is to insert the innovation's vision and goals into those of the larger organization (performance engine). Assuming you've exploited executive sponsors' roles, this should occur naturally and organically through appropriate communication.

Questions to ask

- What is the most appropriate method to do this integration?
- Who needs to review and approve the integration?
- What sort of lessons can be learned about setting and maintaining the vision and goals?

Institutionalize/Motivation

Likewise, your HR team should recognize and integrate the innovation's reward and recognition tools into corporate standards. These might include

- Employee performance goals and objectives
- Impact award (e.g., getting innovations onto product road maps, achieving cost savings through process improvements)
- Innovation sponsor of the year
- Innovation champion of the year
- Innovator of the year
- Idea (innovation) of the year
- Patent filing award
- Patent grant award
- Patent of the year

Questions to ask

- What is the most appropriate method to do this human resources integration?
- Who needs to review and approve the integration?
- What sort of lessons can be learned about defining and executing a motivation program?

Institutionalize/Skills Development

During institutionalization a change refresher course is defined. In addition, all training material developed for the culture change introduced to support the innovation project is incorporated into the organization's formal innovation training curricula.

Define a refresher course covering the unique skills required for supporting and sustaining the innovation project. Then formalize the training required to perform change support and the change (full and refresher). This includes making the training part of the learning organization's standard curricula.

Questions to ask

- Are the training mechanisms offered on a recurring basis?
- Should some of the training be mandatory for all employees?
- Will any new personnel automatically receive the required training?

Institutionalize/Implementation

In this phase we ensure that the infrastructure and maintenance resources are in place to support the changes to help the innovation project become business-as-usual in the organization.

Questions to ask

- Is the appropriate infrastructure in place to sustain the changes made to support this innovation project?
- What level of user support is required?
- Are the appropriate resources available and allocated to support these changes in ongoing operations?

It is important that the relevant culture changes to support the innovation project become accepted as part of the organization's core beliefs or values. Beware of these changes leading to tunnel vision or traditionalism, such that it becomes an unchangeable part of operations. If it does become unchangeable, it will be eliminated as a candidate for continuous improvement or reengineering. It will limit the organization's ability to respond to a rapidly evolving business climate or rapidly shifting technology horizon.

Mainstream the support network. At this juncture, the innovation project should be performing and delivering. It should have all the attributes of a successful product launch. All the unique roles, responsibilities, and temporary assignments created to support the innovation project are assessed for incorporation into ongoing operations or elimination.

- Newly created roles relevant to any future innovation project are added to career ladders.
- Staff is assigned to maintenance of the innovation project, training roles for innovation project skills deployment, rotated onto the next innovation project, or back into assignments in ongoing operations.
- Sponsorship, as well as program and project management are allocated from within ongoing operations, and may no longer be prioritized as a separate or dedicated effort as at the beginning of the change effort.

Collect and analyze data. Collect information on the progress being made during this phase and feed it back into the project management activities.

Questions to ask

- Do the data collection mechanisms cause significant extra work for the group?
- How might we automate some of the data collection activity?
- Do some data collection systems need to be scaled up to allow use by the whole organization?

The analysis of the impact of the change will also be used to identify incremental improvements to the innovation to make it more broadly applicable and useful in ongoing operations. Improvements can be identified for the change process used for this innovation deployment, the innovation project itself, or the organization's culture.

Continue to analyze the daily impact of execution of the changes supporting the innovation project. The analysis of the impact of these changes will be used to identify improvements to make these changes more broadly applicable and useful across the organization. Improvements can be identified for the change process used for this change, the change (innovation project) itself, or the organization's culture.

You can sustain the benefits of the innovation well beyond the completion of the project by continuing to recognize realized benefits and the people who contributed to those achievements. It always has been and always will be about the people who made it work. If you have successfully institutionalized and integrated the change, this, too, will occur organically.

Prosci® ADKAR® Model

Awareness
Desire
Knowledge
Ability
Reinforcement®

FIGURE 42-4   The ADKAR model.

In the words of the blessedly wise Dr. Ram Chillarege,[8] "the successful innovation is the one that survives at least three reorganizations."

Questions to ask

- Will your innovation withstand the test of time?

A Word About Management of Change and the Prosci®ADKAR® Model   Our management of change framework of choice is the Prosci ADKAR model: awareness, desire, knowledge, ability, and Reinforcement.[9] Our choice is a practical one. We first encountered Prosci's ADKAR: *A model for change in business, government, and our community* at a conference a few years ago. Unfortunately, the book lay ignored and forgotten for a full year before the opportunity arose to read it. And what a joy it proved to be! ADKAR's central idea—that change operates at the personal, individual level—resonated with us as a complementary approach to the technology transition framework. Now it was authoritatively written in a book for anyone to understand! We got the traction we needed to personalize our change management efforts, with considerable success. The next three projects at Motorola Solutions were given "the ADKAR treatment" with stunning results and we've been ADKAR advocates ever since.

Here's a brief summary of ADKAR's elements:

ADKAR stands for awareness, desire, knowledge, ability, and reinforcement.

ADKAR is a model for individual change and describes the phases a single person needs to go through in order to successfully change. The below five elements pertain to an individual, but the assessment of those individuals can also be viewed in the aggregate to complement an organizational model such as the technology transition framework.

Awareness focuses on ensuring that each person involved in the innovation project, from sponsors to all organizational functions who must support the deployment of the program, are informed and aware of the need for the change, as well as the unique aspects of the innovation project that are a departure from normal operations. All impacted employees should understand the business, customer, and competitor issues that require the need for the organization to launch a new program.

Desire focuses on ensuring that all impacted employees make a personal decision to support the change. This includes making sure that impacted sponsors, stakeholders, and team members are actively supportive of and participative in the program throughout its life cycle. In order to assess the desire of an organization to change, all the motivational factors, inclusive of compelling reasons for the change, must be captured, as well as the obstacles and anticipated objections to the change or innovation project. Sponsors, especially, must take a personal interest in the program's status over time and ensure any obstacles or risks are removed or mitigated. Sponsors are the passionate catalysts for change in the organization. Managers also play a critical role in understanding barriers to desire in impacted employees and helping to create desire.

Knowledge focuses on deploying new information, training, and education necessary for employees whose work is impacted by the change and the team supporting and implementing the change program or innovation project. This includes information about unique behaviors, processes, tools, systems, skills, job roles, and techniques necessary for successful implementation of the innovation project.

Ability is all about turning the knowledge gained in the previous phase of ADKAR into action. The innovation project team demonstrates their capability to implement the change program successfully. Some innovation projects may require partnering with external resources or hiring new talent for their unique subject matter expertise. Impacted employees who underwent knowledge-building events should also have the chance to demonstrate they can successfully apply the skills needed to work in the future state of the innovation program.

Reinforcement focuses on sustaining the change or innovation program within the individuals of the overall organization. This includes making it part of ongoing operations and integrating any unique incentives into the organization's existing rewards and recognition programs. As well, existing incentives that run counter to reinforcing or rewarding the change or innovation project should be retired or not applied to the innovation project team.

If you would like more information about the Prosci ADKAR model or the licensing options available for applying ADKAR at your organization, please contact changemanagement@prosci .com or +1-970-203-9332.

Sponsorship

Throughout the life cycle of managing an innovation project, from initiation through institutionalization or retirement, sponsorship is the key success factor that must be obtained from the start and continuously recalibrated. Sponsorship ensures that both appropriate initiating and sustaining sponsors are identified and maintained throughout the course of a change program, or innovation project as in our case. These sponsorship activities ensure that management support and resource allocation is maintained. Sponsorship activities every innovation program must include are

Review the innovation business plan with the sponsors. Make sure sponsors commit both budget and personal resources to the innovation project.

Identify sponsors, both initiating and sustaining, and obtain assurance of their continuing sponsorship throughout the innovation project life cycle.

Obtain the commitment from the individual or group who will be the initiating sponsor. This individual or group must understand the organizational needs, the changes from normal operating procedures, and the level of involvement required to be an initiating sponsor for the specific innovation project.

Questions to ask

- Who has the influence and control necessary to commit the affected organization to move forward on the innovation?
- Who can authorize the necessary resources to make the innovation happen (both for the dedicated innovation team and the organizations or employees impacted by the innovation project)?
- Is the potential sponsor willing to commit support to the innovation project publicly?

Identify the sustaining sponsor. Obtain a commitment from the individual or group who will be the sustaining sponsor when the innovation becomes part of ongoing operations (the landing zone). The sustaining sponsor should be a champion with the ability to commit resources and budget. This individual or group must understand the organizational needs, the nature of the innovation project (incremental, adjacent, disruptive), and the level of involvement required to be a sustaining sponsor.

Questions to ask

- Who has the visibility, interest, and control to consistently monitor the progress of the innovation project and its impact?
- Is the sponsor willing to personally commit to the ongoing resource allocation required to sustain the innovation project?
- What are the sponsor's expectations relative to the effects of the innovation project on the organization? Will the selected sponsor be able to meet these? How will you know?

Periodically reevaluate the initiating and sustaining sponsor's change-related activities to assure that they still meet the requirements of the role, and that they are still committed to the innovation project's success.

Questions to ask

- Are the sponsors attending the appropriate meetings?
- Are they providing the agreed-upon levels of support in staffing, facilities, and necessary resources?

- Do the sponsors promote the innovation project to their peers and upper management?

- Has the sponsor continued public and private support for the innovation project?

- Has the sponsor provided the resource support required to effectively continue the innovation project?

- As unexpected issues or problems have arisen, have they worked to resolve them while maintaining the integrity of the innovation project?

- Has the sponsor allowed roadblocks to stall the innovation project's progress?

If during the calibration activity, a problem is uncovered with the initiating or sustaining sponsor, you may have to reeducate or remind the sponsors of their commitments and their expected actions. You may be required to identify a new individual or group to perform the role.

Questions to ask

- What actions do the sponsors need to take or do differently?

- What do you have to provide (e.g., training) so the sponsors can better perform their roles?

- What actions are necessary to reestablish sponsorship?

- Do the current sponsors need additional training, coaching, or reminding?

- Are different sponsors required to continue the innovation project into the next phase?

Strong sponsorship is likely the most important critical success factor for innovation projects in established organizations. Without a senior executive sponsor who can provide necessary budget and resources to protect the innovation project, even with all other critical success factors in place, innovation projects will fail to deliver as the ongoing operations continuously redirect efforts from future deliverables to near-in deliverables for current paying customers. The Harvard Business Review blog offers one successful approach to leveraging the value of sponsorship.[10]

Communication

Here is a matrix Nina used in the past to manage a comprehensive communication plan, informed by her personal knowledge of ADKAR, over the course of the project's phases. This simple tool has been used successfully at three different organizations in different industries; so it might work well for you.

NOTE: *ADKAR is a proprietary model and integration into change management plans requires licensing from Prosci. Contact Prosci for information about how you can integrate ADKAR into your plans.*

Create your own spreadsheet using rows to enumerate specific communication activities tied to ADKAR aspects, project phase, and target audience—adding columns for the types of online and off-line methods and activities available to you. For example,

Online (written, presented, or posted)

- E-mail
- Internal websites
- Newsletters
- Blogs
- Posters and hallway (TV) monitors
- Town hall presentations
- Staff meetings

Off-line (activity-based events)

- Quick polls
- Brown bag meetings

PHASE	ADKAR ASPECT	ITEM / DESCRIPTION	TARGET AUDIENCE	Email	Internal Website	Newsletter	ONLINE Blog	Poster	Table Top	Town Hall Pres	Other	Brown Bag	Quick Poll	Special Event	Video/ Podcast	Kiosk	Hallway Monitors	Other
Pre-launch	Awareness	**Establish prominent visibility among employees,** reflecting program branding. Includes project status, link to application, training path info, downloadable docs	Enterprise		X			X	X	X		X						
Pre-launch	Awareness	**Exec SPONSOR Intro Message stating project goals, WWIFM** and sanctioning broad employee participation.	Enterprise	X	X		X			X					X			
Pre-launch	Awareness	**Exec MANAGEMENT Launch Message including** vision statement, What problem/s are we trying to solve, the ideal solution and timeline, and the possibilities for the company.	Enterprise	X	X		X			X				X	X			
Pre-launch	Awareness	**Manager Pack** - Reusable slide deck for Town Hall & Staff meetings. Includes vision, rationale and goals; with key focus on WIIFM.	Managers & Team Leads							X								
Pre-launch	Awareness	**Print Advertising** - Branded posters and table-top tent cards to promote awareness, point to web site for info.	Enterprise					X	X									
Soft Launch	Desire	**Exec MANAGEMENT Message** - Soft launch schedule and goals.	Managers & Team Leads	X	X		X					X		X	X			
Soft Launch	Desire	**Announce Incentives** - Social/game event to advertise incentives for participation.	Enterprise	X	X		X	X	X	X		X		X	X			
Soft Launch	Desire	**"WII-FM Radio"** - podcast mimicking FM radio persona, offering the What's In It For Me (WIIFM) message.	Managers & Team Leads	X	X										X			
Pilot	Knowledge	**Welcome Deck** - Quick-start instructions for first-time users, includes application tour with screen-shots and basic access and function points.	Enterprise		X					X		X						
Pilot	Reinforcement	**Playbook / Desk Reference** - A hard-copy collection of manager questions and decision points, prompts and cheat-sheet for application usage.	Managers & Team Leads		X													
Pilot	Knowledge	**Exec Mgmt Status Update Message** - Report on Soft Launch outcomes and resulting program changes.	Enterprise	X	X		X			X		X	X	X				
Pilot	Ability	**User Workshops** - One-hour, one-on-one coaching at first system use.	Target Participants	X	X				X	X		X		X				
Pilot	Reinforcement	**Recognition & Rewards** for Early Adopters who have demonstrated most collaborative skills and potential of resulting ideas.	Enterprise	X	X		X	X	X	X				X				

Figure 42-5 ADKAR communication framework.

715

- Contests
- Video or podcasts
- Kiosks
- Receptions
- War room gatherings

During planning, drop an "x" in the column corresponding to each item described on the rows. Replace the "x" with a publication or activity date when it is planned, adding an active link to the document or location of instructions.

Plan for an instant project. All bases should be covered. Keep it real and simple.

Project Management

We make no attempt to prescribe project management tools or methods and instead defer to those common to your organization and environment—with some caveats and recommendations. Specifically, don't confuse program management with project management. Do not encumber innovation projects with the trappings of typical program fixed schedule and resource allocation methods. Start with a standard set of project milestones like those indicated in the technology transition phases (initiate/launch, customization, fan out, and institutionalization), but tailor a unique set of tasks for every innovation project to achieve those milestones.

Task lists are more effective than 10,000-line Gantt charts in the innovation world. Keep it simple. Don't waste your time on excessive detail that isn't well understood outside or even within the core innovation project team. That's why we like the communication framework as a driver to activity. Defer linkage to internal program schedules until the institutionalization phase. Ensure the program office understands this at the outset of the project to avoid distractions later.

Focus instead on APQC's cautionary advice to identify and resolve all potential barriers to change implementations.[11]

Simply stated, failure to address the innovation project's unique vision, skills, incentives, resources, or action plan will result in confusion, anxiety, gradual (rather than definitive) change, frustration, or false starts. This advice on the conduct of innovation projects shouldn't be taken lightly.

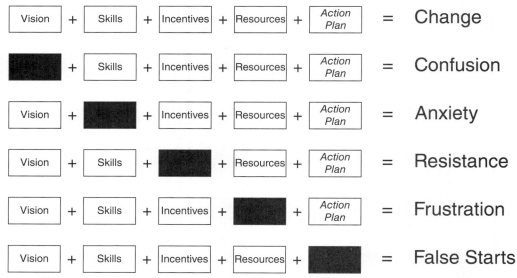

Adapted from Knoster, T., Villa R., & Thousand, J. (2000). A framework for thinking about systems change. in R. Villa & J. Thousand (Eds.), Restructuring for caring and effective education: Piecing the puzzle together (pp. 93-128). Baltimore: Paul H. Brookes Publishing Co.

FIGURE 42-6 Managing complex change.

Conclusion

Our combined experience with Motorola technology transition framework and the ADKAR management of change model tells us a disciplined change management framework is mandatory to effectively manage innovation projects. The Motorola Technology Transition model is complementary to the Prosci ADKAR model for individual change. We believe that ADKAR can be tailored to the needs of any organization's introduction of change, including the culture change required to successfully manage and deploy innovation across an organization.

Having explored the technology transition model and leveraging ADKAR, it becomes obvious that any attempt to optimize an organization's culture to promote innovation requires a personal touch. When we consider skills, accountability, reward, and recognition—we're talking about individuals. When we encourage flexibility, creativity and experimentation – we're talking about individuals. When we digitize environmental support mechanisms and draw focus to the organization's strategy—we're talking about individuals. Therefore, when we communicate to inspire and engage, we do so in pursuit of the hearts and minds *of people individually*. Whatever you take away from the suggestions we've provided, carry it forward with the lightness of creativity. To paraphrase a great catalyst for change, Mahatma Gandhi, "go forth and be the [*culture*] change you want to see in the world."

We've enjoyed having this one-on-one with *you*.

References

1. Trimble, C. The Other Side of Innovation. http://www.amazon.com/Other-Side-Innovation-Execution-Challenge/dp/1422166961.
2. North Carolina State University Innovation Management Maturity Assessment (IMMA). http://cims.ncsu.edu/tools-assessments/im-maturity/.
3. TEDOC (Target, Explore, Develop, Optimize and Commercialize. http://www.accelper.com/innovation.asp.
4. Gibson, R. http://www.imaginatik.com/blog/recap-five-questions-rowan-gibson.
5. Christensen, C. and Dyer, J. The Innovator's DNA. http://innovatorsdna.com/.
6. Creative Problem Solving. http://www.creativeeducationfoundation.org/our-process/what-is-cps.
7. Project Management Knowledge. http://project-management-knowledge.com/definitions/w/war-room/.
8. Chillarege, R. Software Engineering Optimization & Orthogonal Defect Classification. http://www.chillarege.com.
9. Jeffrey, H. *ADKAR: A Model for Change in Business, Government, and Our Community.* Loveland, CO: Prosci Research, 2006. changemanagement@prosci.com.
10. Wessel, M. "How to Innovate with an Executive Sponsor." *Harvard Business Review,* January 2, 2013. http://blogs.hbr.org/cs/2013/01/how_to_innovate_with_an_execut.html.
11. Sustaining Effective Communities of Practice, APQC. http://www.apqc.org/knowledge-base/collections/sustaining-effective-communities-practice-collection.

CHAPTER 43

Measuring for Innovation

Helena Forsman and Serdal Temel

Introduction

Since innovation is regarded as one of the main sources of competitive advantage, companies do their best to invest in innovation aiming at reinforcing their competitiveness. Nevertheless, recent studies suggest that there remains a serious disconnect between what firms are targeting for and what they have achieved through investments in innovation.[1,2] Another challenge is that decision makers and employees seem to have different views on innovation success. For example, a recent report published by the Boston Consulting Group reveals that 59 percent of top executives are satisfied with the return on innovation spending, while only 36 percent of other employees agree with them.[3] The above challenges bring in an important issue, the measurement of innovation in order to support decision-making and innovation activities at company level.

There is a variety of ways in which companies approach measuring the innovation activities. The traditional approach has focused on measuring innovation inputs in terms of RD investments and innovation outputs in terms of the number of patent filings and the number of new products. This is a limited approach. Innovation as a holistic framework, starting from the creation of an idea and then converting it into future business value, demands that the process generation between inputs and outputs should be measured as well. There are, however, challenges. The companies consider themselves the most effective at measuring innovation outputs, while they consider themselves far less successful at monitoring innovation inputs and the quality of their innovation process.[3] All too often companies are measuring what is easy to measure instead of what is important to measure.

The recent trends in measuring innovation appear to be toward a comprehensive view of innovation activities in firms and business networks. These performance measurement systems range from innovation dashboard models and process-oriented systems to the solutions that emphasize the organizational, cultural, and attitudinal aspects to facilitate innovation. The more comprehensive systems allow management to see performance across a variety of innovation facets.

An old adage, "You cannot manage what you do not measure" has been fuelling the discussion on the measurement of innovation performance. This is, however, a controversial topic. Performance management practitioners commonly argue that only the things that get measured are the things that get done. Correspondingly, the common answer from the innovation management practitioners is that innovation is intangible and serendipity-driven, and thus difficult to measure. The measurement of innovation performance is a multidimensional management challenge that raises several questions such as

- What can be measured?
- What indicators are meaningful to use in measurement?
- How to use the measurement information in decision making?

The aim of this chapter is to offer both theoretical and empirical insights for answering the above questions. In the next section, the links between innovation and business success are discussed followed by the identification of the main dimensions for measuring innovation. Finally, a set of potential indicators for each measurement dimension has been introduced.

Innovation and Business Success

Successful innovation should create value for customers, be superior in competition, and create profits for innovators. Thus, it is expected that innovations lead to business success through the better ability of a firm to compete with its commercial rivals.

However, the relationship between innovation and business success is not simple. So far, the links between innovation and success have been studied based on three research streams. In the first stream, the emphasis is on the types of innovations and business success, while the second stream has emphasis on the direct links between innovation development and business success.[2,4,5,6] Finally, the third stream explores the links between innovation and business success affected by moderating or mediating variables.[7,8]

The broad definition of innovation as "the successful exploitation of new ideas" leads us to consider two common typologies that have been used to explore the links between the types of innovation and business success. First, the distinction between "what-innovations" (e.g., new products and services) and "how-innovations" (e.g., process, method, and organizational innovations), have been explored in several studies. The development of what-innovations is commonly connected with competitiveness created through differentiation strategy with an aim to improve the market position or to open up new market segments.[9–11] Correspondingly, the development of how-innovations is linked with competitiveness created through low-cost strategy aiming at improving productivity and cost-efficiency.[12,13]

Another common typology used in innovation studies is the distinction between radical and incremental innovations. The aim of incremental innovations is to enhance processes, make operations more effective, improve the quality, or decrease costs. Correspondingly, the radical innovations, which are characterized by discontinuity in technology and the market, represent entirely new and different offerings through which enterprises aim to get access into new markets or even try to create new markets.[14,15] Nevertheless, McDermott and Prajogo[16] argue that there is an interaction effect between different innovation types. The firms having emphasis on both radical and incremental innovations will perform better than those who only emphasize either innovation.[2]

As regards the second stream, the previous studies have demonstrated mixed results about the direct links between innovation and business success, some positive, some negative, and some have indicated no relationship at all.[2,17–19] It seems that the fuzzy nature of innovation outputs makes it very challenging to measure the change and its impact on business. Also the question of what is the direction of causality has remained unanswered. Prior literature has mostly focused on the impact of innovation on business success, while it tends to overlook the reverse relationship, that is, to what extent innovation is stimulated by past success.[6,20]

Cainelli et al.[6] make a point that there is a two-way cumulative and self-reinforcing link between innovation and business success, and thus the better performing firms are more likely to innovate and to devote their resources to innovation development. This can reflect the fact that high business performance as a result of existing competitiveness enables the allocation of adequate resources for innovation development. A profitable firm has free cash to be used in innovative projects enabling it to foresee and adapt to changing market conditions. The free cash is needed to finance strategies toward building hard-to-copy competitive advantages.[21]

Finally, regarding the third research stream exploring the links between innovation and business success affected by moderating and mediating effects, maybe the most common factors have been linked with learning, capability creation, and knowledge management. Learning may arise from different issues along with innovation process: learning-by-doing, learning-by-using, or learning-by-sharing the internal and external sources of knowledge.[22] Correspondingly, knowledge management is a function that covers the management of explicit and implicit knowledge held by the firm. It is a well-established empirical fact that the accumulation of existing knowledge plays an important role in innovation. With a low level of existing knowledge, a firm is not able to internalize and exploit the external knowledge.

Learning and knowledge have been introduced as a transforming ability between resources and innovation goals. For example, Zott[7] highlights that business success is affected by the abilities of firms to integrate, build, and reconfigure their resources and capabilities. Teece et al.[23] have earlier conceptualized these abilities into a model of dynamic capabilities. The dynamic capabilities will create and shape the resource position of the firm, its capabilities, operational routines, and activities.[24] Timing, cost, and learning effects foster the emergence of robust performance differences among firms with strikingly similar dynamic capabilities. Even small initial differences among firms can generate significant intraindustry differential firm performance.[7] Finally, it should be noted that firms are not isolated islands and thus, learning occurs within and between the firms.[25]

On the basis of the above, it can be summarized that the two-way cumulative and self-reinforcing relationship between innovation and business success is a complex issue for firms. The type of innovation may influence the type of return on innovation investments, new innovations can boost competitiveness, while existing competitiveness may provide a fertile ground to create new competitive advantages through innovations, and finally learning, capability creation, and collaboration are closely connected to the success of innovation efforts, while successful innovation efforts may strengthen the capabilities of a firm as well as its attractiveness as a partner. In the following section, the above issues are holistically explored and the main dimensions for measuring innovation have been identified.

Dimensions for Measuring Innovation

In order to be able to manage the capital and time-intensive innovation process, it has to be measured. Thus, the main purpose of the measurement is to guide decision making. For example, if the aim of the innovation is to create better value for customers so as to gain improved return on investment, then the factors related to this success should be measured. However, innovation often means a new venture to unknown, which makes it difficult to assess the outcome in the early phases. The measurement of the wrong aspects of innovation could even inhibit innovation in firms. Sometimes the indicators are one-sided having focus only on technological innovations, for example, on tracking the patent applications while ignoring nonpatentable innovations. Sometimes the indicators may be misleading, such as accounting the volume of ideas instead of the quality of ideas. Finally, if companies do not know the intended performance results, it is difficult, even impossible, to take suitable managerial actions.

The above challenges suggest that a single measurement dimension, such as revenue growth, the number of patents, or submitted ideas, constitutes a poor innovation performance system. In order to manage innovation successfully, it is needed to combine a series of indicators to provide a more balanced view of innovation activities. This leads to a need of a holistic model that commonly includes at least the input, efficiency, and output indicators.

Some companies have used the balanced scorecard to measure innovation. It is a strategic management and measurement system that looks at the performance of a firm from four perspectives: financial perspective, customer perspective, internal process perspective, and learning and growth perspective.[26] Based on the process-oriented view of innovation, Birchall et al.[27] have studied the current practices of innovation measurement in firms. They extracted five dimensions describing innovation measurement in firms: future focus, market impact, capabilities and image, process, and finally, sustainability and overall effectiveness. These dimensions have some similarities with the balanced scorecard. The future focus dimension underpins strategy and vision, the capabilities and image dimension reflect learning, the market impact dimension is related to customer perspective, the process dimension provides an internal perspective by measuring budgeting, project management, and resource allocation, and finally, the sustainability and overall effectiveness dimension is concerned with long-term sustainability and portfolio management.[27]

Based on a holistic innovation management approach, Adams et al.[28] propose a multidimensional model to be applied to innovation measurement consisting of seven dimensions for measurement: innovation input, knowledge management, innovation strategy, organization, culture, portfolio and project management, and commercialization.

As the above approaches reflect, measuring innovation is a complex operation and identifying innovation performance is at least as complex due to its multidimensional nature. Overall there is no consensus on the definition of innovation performance and quite often the degree of the success of innovation activities has been measured by using indicators that are more suitable for measuring business success, for example, sales growth and profit rate. Birchall et al.[1] define innovation performance as an output of innovation activity, while Alegre and Chiva[29] suggest that innovation performance consists of two different dimensions: innovation efficacy and innovation efficiency. Innovation efficacy reflects the degree of innovation success, while innovation efficiency indicates the efforts made to achieve that degree of success. Firms can no longer focus solely on efficacy; they need to excel simultaneously with efficiency, that is, the speed of innovation process, quality, flexibility, innovativeness, and learning. Thus, innovation performance can also be viewed to reflect the firm's deployment of its overall innovation capabilities, underlying innovation process and seeing innovation from a systemic point of view.[27]

Several scholars have emphasized the importance of measuring performance along with innovation process. The earlier three models can be combined to a general approach that considers innovation process as a system that acquires information, assimilates it, and

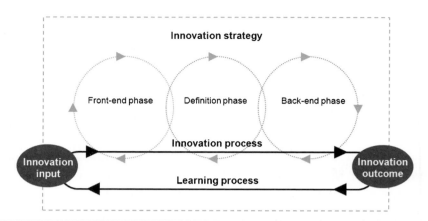

FIGURE 43-1   A system approach of the measurement dimensions. (*Source:* Adapted from Forsman.[31])

converts it to commercial ends.[22] The innovation strategy, culture, and structure of a firm direct this process and thus, create the landscape for innovation. Finally, this system connects innovation and learning so that innovation can be described as an embodiment of learning, and vice versa.[30] The innovation process exposes firms to learning and capability creation, and gives rise to an iterative process that leads to new innovations. This, in turn, gives rise to higher levels of capabilities. Figure 43-1 illustrates the system approach for measuring innovation at firm level.

Five measurement dimensions can be identified from the system approach:

- Dimension 1: innovation strategy
- Dimension 2: innovation input
- Dimension 3: innovation process
- Dimension 4: learning process
- Dimension 5: innovation outcome

The first dimension, innovation strategy, reflects the commitment and ability of a firm to innovation development. The innovation strategy must be consistent with the organizational strategy and the overall business goals implying that management takes well-considered decisions regarding the innovation goals. While innovation strategy provides an answer to the question of why to innovate, the innovation portfolio management selects what to innovate, and innovation process provides an answer to how to make it happen. Thus, innovation strategy directs the innovation activities aiming at optimizing the trade-off between returns and risks through the evaluation and selection of current and future innovation projects.

It has been argued that the innovation process is more efficient if the innovation strategy is embedded in the culture and structure of an organization. Creativity and innovativeness can be promoted by work environment factors. The culture and structure demonstrate the way the staff members are grouped and the way they are working. These have impact on corporate flexibility and responsiveness to change. Thus, the culture and structure of a firm create a landscape that either encourages or hampers innovation.[32] The common aspects for measurement are related to the adaptiveness of staff to changes, their willingness to try new procedures, and the autonomy of individuals in day-to-day decisions.[28,33]

The second dimension, innovation input, consists of tangible and intangible resources to be used for innovation activities. These innovation enablers cover such items as people, physical and financial resources, tools, and idea generation. RD investments and RD staff have been frequently used aspects for measuring innovation. However, these appear not to be very useful for smaller firms that often do not have formal RD activities or for service firms in which innovation is often integrated into customer collaboration. However, adequate funding is a critical input into innovation process and thus, the data demonstrating resources is a commonly used aspect for measurement.

The third dimension, the innovation process, is divided into three phases flowing through the front-end, definition, and implementation phases as well as through the functions of value identification, value development, and value communication.[31] When looking at the project from the point of view of innovation development, the front-end phase focuses on exploring new opportunities and on generating and selecting the best ideas to solve the customer problems, while the definition phase focuses on turning these great ideas into attractive concepts. Finally, the back-end phase converts these concepts into winning new innovations, for example, products, services, processes, or methods. The innovation process turns inputs into exploitable innovations along with this process. Thus, two aspects arise for measuring innovation: the quality of the management of the innovation process and the quality of the process output.

However, the system considers also organizational outcomes such as learning and capability creation. The innovation process provides opportunities for learning by interacting and thus for improving the capabilities of firms to innovate. The firms should learn not only from their successful projects but also from failures and previous mistakes. The quicker the firm will learn, the faster it will succeed. Therefore, measuring the innovation of a firm is also a way to measure the value of what it has learnt. This justifies the fourth measurement dimension: learning process.

A firm's ability to identify, acquire, and utilize external knowledge can be critical to its innovation success.[34] During the front-end phase external information is detected and acquired. Subsequently, the information is assimilated to enable the selection and decision making regarding the potential innovation efforts or projects. During the definition phase external knowledge is combined with internal knowledge, and during the back-end phase the new knowledge combinations are transformed into the competitive advantage of a firm. The aspects that arise for measuring innovation are related to the number and quality of ideas, the knowledge repository of a firm, and its ability to absorb and put to use new knowledge and the linkages with stakeholders and other sources of information.[22,28]

According to Alegre and Chiva[29] innovation performance consists of innovation efficiency and innovation efficacy. The first four measurement dimensions are related to innovation efficiency reflecting the efforts made to improve business success. Correspondingly, the fifth dimension, innovation outcome, reflects the innovation efficacy, demonstrating the impact of innovation on business success. In addition to immediate business impact, this dimension also measures the longer-term impact that may be gained through the changes in the customer perspective and future potentiality of a firm.

Indicators for Measuring Innovation

Success gained through innovations has been measured by a variety of indicators. The research carried out by the Boston Consulting Group reveals that three indicators are considered the most valuable by executives: time to market, new product sales, and return on investment.[3] However, the well-designed measurement system includes indicators for each phase of the innovation process. It should also have a balance between the qualitative (soft) indicators and the quantitative (hard) indicators. The qualitative indicators can be the assessment items or sometimes even the provocative questions aiming at stimulating people to think, reflect, and act. Correspondingly, the quantitative indicators are objective measures based on numeric data. In this section, a set of potential indicators has been identified to each measurement dimension.

Dimension 1: Innovation Strategy

Today's business organizations deal with the increase in ambiguity and turbulence resulting in the change in work toward nonroutine and nonrepetitive in nature. In such a situation, it is hard to find standard solutions, and the innovation process as well as its outcome cannot easily be specified in detail beforehand.[35] In this kind of work environment hierarchical structures become inefficient. Instead, it is needed to give workers authority to make decisions on the issues they are dealing with. The existence of innovation strategy, that directs this shared decision making, is one potential indicator to be measured.[36] Another suggested indicator measures how effective the innovation strategy is in shaping and directing the innovation activities of a firm.[37] In order to be efficient, the structures and systems should be aligned and the innovation goals should be in balance with the strategic objectives.

In addition to strategy, the behavior and attitudes of top management have been found to be influential in innovation success.[38] Top managers should communicate the vision for innovation being supportive to champion the efforts and change in firms. This strengthens the work environment creating culture that encourages innovation. The numbers of innovators in senior positions or CEO commitment in terms of how much the CEO spends their time on innovation are potential indicators for measuring top management commitment in innovation.[39]

Willingness to change is another aspect needed for innovative culture. The following reflective questions can be used as indicators: "How flexible and responsive is the organization to change?" and "To what extent are the people willing to try new procedures and to experiment with change?"[28] With respect to organizational climate, Andersen and West[33] have developed the team climate inventory that highlights the importance of participative safety and task orientation.

In summary, the indicators of the innovation strategy dimension are mainly qualitative in nature, reflecting whether respondents recognize the factors to be present or not. Table 43-1 presents the reflective questions and indicators that can be used to measure the innovation strategy and how well the structure and culture of a firm create a fertile landscape for innovation.

Dimension 2: Innovation Input

Input management allocates resources to innovation activities including factors such as financial, human, and physical resources as well as the capture and generation of new ideas. In the past few years, the emphasis has been on financial input indicators, while the importance of softer inputs, such as skills and knowledge, has been more recently recognized.[28] Therefore, a potential aspect to select the indicators is to focus on different types of inputs, for example, finance, people, equipment, facilities, systems, and tools.

Maybe the most commonly used indicator for measuring financial input has been RD intensity expressed by a ratio between expenditure and the number of employed persons in RD roles. This indicator has received growing criticism for two reasons. First, some researchers have found that the relationship between RD intensity and performance is not linear but

Indicator/Reflective Question	Source
Innovation Strategy	
The existence of an innovation strategy	Adams et al.[28]
The effectiveness of the innovation strategy in shaping and directing	Bessant[37]
The extent to which the structures and the systems are aligned with strategy	Bessant[37]
Do the innovation goals of a firm match with its strategic objectives?	Bessant[37]
Structure	
The extent to which a firm makes it possible for the people working on an innovation to bend the rules	Shane et al.[40]
The extent to which the team's objectives and visions are clearly defined, shared, valued, and attainable	Anderson and West[33]
The extent to which the employees can participate in decision-making procedures	Anderson and West[33]
The extent to which the team has a shared concern with the excellent quality of task performance	Anderson and West[33]
The level of influence that employees have in the decision-making process	Alegre and Chiva[29]
Culture	
The extent to which the top management is committed to innovation, e.g., in terms of the number of innovators in senior management positions or how much the CEO spends her/his time on innovation	Mankin[39]
The extent to which the environment is perceived as interpersonally nonthreatening so that it is safe to present new ideas and improved ways of doing things	Anderson and West[33]
The extent to which new ideas and suggestions are attended to and treated sympathetically	Alegre and Chiva[29]
The degree to which there are expectation, approval, and practical support of attempts to introduce new and improved ways of doing things in the work environment	Anderson and West[33]
The extent to which people are encouraged to take risk in a firm	Alegre and Chiva[29]

TABLE 43-1 Indicators to Be Used to Measure Innovation Strategy

an inverted-U.[41] Second, RD intensity does not appear to be useful in smaller firms nor is it in the service business in which innovation development is often integrated into the daily business operations. For that reason, the expenditure data has become a popular indicator to measure innovation input. The main benefit of using expenditure as an indicator is the availability of data at firm level. The indicators used are total expenditure, expenditure expressed by a share of sales, or expenditure by an assessment item such as project, department, or developed innovation. Also the adequacy of funding for innovation projects belongs to the indicators useful to consider.[28]

The people aspect has been commonly measured by the number of people committed to the innovation tasks. One common indicator has been the number of RD staff. It implies not only the resources but also the importance of innovation in the firm. It is expected that the higher the number of RD staff, the better the innovation results. The innovation studies have also demonstrated that a mix of types of people influence on innovation performance, and the background, behavior and responsiveness of staff are key issues.[42] The higher staff number does not always help to reach expected results. For that reason, also the diversity of employees, in terms of skills, experience, education, or demographic factors, can be a useful indicator.[43]

The number and diversity of people also affect the number and quality of ideas gathered into the innovation process. The formal system that manages the generation of new ideas also provides a useful tool to get the employees to contribute. However, it should be kept in mind that, instead of volume, the quality of ideas matters. Therefore, the firms should incorporate the screening procedures for evaluating and selecting the best ideas, and for killing the ideas that won't work.

With respect to equipment and facilities, it is a broad area with a range of inputs from buildings to computers. These can be measured in terms of the value or adequacy of input.[28] Finally, the indicators related to systems and tools measure whether or not the firm makes use of formal systems and tools for promoting innovation.[42] For example, having a formal RD department in the firm requires continuity in resource allocation for innovation activities, and this forces a firm for sustainability in innovation activities. Finally, it is also a relevant question of whether or not the use of systems and tools is consistent with a firm's innovation requirements.[28]

In the current literature, the focus has been on measuring innovation input by using financial indicators. This system approach expands them to cover also such input aspects as people, ideas, equipment and facilities, as well as systems and tools. Table 43-2 summarizes the potential indicators for measuring them.

Dimension 3: Innovation Process

Innovation process turns innovation inputs into outputs to be exploited in business. Thus, two aspects arise for measuring innovation: the management of innovation process and the process output.

The efficiency of managing the innovation process has traditionally been measured by comparing the realized budget, schedule, and achieved goals with the original plans. This means that the emphasis is on the duration, costs and revenue forecasting, and on the progress of the innovation project. Efficiency has also been operationalized by the indicator of "innovation speed." The innovation speed correlates positively with product quality and the degree to which it satisfies customer requirements. The speed reflects how fast the innovation process is completed, for instance, the length of time between the conception of a new product and its introduction into the marketplace.[49]

In order to achieve the efficiency of managing the innovation process, project-based work is an often-used approach in organizations. It is recommended that firms seeking to innovate should establish formal processes and make use of project management tools and techniques that facilitate innovation activities. The tools, such as the Stage Gate process[50] or Product and Cycle-Time Excellence,[51] commonly offer milestones for the quality control checkpoints and process evaluations leading to a systematic review of innovation projects.

Communication is important for the management of any innovation efforts. The indicators used to measure the extent and diversity of communication are for example the frequency of meetings, the number of contacts, the degree of other stakeholders that are consulted on new ideas, how well the staff of different departments are communicating with one another, and the extent to which decision making is characterized by cross-functional discussions.[28,43,52]

With respect to the output of the innovation process, the rate of launched innovations and the registered intellectual property rights (IPRs) have been commonly used as indicators. Especially, patent filings have been used by many authors to shed light on the output of the innovation process.[53] There, however, has been growing criticism toward the patents as an indicator,

Indicator/Reflective Question	Source
Finance	
RD intensity in terms of a ratio between expenditure and number of employees in RD roles	Adams et al.[28] Flor and Oltra[44]
RD expenditures in terms of total expenditure, expenditure expressed by a share of sales, or by an assessment item such as project, department, or developed innovation	Hall and Bagchi-Sen[45]
Adequacy of funding	Adams et al.[28]
Percentage of projects delayed or cancelled due to lack of funding	Davila et al.[46]
People	
The number of people committed to the innovation task	Adams et al.[28]
The diversity of people committed to innovation development, e.g., in terms of skills, experience, education, or demographic factors	Amabile[42] Damanpour[43]
Staff training in terms of the level of training and the number of training hours	Flor and Oltra[44]
The number of RD staff	Adams et al.[28]
Capture and Generation of New Ideas	
The number and the quality of ideas generated in a period	Adams et al.[28]
The diversity of the sources of new ideas	Inauen and Schenker-Wicki[47]
The use of generative tools and techniques for creating new ideas	Adams et al.[28]
The number of new ideas funded	Mankin[39]
The number of ideas killed	Hempel[48]
Equipment and Facilities	
The value and adequacy of equipment	Adams et al.[28]
The value and adequacy of facilities	Adams et al.[28]
Systems and Tools	
The extent to which the firm makes use of formal systems and tools for promoting innovation	Amabile[42]
The extent to which the use of systems and tools is consistent with a firm's innovation requirements	Adams et al.[28]
The existence of RD department	Flor and Oltra[44]

TABLE 43-2 Indicators to Measure Innovation Input

due to the fact that a patent can measure technological change and technology-intensive innovations, while it ignores the other types of innovations.[28] Mendonca et al.[54] argue that trademarks can be used by a wider set of business firms capturing change also in service activities. The filing of new trademarks reflects the introduction of new offerings aimed at differentiating them from those provided by other firms. In addition, the other forms of registered IPRs, such as the designs and the utility models, can be included in IPR-related indicators.

Regarding the organizational innovations, the potential indicators are the earliness of adoption or the rate of adoption. The earliness of adoption of innovations reflects the timeline of the adoption decision in a firm compared to other firms. It has been measured in terms of the mean time of adoption and the consistency of a firm in adopting innovations either earlier or later than average.[55] Correspondingly, the rate of adoption has been measured as the total number of innovations adopted within a time interval and the percentage of innovations adopted from a pool of innovations within a given time period.[43] The speed of implementation reflects how quickly the innovation is assimilated throughout the organization and becomes a regular part of organizational procedures and behavior after the adoption decision.[56]

In addition to measuring the quality of a single innovation effort and its output, the firm should also measure the innovation portfolio. The portfolio comprises all innovation efforts (or projects) carried out by an organization. In order to translate the goals of innovation strategy into action, the portfolio management selects and prioritizes the projects to be funded. The aim of the portfolio indicators is to measure the diversification of innovation projects and to manage the overall risk in firms. As a measurement challenge there arise such questions as if the portfolio is optimal regarding risks and returns or regarding the radical and incremental innovation projects. The appropriate allocation of resources is also an important aspect for measurement. The firms should design clear selection criteria for optimizing the use of limited resources. Based on the selection criteria, the project proposals can be scored. In the scoring models, the respondents specify the merit of any project proposal according to a set of criteria. Finally, measuring the value of all innovation projects is one way to measure the health of a firm's innovation efforts. An important aspect for measuring the portfolio is also to ensure that there is an optimal balance between the projects aiming at growing existing business, developing new business, and innovating future business.[28,39]

The summary of the indicators and reflective questions to measure the innovation process has been presented in Table 43-3. These indicators for measuring the management of the innovation process, the process output, and the management of innovation portfolio, have

Indicator/Reflective Question	Source
Innovation Process	
The progress of process—comparison between plan and actual The extent to which the planned schedule and costs were kept The extent to which the goals were achieved	Adams et al.[28] Cooper et al.[50]
The speed of innovation process, e.g., in terms of time to market	Adams et al.[28] Kessler and Chakrabarti[49]
The extent and diversity of communication in terms of The frequency of meetings, the number of contacts The degree to which other stakeholders are consulted on new ideas How well the staff of different departments are communicating with one another The extent to which decision making is characterized by cross-functional discussions	Adams et al.[28] Damanpour[43] Parthasarthy and Hammond[52]
Process Output	
The number of launched innovations in a given period	Branzei and Vertinsky[53]
The types of launched innovations in a given period	CIS[57]
Replacement of products being phased out	Allegre and Chiva[29]
The number of IPR filings: trademarks, designs, patents, and utility models	Mendonca et al.[54] Branzei and Vertinsky[53]
The adoption of organizational innovations in terms of: the earliness of adoption the rate of adoption the speed of adoption	Damanpour and Gopalakrishnan[58] Fichman[56] Subramanian and Nilakanta[55]
Innovation Portfolio	
The total number and value of the projects in pipeline	Mankin[39]
The extent to which funding is aligned with the goals of innovation strategy	Adams et al.[28]
The extent to which the resources are allocated to projects to obtain optimal innovation portfolio	Adams et al.[28]
A balance of project portfolio between the high-risk and low-risk projects the radical and incremental projects the short- and long-term projects the phases of front end, definition, and back end the projects enhancing existing, new, and future businesses	Adams et al.[28] Cooper et al.[59] Mankin[39]

TABLE 43-3 Indicators to Measure the Innovation Process

emphasis on the hard aspects. However, the present literature has identified several soft aspects such as learning, capabilities, and collaboration that drive the innovation process success. These are discussed in more detail in the next section (Dimension 4).

Dimension 4: Learning Process

Several studies have emphasized the role of learning and knowledge-related processes in innovation development. Learning is closely connected with innovation, especially in the firms that operate in highly dynamic environments resulting in nonroutine work and the higher level of interdependency among employees. In these kinds of business environments, work is commonly assigned to a team of people that implies knowledge sharing, exploring solutions together, learning from each other, and finding synergy in creative processes. In such a situation, it is difficult to find standard solutions that can be embedded in standardized processes.[35]

In order to survive, the firm must have capabilities to transform knowledge into innovations, and further into competitive advantage. Especially, the creation of radical and highly complex innovations requires such a diversity of capabilities that it is not realistic to expect a single person to have them. The concept of absorptive capacity, which highlights the importance of collaborative learning, has been offered as one framework to assessing capabilities along with the innovation process.

Absorptive capacity has been defined as an ability to recognize the value of new external knowledge, assimilate it, and use it for commercial purposes. Zahra and George[34] have identified four perspectives of absorptive capacity: acquisition, assimilation, transformation, and exploitation. The acquisition expands the future potential of a firm by stimulating the emergence and early development of new knowledge-search routines. Greater efforts to identify and acquire relevant information from external sources foster new connections and increase the speed and quality of learning from external sources, while the assimilation incorporates this new information into the existing production processes of a firm. The transformation fosters the effective cross-pollination transforming the internal expertise into new inventions and improved operational routines, and finally, the exploitation applies the existing know-how to new innovations.[34,53,60]

While the absorptive capacity has been closely connected to innovation success, this knowledge-related process is difficult to measure. Zahra and George[34] and Branzei and Vertinsky[53] provide some indicators for measurement. The acquisition of knowledge can be measured based on the number of sources from which a firm obtains important ideas for the innovation process and based on the direction, intensity, and speed in seeking out the different types of inputs for the innovation process. Correspondingly, the assimilation of knowledge can be measured by a reflective question of to what extent a firm has incorporated these different inputs into its existing production processes. Finally, the transformation of knowledge can be measured based on the output of the innovation process and the exploitation of knowledge reflects how the output of the innovation process has been exploited in business. The indicators for measuring transformation have been introduced in more detail in the previous dimension (innovation process) and the indicators for measuring exploitation are discussed in the next dimension in relation to the innovation outcome.

While the concept of absorptive capacity describes how acquired knowledge is transformed into new sources of competitive advantage, it is evident that also the type of acquired knowledge affects innovation.[61] Therefore, the ability to combine and effectively use different kinds of knowledge bases is crucial to the development of innovation capabilities. Asheim[62] introduces three specific types of knowledge bases needed in innovation development: analytical–scientific-based knowledge, synthetic–engineering-based knowledge, and symbolic–creativity-based knowledge. Activities that require analytical knowledge are geared toward the capabilities to understand and explain. Synthetic knowledge is required while designing solutions to practical problems and while developing innovations by applying existing knowledge or a new combination of knowledge. Activities that draw on symbolic knowledge focus on creating cultural meaning emphasizing the aesthetic attributes and the imagination. Symbolic knowledge becomes very important when competition shifts from the "use-value of products" to the "sign-value of brands".[62] Different knowledge bases should be treated as complementary assets for innovation development and therefore, the use of different knowledge bases is a potential indicator for measuring the diversity of acquired knowledge.

Both the absorptive capacity and the use of different kinds of knowledge sources demand interorganizational and intraorganizational collaboration. From the resource-based point of view, it has been discussed that interorganizational collaboration has potential to deliver a wide range of ideas, resources, and opportunities far beyond the ability of a firm on its own. According to Chesbrough[63], innovative firms have realized that the closed innovation model does not foster the growth of a firm in a dynamic business atmosphere and that interacting with external partners helps them to get access to the different types of new knowledge bases. Thus, the firms that display a high level of collaboration are able to improve their innovative capacity in terms of capabilities and resources.[64] On the other hand, TerWal and Boschma[65] argue that the firms with superior capabilities to innovate and commercialize the innovations are attractive partners for other firms.

The indicators for measuring collaboration are partly related to the measures of knowledge acquisition. The commonly used indicators measure whether a firm has and maintains external linkages with other organizations and the sources of information, for example, through participation in joint research projects, university links, or attendance at trade shows.[28,53] Inauen and Schenker-Wicki[47] highlight the openness and cocreation in the innovation process. This means increasingly participative approaches emphasizing customer involvement, supplier collaboration, and cooperation with competitors in addition, cooperation with private consulting firms and research institutes is commonly used for the creation and commercialization of new products, services, and processes. Nieto and Santamaria[66] summarize that collaborative networks, which include a variety of partners, have a significant impact on innovation novelty.

Nevertheless, the firms must also look for an optimal network structure and the right partners in order to get optimal benefits out of collaboration.[67] For example, collaboration between noncompeting enterprises can strengthen the innovation capacity of collaborators, while networking with competitors can create advantages through shared RD costs and development collaboration in precompetitive areas.[68] Finally, Forsman[69] highlights that the structure of external linkages should also be in balance between the phases of the innovation process, for example, regarding knowledge creation, resource acquisition, and the development of innovations.

While the learning process has been closely connected to innovation success, it is challenging to measure due to its emergent, even unplanned, and unintended nature. This kind of learning process ideally leads to a process of continuous innovation for which new ideas are constantly being identified and tested. This process, however, stays on track feeding opportunities for innovation development only if it is managed and guided.[70] Table 43-4 presents a summary of the potential indicators for managing the process.

Dimension 5: Innovation Outcome

The innovation outcome indicators measure how successfully the outputs of the innovation process have been exploited for creating value for the innovating firm and its stakeholders. Commonly this has been measured based on business success in terms of sales, profits, return on investment, market share, and the stock price of market valuation. As regards the internal innovations, such as process, managerial and organizational innovations, the metrics can

Indicator/Reflective Question	Source
Absorptive Capacity and Knowledge	
The number of sources from which a firm obtains knowledge	Branzei and Vertinsky[53]
The direction of sources from which a firm obtains knowledge	Asheim[62]
The intensity of using sources for obtaining knowledge	Forsman[31]
The extent to which a firm has incorporated different inputs into its innovation process	Branzei and Vertinsky[53]
The speed of incorporating the inputs into existing process	Zahra and George[34]
The extent to which an organization accepts mistakes because such an environment facilitates learning	Alegre and Chiva[29]
The level of absorption capacity	Zahra and George[34]
Collaboration	
The quantity, quality, and diversity of collaboration with external linkages	Inauen and Schenker-Wicki[47] Nieto and Santamaria[66]
The intensity of collaboration with external linkages	Flor and Oltra[44]
The structure of external linkages	Forsman[69] Rowley et al.[67]

TABLE 43-4   Indicators to Measure the Learning Process

Indicator/Reflective Question	Source
Business Success	
Value or percentage of sales from new innovation(s)	CIS[57]
Revenues from sales of licensing the patents	Flor and Oltra[44]
Value or percentage of profits from new innovation(s)	CIS[57]
The change in internal efficiency, e.g., in terms of productivity, cycle time, and quality	Shenhar et al.[71]
Return on investment (ROI)	Mankin[39]
Market share	Alegre and Chiva[29] Shenhar et al.[71]
Impact on Customer	
The extent to which the needs of customers are fulfilled	Shenhar et al.[71]
The speed of customer adoption	Mankin[39]
The change in customer satisfaction	Shenhar et al.[71]
The change in customer loyalty	Mankin[39]
Future Potentiality	
Opening of new markets	Alegre and Chiva[29]
Impact on the reputation of a firm	Enzing et al.[72] Shenhar et al.[71]
Impact on knowledge repository	Shenhar et al.[71] Zahra and George[34]
Impact on staff satisfaction	Kaplan and Norton[26]

TABLE 43-5   Indicators to Measure the Innovation Outcome

include productivity and quality indicators, and the indicators of new process performing time and cycle time.

However, the firms should utilize their resources in such a way that they are efficiently delivering products and services to their current customers while also innovating to serve the future needs of their existing and potential new customers. In line with this, Forsman[31] argues that the innovation outcome should be measured not only based on business success but also on the basis of the impact on customer perspective and future potentiality. The challenge with the business success indicators, if used alone, is that they are lagging indicators demonstrating the result of past innovation efforts, while these indicators are not predictive providing direction for the further innovation activities.[26]

The impact on customer addresses the importance placed on customer requirements and on meeting their needs. This impact can be measured after the innovation has been delivered to the customer and the customer is using it. This relates also to the innovations developed for the internal use of a firm. The common indicators to measure customer impact are the extent to which the solution meets the customer needs, the speed of customer adoption, customer satisfaction, and customer loyalty.[39,71]

While the business success indicators measure immediate and often direct benefits gained through innovations, the future potentiality indicators measure the benefits that will be realized indirectly in a longer term. According to Shenhar et al.,[71] the main questions to be asked are "How does the developed innovation help prepare the organization for future challenges?" and "Did a firm build new skills that may be needed in the future?" The indicators used to measure future potentiality include indicators such as impact on the reputation of a firm, knowledge repository, staff satisfaction, and opening of new markets. All these enhance the abilities of a firm to adapt quickly to external challenges and the unexpected moves of competitors.

Table 43-5 presents a summary of the potential indicators for measuring innovation outcome in terms of business success, impact on customer, and future potentiality.

Conclusion

As growth has become a more and more common strategic goal, it is natural that firms seek to identify the ways to measure the impact of their innovation activities. The critical questions are what to measure, what indicators are meaningful to use in measurement, and how to use the measurement information in decision making. These are not easy questions to answer. For example, based on the survey conducted by the Boston Consulting Group, three out of five respondent firms are dissatisfied with their innovation measurement practices. Customer satisfaction and overall revenue growth are the two main indicators that these firms have used to determine the success of their innovation efforts.[3] The choice of indicators is undoubtedly one part of the problem—especially for the above firms which, based on the same survey, assessed that the most important innovation barriers are a risk-averse corporate culture and lengthy development times. The used indicators are limited providing the managers only with a vague sense of the firm's overall innovativeness, the effectiveness of its innovation process, and the output of it.

The optimal selection of indicators will vary from company to company. One size does not fit all. For example, innovation for a retailer will require different capabilities, resources, and equipment than innovation for a media company. However, no single metric can demonstrate the full meaning in isolation. It is needed to look at several indicators in order to develop a comprehensive view of innovation performance in a firm. This chapter recommends that firms build a set of indicators covering all five dimensions, for example, innovation strategy reflecting the clarity of goals, innovation input demonstrating the adequacy of resources allocated to the activities for achieving these goals, innovation process measuring the effectiveness of the activities, learning process reflecting capability view, and innovation outcome demonstrating the impact of innovation on immediate and future business success. Hempel[48] points out that the top management needs a different set of indicators than do the people executing the innovation strategy. In addition, if a firm uses a performance management system such as the balanced scorecard, the indicators for measuring innovation should be adjusted with that system.

With a comprehensive set of indicators, the firms will more likely recognize the possible problems in their innovation system. However, it is recommended to avoid the temptation to measure every conceivable factor resulting in high measurement costs and a nonmanageable set of indictors. As Neely[73] states, it is not that the wrong things are being measured, but too much is being measured. But what is the optimal number of indicators? Muller et al.[38] recommend that the number of indicators should be no more than 8 to 10, while Andrew et al.[3] argue that 10 to 12 indicators are required to provide the information needed to manage, rather than react to, the innovation process. They continue that the majority of companies use only five or fewer indicators. It is not enough to specify the realistic goals of innovation and to identify and resolve problems hindering progress toward goals, making decisions, managing innovation process, and continuously improving the capabilities to innovate. Kuczmarski[70] adds that firms must also let the existing indicators go when they are no longer relevant.

The measurement also has risks. The firms need to be mindful of unintended consequences that can result from overemphasizing the importance of one single indicator.[38] For example, if the main indicator is the number of new products and the employees are rewarded based on this output, it may accelerate the invention of all sorts of minor product changes resulting in a complex and confusing product portfolio. Also, the indicator measuring the number of ideas can lead to a similar effect resulting in a huge number of suggestions with low quality.

Another potential risk is that the firms tend to focus on the immediate outcomes of innovations expecting that innovation activities result in concrete outcomes all the time. According to Kuczmarski,[70] this can cause projects to be considered failures even though they may have proved to be springboards for other innovations. He continues that innovation should be viewed as a continuous activity that does not go in a straight line. Therefore, it might be needed to allow different time frames to realize the business outcome.

This chapter proposes a five-dimensional conceptualization of the innovation management system and its application to measuring innovation. It also shows the breadth and variety of aspects that should be considered when selecting the measurement areas. Further, a set of potential indicators for each dimension and aspect is provided. A wise selection and use of indicators can improve the innovation success rate of a firm and help it make the right decisions, allocate the resources to right activities, and manage the innovation process efficiently, with less risk of failure.

References

1. Birchall, D., Tovstiga, G, and Gaule, A. *Innovation Performance Measurement, Striking the Right Balance*. London: Grist Ltd., 2004.
2. Forsman, H. and Temel, S. "Innovation and Business Performance in Small Enterprises: An Enterprise-Level Analysis." *International Journal of Innovation Management*, 15 (3): 641–665, 2011.

3. Andrew, J.P., Manget, J., Michael, D.C., Taylor, A., and Zablit, H. *Innovation 2010. A Return to Prominence—and the Emergence of a New World Order.* Report. Boston: Boston Consulting Group, 2010.

4. Abrahamson, E. "Managerial Fads and Fashions: the Diffusion and Rejection of Innovations." *Academy of Management Review*, 16 (3): 586–612, 1991.

5. Acs, Z. and Audretsch, D. "Innovation in Large and Small Firms: An Empirical Analysis." *The American Economic Review*, 78 (4): 678–690, 1988.

6. Cainelli, G., Evangelista, R., and Savona, M. "Innovation and Economic Performance in Services: A Firm-Level Analysis." *Cambridge Journal of Economics*, 30 (3): 435–458, 2006.

7. Zott, C. "Dynamic Capabilities and the Emergence of Intraindustry Differential Firm Performance: Insights from a Simulation Study." *Strategic Management Journal*, 24 (2): 97–125, 2003.

8. Bisbe, J. and Otley, D. "The Effect of the Interactive Use of Management Control Systems on Product Innovation." *Accounting, Organizations and Society*, 29 (8): 709–737, 2004.

9. Ambec, S. and Lanoie, P. "Does It Pay to Be Green? A Systematic Overview." *Academy of Management Perspectives*, 22 (4): 45–62, 2008.

10. Chang, C-H. "The Influence of Corporate Environmental Ethics on Competitive Advantage: The Mediation Role of Green Innovation." *Journal of Business Ethics*, 104 (3): 361–370, 2011.

11. Porter, M. and van der Linde, C. "Towards a New Conception of the Environment-Competitiveness Relationship." *Journal of Economic Perspectives*, 9 (4): 97–118, 1995.

12. Rochina-Barrachina, M.E., Manez, J.A., and Sanchis-Llopis, J.A. "Process Innovations and Firm Productivity Growth." *Small Business Economy*, 34 (2): 147–166, 2010.

13. Roper, S. "Product Innovation and Small Business Growth: A Comparison of the Strategies of German, UK, and Irish Companies." *Small Business Economics*, 9 (6): 523–537, 1997.

14. Dewar, R.D. and Dutton, J.E. "The Adoption of Radical and Incremental Innovations: An Empirical Analysis. *Management Science*, 32 (1): 1422–1433, 1986.

15. Garcia, R. and Calantone, R. "A Critical Look at Technological Innovation Typology and Innovativeness Terminology: A Literature Review." *Journal of Product Innovation Management*, 19 (2): 110–132, 2002.

16. McDermott, C.M. and Prajogo, D.I. "Service Innovation and Performance in SMEs." *International Journal of Operations & Production Management*, 32 (2): 216–237, 2012.

17. den Hertog, P., Gallouj, F., and Segers, J. "Measuring Innovation in a 'Low-Tech' Service Industry: The Case of the Dutch Hospitality Industry." *The Service Industries Journal*, 31 (9): 1429–1449, 2011.

18. Heunks, F.J. "Innovation, Creativity and Success." *Small Business Economics*, 10 (3): 263–272, 1998.

19. Prajogo, D.I. "The Relationship between Innovation and Business Performance—A Comparative Study between Manufacturing and Service Firms." *Knowledge Process Management*, 13 (3): 218–225, 2006.

20. Davidsson, P., Steffens, P., and Fitzsimmons, J. "Growing Profitable or Growing from Profits: Putting the Horse in Front of the Cart?" *Journal of Business Venturing*, 24 (4): 388–406, 2009.

21. Sarkees, M. and Hulland, J. "Innovation and Efficiency: It Is Possible to Have It All." *Business Horizons*, 52 (1): 45–55, 2009.

22. Cohen, W.M. and Levinthal, D.A. "Absorptive Capacity: A New Perspective on Learning and Innovation." *Administrative Science Quarterly*, 35 (1): 128–152, 1990.

23. Teece, D., Pisano, G., and Shuen, A. "Dynamic Capabilities and Strategic Management." *Strategic Management Journal*, 18 (7): 509–533, 1997.

24. Eisenhardt K.M. and Martin, J.A. "Dynamic Capabilities: What Are They?" *Strategic Management Journal*, 21 (10–11): 1105–1122, 2000.

25. Hernandez-Espallardo, M., Molina-Castillo, F-J., and Rodriguez-Orejuela, A. "Learning Processes, Their Impact on Innovation Performance and the Moderating Role of Radicalness." *European Journal of Innovation Management*, 15 (1), 77–98, 2012.

26. Kaplan, R.S. and Norton, D.P. "Using the Balanced Scorecard as a Strategic Management System." *Harvard Business Review*, 74 (1): 75–85, 1996.

27. Birchall, D., Chanaron, J-J., Tovstiga, G., and Hillenbrand, C. "Innovation Performance Measurement: Current Practices, Issues and Management Challenges." *International Journal of Technology Management*, 56 (1): 1–20, 2011.

28. Adams, R., Bessant, J., and Phelps, R. "Innovation Management: A Review." *International Journal of Management Reviews*, 8 (1), 21–47, 2006.

29. Alegre, J. and Chiva, R. "Assessing the Impact of Organizational Learning Capability on Product Innovation Performance: An Empirical Test." *Technovation*, 28 (6): 315–326, 2008.

30. Tran, T. "A Conceptual Model of Learning Culture and Innovation Schema." *Competitiveness Review: An International Business Journal*, 18 (3): 287–299, 2008.

31. Forsman, H. "Improving Innovation Capabilities of Small Enterprises. Cluster Strategy as a Tool." *International Journal of Innovation Management*, 13 (2), 221–243, 2009.

32. Tidd, J., Bessant, J., and Pavitt, K. *Managing Innovation: Integrating Technological, Market and Organizational Change*. Chichester: John Wiley, 1997.

33. Anderson, N.R. and West, M.A. "Measuring Climate for Work Group Innovation: Development and Validation of the Team Climate Inventory." *Journal of Organizational Behavior*, 19: 235–258, 1998.

34. Zahra, S. A. and George, G. "Absorptive Capacity: A Review, Reconceptualization, and Extension." *Academy of Management Review*, 27 (2): 185–203, 2002.

35. Molleman, E. and Timmerman, H. "Performance Management When Innovation and Learning Become Critical Performance Indicators." *Personnel Review*, 32 (1): 93–113, 2003.

36. Kang, K.N. and Lee, Y.S. "What Effects the Innovation Performance of Small and Medium Sized Enterprises (SMEs) in the Biotechnology Industries? An Empirical Study on Korean Biotech SMEs." *Biotechnology Letters*, 30 (10): 1699–1704, 2008.

37. Bessant, J. *High Involvement Innovation: Building and Sustaining Competitive Advantage through Continuous Change*. Chichester: John Wiley, 2003.

38. Muller, A., Välikangas, L., and Merlyn, P. "Metrics for Innovation: Guidelines for Developing a Customized Suite of Innovation Metrics." *Strategy & Leadership*, 33 (1): 37–45, 2005.

39. Mankin, E. "Measuring Innovation Performance." *Research Technology Management*, 50 (6): 5–7, 2007.

40. Shane, S., Venkataraman, S., and MacMillan, I. "Cultural Differences in Innovation Championing Strategies." *Journal of Management*, 21 (5): 931–952, 1995.

41. Arvanitis, S. "The Impact of Firm Size on Innovative Activity—an Empirical Analysis Based on Swiss Firm Data." *Small Business Economy*, 9 (6): 473–490, 1997.

42. Amabile, T.M. "How to Kill Creativity." *Harvard Business Review*, 76 (5): 76–87, 1998.

43. Damanpour, F. "Organizational Size and Innovation." *Organizational Studies*, 13 (3), 375–402, 1992.

44. Flor, M.L. and Oltra, M.J. "Identification of Innovating Firms through Technological Innovation Indicators: An Application to the Spanish Ceramic Tile Industry." *Research Policy Research*, 33 (2): 323–336, 2004.

45. Hall. L.A. and Baghci-Sen, S. "An Analysis of Firm-Level Innovation Strategies in the U.S. Biotechnology Industry." *Technovation*, 27 (1–2): 4–14, 2007.

46. Davila T., Epstein M. J., and Shelton R. *Making Innovation Work: How to Manage It, Measure It, and Profit from It*. Upper Saddle River, New Jersey: Pearson Education, 2012.

47. Inauen, M. and Schenker-Wicki, A. "The Impact of Outside-in Open Innovation on Innovation Performance." *European Journal of Innovation Management*, 14 (4): 496–520, 2011.

48. Hempel, J. "Metrics Madness. Quantifying Innovation Is Key. Here's How to Do It Right and Avoid the Big Mistakes Managers Often Make." *Business Week*, 34–35, 2006.

49. Kessler, E.H. and Chakrabarti, A.K. "Innovation Speed: A Conceptual Model of Context, Antecedents, and Outcomes." *The Academy of Management Review*, 21 (4): 1143–1191, 1996.

50. Cooper, R.G. "Stage-Gate Systems: A New Tool for Managing New Products." *Business Horizons*, 33 (3): 44–54, 1990.

51. Jenkins, S., Forbes, S., Durrani, T.S., and Banerjee, S.K. "Managing the Product Development Process. Part I: An Assessment." *International Journal of Technology Management*, 13 (4): 359–378, 1997.

52. Parthasarthy, R. and Hammond, J. "Product Innovation Input and Outcome: Moderating Effects of the Innovation Process." *Journal of Engineering and Technology Management*, 19 (1): 75–91, 2002.

53. Branzei, O. and Vertinsky, I. "Strategic Pathways to Product Innovation Capabilities in SMEs." *Journal of Business Venturing*, 21 (1): 75–105, 2006.

54. Mendonca, S., Pereira, T.S., and Godinho, M.M. "Trademarks as an Indicator of Innovation and Industrial Change." *Research Policy*, 33 (9): 1385–1404, 2004.

55. Subramanian, A. and Nilakanta, S. "Organizational Innovativeness: Exploring the Relationship between Organizational Determinants of Innovation, Types of Innovations, and Measures of Organizational Performance." *Omega*, 24 (6): 631–647, 1996.

56. Fichman, R.G. "The Role of Aggregation in the Measurement of IT-Related Organizational Innovation. *MIS Quarterly*, 25 (4): 427–455, 2001.

57. CIS: Community Innovation Survey, European Commission, 2010. Online. Available at http://epp.eurostat.ec.europa.eu/portal/page/portal/microdata/cis. Accessed January 25, 2013.

58. Damanpour, F. and Gopalakrishnan, S. "The Dynamics of the Adoption of Product and Process Innovations in Organizations." *Journal of Management Studies*, 38 (1): 45–65, 2001.

59. Cooper, R.G., Edgett, S.J., and Kleinschmidt, E.J. "New Product Portfolio Management: Practices and Performance." *Journal of Product Innovation Management*, 16 (4): 333–351, 1999.

60. Danneels, E. "The Dynamics of Product Innovation and Firm Competencies." *Strategic Management Journal*, 23 (12): 1095–1121, 2002.

61. Verona, G. and Ravasi, D. "Unbundling Dynamic Capabilities: An Exploratory Study of Continuous Product Innovation." *Industrial and Corporate Change*, 12 (3): 577–606, 2003.

62. Asheim, B. "Differentiated Knowledge Bases and Varieties of Regional Innovation Systems." *The European Journal of Social Science Research*, 20 (3): 223–241, 2007.

63. Chesbrough, H. *Open Innovation, the New Imperative for Creating and Profiting from Technology*. Boston: Harvard Business School Press, 2003.

64. Szeto, E. "Innovation Capacity: Working towards a Mechanism for Improving Innovation within an Inter-Organizational Network." *The TQM Magazine*, 12 (2): 149–158, 2000.

65. TerWal, A.L.J. and Boschma, R. "Co-Evolution of Firms, Industries and Networks in Space." *Regional Studies*, 45 (7): 919–933, 2011.

66. Nieto, M.J. and Santamaria, L. "The Importance of Diverse Collaborative Networks for the Novelty of Product Innovation." *Technovation*, 27 (6–7), 367–377, 2007.

67. Rowley, T., Behrens, D., and Krackhardt, D. "Redundant Governance Structures: An Analysis of Structural and Relational Embeddedness in the Steel and Semiconductor Industries." *Strategic Management Journal*, 21 (3): 369–386, 2000.

68. Gomes-Casseras, B. "Alliance Strategies of Small Firms." *Small Business Economics*, 9 (1): 33–44, 1997.

69. Forsman, H. "Innovation Capacity and Innovation Development in Small Enterprises: A Comparison Between the Manufacturing and Service Sectors." *Research Policy*, 40 (5): 739–750, 2011.

70. Kuczmarski, T.D. "Five Fatal Flaws of Innovation Metrics." *Marketing Management*, 10 (1): 34–39, 2001.

71. Shenhar, A.J., Dvir, D., Levy, O., and Maltz, A.C. "Project Success: A Multidimensional Strategic Concept." *Long Range Planning*, 34 (6): 699–725, 2001.

72. Enzing, C.M., Battering, M.H., Janszen, F.H.A., and Omta, S.W.F. "Where Innovation Processes Make a Difference in Products' Short- and Long-Term Market Success." *British Food Journal*, 113 (7): 812–837, 2011.

73. Neely, A. Performance measurement: the new crisis. In: Crainer, S. and Dearlove, D. (eds.). *Financial Times Handbook of Management.* Harlow: Pearson Education Ltd., 2004.

CHAPTER **44**

Intellectual Property for Innovation

Joseph T. Nabor and Steven G. Parmelee

Introduction

There are numerous, and different, intellectual property rights. We discuss four of the more common intellectual property rights (IPRs) here—trade secrets, copyrights, patents, and trademarks. Depending on the IPR, protection may (or may not) exist for mere ideas and half-baked notions as well as fully expressed, physically demonstrated examples, and everything in between. Verbally expressed notions, thoughts captured (literally or virtually) in print, and fully rendered prototypes may all qualify for some protection at least sometimes provided the innovator observes the appropriate protocols. At the same time, however, protection can be similarly waived, lost, or otherwise compromised when the applicable protocols are ignored either willfully or through ignorance.

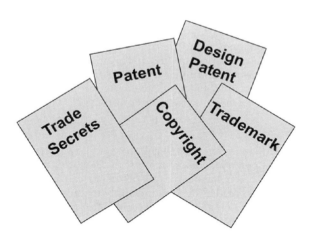

That said, space limitations require that this treatment be more in the way of an introductory primer than an exhaustive exploration of this topic. In this chapter, we do, however, pay particular attention to something that innovators and those who hire or work with innovators, really should understand and understand well—how ownership works in the context of IPRs.

To be sure, there is no monolithic, universally applicable golden rule that applies across all IPR types when it comes to knowing who owns what (and when). Instead, and for a variety of reasons, ownership rights arise (and can be lost) in very different ways depending on which IPR is under the spotlight.

Innovation occurs in a variety of different working contexts, often across various individuals and enterprises, and even spanning distance and time. Can the question "Who owns this?" vary with such differences? Many times, the answer is "yes."

Depending on the IPR (and often the jurisdiction where the innovation occurs), ownership interests can vary depending at least on preexisting written (or implied!) contracts,

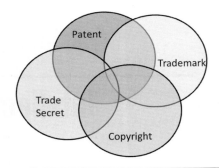

FIGURE 44-1 IPR overlap.

employment status, and whether, where, how, and by whom an idea is made tangible. As a result, the person who would innovate, and those who would prompt such innovation or otherwise work with innovators, are all insufficiently informed if they only have a working knowledge of *what* can be protected by a particular intellectual property right. It is just as important that such persons also understand how ownership rights (and other related rights) arise in the first instance for IPRs and how such rights are transferred.

The innovator should also understand that the various IPRs can and do overlap with one another in various ways (Fig. 44-1). Many patented inventions, for example, begin life as a trade secret, and the original written description of the nascent concept is often protected by copyright. As another example, many designs are potentially simultaneously protectable by copyright, a design patent, and/or trademark law. And a software-based invention often provides a particularly good example in these regards—what the software does in combination with its enabling hardware may be protectable by one or more patents, the compiled, released code may be protected by copyright, the annotated source code may be retained in confidence as a trade secret, and trademark protection may apply to the name of the product and possibly to some aspects of the product's operating appearance, icons, or the like.

Using overlapping IPRs can sometimes be important to achieving an adequate level of protection for a given innovation. To emphasize an earlier point, however, scattered ownership rights as pertain to a variety of IPRs that all apply to a greater or lesser extent to a given innovation can impair or even ruin an otherwise bright and hopeful business plan. To some large extent, ownership and/or other related rights can be controlled by the use of appropriate written agreements between the parties involved.

Trade Secrets

Trade secrets essentially refer to the legal protection often granted to confidential information having at least potential competitive value. That said, the specific details pertaining to what can be protected, to what degree that protection can apply, for how long that protection can persist, and what remedies or penalties apply to transgressors can vary greatly from one jurisdiction to another. Sometimes trade secrets are specifically addressed by statutes (on either a national or regional basis) and sometimes they are protected by courts in the absence of statutes on a common law basis. In the United States, for example, many trade secret protections are handled on a state-by-state basis.

The breadth of subject matter content that qualifies for protection as a trade secret can be broad. Examples include electrical-circuit schematics, mechanical drawings, CAD files, text-based documents of all manner, and variety including both hard copy and digital versions, chemical formulas, prototypes, programming requirements statements, and software code (including annotations when present), manufacturing equipment (including custom jigs, lines, equipment settings, and assemblies), engineering designs and specifications, invention disclosures, and so forth. Trade secret protection can also accommodate a wide variety of nontechnical materials. Examples in these regards include short-term and long-term business plans and strategic plans, financial information, lists of customers and suppliers, employee names and positions, organization charts, and so forth.

Generally speaking, one need not apply to any government or governmental agency in order to perfect a claim to a trade secret (although often one must rely on a local judiciary to enforce such rights when a misappropriation occurs). Instead, at least to a very large extent, the creation, maintenance, and perfection of a trade secret right rests on the party claiming to own that right.

At the bottom line having an enforceable trade secret usually requires that the owner in fact have a "secret." This is typically a fact-dependent question. The more the supposed secret is available to the public without restriction, the less likely the information will be protected as a trade secret. That said, it is usually not a requirement that the trade secret be wholly unshared with others. Instead, it will usually suffice if the owner of the trade secret has taken reasonable steps, commensurate with the value of the information itself, to maintain that information in confidence. As a result, it is usually not a requirement that the information have been successfully protected against any and all unauthorized attempts to gain access to that information.

In a practical sense, then, the existence of a trade secret can depend on the information-handling practices of the party claiming ownership. Locked drawers, clean white boards and clean desks, password-protected accounts, visitor check-in and supervised visitation, and written policies specifying how and when various categories of information can be shared are all typical examples of behaviors that can help to establish the existence of a trade secret. Conversely, the extent to which reasonable safeguards were not observed can negatively impact the likelihood that the information will be protected as a trade secret.

Ensuring the enforceability of a trade secret does not usually require withholding the secret from all other parties. Instead, trade secret laws will typically permit disclosure of the protected information from one person to another provided those disclosures occur in a manner that preserves the confidentiality of the agreement. In other words, a secret remains a "secret" even when disclosed to another person so long as the party learning of the secret promises to preserve the secrecy of the information. To be clear on this point by way of an admittedly extreme example, information can potentially lose its "secret" status upon being disclosed to a single person when that disclosure occurs without imposing an obligation of confidentiality on the receiving party, while information that is disclosed to one million different people can potentially retain its "secret" status so long as each and everyone of those parties is bound to an obligation of secrecy.

The so-called nondisclosure agreement (aka "NDA") is the typical legal vehicle by which one can clearly evidence this kind of control. All nondisclosure agreements are not created equal, however. Keeping in mind the evidentiary value of the document in terms of establishing the trade secret owner's reasonable practices, such a document will preferably identify the parties and the information, will specify the purpose of disclosing the information to the receiving party, will limit the receiving party's use of the disclosed information at least with respect to further dissemination and preferably impose obligations on the receiving party to take reasonable steps to also maintain the confidentiality of that information, and the duration of the follow-on period of confidentiality. It should be noted that many jurisdictions will view *"forever"* with great suspicion as regards the latter requirement. Instead, it can be legally helpful to specify a term up to, for example, 3 years, which many courts and jurisdictions will accept without much question. It is possible to specify longer periods (including *forever*) but it can be very useful to include reasons in the nondisclosure agreement itself to justify a more extensive duration of confidentiality.

Nondisclosure agreements also often include enumerated outs. These are circumstances that can cause the obligations of confidentiality to be released. For example, if the trade secret owners themselves publicize the information without any corresponding obligations of confidentiality, then the nondisclosure agreement can acknowledge that the receiving party is also free to now treat the information as public content rather than per the terms of the nondisclosure agreement itself.

In a business-oriented innovation application setting, virtually all related information that is protected as confidential might be fairly considered a trade secret. For example, the target of innovation, the driving reasons behind wishing to innovate, the perceived business strategy served by the innovation, the innovation team, the innovation plan, and so forth may all, at least initially, be considered to have competitive importance and worth keeping out of a competitor's hands. As the innovation process progresses, at least some of the information will likely leave the domain of trade secrets, although possibly in favor of other legal devices that provide follow-on protection.

Trademarks

Trademarks are most commonly words or logos that are used by someone to identify their products or services and distinguish them from the words or logos of others. Although we are most familiar with trademarks as being words or logos, the laws of many jurisdictions will tolerate trademark status for other devices as well.

In many countries, beyond the usual words and designs, trademarks can be composed of what are considered nontraditional trademarks. These include sounds such as the roar of

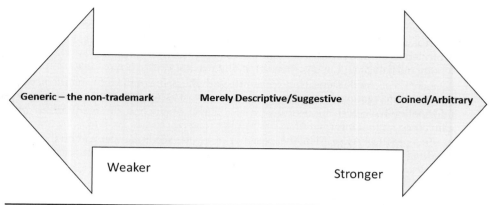

Figure 44-2   Spectrum of distinctiveness.

the MGM lion, or smells such as a scented sewing thread, or motions such as the famed Peabody Duck March at the Peabody Hotel. Essentially all sensory-perceptible indicia have been accepted for trademark protection with the lone notable exception thus far being taste. Since words and logos are the most commonly used form of trademark indicia, we will focus here on those forms while noting that our discussion applies generally to all forms of trademark indicia.

The scope of protection provided to a given trademark depends in part on where the trademark falls on the spectrum of protection. As shown in Fig. 44-2, trademarks can be categorized by the nature of the mark compared to the products or services. Thus a mark having no relevance to the corresponding product can receive a broad scope of protections. An example here is the Apple mark for computer hardware, even though Apple is a commonly used word referring to a variety of fruit that has no bearing as to computer hardware. As a result, the use of the term Apple as a trademark to identify computer hardware is an entirely arbitrary use of that word.

This is also true for coined trademarks such as Pepsi or Kodak. Apart from their use as a trademark to identify brands of beverages or photographic equipment, there is no such word as Pepsi and Kodak. Those words were made up solely for the purpose of serving as trademarks to identify those products. These coined terms share the same space on the spectrum of distinctiveness and are entitled to a broad scope of protection.

At the other end of the spectrum are words that are generic for the corresponding products or services. Thus, if one selects the term Apple as a trademark to identify apples then the trademark does not serve any distinguishing indicia purpose and it is not entitled to any protection. This provides an allowance for all those in the field of selling apples to use that word in a generic manner.

Coming within those extremes on our spectrum of distinctiveness are other categories. Toward the protective side of the spectrum we have suggestive marks, and toward the free use side we have descriptive marks. As the name suggests, descriptive marks are those that as applied to the products instantly tell the consumer information about important features, qualities, or components of those products. In contrast, suggestive marks are those that merely suggest those features, qualities, or components. The essence of a suggestive mark is that it requires the consumer to use a certain mental two-step process to recognize those features and then relate them to the products.

Trademarks are not generally monopolies on words. Instead, the trademark is defined by the products or services that it serves, to distinguish from the products and services of others. These concepts are fundamental to the selection and adoption of trademarks.

In selecting a new trademark to identify your product, it is a best practice to consider existing marks and select a mark that is distinguishable from those that already exist. There are many levels of searching available for selecting a distinguishing mark. A baseline approach simply leverages your knowledge of your particular industry. Some industries are sufficiently small such that those in the industry know the marks used by their competitors.

Another form of searching is to examine the databases of the government authorities for those jurisdictions in which you will distribute products. For example, the Unites States has all of its federally registered trademarks available in its electronic database. This is true of many of the trademark offices throughout the world.

Other search forms include the various Internet search engines. Still, other excellent resources are a number of commercial entities that provide trademark availability search reports. These search companies can provide a search report based on a level of investigation the user selects.

While any of this searching is not required prior to adopting and using a mark, such due diligence is generally advisable in order to avoid as many (potentially costly and attention-diverting) conflicts with other enterprises as possible.

One issue that differs from country to country is the determination of when trademark protection begins. For example, a certain level of protection is available in the United States the moment that one places a trademark on a product and ships that product to an independent consumer. An entirely separate level of protection is available in the United States once a federal trademark application is filed and then ultimately granted. Yet again a different level of protection is available in the United States when a state trademark registration is granted, which is usually issued by the state government office with responsibility for business services in each of the various states.

In contrast, some countries only recognize enforceable trademark rights for marks that are registered in the appropriate governmental office for that national government. To further complicate this picture, some national rights are granted by treaty via supernational trademark registrations such as the European Community trademark registration or the Madrid Agreement.

For the sake of a more specific example we will review trademark registration benefits using the United States as an example. In the United States, an application is filed with the U.S. Patent and Trademark Office. The application has certain minimum requirements that include identification of the mark along with identifying the products and/or services and the owner. As you would expect, there are many other legal issues involved in each of these components as well as the other components of an application.

We have already addressed some of the different forms of indicia that can form the basis of a trademark, but let's look at the mark itself. In the commercial world, trademarks are usually not static. That is, they evolve and change over the course of time. However, those changes need to retain the material essence of the mark. Those changes also have some dependence on the timing involved. For example, a mark created today but changed only a few months later would not likely be able to take much benefit of the original use since a change in such a short time span is often not a market-derived update of an old design. In contrast, a mark created two or three generations ago often is in need of an update to maintain relevance in the market. If such a change maintains the material essence of the original mark, then the updated mark often reaps the benefit of the earlier use. The importance of maintaining the benefit of the earlier use will become clear once we discuss the importance of priority. This also becomes important because unlike patents and copyrights, a well-cared-for trademark can last for an unlimited duration. Most trademark registration systems throughout the world have provisions for the regular renewal of the trademark registration and the rights appurtenant with that registration.

Another aspect of the mark we have identified is the products and services. Again, marks do not exist in a vacuum. The products and services to which the mark is attached go as well to the extent of protection as well as the inherent strength of the mark.

First we should distinguish between the product and the service dichotomy. "Trademark" is a broad term often used to encompass marks used both on products as well as in connection with services. Many of us, however, are familiar with the concept of a service mark as being distinguishable from a trademark. Service marks are simply those indicia that are used to identify the services of one from another. The usual fundamental difference that distinguishes a service is that the service is not a tangible product that can be physically handed from one person to another. Instead, the service is an action or set of actions that are rendered by one for the benefit of another. Many examples of these are quite easy to distinguish. When we go out to dinner and sit down in a restaurant and a meal is prepared and delivered to our table, the dining experience is the result of a series of services. The restaurant, in turn, uses its service mark to distinguish those services from other restaurants.

A stark contrast to the restaurant service is our experience shopping in a grocery store. A walk up and down the aisles of the grocery store bombards the shopper with a vast array of different trademarks, most placed directly on the displayed products. It is here that the importance of trademarks becomes obvious. For example, if you are in the market to buy a canned soup, the choices in most grocery stores will be plentiful. Not only the different flavors of soup, but more importantly here, the different brands of soup. Our shopping experience is enhanced and shorted because of the benefit of trademarks. In many instances, we know the brands that we like and those that have a less than favorable impression on us as consumers. By our retention of this knowledge, we use trademarks to influence our purchasing decisions. This is the primary importance of trademarks to both the consumer as well as the

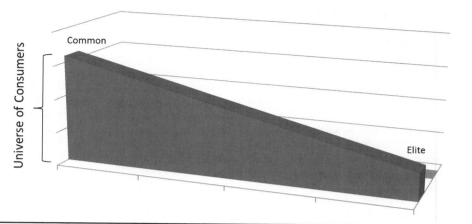

FIGURE 44-3   Elitism of the brand.

company offering the goods and underscores trademarks as an important focal point for the innovator.

When a new product is developed and brought to market, that new product begins a long course in the development of trademark goodwill. The new product developer needs to identify the sort of goodwill they seek to attach to their mark. Is this going to be a luxury branded product offered only to an economically elite group of purchaser, will this be a budget product priced to maximize the volume of sales, or is this product somewhere in the middle of that spectrum (Fig. 44-3). By the development of its reputation among the consumers for the product, the trademark begins to serve a useful purpose to the consumer as well as the owner.

Because of the value of this reputation, it is incumbent upon the owner of the brand to protect that reputation. One of the vehicles for protecting a brand's reputation is to enforce the brand against infringement and dilution. These are two separate issues within the area of trademark law that provide different protections for brands.

Protection against infringement generally prevents the use of confusingly similar brands by others. In that regard the confusing similarity is in the eye of the relevant consumers. Again, the essence of the brand is the reputation that has been cultivated among consumers so that protection from infringement seeks to protect that reputation among consumers.

The issue of what is a confusingly similar brand usually takes into account a whole host of factors. The primary facts, however, are the relative similarity of the marks of the goods and services, and of the channels of trade in which the products or services travel. In examining the similarity of the marks, most courts will look to the degree of similarity in appearance, sound (e.g., the pronunciation of the verbalized mark), and meaning of the marks at issue. This is known as the sight, sound, and meaning trilogy.

In examining the similarity of the products and services, the court will often question whether or not the accused products are the same or related to the protected branded products. This is examined in the grand universe of products and services. As a result, most of us would never confuse a shirt with a pair of pants. However, with both being clothing the degree of similarity of those two pieces of clothing is much more similar than are shirts and automobiles. As the relationship of the products become closer to each other, often the degree to which the other factors need to be more similar diminishes. These all go to the weight to be given the evidence in a particular infringement analysis.

This is true of the other factors as well. It is often stated that the more similar the marks, the less close the other factors need to be. As a result when identical marks are being weighed against each other, even a relatively greater distance between the types of products may still support a claim of infringement. In contrast, the more different the marks at issue, the closer the products likely need to be in order to likely cause consumer confusion.

As these factors are examined, other issues are often taken into account. The nature of the products and the channels of trade often dictate the degree of care to which the consumer will exercise in their purchase of the product. The greater the degree of care demonstrated by the customer, the less likely that the consumer will be confused. As a result impulse products such as a package of breath mints grabbed at the last minute often require a lesser showing of similarity than would a complicated industrial processing machine that requires explicit specifications and study during the purchasing process.

A related issue here is the sophistication of the consumer. The more sophisticated the consumer, the lower the threshold of similarity in the marks needed to support a claim of infringement. Generally the purchaser of a nuclear reactor is more knowledgeable of the vendors in that industry than is the average consumer of an ice cream bar. This sophistication is often highly relevant to the issue of whether or not consumers are likely to be confused. These factors point to the degree to which the actual marketplace for the products at issue is the critical enquiry.

When examining the likelihood that consumers will be confused, the marketplace should be the context of the examination. However, this requires a degree of caution. The marketplaces for different products are often not the same though confusion can still be likely when market overlap occurs. In addition, market change from time to time and the likely expansion of markets in the future is equally a relevant consideration.

As stated previously, another protection afforded the reputation of a brand is protection against dilution. This is the protection of the distinctiveness of the brand. This aspect of trademark laws protects a trademark's distinctiveness from being diluted by the use of similar brands by others. Usually, this relates to products that are not necessarily so similar that a consumer would be confused as to the source of that product. This is not a protection of the immediate reputation held be the consumer but instead protects the distinctiveness of the brand across entirely different product lines and channels of trade. Since this can be a powerful right provided to the brand, this protection is usually afforded to only the most famous or well known of brands. For example, many luxury brands have been afforded dilution protection so that products in a different goodwill spectrum do not diminish the value of that reputation.

Another factor affecting the strength to be afforded the brand is the degree to which the brand has acquired distinctiveness in the marketplace. This acquired distinctiveness is also sometimes called secondary meaning when it is being applied with respect to marks that are primarily descriptive of the goods or services with which the marks are used. Recalling that descriptive brands are not entitled to protection or only the most minimal of protection, that degree of protection to be afforded the brand can be heightened once the brand has acquired a secondary meaning identifying the source of the product. Generally speaking, secondary meaning refers to a meaning and reputation understood by the consuming public that has now been built by use of a given mark as a brand; in a very real sense it is the consumer who imbues a descriptive mark with secondary meaning and when this happens many trademark systems will permit that secondary meaning to be protected.

The consideration of whether or not acquired distinctiveness attaches to a brand is again determined by a number of factors. These factors include the length of time for which the brand has been used, the volume of sales of products bearing the brand, both in units as well as revenue generated, and the volume of advertising as well as the nature of those advertisements. All of these factors provide evidence of the degree to which consumers attach a recognized secondary meaning to a particular brand. Other times, the proof of acquired distinctiveness is gathered through the use of declarations from others knowledgeable of the industry that recognizes the brand. Additional evidence is sometimes gathered through the use of consumer surveys where the recognition of the brand is tested in the simulated marketplace. Again, the ultimate purpose here is to evidence that others attach recognition to the brand as being an identifier of source and not merely as a describer of some attribute of the product or service.

The issue of descriptiveness includes factors other than merely those terms that describe features, qualities, or characteristics of the product. Such additional descriptiveness factors include surnames. Generally, a surname is not given trademark protection unless it has achieved acquired distinctiveness. This allows others to enter the market using their own surname. Whether or not a mark constitutes a surname is often determined based on the number of people in the telephone directory with that name or whether or not anyone with that name is associated with the company.

Another descriptiveness issue is raised by marks that have geographical significance. This usually entails a multipart test that examines the mark in relation to the goods. The first enquiry is whether or not the mark is such that the relevant consumers would expect the goods at issue to come from the identified place. If consumers would likely have such an expectation, then the mark has geographical significance. An example of this might be a mark such a Napa for wine. Because the area of Napa, California is well known for wine, it would be reasonable for consumers to expect wine with the mark Napa to come from Napa, California. The next part of the enquiry then examines whether or not the products actually come from that location. If the wine actually comes from Napa, California, then the mark is geographically descriptive and can only be protected upon showing that the mark has acquired distinctiveness as a brand. In the event the wine does not come from Napa, California, then the mark itself is considered geographically misdescriptive and cannot be protected as a brand under any circumstances.

In the United States, the issue of foreign language can play an important role in determining the protectability of marks. In this case, if the meaning in the foreign language is descriptive, then the protectability of the mark is limited just as if the English translation was used instead of the foreign language mark. Usually the degree to which that language is readily understood is of little significance. That response to a foreign language equivalence is reserved for the most obscure or ancient language that is no longer in use.

In many cases, on day one of a new business or product a newly adopted mark offers little immediate value. As time passes, however, and public goodwill and recognition builds, trademarks can become the single most valuable intellectual property an enterprise may own. Steering clear of conflicts with others and protecting brands against infringement and dilution can seem mundane in the exciting context of innovation, but these behaviors are very much the stuff of long-term value building.

Patents

Patents are often confused with a monopoly. A monopoly typically describes the situation when one enterprise is the only enterprise engaged in the manufacture and sale of a particular product or service. While a patent can sometimes aid in creating such a situation, a patent itself is technically rather the opposite. That is, patents are a government-granted right that (literally and strictly) permits the patent owner to prevent others from practicing the claimed invention while not guarantying the patent owner the right to practice that same invention. Accordingly, on the face of it, a patent does not give the patent owner the clear right to themselves practice the invention. This nuance can be important when considering whether and how to employ patents to protect one's innovations.

Patents are the end result of a formal, complex, and typically expensive acquisition process (Fig. 44-4). Patents are granted by individual countries and, in some cases, by authorized regional authorities (such as the European Patent Office). A patent for a given country typically only provides protection within that country and will not provide the patent owner with rights to prevent parties in other countries from practicing the claimed invention absent a relevant patent in that country as well.

Patents are very much creatures of limited life (Fig. 44-5). For example, patents typically expire at the end of 20 years following the date of filing the application (though numerous circumstances exist that can shorten or lengthen the patent's duration).

Patents typically have two main sections, a technical description of the invention and one or more claims. The legal coverage offered by a patent begins (but does not necessarily end) with those claims. Claim drafting constitutes a fairly arcane art and science complete with its own syntax and burdened (or enlightened, as you wish) by an ever-evolving body of rules, regulations, and court precedent to guide their interpretation. As a result, drafting good patent claims often requires the attention of both the innovator and a patent specialist.

Obtaining a patent requires the preparation and filing of a patent application with the relevant patent office(s) of interest. This step alone typically represents a considerable investment of time and effort and often costs many thousands of dollars. The patent office in turn will eventually examine that application to determine whether the contents meet a number

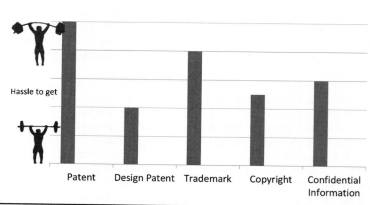

Figure 44-4   Relative comparisons of the difficulty of acquiring and maintaining different IPR's.

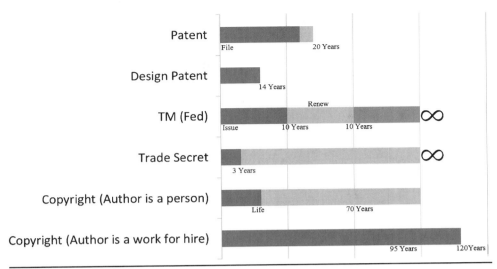

FIGURE 44-5   IPR lifetimes.

of formal and substantive requirements and most especially whether the claims describe an invention that is new enough to warrant the granting of a patent. In many cases the patent examiner will be initially unimpressed to a greater or lesser extent, thereby necessitating a formal dialog to correct or modify the application to the examiner's satisfaction or to otherwise convince the examiner that the application is fine as is. This so-called prosecution activity can stretch out over months or years and can also require an expenditure of considerable sums.

If and when a corresponding patent eventually issues, additional fees are due at that time. Many countries also require an occasional maintenance fee payment to keep an issued patent (or, sometimes, a pending patent application) in force and effect. A failure to pay such a fee leads typically to the early expiration of the patent. And again, these fees over time can represent many thousands of dollars.

There are also any number of ways that a given patent office may reengage with an already-issued patent. Reissue patent applications (typically filed by a U.S. patent owner to correct some mistake in the original proceeding), ex parte reexamination requests, inter partes and post grant reviews (which can involve, to a greater or lesser extent, a third party who seeks to challenge some aspect of the issued patent), supplemental examinations, and opposition-like proceedings are all possibilities in the United States and many other jurisdictions. Any of these proceedings can be fatal to one or more claims of an issued patent and hence can require the full attention of the patent owner. And again, such proceedings can stretch out over lengthy periods of time and can be very expensive as well.

In many jurisdictions the right to seek a patent expires if and when the invention is sufficiently publically disclosed prior to filing a corresponding patent application. As a result, many innovators new to the process discover that they are already too late to file a patent application for their invention as they have already provided a description in some public manner. (The opportunity to communicate and post via the Internet has greatly increased the number of opportunities and temptations for such a thing to happen.) As a result, a healthy patent acquisition program typically goes hand-in-hand with a healthy trade secret program to ensure that innovations remain confidential at least until a corresponding patent application has been filed.

In some countries, and certainly in the United States, the patent ownership interest rests in the first instance with the inventors. Accordingly, and by example, an employer can only claim ownership to an invention of one of its employees via an assignment of those rights. With this in mind, many enterprises utilize employment agreements that arrange for such assignments and follow up with individual assignments of specific inventions to ensure that title rests clearly with the enterprise. In the absence of a written agreement, many jurisdictions will enforce an implied agreement that inventions made by employees who are hired to invent are to be assigned to the enterprise.

In both cases, however, the facts of the situation can sometimes be important. When, for example, the employee invents something that is completely unrelated to their usual business activities for the enterprise and that is unrelated as well to the normal business of the enterprise, and where the work the employee does on the invention is on their own time using their own resources, it may be that the enterprise has no legitimate interest in such an invention.

Further complicating the situation, at least in the United States, is the notion of so-called shop rights. Generally speaking, if an employee invents something that they are not obligated to assign to the enterprise, but they use enterprise resources to reduce the invention to practice, the enterprise may have a shop right in the invention that permits the enterprise to use (though not own) the invention without further payment to the employee.

If ownership is driven by inventorship, this in turn can sometimes present challenging questions. Generally speaking, a person is an inventor if they conceived and contributed to anything that appears in any of the claims of the patent. There is no requirement that each inventor contribute in like substantive manner as compared to one another and there is no requirement that all of the inventors work together or even necessarily know one another. As a result, a given innovator may be an "inventor" of a given patented invention if that patent includes a claim directed to their specific inventive contribution, and may similarly be appropriately denied inventorship status if such is not the case (and regardless of how important their participation might otherwise have been to the project).

Incorrectly identifying the inventors (either through inappropriate inclusion or exclusion of specific individuals) can result, at worst, in rendering the corresponding patent unenforceable. Naming the inventors also can have significant impact on initial ownership of the patent application, however. In the United States, for example, all named inventors share initial ownership equally. A patent application that names two inventors, where one inventor made one small contribution that appears in only one of 20 proffered claims, has its ownership shared completely equally between those two inventors (absent some agreement to the contrary) regardless of the disparity of their relative contributions. The consequences of this result can be significant depending on other circumstances.

Various categories of things can be patented. Many things and processes as well are properly the subject of the so-called utility patent. Ornamental designs can be the subject as well of the so-called design patent. (Design patents are often less expensive than a utility patent to acquire and have a shorter maximum term of life. Unlike a utility patent that expresses its claims in prose, the illustrations of the design patent serve as its claims.) And at least in the United States a so-called plant patent will protect certain nonsexually propagated plants.

That said, patents are not available to protect every idea that every innovator might develop. While the precise standards vary from country to country, to some extent at least the invention must be sufficiently new as compared to past discoveries and endeavors to warrant the granting of a patent. It is well beyond the scope of the space available here to explore such topics as novelty, nonobviousness, inventive step, and so forth.

That said, it is worth noting that, as with trademarks, it is possible to do searching to attempt to identify relevant prior art that can influence whether, and to what extent, a given innovation can be likely patented. Many patent offices (including the U.S. Patent and Trademark Office), many search engines (such as www.google.com/patents), and a number of dedicated sites (such as www.freepatentsonline.com) provide free public access to many records regarding previous published patent applications and issued patents. These free services can vary with respect to their supported search logic as well as the completeness of their offerings for a particular country or region. In addition, there are a number of fee-based online search services that can serve in these same regards.

Sometimes, however, finding relevant prior art feels like the proverbial search for a needle in a haystack, and sometimes things are worse and the activity feels more like searching for just one particular needle in a stack of needles. Numerous professional searchers offer their services in these regards. These service providers range from one-person shops to large operations featuring 100 plus well-trained, highly educated searchers having access to a variety of public, fee-based, and proprietary databases and classification systems.

There is no legal requirement that the innovator conduct a prior art search. And, in fact, there are business settings where conducting such searches may not represent a best use of available resources. The decision to not search, however, like the decision to conduct a search, should be actively considered and not simply happen as a default circumstance resulting from inattention.

In addition to testing the patentability of a given innovation, searching can also help identify whether pursuing the innovation may expose the innovator to the risk of infringing third party patents. Again, the decision to search for potentially infringed patents is not always the right decision for a given enterprise, but the ramifications of infringement are severe enough to justify serious consideration of such due diligence. Infringement (win, lose, or draw) can lead to some of the more costly litigation the U.S. legal system is able to dish up. And a finding of infringement can lead to a serious financial accounting and the risk of having the business involving the innovation shut down.

And things can get worse for a party who willfully infringes a patent. For willful infringement U.S. courts are willing to triple the damages that the infringing party must pay and the infringing party may also be forced to pay the winning party's legal expenses (which, as hinted at earlier, can be considerable).

As with trademarks (but unlike trade secrets or copyright), "innocence" by way of ignorance is not a full defense to a charge of patent infringement (Fig. 44-6). Instead, it is quite

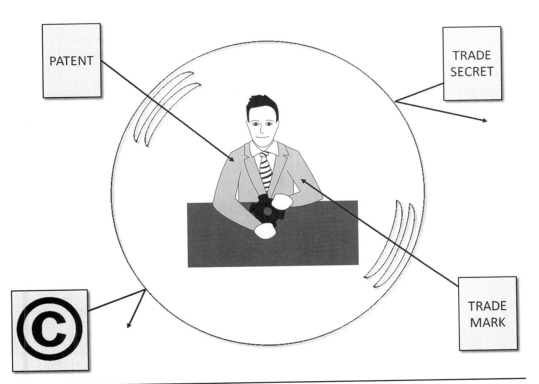

Figure 44-6 Bubble of innocence.

possible for a party to invest in a product they believe to be new and, upon bringing that product to market, discover instead that their efforts are subject to the previously established patent rights of others. Accordingly, many innovators find that any discussion of patents as a means of protecting their innovations is not complete without some consideration of whether those same innovations are somehow subservient to the patent rights of others.

Copyrights

Copyrights protect original works of artistic authorship. These artistic works must be fixed in a tangible media, but various forms of media are accommodated. For example, copyright extends to textual works, recorded-performance works, sculptural, and many others. Protection can also extend to certain technical works such as computer software, technical manuals, and the like. There are also a variety of separate provisions for other creative works such as those that provide copyright protection for architectural works.

The general universal rule is that all of these works are entitled to protection the moment they are recorded in a fixed media. Thus, the author of a book has initial protection of the very first draft of her book.

It is important to understand that what is being protected by copyright is the precise expression contained within the fixed work itself. This is *not* a protection for the idea or concepts contained within the work. Furthermore, a mere recitation of fact, *scenes a faire,* and purely functional components of useful articles are usually not entitled to protection by copyright laws.

One of the issues that often arises with copyright protection is the degree of artistic originality required for protection. Copyright protection requires that the work be original to the creator such that they must have fixed the work in a tangible media from their own creative abilities. That said, the required degree of artistic merit for the work to be entitled to protection is minimal.

Some countries have an established system for recording copyrights. The United States has a copyright registration system (administered by the Library of Congress) but registration is not required to actually establish copyright protection for new works. However, registration is necessary in the United States in order to file civil suits to enforce the copyright claims. A timely registration is also an important requirement for establishing an entitlement to certain monetary damages under the U.S. copyright statute. Many countries in fact have no system for registering copyrights though copyright protection is provided via their courts.

While copyright protection may exist from the moment the creative work is fixed in a tangible media, that right is not endless. There are time limits governing the protectability of that work. For example, for newly created works in the United States the period of protection is limited to the life of the author plus 70 years after the author's death or the last of joint authors to die. If the work is a work made for hire as defined by U.S. law, then the term of protection is the shorter of either 95 years from the year the work was first published or 120 years from its creation (Fig. 44-7). After the expiration of the protectability period, the work is said to fall into the public domain and can then be freely used by others.

Ownership of the copyright in a work can be transferred from one to another by a written assignment. Once the complete copyright interest is assigned, the original owner retains no right to the work. Note, however, that U.S. law provides for a limited one-time exception to claim a reversion of that ownership transfer. That reversionary interest is very limited in nature and application but can act to bring ownership of the copyright back to the original author.

Separate from an outright transfer of ownership of the work, a copyright owner can also license the use of the work to others. Such licenses can be either exclusive or nonexclusive. In an exclusive license, the person receiving the license becomes the only one who can publish the work. While the copyright owner retains ownership of the copyright, they cannot publish the work while the exclusive license is still valid. For a nonexclusive license, the right to publish belongs to all licensees as well as the owner unless otherwise agreed upon. In most circumstances, an assignment of the copyright ownership must be in writing and signed by the owner of the copyright. Licenses, on the other hand, do not need to be in writing although a written document signed by all the parties is generally recommended when practical.

A common misunderstanding related to copyright ownership concerns physical possession. Just because one is in physical possession of the work that is the subject of copyright, that does not also place the legal copyright ownership in that possessor. Thus, when you buy a book or a piece of software, you are in possession of that work and have certain limited freedom to use the work, but that does not include ownership of the copyright in the work. While many of the copyright owner's rights to that particular copy you legitimately possess may be exhausted, the actual copyright in the underlying work still remains with the

AUTHOR

Author is work made for hire   Author is a person

FIGURE 44-7 Author.

copyright owner. This often becomes an important issue when the possessor wishes to use the copyrighted work as the basis for further works. The ability to do so can be dependent on an extensive legal examination of the particular facts at issue for each event.

The ownership of copyright in a work entitles the owner to a certain bundle of rights. These rights include not only the right to transfer ownership or license the work, but also the right to publish the work and the right to create derivative works. Whenever a work appears in public, each iteration of that appearance is considered a publication of the work. If that publication of the work is not authorized by the copyright owner, then it is usually considered an infringement. As for the right to create derivative works, a derivative work is one that is still substantially similar to the original work but recasts, transforms, or adapts the original work into a new work. As such, a derivative without authorization from the copyright holder is often an infringement.

In some manner, a seeming antithesis of copyright protection as it relates to computer software is open source software. Open source software is usually clearly identified as such. It is software in which the source code is unlocked and freely available to others to copy, modify, and redistribute without the restriction of copyright protection. This allows others to freely build on the software for new applications. Much of this software has grown out of the movement that is now directed by a California corporation known as the Open Source Initiative. In return for this free use of the open source software, however, the user usually is obligated to freely license the new use to others as well. And in most cases this obligation is itself enforced via the copyright in the original open source content.

Many parties have found themselves in the unhappy position of permitting others free access to their software products as a result of having built their software innovations using open source content. The "unhappiness" in such a case usually stems from the innovator having been initially unclear (or even completely ignorant) regarding the downstream obligations that often attend the use of open source content.

Internationally, copyright interests are the subject of a number of international treaties. The most significant of these are the Berne Convention and the Universal Copyright Convention. Under each of these treaties the member country must extend copyright protection to works created from other member countries on the same basis as they do nationally created works. These treaties are also significant since most industrialized countries are members of one or both of these treaties.

Conclusion

Intellectual property rights are often a two-edged sword. Used properly, IPRs often help the innovator to preserve at least some of the monetary and sweat-based equity of their innovation efforts and can sometimes provide an opportunity to further leverage the innovation

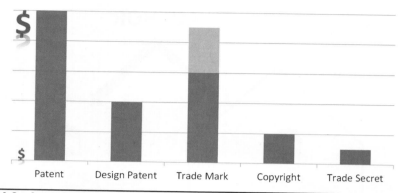

Figure 44-8 Government fee costs.

results. Third-party IPRs, however, can achieve quite the opposite result. Scrupulous behaviors while innovating can often do much to protect the innovator from charges of trade secret misappropriation as well as copyright infringement. Innocence, however, offers no complete defense to either trademark or patent infringement. (That said, knowing intentional infringement of a trademark or patent can result in particularly harsh consequences for the infringer.)

As a result, innovators or those who would seek to invest in the innovations of others sometimes seek comfort by searching for patents or trademarks that might be infringed by the innovation. These search results are assessed to identify potential concerns and opinions of counsel utilized to guide behaviors and inform corresponding business decisions. For example, an identified patent of concern may be targeted for acquisition or licensing. As another example, the innovation activity may be tasked with achieving the desired functionality while avoiding the specific approach of a patent of concern. Many useful advances in technology in fact result through exactly this kind of prophylactic behavior.

As is often the case, the devil is in the details. The book you are presently holding is testimony enough as to the many challenges the successful innovator faces, but overcoming the real-world obstacles to innovation are not, in and of themselves, sufficient to ensure that the successful innovator can reap the rewards of their efforts or even to necessarily possess the required degrees of freedom to practice their own innovations. The details of

Figure 44-9 Different IPRs.

IPR ownership (and the closely related rights to use) must also align with the innovator's business strategy. Unfortunately, in many cases, ownership and rights to use often fall naturally in a less than optimum distribution.

Worse, correcting ownership and rights to use conditions often becomes more difficult as a practical matter over time. It is nearly a cliché that the greater the value of the innovation, the greater the challenges to maintaining one's unchallenged rights to that innovation. The best time (legally, logistically, economically) to deal with ownership and rights to use concerns is at the very beginning.

CHAPTER 45

Product Launch

C. Anthony Di Benedetto

Product Launch as an Academic Field of Study

Leading new product development (NPD) firms depend on successful new products for long-term performance. In a recent study supported by the Product Development & Management Association (PDMA), 49 percent of the sales and profits of top NPD performers is derived from products launched within the past 5 years.[1] Academic research on the NPD process generally shows that executing an effective product launch is strongly related to overall new product performance.[2–13] Indeed, even when all other stages of the NPD process are executed well, if launch is mishandled, the product is likely to fail in the marketplace.[14] (See Fig. 45-1 for a view of the stages of the NPD process.) In addition, launch is often among the most expensive stages, if not the most expensive stage, in the NPD process, due to the costs incurred in ramping up to full-scale production, organizing the supply chain and dealer networks, training the sales force, and advertising.[15,16]

The launch stage is also fraught with risks, in particular when the innovation is of a radical nature. There are many unknowns, including whether a mass market for the product (not including technology enthusiasts or "visionaries") even exists.[17] In addition, other unknowns include likely competitive reaction to the launch, how the competitive framework will change through time, which among the competing technologies will become the dominant form, what levels of product performance will be required by customers, and so on.[18,19]

Despite the importance of the launch decision, product launch did not garner much academic research interest at first. An early meta-analysis on product launch found few empirical studies until the early part of the 1990s.[7] The PDMA published its first *Handbook of New Product Development* about this time; in this volume were practical articles on how to manage a consumer product launch[20] and a business-to-business (B2B) product launch.[21] Academic research began in earnest about this time, with several new empirical data collections initiated in the United States, United Kingdom, and Netherlands.[9,11,22–26] This literature was initially concerned with the strategic and tactical decisions taken at the time of launch, while subsequent articles examined the role of supply chain management,[27,28] competitive reaction to launch,[29] launch signaling,[30] launch of highly innovative products,[31] and the effect of market orientation on launch.[16] This stream of literature is reviewed by Calantone and Di Benedetto,[12] and is summarized in Fig. 45-2. Many of these topics are examined in detail later.

Strategic Launch Decisions

Empirical studies generally distinguish between strategic and tactical launch decisions[9,10,18,23] (see Fig. 45-3). Strategic launch decisions are long term in nature, and involve answering questions such as which types of products to launch and into which markets. Strategic launch decisions are often made early in the NPD process and are in fact "strategic givens."[32] That is, the decisions have already been made, perhaps with senior management involvement, long before the NPD process started, or possibly at the time of product protocol specification (what deliverables should be built into the product's design). Specifically, strategic decisions might include decisions concerning the target market (mass market vs. specific segments or niches), whether to be a leader or fast follower, and whether to aim for radical versus incremental levels of innovation.[7,10,11] Other strategic launch decisions to consider include the timing of the launch[13] and determining the competitive position.[18] These and several other strategic launch decisions are discussed in greater detail later. Keep in mind that, by their nature, strategic launch decisions

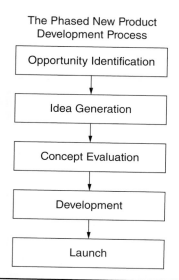

The Phased New Product Development Process

Opportunity Identification

↓

Idea Generation

↓

Concept Evaluation

↓

Development

↓

Launch

FIGURE 45-1 The phased new product development process. (*Source:* Adapted from Carwford and Di Benedetto (2011).)

Author(s)	Type of Study
Calantone and Montoya-Weiss (1994)	Early review article
Hultink et al. (1995, 1997, 1999, 2002)	Empirical studies in UK and Netherlands
Di Benedetto (1999); Lee and O'Connor (2003); Calantone and Di Benedetto (2007)	Empirical studies in N. America
Guiltinan (1999), Bowersox et al. (1999)	Conceptual studies
Thoelke et al. (2001)	Dutch case studies
DeBruyne et al. (2002)	Multi-country empirical study
Langerak et al. (2004)	Dutch empirical study
Calantone et al. (2005)	Later review article
Calantone and Di Benedetto (2012); Calantone et al. (2012)	Empirical studies of launch timing

FIGURE 45-2 Studies of new product launch. (*Source:* Adapted from Calantone and Di Benedetto (2007).)

Strategic Launch Decisions	Tactical Launch Decisions
Type of Demand Sought (primary versus selective)	Promotion (advertising, publicity, sampling, etc.)
Permanence (temporary or permanent position in the market)	Sales and Distribution (distribution structure and incentives, technical support, etc.)
Aggressiveness/Speed of Entry	Pricing (introductory price decisions, price administration)
Competitive Advantage (price, differentiation)	Product/Product Line (breadth or depth of assortment)
Product Line Replacement (terminate old product when the new one is launched?)	Timing (preannouncement, timing of launch, deletion decision)
Segmentation and Positioning	
Branding	
Packaging and Labeling	

FIGURE 45-3 Strategic and tactical launch decisions. (*Source:* Crawford and Di Benedetto (2011); Guiltinan (1999).)

are relatively difficult to change once implemented, so the firm should be quite sure of these long before the launch date so as not to jeopardize the success of the launch.

Strategic and Tactical Launch Decisions

One strategic decision is whether the launch objective should be stimulating primary or selective demand.[32] For radically new products (such as the first iPad, or the first in-line skates), the firm must stimulate demand for the unfamiliar product category by stressing its benefits relative to other products. For example, when Sony introduced the first recordable compact disk system (Minidisk) to the market, it stressed the generic benefit of being able to record music in improved digital sound; the first manufacturers of GPS systems had to convince drivers that they needed an on-board electronic map to provide directions. For more incremental launches, the goal is to draw market share away from competitors, or to hold onto market share. Selective demand refers to a demand objective framed around market share considerations. For example, when Coke or Kellogg's launches a new soft drink or cereal, the goal is often to draw market share from Pepsi or General Mills. The launch plan would therefore focus on generating trial, with the long-term goal of increased adoption and improved market share. As another example, Gillette wants most of the customers of its existing brand (say, Mach III) to upgrade to the newest version (Fusion) and not to migrate away (this kind of selective demand stimulation is sometimes called stimulating replacement demand).

The firm must also consider permanence (whether the new product will be permanently or only temporarily on the market).[32] A new snack product or ice cream flavor may be on the market for a short time only, for example, to tie in to a current movie or sporting event. In fact, its presence there may be strictly tactical: to take shelf space away from competitors. Here, some marketing aspects of the launch may be different (one can take chances with a humorous brand name if the product is only going to be on the market for a short while), as may some operational aspects (if the product is temporary, the firm would rather use contract manufacturing than build a new plant).

One important strategic decision is the desired competitive advantage: cost leadership, differentiation, or focus.[33] Clearly, marketing-mix decisions such as price or promotion will be greatly affected by whether the firm wants to compete as a low-cost alternative or as a differentiated product that offers differential advantage over the competition. A combination of these may also be possible, if the firm strives to offer differentiated products at low prices, such as designer label kitchen appliances for sale at half the regular price at Target stores.[32] Similarly, the firm should also think about how the new product will replace existing ones. Gillette hopes to completely obsolete its existing razor with the newly launched one; when Ford stops making Escorts and begins building the Focus on the same platform, it is doing the same thing. By contrast, electronics manufacturers such as Sony or Samsung frequently simply add a new, higher-performance product to the line while maintaining the existing product on the market at a lower price. Samsung big-screen TVs of 2 or 3 years ago sit side by side with the latest models, and are more affordable now to families that resisted up to now due to the perceived high price.

The speed of product rollout is also a strategic launch decision: a firm may roll out more quickly or more slowly depending on competitive conditions. Overlooked sometimes is the need to choose the right mode of rollout. Many firms choose a simple geographic rollout (e.g., East Coast followed by West Coast followed by Midwest) for their products, yet other options are possible. For high-tech, expensive new games, a video game manufacturer may start in exclusive retailers such as Sharper Image and advertise in airplane magazines, gradually rolling out to lower-end retail stores and discounters. When launching a line of inexpensive toy cars, a toy and game maker may do the opposite: try it out first at Walmart, and if they have no difficulties selling them, roll out to smaller national and regional retailers, then to local toy stores.

Segmentation, targeting, and positioning strategies (sometimes called STP strategies), developed earlier in the marketing plan, will also be strategic givens at the time of launch. One obvious example is geographic segmentation: a product will be launched in a certain region, nationally, or internationally, according to the marketing plan. The same principle applies when benefit segmentation or psychographic segmentation is used. The product is targeted to specific buyers based on apparent benefits sought, or according to activities, interests, or life styles. Perhaps most important at the time of launch is the product positioning statement, which can be thought of a kind of value proposition to the target customer: why should they buy this product as opposed to another competitive offering?[32]

A firm has several options when developing its product positioning statement, as shown in Fig. 45-4. First, it can position its product according to its attributes: its features, functions,

Position	Example
On a feature	A new tire has extra-thick treads (statement of feature only)
On a function	A new tire's extra-thick treads hugs wet roads better (the function that is provided by the feature)
On a benefit	A new tire' extra-thick treads provide a safer ride for your family on wet roads (the customer benefit provided by the feature)
Claim of superiority	The new tire is the best on wet roads (generic claim that it is better than the competition)
Manufacturer	The new tire is manufactured by Goodyear (a trusted name)
Production method	The new tire is produced using new improved rubber technology
Target segment	This is the tire for parents with small children, who are always looking out for their safety (it's the safest tire on the road)
Customer preference	It's the No. 1 selling tire in North America
Endorsement	The new tire has a well-liked race driver as its spokesperson
Satisfaction	It's the tire most families have come to rely on; most families who buy this brand stay with it
Competition	The new tire is just as good as Firestone's version, but it's 15% cheaper
Previous products	You've trusted previous tires by this company so you will also be happy with this one

Figure 45-4   Positioning options at the time of launch. (*Source:* Adapted from Crawford and Di Benedetto (2011).)

or benefits. Miller Lite beer was positioned for years using a slogan that essentially is two product benefits stuck together: "tastes great, less filling."[32] When Kraft Slices are promoted as containing as much dairy as a whole glass of milk, this is a clear example of feature positioning. But a new product may be positioned using surrogates, that is, without making reference to any attributes. These surrogates can include generic claims of superiority (Cascade dishwasher detergent or Perrier water are positioned as "just better"), parentage (Calvin Klein fragrances or Disney movies should be good quality because of the firm's reputation), manufacturing process (BMW and other German cars are supposedly better engineered, and beechwood aging is said to improve Budweiser beer), rank (Honda Accord is often cited as the number one car in North America, while Singapore Airways is often rated the best international airline), and so on. These and many more examples are illustrated in Fig. 45-3.

Another aspect of launch strategy that cannot be overlooked is the choice of brand name. If a poor name is chosen, it is difficult, costly, and risky to try to change it. A name might fail for being unintentionally humorous (especially if launched into a foreign market), or for simply being uninspiring (such as "Fresh and Lite," which was a line of frozen entrees but which could have been almost anything).[32] Sometimes, an unusual but eye-catching brand name (such as Billy Fuddpucker's or Bluetooth) can be a very good choice due to the element of surprise and distinctiveness. Most companies will carefully select their brand names, and seek trademark protection so as to protect the brand's equity. There are numerous guidelines for good brand name selection, some of which are provided in Fig. 45-5. Other associated strategic decisions include labeling and packaging: again, correct decisions here might be vital to launch success, while mistakes would be difficult and/or costly to overcome.

Tactical Launch Decisions

At later stages of the NPD process, when physical product development is nearing completion, tactical launch decisions are usually made. These are shorter term in focus and are familiar marketing program decisions, such as levels of promotion, pricing, personal selling, and distribution, as well as breadth of product line and level of services to offer with the product. It is essential for a successful launch that the front-line sales force be convinced and motivated to sell the new product.[18] Tactical launch decisions are made after, and typically influenced by, strategic launch decisions,[7,11,34] and compared to strategic decisions, are relatively easier to alter during the launch phase.[18]

Another important launch tactic to consider is the preannouncement.[32] The preannouncement is prelaunch communication targeted to customers, distributors, suppliers,

Consider the brand's purpose: a descriptive name such as Everready might be helpful in positioning, but a made-up word like Bluetooth or Kodak might be great for identification.
Consider possible brand extensions: an acceptable name at first might prove limiting (Allegheny Airlines or Western Hotels sound regional, not international).
The expected life of the product is a factor: a temporary brand could be a movie or sporting event tie-in, but the name's popularity will fade.
Avoid anything unintentionally humorous or demeaning: particularly important when entering foreign markets.
Allocate enough time to select a name: the brand name should not be a rush decision as it may be difficult to change later; also allot enough time to get the rights to the name.
Don't be afraid to go out of the comfort zone: Yahoo! or Google might be odd at first, but may really stand out in the crowd.
Assign a knowledgeable team to choose the brand name: don't leave it to popular vote.
Make sure the name is tested: don't fall in love with a name early on; be sure to get feedback.

FIGURE 45-5   Guidelines for brand name selection. (*Source:* Adapted from Crawford and Di Benedetto (2011).)

investors, and journalists.[18] When implemented successfully, it can encourage enthusiasm and interest in the launched product and can also encourage customers to wait for it rather than try competitive products. In one of the most successful computer software launches in history, Microsoft preannounced Windows 95 several months before the targeted launch date. Even though the original launch date was pushed back (from February 1995 to August 1995), customers remained enthusiastic and avoided buying competitive operating systems. Nevertheless, problems can ensue if the launch is delayed, as was the case in Microsoft's launch of Windows 7 several years later.[35]

Another factor affecting launch is network externalities, of which there are two types to consider. Indirect network externalities refer to the sale of a product depending on the sale of complementary products. Video game manufacturers know this, so they first release their newest system (for example, Wii) to game software developers, who develop hundreds of new games that then are launched to the general public several months later at the same time the new console is launched. Nintendo knows that Wii games do not sell without the Wii console, and the console will not sell with only a few available games, so careful coordination at the time of launch is required. Direct network externalities refer to the sale of a product depending on how many other people already have it. Videophones, for example, never caught on, because they were only valuable if others also had a videophone! But the more people on Facebook, the more valuable it is as a greater number of people can be reached.[36]

Other launch tactics are familiar marketing decisions involving pricing, promotion, and distribution. A low launch price may be used to stimulate demand, or to keep potential competitors out (since they may not be profitable at that price point), while a skimming price strategy sets the price high at first, with the intention of selling to the most enthusiastic customers first, followed by the majority as the price is reduced. Promotion and distribution decisions are made in order to stimulate trial (and, eventually, also repeat purchases). All aspects of the familiar communication mix may be used here, from traditional advertising media to the "new media" (web advertising, Facebook or Twitter presence, etc.), to personal selling, and publicity. Distribution must also not be overlooked, as resellers and distributors may not necessarily be highly motivated to add new products to their lines. In fact, problems in the channel may not have been picked up in consumer tests. Parents and children might like the new toy car product line, but retailers may find the display case or in-store promotions to be too large, too ugly, or too difficult to keep clean, or may simply not want to carry yet another line of toy cars at this time. The firm must choose an appropriate strategy by which to motivate distributors to carry the product, such as lowering their costs of doing business, increasing their margins, or promoting directly to them to improve their perception of the new product.[32]

The Lean Launch

Despite the growth of academic interest in product launch, significant questions still remain almost unanswered. Consider that most launch activities involve the efforts not only of the new product development team, but also representatives from logistics and operations.

The contributions of these departments to the successful launch are usually overlooked. Since launch requires coordination of product development or management, logistics, and operations, and the amounts at stake at the time of launch are quite high, a better understanding of the interface among these departments is potentially a major determinant of ultimate success.[13]

One important consideration at this interface is at the handoff from launch strategy to launch tactics. For example, once the launch strategy is in place, ramp-up to full-scale manufacturing production should be ready, and the logistics involved in product distribution and promotion need to be in place as well. In general, a closer consideration of logistics at the time of launch is a critical determinant of a successful launch. Many firms integrate supply chain capability issues directly into their product launch strategy as part of their "lean launch" strategy. This can be briefly described as "the use of a flexible supply chain system to enable the firm to react quickly to emerging customer needs and market demands."[13] A lean launch is characterized by small commitments of resources, slow manufacturing ramp-up, and postponement of commitment to inventory.[37] Firms that use lean launch practices report significant cost-efficiencies, shorter times to market, and ultimately, improved product performance.[27,28]

It is important to retain flexibility in the supply chain so as not to jeopardize the launch. An initially successful product may stumble if out-of-stock problems occur after launch, due to insufficient planning for high demand. Adoption rates may vary by market or geographical segment and stock availability must be adjusted for these differences, otherwise there will be out-of-stock conditions in some locations and unprofitable overstocks in others. Demand may be higher than forecasted, which causes difficulties for both manufacturing and logistics as they struggle to balance unplanned customer demand against previously scheduled lead times and material or component procurement. Conversely, demand may not materialize as expected, leading to overmanufacture, preallocation of inventory, and a situation of overstock in many or all locations. In some cases, additional excess reclamation expenses may be incurred as well.

The role of supply chain logistics and production in the success of a launch cannot be overstated. A new product with high future potential may only do well in limited markets at the time of launch (e.g., in a specific geographic location or distribution channel, or within a limited customer segment). The product may have much potential, but may never be able to cover manufacturing or promotional start-up costs due to slow initial sales that do not generate enough revenue to cover the rollout costs. In an illustrative example, a consumer goods company recently rolled out two flavors of a new cracker product: plain and onion. Both rolled out of the same manufacturing process, with some crackers receiving the onion flavor and others not. The plain cracker proved very popular in the market, while the onion version flopped. However, due to lack of flexibility in the production process, onion cracker inventories mounted up unprofitably, while the company was unable to keep up with demand of regular crackers, resulting in out-of-stock problems in many geographic markets. Consequently, the product was unable to meet financial targets.[13] As a counterexample, clothing retailer Benetton uses sophisticated customer-tracking technology to identify exactly the colors and styles of sweaters that are currently popular. Due to investment in computer-aided design and manufacturing (CAD/CAM), Benetton can move a sweater from design to manufacture in literally a few hours: in-house designs are transferred to computerized garment cutters and knitting machines. In addition, Benetton perfected the ability to add colors to sweaters after they had been knitted. Traditionally, yarn is dyed first then knitted into sweaters, but this can cause overstock in unpopular colors and stockouts in popular ones. Benetton postpones the dyeing process until customer demand is clearly defined, based on information gained from their precise customer tracking technology. In addition, other logistics practices also streamline Benetton's operations. Work processes are standardized: only two box sizes are used and pre-addressed labels and bar codes are used to speed up processing. Benetton claims significant reductions, both in physical distribution costs and in lead times to foreign markets.[38]

The illustrative examples show that successful launch requires careful consideration of the scale of launch, that is, correctly anticipating the "right" size of the market and thereby to calibrate the appropriate scale of production and distribution, as well as the correct amount of marketing investment.

To support the forecast of launch scale and likely product profits, a firm may turn to teardown analysis. For example, a forklift truck manufacturer may decide whether to put a new heavy-duty model into production by buying competitive products and tearing them apart, cost estimating every part. This leads to a good projection of variable materials cost per unit, and also provides clues as to production techniques (i.e., automated or robotic production vs. human intervention), which allows a projection of variable labor costs per unit as well. To complete the teardown analysis, the manufacturer would project total market size for the forklift truck, realistically assess likely market share and total incurred costs per truck at that sales volume, and determine whether the new truck would make money at the forecasted market price and pay back development costs within a reasonable time period.[13]

This kind of analysis requires an accurate launch size projection, and the flexibility of the production process and marketing or distribution resource allocation should the launch size projection be inaccurate. To achieve desired levels of flexibility, the firm would need to closely monitor sales trends. The cracker manufacturer earlier, for example, might have profitably followed point-of-sale information and therefore more rapidly adjusted production levels to sales fluctuations and unpredictability. Firms will sometimes use rollouts at the time of launch (by region, or by type of distributor or retailer) to monitor demand market by market, and ramp up production and distribution as the "real" demand is slowly revealed.[32] This avoids overcommitment to products whose sales are lower than anticipated, and allows time for increasing production on products that are surprisingly high in popularity, so as to avoid out-of-stock situations.[13] Benetton, for example, succeeded at making these adjustments late in the process, while the cracker manufacturer could not.

Lean launch methods are designed to allow great flexibility in the supply chain so as to respond rapidly to sales success and limit production of inventory without evidence of sufficient market demand. Essentially, the principle is to respond to market needs as efficiently as possible, and as soon as possible at the time of launch, rather than risking inappropriately high or low production and inventory levels. Fundamental to a lean launch is the principle of postponement: the firm delays finalizing the product form and production of inventory until the last possible moment in the development process, to ensure it has all the customer and channel information it can gather.[39] Lean launch, therefore reduces uncertainty and increases operational flexibility: production and manufacture is pushed back to the latest possible time, when demand is more accurately known. Another related principle is lead-time reduction, such that the forecasting horizon is shortened as much as possible and the chance of error in demand prediction is minimized.[39]

To be precise, postponement actually can be thought of in terms of both time and form. Time postponement refers to shipping product quantities to specific destinations so as to just meet customer requirements, and conceptually is similar to "just-in-time" inventory procedures. Stocks are then replenished strategically, according to up-to-date sales reports. By delaying shipping until the likely demand is well known, the requirement for large safety stocks is reduced, as is the possibility of stockouts.[13,28] Form postponement refers to avoiding "locking-in" to feature specifications until specific customer information is gathered or orders are received. A firm may manufacture up to a certain point, but delay further customization until customer information emerges (much as Benetton waits to add colors until the fashion trends are known). Other examples might be the paint manufacturer that ships neutral paint bases to retailers, who can mix in colors according to customer needs, or the manufacturer serving several countries where labeling and instruction manuals specific to target countries can be added later.[13]

Advocates of lean launch note that it provides the firm with several types of supply chain management competencies.[39] First, the end goal is to provide the best value to the end user. This might require participation from a downstream partner (e.g., the paint retailer that has the responsibility of adding colors). Therefore, collaboration through the supply chain is enhanced. State-of-the-art information technology is also needed, so that customer buying behavior is accurately tracked and production and inventory replenishment occurs instantly. This is especially important if the firm is operating on a global scale. The firm thereby relies less on potentially inaccurate sales forecasts, but on real-time information emerging from the marketplace. Clearly, to accomplish all of this requires high levels of coordination of internal operations, allowing the firm to find synergies and better meet customer requirements. For example, boundary-spanning employees, such as the sales force or market research personnel, are empowered to focus their efforts and provide information leading to unique products and services that offer customer value in ways competitors cannot match.[39] Increased flexibility can also be achieved through agile manufacture, concurrent engineering and design, or improved transportation methods.[13]

The Importance of Launch Timing

In addition to lean launch activities, several other launch-related topics are only now emerging as academic research topics. For example, few studies have considered the role of launch timing, which is odd given the apparent importance of getting the timing right. Especially in a time of shortening product life cycles, a late launch may result in high competitive barriers to entry as well as a missed market opportunity window.[40] An early launch reduced the amount of time the firm has to prepare all aspects of its marketing strategy and to train its sales force and/or distribution channels; or the technology may not be sufficiently advanced or bugs may still remain in the product. To make the launch decision even more complex, there are many involved parties and their objectives may not all coincide. Senior management may feel that the optimal time to launch is now, while marketing believes a 3-month delay is required to get the distribution channel educated and motivated, while engineering and production may feel they will be unable to ramp up to full-scale production in less than a year. In sum, the launch timing decision is deceptively difficult, yet has up to now received little research attention.

While not attempting to answer all of the earlier questions, a recent study included launch timing as one of the antecedents to new product performance.[13] This study considered launch timing, and also lean launch execution (as discussed earlier) as well as level of marketing effort, and their effects on new product performance. Level of marketing effort, and lean launch execution, were found to have direct impacts on new product performance. Though a direct impact of launch timing on performance was not found, launch timing interacted with lean launch execution to increase the effect of lean launch execution on performance. A firm that executes a lean launch can speed development time, and essentially creates the option of launching right away, or delaying until all interested parties are ready for the launch. That is, a lean launch execution allows the firm to select an early or a later launch option, and to fine-tune the launch to find the best compromise among all parties. Without lean launch, the option to launch early is missed, and the firm loses the flexibility to fine-tune the launch and may be stuck with a suboptimal launch time.

In a related study, Calantone et al.[41] examined the role of launch timing as well as how well launch activities were carried out. Specifically, market orientation, cross-functional integration, and leadership style were hypothesized to be antecedents to both launch timing and to launch activities proficiency, both of which are related to improved product performance. While the results of these studies are intriguing, much more needs to be done to understand the impact of launch timing (with respect to all parties involved in the launch process) on new product performance.

Product Launch by the New Venture

Much of the earlier discussion assumes a medium or large firm, which has the resources to develop and launch a product successfully. The challenges involved in product launch are much different for the entrepreneurial firm or new venture. By definition, the new venture has fewer human and financial resources, probably lacks a track record in the industry (unless the entrepreneurs have prior experience), and lacks the scale and scope of market information available to the larger firm: these are part of what has been called the "liability of newness".[42] As a result, the new venture certainly needs to find outside resources in order to effectively launch its product; this support might come from an experienced supplier who could provide financial resources as well as technical know-how, or from a venture capitalist (VC). It is, however, a classic Catch-22 for the new venture: due to the liability of newness, the new venture is not an attractive investment for the supplier or VC but rather appears highly risky and uncertain, and there has been no time for trust between the prospective partners to emerge.[43-45] Thus, the new venture is dependent on investors for the successful launch of the product, and indeed possibly for its very survival, yet these investors may be reluctant to work with the new venture.[46-48]

The new venture needs somehow to reduce the level of risk it faces when launching its first new product. After all, a failure with the first new product may cause even initially enthusiastic investors to be less supportive in the future—assuming the new venture hasn't gone bankrupt in the meantime. The new venture typically is launching a single new product, which means it has a single large risk to manage. To reduce the risk accompanying this launch, the new venture needs to get smarter—a well-funded launch must inevitably also be a knowledgeable one.[48] The new venture can reduce the chances of launch failure by acquiring knowledge from its suppliers and distributors and otherwise "learning by doing." This knowledge expansion can be at least partially achieved from loosely coupled networks: distributors may be motivated to cooperate because they get some new product offerings, while the VC may see an opportunity to sell the product through distribution channels that have worked in the past.

A research study reported in Di Benedetto and Calantone[48] obtained data from a sample of 334 high-tech and medium-tech entrepreneurial start-ups to determine what factors were associated with better launch capabilities and also better launch execution. It was found that high investments in marketing resources, high levels of marketing skills, and high customer market orientation increased launch capabilities (these included good communications and distribution effort, and the correct choice of target customer), while market orientation increased quality of launch execution (defined as high-quality execution of distribution, channel coordination, and promotion). What is more, however, the presence of a VC permits additional funding of the marketing capabilities, and also increases the knowledge net (the VC brings in its own expertise and experience), which reduces the new venture's risk and increases the "smartness" of the launch. In fact, the study confirmed that VC presence had a significant effect on both launch capability and new product performance, and greatly improved the quality of launch execution.

Summary

There is no question that product launch is a risky and expensive undertaking for most firms. It is strange, therefore, that innovation and marketing academics only comparatively recently began studying strategic and tactical launch decisions in earnest. Recent trends have shown greater interest in not only strategic launch decisions, but also in such topics as launch timing and product launch by the new venture or entrepreneurial firm. Much progress has been made in recent years, but much more still needs to be done to improve our understanding of this complex and risky activity. Certainly, the results all point to the need to do the marketing and supply chain activities prior to launch well. If the market is sufficiently primed (in terms of promotion to end users and through the distribution channel, distributor support, adequacy of supply, etc.), the launch has a much better chance of succeeding.

References

1. Barczak, Gloria, Griffin, Abbie, and Kahn, Kenneth. "Trends and Drivers of Success in NPD Practices: Results of the 2003 PDMA Best Practices Study." *Journal of Product Innovation Management*, 26 (1): 3–23, 2009.
2. Cooper, Robert G. "The Dimensions of Industrial New Product Success and Failure." *Journal of Marketing*, 43 (2): 93–103, 1979.
3. Booz, Allen and Hamilton. *New Product Management for the 1980s*. New York, NY: Booz, Allen and Hamilton, 1982.
4. Maidique, M. and Zirger, B.J. "A Study of Success and Failure in Product Innovation: The Case of the U.S. Electronics Industry." *IEEE Transactions on Engineering Management*, 31 (4): 192–203, 1984.
5. Cooper, Robert G. and Kleinschmidt, Elko J. "New Products: What Separates Winners from Losers?" *Journal of Product Innovation Management*, 4 (3): 169–184, 1987.
6. Calantone, Roger J. and Di Benedetto, C. Anthony "An Integrative Model of the New Product Development Process: An Empirical Validation." *Journal of Product Innovation Management*, 5 (3): 201–215, 1988.
7. Calantone, Roger J. and Montoya-Weiss, Mitzi. Product launch and follow-on. In Souder, W.E. and Sherman, J.D. (eds.). *Managing New Technology Development*. New York, NY: McGraw-Hill, 217–248, 1994.
8. Parry, Mark E. and Song, Michael. "Identifying New Product Success in China." *Journal of Product Innovation Management*, 11 (1): 15–30, 1994.
9. Hultink, E.J., Hart, S., Robben, H.S.J., and Griffin, A. Launching new products in consumer and industrial markets: a multi-country empirical international comparison. In Murray, L.W. (ed.). *Maximizing the Return on Product Development*. Proceedings of the Research Conference of the Product Development & Management Association, Monterey, CA, 93–126, 1997b.
10. Guiltinan, J.P. "Launch Strategy, Launch Tactics, and Demand Outcomes." *Journal of Product Innovation Management*, 16 (6): 509–529, 1999.
11. Di Benedetto, C. Anthony. "Identifying the Key Success Factors in New Product Launch." *Journal of Product Innovation Management*, 16 (5): 530–544, 1999.
12. Calantone, Roger J. and Di Benedetto, C. Anthony. "Clustering Product Launches by Price and Launch Strategy." *Journal of Business and Industrial Marketing*, 22 (1): 4–19, 2007.

13. Calantone, Roger J. and Di Benedetto, C. Anthony. "The Effects of Launch Execution and Timing on New Product Performance." *Journal of the Academy of Marketing Science*, 40 (4): 526–538, 2012.

14. Montoya-Weiss, M.M. and Calantone, R.J. "Determinants of New Product Performance: A Review and Meta-Analysis." *Journal of Product Innovation Management*, 11 (5): 397–418, 1994.

15. Urban, G. L. and Hauser, J. R. *Design and Marketing of New Products*. 2nd ed. Englewood Cliffs, NJ: Prentice-Hall, 1993.

16. Langerak, Fred, Hultink, Erik Jan, and Robben, Henry S.J. "The Impact of Market Orientation, Product Advantage, and Launch Proficiency on New Product Performance and Organizational Performance." *Journal of Product Innovation Management*, 21 (2): 79–94, 2004.

17. Moore, G.A. *Crossing the Chasm. Marketing and Selling Disruptive products to Mainstream Customers*. New York, NY: Harper Business, 1991.

18. Hultink, E.J. Effective launch strategy. In T. Langeler (ed.). *Innopreneur: 101 Appetizers for Innovation and Entrepreneurship*. Pro-Actuate, 55–56, 2011.

19. Utterback, J.M. and Abernathy, W.J. "A Dynamic Model of Process and Product Innovation." *Omega*, 3 (6): 639–656, 1975.

20. Ottum, B.D. Launching a new consumer product. In Rosenau, M.D., Griffin, A., Castellion, G., and Anscheutz, N. (eds). *The PDMA Handbook of New Product Development*. New York, NY: Wiley, 381–394, 1996.

21. Stryker, J.D. Launching a new business-to-business product. In Rosenau, M.D., Griffin, A., Castellion, G., and Anscheutz, N. (eds). *The PDMA Handbook of New Product Development*. New York, NY: Wiley, 363–380, 1996.

22. Hultink, E.J. and Schoormans, J.P.L. "How to Launch a High-Tech Product Successfully: An Analysis of Marketing Managers' Strategy Choices." *Journal of High Technology Management Research*, 6 (2): 229–242, 1995.

23. Hultink, E.J., Griffin, A., Hart, S., and Robben, H.S.J. "Industrial New Product Launch Strategies and Product Development Performance." *Journal of Product Innovation Management*, 14 (4): 243–257, 1997a.

24. Hultink, E.J. and Hart, S. "The World's Path to the Better Mousetrap, Myth or Reality? An Empirical Investigation into the Launch Strategies of High and Low Advantage New Products." *European Journal of Innovation Management*, 1 (3): 106–122, 1998.

25. Hultink, Erik Jan, Hart, Susan, Robben, Henry S.J., and Griffin, Abbie. "New Consumer Product Launch: Strategies and Performance." *Journal of Strategic Marketing*, 7 (3): 153–174, 1999.

26. Hultink, Erik Jan, Hart, Susan, Robben, Henry S. J., and Griffin, Abbie. "Launch Decisions and New Product Success: An Empirical Comparison of Consumer and Industrial Products." *Journal of Product Innovation Management*, 17 (1): 5–23, 2000.

27. Bowersox, Donald J., Closs, David J., and Cooper, M. Bixby. *World Class Logistics: The Challenge of Managing Continuous Change*. Oak Brook, IL: The Council of Logistics Management, 1995.

28. Bowersox, Donald J., Closs, David J., and Stank, Theodore P. *21st Century Logistics: Making Supply Chain Integration A Reality*. Oak Brook, IL: The Council of Logistics Management, 1999.

29. DeBruyne, M., Moenaert, R., Griffin, A., Hart, S., Hultink, E.J., and Robben, H.S.J. "The Impact of New Product Launch Strategies on Competitive Reaction in Industrial Markets." *Journal of Product Innovation Management*, 19 (2): 159–170, 2002.

30. Hultink, E.J. and Langerak, F. "Launch Decisions and Competitive Reactions: An Exploratory Market Signaling Study." *Journal of Product Innovation Management*, 19 (3):199–212, 2002.

31. Lee, Yikuan and O'Connor, Gina Colarelli. "The Impact of Communication Strategy on Launching New Products: The Moderating Role of Product Innovativeness." *Journal of Product Innovation Management*, 20 (1): 4–21, 2003.

32. Crawford, C. Merle and Di Benedetto, C. Anthony. *New Products Management*. 10th ed. Burr Ridge, IL: Irwin/McGraw-Hill, 395–396, 2011.

33. Porter, M.E. *Competitive Strategy*. New York, NY: The Free Press, 1980.

34. Guiltinan, J.P. "The Price Bundling of Services: A Normative Framework." *Journal of Marketing*, 51 (2): 74–85, 1987.

35. Robertson, T.S., Eliashberg, J., and Rymon, T. "New Product Announcement Signals and Incumbent Reactions." *Journal of Marketing*, 59 (3): 1–15, 1995.

36. Le Nagard-Assayag and Manceau, D. "Modeling the Impact of Product Preannouncements in the Context of Indirect Network Externalities." *International Journal of Research in Marketing*, 18 (3): 203–220, 2001.

37. Calantone, Roger J., Di Benedetto, C. Anthony, and Stank, Theodore P. Managing the supply chain implications of launch. In Kenneth B. Kahn, George Castellion, and Abbie Griffin (eds.). *The PDMA Handbook of New Product Development*. 2nd ed. Hoboken, NJ: Wiley, 466–478, 2005.

38. Pepper, C.B. "Fast Forward." *Business Monthly* February, 25–30, 1991.

39. Bowersox, Donald J., Stank, Theodore P., and Daugherty, Patricia J. "Lean Launch: Managing Product Introduction Risk through Response-Based Logistics." *Journal of Product Innovation Management*, 16 (6): 557–568, 1999.

40. Kerin, R.A., Varadarajan, P.R., and Peterson, R.A. "First-Mover Advantage: A Synthesis, Conceptual Framework, and Research Propositions." *Journal of Marketing*, 56 (4): 33–52, 1992.

41. Calantone, Roger J., Di Benedetto, C. Anthony, and Rubera, Gaia. "Launch Timing and Launch Activities Proficiency as Antecedents to New Product Performance." *Journal of Global Scholars of Marketing Science*, 22 (4): 290–309, 2012.

42. Stinchcombe, A.L. Social structure and organizations. In James G. March (ed.). *Handbook of Organizations*. Chicago, IL: Rand-McNally, 142–193, 1965.

43. Wathne, K.H., Biong, H., and Heide, J.B. "Choice of Supplier in Embedded Markets: Relationship and Marketing Program Effects." *Journal of Marketing*, 65 (2): 54–66, 2001.

44. Wathne, K.H. and Heide, J.B. "Opportunism in Interfirm Relationships: Forms, Outcomes, and Solutions." *Journal of Marketing*, 64 (4): 36–51, 2000.

45. Heide, J.B. "Plural Governance in Industrial Purchasing." *Journal of Marketing*, 67 (4): 18–29, 2003.

46. Song, M. and Di Benedetto, C.A. "Supplier's Involvement and Success of Radical New Product Development in New Ventures." *Journal of Operations Management*, 26 (1): 1–22, 2008.

47. Song, L. Z., Song, Michael, and Di Benedetto, C. Anthony. "Resources, Supplier Investment, Product Launch Advantages, and First Product Performance." *Journal of Operations Management*, 29, (1–2): 86–104, 2011.

48. Di Benedetto, C. Anthony and Calantone, Roger J. Venture capital supports innovative product launch. In T. Langeler (ed.). *Innopreneur: 101 Appetizers for Innovation and Entrepreneurship*. Pro-Actuate, 18, 2011.

CHAPTER 46

New Product Launch

Fernando Almeida, José Monteiro, and José Santos

Abstract

The new product launch is a critical stage of any product development process; however, this process is highly risky because new products, either new to the market or the firm or both, are intrinsically associated with a high level of market uncertainty. Therefore, it is vital for a company to properly manage the new product launch process. This work analyzes the several components of the new product launch process in terms of portfolio management, financial impact of new products, and the role of IT technologies in the new product development phase. Additionally, we propose several practices and policies for a successful new product launch strategy.

Introduction

The new product launch phase is a vital point in the entire business process. This is particularly true in the consumer packaged goods arena, where the introduction of new products has significantly increased during the last decade. With this dramatic escalation in the number of new products competing for consumer attention, the quality of launch programs greatly impacts the success of product introductions. Done well, a launch helps a new product rapidly establish itself among its target users, gain market share, and enhance the company's brand position. Done poorly, a launch can negate all the time, money, and human capital that went into developing the new product if it fails to achieve commercial success.

Furthermore, launching new products to market quickly is a prerequisite for acquiring competitive advantage. Today, product development managers face intense pressure to bring world-class products to market in record time. According to Chiu et al, many factors contribute to this pressure, including acceleration in the rate of technological development, improved mass communication, more intense competition due to the maturing of markets and globalization, fragmentation of the marketplace due to changing demographics, shorter product life cycles, and the escalating costs of R&D. This accelerated rate of product obsolescence increases the need to develop new products quickly enough to ensure timely introduction during the product life cycle. To be successful, a company must master product strategy and skillfully navigate through proper development, and application and management of a product strategy that separates enduring success from failure.

Marketing experts estimate that two-thirds of all new products fail within two years.[1] Additionally, Cooper[2] estimates that about half of all resources allocated to "product development and commercialization" in the United States goes to products that a firm cancels, or produces an inadequate financial return. In packaged goods, for instance, less than a quarter of new products introduced in 2008 broke the $7.5 million in sales mark in their year of availability and less than one-half of 1 percent earned more than $100 million in sales. It is also widely accepted that the new product and service failure rate varies widely by company, category, and industry. Many factors contribute to this high failure rate, including a weak positioning strategy, an overoptimism about the marketing plan that led to an unrealistic forecast, an ad campaign that generated an insufficient level of awareness, new products that do not match customer needs, or unforeseen competitive countermoves.[3,4]

Several studies have also examined the determinants to new product launch success and identified many factors that distinguish successful products from unsuccessful ones.

Factors that are necessary and guarantee commercial success are termed as *critical success factors* (CSFs). It is considered imperative to reflect on how one can benefit from each and how one can translate each into an operational aspect of the new product development (NPD) process. The challenge is to design a process for successful product innovation—a process whereby new product projects can move quickly and effectively from the idea stage to a successful launch and beyond.

Companies face the challenge of determining what factors in NPD are essential for success, and how to measure the extent of this success. A metric tracks performance and allows a firm to measure the impact of process improvement over time. In fact, metrics can play an important role in helping companies to enhance their NPD efforts and, at the same time, they can be seen as a fundamentally unique source of competitive advantage. First, metrics document the value of NPD and are used to justify investments in this fundamental, long-term, and risky venture. Besides that, good metrics enable chief executive officers (CEOs) and chief technical officers (CTOs) to evaluate people, objectives, programs, and projects in order to allocate resources effectively. The right metrics have the power to align employees' goals with those of the corporation; wrong metrics are counterproductive and lead to narrow, short-term, risk-avoiding decisions and actions.

Portfolio Management

In current times, customer relationship management has taken priority over the product portfolio review. However, even in this new paradigm,[5] the existence of different products and at different stages of the life cycle is important for the survival of businesses. Thus, in modern marketing we manage clients, but we should not forget to manage the portfolio of current products and developing products. Without customers, the products may not be sold; but with no products, there are no sales.

There are several tools that help to manage the portfolio of existing products. One of the models that allow you to analyze the balance, or not, of the product portfolio is the BCG matrix, created by consulting firm Boston Consulting Group in the 1960s. This template, by itself, does not permit a multifaceted vision and even features some application limitations that in certain sectors of activity should not be neglected. Short product life cycle,[2,6] limited series, and high market fragmentation,[6] are some of the current situations that limit the BCG matrix and influence the development of new products.

A new product for a company, at the time of its release, is either a first mover, fast follower, or delayed entrant,[7] occupying a different dimension in the market. In turn, the own product development will condition the future role in the portfolio of existing products, considering that previous investments were already assumed by the company.

Therefore, in addition to the management of existing products, companies need to manage the portfolio of new products in research and development (R&D),[8] which is crucial for the competitiveness of the companies.[9] "Portfolio management is a dynamic decision process, whereby a business's list of active new product (and R&D) projects is constantly updated and revised."[8]

The management of the portfolio of potential new products is not unconnected to the portfolio management of the products that are being marketed, as the former will be the substitutes of the latter, and this occurrence must take place at the right time.

In a study carried out by Cooper et al,[8] senior managers identified several reasons to focus on portfolio management:

- Maintain the competitive position, and preferably be strengthened by increasing sales and market share.

- Use resources effectively and efficiently.

- Support the strategy.

- Help focus on strategic projects.

- Ensure a balance, between short- and long-term projects, and between low- and high-risk projects.

- Improve organizational communication.

- Encourage objectiveness.

Portfolio management causes the strategic alignment of development projects with business strategy,[10,11] maximizing portfolio value by taking into consideration available resources[11] and load balancing between projects based on various criteria.[12] It is important to have a

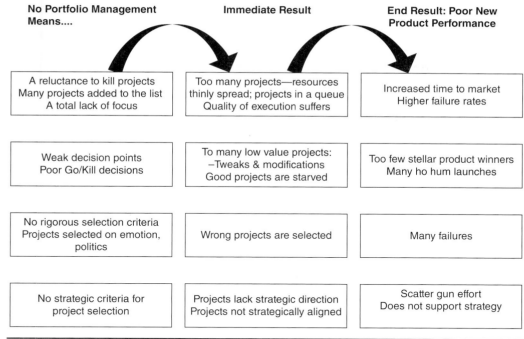

No Portfolio Management Means....

Immediate Result

End Result: Poor New Product Performance

A reluctance to kill projects Many projects added to the list A total lack of focus	Too many projects—resources thinly spread; projects in a queue Quality of execution suffers	Increased time to market Higher failure rates
Weak decision points Poor Go/Kill decisions	To many low value projects: –Tweaks & modifications Good projects are starved	Too few stellar product winners Many ho hum launches
No rigorous selection criteria Projects selected on emotion, politics	Wrong projects are selected	Many failures
No strategic criteria for project selection	Projects lack strategic direction Projects not strategically aligned	Scatter gun effort Does not support strategy

Figure 46-1   Consequences of the lack of a method of portfolio management.

standardized method of managing the portfolio of new product development projects, because its absence has negative results in the short and long term, as can be seen in Fig. 46-1 designed by Cooper et al[11]:

As methods of management of the portfolio of projects that make up product R&D, we highlight the 1 financial methods, 2 strategic buckets model, 3 bubble diagrams, 4 scoring projects, and 5 check lists.[2,13]

The financial methods include various profitability and return metrics, such as expected commercial value (ECV), net present value (NPV), return on net assets (RONA), internal rate of return (IRR), return on investment (ROI) or payback period. It is suggested that three scenarios are used: optimistic, conservative, and pessimistic.

The objective of the ECV metric is to maximize the business value of the portfolio based on the data obtained with a decision tree. Its main advantage is the combination of risk and financial return; its disadvantages are that it does not consider portfolio balancing and requires a very accurate data collection.

NPV considers all the inputs and outputs of cash flow at the current date. The result indicates the lack or excess of investment income to reach the desired minimum rate of attractiveness. If NPV is less than zero, investment isn't attractive; if NPV equals zero, it has the same investment attractiveness rate as the minimum expected attractiveness rate; if NPV is more than zero, investment is economically attractive, surpassing expectations. This method makes it possible to adopt an interest rate that is attractive compared to the risk, but to do this, it is necessary to know the risk and assign it a value.

RONA is calculated by dividing the net income of a company by the sum of its fixed assets and net working capital. It is advisable to use RONA for projects that use capital-intensive and high-volume purchases of assets.

The IRR calculation allows you to determine the minimum interest rate that would not be detrimental to the company, that is, NPV equal to zero. The result obtained and compared with the expected minimum attractive rate of return (MARR) lets you know whether the project is economically attractive (IRR < MARR), does not produce economic advantage (IRR = MARR), or is economically attractive (IRR > MARR).

ROI presents a list of net profit with assets or investments; the value above 100 percent lets you know what recovered the investment and the percentage value of this recovery. It's simple, easy to manage and compare, but does not take into account the risk involved and does not evaluate intangibles.

The payback period calculates the length of time required for the investment of the project to be recovered. The sooner the value becomes positive, the better. This calculation is simple, but does not consider cash flow after the payback period or the value of money over time.

The strategic buckets model suggests the categorization of all budgeted R&D projects into "envelopes" or "buckets" based on selected criteria, leading to the creation of multiple portfolios. Prioritization of projects is then done inside the buckets, thus avoiding projects with different objectives having to compete among themselves for the same resources.

Bubble diagrams are projects plotted on an X-Y plot or map, much like bubbles or balloons. The projects are categorized according to the zone or quadrant they are in. The quadrants are defined as "bread and butter," which includes products in simplified versions, but with high probability to work; "pearl," consisting of products that generate a good reward; "oyster," which includes products with a disruptive technology that would ensure greater profitability; and "white elephant" for products that are not very advantageous economically, but due to other interests, remain in the portfolio. This model shows relative project values using two or three criteria at once, in the two shafts and the size of the bubbles as shown in Fig. 46-2.

The following parameters are suggested for drafting the bubble diagram: alignment with the business or corporate strategies, inventive merit and strategic importance to business, durability of competitive advantage, return based on financial expectations, competitive impact of technology, likelihood of technical and commercial success, costs of research and development through project conclusion, project conclusion time, marketing capital, and investments necessary to the product launch. The bubble size indicates the amount of resources needed (human resources, finance).

In the scoring model, projects are rated or scored on a number of questions or criteria using, for example, a scale of 0 to 5: 0—nonexistent, 1—very poor, 2—poor, 3—reasonable, 4—good, 5—very good. Each predictor of success has a weighting (reflecting the importance of each one), with a total weighting of 100 percent. The sum of weighted factors allows you to obtain the project score.

Using check lists the project is evaluated on a set of yes or no questions. The project must achieve a target number, or all, of positive answers to proceed.

Models used by businesses help in the process of decision making for investment, disinvestment, or withdrawal of existing products, as well as the development and launch of new products. The financial nature always tends to be present, but there are other aspects that cannot be taken lightly in the management of the portfolio, such as the needs and satisfaction of the customer, the strategic component, competition, and the environment. Portfolio management for new product development should consider the current situation, but also be proactive in searching and anticipating possible mutations caused by changes in consumer behavior.

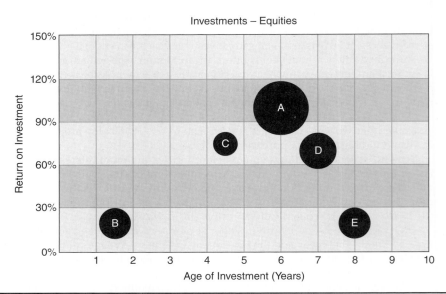

FIGURE 46-2   Example of a bubble diagram.

Launching Strategies

New products are usually associated with uncertainty of market acceptance. The large investment required for new products makes successful product launch even more critical. However, independently of the investment, most products fail at the launch phase. The big and complex question still remains unanswered, why?

Over the years, different approaches have been tried to ensure successful product launch, adjusting focus on product, customer, and communication. Nevertheless, the number of products that fail in the market still remains high.

Literature points out that product performance is influenced by strategic launch decisions followed by tactical launch decisions.[6,14,15] Due to market aggressiveness and the risks of investment in new product development, it is imperative to make more accurate decisions to avoid undesirable results.

Any discussion of strategy necessarily leads to consideration of decision-making methods. Today's managers face frequent changes in decision scenarios, and traditional decision-making methods, by its characteristics, do not always function satisfactorily. The default decision metric is often limited to minimum cost or maximum benefit.[6]

In this section, we present the implications of launch strategies, compare frameworks, and analyze launch strategy categories.

Launch Strategies: Meaning and Implications

Much has been written about launch strategies and much can be learned from this body of knowledge such as how to get competitive advantage, reduce the risks of market introduction, among other things. By definition, launch strategies, can be regarded as decisions and activities necessary to present a product to its target market and begin to generate income from sales.[14]

Using an analogy from the managerial pyramid, three levels of decision making can be applied at new product launch: the strategic level creates the bridge between the firm's objectives and the early stages of new product development; the tactical level embodies marketing mix and is responsible for the methods that enable the introduction of the product to market; and the operational level that is responsible for project and development decisions.

Reasoning about these factors, we may infer that each level cannot be taken in isolation as each has an important role in the strategy of new product launch.

Evaluation of Frameworks

All new product launches are filled with risk. The risk can be minimized by choosing proven new product launch strategies or, eventually, by outsourcing the process to a business management consultancy. In any case, less risk means fewer benefits. Table 46-1 considers the aspects, objectives, and criteria of various new product launch strategies.

To evaluate a strategy, it is necessary to consider the definition of objectives, the selection criteria, and definition metrics. A hierarchical model proposed by Chiu et al[6] defines three levels of concerns, called aspects. Each aspect has a set of objectives with its own criteria. The model favors the adoption of a strategy based on the selection of objectives and the criteria. As negative aspects of the model, we found the absence of recommendations about the metrics. The model is very strict and cannot predict future market conditions.[16] In general, though, the model helps to simplify the process of defining a new product launch.

A more simple approach to defining a new product launch strategy is proposed by Bhuiyan.[17] New product strategy is the first of a seven stage framework. Clear communication, well-documented definition, coordination, and a clear plan are pointed out as critical success factors for new product launch. Another relevant aspect is to take ROI as the preferred metric to determine if the cost to develop a new product exceeds the resulting benefit, or how favorably the payback affects the corporate bottom line.[17] Compared to the hierarchical model, this approach is more simplistic, and puts more focus on overall NPD process, and synthesizes these phases into a single framework.

Another two similar models are offered by Trim and Hultink et al,[14] respectively. Both are based on strategic and tactical decisions to justify product performance.

The main differences are the pharmaceutical-related factors complimented by SWOT analysis to support strategic decisions that Trim adds to his model. In the past, these models have led to verified improvements to Chiu's model.[6]

A more complex approach consists in a dynamic model of new product launch that enables predicting adjustments of launch scale according to market conditions and dynamic

Aspects	Objectives	Criteria
Strategic concern	Product strategy	Innovativeness
		New product development cycle time
	Market strategy	Growth/potential
		Target/position
	Rivalry	Threats of competitors
		Product advantage
	Business strategy	Cost leadership
		Differentiation
		Core competence
		Complementary resource
Marketing concern	Product	Branding
		Breadth of assortment
	Placement	Number of channels
		Distribution expenditures
	Price	Penetration
		Skimming
	Promotion	Promotion expenditures
		Sales force intensity
Organization concern	Structure	Integration
		Differentiation
		Coordination
	New product development organization	Project team
		R&D team
		Product managers
		Separate new product
	Culture	Delegation
		Openness

TABLE 46-1   A Hierarchical Model for New Product Launch Strategy. *Source:* Adapted from Chiu et al. 2006.

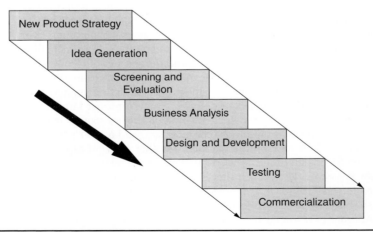

FIGURE 46-3   Stages of new product development.[17,18]

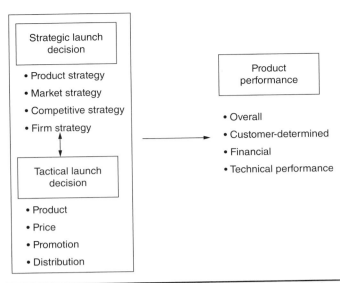

FIGURE 46-4 Impact of launch decisions on new product launch performance.[6,14,19]

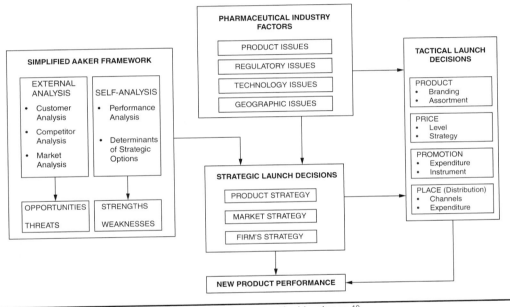

FIGURE 46-5 A new product launch model for pharmaceutical business.[19]

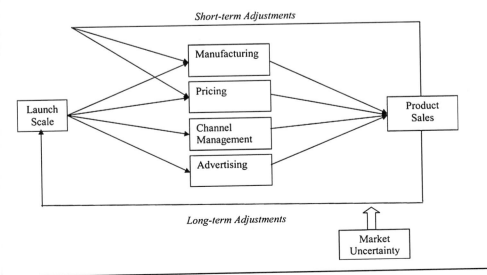

FIGURE 46-6 A dynamic model of new product launch.[16]

interactions.[16] The model follows the classic strategic level and tactical level. Compared to past models, it reacts to the changes of the market.

As a result of our findings, we elected the innovativeness of the system and its attempt to introduce a dynamic system to model new product launch. Less favorably, the model has not been tested in different scenarios to prove its efficiency. Also, it bases the dynamic solely on marketing mix factors.

In any strategy, the definition of critical success factors is strongly recommended. Also, a clear understanding of the objectives for the new product launch and the definition of the ROI is a prerequisite to evaluate the success of a launch strategy.

Categories of Launch Strategies

Despite the differences in studied frameworks, there are decisions that firms need to make. The first decision is related to market timing. Being the first-to-market elevates the risk of market acceptance of the new products. As a fast-follower, firms need to learn quickly from those companies who enter the market first. The fast-follower disadvantage is to be the second with diminished market share. On the other hand, the advantage is leaving the others to take the risks first. A third option is take the position of the delayed entrant. The followers of this category prefer to enter in established markets. The main advantage is to decide to enter when the market is ready for the product, but the disadvantage is adverse market conditions populated by many competitors.

The framework proposed by Cui introduces another type of launch strategy: the dynamic strategies.[16] Different from static strategies, it involves short-term demand and long-term adjustments, where the initial launch scales.

The Financial Impact of New Products

The development and launch of new products entails organizational, strategic, marketing, and financial risks. When a company begins to develop a new product, it assumes that it's technically and economically possible and that the end result will meet market needs. If this assumption proves incorrect when the product is released, the probability of success will be lower and there will be a strong negative impact on the finances of the organization. In addition, the company is unable to recover the investment made, impairing their future capacity to generate revenue that would support research and development of new products.

Studies point to a rate of only 50 percent success in products that are launched to the market.[20] Whether caused by the product itself or by ineffective launch strategies, new product failures lead to the need to launch more products to increase the chances of success, with a consequent increase in overall costs for research and development of new products. Thus, the company has to manage not only the investments of a new product, but also take into account the budget doubling to obtain the desired success.

Ultimately, sales must generate enough profit to justify development costs. As a result, demand forecast accuracy is also important to achieve financial stability. An overly optimistic prediction of market uptake leads to overproduction, unnecessary costs of inventory ownership, and underutilized production capacity. A low forecast fails to take advantage of market opportunities, leading to a negative impact on product and company image, potential loss of economies of scale, and extended financial return cycle.[16]

The type or complexity of the new product also influences the investment and the financial impact on the organization. Major technological innovations rarely occur without significant financial investment.

When a new product is released, it may replace one or more existing products, or occupy a space completely or partially empty, in the portfolio of products that are being marketed. There is the need to predict if or how the sale of the new product will cause an involuntary cannibalization of existing products, because it could lead to a decrease in the volume of sales, margin, and even the reputation of the brand. It is expected that the new product replaces or recoups the old marketing margins, captures new customers from competitors, or finds new customers who haven't previously used the product type in question.

The fact that a new product has more or less variation (i.e., the depth of the line, to be more or less profound) has implications for the number of threads that it could serve. If the company can maintain mass production, but customize in volume for each segment, it can get higher global sales volume, lower unit costs, and consequently a better financial fit. It is expected that during the launch of a new product, the company needs to invest heavily in communication addressed to end-user customers as well as to companies constituting the distribution circuit. Marketing efforts spent at this stage work as a leverage to increase the likelihood of product success. But these efforts will affect cash-flows generated as well as the company's profits. Therefore, the company's financial investment does not end when the product is conceived and developed but throughout the four phases of the life cycle of the product: launch, growth, maturity, and decline. For example, a given new product can lead to the need to educate users so that they can overcome certain psychological barriers to handling.

The financial plan for a new product will depend on margin goals, profitability goals, break-even time, and ROI goals.[21] Whereas fixed costs cannot be changed, the break-even will be less if the new product exhibits a higher selling price and/or a lower variable cost. For the first situation, it is possible to apply a price premium if the product is innovative; and for the second, a market penetration strategy can improve margins and financial results. The state of the economy when the new product is released may also have implications on financial results. In less favorable economic times, the volume of sales may be lower, the price may need to be adjusted, and the company may need to create more favorable payment terms for customers—all of which affect the immediate release of funds.

Maintaining an active relationship with existing customers can facilitate the introduction of a new product and achieve better marketing margins, and also get immersive feedback that will allow for more sustainable future developments of new products, thereby minimizing future investments and maximizing future profitability.

Financial teams face with a number of unique challenges when evaluating new products or services. Management team members may have different ideas about the strategic role of a new product and how its success should be evaluated. Furthermore, there may be little or no historical information, such as sales or cost trends, on which to base a financial model. As a consequence, many assumptions and calculations must be built into a financial model to account for different operation scenarios. Although the financial component is important, Lev[22] argues that financial statements, of themselves, do not show how well the company is doing in a given market or the exact worth of the company. As a consequence, financial statements should not be the primary (or only) documents used to analyze a company's performance, as it is limited in information. Cash-flow charts and income statements only show how much the company is spending and earning on a monthly basis. These elements are limited when analyzing company performance, as they do not reveal what is in demand in the given target market. Income statements show how much the company earns from service and product sales, but these pictures may not be reflective of market trends or demands. As an example, if a company's income is higher than the year before, it does not necessarily means that its performance is good enough, considering that a competitor may earn two times more. In relation to other companies, any given business may not be performing as well as it could be.

To address these limitations, Kaplan and Norton suggest a balance between this area and the internal business perspective (manufacturing process, costs, quality, time to market, segmentation, positioning), innovation and learning perspective (new products development, continuous improvement, learning organization), and customer perspective (service or product performance, complaints, perceived brand, value proposition).

The Role of IT Technologies

The use of social media for new product launch has not received much attention in academic literature. This does not mean that it is completely ignored. Until recently, a new product launch was announced only in magazines, newspaper (publicity), radio announcements, or small spots on television. Nowadays, the means of communication are much more diversified and have become more interactive than ever. The democratization of access to new technologies, in addition with the growth in use of social networks, brought new opportunities for firms to publicize their products and to communicate with customers in more innovative ways. The need to become closer to consumers has led many firms to seek new communication alternatives that complement more traditional means. The growth in use of technology is demonstrated not only in how firms communicate with the market, but also how they work.

In work scenarios, e-mail is the more popular communication technology after mobile phone devices; however, due to the asynchronous characteristics of e-mail, it may not be an adequate solution.

Consider this scenario: A team leader sent a message to team members asking for some ideas to launch a new product. After that, team members provided feedback. Typically, this results in a sequence of iterations that lead to an out-of-phase communication. While one member stays at the first iteration with the leader, one or more other members move to the second, third, or further iterations. To the team leader, this out-of-phase process becomes very difficult to manage, due to the unsynchronized discussion.

Realizing this difficulty, teams having adopted more sophisticated alternatives based on integrated platforms that allow synchronous communications between team members, even when they are in different geographical locations.

Today, e-mail is only one of many diverse tools available to support team work and communication. It has been integrated into hybrid frameworks that combine text, images, audio, or video to facilitate interaction among team members in synchronous or asynchronous mode of communications.

To launch a new product, we have identified two main roles of IT: (1) communications support and (2) development support. In this section we present how IT could benefit firms in communicating with the market as well as help teams to collaborate from development through launch.

The Role of IT in Communication with the Market

Everyday, more and more people are using the Internet. From the perspective of one who wants to communicate using a fast and global solution, we may infer that Internet is the right solution. However, the increasing number of individual and business websites has brought new challenges, like how a business website can be noticed faster than competitors. In fact, the exponential growth of the number of websites offering and announcing things makes it impossible for a human to visit all the sites they need and synthesize whose is more relevant at the right time. Search engines have simplified the process somewhat. Nevertheless, it is necessary to inform search engines what websites exist and what contents they offer. Depending on the quality of the websites metainformation, the process could be automated, based on small programs that "travel" across websites indexing metainformation and informing search engines about the website contents.

To improve a website's relevance for search purposes, multiple techniques have been adopted. In the following paragraphs, four examples are offered: (1) search engine optimization (SEO), (2) recommendation techniques (recommenders), (3) techniques to audit visitor clicks, (4) incentive to site news subscription. Each technique contributes to firms needing to analyze product acceptance, product visibility, and customer preferences, among other metrics.

Using SEO techniques, websites automatically inform search engines about its contents and how they are organized, influencing the relevance of the website to the search engine. When a website refers to another website that is considered a hub authority, it may increase the value of the website to the search engine. Also, if a hub authority points to a website, the value of the site rises, too. Nevertheless, the process to reach the top is slow. To launch a new product using a website to publicize the product could be very slow. The website needs time to get a relevant rank to become more visible and reach the top. On the downside, when a product becomes noticed, it already has competitors.

Recommendation is a technique that is used to recommend a product based on previous preferences of the customer or information preferences previously submitted. Another technique is to audit site clicks as a method of obtaining the preferences of the site visitors relative to the offered products. Finally, incentive to site visitors to subscribe to news is a way of knowing who and how many are interested in the firm's products.

The paradigm of social networks has emerged as a new tier integrated in the main tier, the web. Social networks are the result of a combination of integrated communications and group segmentation. Regarding Facebook or Google Plus, the last improvements provide to users a place to interact with friends using a set of unified communications. Likewise, they make available feeds, e-mail, image, audio, and video. Why this is important to our context?

Companies like Google and Facebook have perceived the potential value of networks as a fast way to disseminate information to segmented receivers. Each company has been developing efforts to offer its own advertising services to firms that want to publicize their products. Due to the increasing number of users connected every day, recommended by a friend, we can easily conclude that social networks have rediscovered the concept of *think global, act local*. In fact, the network concept is global, but people act locally.

Consider this example of how product information could be disseminated through social networks: When someone gets interested in a product and posts their interest in their own area, friends see the same information. If a product is relevant to one or more friends, it will be disseminated more quickly across multiple sharings, reaching a large number of exposures. A technique used in social network to increase product visibility is *helping a friend*. The technique consists of a user submitting a link to other users on their network (friends) claiming help to win a product. The answer to the solicitation leads a product to be quickly popularized.

The benefit of fast propagation of product information across social networks, sometimes, could face adversities. In the event of users' disappointment, viral marketing is easily propagated and could cause severe damage to product and firm reputations. The negative repercussions also transfer to the web because search engines register everything—the good and the bad—that happens on the Internet.

The presentation of a prototype of a product has also experienced changes. IT technologies allow virtualization of products throughout multimedia tridimensional views and personalization. This technique brings the product to the customer and fosters interaction. Beyond fast dissemination of information of a product, technology also provides lower costs. The launch of a new product no longer requires having a physical object to showcase the product to the world.

The role of IT tools has proven valuable to the interaction between firms and costumers or individual users in more closed contexts. Users' feedback can be collected not only on social networks, but also in closed community environments like discussion forums, virtual experimentation, and testing simulators, where suggestions and feedback can be collected for processing.[23]

User feedback collection and management becomes even more valuable in a challenging economy, in which companies are being forced to reduce spending on store expansion and other physical assets, but are focused inward on improving the satisfaction of current customers. Lockwood[24] defends that today's companies do not have an alternative to actively encourage customer feedback and to treat customers as strategic assets in the process to improve operations, marketing, research, product development, and the overall customer experience. In fact, according to a research study conducted by Aberdeen Consulting Group, 41 percent of survey respondents say that they will increase spending on customer feedback initiatives within the next fiscal year, with the goal of retaining existing customers, attracting new customers, and identifying new revenue sources.[25]

The study conducted by Minkara[25] proposes three metrics to track customer feedback:

1) Customer satisfaction—analyzing feedback and responding quickly with necessary changes

2) Customer retention—responding to customer complaints and suggestions

3) Customer-focused innovation—using innovative methods to meet the ongoing needs of consumers.

The most successful companies are actively analyzing, reacting, and responding to customer feedback. Those that have little or no involvement in customer feedback management must start to define the performance metrics, proactively inform customers of any changes resulting from the feedback, and bring on board employees dedicated to customer feedback management. To track customer feedback, companies can use several channels, such as e-mail, corporate ecommerce site, call center, and direct mail.

The Role of IT in the Development of a Strategy for New Product Launch

Team work driven by a collaborative process is not new. When team members are present at the same physical location, communication is easy and each member is available most of the time. Nevertheless, this is not what happens in all cases. The delocalization of company infrastructures and its effect on human resources lead teams to seek better alternatives to lowering the physical barriers to interaction. Popular telephone and e-mail technologies are not sufficient to solve team interaction needs. Even in combination, they are a poor solution for team meetings where valuable "real-time" interaction occurs. New frameworks for team collaboration have put the focus on document revision history, unified communications, and knowledge sharing and preservation.

The discussion of concepts, ideas, or strategies is often a lengthy process that involves a lot of negotiation, commitment, and many hours of meetings and team discussion. Context and circumstances will determine if the process is more or less long, or even more or less difficult. It is perfectly understandable that the discussion process and achievement of consensus are unlike those required for other similar projects. Each new project has its own characteristics and its own difficulties. However, part of those difficulties can be minimized by the use of facilitators to reduce the time to reach the goals. The adoption of IT infrastructures and applications that support team collaboration can be a significant contribution to lowering the barriers to communication across distances.

The taxonomy time or space concept assumes that the needs of synchronous or asynchronous interaction are the basis for choosing collaborative technologies. Table 46-2 shows the recommended categories of technologies that support team interaction. Nevertheless, some issues should be taken into consideration: 1 In face-to-face collaboration, IT technologies may not be strictly necessary, but it does not mean they are useless; 2 Team needs and work processes should be the determinants to choose technologies; 3 Technology is not the solution to all problems, but it helps to moderate them.

Space \ Time	Same	Different
Same	Calendar/agenda File share Collaborative writing Electronic presentations	Calendar/agenda E-mail Wiki Forum File share Document revision history
Different	VoIP Video Audio Chat Whiteboard Collaborative writing Electronic presentations Calendar/agenda	Calendar/agenda E-mail Wiki Forum File share Document revision history

TABLE 46-2 An Example of the Adoption of Technologies to Support Team Collaboration Based on Taxonomy Time or Space[26-28]

Collaborative technologies are available separately or integrated in hybrid solutions, based on highly configurable, on-demand web platforms, or services. One or more options can be enabled or disabled to simplify utilization and reduce usage costs. More complete platforms offer planning tools to manage projects, allocate resources, and define milestones, allowing team managers to intervene when the project is delayed.

IT Usage and Variables

The importance of IT tools is not limited to the launch process. At each of the different activities across the stages of the new product development process, several contributions are observed. IT is used to aid idea generation and product testing as well as for development activities such as process and portfolio management.[29]

As mentioned before, each new product launch project is a new beginning. Nevertheless, experience with IT, the nature of the project and the function of the team are determinants to which and how IT tools should be used:

- Antecedents of IT usage influence the variable "experience."

- Execution of the project influences the type and the number of tools to be used for different activities.

- The location of the team is also a determinant in selecting what type of tool should be used and what software application is the most appropriate.

A conceptual and generalist model presented by Barczac et al[29] will help understand the influence of IT on product performance, as shown in Fig. 46-7.

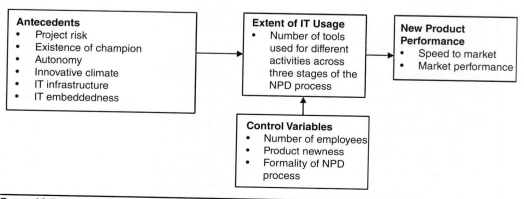

FIGURE 46-7 IT variables and influence on product performance.[29]

In summary, technologies help to get fast results, improve proximity to customers, increase accuracy of predictions, improve team collaboration, extend knowledge sharing and preservation, contribute to reduced investment cost, and, as a result, reduce project risk.

Strategies for Success

The ability to create successful product launch strategies can position a company for survival and growth. In early studies of product development, research focused on the product to be developed rather than the process of product development. During the 1970s, focus expanded to include the study of the *process* of new product development. Product development efforts are typically organized as projects and managed by the classical elements of project components like product performance, schedule, and cost. In the context of product development, these project components are often associated with product performance, speed to market, and development cost.

There have been a significant number of studies to suggest the key success factors used in the process of successful product development including the use of cross-functional teams, support of upper management, and support of the organizational structure.[30] Additionally, private companies and incubation business centers have conducted their own studies and proposed additional key factors for a successful product launch.

In the next section, we will analyze the most popular and successful product launch strategies that aid managers in focusing their efforts appropriately by supporting product launch with the essential resources and techniques.

Assigning Dedicated Resources

Companies come to realize that launch management is not an optional activity. Successful companies dedicate teams exclusively to the challenges of building market momentum for a new product launch. For example, companies such as Amdocs and Abode Systems Incorporated, strive for consistency.[31] They use the same team from launch to launch, leveraging their experience and knowledge of shared processes and organizational decision making.

Dedicated launch teams can include members from the project management office (PMO), public relations, marketing, product management, and events management. It is possible to create such teams with company resources or use a combination of outsourced consultants according to company needs.

Another challenge to consider is the amount of time needed to set up and bring together the whole team, which can take several weeks, especially in big organizations. Therefore, preparation is crucial and the steps should consider the necessary lead time for outside resources and establishing master service agreements, or in-house resource allocation agreements, so the team can start work immediately.

Implementing a Scientific Launch Process

Companies might consider taking the same approach to product launches as they typically apply to product development. This has the advantage of using proven and scientific processes. Product development begins with planning, managing, tracking, and measuring from the conception of a product. As a consequence, management needs to consider the following phases:

- Established end-to-end product development launch processes
- Time-phased tracking
- Product cost capture and measurement
- Workflow management to share information across product development stages
- Process control and component management

Marketing a product launch requires the same foundation as product development. Typically this process starts with a marketing product launch profit and loss (P&L). The main use for preparing P&L projections or statements is to determine how profitable the product launch will be. By comparing P&L forecasts with the actual performance of other market products, we can see where we need to pay attention. Then we should measure marketing ROI with tracking that starts with the marketing bill of materials and follows through to customer use of the product.

This methodology makes it easy to move standard processes and procedures from project to project, resulting in benefits in the following ways:

- Organizational cohesion and consistency across product lines and functions
- Alignment of cross-functional activities
- Effective transitions from engineering to marketing
- A knowledge "legacy" of documented best practices

Investing in Collaboration Tools

Globalization, downsizing, and telecommuting have changed the way enterprises large and small operate. In today's global business environment, the value of working together can have a huge positive impact in a business. Employees and teams need to work together, and share ideas and information across geographic boundaries. This is particularly important for a small business where effective collaboration is essential to improving productivity by empowering their employees to communicate and work more efficiently using the right communications tools. Studies conducted by Schwab[32] demonstrated that small businesses that can work together effectively and execute more aggressively with their competitors.

Collaboration tools generally fall into two categories according to their interaction schema and demands: synchronous and asynchronous.

Synchronous tools are presence-based, used when all parties are working at the same time but in different locations. These tools enable real-time communication similar to a face-to-face meeting, connecting people at a single point in time. These types of tools include audio conferences, web conferences, instant messages, white boards, and application sharing.

Audio conferencing, a telephone connection (typically based in VoIP) between more than three people through audio connection, is useful for verbal discussions and dialogue. Web conferencing is the fastest growing collaboration tool that is being used to promote interaction between remote participants. It connects people together from multiple locations and allows application and video sharing, and can include features such as content, polling, and white boarding. Presence-based instant messaging tools are useful for quick resolution of problems and issues.

On the other side, asynchronous tools enable communication over a period of time, allowing people to connect at each person's own convenience, and not necessarily at the same time. Some examples include messaging (email), discussion boards (forums), blogs, and shared calendars.

The impact of collaboration tools in business performance can be significant, considering that it increases productivity, reduces costs, improves communication, and enhances coordination.

In the first place, collaboration tools enhance business performance by increasing productivity and cutting costs. Collaboration solutions replace face-to-face meetings. For example, collaboration solutions such as audio, video, and web conferencing are reducing the need for travel, eliminating airfare, taxi, and hotel expenses. Eliminating travel also saves many hours and days of downtime away from the office, and reduces employee stress levels.

At the same time, collaboration tools facilitate faster and more effective communication. Adopting them, it is possible to automate activity management and workflow, track task and milestone dates, and build a central repository of launch documents.

Managing Customer Feedback

Most companies understand the value of customer feedback and they collect it for good reasons. These companies realize that how their customers perceive and feel about them, based on their previous experiences, can and do impact their business. Customers have choices of whom to do business with, so if their experiences and perceptions of the firm aren't positive, they will go elsewhere. For that reason, companies collect data across many customer "touch points," including feedback on specific transactions, with the goal of improving the customer experience.

Customer feedback is simply data from customers about their perceptions and experiences. It is typically gathered either directly by companies or outsourced and gathered by market research firms. Feedback can take different forms and can cover a wide range of topics, but is often structured and gathered via surveys conducted by mail, phone, in person, or over the web. It is typically focused on aspects of the customer experience believed to be most critical to customer satisfaction and loyalty.

The acquisition, analysis, and distribution of customer feedback are three of the most challenging areas of customer management for product teams. Success depends on process, people, and technology. Most technology companies have not yet made investments to establish infrastructure for effectively communicating with customers during the development process. According to Shoppel and Davis,[33] this is reflected by industry average participation rates of less than 15 percent during customer feedback programs. For most programs, particularly those that are open to everyone (such as public betas), the overwhelming majority of users who obtain or use the product or service provide no feedback at all.

The challenge consists in promoting the participation of customers employing advanced feedback management systems and providing the support staff necessary for proactively communicating with customers. Customer feedback management (CFM) is the process by which customer feedback is incorporated into operational processes. By managing customer feedback, the company gets inputs to operational processes such as sales, account management, product management, and customer support (among others), allowing those processes to be made more effective, efficient, and customer-centric.

Customer feedback management tools facilitate the CFM process and are designed to make feedback actionable. The number of companies that develop and implement CFM processes is still very low, but it is expected to grow significantly in the coming years due to their potentialities of increasing customer retention. Companies that adopt CFM don't stop at the data collection level, but they also analyze the meaning of feedback and direct it to internal actors who can effectively use it in support of business objectives.

CFM is a process based on the systematic collection of customer feedback data, the analysis of the data, and defined dissemination and follow-up actions. When done correctly, CFM can place individual customer feedback data (both response data and profile data) into the hands of the company. In a corporate environment, the business's objectives drive CFM implementation. As with any business process, business objectives determine what customer feedback to collect, how to analyze it, how to disseminate it, and what actions should be taken with it.

The central goal of effective customer feedback programs is the collection, analysis, and distribution of meaningful data in real time. Collection means actively soliciting feedback from customers, as well as providing mechanisms that make it convenient for customers to give feedback. Analysis means quantifying and qualifying feedback to inform product team and management decision making during development. Distribution means presenting structured, organized, actionable feedback, as it arrives, to feedback stakeholders.

The most important element of being able to process feedback quickly is having pre-defined metrics and categories. The graphical presentation of customer feedback during development can provide a wealth of information to the product team and management, and can objectively represent product performance and customer sentiment at a given point in time.

Additionally, it would be interesting to effectively recognize the value of customer feedback during development. ROI in customer feedback programs can be measured by the following formula:

$$\text{ROI} = \frac{qnt + qlt + tml}{psc + oct} \tag{1}$$

In the formula (1) *qnt* means quantity, *qlt* means quality, *tml* represents the timeliness of customer feedback, *psc* means the program-specific costs, and *oct* means the opportunity costs.

While maximizing ROI for a customer feedback program is a primary objective, minimizing the opportunity costs associated with underexecuted programs is of equal or greater importance. Opportunity costs include unscheduled or costly follow-up releases to fix the product, higher than expected support costs, weaker than expected market acceptance, lost customer feedback to validate and endorse the product, and lost marketing feedback to understand the satisfaction indicators that drive the development of the next release.

Considering that competition, product complexity, and the number of stakeholders in new product introductions are all increasing, while development cycle time is decreasing, there is the need to consider a newer ROI criteria. The measured customer feedback program should also gather feedback for use in marketing and public relations, creating word of mouth among highly valued early-adopter customers and forging stronger working relationships with key customers.

Consumers who participate in the cocreation and colaunching processes show an increased loyalty to the company, a more positive brand image, and an increased likelihood of spreading positive word of mouth. Additionally, this cooperative process creates the opportunity to bring more relevant products to the consumer market and enhances product performance, particularly in terms of sales.

Customer Feedback Objectives	Product Management Objectives
Bugs and showstoppers	Achieving high customer participation
Ease of install and ease of use	Verifying systems
Customer assessment of features	Verifying software compatibility
Functionality, quality, and performance	Testing internal processes

TABLE 46-3   Most Relevant Feedback Objectives

Defining Objectives and the Right Process

The ultimate success of the product launch can almost always be traced to understanding the requirements of the target market. The amount of time initially invested in defining and prioritizing the feedback objectives directly impacts the quantity and quality of information learned. This typically falls into two major categories: customer feedback objectives, and product management objectives. A list with the most relevant objectives organized into the two above categories is presented in Table 46-3.

Currently, companies are continually decreasing product development cycles, which means that there is not enough time for rework. Therefore, without a concrete set of objectives, the process for obtaining and incorporating customer feedback becomes *ad hoc* and unpredictable.

Defining a process for generating the greatest quantity and quality of feedback from customers is arguably the most difficult task, as it requires the greatest ongoing commitment of time to execute successfully. Methods for managing feedback and problem resolution typically include proactive outbound telephone support, e-mail or newsgroup communication, web-based surveys, bug tracking, and on-site interviews.

Most organizations have traditionally employed a combination of telephone and e-mail support. Web-based communication offers a host of new benefits to both product teams and customers, and the case for making the web an integral part of customer feedback programs has become extremely compelling. On-site feedback collection, due to its cost, is typically reserved for the most technically sophisticated, high-end products with few customers.

Working Closely with Lead Users

Selecting the appropriate quantity of target customers with the right characteristics is a crucial step in setting up and carrying out a successful program. Recruiting a sample of customers whose system configurations and usage patterns reflect the target market for the product or service should be the goal of any customer feedback program.

The lead users segment offers great potentialities for any launch programming. According to Hippel,[34] lead user concept is defined as users whose needs significantly anticipate requirements of the broader marker. These users then innovate around the existing products to satisfy their unique requirements. This class of users represents a valuable asset and is becoming recognized by the business public due to the increased prominence of Internet communication channels.

A representation of the lead user segment is presented in Fig. 46-8.

There are several indicators that may help companies identify lead users. The most relevant are identified in Table 46-4.

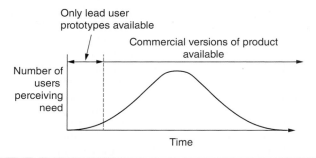

FIGURE 46-8   Identification of the lead user segment.[35]

Criteria/Indicator	Expected Behavior
Being ahead of a market trend	Lead users are expected to be the early adopters of an innovation.
User investment	They are available to invest more money in a product rather than other user segments.
High expected benefit	Considering that they make a higher investment, they expect to have a better return in terms of quality and "status" inside the group.
User dissatisfaction	They tend to resist dissatisfaction when they find a "bug" or a minor "issue" in the product or service. Instead, they look for ways to collaborate with the company to improve the quality.
Speed of adoption	They easily and quickly adopt and use the new, innovative products launched in the market.

TABLE 46-4   Lead User Criteria Identification

Lead users have a higher rate of incremental breakthrough innovation compared to firms due to the natural access to information to their own segment, and the insights that they have for the future trends of their industry. This is also attributed to the fact that they have a tendency to obtain free assistance from fellow community members.

The first step consists in indentifying the market lead users. Knowing where to look, or how to draw in lead users so that the coproduction and colaunching processes can be facilitated, is a critical success factor for harnessing lead-user innovation. According to Moeini et al,[36] lead users can be found in communities, through intermediaries and by using contests.

The first approach suggests that lead users can be found in network communities. These communities form as a result of lead users finding themselves searching for like-minded people with whom they can share their ideas and perhaps find ways to improve their own innovations. There are many ways to facilitate and foster the growth of these pseudo-organizations, thus making it necessary to investigate different network types and how each one can be coerced into revealing its riches in slightly different ways.

When we look to network products, we realize that they tend to foster their own user communities for the very same reasons that lead-user innovation itself fosters communities. Because the users of a network product get more out of it when the network encompasses as much as possible, network product users tend to migrate to common areas in which they can learn and discuss that product. Tapping into these user-formed communities can be great way to find lead users who, after using the product, realize that there was something more that needed to be fulfilled.

Another way is finding lead users through intermediaries. Firms may explore making a connection with lead users by entering relationships with intermediary firms. Companies solicit the help of such intermediaries due to several reasons, including that intermediaries have the resources to efficiently find innovators; the complex nature of the industry requires specific expertise to find users; and intermediaries will likely yield information far more quickly and cost-effectively due to already established networks and relationships.

Finally, it is also possible to look for lead users by using contests. Contests are an effective medium that companies can use to encourage both incremental and breakthrough innovation. The key success factors of contests are the definition of the challenge and the motivation for the reward. Firms need to strike the delicate balance between giving a specific purpose to entrants without limiting them from showing their true potential. Moreover, due diligence still plays a vital role in the attraction of lead users as, without the proper branding of the contest, the segment will not yield optimal results. As result, extensive market research must also be done in order to understand how to incentivize the segment so that ideal applicants can be motivated to participate in these initiatives.

Then, as the next step, the company should define a matrix of target requirements, which includes criteria such as

- Knowledge, experience, and usage level
- System configuration, and usage environment
- Availability to test and provide feedback
- Testimonial potential

The number of target customers to include in this feedback program is a function of the type of product or service, the aggregate amount of product or service usage the company needs for meaningful results, and the selection criteria defined earlier. The objective is to create a sample no larger than the company can adequately manage, and no smaller than the matrix coverage requirements. In general, the company wants to include the largest number of target customers that can be effectively managed and supported.

Education of Sales Staff

The sales staff needs to be educated in the company products and the best people to do this are internal staff. It's critically important to continually educate the sales force about what the product does, not only in terms of features and functions, but in terms of its real business value to customers. In fact, the company should treat the sales force just as attentively as its customers.

The most common approach is the implementation of a face-to-face educational program. However, there are alternative ways to educate the sales staff if the company cannot adopt the first approach. Among them we highlight

- Developing sales material that is easy to read and includes graphical illustrations. It is important to realize that sales staff need clear and simple statements.
- Producing and distributing computer-based training (CBT) on a DVD to the stores (in a retail situation). However, in this case, the stores will need incentives before they would be willing to allocate any time to a CBT program.
- Reserving part of the company website for demonstrations, explanations, and FAQs.

Finally, it is also important to create an internal incentive program. Lovins[37] found that approximately 85 percent of senior managers surveyed believe that managing human capital is the most important means of improving productivity. Additionally, the same study pointed out that engaged organizations have 3.9 times the earning per share growth rate compared to organizations with lower engagement in the same industry.

Tangible incentives encourage healthy competition within a sales team. While current profitability is important, sales performance should be expressed as staffers' projected and long-term competitive positions. The lesson is to measure performance by progress, as opposed to current rigid sales rankings, highlighting team members' sales progress with programs like Salesperson of the Month. Financial bonuses are ideal incentives, but companies can motivate staff with gifts, like participation in sportive popular events or gift vouchers to trendy restaurants.[38]

Involvement of the Company

The involvement of the managerial staff of the company is crucial for success. However, not only the executive team but also everyone within the organization must be involved. Launching a new product in the market is an important step that should elicit pride in all the company staff. Therefore, communication of goals and progress is important. At the same time, the company should use the launch as a way to build morale and unify the team through a common cause.

The dissemination of best practices is also a good way to get the involvement of the whole organization and promote an innovation culture. The company should identify what is working and what isn't, and share this information throughout the organization. Additionally, the company should be flexible and willing to adjust the plan as results are known.

Conclusions

Research and innovation helps create jobs, prosperity, and quality of life. Our future standard of living depends on our ability to drive innovation in products, services, business and social processes, as well as models. For this reason, all companies have tried their best to launch maximum number of products to market. However, the commercial success or failure of a product does not rest solely on the product itself. The launch strategy adopted also determines whether a product succeeds or fails. In fact, the key to success in the launch process often rests in finding the proper strategies.

In the high-tech environment, there is often a small window of opportunity for technology producers to reach their markets. The marketing operations undertaken have to be right first time, this when there are high levels of uncertainly over the new product for both producers and consumers. Because of short lead times, rapid technological change, and high levels of uncertainty, we have argued that the key to understanding the market launch is through actions marketers undertake. Although the actions taken may vary according to the industry or even the producer, there are tactical approaches that have been shown to be consistently used across a wide range of technology producers and industries.

Today's new product projects decide tomorrow's product or market profile of the firm. Recent years have witnessed a heightened interest in portfolio management, not only in the technical community, but also at the CEO's level. Good portfolio management becomes essential. This portfolio management strategy should allocate resources to maximize the value of the portfolio via a number of key objectives such as profitability, ROI, and acceptable risk. Furthermore, the company should obtain the right number of projects to achieve the best balance between pipeline resource demands and the resources available. And last, the company should ensure that the portfolio reflects the company's product innovation strategy and that the breakdown of spending aligns with the company's strategic priorities.

There are a lot of elements to manage simultaneously to achieve a successful product launch. Moreover, it involves pulling together all the internal functions of the business as well as many external entities. New products are also key to remaining a force in the marketplace. However, innovative technologies and first-rate ideas do not automatically become great products, because most new products fail. New products have to meet some criteria to win the sale. In the first place, the product must have highly sought after benefits and enough potential customers to make the offering economically viable. Additionally, a substantial number of perspective customers must be ready and able to purchase and use the product to justify market launch. Furthermore, the new product must offer the customer greater value than alternative offerings.

A product launch strategy defines several stages of the launch including development, internal testing, external testing, objective and goal setting, positioning, excitement building, and event timing. Researchers, companies, and incubation business centers have also conducted their own studies and proposed additional key factors for a successful product launch. Among them, and due to their wide acceptance, we highlight the existence of dedicated resources, the process of managing customer feedback and involving customers in the design process, the capacity to work closely with lead users, and openness to involve all the company members in the product launch process.

Finally, the role of IT technologies should not be overlooked. The Internet is not only the consequence of innovation in IT, but also an enabler for various types of innovation both in the public and private sector, and for broader and more complex innovation processes. It is not the goal of Internet technologies to replace personal communications, but rather to supplement them and to assist in presenting a common data set for all involved parties to operate from. With this in mind, Internet technologies should be used as an additional tool to assure a common foundation from which to communicate. Time and resources can be better utilized through integrating this technology in not only the new product development process, but also in general business processes.

References

1. S&A. "New Product Launch Report." Schneider/Boston University, Schneider & Associates, 2001.
2. Cooper, R. *Winning at New Products*. 3rd ed. New York, NY: Basic Books, 2001.
3. Anthony, S. *The Little Black Book of Innovation: How It Works, How to Do It*. Boston, MA: Harvard Business Review Press, 2011.
4. Kastelle, T. "What We're Talking about When We Talk about Failure." *Innovation Leadership Network*, 2012. timkastelle.org/blog/2012/06/what-were-talking-about-when-we-talk-about-failure/. Accessed August 12, 2012.
5. Santos J.D., CRM offline & online. Vila Nova Gaia: Instituto Superior Politécnico Gaya, 2006.
6. Chiu, Y.C., Chen, B., Shyu, J.Z., and Tzeng, G.H. "An Evaluation Model of New Product Launch Strategy," *Technovation*, 26 (11): 1244–1252, 2006.
7. Barczak, G. "New Product Strategy, Structure, Process, and Performance in the Telecommunications Industry." *Journal of Product Innovation Management*, 12: 224–234, 1995.
8. Cooper, R., Edgett, S., and Kleinschmidt, E. "Portfolio Management for New Product Development: Results of an Industry Practices Study." *R&D Management*, 31 (4): 361–380, 2001.

9.   Kavadias, S. and Chao, R. "Resource allocation and new product development portfolio management." In Loch, C.H. and Kavadias S. (eds). *Handbook of New Product Development Management.* Elsevier/Butterworth-Heinemann, 135–163, 2007.

10. Clark, K.B. and Whelwright, S.C. *Managing New Product and Process Development.* New York, NY: The Free Press, 1993.

11. Cooper, R., Edgett, S., and Kleinschmidt, E. *Portfolio Management for New Products.* Addison-Wesley Publishing, 1998.

12. Cooper, R., Edgett, S., and Kleinschmidt, E. "Portfolio Management in New Product Development: Lessons from the Leaders—I." *ResearchTechnology Management*, 40 (5): 16–28, 1997.

13. Deregowska, D., Grosbois, J., and Kumar, V. Risk integrated framework for R&D portfolio management. In Kumar, U. (ed.). *Technology and Innovation Management.* Proceedings Annual Conference of the Administrative Sciences Association of Canada Technology and Innovation Management, Carleton University, 59–72, 2005.

14. Hultink, E.J., Griffin, A., Hart, S., and Robben, H.S.J. "Industrial New Product Launch Strategies and Product Development Performance." *Journal of Product Innovation Management*, 14: 243–257, 1997.

15. Hultink, E.J., Griffin, A., Robben, H.S.J., and Hart, S. "In Search of Generic Launch Strategies for New Products." *International Journal of Research in Marketing*, 15: 269–285, 1998.

16. Cui, A.S., Zhao, M., and Ravichandran, T. "Market Uncertainty and Dynamic New Product Launch Strategies: A System Dynamics Model." *IEEE Transactions on Engineering Management*, 58 (3): 530–550, 2011.

17. Bhuiyan, N. "A Framework for Successful New Product Development." *Journal of Industrial Engineering and Management*, 4 (4): 746–770, 2011. dx.doi.org/10.3926/jiem.334.

18. Hamilton, A. Booz. *New Product Management for the 1980s.* New York, NY: Booz, Allen & Hamilton, Inc., 1982.

19. Trim, P. and Pan, H. "A new product launch strategy (NPLS) model for pharmaceutical companies." *European Business Review*, 17 (4): 325–339, 2005.

20. Gourville, J.T. "The Curse of Innovation: Why Innovative New Products Fail." MSI Reports—working paper series, Marketing Science Institute, 2005. http://www.google .pt/url?sa=t&rct=j&q=&esrc=s&source=web&cd=1&cad=rja&ved=0CCAQFjAA&url =http%3A%2F%2Fwww.cob.unt.edu%2Fslides%2Fpaswan%2FMKTG4320%2Ffreepdfg rab.pdf&ei=ew6HUPTUEIOXhQf1wIHAAQ&usg=AFQjCNHGMb8v4Mv4QVLFWgZk 9jWVE6fbrA&sig2=dSPbbPZRD0U5E8Q6mOCPBQ. Accessed August 30, 2012.

21. Curtis, T. *Marketing in Practice.* Oxford, UK: Butterworth-Heinemann, 2006.

22. Nadhani, S. "Second Base with Charts: Understanding the Bubble Chart." *FusionBrew* .blog.fusioncharts.com/2009/08/second-base-with-charts-understanding-the-bubble -chart/. Accessed May 4, 2013.

23. Monteiro J.A. and Almeida, F.L. A user-driven approach for service process innovation. In Putnik, G.D., Ávila, P. (eds). *Business Sustainability 2.0.* Guimarães: School of Engineering—University of Minho; Porto: School of Engineering—Polytechnic of Porto, 127–132, 2011.

24. Lockwood, T. *Design Thinking: Integrating Innovation, Customer Experience, and Brand Value.* New York, NY: Allworth Press, 2009.

25. Minkara, O. "Customer Feedback Management: Leveraging the Voice of the Customer to Amplify Business Results." Aberdeen Consulting Group, 2012. boletines.prisadigital .com/2012_voice_of_the_customer.pdf. Accessed October 17, 2012.

26. Ellis, C.A., Gibbs, S.J., and Rein, G. "Groupware: Some Issues and Experiences." *Communications of the ACM*, 34 (1): 39–58, 1991.

27. Grudin, J. "Computer-Supported Cooperative Work: History and Focus." *Computer*, 27 (5): 19–26, 1994.

28. Monteiro, J.A. Uma abordagem para implementação de comunidades de conhecimento em associações empresariais. Dissertação de Mestrado. Porto: Faculdade de Engenharia da Universidade do Porto, 2009.

29. Barczak, G., Sultan, F., and Hultink, E.J. "Determinants of IT Usage and New Product Performance." *Journal of Product Innovation Management*, 24: 600–613, 2007.

30. Schimmoeller, L. "Success Factors of New Product Development Processes." *Advances in Production Engineering & Management*, 5: 25–32, 2010.

31. Sklarin R. and Gee, L. "Four Pillars for Product Launch: Best Practices Form World-Class Companies." Crimson Consulting Group, 2012. http://www.crimson-consulting.com /stageone/files/crimson/collateral/Product_Launch_Article_080206.pdf. Accessed September 15, 2012.

32. Schwab, K. "The Global Competitiveness Report 2011-2013." World Economic Forum, Geneva, Switzerland. http://www3.weforum.org/docs/WEF_GCR_Report_2011-12 .pdf. Accessed September 18, 2012.

33. Shoppel, M. and Davis, P. "The Five Secrets of a Successful Launch." *BetaSphere White Paper*, 2–9, 2011.

34. Von Hippel, E. "Lead Users: A Source of Novel Product Concepts." *Management Science*, 32 (7): 791–805, 1986.

35. Von Hippel, E. *Democratizing Innovation*. Massachusetts: The MIT Press, 2006.

36. Moeini, A., Goldmintz, S., and Alamzir, S. "Lead User Innovation." Master dissertation. Toronto: Schulich School of Business, York University, 2006.

37. Lovins, H. "Engage Your Employees, Educate Your Frontline, Increase Your Profits." Triple Pundit. http://www.triplepundit.com/2012/02/engage-employees-educate -frontline-increase-profits/. Accessed October 17, 2012.

38. Donata, L. "Employee Motivation Techniques: Extrinsic Rewards vs. Intrinsic Rewards." Knoji Consumer Knowledge. leadership-management.knoji.com/employee-motivation -techniques-extrinsic-rewards-vs-intrinsic-rewards/. Accessed October 19, 2012.

CHAPTER 47

Business Innovation Maturity Model

C. Robert Carlson and Praveen Gupta

The fundamental goal for any organization is sustained profitable or positive growth. While making it profitable requires excellence in execution, growth requires thinking of new opportunities. Therefore, development of new products or services is vital to an organization. In today's economy, the proximity of customers and producers creates demand for new products and services even faster. Internet, social media, globalization, and competition for limited resources has created a hunger for doing more faster and better, including even innovation. Customers are experiencing new innovative products and services more frequently, thus raising desire for even more innovations. This cycle of customer demand and innovation is challenging and requires innovation in every aspect of life. Educational institutes are challenged to provide more personalized education to students, governments are required to be more responsive to citizens, health care is being redesigned into more personalized and preventive health care, and businesses are developing new products or solutions more frequently or even on demand, like a cafeteria. Acceleration in innovation has led to experts looking into new ways to innovate and making innovation more pervasive and predictable. Businesses at the front end of the food chain are looking into creating a culture of innovation so all brains in the organization could be deployed in the service of its customers. However, challenges exist. A road map is needed to guide organizations to pursue a simple and robust journey to sustain innovation for profitable or positive growth.

In late 80s, Dr. Watts Humphrey[1] focused on improving software quality, which highlighted the need for a disciplined approach to developing software. His work on a process maturity framework became the basis for the software capability maturity model (CMM)[2] at the Software Engineering Institute (SEI). The software CMM had been initially utilized by a few companies working with the National Aeronautics and Space Association (NASA) and later by the software community. CMM-based education, improvements, and certifications have had an enormous impact on the software industry and software quality overall. Interestingly, the capability maturity model was based on Watts Humphries's process maturity framework that in turn was adapted to Philip Crosby's quality management maturity grid that consists of five stages: uncertainty, awakening, enlightenment, wisdom, and certainty.[3] The software capability maturity model levels include initial or *ad hoc*, repeatable, defined, managed, and optimizing, identifying priorities for an organization in order to develop in-house competency in software creation. Successful deployment of the software CMM has led to many other improvement frameworks, including testing maturity model[4] and the business innovation maturity model (BIMM).

Using Capability Maturity Model

Success of the software capability maturity model led to utilizing the maturity model concept as a roadmap across industries. Some examples include privacy maturity model, online analytics maturity model (OAMM), software assurance maturity model (SAMM), digital asset management maturity model (DAM-MM), open source maturity model (OSMM), process maturity model, smart grid maturity model, cloud computing maturity model, green maturity model, change management maturity model, and testing maturity model (TMM). Now maturity model has become a conceptual model to develop a road map for deploying new methods of achieving sustained and measurable progress and creating a culture of new practices.

The TMM was developed to complement the software maturity model for addressing issues specific to software quality and to define testing process maturity. It consisted of a set of defined testing policies, a test planning process, a test life cycle, a test group, a test process improvement group, a set of test-related metrics, tools and equipment, controlling and tracking, and product quality control. The five maturity levels are initial, phase definition, integration, management and measurement, and optimization and defect prevention and quality control.

The commercially successful software capability maturity model (CMM) was first inspired by Philip Crosby's quality management maturity grid and total quality management (TQM) principles. The capability aspect of CMM is based on the level of consistency and control to achieve high quality according to TQM principles, and maturity is based on Crosby's five stages of management, which are uncertainty, awakening, enlightenment, wisdom, and certainty. One can see that the five stages could be related to confusion or ignorance, education, belief, experience, and intellectual state of mind. The corresponding capability can be understood in terms of probabilities or variation,[5] such random and assignable process variation, statistically stable process, reduced variation or an improving process, minimal variation to reduce or eliminate testing, and measured and managed process to sustain error-free and predictable output.

Surveys conducted by major consulting firms about innovation deployment have found that executives do not believe in their innovation processes due to uncertainties associated with it and new products. Innovation successes have been related to existing as well as new products. The success rate of new products and new businesses or start-ups ranges 5 to 20 percent. Return on investment (ROI) in innovation using conventional methods has been about 10 to 20 percent requiring about five to seven years to recoup the investment, which is too long given shrinking product life cycles. Internet has brought user and producer together, increased dialogue between the customer and supplier, and resulted in demand for more innovation and personalization. The age of mass customization, experience economy, smart innovations, and globally competitive economy has necessitated institutionalization of innovation and a culture of innovation. Innovation is becoming routine on executives' agenda, and new process and tools are being introduced to make innovation more manageable and profitable. Moving beyond policies and strategies for innovation, organizations now need a roadmap or a maturity grid to assess the state of innovation in an organization, and develop a plan for creating a culture of innovation.

Knowledge of innovation methods has been pursued for years in order to guide people to explore new venues for applying gained wisdom to improve life. Innovation has gone through its evolutionary state from a chance, random occurrence, to innovation art, innovation patterns, innovation methodologies, and innovation science. Organizations offer training in various innovation methods, industry certifications and academic degrees in innovation, designate innovation leadership role such as vice president or director of innovation, and even establish measures of innovation. One of the remaining challenges of innovation is the poor success rate and effective measures of innovation. Improving effectiveness of an organization's innovation processes and effectively measuring innovation capability require a framework to assess and guide innovation deployment in an organization.

Business Innovation Maturity Model

The initial drive for developing BIMM is to offer a tool to organizations and its executives for assessing their internal innovation practices and competency. An organization-wide assessment raises awareness of innovation, identifies internal innovation strengths, challenges, and opportunities to accelerate innovation, and makes it more profitable. Recognizing that every organization has some level of innovation capability and innovators, though often insufficient to sustain profitable growth, an innovation capability maturity model is necessary to institutionalize innovation. The following are reasons for the need for an innovation maturity model:

1. Innovation is an evolving business practice.
2. Product life cycles are shrinking.
3. Rate of innovation is increasing.
4. Demand for innovation is growing rapidly.
5. The innovation management process is not well defined.

6. Investment in innovation is becoming a necessity.

7. To ensure ROI, organizations must follow a known process.

8. Need for a roadmap with guidelines to gain more confidence in innovation practices.

BIMM Development

BIMM was developed after the capability maturity model, incorporates aspects of TQM and software capability maturity models. Strategy for innovation is for a long-term commitment to sustained growth. It is a vision to achieve greater successes and create bigger opportunities for growth of stakeholder value. When it comes to innovation, the strategy may look into 5-, 10-, 20-, or 30-year, or even longer time frames to create a path for growth. The path needs to be checked periodically and the roadmap adjusted routinely to achieve desired outcomes. BIMM addresses the strategy, execution, and results at every stage of developing organic innovation competency. It provides a framework to benchmark an organization's innovation system and culture through practical and meaningful stages of innovation deployment, necessary activities or tasks for developing better and faster innovative solutions, and measures of progress at each stage. Ultimately, BIMM offers guidelines for an organization to create a culture of innovation. The model recognizes and facilitates the intellectual potential of everyone in an organization in order to maximize innovation of product or service, as well as business model. It is not the light bulb alone, but the entire ecosystem consisting of power generation, distribution, billing, and related services, that brings electricity home. Moving up through the stages of BIMM, organizations can deploy innovation methods and tools throughout for measurable outcomes. The following summarizes the benefits of BIMM:

Benefits of BIMM

- Higher awareness to practicing innovation management
- Better understanding of the innovation processes
- Measurable and predictable innovation performance
- Common innovation language
- Guidance for creating an innovation culture
- Roadmap for profitable innovations

The BIMM Architecture

BIMM is designed for internal and external assessment of innovation practices and systems in an organization. It is intended for upper management to gain insights of their organization's innovation resources and identify areas to manage innovation for revenue and profit growth. Since innovation is becoming a routine business process, it must be standardized and be continually reviewed for its effectiveness and efficiency.

BIMM is a comprehensive assessment and improvement system developed by IIT's Center for Innovation Science and Applications (CISA). After establishing an ecosystem for deploying innovation, BIMM is a logical next step in managing innovation. It consists of five stages, namely, sporadic, idea, managed, nurtured, and sustained. Each stage builds on the prior stage and advances the culture of innovation, and its contribution to financial objectives of an organization as shown in Fig. 47-1, BIMM ROI. These stages are the milestones on the innovation roadmap for an organization. Each stage is associated with an expected aim to achieve, activities to perform, and symptoms of expected outcomes as shown in Fig. 47-2, BIMM architecture.

The five BIMM stages, corresponding aims, activities, and symptoms are summarized in Fig. 47-3, BIMM stages,[6] and explained along with a list of key levers below.

Sporadic Innovations

The sporadic innovations stage recognizes pockets of innovation and self-motivated, isolated innovators for their contributions to the organization. Sporadic stage affirms leadership's informal support for innovation, random and scattered acts of innovation, or localized deployment of innovation practices. The organization recognizes random employees' innovative contributions formally or informally and is focused on profit increase or cost reduction more than its revenue growth.

Levers: pockets of innovation leadership, unplanned employee innovations, localized recognition, marginal financial performance.

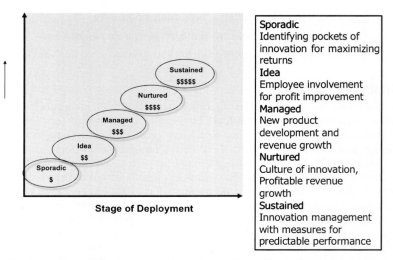

FIGURE 47-1 BIMM ROI.

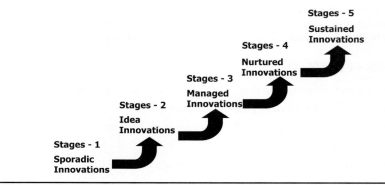

FIGURE 47-2 BIMM architecture.

Idea Innovations

Idea innovations are about inspiring intellectual engagement of all employees. Customers are demanding for more and faster innovations on a continual basis. Today, organizations must develop a dynamic portfolio of innovations that include product, process, services, and business model innovations, thus mandating innovations throughout the organization. In order to engage employees intellectually, they need to be inspired by leadership to support innovation. There may be incentives for recognizing problems in need of solutions, submitting ideas, rewarding and recognizing innovative ideas, and sharing the benefits of one's intellectual output. In the idea innovations stage, our focus is to achieve excellence in idea management, from enlisting people, listening to their ideas, capturing their ideas without any criticism or rejection, evaluating and prioritizing ideas for implementation, and recognizing successful ideas. All of this occurs while ideas are pouring in, being filtered, collaborated, and enhanced for developing innovative solutions. While focusing on excellence in idea management, education in innovation for key or R&D employees begins, and a few pilot projects are initiated with improvement in ideas and innovation methods. Conventional intellectual property protection methods are utilized to ensure proper and sufficient protection of ideas.

Levers: innovation strategy, idea management leader, leadership trained in idea management, rewards for participation at all levels, idea management deployed throughout, innovation projects evolving from ideas, tolerance for experimental failures.

Managed Innovations

The managed innovations are directly related to the new product development process. Too many times, new products are launched and management expects to lose money in its early years. The internal defect rate is high and so is the rework. New product takes away

Stage	Aim	Activities	Symptoms
I **Sporadic innovations**	Innovation Awareness	Leadership's informal support to innovation Scattered employee participations Recognition for innovations Cost focus	Pockets of innovation leadership Unplanned employee innovations Localized recognition Marginal financial performance
II **Idea Innovations**	Employee Intellectual Engagement	Leadership's formal support to innovation Informal organization for Idea Management Incentives or inspiration for ideas Reward and recognition system deployed Some employees are trained in innovation Few pilot projects Intellectual property practices	Short- and long-term innovation strategies Idea management leader Leadership trained in idea management Rewards for participation at all levels Idea Management deployed throughout Innovation projects evolving from ideas Tolerance for experimental failures
III **Managed Innovations**	New successful solutions in the market for revenue growth	Train product developers in excellence and innovation methods for deployment Idea management process is defined and documented Create a portfolio of innovations addressing M&A, open innovations and organic innovations; and market vs. technology-driven innovations Commit resources to the portfolio Establish innovation management plans for open and organic innovations Employees are allowed risk taking and experimentation Formal organization for innovation with Chief Innovation Officer reporting to CEO	Frequent launch of new derivative or variation innovations Idea management software implemented Effectiveness of idea management measured Improved performance of design and development processes Profitable innovations in the marketplace Chief Innovation Officer with innovation function on the organization chart, job description, and a designated visible innovation space
IV **Nurtured Innovations**	Culture of innovation across the organization	Establish all department innovation goals towards profit or growth Provide training in excellence and innovation to all employees for deployment Ensure innovation competency in most departments Establish an Innovation Index CEO award for excellence in innovation	Industry recognition of new innovations Most departments innovating new solutions activity, process, product, or business model level Performance to innovation goals is reviewed, reported and communicated quarterly Higher number of ideas per employee, and more CEO recognitions
V **Sustained Innovations**	Sustained Profitable Growth	Plan and review for sustained profitable growth Budget and achieve desired revenue growth Establish Corporate Innovation Index Annually update portfolio of innovation Identify new growth opportunities and new jobs Introduce new platform every 3-5 years	Top and bottom line impact of innovation internally and externally Corporate recognition in the media Business objectives are being achieved, employees and customers are happy Improved ROI for investors Improve innovation process and its performance including predictability

Figure 47-3 BIMM stages.

production capacity from existing profitable products. Customer returns are excessive, and there is a long list of unhappy customers. It all results in a higher cost of goods sold for the organization and turns excitement about new product into nightmares. Instead of earning premiums on innovative products, the company loses money and cost-reduction programs begin right after its launch.

When a new product launch fails, many things could have gone wrong: questionable customer requirements, ill-defined product development process, insufficient new product evaluation during development, penny-pinching during product development, poor transfer of new product from development to production, and an inefficient internal corrective action process resulting in recurring problems. In order to make new production innovations profitable, one must focus on enjoyable customer experiences, design to target performance, and produce to design targets.

The managed innovations stage emphasizes effective and better management of the new product development process that results in profitable products faster. This includes training product development professionals in excellence (Lean Six Sigma) and innovation methods (TRIZ or TEDOC) for deployment, and idea generation methods. A portfolio of innovative products or services is developed for near-term, short-term, long-term, and future revenue growth. The portfolio includes fundamental or conceptual innovations, platform product innovations, derivative or market-specific innovations, and variation or customized and personalized innovations. The portfolio incorporates various sources of innovations such as organic, open, and acquisition. Finally, the managed innovations stage requires a formal innovation leadership role, such as chief innovation officer, reporting directly to the chief executive of the organization.

Levers: frequent launch of new derivative or variation innovations, idea management software implemented, idea metrics tracked, improved performance of design and development processes, profitable innovations in the marketplace, chief innovation officer with innovation function on the organization chart and job description.

Nurtured Innovations

As the saying goes, developing a light bulb alone would not bring electricity home, and making an iPod alone would not provide its user the experience. One needs the entire system to utilize an innovation. One needs electricity generation, distribution, billing, and payment systems in order to use the light bulb; or one needs easily accessible music, comfortable headphone, and other accessories in order to enjoy the iPod. This requires innovation in all aspects of a business, and the involvement of employees in all departments. In addition to deploying a new innovation methodology, it is necessary to create a culture of innovation where everyone is inspired and every idea is nurtured to accelerate and maximize innovation for profitable growth of an organization. Culture implies that corporate policies, strategies, objectives, procedures, infrastructure, layout, services, incentives, recognition, and performance reviews all are geared to nurture innovation.

In the nurturing innovations stage, we establish all department innovation goals toward profit or growth. All employees receive training in creativity and innovation, giving them confidence that they can be a major innovator at work. Employees also receive training to excel at work, as excellence is a necessity for innovation to succeed. Each department can have resident experts in innovation methods who can ensure efficient and effective innovative solutions. At this stage in an organization there are measures of innovation that provide linkages to financial objectives. One of the simple but powerful ways to nurture innovation is recognizing successful innovations, or acts of innovation. Recognition feeds more innovation.

Levers: industry recognition of new innovations, most departments innovating new solutions activity, process, product, or business model level, performance to innovation goals is reviewed, reported, and communicated quarterly, higher number of ideas per employee, and more CEO recognitions.

Sustained Innovations

Leading an organization through sporadic, idea, managed, and nurtured innovations stages, the challenge remains to sustain the culture of innovation. Often, organizations invest resources in deploying innovation systems and experience sporadic innovations, but are unable to sustain the progress. In the sustain phase the focus is on measuring and managing innovations. Measurements provide feedback and management steers the innovation system in the desired direction of profitable growth. In the sustained innovations stage, organizations develop a plan to sustain profitable growth through the innovation system. Company-level financial objectives are established, investments are made, budgets for the innovation portfolio are planned, innovation objectives and measures are defined, and revenue and profit targets are established. In the final stages of creating a culture of innovation in an organization, all elements of the innovation system are working. Now the challenge is to keep the innovation machine working. With proper measures and management, the innovation system is maintained and a portfolio of innovations is updated and aligned with the organization's objectives and goals.

Levers: top- and bottom-line impact of innovation measured internally and externally, corporate recognition in the media, business objectives are being achieved, employees and customers are happy, improved ROI for investors, improved innovation predictability.

The BIMM Assessment Methodology

For assessing an organization's innovation practice, system, and culture using BIMM criteria, there are detailed requirements for each stage that allows the organization to establish a baseline. There are two aspects to assessing an organization's conformance, BIMM criteria, and the stage level.

Guidelines for Using the BIMM Criteria

The main purpose of the BIMM checklist is to understand and assess the level of conformance and identify opportunities for accelerating innovation. The BIMM checklist contains open-ended statements to allow for flexibility and application to various industries and circumstances, and for benchmarking. The numerical evaluation is an estimated level of implementation of the requirements that can be updated after corrective actions are implemented. There are three components to be considered to assess performance and assign a score. Each statement is rated in percentage for three components: approach, deployment, and results. The approach implies level of maturity and documentation of the methods, deployment (the extent of implementation throughout the organization), and results mean how well the strategy, method, or approach has worked. Figure 47-4 summarizes the scoring guidelines.

Scoring Guidelines

- While listening to responses, the assessor must ask for evidence and review enough samples to gain confidence in assessing the level of approach, deployment, and results.
- Once the level of approach, deployment, and results are understood, a percentage is assigned as applicable for each statement.
- Score for each statement is translated into the average percentage.
- A total score for each stage is calculated, and the average stage score is established.
- At the completion of the assessment, overall level is determined according to guidelines shown in Fig. 47-5, and examples as shown in Figs. 47-6 through 47-9. Examples demonstrate stage determination in different circumstances.

Rating	Approach (A)	Deployment (D)	Results (R)
(0-20)	No formal approach	No awareness of compliance to innovation management processes	Ad hoc innovation outcomes
(21–40)	Reactive approach	Some awareness and sporadic compliance to innovation management processes	Effectiveness of innovation management processes is not assessed regularly
(41-60)	Stable formal system approach	Significant level of compliance to innovation management processes	Effectiveness of innovation management processes exists and evidence is present
(61-80)	Continual improvement emphasized	Full compliance and continual updating of the system	Effectiveness of the entire innovation management system is evident and innovation objectives are achieved.
(81-100)	Best-in-class performance	Full compliance, continual update, highly effective processes	High customer satisfaction, continual innovation improvement is evident, innovation objectives are exceeded

FIGURE 47-4 Scoring guidelines.

Level	Range of Score	Stage	Max	Min. to Qualify
0	1 - 10	Sporadic	15	12
1	11 - 30	Idea	15	12
2	31 - 50	Managed	20	16
3	51 - 70	Nurtured	25	20
4	71 - 90	Sustained	25	20
5	91 - 100			

Step 1. Calculate total score for the assessment.
Step 2. Determine initial qualified BIMM level (QBL).
Step 3. Identify the lowest qualified stage without jump (LQS).
Step 4. Calculate the average of remaining stages above LQS (ARS).
Step 5. Estimate final BIMM stage score: LQS Upper Limit + ARS.
Step 6. Determine the final qualified BIMM Level (QBL).

FIGURE 47-5   BIMM level determination guidelines.

Level	Range of Score	Stage	Max	Min. to Qualify	Actual
0	1 - 10	Sporadic	15	12	5
1	11 - 30	Idea	15	12	5
2	31 - 50	Managed	20	16	5
3	51 - 70	Nurtured	25	20	5
4	71 - 90	Sustained	25	20	5
5	91 - 100				

Step 1. Calculate total score for the assessment = 25.
Step 2. Determine initial qualified BIMM level (QBL) = 1.
Step 3. Identify the lowest qualified stage without jump (LQS) = None.
Step 4. Calculate the average of remaining stages above LQS (ARS) = 5.
Step 5. Estimate final BIMM stage score: LQS Upper Limit + ARS = 10+5 = 15.
Step 6. Determine the final qualified BIMM Level (QBL) = **1**.

FIGURE 47-6   BIMM level determination example A.

Level	Range of Score	Stage	Max	Min. to Qualify	Actual
0	1 - 10	Sporadic	15	12	12
1	11 - 30	Idea	15	12	8
2	31 - 50	Managed	20	16	8
3	51 - 70	Nurtured	25	20	8
4	71 - 90	Sustained	25	20	8
5	91 - 100				

Step 1. Calculate total score for the assessment = 36.
Step 2. Determine initial qualified BIMM level (QBL) = 2.
Step 3. Identify the lowest qualified stage without jump (LQS) = 1.
Step 4. Calculate the average of remaining stages above LQS (ARS) = 8.
Step 5. Estimate final BIMM stage score: LQS Upper Limit + ARS = 30+8 = 38.
Step 6. Determine the final qualified BIMM Level (QBL) = **2**.

FIGURE 47-7   BIMM level determination example B.

Level	Range of Score	Stage	Max	Min. to Qualify	Actual
0	1 - 10	Sporadic	15	12	8
1	11 - 30	Idea	15	12	12
2	31 - 50	Managed	20	16	12
3	51 - 70	Nurtured	25	20	12
4	71 - 90	Sustained	25	20	12
5	91 - 100				

Step 1. Calculate total score for the assessment = 56.
Step 2. Determine initial qualified BIMM level (QBL) = 3.
Step 3. Identify the lowest qualified stage without jump (LQS) = None.
Step 4. Calculate the average of remaining stages above LQS (ARS) = 11.2.
Step 5. Estimate final BIMM stage score: LQS Upper Limit + ARS =10+11.2 =21.2.
Step 6. Determine the final qualified BIMM Level (QBL) = **1**.

FIGURE 47-8   BIMM level determination example C.

Level	Range of Score	Stage	Max	Min. to Qualify	Actual
0	1 - 10	Sporadic	15	12	8
1	11 - 30	Idea	15	12	8
2	31 - 50	Managed	20	16	12
3	51 - 70	Nurtured	25	20	12
4	71 - 90	Sustained	25	20	24
5	91 - 100				

Step 1. Calculate total score for the assessment = 64.
Step 2. Determine initial qualified BIMM level (QBL) = 3.
Step 3. Identify the lowest qualified stage without jump (LQS) = None.
Step 4. Calculate the average of remaining stages above LQS (ARS) = 12.8.
Step 5. Estimate final BIMM stage score: LQS Upper Limit + ARS =10+12.8= 22.8.
Step 6. Determine the final qualified BIMM Level (QBL) = **1**.

FIGURE 47-9   BIMM level determination example D.

BIMM Assessor Qualifications

Assessing to a set of requirements includes some subjectivity by the assessor. In addition to standard assessor qualifications, there are common assessor attributes, and processes to assessing a system irrespective of the requirements. Critical assessor attributes include being objective and respectful of the auditee. The following is a generic list of personal attributes of an effective assessor:

Assessor Personal Attributes

- Ethical—fair, truthful, sincere, honest, and discreet
- Open-minded—willing to consider alternative ideas or points of view
- Diplomatic—tactful in dealing with people
- Observant—actively aware of physical surroundings and activities
- Perceptive—instinctively aware of physical surroundings and activities
- Versatile—adjusts readily to different situations
- Tenacious—persistent, focused on achieving objectives
- Decisive—reaches timely conclusions based on logical reasoning and analysis
- Self-reliant—acts and functions independently while interacting effectively with others
- Creative—able to combine new ideas uniquely and quickly as needed before, during, or *postassessment*

Assessor Expectations and Responsibilities

In addition to possessing personal traits of a good assessor, there are responsibilities that an assessor must dispense appropriately. In order to conduct a value-added assessment of innovation practices, the assessor must be able to understand how the innovation system works in the company, how well it works, and what action is taken to remedy barriers to innovation. The assessor must possess good communication and influencing skills. It is not just opportunities one finds, but how they are addressed that is equally important. The assessor must be knowledgeable and strong enough to ask probing questions to identify deeper issues with the innovation system. Once an assessor understands the strengths and opportunities, one must be able to capture the essence of the shortcoming and document it for the desired response and action. Another critical responsibility an innovation assessor must fulfill is to maintain confidentiality of the information. Risks are much higher auditing for innovation system where information about future products and services is shared with the assessor. Assessors must avoid retaining evidence related to future products or strategy to prevent inadvertent violation of nondisclosure agreements.

Assessor Responsibilities

- Focus on finding facts, but not personal faults
- Follow the assessment guidelines or procedures
- Listen carefully
- Communicate and clarify assessment requirements

- Understand the system
- Document the findings and observations
- Report assessment results
- Retain and safeguard documents
- Ensure confidentiality

Summary

Assessment of innovation competency is a first step in answering a question by the leadership about the state of innovation in order to develop a strategic plan for institutionalizing innovation. The BIMM is a comprehensive approach to assessing the state of innovation based on the proven capability maturity model. Receiving an assessment by a third party, such as IIT's Center for Innovation Science and Applications, will lend credibility to and confidence in the assessment, and can be used to develop strategic execution plans to maximize benefits of internal innovation competency. Return on investment in innovation has a long way to go using conventional methods and models of innovation. The BIMM is a good start to improving ROI in innovation.

References

1. Humphries, Watts. *Managing the Software Process*. Addison-Wesley Professional, 1989.
2. Paulk, C. Mark. *A History of the Capability Maturity Model for Software*. SQP, 12 (1), Quality Press, 2009.
3. Crosby, B. Philip. *Quality Is Free: The Art of Making Quality Certain*. Mentor Books, 1992.
4. Burnstein, Ilene, Suwannasart, Taratip, and Carlson, C.R. "Developing a Testing Maturity Model: Part I." *The Journal of Defense Software Engineering*, 1996.
5. Paulk, C. Mark. The History of Capability Maturity Model for Software, A Power Point Presentation. Software Engineering Institute, 2001.
6. Carlson, C. Roberts, Gupta, Praveen, and Srivastava, Arvind. Certified Innovation Assessor Training Manual. IIT Center for Innovation Science, 2012.

Case Studies

CHAPTER 48

Financial Innovation*

Michael Gorham and Siva Balasubramanian

This case study combines two approaches to the topic of financial innovation. We will, of course, survey the overall issues related to financial innovation, including definitions, classifications, drivers, performance, waves, life spans, success rates, competition, protection, regulation, and the extent to which innovation is good or bad. In addition, we will take advantage of a rich derivatives database containing every new futures product listed on U.S. exchanges from 1956 through 2012.[1] We will use that database to test various ideas about financial innovation and thus put hard numbers to many of the issues that we discuss. So the chapter is really a combination of traditional survey and case study. Finally, we will conclude the chapter with a more focused, derivatives-based case study of how a small, scrappy futures exchange called the Chicago Mercantile Exchange (CME) beat out and eventually acquired a much older, larger, and the more important Chicago Board of Trade (CBOT) through a strategy of aggressive innovation.

What Is Financial Innovation?

A financial innovation creates and introduces something new in finance. Generally, the idea of financial innovation involves not only the creation of a new financial concept, but also its production, diffusion, and popularization. Financial innovation is not an obscure creation that sits on a shelf, but rather a new financial product that is actually used in the marketplace. These products should make life better.

What is a little more difficult is the classification or taxonomy of these various financial innovations. This is important because categorizing the various manifestations that financial innovation can take it helps us to better understand what financial innovation is. There is a large number of innovations out there. Some have to do with products, such as variations on stocks, bonds, and derivatives. Some have to do with processes, such as the ticker tape, electronic trading, algorithmic trading, VWAP or iceberg orders, payment systems like CHIPS, Fedwire, ATMs, and debit cards.

Classifications of financial innovation typically fall into one of two categories: essentially, what the innovation looks like and what the innovation does. The simplest what-it-looks-like classification has only two classes: products and processes. Tufano's definition, "Financial innovation is the act of creating and popularizing new financial instruments as well as new financial technologies, institutions, and markets," does a nice job of both defining and then categorizing financial innovation into four buckets.[2] Very similar to Tufano is the White and Frame[3] classification, which breaks down innovations into

1. **Products**: exchange traded funds, credit default swaps, and mortgage-backed securities
2. **Services**: ATMs, computerized trading, and electronic trading
3. **Production processes**: dematerialization of securities, credit rating of securities, and credit scoring of individuals
4. **Organizations**: electronic exchanges, dark pools, and venture capital firms

---

*The authors would like to thank IIT Stuart research assistants Yang Hu and Kushan Trivedi for their tireless help on this project.

There are several overlapping approaches to the what-it-does system of classifying financial innovation. Merton,[4] for example has six categories of financial innovation based not on their structure, but on what they do. These include:

1. Moving funds across time and space
2. Pooling funds
3. Managing risk
4. Extracting information to support decision making
5. Addressing moral hazard and asymmetric information problems
6. Facilitating the purchase and sale of goods and services through payment systems

Walmsley[5] has a smaller set of classes of financial innovation based on what they do.

1. Transferring risk
2. Enhancing liquidity
3. Broaden credit
4. Broaden equity

Finnerty[6] has still another collection of financial innovation categories based on their function, and these include:

1. Reallocating risk
2. Increasing liquidity
3. Reducing agency costs
4. Reducing transaction costs
5. Reducing taxes or getting around regulations

One other way of categorizing financial institutions is to group them according to the sector in which the innovation has occurred. For example Allen, Yago and Barth[7] organize their discussion around the innovations in business, housing, and environmental finance.

For the purpose of this chapter, we do not have to pick the best possible classification. Each of these helps us to think about financial innovation in slightly different ways and each of these groupings has its own ambiguities and problems with fitting everything neatly into one of the buckets. We list each of these classification systems to both demonstrate how people have thought about the issue and to point out that there is no "Holy Grail" to classify financial innovations.

Financial versus Nonfincial Innovation

Are financial innovations much different from nonfinancial innovations? In terms of the basic definitions, both involve solving existing problems. What is different is that the kinds of problems solved by financial innovations are different from those posed by nonfinancial innovations. When we look at the classifications based on function, the classifications for financial innovations will be quite different from those of nonfinancial innovations. Only the White and Frame classifications could be applied to both and this is because their classification was based on what innovations look like as opposed to what they do.

Another difference between the types of innovation has to do with cost and time-to-market. Physical products can take a significant amount of time and involve significant cost to get to market. Financial products can often be created almost immediately with a much less significant up-front investment. Financial processes, such as a number of the hardware and software components involved in electronic trading, can take substantial time and involve significant expense.

The third difference between financial and nonfinancial innovation has to do with the potential costs of "bad" innovations. While disastrous results exist in both types, and bad nonfinancial innovations (e.g., pharmaceuticals) can actually kill people, we can't think of any nonfinancial innovations that have resulted in widespread trauma to the economy, as financial innovations have been blamed for doing in the financial crisis of 2008.

Deep History

The reason we look at the ancient history of financial innovation is simply to remind us that this is not a phenomenon of just the last few decades, or centuries. People have always looked for ways to get things done more effectively and efficiently. Let's take a look at a few early appearances of financial instruments that we find commonplace today. The very first financial innovation appears to have been the basic loan, along with its ever-present, contrersial companion, interest.

- Loans appeared before money. A form of loan appeared in the first cities in Mesopotamia (today's Iraq, Iran, Syria, and Turkey) about 3200 BC.[8] The loans were not in money, but in agricultural goods. Because these cities, with populations in the tens of thousands, required individuals to specialize in agriculture, fishing, herding, or in the production of cloth or various crafts, they needed a distribution system. It appears that there was a centralized system for gathering staples contributed by each member of society and then redistributing them. The amount received depended on a worker's gender, age, and position. The average adult male received 60 L of barley per month and the average adult female received about half that amount. To make this work, everyone had to make contributions to the central fund. Because of significant uncertainty over the harvest or output of each individual member of the community, there were times when individuals could not make their contribution and would be recorded as owing a debt to the central fund, or alternatively could borrow from another citizen and make the contribution to the fund, making them indebted to the individual from whom they borrowed.

Even 5000 years ago there were financial crises and defaults. At times the amount of indebtedness became so onerous and the prospects of repayment so unlikely, that the king would issue royal decrees that would annul all debts forced upon people because of crop failures. Debt slaves were released and pledged property was returned to the original owners. This debt forgiveness did not affect loans taken out by merchants.

- Equity. For a long time, common stock initially appeared to have had its origin in the financing of projects like sailing expeditions. An investor who fronted some of the money for the expedition would receive a share of the profits of the project. Initially this equity ownership lasted only as long as the expedition itself, but that changed with the creation of the many joint stock companies such as the Russia Company (1553), the British East India Company (1600), the Dutch East India Company (1602), the Danish East India Company (1616) etc. We now think that the concept of equity developed in 5th century BC in Rome.[9] What we know now is that there were these organizations that were essentially government contractors that did things like feeding the Roman geese, building and maintaining roads, outfitting the army, and collecting taxes. These government contractors took on a form very similar to corporations issuing ownership shares that looked very much like equity.
- Bonds.[10] A bond is just a loan that is transferable. A loan is a private contract between two parties. A bond is created as an instrument that can be bought and sold among investors. While bonds were also issued by the Dutch East India Company in the 17th century, the first true bond is considered to be one issued by the French government in 1555.[11]

Let us also take a look at the three ideas that form the building blocks of most financial products.[12] They are the transfer of value or wealth through time, the contingent claim, and negotiability.

- The intertemporal transfer of wealth. You pay something now in order to receive something greater at a later time. You receive something now at the cost of having to pay something greater at a later time. These are the intertemporal transfers upon which finance is built and which is essential to most financial products.
- Contingent claims. A contingent claim is essentially a bet. One party pays the other depending on the occurrence of some event. The wager on the Super Bowl is a contingent claim, but the ones that we talk about in finance typically involve some sort of risk transfer, as in an insurance policy. The party trying to mitigate risk pays premiums over some period of time, and receives a payment if his house burns down, if his ship gets lost at sea, or if he dies (in which case the payment goes to a beneficiary). Options are contingent claims. Inflation indexed bonds are contingent claims. In terms of historical precedents, the stock options traded in 17th-century Holland were probably the most famous.

- Negotiability. If you loan money to someone else, and you find that you suddenly need that money back, private loans generally do not contain the ability to receive the money prior to the day on which it was promised. To get money, you would most likely have to take out a loan yourself. But if you own a bond issued either by a government or corporation, this is a negotiable instrument and can be sold in the marketplace to someone else, allowing you to receive your money. Bonds, notes, equities, and derivatives are typically easily negotiable and can be acquired or disposed of in relatively liquid markets.

Financial Innovation by Embedding Options Inside Existing Financial Instruments

One trick used in financial innovation is to take some existing financial instrument and embed an option inside. The option will give either the issuer or the holder of a financial instrument the right to do something that they would not have been able to do otherwise. And this usually solves some kind of problem. For example, we can take an ordinary bond and embed a call option, giving the bond issuer the right to call back or repurchase the bond from the investor at sometime before the bond matures. The problem this solves for the bond issuer is that if interest rates fall significantly during the period that the bond is outstanding, the issuer can eliminate these high interest rate bonds and replace them with much lower interest rate bonds. Of course this is undesirable to bond investors, so they have to be paid a higher rate of interest initially to put up with this possible inconvenience.

Any time an option is embedded in a financial instrument, the owner or beneficiary of that option has to pay for it. A convertible bond, which gives the holder the right to convert the bond into the common shares of the company issuing the bond, is attractive to the bond holder and thus they would be willing to accept the bond with a lower coupon rate of interest than would be the case with a plain-vanilla bond.

Merton Miller's Drivers of Financial Innovation

There is actually an overlap between the classifications of, and drivers of, financial innovation. The classifications reviewed in the prior section are really lists of the problems solved by, and thus the drivers of, these financial innovations. We would like to focus here on numbers four and five in the Finnerty list of classifications of financial innovation.

Merton Miller has argued that the major driver of financial innovation has been to avoid regulations and taxes.[13] He admits it is with some sadness that he makes this observation. While his focus is on a two-decade period from 1966 to 1986, we know that the drivers of tax and regulation avoidance hold over a much longer span of time than that. Before illustrating this, we note that avoiding taxes and regulations is really a subset of something larger. Financial innovations arise in order to make it easier and more pleasant for individuals and firms to do things of a financial nature (more efficiently, more effectively, with less risk, and with lower cost). William Silber illustrates this with more academic rigor. Think of a linear programming model that attempts to maximize the utility of a firm subject to a set of constraints. The more constraints that we can remove from this optimization problem via innovation, the better our results will be.[14]

For example, during the second half of the last century, the cost of making a significant "calculation" dropped precipitously due to the handheld electronic calculator that debuted in the 1960s with a price tag of $300 and can now be obtained in the corner drugstore for less than $10. Innovations such as integrated circuits allowed those costs to fall precipitously. In the case of financial innovations, avoiding regulations and taxes can often decrease the cost of doing business just as impressively or better. Some examples include:

Preferred stock: Preferred stock is actually a cross between common stock and bonds. It generally gives ownership rights similar to common stocks, but pays a fixed dividend in the same way that bonds pay a fixed amount of coupon interest. It is called preferred, because in the case of liquidation, preferred stockholders would be paid off before common stockholders. While a form of preferred stock existed in the 16th century, it was not until the 1840s that preferred stock became very popular. The driver? The British had a regulation that restricted a company's debt to no more than a third of its equity.[15] Because preferred stock was considered equity, but behaved very much like a loan, it allowed companies to increase their debt-type capital without it really counting as debt. The fact that it counted as equity increased the denominator of the debt-to-equity fraction that companies had to keep below 50 percent, allowing companies to borrow much more than the ceiling that had been set by the British government.

Eurobonds: In the early 1960s, the United States was facing a significant balance of payments deficit. As one measure to help reduce this deficit, President Kennedy in 1963, introduced the Interest Equalization Tax, making it less attractive for U.S. investors to purchase bonds sold by foreign companies into the U.S. market. The bonds that were hit by the tax are those known as Yankee bonds. These are dollar-denominated bonds sold in the United States by companies headquartered in other countries. The tax was successful. Sales of Yankee bonds declined by 90 percent within a year.[16] But yet again, we have a regulation that inspires an innovation.

These foreign issuers still wanted to raise capital in U.S. currency by selling U.S. dollar-denominated bonds, so they simply did so outside the United States in order to avoid the Interest Equalization Tax. The Eurobond is not what it sounds. It does not have to be issued in Europe. A Eurobond is a bond issued in one country by a company based in a second country, in a currency from yet a third country. For example, a bond issued by Japan-based Sony in the U.S. market, denominated in Mexican pesos would be a Eurobond. One of the big advantages is because the currency of issue is not the currency of the country in which the bonds are being issued so there is very little regulation of these bonds, meaning they are much less expensive to issue.

Zero coupon bonds: An unintentional U.S. tax loophole in the early 1980s made it very attractive for corporations to issue zero coupon bonds. A zero coupon bond is sold at a discount and the interest is paid by the issuer of the bond at maturity, when it is redeemed at its full face value. The IRS unwittingly allowed corporations to deduct much more interest than was appropriate in the early years of the bond's life. With the high interest rates of the early 1980s, the value of this loophole was sufficient to stimulate a wave of zero coupon bond issues. The loopholes closed relatively quickly, but not before investors also started to appreciate the value of zero coupon bonds. Investors liked the idea of getting a certain known payment on a specific date in the future and avoiding the reinvestment risk associated with receiving semiannual coupon payments. But with coupon bonds, investors never knew at what rate they would be able to reinvest future coupon payments.

Given this demand, investment banks created a new financial product by stripping a normal Treasury bond that paid regular, semiannual coupon interest. "Stripping" means stripping all of the coupon payments off the bond principal payment, ending up with a number of zero coupon bonds. The banks sold these strips to investors under brands like CATS and TIGRS. Visualize a $1000, six percent coupon, 30-year T-bond. The bond will make a coupon interest payment of $30 ($1000 X .06 X .05) every 6 months for 30 years, and one $1000 payment of principal in 30 years. When the coupons are stripped off the principal, each of those 61 payments becomes a zero coupon bond. The demand was sufficient that the Ttreasury started to let investors have direct ownership of Treasury principal and interest. The new program was appropriately called STRIPS, an acronym for separate trading of registered interest and principal of securities.

Does Financial Innovation Come in Waves?

While there seems to be no study that maps out the waves of financial innovation over very long periods of time, Nobel laureate Merton Miller has argued forcefully that financial innovation had a huge wave between 1966 and 1986. He listed a few of the innovations that squeezed themselves into this brief time span as including "negotiable CDs, Eurodollar accounts, Eurobonds, sushi bonds, floating-rate bonds, putable bonds, zero coupon bonds, stripped bonds, options, financial futures, options on futures, options on indexes, money market funds, cash management accounts, income warrants, collateralized mortgages, home equity loans, currency swaps, floor and ceiling swaps, and exchangeable bonds."[17]

We can test Miller's claim by utilizing the futures product database we referred to earlier. This database includes all new futures products created between 1956 and 2012. Table 48-1 shows the success level of futures products launched in various decades. There are actually six measures, which we will explain later. In all cases, the larger the number, the more successful the products created in that decade. Looking at conventional chronological decades, we find the decade in which new products created the greatest amount of use was the 1980s, with the 1970s and 1990s coming in as close seconds. This supports Miller's belief.

Success Measure	1950–1960	1960–1970	1970–1980	1980–1990	1990–2000	2000–2010	Whole Period
Life span	7.9	6.4	9.7	8.9	6.0	4.2	5.9
Lifetime volume	2064	1212	36,151	90,395	48,775	5990	27,637
PV lifetime revenue	522	546	10,838	27,167	25,385	4816	11,793
PV 10-y revenue	258	416	2495	4994	9288	5760	5530
PV 5-y revenue	182	159	627	1895	2443	3162	2201
5th year volume	34	72	304	707	1101	797	703

TABLE 48-1   New Futures Products: Success by Decade Listed (in 1000s except Life Span)

If we focus on financial derivatives traded on exchanges, this gives us 40 years of history to examine. Because the futures industry has been a highly competitive environment, and because of the first mover advantage, it would seem reasonable to postulate that when a new idea, a new type of product, was introduced by an exchange, other exchanges would be keen to quickly get their own versions of these new products listed. So one might expect to see that when the CME launched seven new foreign exchange contracts, others would be quick to jump on the bandwagon to try to be the first to list their own. We would expect the same when the CBOT listed the first interest rate contract in 1975 and when the Kansas City Board of Trade listed the first stock index futures contract in 1982. If we plotted the number of new contracts annually in any new product category, we might expect to see a quick ramping up over a few years and then as the new ideas were exploited, see the innovation fall off to almost zero.

Waves in foreign exchange futures: Let's first take a look at newly introduced futures contracts on foreign exchange rates, such as the price of the pound or yen in U.S. dollars. The initial impetus to create foreign exchange futures was that the 1945 Bretton Woods Agreement had started falling apart in the early 1970s. For about 25 years, there was little need for protection against fluctuating exchange rates, because exchange rates did not fluctuate. Once there were signs that this quarter century of fixed exchange rates might unravel, futures exchanges started to take notice. The first exchange to list a foreign exchange contract was a New York exchange that listed too early (i.e., before there was really any free movement in exchange rates). In addition, this was a tiny exchange with few resources. The single foreign exchange product we see listed in 1970 in Fig. 48-1 came from the tiny International Currency Exchange. It failed quickly.

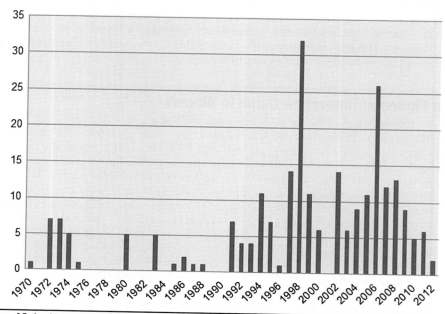

FIGURE 48-1   Number of new currency futures products introduced annually.

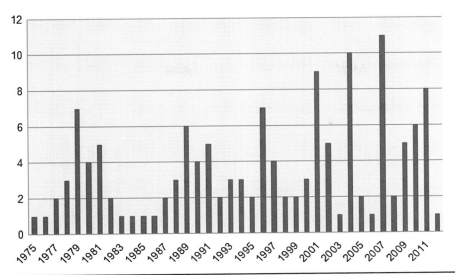

Figure 48-2   Number of new interest rate futures products introduced annually.

It was the much larger Chicago Mercantile Exchange that appeared to have picked just the right time to enter in 1972 when it listed seven different currency products (British pound, Canadian dollar, Deutsche mark, Italian lira, Japanese yen, Mexican peso, and Swiss franc). The choice of currencies was based on both their importance in international transactions and their importance in trade with the United States. The following year, another seven products were listed, mainly clones of the CME's products listed on other exchanges. A similar thing happened in the third year; again, mainly contracts by other exchanges trying to catch up with the same CME products. Then for five years, innovation in this area virtually dried up. What we didn't expect was that in 1980, another exchange would copycat the CME's products and then 3 years later yet another exchange would copycat those same products, though with the twist of making the contracts half the size of the CME products, in hopes of attracting smaller traders.

But an even bigger surprise was that there would be two more large waves of innovative activity in foreign exchange futures. The first of these began in 1991 and lasted for 10 years (including an all-time high of 38 new currency contracts in 1998) and the second wave began in 2002 and also lasted 10 years. Why? A small amount of this was yet other exchanges attempting to list the same star performers that the CME listed in the early 1970s. Much of it was listing new currency pairs that did not include the dollar (such trades are known as cross rates). Because the bulk of interbank foreign exchange trading is largely against the U.S. dollar,[18] naturally all of the early currency futures contracts also had the dollar as the base currency. Some of it was from experimenting with the contract sizes. The initial size of currency futures contracts was about $100,000. This was a compromise between the needs of the wholesale market, in which interbank transactions typically had one million dollar minimums, and the needs of a retail market that needed contracts much smaller. So this last wave of new currency products included both the $1 million size as well as tiny $10,000 contracts referred to as E-Micro's, to directly cater to the retail market. A portion of the last two waves involved emerging-market currencies like the Brazilian real, South African rand, Russian ruble, Malaysian ringgit, Hungarian forint, Czech Corona, Polish zloty, Israeli shekel, Korean won, Colombian peso, Turkish lira, and Chinese renminbi, to allow the management of an entirely new set of risks.

The implications of the pattern that we see here is that not all ideas related to a new product are quickly exploited, even in the presence of winner-take-all markets. The fact that there has been a vibrant competition and innovation in foreign exchange futures over the past 40 years suggests that both changes in the macroenvironment and the strong drive for exchanges to compete, have led to continuing attempts to try something at least a little bit different.

Waves in interest rate and equity futures: While we do not explore these other cases with the same kind of detail, we can see from the figures that innovations in interest rate futures and equity index futures, also demonstrate prolonged periods of innovation and competition, often falling into a series of waves over the past four decades.

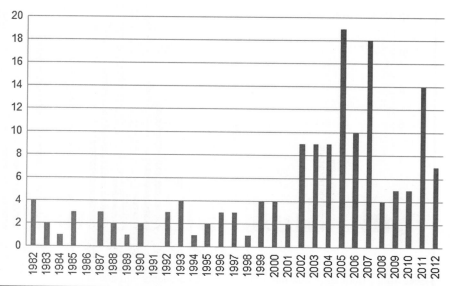

FIGURE 48-3 Number of new equity index products introduced annually.

Modeling Innovation Waves

Moving away from the data-rich futures markets to the over-the-counter (OTC) markets, we notice that there are significant waves as well, along with what Persons and Warther call serious boom and bust patterns.[19] They see these patterns in the horizontal mergers of the 1890s, the vertical mergers of the 1920s, the conglomerate mergers of the 1960s, and the hostile takeovers of the 1980s. The leveraged buyout (LBO) was a huge wave of the 1980s, with the value of LBOs going from $1 billion in 1982 to $60 billion in 1988 and back to $1 billion by 1991. The more recent collateralized debt obligation (CDO) peaked at $300 billion in 1993, falling back to $20 billion by 1997. They note that these kinds of boom and busts are pointed to by critics as examples of irrational excess, but they present a mathematical model of fully rational innovation adopters that is totally consistent with the kinds of boom and bust cycles we see in real financial markets. Specifically, their model finds the same behaviors that we observe in the marketplace:

1. Firms will pass on an innovation in one period, only to adopt the innovation in the next period, even after receiving unfavorable information about the innovation.

2. The last firms to adopt lose money on average.

3. The end of an innovation wave is completely unpredictable, and often ends badly.

4. Most innovations perform worse than expected, even though the distribution of quality across innovations is symmetric.

What Do We Know about the Life Spans of Financial Innovations?

Randall Kroszener describes the life cycle of a new financial instrument as occurring in several phases.[20] First comes the experimentation phase in which market participants try out the new product and give feedback to the product issuer. Then comes the adjustment phase, when the creator-issuer tweaks the product in reaction to this market feedback. Some products simply missed the mark and don't sell. Others, like credit default swaps during the financial crisis, become insanely popular and in some cases, the feedback becomes so intimate that customers are participating significantly in the design of the product. For example, John Paulson, the hedge fund manager who made $5 billion during the financial crisis from buying credit default swaps, sat down with Goldman Sachs and told them specifically which zip codes he wanted included in the mortgage-backed credit default swaps that he wanted to buy.[21]

How long do new financial products last? Luckily we have hard data on the birth, life, and death of exchange traded derivatives in the United States between 1956 and 2012. Figure 48-4 contains two pieces of information. The bar and the left axis tell us the

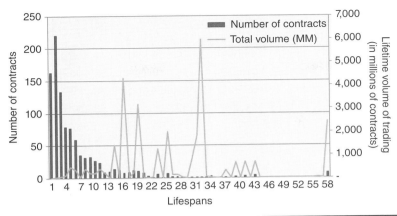

FIGURE 48-4 Number of futures products and lifetime volume for various life spans 1956 to 2012.

number of new futures products that survived only 1 year (there were 163), 2 years (220), 3 years (133), and so on. Even though we are looking at life spans as long as 58 years, we see that most contracts had relatively short life spans. Of the 1026 new futures contracts introduced during this period, 66 percent of them lasted 5 years or less and 83 percent lasted 10 years or less. Two years was the most frequent life span, and 1 year was the second most frequent.[22]

So it appears that the life of the typical new futures contract is "nasty, brutish, and short." Actually the nasty and brutish is an exaggeration because most futures contracts are not competitive situations where multiple exchanges list the same products at the same times and then duke it out for market share. This happens, as is explained in the Competition section later, but the bulk of listings of new futures products are solitary events where no one shows up for the party.

The line and right axis in Fig. 48-5 tells us the lifetime volume of products that lasted for various periods of time. For example, we can see that there was a product with a life span of 31 years that had a huge lifetime volume of almost 6,000,000,000 contracts traded. This is the 3-month Eurodollar contract, the one that carries the LIBOR interest rate. To be clear, the Eurodollar contract did not die in its 31st year of life. It was listed in 1981 and continued to trade through the end of the available data set, 2012. It is still an actively traded product.

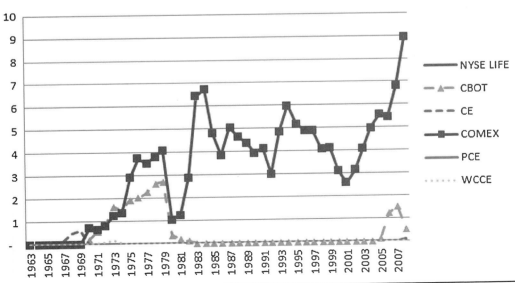

FIGURE 48-5 Annual volume of silver futures contracts 1963 to 2008 (in millions of contracts traded) first mover wins.

Success Measure	Innovations	Imitations	Extensions	All Contracts
Life span	6.0	5.8	5.9	5.9
Lifetime volume	35,743,939	10,859,950	42,498,131	27,636,652
PV lifetime revenue	$14,091,752	$5,330,798	$19,660,245	$11,1792,851
PV 10-y revenue	$6,469,390	$2,917,061	$8,300,627	$5,530,354
PV 5-y revenue	$1,979,929	$2,015,403	$3,161,061	$2,200,996
5th year volume	693,570	553,530	1,038,153	703,137

TABLE 48-2   Success of New Futures Products by Innovation Type

How Successful Are Financial Innovations?

Most financial innovations are not successful. This is an impressionistic statement about financial innovation overall, but we do have some numbers to back this up in the case of futures markets. From 1956 to 2012, there have been almost 1000 new futures contracts created in the United States. The average life of a futures contract is about 6 years. Now to be fair, there are some futures contracts that have lasted well over half a century. The reason for the relatively low average life is because about half of all of new futures contracts die within 5 years, and many of them lasted less than one year.

Does the success of a newly listed futures contract depend on whether it was a totally new innovation, an imitation of the product already created at another exchange, or an extension of the product already listed at the incumbent exchange? It seems like it does. In order to explain the data in Fig. 48-6 we need to describe each of the six metrics that we examined for success. Let us first explain the term *volume*. Volume in futures markets refers to the number of futures contracts entered into, or traded, during a specific period of time—a day, a month, or a year. It is similar to the number of shares traded on a stock market. The volume of trade is an important measure for two reasons. First, the level of trading in a product reflects how important the product is to the trading community, both the commercial users who are trying to manage price risk and the speculators who are willing to take on that risk in hopes of a profit. Second, traders pay exchange fees in direct proportion to the amount of trading that they do. Fees are charged on a per-contract basis. The metrics we used to gauge the success of new products include

- Lifespan
- Lifetime volume
- Present value of lifetime revenue
- Present value of 10-year revenue
- Present value of five-year revenue
- Fifth year volume

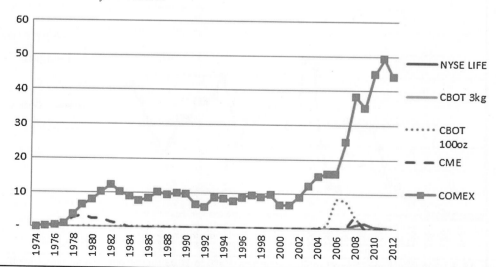

FIGURE 48-6   Annual volume of gold futures products (in millions of contracts traded).

One might expect that true innovations would be the most successful, especially given the first mover advantage that we talk about in the next section. In fact, of the three measures of innovation, true innovations tend to be the second most successful type of product. The most successful product under all five measures of volume or revenue is product extensions. The story that goes with this result is that exchanges list the best designed contract that they can develop at the time. Experience and user feedback cause them to develop product extensions, which either better meet the initial needs or meet other variations on those needs. And it is generally successful product extensions that are the real drivers of trading activity and revenues.

For example, when the CBOT got a license from Dow Jones and announced that they were going to list a small, retail Dow Jones Industrial Average futures contract, the CME felt that it needed to meet that competitive threat and list its own retail size contract. It quickly negotiated a deal with the S&P Corporation and gained a license to list smaller version of its premier S&P 500 contract. The new product was called the E-mini S&P 500® futures contract. It was called "mini" because it was smaller and more accessible to the average trader and had the "E" in the name because it would be traded exclusively on an electronic platform, the first CME product to be available only online. This product extension has become the biggest stock index futures contract in the world, and by 2012, its trading volume has become 94 times the level of the mother product on which it was based.[23]

Competition, Protection, and Monopoly Outcomes

Innovators hate competition. If they create a new invention or business process, they generally want to benefit from this ingenuity by owning exclusive rights to the innovation and selling or leasing the innovation to others. They don't want imitators coming in and reaping the rewards for the difficult creative work they've done. There are two ways of obtaining such a monopoly position. The first is by establishing some sort of legal protection for the intellectual property they have created, specifically through use of a patent, copyright, or trademark. The second is by being first and having a sufficient head start to be able to obtain a first mover advantage that allows them to obtain a dominant market share. The first approach is well known generally, but the second is known only in certain corners of financial innovation. We take each in turn.

In exploring the extent to which financial innovations can be legally protected, we first need to make a distinction between products and processes. For many years, there was a basic assumption that the general category called business processes, and its subcategory, financial processes, could not be protected. Specifically, there was a decision in 1908 that established the *business methods exception* that stated that business methods could not be patented. For the following 90 years, there was very little litigation on this type of intellectual property, even in cases where the U.S. Patent Office had actually issued a patent. It was generally viewed as not worth the cost, since success was very unlikely. There was a sea change in July 1998 in the case of State Street Bank and Trust versus Signature Financial Group. Signature had, in 1993, been given the patent for a software program that allowed them to calculate the value of a mutual fund. State Street Bank sued Signature, arguing that it was a nonpatentable business method. While State Street won the initial case, the Court of Appeals reversed the decision in 1998, thus eliminating the concept of the business method exception. Because the Supreme Court declined to hear the case, this was the beginning of a new era in which business methods became patentable and led to a boom in filing for and receiving business method patents.

Was this good or bad? Both. On the one hand, there was suddenly a greater incentive to create these innovations, as they could now be protected. On the other hand, the diffusion of any of these innovations was slowed down to the rate that was optimal for the holder of the intellectual property, but not for users. The lawyers also did well, as patent litigation typically costs both parties to the contest millions of dollars. The lawsuits are generally filed by individuals and small private companies.

We can make a few simple statements about protecting futures products. No futures contract listed before 1982 required a licensing agreement. All of the futures contracts listed on agricultural products, metals and precious metals, energy products, and foreign exchange rates were all sufficiently generic that there had been no creator who claimed intellectual property for these things. However, beginning with stock index futures and options in 1982, all stock indexes had to be created by some firm, typically a media company or exchange, and any exchange listing a derivative on one of these products created and owned by another, needed to obtain a license to do so from the owner creator. The same holds for OTC products.

But things are more complex than this.[24] For example, futures and options on indexes require a licensing agreement from the creator owner, whereas options on ETFs do not. Also the settlement prices at an exchange are considered to be in the public domain and cannot be claimed as intellectual property. This took the courts some time to work out after Nymex sued the Intercontinental Exchange for using Nymex settlement prices to settle intercontinental markets.

Liquidity-Driven Monopoly

There is a way to establish a significant market share—even a 100 percent market share—without any legal protection. In many cases, being first out of the box can give a company an advantage in establishing a dominant market share. The interesting thing is that in some financial markets, specifically exchange traded derivatives, there is a tendency for exchanges to have portfolios of monopoly products. For example, on the day that the Chicago Mercantile Exchange and the Chicago Board of Trade merged in 2007, the CME had about 75 products and the CBOT had about 30 products; but there was not a single product with any significant turnover that the two exchanges shared in common. Each exchange had its own portfolio of products with 100 percent market share.

How did this happen? There's a principle called *liquidity-driven monopoly*.[25] The idea is that when traders trade, an important thing that they care about is market liquidity. They want to trade at an exchange where there are a lot of buyers and sellers, a narrow bid-ask spread, and very little market impact when they put in a reasonably sized order to buy or sell. Liquidity is a function of the number of participants that are actively trading in the market. So if an exchange launches a product which is successful and attracts a lot of participants, it becomes liquid. Typically, another exchange that lists the same product will find it virtually impossible to attract traders away from the existing liquid market to come over and trade in the new, relatively illiquid market.

One of the classic battles was the competition for silver futures (see Fig. 48-5). In 1963, the New York metals exchange known as COMEX, listed a futures contract based on the price of silver. The first 3 years saw only light trading, but when activity started to take off, it caught the attention of the world's largest exchange, the CBOT, which listed its own silver contract in 1969. With all the resources available at the CBOT, it caught up with the COMEX contract for a few years, but then started falling behind and the CBOT contract fell to zero volume by 1983. Note that in the figure, both exchanges lost significant volume beginning in 1979 due to the piercing of the Hunt brothers silver bubble in that year. So this is a clear case of the first mover advantage and an illustration of the principle of liquidity-driven monopoly.

Let's take a look at a case where there was no first mover, and where several exchanges simultaneously listed identical contracts. In 1974, Congress repealed the Gold Reserve Act of 1934 and made it possible for Americans to own gold for the first time in 40 years. On the day that the law took effect, five different exchanges listed gold contracts. For five years the COMEX and the CME both built significant trading volume at about the same level. In the fifth year, COMEX began to get ahead and the CME's activity fell to essentially zero by 1984. In cases like this where there are multiple competitors starting at the same time, what is it that causes a specific exchange to come out ahead? The reasons vary from case to case. In this case, the CME had much deeper pockets, but the COMEX was the traditional metals exchange with a very well developed distribution network into the metals community. It is likely these key factors emerged as competitive advantages for COMEX. But the point is that exchange traded derivatives are almost always "winner takes all" competitions.

We must point out that, while the principle of liquidity-driven monopoly reigns supreme in exchange traded futures, it has relatively no power in exchange traded equity options or in stocks. For example, in March 2012, there were 3554 different equity options and ETFs listed at the nine different options exchanges.[26] A full 90 percent of these options showed some market share for all nine of the options exchanges. And the dominant exchange had a market share of, not 100 percent as was the case in exchange traded futures, but 29 percent. The two markets are polar opposites. The reason is simple, and it has to do with market structure. Futures markets are much older and were allowed to evolve organically over time. That organic evolution resulted in each exchange having its own clearinghouse. So that if a trader took a long position in gold at one exchange, he could offset that position only at the same exchange since the position was held at that exchange's clearinghouse.

In the case of options, by government mandate there is a single clearinghouse for all options exchanges, known as the Options Clearing Corporation (OCC). This means that traders can buy an option at whichever exchange has the cheapest price at the time of purchase, hold the position at the OCC, and then sell the position back at whichever exchange has the best price at that time.

How Important Are Process Innovations?

Up until now, most of our focus has been on product innovation, which is what generally comes to mind when we think about the world of financial innovation. However, in financial markets, process innovation has been incredibly important in the past few decades and has resulted in a total transformation of the industry.[27] Stock exchanges and derivatives exchanges worldwide were typically floor-based, not-for-profit entities that were owned and governed by the member traders who inhabited the pits. Today, exchanges are almost fully electronic and are generally stockholder-owned, for-profit, publicly traded companies. This occurred in stages.

- In stage I, only exchange members had access to the electronic matching engines, and customers would have to phone in their orders to these members.

- Stage II was where customers had direct electronic access and could enter their own orders, with the orders being routed through the risk and credit filters of their brokerage firms. This could be called the point-and-click stage, as customers would watch price changes on the screen and then when they saw a trading opportunity, they would enter their orders with either a keyboard or mouse.

- Stage III is one in which traders have downloaded their trading strategies from their brains into computer code that is resident either on their own computers or on servers placed in a building adjacent to the matching engine of the exchange. *Algorithmic trading* is the name given to the process in which trading decisions are being made by computer code that has been created by the trader or her assistant.

How did this happen? Interestingly, it did *not* happen because the exchange giants of the day decided to take advantage of the dramatic changes in the sophistication of and rapidly declining cost of new technology. These exchanges were owned by their members who did not want to dump their tried and true trading floors for a life of sitting in offices behind screens. Seasoned floor traders can read faces, body language, and the level of noise and activity in the pit to use to their advantage. So they actually did their best to keep technology at bay. Electronic trading platforms mainly came to the fore through the efforts of newly minted, de novo exchanges (see Table 48-3). The old line exchanges only converted due to competitive pressures.

The world's first partially electronic stock exchange was NASDAQ. It was founded as an electronic exchange by the National Association of Securities Dealers (NASD) because the SEC insisted that it be electronic. When it was created in 1971, to try to bring more transparency to the OTC market in smaller stocks that couldn't meet the listing requirements of the New York Stock Exchange or the American Stock Exchange, it was only the display of bids and offers by OTC dealers that took place on screens. Transactions were accomplished by picking up the phone and calling the dealer with the best on-screen quote.

It was another 13 years before the first fully electronic exchange was created. No one remembers it because it failed. INTEX was an exchange run by Americans, but based in Bermuda, because U.S. regulators advised INTEX's founders that they would be able to get going sooner offshore. If they tried to launch the exchange in the United States, it was very likely that the existing floor-based Chicago exchanges would be able to slow down the

Fully Electronic	Exchange
10/25/84	INTEX
1/20/85	NZ Futures and Options Exchange
6/1/85	OM
4/1/88	Tokyo Grain Exchange
6/15/88	SOFFEX
1/26/90	DTB
6/12/90	Australian Stock Exchange
8/1/90	SAFEX
4/1/91	Tokyo Commodity Exchange
5/28/93	Zhengzhou Commodity Exchange

TABLE 48-3 The First 10 Fully Electronic Derivatives Exchanges

approval process. The problem the INTEX founders were trying to solve was to avoid the trading abuses on the Chicago and New York floors, and they felt that having a transparent electronic trading system would be best for everyone. Despite a reasonable investment and a lot of hard work, INTEX failed because the offshore placement made things more complicated and because they chose the wrong product. Their first product was gold, which was in the middle of a bear market.

Several of the early electronic exchanges were trying to solve a different problem—a problem of location. In the case of New Zealand, there were a number of wool traders spread around in different centers of the country. They had been hedging their wool price on a London exchange, but the hedge was messy because New Zealand and British wool prices did not move together in lockstep. They decided to create a New Zealand-based exchange, but argued over where it should be. Each of the major wool traders wanted the exchange in their own region. The solution was to locate the exchange in cyberspace, thus giving all wool traders all over New Zealand equal access. Partnering with a group of banks gave the exchange the needed capital, and on January 20, 1985, the New Zealand Futures Exchange "opened its servers." It listed both wool and financial products and was a success.

Around the same time, a group of financial leaders in Switzerland decided to start planning for a derivatives exchange. Each of the existing three stock exchanges (Geneva, Zürich, and Basel) wanted to have derivatives listed on the floor of its exchange. Again, a cyberspace solution gave equal access to all and on May 19, 1988, the Swiss Options and Financial Futures Exchange (SOFFEX) was born.

In neighboring Germany, another group of bankers was trying to set up a new derivatives exchange. The bankers were located in Berlin, Hamburg, Stuttgart, Munich, and Hanover and of course each bank wanted the physical exchange in its own city. The innovation of an electronic exchange was again the solution and the DTB was created. This electronic solution was a highly disruptive innovation that would turn the hierarchy of European derivatives exchanges on its head. It also laid the groundwork for the second largest derivatives exchange in the world.

Being a German exchange, DTB decided to list the German government bond as one of its first products. Starting in 1990, it began building up volume as many of the German banks began diverting their business away from London and toward their homegrown German exchange (see Fig. 48-7). Within 2 years, DTB had captured just under a 50 percent market share, though the demand for the Bund was so strong that the London Financial Futures Exchange's (LIFFE) business was still growing. With a few ups and downs, both exchanges continued to grow until 1997 when DTB significantly closed the gap. Then in 1 year, DTB's Bund volume soared from being about three-quarters that of LIFFE to being almost four times LIFFE's volume. By the end of 1998, 100 percent of LIFFE's activity had migrated to DTB. Almost overnight, Europe's largest exchange lost its major product and its position. It was a spectacular victory for electronic trading that frightened floor-based exchanges all over the world. This new technology really was both good and powerful and seemed to be more attractive to customers than the old floor-based mechanism. In that same year, 1998, the electronic German DTB merged with the electronic Swiss SOFFEX to create the world's largest derivatives exchange called Eurex.

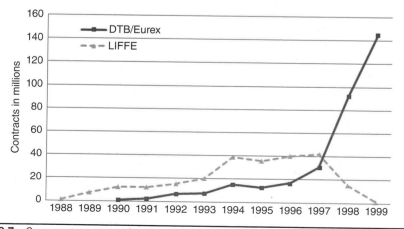

FIGURE 48-7   German government bond volume at LIFFE and DTB.

Who Regulates New Financial Products?

Financial regulation ebbs and flows. In the case of securities regulation, it was the stock market crash in October of 1929 and the abuses that led up to it that resulted in two famous acts of Congress—the Securities Act of 1933 and the Securities Exchange Act of 1934. There was also the Banking Act of 1933, better known as the Glass-Steagall Act, which separated commercial banking activities from investment banking activities, given the conflict of interest when these activities were conducted by the same institution.

As a general matter, derivatives are regulated by the Commodity Futures Trading Commission (CFTC) and securities are regulated by the Securities and Exchange Commission (SEC). So this means that new derivative products traded on futures exchanges come under CFTC oversight, and new securities products traded on securities exchanges come under SEC oversight. One exception to this characterization is that the SEC regulates options based on securities, such as individual stocks and stock indexes. So the SEC regulates all the products traded on the 11 U.S. equity options exchanges. A second exception is that the SEC and CFTC jointly regulate futures on individual stocks that are currently traded only on a single U.S. exchange, OneChicago. This collaborative regulation has been a bit of a disaster. Despite the success of single stock options in other countries such as the United Kingdom, Germany, and India, single stock futures in the United States have been a serious disappointment. A third exception is that the derivative instrument known as a swap, which has been traded in the OTC market over the past several decades, has been essentially unregulated, a problem that was finally corrected by the passage of the Dodd-Frank Act in 2010.

Is Financial Innovation Good or Bad?

In general, all innovations, both financial and nonfinancial, are good and bad. The tractor allowed farmers to plant, cultivate, and harvest a much larger crop than they would have been able to do with a shovel or a mule and plow. But sometimes, on uneven ground, tractors overturned and injured or even killed the farmers who were riding them. And post-Schumpeter, none of us can really even think about innovation without thinking about the destruction of old technology and jobs that is inevitable with the adoption of new technology. But is there something special about financial technology that makes it more likely to be destructive? This question arises because postmortems of the financial crisis of 2008 often point to financial innovation as a cause of the pain that we had inflicted upon ourselves.

There is little doubt that the basic financial product innovations, like loans, stocks, bonds, mutual funds, and exchange traded funds have made it easier for us to move capital from those willing to loan it to those needing it to build companies, to give birth to nonfinancial innovations that would otherwise simply remain as ideas, allow us to invest in a safer, more diversified fashion, obtain advanced educations, and own our own homes.

There were three financial innovations that did play an important role in the financial crisis—mortgages, securitization, and credit default swaps. The first of these is a collateralized loan. The second is a new financial process, and the third is a derivative. All three of these were brilliant solutions to problems that we had. The fixed rate, level payment, self-amortizing mortgage loan made it possible for millions of working class and middle class people to become homeowners with a payment that in real terms declined over the two to three decades of the life of the mortgage. Securitization of mortgages allowed mortgage originators, like S&Ls and banks, to sell the mortgages and obtain new capital to be able to make yet more mortgage loans. The credit default swap enabled those who invested in bonds or in mortgage-backed securities to purchase insurance against a default or other credit event. Let us take a closer look at the innovations being blamed for the financial crisis.

The Credit Default Swap

The credit default swap (CDS) was created in 1994 by J.P. Morgan, and in particular by a British banker named Blythe Masters. A swap is a financial contract in which two parties exchange a series of cash flows over a fixed period of time. The plain-vanilla interest rate swap, for example, involves Party A paying a fixed rate of interest on some specified amount of money, say $1 million, to Party B, and Party B paying a floating rate of interest on $1 million to Party A. The frequency of payments and the duration of the contract can be set in the fashion that is acceptable to both parties. Such an instrument is very useful to a firm that wishes to convert an asset or liability that it holds from floating to fixed rate, or vice versa.

The credit default swap uses that same idea of two parties exchanging cash flows but does so in such a way that essentially mirrors the cash flows that would exist in a traditional insurance contract. If party A held bonds in a company that they feared might default or be downgraded, it could enter into a CDS with party B who would guarantee the value of the bonds against default or downgrade in exchange for a regular quarterly payment that was essentially an insurance premium. If the credit event specified in the contract occurred, then party B would be required to give a sum of money to Party A that would make party A whole.

The Collateralized Mortgage Obligation

The collateralized mortgage obligation (CMO) was created to convert mortgages into different types of bonds addressed to the needs of different types of investors. The plain-vanilla mortgage-backed security is one in which all of the principal and interest payments by mortgage borrowers are simply passed through to the security holders on a *pro rata* basis. Such investors know that there is always a degree of prepayment uncertainty inherent in mortgage-backed securities. People pay off mortgages early for a number of different reasons and when the early payoffs are larger than expected, this means the investors must find a place to reinvest this unexpected capital and the interest rate at which they can reinvest will often be lower than the interest rate on the original mortgage loan. So we have a combination of prepayment risk and reinvestment risk.

The CMO is a much more sophisticated and complex instrument than the simple mortgage pass-through security. With a CMO, the mortgage payments that come into the pool are allocated according to certain rules into one or more of several classes of bonds. For example, a CMO with four classes of bonds might allocate all of the early principal repayments to class I, until all principal due to class I is paid; then the principal repayment flow starts going to class 2 until it is fully paid, and so on. The earlier classes would receive lower interest rates than the later classes because they are much more certain of being paid in full. If there were defaults, this would typically impact the last class that would receive the highest interest rate because of this risk.

So what went wrong? First, housing prices fell because they had been pushed up to an unsustainable level. Second, many mortgages were written for borrowers who were never in any position to pay back the loan. Third, the rating agencies did a poor job of rating instruments such as CMOs and other mortgage-backed securities, with which they had had little experience. And there were many people who simply did not understand the true risk involved. Finally, because OTC swaps were unregulated, there was no reporting on credit default swaps, so neither the government nor the industry realized how many were outstanding. In addition, many of the purchases of CDSs were not the protective hedges they were designed to be, rather speculative bets on a decline in housing prices.

So the bottom line is that there are many great benefits flowing from financial innovation, but there is no doubt that three of these innovations played some role in the financial crisis. Whether the huge regulatory package known as Dodd Frank of 2010 is sufficient to make them safer in the future remains to be seen.

A Case Study or the Role of Innovation in Competition: David Beats Goliath via Aggressive Innovation

One of the most dramatic examples of the role of financial innovation in competitive contexts has to do with the upstart Chicago Mercantile Exchange (CME) overtaking its much larger crosstown rival, the Chicago Board of Trade (CBOT). This is a story of innovation in both products and processes, and how the dogged pursuit of innovation made all the difference.

CBOT Had All the Advantages

Today, CME group is the largest derivatives exchange in the world. But it didn't have to turn out this way, and in fact the CME spent over 90 percent of its 115-year life in the shadow of the CBOT. These are both old exchanges whose major focus has been the listing of futures contracts for trading by hedgers and speculators domestically and eventually worldwide. Initially, the CBOT had all the advantages. It was founded in 1848 and started listing futures contracts in 1865 at the end of the American Civil War. The CME began listing futures contracts in 1898, three decades later, under the name the Butter and Egg Board, reflecting its first two products. With ideas of a broader portfolio of markets, it changed its name in 1919 to the Chicago Mercantile Exchange.

Not only did the CBOT have a head start, it also had much better products. It started with grains, mainly corn and wheat, which had both a national and international market. The CME started with a much more perishable product, butter and eggs, which had only a local market. This means that the potential customer base for the CBOT products extended throughout the grain growing, processing, and distribution area not only in the United States but in Europe as well. Potential customers for the CME's butter and egg futures were much more limited. In fact, if we take a look at the top 20 futures contracts globally in 1955, we would find that CBOT contracts captured three of the top four global spots. The CME on the other hand, had the sixth and fifteenth ranked contracts, and both of these contracts would disappear within a few decades. Out of these top 20 global products, the CBOT had a 30 percent share and the CME had a nine percent share. Looking not just on the top 20 list but at all contracts listed by both exchanges in 1955, the CBOT had almost three times the number of contracts as the CME, eleven versus four.

Had futures markets continued to constrain themselves to agricultural products, the CBOT would still be on the top of the heap and the CME would continue to play the role of the poor cousin. However, the event that changed the game for the entire industry was the birth of financial futures in the 1970s. This was an entirely new playing field, independent of that for traditional futures trading and one that would prove to be much larger and more lucrative. It's difficult to know what the internal thinking was at the world's other exchanges, but in terms of action it was the CME that took the first step into this new world. In the early 1970s, the CME decided that it was going to list a set of foreign exchange futures contracts that would allow businesses and individuals to protect themselves from fluctuations in exchange rates that created uncertainty for importers, exporters, and global investors. Of course, when the CME leadership went to talk to the New York banks to enlist them in the project, they were dismissed as a bunch of pork belly traders who knew nothing about the sophisticated products that were only understood in New York. With the backing of Milton Friedman (a University of Chicago economist who was only a few years away from winning the Nobel Prize), in 1972 the CME launched seven futures contracts on the U.S. dollar price of the German mark, British pound, Japanese yen, Italian lira, Swiss franc, Canadian dollar, and Mexican peso. The Dutch guilder and French franc were also added during the next 2 years.

The driver of this innovation was a change in the global macroeconomic structure. For 26 years, all of the major currencies of the world had been pegged to the dollar and the dollar had been pegged to gold. This was a decision made in the Bretton Woods Agreement in 1944, in which all of the world's major countries came together to plan the post-World War II financial system. The purpose was to make the world safer and more predictable for international trade. It worked for a while, but ultimately couldn't hold because each country needed to pursue its own macroeconomic agenda. The key event was in 1971 when President Nixon revoked the promise to convert dollars into gold at $35 per ounce and devalued the dollar relative to other currencies. This was the beginning of a breakdown of the gold-based system of fixed exchange rates. Suddenly, risks that had been held at bay by an international agreement had to be faced by hundreds of thousands of individual banks, companies, and individuals. And all of those entities now needed some way to help manage the risk.

The Golden Decade of Innovation: 1972 to 1982

The introduction of currency futures ushered in a golden decade of innovation during which four blockbuster financial futures contracts were introduced to the world. Between 1972 and 1982, futures contracts were introduced that helped manage the risk in foreign exchange, long-term interest rate risk, short-term interest rate risk, and stock market risk. The CBOT listed one of these contracts—U.S. Treasury bonds in 1977. The CME listed the other three, which included the seven currency contracts in 1972, the Eurodollar contract in 1981, and a stock index contract based on the S&P 500 in 1982. While no one knew it at the time, this was the springboard for the CME to put itself on a rapid growth path.

If we look at the years 1956 through 2012, we can calculate the average number of new contracts that each major U.S. exchange launched each year (see Table 48-4). What is striking is that when we compare the CME and the CBOT, the CME launched about five new contracts per year while the CBOT launched about two. So the CME was about two and half times more active in product innovation. And this edge of the CME being two to three times more active than the CBOT was true no matter if we focused on the total new listings, or the

Exchange	New Contracts per Year	Innovation	Imitation	Extension
CME	5.09	2.53	1.26	1.3
CBOT	2.05	.84	.68	.54
NYMEX	1.4	.58	.37	.46
ICE	1.76	.8	.9	.06

TABLE 48-4   Average Number of New Contracts per Year by Exchange 1956 to 2012

	Highly Successful	Successful	Moderately	Dead	Total
CME	28%	8%	25%	39%	100%
CBOT	11%	14%	23%	52%	100%
NYMEX	21%	7%	24%	49%	100%
NYBOT	13%	8%	33%	46%	100%
ICE	37%	1%	15%	47%	100%
Others	28%	8%	25%	39%	100%

TABLE 48-5   Success Level by Exchange (% 1956–2006)

actual real innovations, or the cases where one exchange was imitating another exchange, or whether an exchange was creating a product extension of one of its existing futures contracts.

Of course, listing new products is not enough. These new products need a reasonably high level of success if they are going to contribute to the success of the institution. So we also took a look at the degree of success that every new futures contract obtained that was listed between 1956 and 2006 (see Table 48-5). Success was measured by looking at the fifth year of life for each new futures contract and, based on the amount or volume of trading in that fifth year, we assigned contracts to one of four categories: highly successful (more than 1 million traded), successful (between 100,000 and 1 million), moderately successful (between 1 and 100,000), and dead (zero contracts traded). The CBOT excelled in the dead category— 52 percent of all of its new contracts had disappeared by the fifth year compared to 39 percent for the CME. It also excelled in the successful category, which is why it remained a very serious contender for all of these years. The CME beat the CBOT only by a few percentage points in the moderate category. But in the one that counts the most, the highly successful more-than-1-million-contracts traded category, the CME did two and a half times better than the CBOT.

Let us take a look at the battles in each product category. The CBOT did not even compete in the first category of financial products, that is, those based on foreign exchange rates. As you can see in Fig. 48-8, the CME, by being first, was able to bat away competition from

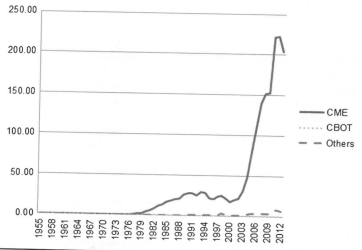

FIGURE 48-8   Volume of FX contracts (in millions).

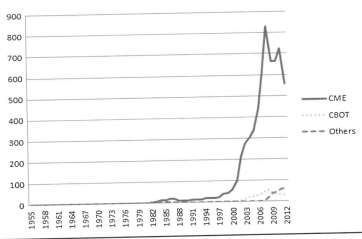

Figure 48-9   Volume of equity index contracts (in millions).

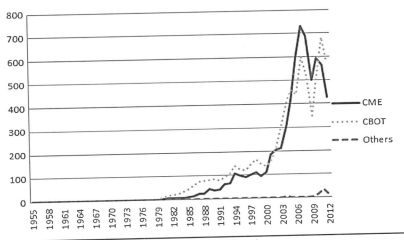

Figure 48-10   Volume of interest rate contracts (in millions).

all of the other exchanges. In the case of futures contracts based on equity indexes, the CBOT did attempt to compete, but not for 2 years after the CME made its entry into this category of product. It was simply too late. In the peak year of 2009, the CME traded over 800 million contracts mainly in S&P 500 futures, while the CBOT and all other exchanges combined traded less than 100 million (see Fig. 48-8).

It was only in interest rates in which the CBOT gave the CME a run for its money. The CBOT was the first innovator and got an early start on long-term debt products, especially U.S. Treasury bonds and U.S. Treasury notes. The CBOT actually stayed ahead of the CME during most of the years since the two exchanges first listed interest rate products in the late 1970s. However, the CME's Eurodollar contract (the interest rate at which large banks in London loan one another U.S. dollars for three months, the rate being referred to as LIBOR) became increasingly important to world financial markets over time and eventually overtook the CBOT's Treasury contracts. As can be seen in Fig. 48-9, the CBOT has stayed right at the CME's heels in this interest rate competition and actually jumped ahead again in 2010 due to a significant drop in the CME's LIBOR business.

So the CBOT had significantly higher trading volumes than the CME for just over a century. Then, because of the CME's more aggressive embrace of financial futures in the 1970s and 80s, we can see from Fig. 48-11, that the CME gradually closed the gap, inched ahead of the CBOT in 2001 and rapidly widened that gap until at one point it was doing twice the volume of its older Chicago brother.

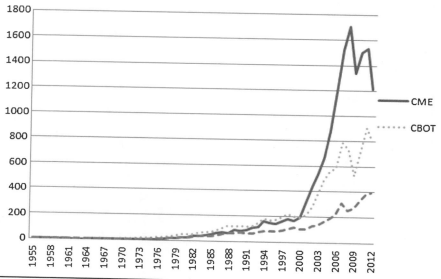

Figure 48-11   Volume of all contracts (millions).

Process Innovation: The Race toward Electronic Trading

The CME's success at besting the CBOT was due not only to innovation in products. It was also due to process innovation, and specifically innovation in technology. One of the most important technological innovations in financial markets was the shift from traditional trading floors to electronic trading. Let's be clear: the member-owned, floor-based exchanges were fearful and dismissive of electronic trading. These owner traders knew how to make a living shouting on floors, but were not so sure how well they would do sitting behind screens, and so they generally went into denial and fought the conversion as long as possible.[28] Both CME and CBOT members disliked electronic trading. The difference was in the leadership of the two exchanges.

Both exchanges knew that they had to solve a problem that could significantly erode their customer base. The problem was that Asia, especially Japan, had awakened to the importance of financial futures trading. The problem was that Japanese customers needed to stay up overnight in order to trade on Chicago markets. At some point the Japanese were likely to list the big volume Chicago products like U.S. Treasuries (at the CBOT) and Eurodollars (at the CME) on exchanges in their own time zone. These products had no intellectual property protection and any exchange worldwide could list them at any time. The objective was to somehow make Chicago markets available during the Japanese day. The two exchanges differed on strategy. The CBOT leadership, reflecting the views of many of its members, decided that the most logical thing to do was to simply have a night shift on the trading floor. It extended the daytime session into the evening, just like a factory would add a second shift when demand dictated. It lasted for a few years but was ultimately unsuccessful. The major brokerage firms were not happy about having to man their booths during evening hours.

CME visionary leader Leo Melamed and Chairman Jack Sander had a more realistic sense of the future and knew that somehow the solution needed to be electronic, but needed to be done in such a way that it did not alienate the members. Member-owned exchanges were democratic in nature. The members either elected both the chairman and the board directly (CBOT), or directly elected the board, which in turn elected the chairman (CME). The two CME leaders sold the membership on the notion of building an electronic trading system by partnering with Reuters which agreed to make available its worldwide network of screen-based customers as well as to cover the costs of implementing the new system. In addition, the members were promised that trading would always be after hours and never encroach on regular trading hours unless the members later voted for it. Finally a major part of the revenues from this new project would be given to CME members.

CME members approved plans for electronic trading in October 1987. While there were many changes in both design and in project partners, the basic vision stayed and 5 years later in June 1992 the new electronic platform, named Globex, went live.

In contrast to this consistent vision at the CME, the CBOT had six different visions of how to handle after-hours trading. The first involved simply adding a night shift on the floor as mentioned earlier. The other five visions involved two projects to build its own system (Aurora and Project A), joining the CME in Globex, and leasing another exchange's electronic platform (Eurex and LIFFE).

So while the CBOT spends over a decade and a half jumping from one solution to another, the CME continues to focus on the single platform, Globex, and continues to build greater quality, list more products on it, and thus substantially increase trading volume.

The endgame is that after years of aggressive product innovation and focused process innovation, the CME ends up with a very attractive product suite actively traded on a state-of-the-art electronic trading system. Trading volume and revenue grow much more rapidly at the CME and in 2007, the CME buys the CBOT. A year later the CME buys the New York energy exchange NYMEX and after the two mergers finds itself with a 97 percent market share in U.S. futures trading.

One final note. While it is clear that the CME was much more successful than the CBOT and it is clear that success was grounded in a culture of innovation, one can't help but ask why. Why does one financial company innovate much more aggressively than another? We can't generalize about all financial companies, but we know from direct observation[29] that the CME was hungrier, more open to ideas, and willing to work much harder because its members and staff and especially its leadership was not content to be number two. The CBOT on the other hand had been number one for 100 years. It's almost impossible to go that long and not become a bit arrogant and complacent. Successful innovation happened, at least in this case, because it was driven by hungry owners and visionary leaders.

References

1. We are grateful to the Futures Industry Association for providing 57 years of data from their annual volume reports.
2. Tufano, P. Financial innovation. In Constantinides, George M., Harris, M., and Stultz, Renée M., eds. *Handbook of the Economics of Finance: Corporate Finance*. Volume 1 a, North Holland, p. 310, 2003.
3. Scott, W. and White, L. "Empirical Studies of Financial Innovation: Lots of Talk Little Action?" *Journal of Economic Literature*, 42 (1): 118, 2004.
4. Merton, R.C. "Financial Innovation and Risk Management." *Journal of Banking & Finance*, 19 (2): 461–481, 1995.
5. Walmsley, J. *New Financial Instruments*. 2nd ed. John Wiley & Sons Inc., p. 4, 1998.
6. Finnerty, J.D. "An Overview of Corporate Securities Innovation." *Journal of Applied Corporate Finance*, 4 (4): 23–39, 1992.
7. Yago, G. and Barth, A. *Financing the Future: Market Based Innovations for Growth*. Pearson Education, 2012.
8. De Mieroop, Marc Van. The invention of interest: Sumerian loans. In Goetzmann, William N. and Rouwenhorst, K. Geert, eds. *The Origins of Value: the Financial Innovations That Created Modern Capital Markets*. Oxford University Press, p. 18, 2005.
9. Malmendier, U. Roman shares. In Goetzmann, William N. and Rouwenhorst, K. Geert, eds. *The Origins of Value: the Financial Innovations That Created Modern Capital Markets*. Oxford University Press, p. 31–42, 2005.
10. The basic features of a bond include: a principal (or par value or face value), a maturity, and a fixed rate of interest payable on a firm schedule. So the bond holder knows in advance all of the payments that he will receive, which is why bonds are referred to as fixed income instruments.
11. Franklin, A. and Douglas, G. *Financial Innovation and Risk Sharing*. Massachusetts Institute of Technology, p. 12, 1994.
12. Goetzmann, William N. and Rouwenhorst, K. Geert. *The Origins of Value: the Financial Innovations That Created Modern Capital Markets*. Oxford University Press, p. 4, 2005.
13. Miller, M. "Financial Innovation: the Last 20 Years and the Next." *Journal of Financial and Quantitative Analysis*, 21 (4), 1986.
14. Silber, William L. "The Process of Financial Innovation." *American Economic Review*, 73 (2): 89–95, 2001.
15. Franklin, A. and Douglas, G. *Financial Innovation and Risk Sharing*. The MIT Press, p. 12, 1994.
16. Interest Equalization Tax Extension Act of 1971, U.S. Senate, 92nd Congress 1st session, Calendar No. 50, Report No. 92–47, p. 12.
17. Miller, M. "Financial Innovation: the Last 20 Years and the Next." *Journal of Financial and Quantitative Analysis*, 21 (4): 459, 1986.

18. The U.S. dollar was on one side of 84.9% of all foreign exchange trading according to the most recent Triennial Central Bank Survey: Report on Global Foreign Exchange Market Activity in 2010, by the Bank for International Settlements. http://www.bis.org/publ/rpfxf10t.pdf

19. Persons, John C. and Warther, Vincent A. "Boom and Bust Patterns in the Adoption of Financial Innovations." *The Review of Financial Studies*, 10 (4): 939–967, Winter 1997.

20. Mester, Loretta J. "Innovation and Regulation in Financial Markets: A Summary of the 2007 Philadelphia Fed Policy Forum." *Philadelphia Federal Reserve Bank Business Review*, Q3, 2008.

21. Zuckerman, G. *The Greatest Trade Ever: the Behind-The-Scenes Story of How John Paulson Defied Wall Street and Made Financial History.* Crown Business, 2010.

22. There are two shortcomings of this analysis. First, our data are annual. This means that a new futures contract we coded as lasting one year, could have lasted only a few days or 365 days. A contract in which we saw volume reported in two calendar years, and which we coded as lasting two years, might have been listed in December of the first year and died in January of the second year, thus having an actual lifespan of only a few days or weeks. Another contract coded with a two-year lifespan might have started in January of the first year and died in December of the second year, thus having two full years of life. We trust that for most purposes of analysis, the actual life spans are randomly distributed and the errors balance out.

23. In 2012, the product extension E-mini S&P 500 traded 474 million contracts, while the original, large S&P 500 product traded only 5 million contracts.

24. Pokotilow, Stephen B. and DiBernardo, Ian G. "Protection for Financial Indexes, ETF's, Other Products." *New York Law Journal*, 236 (63), 2006.

25. Gorham, M. and Singh, N. *Electronic Exchanges: the Global Transformation from Pits to Bits.* Elsevier, p 52, pp. 61–62, 2009.

26. Data from the Market Data section of the OCC's website. http://www.optionsclearing.com/market-data/.

27. Gorham, M. and Singh, N. Liquidity driven monopoly. In *Electronic Exchanges: the Global Transformation from Pits to Bits.* Elsevier, 2009, and Gorham, M. "Product Innovation, Clearing and Competition among U.S. Derivatives Exchanges." *Global Markets Law Journal*, 1 (1), Fall 2012. globalmarkets.jmls.edu/articles/show/53.

28. This denial, fear and anger are beautifully portrayed in the 2010 documentary film called *Floored*, that premiered to sold-out audiences in Chicago. One trader in the film is heard saying "I hate email. I hate computers…the computer is the worst and most evil thing of trading that I've ever seen….It's more devious…trust me when I say these people are cheating on the computer…The people who are making money are the programmers."

29. One of the authors, Michael Gorham, held a variety of research and marketing positions at the Chicago Mercantile Exchange from 1979 through 1997, was director of the Division of Market Oversight at the Commodity Futures Trading Commission, from 2002 to 2004, and has continued to be involved with the derivatives industry up until the present.

CHAPTER 49

Case Study: Technology Innovation within Government

Suzanna Schmeelk

E-Government Movement across the Globe

As we know, technology is revolutionizing communication, activities, and knowledge on our planet. We have been witnessing over the past two decades a slow government adoption of technology. Innovations, affecting the public sector, in government can be categorized within the e-government movement. Almarabeh and AbuAli, define e-government as, "Like other concepts of contemporary there are multiple definitions of e-government among researchers and specialists, but most of them agreed to define electronic government as government use of information communication technologies to offer citizens and businesses the opportunity to interact and conduct business with government by using different electronic media such as telephone touch pad, fax, smart cards, self-service kiosks, e-mail or Internet, and EDI. It is about how government organizes itself: its administration, rules, regulations, and frameworks set out to carry out service delivery and to coordinate, communicate, and integrate processes within itself."[1] We use this definition throughout the chapter.

There are many motivating factors to the gradual transition to e-government. Northrup and Thorson (2003) discovered three positive claims used to support the e-government initiative: increased efficiency, increased transparency, and transformation.[2] Capacity[3] and system architecture are important aspects of e-government development. Capacity, security—confidentiality, integrity, and availability—and performance[4] should be underlying concerns to e-government innovation.[5–7]

This research chapter describes recently published case studies and technology innovations to help the reader understand the current state of public sector government technology innovation. The research first explores the global state of e-government and, then, broadens the research to explain e-government frameworks.

The Global E-Government Movement

In 2012, the Department of Economic and Social Affairs (DESA) within the United Nations published its e-government survey.[8,9] The survey assessed government web portals of all 193 United Nation's entity countries.[9] The survey examined interactive, transactional, and e-participation e-service features of e-government portals. The DESA accordingly ranked each country's e-government models with an e-government development index (EGDI), "a weighted average of three normalized scores on the most important dimensions of e-government: scope and quality of online services, development status of telecommunication infrastructure, and inherent human capital."[8] The United Nations found the top ranking country to be North Korea, as it did in 2010.[10] Following Korea was the Netherlands and, then, the United States. Other United Nations surveyed e-government top countries were France, Sweden, United Kingdom, Denmark, Israel, Australia, Norway, and New Zealand.[8]

Canada, like the United States of America, has introduced e-government and web-mediated citizen-to-government interaction. Parent et al.'s research, "building citizen trust through e-government,"[11] explored public views, beliefs, and feeling s about the Canadian government's electronic movement and found that using the government's electronic points-of-access had a positive impact on trust and external political efficacy.

Similar to the United States of America, Canada's e-government movement started with initial one-way information-based websites. Examples of such information-based sites include posting contact and program information. The next big step was to offer a one-way point-of-access for delivering public information, which includes such sites as those for tax filing and bill payments. Finally, gradual shifts toward web-based public services and "customer-centered"[11] websites are transpiring.

China is another country moving steadily to e-government. Wenhu and Jian's (2010) research paper, "e-government and the change of government management mode"[12] and Chen Jianmei's (2011) research paper, "the construction of efficient e-government to establish brand-new service-oriented government management model,"[13] explore the transition from internal private sector e-government to public citizen-based service-oriented e-government. The Wenhu and Jian's research paper first explores key issues in China's e-government management systems, including administration, social development, legislation, planning, and security, that limit quick e-government development. The research then explains next-step policy issues that need to be overcome for future e-government development. Jianmei's research paper explores the necessity of service-oriented government as it pertains to understanding customer demand, preferences, and service delivery. Both papers help distinguish motivating outcomes of a service-oriented e-government. Additionally, China has published quite a few articles on how they are (contemplating) using e-government systems to manage electric power,[14] accounting,[15] water crisis management based on water pollution,[16] urban traffic,[17] and public crisis management.[18] All applications show an increasing trend toward e-government.

Australia is at the forefront of e-government initiatives. Shackleton et al.'s (2004) research, "evolution of local government e-services: the applicability of e-business maturity models,"[19] used a two-stage research approach to evaluate federal and state websites. At the time of their study, they found that most e-government websites fell into one of four general categories: e-management, e-service, e-commerce, and e-decision. The second stage of their research quantitatively evaluated sites within the same category. The evaluation found that the strength of e-management sites, (sites based on branch-related public information to assist a citizen and/or resident), lay in providing information; their weakness lay in providing news and coming events. It found that the strength of e-service sites, (sites based on assisting a citizen and/or resident to provide information about a service or perform the service itself), lay in service details, while the sites' weakness lay in service tracking, FAQs, and e-mail support. In addition, it revealed that e-commerce sites, (sites based on transaction handling for placing orders for services or products over the web), had slight strengths in online payments but weaknesses in ordering facilities and e-mail. Finally, the evaluation found that e-decision sites, which are set up to provide broad issues about the operation of the council, were strong in links to other organizations or businesses but weak in community information and bulletin boards.

India is yet another country that is transitioning to e-government. In 2010, the government announced that it was going to create a unique identity number for every resident in India.[20–23] The magnitude of the project is to enroll *over 2 billion people*, as India is the world's fourth largest economy.[24] The identity number given to citizens is a 12-digit number based on the 10 fingerprints, face, and both iris scans. The enrollment data is multimodal to alleviate discrepancies based on eye colors (such as dark eyes and missing irises that are hard to distinguish), religion (when the individual wears a religious head covering that affects the facial image), and hand-laborers that may not have lingering fingerprints. The Unique Identification Authority of India (UIDAI) can then track patterns and provide services to citizens.[20] To date, the unique identity number has been used in health programs,[22] school attendance,[22] legal court cases,[22] financial services,[22] public distribution,[25] and signature identification.[26–28]

E-Government Innovation in the United States of America

In the United States of America, e-government became an official direction when the Senate passed the E-Government Act of 2002.[29] The Act's motivation was to improve management, improve citizen access, and reduce long-term, growing costs. It helped agencies establish performance measures, demonstrating how the e-government initiative enables progress toward the agencies' goals, including customer satisfaction, productivity, adoption of innovative information technology (IT), and performance improvement in delivering programs to constituencies.[30]

The Act created a position within the Office of Management and Budget to unify e-government management. The position was given the title of Federal Chief Information Officer (CIO). The United States now had a CIO to help manage its e-information.

In March 2009, President Obama appointed the first United States CIO, Vivek Kundra.[31] Kundra, prior to becoming the US-CIO, had worked on or inspired many e-government technologies including D.C. Data Catalogue,[32] Apps for Democracy,[33] New York City's Big-Apps,[34] San Francisco's Data Portal,[35] and a project management system using Stock Market Portfolio Management tools, among others. As the United States CIO, Kundra made it a priority to develop open, transparent, cost-effective, innovative, secure, democracy-driven technologies. He launched the Federal IT Dashboard[36] and Data.gov[37] in 2009.

Data.gov, (Fig. 49-1) launched in spring 2009, was developed to provide public access to raw Federal Executive Branch datasets. Data.gov provides federal data in both raw format and geographic format. The site currently supports over 200 datasets from over 40 government agencies, including cross-continent wages, retirement benefits, White House visitor requests, medical facility information, and others. The data can be manipulated and downloaded as discussed on the website. Cities across the United States and the world have launched conceptually parallel websites providing public access to government datasets.

Data.dc.gov (Fig. 49-2) is an information-based website providing District of Columbia operational data. The website provides the public with over 497 datasets including crime, property, public notary, school attendance, metro stations, zoning, snow removal, and transportation features. The Java-based site provides a link to an IBM visualization tool, Many Eyes, to help graphically interpret the data.[38] Additionally, the public can interface with various forms of the datasets, such as web-feed data in Atom and Rich Site Summary (RSS) format, text-based format, xml-based format, Environmental Systems Research Institute (ESRI) ASCII raster format [a language used by Geographic Information System (GIS) software], and Keyhole Markup Language (KML), a language used by Google Earth and Google Map documents.

AppsForDemocracy.org (Fig. 49-3) was launched in 2009 and 2010 to spur District of Columbia area public development of applications that use open data given by Data.dc.gov. The competition cost approximately $50,000 and returned 47 iPhone, Facebook, and web applications to the general public. Example applications created by the competition were bike maps, a red light watch to add insight into which red lights have cameras watching and ilive.at, an application that adds insight into life-style components around a selected geographic location.

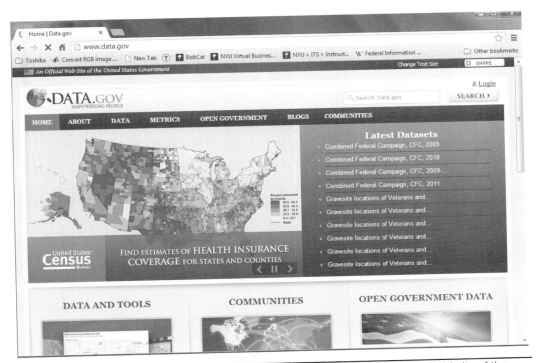

Figure 49-1   Data.gov United States Government Open Access Datasets: An Official Website of the United States Government.

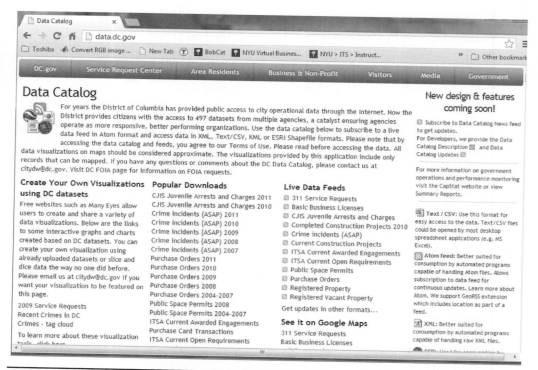

FIGURE 49-2   District of Columbia Open Access Datasets.

FIGURE 49-3   Apps for Democracy.

In New York City (NYC) (Fig. 49-4), a Big Apps 3.0 event held in 2011 created 96 public applications that used government provided NYC public information.[39] The event was sponsored by New York City Economic Development Corporation and the New York City Department of Information Technology & Telecommunications. The motivation behind the event was to increase both government transparency and access as well as NYC life-style services.

FIGURE 49-4   New York City Big Apps Challenges.

The winner apps included *NYCFacets*, an application that provides access to standardized NYC public data; *Work+*, an application to help find a great place to work based on personal requirements; and *Funday Genie*, an application to help plan a fun day-off.

Data.sfgov.org (Fig. 49-5) is an information-based website providing links to San Francisco's public-domain government data. Similarly to the District of Columbia, its public-domain data includes crime, film locations, political boundaries, neighborhood planning, and registered

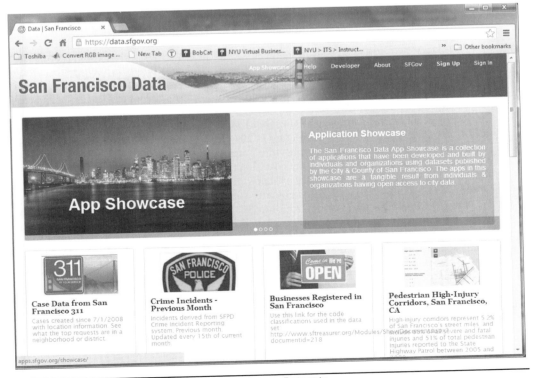

FIGURE 49-5   San Fransisco in California Open Access Datasets.

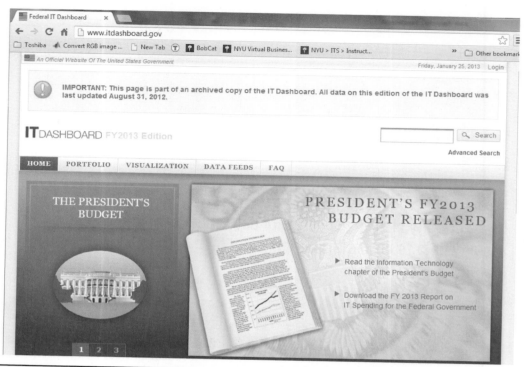

FIGURE 49-6   IT Dashboard: An Official Website of the United States Government.

businesses. Additionally, the site sponsors links to applications developed for traffic, crime, parks, playgrounds, tree guides, recycling facility locations, city parking information, and transportation maps and times, among others.

The it.usaspending.gov (Fig. 49-6) dashboard,[36] launched in summer 2009, provides a methodology for tracking federal IT spending within the larger government spending tracking, USASpending.gov.[40] The IT dashboard provides the public with a portfolio of federal IT spending, visualization tools to interpret the data and direct data feeds to export the data. Recently, the government provided the dashboard source code to the general public. Download statistics show top countries where the dashboard has been downloaded. For the year 2013, the United States of America is the top download country followed by India, Japan, Brazil, Spain, France, and Indonesia.[36]

The USASpending.gov[40] (Fig. 49-7)site is a wider initiative by the federal government to publically track federal spending. Data presented on the website include data presented by over 20 agencies on museum spending, oil spill contracts, construction contracts, housing assistance, student assistance, and higher education grants. In winter 2010, Techstat was launched for accountability sessions regarding IT spending.[41]

Apps.gov, now Info.Apps.gov,[42] (Fig. 49-8) was launched by the General Services Administration (GSA) under Kundra's direction in 2009 to provide cloud services to federal agencies. The website was created to provide a unifying website where agencies could find pertinent information about potential cost-effective cloud computing service solutions. The information-based website currently explains cloud computing, federal and state case studies and how to order services. Apps.gov also interfaces with a federal cloud security initiative, the Federal Risk and Authorization Management Program (FedRAMP).[43] FedRAMP (Fig. 49-9) must be used by federal agencies when authorizing cloud services. The site explains that "FedRAMP is a government-wide program that provides a standardized approach to security assessment, authorization, and continuous monitoring for cloud products and services."[43] Currently, FedRAMP has a three-process model to allow agencies to use cloud services.

In 2011, President Obama appointed the second federal CIO, Steven VanRoekel.[44] VanRoekel (Fig. 49-10) was appointed partly because he had already adapted, developed, managed, and updated many e-government technologies while managing director of the Federal Communications Commission (FCC). At the FCC, he directed the relaunch of the agency's public web presence,[45] began a hugely followed agency twitter account,[46] initiated an agency "developer" website,[47] which provides standard application programming interfaces (APIs) to access FCC public data, launched the agency's National Broadband Map,[48]

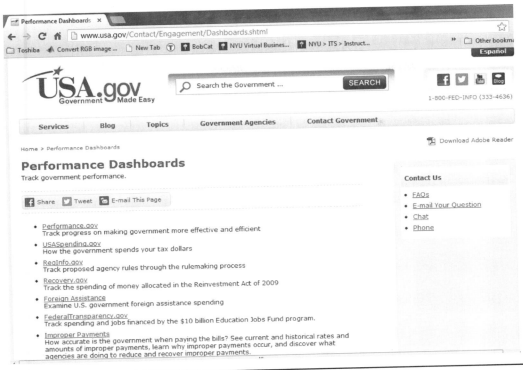

FIGURE 49-7 Usa.gov The United States Government Official Web Portal.

FIGURE 49-8 Formerly apps.gov, currently cloud.cio.gov.

used social media tools to accept public comments, and created an agency-specific employee technology experience center (TEC).

The National Broadband Map (Figs. 49-11 and 49-12) provides the public with an option to view or analyze federal broadband data. Its maps are based on statistics about broadband availability around the country. Map visualization data include maximum advertised broadband speeds, broadband technologies, service areas and providers, and broadband availability

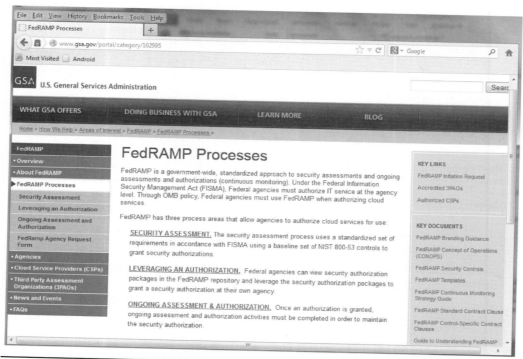

FIGURE 49-9   United States General Services Administration FedRAMP Process.

FIGURE 49-10   United States CIO Twitter Account.

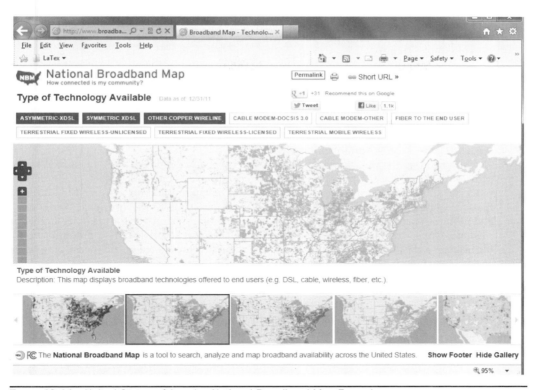

FIGURE 49-11   United States of America National Broadband Map Example.

FIGURE 49-12   United States of America National Broadband Map.

at community anchor institutions (schools, libraries, medical facilities, and public safety enti-ties). Federal data also can be analyzed, ranked, and summarized based on geographic loca-tions. The website provides the public with current knowledge of the state of broadband throughout the country.

Recently the United States Government Accountability Office (GAO) published its report to the Committee on Homeland Security and Governmental Affairs, U.S. Senate (2012) on "Electronic government Act: agencies have implemented most provisions, but key areas of attention remain."[30] The study examined federal government efforts to participate in and to comply with the E-Government Act of 2002. The GAO examined 24 executive branch agencies as well as the Office of Management and Budget (OMB) and the National Archives and Records Administration (NARA).

According to the GAO report, "The E-Government Act codified the CIO council as the principle interagency forum for improving agency practices on matters such as the design, modernization, use, sharing, and performance of agency information resources. In addition, the act requires the CIOs of each of the 24 agencies to participate in the functions of the council and monitor the implementation of information technology standards for the federal govern-ment developed by the National Institute of Standards and Technology and promulgated by the Secretary of Commerce, including common standards for interconnectivity and interoper-ability, categorization of federal government electronic information, and computer system effi-ciency and security."[30] One website developed by the CIO council (Fig. 49-13) facilitates the exchange of IT management best practices through its website, cio.gov.[49] Through this public website, the CIO council provides guidance for the IT community (from the OMB) and docu-ments and presentations created by the council committees. Additionally, the website includes pertinent information for both committee members and the general public.

In January 2007, USA.gov[50] (Fig. 49-14) became the federal Internet portal prescribed by the E-Government Act of 2002.[30] Until 2007, the portal had been known as *FirstGov.gov* and the general public had requested an easier to remember uniform resource locator (URL). The web-site features a partial list of government agencies and bureaus organized by services rather than agencies, access to government information and services, expanded search functionality, and mobile applications. The website (Fig. 49-15) features a guest blog, Blog.USA.gov.[51]

Three innovative e-government websites produced as a response to the E-Government Act of 2002 are publications.gov (Fig. 49-16), science.gov (Fig. 49-17), and grants.gov (Fig. 49-18). Each website conveys government activities and funding opportunities to the general public. For example, the publications website[52] features the most popular government publications, and the science portal and repository houses information on federal research and provides links to science websites and scientific databases.[53] Interestingly, the GAO found that as of 2012, 11 of 24 government agencies provided research information to Science.gov.[30]

FIGURE 49-13   The website of the U.S. Chief Information Officer and the Federal CIO Council.

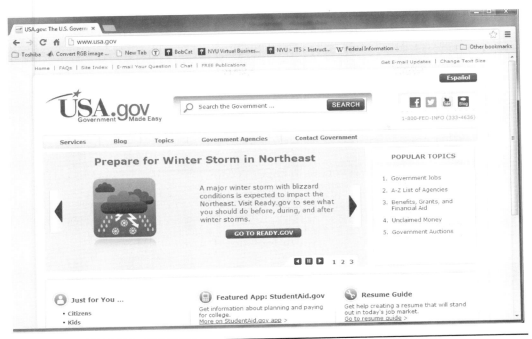

FIGURE 49-14   The United States Government's Official Web Portal.

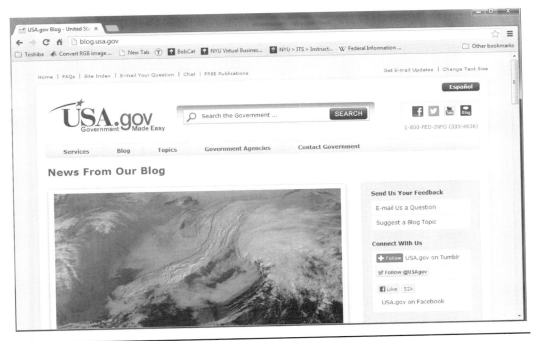

FIGURE 49-15   United States Government Blog.

The information and service-based portal, Grants.gov,[54] provides information "on over 1000 grant programs awarded by 26 grant-making agencies and other federal grant-making organizations."[30] Agencies reported that the grant-making web-portal reduced application submission error rates by performing data validation checks, reducing respondent burden, and improving internal efficiencies.

Another requirement of the E-Government Act of 2002 "required federal agencies to electronically accept submissions of comments on proposed rules, among other things, and to make electronic dockets publicly available online."[30] Many agencies took actions to address the requirements by participating in regulations.gov (Fig. 49-19). The website allows the general public to search proposed regulations and comment on proposed regulations, and it provides public access to electronic dockets.[55] As an example of the government's

FIGURE 49-16   Publications based on categories supported by the United States Government.

FIGURE 49-17   Science.gov is a gateway to United States Federal Science.

websites' popularity, the GAO cited that the general public in 2011 posted over 500,000 public comments on the Environmental Protection Agency (EPA) through the website.

Ready.gov[56] (Fig. 49-20) is a web-portal sponsored by the Federal Emergency Management Agency (FEMA). The website provides information in approximately 10 languages on what to do during an emergency. It helps the general public make a plan to prepare for emergencies, stay informed during emergencies, and get involved with various emergency components. Additionally, it helps adults and children to build necessity kits for different types of emergencies. Finally, the portal provides e-mail notifications, lists of publications, and volunteer opportunities.

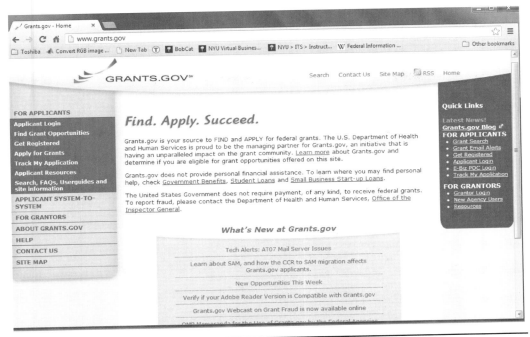

FIGURE 49-18   Grants.gov is a place to find and apply for federal grants.

FIGURE 49-19   Regulations.gov is a source for information on the development of Federal regulations and other related documents issued by the U.S. government.

The United States Department of Education also has an innovative portal for the public to access, Ed.gov.[57] The website (Fig. 49-21) provides a plethora of information for teachers, parents, families, schools, and students as well as links to federal funding, accreditation,[58] policies (Fig. 49-22), research, and news.

E-Governance Frameworks

The federal government models of e-government have been categorized in many ways. The Government Accountability Office (GAO) in 2001,[59] categorized four high-level models of electronic government: government-to-citizen (G2C), government-to-employee (G2E),

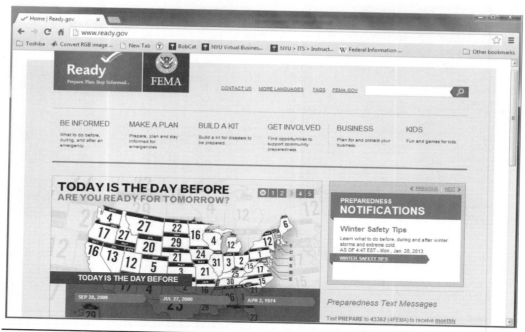

FIGURE 49-20   Ready.gov is sponsored by Federal Emergency Management Agency (FEMA).

FIGURE 49-21   United States Department of Education Website.

government-to-government (G2G), and government-to-business (G2B). G2C government encompasses citizen-based services. Examples of such services are Savings Bond Direct and bill payment. G2E comprises ways for federal employees to interact with the government. Examples of such services are payroll, timesheets, savings plan, and overarching employee portals. G2G government involves internal government interaction. An example of such interaction is the National Environmental Information Exchange Network that connects different state systems with the Environmental Protection Agency's system using a common secured Internet connection. G2B provides point-of-access to business-related interactions.

FIGURE 49-22 United States Department of Education database of accredited postsecondary institutions and programs.

An example of a G2B model is the federal business opportunities website, fbo.gov,[60] which is a point-of-access to federal business opportunities involving more than $25,000.

The Office of Management and Budget (OMB) classified e-government service models using all but one of the terms used by the GAO. The OMB classified the fourth model G2E as internal efficiency and effectiveness (IEE). The IEE, "brings commercial best practices to key government operations, particularly supply chain management, human capital management, financial management, and document workflow."[61]

Hiller and Belanger's (2001) research, "privacy strategies for electronic government,"[62] categorized e-government into six categories. They retained G2G and G2E categories used by the GAO and expanded the G2C and G2B categories into four additional categories: government delivering services to individuals (G2IS), government to individuals as a part of the political process (G2IP), government to business as a citizen (G2BC), and government to business in the marketplace (G2BMKT). The two e-government categories, G2IS and G2IP, distinguish between the services the government provides to its citizens, and communications as part of the political process. The two e-government categories, G2BC and G2BMKT, expand the GAO and OMB's G2B, as the former includes financial services (e.g., paying taxes and filling SEC reports), and the latter includes e-procurement.

Carter and Belanger's (2004) research, "citizen adoption of electronic government initiatives,"[61] showed that perceived usefulness, relative advantage, and compatibility were indicators of a citizen's intention to use e-government services. The researchers sampled 140 undergraduate students with an average age of 19 at a southeastern research university. Subjects had an average of 9 years' experience using a computer as a tool. The research found that federal marketing for agency-specific services could lead to an increased perception of usefulness.

Klischewski's (2003) research paper, "semantic web for e-government,"[63] brought fourth four important categories for e-government website development: organizational cost and benefit, user involvement, technical integration, and implementation strategy. They suggest that following e-government innovation in these four categories can help guide and support the future application development of semantic web e-government technologies.

This research chapter maps out e-government innovation around the globe. As we saw with the United Nations study, many countries are delving deep into e-government innovation. We then presented an overview of public sector e-government innovation in the United States of America. The end of the chapter helped summarize ways to standardize e-government terms and access models. We hope you enjoyed reading the chapter as much as we enjoyed writing it!

References

1. Almarabeh, T. and A.A. "A General Framework for E-Government: Definition Maturity Challenges, Opportunities, and Success." *European Journal of Scientific Research*, 39 (1), 2010.

2. Northrup, T.A. and Thorson, S.J. "The Web of Governance and Democratic Accountability. in System Sciences." *Proceedings of the 36th Annual Hawaii International Conference*, 2003, IEEE.

3. Kim, H.J. and Bretschneider, S. "Local Government Information Technology Capacity: An Exploratory Theory." *Proceedings of the 37th Annual Hawaii International Conference on System Sciences (HICSS'04)—Track 5—Volume 5* 2004, IEEE Computer Society, 50121.2.

4. Xiao-Lan, Z., et al. Performance evaluation index system of government management information construction. *Machine Learning and Cybernetics*. International Conference, 2009.

5. Whitman, M.E. and Mattord, H.J. *Principles of Information Security*. Course Technology Press, 2004.

6. Wang, J.F. E-government security management: key factors and countermeasure. *Information Assurance and Security IAS '09*. Fifth International Conference, 2009.

7. Wang, K., et al. E-government safety management based on data encrypt. *Industrial Control and Electronics Engineering (ICICEE)*. International Conference, 2012.

8. Affairs, U.N.D.o.E.a.S. United Nations E-Government Survey 2012: E-Government for the People, 2013.

9. Nations, U. E-Government Survey 2012: E-Government for the People, 2012.

10. Government, K. Korea.net Gateway to Korea, 2013.

11. Parent, M., Vandebeek, C.A., and Gemino, A.C. "Building citizen trust through e-government." *Government Information Quarterly*, 22 (4): 720–736, 2005.

12. Xu, W. and Ye, J. E-government and the change of government management mode. *E-Business and E-Government (ICEE)*. International Conference, 2010.

13. Chen, J. The construction of efficient e-government to establish brand-new service-oriented government management model. *Management Science and Industrial Engineering (MSIE)*. International Conference, 2011.

14. Zhu, B., et al. Research on the government management pattern of the Chinese electric power build-operate-transfer project application. *Management Science and Engineering, 2006, ICMSE '06*. International Conference, 2006.

15. Yue, Q., Yu, C., and Sun, Y. Accounting information sharing in government management. *Multimedia Technology (ICMT)*. International Conference, 2010.

16. Dajin, Y. Study on the innovation of China government's water crisis management based on water pollution occurred frequently. *Mechanic Automation and Control Engineering (MACE)*. International Conference, 2010.

17. Hong, Z., Rui-ping, Y., and Jun, L. Research on construction of e-government platform based on urban traffic management. *Information Management, Innovation Management and Industrial Engineering (ICIII)*. International Conference, 2011.

18. Liu, C.Z. and Gao, J. Public crisis management and government responsibility. *Information Systems for Crisis Response and Management (ISCRAM)*. International Conference, 2011.

19. Shackleton, P., Fisher, J., and Dawson, L. "Evolution of Local Government E-Services: The Applicability of E-Business Maturity Models in System Sciences." *Proceedings of the 37th Annual Hawaii International Conference*, 2004, IEEE.

20. (UIDAI), U.I.A.o.I. UIDAI Strategy Overview: Creating a Unique Identity Number for Every Resident in India, April 2010.

21. Turbeville, B. Cashless Society: India Implements First Biometrics ID Program for All of Its 1.2 Billion Residents, 2012.

22. Paik, M., et al. "A Biometric Attendance Terminal and Its Application to Health programs in India." *Proceedings of the 4th ACM Workshop on Networked Systems for Developing Regions*, ACM: San Francisco, California, 1–6, 2010.

23. Romero, J.J. "Fast Start for World's Biggest Biometrics ID Project." *Spectrum, IEEE*, 48 (5): 11–12, 2011.

24. Ricanek, K. "Dissecting the Human Identity." *Computer*, 44 (1): 96–97, 2011.

25. Krishnan, A., Raju, K., and Vedamoorthy, A. Unique Identification (UID) based model for the Indian Public Distribution System (PDS) implemented in Windows embedded CE. *Advanced Communication Technology (ICACT)*. 13th International Conference, 2011.

26. Singh, A., et al. The ultimate signature identifier? *Electronics Computer Technology (ICECT)*. 3rd International Conference, 2011.

27. Kore, S. and Apte, S. "The Current State of Art: Handwriting a Behavioral Biometric for Person Identification and Verification." *Proceedings of the International Conference on Advances in Computing, Communications and Informatics*, ACM: Chennai, 925–930, 2012.

28. Govinda, K. and Ravitheja, P. "Identity Anonymization and Secure Data Storage Using Group Signature in Private Cloud." *Proceedings of the International Conference on Advances in Computing, Communications and Informatics*, ACM: Chennai, 129–132, 2012.

29. Turner, R.J. *H.R.2458: E-Government Act of 2002*, 2001.

30. (GAO), U.S.G.A.O. Electronic Government Act: Agencies Have Implemented Most Provisions, but Key Areas of Attention Remain. Report to the Committee on Homeland Security and Governmental Affairs, U.S. Senate, September 12, 2012. GAO-12-782: 50.

31. Government, U.S.F. Our Nation's First Federal CIO. Whitehouse.gov, 2011.

32. Government, D.o.C. District of Columbia Public Data Repository, 2013.

33. Community, D.o.C. Apps for Democracy, 2009.

34. Community, N.Y.C. NYC Big Apps, 2011.

35. Government, S.F.C. San Fransisco City Public Data Repository, 2013.

36. (OMB), U.S.O.o.M.a.B. Federal IT Dashboard, 2013.

37. (OMB), U.S.O.o.M.a.B. Federal Datasets, 2013.

38. IBM. Many Eyes: Data Interpreter, 2013.

39. York, C.o.N. NYC Open Data, 2013.

40. (OMB), U.S.O.o.M.a.B. Federal Government Spending, 2013.

41. Government, U.S. PortfolioStat: Saving Billions in IT Spending, 2012.

42. (GSA), U.S.G.S.A. A Place Where Agencies Can Gather IT Information, 2013.

43. (GSA), U.S.G.S.A. Federal Risk and Authorization Management Program (FedRAMP), 2013.

44. Government, U.S.F. Office of Management & Budget Members: Steven VanRoekel. cio.gov, 2013.

45. (FCC), F.C.C. FCC Website, 2013.

46. VanRoekel, S. Twitter Account, 2013.

47. (FCC), F.C.C. FCC Developer Website, 2013.

48. (FCC), N.T.a.I.A.N.a.F.C.C. National Broadband Map, 2013.

49. Council, U.S.C.I.O.C. Federal Chief Information Officer Website, 2013.

50. (OMB), U.S.O.o.M.a.B. Federal Internet Portal, 2013.

51. (GSA), U.S.G.S.A. Federal Internet Blog, 2013.

52. (GSA), U.S.G.S.A. Federal Most Popular Publication List, 2013.

53. (OMB), U.S.O.o.M.a.B. Federal Research, 2013.

54. (OMB), U.S.O.o.M.a.B. Federal Grants Website, 2013.

55. Agencies, P.F.G. Proposed Federal Government Regulations, 2013.

56. (FEMA), F.E.M.A. Federal Emergency Plan, Prepare and Information Website, 2013.

57. Education, U.S.D.o. US Department of Education Webiste, 2013.

58. Education, U.S.D.o. The Database of Accredited Postsecondary Institutions, 2013.

59. (GAO), U.S.G.A.O. Electronic Government: Challenges Must Be Addressed with Effective Leadership and Management. Testimony Before the Committee on Governmental Affairs, U.S. Senate, 39, 2003.

60. Government, U.S.F. FedBizOpps.gov, 2013.

61. Carter, L. and Belanger, F. "Citizen Adoption of Electronic Government Initiatives in System Sciences." *Proceedings of the 37th Annual Hawaii International Conference*, 2004, IEEE.

62. Hiller, J.S. and Belanger, F. "Privacy Strategies for Electronic Government." *E-Government*, 173, 2001.

63. Klischewski, R. Semantic web for e-government. *Electronic Government*. Springer Berlin Heidelberg. 288–295, 2003.

CHAPTER **50**

Case Study: Technology Innovation within Education

Suzanna Schmeelk

Technology Innovation in Education

Education is radically evolving—predominantly due to ongoing technology advancements. Traditionally, education has been a top-down linear process where a student receives previously selected material. The student is, then, expected to read it, perform some exercises, and perhaps be "evaluated" on the material. In the traditional sense, a student might need to borrow a car to visit either a local library or a tutor so that he or she could look for additional resources to either find additional material or to look for a different discussion of the same material. This is the traditional educational model, which is slowly, but surely, dynamically changing.

This paradigm was my personal experience until I entered my second graduate school program at Rutgers University in New Jersey. At Rutgers, I was offered many additional resources—centers for learning, a plethora of computer labs, software, consortiums, seminars, smartphone apps, and extended inner-library loan resources from universities around New York City. Suddenly, my learning experience became dynamic and from many, many sources. My personal learning was no longer confined by a top-down linear process; it was organic and increasing at an alarming rate. The school gave me the power to tap human intelligence from all over the world. It was a truly amazing personal revelation. Technology was not only revolutionizing my educational experience, but also I began to realize that it was, and would continue to, globally change the learning, teaching, and understanding paradigms as we had previously known them on our planet.

Today, students of all ages can personally experience a more democratic educational experience simply through the power of the Internet resources. For example, an older student who might want affirmation in converting *amps* to *ohms* can simply get an *app* for less than $10 for their smartphone or tablet or try to find answers using a search engine rather than to sign up for a 16-week, $4000 electrical engineering basics course that forces students to meet arbitrary deadlines and that may even penalize the student for intuitive, innovative ideas. The Internet resources results can, for example, serve as a benchmark for comparison with the student's personal, handwritten calculations. Online reviews of found information, which can be either *valid* or *invalid*, can serve as a standard for the information's validity.

This chapter discusses an overview of many types of current, innovative technology resources available for students, of all ages, to use to aid their increasing understanding and knowledge expansion.

Technology Innovation for Education Using Internet-Based Services

Websites

One of the first open-access teacher, teacher educators, students, and researcher video repositories is the Video Mosaic Collaborative (VMC).[1] The VMC preserves a major video data collection on student reasoning.[2] The collection, consisting of over 4500 hours of video data and related metadata, has been collected from longitudinal and cross-sectional studies from diverse settings, making available new tools for practicing teachers, teacher educators, and researchers.[3] The VMC "is a unique compilation of video clips assembled from a longitudinal study that followed the mathematical thinking of a cohort group of students doing mathematics in and out of classrooms, from grade school through high school. The study is now beginning its 22nd year and tracking the subjects as young professionals".[4] Support for the research and building of the repository has come from the National Science Foundation and private foundation of over $15 million research dollars that have resulted in "new discoveries and innovations in understanding how students develop critical thinking skills and how to practically apply [the knowledge]."[4] Components of the first 12 years of research were featured in the *Private Universe Project in Mathematics*, "a 1-hour documentary produced by River Run media and a series of six video workshops for teacher professional development that was produced by the Harvard-Smithsonian Center for Astrophysics."[4]

The VMC makes available video clips of students doing mathematics. Currently, the VMC illustrates three strands of mathematical learning: fractions and rational numbers, counting, and early algebra. With video clips as session transcripts, student work and other metadata from over two decades of videos collected of students working on mathematical strands in rural, suburban, and urban settings. The repository currently stores video clips of students working on over 20 mathematics problems and can provide up to three camera views of classroom sessions, small group work, and individual task-based interviews. Joining the VMC enables users to watch the videos of sessions where students are learning and producing analytics. *Analytics* are an online video editing and annotation tool that can be used further for classroom analysis and for analysis of research evidence.[5] In many cases, mathematics educators do not have the funds or resources for collecting valid student data. Collecting this information is an expensive proposition. For example, *triangulating data*, a concept published by Francisco, Powell, and Maher, is a concept to link student written work, verbal and visual work when analyzing data.[6] Collecting three different representations of session information—written, verbal, and visual—over a very long period of time, is extremely challenging. This open portal provides NSF sponsored data to researchers around the globe on terms that the information is used in accordance to VMC policies.[7]

The Open University (OU) started in Milton Keynes, United Kingdom, in 1971 offering "a new type of learning."[8] The school opened with a radical open admissions policy: "it did not insist on any prior educational qualifications—the university required students to take two foundation courses before moving on to higher level study and, eventually, a BA Open degree."[8] In 1983 the university opened, "the OU Business School, whose worldwide success has seen it become the largest business school in Europe."[8] In 1986, the school staff started using text-based communication applications and by 1988 the university had a personal computing policy. By the mid-1990s, the school's use of the Internet "made the OU the world's leading e-university."[8]

The university's website states that, "the OU's mission is to be open to people, places, methods, and ideas."[8] In the 2011–2012 annual report,[9] the school reported over 250,000 students' worldwide with approximately 86,000 full-time students. The school claimed over 1000 full-time academic staff members and more than 3500 support staff. In 2011, BBC announced that the Open University full-time tuition is less than $10,000 annually.[10]

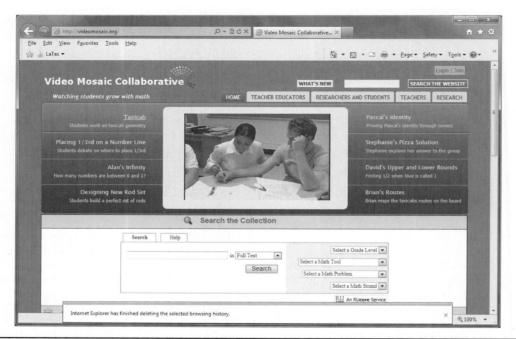

Figure 50-1   The Video Mosaic Collaborative: Watching Students Grow with Mathematics. http://www.videomosaic.org.

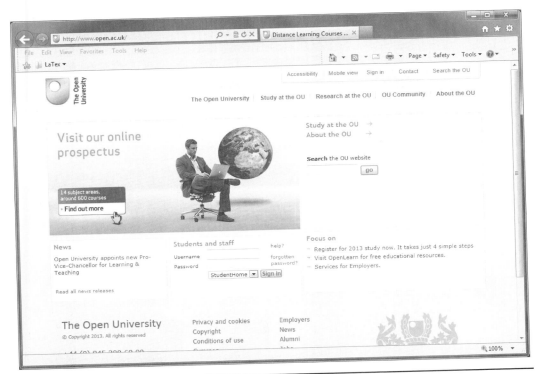

FIGURE 50-2   The Open University. http://www.open.ac.uk.

The African Virtual University (AUV) piloted in the 1990s as a vision, "to be the leading Pan-African open, distance, and e-learning network."[11] The International Review of Research in Open and Distance Learning published an article by Rashid Aderinoye in which he reviews a 2002 book on the development of the university entitled *African Virtual University: The Case of Kenyatta University, Kenya*.[12] The report states that the university "was established to serve the Sub-Saharan Africa. Funded by the World Bank, the University's mission is to use the power of modern information technology to increase access to educational resources throughout Sub-Saharan Africa."[12] The school has some paying students and some scholarship-sponsored students.

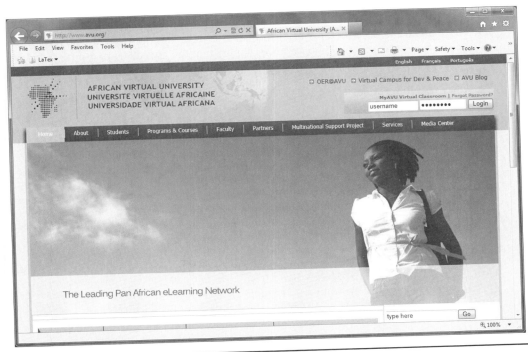

FIGURE 50-3   The African Virtual University (AVU). http://www.avu.org.

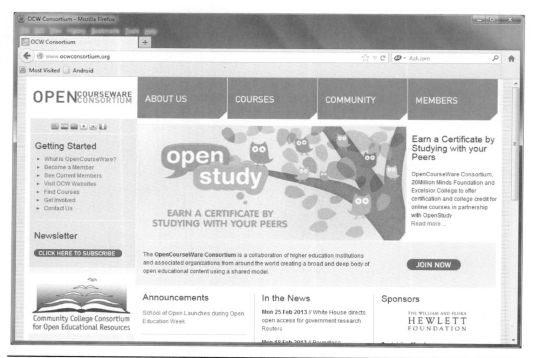

FIGURE 50-4 The Open Courseware Consortium. http://www.ocwconsortium.org/.

The AUV portal, myAUV, is a virtual classroom "that provides individualized resources, information, and services from one simple and secure login point. Once you successfully log in, myAVU Virtual Classroom will allow you to view your lessons, resources, assignments, quizzes, and marks; access live sessions; access a digital library; engage in learning activities; communicate and interact with lecturers and classmates; plan by managing your calendar; and, ask for academic and technical support when needed." The site is hosted in English, French, and Portuguese.

Currently, the website states that there are over 50 African tertiary education institutions in its partner network.[13] It currently offers a plethora of programs and courses ranging from self-study to certificate and degree programs. Topics range from Renewable Energy courses to Education courses. The website claims to have had over 4500 enrolled students in 2010.[14]

OpenCourseWare (OCW) Consortium, http://www.ocwconsortium.org/, is a collaboration of higher education institutions and associated organizations from around the world, creating a broad and deep body of open educational content using a shared model. The courseware is free and open "digital publication of higher high-quality college and university-level educational materials."[15] Current members as of this publication consist of institutions and organizations from over 20 countries and include African Virtual University, China Open Resources for Education, Japan OCW Consortium, Johns Hopkins Bloomberg School of Public Health, The Open University of Israel, Tufts University, University of California at Irvine, University of Michigan, among others.[16]

The Stanford University Engineering Everywhere (SEE) initiative offers some of the most popular engineering classes free of charge to people around the globe. The content includes video lectures, access to reading lists, course handouts, syllabi, homework, quizzes, tests, and ways to communicate with other SEE students. The courses selected fall into four general categories: Introduction to Computer Science, Artificial Intelligence, Linear Systems and Optimization, and Additional School of Engineering Courses.[17] The additional courses offered by SEE include iPhone Application Programming and Programming Massively Parallel Processors and Seminars. SEE students can pick and choose which courses and materials "best meet their needs and interests."[18] At the time of this publication, educators can use complete or partial course material "free of charge under the Creative Commons Attribution-Noncommercial-Share Alike 3.0 Unported License."[18,19]

The Massachusetts Institute of Technology (MIT) Open Courseware project was launched in 2001. Already by 2002, over 50 courses were published openly online. Currently in 2013, over 2150 courses have been published, including multilingual course translations. The website has had over 125,000,000 visitors[20,21] where visits per month are approximately 1,000,000 and monthly translation requests are approximately 500,000.[22,23]

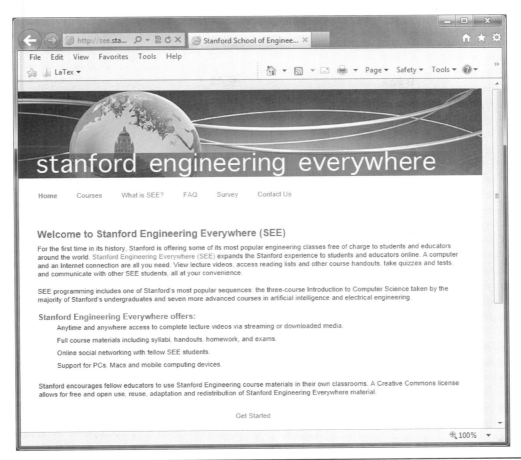

FIGURE 50-5   Stanford Engineering Everywhere (SEE). see.stanford.edu.

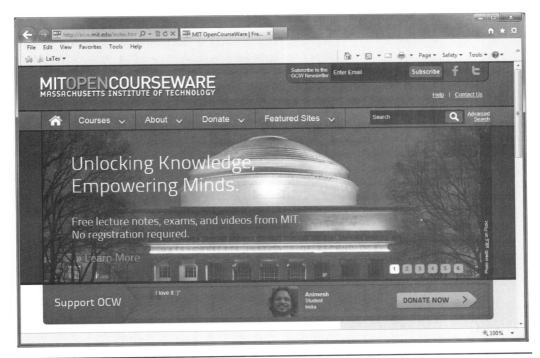

FIGURE 50-6   Massachusetts Institute of Technology Open Courseware. ocw.mit.edu.

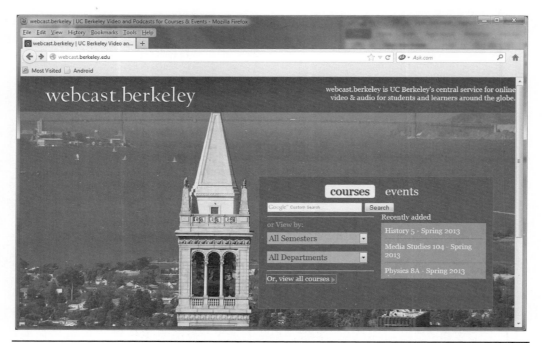

FIGURE 50-7   University of California at Berkeley Open Online Video and Audio. webcast.berkley.edu.

Courses can be viewed with respect to MIT's current schools' offerings: School of Architecture and Planning; School of Engineering, School of Humanities, Art, and Social Sciences; School of Science; Sloan School of Management; and Other (including Athletics, Experimental, and Special Programs).[24] For example, examining offered courses within the Sloan School of Management shows a listing of over 160 courses.[25]

University of California at Berkeley Webcast is "Berkeley's central service for online video and audio for students and learners around the globe."[26] Currently the website hosts videos and audio services for courses and topics from over 20 departments. The videos are hosted on the content distribution network YouTube; viewers can download videos to their desktop. Additionally, Berkeley hosts iTunes audio recordings of courses that can be downloaded

FIGURE 50-8   The Open Information Sciences Institute (ISI) DeterLab—A Cyber-Security experimentation and Testing Facility. http://www.isi.deterlab.net.

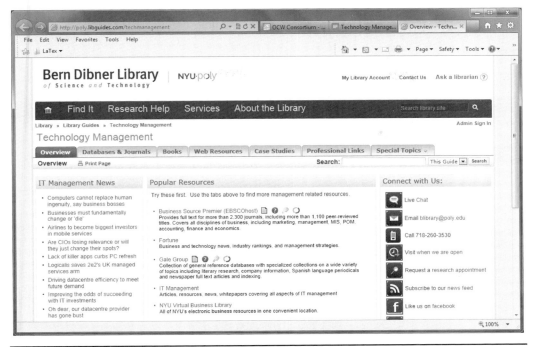

FIGURE 50-9   Online Library Interface. http://www.poly.libguides.com.

free-of-charge for both events and lectures.[27] For example, iTunes shows a list of 86 Spring 2013 courses that can be freely subscribed to via iTunes. The iTunes selection is quite extensive.

Centered at University of Southern California Information Sciences Institute (USC-ISI) is the shared experimentation research facility, DETER.[28] "Designed specifically for large-scale cyber-security research, DeterLab provides an open, remotely accessible, shared network research lab. Facilities include networking and computing resources, and an expanding set of tools for using them to construct and operate experiments. DeterLab's users are cyber-security researchers and experimenters who typically work in project teams."[29] The website statistics have showed that hundreds of research projects, thousands of students trained, and at least a hundred research papers have sprung from the lab.[30]

Online library portals are an incredible source of dynamic information for learners. For example, the New York University (NYU),[31] The Polytechnic Institute of New York[32] University (NYU-Poly), and Rutgers University[33] have specialized portals depending on subject, department, and school. Every portal can be different. For example, some portals allow students to directly chat with library reference desk help. Some libraries have online book access for students. Some libraries have access to different collections.[34] For example, the NYU library has access to United Nations documents.[35] One of the best resources that online libraries, or their subscribing journals and databases, can provide are formatted bibliography downloads. Downloadable bibliographies help to remove citation error.

GEAZLE, launched to the public in 2009,[36] is a website for Science, Technology, Engineering, Mathematics (STEM) learning. The acronym stands for *G*row intellectually, *E*nhance your thinking, *A*pply your knowledge, *Z*en your world, *L*earn something new, and *E*levate your aptitude and altitude. The mission of GEAZLE is for everyone to acquire new knowledge, share new ideas, grow intellectually, and advance the purposes of humanity within the STEM disciplines.[37] The focus of the website is on opening lines of communication among individuals in the STEM disciplines, employed or academia, providing a forum for collaboration, and establishing a web-based user controlled environment for a better future."[37] The website provides a portal for communications. It is accessible in over 250 countries and users can chat in real time with other users and viewers.

Moodle "is a Course Management System (CMS), also known as a Learning Management System (LMS) or a Virtual Learning Environment (VLE). It is a free web application that educators can use to create effective online learning sites."[38] The Moodle acronym started as, "Modular Object-Oriented Dynamic Learning Environment."[39] Moodle statistics show that over 75,000 websites host the Moodle application from over 220 countries.[40] The website statistics estimate over 7,000,000 courses that are being taught using the LMS.

Figure 50-10   Geazle website for Science, Technology, Engineering, Mathematics (STEM) Learning. http://www.geazle.net.

The application needs to reside on a web server that supports PHP and SQL. The Moodle website links website visitors to demonstration sites, if interested.[41]

The Khan Academy, https://www.khanacademy.org/, was created in 2006 by Salman Khan. The mission of the website is as a nonprofit organization, "with the goal of changing education for the better by providing a free world-class education for anyone anywhere."[42] The site hosts videos and lectures for a plethora of topics for all ages. The site also hosts both coach, or teacher, material and a dashboard where people can sign up to learn how to become

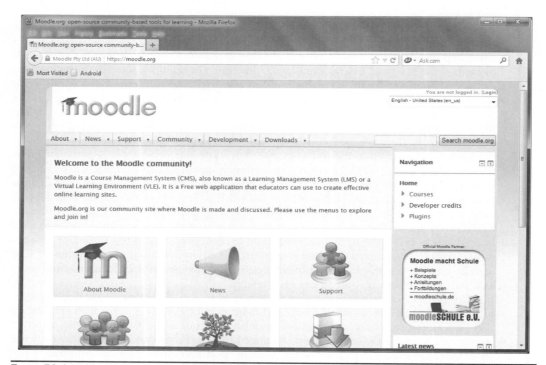

Figure 50-11   Moodle Course Management System (CMS). http://www.moodle.org.

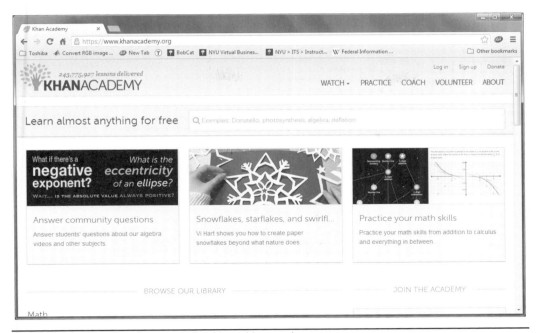

Figure 50-12   The Khan Academy. http://www.khanacademy.org.

and/or improve coaching or teaching.[43] Additionally, the Khan Academy is interfacing with some pilot schools—Los Altos School District K-8 Public School, Oakland Unity, Summit Public Schools, KIPP Bridge, Eastside College Prep, and Marlborough—to include the online resource in the classroom.[44]

Yahoo!,[45] Google,[46] DogPile,[47] Ask,[48] and DuckDuckGo[49] are all search engines which can help pupils of all ages to find information. The review of information for accuracy is one of the most difficult challenges that a young pupil can encounter, and a safe solution is still in development. Typically, a person can see if information is given in a blog, and then the person can check for positive and negative reviews. Blogs have been proven to be effective learning tools.[50] Unless the website is from a bona fide source, be wary of found information. Google offers additional online free resources with Google documents[51] and hangouts.[52] These services encourage distance collaboration. Similarly, Dropbox[53] is another free utility

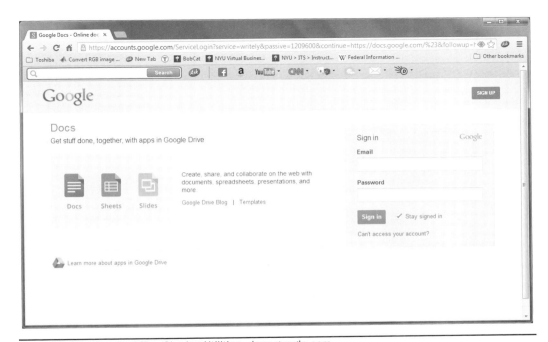

Figure 50-13   Google Office Sharing Utilities. docs.google.com.

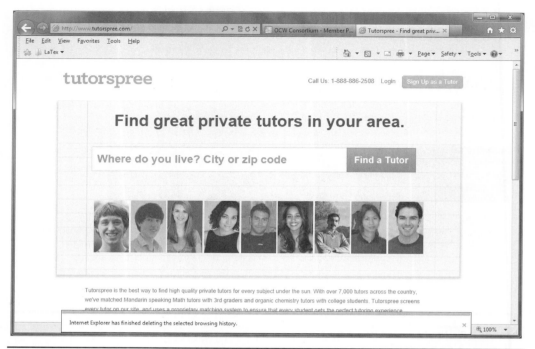

FIGURE 50-14   Tutorspree an online tutor locater. http://www.tutorspree.com.

that allows distributed collaboration. Google docs and Dropbox are especially important and useful when documents exceed e-mail limits.

Technology Innovation for Education Using the Cloud

Tutorspree[54] is one of the first online tutoring services that screens tutors and matches prospective pupils with tutors using a proprietary mechanism. Tutorspree helps pupils find private tutors for a plethora of subjects including Math, Language, Finance, Science, Economics, Programming, English, History, and Test Preparation. The site claims it has

FIGURE 50-15   Apple Apps. http://www.apple.com.

Figure 50-16   Google Apps. play.google.com.

"over 7000 tutors across the country."[54] Tutorspree states that they have matched Mandarin speaking Math tutors with third graders and organic chemistry tutors with college students. The site is uniquely marketed. It has a phone number to directly contact tutoring schedulers, a way to browse particular tutors, ways to submit feedback on sessions on the website, ways to help pupils coordinate meeting tutors in valid locations, and a history methodology so that all Tutorspree website-scheduled meetings statistics are stored on the portal for security mechanisms. Additionally, the website supports a TRUSTe certified privacy policy to help protect collection and use of pupil information.

Technology Innovation for Education Using Smartphone and Tablet Apps

Apple[55] is a popular technology brand for both smartphones and mini tablets. These devices host a proprietary operating system. The iTunes store hosts apps for download. An example of a useful education iTunes application is the *U* app which can host an entire course in one application. Additionally, "students can play a video or audio lecture and take notes that are synchronized with the lecture. You can read books, play presentations, see a list of tasks for the course, and they tick off when they are done. And when a teacher sends a message or presents a new challenge to give the user a push, notification with the new information [is sent to the user]."[56] Additionally, the *U* application can interface with many of the world's open-source repositories. In addition to *U*, iTunes has a wide assortment of helpful learning applications.

Google has a repository for Android Smartphone and Tablet Applications.[57] The application contents consist of both free and paid music, books, movies, and android applications. It is also possible to buy Google's Nexus devices, smartphone, mini tablet, and 10-in tablet, from this location.

NASA is an example of a government institution that features free apps for your smartphone—currently android-based. A visit to the NASA App page gives you a scanable Google-wallet image which can be scanned directly by your smartphone from the viewing-computer screen.[58] The image opens a list of free downloadable apps for Android.[59]

Samsung, a company based in South Korea, is another popular technology brand for smartphones [four from marketing]. Samsung sponsors a proprietary operating system, Bada. A search on the applications sponsored for Bada shows a plethora of education games applications. Example of useful educational applications include "Learning Hub,"[60] "Basic Trigonometry,"[61] and "Miracle Math."[62]

Technology Innovation in Education Considering Influencing Organizations

This chapter briefly touches on the ethics of technology in education by exploring places where e-education can be reviewed or ideas conceived. There are open areas of concern such as making sure that students know how to parse information, find incubators for developing safe technology, keep privacy, and practice security measures.

FIGURE 50-17   NASA Apps. http://www.nasa.gov.android.apps.

FIGURE 50-18   Samsung Apps. http://www.samsung.com.

The United States Department of Education[63] hosts a website, http://www.ed.gov. The website discusses public schools around the country at large and provides resources to teachers, parents, families, schools, and students. It is an excellent place to turn to for advice. Additionally, organizations such as the National Council of Teachers of Mathematics (NCTM)[64] or the National Council of Teachers of English (NCTE)[65] all host online websites full of resources for teachers, schools, students, and parents. At a local level, school districts and states sponsor their own websites for local policies and resources.

Western Interstate Commission for Higher Education (WICHE) is a consortium of 15 western states including Hawaii and Alaska. In 1989, the commission proactively started the WICHE Cooperative for Educational Technologies (WCET), an organization that "accelerates the adoption of effective practices and policies, advancing excellence in technology-enhanced

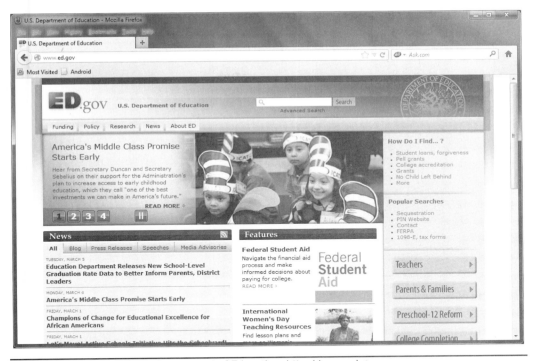

FIGURE **50-19**   United States Department of Education. http://www.ed.gov.

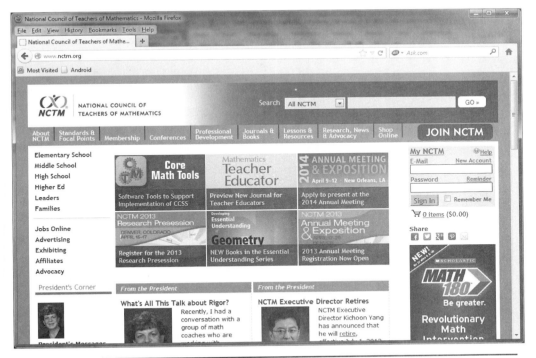

FIGURE **50-20**   National Council of Teachers of Mathematics (NCTM). http://www.nctm.org.

teaching and learning in higher education."[66] Currently the organization has nearly 2000 subscribing members who regularly share information about the state of e-education.[66]

StartingBloc,[67] http://www.startingbloc.org, started in 2005 to "educate, connect, and inspire emerging leaders to drive social innovation across sectors. StartingBloc focuses on activating individuals, while simultaneously teaching a community approach to lasting change."[68] Typically, the innovation involves use of technology whether via development of a website or standalone software.

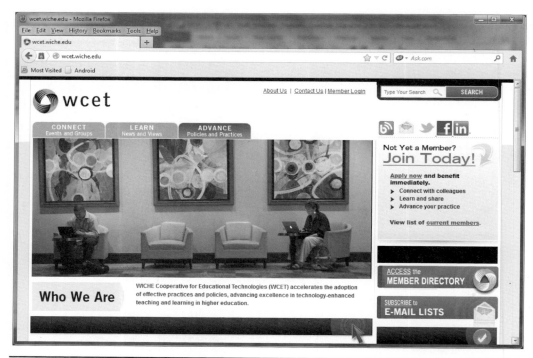

FIGURE 50-21   Western Interstate Commission for Higher Education (WICHE). wcet.wiche.edu.

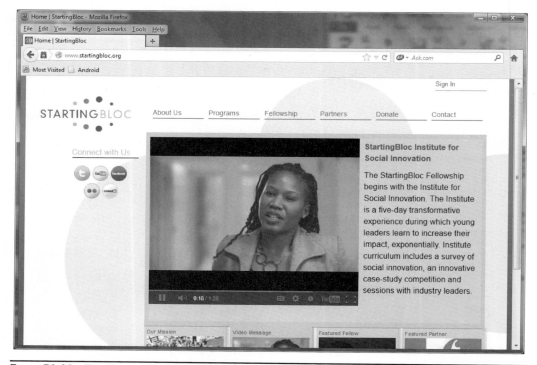

FIGURE 50-22   The Starting Bloc Institute. http://www.startingbloc.com.

Conclusion

This chapter introduced examples of educational technology resources that are revolutionizing how the world learns, teaches, and conveys information. There is a plethora of software, both open-source and proprietary, for education. New York City is currently sponsoring iZone[69,70] to boost K-12 technology in city public schools. As mentioned earlier,

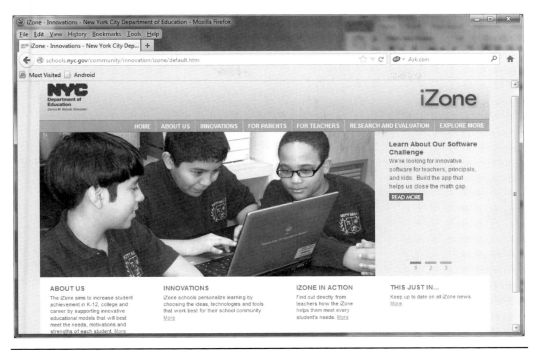

FIGURE 50-23   New York City Schools. schools.nyc.gov.

the Khan Academy is being piloted in San Francisco Bay Area schools and Moodle is being used regularly in K-12 schools such as Rutgers Prep.[71,72] Technology is quickly changing and challenging our earlier and traditional learning models.

References

1.   University, R. Video Mosaic Collaborative, 2013.
2.   Maher, C.A. "How Students Structure Their Investigations and Learn Mathematics: Insights from a Long-Term Study." *The Journal of Mathematical Behavior*, 24 (1): 14, 2005.
3.   Maher, C.A. "The Longitudinal Study." *Combinatorics and Reasoning: Representing, Justifying, and Building Isomorphisms*, 5, 2010.
4.   University, R. Video Mosaic Collaborative, 2013.
5.   G. Agnew, C.M.a.C.M. VMC Analytic: Developing a Collaborative Video Analysis Tool for Education Faculty and Practicing Educators. Hawaii International Conference on System Sciences (HICSS-43), 2009, IEEE.
6.   Powell, A.B., Francisco, J.M., and Maher, C.A. "An Analytical Model for Studying the Development of Learners' Mathematical Ideas and Reasoning Using Videotape Data." *The Journal of Mathematical Behavior*, 22 (4): 405–435, 2003.
7.   University, R. Terms of Use for the Video Mosaic Collaborative, 2013.
8.   University, T.O. History of the Open University, 2013.
9.   University, T.O. The Open University Annual Report, 2012.
10.   Coughlan, S. Open University Sets £5,000 Tuition Fees, *BBC News*, 2011.
11.   University, A.V. African Virtual University (AVU) Website, 2013.
12.   Aderinoye, R. Book Review—African Virtual University: The case of Kenyatta University, Kenya. *The International Review of Research in Open and Distance Learning*, 2003.
13.   University, A.V. African Virtual University Partners, 2013.
14.   University, A.V. The AVU Scholarship Fund: More than 600 Beneficiaries in 2009 and 2010. *The Leading Pan African eLearning* Newsletter, 2011.
15.   Consortium, O.C. What is Open Courseware, 2013.
16.   Consortium, O.C. Open Courseware Consortium Members, 2013.
17.   University, S. Stanford Engineering Everywhere Courses, 2013.
18.   University, S. What is Stanford Engineering Everywhere?, 2013.
19.   Commons, C. Attribution-Noncommercial-ShareAlike 3.0 Unported License, 2013.
20.   (MIT), M.I.o.T. Our History of MIT Open Courseware, 2013.
21.   (MIT), M.I.o.T. About Open Courseware, 2013.

22. Technology, M.I.o. MIT Open Coursware Site Statistics, 2013.
23. (MIT), M.I.o.T. MIT Open Courseware Monthly Reports, 2013.
24. (MIT), M.I.o.T. MIT Open Courseware Course Finder, 2013.
25. (MIT), M.I.o.T. MIT Open Courseware Sloan School of Management, 2013.
26. Berkeley, T.U.o.C.a. Berkeley Webcast, 2013.
27. Berkeley, U.o.C.a. UC Berkeley Events on YouTube and iTunes, 2013.
28. Project, T.D. DETER Project History, 2013.
29. Project, T.D. DeterLab: Accelerating Cyber-Security Advances, 2013.
30. Project, T.D. DeterLab: Resources, Capabilities, and Benefits, 2013.
31. University, N.Y. New York University Library, 2013.
32. University, P.I.o.N.Y. Bern Dibner Library of Science and Technology: Guides and Tutorials, 2013.
33. University, R. Rutgers University Libraries, 2013.
34. University, N.Y. New York University Collections, 2013.
35. University, N.Y. NYU United Nations and International Documents Collection, 2013.
36. GEAZLE. About GEAZLE, 2013.
37. GEAZLE. GEAZLE: About Us, 2013.
38. Moodle. Welcome to the Moodle Community!, 2013.
39. Moodle. About Moodle, 2013.
40. Moodle. Moodle Statistics, 2013.
41. Moodle. Moodle Demonstration Site, 2013.
42. Academy, K. About the Khan Academy, 2013.
43. Academy, K. Khan Academy Resources, 2013.
44. Academy, K. Khan Academy in your Classroom, 2013.
45. Yahoo!. Yahoo! Search, 2013.
46. Google. Google Website, 2013.
47. Dogpile. Dogpile Website, 2013.
48. Ask. Ask, 2013.
49. DuckDuckGo. DuckDuckGo Website, 2013.
50. Christine O'Connor, Sue Bond, D.M. "Blended Learning: Issues, Benefits and Challenges." *International Journal of Employment Studies*, 19 (2): 20, 2011.
51. Google. Google Documents and Calendar, 2013.
52. Google. Google Hangouts, 2013.
53. Dropbox. Dropbox Website, 2013.
54. Tutorspree. Tutorspree Website, 2013.
55. Apple. Apple Website, 2013.
56. Apple. iTunes-U, 2013.
57. Google. Google Play, 2013.
58. NASA. NASA App for Android, 2013.
59. NASA. NASA App on Google Play, 2013.
60. Samsung Electronics Co., L. Learning Hub Samsung App, 2013.
61. Singh, V.S. Basic Trigonometry Samsung App, 2013.
62. Soft, W. Miracle Math Samsung App, 2013.
63. Education, U.S.D.o. United States Department of Education Website, 2013.
64. Mathematics, N.C.o.T.o. National Council of Teachers of Mathematics Website, 2013.
65. English, N.C.o.T.o. National Council of Teachers of English, 2013.
66. Education, W.I.C.f.H. WICHE Cooperative for Educational Technologies Website, 2013.
67. StartingBloc. StartingBloc Website, 2013.
68. StartingBloc. About StartingBloc, 2013.
69. Education, N.D.o. NYC DOE iZone Schools, 2013.
70. Coughlan, S. New York schools enter the iZone, *BBC* News, 2011.
71. Prep, R. Rutgers Prep Website, 2013.
72. Thibault, J. Moodle Website: Gone Moodle: Book clubs, Snow Days and More, 2011.

Index

Note: Page numbers followed by *f* denote figures; page numbers followed by *t* denote tables.